THE
BIOLOGY OF MAYFLIES

CAENIS SIMULANS
HEPTAGENIA HEBE — BAETISCA RUBESCENS
HEXAGENIA ATROCAUDATA
SIPHLOPLECTON BASALE — SIPHLONURUS MARSHALLI
CALLIBAETIS FERRUGINEUS

THE BIOLOGY OF MAYFLIES

WITH A SYSTEMATIC ACCOUNT OF NORTH AMERICAN SPECIES

BY

JAMES G. NEEDHAM, PH.D., SC.D., LITT.D.
Professor of Entomology and Limnology, Cornell University

JAY R. TRAVER, PH.D.
Honorary Fellow in Entomology, Cornell University

YIN-CHI HSU, PH.D.
Professor of Zoology, Yenching University

ASSISTED BY SPECIALISTS
IN CERTAIN SUBJECTS

CLASSEY PUBLICATIONS

**Reprinted 1972 by
E. W. CLASSEY LTD.
353 Hanworth Road
HAMPTON, MIDDLESEX
ENGLAND**

ISBN 0 900848 06 5

Sole Agent in USA
Entomological Reprint Specialists
P. O. Box 77971, Dockweiler Station
Los Angeles, California 90007

ORIGINALLY PRINTED IN NEW YORK BY
COMSTOCK PUBLISHING CO. INC., 1935
REPRINTED 1969
REPRINTED 1972
HAIN – MEISENHEIM
ECM

PREFATORY NOTE BY THE SENIOR AUTHOR

SOMETHING should be said about the part that each of us has taken in the preparation of this book. Dr. Hsu has contributed chapters II, III, IV, V, and VI, and many of his notes on Stenonema and other Heptageninae are incorporated elsewhere in the text. Nearly the whole of the systematic portion has been done by Dr. Traver, who has also contributed largely, out of her extensive field experience, to our account of methods of collecting and rearing, and of mayfly ecology in general. I have gathered materials and have written introductions, and have directed the work throughout.

Something should also be said about the way in which materials for this book have been accumulated through the labors of others than those named on its title page. First of all, I would make mention of the chapters that have resulted from special studies recently made in my laboratory: the chapter on eggs and egg-laying prepared by Dr. Osgood Smith, and the two chapters on the mayfly thorax and its musculature prepared by Dr. Velma Knox. Dr. Knox has also contributed the colored frontispiece and numerous drawings.

Work on mayflies has employed the hands of a good many of my graduate students. Their names and papers make up a considerable part of our bibliography. I would here make grateful acknowledgement of their useful labors and generous cooperation.

A beginning was made by Dr. Cornelius Betten and myself when in the Adirondacks we together worked out the first life histories of North American mayflies. The results were published in four successive bulletins of the New York State Museum.

The book itself was first planned, and in part written, when Dr. Anna H. Morgan was a graduate student in my laboratory, at the time of the publication in the *Annals of the Entomological Society of America* of her first three excellent papers. Her contributions will be noted on many pages. After her departure from Ithaca, to assume elsewhere important work that precluded her further participation in this task, the following persons worked as graduate students in my laboratory, preparing the papers noted in the bibliography and providing materials that we have drawn upon freely in the preparation of this volume:

Dr. Wilbert A. Clemens made a careful ecological study of *Isonychia albomanicata.*

Miss Emily Reed (now Mrs. Harold Morrison) studied the rudimentary ovipositor as developed in Paraleptophlebia and Hagenulus.

Dr. Helen E. Murphy studied mayfly mouthparts, their embryonic development, and their metamorphosis; also she studied the local species of the genus Baetis and their life cycle. We have drawn largely upon her work in the following pages.

Dr. R. N. Chapman studied the tracheation of the thorax and of the wing bases.

Mrs. Theresa Robinson Seeman made a preliminary study of the mayflies of Southern California.

Mr. Virgil Argo studied the regional species of Potamanthus.

Dr. Louise Steger made a careful study of biology of the genus Ephemerella in the northeastern United States, and Miss Eva Gordon made a parallel study of the genus Paraleptophlebia.

Meanwhile Dr. Traver had been collecting and rearing mayflies in North Carolina, and in 1931 she began publishing in the *Journal of the Elisha Mitchell Scientific Society* her excellent monograph of the mayfly fauna of that state. Then she joined me in the work of completing this volume. The largest single contribution to it is hers.

We have also been generously aided by two entomological colleagues in other institutions, Dr. James McDunnough of Ottawa and Dr. Nathan Banks of the Museum of Comparative Zoology at Cambridge, Massachusetts. Doctor McDunnough, who has made the greatest additions to the knowledge of our mayfly fauna, has aided us by gifts of specimens, by loan of rare material, and by much good advice. It fell to him to make the first thorough-going effort to identify the poorly described species of the older authors, and to properly attach the names they gave. This was a task of enormous difficulty, owing to the loss of many types, the fragmentary condition of others, and the immature and ill-preserved condition of still others. Absolute certainty as to the species that the older authors had before them is now impossible in some cases; but we believe that Doctor McDunnough's results are not likely to be bettered, and we have in nearly all cases followed his usage. It provides a new basis from which future work may proceed.

To Dr. Nathan Banks, who was for so long the only American working with mayflies, and whose collection contains so many types of species, we have had to appeal again and again, and always with generous response from him. He has given us specimens for study, whenever a series of paratypes was available. He has loaned us others, has spent much time and care making comparisons of our material with his own, and we owe him a large measure of thanks.

Mr. William D. Sargent has furnished a number of drawings for our pages and Mr. Chin-jen Luh has contributed the plate of the Ametropus nymph from New Mexico, that was obtained for us by special effort of Mr. Philip N. Albright.

For the loan of plates and cuts we are indebted to various persons as follows: To Dr. John T. Lloyd of Cincinnati for plates from *Bulletin 22 of the Lloyd Library;* to Dr. C. H. Kennedy for those from the *Annals of the Entomological Society of America;* to Dr. H. B. Hungerford for the one from *Kansas University Studies;* to Mr. Harry B. Weiss for the one from the *Journal of the New York Entomological Society;* to Dr. James

McDunnough for the one from the *Canadian Entomologist,* and to the Comstock Publishing Company of Ithaca for a number of cuts from Comstock's *Wings of Insects* and from his *Introduction to Entomology.* We are indebted to the Academy of Natural Sciences of Philadelphia for the loan of the original from which our portrait of Thomas Say was made.

As senior author I wish to assume responsibility for the omission from the text of names of first describers after the names of species. These are no part of the names of species. They are retained only where needed for systematic reference. Where so retained I omit also the parentheses, now commonly used to indicate a shift of the species from the original genus. For me, a name is a name, and not an historical sign board.

We have endeavored to bring together and make available for the use of students the little that is known concerning mayflies, and to present a condensed systematic account of our North American species with means of identifying both immature and adult stages. This is only a beginning of the study of this long neglected order. We hope that the materials here assembled will form a substantial basis for much future work on these interesting insects.

James G. Needham

Cornell University,
September 1, 1935.

CONTRIBUTORS OF SPECIMENS

The following persons have contributed specimens for study, from the regions and localities set opposite their names; those marked with an asterisk have contributed especially valuable materials:

Abbott, C. H., Cal.
Ainslie, G. G., Tenn.
Albright, C. C., Mont.
*Albright, P. N., N. Mex., Colo.
Alexander, Chas. P., Mass., Conn.
Baerg, W. J., Ark.
Bailey, J. W., N. Car.
Bennett, Geo., Neb.
Bibby, F. F., Tex.
*Bird, R. D., Okla.
Birge, W. I. (Miss), Tex.
Bishopp, F. C., Tex., Cal.
Blake, I. H., Wyo.
Bowman, H. B., N. Y.
Bowen, J. D., Col.
Brimley, C. S., N. Car.
Bromley, S. W., Conn.
Brower, A. E., Me.
Canova, M. F., Ore.
Carruth, L. A., N. D., Man., Minn.
Christenson, R. O., Minn.
Clarke, S. W., Tex.
Clemens, W. A., N. Y., B. C.
Coker, R. E., N. Y.
*Crosby, C. R., N. Y., Que.
Davis, D. W., Va.
DeLeon, D., Wis.
Didlake, M. L., Ky.
Dietrich, H., Miss., N. Y.
Dodds, G. S., Colo.
Dunavan, David, S. C., Ont., Minn., Tenn., Ia.
Eastburn, I. K., Ga.
Engelhardt, G. P., Fla.
Eno, Esther W., N. Y.
*Fattig, P. W., Ga., Ala.
Farrell, M. A., N. Y.
Fisher, Elizabeth G., Alta.
Fisher, I. W., Okla.
French, Anna May, Ia., Mont.
Friedrich, George, Minn.
Gaines, J. C., Tex.
Gilmore, R. J., Wyo.
*Gordon, Eva., N. Y.
George, Evelyn (now Mrs. A. Grieve), Ont., Que.
Hansberry, R., Wash.
Hardy, Christine N., N. Y.
*Harper, Lawrence, N. Y.
Hazzard, A. S., Utah, Mont.
Heath, H., Cal.
Hodson, A. C., Minn.
Ingles, L. G., Cal.
Jennings, P. S., N. Y.
Jones, S. E., Tex.
Klots, Mrs. E. B., Wyo.
Knox, Velma, N. Y., Cal.
*Leonard, M. D., Wash., Ore., Fla., N.Y.
Lowry, P. R., N. H.
Leim, A. H., N. S.
Little, V. A., Tex.
Marshall, Byron, Ark.
Marshall, Ruth., Ill.
Marshall, W. S., Wis.
*Michener, Chas., Cal.
Mickel, C. E., Minn.
Miller, E. D., Va.
Morrison, H., La.
Musgrave, P. W., W. Va., Fla., Ala.
Neave, F., Man.
Needham, P. R., N. Y., Cal.
Nevin, F. R., N. Y.
Parks, H. B., Jr., Tex.
Parsons, T. M., Iowa, Minn.
Pate, V. S. L., N. Y., Mo.
Reinhard, H. J., Tex.
Rice, P. L., Idaho
Ries, D. T., Fla.
Rodock, R. E., Idaho
Rogers, J. S., Fla.
Sandoz, A., Okla.
Sargent, W. D., Mass., Mo.
Schradieck, H. E., Ia.
*Scullen, H. A., Ore., Ia., Tex.
Shadlack, J., Minn.
Shaver, J. M., Tenn.
*Shepherd, D., Cal.
Sherman, F., S. C.
Shipe, Lula, Tenn.
Smith, O. R., Cal.
Smith, R. C., Kan.
Spieth, H. T., Ind.
Steger, A. Louise, N. Y.
Sumner, F., Cal.
Surber, E. W., W. Va.
Tasker, R. C., N. Y.
Thomas, F. L., Tex.
Thomsen, Lillian, Mo.
*Townes, Henry, Jr., S. C., N. C., N. Y.
Van Duzee, E. P., Cal., Utah.
Vial, Mrs. E. E., N. Y.
Wehrle, L. P., Ariz.
Whelan, D. B., Neb.
Williams, Inez, Mass.

CONTENTS

PART I. MAYFLIES IN GENERAL

PART II. NORTH AMERICAN MAYFLIES

A systematic account of North American species in both adult and nymphal stages

Under each genus the species are in alphabetic order.

PRINCIPAL ANALYTICAL KEYS

Keys to the species will be found under their respective genera, and these are named in the right-hand running page-heads of part II.

SOLILOQUY OF A VENERABLE EPHEMERA WHO HAD LIVED FOUR HUNDRED AND TWENTY MINUTES*

By Benjamin Franklin

"It was," said he, "the opinion of learned philosophers of our race, who lived and flourished long before my time, that this vast world could not itself subsist more than eighteen hours; and I think there was some foundation for that opinion, since, by the apparent motion of the great luminary that gives life to all nature, and which in my time has evidently declined considerably toward the ocean at the end of our earth, it must then finish its course, be extinguished in the waters that surround us, and leave the world in cold and darkness, necessarily producing universal death and destruction. I have lived seven of those hours, a great age, being no less than 420 minutes long. How very few of us continue so long! . . . By the course of nature, though still in health, I cannot expect to live above seven or eight minutes longer. What now avails all my toil and labor! What the political struggles I have been engaged in, for the good of my compatriot inhabitants of this bush, or my philosophical studies for the benefit of our race in general! for, in politics, what can laws do without morals! Our present race of ephemerae will in a course of minutes become corrupt, like those of other and older bushes, and consequently be as wretched. And in philosophy how small our progress! Alas! art is long and life is short! My friends would comfort me with the idea of a name, they say, I shall leave behind me; and they tell me I have lived long enough to nature and to glory. But what will fame be to an ephemera who no longer exists? And what will become of all history in the eighteenth hour, when the world itself shall come to its end, and be buried in universal ruin?

"To me, after all my eager pursuits, no solid pleasures now remain, but the reflection of a long life spent in meaning well, the sensible conversation of a few good lady ephemerae, and now and then a kind smile and a tune from the ever amiable Brillante."

Smyth's Writings of Benjamin Franklin, Vol. X, p. 408.

*Part of a bagatelle addressed to Madam Brillon, "a lady of most respectable character and pleasing conversation."

The sun comes forth, and many reptiles spawn;
He sets, and each ephemeral insect then
Is gathered into death without a dawn,
And the immortal stars awake again.

—Shelley (in *Adonais*).

THE BIOLOGY OF MAYFLIES

INTRODUCTION

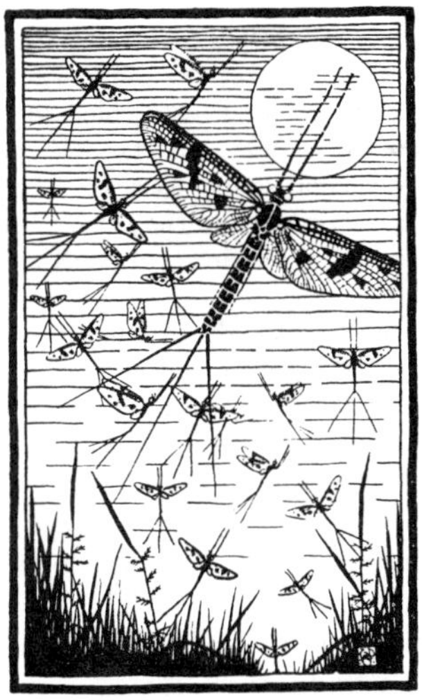

MAYFLIES are fragile insects of singularly elegant form and of very peculiar and interesting habits. In their immature stages they are all aquatic, and the adults are to be met with only near the water. As a group they have received scant attention from collectors of insects. They are well known to all fishermen, and even philosophers have heard of them.

Certain mayflies appear on the wing but once a year, and vanish again completely after a few hours flight. These have served the moralists as examples of the brevity of earthly existence. Many of them hover over the surface of the water, where fishes leap to catch them. These are the ones that have chiefly interested the fisherman.

The fisherman has studied them in his own way. He has learned their times and seasons, and their favorite haunts. He has captured them when they hovered near enough, and has used them to bait his line. Finding them too delicate for rough handling, he has invented effigies of them for his hook, crudely fashioned out of feathers and wire and silk. He has given them his own names—soft, tripping, facile names, such as grow readily outside the Nomenclatural Garden. In his mother tongue he has named them "dotterels" and "mackerels," "cocktails" and "spinners," "drakes "and "duns," "cob-flies" and "dayflies," as well as mayflies. These are the names given for the fourteen British mayflies, listed in Ronalds' *Fly Fishers Entomology* (1877). Some of these names are not in use in America, and others, such as "shad flies," "eel flies," "willow flies," and "cisco flies" are more often heard. The two names, "dun" and "spinner," applied in England to the two successive winged stages of the same insect, are practically unknown among us.

Some mayflies suddenly appear in flight in countless numbers and form vast swarms along the waterside. Such swarms compel the notice of everyone. They are harmless, of course, but they settle all over one's clothing, and get into one's mouth and ears, and are sadly indifferent to the blow that threatens to crush them. The earliest published scientific records concerning American mayflies appear to be descriptions of such swarms.

HISTORICAL

In 1749 John Bartram, botanist, of Philadelphia made, near his home on the Schuylkill River, some observations that were published a year later in the Transactions of the Royal Society of London, being communicated by Mr. Peter Collinson, as follows:—

"May the 4th 1749, Mr. Bartram perceived many Mayflies had attained wings, and were very thick spread on the bushes and grass, and by the river side. The second day after their leaving their aquatic abode they cast another skin, after which their tails are longer, and their wings drier, and more transparent. The 5th and 6th was rainy, the 7th windy; so very few came out. The 8th was cool; so few were seen: but the 9th and 10th, being warm, many swarmed late in the evening; and the 11th, 12th, 13th, they swarmed abundantly. What he calls swarming, was their gathering thick as bees, near the rivers to lay their eggs in the water.

"In their flight they mount to the tops of trees, 20 or 30 feet high; their motion is surprising, hovering up and down, rising and falling, 7 or 8 feet at a time: this he takes to be the time and manner of their impregnation. After which they fly to the brooks, cast out their eggs, and perish immediately: their eggs sink directly to the bottom, and lodge among the mud and gravel, and may be food for some minute water animal."

Some forty years later his son, William Bartram, ornithologist and Professor of Botany in the University of Pennsylvania (1791), published a description of a swarm he observed on St. John's River, Florida. This description, however, is little more than a flood of rhetoric.

While Thómas Jefferson was still President of the American Philosophical Society, Dr. Hugh Williamson, physician and statesman, read before that society his description of *Ephoron leukon*. This appears to have been the first American mayfly to be made known to science. This paper (1802)* contains a recognizable description, and adds the following interesting observations on the habits of the species:

"They begin to rise out of the [Passaic] river 35 or 40 minutes after the sun sets and continue rising about 15 minutes. . . . The chrysalis deposits a thin white pellicle or skin on the surface of the water and rises a perfect insect. It continues on the wing an hour and perishes.

"Some of them, not one in a hundred, rise from the water in the form of a chrysalis. They fly immediately to the shore and in less than a minute they creep through the white pellicle that covered the trunk, abdomen and appendices, and rejoin their companions on the wing.

"In their flight they seldom rise more than six or eight feet above the water, but they usually skim or play near the surface.

"The female drops two clusters of eggs upon the water and perishes immediately.

"The eggs are yellow. Each cluster is nearly one quarter of an inch in length and the thickness of a common pin, resembling the roe of a fish and containing about 100 eggs. They sink in the water.

"These flies are so numerous that they appear some evenings like thick driven snow in a cloud that is hardly transparent.

*Linnaeus and Müller had each described earlier in Europe one holarctic species that has since been reported from Arctic America.

"Their first appearance every year is about the 20th of July, and they continue rising every evening, more or less, for about three weeks.

"They seek the light, for they fly in crowds to a lamp or candle, but they are supposed to be the only genus of winged insects that never see the sun.

"The insect of an hour, that is never at rest, might serve for a strong figure in the hands of a peevish philosopher."

Descriptive work on American mayflies may fairly be said to begin with the Quaker naturalist, Thomas Say, "the Father of American Entomology" (born in Philadelphia, 1787). Say started in life as a drug clerk and followed mercantile pursuits. At the age of 25 he joined the Philadelphia Academy of Science and soon found his bent. He visited Florida, and later was naturalist of Long's expeditions, the first to the Rocky Mountains, and the second to the sources of the Mississippi River. He discovered and described many insects new to science. Among these were about a dozen species of mayflies, described in three papers with dates ranging from 1823 to 1839.

Say had a genius for description. Despite his lack of books and traditional knowledge, he could select and briefly state the salient characters by which species are recognized. The species he described are today recognized in a larger proportion than are those of almost any other author.

Say was an observer of habits, also; witness the following concerning his *Baetis alba* (Ephoron) in the narrative of Long's second expedition:

"Among the objects which chiefly attracted our notice, were the interesting ephemera which we had seen on Winnepeek [Winnipeg] River. They became so abundant on Rainy R. toward sunset, that they presented the appearance of a snow storm. They continued for some time, until they were driven by the wind into a small tributary valley where they formed white clouds, beautifully relieved against the dark green of the forest, deepened in its shade by the approach of night. The ensuing morning their dead bodies were seen floating on the stream, and drifted by the wind into small coves near the shore. From their great abundance, Mr. Say was led to believe that this short-lived insect never witnesses a rising sun, but that after each performance, in a short time, all the duties assigned to it in its perfect state, it deposits its eggs and expires in the night, a few hours after it has been evolved from the chrysalis. The next evening the ephemera were again seen very abundantly, but it was evident that this was a new swarm, and not a part of that previously observed." (Keating, 1824).*

The long interval of silence that followed Say's descriptive work was broken when Dr. H. A. Hagen of Königsberg, Prussia, at the request of the Smithsonian Institution of Washington prepared his *Synopsis of the Neuroptera of North America.* This was published in 1861. It was for many years the chief manual for American forms in all the Neuropteroid groups.

---

*For additional data concerning Thomas Say, see his biography by Harry B. Weiss and Grace M. Ziegler.

Say's widely scattered descriptions were all made accessible when Dr. John L. LeConte gathered them up in two volumes under the title *The Complete Writings of Thomas Say.*

H. A. HAGEN

B. D. WALSH

THOMAS SAY

A. E. EATON

A. VAYSSIERE

The *Synopsis* included descriptions of 45 nominal species of mayflies. Say's species were all included. Ten species were described as new; the remainder were species that had been described from time to time in Europe (mostly by Mr. Francis Walker for the British Museum), all of which Hagen redescribed. The *Synopsis* was thus a careful and complete compilation of the existing knowledge of our species by one who was "the highest living authority" in the knowledge of the group, and who was also a most careful and learned bibliographer.

The *Synopsis* was but the beginning of Dr. Hagen's services to America. In 1867 he accepted the invitation of Professor Louis Agassiz to take charge of the entomological department of the Museum of Comparative Zoology at Cambridge, Mass., and came to America to live. He set a new standard for entomological collections. Not fully satisfied with fine dried specimens set on pins, he planned a collection less for show, and more for real knowledge. He insisted on the preservation of full life-history materials. He entered with zeal and unstinted labor upon the preparation of what he called "a biological collection,"* covering the whole field of entomology. He did not live to make use in his writings of the mayfly materials of this collection, but how rich it became may be learned by consulting Eaton's great Monograph, in the preparation of which it was later in part used, and to the biological part of which it so largely contributed.

Dr. Hagen was stricken with paralysis in 1890, died in 1893, and was buried in the grounds of Harvard University at Mount Auburn.

When the far-sighted Secretary of the Smithsonian Institution, Professor Joseph Henry, sent the little used collection of the Institution to Dr. Hagen in Königsberg for study and received back from him the manuscript of his *Synopsis* and published it, Hagen said he hoped that "the publication of this work and its distribution through the country would call attention to the insects of this order, and result in the collection of fuller materials." That hope was speedily realized. A copy of the *Synopsis* fell into the hands of Benjamin D. Walsh of Rock Island, Ill., and he was seized with a zeal to study the Neuroptera of his own home neighborhood. In 1862 and 1863 he published two important papers on the Neuroptera of Illinois, which added vastly to the knowledge of our fauna and made Rock Island, Ill., for a long time and for these orders, the most carefully studied spot on the continent. Of mayflies he added eighteen nominal species, nearly all of which are still recognized.

Walsh also made a beginning in the study of the immature stages of mayflies in this country. On the stones in the rapids of the Mississippi

*"Dr. Hagen's influence upon the formation of such biological collections has been very great; few were in existence at the time when, almost unaided, he created that at Cambridge; and the care and elaborateness with which the whole is labelled makes it not only a worthy model, but most truly a monument to persistent and well directed industry." From Mr. Samuel Henshaw's biographical sketch of Hagen, published in the Proceedings of the American Academy of Arts and Sciences, Vol. XXIX, p. 422.

River near his home he found the curious nymph of *Baetisca obesa*, described it, and published (1864) a figure of it that has been widely copied.

Walsh was born in England in 1808. He was a graduate of Trinity College, Cambridge, and a schoolmate of Charles Darwin. He came to Illinois at 30 years of age, was first a farmer, then a lumber merchant, then a reformer of ward politics, and then an entomologist, and he devoted himself whole-heartedly to each of these occupations in succession. Throughout, he was a philosopher. Only the last twelve years of his life were devoted to the study of insects, but in that brief time he made a name and a place for himself in entomology. He was an editor of two of our early entomological journals, and was the first State Entomologist of Illinois. He suffered an accidental injury on a railroad near his home and died while still in the full enjoyment of his powers.*

The great work upon mayflies is Eaton's *Monograph of the Recent Ephemeridae*, which was issued in 1883–1888. This Monograph covered the then known fauna of the world and included descriptions of 55 genera and 270 species. Immature stages were also described and adequately illustrated.

Important studies of mayfly nymphs.—The main features of the life of mayflies have long been known. As long ago as 1737 the great Dutch naturalist, Swammerdam, published an account of the structure, habits, and life history of the European species, *Palingenia virgo*, in remarkable fullness of detail. He figured the nymphal burrows in the mud of the stream bed, the nymphs of various sizes, the first winged stage (subimago), the expansion of its crumpled wings, the final molting, and the male adult (imago). He figured also with great fidelity the internal anatomy of the nymph. He studied the deposition and the dispersal of the eggs. He endeavored to determine the duration of nymphal life. He observed the habits of the nymphs, noting that

> "When these insects are expelled out of their holes, so that they must creep on the plain or smooth bottom, which does not support every part of their bodies, they immediately lose their ability to go forward. . . . They always fell on their back, and, as if they were in a swoon, could not turn themselves again."

He quaintly describes the inoffensive habits of the nymph thus:

> "Of all species of insects, I never saw any one more mild, gentle or innocent. For, in whatever manner it is treated, it is always calm and peaceful; and when left to itself, it immediately goes to work and begins to dig a cell for its habitation."

By direct dissection Swammerdam studied the food of the nymphs and found it to consist largely of clay and bottom mud. Concerning the use to which anglers put the nymph Swammerdam said:

---

*"Un naturaliste Américain, que la science a perdu, malheureusement trop tôt." Vayssiere (1882, p. 14).

A sketch of Walsh's life by Dr. C. V. Riley, with a portrait will be found in the *American Entomologist* (1870).

"When the fishermen use this worm, they fix the hook in its head, which is the hardest and most firm part. And as it is very long lived, by the continuous motion it makes in the water, it becomes a very proper bait to allure the fish to swallow the insidious hook."

Much more that Swammerdam wrote is worthy of a place here had we space to quote it. Almost every important phase of the life of the mayfly was seriously studied by this great pioneer naturalist.

It is in France that the habits and structure of nymphal mayflies have been most diligently studied. Reaumur (1742) gave an account of their habits, and especially of the peculiarities of mayfly metamorphosis. Pictet (1843) pointed out the main lines of nymphal specialization. During the latter half of the nineteenth century Messrs. N. and E. Joly and A. Vayssiere published a long series of brilliant researches on the anatomy of mayfly nymphs. These studies culminated in Vayssiere's *Recherches sur l'Organisation des Larves des Ephemerines* (1882), which contains the best existing general account of nymphal structures. The reader will find in it also an excellent review of earlier anatomical work.

Vayssiere's interest and splendid activity in this field have continued down to the present day, and at the age of eighty years he has recently published an excellent memoir on the internal anatomy of nymphs of the American genus Baetisca (1934).

One general listing of the described species of the world has been published since Eaton's *Revision*. It is Ulmer's *Uebersicht ueber die Gattungen der Ephemeropteren* (1920). This has greatly stimulated local studies. Regional monographs have since been published by Schoenemund in Germany, by Chopra in India, by Ueno in Japan, and by Phillips in New Zealand.

The recent work upon American mayflies has been presented, mostly in small and scattered papers. A considerable part of it has been devoted to the study of habits and life histories. Such ecological data as the authors have been able to accumulate will be found in the various chapters of this book.

In Part II will be found systematic descriptions of the North American species of mayflies in all stages, so far as known to us.

CHAPTER I

THE LIFE OF A MAYFLY

The Adult

Our first acquaintance with mayflies is likely to be made when we come upon a little social company of them swarming. Suppose the species be the common band-wing, *Callibaetis fluctuans*. As we walk on a sunshiny forenoon through openings in a wood near a pond, our eye is caught by glimmers of light from swiftly flitting wings. We stop and look, not quite sure whether we have seen anything, and then we see them plainly: a little company of delicate "spinners," rising and falling on transparent gauzy wings, light as thistle-down. We approach cautiously and fix our attention on individuals. We see each one beating its way upward in slowly rising flight, its slender fore legs extended forward like outriggers, its long tails trailing in the air behind. Then we see it shift to a horizontal position and fall upon its outspread wings, slowly descending to the lower level of the company. Flying upward and falling again in endless repetition, zig-zagging now and then, or drifting a little on gentle air currents, passing and repassing its companions—this is its behavior in the "nuptial flight" that has been celebrated by so many fascinated writers. It is the embodiment of light and airy motion. It comprehends the chief—almost the only—locomotor activity of adult life.

When we have observed the company closely enough, we begin to see that the individuals comprising it are all alike. They are all males. All have clear wings and very long white tails, whereas females would have brownish costal wing bands and shorter tails. Is this dance then a masculine performance merely? If we watch closely enough, sooner or later we will see that it is not. We will see a solitary female, recognizable even on the wing by slightly larger size and heavier flight, as well as by her wing bands, slip into the swarm from some unobserved quarter and take her place among the others, getting in step, so to speak, and rising and falling with the company. And soon we will see some elect male leave his place in the swarm, seize hold of her, and fly away with her, while the remainder of the company continues undisturbed in its rhythmic and tireless flight. We have then witnessed the ancient ceremonial by which nature provides for the selection of mayfly mates. The pair leaves the swarm, flying together to some resting place among the trees or in the grass.

Swarming, in this genus of mayflies (Callibaetis) occurs oftenest, so far as we have observed, on calm sunshiny days, over grassy levels near the waterside. It recurs throughout the season of warm weather, swarms appearing whenever the weather is fit, if there are enough adult males to form a company. The winds greatly interfere with it, even a strong breeze tending to scatter the fragile "spinners." The flight is sometimes so low that one may walk directly into the midst of a swarm. And if their flight be thus interfered with, they scatter and settle upon the low herbage, suddenly vanishing ghostlike from view. If one keep his eye constantly upon a flying specimen, he may perhaps follow it to its resting place in the grass; but to find its mates, as they sit motionless all about, is exceedingly difficult, so inconspicuous are they.

Other mayflies fly at a higher altitude in swarming; some so high as to be observed with difficulty from the ground. Others swarm for but a few days out of an entire season, being more synchronous in habits. These form much larger companies and are, therefore, much more celebrated in popular literature. Others swarm in late afternoon, or at evening twilight, or even after darkness falls. And for many mayflies the swarming habits are quite unknown.

If we would see what mayflies are doing when not swarming, we must search the vegetation near the water. On the leaves we will find them sitting, motionless; or on the trunks of trees; or on the vertical walls of buildings; or occasionally, on the windowpane, where most easily seen. Here we are likely to find males and females about equally

FIG. 2—A subimago of *Callibaetis fluctuans*, just emerged from its nymphal skin, and resting on the surface of the water before taking flight (from Morgan).

numerous. And here we will find two forms of both sexes: *imagos* and *subimagos*. The former are the fully adult males and females that we have noted swarming; they are smooth and shining, have longer legs and tails, and more transparent wings. The latter are the immature winged forms, which, after a short resting period and another partial molt, will become adults. They have a dull appearance on all surfaces, with more or less pubescence about the borders of the wings and on the tails. Usually they are darker in color—often, so much darker on the wings as to appear to be different species.

The adults are inactive the greater part of the time. They sit with wings erect and fore feet extended, silent and immobile as sphinxes. They do not eat: their jaws are vestigial. They do not fight: they have no weapons. They do not play. They scarcely move to escape enemies. Such enemies as can find them may have them. If removed from one resting place, they will fly to another and rest again, the subimagos doing nothing at all until the final molting, and the imagos thereafter resting again until moved by their reproductive instincts to enter upon the mating flight.

Reproduction is the one schedule of unfinished business on the program of the winged mayflies. Feeding and growth are ended before leaving the water at the close of nymphal life. The inactive hours of later life are a time of maturing the sexual products and perfecting the powers of the winged creature in preparation for its grand finale of mating flight and subsequent oviposition. With these over their careers are ended.

Beside the pond or the still bay filled with water plants we may often see the closing event of mayfly life. Thither the females return singly to deposit their eggs. We may see a female come from an unknown quarter, and after darting about over the water a few times in short irregular flights, descend and touch the shining surface in one of the more open spaces between the floating lily pads. Instantly she rises and descends again, lightly touching the surface at each descent. Finally she falls flat upon the surface with wings outspread and remains there, floating. The eggs are liberated at the surface. They fall to the bottom, and in descending, tend apart, and so are scattered. They are very minute and of a pale yellowish color. Their surface coating being slightly adhesive, they gather to themselves in falling a protective covering of the finest suspended particles that are in the water, and become quite undiscoverable when they lie upon the bottom

Comprised between the emergence from the water of the subimagos and the deposition of the eggs by the imagos is a period of terrestrial life so brief that it has often served as a figure of fugitive existence. It is the briefest existence known in the winged state among insects. Callibaetis leaves the water one day, lives as a subimago for about 24 hours, transforms to an imago, mates, oviposits, and dies on the second day. A few other mayflies live a day or two longer; a number of others have a very much shorter period of terrestrial life.

The perils of this brief period are manifold. Robins and other birds search the shore—a reception committee for them at their arrival upon land! We have seen robins at the waterside with bulging crops and half a dozen mayfly tails hanging out of the angles of their beaks. Other birds search the foliage for them; others harvest them when they swarm. Spiders lay snares that capture them by thousands. Let any one examine the spiders' webs spun in the outer angles of any inland bridge he crosses over a clean stream, and he will find mayflies among the things ensnared. Collectors form the habit of watching spiders' webs as a means of tracing the seasonal appearance and successions of the species. Dragonflies of many kinds hover about the swarms, capturing them at will, as hawks follow coveys of pigeons and quail. Bullfrogs sit by the waterside, alert to catch them whenever they fly past, and the return of the females to the water for egg-laying marks the renewal of attack by carnivorous fishes. Trout, bass, pickerel, and many other fishes snap them up at the surface, or even spring into the air to seize them while in flight.

2. The Nymph or Larva

From the egg of Callibaetis there comes forth a nymph or larva. It lives in the midst of the submerged weeds in shoal water. It shuns the

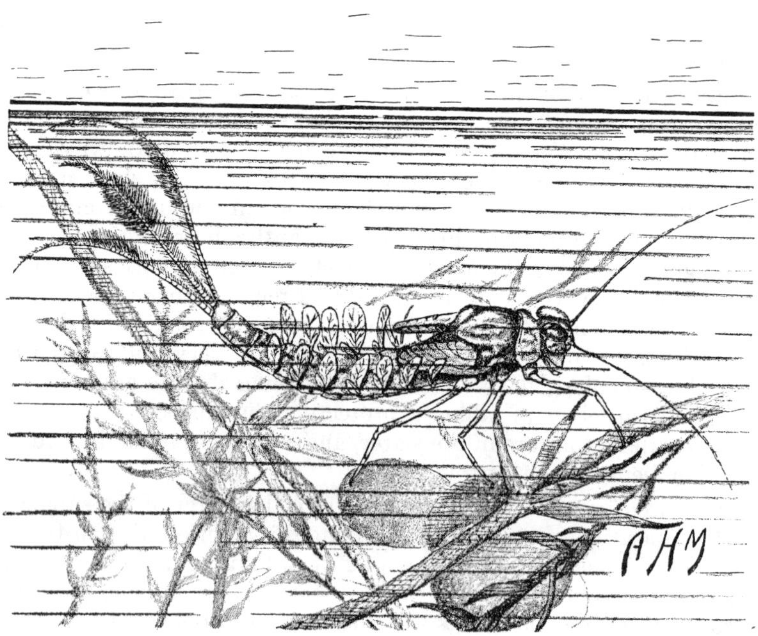

Fig. 3—Nymph of *Callibaetis fluctuans* (from Morgan).

dense mats of filamentous algae, and seeks the more open tangles of the seed plants. Beds of Marsilea, Elodea, etc., afford the right conditions. Except in newly formed pools, of which the rapid rate of increase of its species has enabled it to come into quick possession, it is rarely very abundant, for its enemies are many. It is subject to all the perils of the shoal waters it inhabits: perils of evaporation, when the pools dry up; perils of hopeless and fatal entanglement among the filaments of "blanket algae" (Spirogyra, etc.), and most commonly of all, the ever present peril of being eaten. Everything that has a taste for flesh seems to prey upon these nymphs. All the lesser fishes pursue them eagerly, so that in all the more open fish-inhabited waters no nymphs are to be found. They can survive only in sheltering vegetation where the fishes cannot readily find them all. Frogs, salamanders, and shore birds eat some of them; but carnivorous insects are doubtless their most constant and ever-present enemies. Diving beetles and damsel-fly nymphs, backswimmers and other water bugs pursue them relentlessly through the water meadows. Ranatras and diving beetle larvae seize them when they approach the surface, and dragonfly nymphs and crawfishes lie in wait for them on the bottom.

Against this ever-besetting horde of bloodthirsty enemies the nymph has no weapon of defence. It cannot fight its enemies; it can only escape them. Its success, therefore, depends on its alertness, its agility, and its protective coloration.

The ordinary activities of Callibaetis nymphs may readily be seen by watching specimens in a small aquarium. Their clambering about among water plants is strikingly intermittent: a run here, a dash yonder, a sprint across the bottom, and then a sudden halt, in an attitude of alertness, body gracefully arched, head and tail up, antennae extended, and caudal plumes outspread. Such sudden starting and stopping is common to the behavior of many defenceless herbivores and is well calculated to throw the eye of a pursuer off the track. The stopping is so sudden, and the nymph sits so tight among the protective greens and grays and browns of its environment, that the eye with difficulty locates it while at rest. So it sits still for a time, lightly poised, facing an opening between the stems, ready for another instant spring, its banded caudal filaments gracefully arched and drooping, displayed like the plumes in a proudly decorated rooster's tail. Very slowly the antennae, long and slender, and so transparent as to be well nigh invisible, are set swinging up and down. They are moved alternately, one being lowered while the other is lifted.

The first motion of a resting nymph that is likely to catch the attentive eye is the shuttle-like vibration of the gills. These are moved at will, set going for little intermittent runs of a few strokes, or run with rhythmic, machine-like continuity, according to the mood of the nymph, or according to its needs of fresh oxygen.

The seven pairs of gills in Callibaetis are articulated upon the side of the first seven abdominal segments and moved by internal abdominal muscles. Each gill is pulled forward with a quick stroke and swings back again a bit more slowly, rotating a little on its axis as it swings. The strokes are too swiftly executed for the eye to follow the separate movements—only the slight pause between strokes allows them to be counted. They are made commonly at the rate of 5 or 6 complete movements per second.

The seven pairs of gills, thus moved in harmony, set up currents in the water, which doubtless renew the oxygen supply. By their more rapid forward stroke, the water is driven toward the mid-dorsal line of the abdomen, and forward along this line to the hump of the thorax, where the current divides and passes outward to right and to left. To facilitate such passage, front and rear gills are set at somewhat different angles from the body. As the water in contact with the gills is driven forward, a new supply flows over them, bringing fresh oxygen.*

Although the nymph "sits tight" and waits long in one position when undisturbed, it does not submit passively to being eaten when an enemy approaches. Eternal vigilance is indeed the price of its safety. It sits ever alert, clinging lightly with outspread feet, or making quick little runs in search of food, and when disturbed it darts forward or dodges aside with surprising suddenness and agility. Let any one who would test the powers of escape try picking an active nymph from a dish of water with a forceps. Yet it is not distance-covering, but dodging and darting into shelter that is its mode of escape. Its longest single leaps appear to cover less than an inch of distance, and when it makes a rapid circuit of several inches across an open space this movement is chiefly effected by vigorous undulations of the body and by sculling with the fringed tails. A nymph which has lost its tails has its jumping powers very greatly limited. One that has lost its gills is retarded nearly as much, while one that has lost both tail and gills has left little power of locomotion. The violent swings and undulation of its bare body result in very little progression. The slender legs assist but little, being chiefly used in poising.

Nymphs, left undisturbed in an aquarium containing water plants, may often be observed feeding. They commonly browse upon the stalked diatoms and other minute sessile algae that grow upon the stems of water weeds. A nymph will be seen clinging to a weed stem with its head lowered and its jaws swinging. Under a lens it presents a singularly grasshopper-like face. Its mouth is pressed down to the stem in the midst of the miniature herbage and is slowly drawn backward while the jaws swing laterally, leaving the close-cropped diatom stubble when it has

*The water currents may readily be observed under a pocket lens, if one places a nymph in a vial or watch glass of water that is turbid with suspended particles, say, of powdered carmine or of India ink, or even of mud.

passed. A step forward or a turn about the stem, and another little strip is cropped close. Perhaps if the stem be one long dead, and its tissues softened with decay, the jaws may be seen to tear up strips of the soft tissues and chew them. Often a loose algal filament will be picked up in the middle, bent double, and drawn in length by length with each swing of the jaws, until consumed. Thus the nymph, when undisturbed, goes about gathering its food from the surface of the stems, and also from the silt of the bottom.

Molting. During the growth of the nymph of Callibaetis, molting frequently occurs. The thin chitinous skin with which the nymph is clothed splits across the head and down the back of the thorax, and the pale nymph slips out of it and grows a new skin of larger size. On first emerging from the old skin, the nymph is very pale and transparent; its markings are lost, and the hairs that fringe its caudal plumes are white. It is then most soft and defenceless, and remains inactive until the hardening of the chitin in its new skin has taken place. Just after molting the gills remain motionless for a time, their aid in respiration being presumably less necessary while the skin is soft and permeable all over the surface of the body. After a few minutes their shuttle-like vibration begins, at first slowly (once or twice a second) and intermittently, but soon as rapidly as ever.

How many molts occur during the life-time of a Callibaetis nymph has not been determined, but they are many—probably a score or more.

Near to the close of nymphal life signs of impending transformation begin to appear. The wing pads, that have been increasing in size with each successive molt, become thickened and dark colored. The wings that are forming inside are expanding and becoming pigmented and the pigment shows through. Sometimes the folds formed by the crumpling of the rapidly expanding wings may be seen through the transparent sheaths. The eye cases also are bulging with the pressure of the huge eyes of the adult developing within, especially in the male. And in the male, at the other end of the body the short rudimentary sheaths of the developing forceps appear. On the other hand, certain organs that are destined to disappear at transformation, such as the middle tail and the jaws, come to appear thin and hollow as the contained organs atrophy.

This is a time of great peril in the life of the mayfly, for the nymph, when nearly ready for transformation, comes to the surface of the water and remains more or less quiescent there, an easy prey to water striders and to many other predatory enemies.

If it escapes being eaten at this time, it sheds its last nymphal skin and emerges from the surface of the water as a subimago. It immediately takes a short, unsteady flight and alights upon the leaves of shrubs or trees near the shore. There it remains in a shady place, sitting quietly upon the under side of the leaves. If disturbed, it either moves over the surface in short hops or flies to some retreat higher in the tree.

Next day, as the time for final molting approaches, the insect becomes a duller gray, due to the loosening of the outer layer of its skin. Finally this splits upon the head and thorax, and the fully developed adult with its translucent wings and shining body steps out. The legs and tails are much longer than those of the subimago, and if it be a male, the eyes are twice as large. With the completion of the subimago stage the sex organs attain their full development and the insects are ready to enter upon their reproductive activity.

Such in brief is the life of one species. Fuller details and comparative notes on many other mayflies will be given in subsequent chapters.

CHAPTER II

THE STRUCTURE OF A MAYFLY

In this chapter and in the three next following an attempt has been made to prepare a detailed account of the structural characters of a representative mayfly, and to homologize them with those of more generalized insects as described by recent investigators. The genus Stenonema* has been selected because of the abundant material available for study, and because it is relatively generalized and, therefore, fairly typical.

In plate I we present whole figures of adults and nymphs of *Stenonema vicarium* for comparison with the figures of Callibaetis in the preceding chapter. These show a very different type of mayfly. In the adult the eyes are smaller and simpler. The hind wing is larger. The venation is much more abundant. The joints of the tarsi are more freely movable.

The nymph is strikingly different. It is flat, sprawling, dorsally depressed, adapted to life under stones in rapidly flowing water; whereas, the nymph of Callibaetis is light on its feet, laterally compressed, and adapted to life in the still water ponds.

1. The Head

The head of Stenonema (Pl. II, fig. 1), like that of other mayflies, appears triangular in shape when viewed from above. The vertex, the occiput, and the genae are closely fused into a single large piece, the epicranium, which forms the main portion of the head capsule. The epicranial sutures between these constituent sclerites are entirely obscured.

The face is a fusion of the front and the vertex. It is triangular in form in the male, being longer than it is wide, and is more or less quadrangular in the female, being wider than it is long. It is bounded on the sides by the inner margins of the compound eye, behind by the posterior margin of the head, and in front by the transverse clypeo-frontal ridge.

The clypeus is a thin triangular flap, held roof-like with a distinct longitudinal carinated ridge along the median longitudinal line. It is situated in front of the face and is separated from it by the clypeo-frontal ridge.

At the base of each antenna on the front border of the clypeo-frontal ridge and on the sides of the carinated ridge of the clypeus, there is an annular antennal sclerite.

*This genus was formerly included in Heptagenia, which was then the equivalent of what we now consider a sub-family, the Heptageninae. This subfamily includes Stenonema, Iron, etc. and Heptagenia proper, a group of rather closely allied modern genera, some of which will be used to furnish additional structural data.

PLATE I.

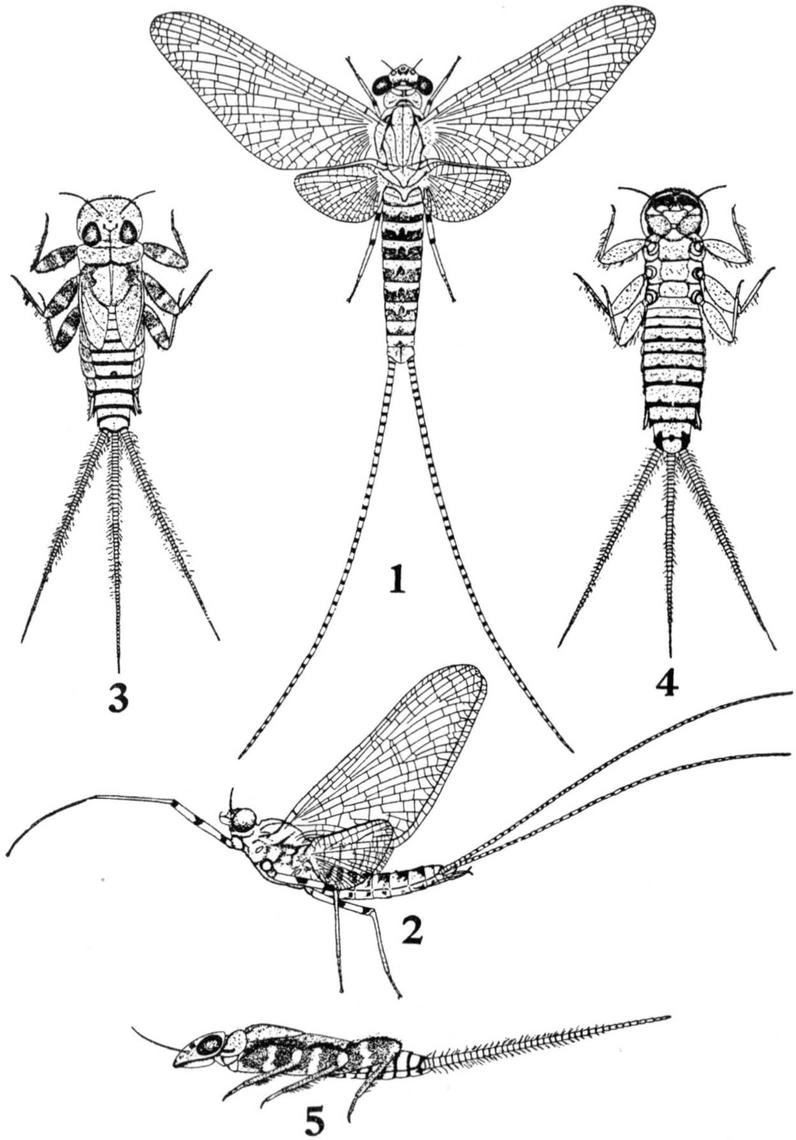

A HEPTAGENINE MAYFLY

Stenonema vicarium. Fig. 1. Female imago. Fig. 2. Male imago. Figs. 3-5. Nymph.

PLATE II.

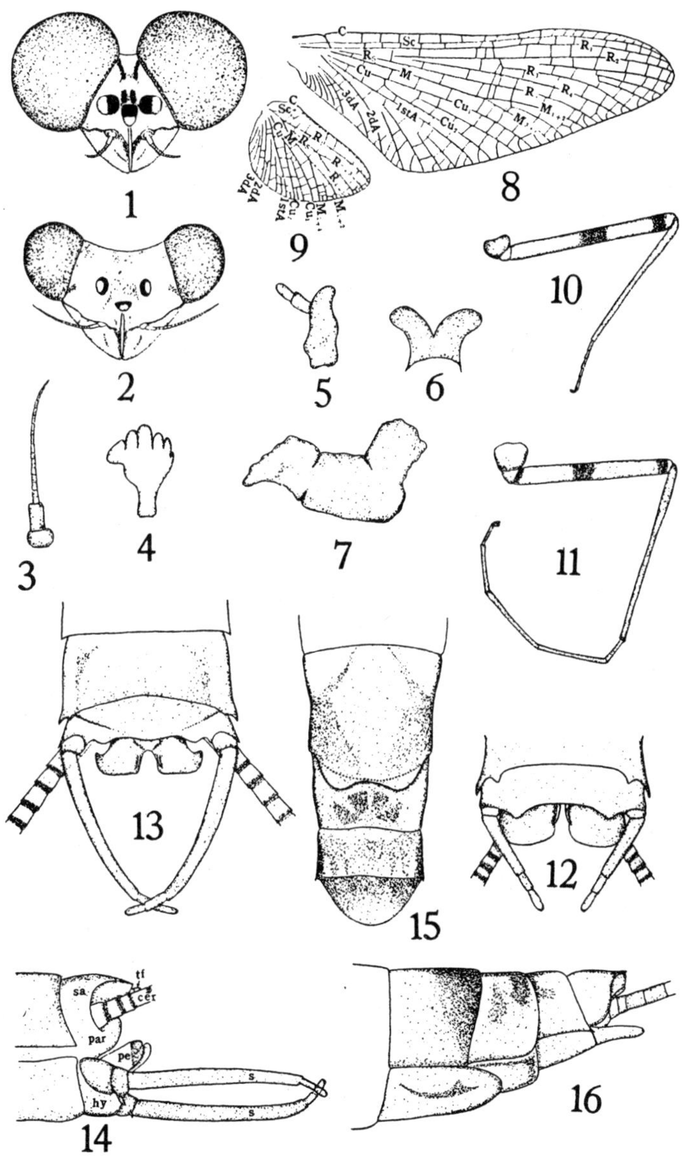

EXTERNAL MORPHOLOGY OF THE IMAGO

Fig. 1. Head of male; Fig. 2. Head of female; Fig. 3. Antenna; Figs. 4-7. Atrophied mouthparts; Figs. 8 and 9. Fore and hind wing; Fig. 10. Hind leg; Fig. 11. Fore leg of male; Fig. 12. Genitalia of male subimago; Figs. 13 and 14. Same of male imago; Figs. 15 and 16. End of abdomen of female.

The Compound Eyes. In correlation with the extraordinary development of the sense of sight, the eyes are greatly developed, especially in the males. They occupy the largest part of the head. They are more or less rounded, with their inner margins straight. In *Stenonema vicarium*, as well as in most other species of this group, the eyes are approximated very closely behind in the males, while widely apart in the females. In *Stenonema interpunctatum* and *S. canadense* the eyes are quite far apart in the males.

Each eye is divided transversely into two parts. The upper portion has larger facets and is a little darker, while the lower portion is composed of smaller facets and is lighter in color. The line of demarcation is sometimes very well marked and sometimes not. In *Stenonema interpunctatum* and *S. canadense* the two parts are very distinct and the upper part occupies half of the eye. In *S. vicarium* the upper part occupies three-fourths of the eye. In the female a line of demarcation is never clear.

The Ocelli. On the face between the two prominent compound eyes there are three ocelli, one of which is median and anterior in position, while the other two are lateral and posterior. They are arranged in a triangle. In the males they are strongly elevated and are close together. In the females they are only slightly elevated and wide apart. They are black at their bases and pale distally.

The Antennae (Pl. II, fig. 3). These are filiform and smooth. Each is inserted into a small pit or hollow cavity surrounded by the antennal sclerite beside the carinated ridge of the clypeus and anterior to the clypeo-frontal ridge. The antennae are shorter than the head. The number of segments is not constant, but it is around twenty-five in each. The basal joint, or scape, is short and thick. The second joint, or pedicel, is longer. The remaining joints are very slender and form the setaceous flagellum.

Mouthparts. The mouthparts (Pl. II, figs. 4 to 7) are vestigial, but they are externally visible. They are reduced in size, asymmetrical in form, and lacking in chitinization. Internally there is a degeneration of their musculature. The molar surface, canines, and lacinia of the mandible are not differentiated. The maxillae show no segmentation, except that the palpus is still two-segmented. The hypopharynx shows two wing-like structures. The labium does not contain any traces of the usual parts. All are shrunken very closely together in a single whitish mass underneath the clypeus.

2. The Thorax

In accordance with the requirements of flight and reduced functions of the legs, the thorax is very highly specialized. It is little like that of other orders of insects, except that in the Odonata there is some correspondence in structures for wing articulation. The prothorax is reduced in size. The meso- and metathorax are strongly developed and fused together into a solid synthorax, forming a box to enclose the strong flight muscles. Each thoracic segment presents little indication of boundaries of constituent sclerites.

The prothorax is the smallest of the three thoracic segments. The pronotum is deeply concave behind, with its two sides extended in wing-like processes. The pleura are greatly reduced by the pushing down of the pronotum on the two sides. The sternum is a more or less narrow and constricted sclerite, broader at the anterior margin and bounded by the insertion of the fore legs on the sides.

The mesothorax is the most strongly developed, for the enclosure of the muscles of the large fore wings. The mesonotum is deeply furrowed, oval in outline, and strongly convex dorsally. On the mesopleurum the wing processes and pleural suture are present, but the episternum and epimeron are not well defined. The sternum is also a single piece, which is constricted at the middle, with the portion posterior to the constriction broader than the anterior portion.

The metathorax is smaller than the preceding segment. The metanotum is quadrangular and transversely ridged. The metapleurum is provided with distinct wing processes, but without definitely distinguished pleurites or pleural suture. The sternum is a small triangular piece, situated between the hind legs.

There are two pairs of thoracic spiracles. They are of the same size and are oblique elongated slits. The anterior spiracles or mesostigmata are situated a little distance antero-dorsally to the base of the middle leg. The posterior spiracles or metastigmata are situated a little lower and closer to the base of the hind leg.

The Legs. The legs (Pl. II, figs. 10 and 11) are very long and slender and are unarmed. Each is composed of the five usual divisions. The coxa is short and stout, more or less conical in shape. The trochanter is small and triangular. The femur is uniformly slender. The tibia is about the same length as the femur, but much more slender. The tarsus is composed of five segments. In the fore legs of the male it is much longer than in the female, and it is a little longer than the tibia. The relative lengths of the tarsal segments in the fore tarsi of the male are used for the separation of the genera in the subfamily Heptageninae. In the middle and hind legs of male and female, and in the fore legs of the female as well, the tarsal segments are all very short and subequal. They are together about half the length of their respective tibiae. The fifth or last tarsal segment bears a pair of tarsal claws (ungues); one terminates in a sharp hook; the other is just a thin oval flap.

The Wings. The wings (Pl. II, figs. 8 and 9) are somewhat triangular in outline, and when at rest both pairs are held rigidly upright. In *Stenonema vicarium* they are entirely hyaline, without any color patches, but in other species of this genus there are distinct color patterns on the wings which are useful for distinguishing the species. The veins along the costal margin are strong and heavy, so the costal area appears darker. The anal area is very transparent with thin, weak veins.

The venation of the wings will be more fully discussed in Chapter X, wherein the system of names used in this book and the reasons therefor are set forth. But for purposes of comparison with current mayfly literature, the system most widely used hitherto in taxonomic work is shown in Plate II, figures 8 and 9.

3. The Abdomen

The abdomen consists of eleven segments, the last of which is vestigial and completely fused with the tenth. The epiproct is the tergal part of the eleventh segment. It is truncate along its posterior margin. It is also called supra-anal plate or pygidium (Crampton '18). It bears the stump of a middle tail or "telofilum." The paraprocts are the lateral portions of the vestigial eleventh segment. They bear the two lateral tails, or cerci. These are little more than twice the length of the body. They are smooth and are banded at the joints. The joints are short and wide at the base, and the following segments gradually increase in length to the distal end, where they attain their maximum length and slenderness. The hypandrium or hypoproct is the tenth sternite which is modified into a transverse plate. It is truncate at the posterior margin.

Each abdominal segment is ring-like and is composed of a tergite (the dorsal half) and a sternite (the ventral half), which meet on the sides. The tergite is wide and convexly arched. It occupies not only the dorsal surface, but also the larger portion of the lateral regions of the segment. The sternite is a transverse, quadrangular piece. In the first abdominal segment of both sexes the tergite and the sternite are for the greater part separated by the insertion of the hind coxae on the sides.

There are eight pairs of abdominal spiracles or stigmata on the sides of the abdomen from segments one to eight. Each lies at the middle of its segment near the lateral margin of the tergite. The first and the last pairs are smaller than the others.

Genitalia. The male forceps (Pl. II, figs. 12 to 14) is a pair of four-jointed, filiform appendages arising from the postero-lateral corners of the sub-genital plate. They are called by Crampton "styli." The basal segment is very short. The second segment is the longest. The distal two segments are equal in length, and the two together are about one-fourth of the length of the second. They are used to grasp the abdomen of the female at the time of mating.

Two penes arise from the antero-dorsal border of the sub-genital plate. They are chitinized, flattened tubes which lie very close together, with one spine, the titillator, arising from the middle of the inner margin of each. In each penis the seminal duct can be clearly traced to its termination at the tip. The form of the male genitalia varies in different species and is, therefore, much used for determining them.

The male genitalia of the subimago are very much like those of the imago, except that the whole structure is soft and very weakly chitinized. The forceps limbs are short, and the penes are more or less club-shaped. The two titillators are already well developed.

The Female Genitalia. A simpler condition exists in the female (Pl. II, figs. 15 and 16). Each oviduct lies well to the side of the abdomen and opens between the seventh and eighth sternites. Each opening is perfectly distinct. The seventh sternite elongates and projects posteriorly beneath the eighth sternite to its middle. This prolongation is called the egg valve. The under surface of the egg valve is armed with heavily chitinized compound mating spines.

CHAPTER III

THE STRUCTURE OF A MAYFLY NYMPH

The nymph of *Stenonema vicarium* lives in running water underneath stones. In accordance with its need to stand fast in rapid running waters, the body of the nymph has been greatly modified. The entire body is dorso-ventrally flattened. The femora are also flattened and laid down against the stones.

1. The Head

The head (Pl. III, fig. 1) is strongly depressed, convex on the dorsal side and concave on the ventral surface, where the mouth parts are concealed. In shape it is more or less quadrangular, with rounded anterior margin and truncate posterior border. The anterior and lateral margins are fringed.

The larger part of the head capsule, consisting of the fused vertex, occiput, and genae, appears to be a single large piece surrounding the compound eyes and bearing the ocelli. This entire flattened top of the head capsule is termed the epicranium. The coronal suture or the stem of the Y-shaped epicranial suture extends from the posterior margin of the head forward along the median line to near the center of the dorsal surface of the head, where it branches into two frontal sutures or the arms of the epicranial suture. These frontal sutures proceed antero-laterally, passing the two lateral ocelli, and can be distinctly traced as far as the anterior end of the compound eye. The vertex is the dorsal part of the epicranium behind the frontal sutures, and is divided by the coronal suture. The fronto-clypeus is an unpaired and large sclerite occupying the remaining dorsal part of the epicranium, bounded behind by the diverging frontal sutures and antero-laterally by the margins of the head. It bears the median ocellus and two antennae. Each compound eye is surrounded by an annular sclerite, known as the ocular sclerite. Surrounding the base of each antenna there is also a small annular sclerite, known as the antennal sclerite, which is light in color and very weakly chitinized. The ventral side of the head presents a concave surface and is separated from the dorsal part by the sharp and thin edges of the head. There is a large opening, the occipital foramen, through which the internal organs pass from the head backward into the thorax. The parts on the sides of the foramen are the genae. The occiput is the narrow transverse part, lying immediately behind and above the foramen.

The Eyes. On the postero-lateral portions of the dorsal surface of the head there are two large, prominent, elevated, and more or less oval-shaped compound eyes. There are three ocelli arranged in a triangular position on the dorsal surface of the head between the two compound

PLATE III.

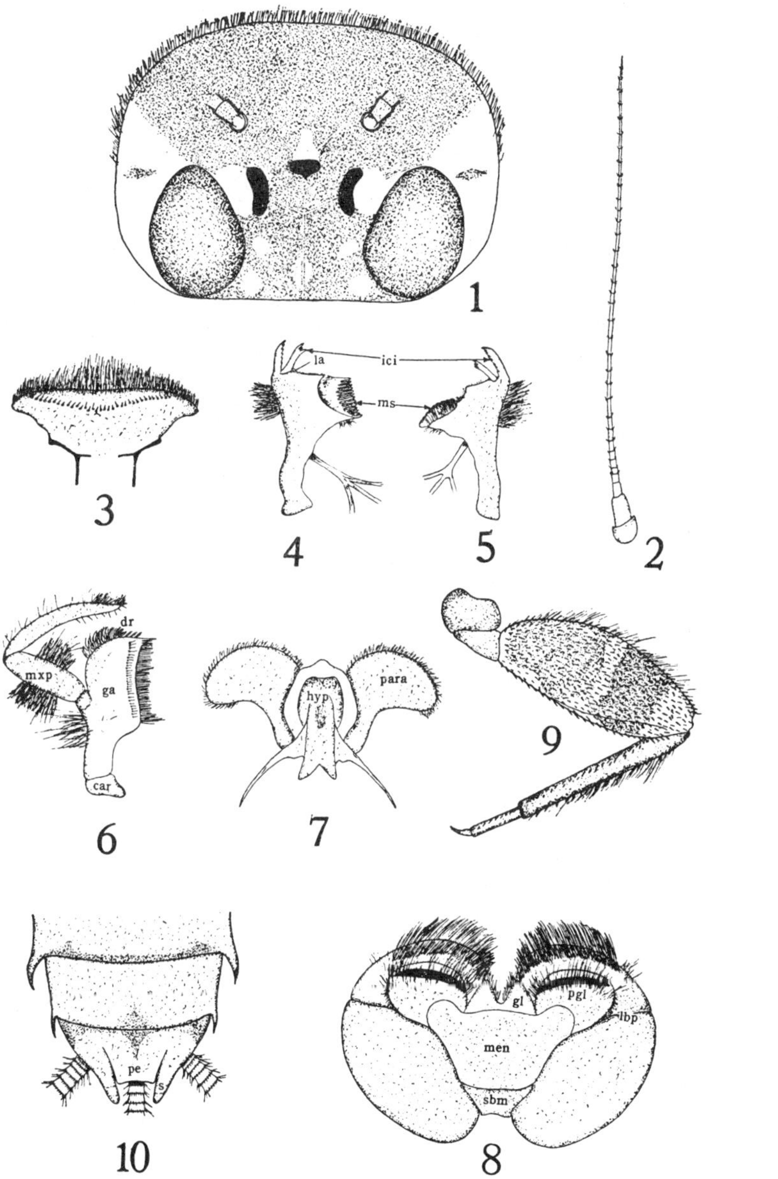

EXTERNAL MORPHOLOGY OF NYMPH

Fig. 1. Head; Fig. 2. Antenna; Fig. 3. Labrum; Figs. 4 and 5. Left and right mandibles: *ici*, incisors; *la*, lacinia; *ms*, molar surfaces; Fig. 6. Maxilla: *dr*, crown of galea-lacinia; *ga*, galea-lacinia; *car*, cardo; *mxp*, maxillary palp;Fig. 7. Hypopharynx: *hyp*, hypopharynx proper; *para*, parapsides; Fig. 8. Labium: *gl*, glossa; *pgl*, paraglossa; *lbp*, labial palp; *men*, mentum; *sbm*, submentum; Fig. 9. Leg; Fig. 10. End of abdomen of male: *pe*, penes; *s*, forceps (stylus).

eyes, a single median ocellus more or less triangular, situated at the center of the dorsal surface and anterior to the diverging frontal sutures, and two kidney-shaped lateral ocelli, situated one on each side, at the middle of the frontal sutures.

The Appendages of the Head. The antennae (Pl. III, fig. 2) are a pair of very mobile, many-jointed appendages which are attached to the dorsal surface of the head a little in front of the compound eyes. They are long, slender, and setaceous, with very short bristles at the joints. They are light brown in color. Their length is a little less than the width of the head. The number of segments is not constant, about twenty-four, more or less. The scape, or basal segment, is short and thick. The pedicel or the second segment is quite long and is about twice the diameter of the segments immediately following. The remaining segments are slender and gradually increase in length as they near the distal end.

The mouthparts are not visible from the dorsal surface of the nymphal head. They are completely concealed by the dorsal wall of the head and are hidden in the ventral concavity. They consist, as in other insects, of labrum, mandibles, maxillae, hypopharynx, and labium.

The labrum (Pl. III, fig. 3) is a more or less triangular, flap-like, and chitinized piece, with long fringes on its anterior margin. It is articulated with the head on the ventral side of the fronto-clypeus. On the ventral surface of the labrum there is a single transverse curving row of short sensory spines, each of which arises from an annular pit.

The two mandibles (Pl. III, figs. 4 and 5) are hidden in part by the labrum and the maxillae, and directly opposed to each other. They are narrow in the basal half, broad in the distal half, and heavily chitinized. They are articulated on the two sides of the mouth at their narrow base. Each is doubly equipped for feeding. Arising from the outer anterior corner of each, there are two heavily chitinized spines called canines, adapted for holding food, and at the outer interior angle there is a flattened and corrugated molar surface, adapted for grinding food.

The outer canine is immovable and toothed along its inner margin, with the number of teeth varying from five to ten. The inner one is movable and toothed only at the tip.

The molar surface is different in the two mandibles. On the left mandible it occupies the inner anterior region in the form of a chitinized depression with heavily chitinized ridges, which are toothed along their two sides. The number of ridges varies from twelve to seventeen. On the right mandible the molar surface occupies the inner third of the anterior margin and is armed with a row of plate-like and heavily chitinized ridges with smooth margins. Lateral to the ridges is a depression with a row of bristles.

The mandibles are further unsymmetrical with respect to the lacinia. There is, in the left mandible, a single, strong bristle that arises from the inner base of the inner canine. It is absent from the right mandible.

In both mandibles, along the anterior margin between the canines and the molar surface there is a row of short bristles. Along the outer lateral margin of each mandible there is a thick row of long bristles extending from the base of the canines to the middle. Beneath each molar surface there is a small tuft of bristles. A chitinized tendon for the attachment of mandibular muscles arises at the middle of the inner margin of each mandible.

The two maxillae (Pl. III, fig. 6) are thin, flap-like appendages borne on the sides of the mouth just behind the mandibles, between them and the hypopharynx. The fundamental divisions which make up the maxilla are so closely fused with each other that boundaries of component parts are very difficult to determine. The principal sclerite, here called the galea, is formed by the fusion of stipes, galea, subgalea, lacinia, and palpifer. Attaching to the base of the galea there is a small triangular sclerite, the cardo, which articulates the maxilla to the head. The maxillary palpus is three-segmented and arises from the middle of the outer lateral border of the galea. Its basal segment is short and is fused with the galea. Its middle segment is long and stout with a row of bristles on each side. Its distal segment is elongated, recurved, and terminated at the tip by a claw-like spine with a row of long hairs along its dorsal margin and a row of a few scattered bristles along its ventral edge.

The vestiture of the galea is highly differentiated. Along the anterior half of its inner edge there are two rows of plumose bristles. The inner compact row is situated right along the edge; the outer loose row is a little away from the inner margin, and each of its bristles is inserted in a small annular pit. Along the distal margin of the galea there is a row of diatom rakers, composed of two kinds of structures: bristles and pectinate spines. The latter are borne in saw-toothed notches along the anterior margin of the galea. The number of these spines differs in different species. The character of the diatom rake is, therefore, of very much use in the classification of the nymphs. The maxillae appear to be used chiefly to scrape the thin films of green diatoms from the surface of stones for food.

A tuft of bristles is borne on the outer lateral border of the galea, below the insertion of the palpus. At its apex the galea terminates in three prominent teeth.

The hypopharynx (Pl. III, fig. 7) is a relatively large, tongue-like process on the dorsal surface of the labium, lying behind and between the bases of the maxillae and forming the floor of the mouth opening. It is articulated with the head by the two chitinized tendons arising at its base. It is composed of a median, more of less oval-shaped organ, the hypopharynx proper, and two diverging wing-like processes at its sides. The latter are provided with rows of hairs along their margins. These are called paragnatha by Crampton, and are considered as the appendages of an extra segment of the head.

The labium (Pl. III, fig. 8) underlies the other mouthparts on the ventral side. Its constituent parts are for the most part distinct. Basally there is a small semi-membranous piece, the submentum, followed by a large mentum bearing the ligula in front, and there is a pair of labial palpi.

The labial palpi are very short and stout. Each is composed of two segments. The basal segment is more or less elliptical. On the outer margin of the distal segment there is a dense brush of rigid bristles. A crescent-shaped, heavily chitinized, transverse bar is situated at the center of the segment.

The four-parted ligula consists of a pair of inner triangular lobes, the glossae, and a pair of outer club-shaped lobes, the paraglossae. The glossae are fused at their base and are fringed along their inner margins, while the paraglossae are fringed with longer bristles along both margins.

2. The Thorax

The short but well formed prothorax of the nymph is larger than in the imago. On the dorsal side the pronotum is a transverse piece, which has thin and sharp lateral edges. The meso- and metathorax are closely fused with each other on the dorsal side, and each bears a pair of wing pads. One or two days before transformation the larger anterior pair of these becomes black with the pigment of the developing wing.

The thorax is heavily chitinized above and very weakly so below. Along the middorsal line there is a longitudinal suture, which continues in line with the coronal suture of the head. It is along this suture that the nymph splits its old skin at the time of molting.

The legs of the nymph (Pl. III, fig. 9) are shorter than those of the adult, and are much stouter, especially the femora. The entire leg is flattened and fringed along the outer margin, and flexed downward so as to lie close against the stones. Coxa and trochanter are short, unarmed, basal segments. The femur is large, oval in outline, and much flattened. Both its outer surface and its upper and lower margins are armed with short spines. The upper margin is fringed with a row of long, stream-line bristles. The outer surface is banded with two transverse dark patches. The inner surface is unarmed, smooth, and uncolored. The slender tibia is about the same length as the femur. It is armed with two rows of short spines on the outer surface, one along either margin. The outer margin is also fringed with a row of long bristles. The tarsus of the leg is composed of only one segment and is armed with a row of spines along its inner margin on the dorsal surface. It bears at its distal end a single prominent claw.

3. The Abdomen

This last division of the body is depressed and broad and tapers to rearward (Plate I). It bears seven pairs of gills on the back and three tails at the end. It consists of ten distinct, visible segments. The dorsal tergites are convexly arched in the middle and much flattened at the sides, where they meet with the ventral sternites, forming a sharp edge. The sternites are transverse rectangular plates which are entirely flat. The posterior margin of each abdominal tergite is armed with a row of short spines. The postero-lateral angles of the more posterior segments are produced backward into sharp lateral spines, those of the seventh and eighth segments being most prominent. The others are very small.

The gills are seven pairs of plate-like structures borne on the sides of the abdomen from the first to the seventh segment. The last pair is thread-like or filamentous in this genus. The gills arise from the postero-lateral angles of the dorsum. They are ramified with tracheal tubes for the function of respiration.

External genitalia become visible only toward the end of nymphal life, and then only in the male (Pl. III, fig. 10). The ninth sternite of the female is simple and its posterior margin is rounded, but in the male it is greatly modified. At the sides of the sternite two short slender appendages appear, with short, barely visible segments at their apex. These later develop into the two four-segmented forceps, or styli, of the adult male. Between these two appendages there is a more or less rectangular lobe, which is truncate at its posterior margin. Within this lobe later arise the two penes.

Three long, filiform tails terminate the abdomen, one median and two lateral. They probably represent the appendages of the degenerated eleventh abdominal segment. They are about the same length as the body. The median one disappears when the nymph transforms into the subimago. The basal segments are short, and the following segments gradually increase in length toward the tip. There is a whorl of very short spines at each joint and a few long bristles on each side. The bristles gradually decrease in length toward the tip of the tail.

CHAPTER IV

INTERNAL ANATOMY

This study of the internal organs is based on *Stenonema vicarium* for both nymph and imago. Four other local species, namely, *S. interpunctatum*, *S. ohioense*, *S. ithaca*, and *S. fuscum*, have been used for comparison and verification.*

1. The Body Wall

The body wall, or integument (Pl. IV, figs. 2 and 3), as in other arthropods, is composed of cuticle, hypodermis, and basement membrane. The cuticle, or cuticula, is the outermost investment of the body and appendages and is the product of the cells of the underlying hypodermis. It forms the hard inelastic exoskeleton of the animal, which protects the internal delicate organs. The cuticle of the nymph is much thicker and more rigid than that of the subimago and imago. It is differentiated into two distinct layers, the outer primary cuticle and the inner secondary cuticle. The primary cuticle of the subimago bears numerous minute bristles (microtrichia). In the nymph it is entirely smooth. The secondary cuticle is thicker in the nymph.

The hypodermis forms a continuous layer of columnar epithelial cells with quite distinct cell walls. Each hypodermal cell has an oval nucleus.

The basement membrane is a continuous, apparently structureless layer, bounding the inner surface of the hypodermis. It is extremely thin.

The Skeletal System. The internal skeleton of the head is the tentorium (Pl. IV, fig. 1). This is easily studied in the flat nymphal head. After the mouthparts on the ventral side of the head have been dissected out and all soft parts removed by boiling for a few minutes in 2% caustic potash (KOH), the entire tentorium shows nicely from the ventral side. It is composed of three pairs of ingrown arms, or apodemes: 1. the anterior arms, 2. the dorsal arms, and 3. the posterior arms. The tips of these arms expand and coalesce to form a plate, the body of the tentorium. The posterior arms form the anterior boundary of the occipital foramen, through which the alimentary canal passes into the thorax.

The endothorax of the imago consists of all invaginations of the thoracic segments for the attachment of muscles. Those that are derived from the tergite are called the phragma; those from the pleurite, the lateral apodemes; and those from the sternite, the furcae. Further details concerning these will be given in a subsequent chapter.

*The best preservation of materials obtained for dissection and for biological study was with Dietrich's solution (40 per cent formalin, 12 cc; 95 per cent alcohol, 30 cc; glacial acetic acid, 2 cc; distilled water 60 cc) by injecting the solution into the body cavity with a syringe. The materials were left in the solution for twenty-four hours and then transferred to 95 per cent alcohol for keeping.

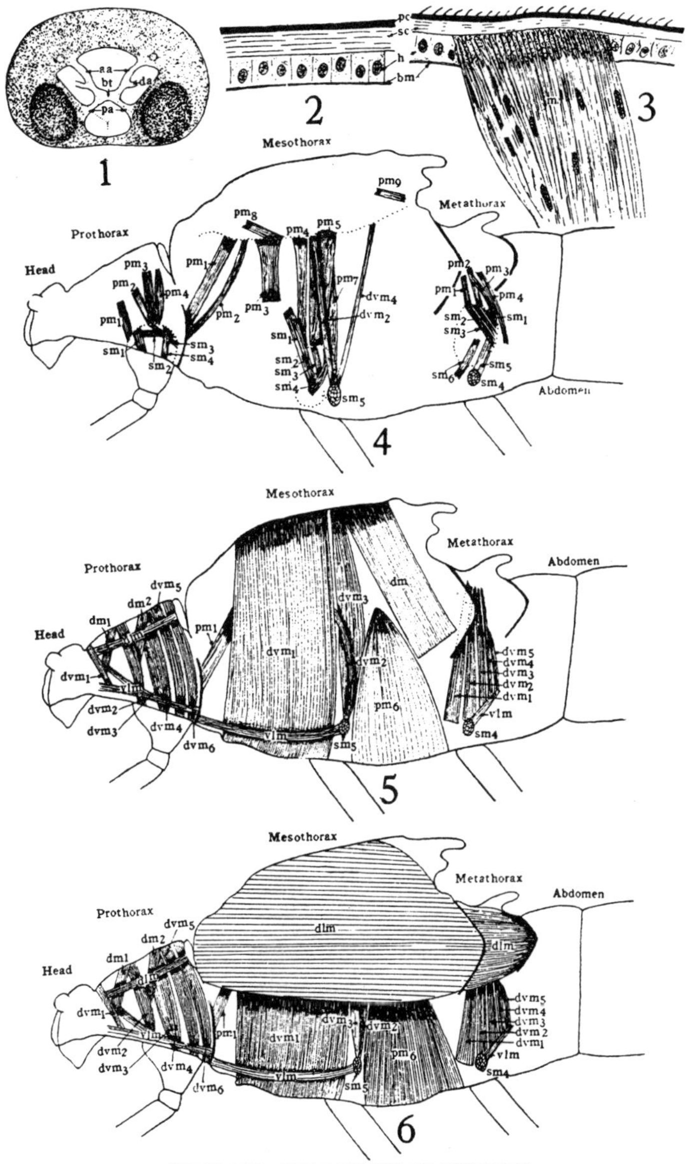

SKELETAL AND MUSCULAR SYSTEMS

Fig. 1. Ventral view of nymphal head showing the tentorium: *aa*, *da*, and *pa*, anterior, dorsal, and posterior arms respectively, and *bt*, body formed by their fusion. Fig. 2. Section of nymphal body wall, and Fig. 3, section of imaginal body wall showing attachment of body muscles: *pc*, primary cuticle; *sc*, secondary cuticle; *h*, hypodermis; *bm*, basement membrane; Fig. 4. Outer layer of thoracic muscles lying immediately within the body wall on the side; Fig. 5. Middle layer of thoracic muscles; Fig. 6. Inner layer of thoracic muscles.

The designations used for the muscles seen at these three levels are as follows: *dlm* and *vlm*, dorsal and ventral longitudinal muscles, one pair of each in each segment; *dm*, dorsal muscles; *dvm*, dorso-ventral muscles; *pm*, pleural muscles; *sm*, sternal muscles. Two of these (*sm*5 in mesothorax, and *sm*4 in metathorax) are unpaired transverse muscles, extending from coxa to coxa across the ventral nerve cord.

See Chapter XII for more detailed treatment.

2. The Digestive System

The very short adult life of mayflies is concerned only with reproduction. They are aerial fliers, and the alimentary canal in the adult stage is greatly modified for aerostatic purposes. It no longer functions as a digestive tract. On the other hand the alimentary canal of the nymph remains normal. This system is discussed, therefore, under the following two headings: 1. normal digestive tract of the nymph; 2. modified alimentary canal in the subimago and imago.

Digestive Tract in the Nymph (Pl. V, fig. 1). The alimentary canal of Stenonema is almost a straight tube, extending throughout the length of the body from the mouth at the anterior end to the anus at the posterior end. In it are the usual three principal divisions, the stomodaeum, the mesenteron, and the proctodaeum. The stomodaeum, or fore-intestine, is the anterior division of the alimentary canal and is formed as an invagination of the ectoderm in the region of the mouth. Its epithelial lining is therefore continuous with the hypodermis around the mouth and is lined internally by a fine chitinous intima, corresponding to and continuous with the cuticle. Likewise, the proctodaeum or the hind intestine, which is the posterior division of the alimentary canal, is a similar invagination of the ectoderm, and its epithelial lining is similarly covered by a fine internal chitinous intima. Between these two divisions lies the mesenteron, or mid-intestine, or mid-gut, or stomach, or chylific stomach, or ventriculus, formed from the endoderm.

The stomodaeum consists of the buccal cavity, pharynx, and oesophagus. The constriction at the posterior end of the oesophagus marks the boundary between the fore- and mid-intestine.

The mouth, on the ventral surface of the flattened head, is the foremost part of the digestive tract. It is a wide undivided cavity, bounded above by the labrum, below by the labium and hypopharynx, and laterally by the mandibles and maxillae. Its walls converge into the pharynx, which passes posteriorly into a very narrow tube, the oesophagus. This passes upward and backward between the brain and suboesophageal ganglion and projects some little distance into the anterior end of the mesenteron at the level of the anterior margin of the prothorax, forming with the anterior part of the mesenteron a single oesophageal valve. The oesophagus is therefore confined within the head. Its posterior part, behind the brain, is wider.

The mesenteron, which is much the longest division, extends from its junction with the stomodaeum to about the middle of the seventh abdominal segment at the insertion of the Malpighian tubules. The mesenteron does not possess any diverticula or caeca. It attains its greatest diameter in the mesothorax. The anterior four-fifths of the mesenteron are greater in diameter than the posterior fifth. The narrowing begins about the middle of the fifth abdominal segment.

PLATE V.

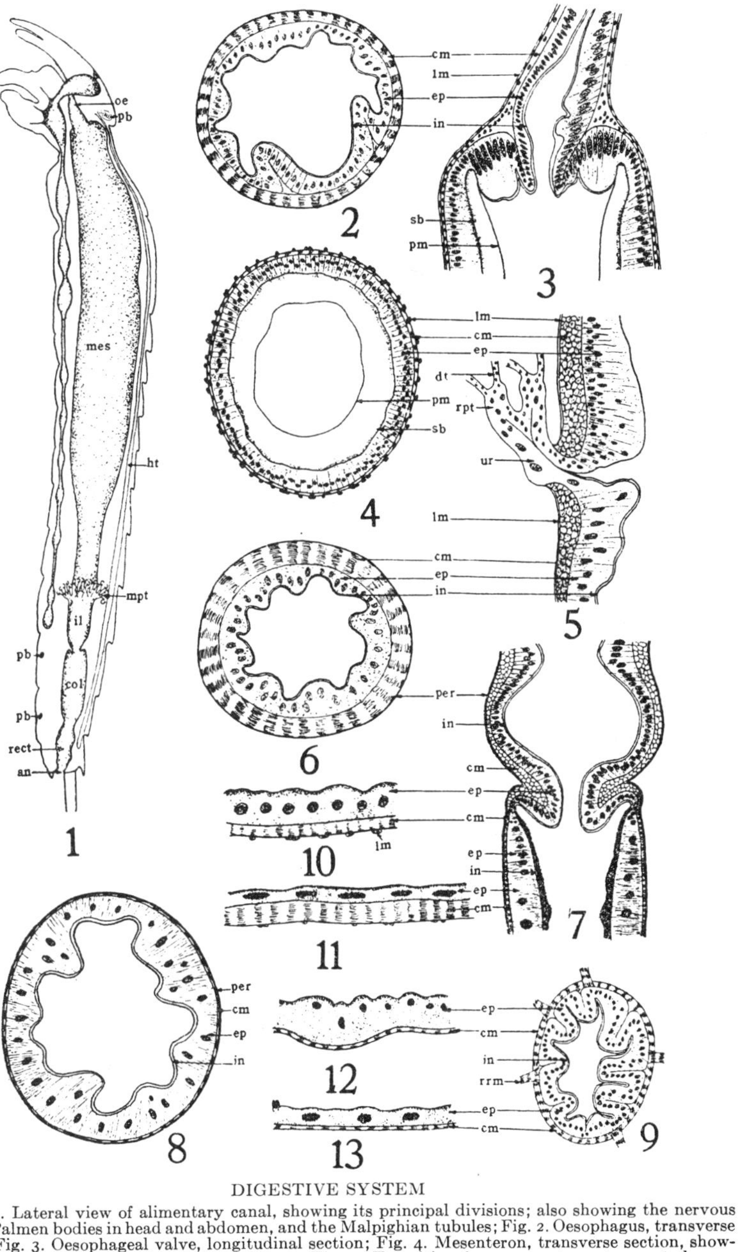

DIGESTIVE SYSTEM

Fig. 1. Lateral view of alimentary canal, showing its principal divisions; also showing the nervous system, Palmen bodies in head and abdomen, and the Malpighian tubules; Fig. 2. Oesophagus, transverse section; Fig. 3. Oesophageal valve, longitudinal section; Fig. 4. Mesenteron, transverse section, showing striated border and peritrophic membrane; Fig. 5. Transition from mesenteron to proctodaeum, longitudinal section, showing insertion of ureter of Malpighian tubules; Fig. 6. Ileum, transverse section; Fig. 7. Ilial valve, longitudinal section; Fig. 8. Colon, transverse section; Fig. 9. Rectum, transverse section, showing radial muscles and deep epithelial folds; Fig. 10. Mesenteron, transverse section, at transition between nymphal and subimaginal stages; Fig. 11. Mesenteron, transverse section, during subimaginal stage; Fig. 12. Colon, transverse section, during subimaginal stage; Fig. 13. Colon, transverse section, during imaginal stage. Explanation of abbreviations: *oe*, oesophagus; *pb*, Palmen's body; *mes*, mesenteron; *ht*, heart; *mpt*, Malpighian tubule; *il*, ileum; *col*, colon; *rect*, rectum; *an* anus; *cm*, circular muscles; *lm*, longitudinal muscles; *ep*, epithelium; *in*, intima; *sb*, striated border; *pm*, peritrophic membrane; *dt*, duct; *rpt*, receptaculum; *ur*, ureter; *per*, peritoneum; *rrm*, radial muscles.

The proctodaeum is the slender posterior division of the digestive tract, which extends from the insertion of the Malpighian tubules to the posterior opening, the anus, in the tenth segment. It exhibits three well-marked divisions: the ileum, the colon, and the rectum.

The ileum, which is the anterior third of the proctodaeum, extends from the insertion of the Malpighian tubules to the strong constriction of the ilial valve at the anterior end of the eighth segment. The colon, which is the middle third of the proctodaeum, extends from the constriction of the ilial valve to a weak constriction at the posterior end of the ninth segment. The rectum is the posterior third of the proctodaeum. It is more slender than the anterior regions. It terminates at the anal opening.

Histology of the Digestive Tract. The oesophagus (Pl. V, fig. 2) is lined internally by a distinct and well developed chitinous layer, the intima. Outside of the intima there is an epithelial layer, which is thrown into large longitudinal folds. The cells of the epithelial layer are short and columnar, with oval nuclei. A layer of circular muscles surrounds the epithelium on the outside. A sheath of longitudinal muscles forms the outermost layer of the oesophageal wall in its posterior half.

The narrow anterior portion of the oesophagus, below and before the brain, has much thicker circular muscles and a more strongly folded epithelial layer than its wider posterior portion behind the brain.

The Oesophageal Valve (Pl. V, fig. 3). The oesophagus sinks slightly into the mesenteron to form an oesophageal valve, which is formed by a complete circular fold of the epithelial and intima layers. The epithelium of the oesophagus joins that of the mesenteron without a break. The inner part of the fold is composed of short columnar cells, while in the outer part of the fold the cells begin with short columnar ones and then suddenly increase in size into a group of very long columnar cells. The intima lines the inner surface of the epithelium as far as to the outer end of the group of long columnar cells. At this level the peritrophic membrane arises, and the wall begins abruptly with columnar cells, having a striated border. The circular muscle layer continues into the fold of the oesophageal valve, and is continuous with the circular muscle of the mesenteron; the longitudinal muscles, on the other hand, do not enter into the fold of the valve, but bridge over it and continue directly as the longitudinal muscle sheath of the mesenteron.

The histological structure of the mesenteron (Pl. V, fig. 4) is markedly different from that of the stomodaeum in that there is no chitinous intima lining. The epithelial cells are columnar. Immediately outside of the epithelium there is a layer of circular muscles, and outside of that there is a sheath of scattered, longitudinal muscle fibers. An exceedingly delicate peritoneum forms the outermost layer.

The muscular layer of the mesenteron is thin in front and most strongly developed at its posterior fifth, where it attains the same thickness as the epithelial lining. The epithelial layer is thicker at the front. The inner free margin of the epithelium is covered by a thick, striated border, which gives it a ciliated appearance. The peritrophic membrane, which originates at the oesophageal valve, forms a tube inside the epithelium enclosing the food, and is freely suspended in the mesenteron. It is very thin and sometimes it may not be distinguished in sections.

The food material observed in the mesenteron consists largely of filamentous algae and diatoms.

Transition from the Mesenteron to the Proctodaeum (Pl. V, fig. 5). In the seventh abdominal segment, the epithelium of the mesenteron comes to an abrupt end just in front of the origin of the ureters of the Malpighian tubules. Posterior to the insertion of the Malpighian tubules the ileum or the small intestine begins suddenly with an epithelial layer of scattered nuclei, covered by an intima which continues into the conducting tubes of the Malpighian tubules. The ureters of the Malpighian tubules thus open directly into the alimentary canal through the narrow break between the epithelium of the mesenteron and that of the ileum.

The epithelium of the ileum (Pl. V, fig. 6) is thrown into several weak longitudinal folds and is composed of large and granular nuclei. It is lined with an intima. A thick circular muscle layer surrounds the epithelium on the outside, and outside of this there is a very thin peritoneum.

The Ilial Valve (Pl. V, fig. 7). The posterior end of the ileum constricts and sinks slightly into the colon to form the ilial valve. This is like the oesophageal valve in that it is formed by a complete circular fold of the epithelial layer and intima of the ileum. The inner part of the fold is composed of columnar cells, while in the outer part of the fold the cells begin with columnar cells and gradually decrease in thickness to merge into a homogeneous epithelial layer of the colon. The muscular layer continues into the fold of the ilial valve and is continuous with the very thin muscular layer of the colon.

The Colon (Pl. V, fig. 8). The epithelial lining of the colon is a homogeneous, thick, and weakly folded layer with large granular and oval nuclei (the "grosskernige Epithel" of Fritze). Its outer surface is surrounded by a thin layer of circular muscles. A very thin peritoneum encloses the muscular layer on the outside.

The Rectum (Pl. V, fig. 9). The epithelial layer of the rectum is thrown into several deep folds with columnar cells, having small nuclei. The layer of circular muscles is of moderate thickness. Radial muscles are also present, attaching to the body wall and holding the posterior end of the digestive tract in place.

3. The Circulatory System

The organ of circulation (Pl. VI, fig. 3) is a single dorsal vessel which lies just beneath the dorsal body wall, above the alimentary canal, and along the median line. It extends from the tenth abdominal segment to the head.

The posterior, wider portion of the vessel in the abdomen, the so-called heart, consists of ten chambers, corresponding to the ten abdominal segments in which it lies. Each chamber is provided with a pair of slit-like openings (ostia) guarded by a pair of internal valves. Zimmermann ('80) found that in *Cloeon dipterum* the heart divides into three branches at the posterior end of the tenth segment in the nymph, and each branch goes to a tail. The branch has ostia in the tail. This condition has not been noted in Stenonema.

The aorta is a very delicate, slender, straight, and non-contractile tube, which carries the blood stream from the heart forward through the thorax to the head. It is not divided into chambers. It opens in the head above the brain.

The general body cavity in which blood flows is called the haemocoele. Owing to the position of the various internal organs in the body, the haemocoele forms a kind of imperfect sinus along either side of the middle line but chiefly ventral. It extends into the legs and other appendages. It is filled with blood plasma.

The blood is a lymph-like fluid, in which corpuscles (Pl. VI, fig. 4) of various kinds float. The blood corpuscles are of various forms; some are spindle-shaped, some rounded, and others ameboid as will be shown in more detail in Chapter IX.

4. The Excretory System

In Stenonema there are eight ureters, which open into the digestive tract between the mesenteron and the proctodaeum (Pl. V, figs. 1 and 5). Soon after origin, each ureter branches into two or three short stub-like structures, the receptacula, which receive ducts from the Malpighian tubules proper. The number of Malpighian tubules varies in different individuals around one hundred and fifty. Drenkelfort ('10) found that there were more than a hundred tubules in *Siphlonurus lacustris*.

Each Malpighian tubule (Pl. VI, figs. 5 to 10) is divided into two main portions: the distal, secretory glandular part, and a slender tube or duct which leads to one of the receptacula. The distal glandular portion can again be divided into two parts: a long, distal, curved, and bent portion, which leads into the swollen central coiled body. From the end of this the duct leads. The distal portion is blind and blunt at its tip. Its widest part is near the apex, and its diameter gradually decreases proximally about one-half. This narrow portion continues for a short distance, making a second abrupt curve, and then the tube starts to make a spiral-like loop. It suddenly becomes very much widened to form the coiled central body. The central body suddenly decreases in diameter to form the beginning of the duct, which forms a nearly circular loop on the circumference of the central body and connects with one of the receptacula.

The Malpighian tubules together with the ureters are formed from

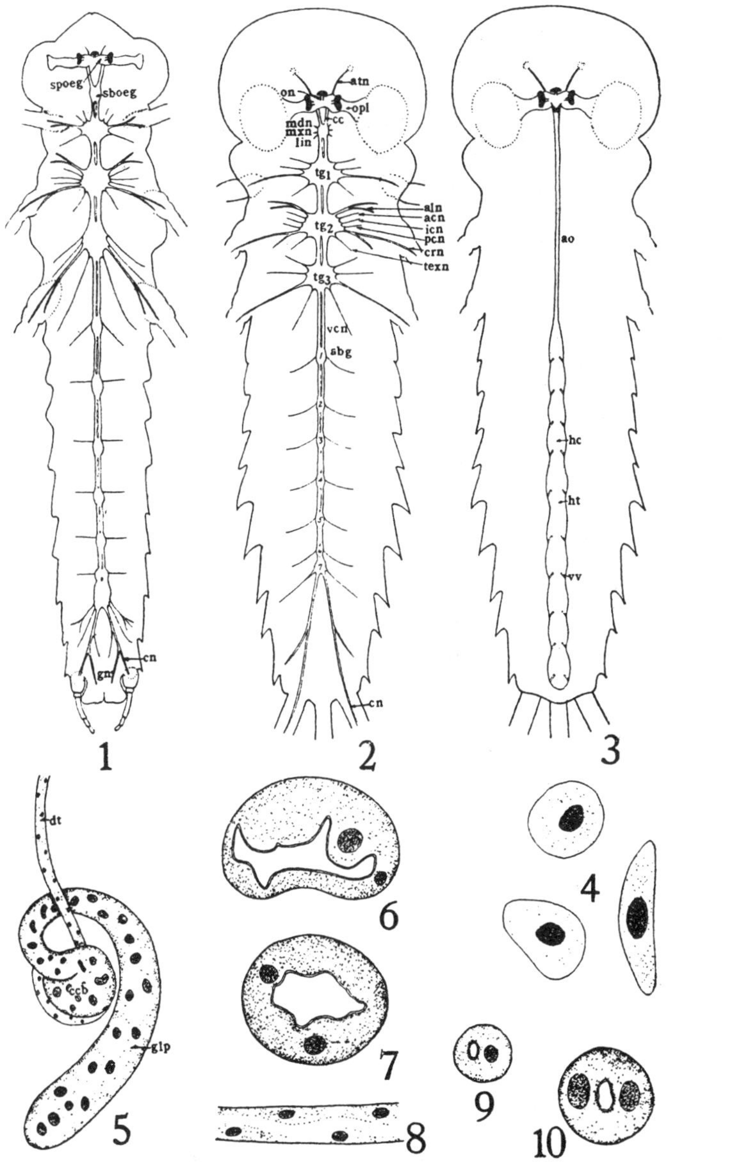

NERVOUS, EXCRETORY AND CIRCULATORY SYSTEMS

Fig. 1. Dorsal view of central nervous system of imago; Fig. 2. Dorsal view of central nervous system of nymph; Fig. 3. Ventral view of dorsal blood vessel in nymph, showing the aorta opening on the ventral side of the supra-oesophageal ganglion, and the ten chambers of the heart; Fig. 4. Forms of blood cells; Fig. 5. Malpighian tubule, showing its different parts; Fig. 6. Central coiled body of Malpighian tubule, sagittal section; Fig. 7. Central coiled body of Malpighian tubule, transverse section; Fig. 8. Duct of Malpighian tubule, longitudinal section; Fig. 9. Duct of Malpighian tubule, transverse section; Fig. 10. Glandular portion of Malpighian tubule, transverse section.

Explanation of abbreviations: *spoeg*, supraoesophageal ganglion; *sboeg*, suboesophageal ganglion; *atn*, antennal nerve; *on*, optic nerve; *mdn*, mandibular nerve; *mxn*, maxillar nerve; *lin*, labial nerve; *tg1*, prothoracic ganglion; *tg2*, mesothoracic ganglion; *tg3*, metathoracic ganglion; *vcn*, ventral nerve cord; *abg*, abdominal ganglion; *opl*, optic lobe; *cc*, crura cerebri; *cn*, cercal nerve; *gn*, genitalic nerve; *ao*, aorta; *hc*, heart chamber; *ht*, heart; *vv*, valve; *dt*, duct; *ccb*, central coiled body; *glp*, glandular portion.

the invaginations of the proctodaeum, and are therefore composed of three layers: a peritoneal membrane on the outside, an intima lining, and a layer of large epithelial cells on the inside. The distal glandular portion of the tubule has thicker epithelium with large nuclei. The lumen is largest at its distal end, and gradually decreases in diameter. The lumen in its course alternates first towards one side of the tube and then towards the other. The cells of the central body are larger than in any other part of the tubule, but their nuclei do not show a corresponding increase in size.

The duct is a very slender tube with its lumen swinging from side to side of the tube. It shows more or less uniform width throughout its entire length. From a surface view the nuclei are seen to be circular in outline and to alternate in position along the walls of the duct.

The Fat Body (Pl. VII, fig. 2). The fat body is more conspicuous in the nymph than in the imago. It forms large masses in the body cavity around the digestive tract and other internal organs. These masses are whitish or yellowish in color. The adipose cells forming the fat body are very numerous. They are closely packed together, so as to assume an irregular polygonal form, but appear round and distended wherever their surfaces are free. Their nuclei are small and more or less oval, and the cells themselves are filled with a store of fatty matter, held between the meshes of a fine network of cytoplasm.

5. Respiratory System*

The nymph breathes by means of seven pairs of plate-like gills on the sides of the abdomen. Each gill is traversed by numerous branches of tracheal tubes. All the gills are connected inward by the spiracular trunks with two longitudinal trunks (Pl. VII, fig. 1) which extend from the anterior border of the mesothorax to the last abdominal segment. These two longitudinal trunks run parallel to each other along the dorsolateral borders of the alimentary canal. There are tufts of short tracheal tubes arising along the inner margin of the two longitudinal trunks, but they are not connected transversely in the first seven adominal segments. In segments eight and nine, they connect transversely. Farther to rearward a branch from the outer side of each longitudinal trunk in the tenth segment extends to one of the lateral tails.

In the mesothorax each longitudinal trunk breaks up into two branches leading forward into the head, one branch extending anterolaterally into the fore leg, and another branch running postero-laterally to the middle leg. In the anterior border of the metathorax there is one branch coming out from the longitudinal trunk and running into the hind leg. Each of the leg branches divides into two in the base of the femur.

*The details of tracheal distribution, so difficult of dissection, may easily be made out in whole preparations if nymphs be collected immediately after molting, cleared in glycerine, and studied while all the tubes are still filled with air. The body is thus rendered transparent, while the whole system of tracheal tubes is sharply defined.

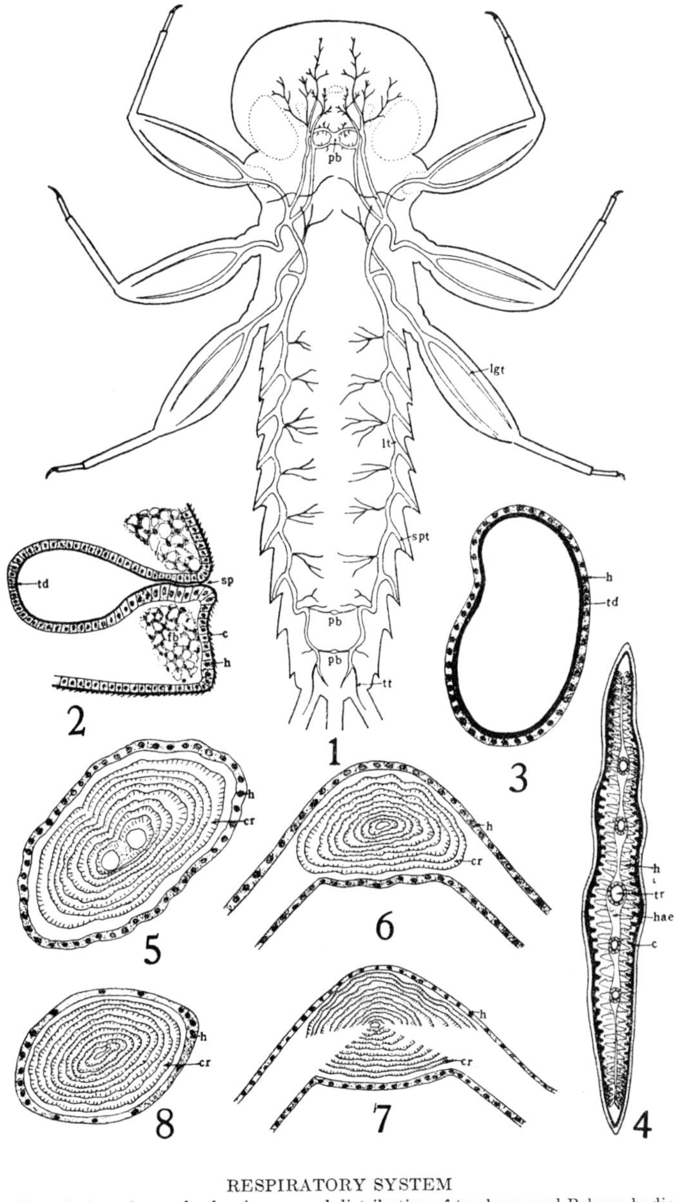

RESPIRATORY SYSTEM

Fig. 1. Dorsal view of nymph, showing general distribution of tracheae, and Palmen bodies in head and abdomen; Fig. 2. Transverse section of a portion of abdomen through the spiracle; Fig. 3. Trachea, transverse section, showing the histological structures; Fig. 4. Gill, transverse section, showing the histological structure; Fig. 5. Section through Palmen body of head; Fig. 6. Section through Palmen body in head; Fig. 7. Section through Palmen body in head; Fig. 8. Section, Palmen body in abdomen.

Explanation of abbreviations: *hae*, haemocoele; *tr*, trachea; *c*, cuticle; *h*, hypodermis; *cr*, concentric rings of intima; *td*, taenidium; *spt*, spiracular tracheal trunk; *lgt*, leg trachea; *pb*, Palmen body; *tt*, tail trachea; *sp*, spiracle; *fb*, fat body; *lt*, longitudinal trunk.

"Palmen bodies" or **induvia** (Pl. VII, figs. 1 and 5 to 8). A chitinous, concentrically-layered structure in the head of mayflies that has been much discussed by students of tracheation, is the "Palmen body."* It is symmetrically situated between the compound eyes and a little posterior to the brain, above the oesophagus. It is formed by the inner ends of the four large tracheal tubes coming together at a common point. These four tubes are arranged in an hour-glass fashion or like a horizontal 8, two on each side. The anterior tube is a branch from the outer longitudinal tracheal trunk and the posterior tube from the inner tracheal trunk. The entire structure is surrounded by a continuous hypodermal layer. It consists of several concentric rings of intima, or tracheal lining, retained at each successive molt. In the center of the organ is an irregular mass which does not show any ring-like structure and which is apparently secreted before the larva casts the first lining of its tracheal system. This central mass is larger than the opening in any of the four tubes; therefore, it is impossible to cast it out of the body. The different conspicuous rings, which are sections through the concentric layers of the organ, represent successive ecdyses. The number of rings increases with each instar. The hypodermal cells surrounding the organ are larger than those surrounding the tracheae and this accounts for the greater thickness of the chitinous layers of the organ, which are secreted by these larger hypodermal cells.

On the ventral side of the eighth and ninth abdominal segments there are two smaller induvia that evidently have been overlooked by the previous workers (Hsu 1933). Each of these is about one-fourth the size of the one in the head and is formed by the union of the inner ends of the two transverse tracheal branches from the longitudinal trunks. Each is also composed of a series of concentric chitinous layers or rings of intima, left behind at successive molts, and each is surrounded on the outside by a layer of larger hypodermal cells.

Gills (Plate VIII). The gills are plate-like and double on abdominal segments 1 to 6. Those on segment 7 are rudimentary and thread-like or filamentous in this genus. The upper division is plate-like and large, and the lower part is small and fimbriate-lamelliform. The former may be either pointed or rounded or truncate at the posterior margin and may either be fringed along the margins or not, according to species.

The gill plate (Plate VII, fig. 4) has a rather thin cuticula covering the outside. A continuous layer of pigmented hypodermal cells lies beneath the cuticle. They tend to assume a pyramidal form with clear spaces interposed along their inner ends. The pigment is nearly all deposited in the layer of protoplasm lying closest to the cuticle, where the nuclei are also situated. The inner ends of the hypodermal cells rest on a definite basement membrane. In the interior of the gill there are several branches of the tracheae, and between them are blood-spaces of the hæmocœle.

*Named for J. A. Palmen, who discovered it, (1877).

PLATE VIII.

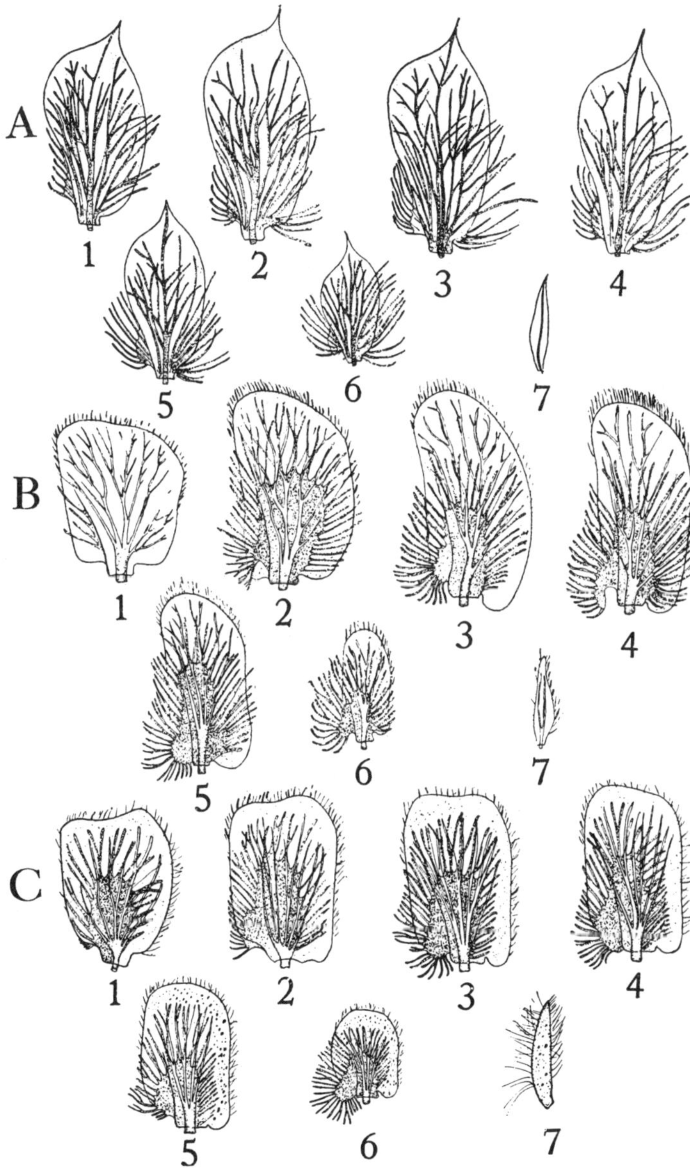

NYMPHAL GILLS

A, the pointed gills of *Stenonema sp.*, *interpunctatum* group; *B*, the obliquely rounded gills of *S. tripunctatum; C*, the truncate gills of *S. vicarium.*

In the imago the whole system of tracheal tubes opens to the outside by ten pairs of spiracles, two belonging to the thorax and eight to the abdomen. They are situated on the last two thoracic and the first eight abdominal segments.

Each thoracic spiracle is a rather large oblique elongated opening guarded on two sides by weakly chitinized lips with no special closing device between the lips. The abdominal spiracles are very small and simple.

The histology of the spiracle (Plate VII, fig. 2) is very simple. The body wall invaginates to form the spiracular trunk. The cuticle and hypodermis of the body wall continue in those of the trachea. The mouth of the invagination is the spiracle.

The tracheae in transverse section (Plate VII, fig. 3) are seen to have a hypodermal layer composed of flattened cells with oval nuclei. The lining cuticular layer is strengthened by peculiar annular thickenings known as taenidia.

6. The Muscular System

The muscles of mayflies (Plate IV, figs. 4 to 6) are striated as in all other insects without exception. The muscle fibers are bound together in bundles or fasciculi, with or without an enveloping sarcolemma. In the nymph the developing muscles are very different from the functional ones; they contain numerous nuclei.

Most of the muscles are of a white, creamy, or pale greyish color. They may be attached either directly (as in the case of the body muscles to the hypodermis) or by means of a tendon, which is a chitinous ingrowth from the cuticle.

In the accompanying plate is shown the musculature of the three thoracic segments. All the muscles are paired except two small ones that extend transversely between the coxae of the synthorax. Hence those of one side only are shown in the figures. The thorax is divided in its median plane, and the muscles are shown as seen from within at three levels, the uppermost being nearest the body wall, the lowermost next the median plane. It will be seen that the huge muscles that move the wings are far less numerous than are those that move the legs.

CHAPTER V

INTERNAL ANATOMY (continued)

1. The Nervous System

The central nervous system (Pl. VI, fig. 1) consists, as in other insects, of the following principal parts: 1. the brain or the supra-oesophageal ganglion, situated in the head, above the oesophagus, 2. the suboesophageal ganglion, also situated in the head, but below the oesophagus, 3. three thoracic ganglia in the thorax, below the digestive tract, and 4. seven abdominal ganglia in the abdomen, also below the digestive tract.

The bilobed supra-oesophageal ganglion consists of two hemispheres which are closely fused together on the meson. It is situated transversely in the anterior portion of the head above the alimentary canal and between the eyes. Laterally are the optic lobes, which innervate the compound eyes. Anteriorly each hemisphere of the brain sends out a very small nerve which arises from the anterior border and extends antero-laterally to innervate the antenna. The two crura cerebri arise from the antero-ventral border and run parallel, one on each side of the oesophagus, extending postero-ventrally half way around the oesophagus to join the suboesophageal ganglion at its anterior border. Arising from the dorsal surface of the brain there are three, short, club-shaped ocellar nerves which extend directly dorsally to innervate the ocelli.

The suboesophageal ganglion is located below the oesophagus and above the labium in the nymph. It sends out three pairs of nerves: 1. the mandibular nerves, arising from its antero-lateral borders and extending antero-ventro-laterally to the mandibles, 2. the maxillary nerves, arising from its lateral borders and extending latero-ventrally to the maxillae, and 3. the labial nerves, arising from its postero-lateral borders and extending ventrally to the labium. Posteriorly the suboesophageal ganglion is connected with the prothoracic ganglion by a pair of longitudinal nerves, the ventral connectives.

In the subimago and imago these three pairs of nerves are obsolete following the degeneration of the mouthparts.

There are three large ganglia in the thorax, (Pl. VI, figs. 1 and 2) the prothoracic, mesothoracic, and metathoracic ganglia, which are situated on the ventral side of the corresponding segments between the alimentary canal and the ventral muscles. They are each composed of two hemispheres and are connected with each other by the paired ventral connectives. This bilateral condition shows very nicely in the nymph, but not so well in the imago.

Since they are quite similar in shape and arrangement of nerves, the mesothoracic ganglion is herewith described in detail to illustrate the general condition, and the differences in the other two ganglia are then mentioned. In the following description the terms used are adopted from the terminology of Holste ('10).

The mesothoracic ganglion is large, dorso-ventrally depressed, and is connected with the prothoracic ganglion in front and metathoracic ganglion behind by the paired ventral connectives. Two strong nerves arise from the antero-lateral borders and extend antero-latero-dorsally to innervate the muscles of the fore wing. They are branched in both nymph and imago. These nerves are absent in the prothoracic ganglion and are very slender in the metathoracic ganglion, where they innervate the smaller muscles of the reduced hind wing. From its postero-lateral borders two crural nerves extend postero-latero-ventrally into the coxae and thence into the legs. They are branched in the coxae in both subimago and imago, and single in the nymph. On each side of the ganglion between the alar nerve and crural nerve three slender nerves extend laterally to the muscles. Beginning with the anterior one, they are: 1. the anterior coxal nerve, 2. the inferior coxal nerve, and 3. the posterior coxal nerve. The trochanter-extensor nerve is a slender nerve which arises from the posterior border of the ganglion near the base of the crural nerve on each side and extends postero-laterally to the region of the coxa.

There are seven abdominal ganglia (Pl. VI, figs. 1 and 2). The first five are similar in size and shape. Each of them gives off one pair of nerves to innervate the abdominal segments two to six. The sixth or the next to the last abdominal ganglion, which is situated in the seventh segment, is small and closely united with the last one. It also sends out one pair of nerves to innervate the seventh abdominal segment. The seventh or last abdominal ganglion is much larger in size and is situated between segments seven and eight. It sends out two pairs of nerves in the nymph, an anterior slender pair which innervates the eighth segment, and a posterior heavy pair of cercal nerves, which branch in the ninth segment and innervate the ninth segment and the two lateral tails. In the subimago and imago the last abdominal ganglion sends out three pairs of nerves, two anterior slender pairs, innervating segments eight and nine, and a posterior heavy pair of cercal nerves, which branches twice; the anterior branches innervate the tenth segment and the posterior branches innervate the genitalia in the male. The cercal nerves themselves innervate the two cerci. All the abdominal ganglia are connected with each other and also with the thoracic ganglia by the two longitudinal nerve cords, the ventral connectives.

The foregoing description shows that a slight cephalization of the abdominal ganglia has taken place and that the original first abdominal ganglion has fused with the metathoracic ganglion. This is shown by the facts that the metathoracic ganglion sends out one extra pair of nerves posteriorly to innervate the first abdominal segment and that the apparently visible first abdominal ganglion now sends its nerves to the second abdominal segment. It is also to be noted that the last abdominal ganglion is a compound one and is formed by the fusion of the eighth, ninth, and tenth abdominal ganglia.

Histology—The brain (Pl. IX, fig. 1) is completely surrounded by a thin membrane, the neurilemma, and consists of two kinds of cellular elements, the ganglionic cells and the medullary substance. Fundamentally it is composed of three principal divisions: 1. the protocerebrum, or ganglion of the first head segment, which forms the greater portion of the whole brain, 2. the deutocerebrum or ganglion of the second head segment, which gives off nerves to the antennae, 3. the tritocerebrum or ganglion of the third head segment. They are closely fused together to form the single brain.

The protocerebrum is composed of two large lateral lobes, the protocerebral lobes, formed of an immense number of ganglionic cells (the cortex) enclosing a mass of nerve-fibers (the medulla). The cortical cells are most abundant dorsally and are of the usual unipolar form, with large nuclei. The two protocerebral lobes are connected across the middle line by two commissures: 1. the central body and 2. the bridge. The central body is small and forms merely a long, oval, thick, rod-like mass of transverse nerve fibers. The bridge is a thicker commissure, appearing as the fusion of two transverse bodies, and is situated above the central body.

On the sides of the central body and the bridge there is a pair of mushroom bodies which are not so peculiarly and highly developed in this group as in other insects.

The deutocerebrum consists of two small masses of medullary substance situated in front of the protocerebrum in the antero-dorsal portion of each hemisphere. It sends out the nerves to the antennae.

Below the deutocerebrum in the antero-ventral portion of each hemisphere there exist two lobes of the tritocerebrum, composed of medullary granules and fibers.

The optic lobe (Pl. IX, fig. 3) is the most complicated portion of the anterior nerve-mass. This is chiefly due to the enormous development of the compound eyes. The present discussion follows Viallanes ('84) in dividing the optic lobe into six main parts. They are discussed in order, from the compound eyes inwards to the protocerebral lobe.

1. First comes the layer of post-retinal fibers or terminal anastomosis. All the ommatidia of the compound eye rest upon the fenestrate membrane. Each ommatidium sends delicate nerve-fibers through the membrane, and these fibers collect into bundles which pass into a narrow layer of nerve cells to form a plexus. From this plexus the bundles emerge irregularly. They next enter another small layer of nerve cells, forming a second plexus, lying just outside of the ganglionic plate. This whole structure is called the terminal anastomosis.

PLATE IX.

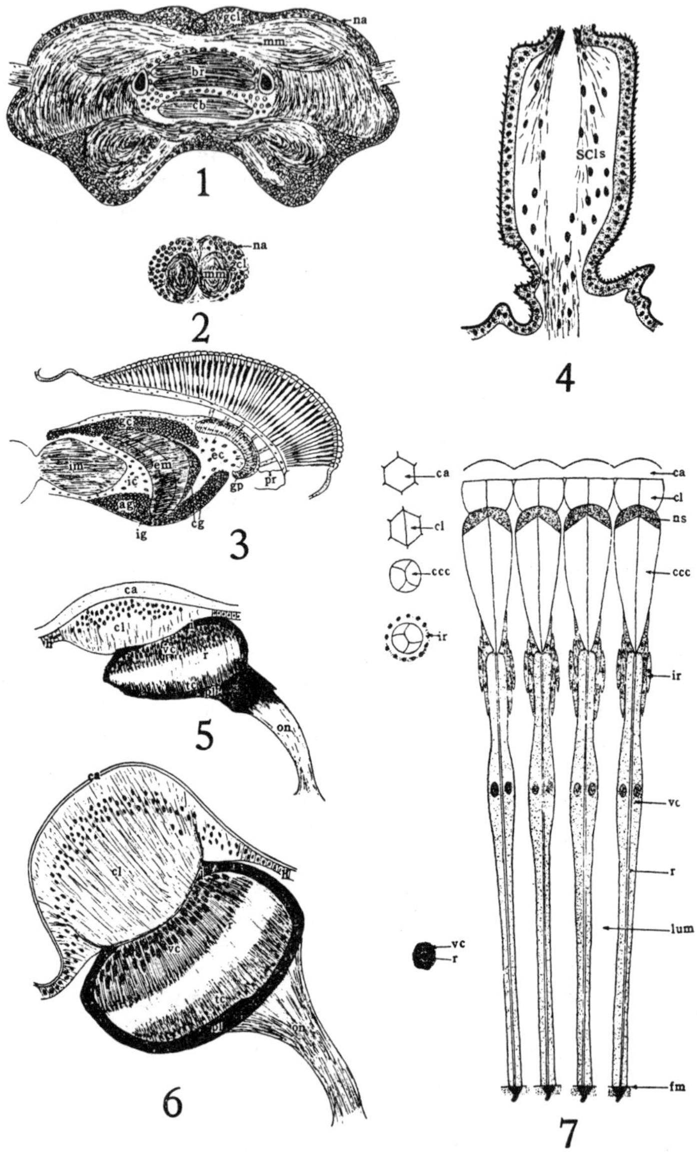

NERVOUS SYSTEM AND SENSE ORGANS

Fig. 1. Supra-oesophageal ganglion, horizontal section through center; Fig. 2. Ventral connective, transverse section; Fig. 3. Optic lobe, horizontal section through center; Fig. 4. Pedicel of antenna, longitudinal section, showing the Johnston's Organ; Fig. 5. Section of nymphal ocellus; Fig. 6. Section of imaginal ocellus; Fig. 7. Ommatidia of compound eye, longitudinal and transverse sections.

Explanation of abbreviations: *br*, bridge; *gcl*, ganglionic cell; *na*, neurilemma; *mm*, medullary mass; *cb*, central body; *ec*, external chiasma; *gp*, ganglionic plate; *pr*, postretinal fibers; *gc*, ganglionic crown; *im*, internal medullary mass; *ag*, anterior ganglionic mass; *ic*, internal chiasma; *ig*, internal ganglionic mass; *cg*, corner ganglion; *em*, external medullary mass; *ca*, cornea; *cl*, corneagen layer; *vc*, visual cells; *r*, rhabdom; *on*, ocellar nerve; *tc*, tapetum cells; *ccc*, crystalline cone cells; *ir*, iris; *ns*, Sempers nuclei; *lum*, lumen; *fm*, fenestrate membrane; *SCls*, sense cells; *pl*, pigment layer.

2. The ganglionic plate is a kind of screen placed across the path of the post-retinal fibers. It is composed of a layer of cylindrical columns of nervous matter with many nuclei on the outer margin and a single row of nuclei near its inner margin.

3. The external chiasma comes next. The nerve-fibers from the eye cross one another very completely after leaving the ganglionic plate, forming a distinct chiasma known as the external chiasma.

4. The external medullary mass is next. It consists chiefly of a mass of granular nerve matter, which receives the nerve-fibers from the external chiasma and passes them on again to the internal chiasma. It is, in section, a crescent-shaped body, with its concavity towards the brain. There are four ganglionic masses attached to the medullary mass, namely: 1. the ganglionic crown, 2. the corner ganglion, 3. the anterior ganglionic mass, and 4. the internal ganglionic mass. Each of these four masses is formed of small unipolar nerve cells with prolongations passing into the main medullary mass.

5. The nerve-fibers cross again completely to form a second chiasma after passing out from the external medullary mass. This is known as the internal chiasma.

6. The internal medullary mass is a large, more or less ovoid mass which receives the nerve-fibers from the internal chiasma.

The Ventral Nerve Cord.—As in the brain, the ventral nerve cord (Pl. IX, fig. 2) is completely covered on the outside by neurilemma. Immediately below this peripheral layer of neurilemma there is a second or inner envelope of cortical cells similar to those of the brain in form and structure. Inside of this second envelope there is in the middle of each hemisphere a round mass of medullary granules and fibers.

2. The Sense Organs

The sense of sight is highly developed in mayflies, as in dragonflies, especially in the males, whose compound eyes occupy the greater part of the head. The three ocelli in the nymph are flat, while in the subimago and imago they are strongly elevated, especially in the male. Each ocellus is innervated by a small ocellar nerve, which arises from the dorsal surface of the brain. The nerves of the lateral ocelli pass dorsally and laterally to the ocelli, while the nerve of the median ocellus passes directly dorsally to the ocellus.

The histology of the various parts of the ocellus passing from without inward is as follows: (Pl. IX, figs. 5 and 6). The cuticula or the corneal layer over the ocellus is the continuation of the cuticle of the body wall. In the nymph the cornea is thick, while in the imago the cuticle is very thin. No sign of a lenticular structure is seen in the cuticle.

Immediately inward from the corneal layer there is a layer of elongated and loosely arranged cells, known as the corneagen cells, which are continuous with the hypodermis. This is called by Seiler ('06) the lens. The corneagen cells are colorless transparent cells and form the vitreous body which supplants the lens in function. There is a layer of small nuclei near its outer end. In the nymph this layer is not very thick, while in the subimago and imago it is very thick, especially in the males, less so in the females.

Inside this layer is a layer of retinal or visual cells which are elongated with oval nuclei at their outer end. Near its center there is a row of longitudinal optic rods, known as the rhabdom, arranged transversely. Seiler called this layer the corneagen layer.

Inside the visual layer there is another layer of spindle-shaped cells with very small nuclei. They are the tapetum cells which are said to serve as a reflecting layer. They are called by Seiler the sense cells. The visual layer and the tapetum are not very thick in the nymph, while they are very thick in the subimago and imago.

The pigment layer is a layer of accessory cells loaded with pigment and surrounding the visual and tapetum layers. Between the corneagen cells and the visual cells the pigment layer extends a little inward to form the iris.

The pigment layer lies against the inside wall of the ocellar structure, and between them pass the nerve fibers of the visual cells. The fibers then pass directly into the ocellar nerve.

The compound eyes are the two most prominent organs on the head. They are situated on its sides. In the nymph they are only a little convex and not divided, while in the subimago and imago the eyes are greatly bulged out, especially in the males, where the two eyes almost touch each other dorsally. In the males the eyes are distinctly divided transversely into the upper large-faceted portion and the lower small-faceted portion, while in the female they are not so well divided.

Each compound eye (Pl. IX, fig. 7) is composed of hundreds of units, each known as an ommatidium. These are all similar in structure, so a detailed description of one ommatidium only will be needed.

A longitudinal section of the compound eye shows the various parts which enter into the composition of an ommatidium, passing from without inwards, to be as follows:

The cornea is the outer transparent hexagonal area of the cuticle, forming the facet or lens of an ommatidium. It is convex on the outside. It is cast off during each nymphal molt. In the nymph the cornea is very thick, about twice as thick as in the subimago and imago.

The part of the hypodermis which extends beneath the cornea is known as the corneagen layer. It is composed of two large cells in the subimago and imago, and two very small cells in the nymph. The cornea is secreted by these cells.

Beneath the corneagen cells there is a beautiful structure, the crystalline cone. It is formed of four elongated transparent cells which are highly refractive and vitreous in appearance. Together they form a cone with the apex inwards and the broad outer end convexly rounded. Their nuclei are situated close to their outer ends and are known as Semper's nuclei.

Surrounding the inner, tapering end of the crystalline cone cells and the distal end of the retinal cells are arranged numerous slender, spindle-shaped pigment cells. These are the iris cells.

The retinula forms the basal portion of an ommatidium and is composed of a group of seven slender pigmented visual cells. These visual cells collectively give rise to an internal optic rod, known as a rhabdom. They contain pigment granules throughout their entire length.

The bases of these visual cells rest upon the fenestrate membrane, which forms the inner boundary of the eye proper.

The Johnston's Organ. The Johnston's organ (Pl. IX, fig. 4) is located in the second segment of the antenna, the segment commonly known as the pedicel. Hitherto known to occur in Odonata, Orthoptera, Hemiptera, Anoplura, Neuroptera, Coleoptera, Trichoptera, Lepidoptera, Diptera, and Hymenoptera (Snodgrass '26), it has recently been reported by one of us (Hsu, 1933) for the Ephemeroptera. It is a simplified chordotonal organ. Some writers say that it is an auditory organ, but there is little evidence to support this view. According to Snodgrass it may be perceptive of slight air motions, but whether of waves of sound or not is still undetermined.

It consists essentially of a ring of large bipolar nerve cells, whose fibers run back and are continuous with the fibers of the antennary nerve. Their distal extremities are attached to the conjunctiva between the second and third joinings of the antenna.

The Gustatory Organ (Pl. III, fig. 3).—On the under side of the triangular labrum in the nymph there is a row of sensillae, which are probably organs of taste.

3. The Reproductive System

The most remarkable thing about the reproductive organs of mayflies is their primitive nature. There are no accessory glands, and the genital ducts retain their paired condition in both sexes, each duct opening to the exterior separately.

In the adult male the two testes (Pl. X, fig. 7) are elongate and subcylindrical whitish organs. They are situated dorsally at the sides of the alimentary canal. From the posterior end of each testis a very slender duct, the vas deferens, extends posteriorly for a short distance and suddenly enlarges into a large sac, the seminal vesicle. The latter narrows to form the ejaculatory duct, which divides into two parts: an anterior, wider, and coiled part, and a posterior, slender, non-coiled portion. The latter opens out through the male genital opening on the posterior end of the penis of the same side.

PLATE X.

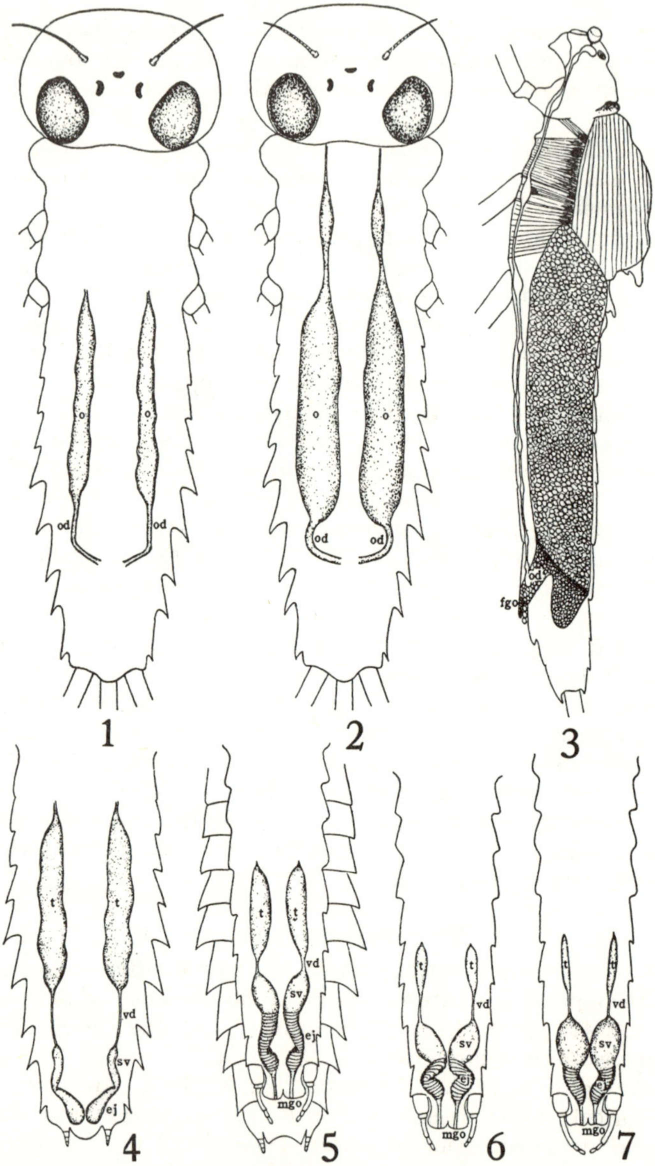

REPRODUCTIVE SYSTEM

Fig. 1. Dorsal view of female early nymphal reproductive system; Fig. 2. Dorsal view of female late nymphal reproductive system; Fig. 3. Lateral view of female subimaginal and imaginal reproductive systems; Fig. 4. Lateral view of male nymphal reproductive system; Fig. 5. Dorsal view of male reproductive system between nymphal and subimaginal stages; Fig. 6. Dorsal view of male subimaginal reproductive system; Fig. 7. Dorsal view of male imaginal reproductive system.

Abbreviations used are as follows: *o*, ovary; *od*, oviduct; *fgo*, female genital opening; *vd*, vas deferens; *sv*, seminal vesicle; *ej*, ejaculatory duct; *t*, testis; *mgo*, male genital opening.

In the full grown male nymph (Pl. X, fig. 4) the two testes extend from the anterior border of the second abdominal segment to the anterior end of the seventh. Histological sections show that they enclose large masses of mature spermatozoa. A long, slender vas deferens leads from the posterior end of the testis to the middle of the eighth abdominal segment, where it enlarges into a somewhat wider tube, the seminal vesicle. In the middle of the ninth segment the seminal vesicle twists inward and extends postero-mesally to form a flattened and expanded ejaculatory duct which has not yet differentiated into two parts. It is not coiled and is closed at the posterior end.

In the transition from nymphal to subimaginal stages (Pl. X, fig. 5) important changes occur. The testes shrink and are reduced in size, due to the flow of the spermatozoa into the seminal vesicles. The vasa deferentia are greatly reduced in length. The seminal vesicles swell up to form an oval sac containing the spermatozoa. The ejaculatory ducts are differentiated into two parts, and the posterior, slender, and uncoiled part extends as a tube to rearward.

During the subimaginal stage (Pl. X, fig. 6) the testes are reduced considerably in size, the seminal vesicles swell still more with the inflow of the spermatozoa, and the coiled part of the ejaculatory duct becomes shorter and wider.

In the female the two ovaries (Pl. X, fig. 3) are very large, very elongate, subcylindrical organs. They are full of eggs and occupy nearly the whole body cavity in the subimago and imago. Each ovary leads into a fairly wide muscular tube at its postero-latero-ventral border, the oviduct, which extends downward to open through the female genital opening above the egg valve.

In a young female nymph (Pl. X, fig. 1) the two ovaries are elongate, slender, and subcylindrical, and extend from the anterior end of the first abdominal segment to the anterior portion of the sixth abdominal segment, lying dorsally at the sides of the alimentary canal. The slender oviducts extend from the posterior ends of the ovaries downward to the middle of the posterior margin of the seventh abdominal segment.

As the nymph approaches maturity, the ovaries (Pl. X, fig. 2) increase tremendously in size by reason of the growth of eggs that they contain. They extend forward quite through the thorax and rearward into the seventh abdominal segment.

In the later, winged stages there is further increase in the ovaries until they occupy practically the entire body cavity of thorax and abdomen, and the oviducts become much shortened and dilated.

Histology of the Reproductive Organs. The vasa deferentia (Pl. XI, fig. 6) of the male are enclosed in a thick layer of circular muscles and are lined with an epithelial layer. The epithelial cells are more or less glandular in structure, with large nuclei but with no trace of cell walls.

Each seminal vesicle (Pl. XI, fig. 4) is a large thin-walled sac. The

PLATE XI.

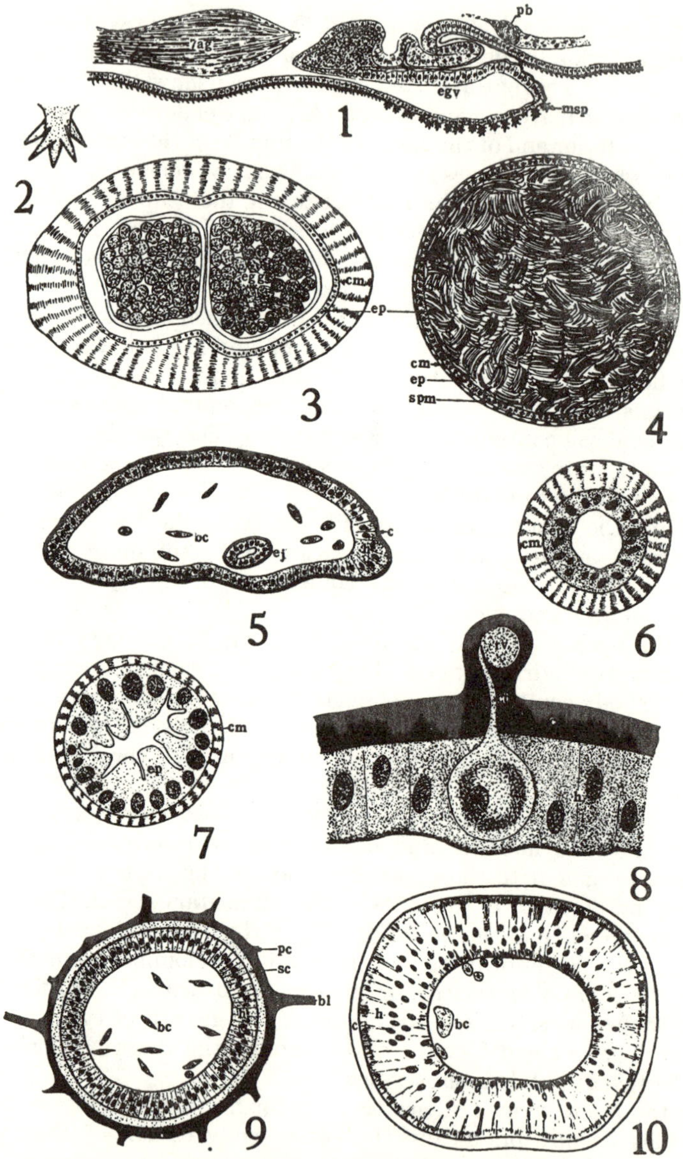

HISTOLOGY OF REPRODUCTIVE SYSTEM

Fig. 1. Longitudinal section of ventral half of abdomen at seventh and eighth segment, showing the folds of body wall to form the female genital opening, the mating spines, and the Palmen body in abdomen; Fig. 2. A mating spine, greatly enlarged; Fig. 3. Oviduct, transverse section; Fig. 4. Seminal vesicle, transverse section, showing bundles of spermatozoa; Fig. 5. Penial lobe, transverse section, showing the lumen with blood cells, and the ejaculatory duct; Fig. 6. Vas deferens, transverse section; Fig. 7. Ejaculatory duct, transverse section; Fig. 8. Section of a mating gland; Fig. 9. Nymphal cercus or tail, transverse section; Fig. 10. Imaginal cercus, transverse section.

Explanation of abbreviations: *pb*, Palmen body; *7ag*, seventh abdominal segment; *egv*, egg valve; *msp*, mating spine; *ep*, epithelium; *cm*, circular muscles; *spm*, spermatozoa; *c*, cuticle; *pc*, primary cuticula; *sc*, secondary cuticula; *bl*, bristle; *h*, hypodermis; *bc*, blood corpuscle; *ej*, ejaculatory duct.

wall is composed of an outer very thin layer of circular muscles and an inner very thin layer of epithelium with no trace of cell walls, and with very small nuclei. Large bundles of spermatozoa are massed inside.

The ejaculatory ducts (Pl. XI, fig. 7) have a circular muscle coat on the outside. The inner epithelium is thick and more or less glandular, with very large nuclei. It is thrown into deep irregular folds.

In transverse section, the penes (Pl. XI, fig. 5) have a bean-shaped outline. Outside there is hypodermis and cuticle, as elsewhere in the body wall. The lumen is full of blood corpuscles, and among them lies the slender uncoiled portion of the ejaculatory duct.

The oviducts (Pl. XI, fig. 3) are large tubes with a thick circular muscle on the outside and a very thin layer of epithelial cells lining the inside.

The paired female genital openings (Pl. XI, figs. 1 and 2) lie in the ventral body wall between the seventh and eighth abdominal segments among several folds of the smooth cuticle. Below them extends the protective egg valve, whose lower surface is covered with mating spines of very peculiar compound conical form. These evidently are used to hold the male genitalia to their place at the time of mating.

The histology of the tails we may mention here, although in this order they are not accessory organs of reproduction. In the tails (cerci) of the nymph (Pl. XI, fig. 9) the thick cuticular layer on the outside is beset with large bristles and more numerous small spinules. The inner secondary cuticle is soft. The inner hypodermal layer is a little thicker than the cuticular layer. It is composed of columnar epithelial cells with oval nuclei and distinct cell wall. The lumen of the cerci contains many, mostly spindle-shaped, blood corpuscles.

The histological structure of the tails of the imago (Pl. XI, fig. 10) is very different. The cuticular layer is very thin, unarmed, smooth, and undifferentiated. The inner hypodermal layer is very thick, about five times as thick as the cuticular layer. The hypodermal cells are entirely modified. They have no definite cell walls. Their deeply-staining protoplasm is concentrated at their ends. Their middle portion is composed of a network of entangled protoplasmic strands or threads with oval nuclei suspended on the network. The lumen contains blood corpuscles mostly of ameboid form.

The inner surface of the male forceps is covered with minute unicellular mating glands (Pl. XI, fig. 8) that are distributed from the base of the main segment to the tip of the distal segment. Externally they appear as very minute club-shaped protuberances that are barely visible to the unaided eye. The forceps is lined on the outside by a thick layer of cuticle. A layer of columnar hypodermal cells with large oval nuclei is situated inside the cuticle. The cell walls are not well defined.

A mating gland is a large unicellular globular gland embedded in the hypodermal layer, and is derived from the hypodermal cells. The nucleus of the gland is eccentric. The cytoplasm forms a dense ring around the center. The globular cell elongates into a very slender capillary tube, penetrating through the cuticle. At the tip of the tube, it swells into a round reservoir, where the mating fluid is temporarily stored. The cuticle of the wall of the forceps projects out and encloses the reservoir, forming a club-shaped protuberance, which is visible on the inner surface of the male forceps. At the time of mating the thin cuticle surrounding the reservoir probably bursts open, thus liberating the adhesive fluid.

CHAPTER VI

POSTEMBRYONIC DEVELOPMENT

In this chapter we will endeavor to trace the changes of external form that are undergone by Heptagenine mayflies after hatching, with special reference to tracheal gills. Representations of three genera will be discussed and illustrated.

1. *Stenonema interpunctatum*

First instar nymphs of this species were obtained by hatching the eggs in the laboratory. The newly hatched nymphs are very small, and they are difficult to find in their natural surroundings. They are hardly visible to the naked eye. They cannot be collected in the field in nets used for the large nymphs. We used a wide-mouth glass bottle with a fairly deep tin cap. The stones along the creeks were picked up, and the underside of each stone was washed several times by using the tin cap to pour water over the surface into the mouth of the bottle. All the insects, large and small, were thus washed into the bottle. The materials thus collected in the bottle were brought into the laboratory and poured into a white enameled dish. After the sediment settled for a few minutes, the different stages of the tiny nymphs were sorted out under the binocular microscope. Many specimens were collected, and a series of eleven successive stages was obtained. The series was complete up to the full development of the gills.

The newly hatched nymph (Pl. XII, fig. 1) measures .5 mm. in length,* not including the antennae and tails. The head is almost pentagonal in shape, except that its anterior angle is more or less rounded off. From the anterior border of the head between the antennae arises a small rounded projection, which is called by Lestage the "clypeo-cephalic prolongment." The head is almost as wide as the prothorax and a little wider than the meso- or the metathorax. The dorsal surface of the head bears five ocelli: an anterior pair, a posterior pair, and one unpaired median ocellus in front. The anterior three are ocelli proper, and the posterior two will develop into the compound eyes. These latter are now each composed of only a single ommatidium. Their size is the same as that of the true ocelli, and they are situated at the postero-lateral angles of the head.

Each antenna is composed of five segments. The proximal segment is the shortest. The second is about twice as long as the first, and both increase in thickness towards the apex. The third segment is a little shorter than the second and a little longer than the first, and of uniform thickness throughout. It bears a long bristle on its inner margin at the

*For further measurements of parts see table at the end.

PLATE XII.

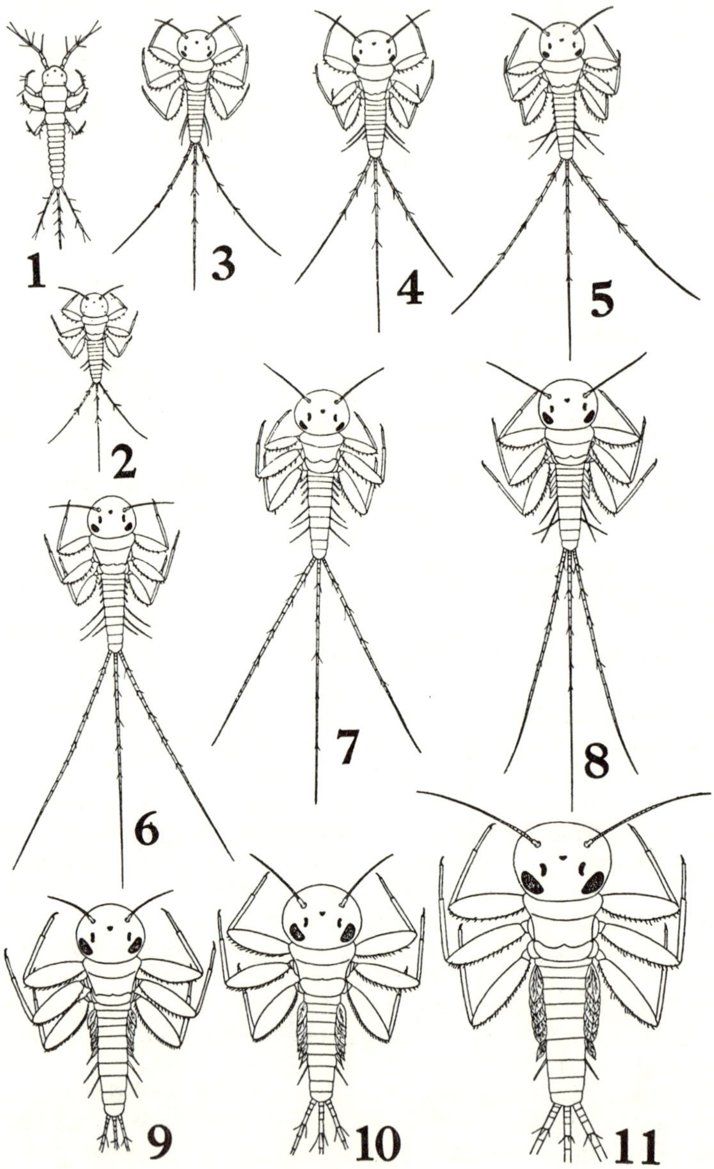

DEVELOPMENT OF GILLS IN

Stenonema interpunctatum. Figs. 1 to 11, successive stages; Fig. 1 somewhat more enlarged than others

distal end. The fourth and the fifth segments are approximately equal in length and each about two and a half times as long as the first one. The fourth segment bears three bristles at its distal end, and the fifth segment carries two long bristles at its apex.

Each thoracic segment is more or less quadrangular, with rounded lateral margins. It is wider than long. The three pairs of legs are all similar. They have not yet become flattened. Each of the fore and middle femora bears a bristle, and the hind femur bears two bristles. All tibiae bear two bristles each.

The abdomen is composed of ten distinct segments. All the segments are wider than long, except the last one, which is longer than wide, with its posterior margin rounded. Gills are entirely wanting in this stage.

The last abdominal segment bears three tails. The lateral tails are each composed of five segments. The first segment is the shortest; the second, third and fourth segments are equal in length; and the fifth segment is the longest. The second, third, and fourth segments bear a bristle each, and the fifth segment terminates in two bristles at its apex. The median tail is composed of six segments. The first and the sixth segments are equal in length and shorter than the others, and do not bear any bristles. The remaining four segments are equal in length. Each of them bears one or two bristles.

Second instar nymphs (Pl. XII, fig. 2) are much more like the later nymphs in appearance than like those of the first instar. The head is more rounded in outline. The five ocelli are as before. The two posterior ones forming the compound eyes are still composed each of one ommatidium. The antennae have increased in length and in the number of segments, and have lost their bristles. The legs are now flattened, especially the femora, which are strongly developed and armed with four to seven spines along the hind margin. They are very much like the legs of later nymphs, except that there are fewer spines and no fringes on the femora and tibiae.

The most important change has taken place in the abdomen, and this is the beginning of the development of gills. Two pairs arise as simple filaments from the postero-lateral angles of segments five and six, each with a single tracheal tube inside. The fifth pair measures .192 mm. and the sixth pair .184 mm. in length.

The three tails have grown in length to about .75 mm. The number of their component segments has increased, and each tail bears two whorls of bristles.

In the third instar (Pl. XII, fig. 3) the five ocelli have differentiated into two compound eyes and three true ocelli. Beginning with this stage the compound eyes increase rapidly in size and in the number of ommatidia. Each eye of this stage is composed of more than ten ommatidia. The number of spines on the femora is now seven to nine. The tails are longer.

The fifth and the sixth pairs of thread-like gills, developed in the previous stage, have increased in length to .273 and .262 mm. respectively, and another pair has appeared as little thread-like stubs arising from the postero-lateral angles of the seventh abdominal segment. They measure .026 mm. in length.

In the fourth instar (Pl. XII, fig. 4) the most important changes are in the growth of the compound eyes and in the length of the three pairs of thread-like gills on segments five, six, and seven. These now measure .299 mm., .309 mm., and .055 mm. respectively.

The fifth instar (Pl. XII, fig. 5) marks the first appearance of all seven pairs of gills and the maximum increase in relative length of the fifth and the sixth pairs. The first four pairs are developed as short bud-like out-pocketings of body wall. They do not contain tracheal tubes. They measure about .019 mm. in length. The fifth and sixth and seventh pairs measure .381 mm., 390 mm., and .179 mm. respectively.

In the sixth instar (Pl. XII, fig. 6) there occurs a reduction in relative length of the fifth and the sixth pairs of gills. The fifth, sixth, and seventh pairs measure .351 mm., .378 mm., and .214 mm. respectively, while the first four pairs of bud-like gills increase to about .045 mm.

In the seventh instar (Pl. XII, fig. 7) the eyes have become about twice as large as the ocelli; the first four pairs of gills have increased in length to about .130 mm. The fifth, sixth, and seventh pairs of gills now measure .348 mm., .364 mm., and .247 mm. respectively.

In the eighth instar (Pl. XII, fig. 8) the first four pairs of gills change from bud-like into plate-like organs. They are not yet penetrated by tracheal branches. Each is a small oval structure. The length of the fifth, sixth, and seventh pairs of thread-like gills is .325 mm., .338 mm., and .279 mm. respectively.

In the ninth instar (Pl. XII, fig. 9) the last three pairs of thread-like gills begin to develop into plate-like organs. Those of the fifth segment are shortened and widened in their basal half, while the distal half is still thread-like in form. The sixth and the seventh pairs remain thread-like in form. The three measure .169 mm., .195 mm., and .208 mm. respectively. The first four pairs of gills have become entirely flat and leaf-like with tracheal ramifications inside. They are pointed at the tip.

The tenth instar (Pl. XII, fig. 10) marks the completion of the modification of the fifth pair of gills from thread-like filaments to leaf-like plates, similar to those of the first four segments, and the beginning of the change in the sixth pair. The seventh pair still remains filamentous. The last three pairs of gills now measure about .117 mm., .156 mm., .208 mm., respectively.

The eleventh instar (Pl. XII, fig. 11) marks the final stage of the development of gills to plate-like form. The gills of the sixth pair become leaf-like plates similar to the first five pairs, except for smaller size. The seventh or the last pair (.208 mm. in length) always remains filamentous in this species.

The remaining nymphal instars show only increase in size of the various parts of the body.

The study of the eleven early instars of the nymph *of Stenonema interpunctatum* shows two very interesting features of development. In the first place, the gills are developed in two entirely different ways. The first four pairs appear later, but are completed earlier than the others and are developed from buds to plates. The last three pairs appear earlier, but are completed later and are developed from filaments to plates. In the second place, compound eyes and ocelli begin with the same size and structure. They look like five ocelli. As development goes on, the compound eyes grow much more rapidly than the ocelli and they increase the number of their ommatidia, while those of the ocelli remain single.

DEVELOPMENT IN STENONEMA INTERPUNCTATUM

Table of measurements for eleven instars

Instars	Body length	Head length	Head width	Antenna length	Tail length	No. of spines on femora
1st	0.5	.082	.091	.182	0.2	1–2
2nd	0.8	.184	.234	.273	0.75	4–7
3rd	1.0	.227	.299	.390	1.5	7–9
4th	1.2	.266	.348	.455	1.6	7–9
5th	1.25	.275	.358	.494	2.0	8–11
6th	1.4	.292	.377	.533	2.0	9–12
7th	1.5	.331	.416	.637	2.4	10–12
8th	1.6	.361	.461	.689	2.6	11–16
9th	1.95	.435	.539	.754	3.0	13–23
10th	2.1	.481	.624	.806	4.0	14–26
11th	2.6	.559	.721	.975	5.0	19–27

All measurements are in millimeters.

2. *Heptagenia hebe*

The tiny nymphs of this species were collected as before from the streams at Ithaca, New York, during the latter part of August and in September. Ten instars were found. The fertilized eggs were not secured; so the first and probably also the second instar nymphs are missing in this series.

The first figure (Pl. XIII, fig. 1) is probably that of a third instar nymph. It is very small.* The head is more or less rounded in outline. The compound eyes are only about twice the size of the ocelli, which appear as three small dark spots on the dorsal side of the head. The legs are flattened, especially the femora which are strongly developed and armed with five to nine spines along the hind margin.

*For measurements see table on page 54.

PLATE XIII.

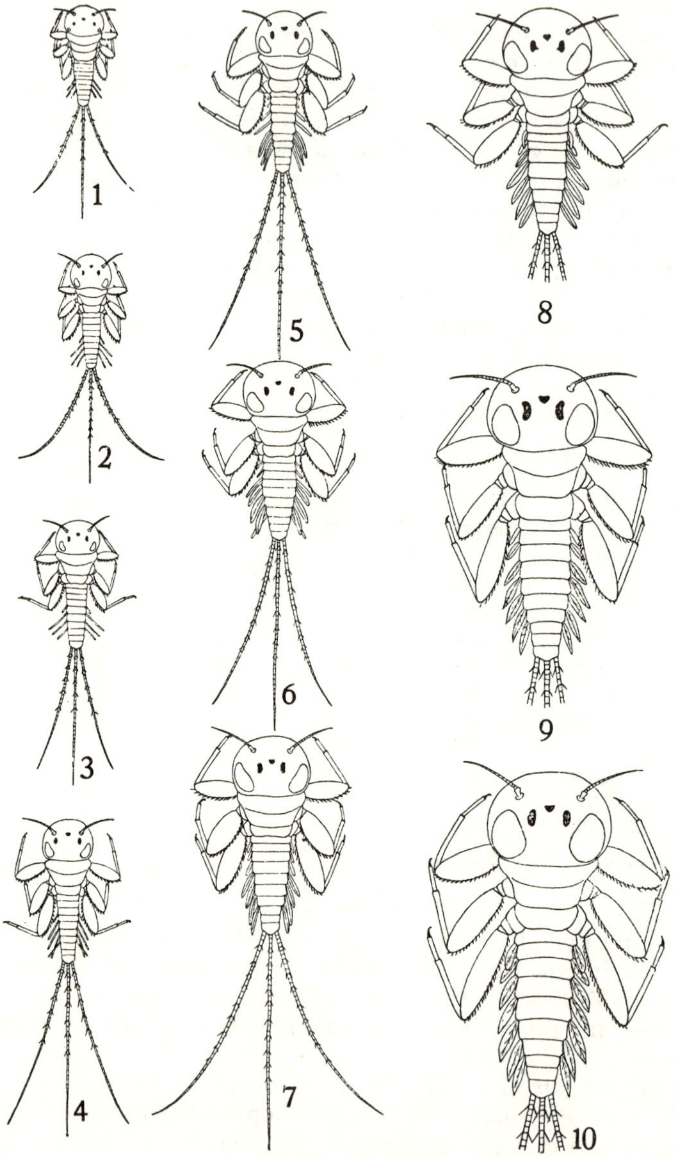

STAGES IN THE DEVELOPMENT OF GILLS

Heptagenia hebe. Figs. 1 to 10, successive stages.

Three pairs of thread-like gills arise from the postero-lateral angles of segments five, six, and seven, a pair on each segment. No traces of gills are found on the first four abdominal segments. The three tails are composed of many segments, and each tail has three whorls of bristles.

In the next instar (Pl. XIII, fig. 2) the compound eyes have increased in size and in the number of ommatidia, and are now about three times the size of the ocelli. Antennae and tails are longer, and the number of spines on the femora has increased in number, ranging from six to eleven. The fifth, sixth, and seventh pairs of gills, which were developed during the previous stage, have increased their length, and another small pair has developed on the fourth segment. These are also thread-like, resembling the posterior pairs.

In the next instar (Pl. XIII, fig. 3) there are no changes of form, but only the increases in size shown in the table at the end. In the next instar (Pl. XIII, fig. 4) these increases in size continue, and another pair of thread-like gills appears on the third abdominal segment. The next instar (Pl. XIII, fig. 5) marks the completion of the set of seven pairs of gills. Two minute filamentous pairs arise on the first and second abdominal segments. The fifth and the sixth pairs of gills have expanded to plate-like or leaf-like form with ramifying tracheal branches, while the third, fourth, and seventh pairs remain thread-like.

In the next instar (Pl. XIII, fig. 6) the first and second pairs of gills increase in length, and the fourth pair also becomes expanded into plate-like form. The others remain thread-like.

In the next instar (Pl. XIII, fig. 7) the seventh pair of gills has also expanded into plate-like form, while the first three pairs remain thread-like.

In the next three instars (Pl. XIII, figs. 8, 9, and 10) growth steadily increases, and the three remaining pairs of thread-like gills become plate-like, one pair changing form at each instar proceeding forward from the rear, the third, the second, and the first pairs successively. All seven pairs of gills are now plate-like or leaf-like with tracheal ramifications, and are well developed.

The mode of the development of gills in *Heptagenia hebe* is distinctly simpler than in *Stenonema interpunctatum*. The gills of the latter first appear in two forms, some as buds and some as filaments, while those of the former all begin alike as filaments and later expand to plate-like form. The posterior gills form first, and the development gradually extends toward the anterior end. The seventh or last pair of gills is plate-like, similar to the others, instead of remaining thread-like as in *Stenonnema interpunctatum.*

3. *Iron humeralis*

The materials for this series were collected under the stones along the Fall Creek, Ithaca, New York, during the later part of September. From their study six distinct instars were recognized, with probably the

DEVELOPMENT IN HEPTAGENIA HEBE

Table of Measurements

In-stars	Body length	Head length	Head width	Antenna length	Length of gills 1st	2nd	3rd	4th	5th	6th	7th	Tail length	Spines on femora
3	.8	.204	.264	.143					.117	.130	.078	1	5–9
4	1	.234	.286	.156				.067	.156	.178	.128	1.4	6–11
5	1.2	.247	.312	.234				.067	.182	.182	.134	1.5	9–12
6	1.3	.305	.377	.247			.067	.136	.199	.208	.156	1.6	10–15
7	1.4	.338	.416	.264	.036	.034	.1	.162	.208	.208	.182	1.7	13–18
8	1.5	.377	.514	.299	.065	.054	.130	.195	.243	.243	.195	1.8	16–20
9	1.7	.416	.554	.390	.097	.104	.160	.195	.238	.247	.201	2	20–22
10	2	.455	.637	.410	.143	.119	.182	.247	.260	.273	.240	2.2	22–26
11	2.5	.559	.780	.468	.182	.169	.221	.240	.286	.286	.260	2.7	30–24
12	2.7	.611	.806	.494	.182	.169	.234	.292	.312	.312	.260	2.8	31–39

DEVELOPMENT IN IRON HUMERALIS

In-stars	Body length	Head length	Head width	Antenna length	Length of gills 1st	2nd	3rd	4th	5th	6th	7th	Tail length Median	Lateral	Spines on femora
3	.75	.188	.267	.130					.143	.143		.071	.546	7–9
4	.9	.227	.329	.169					.143	.143		.052	.8	11–16
5	1.1	.266	.397	.234					.156	.156		.049	1	14–20
6	1.4	.299	.407	.312					.169	.169		.045	1.4	17–24
7	1.8	.403	.533	.325	.007	.026	.039	.041	.208	.186	.048	.019	1.7	24–34
8	2.2	.435	.617	.360	.013	.052	.072	.084	.221	.260	.091	.013	2.6	30–40

All measurements are in millimeters.

first and second missing in this series. The fertilized eggs were not secured.

The first figure (Pl. XIV, fig. 1) is probably that of a third instar nymph. It is very small (see table). The head is more or less rounded in outline. The compound eyes are about twice the size of the ocelli. The legs are flattened, especially the femora, which are strongly developed.

The development of gills has already begun. Two pairs of filaments arise from the postero-lateral corners of segments five and six. No traces of gills are found on the other abdominal segments. Both pairs measure .143 mm. in length.

The last abdominal segment bears three tails during the early nymphal stages. The middle tail is short and degenerates as the nymph grows. It is now composed of five to six segments. The lateral tails are long and are composed of many segments. They are not armed with whorls of bristles.

The next three figures (Pl. XIV, figs. 2, 3, and 4) probably represent instars four, five, and six. They show the eyes enlarging, the antennae and lateral tails steadily increasing in length while the middle tail is dwindling, the spines on the hind margin of the femora increasing in number. Numerical changes are shown in the table.

The next figure (Pl. XIV, fig. 5) is probably of the seventh instar. The most important change is the simultaneous appearance of the first four pairs and the seventh or last pair of gills. These are very short thread-like gills. The seven pairs measure in succession .007 mm., .026 mm., .039 mm., .041 mm., .208 mm., .186 mm., and .048 mm. respectively.

The last figure of this series (Pl. XIV, fig. 6) perhaps represents the eighth instar. The gill condition remains the same as that of the preceding instar, except that every pair has increased in length. The later instars of nymphs of this species were not secured.

From the foregoing it is evident that in this species, as in *Heptagenia hebe*, all the gills first appear in thread-like form and gradually expand into plates with tracheal ramifications. The fifth and the sixth pairs come out first, and the other five pairs come out later and simultaneously. Gros ('23) found that in *Ecdyonurus forcipula* the gills appeared on the fifth and sixth segments in the second stage and on the fourth and seventh in the fourth stage. It was not until the sixth stage that six pairs of gills were present. Lubbock ('66) found in *Cloeon dimidiatum* no traces of gills until the third stage, when they were present on segments two to six.

In *Iron humeralis* the middle tail is present and well developed during the first few instars of the nymph, but it is much shorter than the lateral tails. It gradually degenerates, to become completely invisible during the later nymphal instars. We have seen that lateral and median tails are equally well developed in the first stage in *S. interpunctatum*,

PLATE XIV.

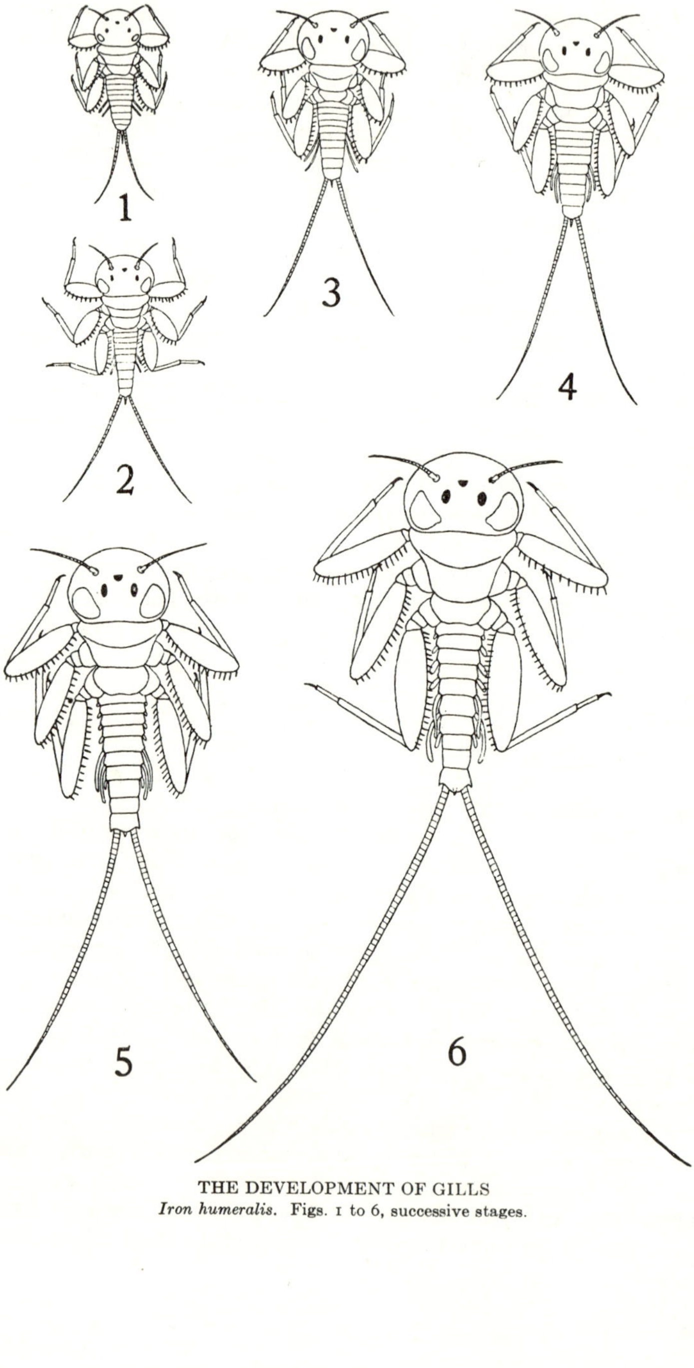

THE DEVELOPMENT OF GILLS

Iron humeralis. Figs. 1 to 6, successive stages.

and this agrees with conditions found in *Hexagenia bilineata* by Wiebe ('26). Lubbock ('66) stated that the lateral tails of *Cloeon dimidiatum* were each composed of nineteen segments in the first stage. The median tail appeared only as a minute knob in the second stage and did not show segmentation until the sixth stage. Vayssiere ('82) found that the median tail is much shorter than the lateral ones in *Heptagenia longicauda.* Gros ('23) found just the opposite relation in *Ecdyonurus forcipula.* Murphy ('22) found that the median filament does not appear until the fifth stage in *Baetis vagans* (*posticatus*).

CHAPTER VII

THE EGGS AND EGG-LAYING HABITS OF NORTH AMERICAN MAYFLIES

By Osgood R. Smith, Ph.D.

This chapter is made up of three parts. The first is a record of observations on the ovipositing habits of a few common species. The second contains descriptions of the eggs of most of the known genera of North American mayflies, and a key to the eggs described. The third is an account of the development of the ovaries and eggs.

I. THE EGG-LAYING HABITS

Mayflies have at least three distinct methods of laying eggs, and there are a few minor differences within the three main groups. Some species alight on the water many times and let the current wash off a few eggs at a time; some lay their eggs all at once by dropping them into the water in a clump; and some crawl down under water to lay their eggs on a stone. Eaton in 1888 (*Rev. Monogr.* p. 11) mentions these three types of oviposition.

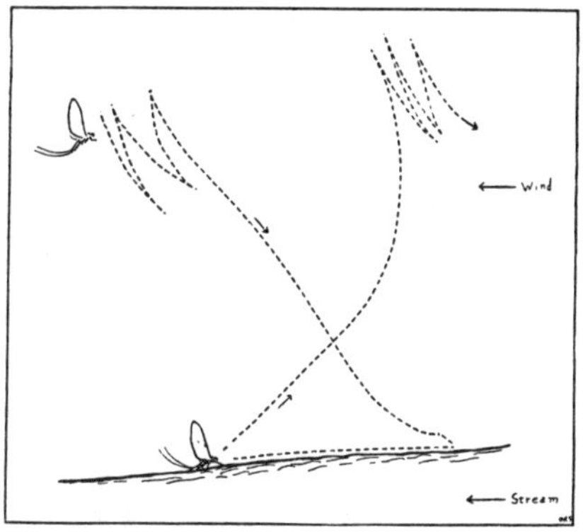

FIG. 4—Diagram of ovipositing flight of *Heptagenia hebe.*

In the first class belongs *Heptagenia hebe* which scatters its eggs a few at a time. Oviposition occurs in the evening, just after sundown. The females, sometimes singly, but usually in small swarms, dance up and down over the riffles of a stream, from two to ten feet above the water. This performance is much like the well-known mating flight of males. Each female, flying against the gentle evening breeze, which generally flows down a valley, flies down and forward a foot or so, and then is carried back again as she flies up, thus staying in nearly the same place. During this dance the abdomen is bent down by the weight of its load of eggs.

After one individual dances up and down in this way many times, she may be seen to continue one of her downward flights till she lands on the running water, where she stays long enough to be carried downstream about a foot. She then flies up again to dance awhile, before making another flight down to the water.

PLATE XV.

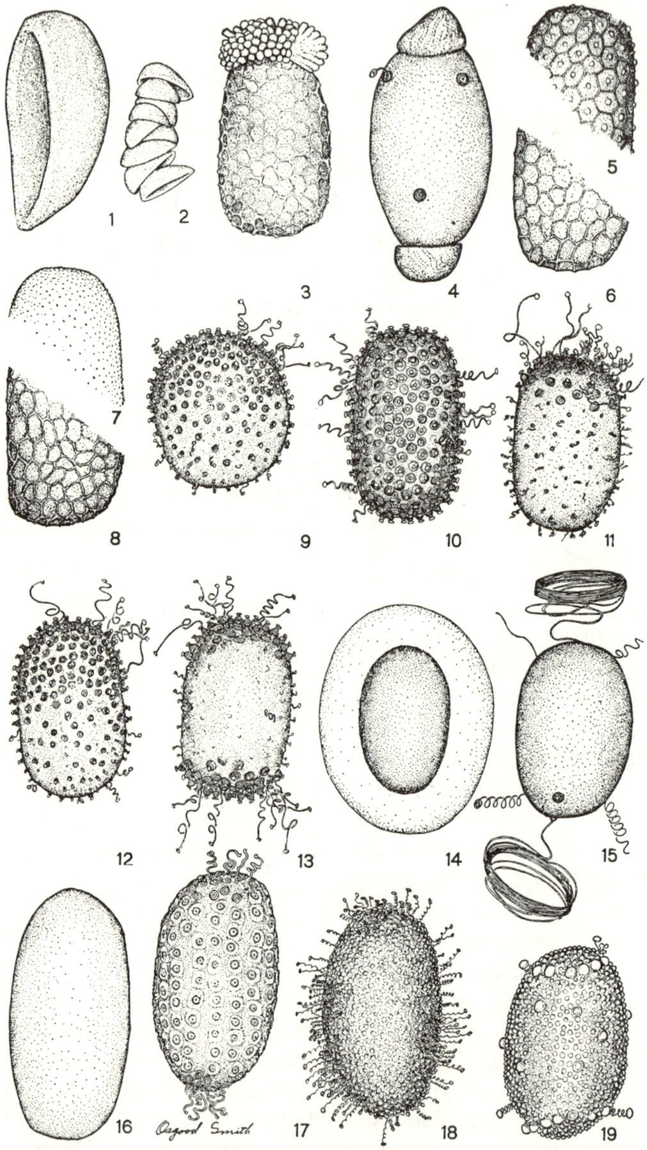

MAYFLY EGGS

Fig. 1. Egg of *Campsurus sp.* (in alcohol); Fig. 2. Several eggs of *Campsurus sp.*, showing how they are packed together in the oviduct (in alcohol); Fig. 3. Egg of *Ephoron album*. Part of the polar cap is shown in cross-section (fresh, in water); Fig. 4. Egg of *Potamanthus rufus* (in alcohol); Fig. 5. One end of *Hexagenia bilineata* egg (in alcohol); Fig. 6. One end of *Hexagenia atrocaudata* egg (in alcohol); Fig. 7. One end of *Hexagenia recurvata* egg (in alcohol); Fig. 8. One end of *Hexagenia elegans* egg (in alcohol); Fig. 9. Egg of *Heptagenia marginalis* (in alcohol); Fig. 10. Egg of *Heptagenia hebe* (in alcohol); Fig. 11. Egg of *Heptagenia flavescens* (in alcohol); Fig. 12. Egg of *Heptagenia aphrodite* (in alcohol); Fig. 13. Egg of *Heptagenia pulla* (in alcohol); Fig. 14. Egg of *Stenonema ithaca*, with the thick gelatinous coat (fresh, in water); Fig. 15. Egg of *Stenonema interpunctatum* (in alcohol); Fig. 16. Egg of *Ironodes nitidus* (fresh, in water); Fig. 17. Egg of *Pseudiron sp.* (near *centralis*) (in alcohol); Fig. 18. Egg of *Rhithrogena doddsi* (in alcohol); Fig. 19. Egg of *Siphloplecton signatum* (in alcohol).

(The figures are not drawn to scale).

If one of these females is interrupted in the midst of her dance and placed gently on the surface of the water in a glass, the small whitish eggs may be seen dropping to the bottom—a few at a time; if she cannot escape, the process will continue for about 20 minutes, or until she is spent. Contact with the water seems to be necessary to start the flow of eggs, and if the female is taken while in an ovipositing dance the contact with the water always results in a few eggs being laid. It therefore seems obvious that a few eggs are laid each time the female alights on the water.

The eggs are covered with many fine strings coiled like springs, each string with a terminal disc which will adhere to pebbles or even to glass. (See Pl. XV, fig. 10). The eggs are probably carried along by the current until they attach to some stone. This is an admirable adaptation to rapid water conditions because the method of oviposition insures a wide dispersal of the eggs, and the eggs themselves stand a very good chance of sticking to some stone where they will be aerated and kept clean by the running water. Their adhesive quality also prevents the eggs from being carried down to some deep pool where the nymphs would hatch in unfavorable conditions.

The ovipositing flight of *Iron humeralis* is like that of *Heptagenia hebe.* The adult generally emerges earlier in the season. It lays its eggs in or just above a riffle, and in the same manner. I have watched one individual land on the water to oviposit and then resume its aerial dance seven times. Continuous observation is very difficult, and I have never been able to follow one any further.

When holding a female by the wings, I have seen a few of the pink eggs ooze out of the genital opening, but these will not fall off until they come in contact with the water. The female, therefore, evidently washes off a few eggs each time she alights on the stream. The pinkish, ellipsoid eggs are smooth, without either strings or an adhesive gelatinous coat. They will not adhere to stones, and it is difficult to see how such eggs hatch safely in running water. A few of them probably manage to lodge in some sheltered spot in the stream bed.

Mayflies belonging to the genus Ephemera apparently also scatter their eggs, laying a few at a time, but this occurs after dark. In the slower parts of Fall Creek at Ithaca, N. Y., females of *Ephemera varia* were frequently seen to start dancing over the water just as it was getting dark, long after sundown. They would dance up and down a few feet over the water, and occasionally land on it. Females caught at this time were always spent; yet they had not been ovipositing early in the same evening. It seems likely that they had oviposited on the preceding night and were vainly trying to lay a few more eggs, because females full of eggs were frequently found at lights an hour or more after dark, and these females, if placed on the water in a glass, always laid all their eggs in a steady stream. The eggs settled to the bottom in little bunches held together by a stringy gelatinous material (Pl. XVIII, fig. 50).

The lake-inhabiting species, *Ephemera simulans*, seems to lay its eggs in much the same way. Just as it is getting dark, females may often be seen flying irregularly over the water, from four to six feet above its surface. There usually is no regular dance, such as the males have, but females often settle on the water a moment and then fly up again. While they are on the water, the abdomen is held flat on the surface, but the wings are held up. Subimagos also may alight on the water, but they always hold the abdomen up above its surface. Imagos which were caught at the time were full of eggs, but they did not lay them in a glass of water. It therefore seems likely that oviposition occurs later in the evening, after dark.

Three species of Ephemerella, *E. rotunda*, *E. dorothea*, and *E. needhami* were observed by me. They all lay their eggs in riffles, but their method of meeting these conditions is very different from that of *Heptagenia hebe*. Instead of having an ovipositing dance, allowing wide dispersal of the eggs, they lay their eggs all at once.

When ready to oviposit, the female carries the eggs in a compact spherical mass attached to the underside of the abdomen, just back of the genital opening and the abdomen is bent over the top of the mass of eggs—the ventral side being concave to receive and retain it. Both *E. rotunda* and *E. dorothea* have a yellowish mass of eggs, but in *E. needhami* the egg-mass is black.

Carrying this load of eggs, the female will fly back and forth over a riffle many times, seldom going either above or below the rapid part. Occasionally she will dip down near the water, as if to look over a landing place, but she usually flies strongly back and forth in a straight line about five or six feet over the water. I have never actually seen an Ephemerella drop its load of eggs, but a species of Stenonema which has similar habits, alights on the water, drops its load of eggs, and then, having completed its life work, flies away. It seems probable that Ephemerella frees its eggs in the same way. If a female is captured carrying her load of eggs, the abdomen may be seen to be empty; hence the eggs are obviously laid all at once.

The mass of eggs will usually fall off in the net when the female is captured, and it will not stick to dry cloth, but if it is placed in a vial of water, remarkable changes take place almost at once. The spherical mass sinks to the bottom and adheres to the glass or a pebble, at the same time swelling a little and becoming fuzzy in appearance. If the mass is shaken slightly, as it would be in a riffle, each part that touches anything will stick to it until finally the mass spreads out and becomes very irregular in shape. Under a lens it may be seen that each egg is sticking to the pebble or to its neighbor by its adhesive terminal cap, and in addition the whole mass is bound together by a tangle of the long lateral strings on each egg, so that the whole mass looks very much like yeast (Pl. XVIII, fig. 49). In this way the eggs are able to adhere to a stone or a pebble in the rapidly running water and to be aerated until they hatch.

I have found similar masses of shells of eggs which had already hatched in the stream bottom still firmly attached to a pebble, showing that the eggs are well able to withstand the current long enough for the nymphs to hatch safely. In escaping from the shell, the nymph does not disturb the polar cap or the arrangement of the mass of eggs, but the shell of each egg splits neatly longitudinally on one side, or a little more than halfway around.

The genus Siphlonurus is much like Ephemerella in its ovipositing habits. A large number of females of *S. quebecensis* were observed ovipositing in Cayuta Creek, N. Y., on June 16, 1933. They flew back and forth over a riffle, just after sundown, carrying a spherical mass of eggs under the abdomen in the way just described. The egg mass itself, however, is very different. The green eggs have no caps or chorionic strings, but they are all imbedded in a mass of green gelatinous material. This green egg mass falls off the female when she is placed in a glass of water, and sinks to the bottom, where it flattens out and sticks to pebbles or to the glass. If a female containing eggs but not carrying a mass of them under her abdomen be placed in water, no eggs are laid. Evidently they must first be extruded into a ball, and this seems to take place during flight (Pl. XVIII, fig. 51).

Females of the genus Isonychia also lay their eggs all at once, in a spherical mass, but oviposition seems to occur after dark. At any rate, females carrying dark brown spherical masses of eggs under the abdomen were often found at lights about ten o'clock in the evening. When these females were placed on water, the eggs would fall off and sink to the bottom, not in one mass like Ephemerella or Siphlonurus, but falling apart as they descended. Each spherical egg is thickly covered with adhesive nodules, so that it sticks to the bottom of the glass. If the water in the glass is washed about, the eggs do not come loose, but the nodules stretch out into strings, as shown in Pl. XVII, fig. 42.

The female of *Ephoron album* also lays her eggs all at once. I have not seen the act of oviposition, but Needham (1918) gives a figure of a female in flight with two packets of eggs extruding. Oviposition probably occurs at night; at any rate, a female carrying two white sausage-shaped packets flew in a window in Ithaca, N. Y., on the evening of July 28, 1933. When these packets were placed in water, the eggs settled singly to the bottom and stuck to the glass by their white caps. They did not stick to each other. These eggs looked like the preserved specimen shown in Plate XV, fig. 3, except that the caps had swelled a little.

The genus Stenonema is divisible into two groups, and both the eggs and the ovipositing habits of each group are so distinct that they must be described separately.

1. *The pulchellum-ithaca group.* Three members of this group, *S. ithaca*, *S. fuscum*, and an undetermined species, have an ovipositing flight much like that of *Heptagenia hebe*. They oviposit in the same sort

of place and at the same time of day. The ovipositing dance, however, usually seems to be more irregular, and once I saw *S. fuscum* alight on a quiet pool below a riffle. When placed on water in a glass, this individual continued to lay eggs, so she must have been ovipositing, but this species usually oviposits by repeatedly alighting in a riffle for a moment, washing off a few eggs at a time.

Each egg, after being laid in water, is surrounded by a thick gelatinous coat by which it will adhere to pebbles or even to glass. It is therefore well adapted to hold fast to any stone in a riffle (Pl. XV, fig. 14).

Another member of this group is *Stenonema tripunctatum.* It is typically a lake species, whereas the others mentioned are stream species; thus it is not surprising to find that the ovipositing habits are a little different. Like their stream-dwelling relatives, the nymphs of this species prefer stony bottoms, and the adults were always found skimming along the shores when bottoms are stony. On several occasions in June 1933 both males and females were seen skimming rapidly about in a sheltered cove on Skaneateles Lake, New York, one or two feet above the water. There was no up and down dance, but there were occasional pauses and then sudden starts, somewhat after the manner of a dragonfly. Once in a while a female would slow up a bit and dip her abdomen in the water without actually alighting, and sometimes she would alight on the water for just a moment. Probably the females were ovipositing in both cases, although the few females caught did not lay eggs when placed in a glass of water. Unfortunately it was difficult to make good observations because of the rapid flight and the fact that my insect net was wielded from a canoe. The eggs were plain ellipsoid, with gelatinous coat. Males and females were seen clasped together, but the females did not oviposit as soon as they left the males; therefore some time must elapse between fertilization and oviposition.

2. *The frontale-interpunctatum group.* Members of this group have ovipositing habits much like those of Ephemerella. I have observed the act of ovipositing by a mayfly in this group only once, on the west branch of the Sacandaga River, New York, on July 5, 1932, but I had a clear view of it and caught the individual for identification. Since all the eggs known from species in this group are alike, I think it is safe to expect that the methods of ovipositing are alike also.

This individual, when I first saw her, was flying straight upstream, about five feet above the water, and was barely making headway against the evening breeze. On the underside of the abdomen she was carrying a fairly large mass of eggs, just as Ephemerella does. The eggs were of an orange or reddish color. Finally she settled down to the water and was carried downstream about a foot. She then flew up again, but without the eggs. This individual was captured. She was spent, and the tip of the abdomen was still curved down, where it had fitted over the egg mass.

Unfortunately no other females carrying egg masses were captured; so I could not find out what happens to the mass when it is laid in water, but the eggs are equipped with two very long coiled threads, which probably adhere to the bottom and to other eggs.

Mayflies belonging to this group are seldom seen in the early evening, but may often be found later at lights. It therefore seems likely that oviposition usually occurs after dark.

Baetis. The oviposition by Baetis has been described by Eaton, 1888; Morgan, 1911; and by Murphy, 1912. According to them, the female lands on a partially submerged stone in a rapidly flowing stream and then folds her wings and crawls down into the water. Having once gotten through the surface film, she wanders over the stone looking for a suitable place, and when she finds it, she presses the openings of her oviducts against the stone. The abdomen is then swung from side to side, leaving a row of whitish eggs stuck to the stone, and at the same time the female crawls slowly forward so that the eggs lie in a compact mass made up of consecutive rows. When all the eggs have been laid, she may crawl out and fly away, or she may be carried down stream to drown.

II. THE EGGS OF NORTH AMERICAN MAYFLIES

It has been known for a long time that mayflies have very peculiar eggs, some having polar caps or various kinds of chorionic strings. Bengtsson (1913) in Europe, and Morgan (1913), in America, were the first to attempt systematic descriptions of the eggs of various species. Miss Morgan described and figured the eggs of thirteen American species. Except for her account, very little has been known about them.

The type of egg is quite constant for a species, and frequently for subgeneric groups and genera, but there is no correlation with families. There also is surprisingly little relation between the types of eggs and the places where they are laid. Eggs both with and without special anchoring structures are laid in rapid streams and in quiet ponds.

A KEY TO THE KNOWN EGGS OF NORTH AMERICAN MAYFLIES

1—Ovoviviparous mayflies. . . . . *Callibaetis sp.**
—Oviparous mayflies. 2
2—Eggs with polar caps, or having at each pole coiled strings which are several times as long as the egg. . . . . 3.
—Eggs not as above. . . . . 9.
3—Polar caps not made up of coiled strings. . . . . 4.
—With long coiled strings at each pole, which when compacted together may look like a solid cap. . . . . 8.
4—Two caps, one at each pole. . . . . *Potamanthus rufus*
—A cap at one pole. . . . . 5.

*Of McLean, New York (also *C. vivipara* of S. America and *Cloeon dipterum* of Europe).

5—With chorionic filaments or strings .... 6.
—Without chorionic strings .... 7.
6—Strings with terminal knobs ....
.... *Ephemerella*, except *E. maculata* and the *temporalis* group
—Strings without terminal knobs .... *Tricorythodes allectus*
7—Longitudinal ridges on chorion .... *Brachycercus nitidus*
—Reticular pattern on chorion .... *Ephoron album*
8—One or two long strings wound around each pole many times, like a coil of rope .... *Stenonema*, of the *interpunctatum* group
—Many long strings at each pole. In preserved material they are compacted together and look like a solid cap .... *Caenis*
9—Chorion smooth .... 10.
—Chorion with appendages or sculpturing .... 13.
10—Eggs spherical .... 11.
—Eggs not spherical .... 12.
11—Diameter about .12 mm., pond species .... *Callibaetis*
—Diameter about .18 mm., stream species .... *Baetisca*
12—Ellipsoid eggs .... See list No. 1.
—Concavo-convex eggs, like a shallow bowl .... *Campsurus segnis*
13—Chorion with reticular sculpturing .... 14.
—Chorion without reticular sculpturing .... 18.
14—Relatively large eggs, .26 to .36 mm. long .... 15.
—Smaller, .17 to .20 mm. long .... 16.
15—Eggs flattened ellipsoid .... *Ameletus*
—Eggs truely ellipsoid .... *Hexagenia*, except *H. recurvata*
16—Mesh of reticular pattern large, 5 or 6 polygonal areas across width of egg .... *Ephemerella temporalis* group
—Mesh of pattern much smaller .... 17.
17—With finger-like translucent projections .... *Leptophlebia gracilis* and *johnsoni*
—Without such projections .... *Centroptilum conturbatum*
18—Chorion covered with shallow pits, egg nearly spherical .. *Siphlonisca aerodromia*
—Not as above .... 19.
19—Chorion, in preserved material, with scattered crater-like elevations, each with a corona of radiating ridges .... *Choroterpes basalis*
—Chorion with small nodules, strings, or adhesive projections .... 20.
20—Chorion partly or wholly covered with strings, each of which has a terminal knob and is coiled up like a bed spring when in alcohol but uncoils in water .... *Heptagenia*
—With chorionic strings or nodules, but not as above .... 21.
21—Either spherical or disc shaped eggs .... *Isonychia*
—Ellipsoid eggs .... 22.
22—With longitudinal rows of low bosses, each with a knob at its summit ....
.... *Cloeon rubropictum*
—Not as above .... 23.
23—With several short thick threads at each pole, remainder of chorion with raised circles arranged in longitudinal rows .... *Pseudiron centralis*
—Not as above .... See list No. II

OTHER GENERA

The eggs of the genera listed below could not be separated by a key. The reader should refer to the descriptions and illustrations.

I. Ellipsoid eggs, without polar caps, chorionic appendages, or sculpturing.

Pentagenia	*Habrophlebiodes*
Ephemera	*Siphlonurus*[2]
Oreianthus	*Baetis*[3]
Stenonema (*ithaca* group)[1]	*Cloeon*
Iron (not adhesive)	*Centroptilum*

[1] With thick gelatinous adhesive coat when fresh in water.
[2] Green, many eggs laid in a gelatinous mass.
[3] Attached to stones under water in flat patches.

II. Ellipsoid eggs, covered with small nodules or bosses when in alcohol. When laid naturally in water some of the eggs are covered with short strings or translucent adhesive projections.

Hexagenia recurvata	*Blasturus*[2]
Rhithrogena	*Siphloplecton*
Cinygma	*Pseudocloeon*
Cinygmula	*Neocloeon*
Paraleptophlebia[1]	

[1] With very fine translucent hair-like projections when fresh in water.
[2] With short finger-like translucent projections when fresh in water.

The measurements given with the following descriptions are in millimeters and indicate length and width.

EPHEMERIDAE

1. *Campsurus.* The eggs of *C. segnis,* a neotropical species, and of *C. incertus,* from Macon, Georgia, look like hollow, rubber, ovoid balls that have been pushed in on one side. The longest diameter is usually .48 (Pl. XV, fig. 1).

Although no fresh specimens were seen, I feel safe in saying that this is the normal shape of the egg because in the several females examined the eggs were all of the same shape, and they fitted very neatly against each other like a pile of caps. (Pl. XV, fig. 2). If this condition had been due to shrinking there would have been surplus space in the abdomen and probably some other evidence of distortion.

2. *Ephoron.* The eggs of *E. album* are broadly ovoid, with polar caps, and about .28 long. The chorion is sculptured with a large meshed reticular pattern of ridges. The white polar cap is made up of a mass of conical structures radiating from a common center, probably the micropyle (Pl. XV, fig. 3). When these eggs are laid in a glass of water by a live mayfly they settle to the bottom, where they become attached by the adhesive conical structures of the cap.

3. *Potamanthus.* The eggs of *P. rufus* are elongate ellipsoid, about .22 by .10, including a cap at each pole. The chorion is smooth, but has scattered over it a few, comparatively thick, coiled threads with terminal

knobs (Pl. XV, fig. 4). In alcohol the eggs are yellowish with white polar caps.

A Potamanthuss egg was described and excellently figured by Grenacher (1886) (as an Ephemera egg). According to him, the two caps swell up when placed in water. Then it may be seen that each cap is made up of a solid basal portion and an outer portion consisting of a compact bundle of fine rod-like structures, which radiate slightly from the basal part. He also says that each coiled "anchor line" is made up of four to six strands.

Probably both the caps and the "anchor lines" serve as adhesive structures when the eggs are laid in water.

4. *Pentagenia.* The eggs of *P. vittigera* from preserved material are slightly flattened ellipsoids with blunt ends, about .22 by .13 by .09. No chorionic structures or markings could be seen.

5. *Ephemera.* The eggs of *E. simulans*, *E. varia*, and *E. blanda* are all plain ellipsoid, about .22 by .12, and they have no chorionic patterns or appendages. When they are laid by the female in a glass of water the eggs scatter over the bottom but they are held together in small bunches by a more or less stringy gelatinous material (Pl. XVIII, fig. 50).

6. *Hexagenia.* The species of this genus all have relatively large ellipsoid eggs, and of those examined all but *H. recurvata* have reticular patterns of ridges on the chorion.

H. recurvata eggs (.28 by .15) have many small nodules rather sparsely scattered over the surface (Pl. XV, fig. 7).

H. bilineata eggs (.27 by .15) are covered with a reticular pattern of ridges. The "strands" of the net are straight. At the center of each of the polygonal areas of the pattern is a small boss-like nodule which refracts light when the microscope is focused up and down. These nodules are not dissolved by KOH (Pl. XV, fig. 5).

H. atrocaudata (.36 by .19), *H. affiliata* (.34 by .18), *H. weewa* (.31 by .18), *H. mingo* (.28 by .15), *H. occulta* (.28 by .15), and *H. viridescens* (.28 by .15) all have eggs with simple reticular patterns, and the strands of the nets are straight (Pl. XV, fig. 6). The eggs of *H. atrocaudata* are not only larger than those of any other Hexagenia examined, but the strands of the net are wider.

The eggs of *H. elegans*, *H. venusta*, and *H. rigida*, (all about .28 by .15) have reticular patterns in which the strands of the net are sinuous (Pl. XV, fig. 8).

The eggs of *H. orlando* (.30 by .17), and *H. carolina* (.28 by .15) have a reticular pattern in which the strands may be either straight or sinuous. Different eggs from one individual may have either straight or sinuous strands, or both types may blend into each other on one egg. The chorionic patterns of these three species seem to show that the nature of the strands of the net is not always constant for a given species, but F. Neave (1931), working with an abundance of material, found that *H.*

occulta and *H. rigida* could be identified by the patterns on the eggs, so this character must be constant for some species.

7. *Oreianthus.* The eggs of *O. purpureus* are ellipsoid, about .24 by .15, and they have no chorionic appendages or sculpturing.

HEPTAGENIIDAE

8. *Heptagenia.* The eggs of the mayflies belonging to this genus all have chorionic appendages consisting of coiled strings, each with an adhesive terminal knob. In material preserved in alcohol these strings are coiled up in the form of a cone which stands out prominently from the surface of the egg like a coiled bed-spring. The terminal knobs are on the tip of each conical "spring."

Some of the strings will uncoil in preserved material, but when freshly laid in water they all uncoil and the terminal knobs adhere to anything they touch, thus holding the egg firmly in place even on glass.

The spaces on the chorion between the coils are covered with small nodules, set closely together and too small to be shown in the illustrations.

Although the eggs of this genus cannot all be identified to species there are a few fairly well marked differences within the genus.

In *Heptagenia hebe*, *H. rusticalis*, and *H. lucidipennis* the eggs are ellipsoid about .17 by .08, and the prominent coiled "springs" are thickly set over the entire surface of the chorion (Pl. XV, fig. 10).

In *H. aphrodite*, *H. maculipennis*, and *H. elegantula* the eggs are ellipsoid, from .12 to .16 mm. long, and about .07 mm. across. The "springs" appear a little smaller in proportion to the egg than in *H. hebe*, and they are more thickly set around one pole. The other half of the egg has smaller "springs" set farther apart (Pl. XV, fig. 12).

The eggs of *H. marginalis* are like the above three, except that they are spherical instead of ellipsoid. They are about .12 mm. in diameter (Pl. XV, fig. 9).

The eggs of *H. pulla* are ellipsoid, about .16 by .09. The "springs" are large like those on *H. hebe* eggs, but they are limited to two patches —one at each pole. The middle of the egg has a few very small "springs" scattered over it (Pl. XV, fig. 13).

The eggs of *H. flavescens* are similarly shaped but smaller, being about .12 mm. long. The "springs" are large like those on *H. hebe* eggs, but they are limited to an area at one pole (Pl. XV, fig. 11).

9. *Stenonema.* There are two very different types of eggs in this genus.

The eggs of the *interpunctatum* group are ellipsoid, about .20 by .15, and they have a single long thread attached at each pole. In preserved material this thread is neatly coiled like a rope around each pole. These threads may be pulled out with needles to a length of over two inches, and when so pulled out they appear to arise from a small protuberance on the chorion. In addition to the two long threads there are a few small threads which, in preserved material, are usually tightly coiled up against the chorion (Pl. XV, fig. 15). Miss A. H. Morgan (1913) figures the egg of a species allied to *S. interpunctatum* showing the long threads, and she says that they uncoil when placed in water.

The eggs of *S. vicarium, S. ithaca, S. fuscum,* and *S. tripunctatum* have no chorionic "threads" or patterns of any kind. In preserved material the eggs are plain ellipsoid about .18 to .20 by .14. When these eggs are laid by the living mayfly in a glass of water it may be seen that each is surrounded by a gelatinous adhesive coat, which quickly swells up to twice the size of the egg (Pl. XV, fig. 14).

10. *Rhithrogena.* The eggs of *R. doddsi* and *R. brunnea* are ellipsoid, about .18 by .10, and they are thickly covered with many small coiled threads, each with a small translucent terminal knob (Pl. XV, fig. 18). In preserved material these threads do not always uncoil, in which case they appear like many small boss-like protuberances covering the surface of the egg. In *R. brunnea* the threads seem to be slightly smaller at one end of the egg.

11. *Cinygmula.* The eggs of *C. mimus* and *C. atlantica* from preserved material are ellipsoid, about .16 by .10, and the chorion is covered with small threads, like the eggs of *Rhithrogena doddsi* (Pl. XV, fig. 18).

12. *Cinygma.* The eggs of *C. integrum* are ellipsoid, about .14 by .10. In preserved material there are no chorionic strings, but the chorion is thickly covered with very small bumps, giving it a granulated appearance, like shagreen. Possibly these bumps are adhesive when laid naturally in water.

13. *Iron.* The mayflies of this genus all have plain ellipsoid eggs, but the eggs differ in size and shape. The eggs of *I. longimanus* are about 16 by .09; *I. fraudator* .15 by .09, and *I. fragilis*, .14 by .07.

These eggs have no chorionic sculpturing or appendages. When laid in water by the live female they are pinkish, have no gelatinous coat, and do not adhere to pebbles or to the bottom of the glass.

13a. *Ironodes.* Eggs of *I. nitidus* are similar to those of Iron, being about .20 by .10 mm. in size (Pl. XV, fig. 16).

BAETIDAE

METRETOPINAE

14. *Siphloplecton.* The eggs of *S. signatum* and *S. basale* are ovoid, about .22 by .14. In preserved material they are covered with many small boss-like protuberances. In addition there are a few much larger protuberances scattered over the surface and others arranged in a ring around each pole (Pl. XV, fig. 19). These larger protuberances sometimes comes off and then it may be seen that they are knobs on the ends of coiled "strings." Probably both the large and the smaller knobs spring out on the end of strings when they are laid in water.

The eggs of *S. speciosum* are like the above two except that there are fewer of the large knobs, and these are all at one pole (Pl. XVI, fig. 20).

14a. *Pseudiron.* The eggs of *P. centralis* are ellipsoid, about .30 by .19. Eggs from preserved material have a number of short thick chroionic strings grouped around each pole. Before it is pulled out, each string forms a short compact spiral, appearing more like a gastropod shell with three whorls than like the "bed spring" coils on Heptagenia eggs. The remainder of the surface of the egg is covered with rows of peculiar, raised, nearly square areas, on each of which is a circular ridge with a small knob in its center. (Pl. XV, fig. 17).

BAETISCINAE

15. *Baetisca.* The eggs of *B. obesa* and *B. carolina* are nearly spherical, about .18 mm in diameter, and the chorion is smooth without sculpturing or appendages (Pl. XVI, fig. 21). In preserved material the eggs have flattened surfaces on them, apparently caused by the compact way in which they are crowded in the abdomen of the adult mayfly.

LEPTOPHLEBIINAE

16. *Paraleptophlebia.* The species *P. moerens*, *P. debilis*, *P. adoptiva*, *P. mollis*, and *P. pallipes* all have ellipsoid eggs, .18–20 by .10. In preserved material the chorion is thickly covered with many small bosses, and by focusing carefully, a few small finger-like projections may be seen (Pl. XVI, fig. 22). When these eggs are laid in water each of the bosses must spring out in some way to form a narrow, finger-like, adhesive projection, because Morgan (1913, p. 399) states that the eggs when freshly laid in water are thickly covered with short "hairs," and look like ciliated protozoans. She also says that the eggs are adhesive.

17. *Blasturus.* The eggs of *B. austrinus*, *B. grandis*, *B. nebulosus*, and *B. cupidus* are ellipsoid, about .18 by .09. In preserved material the chorion is thickly covered with minute bosses, and in addition there are a few large translucent projections scattered over the surface in groups of twos and threes. These are more or less finger-like but they frequently are pressed down against the shell (Pl. XVI, fig. 23). *B. cupidus* eggs usually have a few more of these projections than do the eggs of the other three species. When the live eggs of *B. cupidus* are placed in water these projections all stand out at right angles to the surface of the egg and adhere even to glass. These adhesive structures are evenly scattered over the entire ellipsoid egg. Morgan (1913) says that the eggs of *B. cupidus* are slightly flattened and "about the longitudinal circumference is a shelf-like extension which bears a large number of strap-shaped pegs"; but this appearance seems to me to be merely the effect of using too high a magnification.

PLATE XVI.

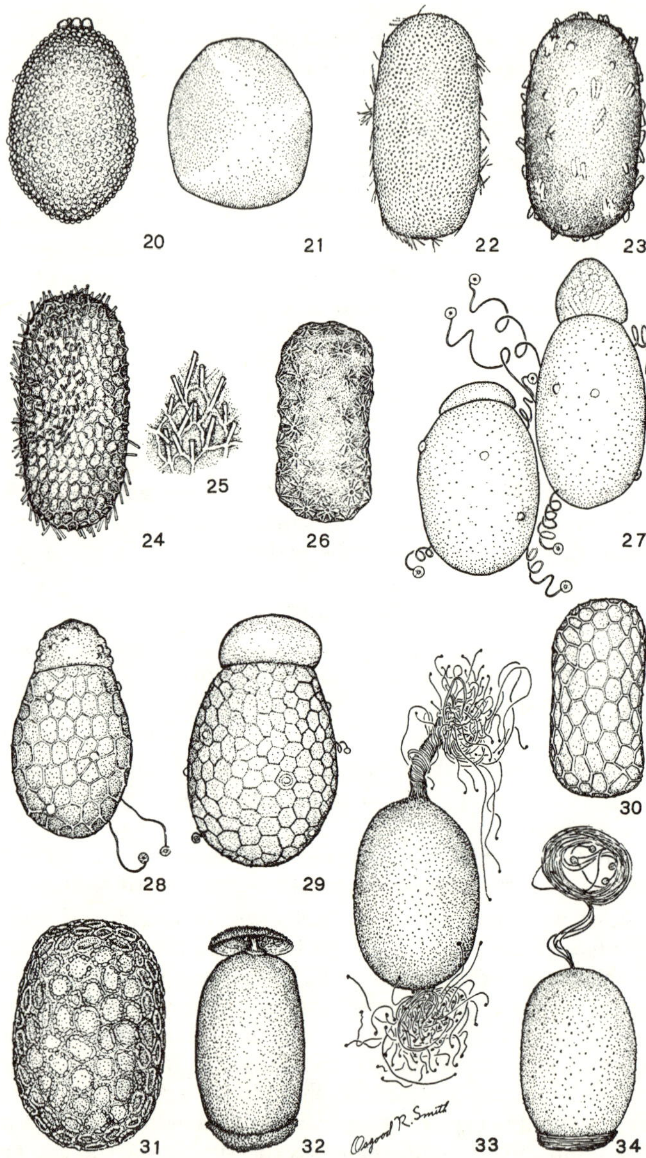

MAYFLY EGGS

Fig. 20. Egg of *Siphloplecton speciosum* (in alcohol); Fig. 21. Egg of *Baetisca obesa* (in alcohol); Fig. 22. Egg of *Paraleptophlebia moerens* (in alcohol); Fig. 23. Egg of *Blasturus austrinus* (in alcohol); Fig. 24. Egg of *Leptophlebia gracilis* (in alcohol); Fig. 25. An enlarged portion of the same (in alcohol); Fig. 26. Egg of *Choroterpes basalis* (in alcohol); Fig. 27. Two eggs of *Ephemerella rotunda*, showing variation in shape (in alcohol); Fig. 28. Egg of *Ephemerella septentrionalis* (in alcohol). Fig. 29. Egg of *Ephemerella hecuba* (in alcohol); Fig. 30. Egg of *Ephemerella temporalis* (in alcohol); Fig. 31. Egg of *Ephemerella maculata* (in alcohol); Fig. 32. Egg of *Caenis hilaris* (in alcohol); Fig. 33. Egg of *Caenis sp.* (laid naturally in a pond); Fig. 34. Egg of *Caenis perpusilla* (in alcohol).

18. *Leptophlebia. L. gracilis* and *L. johnsoni* have elongate eggs that are just alike. They are ellipsoid, about .20 by .10. The chorion is sculptured with a reticular pattern of sinuous ridges, and translucent projections stand up like fence-posts wherever these ridges branch (Pl. XVI, figs. 24, 25).

19. *Choroterpes.* The eggs of *C. basalis* are about .18 by .09. They are not ovoid but are more like a cylinder with rounded ends and usually slightly dumb-bell shaped. In preserved material the chorion has shallow depressions scattered over its surface, and from each of these depressions radiates a corona of ridges making the egg look as though it were decorated with sculptured suns. In this dainty little species each of these depressions is at the summit of a low cone, like a low volcanic crater (Pl. XVI, fig. 26).

20. *Habrophlebiodes.* The eggs of *H. americana* are plain ovoid, about .17 by .09, and they have no chorionic patterns or appendages.

EPHEMERELLINAE

21. *Ephemerella.* There are at least three distinct types of eggs in this genus.

The eggs of *E. rotunda* are typical of the first group. They are ovoid, about .19 by .10, including a prominent cap at one pole. The chorion is smooth but it has on it six or more translucent bumps which form long threads with adhesive terminal knobs when the egg is laid in water (Pl. XVI, fig. 27). The polar cap also is adhesive. When laid naturally in water, the eggs adhere to each other and to stones by the caps and the tangled strings (Pl. XVIII, fig. 49).

The eggs of the following species differ from those of *E. rotunda* only in size. The size is not very constant but I have attempted to give the usual measurements in millimeters.

Species of *Ephemerella* with eggs similar to those of *E. rotunda*

angusta	(from nymph)	*euterpe*	.19 × .14
attenuata	.24 × .15	*fuscata*	.18 × .10
aurivillii	.19 × .15	*grandis*	.19 × .12
cognata	.19 × .12	*hystrix*	(from nymph)
concinnata	(from nymph)	*micheneri*	.19 × .11
cornuta	.20 × .12	*needhami*	(from nymph)
cornutella	.20 × .12	*serrata*	(from nymph)
deficiens	.18 × .11	*simplex*	.23 × .09
doddsi	.20 × .13	*sp.* (near *tibialis*)	.19 × .13
dorothea	.18 × .11	*temporalis*	.24 × .12
		tuberculata	.18 × .11

The eggs of *E. septentrionalis* (.20 x .11) are similar to those of *E. rotunda* except that the chorion is covered with a large meshed reticular pattern of ridges (Pl. XVI, fig. 28).

The eggs of *E. hecuba* might also be put in the *E. rotunda* group, but they are much larger, being .27 x .17, and the chorion is covered with a reticular pattern having a finer mesh than that of *E. septentrionalis* eggs (Pl. XVI, fig. 29).

The eggs of *E. temporalis*, *E. prudentalis*, and *E. verisimilis* belong to the second group. All of them are ellipsoid or slightly dumb-bell-shaped, about .17 by .08, and there are no polar caps or adhesive strings. The chorion is covered with a large-meshed reticular pattern of ridges (Pl. XVI, fig. 30).

The eggs of *E. maculata* are distinctly of a third type. They are broadly ellipsoid, about .20 by .13, and there are no polar caps or strings, but the chorion of preserved material has a very peculiar pattern of irregular bumps arranged in circles (Pl. XVI, fig. 31). This egg was taken from a mature nymph.

CAENINAE

22. *Caenis. C. hilaris*, *C. jocosa*, and *C. amica* eggs, from preserved material are ellipsoid with a cap at each pole, and measure about .14 by .08. The eggs of *C. simulans* are similar but larger, being .16 by .09.

Each polar cap consists of a mass of strings with terminal knobs, arranged spirally, tightly cohering; but when the eggs are laid naturally in water the strings uncoil and become entangled with each other, with the strings on other eggs, and with the algae which are usually present in the ponds where Caenis lives. The chorion is smooth (Pl. XVI, figs. 32, 33).

Five undetermined species of Caenis were examined and they all had eggs similar to those described above, but differing in size. These were isolated females, and in absence of males, the females of this genus are difficult of determination.

A species of Caenis from India, *C. perpusilla*, has eggs which are interesting because they are much like the North American Caenis eggs except that there are clearly just four threads in each polar cap or coil (Pl. XVI, fig. 34).

23. *Tricorythodes.* The eggs of *T. allectus* are ovoid with polar caps about .18 by .12. The chorion is sculptured, not with "shingles" as figured Morgan (1913), but with a reticular pattern having the ridges interrupted in such a way that the egg appears to be covered with overlapping shingles or scales. The chorion also has on it several strings arranged in pairs. These strings are thicker than those of Ephemerella, and they have no terminal knobs. Morgan says that the fresh eggs are green with yellowish caps (Pl. XVII, fig. 35).

24. *Brachycercus. B. nitidus* has elongate ovoid eggs with polar caps, about .24 by .09. The chorion is sculptured with prominent sharp-edged longitudinal ridges. In the preserved material examined no other chorionic structures could be seen (Pl. XVII, fig. 36).

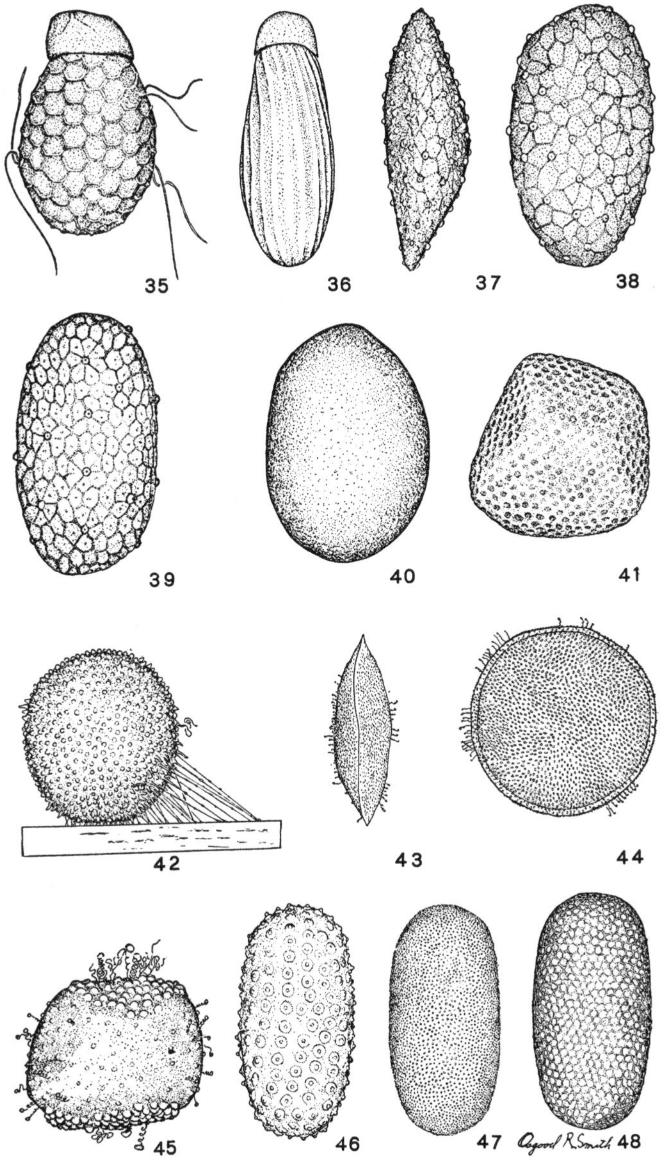

MAYFLY EGGS

Fig. 35. Egg of *Tricorythodes allectus* (in alcohol); Fig. 36. Egg of *Brachycercus nitidus* (in alcohol); Fig. 37. Egg of *Ameletus lineatus* (in alcohol, edgewise view); Fig. 38. Egg of *Ameletus lineatus*, view of flat surface (in alcohol); Fig. 39. Egg of *Ameletus sp.*, view of flat surface (in alcohol); Fig. 40. Egg of *Siphlonurus rapidus* (in alcohol); Fig. 41. Egg of *Siphlonisca aerodromia* (in alcohol); Fig. 42. Egg of *Isonychia christina*, showing how it becomes attached to a glass slide when freshly laid in water; Fig. 43. Egg of *Isonychia aurea*, edgewise view (in alcohol); Fig. 44. Egg of *Isonychia aurea*, view of flattened surface (in alcohol); Fig. 45. Egg of *Isonychia rufa* (in alcohol); Fig. 46. Egg of *Cloeon sp.* (in alcohol); Fig. 47. Egg of *Neocloeon alamance* (in alcohol); Fig. 48. Egg of *Centropilum conturbatum* (in alcohol).

SIPHLONURINAE

25. *Siphlonurus.* The mayflies of this genus all have plain ellipsoid eggs. The eggs of *S. rapidus*, *S. decorus*, and *S. quebecensis* are about .22 by .15; those of *S. mirus*, .19 by .13, and those of *S. berenice* about .26 by .18 (Pl. XVII, fig. 40).

When the live eggs are laid in water they are green and are held together in a gelatinous mass which will adhere to pebbles (Pl. XVIII, fig. 51).

26. *Siphlonisca.* The eggs of *S. aerodromia* are large, and subspherical, about .44 in diameter. In preserved material the eggs are not perfect spheres but are more or less polygonal, probably due to crowding in the oviducts. The chorion is covered with broad, shallow pits, set fairly closely together (Pl. XVII, fig. 41).

27. *Ameletus.* The eggs of *A. lineatus* are slightly flattened ovoids, about .32 by .19 by .10. The chorion is sculptured with a reticular pattern, and scattered over the surface are translucent boss-like structures which are probably adhesive when laid fresh in water. Each of these boss-like structures is located where three or more strands of the net join, but there is not a boss at every one of these joinings (Pl. XVII, figs. 37, 38).

The eggs of *A. subnotatus* are of the same shape, but smaller, being about .26 by .15 by .08. The reticular pattern has a slightly finer mesh than that of *A. lineatus,* but otherwise the sculpturing on the eggs of these two species is the same (Pl. XVII, fig. 39).

28. *Isonychia.* There are at least three very different kinds of eggs in this genus.

In the first group, to which *I. similis, I. annulata, I. serrata, I. georgiae, I. aurea,* and *I. notata* belong, the eggs are shaped like bi-convex lenses, from .32 to .36 in diameter and about .10 thick. When they are dissected out of the abdomen of the mayfly they look much like a pile of pennies. In preserved material the chorion is covered with many small bosses which may often be stretched out into small threads (Pl. XVII, figs. 42, 44). This may easily be demonstrated by allowing fresh eggs to fall in a watch glass of water, and after a few minutes stay, trying to pour them out.

A second group, consisting of *I. pacoleta, I. albomanicata, I. sicca,* and *I. christina* has spherical eggs about .2 in diameter. The chorion is covered with adhesive bosses which are slightly larger than those on the eggs of the first group. When laid naturally in water these bosses are adhesive, and may become stretched out to form strings (Pl. XVII, fig. 42).

In the third group, consisting of *I. rufa* and *I. manca,* the eggs are slightly flattened spheres, about .19 mm. on the longest diameter. At each more or less flattened pole there is a group of large bosses, the remainder of the chorion having only a few very small bosses scattered over it (Pl. XVII, fig. 45).

BAETINAE

29. *Baetis*. Several preserved specimens belonging to this genus were examined, and they all had smooth-shelled ellipsoid eggs, about .13 by .11.

The eggs of the *Acentrella* group of Baetis are plain ellipsoid, about .15 by .08. In the preserved material examined the chorion appeared to have a rough, more or less granular surface.

30. *Callibaetis*. *C. fluctuans* and *C. hageni* have plain, nearly spherical eggs, about .12 mm. in diameter.

A number of undetermined specimens were examined, and all but one had eggs like *C. fluctuans*. The exception, from McLean, New York, July 27, 1924, was ovoviviparous, and first-stage nymphs .6 mm. long were dissected out of the abdomen of the imago. Unfortunately these nymphs were poorly preserved, and I did not attempt to make a drawing of them.

31. *Cloeon*. The eggs of *C. rubropictum* from Ithaca, New York, from preserved material, are ellipsoid, about .13 by .08. The chorion is covered with rows of flat bosses, each with a small translucent knob at its summit. There are ten or twelve rows running longitudinally with about twenty bosses in each row. Even in preserved material some of the small translucent knobs are lengthened out into irregular cones, so it seems likely that these serve as adhesive structures when the eggs are laid in water (Pl. XVII, fig. 46).

Bengtsson found that *C. bifidum* has eggs much like this, but that *C. dipterum* has smooth eggs and *C. simile* has eggs with a reticular pattern; so it is not at present possible to make any generalizations for the genus.

32. *Pseudocloeon*. The eggs from several unidentified specimens were examined, and all were broadly ellipsoid, about .12 by .9. In preserved material the chorion is thickly covered with many small round bumps, much like those on the eggs of Neocloeon, but smaller and set a little closer together (see Pl. XVII, fig. 47, of Neocloeon egg). These bumps are probably adhesive, and stick to the stones when the eggs are laid in water.

33. *Neocloeon*. The eggs of *N. alamance* are rather elongate ellipsoid .16 by .08. In preserved material the chorion is covered with many small boss-shaped bumps, which probably are adhesive (Pl. XVII, fig. 47).

34. *Centroptilum*. The eggs of *C. conturbatum*, from preserved material, are ellipsoid, about .15 by .08. The chorion has a reticular pattern of ridges, with an unusually fine "mesh" (Pl. XVII, fig. 48). Another unidentified species has eggs similar but slightly larger, being .18 by .10. Another unidentified species has ellipsoid eggs, .14 by .09, without any pattern on the chorion. Except for a slight difference in size this is like the egg of *C. diaphanum* described by Bengtsson.

PLATE XVIII.

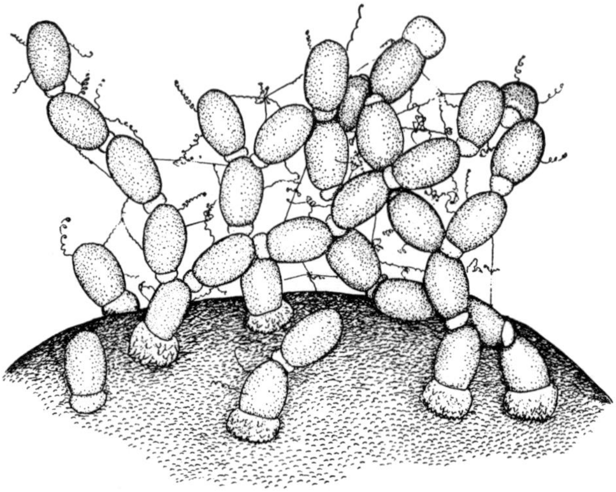

49

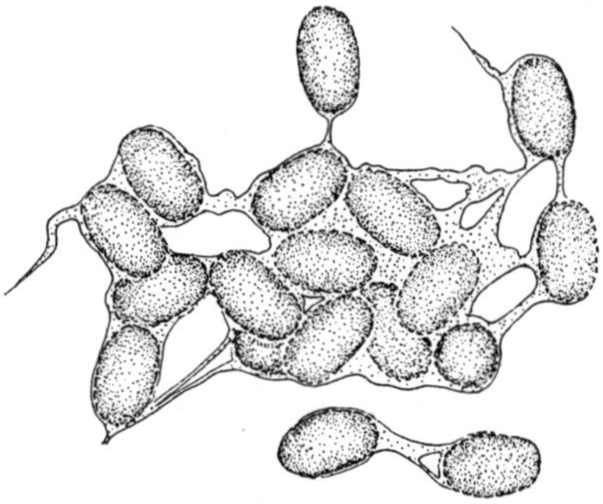

50

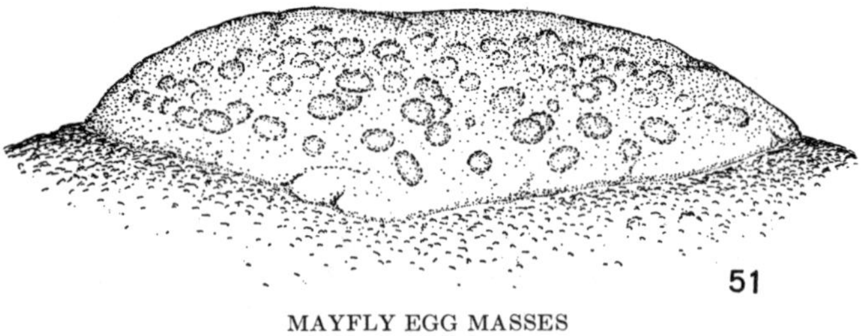

51

MAYFLY EGG MASSES

Fig. 49. A portion of a mass of live eggs of *Ephemerella rotunda* attached to a pebble in water; Fig. 50. Live eggs of *Ephemera sp.* held together by gelatinous material, in water; Fig. 51. Live eggs of *Siphlonurus sp.* in gelatinous mass on a pebble in water.

III. EGG DEVELOPMENT

Most mayflies have a long nymphal life extending over one or two years and a very short adult life of a few days. Therefore the reproductive system must be practically mature and the eggs ready to be fertilized and laid almost as soon as the subimago can fly. Because of these facts the development of the ovaries and of the eggs must be studied in the nymphs. This was done in an undetermined species of Isonychia by sectioning nymphs of various sizes which were collected in winter, in spring, and in early summer.

In a nymph 7.5 mm. long each ovary is merely a band of loosely fitting, irregularly shaped cells, surrounded by a peritoneal sheath (fig. 5). These cells stain a dark blue with Delafield's haematoxylin. These

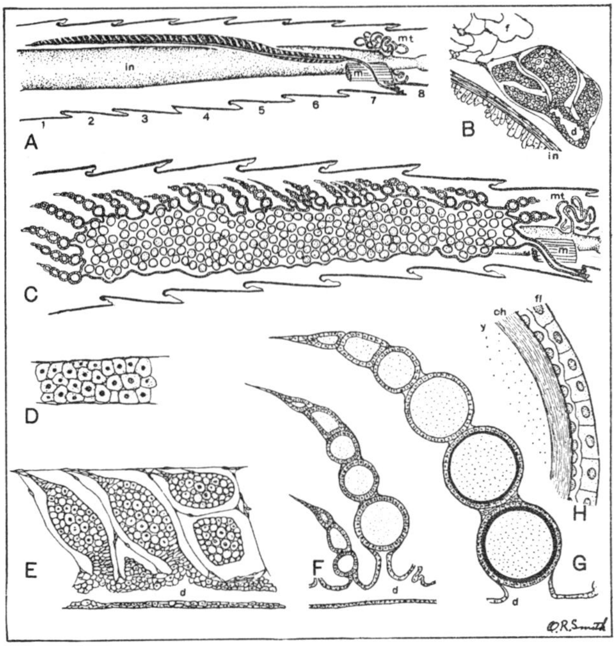

FIG. 5—Egg development in Isonychia. A. Diagrammatic longitudinal section of the abdomen of a 12 mm. nymph, showing position of ovary and oviduct; B. Cross section of the ovary, part of the midinitestine, and part of the dorsal fat body in a 12 mm. nymph. C. Diagrammatic longitudinal section of the abdomen of a mature nymph, about 17 mm. long, showing oviduct full of eggs and the egg tubes on the distended oviduct; D. Longitudinal section of the developing ovary of a 7.5 mm. nymph; E. Enlarged longitudinal section of part of the ovary of the nymph shown in A; F. Longitudinal section of two egg tubes and part of the oviduct of a 17 mm. nymph that was not quite ready for transformation; G. An enlarged section of one of the egg tubes shown in C; H. Part of a section showing the adhesive nodules on the chorion of the egg and the follicle cells.

Arabic numerals indicate the abdominal segments: *in*, intestine; *m*, muscle; *mt*, Malpighian tubules; *f*, fat body; *d*, oviduct; *y*, yolk; *ch*, chorion; *fl*, follicle cells.

bands of germ cells and ovarian cells extend from the first abdominal segment about to the seventh. Anteriorly, the ovaries are dorsal to the mid-intestine, but the posterior portions curve down and are lateral to the intestine.

The cells of the rudimentary ovary proliferate, and by the time the nymph is 12 mm. long, lobes have been formed all along it, in at least two rows, and the hollow oviduct has been formed. Each of these lobes is a rudimentary egg tube, and probably the larger, rounded cells in its center are the germ cells. The smaller peripheral cells would in that case form the epithelial lining of the egg tube. Every part of the ovary is of course covered by a peritoneum, and each lobe already has a terminal filament which is attached to a thin connective tissue sheath surrounding each lobe and the ovary as a whole. The connective tissue sheath is attached to the fat body dorsal to the midintestine (figs. 5B and E).

At this stage the ovaries may be traced from the first abdominal segment to their separate endings between the seventh and eighth sternites. The anterior portions of the ovaries lie dorsal to the mid-intestine, and the lobes or rudimentary egg tubes are on the dorsal side of the oviduct. As the ovaries proceed posteriorly, they come to lie on each side of the intestine, and each twists so that the oviduct is on the outside, and the lobes point inward and ventrally. In the seventh segment the oviducts turn still farther outward and pass down outside, or lateral to the large, ventral, longitudinal, intersegmental muscles. Then the oviducts curve toward the midventral line, and passing posteriorly, each ends blindly against the conjunctiva between the seventh and eighth abdominal sternites. (fig. 5A).

By the time the nymph has grown to 17 mm. and is nearly mature, the lobes of the ovaries have developed into definite egg tubes of the "panoistic" type (fig. 5F). The eggs have not yet reached their final size, and there is no chorion. The oviducts are still empty tubes, but the lumen has become enlarged.

Nymphs of about the same size, but with darker wing pads and therefore more mature, have fully developed egg tubes. These have already produced and extruded into oviducts so many mature eggs that the oviducts are swollen up like sacks of potatoes. The walls of the distended oviducts are thickly covered with short egg tubes, almost everywhere except on the ventral surface. The oviducts and egg tubes are crowded into every available space in the abdomen between the intestine and the muscles, and they have largely replaced the fat bodies of the smaller nymphs. (fig. 5C).

Eggs in all stages of development may be seen in the egg tubes. In the terminal portion there are usually two or three egg cells of various sizes, each surrounded by follicle cells. Near the mouth of the tube the eggs have reached full size, and the chorion has been formed. The egg nearest the mouth is usually mature and has a thick chorion covered with small nodules (fig. 5G).

Figure 5H shows how the nodules are formed under the follicle cells. Evidently the follicle cells secrete a homogeneous substance on the surface of the chorion, and then this substance draws up into nodules or droplets between the chorion and the follicle cells.

When the eggs pass out of the egg tubes into the oviduct, they lose their follicle cells.

The adhesive nodules take the blue Delafield's haematoxylin stain, while the chorion itself takes the eosin.

No structures could be seen within an individual nodule, so they evidently form the strings shown in Plate XVII, figure 42 by adhering to some object and then being stretched out.

CHAPTER VIII

THE LIFE CYCLE

1. Embryonic Development

Mayflies hatch from eggs, and we have already seen in Chapter VII how very diverse the eggs are in size, in form, in sculpturing, in adaptations for attachment, and in numbers produced. The first period of the life cycle is the period of embryonic development. It varies with the species, with the store of food material supplied in eggs of different species, with temperature, and with environing conditions. Doctor Murphy found that the eggs of *Baetis vagans* hatch in eleven days in a warm laboratory, but require twenty-eight days in the cooler water of the creek, and similar results are reported by other workers.

A striking example of adaptation of the embryonic period to environment was reported by Dr. W. A. Clemens in 1922. In *Ameletus ludens*, which lives in the temporary pools of small brooks, the nymphs develop in late winter and spring while the pools are full of water. The eggs are laid in May and early June. The pools dry up, and, except for occasional showers, remain dry until the time of autumnal rains. Then the eggs hatch after a period of five months' incubation.

Doctor Clemens found that the eggs of other mayflies at Ithaca, kept under similar laboratory conditions, hatched as follows:

Heptagenia hebe	in 12 days
Hexagenia recurvata	in 14 days
Ephemera varia	in 15 days
Stenonema tripunctatum	in 11 to 23 days

One of us (Hsu) found that the eggs of *Stenonema interpunctatum* hatched in the laboratory in from 13 to 15 days. Heymons (1896) found that the eggs of *Ephemera vulgata* kept at a temperature of 20° to 25° C. would hatch in ten to eleven days. Wiebe fertilized eggs of *Hexagenia bilineata* by mixing them with spermatozoa in normal salt solution, and they hatched in the laboratory in nine days. From nine days to five months thus appears to be the known range of the incubation period.

During this period antennae, mouthparts, legs, tails, and all necessary internal organs are developed (fig. 6) and made ready to function on eclosion.

We have already seen in Chapter VI that these nymphs emerge from the egg in a very simple form, and so far as known, in all mayflies very much alike. They have elongate bodies made up of similar segments, and they lack both wingbuds and gills. They are not very different in the beginning from the young of generalized insects of other orders, especially those of Plecoptera and Odonata-Zygoptera.

2. Nymphal Life

The form of the nymph slowly changes with many successive castings of skin. The manner of molting as observed by one of us (Hsu) in Stenonema is as follows: The old skin splits along the mid-dorsal line of the thorax and along the epicranial suture of the head. The head is first bent a little downward and the thorax is elevated a little and forced out

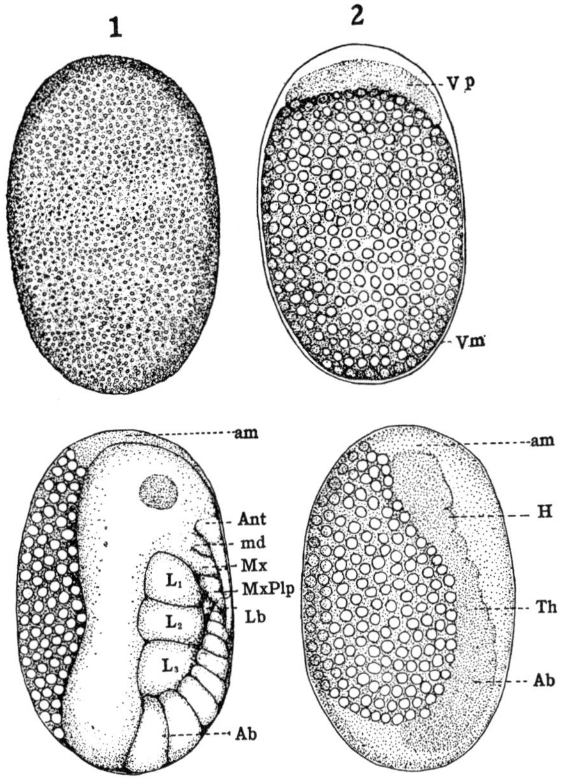

Fig. 6—Egg development in *Baetis vagans* (from Murphy): 1. The ripe egg; 2. The same after two days development, showing the ventral plate (*v.p.*) already formed upon the surface of the yolk; 3. The same after four days development, showing the amnion (*am*), covering the elongate body in which head (*H*), thorax (*Th*), and abdomen (*Ab*), are but little differentiated; 4. The same after nine days development, showing antennae (*Ant*), mandible (*md*), maxillae (*Mx*), maxillary palpus (*MxPlp*), labium (*Lb*), legs (*L1*, *L2*, *L3*), and sharply segmented abdomen (*Ab*).

through the split. The legs and the mouthparts are gradually pulled out. Finally when the head, mouthparts, and legs are completely free, the legs are used to grasp the support and pull out the hind part of the abdomen. The whole process of molting is completed in about five to ten minutes.

The number of moltings (instars) appears to have been determined hitherto for only a single species of mayfly, *Baetis vagans*. That species was studied by Dr. Helen E. Murphy at Ithaca, N. Y., in 1918-20. She reared from egg to adult two successive broods and published (1922) the following record:

LIFE HISTORY OF *Baetis vagans*

Two broods of laboratory reared specimens

	Days duration		*Length of body*	
Instar	*Brood A**	*Brood B***	*Brood A*	*Brood B*
1	2	2	.30	.30
2	2–3	2	.40	.40
3	4–6	3–5	.50	.50
4	4–6	4–5	.60	.60
5	6–7	4–6	.71	.70
6	7–8	4–7	.84	.82
7	6–10	5–7	1.3	.95
8	6–10	6–8	1.7	1.2
9	6–10	6–8	2.1	1.5
10	6–11	6–9	2.5	1.8
11	7–10	8–11	2.9	2.1
12	7–11	8–10	3.3	2.4
13	6–10	8–11	3.7	2.7
14	7–10	8–11	4.0	3.0
15	7–10	7–10	4.3	3.3
16	7–10	7–10	4.6	3.7
17	7–10	8–10	4.9	4.0
18	7–10	8–10	5.1	4.3
19	7–10	8–10	5.4	4.7
20	7–10	8–10	5.6	5.0
21	7–10	8–10	6.0	5.3
22	7–10	8–10	6.3	5.7
23	7–10	8–10	6.6	6.1
24	7–10	8–10	6.9	6.3
25	7–10	8–10	7.2	6.5
26	7–10	8–10	7.5	6.8
27	7–10	8–10	7.8	7.0

These were reared in the laboratory. In order to determine how the species develops under natural conditions, she also reared four successive broods in the cool water of the creek. This revealed a striking relation of length of nymphal life to temperature, as shown by her in the following table:

* May 1918 to October 1918: 376 specimens.

**October 1918 to May 1919: 237 specimens.

SUMMARY OF CONTINUOUS CYCLES, *Baetis vagans*

Cycles	*Periods*	*Duration in months*	*Average temperature*
Brood 1	May 1918 to October 1918	6	60.7° F.
" 2	October 1918 to Aug. 1919	9	47.9° F.
" 3	August 1919 to May 1920	9	41.4° F.
" 4	May 1920 to October 1920	6	62.2° F.

This also explained the three periods of emergence of this species in May, August, and October of each season at Ithaca, N. Y. The succession of broods is shown by Doctor Murphy in the accompanying diagram (fig. 7), which is to be read from the outside inward. Three cycles are completed in two years.

The length of nymphal life is probably greater than this in most mayflies. It is certainly greater in some of the large burrowers,* for adults appear at one short period each season and half-grown nymphs are then present in the beds from whence they come. It may be inferred from uniformity in size of nymphs of most species, coupled with regular seasonal appearance of the adults, that one brood a year is the rule. Yet such is the rapidity of development of Callibaetis nymphs (five or six weeks from the egg) that a succession of broods in a single season would seem possible.

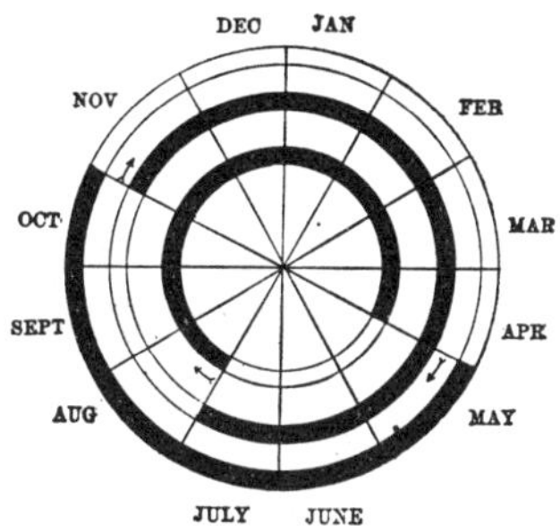

FIG. 7—Diagram of the seasonal cycle of development in *Baetis vagans*. The small arrows indicate successive broods (from Murphy, 1922).

It goes without saying that the winter season is passed in the water, and that as a rule hibernation is in the nymphal stage.

We have already seen in Chapter VI how and when the gills arise in three genera of Heptageniidae. There are slight differences in other genera. For example, Wiebe (1926) found gills present on nymphs of *Hexagenia bilineata* in their first instar. Murphy (1922) says that in Baetis, gills first appear in the fifth instar as tiny outpocketings of the body wall, and that in the seventh "the gills reach their full relative size and are freely moved."

The wings begin as folds of skin at the postero-lateral angles of meso- and metanota. They increase in size at each successive molt. Tracheae from the body trunks grow out into them and in the more primitive forms veins develop about the tracheae. For several instars before the final transformation the venation is clearly visible, and it is identical with the adult vein pattern.

*Average for the period.

3. Internal Metamorphosis

The change from a heavy, sure-footed aquatic nymph to a light, winged, aerial adult consists in far more than the shedding of a skin. It has to be prepared for: the preparation occupies the last days of nymphal life. The nymph ceases feeding and becomes quiescent unless disturbed. The sexual products, eggs or sperms, mature rapidly. The eyes and forceps, wings and feet, grow and press for release, while other parts that are destined to disappear begin to degenerate.

These changes have been studied in *Hexagenia recurvata*, first by Dr. Anna H. Morgan and later by Dr. Helen E. Murphy. We present herewith (fig. 8) Miss Morgan's unpublished parallel drawings of nymph and adult. The most extraordinary changes are in the head, where there is loss of mouthparts, reduction of antennae, and great increase in size of eyes.

Fig. 8—Adult and nymph of *Hexagenia recurvata* showing losses and gains of parts. Note especially mouthparts, antennae and gills; wings of adult not shown (drawings by Anna H. Morgan).

The loss of jaw muscles has been carefully studied by Doctor Murphy, whose account (1922) we in part reproduce below, together with her plate showing the normal musculature, and two figures showing internal changes. She found that portions of these muscles began to degenerate five days before emergence of the subimago. She studied the process by means of both sections and smears, staining both with Giemsa's blood stain.

The first figure (fig. 9) is of normal plasma cells. She says of them:

"In the body plasma from the first instar and through imaginal life there are five varieties of cells, which intergrade with no sharp lines of demarcation:

"A. Little cells with a small amount of deep-blue staining, homogeneous cytoplasm; nuclear chromatin evenly distributed, deep-purple staining, and coarsely reticular. These cells show very slight ameboid movement.

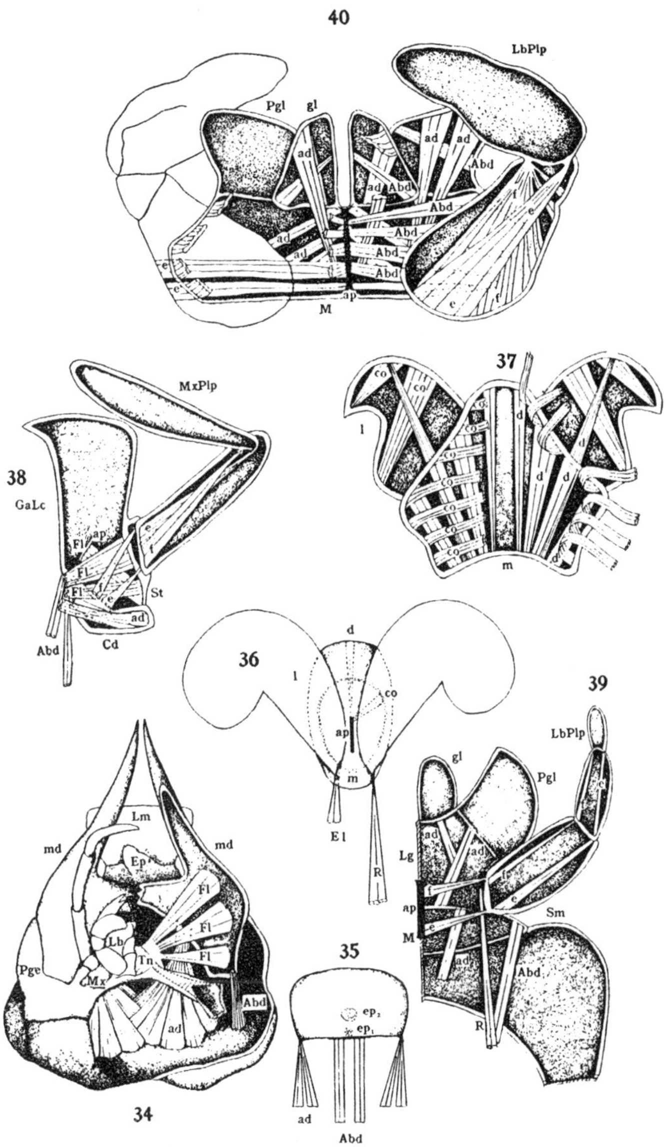

NORMAL MOUTHPARTS AND THEIR MUSCULATURE

Fig. 34. Head of Hexagenia from beneath, with parts on one side removed to show labrum (*Lm*), epipharynx (*Ep*), tentorium (*Tn*), mandible (*md*), cut open to show muscles; Fig. 35. Labrum of Ephemerella; Fig. 36. Hypopharynx of Ephoron; *m*, median, and *l*, lateral lobes; Fig. 37. Hypopharynx of Heptagenia cut to show musculature; Fig. 38. Right maxilla of same: *Cd*, cardo, *GaLc*, blade (galea-lacinia), *ap*, apodeme; *MxPlp*, maxillary palpus; Fig. 39. Half of labium of Ephemerella: *gl*, glossa; *Pgl*, paraglossa; *Lg*, ligula; *LbPlp*, labial palpus; *M*, mentum; *Sm*, submentum; Fig. 40. The labium of Heptagenia; *ap*, apodeme; *St*. Stipes.

The following abbreviations designate muscles on the several figures: *Abd*, abductor; *ad*, adductor; *co*, compressor; *d*, depressor; *e*, extensor of palp; *f*, flexor of palp; *Fl*, flexor; *R*, retractor.

"B. Cells 1.5 to 3 times the size of variety A, with a large amount of homogeneous cytoplasm varying from dark to light blue; nuclear chromatin deep-purple staining, coarsely reticular and evenly distributed. These cells show mitotic division and active ameboid movement.

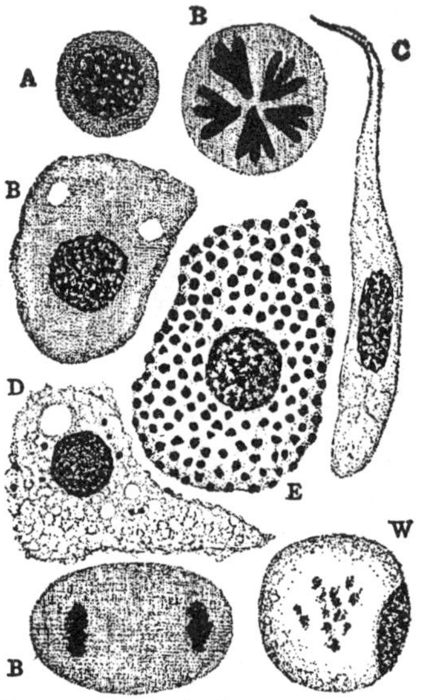

FIG. 9—Typical plasma cells from a blood smear. For explanation see text (fromMurphy, 1922).

"C. Cells 2 to 3 times the size of variety A; with light blue staining vacuolated cytoplasm; nuclear chromatin finely reticular and light-purple to dark-blue staining. Spindle forms are common. Ameboid movement is active.

"D. Cells 2 to 3 times the size of variety A; with small pink staining areas or a few pink granules; cytoplasm light blue staining and vacuolated, similar to variety C. Spindle forms are common. Ameboid movement is active.

"E. Cells 1 to 4 times the size of variety A; with many deep-pink staining granules; cytoplasm staining a very light reticular blue, or not at all; nucleus light purple, with coarsely reticular chromatin, and light staining ground substance.

"Smears made from the abdominal cavity of the nymph and subimago just after molting show many degenerating plasma cells. The most common form is vacuolated, with the nucleus crowded to the periphery (fig. 9, w). Some of the vacuoles contain pink granules or diffuse pink areas."

Her second figure (fig. 10) is of degenerating muscle fibers, with these same kinds of plasma cells interspersed. She says:

"In all the specimens examined cross-striations were visible in portions of the muscle just before transformation.

"Early in degeneration the dark bands stain lighter and irregularly (I). The light bands increase in width, and Krause's membrane is no longer evident. The muscle nuclei (n) are swollen and stain faintly. There is also slight evidence of fibrillation. Then the muscle substance changes from blue to pink in color, and the cell contents (II), begin to liquify. The nuclei (n) assume a spherical shape and their chromatin becomes diffuse. Plasma cells of the five varieties (A, B, C, D, E) increase in the proximity of the degenerating muscle. In cross section (IV) Cohnheim's areas are not evident, and many of the fibers are vacuolated. As degeneration increases, the dark bands disappear (III), and the muscle substance stains more and more faintly. The sarcolemma is no longer evident, and the nuclei (n) rupture. Finally nothing remains but a formless mass (V), and that soon ceases to stain. There is no evidence of phagocytosis.

"During the muscle degeneration no plasma cells are present that do not occur in a first instar nymph. The following differential count from smears of the plasma of nymph, subimago and imago shows a proportional increase in varieties D and E after transformation, and a decrease in varieties A and B.

Percentage of Plasma Cells in Different Count

Variety	*In Nymph*	*In Subimago*	*In Imago*
A	25.0	1.3	1.5
B	12.5	2.3	1.5
C	27.5	16.7	22.0
D	5.0	11.7	26.0
E	30.0	68.0	49.0

"Tracheae (Tr) and tracheoles (tr) persist after the muscles have entirely lost their identity. These cells have a finely granular dark purple staining cytoplasm, and a densely staining finely reticular nucleus.

"In the metamorphosis of the muscles of mayfly mouth-parts, I find no evidence of a proliferation of free tracheal cells. There is a rapid increase in tracheal cells just after the various nymphal molts, but they form tracheoles."

Since mouthparts degenerate, and food is not taken after metamorphosis begins, the stomach ceases to be digestive and becomes aerostatic in function. The nymph just before emergence fills it with gulps of water, which is later replaced by air. The swallowing of air distends the stomach, lightens the specific gravity of the body as a whole, and probably aids in the expulsion of eggs and sperms.

The metamorphosis of the alimentary canal is chiefly degenerative, reducing both epithelium and muscle. The changes are rather slight in stomodaeum and proctodaeum (more extensive in the latter during the subimaginal period), but in the mesenteron they are very great. One of us (Hsu) has traced them in Stenonema. Before the emergence of the subimago the mesenteron is reduced in diameter. The epithelial cells become somewhat distorted. The cells also lose their well-defined boundries and are no longer columnar in shape. The protoplasm becomes very coarsely granular, and the nuclei are reduced in number and degenerate. The whole epithelium is reduced very

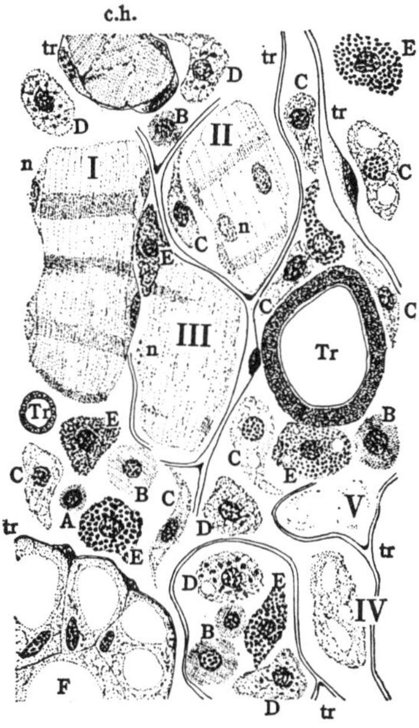

Fig. 10—Degeneration of jaw muscles in *Hexagenia recurvata* in a subimago of 24 hours. A to D represent the same types of plasma cells shown in figure 9. I, II, II, and IV indicate progressive changes in the muscles. *Tr*, trachea and *tr*, tracheole; *c.h.* Cohnheim's area; *n*, nucleus (from Murphy, 1922).

much in thickness. The muscular layer also becomes very thin. The striated border of the mesenteron is entirely lost (Pl. V, figs. 10 to 13).

In the subimago the protoplasm of the epithelium condenses still more towards the base of the cells, and still remains very coarsely granular. The nuclei disintegrate entirely, leaving only the oval-shaped, deeply stained spots in the section as the indication of their place. The entire wall of the mesenteron becomes very thin.

In the imago the epithelium and muscular layers of the mesenteron become reduced and form an extremely thin membrane.

The details of these changes have been described by Pickles (1931) for Ephemera.

The Subimago

When the course of nymphal life is fully run, and transformation is about to take place, all the parts of the body of the adult are already present and fairly well developed. They are hidden beneath the nymphal

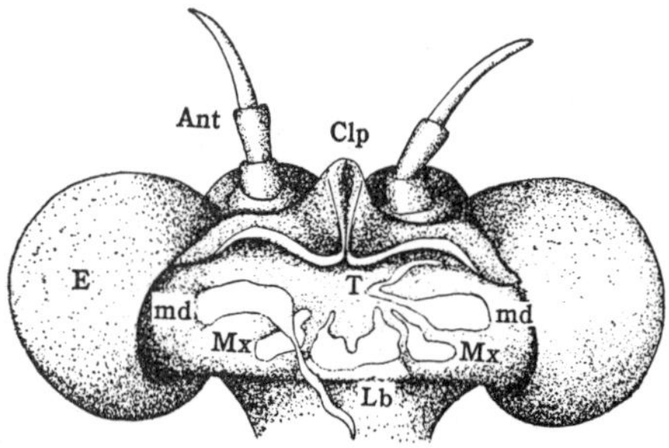

Fig. 11—Head of a subimago of *Hexagenia recurvata*, viewed from beneath showing vestigial mouthparts: *Ant*, antenna; *Clp*, clypeus; *E*, eye; *md.*, *md*, mandibles, with *T*, the vestigial tusk of one of them; *Mx*, maxilla; *Lb*, labium (from Murphy, 1922).

skin, waiting to be released when that skin is cast off. The wings are ready when released to expand to many times the area of the sheaths. The slender legs within the stout nymphal leg-sheaths have now two claws at the tip of each, where in the nymph was but one. Five joints are present within the one-jointed tarsus. The eyes are pressing for release, especially in the male. The eggs in the ovary of the female are so nearly mature as to exhibit their final sculpturing. The forceps of the male is fairly well formed, and the tails in both sexes are ready for a sudden linear extension.

On the other hand, the antennae inside the nymphal antennal sheaths are dwindled and reduced in size, as is the middle tail in many forms, and as are the middle and hind legs in a few (Campsurus, etc.). The mouthparts are so reduced as to be scarcely recognizable, and the gills (save for vestigial filamentous tufts beneath the head and prothorax in a few forms like Isonychia) are altogether lost.

Callibaetis emerges from the nymphal skin at the surface of the water and leaves it floating, as do most mayflies. The nymph of Siphlonurus clambers from one to four inches above the water up vertical stems, and leaves its skin clinging to the stems after the manner of the damselflies. That of Isonychia crawls out on top of a partly submerged stone in the riffles and leaves its stiff and shining brown parchment-like skin there, usually alongside the skins of stoneflies. Occasional nymphs of almost any species will crawl part way out of the water upon a sloping bank before transforming.

The time of day at which transformation takes place varies with the species and with individuals. We have personally observed *Siphlonurus alternatus* transforming at 10:00 A. M.; *Blasturus cupidus*, at 11:00 A. M. to 12:30 P. M.; *Iron fragilis* at 7:00 to 8:00 P. M.; *Ameletus ludens* at 2:00 to 5:00 P. M.; Caenis at the fall of darkness; Hexagenias all day long. Some species that have not been seen emerging are supposed to transform at night. Evidence for this is found in their abundance in spiders' webs in the early morning.

The cast skins of the more delicate nymphs quickly sink, or break up and disappear; the stouter husks of the big species that frequent the beds of larger bodies of water often drift in windrows upon a lee shore.

The first flight of the subimago is a brief one in some species, and a very long one that cannot be followed to its end in others. Everyone who is observant of such things has seen the big Ephemeras rise from the surface of the water and go fluttering slowly upward, always as if about to alight, but never alighting, and all the time going higher and farther until lost from view. Little Caenis, on the other hand, alights on the first solid support that offers, even if it be only a few feet distant from the water. Callibaetis and Blasturus and the commoner Heptagenines fly rather directly to nearby shrubs and settle on their outer leaves; but *Iron fragilis*, although a delicate little fellow, rises from the stream in the bottom of the deep gorge, up and up until lost among the tree-tops that crown the bordering cliffs.

Finding a suitable place, the creature rests during a period that may normally extend to 48 hours or more in some forms, such as Siphlonurus, Stenonema, and Iron, or may be but momentary in others. Caenis, flying to the first perch that offers, settles and molts again almost immediately, and its life as a subimago is over. This stage for most of the commoner mayflies lasts about twenty-four hours.

Subimagos are very inactive. They prefer to spend the daytime resting in the shade. Usually they sit quietly upon the under side of leaves near the stream from whence they emerged. When disturbed they fly away. During warm summer nights they are active, and they fly to the strong lights. The warmer the weather and the stronger the lights, the more they are attracted.

5. The Imago

Mayflies are peculiar in having a partial shedding of skin after the adult form has been assumed. This has been the subject of much speculation—and over-emphasis. It has even been looked upon as the equivalent of a pupal stage. It is only a casting of the delaminated, hairy outer cuticle. It does not occur in all mayflies (Palingenia is an exception). It is partial in some. It occurs only in the male in some. It does not represent a growth period, and is therefore not a true instar. Growth ceases with the nymph. After the wings expand and harden no further growth is possible. There is, however, extension of other appendages at the final molting, due to pressure from within the turgid body against the soft and newly released inner cuticle. Thus the fore legs and outer tails, and the genital appendages of the male subimago, are lengthened to their imaginal proportions.

It has been suggested by Lameere (1917) that this skin-casting (which he recognized as a mere delamination) is advantageous because it lightens the body weight for flying. In order to learn how much loss of weight there is, Mr. E. V. Welch made for us a large number of weighings of living specimens of *Isonychia albomanicata*. He collected several dozen fully grown nymphs, and reared them, weighing the same specimens before and after each of their last two castings of skin. He found that the newly emerged subimago weighs on an average 24.8 milligrams. After two days in a favorable moist atmosphere, the skins were shed, and these skins weighed when dry only .425 mg. Yet the living imagos emerging from them weighed on an average but 19.42 mg. Thus it would seem that the weight discarded in the skins is only about 1½%, while that lost through evaporation is about 20%. Doubtless the removal of the hardened outer cuticle facilitates loss of water by evaporation.

The multitudinous fine hairs of the skin are discarded with the outer cuticle, leaving the surface smooth and shining. This doubtless favors flying by diminishing friction against the air.

It is perhaps significant that those mayflies that do not shed a subimaginal skin do not have a prolonged mating flight.

The casting of the subimaginal skin during flight has been reported for a number of species. Dr. C. V. Riley discussed this matter (Amer. Nat. 15:395, 1881) as follows:

"*Polymitarcys alba* is extremely common on the Red River of the North, where in August, 1877, we found it actually giving the river a white appearance, and falling upon vessels like snow. The submaginal skin was shed in an incredibly short space of time (less than a minute), but was almost invariably preceded by a brief period of rest. The impatience to fly off after the true wings were withdrawn, however, was such that in the large majority of cases the insect took wing before the subimaginal skin was fully cast, in which case exuviation would be completed on the wing. We cannot conceive of the beginning of the process taking place on the wing, for there is a period, however short, from the bursting of the skin on the thorax to the extraction of the wings from their covering, when the use of the wing, it seems to us, is impossible, and we can conceive of full exuviation in the air only on the hypothesis that the insect during the process descends from a sufficient altitude to afford time for the extraction of the new wings."

The manner of emergence has been described by one of us for Blasturus (Traver, 1925) as follows:

"The subimago remains more or less quiet for nearly 24 hours and is then ready for the last moult. This moult again resembles in many respects the normal moulting of a growing nymph and the transformation of the nymph to the subimago. In this case, however, the wings must also be pulled out from the tissue enclosing them, and any slightest injury to these delicate organs before the time for this last change usually causes disaster.

"A subimago would take a firm grip on some convenient object, wave its wings up and down a few times, then gradually lower them till they were at right angles to the body, and their tips were as low as the body. The characteristic wave-like contortions of the abdomen were observed at this moult also, beginning as always at the caudal [end] and working forward. The wings were moved backward little by little, gradually decreasing the angle between them and the body, till they were held flat against the abdomen. The thorax then split along the median dorsal line; the head and thorax of the adult at once pushed through the opening, followed immediately by the upper portion of the wings and the abdomen and the legs. The legs were held seemingly rigid for several seconds after the adult had emerged. Then the adult grasped the support with its feet and pulled out the tails and the tip of the abdomen from the subimaginal skin. The whole proceeding occupied 6 or 7 minutes. The adult is at once very active and immediately flies away."

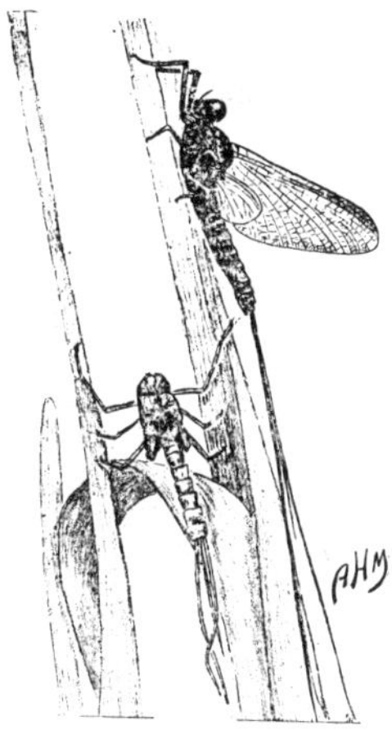

FIG. 12—A male imago of *Blasturus cupidus* standing above its cast-off subimaginal skin. The crumpled wings of the latter have carried off with them most of the black wing pigment (from Morgan, 1913).

The final molting reveals an adult even more delicate in form and of lighter but livelier tints, with jaws and other waning parts more completely atrophied and with longer appendages and larger eyes.

After the final molt there is another brief period of waiting until everything is ready and conditions are right for the "nuptial flight." Caenis is ready at once and takes to the air again almost as soon as the last skin is cast off. But most of the commoner species remain quiescent a while longer, and some of them may live for several days waiting for the passing of unfavorable conditions of weather.

During this waiting period mayflies are little noticed except when they become very abundant. When settled thickly all over the vegetation, they are disturbed by every passing breeze and brushed off by pedestrians, and in fluttering about they disturb and dislodge each other.

Thus the air is kept filled with them. Sometimes at waterside pleasure resorts they are counted a nuisance, for they desire not only to alight on the first available spot, even though it be the rim of one's ear, but to remain there indefinitely.

The mating flight. We have already described in Chapter II the mating flight of Callibaetis. Other mayflies have a different manner of flight, or fly at a different height or at other times of day. Males chiefly compose the familiar swarms. They gather at the waterside in companies great or small, according to the number in the neighborhood at the time, and fly persistently, waiting for the females to appear. The latter enter the swarm singly and intermittently, and usually remain only long enough to find mates. The coming and going of the females scarcely disturbs the rhythm of the great aerial performance.

The place selected is convenient to some water favorable for the deposition of eggs, where, flying in the open, the swarm can easily be found. The Hexagenias of the Great Lakes form swarms of vast extent over shoal waters alongshore. Some of the larger, creek-inhabiting species swarm over open upland fields at some distance from the stream. Some of the lesser, rivulet-inhabiting species, such as *Habrophlebia vibrans*, swarm in compact little companies in forest openings by the brookside, the individuals dancing up and down so close together that a single stroke of a large net may capture scores of them and may largely depopulate the swarm.

The time of the mating flight varies with the species, possibly also with the weather conditions, since high winds or rains may be a great hindrance. In favorable weather some forms, such as Tricorythodes, fly in the open sunshine at midday. Baetis appears to prefer the early forenoons. On almost any calm, bright morning in summer they may be seen from the high bridges over Fall Creek at the Cornell University campus, filling the air of the gorge like bits of floating thistledown. Callibaetis, Habrophlebia, Choroterpes, and Paraleptophlebia fly at any time of day, but seem to prefer the afternoon. Siphlonurus was once observed by the senior author swarming in mid-afternoon over the shoal margin of an Adirondack lake, and at another time in late afternoon, flying over a streamside highway in Pennsylvania. In the latter case, mating pairs settled all over the top of the automobile when a stop was made to observe the swarm. Potamanthus and Ephemera and all their near allies come out in the evening twilight, for their fine performance; Caenis and certain species of Ephemerella and the Heptagenines appear just afterward, as the darkness gathers, putting an end to observation. Before any of them appear it is so dark they can only be seen against the lighted background of the sky. The compact little swarms of Caenis are hardly visible at a distance of more than a few feet.

Manner of flight. Each species has its own characteristic mode of flight in swarming. The habits of a few that are typical and that have been most carefully observed will be here described. Most fascinating of all is the flight of Ephemera, which flight the senior author described in (1908) in the following words:

"After sundown the beautiful mayfly, *Ephemera simulans*, appears in companies of males over the edge of the water. The flight of one of these companies is a most delightful performance to witness, it is so light and graceful, and appears, withal, so exhilarating. Yet it is all up and down in vertical lines. With upturned head each individual flies rapidly upward, mounting quickly to a height of 10 or 15 meters; then, spreading its wings out horizontally, it falls upon them, with long forelegs extended forward and longer tails extended backward full length, rudder-like, keeping it always head to the wind. Thus it descends, floating on the air, yet not drifting, until at the lower level of the swarm (4 or 5 meters above the water), it lifts its head and rises rapidly again in flight. And the whole company flying and falling thus, weaving up and down in vertical lines, and passing and repassing each other creates a scene of great animation."

Ascending and descending lines of flight are long and vertical in Siphlonurus also; but in most other high-flying forms there is a considerable swaying from side to side, especially in descending.

The flight of the little *Baetis flavistriga* is even more irregular, and the position in descending is different. The upward flight is made with such quickness that the eye with difficulty follows it. When the little creature settles on its outspread wings and trails its uptilted reversible-jointed front feet and its long tails at the tip of its upcurved abdomen, it gives an excellent imitation of a dogbane seed floating downward on its crown of drifting hairs. Wings at the sides, feet in front, and tails behind slope down to where the center of gravity lies in the compact little body.

There are many degrees of concentration of flight. Along the west shore of Lake Michigan certain Heptagenias that have a rather long season often show a maximum abundance for a single week in July. An interesting account of such a maximum was published in the *American Agriculturist* for 1884, page 429.

"We were not a little surprised, on alighting from the carriage at nine o'clock on a last July evening, at the Leland Hotel, Chicago, located close by the lakeshore, to find the air was filled with snow flakes—so it appeared. It was certainly a phenomenon—a snow-storm in midsummer! The air was filled with these apparent snow-flakes. The porters were sweeping from the sidewalks around the hotel the two or three inches of gathered 'snow.' The Brush lights were flickering as if about to go wholly out, submerged by the flakes which were rapidly filling the glass globes surrounding them. One light had already been quite suffocated and the globes filled to the top. Rifts of 'snow' swept into the passage-ways leading to the rotunda, and the verandas along the lake-side of the hotel, were fairly flecked from one end to the other with the whitening shower.

"It was indeed a most astonishing sight with the thermometer at 80° F. But a moment, however, was sufficient to dispel the illusion. These were not snow-flakes which whitened and covered roof and pavement, but sand-flies—countless myriads—coming up like the locusts of Egypt from the sands of the lake shore to harmlessly fly and flutter for a day and die. The next morning bushels of them (so the head porter informed us) had been swept up during the night and carried away. They lay dead about the rotunda, through the hallways, in the dining hall, and in guests' chambers whose windows had remained open."

The numbers of mayflies in some of these swarms is almost unbelievably vast. A few examples must suffice. We quote the field notes of Mr. R. C. Lange, formerly of the U. S. Bureau of Entomology, and kindly sent us by Dr. Anna H. Morgan.

"While camping on the south shore of the Meramec River, across from Fern Glen, Missouri, July 22, 1909, two large swarms of mayflies were observed that night, or maybe the same swarm twice. . . . About 8:30 P.M. our campfire was just replenished with fresh wood and was burning brightly when things began to drop into the fire with a noise like that of a lighted match dropped into water. Faster and faster it came. My partner looked up, 'What is it? It is not raining. The stars are out!' Well, the fire was completely extinguished in less than five minutes by the largest swarm of insects that I ever noticed. The stench, like that of old burning grease, was awful. We decided to move, and stepping into our boat we had the second shower, the swarm still hovering above us. We lighted a coal oil torch and before I could put out the flame, the bottom of the boat was literally covered with mayflies to a depth of about two inches. No more flies were noticed after that, except a few clinging to the willows next morning. We stopped at the sand bar where fishermen were camping and they related to us how bugs had put out their fire about 9:00 P.M. One of the fishermen showed me some of the bugs. I think it was the same swarm that put out our fire."

Mating occurs in flight. Probably the necessary positions can only be assumed in the air. The male flies underneath the female and reaches upward with his long reversible fore feet and lays hold of her in front. Then he tips the rear end of his abdomen forward and seizes her with his forceps about the ninth segment of the abdomen. Thus the ventral surfaces of both at the opening of the ducts from the sex organs are brought into juxtaposition (that of the male being turned upward) and copulation is effected.

Apparently in many of the commoner mayflies there is little concentration of period of flight beyond that imposed by the yearly cycle of summer and winter conditions. One finds nymphs of all sizes of Callibaetis and Stenonema all the summer through, and one finds no great swarms, but always a few adults throughout the season of warm weather.

The production of living young. Viviparous reproduction among mayflies has long been known for the European species *Cloeon dipterum*. Von Siebold observed the birth of living mayfly nymphs in 1837. Calori observed it again in this species in 1841. Causard rediscovered the phenomenon independently in 1896. His interesting account of it is in part as follows:

"Desiring one day to study the circulation of the blood in one of these insects in the living state, which I supposed to be sufficiently transparent for the purpose, I took one of them and placed it between two plates of hollowed glass. A quantity of greyish matter exuded from the insect caused by the pressure of the plates. On examination with the microscope, I observed that the extruded matter was formed of a large number of little ovoid bodies, which immediately began to move about and unroll themselves. Each of them was a little larva which was very active and began to swim about. Were these Ephemerae viviparous? Then I examined a large number of individuals and found the eggs in all stages of development: in some the segmentation was but slightly advanced, but a commencement of evolution was very distinct; in others, the extruded larvae showed segmentation; the most advanced enclosed completely developed larvae, but still enclosed in the transparent membrane of the egg. I have since been able to observe the female emitting their larvae freely without any pressure, so that their viviparity is a proven fact."

In certain species of the American genus Callibaetis the production of living young has since been found. Doctor Murphy, while dissecting a specimen from Brazil, found advanced embryos and fully developed first instar nymphs in the ovaries; wherefore the species was named *C. vivipara* (Needham and Murphy, 1924). Dr. Osgood Smith made a similar discovery in an undetermined female of this same genus collected in the wild life preserve at McLean, New York. It is not known at the present time (1934) whether this female belongs to any of our numerous described species.

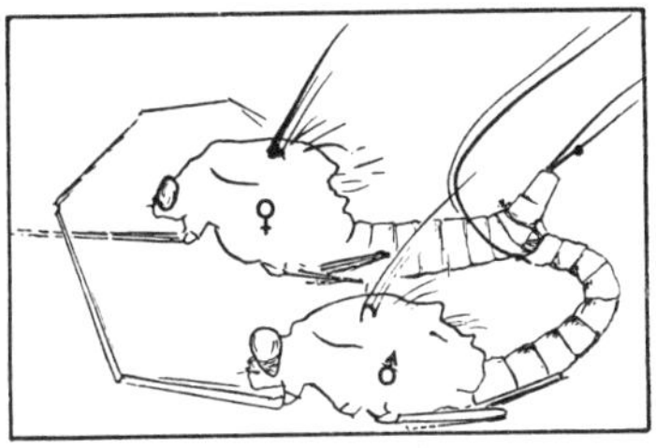

FIG. 13—Diagram showing the copulatory position of a pair of mayflies in flight; based on observations made on Siphlonurus, Paraleptophlebia and Cinygmula.

Parthenogenesis. Very little is as yet intimately known concerning parthenogenesis in mayflies, and that little is somewhat puzzling. At Ithaca, New York, *Ameletus ludens* is a very common species, and several good collectors have gathered specimens by the score, seeking males, and finding only females. Scores more have been reared from nymphs, all females. Yet, two males have been found, and one of them (the male type specimen) is in the Cornell University collection. In this species parthenogenesis seems to be partial. Clemens (1922) demonstrated parthenogenesis by taking the eggs from females reared in isolation and hatching them.

Somewhat different is the case of *Ephemerella rotunda*, also common at Ithaca. Doctor Morgan reared only females when she described the species in 1910. Scores more were reared by Dr. C. K. Sibley in 1923, all females. Yet in 1933, in preparation for this book, scores more were reared by us, and about half of these were males. Is parthenogenesis in this species intermittent? Dodds (1923) wrote concerning an unnamed species of Ephemerella that he collected in the mountains of Central Colorado:

"About thirty nymphs of this species in our collection include no males."

The most remarkable thing about the life cycle in mayflies is the sharp line of demarcation between periods of growth and reproduction. The nymph eats and grows. All that it has built up and stored, the imago uses up in reproduction. The body of the female mayfly has become hardly more than a scaffolding for carrying a mass of eggs. Her mouth parts are atrophied; her alimentary canal is an air reservoir; her large muscles are all muscles of flight; her chief appendages are outriggers for control of flight; and her body is filled with eggs from end to end, even up into the rear of the head. To produce this great mass of eggs and to get them fertilized and back safely into the water is her great end in life.

CHAPTER IX

TAXONOMIC CHARACTERS OF THE ADULT

In this chapter and in several which follow we propose to review the characters commonly used in taxonomic work, setting forth something of the range of their variation within the order. We will deal first with the adult and then with the nymph. In general we will not consider the subimago separately, for structurally the subimago is like the imago: the heavily chitinized parts with which we are dealing, have already taken on their final form.

The largest known mayflies are fossils of the genus Protereisma (see Chapter XIV) some of which had a wing expanse of nearly two and a half inches. The largest recent North American forms are the well known members of the genus Hexagenia, which attain a length of body of about an inch, with tails in the male that are more than twice as long. The smallest of our mayflies is the little white-winged crepuscular *Caenis hilaris* (*diminuta*) which has a length of about a twelfth of an inch (2 to to 2.25 mm), with tails in the male three times as long.

The form may be long and slender, as the generic name indicates for Leptophlebia; or short and thick, as the specific name indicates in *Baetisca obesa*.

There is in general likeness in form throughout this order. Adult mayflies have a brief aerial existence, and the air is the same for all. There is, therefore, little chance for changes due to the impress of environment.

1. The Head

The head of the mayfly nymph is of the usual form; but it is safe to say that there is hardly another order of insects in which the head of the adult has undergone such extraordinary modifications. These are due to the loss of mouthparts with their accompanying musculature and to the extraordinary development of the eyes (fig. 8). The face is reduced to a shrunken shelf, projecting above the vacant space where mouthparts should normally be, and carinated in front of the middle ocellus. The longitudinal ridge, the *nasal carina*, is often nose-like when viewed in profile. The transverse shelf below it is often bilobed and mask-like on its flaring margin. It expands downward to grotesque proportions in the South American genus Alloydia (fig. 14). In the more specialized genera, ocelli and eyes well nigh cover the top and the sides, especially in the males, and the occiput is reduced to another, lesser, shrunken shelf of variable form in the rear. The antennae are correspondingly reduced. They are minute, with a setaceous, almost vestigial flagellum.

Three ocelli are present, though the middle one tends to atrophy. Often they are larger in the male than in the female. The median ocellus is in most genera about half the size of the others. In Caenis and Tricorythodes they are farthest apart with a wide, flat area on the top of the head between them. Here the lateral ocelli lie against the compound eyes and are directed laterally and obliquely outward. In Heptagenia they lie closer together on the top of the head and are directed upward. With the expansion of the compound eyes the ocelli are crowded forward and closer together. In such forms as Blasturus, Siphlonurus, and Isonychia they become approximated, subpedunculate, and point forward.

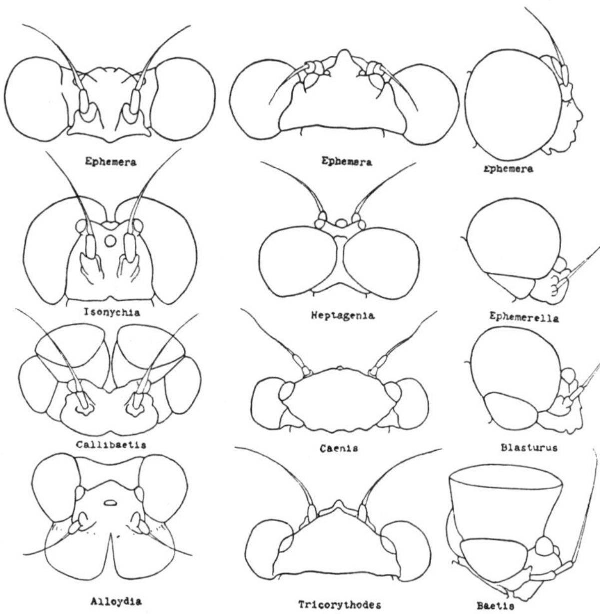

FIG. 14—Eyes of males: first column, face views; second column, top views; third column, side views.

The compound eyes of the imago exhibit a very great range in size and in degree of differentiation between the sexes. With a few exceptions (such as the Old World genus Palingenia) they are larger in the male than in the female (fig. 15). They expand on the dorsal side until in certain genera they entirely cover the top of the head. In Ephemera there is only a moderate enlargement of the eyes of the male. In such genera as Heptagenia, Baetisca, Siphlonurus, and Isonychia, they are enlarged and somewhat lighter in color on the expanding upper portion. In such as Blasturus and Ephemerella a groove appears between lower and upper portions, and there is further differentiation in size and color, the small lower division being blackish. The expanding upper division quite overspreads the top of the head. In all the genera of the Baetinae the eyes are double, the upper division having become pedunculate upon the lower, growing upward in a more or less mushroom-shaped cap, with the eye elements (ommatidia) greatly elongated (fig. 16).

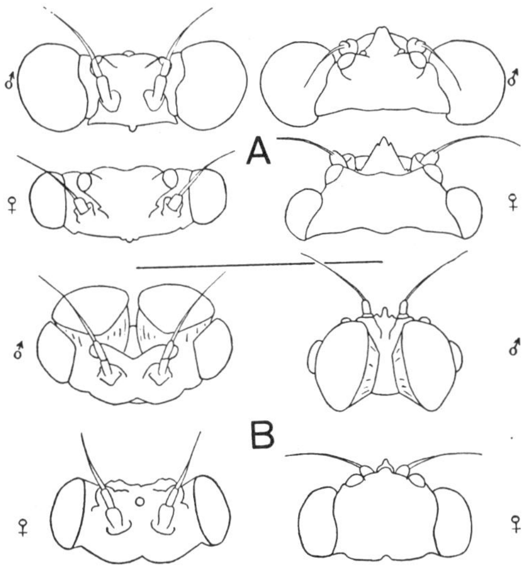

FIG. 15—Eye development in the sexes in *A*. Ephemera and *B*. Callibaetis; male above, female below; face view at left, top view at right.

The compound eyes of the male subimago are somewhat smaller than those of the imago. They are differentiated in the nymphal stage and made ready for successive expansions at the last moltings. On the inner side of the eyes of the male nymph they become externally visible as swellings on the top of the head. When the nymphal skin breaks at ecdysis, they suddenly spring up to mushroom-like form through the gaping sutures in the old skin. And again later, on the breaking of the skin of the subimago, they immediately push through the rent and expand to final size, swelling almost to bursting with liquid.

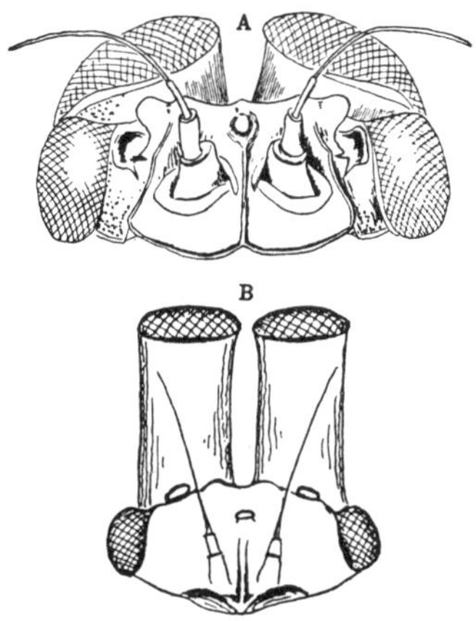

FIG. 16—Divided eyes of the male in two Baetinae; *A*, *Callibaetis fluctuans* (Drawing by Anna H. Morgan); B, *Pseudocloeon binocularis* (from Peru).

This extraordinary development of upper eyes in the male has evolved along with the habit of mating in the air, in a nuptial flight in which the male approaches the female from beneath. When, on the other hand, the female is

valve has remained rather short, while the long egg guide has developed along the venter of the eighth sternite.

The tails* of adult mayflies are either two or three. The two lateral tails (cerci) are always present. These are of ancient ancestry and wide currency among insects. The middle tail (telofilum) may be present, or vestigial, or absent. It may be present in one sex (the female) and absent in the other. The tails vary remarkably in length as between different families, genera, species and sexes, and their length is sometimes correlated with the length of the fore legs in the male.

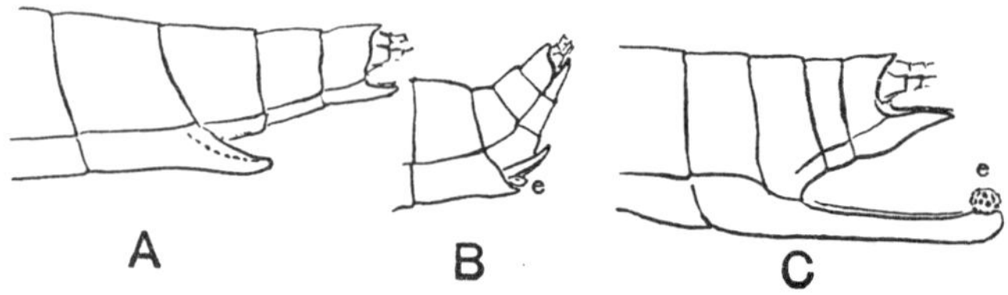

FIG. 20—Ovipositors (after Emily Reed Morrison); *A*. that of *Paraleptophlebia praepedita; B*. in that of *Habrophlebiodes betteni; C*. that of *Hagenulus caligatus* (Cuba). *e.,e,* eggs.

Flaring lateral margins that terminate to rearward in a sharp angle or spine are developed on some of the abdominal segments. These are generally moderate in size and prominent only on the segments toward the rear of the abdomen; but in two genera they are very wide and give anonisciform appearance to the abdomen. In *Siphlonisca aerodromia* (fig. 21) they occur on segments five to nine; in *Oniscigaster wakefieldi* (New Zealand), on six to nine in the female.

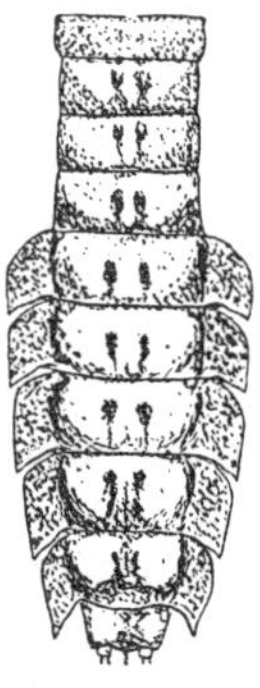

FIG. 21—The abdomen of *Siphlonisca aerodromia,* from above.

Gynandromorphism is not wholly unknown in Ephemeroptera. There is a specimen in the Cornell University collection, taken by one of us (Traver) on Lake Erie, July 7, 1934, that curiously combines male and female characters. It is one of the *interpunctatum* group of Stenonema. It is predominantly a female, with the body full of eggs; but the left side is strikingly masculine in the deep brown color of the top of the thorax (which in the female is pale yellow), and in the well developed half forceps and single penis at the apex of the ninth segment. Head, legs, and tails are those of the female on both sides.

*The term "setae" commonly used for these is very inexact. Many (including ourselves!) have been guilty of its use. The tails bear setae.

CHAPTER X

WINGS AND LEGS

The appendages of the thorax in adult mayflies are so important for purposes of classification that we deem it worth while to devote a special chapter to them.

1. Wings

The chief activity of adult Ephemeroptera is flight, and mayflies have good wings. These are less fragile than are the other appendages of the body. They are firmly attached to the thorax and rarely become detached or broken. There are generally two pairs of very unequal size. The hind wings are small and become progressively smaller as specialization proceeds, until in a few genera they disappear altogether.

The two pairs are coadapted to function as a single pair in flight. With compacting of the synthorax, the hind wings have been drawn forward at the base. An expansion of their front margin near the base underlies and supports the axillary region of the fore wing, and their broad expanse is directed to rearward. The hind wings in the oldest fossil mayflies are as large as the fore, but in all recent ones they are greatly reduced. They are relatively longest in Siphlonurus and its allies, where they are fully veined. Both pairs are traversed by long straight veins that radiate outward from the base, the middle veins widely forking. They are fan-like in shape, and in the regular fluting of their surface, high and low ("convex and concave") veins regularly alternating except near the hind margin. In completeness of fluting they are unique among winged insects.

The fore wings are roughly triangular in shape, with a stiff, straightish front margin, a rounded apex, and a broad hind angle. As in other insects, the front of the wing is stiff and the hinder part thin and pliant, as befits their sculling action in flight. The flexibility of the wing tip is increased by weak spots called *bullae* in some of the principal veins midway their length.

The fore wing on first inspection naturally falls into three principal areas (fig. 22), of which the foremost is the costal strip strongly supported by three nearly parallel veins; the second is the broad, fan-shaped middle field of the wing, traversed by widely forking veins weakly attached at base, and the third is a narrow field at the rear traversed by a graduated series of rapidly shortening veins that are conjoined and bent rearward at their bases.

A deep, cup-shaped, thick-walled cavity in the wing base, opening downward, is another peculiarity of the mayfly wing. The origin and significance of this pleural wing recess are as yet unexplained.

Venation

To the venation of the wings the student who hopes for acquaintance with the genera of mayflies should pay special heed. Venation is definite and easily observed. It suffers least distortion in preserved specimens. It remains the same through the different developmental stages. It is available even in broken specimens, for the wings are usually retained after other appendages are lost. Its main features are easily learned, and they afford a ready clue to relationships. Eaton says of it (*Rev. Monograph*, p. 4), "Unstable in minutiae, so closely is the essential plan of the neuration adhered to by nearly related mayflies that the general facies of the wing is an important aid to their classification, affording characteristics as easily recognized as the style of branching in the case of trees."

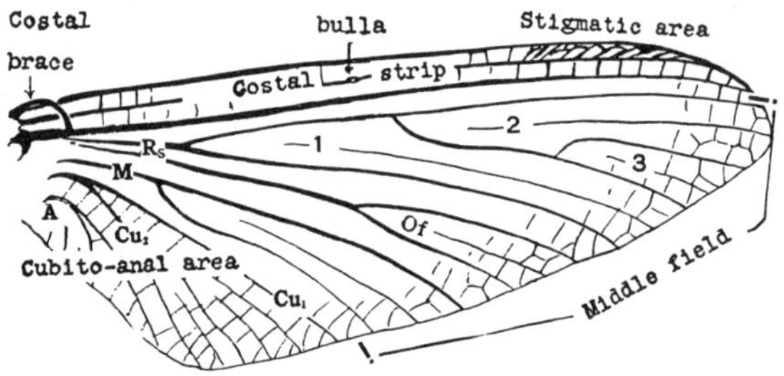

FIG. 22—A diagram of the main topographic features of a mayfly wing (based on a fore wing of *Ironopsis permagnus*), cross veins of the middle field mostly omitted. The numerals 1, 2 and 3 designate successive forks in the triple triad of the radial sector; forks 2 and 3 are not present in the hind wing of recent mayflies. *Of*, outer fork of Rs.

The veins of mayfly wings are the same as in other insects:

Costa	(C)	Media	(M)
Subcosta	(Sc)	Cubitus	(Cu)
Radius	(R)	Anal Veins	(A)

Three nearly parallel veins extend along the front or costal margin of the wing: costa (C), subcosta (Sc), and radius (R1). The subcosta lies in a groove between the other two and is braced against them by numerous weak cross veins, and by a strong costal brace at the wing base. These three thus joined together support the wing as a mainmast supports a sail.

Two widely forking veins occupy the outspread middle area of the wing: the detached radial sector (Rs) and media (M). Both are forked almost to the wing roots, and the former is deeply forked again dichotomously farther out. The remaining small area at the inner margin of the wing contains one long vein, the cubitus (Cu), and several very short anals (A). The cubitus is divided to its very base, and its anterior division (Cu 1) is often branched, with the branches pinnately arranged on its posterior side. There is much variability in this region in the different genera.

This venation pattern holds throughout the order. It differs from that of other recent orders in the straightness of the veins; in lack of differentiation among the numerous cross veins; in the basal detachment of the radial sector (Rs) from the radial stem (R 1); in the expanded area occupied by the radial sector, and in the possession of a distinctive costal brace.

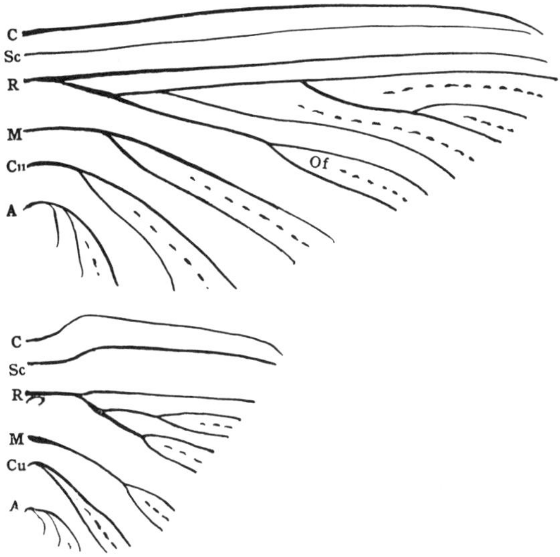

FIG. 23—A diagram showing manner of branching of the principal veins in the wings of an hypothetical primitive mayfly.

The anterior division of the radial sector is a low-lying or concave vein; the posterior is a convex vein. The fork in the latter is one of the best venational landmarks, for it is usually strongly developed, stands well out in the middle of the wing beyond the level of the other primary forks. For brevity and convenience of reference we will call this the *outer fork* and label it **Of**. In Leptophlebia and its near allies this fork is sagged to rearward, becoming strongly unilateral.

In the large middle fan-shaped portion of the wing extra veins occur toward the margin. They are arranged in threes (triads). The fork of the median vein and that of the posterior division of the radial sector are single triads, but the anterior wide-spreading fork of the radial sector is a triple triad, being subdivided as indicated in figure 23.

Throughout this area high and low veins alternate, and this regular furrowing of the wing surface is a further aid to the identification of veins. The aid is needed when forks become detached, and main branches appear as intercalaries or independent veins.

Longitudinal veins. The costa forms the front border of the wing and extends from base to tip. It is a stout vein that widens at the base into an articular callosity. The subcosta is a low-lying, weaker vein that generally parallels the costa almost to the wing tip. Midway in its length it is interrupted by a bulla—a weakly chitinized dilation that is supposed to be a point of special flexion. Similar bullae at the same level may occur in several other low veins to rearward.

The radius in mayflies is unique and remarkable for the depth of its forks, for the great development of its sector, Rs, and for the basal detachment of the sector from the main stem, R_1. In the generalized wings of Siphlonurus and its near allies, the wide-spreading anterior division of

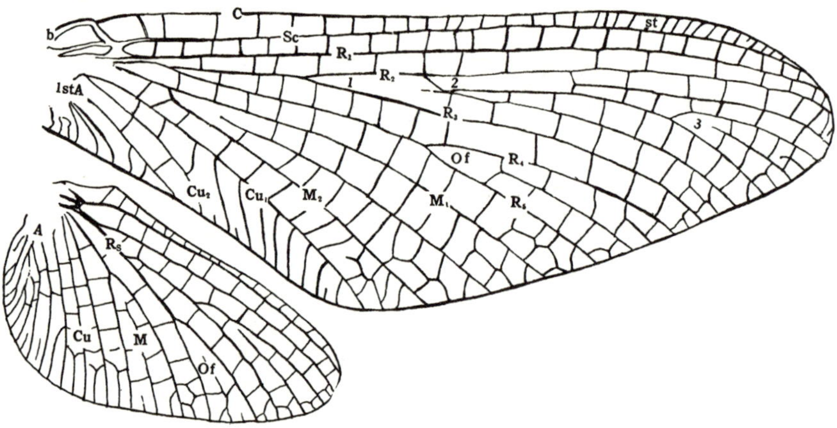

FIG. 24—Wings of *Siphlonurus berenice.*

the radial sector is twice subdivided by a fork within a fork. The first added fork springs from the rear of vein R_2; the second from the front of the first. All these are separated by intercalaries as shown in our figures 22, 23, and 24. With specialization, these extra branches tend to become detached. They then appear as intercalaries. They may even become shifted at base and reattached to the opposite side of their forks. This will be confusing unless one carefully notes that they remain constant in number and in fluting; high and low veins regularly alternate as before and continue to occupy the same relative areas.

The median vein in this order has a single deep fork that lies proximal to or nearly at the level of the first fork of the radial sector. This is typically a complete and symmetrical fork, but in the Ephemeridae it lies close to the wing base and is sagged to rearward, becoming strongly unsymmetrical. In the Baetinae and in others with reduced venation the posterior branch becomes detached and appears as an independent vein.

The cubital vein is divided almost or quite to the wing base and (except in Baetisca) into two very unequal and unlike parts. The anterior branch, Cu 1, is smoothly curved and often bears short branches or long marginal veinlets on its posterior side. It ends on the outer margin of the wing near the hind angle and is a high vein. The other branch, Cu 2, is generally much shorter, ends on the inner margin and is a low vein.

The anal veins are weak and rather irregular (so far as present study has revealed). They are crowded into the little remaining inner corner of wing that is bounded externally by vein Cu 2.

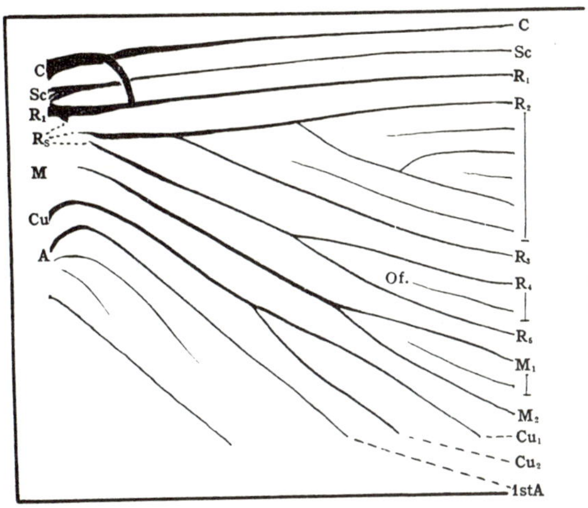

C	M	T	E	R	
C	C	C	1	i	+
Sc	Sc	Sc	2	ii	−
R_1	R_1	R_1	3	iii	+
R_2	M_1	R_2	4		−
R_3					+
		R_3			−
					+
	Rs			iv	−
R_4				v	+
R_5	M_2	R_{4+5}	5	vi	−
M_1	M_3	MA_1	6	vii	+
M_2					−
M_3	M_4	MA_2			+
Cu_1	Cu_1		7	viii	−
					+
Cu_2	Cu_2	M_{3+4}			−
1stA	1stA	Cu_1	8	ix	+
2dA	2dA	Cu_2	9		−
3dA	3dA	1stA			+

FIG. 25—Diagram showing the system of vein designations used in this book, and in parallel columns, their equivalents in other systems. The column headed *C* is that of Comstock and Needham (1899), used hitherto by us, by McDunnough, Lestage, Ulmer, Schoenemund, Chopra, Ueno, Phillips, and others. It is illustrated in Plate II. The *M* column shows that of Dr. Anna H. Morgan (1912); it was adopted by Professor J. H. Comstock in his *Wings of Insects*. It is illustrated in fig. 28. The *T* column shows Tillyard's latest system, rejected by us for reasons stated in the text. The *E* column shows Eaton's system in Arabic numerals; and the *R* column, that of Redtenbacher in Roman numerals. One of Eaton's labeled figures is copied on p. 123. The plus and minus signs in the last column indicate high and low veins respectively.

Having set forth in the preceding pages the system to be used in this book, we now place side by side in a diagram (fig. 25) the other systems that have attained some currency. These will aid in using the literature of the group.

Eaton numbered the veins without attempting to homologize them with the veins in other orders. Within the order, for the fore wings at least, he indicated homologies correctly. The system proposed by Comstock and Needham in 1899 is the one that has hitherto been most widely used. It appears to have been in error in locating the median vein and all that follow it one triad too far forward in the wing. The venation pattern of the hind wing is composed of these same elements in somewhat altered and abbreviated form. The costal border is dilated just beyond the costal brace into a thin flat lobe or angulation of very variable form, that lies beneath the anal area of the fore wing. Beyond this the costa and subcosta are often fused. The radius is more typical of insects in general than in the fore wing. In generalized forms its sector is attached normally to R 1, and only twice forked, both divisions of it with their single intercalaries being simple triads. Media is often forked farther out

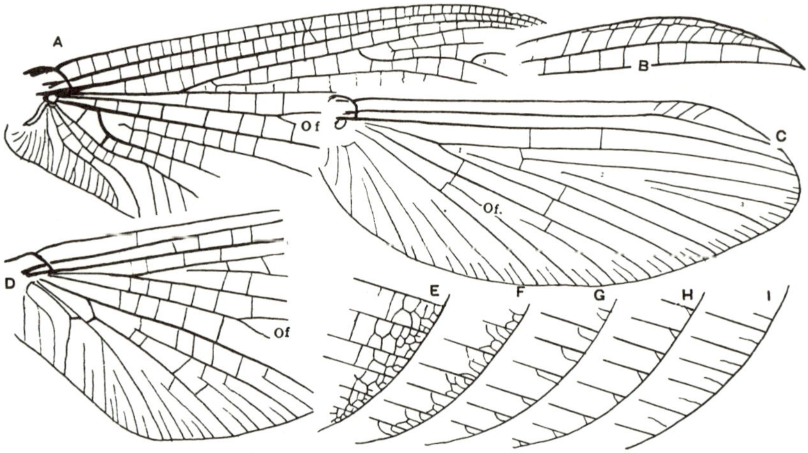

Fig. 26—Specializations in the mayfly fore wing. *A*, the rather generalized wing of *Hexagenia sp.*?, showing a deep median fork, heavily sagged to rearward and strongly braced against a sigmoidly curved Cu2. *B*, the stigmatic area in *Cinygma integrum*. *C*, highly specialized wing of *Pseudocloeon sp.*?, with all forks but one detached, and with crossveins greatly reduced. *D*, The wing base of *Ephemerella rotunda* showing the median fork strongly braced against an angulated Cu2. *Of*, outer fork of the radial sector. *1*, *2*, *3*, successive forks of the triple triad. *E* to *I*, a reduction series showing the evolution of regular marginal veinlets from an irregular archedictyon. *E*, as in Ephoron; *F*, as in Hexagenia; *G*, as in Leptophlebia; *H*, as in Paraleptophlebia; *I*, as in Cloeon.

in the wing. The cubital fork is extended to the wing base. Its intercalary is extended to the base also, so that the cubital triad has come to appear as three complete, subparallel, independent veins. The weak anal veins when present are even more variable than in the fore wing.

The evolution of the hind wing followed a different course from that of the fore wing. After the two wings came to act as one, and after the function of the hind wing became in part that of support for the anal angle of the fore wing, the costal angle, which assumed that function, became strengthened and expanded, while the exposed hind portion became very thin and tended to atrophy. There was non-selective reduction of chitin toward the rear, and when numerous veins persisted they became long, closely parallel, weak, and thin, as in Baetisca, or, in others, faint unto complete disappearance.

The forks of the radial vein did not deepen to the wing base, as in the fore wing, but retained generally the normal attachment of Rs to R 1. At this point a strong triradiate brace developed, with reduction of superfluous crossveins. The outer fork (Of) of Rs disappeared in several families by the fusion, R4+5 (fig. 27, B and C).

In the hind wing of Ephoron occurs the only long vein known to us in this order that has not been reduced to regular longitudinal fluting (Fig. 31, Vx). It is a high vein lying behind another high vein, R 1. It is unattached basally and somewhat zig-zagged in its course, and it appears to be a stranded relic of the old archedictyon, that has not been reduced to the system.

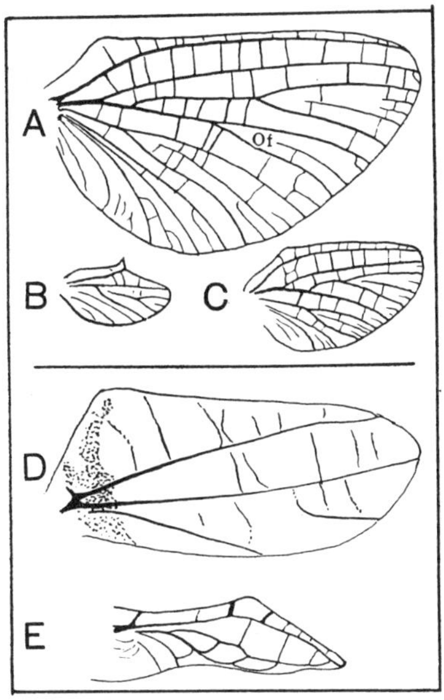

FIG. 27—Hind wings. *A, Ameletus subnotatus; B, Thraulus presidianus; C, Ephemerella dorothea; D, Callibaetis skokiana; E, Dipteromimus tipuliformis* (Japan). *D* and *E* are more enlarged.

The hind wing of Ameletus, shown herewith at **A**, clearly exhibits certain trends of venation within the order. The radial vein is dichotomously five branched, with the branches all in relations that may be considered normal for insects in general. Of the triple triad of vein Rs that was present in the hind wing of Permian fossil mayflies, only the first fork and a vestige of the second (2, of figure 22 on page 116) remains. This vestige recalls that found in the reduced fore wing venation of certain species of Campsurus.

At the first fork of the radial sector the anterior branch is slightly askew forward. It becomes weaker and more askew in Ephemerella (as shown at **C**), and entirely detached in Thraulus (as shown at **B**). With further losses the reduction is so great that homologies are not readily traceable.

Cross veins. Mayflies have made little use of cross veins for special braces in the framework of the wing. Instead, they have developed fluting of the whole wing surface and have used cross veins mainly to hold the furrows and ridges in place. Long ridgings of slender veins require support at many points. Hence cross veins are numerous and for the most part undifferentiated.

There is no stigma in the mayfly wing, but there is in the fore wing a corresponding stigmatic area in which the cross veins are generally heavier, more numerous, more irregular, and more aslant than elsewhere. Often they are conjoined or forking, and in a few genera they divide the costal space into a double row of more or less pentagonal cells.

Clearance of cross veins occurs in certain areas; most frequently, in the costal strip before the bulla. Here the very strong costal brace holds the three long veins securely at the base; and it may well be that this clearance beyond the base allows freer torsion of the wing around vein R 1 as an axis and makes for sculling efficiency.

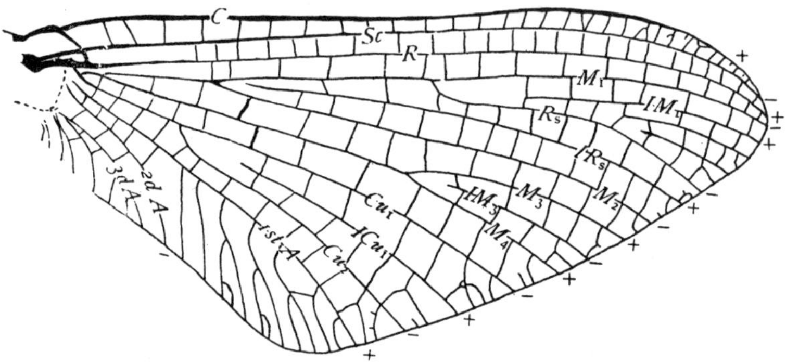

Fig. 28—The wings of *Isonychia albomanicata* showing the Morgan system of vein designation (from Comstock).

Clearances also occur wherever the base of a principal fork becomes strongly unilateral. The offset branch becomes thickened. Perhaps the material of the missing cross veins is used to strengthen it. Adjacent cross veins that stand in opposition to the sagging fork then become strengthened. This is the only kind of special bracing found in mayfly wings. It is illustrated in fig. 26 A and D.

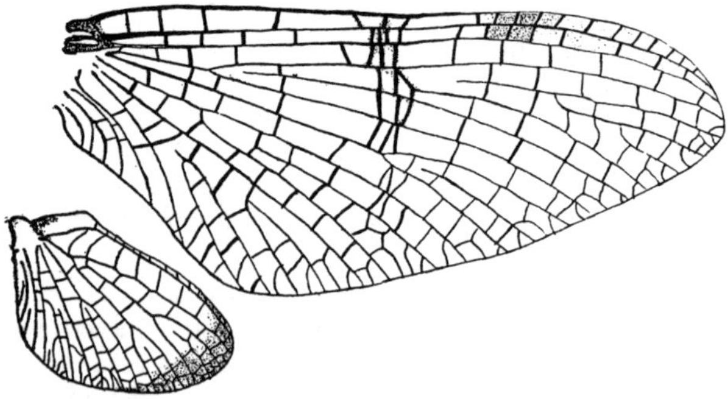

Fig. 29—Wings of *Stenonema pudicum*.

Sometimes the cross veins of the wing disc tend to cluster in a cross band behind the bulla as in the *Stenonema pudicum* (fig. 29) and in those species of Heptagenia and Stenonema that have the more heavily pigmented wings. The cross veins are clustered in a transverse band, with moderate clearances on either side of the band. Further condensation of vein-forming materials may result in fewer cross veins arranged in broken transverse lines as in the Oriental genus Compsoneuria (fig. 30).

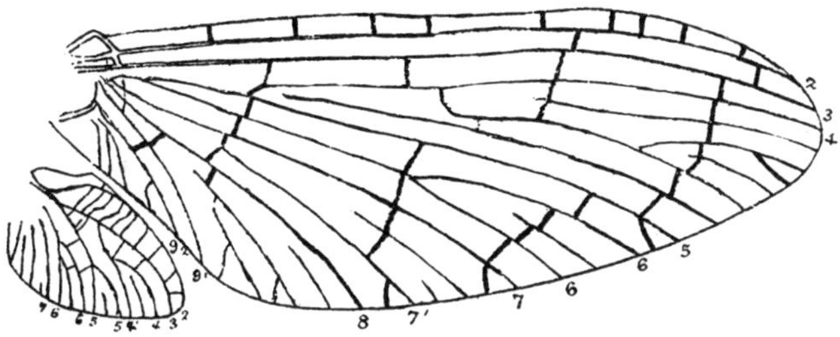

FIG. 30—The wings of *Compsoneuria spectabilis* (from Eaton)

Specialization is plainly seen in the reduction of the number of cross veins. The primitive original net work of them (archedictyon) that characterized the wings of many of the most ancient fossil insects is practically wanting in recent mayflies. There are vestiges of it near the outer wing margin in Ephoron and Siphlonurus and other generalized forms, where a border of irregular cells remains. Selective segregation of vein-building material is evident, as in other orders of winged insects.

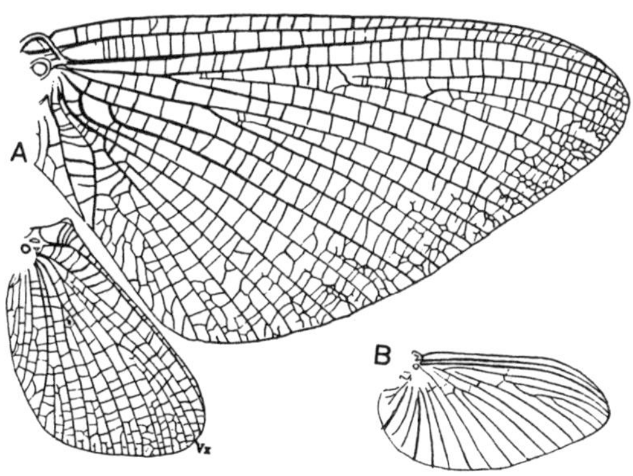

FIG. 31—Extremes in cross-venation: *A. Ephoron album; B, Caenis hilaris.*

For the most part cross veins stand in regular rows perpendicular to the bordering veins. As reduction proceeds their number diminishes markedly. In Ephoron (fig. 31), for example, the cells into which they divide the surface of the fore wing may number between 350 and 400; in Caenis, 10 or 12.

Marginal veinlets. With the complete disappearance of the irregular network there emerges out of it a series of marginal veinlets. These occupy the interspaces between the outer ends of the veins. They vary in number, in length, in regularity, in slant, in straightness, by genera and by species, according to the degree of their reduction. In the Baetinae they form a distinct and regular marginal band, occupying the interspaces singly or by twos. In a number of genera (Campsurus, Choroterpes, Caeninae, and all Oligoneurinae) they are wanting. A few of the commoner types are shown in figure 26.

The mistaken idea that mayflies are generalized insects has twice led to proposals of theoretical venation systems based on the longitudinal fluting of the veins and on their number. This fluting is more complete in mayflies than in any other insects; but, unfortunately for these theories, it is most complete in the most highly specialized mayflies.

Adolph (1879) saw the regular alternation of high and low veins and he thought them to be independent veins, as they appear to be in the most specialized mayflies. He numbered them independently from front to rear, with odd numbers for the high veins and even numbers for the low ones.

Adolph's theory was carefully studied by Brauer and Redtenbacher (1888), Haase (1891), Spuler (1892), Comstock (1893), Packard (1895) and Kellogg (1895), and was rejected by them all. It is now of only historic interest. It recognized too many veins and could not be applied to insects generally.

Lameere (1917) also was impressed by this fluting and by the depths of the forks, but had a different notion. He conceived a primitive wing with half as many principal veins, each with a high anterior and a low posterior division. The radius with its high stem and its low sector was for him a typical vein. He believed that media and cubitus had originally been formed on the same plan—possibly the costal and anal vein as well (costa, with Sc as its low member).

This was an improvement over Adolph's system in that it named as principal veins only those that are found in the wings of insects generally. It was like that system in that vein homologies were to be determined by alternating high and low position and by that only: and strangely enough neither Adolph nor Lameere nor any of their few followers considered it necessary to show any reason why in all the vast diversity of insect wings high and low should always fall in the same place. Whenever two high or two low veins occur side by side, they gratuitously assume that a vein of the opposite character has disappeared between, leaving no trace.

Vein fluting as a basis for determining homologies is attractively simple and easy of application—too easy in fact. The systems based upon it have given no heed to functions of crossveins or to types of bracing and they have found it necessary to ignore the vast body of evidence from wing tracheation now available. They have assumed an unusual consistency on the part of nature and have used the phenomenon of fluting in a wholly uncritical way. Anyone who carefully studies the fluting of mayfly wings may see that it has had an evolution within the order and is most complete and perfect in the most specialized forms. As fluting has progressed forks have either deepened to the wing roots or else they have become detached, and appear as independent veins.

When fluting begins forks are in the way, for they stand at the junctions, and there the vein cannot be both high and low at the same time. There are no independent veins in the older fossil mayflies, and the forks are farther out in the wing than in recent forms. In this they resemble the ancient Palaeodictyoptera. The primitive wing was flat, as the nature of its origin necessitates.

2. Wing Tracheation

Wings, like other organs, are best understood when studied in the light of their origin. We have already seen how they develop in mayfly nymphs as outgrowing folds of the dorsal integument of meso- and metathorax. At first they are flat and veinless, and the two pairs are alike in size and form. After their first appearance they expand with each successive moult, and tracheae grow out into them, forking as they go.

In primitive insects, as growth proceeds, the two layers of the flattened wing sac gradually become conjoined between the tracheae, enclosing them in channels which later will be the hollow tubular wing veins. Heavy scleritization of the walls of these channels, together with expansion and thinning and final stretching of the integument that lies between, result in the differentiation of wing veins from wing membrane.

In recent insects of the more generalized forms—those that have the largest wings in their immature stages—the longitudinal veins are all formed around tracheae that are fairly definite in number and in order of arrangement. In the stoneflies (Plecoptera) the coincidence between long veins and tracheae is exact. There are differences in detail in the different orders, but there is evident similarity, and it provides the best basis yet known for determining vein homologies (fig. 32).

However, in wing tracheation, as in so many other particulars, mayflies have departed far from primitive conditions. The wing folds of the nymphs are partially obstructed at the base by a cup-shaped cuticularization that appears to have thrown certain tracheae out of their normal course, and that holds the walls apart, even in the later instars, allowing the tracheae to stray and to compete for territory. Hence, primitive conditions have been hard to find.

Comstock and Needham were the first to try to interpret the venation of this order with the aid of tracheation. Having demonstrated for other orders of insects the uniformly double origin of the tracheal supply of the wing, they were at a loss to account for the single entering tracheal trunk that they found in this order. Their materials were wholly inadequate.

A much more thorough-going examination of the tracheae in nymphal

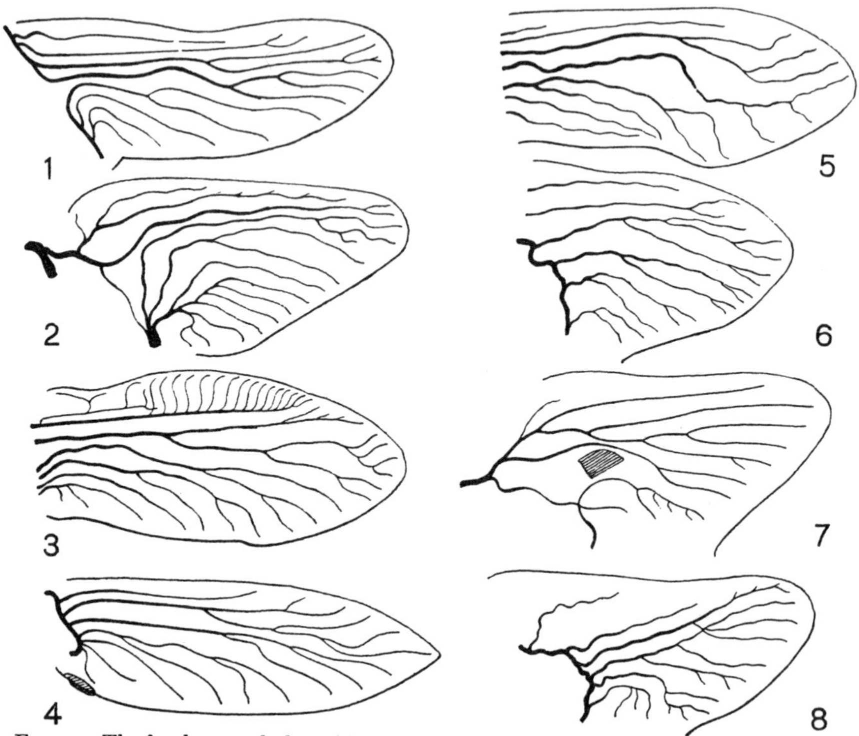

FIG. 32—The fundamental plan of insect wing tracheation, as seen in rather generalized representatives of eight orders.

1. Taeniopteryx (Plecoptera)
2. Xiphidium (Orthoptera)
3. Sialis (Neuroptera)
4. Eriocrania (Lepidoptera)
5. Psocus (Corrodentia)
6. Cicada (Homoptera)
7. Ameletus (Ephemeroptera)
8. Gomphus (Odonata)

Number 7 shows the most generalized tracheation yet found within the Ephemeroptera. The shaded area in the wing base is the developing pleural wing recess. Its chitinization seems to be the cause of the departure of tracheae Rs and M from this pattern in later stages.

mayfly wings was made by Dr. Anna H. Morgan. She studied some fifteen North American genera. She found great diversity among them with a strong tendency for marginal tracheae to disappear, and alternate intermediates as well, with the intervening trunks assuming dominance and filling the open wing sac with their branchlets. But she found one relatively primitive wing of great significance, as her figure which we copy here (fig. 33) will show. She clearly showed the separate origin of two groups of wing tracheae, costo-radial (*c-r*) in front and cubito-anal (*cu-a*) in the rear. She found the rearward attachment of the latter group almost atrophied, its burden shifted forward on a connecting basal trunk (tr). In this figure she correctly interpreted the costal strip of the wing in front and the cubito-anal area in the rear, but did not finally resolve the puzzle of the middle area, and was in a degree misled by the traditional error of belief that the Ephemeroptera and Odonata are closely allied orders.

For further light on tracheal origins Dr. R. N. Chapman (1918) traced the two groups of wing tracheae back to their origin in the thorax. We copy here (fig. 34) his diagram of their normal arrangement. It will be seen that the two groups of wing tracheae come from front and rear, springing from the two arms of a Y-shaped trachea that extends transversely between the leg and one of the main longitudinal trunks of the body. The arms of the Y reach out to two spiracles; thus each wing has two sources of air supply.

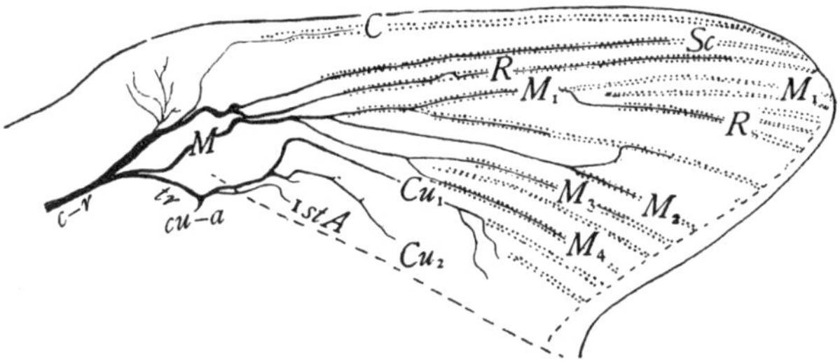

FIG. 33—The tracheation of a nymphal fore wing of *Iron humeralis* (after **Morgan**).

Chapman found that in the nymph of the mayfly *Iron humeralis* (fig. 35) the posterior arm of the Y is atrophied and with it the connection of the cubito-anal group of tracheae with the rearward spiracle. The air supply is then shifted forward on the basal trunk (*tb*). Chapman was the first to call attention to the saucer-shaped disc in the base of the nymphal wing as a possible cause for the diversion of tracheae from their normal course.

It fell to Dr. R. J. Tillyard to complete the explanation of the shift and to resolve the puzzle of vein homologies. Working in New Zealand with forms more primitive than had been available to other investigators, he found the wing tracheae developing after the normal insect pattern, with the radial sector springing from the main radial stem, and with the cubito-anal group of tracheae still attached to rearward. (fig. 36). This however, was in a nymph in its antepenultimate instar. He found that in the next instar the attachment of the radial sector was shifted to media at the point where the cup-shaped cavity ("callosity") in the wing base was developing, and he explained the shift as due to crowding. This is a reasonable explanation. Tracheae must remain open in order to function as passageways for air.

Such a transfer of branches is possible only at the time of ecdysis, when the chitinous lining of the trachea is withdrawn, together with all the air contained in it. The walls are then soft, and their cavity is wholly filled with liquid. As is well known, the refilling of the tracheal system with air begins in the larger trunks of the body and pushes thence out into all the lesser branches. On reaching the plastic and closely crowded tracheae of the wing base, and meeting with a growing obstruction in its

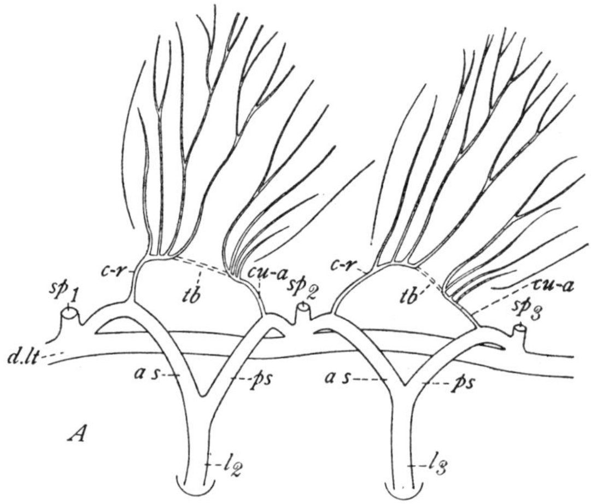

FIG. 34—A diagram showing the typical basal relations of the tracheae of the wings of insects (after Chapman). *d.lt.*, dorsal longitudinal tracheal trunk; *sp*, spiracles 1, 2, and 3; *l2* and *l3*, legs of meso- and metathorax respectively; *as*, anterior, and *ps*, posterior stem of each leg trachea, from which spring separately *c-r*, costo-radial, and *cu-a*, cubito-anal groups of wing trachea; *tb*, (in broken lines) the tracheal bridge that comes to connect these two groups of tracheae in the wing base.

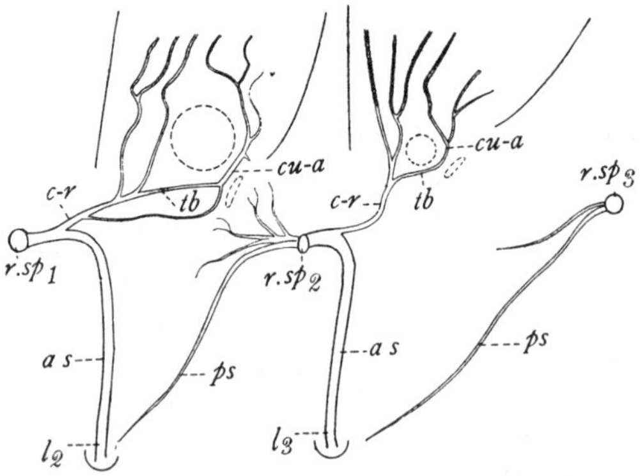

FIG. 35—The basal connections of the tracheae in the nymphal wings of *Iron humeralis* (after Chapman). Lettering as in figure 34. The rearward connection with a spiracle is interrupted in both wings, and the air supply for the *cu-a* group comes from the front over the tracheal bridge. Circle of broken lines is the developing pleural wing recess of the wing base, and the oval is a small apodeme.

old course, the air might force a new passageway, altering the conjunction of branches. That, apparently, is what has happened. That, at least, is the best available explanation of the detachment of the radial sector and of the cubito-anal group of tracheae from their normal bases.

That explanation is in accord with the usual evolutionary trend, in that more primitive conditions are found in the early than in the late stages of development. Furthermore, all lines converge backward toward a common starting point for the order, bringing the generalized members of the several mayfly families into accord with each other and with other orders of insects.

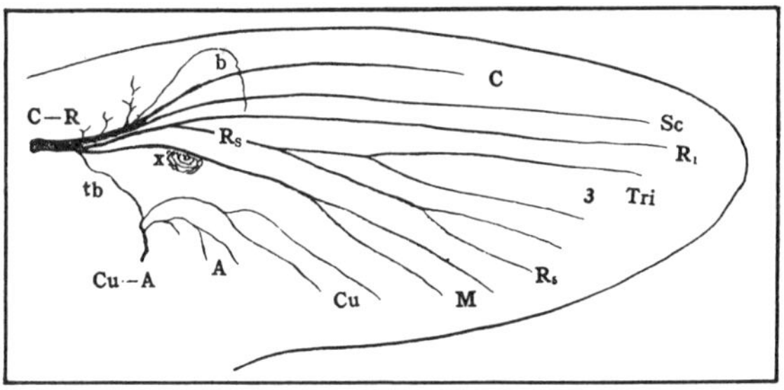

FIG. 36—Diagram of primitive tracheation of a nymphal mayfly wing. The two primary basal groups of tracheae, *C-R* and Cu-A, conjoined by a transverse tracheal bridge, *tb*, with vein M attached to the former group: *b* is a little precostal trachea about which the costal brace develops; *3 Tri* is the area in which the triple triad develops in the fore wing: the forks of Rs become more nearly equal in depth in the hind wing, more unequal in the fore wing; *x* is the developing pleural wing recess.

It is regrettable that the set of names for wing veins now most widely used in this order (fig. 25 C) is found in error and must in part be changed. It is quite clear that the tracheae of the cubito-anal group do not lie in the veins designated for them in that system. This anyone may see for himself if he will take the trouble to examine the tracheae and veins as they appear together in the penultimate nymphal instar of any of our more generalized mayflies, such as Siphlonurus, Isonychia, or Iron. Correct determination of homology is the only sure basis for a uniform system of nomenclature.

We regard Tillyard's 1923 paper as a good and satisfactory exposition of vein homologies in this order. We cannot follow him in his later (1926) abandonment of facts in favor of an untested theory.

During the last nymphal instar the wings expand enormously within their sheaths. In mayflies they become fluted lengthwise and crumpled crosswise and finally folded edgewise, pressing for release. After the fluting begins, the thickening of the veins proceeds rapidly, and mainly on the freer outer edges of the folds; hence the veins acquire walls that are very unequal in thickness above and below, the high veins being thickened in the upper integument layer, the low veins in the lower layer. The unilateral thickenings of the veins follow and probably result from the flutings of the wings.

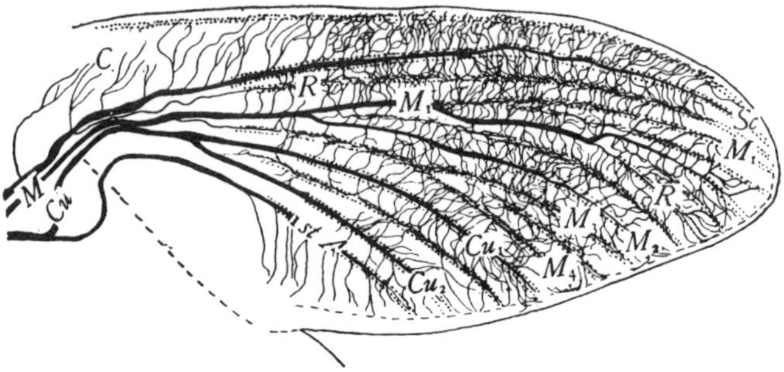

FIG. 37—Tracheation of a nymphal wing of Siphlonurus (after Morgan). Note the dwindling of alternate high-vein tracheae.

The dominance of tracheae over vein-development is shown in this order of insects in two ways: 1. In the development of veins about strong tracheae that, in the loose nymphal wing sac, have strayed out of their normal course. This is illustrated by Eaton's figure of an aberrant wing of Hexagenia (Monogr. Pl. VII, fig. 11c) and by a number of Adolph's (1879) published photographs. 2. When extensive reduction of venation occurs, as in the exotic Oligoneurinae, it follows the reduction of alternate tracheae herein previously noted (fig. 37).

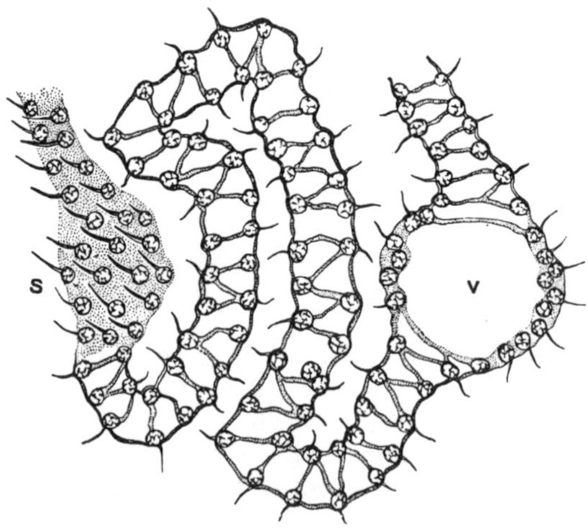

FIG. 38—A section of a crumpled wing of a subimago just withdrawn from the nymphal wing-sheath and not yet expanded; *v*, a developing vein; at *s* is shown a surface view. Note the microtrichiae arising singly from each cell (drawing by Velma Knox).

The wings of adult mayflies are predominantly hyaline, but those of most subimagos are colored in dull tints or in broad patterns. The pigment is mostly superficial, and as a rule it is shed at the final molt along with the small hairs (microtrichia) and primary cuticle. The color is deeper, however, and persists in the imagos of many species; oftenest in the hind wings, sometimes in the fore, more rarely in both. In these the colors are no longer dull. They are bright tints of light yellow and rich brown. The surfaces of hyaline wings are often beautifully iridescent.

FIG. 39—A resting pair of Hexagenia subimagos: note the position in which the fore legs are held. The upper larger figure is the female (photo by Dr. Miles D. Pirnie).

The sexes are generally alike in wing coloration, but in most species of Callibaetis the costal strip of the fore wing in the female imago is colored much more deeply than in the male, while as subimagos the sexes are similar. In Hexagenia, on the contrary, the veins of the male are more deeply colored, having dark spots on the wing and brown margined cross veins that the female generally lacks.

3. Legs

Adult mayflies do not walk or run or climb: they only stand; and some do not even that. Their legs are weak, and they grow weaker in certain groups as specialization proceeds until in a few genera they become useless rudiments. Nevertheless, though weak, they have undergone a high degree of differentiation. The fore legs differ in both form and function from the middle and hind pairs. They are lengthened and directed forward. They balance the tails in the rear and serve as outriggers in flight.

In the males the fore legs also serve for grasping and holding the female in copulation. They are extremely lengthened, and provided with a reversible joint at the base of the tarsus. This joint permits the tarsi to turn backward and take hold of the top of the female thorax when the male approaches her in flight from beneath. The fore legs are always longer in the male than in the female, and (except in Oligoneuria) longer than either of the other pairs.

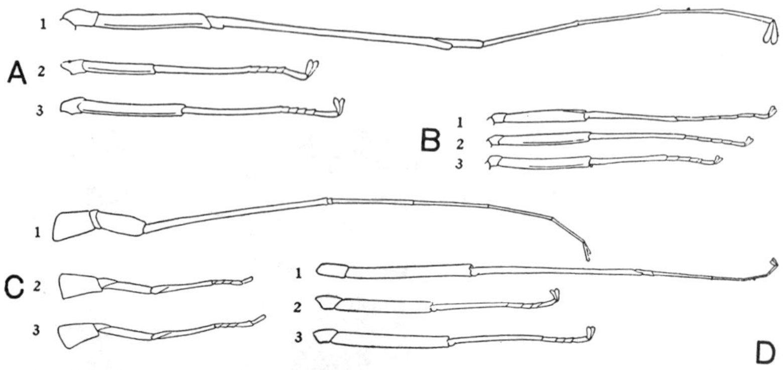

Fig. 40—Legs of adult male mayflies, showing proportionate length of segments. *A*. Ephemera; *B*, Isonychia; *C*, Ephoron; *D*, Potamanthus; in each *1*, *2* and *3* indicate fore, middle and hind legs respectively (from Kennedy).

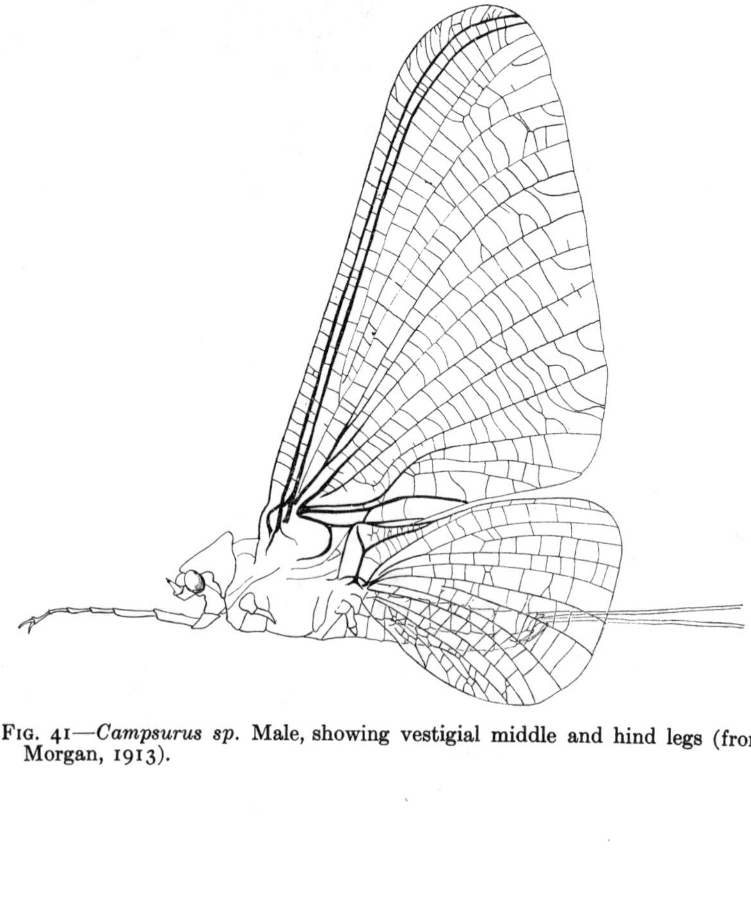

Fig. 41—*Campsurus sp.* Male, showing vestigial middle and hind legs (from Morgan, 1913).

The legs are longer in the imago than in the subimago. The middle and hinder pairs of legs do not differ greatly in length. In certain Old World genera (Palingenia, etc.) they become excessively weak and flabby, and in one American genus (Campsurus) they are reduced to

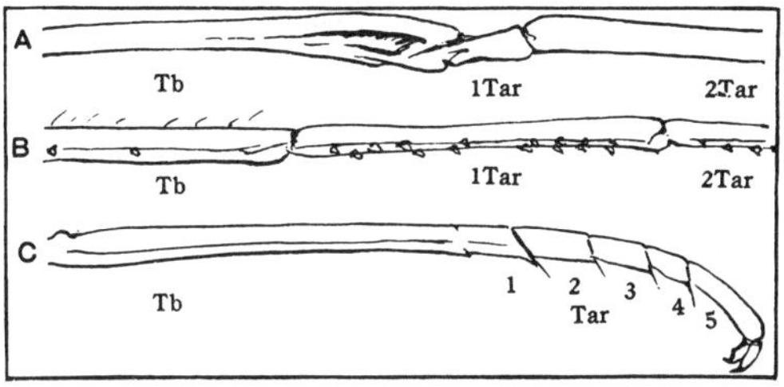

FIG. 42—Leg armature in adult mayflies. *A* shows the reversible joint at junction of tibia and tarsus in fore leg of the male *Ephemera simulans*. *B* shows the short blunt spines in the same parts in *Parameletus midas*. *C* shows the hind tibia and tarsus of the latter. Note the partial fusion of the first tarsal segment with the tibia.

mere stubs. All that remains of them is a small pointed basal portion consisting of coxa, trochanter, and a stub of the femur, chitinized and fixed in the wall of the synthorax and quite functionless. In Campsurus the fore legs are weakly developed and greatly reduced in size, but all the usual parts are still present (fig. 41).

In the male the fore tarsus is generally longer than its tibia, but not always. It is somewhat shorter in the Caeninae, in Anepeorus, and Heterocloeon; equal in a few others. In Siphlonurus and Baetisca it is two and a half times as long, and in Ametropus it is nearly five times as long as its tibia. The varying proportions in length of the tarsal segments serve to distinguish the genera of Heptageniidae.

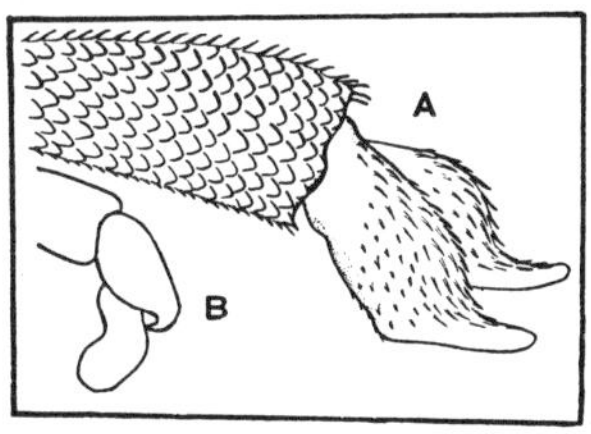

FIG. 43—Tarsal claws: *A*. In *Isonychia pictipes* adult male; note the scaly tarsus and the similar claws with reversed curvature. *B*. In *Ephemera Sp?*; note the unlikeness of the two claws.

A peculiarity of mayfly tarsi is their tendency to eliminate the ankle joint by fusion with the tibia. In the more generalized mayflies all tarsi are five-jointed, but one or two basal joints may be more or less completely fused with the tibia. This occurs less frequently in the fore legs. Degenerate legs may have only one or two tarsal joints.

Each tarsus bears at the tip the pair of claws that is common to insects generally, and in Siphlonurus these are alike, pointed, and hooked at the tip; but in most mayflies one claw of each pair is a degenerate, blunt affair, quite as large as the other but not claw-like at all. Often both are so. In some of the weak-legged Oligoneurines both claws are elongated and pointless.

There are short spines on the legs of a number of Siphlonurine genera (fig. 42B), mostly toward the tip of the tibia and underneath the basal tarsal segments, and terminating the obliquely truncate tips of all tarsal segments. There is sometimes very beautiful and delicate microscopic sculpturing and scaling of their apparently smooth surfaces (fig. 43A). But for the most part, mayfly legs are slender, carinated along their margins, and bare.

CHAPTER XI

THE BODY-WALL OF THE THORAX

By Velma Knox, Ph.D.

Much of the active life of mayflies has its mechanical center in the thorax. Organs of locomotion, the legs and wings, which play such an important rôle in the activities of the individual, are outgrowths of the thorax. Modifications of the nymphal legs to comply with varied conditions in the water, and the development of wings to make possible adult life in the air, necessitate a complex and highly developed thoracic mechanism. The discussion of this chapter deals only with the framework of the structure, with the outer skeleton which lends rigidity and support to the stress and strain of muscular action.

The importance of structure cannot be overemphasized in the interpretation of function. It was the purpose of this study therefore to work out in detail the structure of the mayfly thorax in order that a better interpretation might be made of its function. Since the early part of the nineteenth century much work has contributed towards a knowledge of the insect thorax. However, throughout the literature there is practically no information on the thoracic sclerites of the mayfly.

A true interpretation of the sclerites cannot be made without determining the muscle attachments, and likewise an adequate interpretation of the muscles is not possible without a knowledge of the integument to which they attach. Therefore the following two chapters deal with the sclerites as well as with the musculature in the thorax of the nymph and adult of a representative species, *Hexagenia recurvata*.

For the study of the sclerites the abdomens were clipped off and the specimens boiled for ten minutes in a ten per cent solution of potassium hydroxide. With a small pipette the muscles were washed away as much as possible. The integuments were rinsed in water, transferred to a concentrated solution of acetic acid and rewashed until quite clear. Picro fuchsin was used as a stain.

THE THORACIC SCLERITES OF THE ADULT

A. THE TERGUM

Mesothorax (Pl. XX): The tergum of the mesothorax is divided into the following regions: the antecosta (*Ac*)[1] or the pretergite[2], the prescutum (*Ps*), the scutum (*Sc*), the scutellum (*Sct*), and the post-scutellum (*Psct*).

[1] Snodgrass, 1927. Morphology and mechanism of the insect thorax. Smithsonian Miscellaneous Collections, 80:1–108.

[2] Crampton, 1914. Ground plan of a typical thoracic segment. Zoological Anzeiger, 44:56–67.

Antecosta (Pl. XX, Pl. XXII, fig. 1, *Ac*).—The antecosta is a narrow, transverse region well developed on the anterior margin of the tergum. The anterior margin of the antecosta bears two small lobes which project upward and serve for the attachment of the dorsal longitudinal muscles of the prothorax. The antecosta bears a pair of internal lobes, the anterior phragma (*Ph I*) which project into the cavity of the mesothorax and serve for the attachment of II Dlm.

Prescutum (Pl. XX, Pl. XXII, fig. 1, *Ps*).—The prescutum is present on the anterior lateral part of the tergum as an elongated triangular sclerite. It is deflected anteriorly into the antecosta. The prescutum is limited on the inner margin by a suture which terminates in an unsclerotized area separating the sclerite from the scutum. Posteriorly it articulates with the basalar sclerite (*Ba*) which supports the humeral plate (*HP*) of the wing (Pl. XX).

From the outside, in a specimen which has not been treated with potassium hydroxide, the prescutum appears to be a sclerotized groove on either side around the anterior base of the tergum. The groove becomes deeper anteriorly to fuse with the antecosta, forming the deep invaginations or phragmata.

The prealar bridges which serve as anterior supports of the tergum on the pleura are extensions from the anterior part of the prescutum and also from the antecosta (Pl. XX, extended outward; Pl. XXI, fig. 2, Pl. XXII, fig. 1, *PrAl*). The prealar bridge is very heavily sclerotized, forming a rigid support between the unsclerotized membrane of the prothorax and the pleuron. It fuses with the episternum (Pl. XXI, fig. 2, *PrAl*).

Scutum (Pl. XX, Pl. XXII, fig. 1, *Sc*).—The scutum is the largest sclerite of the tergum, and forms the characteristic hump. This shield-like structure which serves as a covering for the large dorsal longitudinal muscle is deeply rounded from side to side, and its surface is increased by a pronounced fold, the notal furrow, which could easily be mistaken for the anterior lateral margin of the scutum. This margin is made, however, by the prescutoscutal suture. The surface of the scutum is divided by two pair of ridges, (Par_1 and Par_2) the innermost of which is very indistinct. These ridges are homologous with the parapsides of Snodgrass (1927).

The lateral margin bears the anterior notal wing process (*AWP*) which projects outward to support the articular sclerites of the axillary region of the wing (Pl. XX). A process or invagination just behind the articulation forms a support for the surrounding muscle attachments. Anterior to this articulation the tergal portion of the basalar sclerite articulates with the scutum, (*Sc*) and also at that point with the prescutum (*Ps*). Posterior to the anterior process is a deep notch or lateral emargination of the tergum (*LE*). The scutum itself does not bear a posterior notal wing process in this species, but that which corresponds to a posterior process is a projection of the scutellum.

PLATE XX.

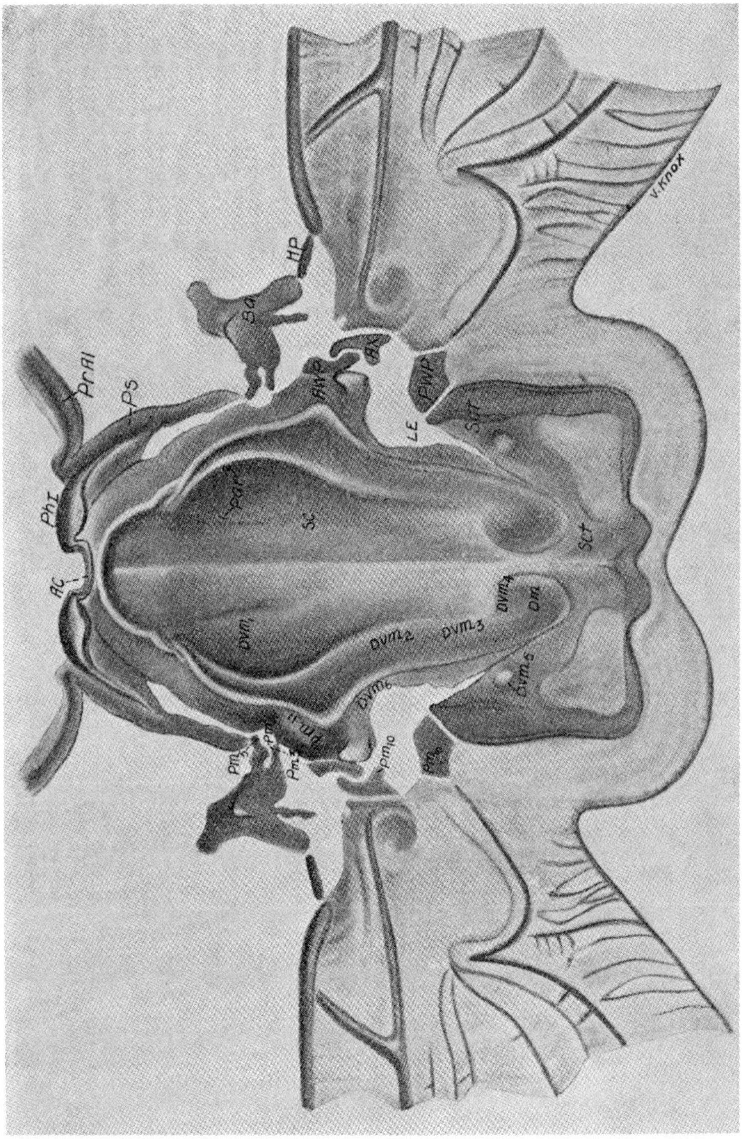

THE MESOTERGUM DRAWN FROM THE INSIDE

PLATE XXI.

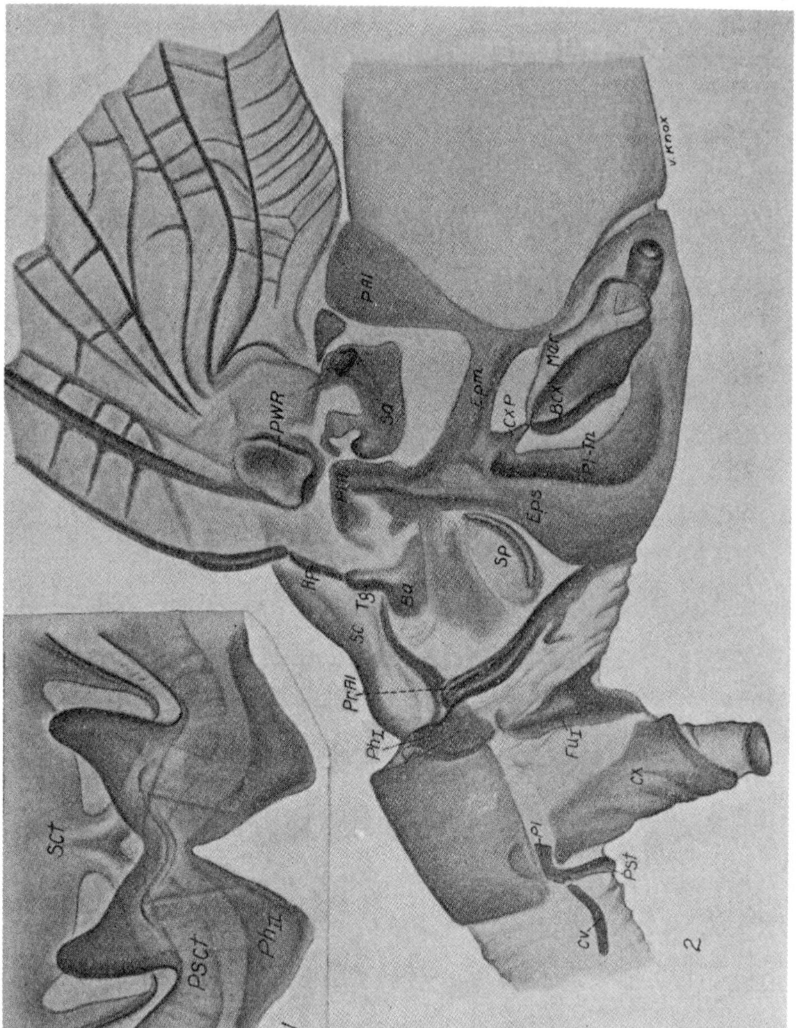

Fig. 1. The scutellum and postscutellum of the mesotergum; Fig. 2. The pleuron of the prothorax and mesothorax.

Scutellum (Pl. XX, Pl. XXI, fig. 1, *Sct*).—The posterior area of the tergum known as the scutellum is characterized by the prominent posterior hump as seen in the lateral view. It is marked off from the scutum by the deep narrow notch of the lateral emargination, and by a suture, an extension of the lateral fold of the scutum, which makes a curve upward close to the mid-line, but does not entirely set off the scutellum from the scutum in the mid-dorsal region.

Laterally the scutellum bears two large plates to which the notal surface of the wing membrane attaches. A strongly sclerotized strip along the lateral margin terminates anteriorly in an articulation with the wing, thus forming a lobe which corresponds to the anterior notal wing process of the scutum (*PWP*). The large sclerite articulating with the wing is ordinarily bent back to form a reduplication of the tergum. In the figure it is extended to show the articulation better (Pl. XX). Medial to this sclerotized area to which the wing membrane attaches laterally, is a triangular area which appears to be thinner (Pl. XX). This region is reinforced in the specimen which has not been treated with potassium hydroxide by a sclerotized triangular process which is a part of the posterior phragma, and will be described in connection with the postscutellum. The posterior border of the scutellum is notched and the lobes thus formed project steeply downward on either side of the mid-line.

Postscutellum (Pl. XXI, fig. 1, Pl. XXII, fig. 1, *Psct*, Pl. XXIII, fig. 6).—The postscutellum of the mesothorax consists of a well developed marginal process, the precosta (Pl. XXIII, fig. 6, *Pc*). The precosta is widened laterally to form two large triangular lobes which are firmly attached to the under surface of the scutellum in the untreated specimen; (i.e., one which has not been boiled in potassium hydroxide). The dorsal surface of the wing is continuous with the lateral margin of the tergum. The dorsal membrane is attached along the lateral and posterior borders of the scutellum. The ventral surface of the wing is continuous with the pleural region and the membrane is attached along the margin of the subalar plates (Pl. XXI, fig. 2). This subalar membrane is reflected over and attached to the postscutellum along the base of the triangular process of the precosta. The attachment of the ventral membrane to the postscutellum corresponds in position to the attachment of the dorsal membrane to the scutellum. In a specimen which has not been treated with potassium hydroxide the postscutellum is drawn under the scutellum and the lateral processes of the precosta are attached to the scutellum; thereby the dorsal and ventral wing membranes are held together. The lateral processes of the postscutellar precosta appear to serve as a seal reinforcing the fusion of the membranes of the wing.

In specimens treated with potassium hydroxide and acetic acid the lateral processes of the precosta are detached sclerotized flaps separated

PLATE XXII.

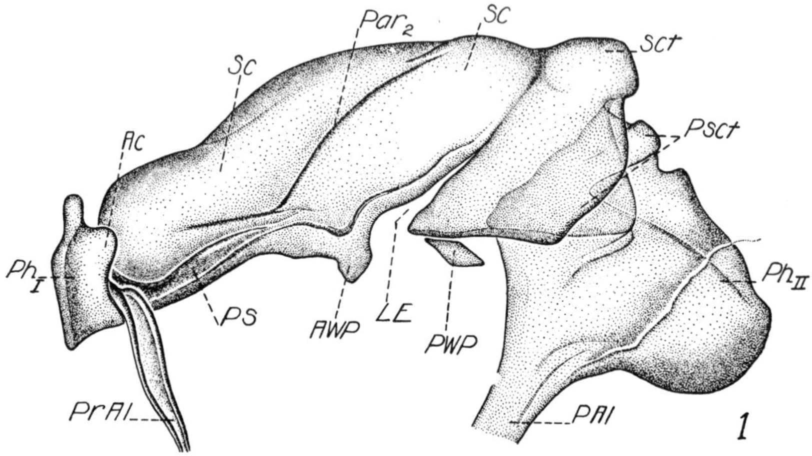

1

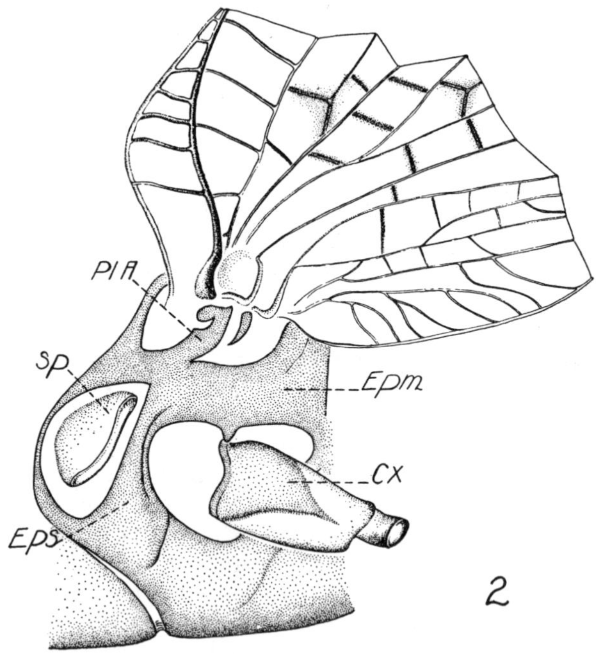

2

SKELETAL PARTS OF THORAX

Fig. 1. Lateral view of the mesotergum; Fig. 2. Pleuron of the metathorax.

from the scutellum. Likewise the two membranes are separated and the wings appear as sack-like evaginations of the body wall.

The precosta is separated from the antecosta and phragmata medially by unsclerotized chitin and laterally by a distinct suture. The second phragmata of the mesothorax are very large outgrowths of the antecosta. They project into the thoracic cavity as two distinct plates. Each plate is made up of two separate surfaces; the ventral surface is a prolongation of the antecosta and is part of the mesotergum, but the dorsal surface is a continuation of an infolding of the metatergum; hence there is no sharp division between the terga of the meso- and metathorax. The phragmata serve as attachments for the dorsal longitudinal muscles. The antecosta fuses laterally with the mesopleuron.

METATHORAX (Pl. XXIII, figs. 5a and 5b):

The tergum of the metathorax is much less complex than that of the preceding segment. The first two sclerites present in the mesotergum, namely, the antecosta and prescutum, are not differentiated from the anterior margin of the scutum. The scutum (*Sc*), scutellum (*Sct*) and postscutellum (*Psct*) are readily discernable. There is no anterior phragma which belongs strictly to the metathorax, but the small metathoracic dorsal longitudinal muscle takes origin from the posterior surface of the mesophragma which is fused with the metatergum.

Scutum (Pl. XXIII, fig. 5a, *Sc*).—The scutum is not subdivided by ridges or parapsides. Anteriorly the margin is invaginated to form the dorsal surface of the phragmata. The anterior notal wing process (*AWP*) on the lateral margin is present as a very small rounded piece which articulates with a small sclerite of the axillary region.

The lateral emargination is distinct and partly cuts off the scutum from the scutellum.

Scutellum (Pl. XXIII, fig. 5a, *Sct*).—There is a more complete division between the scutum and scutellum than that occurring between those sclerites of the mesotergum. The scutellum consists of a sclerotized median lobe and lateral plates to which the dorsal membrane of the wing attaches. The lateral margins of the scutellum are nearly straight. Anteriorly the process corresponding to the anterior notal wing process of the scutum articulates with the wing (*PWP*).

Postscutellum (Pl. XXIII, fig. 5b *Psct*).—The postscutellum is similar to that of the mesotergum. A narrow precosta (*Pc*) bears lateral lobes, or rectangular plates which are firmly attached to the surface of the scutellum in a specimen which has not been treated with potassium hydroxide. The precosta is drawn under the posterior margin of the scutellum as was the case on the mesotergum. The ventral membrane of the wing is attached to the margin of the precostal lobes which are free plates in specimens which have been boiled in potassium hydroxide. The two membranes of the wing are free, forming a membranous sack.

The antecosta is well developed as a straight sclerotized piece posterior to the precosta. There is no true phragma here, but the surface of the antecosta itself serves for the insertion of IIIDlm. Laterally the antecosta fuses with the epimeron of the metapleural region.

PROTHORAX (Pl. XXI, fig. 2).

The tergum of the prothorax consists of a perfectly simple rectangular plate lightly sclerotized which is continuous anteriorly with the cervical membrane and posteriorly with the membrane which attaches to the margin of the antecosta and the prealar bridge of the mesotergum.

B. THE PLEURON

MESOTHORAX (Pl. XXI, fig. 2):

The sclerites of the mesopleuron of *Hexagenia recurvata*, namely, the episternum (*Eps*), and the epimeron (*Epm*) are more or less fused into an irregular plate.

There is no pleural suture marking the division between the episternum and epimeron. The coxal process (*CxP*) of the pleuron is a prominent triangular piece articulating with the basicoxal sclerite of the coxa (*Bcx*). From the point of articulation with the coxa to the pleural wing process a division can be made whereby that part of the pleural plate lying anteriorly constitutes the episternum; the posterior part is the epimeron.

Episternum (Pl. XXI, fig. 2, *Eps*).—The episternum is a large, irregular chitinized plate extending from the pleural wing process to the sternum below where it fuses with the basisternum. Anterior to the coxa the episternum is extended to form the precoxal bridge. This is a wide extension of the episternum which borders the spiracle (*Sp*) and fuses with the prealar bridge (*PrAl*) of the tergum. Posteriorly it is continuous with the basisternum.

The pleurotrochantin (*Pl-Tn*) is marked off from the precoxal bridge by a suture which encloses a heavily sclerotized area. The term pleurotrochantin has been applied by Crampton (1914) to a combination of sclerites in which the true trochantin has become fused with the episternum. In this species the pleurotrochantin extends ventrally to the basisternum and is separated from the latter by a continuation of the episternal-trochantinal suture.

There is an extension of the episternum between the basalar plate (*Ba*) and the spiracle (*Sp*).

The episternum terminates in the pleural arm which acts as a support to the pleural surface of the wing. There are several dorso-ventral folds in the episternum.

Epimeron (Pl. XXI, fig. 2, *Epm*).—That portion of the pleural plate lying posterior to a line drawn from the coxal articulation to the pleural wing process corresponds to the epimeron. An extension of this sclerite posterior to the base of the coxa forms the postcoxal bridge which fuses with the furcisternum. The postcoxal bridge, always an extension of the epimeron, corresponds to the precoxal bridge which is an extension of the episternum. A pleural arch is thus formed above the base of the coxa.

A second reason for maintaining that this sclerite corresponds to the epimeron is, that posteriorly it fuses with a portion of the post scutellum forming the postalar bridge (*PAl*). This corresponds to the prealar bridge (*PrAl*) which serves as the anterior support of the tergum on the pleuron and fuses with the episternum.

Wing Articulation with the Pleuron.—Directly above the pleural arm (*PLA*) is a distinct cup formation, here designated as the pleural wing recess (*PWR*), an infolding of the membrane, the edges of which are slightly sclerotized forming a conspicuous rounded depression in the pleural surface of the wing. On the dorsal surface, the base of this depression is marked by the axillary region.

Posterior to the pleural arm and directly beneath the wing base, is a large sclerite, the subalar plate (Pl. XXI, fig. 2 *Sa*). This sclerite corresponds to the basalar plate (*Ba*) which is anterior to the process and above the episternum. A deep invagination in the posterior region of the subalar plate furnishes a strong finger-like process or apodeme to which the very large direct wing muscle, Pm7, attaches. Anterior to this the anterior margin terminates in a smaller rounded apodeme which is formed by an infolding of the corner of the sclerite. Between these two apodemes is an extension of the plate just beneath the wing base.

Basalar Process (Pl. XX, Pl. XXI, fig. 2 *Ba*).—The basalar process (*Ba*) is a large sclerite which lies anterior to the pleural arm, and articulates with both the tergum and the pleuron. That part of the sclerite articulating with the tergum is characterized by two small processes, one anterior which articulates with the prescutum and scutum, the other slightly posterior which articulates with the scutum near the anterior notal wing process. The base of the sclerite which articulates with the episternum is widened lengthwise. Between the articulations with the tergum and pleuron the sclerite is evaginated to form a large projection, the tegula (*Tg*). To the tegula the costal margin of the wing articulates by means of the humeral plate (*Hp*).

Posteriorly a small sclerite, a second subalar piece, articulates with the wing at the base of a small apodeme which projects downward over the subalar plate (Pl. XXI, fig. 2). The subalar plate which is posterior to the pleural arm is homologous with the basalar plate which is anterior to the arm and above the episternum. The subalar piece posterior to the plate is triangular and almost straight along the upper surface which serves as attachment for the pleural wing membrane. The pleural membrane of the wing is reflected from the precosta of the postscutellum to the margin of the subalar processes.

Meatathorax (Pl. XXII, fig. 2):

The pleuron of the metathorax shows less specialization than that of the mesopleuron. The coxal process of the metapleuron which serves as the point of articulation of the pleuron with the coxa, is directed almost vertically and is considerably smaller than the corresponding sclerite of the mesopleuron. As in the mesopleuron, that part of the pleural plate lying anterior to the coxal process is the episternum, and likewise that which lies posterior to the point of articulation is the epimeron.

Episternum (*Eps*).—The metaepisternum is less complex than the mesoepisternum. The episternum surrounds the spiracle (*Sp*). Anteriorly it forms a thin margin around the surrounding membrane of the spiracle. A distinct suture dividing the mesoepimeron from the metaepisternum marks the division of the two segments. Dorsally this suture determines the margin of the prealar bridge of the metathorax, an extension of the metascutum which fuses with the episternum. Posterior to the spiracle in a nearly vertical line the episternum is extended ventrally to form the precoxal bridge which fuses with the furcisternum. An indistinct suture near the border of the coxal membrane marks the position of the pleurotrochantin.

The extension of the pleural arm (*PlA*) at the base of the wing is pronounced. A basalar plate anterior to this is lacking. There is a small sclerite posterior to the wing support.

Epimeron (*Epm*).—The epimeron is a wide, straight sclerite, extending up behind the wing where it forms by fusion with the antecostal piece of the postscutellum, the postalar bridge. Below the postalar bridge the epimeron forms an extension around the base of the metacoxa, the postcoxal bridge.

Prothorax (Pl. XXI, fig. 2):

The pleuron of the prothorax presents quite a different structure from that of either the meso- or metapleura. The pleural plate (*Pl*) in the anterior part of the segment is very simple. The small sclerite articulates at the base with the coxa. There is no coxal process, but the basicoxal piece of the coxa fits into an angle formed at the base of the pleural plate. Anterior to the coxal articulation a small vertical extension of the pleural plate fuses with the lateral extension of the basisternum. An elongated cervical sclerite (*Cv*) articulates with the fused portion of the pleural plate (*Pl*) and presternite (*Pst*). The pleural plate extends beyond the protergum as an apodeme. The remainder of the pleuron consists of unsclerotized chitin which is attached to the rectangular plate of the protergum and to the prealar bridge of the mesotergum posteriorly.

C. THE STERNUM

Prothorax (Pl. XXIII, fig. 1):

The prosternum may be subdivided into four sclerites, the presternite (*Pst*), the basisternum (*Bs*), furcisternum (*Fs*), and postfurcisternum (*PFs*). The anterior margin of the segment is marked by a distinct, narrow sclerite, the presternite. This sclerite is present in front of the prothorax alone. To its margin is attached the cervical membrane.

The basisternum is represented by a sclerotized strip between the bases of the coxae. Anteriorly it widens into a narrow transverse strip which extends laterally to fuse with the pleuron. The margin is marked by the presternal suture. Between the posterior margins of the coxae the basisternum is marked off from the furcisternum by a distinct ridge which extends upward in the mid-line to form an inverted V shaped suture.

Heavy sclerotization of the furcisternum along the mid-line marks the position of the well developed furcae (*FuI*) or entosternal arms. From the basisternal suture the furcisternum slopes steeply inward. The furcae arise independently on either side of the mid-line, where their bases are marked by two large external pits. The furcal arms of the prosternum are much longer than those of the other segments. They extend into the prothoracic cavity, almost touching the anterior phragma of the mesothorax.

The postfurcisternum extends as a fold laterally where it bears a pair of small processes, the furcillae (*fu*). These are marked externally by a slight infolding of the chitin. Between the furcillae and the prealar arms of the mesotergum which fuse here with the prosternum, there is a slight sclerotization. Aside from the sclerotized strip of the basisternum and furcisternum along the mid-line and the slightly sclerotized area posterior to the furcillae, the entire prosternum is unsclerotized.

Mesothorax (Pl. XXIII, fig. 2):

The sternum of the mesothorax is made up of two regions: the basisternum (*Bs*), lying mostly anterior to the bases of the coxae, and the furcisternum (*Fs*), posterior to the coxae. The division of the two sclerites is marked by the position of the furcae (*Fu II*). A suture from each sternal arm extends toward the mid-line, widens slightly laterally and passes anteriorly where it terminates in the chitin of the basisternum. On either side of the mid-line of the furcisternum the suture forms an internal ridge for the attachment of the large subalar muscle, *Pm*7.

The basisternum is widened laterally where it fuses with the precoxal piece of the pleuron. The pleurotrochantin which terminates ventrally on the basisternum is prominent as a heavily sclerotized area above the coxal membrane.

The surface of the furcisternum between the coxae is convex. It slopes laterally to form the invaginations of the mesofurcae. The basicoxal sclerite of the coxa terminates in this invagination. The furcae are marked externally by small pits at the base of the coxae, which are much less prominent than those of the prosternum. The posterior margin of the furcisternum is convexly rounded where it is attached by intersegmental membrane to the metasternum.

METATHORAX (Pl. XXIII, fig. 3):

There is no division between basisternum and furcisternum of the metathorax. The anterior portion which corresponds to the basisternum of other segments is extended laterally where it fuses with the pleuron.

The furcae of the metasternum are smaller than those of the other segments. Each furca (*FuIII*) is strengthened on the lateral surface by a sclerotized piece which extends the length of the arm.

In the region between the furcae the furcisternum fuses with the first abdominal segment. Laterally a suture marks a division between the thorax and abdomen which extends from the furcisternum posterior to the furca along the posterior base of the coxa for a short distance.

At the base of the coxa is a basicostal suture (Pl. XXIII, fig. 4) which forms internally a distinct narrow sclerite, the basicoxite (*Bcx*). The outer wall of the basicoxite is inflected to form the pleural articular surface, which divides the basicoxite into two parts. The posterior of these is the meron (*Mer*). In this species the meron is much enlarged, extending distally on the posterior part of the coxa almost to the end of the segment. The suture limiting the meron from the anterior part of the coxa is a continuation of the basicostal suture. The meron is unsclerotized in contrast to the anterior part of the coxa. Arising from the anterior trochanteral articulation is the anterior coxal suture (*Acx*) which extends to the basicoxite.

In each coxal segment the wall of the basicoxite is inflected to articulate with the coxal process of the pleuron (*CxP*). There are two articulations with the trochanter, one anterior and one posterior. In addition to these three constant points of articulation the coxae of the meso- and metasternites have a fourth, ventrally with the furcisternum. It is supported at the other end of the basicoxite which articulates with a sclerotized piece in the lateral wall of each furcal arm.

THE THORACIC SCLERITES OF THE NYMPH

In the nymphal stage most of the sclerites characteristic of the adult have not developed. With the exception of the sternum which is represented in figure 1 of Pl. XXVI the integument is practically unsclerotized.

In the sternum the furcae are well developed (Pl. XXVI, fig. 1). Those of the prothorax are greatly elongated and at the base of each furca is a sclerotized projection or plate (*a*) to which the ventral longitudinal muscles attach.

PLATE XXIII.

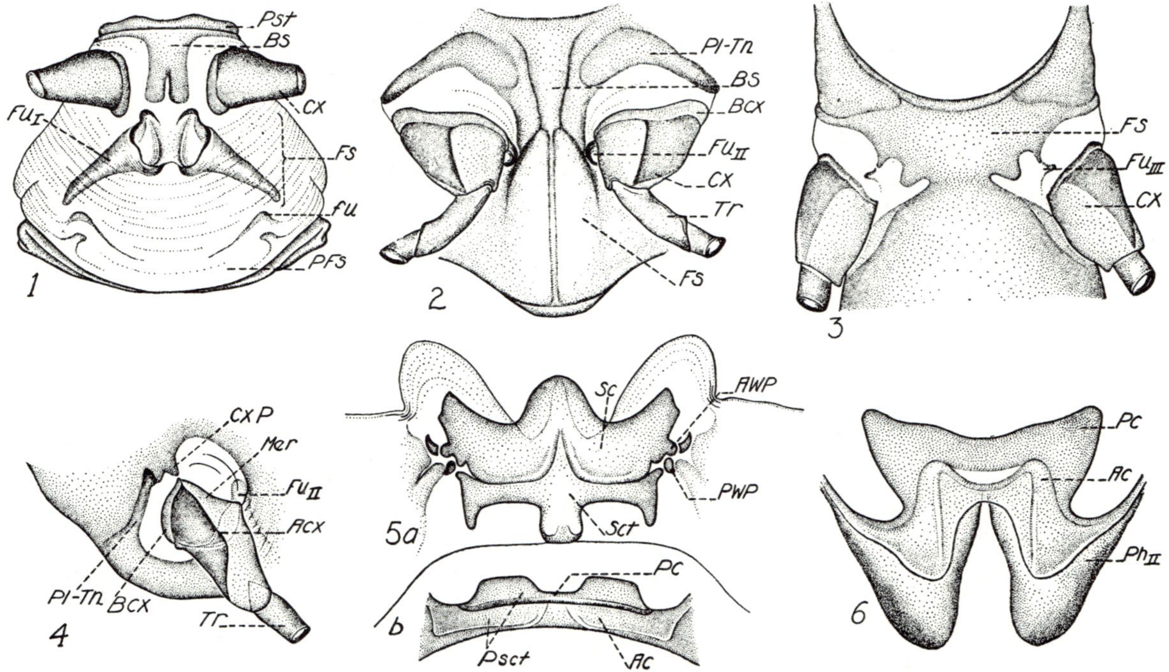

Fig. 1. Prosternum; Fig. 2. Mesosternum; Fig. 3. Metasternum; Fig. 4. Mesocoxa; Fig. 5a. Tergum of the metathorax; Fig. 5b. Postscutellum of the metatergum; Fig. 6. Postscutellum of the mesotergum.

Opposite the furcae at the point which might be termed the pleural coxal articulation, the presternite is invaginated to form well developed apodemes (*Ap*). These are peculiar to the nymph and form a strong support for the leg muscles. Likewise in the metasternum there is a well developed apodeme (*Ap*) at the pleural coxal articulation. In the mesosternum however, there is only a small projection corresponding to the apodemes in the other two segments.

The pleural sclerites consist only of a simple sclerotized strip in each segment. In the prothorax the pleuron is a continuation of the presternite. In the other two segments the sclerite widens at the pleural coxal articulation to form the pleural ridge (*PlR*)—a precursor of the adult pleural arm.

The terga are simple plates. The meso- and metaterga are evaginated posteriorly to form the wing pads.

Further details of the nymphal integument may be found with the description of the muscles of the nymph in the following chapter.

CONCLUSION

The thoracic sclerites of *Hexagenia recurvata* resemble those of higher forms rather than those of the more primitive insects. It is believed that there is a similarity between the Ephemeroptera and certain Neuroptera, the Trichoptera and the Lepidoptera. The following is a brief comparison of *Hexagenia recurvata* with several forms.

The region of the mesothoracic tergum and wing bases of *Hexagenia recurvata* bears a close resemblance to that of the higher forms. In the Trichopteron, Neuronia, and the Lepidopteron, Phassus, the region designated by Crampton (1919) as the subalar is well developed and similar in form to the region in *Hexagenia recurvata* here designated as *AWP* (Pl. XX). The lateral emargination (Pl. XX, *LE*) is enlarged somewhat as in the Trichoptera and Lepidoptera. The sclerites labelled *Ax* and *PWP* (Pl. XX) of *Hexagenia recurvata* are well developed as are similar sclerites in the higher forms. In the postscutellum of *Hexagenia recurvata* there is an unsclerotized area between precosta and antecosta which corresponds to a similar region in the postscutella of Neuronia, Phassus, and to the Neuropteron, Nemoptera.

Dürken's work (1907) on the Ephemeroptera deals principally with the musculature of the tracheal gills with special reference to the structure of the thorax of *Centroptilum luteolum* and *Ephemerella ignita*.

Discussion of the prothorax was not included by Dürken except for the fact that he described the first spiracle as belonging to that segment. Likewise the second spiracle he placed in the mesothorax rather than in the metathorax. In the mesopleuron Dürken designates as the episternum that portion which has here been termed the episternum and epimeron together. The epimeron he believes to be what is here described as the subalar plate below the wing base. Taking into consideration the muscle attachments from this sclerite to the coxa and sternum, there seems to be no doubt but that it is a subalar sclerite which corresponds to the basalar anterior to the pleural arm. Further comparison of this region with that of allied forms is discussed later. The sclerites of the tergum and sternum are not described in detail by Dürken.

ABBREVIATIONS

Abdominal segment—AB
Antecosta—Ac
Anterior coxal suture—Acx
Anterior notal wing process—AWP
Apodeme—Ap
Axillary sclerite—Ax
Basalar sclerite—Ba
Basicoxite—Bcx
Basisternum—Bs
Cervical sclerite—Cv
Coxa—Cx
Coxal process—CxP
Epimeron—Epm
Episternum—Eps
Furca—Fu
Furcisternum—Fs
Furcilla—fu
Humeral plate—HP
Meron—Mer
Lateral emargination—LE
Parapsides (inner)—Par 1
Parapsides (outer)—Par 2
Phragma—Ph
Pleural arm—PlA
Pleural coxal articulation—Px
Pleural ridge—PlR
Pleural wing recess—PWR
Pleurotrochantin—Pl-Tn
Pleuron—Pl
Postfurcisternum—PFs
Postoccipital ridge—PoR
Postscutellum—Psct
Postalar bridge—PAl
Posterior notal wing process—PWP
Prealar bridge—PrAl
Precosta—Pc
Prescutum—Ps
Presternite—Pst
Scutellum—Sct
Scutum—Sc
Spiracle—Sp
Sternum—St
Subalar apodeme (anterior)—SaAp 1
Subalar apodeme (posterior)—SaAp 2
Subalar sclerite—Sa
Tegula—Tg
Tentorium—Tt
Tergum—T
Trochanter—Tr

CHAPTER XII

THE MUSCULATURE OF THE THORAX

By Velma Knox, Ph.D.

The integument of the thorax as discussed in the preceding chapter serves only as the framework for the complex system of muscles beneath. It is the muscles themselves which furnish the locomotive power. They are by no means primitive in the mayfly. The muscles as described in this chapter are grouped as cervical (*Cv*), dorsal longitudinal (*Dlm*), dorsal ventral (*Dvm*), intersegmental (*Im*), pleural (*Pm*), sternal (*Sm*), and ventral longitudinal (*Vlm*).

There are two groups of muscles in the imago which produce flight, the direct and indirect wing muscles. The direct muscles are attached to the basalar and subalar sclerites at the wing base and below to various parts of the sternum. The indirect muscles are the dorsal longitudinal. By artificial contraction of the large dorsal longitudinal muscle the downward stroke of the wings is easily demonstrated. It is very difficult to determine the exact function of other individual muscles however, and in the following discussion only theoretical explanations of function are possible.

The best results in the study of the musculature were obtained from specimens which had been preserved in Dietrich's well known fixative. The specimens were cut in sagittal section, and the muscles dissected by layers under a binocular microscope. Microtome sections were also used, and thick sections cut by hand were found to be most useful as a means of orientation of the muscles in the preliminary study.

MUSCULATURE OF THE NYMPH

Prothorax:

The habit of burrowing in the mud which is characteristic of the nymphs of Hexagenia, necessitates the development of very strong leg muscles and as a result the musculature of the nymphal prothorax is more highly developed than that of the imago.

Dorsal Longitudinal Muscles:

Dlm 1 (Pl. XXIV, figs. 1 and 5).—The entire median portion of the tergum serves as attachment for the fibers of *Dlm 1*. They converge toward the occipital ridge and attach to the phragma formed by that ridge and the cervical membrane.

Dlm 2 (Pl. XXIV, figs. 1 and 5).—*Dlm 2* is a wide, thick muscle, with a large attachment on the mesotergum. The area of attachment is marked off by a suture extending from the lateral anterior margin. The longer fibers attach on the postoccipital ridge, and the shorter fibers to the anterior part of the tergum.

PLATE XXIV.

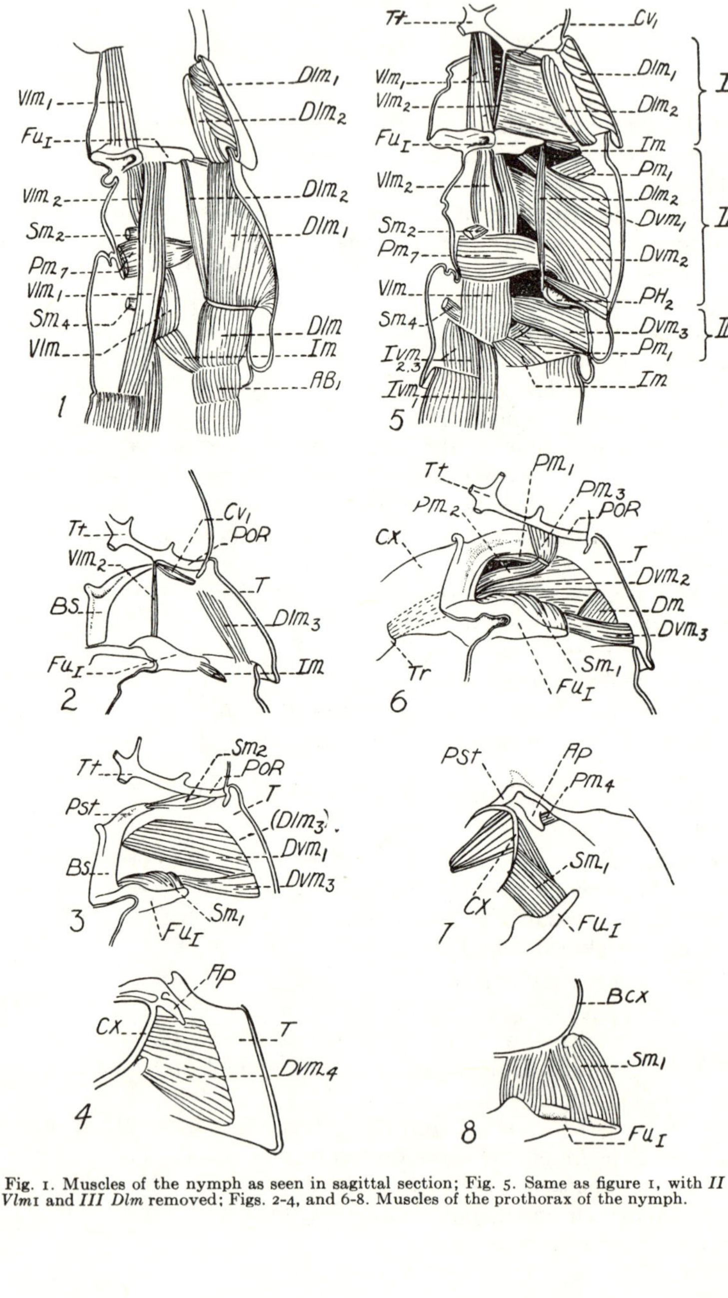

Fig. 1. Muscles of the nymph as seen in sagittal section; Fig. 5. Same as figure 1, with *II Dlm* 1, *II Vlm*1 and *III Dlm* removed; Figs. 2-4, and 6-8. Muscles of the prothorax of the nymph.

Dlm 3 ((Pl. XXIV, fig. 2).—*Dlm 3* is attached to the anterior margin of the mesotergum, dorsal to *Dlm 2*. It is inserted near the anterior margin of the protergum by a wide, oval attachment which is indicated externally by a lightened area in the color pattern on the integument.

Dorsal Muscle, Dm (Pl. XXIV, fig. 6).—*Dm* lies beneath *Dvm 1* and *Dvm* 3 and above *Dvm 2*. It cuts diagonally across the posterior corner of *Dvm 2*. It attaches on the protergum in line with the attachment of *Dvm 2*, and posteriorly to the margin of the mesotergum. The short, wide muscle narrows slightly towards the attachment on the mesotergum.

Ventral Longitudinal Muscles:

Vlm 1 (Pl. XXIV, figs. 1 and 5).—*Vlm 1* is a composite of groups of fibers which attach to the evaginated plate (*a*) (Pl. XXVI, fig. 1) at the base of the profurca and anteriorly to the tentorium (*Tt*). The thickness and apparent strength of this muscle offers a striking contrast with the ventral longitudinal muscle of the adult. The fibers originate on the medial surface of the plate (*a*) by a wide attachment. Cephalad the muscle inserts on the medial border of the posterior arm of the tentorium (*Tt*).

Vlm 2 (Pl. XXIV, figs. 2 and 5).—*Vlm 2* is a very small, ribbon like muscle which attaches with *Vlm 1* on plate (*a*) of the furca. It inserts anteriorly on a small apodeme at the end of the posterior tentorial arm (*Tt*). This is drawn down close to the large apodeme.

Dorsal Ventral Muscles:

Dvm 1 (Pl. XXIV, fig. 3).—*Dvm 1* is a very strong, fan-shaped muscle which attaches to the basicoxite and to the tergum lateral to the mid-line. The attachment extends from that of *Dlm 3* on the anterior part of the tergum, nearly to the posterior margin of the segment. Posteriorly the area of attachment widens.

Dvm 2 (Pl. XXIV, fig. 6).—*Dvm 2* attaches below to the posterior trochanteral articulation and dorsally to the tergum lateral to *Dvm 1*. *Dvm 2* is very long and, as in the case of *Dvm 1*, the attachment on the tergum (*T*) is very much wider than that on the leg.

Dvm 3 (Pl. XXIV, figs. 3 and 6).—*Dvm 3* is attached to a small sclerotized area near the base of the coxa (Pl. XXIV, fig. 8). Dorsally the muscle attaches on the tergum medial to the posterior corner of *Dvm 1*. The attachment is continuous with that of *Dvm 1* on the tergum. This is probably homologous with *Dvm 3* of the adult.

Dvm 4 (Pl. XXIV, fig. 4).—*Dvm 4* lies next the integument. The muscle is somewhat fan-shaped and attaches in a nearly straight line on the tergum. The attachment is about halfway between the mid-dorsal line and the lateral margin of the tergum. Ventrally the muscle attaches close to the rim of the coxa and to the edge of the sclerite near the coxa which serves as attachment for *Dvm 3*.

Cervical Muscles:

Cv 1 (Pl. XXIV, fig. 2).—*Cv 1* attaches to a small apodeme at the base of the posterior tentorial arm (*Tt*). Dorsad and slightly posteriorly the muscle attaches to the anterior margin of the protergum. *Cv 1* is a thick, strong muscle which is homologous to the group of cervical muscles in the adult.

Pleural Muscles:

Pm 1 (Pl. XXIV, fig. 6).—*Pm 1* attaches by a small base to the lateral rim of the coxa and to the arm of the apodeme at the pleural coxal articulation. It is homologous with *Pm 1* of the adult. Contraction of this muscle would draw the coxa in.

Pm 2 (Pl. XXIV, fig. 6).—*Pm 2* originates by a wide attachment on the lower surface of the apodeme. The muscle is attached below to the trochanter (*Tr*). It is homologous with *Pm 2* of the adult.

Pm 3 (Pl. XXIV, fig. 6).—*Pm 3* which corresponds to *Pm 3* of the adult, is a wide, short muscle originating on the anterior surface of the apodeme. The fibers extend cephalad and insert on the postoccipital ridge (*POR*) by a wide attachment.

Pm 4 (Pl. XXIV, fig. 7).—*Pm 4* is a very small muscle which attaches to the anterior surface of the apodeme and extends to the ridge of the pleuron where it attaches.

Sternal Muscles:

Sm 1 (Pl. XXIV, fig. 8).—In the nymph the muscle attachments between the coxa and the profurca are much stronger than in the adult. Muscles attach along the whole length of the furca to the inner ridge below the projection (*a*), and insert on the lateral rim of the coxa (*Cx*) from the pleural coxal articulation halfway around the leg.

Sm 2 (Pl. XXIV, fig. 3).—When *Vlm 2* and *Cv 1* are lifted, *Sm 2* is uncovered. It attaches to the presternite or lateral extension of the basisternum about midway between the apodeme (*Ap*) and the mid-line (Pl. XXVI, fig. 1). Dorsally the muscle attaches to the postoccipital ridge (*POR*). The attachment is very wide and extends from the projection on the occipital ridge nearly to the mid-dorsal line. *Sm 2* of the nymph is homologous with *Sm 4* of the adult.

Intersegmental Muscle (Pl. XXIV, fig. 5).—*Im* is a small muscle which attaches to the tip of the profurca and extends dorsally where it attaches to the mesotergum. The attachment is on the anterior portion of the segment just posterior to the attachment of *Dlm 2* of the prothorax.

MESOTHORAX:

Dorsal Longitudinal Muscles:

Dlm 1 (Pl. XXIV, fig. 1).—*Dlm 1* of the nymph is very similar to that of the adult. There is no anterior phragma in the nymph, but the muscles are attached to the scutum itself. and the intersegmental membrane is folded back externally over the anterior portion of the mesotergum (Pl. XXIV, fig. 5). The longer fibers attach to the anterior margin and the shorter fibers to the entire inner surface of the scutum on either side of the mid-line. The fibers converge to attach on the well developed posterior phragma.

Dlm 2 (Pl. XXIV, figs. 1 and 5).—*Dlm 2* is a narrow band of fibers attaching anteriorly to the tip of the profurca. Posteriorly the muscle attaches to the lateral portion of the phragma or the under surface.

Dm (Pl. XXV, fig. 5).—*Dm* is homologous with *Dm* of the adult mesothorax. It attaches to the posterior part of the scutum and to the intersegmental membrane between the meso- and metathorax close to the lateral base of the phragma.

Ventral Longitudinals:

Vlm 1 (Pl. XXIV, fig. 1).—The muscle is a wide ribbon-like band. It attaches to the base of the posterior surface of the profurca and posteriorly to the anterior margin of the first abdominal segment. This muscle is peculiar to the nymph.

Vlm 2 (Pl. XXIV, figs. 1 and 5).—*Vlm 2* is a wide muscle which attaches to the base of the profurca lateral to *Vlm 1*, and to the anterior basal portion of the mesofurca. *Vlm 2* is very wide and flattened laterally. It is homologous with *Vlm* of the adult.

Dorsal Ventral Muscles:

Dvm 1 (Pl. XXIV, fig. 5 and Pl. XXV, fig. 1).—*Dvm 1* is attached to the anterior point of the basisternum which is invaginated to form a small apodeme (Pl. XXVI, fig. 1). The attachment appears to be pointed but the fibers turn under slightly to attach along the lateral margin of the basisternum. Dorsally the muscle attaches to the scutum lateral to *Dlm 1*. *Dvm 1* is homologous with *Dvm 1* of the adult.

Dvm 2 (Pl. XXV, figs. 1 and 2).—*Dvm 2* corresponds to *Dvm 2* of the adult. It originates on the medial border of the coxa. This attachment is relatively much more narrow than the corresponding attachment of *Dvm 2* of the adult. Dorsally the anterior portion of the muscle attaches lateral to *Dvm 1* on the scutum. The attachment is in line with that of *Dvm 1* and equal in width.

Dvm 3 (Pl. XXV, figs. 1 and 5).—*Dvm 3* attaches to the lateral rim of the coxa by a narrow base. The muscle widens considerably at the attachment on the scutum which corresponds in position to the attachment in the adult of *Dvm 3*.

Dvm 4 (Pl. XXV, figs. 1 and 5).—*Dvm 4* lies posterior to *Dvm 3*. It attaches to the coxa close to *Dvm 3* and attaches on the scutum closer to the mid-line than *Dvm 3*. Contrary to the case in the adult, *Dvm 4* is a larger muscle than *Dvm 3*.

Dvm 5 (Pl. XXV, fig. 5).—*Dvm 5* is homologous with *Dvm 6* of the adult. The muscle attaches to the trochanter below, and dorsally to the lateral border of the scutum by a wide attachment.

PLATE XXV.

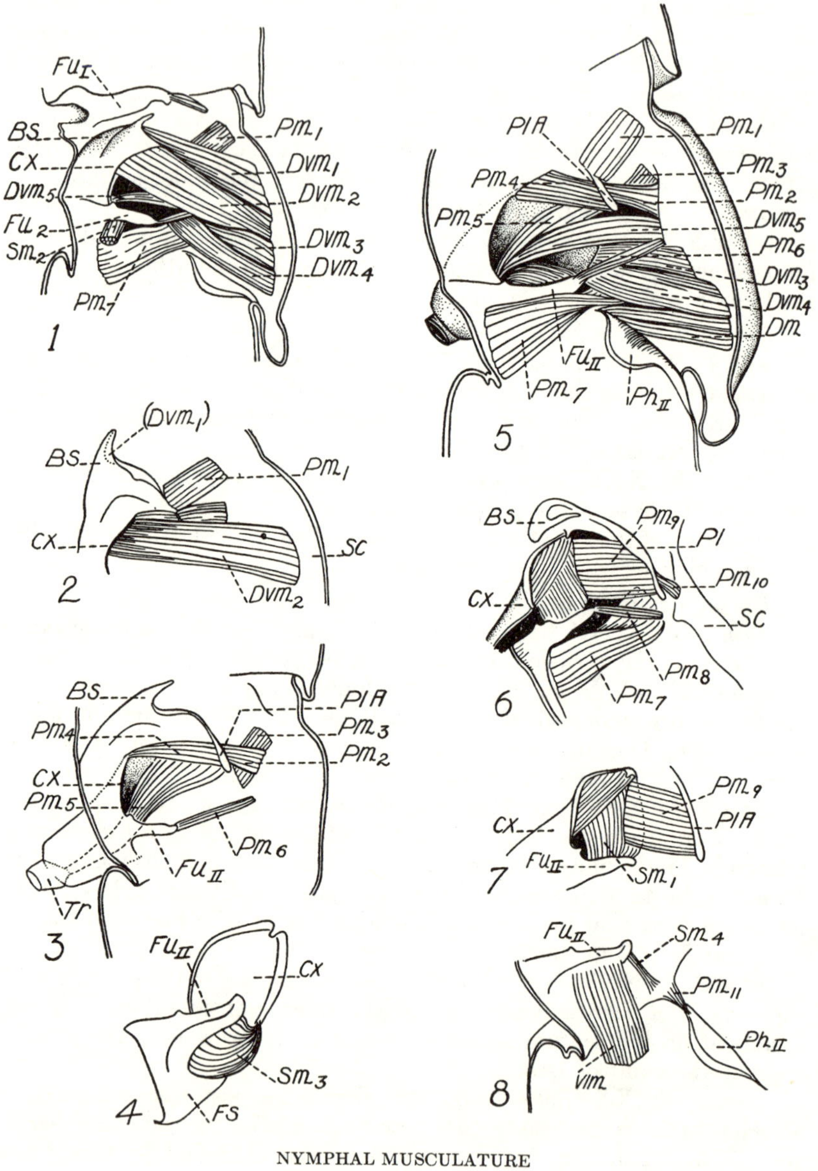

NYMPHAL MUSCULATURE
Muscles of the mesothorax.

Pleural Muscles:

There are several variations from the adult type in the pleural muscles of the immature nymph. Since the wings are not functional the basal wing sclerites are not developed and the muscles necessarily function as motors of the legs. There are however, definite homologies which are indicated in the descriptions, but the numbers do not correspond throughout to those of the adult pleural muscles.

Pm 1 (Pl. XXV, figs. 1, 2 and 5).—*Pm 1* is a wide, short muscle, the fibers of which run obliquely to those of *Dvm 1* and *Dvm 2*. The muscle attaches to the anterior medial portion of the scutum just posterior to the attachment of *Dlm 2* of the prothorax. Below it is attached to a thickening of the pleural ridge which is opposite the pleural coxal articulation. Contraction of *Pm 1* would lower the anterior part of the scutum.

Pm 2 (Pl. XXV, figs. 3 and 5).—*Pm 2* attaches to the pleural ridge close to the termination which forms the wing support or fulcrum in the adult. The muscle lies almost at right angles to *Pm 1*, and extends directly dorsad where it attaches to the lateral part of the scutum. The attachment is marked by a darkened line in the integument which corresponds to the notal furrow of the adult tergum. *Pm 2* is homologous with *Pm 11* of the adult.

Pm 3 (Pl. XXV, figs. 3 and 5).—When *Pm 2* is lifted, *Pm 3* is exposed. The fibers are parallel to those of *Pm 1*. *Pm 3* is attached to the scutum posterior and slightly ventral to the attachment of *Pm 1*. It attaches to the pleuron lateral to *Pm 2* slightly dorsal to the pleural ridge.

Pm 4 (Pl. XXV, figs. 3 and 5).—*Pm 4* attaches to the pleural ridge directly opposite *Pm 2*. It inserts by a wide attachment on the medial rim of the coxa at a place made prominent by the anterior coxal suture. *Pm 4* is homologous with the adult *Pm 12*.

Pm 5 (Pl. XXV, fig. 5).—*Pm 5* attaches to the pleuron lateral to *Pm 4*, and posterior to the pleural coxal articulation. Below it attaches to the trochanter. *Pm 5* is homologous with *Pm 13* of the adult.

Pm 6 (Pl. XXV, figs. 3 and 5).—*Pm 6* is a very narrow muscle which attaches to the tip of the mesofurca and to the scutum ventral to the line marking the attachment of *Pm 2*. Although the axillary sclerites have not yet developed, and the attachment is therefore not identical with that of the adult, the position of *Pm 6* on the scutum indicates that it is homologous with *Pm 10* of the adult.

Pm 7 (Pl. XXIV, figs. 1 and 5, Pl. XXV, figs. 1, 5 and 6).—*Pm 7* is unquestionably homologous with the large adult subalar muscle, *Pm 7*. It attaches to the mid-line of the sternum and to the area bounded by the posterior ridge on the furca. The muscle attaches to a sclerotized area below the pleural ridge. This sclerite will become the subalar sclerite of the adult and already a small invagination forms an apodeme to which the posterior fibers of *Pm 7* attach.

Pm 8 (Pl. XXV, fig. 6).—*Pm 8* is homologous with *Pm 8* of the adult. It is a narrow muscle which attaches to the tip of the mesofurca and close to the posterior attachment of *Pm 7*. In a more mature form where the apodeme on the subalar sclerite has formed, *Pm 8* attaches to the base of the arm.

Pm 9 (Pl. XXV, figs. 6 and 7).—*Pm 9* attaches to the lateral rim of the coxa, from the pleural coxal articulation to a point about midway between the coxal articulation and the furca. The muscle is sufficiently wide to be almost square. It attaches to the pleural ridge directly opposite the attachment on the coxa. *Pm 9* lies next the integument and aside from its function as a leg muscle it probably serves as a support between coxa and pleural ridge. It is believed to be homologous with *Pm 9* of the adult. The adult characters of the subalar sclerite are not distinguishable in the nymph and hence the apodeme to which *Pm 9* attaches in the adult is not developed in the immature form. The attachment from the pleural ridge might easily be shifted ventrally in development and form the inturning of the corner of the sclerite.

Pm 10 (Pl. XXV, fig. 6).—Attaching to the upper surface of the wing pad under the pleural ridge near the tip is a very small muscle which inserts on the notum at the line corresponding to the adult notal furrow. *Pm 10* is homologous with *Pm 6* of the adult. In the mature nymph, one in which the subimaginal cuticula has formed and the sclerites already bear a close resemblance to those of the mature form, this minute muscle may be detected by pulling the pleuron from the wing pad. The posterior notal wing process is formed and is folded directly inward at right angles to the notum. It is to the inner surface of this sclerite that *Pm 10* attaches, the attachment having been shifted from the notum proper in the less mature stage.

Pm 11 (Pl. XXV, fig. 8).—In the nymph there is a wide infolding of the integument between the segments. To the border of the fold a very small muscle *Pm 11* attaches, just posterior to the attachment of *Sm 4*. *Pm 11* has a pointed attachment on the medial base of the phragma.

Sternal Muscles:

Sm 1 (Pl. XXV, fig. 7) attaches to the inner, anterior surface of the mesofurca. The attachment extends the entire length of the furca. The muscle attaches to the inside of the coxa along the anterior coxal suture. The fibers which originate on the tip of the furca are parallel to the rest of the muscle, but attach to the rim of the coxa near the pleural coxal articulation. This part of the muscle lies over the attachments of *Dvm 3* and *Dvm 4*.

Sm 2 (Pl. XXV, fig. 1) is identical to that of the adult. The muscle attaches to each mesofurca anterior to the attachments of *Pm 7*, and crosses the mid-sternal line.

Sm 3 (Pl. XXV, fig. 4) attaches to the posterior border of the mesofurca, extending from near the tip down to the base of the sclerite. The fibers curve under abruptly and attach to the lateral rim of the coxa close to the tip of the furca. The attachment on the coxa is much less wide than that on the furca.

Sm 4 (Pl. XXV, fig. 8) is a short, strong muscle which attaches close to the tip of the furca and extends almost at right angles to the furca. It attaches to the sternum very close to the posterior border of the segment.

METATHORAX:

Dorsal Longitudinal Muscles:

Dlm (Pl. XXIV, fig. 1) corresponds to *Dlm* of the adult. Like *Dlm* of the mesothorax, it is a very large muscle. It attaches anteriorly to the entire posterior surface of the phragma. The fibers extend directly caudad and attach to the intersegmental membrane and anterior border of the first abdominal segment.

Vlm (Pl. XXIV, fig. 5).—This muscle is not present in the adult. The large muscle is attached anteriorly to the posterior surface of the mesofurca below the posterior ridge. On the metasternum it is attached to the flattened anterior surface of the furca close to the base.

Dorsal Ventral Muscles:

Dvm 1 (Pl. XXVI, fig. 5) is homologous with the adult *Dvm 1*, but relatively it is very much smaller. *Dvm 1* attaches to the border of the intersegmental fold medial to the attachment of *Pm 11* of the mesothorax. The attachment on the scutum is covered by the lip of the phragma, and is situated about midway between the mid-scutal line and the lateral tip of the phragma.

Dvm 2 (Pl. XXVI, fig. 5) is attached to the medial rim of the coxa by a narrow rounded base. The muscle widens and inserts by a wide flattened attachment to the lip of the phragma.

Dvm 3 (Pl. XXIV, fig. 5 and pl. XXVI, fig. 5).—Beneath *Dvm 2* this muscle attaches to the medial rim of the coxa. The unsclerotized chitin around the leg base is folded back over the coxa, so that the rim projects into the body cavity. *Dvm 3* is a large, thick muscle. Towards the attachment on the scutum it narrows somewhat, and the anterior border follows the curvature of the phragma. The insertion on the scutum is close to the mid-line and just below the phragma. *Dvm 3* is homologous with *Dvm 2* of the adult.

Dvm 4 (Pl. XXVI, fig. 2) is a long, wide muscle which attaches to the trochanter and dorsally to the scutum. The insertion on the scutum is lateral to that of *Dvm 3* and posterior to the medial portion of the phragma.

Pleural Muscles:

Pm 1 (Pl. XXVI, fig. 3).—Corresponding to the condition in the adult in which a small invagination of the basicoxite forms a process at the pleural coxal articulation for the attachment of pleural muscles,

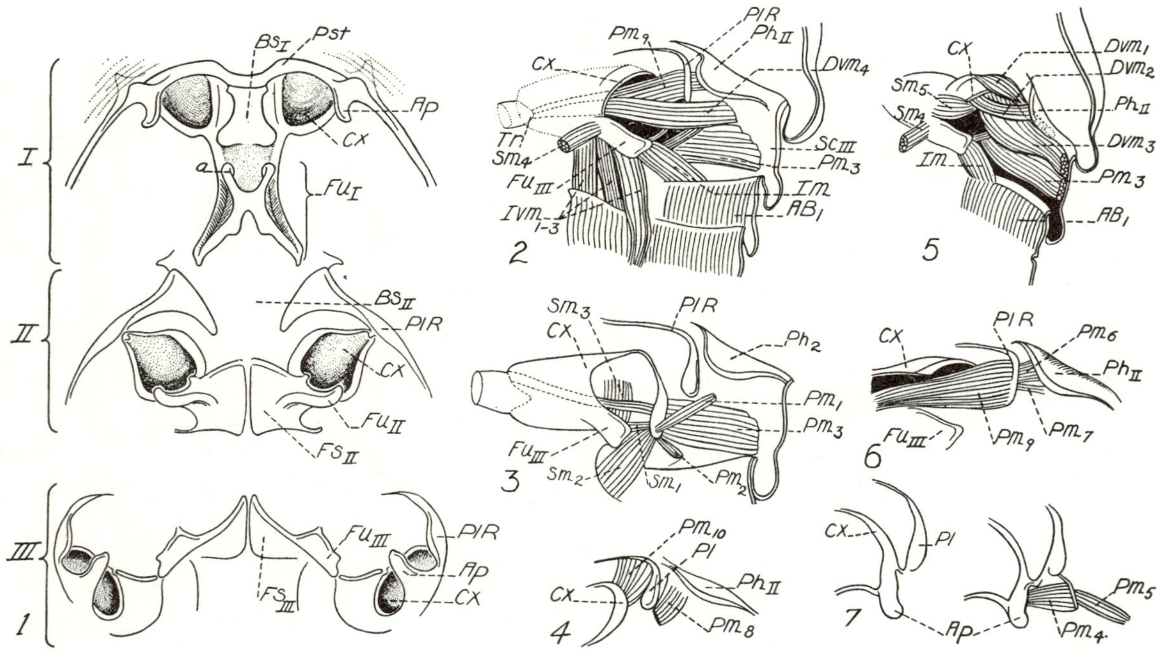

Fig. 1. Sternal and pleural sclerites of the nymph; Figs. 2-7. Muscles of the metathorax of the nymph

is the large apodeme of the nymph. The origin of the apodeme from the rim of the coxa at the pleural coxal articulation is similar to that in the adult. The apodeme is longer and more prominent than the metafurca. At the base of the apodeme is a small projection which articulates with the pleuron. *Pm 1* is a very narrow muscle which attaches near the tip of the apodeme to the upper surface. It extends across *Pm 2* and attaches on the scutum just above the pleural ridge. In the adult the homologue of *Pm 1* attaches to the anterior notal wing process and in the immature nymph the area of attachment probably corresponds to the notal wing process.

Pm 2 (Pl. XXVI, fig. 3) attaches to the tip of the apodeme posterior to *Pm 1*. The fibers extend almost at right angles to those of *Pm 1*, and attach to the dorsal border of the epimeron near the first abdominal segment.

Pm 3 (Pl. XXVI, figs. 2, 3, and 5) is a large, wide muscle attached to the under surface of the apodeme. Dorsally it is inserted on the notum near the mid-line approximately in line with the median end of the phragma.

Pm 4 (Pl. XXVI, fig. 7) is a small flat muscle which attaches lateral to *Pm 2* to the under surface of the apodeme near its base. There is a heavy ridge, a continuation of the pleural arm, which serves for the attachment of both *Pm 4* and *Pm 5*. This extension of the pleuron furnishes rigidity and support to the base of the wing pad. In a specimen treated with potassium hydroxide the two layers of integument forming the wing bud can be pulled apart, and the medial surface of the opening is bordered by the ridge. A similar condition occurs in the mesothorax of the nymph, but there are no muscles corresponding to *Pm 4* and *Pm 5* of the metathorax.

Pm 5 (Pl. XXVI, fig. 7) is opposite the attachment of *Pm 4* and arises on the other side of the ridge. The insertion on the scutum is ventrad to the attachment of *Pm 2*.

Pm 6 (Pl. XXVI, fig. 6) is a wide, thick muscle which attaches to the pleural arm and to the under side of the rim of the phragma.

Pm 7 (Pl. XXVI, fig. 7) attaches to the pleural ridge posterior to *Pm 5* and to the scutum at a point near the deepest width of the phragma. *Pm 7* is a very small, narrow muscle.

Pm 8 (Pl. XXVI, fig. 4) is exposed when *Pm 6* and *Pm 7* are lifted. The muscle attaches to the anterior portion of the ridge which forms a support on the base of the wing pad. The attachment of *Pm 8* is anterior to that of *Pm 5*. The fibers spread towards the termination on the scutum where they attach between the pleural arm and the phragma.

Pm 9 (Pl. XXVI, fig. 6) attaches to the trochanter and to the pleural arm opposite the attachment of *Pm. 6*. *Pm 9* is a thick, strong muscle.

Pm 10 (Pl. XXVI, fig. 4).—When the insertion of *Pm 9* is lifted, *Pm 10* is uncovered. It is a thin muscle lying next the integument. It attaches to the rim of the coxa anterior to the apodeme, and to the pleuron where the latter turns nearly at right angles just ventral to the phragma. *Pm 10* is homologous with *Pm 9* of the nymphal mesothorax.

Sternal Muscles:

Sm 1 (Pl. XXVI, fig. 3) is homologous with *Sm 1* of the adult. It is a very small muscle extending from the anterior corner of the metafurca to the ventral border of the apodeme close to the attachment of *Pm 1.*

Sm 2 (Pl. XXVI, fig. 3) is a strong muscle which attaches to the tip of the apodeme near *Pm 2*. It attaches to the lateral portion of the sternum. Just below the furca the sclerite curves medially and it is to this curved border that *Sm 2* inserts by a wide attachment.

Sm 3 (Pl. XXVI, fig. 3).—To the anterior edge of the furca *Sm 3* attaches by a wide base and to the inside of the coxa along the ridge formed by the anterior coxal suture. *Sm 3* is homologous with *Sm 2* of the adult.

Sm. 4 (Pl. XXVI, figs. 2 and 5) corresponds to *Sm 2* of the mesothorax and also with *Sm 3* of the adult metathorax. It attaches to the base of each metafurca and extends across the mid-sternal line.

Sm 5 (Pl. XXVI, fig. 5) is homologous with *Sm 4* of the adult. It is a wide short muscle which attaches to the furcisternum close to the base of the furca. It extends into the coxa where it attaches to the rim under the attachment of *Dvm 2.*

Intersegmental Muscles:

Ventrally the metafurca is on the border of the first abdominal segment and therefore the muscles which attach to the posterior border of the furca can be designated as intersegmental.

Im (Pl. XXVI, figs. 2 and 5) arises on the tip of the metafurca, by an attachment which extends the width of the arm and the fibers pass dorsocaudally to attach to the anterior border of the tergum of the first abdominal segment.

Ivm 1 (Pl. XXIV, fig. 5 and Pl. XXVI, fig. 2).—Medial to *Im* and attaching to the posterior surface of the furcal arm are the ventral intersegmental muscles. *Ivm 1* which arises nearest the tip of the furca, attaches to the posterior border of the first abdominal segment and continues directly caudad where it inserts to the posterior border of the second segment. Medial to *Ivm 1* are *Ivm 2* and *Ivm 3* which attach to the posterior surface of the furca and to the anterior border of the second abdominal segment.

MUSCULATURE OF THE ADULT

PROTHORAX:

Dorsal Longitudinal Muscles:

Dlm 1 (Pl. XXVII, fig. 1) attaches with *Dlm 2* on the antecosta of the mesotergum. The attachment is lateral to the mid-line close to the point where the prealar bridge fuses with the antecosta. Both dorsal muscles are very narrow and not at all comparable to the large dorsal longitudinal muscle of the mesothorax which fills nearly two thirds of the segment. *Dlm 1* extends cephalad and attaches to the tergum near the anterior border.

Dlm 2 (Pl. XXVII, fig. 1) attaches beneath *Dlm 1* on the antecosta of the mesotergum. It extends beyond the anterior border of the tergum and attaches to the postoccipital ridge near the mid-line.

Ventral Longitudinal Muscles:

Vlm 1 (Pl. XXVII, fig. 5) which is the same width and thickness as *Dlm 1*, passes directly cephalad and inserts on the tentorium (*Tt*) by a side attachment to the medial base of the posterior arm.

Vlm 2 (Pl. XXVII, fig. 5) is exceedingly narrow, almost thread-like, and attaches with *Vlm 1* on the profurca slightly laterad. It passes obliquely forward and attaches near the middle of the cervical sclerite (*Cv*).

Dorsal Muscle. Dm (Pl. XXVII, fig. 3) is a wide muscle attaching to the posterior part of the tergum lateral to *Dlm 1* and *Dlm 2*. The fibers extend posteriorly and attach to the posterior border, slightly ventral to the origin of *Dlm 1* and *Dlm 2*. The muscle passes over the longer fibers of *Dvm 5*.

Dorsal Ventral Muscles:

Dvm 1 (Pl. XXVII, fig. 2) which is flattened laterally, attaches to the basicoxite near the pleural coxal articulation. The muscle extends obliquely caudad and attaches dorsally to the tergal plate. There is no internal ridge for the insertion of the dorsal muscles, but the attachment is indicated by a pigmented line lateral to the mid-line. The muscle widens towards the dorsal attachment which is about twice the width of the ventral. Contraction of *Dvm 1* would tend to pull the coxa up and forward.

Dvm 2 (Pl. XXVII, fig. 2) is approximately the same width as *Dvm 1*. It is attached ventrally to the coxal trochanteral articulation. It widens and attaches to the tergum lateral to *Dvm 1*. The attachment is about midway between that of *Dvm 1* and the pleural sclerite.

Dvm 3 (Pl. XXVII, fig. 6) is attached to the rim of the meron of the coxa about midway between the pleural and trochanteral articulations. It attaches to the tergum near the posterior border of the sclerite, in line with the attachment of *Dvm 1*.

Dvm 4 (Pl. XXVII, fig. 3) is lateral to both *Dvm 1* and *Dvm 2* and may be seen when the latter are lifted. *Dvm 4* is attached ventrally to the basicoxite posterior to the attachment of *Dvm 1*. The attachment is narrow, but the muscle widens considerably, and the attachment on the tergum, which is about four times the width of that on the coxa, is very thin. The fibers seem to "fade out" on the integument.

Dvm 5 (Pl. XXVII, fig. 3) is a delicately thin muscle attached to the rim of the meron posterior to *Dvm 4*. The fibers converge here to a nar-

PLATE XXVII.

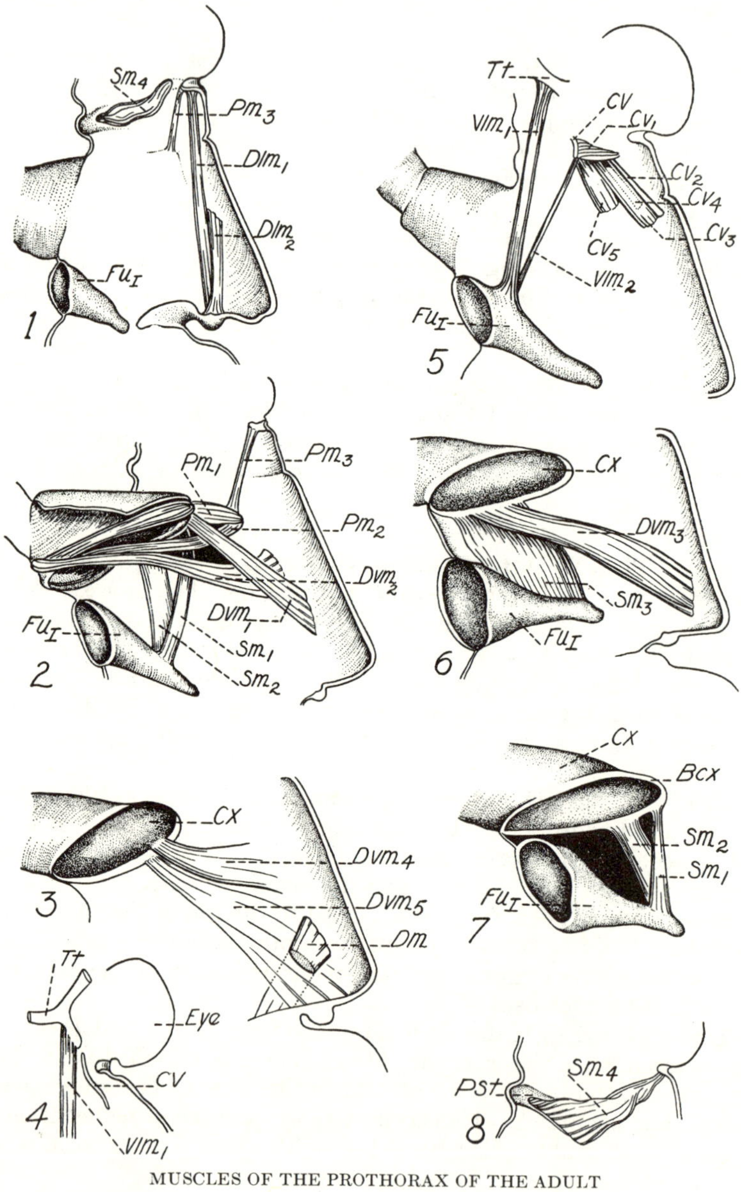

MUSCLES OF THE PROTHORAX OF THE ADULT

row flattened attachment, while those on the tergum are spread out over a wide area. The anterior fibers are shortened and follow the course of those of *Dvm 4*. The fibers gradually lengthen and the longer ones attach to the posterior border of the tergum. *Dvm 5* lies next the integument.

Pleural Muscles:

Pm 1 (Pl. XXVII, fig. 2) is a strong, short muscle attached to the coxa at the pleural coxal articulation, and dorsally to the anterior margin of the pleural plate.

Pm 2 (Pl. XXVII, fig. 2) is a long narrow muscle attached to the anterior portion of the pleural plate and below to the trochanter.

Pm 3 (Pl. XXVII, figs. 1 and 2) is attached anteriorly to the pleural plate near *Pm 1*. It is a narrow muscle which extends cephalad to attach to the postoccipital ridge, close to the eye and slightly lateral to the median line.

Cervical Muscles:

Cv 1 (Pl. XXVII, fig. 5) attaches by a wide base to the anterior half of the cervical sclerite. The fibers converge to a pointed attachment in the cervical membrane directly dorsad. The muscle is not very long, but judging from the width of the base and thickness is probably strong.

The remaining cervical muscles are grouped together and are attached lateral to *Pm 1* on the cervical sclerite.

Cv 2 (Pl. XXVII, fig. 5) which is the most anterior of the group is a very narrow muscle. It extends from the anterior border of the cervical sclerite up to the cephalic part of the tergum where it attaches lateral to the mid-line and posterior to the anterior margin of the sclerite.

Cv 3 (Pl. XXVII, fig. 5) is a wide, strong muscle, the fibers of which run parallel to those of *Cv 2*. It inserts by a wide attachment to the tergum just below *Cv 2*.

Cv 4 (Pl. XXVII, fig. 5) is a small muscle lateral to *Cv 3* and is completely covered by the latter. It attaches very close to the anterior border of the tergum.

Cv 5 (Pl. XXVII, fig. 5) is a short, wide muscle which attaches near the middle of the cervical sclerite, and to the tergum just posterior to the margin, and ventral to *Pm 3*.

Sternal Muscles:

Sm 1 (Pl. XXVII, figs. 2 and 7) attaches near the tip of the profurca, and to the basicoxite close to the posterior border of the pleural sclerite. The attachment on the furca is slightly wider than that on the coxa. The muscle appears to be strong and its contraction would pull the anterior portion of the coxa caudad, thus elevating the posterior part of the segment.

Sm 2 (Pl. XXVII, figs. 2 and 7) is a wider muscle which attaches lateral to *Sm 1* near the tip of the profurca. *Sm 2* extends obliquely forward and by a wide attachment inserts to the basicoxite ventral to the attachment of *Sm 1*. The contraction of this muscle would strengthen the backward pull of the anterior portion of the coxa, for the direction of the force would balance that of *Sm 1*.

Sm 3 (Pl. XXVII, fig. 6).—Beneath *Sm 1* and *Sm 2* and extending nearly to the base of the furca, very delicate fibers connect the furca with the coxa. They take a slightly oblique course and attach on the basicoxite close to *Dvm 3*.

Sm 4 (Pl. XXVII, figs. 1 and 8) arises by a very wide base on the lateral extension of the basisternum. The muscle is one of the strongest in the segment. The fibers converge to a pointed attachment on the evagination on the lateral portion of the head just at the base of the eye. Contraction of the muscle would draw the head up and back.

Intersegmental Muscle (Pl. XXVIII, fig. 1) like that of the nymph, attaches to the tip of the profurca and dorsally to the phragma of the mesothorax.

MESOTHORAX:

Indirect Wing Muscles:

Dlm (Pl. XXVIII, fig. 1) is the largest in the mayfly thorax. It is so wide that viewed from the median plane it appears to fill nearly two-thirds of the thorax. It attaches anteriorly to the mesophragma and to the scutum. The attachment on the scutum extends laterally from the mid-line to the inner parapsidal ridge. The fibers converge slightly and attach posteriorly to the mesophragma. The contraction of this muscle produces the downward stroke of the fore wings.

Dm (Pl. XXVIII, fig. 2) is a large thick muscle which lies lateral to the dorsal longitudinal. It attaches by a wide base close to the suture between the meso- and metathorax above the metathoracic spiracle. The muscle passes obliquely forward, across the lateral plate of the scutellum and attaches on the scutum in the rounded posterior hump near the mid-line.

Ventral Longitudinal, *Vlm* (Pl. XXVIII, figs. 1 and 5) is a narrow band of fibers flattened laterally. The anterior attachment is at the posterior base of the profurca. The ribbon-like muscle extends to the mesofurca to the base of which it is attached.

Dvm 1 (Pl. XXVIII, figs. 1, 3 and 5) attaches to the basisternum where the latter widens laterally to fuse with the precoxal piece of the pleuron, and to that portion of the pleural trochantin which terminates ventrally on the basisternum. The attachment covers the entire basal portion of the pleural trochantinal sclerite and extends laterally along the basisternal ridge close to the furcilla of the prosternum. The muscle widens somewhat towards its attachment on the scutum. The medial anterior fibers insert along the inner parapsidal suture of the scutum. The lateral posterior fibers insert along the heavy outer parapsidal ridge.

PLATE XXVIII.

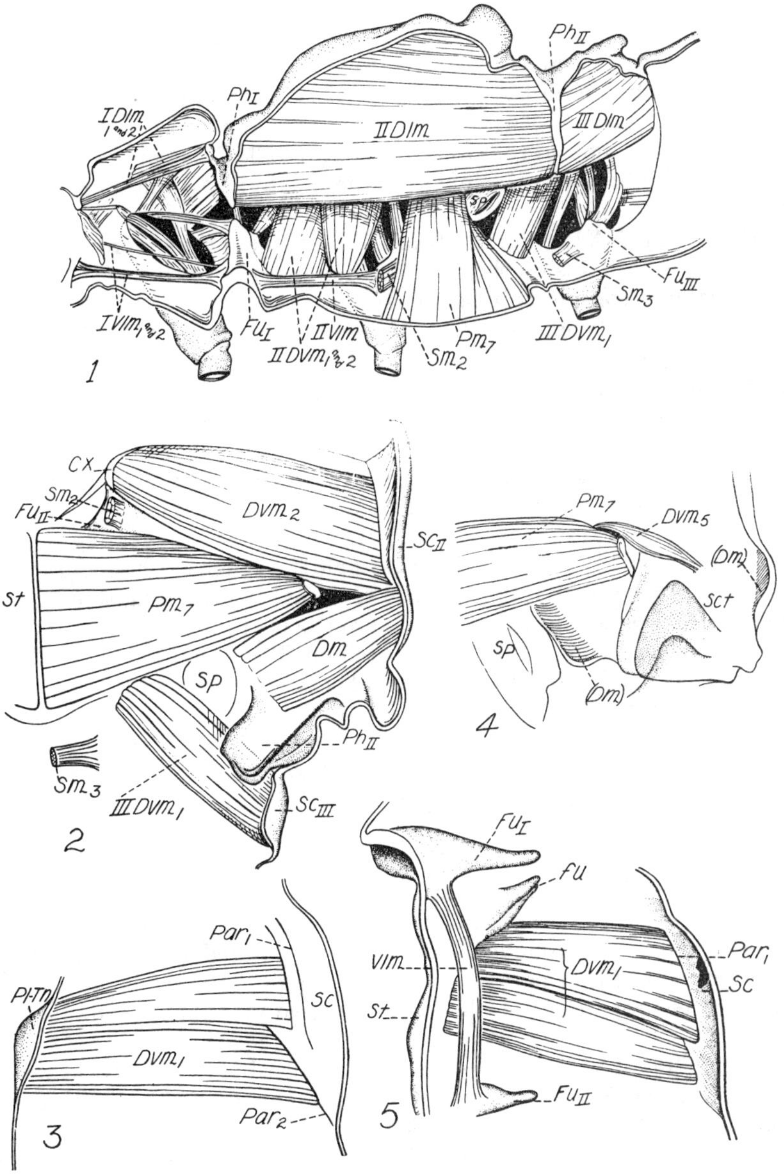

ADULT MUSCULATURE

Fig. 1. Muscles as seen in sagittal section; Figs. 2-5. Muscles of the mesothorax

The area of insertion lies therefore, between the two parapsidal ridges on the scutum (Pl. XX).

Dvm 2 (Pl. XXVIII, figs. 1 and 2; Pl. XXIX, fig. 1) is very thick although somewhat more narrow than the preceding (*Dvm 1*). It attaches to the medial border of the basicoxite. The attachment extends from the coxal articulation (*Cx*) with the pleuron to the mesofurca. The muscle attaches on the scutum (*Sc*) to the posterior half of the outer parapsidal ridge and the area of attachment extends laterad to the notal furrow (Pl. XX).

Dvm 3 (Pl. XXIX, fig. 3) attaches to the rim of the meron. The attachment of *Dvm 3* is on the opposite rim of the coxa to that of *Dvm 2*. The muscle is considerably smaller than either *Dvm 2* or *Dvm 1*. It is narrow, but thickened between attachments so that the body of the muscle is oval in shape. The fibers converge and form flattened attachments. The attachment on the scutum is opposite the termination of the outer parapsidal ridge, and just lateral to the posterior part of the attachment of *Dvm 2*.

Dvm 4 (Pl. XXIX, figs. 3 and 6) attaches to the rim of the meron posterior to *Dvm 3*. The muscle at its attachment is about one-third as wide as *Dvm 3*. It widens slightly, then narrows considerably at the attachment on the scutum which is slightly medial to that of *Dvm 3*.

Dvm 5 (Pl. XXVIII, fig. 4; Pl. XXIX, fig. 7) is a strong thin muscle which attaches to the rim of the meron beneath *Dvm 3*. The muscle attaches to the lateral plate of the scutellum. The area of attachment lies between the notch formed by the lateral emargination and the triangular area to which the postscutellum is fused (Pl. XX).

Dvm 6 (Pl. XXIX, figs. 4 and 6) is a wide muscle attaching to the trochanter base which projects into the coxa. The muscle inserts by a wide attachment on the scutum just posterior to the apodeme on the anterior notal wing process.

Direct Wing Muscles:

Pm 1 (Pl. XXIX, fig. 5) attaches to the furcilla in the postfurcisternum of the prothorax. The furcilla is drawn up near the prothoracic furca. The muscle is rounded and tapers to a pointed attachment on the anterior point of that part of the basalar sclerite (*Ba*) which articulates with the pleuron. *Pm 1* is the largest muscle attaching to the basalar process.

Pm 2 (Pl. XXIX, fig. 5) is a very small ribbon-like muscle which attaches to the furcilla posterior to the attachment of *Pm 1*. It inserts opposite *Pm 1* on the posterior portion of the basalar sclerite above the episternum. *Pm 1* and *Pm 2* attach to the two extremities of the pleural portion of the sclerite and between these is the evagination forming the tegula (*Tg*).

Pm 3 (Pl. XXIX, figs. 4 and 6) is a thin muscle attaching to the trochanter at the point where the trochanter base projects into the coxa,

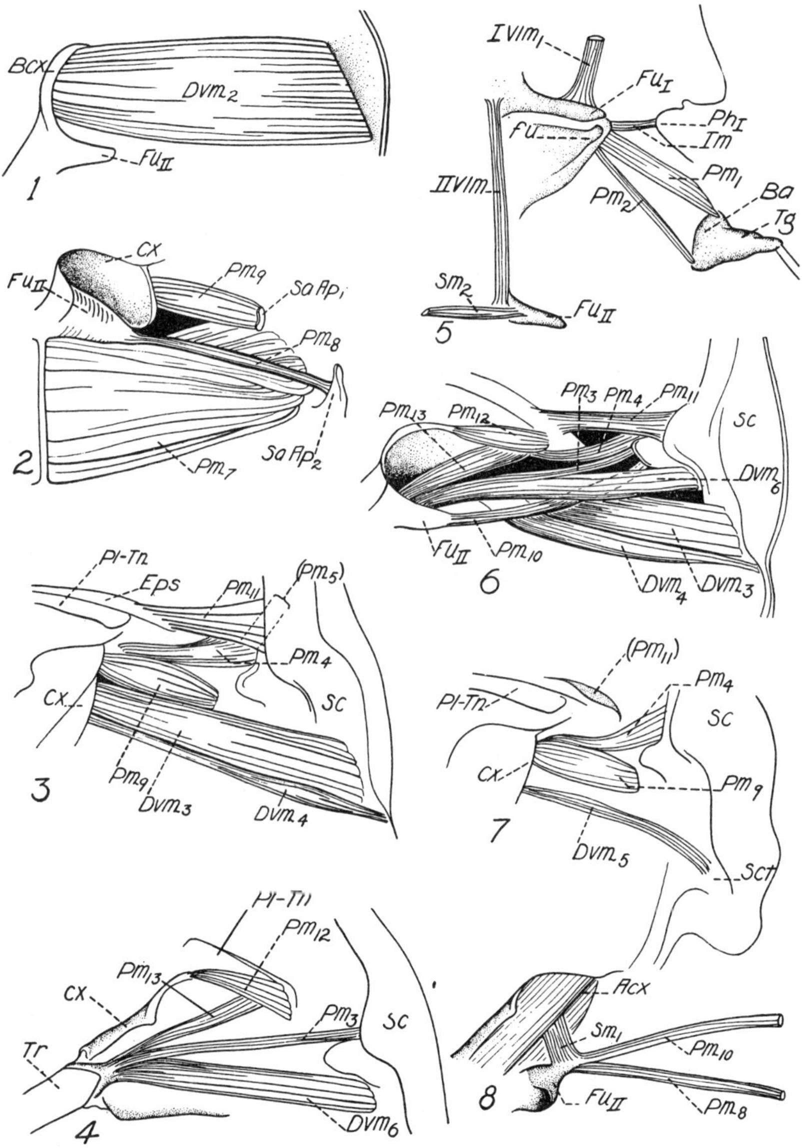

ADULT MUSCULATURE
Muscles of the mesothorax.

forming a strong process for the attachment of *Pm 3*, *Pm 13*, and *Dvm 6*. *Pm 3* attaches to the basalar sclerite on that process which articulates with the scutum near the anterior notal wing process.

Pm 4 (Pl. XXIX, figs. 3, 6, and 7) attaches to the coxa at the pleural coxal articulation. The point of attachment is very small. The muscle is thin and narrow for a short distance until it turns slightly, thickens considerably, and widens towards the insertion. The attachment is above that of *Pm 3* on the basalar sclerite near the articulation with the tergum.

Pm 5 (Pl. XXIX, fig. 3) is a short wide muscle which attaches to the pleural wing process close to the top just beneath the wing base. The muscle extends obliquely forward and attaches to the basalar process. The tergum folds down over *Pm 5*, nearly obscuring it. It serves as a very strong connection between the basalar process and the pleural wing process.

Pm 6 (Pl. XX, *Pm.*) attaches to the base of the sclerite forming the posterior notal wing process which folds down like a hinge over the pleuron at the wing base. The muscle attaches on the pleural surface of the wing to the apodeme on the base of the pleural wing recess. The muscle is exceedingly small but forms a firm connection between the tergum and the pleural wing membrane.

Pm 7 (Pl. XXVIII, figs. 1 and 2: Pl. XXIX, fig. 2) is a very large triangular shaped muscle which arises on the furcisternum at the mid-line. The attachment extends the entire length of the furcisternum and is marked exteriorly by a distinct suture on either side of the mid-line. The muscle attaches to the subalar plate by two heads.

The innermost fibers, or those viewed on top in a mid-sagittal section, converge from the wide origin on the furcisternum to attach to the entire length of the finger-like apodeme on the posterior part of the subalar sclerite (*Sa*). The insertion is approximately one-fourth as wide as the origin. Beneath the attachment to the subalar apodeme the fibers extend obliquely from the furcisternum to the plate of the subalar sclerite extending between the two apodemes. Here the second head of the muscle attaches between the bases of the anterior and posterior apodeme.

Pm 8 (Pl. XXIX, figs. 2 and 8) is a very strong narrow muscle which attaches to the posterior part of the tip of the mesofurca (*Fu II*) and extends to the posterior subalar apodeme (*Sa Ap* 2). The attachment on the base of the apodeme is posterior to that on the mesofurca. The fibers, therefore, cross those of the second head of *Pm 6*.

Pm 9 (Pl. XXIX, figs. 2, 3, and 7) attaches to the border of the meron anterior to *Dvm 3*, and to the anterior subalar apodeme which is formed by the infolding of the corner of the sclerite.

These three, *Pm 7*, *Pm 8*, and *Pm 9* (Pl. XXIX, fig 2) comprise the group known as the subalar muscles. The usual attachment of the subalar muscle to the coxa is supplemented in this case by the attachment on the furcisternum and mesofurca.

Pm 10 (Pl. XXIX, figs. 6 and 8) is a very strong thin muscle which attaches to the tip of the mesofurca (*Fu II*) anterior to the attachment of *Pm 8*. It passes over *Dvm 3* and *Dvm 4*, under *Dvm 6*, and attaches to the tip of the first axillary sclerite. There is a projection of the sclerite below the anterior notal wing process which serves as the point of attachment for *Pm 10* (Pl. XX).

Although the wings of the Ephemeroptera are not flexed, it is possible that *Pm 10* may represent a flexor muscle. Ordinarily the flexor is inserted to the third axillary sclerite, but in this case the third axillary is not distinguishable.

Pm 11 (Pl. XXIX, figs. 3 and 6) originates on the episternum, attaching to the ridge formed by the infolding of the chitin. The longest fibers attach in line with the pleural coxal articulation and the head of the pleural trochantin. The fibers attach in an oblique line above that point on the episternum. The muscle is attached to the anterior notal wing process of the scutum, to the notal furrow (Pl. XX).

Pm 12 (Pl. XXIX, figs. 4 and 6) is a short, wide muscle which attaches to the medial border of the coxa beneath the anterior part of the attachment of *Dvm 2*. The attachment of *Pm 12* is approximately opposite the furca on the rim of the coxa (*Cx*). The muscle is narrow at the base, but widens, becomes slightly fan-shaped and is inserted by a wide attachment to the anterior part of the pleural trochantin (*Pl-Tn*).

Pm 13 (Pl. XXIX, figs. 4 and 6) attaches to the trochanter base (*Tr*) which projects into the coxa. The muscle widens towards the attachment which is at the head of the pleural trochantinal sclerite beneath *Pm 12*.

Sternal Muscles:

Sm 1 (Pl. XXIX, fig. 8) is a wide short muscle which attaches to the anterior part of the mesofurca (*Fu II*) and to the inside of the coxa to the ridge formed by the anterior coxal suture (*Acx*).

Sm 2 (Pl XXVIII, figs. 1, 2: Pl. XXIX, fig. 5) is a short muscle which attaches to the median base of each mesofurca (*Fu II*) and extends across the mid-line.

METATHORAX:

The more specialized function of the fore wings as compared to that of the hind wings probably accounts for the more complex mechanism of the mesothorax. Added stress on the wing muscles necessitates not only a more complex musculature, but increased strength in the entire skeletal mechanism. The sclerites and components of the endoskeleton of the metathorax are much less specialized than those of the mesothorax. It is difficult therefore to ascertain exact homologies in certain cases between the muscles of the two segments.

Dlm (Pl. XXVIII, fig. 1) of the metathorax corresponds to that of the mesothorax in position and relative size. It is attached anteriorly to the phragma (*Ph II*) and metascutum. There is no posterior phragma in the metathorax, but the antecosta is drawn down to serve for the attachment of the dorsal longitudinal muscle posteriorly.

Dvm 1 (Pl. XXVIII, fig. 1, and Pl. XXX, fig. 1) is a large, thick muscle which attaches to the anterior lateral portion of the furcisternum. The area of attachment is nearly round. The muscle is attached to the scutum nearly covering the entire lateral portion of the sclerite. The attachment on the scutum is posterior to that on the sternum, so that the muscle lies in an oblique position. *Dvm 1* of the metathorax is homologous with *Dvm 1* of the mesothorax. Contraction of the muscle draws the scutum down towards the sternum and in so doing helps to elevate the wings. This is by far the largest dorsal ventral muscle in the metathorax, hence its function as wing elevator is probably much greater than that of the other muscles of the group.

Dvm 2 (Pl. XXX, fig. 1) is attached to the medial border of the coxa, and to the scutum close to the suture between the scutum and scutellum where the suture begins to curve anteriorly towards the mid-line.

Dvm 3 (Pl. XXX, fig. 1) is attached to the coxa at the border of the meron between the pleural coxal articulation and the metafurca. A short distance from its attachment the muscle is divided and the two heads are inserted together on the scutum posterior to *Dvm 2*. *Dvm 3* is homologous to *Dvm 3* of the mesothorax.

Dvm 4 (Pl. XXX, fig. 1) is a small muscle which is attached to the tip of the metafurca. The muscle passes over *Dvm 3* and attaches to the lateral portion of the scutellum just below the anterior margin. Towards the insertion the muscle turns and attaches on its upper surface.

Dvm 5 (Pl. XXX, fig. 4) is a strong muscle attaching to the head of the trochanter which projects into the coxa. The muscle is fan-shaped and widens considerably towards the attachment on the lateral border of the scutum at the wing base.

Pleural Muscles:

Pm 1 (Pl. XXX, fig. 2) is attached to the coxa at the pleural coxal articulation. There is an evagination of the basicoxite forming a process at the pleural coxal articulation to which *Pm 1* attaches by a small base. The muscle inserts on the anterior notal wing process, which is pulled down by the attachment. *Pm 1* is very narrow at both attachments but widens considerably between.

Pm 2 (Pl. XXX, fig. 2) is a very small muscle which attaches to the pleuron anterior to the pleural coxal articulation. It attaches dorsally on the lateral oblique margin of the scutum which is anterior to the notal wing process. A possible homology may be drawn between *Pm 2* of the metathorax and *Pm 1* of the mesothorax. *Pm 1* is a basalar muscle attaching to the basalar sclerite near the point of articulation of the latter with the scutum. There is no basalar sclerite in the metathorax, hence the direct wing muscles differ in attachments. It would seem logical, however, that the contraction of *Pm 2* of the metathorax would deflect

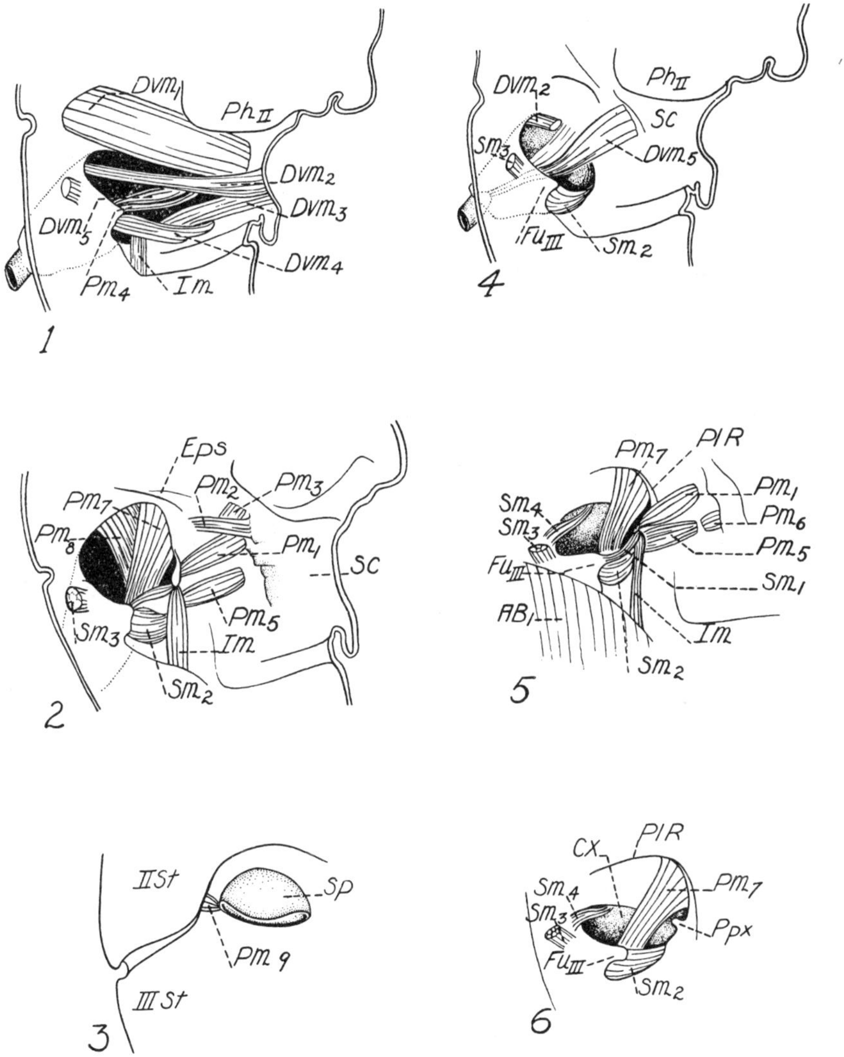

ADULT MUSCULATURE
Muscles of the metathorax.

the costal region of the wing. The function would be similar to that of the basalar muscle, *Pm 1* of the mesothorax.

Pm 3 (Pl. XXX, fig. 2) is an exceedingly small muscle beneath *Pm 2*. It is attached to a heavy ridge in the pleural wing process close to the wing base. It widens somewhat and attaches to the sclerotized piece at the base of the pleural wing recess.

Pm 4 (Pl. XXX, fig. 1) is attached to the metafurca along the anterior edge close to the tip. The narrow muscle passes over *Dvm 5*, under *Dvm 2* and attaches to the small projection of the sclerite behind the anterior notal wing process of the scutum. *Pm 4* may be homologous with *Pm 10* of the mesothorax.

Pm 5 (Pl. XXX, fig. 2) has a wide attachment to the meron of the coxa posterior to the pleural coxal articulation. It attaches to the unsclerotized chitin between the epimeron and the scutum. An homology might be drawn between *Pm 5* of the metathorax and *Pm 9*, a subalar muscle of the mesothorax.

Pm 6 (Pl. XXX, fig. 5) is smaller than *Pm 3*. It is homologous with *Pm 6* of the mesothorax. It attaches to the sclerite corresponding to the posterior notal wing process which it pulls down. *Pm 6* attaches to the pleural surface of the wing.

Pm 7 (Pl. XXX, figs. 2 and 5) which is homologous with *Pm 13* of the mesothorax, attaches to the trochanter. The muscle is wide and broadens slightly at the attachment on the inner border of the episternum. The wide origin extends from the pleural coxal articulation past the curve of the episternum around the coxal membrane at the pleural ridge (*PlR*).

Pm 8 (Pl. XXX, fig. 2) inserts on the rim of the basicoxite anterior to the pleural coxal articulation. The muscle is extremely thin. The attachment on the inner margin of the episternum is about twice as wide as that of *Pm 7*. *Pm 8* lies next the body wall. Its contraction would draw the coxa forward towards the episternum.

Pm 9 (Pl XXX, fig. 3).—To the posterior portion of the lip of the metathoracic spiracle is attached a small muscle, *Pm 9*, which originates on the intersegmental ridge at the base of the spiracle.

Sternal Muscles:

Sm 1 (Pl. XXX, fig. 5) is a very narrow muscle which attaches under *Pm 4* on the anterior margin of the metafurca at the tip of the arm. The muscle attaches to the process formed by the evagination of the basicoxite at the pleural coxal articulation posterior to the attachment of *Pm 1*.

Sm 2 (Pl. XXX, figs. 2, 4, 5, and 6) is a wide, thin muscle which attaches to the inside of the metafurca. The fibers run parallel to the anterior margin of the furca and the muscle attaches to the inside of the meron of the coxa. The muscle fills up the gap between the furca and the rim of the meron.

Sm 3 (Pl. XXX, fig. 2) is homologous with *Sm 2* of the mesothorax. *Sm 3* attaches at the base of each metafurca and stretches across the mid-line. The muscle, like that of the mesothorax (*Sm 2*), is wide and thick. The areas of attachment are nearly round.

Sm 4 (Pl. XXX, fig. 5) is a very small muscle which attaches to the sternum at the base of the metafurca slightly anterior to *Sm 3*. The muscle extends into the coxa, attaches near the rim just posterior to the attachment of *Dvm 2*. The attachment of *Sm 4* is on the anterior border of the meron.

Intersegmental Muscles:

Im (Pl. XXX, fig. 2) is a fairly wide muscle attaching to the evagination of the basicoxite at the pleural coxal articulation. It extends directly caudad and by a wide attachment inserts to the anterior ridge of the first abdominal segment.

Ventrally the muscles of the first abdominal segment attach to the posterior surface of the metafurca and along a line on the sternum between the two furcae, at the fusion of the thoracic sternum and the abdomen. These muscles, three on each side, insert posteriorly to the anterior border of the second segment.

Conclusion

The musculature of the immature nymph of *Hexagenia recurvata* is similar in many respects to that described by Weber for *Psylla mali* (1929). The musculature of the adult of *Hexagenia recurvata* resembles that of *Papilio machaon* as described by Weber (1928).

The ventral longitudinal muscles of the nymph of *Hexagenia recurvata* bear a closer resemblance to those of *Papilio machaon* than do the adult muscles. *Vlm* of the adult (*H. recurvata*) is comparable to *I Vlm 1* of *Papilio machaon*. The muscle between the mesofurca and metafurca (*II Vlm 1* of Papilio) is lacking in the adult mayfly, but is present in the nymph (*Vlm* Pl. XXIV, fig. 5). There is no muscle in *H. recurvata* comparable to *II ism*, a short muscle of Papilio which attaches to the mesofurca and postnotum. In the nymph of *H. recurvata* there are two intersegmental ventral muscles (*Ivm 1* and *Ivm 2*) which are similar to *III Vlm 1* of Papilio. They extend from the metafurca to the anterior border of the second abdominal segment. A third muscle, *Ivm 3* of the nymph, extends from the metafurca to the posterior border of the second abdominal segment, a condition which does not occur either in the adult *H. recurvata* or in Papilio. According to Snodgrass the absence of ventral muscles from thorax to abdomen is common in adults although present in the nymphs (1929). A condition comparable to that of *H. recurvata* occurs in *Psylla mali* as recorded by Weber. Snodgrass describes ventral muscles in the Cicadidae extending from the metathorax to the second abdominal segment. Likewise in the adult honey bee there are two pairs of muscles going from the metafurca to the anterior margin of the second abdominal segment (Morison 1927). The presence of ventral muscles extending from the metafurca to the second abdominal segment is not recorded by Dürken for *Centroptilum luteolum* and *Ephemerella ignita* (1907).

Pleural Muscles:

The pleural muscles of *Papilio machaon* are considerably larger than those of *H. recurvata.* The two basalar muscles of Papilio attach to the midsternal line and to the basicoxite. They are more prominent than the basalar muscle of *H. recurvata,* although the basalar sclerite (*die Episternalgelenkplatte,* Weber) is not as well developed as that of *H. recurvata.*

On the other hand, the enlarged subalar sclerite of mayflies bears a marked resemblance to that of the Lepidoptera (see the figures of Shepard 1930.) However, the subalar muscles of the two groups vary. The three subalar muscles of the Lepidopteron *Papilio machaon* attach to the meron of the coxa and to the postnotum. The corresponding muscles of *H. recurvata* attach to the meron, mesofurca, and mid-line of the mesofurcisternum. There is no connection between the subalar sclerite and the postnotum in *H. recurvata.* The sternal attachment of these muscles is more widespread than those of the Lepidoptera. The small pleural muscle, *II Pm 6* of Papilio, corresponds to *Pm 12* of *H. recurvata.* Both attach to the basicoxite and to the episternum.

The metathorax of the Lepidoptera is more highly specialized than that of the Ephemeroptera. The dorsal longitudinal muscle of *H. recurvata* is larger, and the fibers attach to the entire posterior surface of the phragma rather than to the scutum as is the case in Papilio. *III Dvm 1* of *H. recurvata* is greatly enlarged and more nearly comparable to the dorso-ventral muscles of Papilio than the other muscles of the metathorax, which are very small.

A sagittal section of *Papilio machaon* is strikingly like that of *H. recurvata.* The large dorsal longitudinal muscle of the mesothorax is similar to that of the mayfly although not quite as large in comparison.

Summary

Prothorax: In the prothorax of the nymph the muscles are more highly developed than those of the adult. There are three dorsal longitudinal muscles in the nymph comparable to two in the adult (Pl. XXIV, figs. 1 and 5). *Dvm 4* of the nymph is comparable to *Dvm 4* and *Dvm 5* of the adult (Pl. XXIV, fig. 4. Pl. XXVII, fig. 3).

The five small cervical muscles of the adult are represented in the nymph by one muscle (Pl. XXVII, fig. 5: Pl. XXIV, fig. 2). *Pm 4* of the nymph is not present in the adult (Pl. XXIV, fig. 7). The adult muscles *Sm 1* and *Sm 2* are represented in the nymph by *Sm 1* (Pl. XXIV, fig. 8: Pl. XXVII, fig. 7).

Mesothorax: The following muscles of the nymph are not present in the adult: *Dlm 2* (Pl. XXIV, figs. 1 and 5), *Vlm 1* (Pl. XXIV, fig. 1), *Pm 1* (Pl. XXV, figs. 2 and 5), *Pm 3* (Pl. XXV, fig. 3), *Pm 11* (Pl. XXV, fig. 8), *Sm 4* (Pl. XXV, fig. 8).

The muscles corresponding to the basalar group of the adult are not developed in the immature form. The subalar group of muscles is distinguishable in the nymph.

Metathorax: The principal variations in the muscles of the metathorax of the adult and immature forms are in the ventral longitudinal muscles. It is difficult to make accurate comparisons, particularly of the pleural muscles, because of the indistinct scleries. *Vlm* of the nymph is not present in the adult (Pl. XXV, fig. 8). The attachment of the intersegmental muscle, *Ivm 3* of the nymph, to the posterior border of the second abdominal segment is lacking in the adult (Pl. XXVI, fig. 2).

The thorax of the Ephemeroptera is more highly specialized than that of the Odonata. There is not a close relationship between the thoracic segments of the two orders.

SELECTED BIBLIOGRAPHY

Bauer, Albert, 1910. Die Muskulatur von *Dytiscus marginalis.* Zeitschrift für Wissenschaftliche Zoologie. 95:594–646.

Börner, Carl, 1909. Die Tracheenkiemen der Ephemeriden. Zoologischen Anzeiger, 33: 806–823.

Crampton, G. C., 1909. A Contribution to the comparative morphology of the thoracic sclerites of insects. Proceedings of the Academy of Natural Sciences of Philadelphia, 61: 3–54.

Crampton, G. C., 1914. The Ground plan of a typical thoracic segment in winged insects. Zoologischen Anzeiger, 44: 56–67.

Crampton, G. C., 1914. Thoracic sclerites of winged insects. Entomological News, 25: 15–25.

Crampton, G. C., 1917. A Phylogenetic study of the lateral head, neck and prothoracic regions in some Apterygota and lower Pterygota. Entomological News, 28: 398–412.

Crampton, G. C., 1918. A Phylogenetic study of the terga and wing bases in Embiids, Plecoptera, Dermaptera, and Coleoptera. Psyche, 25:4–12.

Crampton, G. C., 1918. Thoracic sclerites of immature Pterygoten insects, with notes on the relationships indicated. Proceedings of the Entomological Society of Washington, 20: 39–65.

Crampton, G. C., 1926. A Comparison of the neck and prothoracic sclerites throughout the orders of insects from the standpoint of phylogeny. Transactions of the American Entomological Society. 52: 199–248.

Dürken, B., 1907. Die Tracheenkiemenmuskulatur der Ephemeriden unter Berücksichtigung der Morphologie des Insektenflugels. Zeitschrift für Wissenschaftliche Zoologie. 87:435–550.

Morison, Guy D., 1927. Muscles of the adult Honey Bee. Quarterly Journal of Microscopical Science. 71:395–463. 72:511–526. 1928–29.

Shepard, H. H., 1930. The Pleural and sternal sclerites of the Lepidopterous thorax. Annals of the Entomological Society of America. 23: 237–260.

Snodgrass, R. E., 1927. Morphology and mechanism of the insect thorax. Smithsonian Miscellaneous Collections. 80: 1–108.

Snodgrass, R. E., 1928. Morphology and evolution of the insect head and its appendages. Smithsonian Miscellaneous Collection. 81: 1–158.

Snodgrass, R. E., 1929. How insects fly. Smithsonian Report for 1929, 383–421.

Snodgrass, R. E., 1929. The thoracic mechanism of a grasshopper, and its antecedents. Smithsonian Miscellaneous Collections. 82: 1–111.

Traver, J. R., 1931. Seven new southern species of the mayfly genus Hexagenia, with notes on the genus. Annals of the Entomological Society of America. 24: 591–621.

Voss, F., 1905. Über den Thorax von *Gryllus domesticus* (Pts. 1–4) Zeitschrift für Wissenschaftliche Zoologie, 78: 268–521, 645–759. Pt. 5, 100: 589–834.

Weber, H. 1924. Das Grundschema des Pterygotenthorax. Pt. 1, Zoologischer Anzeiger, 60: 17–37. Pt. II, 57-83.

Weber, H., 1928. Die Gliederung der Sternopleural-region des Lepidopterenthorax. Zeitschrift für Wissenschaftliche Zoologie, 131: 181–254.

Weber, H., 1929. Kopf und Thorax von *Psylla mali* Schmidt. Zeitschrift für Morphologie und Ökol. der Tiere. 14: 59–165.

Weber, H., 1933. Lehrbuch der Entomologie, Verlag von Gustav Fischer in Jena, 1933. Der Thorax, pp. 116–197.

Young, B. P., 1921. Attachment of the abdomen to the thorax in Diptera. Cornell University Agricultural Experiment Station. Memoir 44: 255–282.

CHAPTER XIII

NYMPHAL STRUCTURES AND ADAPTATIONS

1. BODY-FORM IN GENERAL

In contrast with the uniformity of the adults, the nymphs of mayflies are remarkably differentiated in adaptation to diverse environments. While there are intergradations, the forms of body fall into categories that match their habitat as follows:

Still water (static or lenitic) forms	a. Climbers amid vegetation. b. Sprawlers upon the bottom. c. Burrowers in the bottom.
Rapid water (lotic) forms	d. Agile, free-ranging, stream-lined forms. e. Close-clinging, limpet forms found under stones. f. Stiff-legged, trash and silt inhabiting forms.

I. Still-water forms

a. *The mayfly nymphs that clamber amid green vegetation* or that dart about rapidly over the bottom in still waters are streamed-lined, smooth-bodied forms with high heads, plate-like gills, and heavily fringed tails such as Siphlonurus, Blasturus, and Callibaetis. They have better eyes than have other nymphs, better locomotor muscles, and a better developed tail fin; they have gill plates that stroke backward synchronously and that wonderfully assist in jumping.

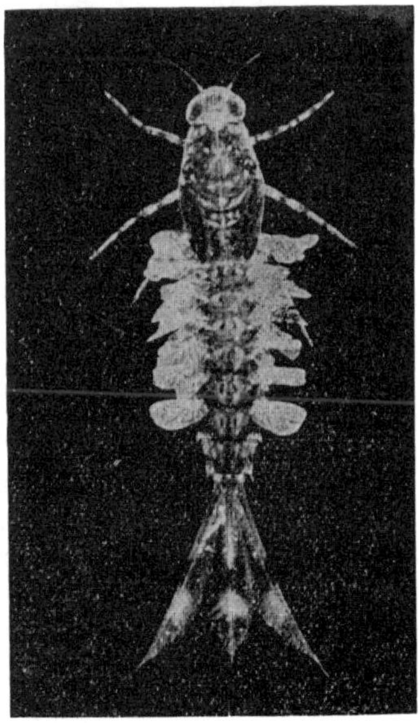

Fig. 44—Nymph of *Siphlonurus alternatus* (photo by Anna H. Morgan).

For taking care if itself in such situations, the nymph of Siphlonurus (fig. 44) is one of the best equipped. It skips about in the water like a minnow, dashing this way and that, making long sweeping runs with most graceful undulations of its lithe body and long tails. It cannot be picked up with a forceps. Indeed, it readily eludes a small net and is most easily taken with a fine-meshed seine. Its double gills are provided with a fine set of muscles and are under perfect control. Its tails bear dense lateral fringes of swimming hairs. The middle tail is fringed on both sides; the lateral tails are fringed on the inner side only. The fringes of the three together compose the webbing of an excellent tail fin; the half-bare laterals are its cutting edges, and the ready adjustment of its position gives wonderful control of the direction of movement.

The nymphs of Ameletus (fig. 45), Callibaetis, and Cloeon are similar. In all these the body as a whole is smoothly contoured; the thorax is high, compressed; the face is long and the large eyes are set high up on the head. The gills are held erect and swing forward in respiratory movements; the water currents flow from rear to front along the mid-dorsal line of the abdomen.

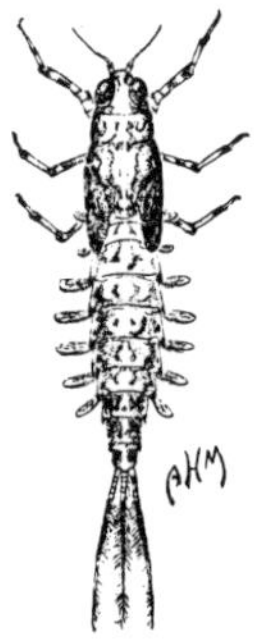

FIG. 45—The nymph of *Ameletus ludens* (from Morgan).

b. *The sprawlers amid the silt* in still waters are stiff-legged, hairy, dirt-covered forms with stout bodies and thinly fringed tails. These nymphs settle down amid the silt, either on the bottom or amid plant stems, moving about but little, and that little, slowly. Their safety lies in escaping observation—not in flight. Typical of this group are the nymphs of Caenis (fig. 46) and the flatter Ephemerellas. The legs are spread out laterally. The body is more or less depressed; the tails are thinly hairy on all sides and of little use in swimming. Body and legs are more or less hairy all over, and silt adheres copiously to the hairs, giving excellent concealment. It is hard to find such nymphs when collecting. Scraped up with the bottom ooze or buried in the dirt covering plant stems, they cannot be seen until they move.

In such places the gills need protection from the dirt which would clog them and interfere with air taking. They exhibit a high degree of gill differentiation. One pair of gills in front has become thickened and broadened and has developed into large opercula or gill covers, overlapping and protecting the gills behind them from falling silt. This cover is alternately lifted and lowered again in breathing. Interlacing fringes upon the margins of the thin underlying gills strain the silt out of the water as it enters the chamber beneath the gill cover, keeping the inner area clean. Thus Caenis gets along comfortably in waters so dirty that they would quickly smother the nymphs with exposed gills. The opercula of Tricorythodes are less perfect.

c. *The burrowers in the lake or pond bed* are pale, more or less thinly hairy forms, with narrow heads, tusked mandibles, digging front feet, and plumose gills, close-laid over the back. They are such as Ephemera, Hexagenia (fig. 47) and Ephoron. These live in the mud and sand of the bottom, more or less out of harm's way. They are highly specialized for digging passageways through the mud. The front feet are flattened like scrapers, and the head is narrowed forward. The short legs are drawn close up under the body. Most remarkable of all, the mandibles are produced forward into long, strong, tapering, horn-like tusks that may be longer than the head. These are used for lifting the roof of the burrow.

If a nymph of Ephemera be thrown out of its burrow, it immediately sets about with the persistence of a mole to make a new one. It thrusts its shovel-like feet forward, close together, edge upward into the mud and then spreads them apart, removing the mud to right and left. Into the cleft thus opened, it then thrusts its head and gives it an upward toss, lifting the earth far forward on the tops of its long mandibular tusks. Then, lowering its head, it lunges forward into the opening, pushing with its legs and propelling itself also by means of undulations of its long abdomen. It repeats these movements, each time going a little deeper, until it has reached a satisfactory level. The gills bear dense internal fringes of filaments, which meet on the mid-dorsal line of the abdomen when swung inward, and separate widely when swung out. They are moved together and apart in pairs, a wave of motion running along the back. This motion sets up a current of water and renews the oxygen supply.

II. Lotic forms

d. *The free-ranging stream-lined forms of rapid waters* are much like those of group *a* above. Stream-lining diminishes resistance to the flow of the water past them as they cling to the stones. It also facilitates their passage through it when they dart from place to place. Here belong Isonychia, Ameletus, Paraleptophlebia, Baetis, *et. al.* These forms also are smooth of body and prettily colored.

Some of these mayfly nymphs live so openly in running water that one may look down into it and see them skipping about. Of necessity they stand with head upstream, facing the current and swaying with its eddies, while hanging on with their feet. One marvels that they are not washed away.

The nymphs of Baetis and Paraleptophlebia live on the pebbly riffles in many a brook-bed, on the shifting sands of the bars, on the flat beds of trailing, brown silt in streams flowing from peat bogs, and in many other situations in the lesser streams. They are little, slender, agile, and protectively colored. Their enemies here are numerous—all the lesser fishes that roam the shoals and many carnivorous dragonfly nymphs are ever ready to grasp a nymph, should it wander within reach.

The nymph of Isonychia (fig. 48) dwells on the stream bed and is a fine, big fellow, trim as a race horse. It lives mainly among the rubble on the flat rock ledges of creek beds, over which the water glides in broad thin sheets, or under the sheltering edges of large half-submerged stones. The ease and precision with which it darts about from place to place in the current are remarkable. This nymph anchors itself to the rock bed with its claws and sits erect with the current sweeping the sides of its shapely body, its head up, its stout gill plates outspread, its caudal plumes trailing, its long fore feet extended forward.

Most curious and interesting is the food gathering equipment of these big fore feet. They bear two rows of long, stiff, brownish bristles along the anterior side. These rows of bristles are set at an angle with each other, opening forward, upstream. Together they constitute a collecting basket that is held ever ready to receive such contributions of food as the current may sweep into it. The bottom of the basket lies beneath the head, convenient to the jaws.

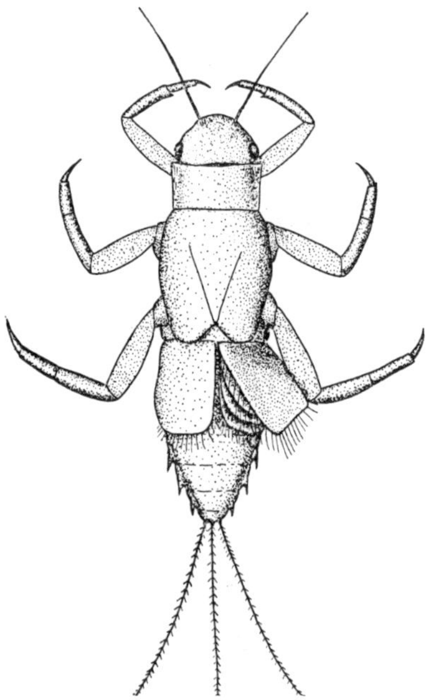

Fig. 46—Nymph of Caenis.

e. *The close-clinging, stone-loving forms* of rapid waters are very flat, widened toward the front, with eyes and antennae dorsal in position. They have broad, depressed femora, slender tibiae, and wide-spreading tails. Here belong Iron (fig. 49), Stenonema, Heptagenia, and all the Heptageninae, *s. str.*, and a few (in Tropical America, many) Baetinae.

Under the stones we may find representatives of other ecological groups. Some dwell there; others are stranded there; others run in for shelter there. Every creature has a right to such shelter as he can find from the vicissitudes of life. In havens of refuge many strangers meet. So it may happen that representatives of any of the foregoing groups may sometimes be found under stones in running streams, even of the

Fig. 47—Nymph of *Hexagenia recurvata* (from Morgan).

still-water forms; for in the lee of the stone, or in the little bay it shelters, there is quiet water. But there is a group of very characteristic flat-bodied nymphs that live on the under surface of the stones, where they are sheltered from the light and habitually exposed to the wash of run-

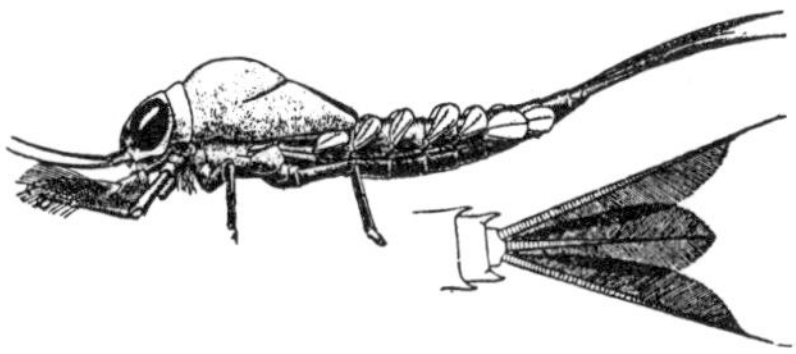

FIG. 48—Nymph of *Isonychia sicca* (after Kennedy).

ning water. Most typical are the Heptagenines, which abound in every stony rapid and on every wave-washed shore. Lift almost any stone or block of wood from running water, turn it over, and there they are, sticking to it so closely and so quietly that they look as though engraved on its surface. They cling tightly, with legs and tails laid flat, until the water drains off. They then shift about a step at a time over the surface, loosing their foothold to clutch again quickly and securely, going sidewise and backward as readily as forward, over the edge and down into the sheltering shadow of the underside. Between the rock and the surface on which it lies is always a sheltering crevice where there is safety from the powerful enemies of the open waters.

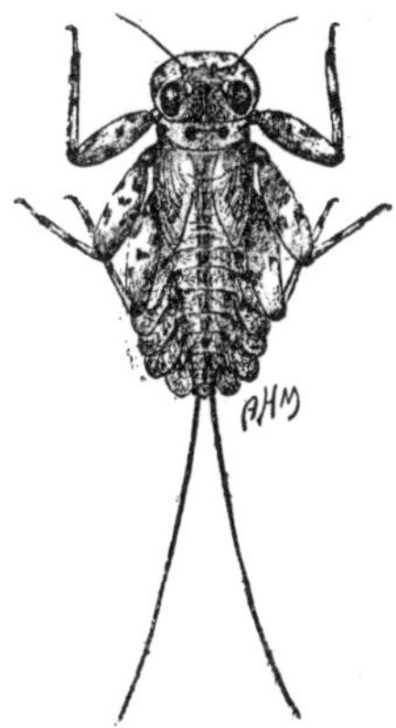

FIG. 49—The nymph of *Iron humeralis* (from Morgan). The outline of the hind wings is dimly seen through the transparent fore wings.

The nymph of Choroterpes is a dainty, little, yellow-legged, brown-bodied creature that lives under the edges of the alga-draped stones in beds of streams, usually in clear-flowing creeks. It is not quite as flat as are the Heptageninae, nor is its head as broad, and it is a much more graceful creature, with beautifully scalloped gills. The gills lie prone upon the back, or, when set in vibration, are lifted only a little. The long front legs lie flatwise on the stone with elbows akimbo. The nymph appears to run backward more readily than forward.

Some of the more specialized of these flat nymphs are torrent-inhabiting forms that are able to cling limpet-like to the face of the bare rock. A few of them have gone farther than any other insects save the sucker-bearing larvae of the net-veined midges (Blepharoceras) in braving the strength of the torrents. Such are the nymphs of Rhithrogena (fig. 50), that have added to the depressed body and legs and the flaring margins of other Heptagenines a disc-like mechanism for attachment to stones. The gill plates are pressed into service for a new function. They are expanded and laterally outspread and fitted together in such a way that their roughened outer margins form a complete oval ring about the abdomen, meeting closely the surface on which the nymph rests. The ring is completed in front by the turning forward beneath the thorax of the foremost pair of expanded gill plates until they meet there, and in the rear, by the reverse rotation of the hindmost pair beneath the tip of the abdomen. Thus the whole abdomen becomes remarkably limpet-like. With the edges appressed to the stone, any elevation of the middle part

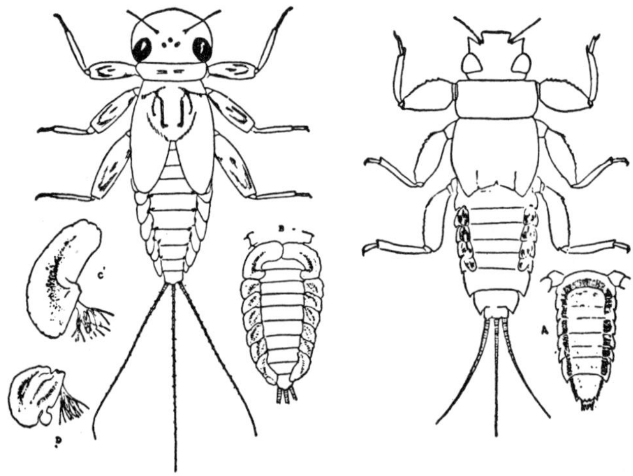

FIG. 50—Parallelisms in hold-fasts in lotic nymphs. At left *Rhithrogena doddsi* with a ventral attachment disc beneath the abdomen that is formed out of expanded gill plates. At right *Ephemerella doddsi*, with disc formed from the abdomen itself.

creates a suction which holds the body down tightly. This doubtless relieves the legs of the greater part of their task of holding against the current. They need to do all of the holding only when a shift of position is desired, when new algal pasturage is sought on the rock face. Doubtless there are few enemies that can follow these nymphs into such an environment.

The same result is attained by an entirely different arrangement of parts in *Ephemerella doddsi*—a most striking parallelism. An adherent disc is formed beneath the abdomen, but the gills are not involved in the disc. Instead, the margins of the wide abdomen have developed soft fringes of hairs for close contact with the stone. The fringe line is continued across the venter between the two basal abdominal segments in a forward-arching curve to complete the disc.

f. *The trash-, silt-, and moss-inhabiting* nymphs of the streams are stiff-legged, more or less spiny forms, among which are Oreianthus and many members of the great genus Ephemerella (fig. 51). Old leaf drifts that accumulate on the upstream side of obstructions in the current are full of them. Some hide in close-growing mosses whose branches are commingled with small rubbish and silt. Others live among grass roots that trail in the edges of the brook from under-cut banks. Typical of these is the sprawling, stiff-legged *Ephemerella septentrionalis*, a regular inhabitant of grass roots, where it is associated with the common black-wing damselfly, *Agrion maculatum*, of similar elongate form. Many of the Ephemerellas are trash dwellers.

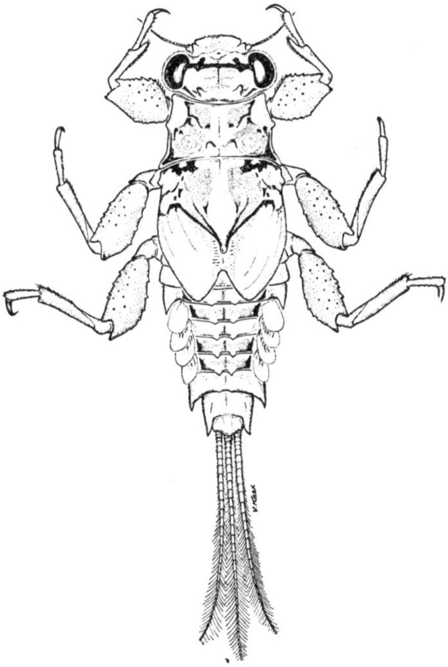

FIG. 51—The nymph of *Ephemerella allegheniensis* (drawing by Velma Knox).

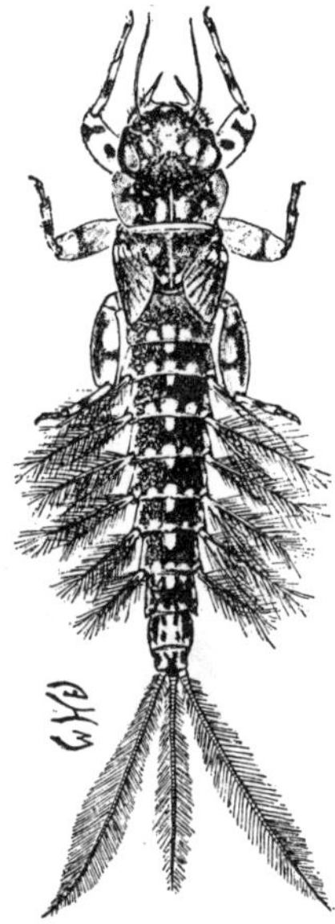

FIG. 52—Nymph of *Potamanthus sp.* (from Morgan).

The flatter, smoother nymphs of Potamanthus (fig. 52) live on the bottom amid rather clean dirt, silt, and sand, and they possess a rather pretty, highly protective color-pattern.

2. THE FORM OF THE PARTS

THE HEAD

Let us now consider the parts of the body more in detail. As indicated in the foregoing, the head conforms to situation in a very important way. It reaches its maximum of breadth and flatness in the close-clinging Heptageninae; its maximum of relative narrowness in the burrowing Ephemerinae; its maximum of height in the Siphlonurinae. In the last-named it is grasshopper-like in form, with ocelli and antennae directed forward. In all these the head is smooth above and at the sides; but in the Ephemerellas it may be variously armed with spines (fig. 53).

The mouth parts of mayflies are always of the mandibulate type already set forth in Chapter IV, as exemplified in the nymph of Stenonema. Their variations in families and genera offer some of the best of group characters.

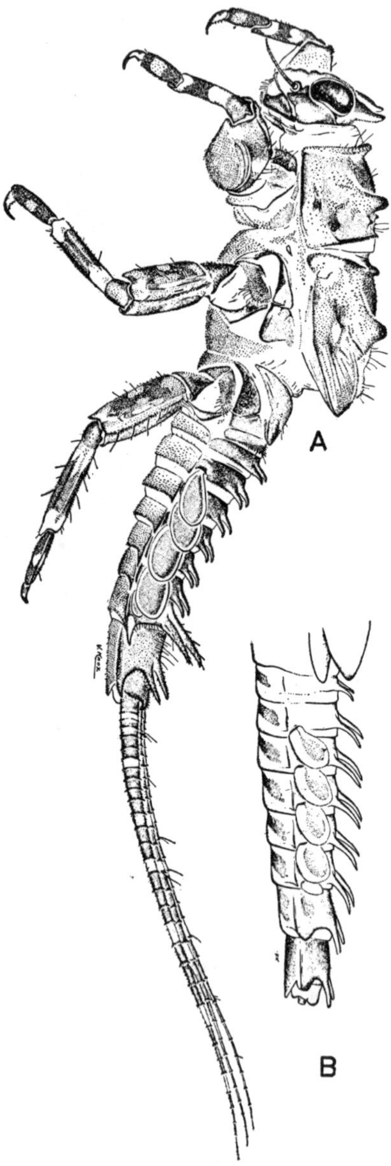

FIG. 53.—Spinous nymphs of Ephemerella (drawings by Velma Knox). *A*, Side view of *E. yosemite*; *B*, abdomen of *E. hystrix*.

The labrum is typically a simple, more or less quadrangular plate covering the jaws. Normally it hangs down in front, but in Heptagenia where it is reduced in size and pointed toward the sides, it is withdrawn under the projecting front margin of the clypeus. In Iron the head is widened at the front; but in the Baetine genus Hagenulus (Pl. XXXI, fig. 47), the labrum itself becomes transversely elongated until its rounded ends project far beyond the head. Thus the advantage of a broad front and tapered rear for more perfect stream-lining, with diminished resistance to the drive of the water, is gained by two entirely different means.

A squarish notch in the middle of the front margin has come to be characteristic of the Baetinae (*s. str.*, Pl. XXXI, fig. 43). Spines and hairs are variously disposed over the surface in most forms: spines for the most part in the swift-water forms, and hairs in the silt-dwellers.

The mandibles are toothed at the tip and bear on their inner opposed surfaces (ex-

PLATE XXXI.

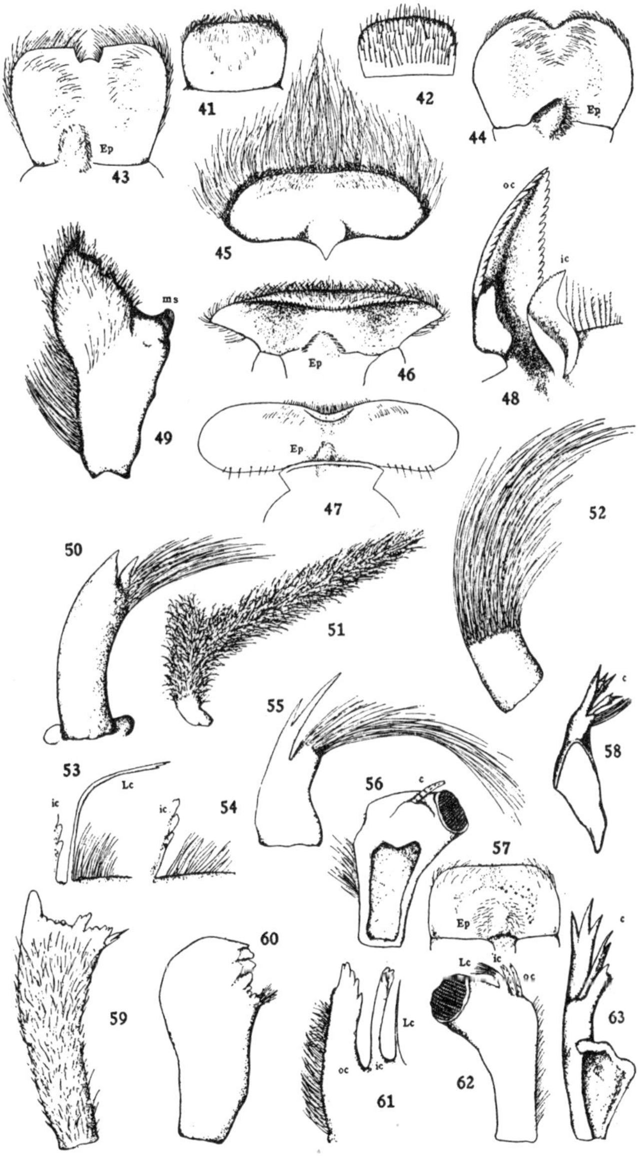

NYMPHAL MOUTHPARTS

Fig. 41. Labrum of Ephoron, dorsal aspect; Fig. 42. Labrum of Isonychia, dorsal aspect; Fig. 43. Labrum of Callibaetis, ventral aspect; Fig. 44. Labrum of Baetis, ventral aspect; Fig. 45. Labrum of Oligoneuria, dorsal aspect; Fig. 46. Labrum of Heptagenia, ventral aspect; Fig. 47. Labrum of Hagenulus ventral aspect; Fig. 48. Canines of Rhithrogena; Fig. 49. Right mandible of Palingenia (after Eaton); Fig. 50. Lacinia of mandible of Ephoron; Fig. 51. Lacinia of mandible of Siphlonurus; Fig. 52. Lacinia of mandible of Leptophyes; Fig. 53. Lacinia of left mandible of Stenonema; Fig. 54. Lacinial region of right mandible of Stenonema; Fig. 55. Lacinia of mandible of Choroterpes; Fig. 56. Right mandible of Lachlania; Fig. 57. Labrum of Ephemerella, ventral aspect; Fig. 58. Right mandible of Prosopistoma (after Eaton); Fig. 59. Lacinia of mandible of Euthyplocia; Fig. 60. Lacinia of mandible of Baetis; Fig. 61. Apex of right mandible of Ameletus; Fig. 62. Left mandible of Homoeoneuria; Fig. 63. Left mandible of Chiloporter (after Eaton).

On all figures: *Ep*, epipharynx; *c*, canine; *ic*, inner canine; *oc*, outer canine; *Lc*, lacinia mobilis; *ms*, molar surface.

cept in a few aberrant carnivorous genera) a grinding device consisting of alternating parallel grooves and ridges. The mandibles are unsymmetrical, and these surfaces rub obliquely against each other in comminuting the food. The teeth at the tip tear off shreds of plant tissues, and these molar surfaces grind it.

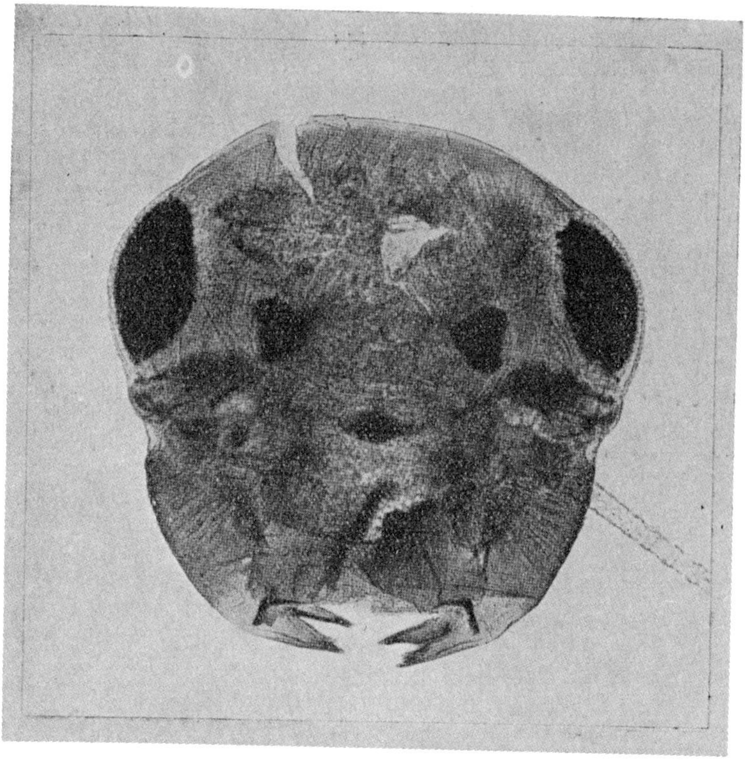

FIG. 54—Head of the nymph of Paraleptophlebia showing by transparency the overlap of the obliquely-grooved molar surfaces of the unsymmetrical mandibles (microphotograph by Eva Gordon).

The most careful study of the mandibles is that of Dr. Helen E. Murphy (1922), who has given a key to the genera of North American mayflies, based on mandibles alone. This key, like many another, admits of some exceptions; but it yet shows the remarkable constancy of these small details.

The teeth at the tip of the mandibles, called canines in most descriptions, are arranged in two groups, an outer and an inner group, with the number of free teeth in each varying by genus and species. In the genus Baetis they become flattened, truncated in an even row, and incisor-like.

The molar surface is for grinding food. Doctor Murphy says of it: "Among the Ephemerinae there is one prevailing type. The left molar surface of Ephemera bears eight transverse ridges, the anterior edge of each provided with teeth. Between these ridges are deep gutters. The right molar surface has seven transverse ridges toothed at the anterior edge. The molar surfaces move in a dorso-ventral plane, rubbing together the transverse ridges. The teeth extending out into the gutters doubtless serve as strainers for the food. Cast skins show many of the transverse ridges worn smooth, but the teeth still sharp, and clogged with ground food. Among the Baetinae and Heptageninae are two main types of molar surface. In one, the transverse ridges are irregularly serrated with stiff bristles acting as strainers across the gutters. In the other, the transverse ridges are rounded and knobbed. They are broad at the posterior portion, but fray out anteriorly. This is the prevailing type among those forms inhabiting streams with gravel or sand bottom."

Between the canines and the molar surface on the distal end of the mandible there is a little appendage that has been called by several names: lacinia mobilis, mandibular palp, and prostheca. Having no muscles, it is movable only as a pedunculate spine is movable, having generally (not always) a basal articulation. Though much discussed, its significance has not yet been disclosed. Doctor Murphy's figures of Baetis embryos (fig. 55) show that in ontogeny it has an early origin.

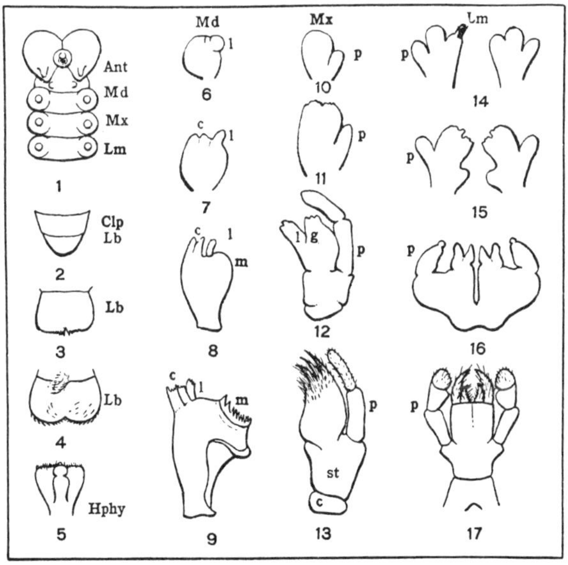

FIG. 55—Development of the mouthparts in *Baetis vagans* (after Murphy). *1.* Head region of an embryo at five days' development: antennae, mandibles, maxillae and labium as indicated by abbreviations. *2, 3, 4,* development of labium at eight, ten, and eleven days; *Clp*, clypeus (indistinguishable after the ninth day); *Lb*, nymphal labium. *5.* Hypopharynx at eleven days. *6, 7, 8, 9,* mandible at 6, 7, 9 and 11 days; *l*, lacinia mobilis; *c*, canines; *m*, molar surface; *10, 11, 12, 13,* development of the maxilla at six, seven, eight, and eleven days; *l*, lacinia; *g*, galea; *p*, palpus; *14, 15, 16, 17,* development of labium at six, seven, nine, and eleven days; *p*, palpus.

It is a little, elongate appendage lodged in the angle at the base of the inner canines, often curved, sometimes blunt at the tip or minutely toothed there, but more often frayed out in a long brush of fine hairs. We copy herewith Doctor Murphy's figures of it (Pl. XXXI). Doctor Spieth (1933) has published many more figures. When unsymmetrical, the heavier chitinization appears to be on that of the left mandible.

There is developed on the outer side of the mandibles of nymphs of all the Ephemeridae except Oreianthus, a long tusk (fig. 56), which may assume grotesque size and proportions. In the burrowers (Hexagenia *et. el.*) it is upcurved and is useful for digging. In the sprawlers (Euthyplocia) it is blade-like and incurved and is laid out flat upon the bottom. What good purpose it then serves, if any, is not known.

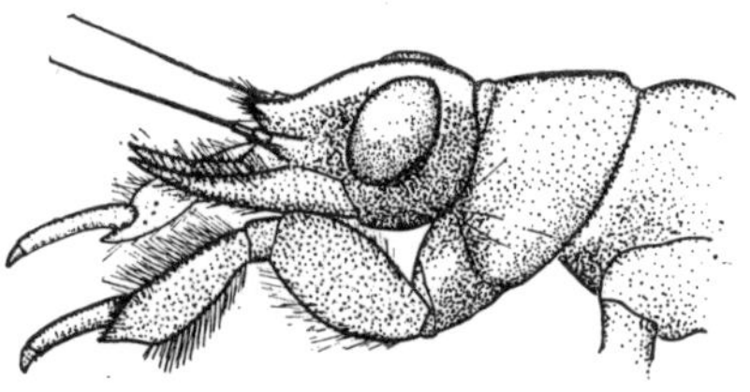

FIG. 56—Head of the nymph of Ephemera, showing tusked mandibles, and digging fore feet.

In the genus Paraleptophlebia the outer group of canines becomes greatly enlarged and consolidated. In certain of them (*P. packii* fig. 57, *et. al.*) a tusk is developed longer than the head, paralleling that of the Ephemeridae but differing in structure as well as in origin. It is blunt at the tip, bare, and inclined obliquely upward.

The maximum development of the molar surface together with the minimum development of the canines is shown by the Neotropical genera, Homeoneuria and Lachlania (Pl. XXXI, figs. 62 and 56). Conversely, in two Antipodean genera, Chiloporter from Chili and Ameletopsis from New Zealand, and also in the famous Old World genus Prosopistoma, there are only canines, no molar surface being developed. These are long-pronged canines. These nymphs are carnivorous.* Phillips (1930, p. 333) says of Ameletopsis that it feeds on "small specimens of Leptophlebiid nymphs and other organisms."

The maxilla of mayfly nymphs is typically of the form shown in fig. 58A. Cardo and stipes are separate, but galea and lacinia are fused lengthwise and consolidated at base with the stipes. The resulting plate may be conveniently designated as the blade of the maxilla. The blade often terminates in two hooks or teeth, or in two groups of teeth, that seem to stand on either side of the original boundary between galea and lacinia. In Ephemera a line of bristles that once may have been marginal on the back of the lacinia runs lengthwise of the blade, as if to mark such a lost boundary.

*In the records of studies in tropical medicine there are a number of references to mayfly larvae as predators on mosquito larvae. Dr. E. H. Hinman has assembled these references in the *Journal of Tropical Medicine and Hygiene* for May 1st 1934; but one who investigates them will find them so vague it is impossible to tell what kind of predatory larvae are intended, and it is highly probable that they do not refer to mayfly larvae at all.

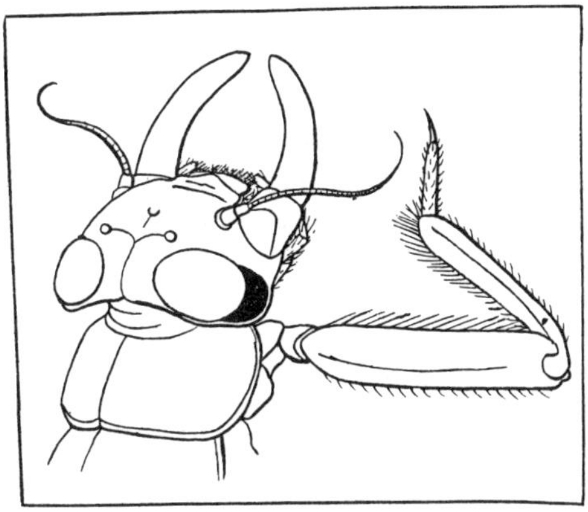

FIG. 57—Head of a male nymph of *Paraleptophlebia packii*, showing the blade-like elongation o the outer canines of the mandibles. Note heavier pigment of lower division of the compound eye.

The maxillary palpus, normally 3-jointed, slender, and thinly clothed with sensory hairs becomes highly modified in several ways. It is reduced to two short, unarmed segments in the Baetinae; to two long, flattened segments, with the last pointed at the tip and furnished distally with a dense patch of short stiff hairs, in the Heptageninae; and to two robust segments, the basal one naked, the second one densely clothed with long, soft hair in Isonychia (also in certain Neotropical genera, Hermanella, Murphyella, and Lachlania, and in the European Oligoneuria).

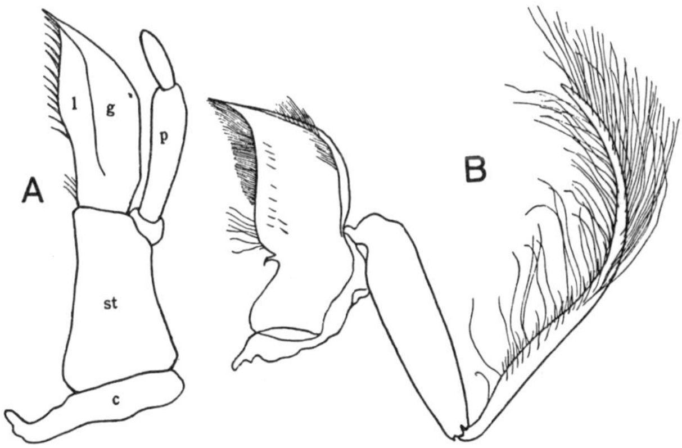

FIG. 58—The maxilla of the mayfly nymph. *A*, typical, to show the parts; *c*, cardo; *st*, stipes; *l*, lacinia; *g*, galea; *p*, palpus. *B*, a maxilla of *Arthroplea bipunctata* (after Ide; for other types see Pl. XXXII).

In Hermanella (Peru) the maxillae attain to remarkable size. They project laterally beyond the sides of the head, and forward farther than the front legs. They thus hold their maxillary brushes outspread to the currents. In this genus maxillary and labial brushes probably function as plancton strainers, as do the tibial brushes of Isonychia. In Ephemerella and in Tricorythodes the palpus may become very small or altogether wanting, as in *Ephemerella deficiens* (Pl. XXXII, fig. 82) and in all the species of the *E. bicolor* group.

But, on the contrary, the palpus is many-jointed in certain aberrant Antipodean genera; 15-jointed in Ameletopsis, according to Phillips (1930); 11-jointed in Chiloporter according to Eaton (Monogr. pl. 53, fig. 7x). This and other forms of maxilla are shown in Plate XXXII, which we copy from Murphy (1922).

The labium is of the usual form of mandibulate insects. The submentum is distinct, the palpus is three-jointed, the glossae and paraglossae are long and separate. The commonest variations are slight differences in length of the palpal segments, especially of the last segment, and in length and width of glossa and paraglossa.

Two-jointed palpi are met with among lotic species. These show remarkable adaptations to feeding under stream conditions. They follow two lines. First there is the homogeneous Heptagenine series of genera in which the second palpal joint (and the tips of both pairs of glossae as well) bear marginal brushes of stiff bristles. These brushes are supposed to be used for sweeping slime-coat algae from the surface of the rocks under which these flat nymphs live. Then there are scattered genera, Isonychia, Ephoron, and Pentagenia, in which the second palpal joint is greatly elongated and densely clothed with long soft hairs. These brushes are strainers and serve to gather food out of the passing current. However, as in the case of the maxillae, the palpi are multiarticulate in the two carnivorous Antipodean genera. Chiloporter and Ameletopsis. The two terminal segments of the palpus acquire a somewhat cheliform adjustment in Baetisca (Pl. XXXII, fig. 74), Parameletus and Metretopus.

Form and function of mouthparts were studied together by Dr. Anna H. Morgan in the Heptagenine mayfly, *Iron fragilis*. She watched them in operation under a binocular microscope. Her account of them (1911, pp. 102–103) is in part as follows:

"The mouth parts are completely hidden from above by the flaring margin of the head with its bordering fringe of soft hairs. Viewed from beneath the small labrum may be seen curving downward and backward over the tips of the mandibles and maxillae to meet the median flaps of the broad labium. The labium is flat and its outer surface fits close down to the surface upon which the insect is foraging. Along the anterior margin of the outer surface of the labial palp is a series of incurving hairs, behind these, a set of overlapping plates, and still farther backward a single strongly chitinized scraper. On the inner surface of the palp is a semi-circular patch of inwardly directed hairs. Closely apposed to the inner surface of the labium is the outer

PLATE XXXII.

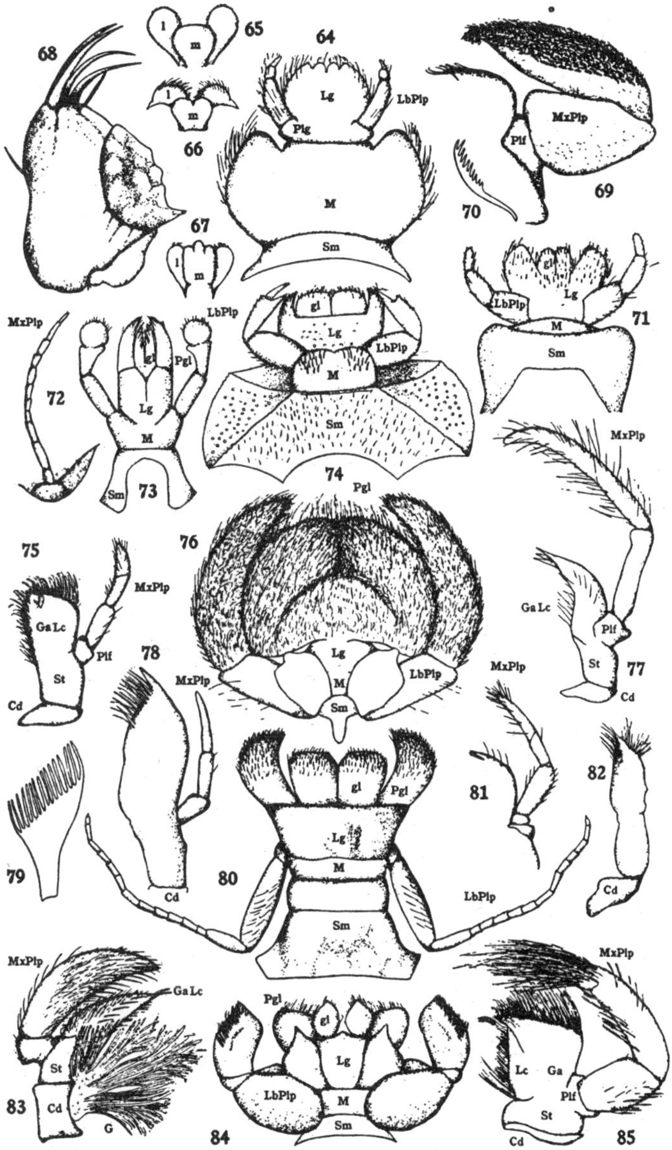

NYMPHAL MOUTHPARTS

Fig. 64. Labium of Tricorythodes, ventral aspect; Fig. 65. Hypopharynx of Ameletus; Fig. 66. Hypopharynx of Habrophlebia; Fig. 67. Hypopharynx of Callibaetis; Fig. 68. Right maxilla of Chiloporter (after Eaton); Fig. 69. Left maxilla of Rhithrogena in part; Fig. 70. Diatom raker from maxillary palp of Rhithrogena; Fig. 71. Labium of Ephemerella, ventral aspect; Fig. 72. Maxillary palp of Chiloporter (after Eaton); Fig. 73. Labium of Baetis, ventral aspect; Fig. 74. Labium of Baetisca, ventral aspect; Fig. 75. Left maxilla of Blasturus, ventral aspect; Fig. 76. Labium of Oligoneuria, ventral aspect; Fig. 77. Left maxilla of Hexagenia, ventral aspect; Fig. 78. Left maxilla of Baetisca, ventral aspect; Fig. 79. Diatom raker of maxilla of Rhithrogena (shown *in situ* in figure 69); Fig. 80. Labium of Chiloporter (after Eaton); Fig. 81. Left maxilla of Caenis in part, ventral aspect; Fig. 82. Left maxilla of *Ephemerella deficiens*, ventral aspect; Fig. 83. Right maxilla of Lachlania, ventral aspect; Fig. 84. Labium of Iron, ventral aspect; Fig. 85. Left maxilla of Hagenulus, ventral aspect.

On all figures: *l*, lateral and *m*, median lobe of hypopharynx; *M*, mentum and *Sm*, submentum; *gl*, glossa and *Pgl*, paraglossa; *LbPlp*, labial palp on all labia; *Cd*, cardo and *St*, stipes; *lc*, lacinia and *Ga*, galea; *Plf*, palpifer and *MxPlp*, maxillary palp (all from Murphy 1922).

surface of the hypopharynx made up of two lateral, and one median portion, distinct except at their bases. The separation of the two lateral portions from the median portion leaves a gutter-like trough between them on the inner side. The labial palpi are freely movable horizontally. They are moved outward, then pressed slightly downward and inward when feeding. Sometimes the labial palp is slipped between the median and the lateral flaps, sometimes over the inner sides of both. Directly above the median portion of the hypopharynx are the teeth of the maxillae with the grinding surfaces of the mandibles directly above them. Food scraped inward by the labial palps is evidently deposited beneath the hypopharynx. From here it is probably sucked up into the mouth cavity through the slits between the median and lateral folds. The maxillary palps act in a manner similar to those of the labium, but because

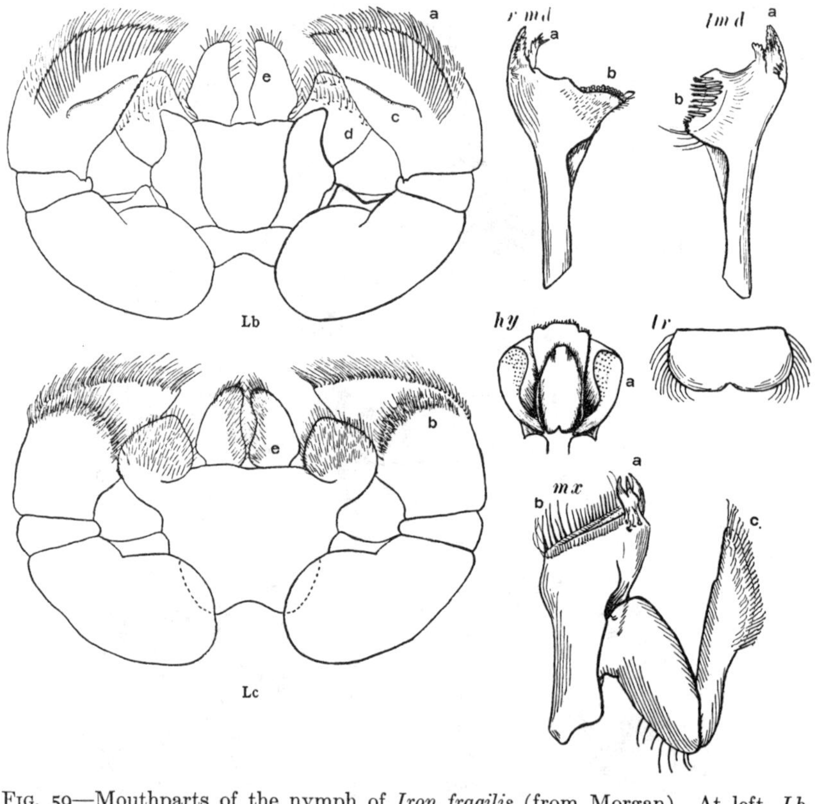

FIG. 59—Mouthparts of the nymph of *Iron fragilis* (from Morgan). At left, *Lb*, labium from beneath, and *Lc*, the same from within. *a*, incurving hairs, with shorter overlapping plates; *b*, brush of shorter bristles; *c*, chitinized scraper; *d*, paraglossa; and *e*, glossa.

At right, *rmd* and *lmd*, right and left mandibles respectively; *a*, canines, and *b*, molar surface: *hy*, hypopharynx (the gutter between the outer chitinized portion, *a*, and the median part is swept by the crescentic brush, *b* of the labial palp when that is swung inward); *lr*, labrum; *mx*, maxilla; *a*, the terminal trident; *b* straight bristles in three graduated series; *c*, hooked bristles on the end of the palp.

of their position above the hypopharynx, they must be able to place the food directly in the mouth cavity. The inwardly curving hairs on the lateral borders of the labrum help to keep the food in the mouth while it is being chewed."

Eyes and ocelli are lateral in the burrowers, and dorsal as the head becomes flattened in the swift-water inhabiting species. In the males of most species they grow very rapidly in late nymphal life by increments added on the inner side. These do not become functional until after transformation. In forms with divided eyes these will form the upper division. The ocelli are directed forward in the free-ranging forms but upward in both burrowers and sprawlers. Before the middle ocellus a nasal carina is already developed in the nymphs of Oniscigaster.

The antennae are of the normal filiform or setaceous type, with short, thick scape, longer and thinner pedicel, and long, rapidly tapering, many-jointed flagellum, varying moderately in length and in hairiness. But the antennae become shortened to seven segments in *Baetisca obesa* and to five in Prosopistoma. In Lachlania they are stouter and 12-jointed, and each of the joints of the distal half has a large thumb-like branch on the inner side (fig. 60).

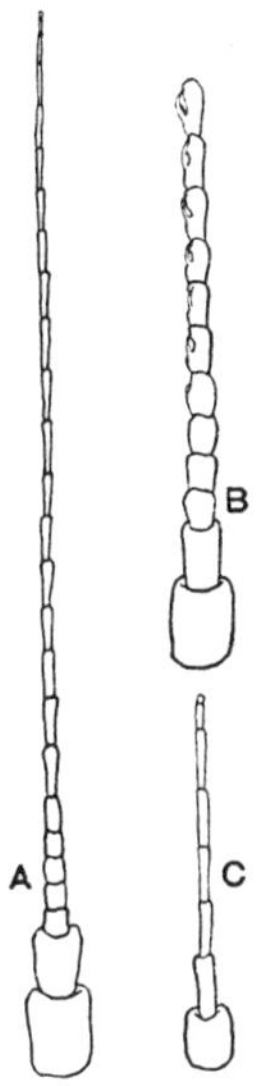

FIG. 60—Nymphal antennae. *A*, Callibaetis; *B*, Lachlania; *C*, Baetisca.

THE THORAX

The mayfly nymph begins its free existence with an elongate thorax, consisting of three nearly equal segments, rather loosely put together. As it grows, these become consolidated and gradually take on more of the adult form. The mesothorax expands, especially in late instars, as the muscles of the fore wings develop within it. The prothorax remains a little detached.

The thorax varies greatly in form, becoming compressed in the agile swimmers and depressed and wide in those that cling to current-washed stones. The metathorax is reduced in size and consolidated with the rear of the mesothorax. The terga of these two segments together form a sort of carapace. As they elongate to rearward, they carry the wings backward over the gills in Hermanella, but in Baetisca and in Prosopistoma the mesothoracic tergum itself overgrows all of the metatergum, and covers both wing sheaths and gills. It thus forms a rather capacious respiratory chamber. In Baetisca (fig. 61) this chamber is completed in the rear by a hump on the dorsum of the middle abdominal segments.

In several genera of burrowers the prothorax buttresses the rear of the head by means of supporting processes apposed to its hind angles. It expands to rearward, overspreading the shoulders of the mesothorax with a pair of very wing-like processes in Ecdyonurus. In the trash-

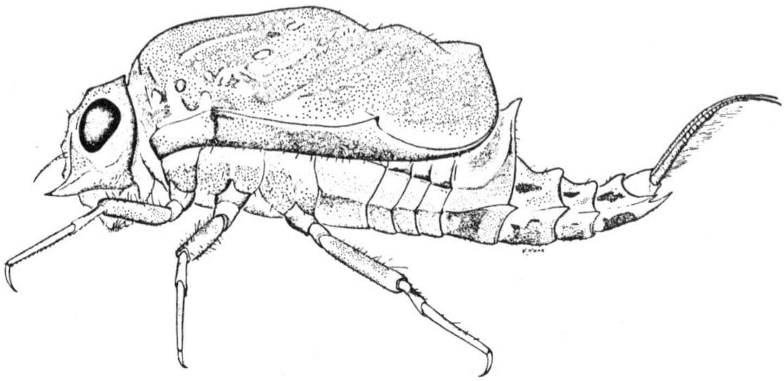

Fig. 61—The nymph of *Baetisca carolina* (drawing by Velma Knox).

inhabiting Ephemerellas, tubercles develop on the back of the thorax in considerable number and variety according to the species.

The legs of mayfly nymphs bear always a one-jointed tarsus with a single claw at its tip. Usually the fore legs are a little shorter than the others, but in a few genera of tusk-bearing sprawlers (Potamanthus, etc., fig 62), and in a few others that collect plancton with them (Isonychia, etc.), they are longer. In the burrowers the front tibiae are flattened and widened to a shovel-like form that fits them for digging, while the hind foot is shaped for pushing.

In many of the Ephemerellas, especially those that live in swift-flowing waters, the front femora are thickened and heavily chitinized, especially along their front margin and on the upper surface. This

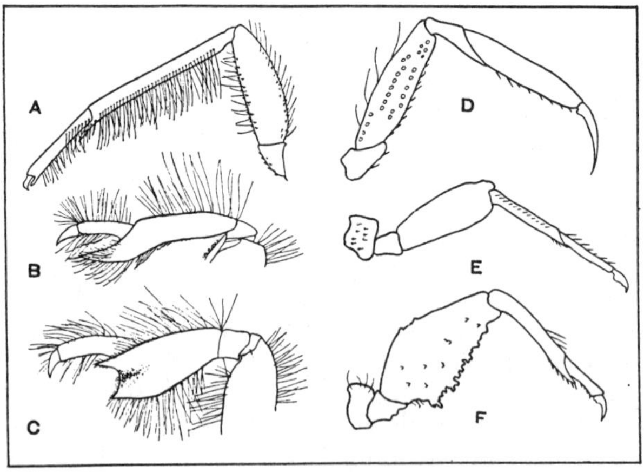

Fig. 62—Nymphal legs. *A*, fore leg of Potamanthus; *B*, hind leg and *C*, fore leg of Hexagenia; *D*, fore leg of Baetisca (note the shortness of the tibia); *E*, fore leg of *Ephemerella bicolor; F*, fore leg of *Ephemerella tuberculata*.

perhaps fits them to withstand the impact of the current and whatever solids (pebbles, etc.) it may carry. The marginal thickenings take on a variety of forms.

In Heptageniidae (fig. 63) the legs are laid down flatwise on the stones, knees out to rearward instead of in the usual upward position. The femora are flat and broad; the tibiae are long and slender, and the

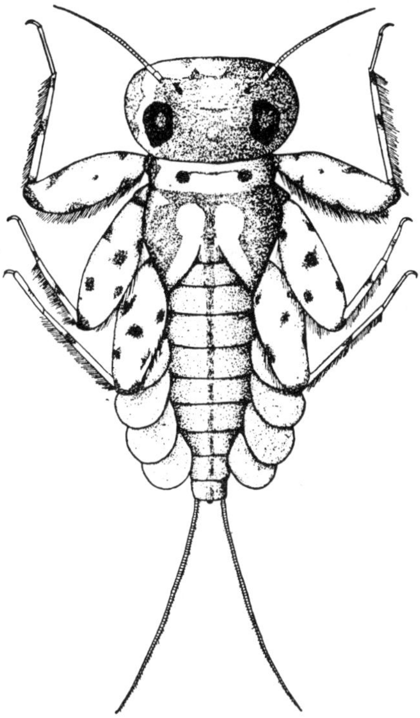

FIG. 63—A nymph with flattened femora, *Iron humeralis* (from Murphy).

tarsal claws have denticles placed asymmetrically on the lower (contact) side. The femora have to meet the flow of water. They are stream-lined in cross section, like the struts on an aeroplane, thick in front and tapered to rearward, with the taper continued by a fringe of long hairs on the hind margin. The tibiae take a fore-and-aft position, more or less in line with the current.

The claw on the tarsus varies greatly in form (fig. 64). It is long and slender and unarmed in Centroptilum, short and thick and coarsely toothed in Isonychia. It is cleft at the tip in Siphloplecton. Its asymmetry is a marked characteristic of the Heptageniidae.

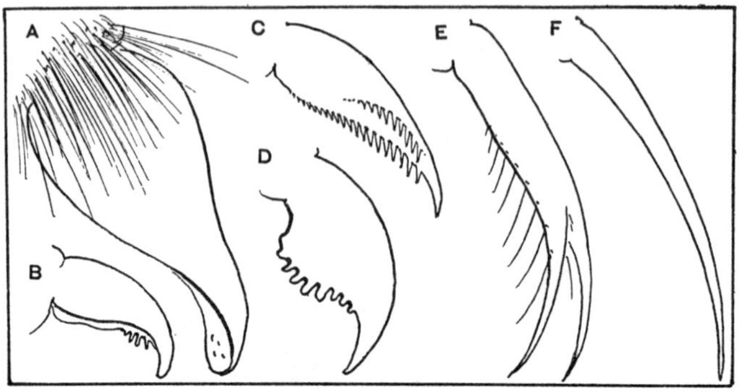

Fig. 64—Claws of mayfly nymphs: *A*, of middle foot of *Hexagenia carolina; B*, same of *Iron fraudator; C*, same of *Blasturus grandis; D*, of hind foot of *Isonychia albomanicata; E*, of front foot, and *F*, of hind foot of *Siphloplecton basale.*

GILLS

The gills of mayfly nymphs are of two types (fig. 65). The prevailing type is that of abdominal gills, segmentally arranged and bearing some evidence of having been derived from ancestral locomotor appendages (probably legs: see Borner 1908 and Snodgrass 1927). These are generally dorsal in position. The other type is that of thoracic or maxillary gills, adventitious in their nature and ventral in position. The former are highly developed as lamellae or filaments or clusters, or combinations of both, and are provided with a musculature of their own. The latter are single filaments or simple clusters of filaments without muscles and without definite serial arrangement.

1. Abdominal gills now occur only on segments one to seven. In the Jurassic genus Mesobaetis they were present on one to eight, and in the Permian genus Phthartus on one to nine (see Chapter XIV). They are situated on the dorsal side close to the lateral margin, except that in Oligoneuria (European) and Lachlania (Neotropical) the foremost pair is ventral in position. In most genera they are present on all of these seven segments, but they may be lacking from certain segments at either end of the series. In the great genus Ephemerella they are always wanting on the second segment and often also on the first and third (on the first, if present, rudimentary). In Leptohyphes (Neotropical) they are present on segments two to six. In Baetisca and in the Brazilian genus Baetodes they are present on one to five; in Siphlonella (Neotropical) on one to four. They originate in very simple form as buds or filaments that later become double, as shown in Chapter VI. Alike on the seven segments, they may become highly differentiated in relation to each other and highly modified in relation to surrounding parts. They are respiratory organs (necessary ones, as the work of Morgan and Grierson 1932 has shown), but they may also take on other functions.

Strong plate-like gills equipped with powerful muscles may render efficient aid in locomotion, as in Isonychia and Siphlonurus, which are able by concurrent lashing of gills and tail to dart about like minnows.

They may, with a shift of position, serve as hold-fasts in swift water. This is well illustrated by the Heptageninae. Thus in Iron the laterally extended gill plate, provided on the front margin with a contact ridge of strong chitinization and roughened with prickles on the lower side, serves as an outrigger when appressed to the stones to hold the nymph against the current. In Rhithrogena, as already noted, a deflection and overlapping of all the gills together with a widening and thinning of those at front and rear, with approximation of them underneath the body, makes of the whole abdomen one large sucking disk that adheres to the stones very strongly.

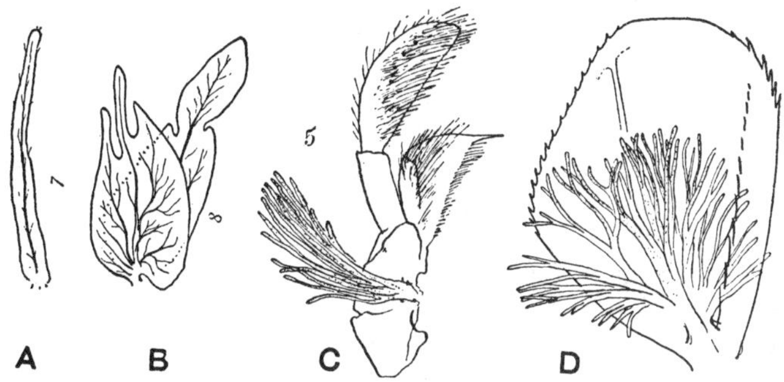

FIG. 65—Gills of mayfly nymphs: *A*, a single and *B*, a double lamelliform gill of *Choroterpes basalis; C*, filamentous gills attached to base of maxilla, and *D*, a compound cluster of same attached beneath the protective gill lamella of the fourth abdominal segment of *Isonychia albomanicata.*

They may be modified for silt avoidance. One pair of gill plates may become enlarged and thickened to form opercula covering the others, and the edges of the others may become fringed with out-pointing bristles to strain the silt from the inflowing water when the opercula are lifted. In such case the segments become shortened to bring the gills closer together under the opercula. This device has been developed independently, several times. In Siphlonella (Neotropical) it is the first pair of gill plates that has become elytroid or operculate; in Caenis, Tricorythodes, and Oreianthus, it is the second, and in Ephemerella it is the fourth pair. In the genus Oligoneuria each gill has developed its own protective shield, the hard anterior half becoming shaped so as to cover and enclose the filaments of the soft posterior part as in the hollow of a hand.

There are two ways in which other parts of the body have been strangely modified to protect the gills. (1) In Baetisca and in Prosopistoma (Old World) the thorax becomes enlarged and extended to rearward, overarching the gills and forming a protective chamber for them. In Baetisca the rear of that chamber is closed by a hump on the middle abdominal segments, and when closed the gills are shut inside. (2) In Hermanella (Neotropical) the abdomen is shortened, and the wing cases are extended like a roof over the gills.

2. Ventral gill filaments on the forward parts of the body are less common. In Isonychia and in Oligoneuria (European), both of which have full series of plate-like abdominal gills, they are restricted to a cluster of filaments on the base of each maxilla and an additional tuft at the base of each fore leg.

In nymphs of Ametropus (Pl. XXXV) there is a single flattened gill plate or gill sac appended to the inner side of each front coxa. The nymph has seven pairs of lamelliform dorsal abdominal gills.

The most remarkable disposition of gills yet found in the order is that of the Chilian genus Murphyella, which lacks gills altogether on the dorsum of the abdomen, but has them along the mid-ventral line of the thorax. There is a single, long, finger-like median gill on each thoracic segment, with an added pair close beside the one on the prothorax. There are similar thick gill filaments attached to the base of the mouth parts, two pairs on the maxillae and one pair on the labium (Needham and Murphy, 1924).

In form the gills of mayflies are of two principal types, lamelliform or plate-like, and filiform or thread-like, with many combinations of these and endless variations in form and size.

VESTITURE

The hairs that grow upon the integument of mayfly nymphs are of great diversity. They vary in form according to the functions they serve, and they vary with genera and species. They may be simple or plumose or serrate or pectinate or forked, or even compound. Bogoescu (1932) has made a careful study of the vestiture of Roumanian mayfly nymphs, and he has described and figured fifty-two different types of hairs.

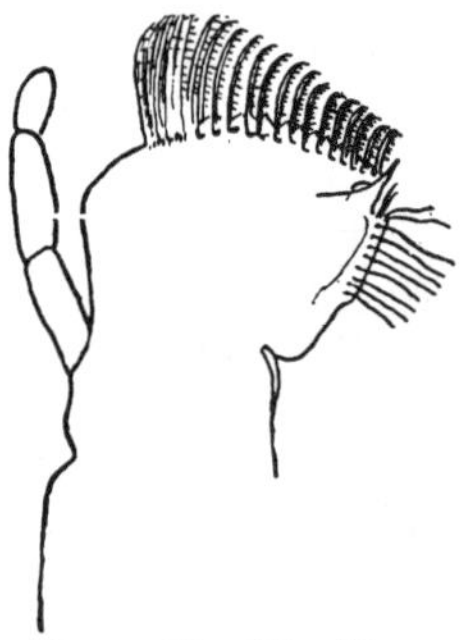

FIG. 66—Maxilla of *Ameletus ludens*, showing "diatom rake."

Those of like functions are alike in form even when found on different parts of the body, and the greatest development of any type of hair is at the point where its function is at the height of intensity. Thus the plancton gathering hairs are longest on the front legs of Isonychia nymphs (fig. 48) and on the tips of the maxillae in Hagenulus (Pl. XXXII, fig. 85).

The hairs that are fitted for scraping algae from the stones in torrential waters are thin and flat and strong with pectinate edges directed forward. Such hairs are found on the maxillae of Rhithrogena (see Pl. XXXII, figs. 69 and 70). They are not found on nymphs that live in quiet or slowly flowing waters. The loose flocculent algae that lie on the bottom in such places do not require such highly specialized tools for their gathering. But the algae on the rocks in the rapids are close laid and must be dug out from the slime-coat in which they develop. These short stiff pectinate spines are the digging tools needed. A lighter implement for use in streams of less velocity and with looser-growing algae is the so-called "diatom rake" (fig. 66) that forms a terminal border to the blade of the maxilla of *Ameletus ludens*.

CHAPTER XIV

MAYFLY PHYLOGENY

1. The Fossil Record

This is the story of the evolution of mayflies. It is not a complete story, for, owing to gaps in the fossil record, some of the earlier chapters are missing. Moreover, there are different versions of it; for like other stories, how it is told depends on who is telling it. This is our version of it.

The earliest known fossils that are well enough preserved to be certainly identified as mayflies occur in Permian shales, where they have been found in considerable variety. Some of these specimens are beautifully preserved, the venation of the wings showing almost as well as in recent specimens. They were first studied by Sellards (1907) in material coming from the now famous beds at Elmo, Kansas. Additional material from the same source was later studied, first by Tillyard (1925, 1932) and then by Carpenter (1933). Other Permian fossils from the Archangel District, North Russia, were studied and reported upon by Martynov (1932). To the excellent work of these four authors our present knowledge of Permian mayflies is due.

The best specimens obtained by Carpenter from the Kansas shales show the body well enough to give a good idea of what these ancient mayflies were like. Carpenter has drawn a restoration of one of them, *Protereisma permianum*, which we copy in figure 67. His figure shows a long-winged, long-bodied insect with slender legs and three long tails. It is much more primitive than modern mayflies in many characters, among which we may note:

1. The nearly equal, little consolidated divisions of the thorax.
2. The nearly equal legs, each with five-jointed tarsus and two good claws.
3. The nearly equal eleven joints in the antennae.
4. The nearly equal *oblong* wings, very moderately narrowed at the base, and lacking a prominent hind angle.
5. The connected intercalaries, all branches and intercalaries being attached at their bases.
6. The radial sector attached at base to R 1, the ancestral relation, rather than to vein M.
7. The three equal tails of moderate length.
8. The long cubital vein and the extensive anal area behind it.

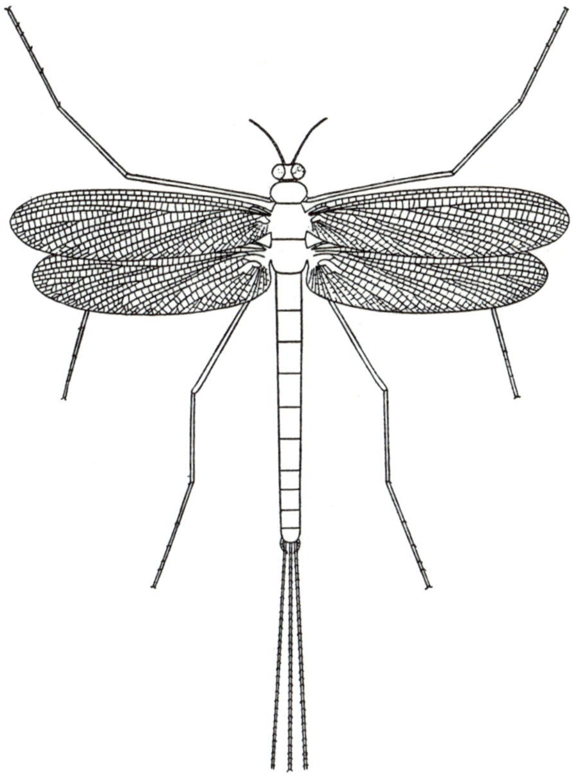

FIG. 67—Carpenter's restoration of mayfly, *Protereisma permianum.*

Nevertheless, Protereisma was far from primitive in these respects:

1. The ancient irregular network (archedictyon) had already been replaced throughout the wing membrane by regular series of straight cross veins.

2. The originally flat wings had become fluted; deeply, between principal veins; moderately in the shallower forks.

3. The modern specialized mayfly type of male accessory genitalia was already present.

Of such a mayfly as this Doctor Tillyard rightly says that there is not a single character which fails to agree with our conception of an archetype for modern mayflies, before the reduction of the hind wing began. "All the characters of the head, thorax, abdomen, legs, and tail filaments, so far as revealed in the fossil Permoplectoptera agree admirably with what is required in such an ancestor."

Another Permian genus that is well represented in the material from the Kansas beds is Misthodotes. Several species have been carefully described by Doctor Tillyard, one of whose figures of venation we copy herewith (fig. 68). These were somewhat smaller mayflies, stouter of body and shorter of legs, with the venation of the relatively broader wings somewhat reduced. The cross veins were fewer and stronger; the wings were shaped more like those of recent mayflies.

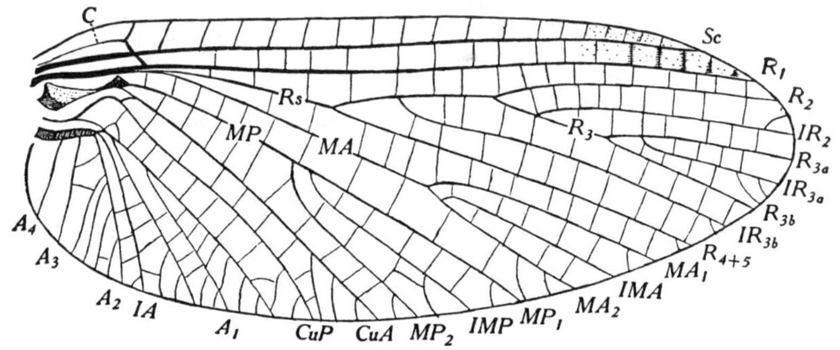

FIG. 68—*Misthodotes obtusus* (from Tillyard). Note that the outer fork of the radial sector (*Of*) is here labelled MA.

Another Permian genus, Palingeniopsis, is known from Russia, and we present a copy of Martynov's figure of it herewith (fig. 69). The single known specimen is not so well preserved. The wing apex is missing and few cross veins remain. The longitudinal veins show its likeness to Protereisma; but it differs markedly in the strongly sigmoid curvature of vein Cu2, in the convergence of branches of the radial sector in two pairs, and in the depth of the forks.

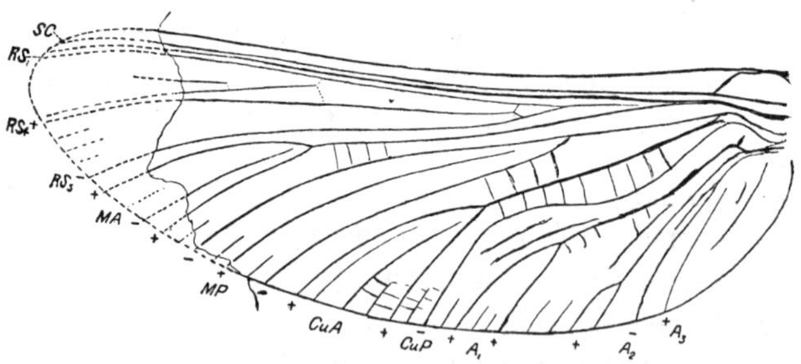

FIG. 69—*Palingeniopsis praecox* (after Martynov).

Mayfly nymphs, also, are known from the Permian of Russia and we copy Handlirsch's figure of one of them, *Phthartus rossicus* (fig. 70). At first glance it looks singularly like a modern Baetine nymph, in head, in tails, and in legs with their one-jointed tarsi and single claws; but on closer inspection one sees that the wings were larger and the two pairs of equal size and wide spacing; gills were well developed on abdominal segments two to nine, not stopping at the seventh segment as in modern mayfly nymphs. It will be noted that these gills were of the simple type that appears first in the ontogeny of the nymphs of various modern groups, whatever form they may later take on. This may fairly be called the ancestral form of the mayfly nymphal gill.

These Permian mayflies show that the group was already well on its way. The gills show that the nymphs were aquatic. The wings of the adults have since been altered mainly by coadaptation of the two pairs, and reduction of veins. The large copulatory forceps of the male indicates that already the chief concern of adult life may have been with reproduction. It is probable that the segregation of the life cycle into aquatic feeding and aerial reproductive phases was already established, and that these oldest well known fossils are far from the beginning of the evolution of the group.

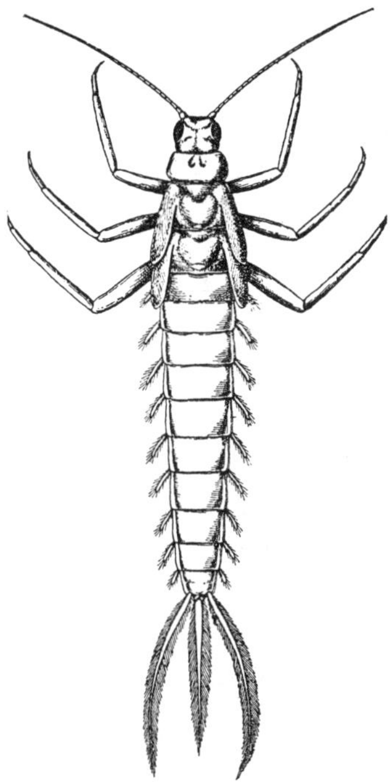

FIG. 70—The nymph of *Phthartus rossicus* (from Handlirsch).

The next good fossil mayflies as yet known are from the Solenhofen beds of Bavaria (Jurassic age of the Mesozoic era). A number have been found but few of them are well preserved.

In one genus, Mesephemera, the hind wings were as long as the fore, but much narrower. In the known specimens of this genus, however, there is little venation preserved.

The venation of two species of Paedephemera is figured by Handlirsch, and we copy his figure herewith (fig. 71). In this genus the hind wings were reduced in size and more or less coadapted to the rear of the fore wings. They were two-thirds to three-fourths as long as the fore wings. They were very like the hind wings of the modern genus Siphlonurus. The fore wings had a somewhat triangular form, especially in *P. schwertschlageri*, with a well developed hind angle, and a narrowed base. About the wing margin, there were free intercalaries as in modern mayflies.

The best known nymphs of this era are from Jurassic beds in Siberia. One genus, Mesobaetis, seems to have gills on nine abdominal segments as in Phthartus. Two others, Ephemeropsis and Phacelobranchus, have them on the first eight segments, and others have them only on the first seven segments as in modern forms. Phacelobranchus shows marked compacting of the synthorax with reduction (possibly, suppression) of the hind wings. The tails have a modern Baetine aspect, being stout and strongly tapered and densely fringed, the laterals being fringed only on the inner side.

Tertiary mayflies so far as known have a very modern aspect. Huge eyes that meet on top of the head in male specimens, differentiated claws, and reduced hind wings are among their signs of specialization. Nearly all have been referred to modern genera. Singularly enough, they come from two different sources, one in the old world and one in the new, adults from one and nymphs from the other.

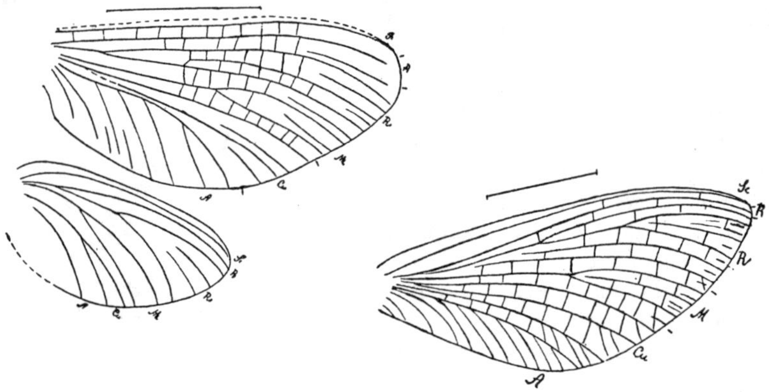

FIG. 71—Mesozoic mayflies: left, *Paedephemera multinervosa;* right, *P. schwertschlageri* (from Handlirsch).

The adults come from the old world and are found in Baltic amber. Since this is fossil resin formed as an exudate from the trunks of coniferous trees, it is only the winged stages of mayflies that could become embedded in it in its soft condition, to be fossilized with it. Both subimagos and imagos have been found (along with hosts of other small insects) in the transparent lumps; also an imago half out of its subimaginal skin.

The known American Tertiary fossil mayflies are nearly all nymphs. They come from the beds in the Florissant basin, high in the Rocky Mountains of Colorado. Certain shales that were formed as deposits in the edge of what was once a shallow lake hold the abundant insects of this famous locality. The mayfly nymphs are not numerous, nor are they very well preserved; but they seem to have been very like present day forms. Because of their wide tibiae, S. H. Scudder, who described five species of them (Tertiary Insects of N. Amer. pp. 120-124, 1890), considered them burrowers and referred them all tentatively to the genus Ephemera.

Thus it appears that the fossils from the Permian, the Jurassic, and the Tertiary form a consistent and progressive series leading up to modern mayflies and showing trends in specialization that have been carried further in recent times. What lies back of the Permian, we do not yet know. The efforts that have been made to link up mayflies with some older Palaeozoic forms have not been very successful. Names that indicate ancestry have been rather inappropriately attached to two Palaeozoic orders: Protephemerida and Protodonata. The nearest ally in the former group, Triplosoba, differs from the Permian fossils in having the hind wing larger than the fore, in lacking the triple triad in the first fork of the radial sector (having the sector pinnately branched instead), and it could not have been ancestral to them.

Furthermore, the Protodonata (Meganeura, etc.) have as many characters in common with mayflies as they have with dragonflies, although they could not have been ancestral to either. They have a great development of the cubito-anal area of the wing. It occupies nearly half the wing and has its own peculiar sigmoid curvature of veins toward the base, and its own pattern of branches supporting the anal field. The Meganeuridae entirely lack the distinctive characters of the Odonata: nodus, stigma and triangle braces, a crossed radial sector in the wing, and male accessory genitalia on the second abdominal segment.

All these are independent orders. None is derivable from any other, nor very closely related to any other. All have the wing surface deeply and extensively fluted and the fluting is different from that of all other insects.

As a group, mayflies appear to have reached their maximum relative abundance in the Permian. In 1930 Carpenter summarized the number of known species, age by age, for mayflies as compared with the total of known species for all insects, in the following percentages:

In the Permian they numbered 3.5% of the total.
In the Mesozoic they numbered 2.0% of the total.
In the Tertiary they numbered .3% of the total.
In recent times they number .095% of the total.

They are weaklings; and in the race for possession of the earth they appear to have steadily fallen behind.

2. More Recent Trends

Such is the fossil record in so far as yet made known. It clearly shows the main trend of evolution within the order, which is toward enlargement of eyes in the male; toward consolidation of thorax and loss of hind wings; and toward complete fluting of the wing surface with deepening of the forks, basal detachment of veins, and non-selective reduction of cross veins. We have now to trace in recent forms the principal lines along which further modifications have been made.

The group as a whole, while retaining a good many primitive characters, has gone a long way in specializing on lines of its own. There are no very primitive mayflies. There is no one genus or species that can be set down as nearest to the ancestral form in all particulars. The lines run criss-cross here as elsewhere, and do not lend themselves to a serial arrangement.

If we seek among living mayflies for a generalized form to compare with the fossils we are somewhat at a loss to choose. One primitive character is preserved here and another there, and that in distantly related genera such as Ephemera and Siphlonurus. In Ephemera, for example, three tails are preserved, the eyes of the male are but little larger than those of the female, and the genitalia are comparatively simple; but the basal joint of the tarsus is fused with the tibia, the claws are unlike on each tarsus, and the median vein is askew at its basal fork, with the posterior branch stiffened and arched to rearward in a way that is certainly not primitive. Also, in the hind wing veins R_4 and R_5 are completely fused. In Siphlonurus the tarsi all have five freely movable joints, the claws are all hooked, the hind wing is large and the venation is rather primitive; but the middle tail is lost, the eyes of the male are very large, and the genitalia are very elaborate.

These two stand as primitive members of their respective families, and they agree in the following generalized characters: both have preserved relatively large hind wings and abundant cross-venation, both have the eyes of the male undivided, and both have the sub-imaginal period not greatly abbreviated. Also, their nymphs have gills preserved on seven abdominal segments.

However, one may find equally primitive characters in the other more specialized families. In Caenis, for example, along with compacted thorax, a single pair of wings and almost no crossveins, one finds three tails, and eyes in the male hardly enlarged at all. In Baetisca, along with extraordinary consolidation of the thorax and (as Vayssière has shown) of all the ganglia of the ventral chain within it, there is preserved a relatively primitive venation, a large hind wing, and in the fore wing the best developed cubito-anal area within the order. And in Campsurus, small eyes, large hind wings, and three tails go with almost complete loss of legs.

It will not do to formulate our ideas of mayfly evolution on the basis of adults alone, for in this order the struggle for existence has fallen largely upon the nymphs, which are better equipped to meet competitors. These show greater divergences in adaptation to their several types of habitat. These have differentiated on lines of their own, independently of the adults, and tell their own story. It goes without saying that our interpretations of nymphal and of adult evolution will, when correct, be in agreement. One must corroborate the other; for nature preserves or eliminates species as wholes.

This corroboration will be found in the tables and keys of Part II of this book. Group characters for nymphs and for adults go well together, in the main. The structural characteristics of both will be given in detail under the appropriate headings. Here we are noting merely the general trends in the evolutionary history of the order.

Although the mayflies of the present time are the isolated remnants of a vanishing race of insects, once far more abundant relatively than now, they may not have been actually more numerous, for a few types of modern mayflies have found situations that they seem to fit better than do any other animals. There are not a few places in the world of fresh waters where a few forms are numerically dominant, as for example, the Hexagenias of the beds of our large rivers, and the Stenonemas of the boulder-strewn rapids of our creeks. A few genera appear to be flourishing in our own time, if the great number of closely similar species is to be taken as evidence. In the great Ephemerella complex species-making is perhaps still going on.

The dominant modern group is that of the Baetidae, in which all lines of specialization traceable among the known fossils seem to culminate. Among the Baetidae the two lines of extremest specialization that are represented by the subfamilies Caeninae and Baetinae stand in striking contrast. The Caeninae have preserved all the forks of the triple triad while losing nearly all crossveins and all marginal veinlets, and the eyes of the male have remained small. The Baetinae have detached these forks while preserving a fine series of marginal veinlets; eyes in the male have become enlarged, then divided, and then the upper division has become pedunculate upon the lower. In both groups hind wings dwindle and disappear; cross veins vanish; forks at the rear of the wing become detached or deepened to the wing base; one or more joints of the tarsus become solidly joined to the tibia; size diminishes, and the period of adult life becomes remarkably short.

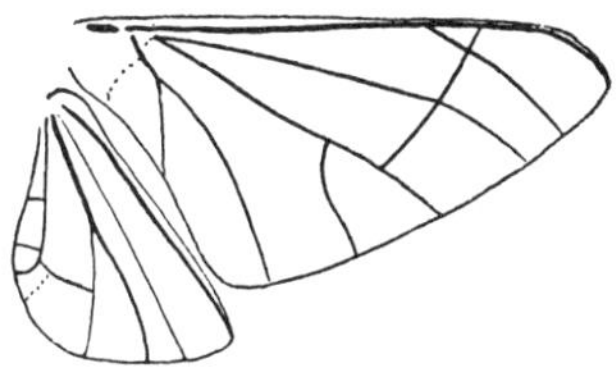

FIG. 72—Wings of *Lachlania abnormis* (after Eaton).

Aeroplane traps capture in the upper air the smaller members of both groups in some numbers, especially Caenis, of which specimens taken in Louisiana at above 1000 feet were sent us by Dr. Harold Morrison.

The Oligoneurine group has followed its own peculiar trend in wing evolution. The subcostal vein has disappeared; also, the second and third forks of the normal triple triad in Rs; also, all marginal veinlets. Other long veins have disappeared by atrophy or by fusion. Cross veins have dwindled; they have been reduced to three in a single cross-line in Lachlania (fig. 71) and in Homeoneuria have disappeared altogether. This is a group of remnants.

Other more isolated remnants are Oreianthus and Baetisca in our own fauna, Prosopistoma in the old world, Ameletopsis in New Zealand and Murphyella in Chile.

Baetisca and Prosopistoma* are an aberrant pair. Both are characterized by a consolidated, carapace-like thorax that is elongated to cover the gills in the rear, and by an abbreviated abdomen. Vayssière in 1882 (Ann. Sci. Nat., Zool. 13: 129) made the prediction that they would be found to be much alike in cephalization of the central nervous system. This prediction is brilliantly fulfilled by his recent works, as one may see by comparing his figure of Baetisca, copied herewith, (fig. 73) with his earlier one of Prosopistoma (*l.c.* Pl. 10, fig. 109). The complete consolidation of all thoracic and abdominal ganglia in both is very remarkable. In internal anatomy they depart from the mayfly norm as widely as they do in external form. Baetisca is unique, however, in the possession of a large intestinal caecum (fig. 74).

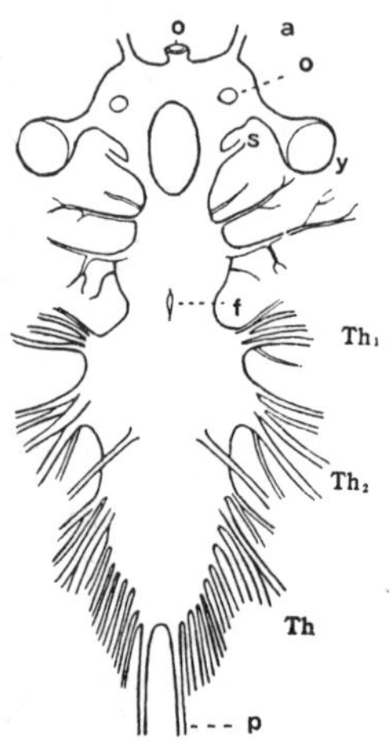

FIG. 73—The central nervous system in a nymph of *Baetisca obesa* (after Vayssiere): *o,o,* ocelli; *a,* base of antennal nerve; *y,* eye; *s,* "sympathetic," that innervates the wall of the alimentary canal; *f,* foramen between the longitudinal trunks, lying between subesophageal and thoracic ganglia; Th1, Th2, Th3, fasicled nerves springing from the pre- meso- and metathoracic ganglia respectively; the latter ganglion, consolidated with all the abdominal ganglia; *p,* a large nerve at posterior end of the latter.

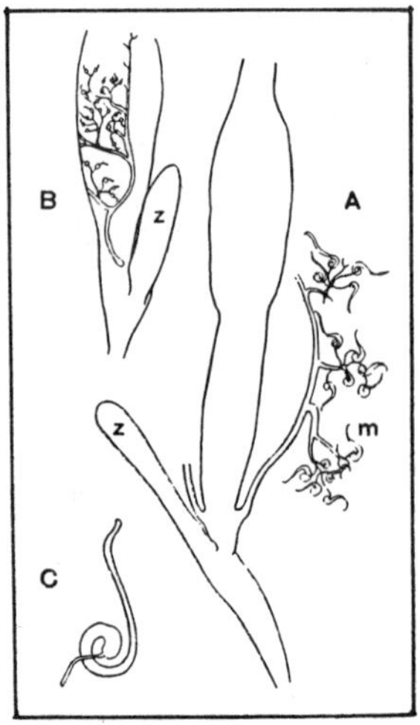

FIG. 74—Some pecularities of the alimentary canal and its appendages in *Baetisca obesa* (after Vayssiere). *A* and *B,* two views of the middle portion of the alimentary canal, with one cluster of Malpighian tubules, *m,* and an intestinal caecum, *z; C,* the secretory end of one Malpighian tubule, more enlarged.

*An elusive old world form, its nymph known since Geoffroy described it in 1764 as a crustacean! After 171 years of discussion and picturization its imago is still unknown!

The supposed affinity of Ephemeroptera and Odonata has recently been discussed by the senior author (N. Y. Ent. Soc. Jour. 43:122, 1935) in the following words:

"Ever since Latreille put mayflies and dragonflies together in one order Subulicornia, it has been commonly held that they are closely related groups. I formerly shared in this opinion, but further study has convinced me that both are very isolated groups, well apart from other insects and from each other.

They have a few very well known common features. Both groups have reduced antennae, as Latreille's name for them indicated, but this is a departure from primitive conditions that may well have been a parallel development. Both groups have enormously developed compound eyes but the eyes have little in common save what is common to insects generally. Both groups have a ten-segmented abdomen, but so had many unrelated fossils. Both groups hold their wings outspread and not folded on the back.

"This last common feature on close inspection will reveal little affinity; for the wings of the two groups differ vastly in development, in mechanical arrangements, and in general evolutionary trends. The nymphal wings lie flat on the back in Ephemeroptera, inverted in Odonata. The hind wings tend to be reduced in Ephemeroptera, expanded in Odonata. They are coadapted to the fore wing in Ephemeroptera but remain free and independent in the Odonata. Huge dorsal longitudinal muscles propel the wings downward in Ephemeroptera by uplift of the tergum, while these muscles are lacking in Odonata and the down stroke of the wing is effected by direct pleural wing muscles. The base of the nymphal wing bears ingrowing callosities in Ephemeroptera that are lacking in Odonata. The stiffening of the wing has been chiefly by means of fluting in Ephemeroptera with hardly any use of special crossveins, while in the Odonata three strong cross bracings of the veins at arculus, nodus, and stigma are characteristic of the entire order. The area of the radial vein is expanded and that of the median reduced in Ephemeroptera, while these conditions are reversed in Odonata.

"Other significant contrasts are seen in the different type of metamorphosis and life cycle; in mouthparts and feeding habits; in form and armature and segmentation of the tarsi; in gills and abdominal appendages; in hind gut and Malpighian tubules; in sex organs and copulatory apparatus in the male, etc."

These contrasts may be further indicated by a parallel statement of the principal differences, as follows:

Character	*Ephemeroptera*	*Odonata*
I. Adult		
Head	Rigid	Versatile
Mouthparts	vestigial	strong, raptorial
Fore legs of male	copulatory	raptorial
All tarsi	five-jointed	three-jointed
Lateral sutures of synthorax	erect	askew
Metathorax	reduced	enlarged
Hind wing	reduced or wanting	often enlarged
Wing fluting	deep and regular	irregular
Crossveins	all similar	a few are highly differentiated

Character	*Ephemeroptera*	*Odonata*
Special braces	almost wanting	highly developed
Area of vein Rs	enlarged	reduced
Dorsal longitudinal wing muscles	very large	wanting
Subimago	present	absent
Copulatory apparatus of male	forceps on ninth abdominal segment	hamules on second abdominal segment
Genital aperture of female	between segments 7 and 8	between segments 8 and 9
II. NYMPH		
Labium	Normal	Large raptorial
Paired appendages attached to hypopharynx	large	wanting
Tarsal claws	single	in pairs
Wing pads	normal	inverted
Malpighian tubules	complex and convoluted	simple
Abdominal gills	generally dorsal	internal or caudal
Cerci	long tails	short spines
Replacement cells of stomach epithelium	scattered	in definite nidi
Thoracic nephrocyte clusters	mainly dorsal	about leg bases

Adult mayflies are specialized in their own ways far beyond any other group of insects. They have departed so far from the insect norm that they may be characterized in good part by the losses they have sustained: loss of mouthparts and of all capacity for feeding; loss of legs and of the walking habit; loss of all defensive equipment, and of all instincts to defend themselves. It seems a marvel that after such losses they still survive and flourish, such fragile creatures in so rough a world!

Their reproductive capacity saves them. Vast numbers of offspring allow for great losses and yet permit a few to survive. The few find their niches and remain as examples of the survival of the unlike.

NOTE—Concerning the fossil *Platephemera antiqua* Scudder, be it said that it is not a mayfly, and that it is, as Handlirsch thought, Palaeodictyopterous. Dr. F. M. Carpenter has sent us a fine photograph of the type specimen. Comstock's figure (*The Wings of Insects* p. 86, fig. 75) is inaccurate. Scudder's figure published by Handlirsch in *Die Fossile Insekten,* (Pl. 10, fig. 22) is better.

CHAPTER XV

CASUALTIES AND ENEMIES

Over against the remarkable reproductive capacity of mayflies stands a host of eliminative agencies: casualties and enemies. There are casualties of storm and flood, of drought and winds. There are enemies both predatory and parasitic. Unfortunately, we have at the present time hardly any positive data for measuring the effects produced by any of these agencies. We can therefore give only the results of chance observations. We have always to remember that if, in each bisexual generation, a single pair survives to reproduce in its turn, the natural balance is preserved; the species is holding its own, even though all the other individuals of the great brood perish without issue.

The losses begin with the eggs and are probably heaviest in the egg stage. Floods carry them away and bury them in bars of mud, or leave them stranded on the shores to die of desiccation when the waters recede. Small egg-eating enemies, caddis worms, snails, and other lesser invertebrates, doubtless devour many of them.

The nymphs in the water are subject to many perils. Even the smallest nymphs have predatory foes lying in wait for them. An interesting unpublished observation reported to us by Dr. Lillian Thomsen is that the microscopic punkie larvae (larvae of Bezzia: Ceratopogoninae: Diptera), that twine their sinuous bodies among the tangled, slender filaments of algae in the pond, feed habitually on the infant nymphs of Callibaetis.

Every observant collector has seen the swarms of very minute nymphs on the stones in a quietly flowing stream and has noted the rapid decrease in numbers as the individuals increase in size. No one knows what becomes of them, but the elimination of large numbers is obvious. Doubtless in quiet waters carnivorous enemies are chiefly responsible.

The destruction wrought by flood waters is often very great. Stones and gravel are swept along by the current, and, in the spring, blocks of outgoing ice scrape the banks and the bottom. The pounding of these solids together can be distinctly heard at a distance from the water. The nymphs are in danger of being crushed by the stones that in quiet times furnish them shelter. If not killed outright, they may be carried out of their proper lotic environment into pools or basins where they cannot live.

Conclusive evidence of this destructiveness was obtained during a high March flood in Six Mile Creek near Ithaca, New York, in 1927 by Dr. Paul R. Needham. He set a drift net (1928) in the flood and strained out the drift. He writes of it (1930, p. 5):

"The results were most illuminating. Practically every kind of aquatic organism which had been collected from this stream during the previous summer was taken in the net. The great majority were dead or injured by the grinding action of rocks and gravel which were being carried downstream. Many parts of insect larvae, such as heads, legs, tails and abdomens, offered evidence of the destructive action of high waters. Many aquatics such as black-fly larvae (Simulium) which are never taken drifting free in the current under normal circumstances, were collected, bruised and battered, as they were carried downstream. Bottom studies made after the flood had subsided showed but a fraction of a gram of organisms remaining per unit area."

Fishes are probably the chief enemies of mayfly nymphs, although water birds and carnivorous salamanders doubtless eat them also. Among invertebrate enemies, nymphs of dragonflies in the ponds and of stoneflies in the streams stand preeminent. One of us (Traver, 1925, p. 215) has recorded observing nymphs of the stonefly *Perlinella drymo* capture and devour nymphs of *Blasturus cupidus*. Nymphs of Ephemerella were found by the senior author to be the principal food of the nymphs of the dragonfly *Dythemis velox* in San Dimas canyon, California. In the aquaria of our laboratory mayfly nymphs of many sorts have been used as food for dragonfly nymphs. All seem to be equally acceptable.

These lesser carnivores can pursue the nymphs among their sheltering rocks and vegetation where the larger vertebrate enemies cannot go. Fishes cannot thoroughly search the stones in the riffle, but in the pools below they lie in wait for any dislodged unfortunates that may come drifting down. Some data showing in what proportions they enter into the diet of fishes will be given in the next chapter.

In Chapter I we have already spoken of some of the perils of the winged mayflies—of the spiders that ensnare them in their webs, of the fish that snap them up at the water's surface, of the robins that await their emergence from the water. Dr. Osgood Smith, while collecting mayfly eggs along Fall Creek at Ithaca, made an observation that is worth recording here. It was on the 31st of May. The dainty little mayfly *Stenonema ithaca* was emerging from the riffles, and cedar waxwings were catching them. The birds sat on convenient perches waiting and watching for the subimagos to come out. Only the few that emerged near shore and flew at once into nearby vegetation escaped capture by the waxwings. When one flew up from the water, a waxwing would dash after it. Doctor Smith was taking imagos also, to obtain their eggs, and a waxwing took one of them almost out of the mouth of his net.

Everyone who has watched swarms carefully has seen swifts and swallows following them and darting through them, foraging. Indeed, the easiest way we have found to locate mayfly swarms at dusk is to watch the flight of these birds. At Ithaca the swallows come down out of the sky at evening twilight and skim over the streams through the lanes of the gorges, making long sweeps until a swarm of insects is found. Then they make numerous short turns and dash through it.

The twilight-flying dragonflies, chiefly the big blue and green darners of such genera as Anax and Aeschna, also take quick advantage of mayfly swarms for easy foraging. On Cayuga Lake in midsummer the Aeschnas may be seen all day long, coursing the rocky shore on tireless wing near the edge of the water, following its curves in and out at a rather constant level of a few feet; they make quick sallies now and then into the upper air to chase a fly or to greet another darner, but they quickly return to the regular beat, where they may be seen to pick up a stray midge now and then, or perchance a mosquito. So they fly through the long hours of sunshine. But suddenly, when the deepening twilight has brought the Heptagenias out for their nuptial flight, the darners rise from the edge of the shore and begin darting through the swarms. Standing on the shore, we lately saw them just beyond our feet. Now we see them overhead. It is too dark to see the actual capture of Heptagenias; but no doubt that is what is going on. The darners continue to fly through the swarm as long as visible, without once returning to the shore line beat.

The parasitic enemies of mayfly nymphs are numerous but they have been little studied. A large external Chironomid parasite, *Trissocladius equitans* (Pl. XXXIII), was found by Dr. P. W. Claassen infesting the nymph of a species of Rhithrogena at Estes Park, Colorado. We reproduce here a part of his account of this species together with one of his plates.

"The parasite attaches itself to the posterior margin of the mesothorax underneath the wing pads, where it imbeds its head in the softer tissue of the nymph and where it draws its nourishment from the host. The larva spins a sheet of silk, which completely invests its body, and by means of which it attaches itself firmly to the body of the nymph. At first, stretched to its full length, the larva lies across the body of the nymph, but as it grows larger the posterior end of the body doubles under, and later, when the larva becomes full grown, the middle part of its body projects backward over the abdomen of the nymph in the form of a letter U. When ready to pupate the larva releases its hold at the head end and turns back over the abdomen of the nymph. The pupa thus extends over about half the abdomen of the host. The transparent sheet of silk invests the pupa as well as the larva. Just before the adult is ready to emerge the pupa breaks through this investing membrane and rises to the surface of the water, where the adult emerges in a manner similar to that in other chironomids.

"That the larva is a parasite upon the nymph is apparent from the following facts: First, the silk membrane completely surrounds the larva. There is no opening in this membrane which would make it possible to obtain food from the outside. Second, an examination of the stomach contents of a number of larvae failed to reveal the presence of any vegetable material, but did contain animal matter, especially fat bodies. This fact was also borne out in a study of cross sections of the larvae. Third, a study of the head and mouth parts of the parasite reveal their reduced condition. The head is very small in proportion to the size of the body and the mouth parts are much reduced.

PLATE XXXIII.

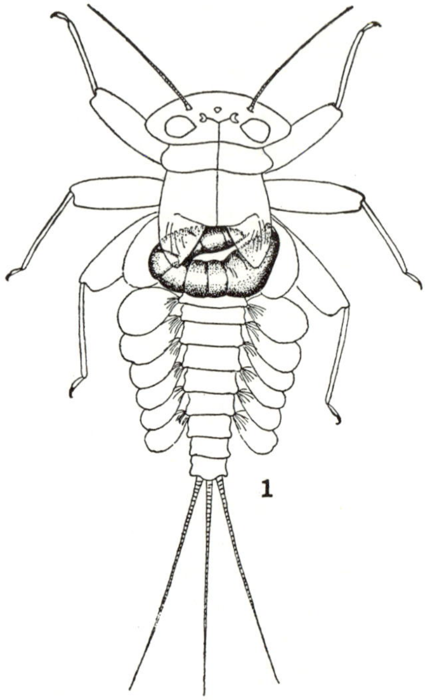

1

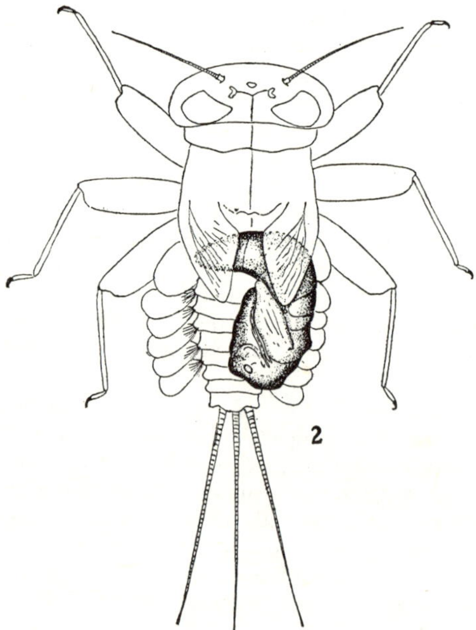

2

AN EXTERNAL PARASITE OF RHITHROGENA
Trissocladius equitans (from Claassen). 1. Larva; 2. Purpa.

"There appears to be no relationship between the relative ages of the host and parasite. Pupae of the parasite were found upon medium-sized nymphs, while, on the other hand, young larvae occurred upon nymphs that were evidently in the last nymphal instar."

Doctor Claassen found that more than ninety per cent of some 300 Rhithrogena nymphs from this particular locality carried either larvae or pupae of the parasite.

We have found larvae of similar form attached ventrally to nymphs of *Ephemerella flavilinea* from the Glacier National Park.

In Europe two species of Phaenocladius have been found to be parasitic on nymphs of Rhithrogena and Ecdyonurus, while another Chironomid larva of the genus Camptocladius has been found attached to nymphs of Ephemera. These, however, are commensals and not parasites. They feed upon diatoms and other algae that grow upon the skin and hairs of the nymphs. They attach their dwelling tubes to the nymphs (as other Camptocladius larvae do to plants) and forage round about them over the nymph's body, keeping its back clean. Thus they perhaps keep clear the channel through which fresh water flows over the nymph's gills. Sulc and Javrel (1924, p. 389) have remarked that these species of Chironomids appear to have specific mayfly hosts on which alone they may be found.

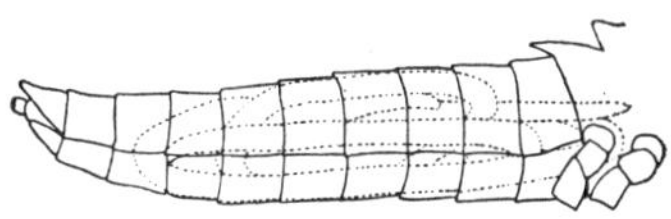

FIG. 75—A nematode parasite coiled within the abdomen of a female imago of the mayfly, *Baetis vagans* (from Murphy).

Many workers have noted the presence of nematode worms in the abdomen of nymphs. We have seen very small nematodes embedded in the fat tissue of Stenonema nymphs, often ten or more of the worms in a single nymph. These are separately encased in a thin-walled cyst with the worm coiled inside. Occasionally two of the parasites are found in a single cyst.

A larger species of nematode is found in the body cavity of both nymphs and imagos, generally coiled around the digestive tract, often filling the space from which both fat and gonads have been eaten. Doctor Murphy (1922, p. 42) records this for *Baetis vagans* as follows:

"Four females captured during a mating flight October 28, 1918, were found to be parasitized. Each imago contained a single nematode coiled up in the body cavity (fig. 75). The hosts appeared natural in flight, and it was not until the specimens were preserved that the presence of the parasite was observed. The abdomen, normally conspicuously filled with eggs, seemed to contain nothing except the nematode."

Mayfly nymphs serve as intermediate hosts for a number of flukes (parasitic trematode worms) that infest fishes and salamanders. For example, the fluke, *Crepidostomum farionis** lives as an adult in the alimentary canal and gall bladder of trout and grayling in England. Its

*From the account of F. J. Brown in *Parasitology* 19:86–99, 1927.

complicated life history begins when the eggs of the adult fluke, passed out into the stream, find lodgment in the gills of little white clams of the genus Pisidium or Sphaerium. The eggs hatch into little spindle-shaped rediae, which attach themselves by the pharynx to the gills of the clam. Each redia later divides into a large number of daughter rediae, and each of these transforms into a cercaria, with a minute stylet or boring spine at the front end and a long tail at the rear. The cercariae leave the body of the clam and swim freely about in the water of the stream, until they encounter a nymph of Ephemera. They make their way into the body of the nymph, and encyst themselves in the fat or in the muscle tissue. There they remain for a time, undergoing another change of form, losing the long tail and taking on the form of the adult fluke. When the mayfly nymph is eaten by a trout or grayling, the cysts are abandoned and the liberated flukes attach themselves to the wall of the alimentary canal, feed there and grow to sexual maturity.

Thus three successive hosts are necessary for the complete development of this parasite, and the mayfly nymph is the intermediate host. The following list of helminth parasites now known to infest mayflies has been kindly furnished us by Dr. Maurice C. Hall* from his records at the Bureau of Animal Industry of the U. S. Department of Agriculture.

HELMINTH PARASITES FROM EPHEMEROPTERA

Mayfly Host **Parasite**

Blasturus cupidus (*Leptophlebia cupida*); *Allocreadium commune* (Trem.; Allocreadiidae)
Cloeon sp.; Cercaria of *Distomum sp.* of Sinitsin, 1905 (Trem.; Fam. incert.)
Cloeon dipterum; ?*Opisthioglyphe endolobum* (Trem.; Plagiorchiidae)
Cloeon sp.; *Plagiorchis sp.*; (Trem.; Plagiorchiidae)
Ephemera danica; *Crepidostomum farionis* (Trem.; Allocreadiidae)
Stephanophiala farionis (Trem.; Allocreadiidae)
Ephemera sp.; *Lepoderma muris* (Trem.; Plagiorchiidae)
Lecithodendrium ascidia (Trem.; Lecithodendriidae)
Ephemera vulgata; *Allocreadium isoporum* (Trem.; Allocreadiidae)
Opisthioglyphe endolobum (Trem.; Plagiorchiidae)
Opisthioglyphe rastellus (Trem.; Plagiorchiidae)
Filaria ephemeridarum (Nem.; Fam. incert.)
Spiroptera ephemeridarum (Nem.; Fam. incert.)
Ephemerid; *Cercaria secunda* (Trem.; Plagiorchiidae)
Lecithodendrium lagena (Trem.; Lecithodendriidae)
Pleurogenoides medians (Trem.; Lecithodendriidae)
Hexagenia limbata; *Crepidostomum cooperi* (Trem.; Allocreadiidae)
Hexagenia sp.; *Crepidostomum canadense* (Trem.; Allocreadiidae)
Crepidostomum cooperi (Trem.; Allocreadiidae)
Crepidostomum cornutum (Trem.; Allocreadiidae)
Crepidostomum isostomum (Trem.; Allocreadiidae)

*In part published by him in 1929 in Smithsonian Misc. Collections Vol. 81, No. 15, p. 54.

Crepidostomum laureatum (Trem.; Allocreadiidae)
Megalogonia ictaluri (Trem.; Allocreadiidae)
Mayfly nymphs; *Crepidostomum isostomum* (Trem.; Allocreadiidae)
Oligoneuria rhenana; *Filaria ephemeridarum* (Nem.; Fam. incert.)
Spiroptera ephemeridarum (Nem.; Fam. incert.)

It will be noted that the parasites listed are mainly trematodes, only four being nematodes. The names of but two American mayflies, *Hexagenia limbata* and *Blasturus cupidus*, appear in the list; the others are all European. Clearly there is need of much further study of mayfly parasites in America.

CHAPTER XVI

VALUES AND USES OF MAYFLIES

1. Economic Considerations

We will not entitle this section *economic values*, because the evaluation of mayflies in the economy of the water is not yet possible. We know only that fishes eat many of them and that fishes are the most valuable product of our lakes and streams.

Mayflies are important herbivores in all shoal fresh waters. They are hardy. They are especially adapted to the utmost diversity of aquatic situations. They eat food stuffs that are everywhere abundant in shoal waters, and they have great powers of reproduction. Like prolific, defenceless herbivores of other groups, from waterfleas to rabbits, they represent the producing class among animals. They turn grass into flesh. They feed the young game fish.

We have already mentioned a few groups of aquatic animals that feed upon them. Most important of these to us are the fishes. All carnivorous fresh-water fishes eat mayflies when they can get them, and in all stages.

Food studies have everywhere demonstrated how generally the nymphs of the big, burrowing mayflies are eaten by fishes. Forbes' report (1888) was one of the earliest. He found that these larvae constitute nearly a fifth of all the food taken by the fishes that he studied. He says (pp. 484–485):

"From the (Linnaean) order Neuroptera fishes draw a larger part of their food than from any other single group. In fact nearly a fifth of the entire amount of food consumed by all the adult fishes examined by me consisted of aquatic larvae of this order, the greater part of them larvae of dayflies (Ephemeridae), principally of the genus Hexagenia. These neuropterous larvae were eaten especially by the miller's thumb, the sheepshead, the white and striped bass, the common perch, 13 species of the darters, both the black bass, 7 of the sunfishes, the rock bass and the croppies, the pirate perch, the brook silversides, the sticklebacks, the mud minnow, the top minnows, the gizzard shad, the toothed herring, 12 species each of the true minnow family and of the suckers and buffalo, 5 catfishes, the dogfish, and the shovelfish—70 species out of the 87 which I have studied.

"Among the above I found them the most important food of the white bass, the toothed herring, the shovelfish (51 per cent) and the croppies; while they made a fourth or more of the alimentary contents of the sheepshead (46 per cent), the darters, the pirate perch, the common sunfishes (Lepomis and Chaenobryttus), the rock bass, the little pickerel, and the common sucker (36 per cent). . . .

"The larvae of Hexagenia, one of the commonest of the "river flies," was by far the most important insect of this group, this alone amounting to about half of all the Neuroptera eaten. They made nearly one-half of the food of the shovelfish, more than one-tenth that of the sunfishes, and the principal food resource of half-grown sheepshead; but were rarely taken by the sucker family, and made only 5 per cent of the food of the catfish group."

Forbes' studies were made on such material as happened to be at hand, without regard to times or seasons or conditions under which the fishes he studied had obtained their food. Subsequent studies of the food of the shovel-fish have made it necessary to qualify his statement as regards that fish. Wagner (1908, p. 28) says:

"The food of Polyodon consists, in Lake Pepin, entirely of plankton material, in largest part of entomostraca, but not unmixed with algae. There is one seeming, but only seeming, exception to this. Occasionally one finds the specimens of a morning's catch largely gorged with larvae of Ephemerids. But in every such case it was found that ephemerid imagines appeared in vast numbers the same evening. It appears plain, therefore, that the larvae taken by Polyodon were captured on their journey to the surface of the water."

In the matter of food, fishes are opportunists, like ourselves. They eat what nature sets before them. In streams they find mainly Ephemerellas, Leptophlebias, Baetids, and Heptagenias; in still waters, mainly Callibaetis in ponds, and Ephemeras and Hexagenias and midge larvae in lakes.

Wagner (1908, p. 31) says concerning the valuable rock sturgeon: "The food of this fish in Lake Pepin, in the summer at least, consists entirely of the larvae of the Ephemerids"; but he does not specify what sort of mayflies.

Pearse (1915) records that 7 of the 16 species of small, shore fishes studied by him at Madison, Wis., had eaten mayflies, the percentages of this sort of food averaging as high as 58.3 per cent in the large-mouth black bass and 40 per cent in the bluegill sunfish. In regularity of consumption and in amount consumed, the young of mayflies were second only to midge larvae.

One of us (Traver, 1925, p. 215) found that in a small stream near Ithaca, New York, the black-nosed dace, *Rhinichthys atronasus*, fed largely upon mayfly nymphs, among which were many *Blasturus cupidus*.

As a primary food mayflies are of obvious importance, but as a secondary food supply we believe them to be of greater importance. They are the natural food of the lesser fishes which constitute the greater part of the food of the larger game fishes. These lesser fishes invade the shoals where vegetation grows more or less thickly and where the mayflies thrive. The relation between the size of fishes and of mayfly nymphs is indicated by studies made by the senior author (1908) for sunfishes of different sizes. The fishes were taken at one time and place from a muddy weed-grown pond where they all had the same choice of food. In the weed beds seined to obtain these sunfishes, the only mayfly nymphs commonly present were the minute nymphs of Caenis. With the increase in size of these fishes the principal element of their diet shifted from Entomostraca to midge larvae, then to mayfly nymphs, as shown by the following figures:

The food of sunfishes (Eupomotis gibbosus) in Old Forge Pond

Length of Fishes	No. of fishes	Mayfly nymphs	Midge larvae	Other insects	Entomos-traca
¾ to 1 inch	8	0	6	0	13
2 to 3 inches	9	6	24	1	13
3 to 4 inches	8	16	12	9	2

Further and more complete evidences of change of diet with age is furnished by Clemens' (1928) studies of the food of brook trout in the streams of Oneida County, New York. His results in so far as they relate to mayflies are shown in the following table:

Trout		*Percentages of food*		
No.	*Length*	*Nymphs*	*Adults*	*Both*
78	1–2 inches	11	1	12
77	2–4 "	22	10	32
64	4–6 "	12	4	16
13	6–8 "	2	8	10
10	8–10 "	10	0	10

Clemens also found that the food of seven rainbow trout two to three inches long was 34% mayflies, while of nine others, five to twelve inches long, it was only 7% mayflies.

As to the effect of such diet on fish there is little experimental evidence, but there is much personal conviction on the part of fishermen. They all agree that the natural food is the best for developing both gaminess and a fine flavor. Isaac Walton (1863) stated the case for them long ago:

"And then as he (the trout) grows stronger, so he gets him into swifter and swifter streams, and there he lies at the watch for any fly or minnow that comes near to him; and especially he loves the May-flie; and these make the trout bold and lustie, and he is usually fatter and better meat at the end of that month than at any other time of the year.

"Only during the last days of May or early ones in June, when the May-fly is on the wing, is the cisco seen. Then the anglers go in boats out on the lake where the water is 50 to 100 feet deep and where experience has shown the cisco may be found. Until casting begins not a fish can be seen, nor the slightest ripple upon the water; but no sooner have a few impaled Ephemeras dropped upon the surface than the ciscos begin to appear. They can be seen coming up from the depths, 'their pearly sides burnished by the gleam and glint of the afternoon sun.' In a moment the water all about the many boats is a ripple of eager fish, every hook has been taken, and the happy anglers are busy removing the catch and dropping it into their boats."

The fisherman's name "cisco fly" is derived from the well known attractiveness of some of the larger mayflies as food of the fish called cisco in our deeper freshwater lakes. Jordan and Evermann (1904, p.

135) give an account of the lure of the mayflies for the cisco of Lake Tippecanoe (Indiana).

But the best observations on the relations between fishes and adult mayflies appear to be those of that good naturalist Henry D. Thoreau. We quote from his "Journal":

"June 2. Friday P.M.—Up Assabet to Castilleja and Annuasnack. . . . When we returned to our boat at 7:00 P.M., I noticed first, to my surprise that the river was all alive with leaping fish, their heads seen continually darted above water, and they were large fish, too. Looking up I found that the whole atmosphere over the river was full of shad-flies. It was a great flight of ephemerae. This is what the French fishermen call 'manna.'

"June 6. Tuesday P.M.—To Assabet Bathing-Place and return by stone bridge. The black oaks, birches, etc. etc., are covered with ephemerae of various sizes and colors, with one, two, three, or no streamers, ready to take wing at evening, i.e., about seven. I am covered with them and much incommoded.

"June 8, P.M. To Cedar Swamp. . . . When I returned to my boat, about five, the weather being mizzling enough to require an umbrella, with an easterly wind and dark for the hour, . . . I noticed a great flight of ephemerae over the water. . . . They advanced against the wind faster than I floated along. From time to time each one descended to the water and touched it, or rested on it a second or two, sometimes several minutes. . . .

"At one place, against Dodge's Brook, where they were driven back by a strong head wind at a bend, more than usual were wrecked on the water and the fishes were leaping more numerously than elsewhere. The river was quite alive with them, and I had not thought there were so many in it, great black heads and tails continually thrust up on all sides of my boat. You had only to keep your eye on a floating fly for a minute to see some fishy monster rise and swallow it with more or less skill and plashing. Some skillfully seized their prey without much plashing, rising in a low curve and just showing their backs; others rose up perpendicularly, half their length out of water, showing their black backs or white bellies or gleaming sides; others made a noisy rush at their prey and leaped entirely out of water, falling with a loud plash. You saw twenty black points at once. They seemed to be suckers; large fish at any rate, and probably various kinds. What a sudden surfeit the fishes must have!"

Little is known about the relative abundance of mayflies of the different species. A bit of census-taking concerned with the bottom fauna in rapid streams at Ithaca, New York, done by Dr. Paul Needham in 1927, 1928 revealed the following numbers and percentages of mayfly nymphs.

	Genus	*Total No.*	*Percentage*		*Genus*	*Total No.*	*Percentage*
1.	Baetis	1512	29.07	8.	Isonychia	244	4.69
2.	Ephemerella	833	16.02	9.	Heptagenia	143	2.75
3.	Paraleptophlebia	761	14.63	10.	Caenis	67	1.29
4.	Stenonema	583	11.21	11.	Iron	59	1.13
5.	Hexagenia	366	7.05	12.	Ephemera	48	.92
6.	Epeorus*	295	5.67	13.	Blasturus	15	.29
7.	Tricorythodes	272	5.22	14.	Ameletus	3	.06
					Total	5,201	100.00

*Included by us in the genus Iron in Part II: these are the Eastern species of Iron, having gills of the first and last pairs extended outward.

That mayflies are a choice food for many kinds of fishes is evident. That they are a staple food in nature is equally evident. As such they rank with caddis-worms and midge larvae and water-fleas and other small, quick-growing, prolific herbivores. Hardly any effort has as yet been made to increase the supply of them or to control them in any way in the interest of fish culture.

Such control will demand knowledge of their ways. Intimate and detailed knowledge of life histories and habits is the very basis of all economic procedure with animals. Let us now briefly review the little that is known in this neglected field.

The food of mayfly nymphs is mainly vegetable substances. One New Zealand genus, Ameletopsis, is positively known to be carnivorous, feeding in the main on nymphs of smaller mayflies. Several other exotic genera (Chiloporter, Prosopistoma, etc.) may be inferred to be the same, judging from the armature of their jaws. Those that strain the water by means of plancton-gathering brushes obtain with the algae and seed-plant-fragments of the drift an admixture of plancton animals and of very small insects. Among the latter there may occur young mayfly nymphs of such genera as Heptagenia and Caenis (Needham 1905, pp. 35 and 36).

A careful study of the food of nymphs of half a dozen local genera was made by Doctor Morgan (1913, pp. 386–7). She found that the principal food of Callibaetis is filamentous algae. That of Blasturus and Hexagenia is diatoms. Stenonema adds to the diatoms a large proportion of seed-plant fragments. Siphlonurus and Isonychia are more general herbivorous feeders. All eat diatoms in varying proportions.

One of us (Traver, 1925, p. 214) determined the preference of Blasturus for diatoms, and in an examination of the stomach contents of 57 nymphs of *B. cupidus* found diatoms of different genera in the following numbers of stomachs: Navicula in 51 stomachs; Cymbella in 38; Meridion in 28; Diatoma in 28; Synedra in 27; Gomphonema in 10; Cocconeis in 7; Stauroneis in 6. These nymphs were collected during the winter and spring from the pools in a small creek. She says:

"Brushes of hairs upon nearly all of the mouthparts are probably useful as diatom rakes, particularly those of the maxilla, many of which are barbed and bent slightly toward the front. In the case of one medium-sized nymph which was observed feeding upon a spray of Cladophora, the action of the maxillary and labial palpi could be plainly observed sweeping the food toward the mouth. The action here seemed to be mainly one of scraping diatoms from bur-reed stems or from sprays of Cladophora. Some sprays of the latter, examined under the microscope, were seen to be well coated with the diatom, Cocconeis. These were given the nymphs, which fell to stripping them eagerly. The same sprays were later re-examined and it was found that the nymphs had done the stripping work with considerable thoroughness. Bur-reed stems seemed on the whole acceptable, and were also easy of access, so these were used to furnish the bulk of the food for the nymphs in captivity. The smaller nymphs confined themselves mainly to stripping these stems of their thick coating of diatoms, while the larger ones fed also upon the actual tissues of the bur-reed, skeletonizing them in a short time."

Concerning mayfly forage in general Doctor Morgan says:

"The kinds of algae and diatoms found in the stomach varied a good deal with the locality and date of collection. In certain parts of Cold Brook during March, 1911, every available object was brown with Meridion and the stomachs of nymphs collected there contained little else. Nymphs taken in the same place a month later contained no fragment of Meridion. May-fly food is most abundant in April and May, especially for the running water forms. Later the thick mats of Meridion, Cladophora and Spirogyra begin to decay, there is a diminished supply of water and consequently greater competition for food."

The total length of life profoundly affects the reproductive capacity of a species, since the number of young that may be produced in a given period is more largely determined by the recurrence of reproductive periods than by the number of eggs laid. A mayfly like Callibaetis, for example, laying a thousand eggs, with several possible generations in a season, might have (if all survived) millions of descendants, while a single nymph of Hexagenia is merely getting its growth. Length of life also determines the ability of a species to take advantage of transient opportunities for livelihood. A short-lived species may raise a generation in a pond that evaporates in late summer, and then may fly away to oviposit in other more permanent waters; but a species requiring a year for growth is doomed to extinction in such a pond. Such forms as Callibaetis and Caenis thus quickly establish themselves in freshly dug ponds, as quickly growing annual weeds invade a new-made garden.

2. Educational Uses

Mayflies, by reason of their abundance and the ease with which they may be collected, kept, fed, and reared, are ever available aids to instruction in the biological laboratory. The forms that inhabit still waters make excellent aquarium specimens. They will eat almost any green algae* or the disintegrating remains of the higher plants, and so are easily provided with food. They must be protected from carnivorous enemies, for to all these they are dainty and delectable morsels. Swammerdam found the burrowing nymphs very interesting, and so has every one who has observed them. The climbing nymphs, if placed among submerged plants, will add life and graceful motion to any aquarium. The nymphs of Siphlonurus, for example, are of remarkable agility. The shuttle-like motion of their gills when the body is at rest is fascinating. Some of their antics in the water are the embodiment of graceful motion; their coloration also is beautiful.

Mayflies illustrate an unique type of metamorphosis, for the molting

*Preferably non-filamentous algae, for in masses of long filaments they become entangled and die. The diatoms that fringe the coarser stems of the higher plants are constant items in the food of most species.

of the fully-winged insect does not occur in other groups. The teacher will find elsewhere no such illustration of the brevity of adult life and of the independent specialization of immature stages.

Satisfactory observation of actual metamorphosis can be made by large classes, using mayflies for illustration. From the time when the ice goes out in spring until the heavy frosts of autumn, grown nymphs may be found in many a pond. And their actual transformation may be witnessed in any laboratory with a modicum of trouble. Blasturus nymphs are ready when the season opens; Callibaetis nymphs are still transforming when it closes, and others fill the intervening time. If grown nymphs be placed before a closed window in flower-pot saucers filled with water they will transform, fly to the window, and remain sitting there. If a considerable number of fully grown nymphs* be collected, some of them will be found transforming at various times.

The transparency of the nymphs, especially of young nymphs, renders them especially available for the study of internal organs under low magnification. The tracheae and the blood vessels—organs that in insects are studied with great difficulty by dissection—may, in the more transparent nymphs, be observed with the greatest facility and satisfaction. The tracheae in the living insect, filled with air, are distinctly visible to their finest ramifications. And as for the blood vessels, let the bewildered student who has tried to get some idea of the insect's heart and circulation by dissecting a grasshopper try looking for these things with the microscope in the transparent body of a small, living, mayfly nymph, and he will find relief. He will see the heart in action with its valves working, and the blood currents sweeping around the entire circuit of the body. The blood is transparent, of course, but its flow is traceable by the minute white corpuscles (leucocytes) swept along in its currents. It is now a century since Mr. James Bowerbank (1833) called attention to the availability of mayfly nymphs for this purpose, showed how the corpuscles could be followed about the body in complete circulation, and explained and diagrammed the action of the valves of the heart.

All that is necessary in order to see these things is to select a small nymph, the smaller the better, with empty food canal, preferably one that has recently molted and that is, therefore, very transparent, to mount it on a slide in water, supporting the coverglass a little to prevent crushing, and to examine it at once with the microscope.

*Signs of approaching transformation unmistakably appear in the wing pads of the nymphs. These become dark colored shortly before transformation, owing to the pigmentation of the developing wings inside them.

CHAPTER XVII

COLLECTING, PRESERVING, AND REARING MAYFLIES

Any one who has read the preceding pages will know where to look for specimens. The chief equipment of a good collector is accurate knowledge of habits. Little apparatus is necessary for getting mayflies in either winged or nymphal stages.

1. The Winged Stages

Fingers alone will do a part of one's collecting, since mayflies will allow themselves to be picked from leaves, fences, or window-panes. A killing bottle* is necessary, as with other insects, to receive the captures and quickly quiet them. An ordinary insect net is most useful and generally necessary.

Many forms are hard to find even when common and within reach; "sweeping" vegetation by the waterside will reveal many specimens which would otherwise be overlooked. The mayflies are very fragile. They should be taken with gentle strokes of the collecting net, and introduced into the cyanide bottle with as little handling as possible.

A net is most necessary when swarms are seen, for often with a few strokes of it, dozens of specimens may be secured. These will be adult males, with rarely a female or a subimago. At times, however, hand-picking with only a cyanide bottle as equipment is very satisfactory, yielding many fine specimens of such species as spend their resting hours where they can be found. It yields both winged stages and both sexes. Good specimens may often be found held fast in spiders' webs, yet quite uninjured. These may easily be released and introduced into the cyanide bottle.

Many mayflies are strongly attracted to light at night, and collecting beside street lamps is often excellent on warm summer evenings. Those obtained are mainly subimagos, just come up from the water.

A good light lure is a great aid to the collector. Two lures that the authors have found most useful consist alike of a strong light and a white sheet. In the one, the sheet is supported on a collapsible wire frame over a gasoline lantern. This may be carried and set down anywhere. In the other the white sheet is spread, apron-like, over the front of an automobile. There are two holes in the sheet for the lamps. The bottom of the sheet is outspread to catch such mayflies as fall. When darkness has fallen and the lights are turned on, then the collecting begins by hand picking.

*A wide-mouthed bottle, most easily prepared with a spoonful of dry cyanide of potassium in the bottom, held down under a layer of dry sawdust by means of some close fitting discs of thick blotting paper, glued to the glass at their edges.

It is easy by such means to collect all the more common and important species of a neighborhood; but there are sure to be a few species of more elusive habits that will only be obtained by finding the nymphs and rearing them—inconspicuous species that, on leaving the water, fly to resting places in the tree tops. Some species which have not been seen to swarm or oviposit are supposed to perform these functions at night. A few of the most ephemeral, Caenis, for example, whose winged existence is compassed in a few hours, must be collected in the evening of their emergence if they are to be obtained alive. Dead specimens of such may often be found in spiders' webs the following morning.

Specimens should not be allowed to remain more than a few hours in the killing bottle. They may be pinned for the cabinet, or put in alcohol, or placed for drying between sheets of glazed cotton cut to fit small temporary storage boxes, or enclosed in envelopes of thin, soft paper. When thoroughly dry they may be transferred to cellophane envelopes, labeled, and filed like cards in a pestproof filing cabinet, with great economy of space and convenience of access for examination. The colors are somewhat better preserved in dry specimens than in alcoholics.

2. The Immature Stages

Nymphs may also be collected by hand. Many of them cling to submerged stones or to trash or water plants when they are lifted from the water. Fingers are rather clumsy, however, even the daintiest of them, and may be supplanted with advantage by a forceps, or, for lifting the smaller, more closely clinging types, by a knife blade slipped beneath them.

For stream collecting a hand screen is of great assistance. It is held against the bottom while the stones on the upstream side are stirred to dislodge the nymphs. These are washed by the current upon the screen, from which, when it is lifted, they may be picked by hand.

A water net will be especially useful for obtaining pond forms, and sweeping it through beds of submerged plants will usually obtain many specimens. There are certain little, close-clinging nymphs, however, such as Caenis or small Ephemerellas that are scarcely to be dislodged with a net, but are best picked off by a forceps or induced to leave their resting places by submerged light traps. For obtaining the agile nymphs of the many species of Baetinae from beds of aquatics, an apron net, designed by the senior author, has been found to be very efficient. If the bottom be of proper mesh, it retains nymphs of all sizes. Where nymphs are at all common, it gathers them in great numbers and very quickly.

The alert collector will obtain hints as to when and where to find grown nymphs from the empty nymphal skins he finds floating on the surface of the ponds or clinging to stones in the stream or gathered in the drift line of the beach. In the water nearby, grown nymphs of the same species are likely to be found. However, it is the freshly emerged subimagos that usually first attract notice. When dangling in a spider's web beside one's path, they furnish excellent reminders that the collecting season for their species is at hand.

3. Collecting Eggs

The easiest way to get living mayfly eggs for study is to visit streamside lights on warm nights at the proper season. Among the many subimagos attracted to the lights there will generally be found a few females carrying under the tip of the abdomen a clump of extruded eggs. The clump will often fall off and be found in the bottom of a collecting net. If not, the female bearing them may be gently taken from the net by the wing tips and dipped to the surface of a beaker of water, whereupon the eggs will be released. For such mayflies as Isonychia and Stenonema we have found at Ithaca, New York, that the best hour for egg-gathering is between ten and eleven o'clock at night.

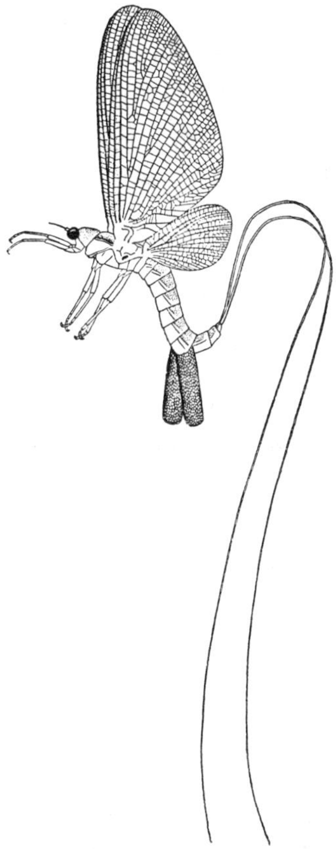

Fig. 76—Female imago of *Ephoron album*, carrying egg packets (drawing by C. H. Kennedy).

In figure 76 we show how the female Ephoron carries her eggs in flight. Large, cylindrical packets like these are characteristic of the burrowing Ephemerinae. Smaller oblong clumps held in the hollow of the overarching eighth and ninth abdominal segments are characteristic of the Heptageninae. The beautiful greenish eggs of Isonychia are carried in a big, round ball. Little rounded balls of some two hundred eggs each are carried by females of Caenis. These are sometimes deposited thickly on the windshield of an automobile when the car passes through a swarm at nightfall. This may be much to the motorist's discomfiture, for they adhere strongly to the glass on drying.

The eggs of Hexagenia may be obtained from a subimago by placing her on the surface of a beaker of water and snipping off her head. She will pour them all out immediately from both oviducts—in the case of *H. occulta* some 5000 of them (Needham. 1908, p. 260). These eggs may then be fertilized by adding sperms dissected from a male. Either subimago or imago will do for this, for the sex products are already mature. The eggs of many species are shed so quickly on contact with the water that, if one attempts to get them by taking floating females from the surface, he will generally find he has taken only spent individuals.

4. Life History Material

The easiest way to get the life history material that will connect up nymph and imago is to collect specimens of all the later stages together: nymphs from the water, cast skins from its edges, and subimagos from the shore. It will not do to assume that stages found together are of the same species. This is to be determined only by rearing isolated specimens. Careful life-history work demands that individual specimens and the skins from which they come be kept separately for each individual and all stages kept together. The adults may be had in a day or two by merely keeping some of the subimagos alive until the final molting. Thus the principal developmental stages will all be speedily obtained.

Adults are often necessary for specific determination. Subimagos are not of much value as specimens; hence they should be kept through their final molting. This is not difficult if they be kept in a moist atmosphere. In a dry place they will die of evaporation. Grocers' small paper bags are good containers for them, and enough moisture may be provided by placing in each bag a leaf of some green plant. If subimagos, however obtained, be handled carefully by the wing tips and introduced uninjured into the paper bags, they will generally complete their final transformation.

5. Rearing Mayflies

The secret of success in rearing mayflies, as in other things, consists in imitating natural conditions. Pond species are the easiest to manage in ordinary aquaria. Running water forms require better aeration of the water in which they live. The most successful cages are those which simply enclose the nymphs in their natural environment, keep out enemies, afford a place of transformation, and retain the winged forms. We recommend a pillow cage*, as being efficient and inexpensive. It should be set aslant in still or running water with only the lower end immersed. The gently sloping side affords a place for emergence and also a place where any subimago that may have fallen into the water can readily crawl out.

Such cages when set in public waters usually need to be hidden from the eyes of the untutored and irreverent, and are always much safer at home. A trough of running water in one's dooryard, filled with such cages, makes an excellent equipment for raising mayflies.

*Made from a square of wire cloth by first twice folding together two opposite edges (in a tinker's "hem") to form a cylinder, and then cross folding its ends. It opens easily and closes securely with the fingers. Woven edges should be placed at the top to avoid sharp wire-ends and consequent pricked fingers. A square yard of wire cloth will make four cages of suitable size for rearing mayflies.

6. Preservation

The simplest, most efficient, and most permanent preservation of mayflies in all stages is had by putting them in alcohol of 70 to 80 per cent strength. No other fluids used by us have proved so satisfactory. Nothing further is necessary than to put the specimens directly into the alcohol and to keep the container full. If shaken about in partly filled bottles, they will be broken. If the alcohol be allowed to evaporate, they will be spoiled.

For study, and for ready comparison of the structures that are used in the determination of specimens, the alcoholic collection should be supplemented by a series of slides with the part concerned permanently mounted in Canada balsam.[1] Legs, tails, genitalia, and wings are thus made most easy of access for comparison. The wings should be mounted dry,[2] for if immersed in balsam some of the weaker veins will become invisible.

For exhibition purposes mayflies may be pinned and spread like other insects, and even a few of the more strongly chitinized nymphs may be set on pins for the insect cabinet; but dry, pinned specimens prepared in the ordinary way are very hard to keep whole. Legs and tails are so delicate and fragile that they break off with a breath of air, and even the bodies of the slender ones fall away without apparent cause. In some of our museum collections the most important type specimens are represented now by bare pins. As stated above, the use of cellophane envelopes as containers gives better preservation of dry specimens, with retention of any parts that may chance to become detached.

If precise arrangement of wings, legs, and tails be desired, a smooth board with small holes in it a quarter of an inch deep makes an excellent spreading board for mayflies. The insect, pinned and placed in an inverted position on the board with the head of the pin in the bottom of the hole, is automatically adjusted to the proper height on the pin. Its wings are then spread and pinned down under strips of paper in the usual way and left until dry. Its long legs and tails are much more easily arranged and kept in position than with the more complicated spreading boards commonly used by entomologists.

In shipping specimens of mayflies by mail great care must be used in packing them. The containers of dry specimens must be *loosely* packed inside a stout box. The packing must ease the bumps. Vials containing

---

[1] Our usual method: dissect off the part to be mounted and clear the chitin by soaking it in caustic potash (KOH) for a few hours if small, over-night if large species. Then wash, and transfer to 35%, then 70%, then to 95% alcohol for a few minutes each. Immerse in carbol-turpentine (turpentine, 3 parts; carbolic acid crystals, 1 part) until clear. Transfer to a small drop of balsam on a slide and arrange in the desired position. Adjust a clean, dry, cover glass on it, and then label the mount accurately and fully.

[2] Dry mounts may be held in place by binding each end of the cover glass down to the slide with a strip of gummed paper. Wet wings may be floated into place on a slide in alcohol, drained a little, covered and bound, and left to dry.

alcoholics must be filled full to the corks, else their fragile contents will be dashed about and broken.

7. General Suggestions

The practical student of mayflies will accumulate much material in vials of alcohol, and it may be of help to some one if we suggest that in our experience we have found several very minor matters to be of some importance. It is best to use one size of vials as far as possible, and that *not too small.* Specimens are easily broken in pulling them out of a narrow aperture, and small corks do not last as long as larger ones of the same quality. Four dram homeopathic vials are our choice. The "life" of the corks is longer when not in contact with the alcohol; so, the vials should stand erect in trays or hang by wire hooks on screen racks.

Uniform labeling will save time. The labels should be of stiff white paper and should conform to the size of the vial. Small pin labels in vials are a nuisance, and big labels folded for insertion are an abomination. The labels should be wide enough for two or three lines of fine writing and long enough to reach from the bottom of the vial up to the shoulder. On a long strip of suitable paper, cut to the proper width, name, place, and date may be written crosswise; then a single cut with scissors makes a label. It should be inserted facing outward between the specimens and the glass. It should read from the bottom up, to avoid lost motion in hunting for something. Verily, it should be legible; hence ordinary inks, which fade, and hard pencils, which mark faintly, should be avoided. Soft black lead pencils are good; waterproof India ink (dried on the label before insertion in the alcohol) is best of all.

CHAPTER XVIII

THE CLASSIFICATION OF MAYFLIES

Plectoptera, Ephemerida, or Ephemeroptera

Linnaeus put the six species of mayflies that were known to him in the original genus Ephemera, and included that genus in his heterogeneous order Neuroptera. This ordinal name was later restricted to net-veined insects that have complete metamorphosis, and the mayflies then reposed for a time, along with dragonflies, stoneflies, termites, psocids, etc., in an almost equally heterogeneous order Pseudoneuroptera. In the course of subdivision of the latter group, the mayflies have been shuffled about considerably. For example, Latreille grouped them along with dragonflies in an order Subulicornia; and Banks catalogued them along with stoneflies in an order Archiptera.

Many names have been proposed for this order of insects. Three are now in common use by different authors. Brauer's name Plectoptera has priority. It also conforms in its ending (*ptera*, from *pteron*, wing) with the Linnaean names for other insect orders. But it is too near in form and quite identical in significance (*plecos*, folded) with Plecoptera, the generally accepted ordinal name for stoneflies. For these reasons Comstock rejected it and proposed in its stead the name Ephemerida (based on the original Linnaean genus Ephemera). This name has been used by us hitherto. It is open to the objections of nonconformity of ending and of accent. The shift of the accented syllable introduces a difficulty of pronunciation and also tends to obscure its significance. The name Ephemeroptera, now rather widely used, does not have these objections but adds a syllable in length. Since it may be freely rendered as signifying *ephemeral winged insects,* it seems on the whole to be the preferable name for the order. Martynov's more recently proposed name Agnatha (meaning lacking jaws) is in its significance descriptive but not distinctive; it does not conform in ending, and it is not euphonious.

There have been three listings of the world fauna in this order: those of Pictet (1843), Eaton (1883–1888), and Ulmer (1920, with a supplement on newly described genera in 1932). Each of these authors has given a comprehensive review of antecedent work leading up to the then current classification, and there is no need that we should repeat this taxonomic history here.

The basis of our classification of the order was laid by Eaton, and not much progress has been made since the publication of his *Revisional Monograph.* Many new family names have been applied to his subdivisions ("series" and "sections") of the one family (Ephemeridae) that he recognized; but merely adding names is surely no great contribution to science. In a series so consistent and so clearly monophyletic as are the mayflies of the present time, we can see no justification for the division of the group into suborders, especially along the vague lines where such have recently been proposed. We do, however, recognize three families, separated along the lines proposed for subfamilies by the senior author in 1905.

Recent mayflies may for convenience be grouped as follows:

Order Ephemeroptera

I. Family Ephemeridae, with subfamilies as follows:
   1. *Palingeniinae*—Palaearctic and Oriental
   2. *Ephoroninae*—Holarctic and Neotropical
   3. *Ephemerinae*—Cosmopolitan
   4. *Potamanthinae*—Holarctic and Oriental
   5. *Campsurinae*—Nearctic and Neotropical
   6. *Neoephemerinae*—Holarctic

II. Family Heptageniidae
   7. *Heptageninae*—Cosmopolitan

III. Family Baetidae
   8. *Oligoneurinae*—Cosmopolitan except Nearctic
   9. *Ametropinae*—Holarctic
   10. *Metretopinae*—Holarctic
   11. *Siphlonurinae*—Holarctic
   12. *Baetiscinae*—Nearctic
   13. *Ephemerellinae*—Holarctic
   14. *Leptophlebiinae*—Cosmopolitan
   15. *Caeninae*—Cosmopolitan
   16. *Baetinae*—Cosmopolitan
   17. *Prosopistomatinae*—Palaearctic and Oriental

Of these seventeen subfamilies it will be noted that three (Nos. 1, 8 and 17) are not as yet known from North America, and only one (No. 12) is peculiar to our fauna. That one, with a single genus, Baetisca, is so different from the others in many ways, and in both external and internal parts, that it might perhaps be given family rank.

The remainder of this book will be devoted to an account of North American species occurring north of Mexico. As narrated in chronologic order in our introductory chapter, Hugh Williamson (1802) described the first one, Thomas Say (1823–39) a dozen more, Hermann A. Hagen (1861) forty-five in all, and Benjamin D. Walsh (1863–4) eighteen more. Meanwhile a few species were being described in Europe by Pictet, Burmeister and Walker. After Hagen's *Synopsis* there was no general listing of North American species until the publication of Banks' *Catalogue of Neuropteroid insects.* His first catalogue in 1892 contained 86 species of mayflies; his second in 1907 contained 102 species. In the following pages are described 507 species. Among them 48 species are described as new to science.* These all may be grouped by genera as follows:

EPHEMERIDAE	
Ephoroninae	
Ephoron	2 species
Ephemerinae	
Ephemera	6 "
Pentagenia	2 "
Hexagenia	18 "
Potamanthinae	
Potamanthus	8 "
Campsurinae	
Campsurus	4 "
Neoephemerinae	
Neoephemera	1 "
Oreianthus	1 "
HEPTAGENIIDAE	
Heptageninae	
Stenonema	35 "
Heptagenia	34 "
Cinygma	3 "
Rhithrogena	19 "
Cinygmula	11 "
Iron	20 "
Ironodes	6 "
Ironopsis	2 "
Anepeorus	2 "
Arthroplea	1 "
BAETIDAE	
Ametropinae	
Ametropus	2 "
Metretopinae	
Metretopus	1 "
Siphloplecton	1 "
Pseudiron	2 "
Siphlonurinae	
Ameletus	19 species
Siphlonurus	19 "
Isonychia	26 "
Siphlonisca	1 "
Parameletus	2 "
Leptophlebiinae	
Leptophlebia	2 "
Blasturus	9 "
Paraleptophlebia	29 "
Choroterpes	7 "
Habrophlebiodes	3 "
Habrophlebia	2 "
Thraulodes	1 "
Thraulus	2 "
Baetiscinae	
Baetisca	6 "
Ephemerellinae	
Ephemerella	67 "
Caeninae	
Caenis	12 "
Brachycercus	5 "
Tricorythodes	8 "
Baetinae	
Callibaetis	19 "
Baetis	41 "
Centroptilum	19 "
Heterocloeon	1 "
Pseudocloeon	12 "
Neocloeon	1 "
Cloeon	10 "
Total	507

A number of species are as yet known only in their immature stages, and a few new species are hereinafter described from nymphs alone, but only in those genera in which the nymphs are highly differentiated (Baetisca and Ephemerella). In such genera differential characters are often better developed in the nymphs than in the adult. In general, we believe that species are best described in the stage that is most highly differentiated, and therefore, most easily recognized.

*The descriptions of these new species are all written by Dr. Jay R. Traver, and in all future references to them they are to be accredited only to her. J.G.N.

Keys and Tables.—In Part II of this book we have tried to make access to a knowledge of the North American mayfly fauna as easy as possible, first by means of keys, and then by verification tables supplementing the keys. We have cast the keys in the form that our experience has found most convenient, and have endeavored to make them concise, using by preference single characters. In the tables which follow the keys half a dozen characters are conveniently placed in columns for ready comparison. These should resolve most of the user's difficulties and uncertainties, and should save much time in finding the desired description.

Size and distribution are given in every key to species. The latter is broadly stated by abbreviations, as follows: **E**, eastern; **W**, western; **N**, northern; **S**, southern, **C**, central; **G**, general; and these in combination, as **NE**, northeastern, etc. When the known distribution is more limited it is indicated by the usual abbreviations for states and provinces. Much additional abbreviating has been necessary to economize space in the tables. Arabic numerals in the column headings designate abdominal segments unless otherwise stated, and Roman numerals, thoracic segments. The fore wing is always meant unless the hind one is specifically mentioned. **Var.** under any heading means variable. Measurements are all in millimeters. Under each genus the species are arranged in alphabetic order.

PART II
SYSTEMATIC

NORTH AMERICAN MAYFLIES

ORDER EPHEMEROPTERA

These are winged insects of moderate size and incomplete metamorphosis whose immature stages are aquatic, whose adult life is very brief, and whose characteristics are more fully set forth in the preceding pages of this volume. They are found in all fresh waters in every part of North America, and are among the most important of herbivorous invertebrate aquatic animals.

They may be grouped in three main families, distinguished as follows.

KEY TO THE FAMILIES OF EPHEMEROPTERA

Adults

1—Veins M and Cu 1 strongly divergent at base, with M 2 strongly bent toward Cu basally. Outer fork (Of) in hind wing wanting. Hind tarsi 4-jointed...EPHEMERIDAE

—Veins M and Cu 1 little divergent at base and fork of M more nearly symmetrical. Outer fork (Of) of Rs in hind wing present or absent. Hind tarsus 4- or 5-jointed...2

2—Hind tarsus with five freely movable joints. Cubital intercalaries in two parallel pairs, long and short alternately. Venation never greatly reduced. Eyes of the male simple..HEPTAGENIIDAE

—Hind tarsi with three or four freely movable joints. Cubital intercalaries not as above (except in Metretopinae). Venation sometimes greatly reduced. Eyes of the male often divided..BAETIDAE

Nymphs

1—Mandible with an external tusk projecting forward and visible from above the head (except in Neoephemerinae)...........................EPHEMERIDAE

—Mandible with no such tusk, except in some species of Paraleptophlebia.......2

2—Head strongly depressed; eyes dorsal*.......................HEPTAGENIIDAE

—Head not strongly depressed; eyes lateral..........................BAETIDAE

* Certain Leptophlebiinae, of family Baetidae, which may seem to fall here, have gills of each pair filiform or lamelliform, while in Heptageniidae the upper member of each gill pair is plate-like, the lower member fibrillate.

Family **EPHEMERIDAE**

Mostly large mayflies, the wings with abundant cross venation. Median vein and anterior branch of cubitus of fore wing strongly divergent at the base. Posterior fork of media very deep, its basal portion strongly sagged toward cubitus. Outer fork (Of) in hind wing wanting, R_4 and R_5 being fused. Stigmatic area of fore wing never widened, often becoming somewhat narrowed; no sag in the costal margin. Male fore tarsus five-jointed, the basal joint very short. Middle and hind tarsi four-jointed, the additional basal joint being fused with the tibia. Legs rather weak throughout, becoming very weak in certain genera; in Campsurus the middle and hind legs dwindle to mere vestiges.

Among the members of this family there is a very great diversity of form and structure, as will appear from the characters stated in the following keys and tables.

KEY TO THE SUBFAMILIES OF EPHEMERIDAE

Adults

1—Costal cross veins normally developed 2
—Basal costal cross veins vestigial or wanting . . . . **Neoephemerinae**
2—Marginal veinlets present. 3
—Marginal veinlets absent. . . . . **Campsurinae**
3—Cubital intercalaries widely forking. 4
—Cubital intercalaries simple, four long. . . . . **Ephoroninae**
4—Cubital intercalaries one or two, long and deeply forked . . . . **Potamanthinae**
—Cubital intercalaries shorter and with shallow forks, more like marginal veinlets, decurrent from vein Cu 1. . . . . **Ephemerinae**

VERIFICATION TABLE

			Fore wing		Hind wing		
	♂ eyes, above	*Wing membrane*	*Cu. intercalaries*	*Marg. veinl.*	*Costal angul.*	*Legs*	*Forceps joints.*
Campsurinae	remote	transluc.	1 fk.	none	low	weak	2
Ephemerinae*	variable	hyaline	2–4 fks.	irreg.	low	normal	4
Ephoroninae*	remote	transluc.	4, long	irreg.	low	weak	4
Neoephemerinae	approx.	hyaline	1–2 fks.	few	acute	normal	4
Potamanthinae	remote	hyaline	2 fks.	irreg.	mod.	normal	3

* Cross veins very numerous.

KEY TO SUBFAMILIES OF FAMILY EPHEMERIDAE

Nymphs

1—Gills present on segments 1–7, rudimentary on segment 1; mandibular tusk present. 2
—Gills present on segments 1–6 only, rudimentary on 1, operculate on 2, so that gills on segments 3–6 are under opercula; mandibular tusk absent. . . . . **Neoephemerinae**
2—Gills dorsal in position, curving up over the abdomen; fore tibiae more or less flattened and widened (burrowing forms). 3
—Gills lateral in position, outspread at sides of abdomen; fore tibia long and slender, not as above (sprawling forms). . . . . **Potamanthinae**
3—Head with a conspicuous frontal process. 4
—Front of head rounded, lacking frontal process. **Campsurinae**
4—Gills narrow, tapering, with no distinct lateral tracheal branches, margined on each side by a fringe of hairs at least twice as long as width of gill; mandibular tusks upcurved apically, in lateral view. . . . . **Ephemerinae**
—Gills much wider, laminate, tapering only near tips, with many prominent lateral tracheal branches; marginal fringe of hairs short, inconspicuous; mandibular tusks curve downward apically in lateral view. . . . . **Ephoroninae**

VERIFICATION TABLE

Subfamilies	*Mdb.*[1] *tusks*	*Frontal process*	*Legs*	*Gills on segments*	*Gills, position*	*Distr.*
Campsurinae	Converg.	Absent	Digging	1–7	Dorsal	S
Ephemerinae	Diverg.	Present	Digging	1–7	Dorsal	G
Ephoroninae	Converg.	Present	Digging	1–7	Dorsal	G (S)
Neoephemerinae	Absent	Absent	Sprawling	1–6[2]	Dorsal[3]	E, S
Potamanthinae	Converg.	Absent	Sprawling	1–7	Lateral	E, C, S

1 Tips of mandibular tusks, in dorsal view.
2 Elytroid on segment 2.
3 Gills on 3–6 under opercula.

Subfamily EPHORONINAE

Wings translucent, especially in female. Cross veins very numerous in both wings. Fore wing with a wide cubital fork; usually four long intercalaries in this area. In hind wing, an extra branch in the radial sector, directly behind R_1; costal angulation obtuse. Legs weak, except fore leg of male, which is usually long and normally developed. Claws weak, slender, blunt. Forceps 4-jointed. Tails 2 in male, 3 in female.

Genus Ephoron Williamson

1802—Williamson—Trans. Amer. Soc. Phil. 5: 71.
1868—Eaton—Ent. Mo. Mag. 5: 84 (Polymitarcys).
1883—Eaton—Revis. Monogr. p. 43 (Polymitarcys).
1920—Ulmer—Stett. Ent. Zeit. 81: 107 (Polymitarcys).
1920—Needham—Bull. Bur. Fish. 36: 285 (Polymitarcys, *nec* nymph).
1926—McDunnough—Canad. Ent. 58: 184 (Polymitarcys =Ephoron).
1933—Spieth—J. N. Y. Ent. Soc. 41: 347 (Polymitarcys).

Pale whitish mayflies of moderate size; wing 11 to 12 mm., in our species. Eyes of male rather small, remote, separated by distance somewhat greater than diameter of eye. Ocelli large, prominent; median one slightly smaller than laterals. Posterior margin of head in female almost straight. Pronotum slightly wider than long, somewhat narrower anteriorly; posterior portion of pleura behind coxa slightly expanded. Fore femur of male short; tibia long, 4 to 4½ times length of femur; tarsus very slightly shorter than tibia. Fore leg about as long as body. Claws very slender, weak, one very slightly shorter than the other. All legs of female, and middle and hind legs of male, weak and considerably reduced in size; coxae and trochanters rather well developed, all other joints weak; a long apical flange on middle and hind trochanters. Femur of hind leg slightly longer than tibia, tarsus shorter than tibia.

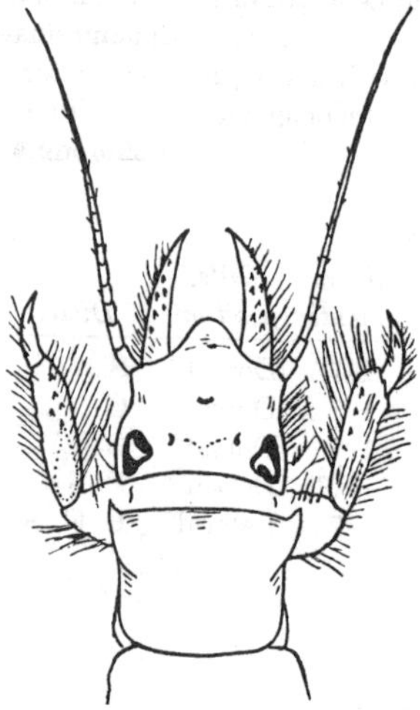

FIG. 77—Head of nymph, *Ephoron sp.*

Wings slightly translucent in male, distinctly so in female; main veins of costal margin purplish, all others pale. Posterior branch of radial sector forked about ¼ the distance from base to margin. Between the two branches of cubitus are three or four long straight intercalaries, joined together by cross veins, and converging slightly toward base: one nearest posterior branch joined to hind margin by a series of shorter veins, often forked, somewhat sinuate. Hind wing about 1½ times as long as its greatest width. An extra branch in radial sector, immediately behind R_1. Median vein forked about ¼ of distance from base. Venation of both wings copious; cross veins very numerous, especially near outer margin of fore wing, where they form more or less of a network with the numerous marginal intercalaries which are joined to each other and to the cross veins basally (see fig. 31).

Tails 2 in male, 3 in female; a very short third rudiment in male. Pleural fold slightly expanded, at posterior margins of segments. Forceps base very narrow, apical margin almost straight. Forceps four-jointed; basal joint short, second one longest; fourth shorter than third, these two together subequal to first. Penes united on median line, apical outer margin of each prolonged into a blunt horn-like process; penes covered by forceps base only at extreme anterior margin. Posterior margin of 9th sternite of female almost straight.

Nymph fossorial. Head with a rounded median frontal prominence (see fig. 77). Mandibular tusks long, down-curved, convergent apically, irregularly toothed on outer margin, bearing short spines on dorsal and lateral surfaces. Long hairs at base on inner margin, a fringe of hairs of graduated length basally on outer margin, longer hairs apically. Labrum almost as long as broad. Maxillary and labial palps each two-jointed. Fore leg short and stout; tibiae and tarsi of middle and hind legs slender, femora moderately stout. Gills dorsal in position; single on segment 1, double on segments 2–7, each member of a pair fringed on each side with short hairs; lateral tracheae numerous, distinct.

Wings and genitalia of the imago and structures of the nymph are figured by Eaton (1883) and Spieth (1933); wings and genitalia by Needham (1920; nymphal figures indicated for this genus belong instead to genus Potamanthus).

Genotype—*E. virgo* (in Ephemera), Olivier.

Distribution—Holarctic; in Nearctic Realm, eastern, northern, western.

KEY TO SPECIES

Male adults

1—Mesonotum dark reddish brown; cross veins of fore wing darkened except in anal area (blackish at costa, purplish grey beyond)........................**leukon**

—Mesonotum yellow shaded with yellowish brown; cross veins of fore wing not darkened beyond anterior branch of media..........................**album**

Ephoron album Say

1823—Say—West. Quart. Rep. 2; 305 (Baetis)..Man.
1859—LeConte—Complete writings of T. Say, 1: 204 (Baetis).
1861—Hagen—Syn. Neurop. Ins.: p. 40 (Palingenia).
1863—Hagen—Proc. Ent. Soc. Philad. 2: 170 (Baetis).
1863—Walsh—Proc. Ent. Soc. Philad. 2: 191 (Cloe A).
1871—Eaton—Trans. Ent. Soc. London: p. 124 (Baetis).
1873—Hagen—Trans. Ent. Soc. London: p. 390 (Baetis), p. 391 (Palingenia).
1883—Eaton—Revis. Monogr. p. 47 (Polymitarcys).
1894—Banks—Ent. News 5: 178 (Polymitarcys) ....Kans.
1905—Howard—(in Needham, *et al.*) N. Y. St. Mus. Bull. 86: 60 (Polymitarcys)................Ill.
1920—Needham—Bull. Bur. Fish. 36: 285 (in part)—Figs. (Polymitarcys).
1926—McDunnough—Canad. Ent. 58: 184...O., Wash.
1932—Neave—Canad. Field-Nat. 46: 54.

Length: body 12; wing 11–12 mm.

Vertex of head yellowish brown with purplish markings; mesonotum light yellowish brown; veins of costal margin of fore wing purplish grey, all others pale.

Face pale yellowish; second basal joint of antenna dusky at apex, filament whitish. Eyes black. Vertex of head yellowish brown, marked and sometimes almost obscured by purplish shading; darkest between ocelli, which are heavily dark-ringed at base. Pronotum yellowish; pleura with purplish grey streaks above leg base. Mesonotum yellow, tinged with yellowish brown on each side of median area; scutellum pale yellowish. Metanotum yellowish brown. Pleura and sternum pale yellowish, the former often tinged with yellowish brown. Fore femur and tibia purplish; femur marked with yellowish white along ventral edge; ventral edge of tibia whitish at extreme base; tarsus smoky grey at base, becoming silvery white apically. Middle and hind legs pale yellowish white. Wings rather semi-hyaline, with faint purplish tinge along the costal border. Longitudinal and cross veins of costal margin purplish grey, cross veins somewhat darker than longitudinals; as far back as the anterior branch of media, the cross veins are faintly greyish; all other veins of both wings colorless.

Abdomen pale yellowish white, basal and middle segments often semi-hyaline; apical tergites yellowish. Genitalia pale (see fig. 78). Tails pale smoky at base, becoming whitish apically; joinings opaque whitish.

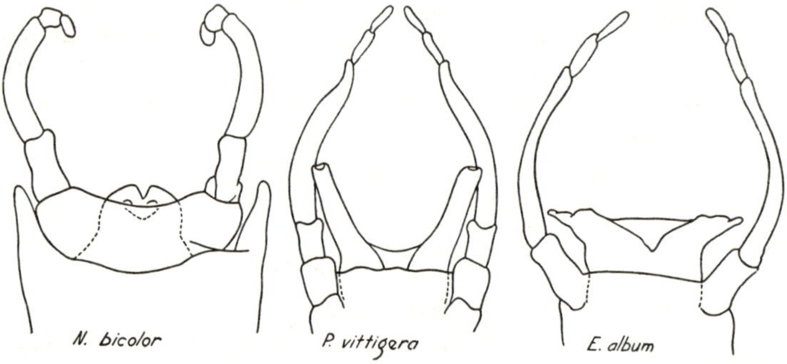

FIG. 78—Male genitalia of Neoephemera, Pentagenia, and Ephoron.

Above description drawn from specimens taken at L. Winnipeg, Manitoba, by Mr. F. Neave. A paler species than *E. leukon*, especially as to the thoracic notum; fore legs, pleura and sternum likewise paler; area of dark across veins of fore wing more restricted; no dusky markings on abdominal tergites; tails somewhat paler. The size of the body is slightly smaller, in specimens measured. Specimens of this species in the Cornell collection are from L. Winnipeg, Man., as mentioned above; also from Fairport, Ia. (H. E. Schradieck); Parks, Neb. (G. Bennett); and the Jordan River, Utah, (J. G. Needham).

Ephoron leukon Williamson

1802—Williamson—Trans. Am. Phil. Soc. 5: 71 .... N. J.
1911—Morgan—Ann. Ent. Soc. Amer. 4: 98 (*album*, in Polymitarcys) .... N. Y.
1920—Needham—Bull. Bur. Fish. 36: 285 (in part) (Polymitarcys).
1926—McDunnough—Canad. Ent. 58: 184 .... Ohio, Ont.

Length: body 13–14; wing 11–12 mm.

Vertex of head yellowish brown, heavily shaded with purplish black between ocelli; mesonotum largely dark reddish brown; veins of costal margin of fore wing purplish, cross veins of median and cubital areas greyish lavender.

Face yellowish; antennae pale whitish, faint grey shading at apex of second joint; eyes black. Vertex yellowish to pale reddish brown, heavily shaded with purplish black between ocelli, often with faint brown shading near posterior margin; bases of ocelli heavily black-ringed. Pronotum yellowish brown, tinged distinctly with purplish grey in median area; reddish brown shading laterally; purplish grey streaks and shading above base of fore leg. Mesonotum dark red-brown in anterior and median portions; posterior portion anterior to and laterad of scutellum pale yellowish; scutellum smoky; sutures and a narrow median line blackish, except in anterior portion, where median line is yellow. Metanotum red-brown, scutellum dark smoky. Pleura yellowish with red-brown areas anterior to middle leg and below wing roots;

sternum yellow, washed with red-brown in median area. Fore femur and tibia deep purple, the femur with whitish areas along ventral edge and near middle of outer surface; tarsus paler, purplish grey. Middle and hind femora yellowish, tibiae and tarsi whitish. Wings as in *E. album* Say, but both longitudinal and cross veins of costal margin are darker, cross veins appearing blackish in some lights; cross veins as far back as the anal region are purplish grey, becoming fainter and paler posteriorly.

Abdomen yellowish white; basal and middle segments often semi-hyaline; segments 8–10 yellowish with tinge of red, the tergites washed with purplish grey except on lateral margins, this shading deepest in median area and anterior margin. Middle tergites with a greyish median streak, sometimes faintly seen also on basal tergites; occasional traces also of grey shading laterally. Genitalia pale; tips of penes and tip of long joint of forceps dusky. Tails rather deep purplish at base, becoming pearly grey in median area, tips whitish; joinings opaque white.

Above description drawn from specimens taken in the Potomac River near Harpers Ferry, Md. by Mr. V. Argo. This species differs from *E. album* in the deep red-brown mesonotum; smoky pronotum and reddish brown markings on pleura and sternum; darker fore legs; area of dark cross veins in fore wing more extensive; tails darker; tergites shaded with grey. The body, in individuals measured, is slightly longer. Specimens of this species in the Cornell collection are from Harpers Ferry and New Windsor, Md., (V. Argo); Amherst, Mass., (C. P. Alexander); Trenton, N. J., (L. Harper); Monteagle, Tenn., (A. G. Richards); and from the mouth of the Lievre River, Que., (F. P. Ide).

The specimen from Quebec is considerably smaller than the Maryland forms, the body not exceeding 7 mm.; thorax more uniformly dark red-brown; wings as pale or paler than in *E. album*, especially cross veins; abdomen wholly pale except apical ter-, gites; tails pale. There are also pale specimens among the females from Trenton, N. J. but we do not know whether this indicates the presence of two species. Williamson's description calls for an insect ½ inch in length. He further states, "wings, abdomen and legs are perfectly white; eyes black; the trunk of a brownish color." This does not agree in all particulars either with the Maryland specimens or the specimen from Quebec. For the present, we retain the larger specimens as *E. leukon.*

Addendum

The mature nymph of *Ephoron leukon* has been described recently by Dr. Ide, likewise nymphs of the first and second instars (Canad. Ent. 67: 114, 1935). Figures of the egg, and of nymphs of the first and last instar, are presented.

Subfamily **Ephemerinae**

Eyes of male of moderate to large size, more or less approximated apically. Wings hyaline. Marginal intercalaries irregular. Costal cross veins well developed. Cubital intercalaries somewhat resemble marginal veinlets, consisting of two to four forks decurrent from anterior branch of cubitus. Costal angulation of hind wing low. Legs normally developed. Claws on fore leg of male similar, blunt; on middle and hind legs of male and all legs of female, dissimilar on each tarsus. Forceps four-jointed. Nymphs fossorial. Mandibular tusks present, divergent at tips. Head with a distinct frontal process. Gills dorsal in position.

KEY TO GENERA

Adults

1—Cross veins somewhat crowded at and below bulla; wings with a variable pattern of dark spots.......... **Ephemera**

—Cross veins not crowded at bulla; wings with no such distinct pattern of dark spots*.......... 2

2—Penes tube-like, tips not incurved.......... **Pentagenia**

—Penes broader at base than at apex, tips more or less incurved.......... **Hexagenia**

Nymphs

1—Frontal process of head bifid.......... 2

—Frontal process of head entire; truncate, conical or rounded.......... **Hexagenia**

2—Mandibular tusks crenate on outer (upper) margin; labial palp 2-jointed.......... **Pentagenia**

—Mandibular tusks smooth on margins; labial palp 3-jointed.......... **Ephemera**

Genus **Ephemera** Linnaeus

1758—Linnaeus—Syst. Nat. ed. 10: 546.
1815—Leach—Brewst. Edinb. Encycl. 9: 137 (restricted).
1883—Eaton—Revis. Monogr. p. 58.
1920—Ulmer—Stett. Ent. Zeit. 81: 109.
1920—Needham—Bull. Bur. Fish. 36: 283.
1933—Spieth—J. N. Y. Ent. Soc. 41: 347.

Mayflies of moderate size, wing from 10 to 15 mm. in length; wings more or less dark-spotted. Eyes of male rather small, separated apically by a distance at least as great as diameter of one eye. Posterior margin of head of female almost straight between the eyes. Fore leg of male almost as long as body. Tibia 2 to 2½ times as long as femur; tarsus subequal to or slightly longer than tibia. Second joint of fore tarsus longest, equal to fourth and fifth combined; third and fourth progressively shorter; fifth slightly more than ½ of fourth. In hind leg of both sexes, tibia 1⅓ times length of femur; tarsus slightly shorter than femur. First three tarsal joints shorter, more or less subequal; fifth as long as these three combined. In fore tarsus of female, joints two to four are relatively longer and distal joint shorter than above. In fore wing, cross veins at and below bulla tend to be distinctly crowded, in area of dark band which may extend backward as far as vein M. Other clouds, variable in extent, may be present in both wings. In hind wing, M_2 is often detached from M_1 at base, and directed downward toward Cu. Second joint of forceps more or less strongly bowed, with an extension on inner margin near base. Penes irregularly rod-like, separated near base, almost parallel; a long spine on inner margin of each division. A median V-shaped cleft on apical margin of subanal plate of female. Tails three, and of equal length in both sexes.

---

*In Pentagenia, four small dark dots may be present at or below bulla.

Frontal process on head of nymph bifid. Mandibular tusks slender, their margins smooth. Maxillary palp 3-jointed; slender, quite long, second and third joints fringed with long hairs. Galea-lacinia slender, pointed at apex. Labial palp 3-jointed; apical joint widened toward apex, its apical margin truncate; hairs and spines less numerous than in Pentagenia.

Wings and genitalia of the imago and structures of the nymph are figured by Eaton, Needham and Spieth.

Genotype—*E. vulgata* Linn.

Distribution—Holarctic; in Nearctic, generally distributed.

KEY TO SPECIES

Male adults

1—Abdomen largely creamy white, no dark markings; wings heavily blotched, appearing blackish . . . . **guttulata**

—Abdomen pale yellowish to brown, with prominent blackish markings; wings not as above 2

2—Abdomen yellow-brown to pale red-brown; small dark spots on disc of wing 3

—Abdomen whitish to yellowish, in one species tergites largely purplish grey; no dark spots in disc of wing 4

3—Abdominal tergites with a pair of black longitudinal streaks on each side; a single lateral black streak on each sternite . . . . **compar**

—Abdominal tergites each with a large irregular dark blotch on each side; a dark lateral triangle on each sternite . . . . **simulans**

4—Abdominal tergites largely purplish grey; a wide dark band at bulla, two other small clouds . . . . **varia**

—Abdominal tergites paler; dark areas in fore wing less prominent 5

5—Band at bulla not prominent, extends back 3 or 4 spaces only; no cloud on media beyond outer fork of radial sector . . . . **blanda**

—Band at bulla prominent, extends back to media; small cloud on media beyond outer fork of sector . . . . **triplex**

VERIFICATION TABLE

Adults

Species	*Wing Length*	*Spots, f-wing*	*Spots, h-wing*	*Abd. terg.*	*Distr.*
blanda	12–13	Few	None	Yellowish	S E
compar	15	Many	Few	Br-yellow	Colo.
guttulata	13–15	Many	Many	Yellowish	E
simulans	12–14	Many	Many	Pitch-br.	N. C.
triplex	10–11	Three	None	Whitish	W. Va.
varia	12	Few	None	Violet	E

KEY TO THE SPECIES OF EPHEMERA

Nymphs

1—Outer curve of mandibular tusk, in dorsal view, visible between antenna and frontal process . . . . **guttulata**

—Outer curve of mandibular tusk, in dorsal view, visible outside antenna 2

2—Frontal notch quite regularly rounded; mandibular tusk projects beyond frontal process almost twice length of latter . . . . **simulans**

—Bottom of frontal notch squared off; mandibular tusk shorter, projects 1½ times length of frontal process 3

3—Semi-lunate grey lateral patches on abdominal tergites, and black streak near pleural fold . . . . **varia**

—No such lateral grey patches nor black streak near pleural fold . . . . **blanda**

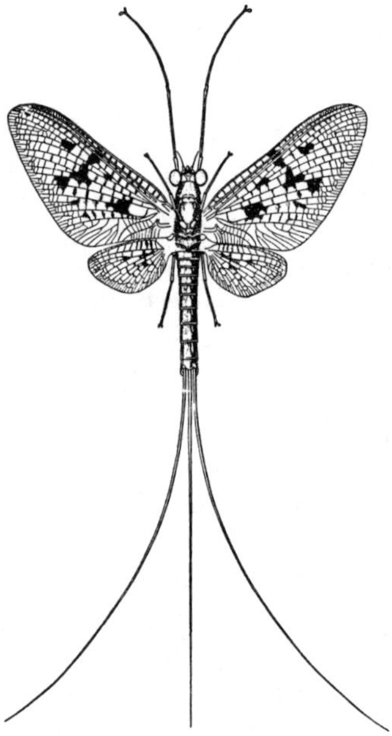

FIG. 79—Male imago of *Ephemera simulans* (drawing by C. H. Kennedy).

FIG. 80—Nymph of *Ephemera varia* (drawing by C. H. Kennedy).

VERIFICATION TABLE

Nymphs

Species	*Body length*	*Tails length*	*Frontal notch*	*Outer curve*[1]	*Tusk projects*[2]	*Gills*
blanda	17–21	6–7	Squared	Between	1½	Yellow
guttulata	20	11	Rounded	Outside	1	Purp. grey
simulans	18	8	Rounded	Outside	2	Purp. grey
varia	18	8	Squared	Outside	1½	Purp. grey

[1] Outer curve of mandible visible from above, *between* antennae and frontal process, or *outside* of antennae.

[2] Tusk projects beyond frontal process as far as length of process, 1½ times or twice length of process.

Ephemera blanda Traver

1932—Traver—J. Elisha Mitchell Sci. Soc. 47: 104. Figs., nymph described.................N. C.

Length: body 13–14; wing 12–12½ mm.

A yellowish brown species, the wings having very small dark spots only, in place of the usual wide dark blotches.

Head yellowish brown; vertex dark red-brown; eyes slate-grey, upper portion paler than lower; ocelli pearl-grey, widely black-ringed at base. Basal joints of antenna cream-colored, black at apex; filament pale grey. Pronotum yellow in median area; posterior and lateral margins creamy; a longitudinal black stripe on each side, beyond which is an outer red-brown area. Mesonotum dark red-brown, its antero-lateral margins narrowly pale; median line may be narrowly pale; other margins darker brown. Metanotum deeper red-brown. Pleura yellowish with brown markings. Sternum light yellowish brown; posterior of prosternum dusky, areas above antero-lateral portion of mesosternum brown. Coxa, trochanter, femur and tibia of fore leg yellowish to yellowish brown, tibia often paler in basal half; apex of femur, and base and apex of tibia, rather widely black-brown. Tarsus greyish, joinings blackish grey. Middle and hind legs creamy white, claws greyish.

Costal margin of fore wing tinged with purplish brown, rather deeper in subcostal space. Cross veins in basal costal and subcostal spaces margined with dark brown, but less prominently than in *E. varia.* About 10 basal costal cross veins, about 20 beyond the bulla; the latter considerably anastomosed in area halfway to apex of wing. Wing clear except for a very small purplish cloud extending downward from bulla for 3 to 6 spaces, very narrow, often not covering two cross veins; another small cloud in fork of posterior branch of radial sector, a third at basal end of long median intercalary. No other veins margined at all prominently, no other clouded areas. Main longitudinal veins of costal margin yellowish brown, other longitudinals brownish, very pale in anal area; principal cross veins brown. Veins on costal margin of hind wing yellowish, all others colorless. Outer margin of hind wing narrowly yellowish.

Abdomen pale yellowish, segments 8–10 opaque, and tinged with reddish yellow. Irregular purplish grey lateral spots on tergites 1 and 2. On tergites 3–9, a pair of longitudinal curved black lines on each side of median area, coalescing on 9; between these and pleural fold, a small brownish blotch; on 7–9, a pair of submedian streaks between the lateral pairs. A black spot on pleural fold at anterior margin of each tergite, and an irregular blackish line above it from posterior margin. Sternites 3–9 each with a single black penciling on each side of median line; 2 small median dots on sternite 1; mid-ventral line largely blackish, on sternite 9, and a small round dark dot on each side of this, anteriorly. Forceps base brown; forceps greyish; second joint of forceps relatively longer than in *E. simulans.* Tails olive brown, joinings dark brown; in some areas, alternate joinings only are darker.

Distinguished from other species by the lack of dark blotches in fore wing; otherwise, rather close to *E. varia.* Besides the type material from North Carolina, there are specimens of this species in the Cornell collection from the Chestate River near Blairsville, Ga. (P. W. Fattig), and from Greenville, S. C. (H. K. Townes, Jr.).

Ephemera compar Hagen

1875—Hagen—Ann. Rep. U. S. Geol. and Geogr. Survey of Colo. Terr. 1873, pt. 3, Zool. p. 578.
1883—Eaton—Revis. Monogr. p. 65.

Length: body ?; wing 15 mm.

A brownish yellow species, marked with black and dark reddish brown.

Head pitch black, antennae somewhat yellowish. Lateral stripes on pronotum pitch black, the stripes widest posteriorly. Mesonotum light rufo-piceous with a pale translucent yellowish brown median stripe. Wing roots pale; a black streak extends forward from them along the margin of the mesonotum. A small black dot on outer margin of each trochanter. Fore femur rufo-piceous, greyish basally on upper edge; tibia and tarsus piceous. Middle and hind femora yellowish brown; tibiae and tarsi lighter, claws brownish. Wings hyaline, tinged with pale dull yellowish. The three main costal veins of fore wing rufo-piceous, other longitudinal veins light yellow; cross veins black. Stigmatic portion of subcostal space, the half of the costal space which adjoins it, and the extreme apical portion of the space below the subcosta, burnt umber brown. Basal portion of subcostal space to a little beyond the humeral cross vein, and the bases of subcosta and radius, same color, as are the margins on the cross veins and the usual dark spots; the latter are small and distinct. Longitudinal veins of hind wing yellowish, subcosta brownish at base; cross veins mostly black, several margined faintly, and a few with small dark brown spots; outer margin brownish grey.

Abdomen yellowish brown. Tergites marked on each side with black streaks; 3–8 with "a pair of longitudinal subparallel black linear stripes, gently curved outwards, and every stripe at its proximal extremity slightly enlarged on the inner side; between these stripes, in segments 6–8, is interposed a single pair of divergent black streaks from the base, longest in the hindermost; in segment 9, the corresponding markings are confluent across the back" (Eaton). Lateral lines on tergites 1 and 2 coalesce to form blotches, small and indistinct on 1; tergite 10 pale, "with a triangular greyish cloud on each side in place of the stripes." On each tergite a short black streak from the anterior margin lies next to the pleural fold; on 5–7, a similar streak arises from the posterior margin. Joinings of tergites somewhat opaque. Sternites 2–8 with a longitudinal black streak on each side of median line; on 9, "a quadrangular black blotch in the midst of the disc, narrowed somewhat anteriorly; in segment 10 a large rounded black spot on each side at base" (Eaton). Tails light brown, joinings subopaque. Forceps yellowish brown.

We have seen no specimens of this species, which is apparently close to *E. simulans*, but slightlv larger.

Ephemera guttulata Pictet

1843—Pictet—Hist. Nat. Neurop. 2, Ephem. p. 135.
1853—Walker—List Neurop. Ins. Brit. Mus. p. 536.
1863—Hagen—Proc. Ent. Soc. Philad. 2: 177 (n. sp?).
1871—Eaton—Trans. Ent. Soc. London, p. 71 (*myops var.?*).
1873—Hagen—Trans. Ent. Soc. London, p. 393 (*myops*).
1875—Hagen—Ann. Rep. U. S. Geol. and Geogr. Survey of Terr. 1873, pt. 3, Zool., p. 579.
1876—Provancher—Nat. Canad. 8: 265 (*simulans*)..Que.
1877—Provancher—Fr. Ent. d. Canad. 2: 81 (*simulans*).
1920—Needham—Bull. Bur. Fish. 36: 283, 285. Fig..N. Y.
1925—Kennedy—Biol. Bull. 48: 393.
1925—McDunnough—Trans. R. S. C. 19: 208..Ont.
1926—Kennedy—Canad. Ent. 58: 61. Nymph figs..Tenn.

Length: body 13: wing 13½–15 mm.

Abdomen largely creamy white, no black pencilings; contrasted strongly with wings, which appear largely blackish.

Head red-brown; frontal portion of median carina blackish; basal joints of antenna whitish, filament pale red-brown; heavy black rings at bases of ocelli. Pronotum light red-brown, with darker lateral areas; posterior margin whitish at median line.

Remainder of thorax very dark red-brown; large pale areas on pleura, anterior to wing roots, at bases of legs, between wing roots and legs. Fore legs dark reddish brown; a white area apically on dorsal edge of coxa; apical half of tibia and all joinings deeper brown; tarsus slightly paler than femur and tibia. Claws marked with whitish. Middle and hind coxae and trochanters red-brown; femora yellowish with faint tinge of red; tibiae and tarsi whitish; apical margin of femur red-brown, claws largely pale purplish brown (Eaton says, blackish spot at apex of tibia, tarsus sometimes light red-brown).

Wing so heavily blotched with dark purplish or blackish brown as to appear largely blackish. Longitudinal veins in anterior half of fore wing dark red-brown, elsewhere yellowish amber. Cross veins blackish or dark red-brown. Majority of cross veins in fore wing widely margined with dark purplish brown, forming many large blotches of same color, arranged much as in *E. simulans*, but more extensive and with additional dark areas. Bullar stripe may extend to cubital vein; very wide at fork of posterior branch of radial sector, covering 5 to 6 veins within this fork. A partial second stripe on some wings, at first apical blotch beyond bulla; in others, this and next blotch beyond it may be partially fused; several other small blotches near margin of wing, from apex to cubitus; elongate blotches between branches of cubitus; other dark areas larger than in *simulans*, cross veins more widely margined. In hind wing, majority of cross veins rather widely margined, those of anal region only narrowly dark-edged. A large dark blotch near base of subcostal space; another in the two spaces just below it, at basal and second fork of radial sector; below the latter, cross veins very widely margined for three spaces; smaller blotches nearer outer margin, which is widely tinged with pale reddish brown.

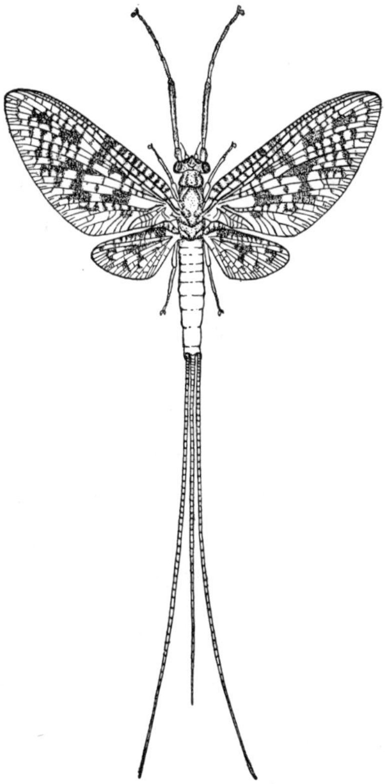

FIG. 81—Male imago of *Ephemera guttulata* (drawing by Evelyn George).

Abdominal segments 2–8 wholly pale yellowish white, except for reddish pleural streak on 8; median apical area of tergite 1 also pale, remainder of 1 red-brown; segment 10 largely bright red-brown, two yellow lateral streaks on tergite; lateral reddish-brown stripes on sternite 9, and pleural margin same color. Forceps reddish-brown, second joint paler; third joint relatively longer than in others of this genus. Penes distinct structurally from other species in our fauna (fig. 82). Tails olive to pale reddish brown, joinings blackish brown.

The very pale abdomen, wholly lacking the usual blackish pencilings, combined with the very dark-appearing wings, distinguishes this from other species. There are specimens of this species in the Cornell collection from McLean and Chicago bogs, near Ithaca; Oswegatchie, and Beaver Kill, N. Y. (P. R. Needham); Sport Is., Sacandaga R., N. Y. (C. P. Alexander); Ausable River, N. Y. (P. S. Jennings); Ile Bizard, Que. (Mrs. E. G. Grieve).

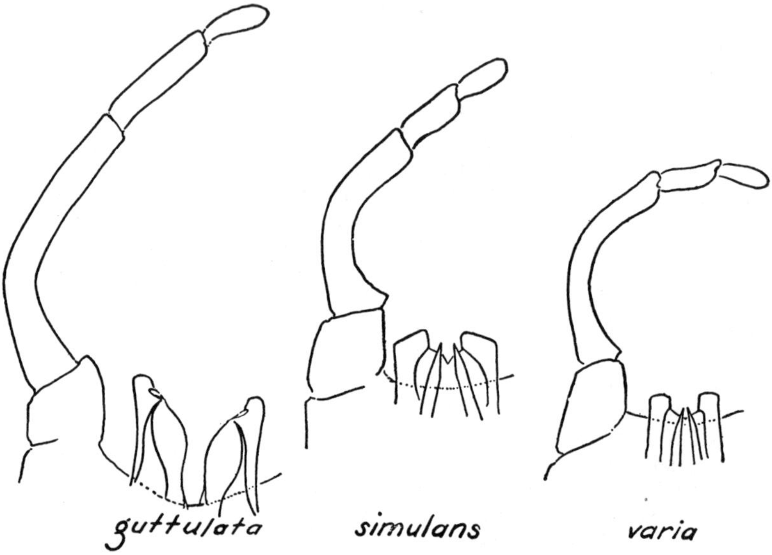

FIG. 82—Male genitalia of three species of Ephemera (drawing by C. H. Kennedy).

Ephemera simulans Walker

1853—Walker—List Neurop. Ins. Brit. Mus. p. 536 (*simulans*).
1862—Walsh—Proc. Acad. Nat. Sci. Philad. p. 376 (*decora*) .... Ill.
1863—Hagen—Syn. Neurop. Ins. N. Amer. p. 38 (*simulans*), 39 (*natata*).
1863—Hagen—Proc. Ent. Soc. Philad. 2: 177 (*natata*).
1863—Walsh—Proc. Ent. Soc. Philad. 2: 177 (*decora*).
1871—Eaton—Trans. Ent. Soc. London: p. 69. Figs. (*guttulata*).
1873—Hagen—Trans. Ent. Soc. London: p. 393 (*natata*).
1875—Hagen—Ann. Rep. U. S. Geol. and Geogr. Survey of Terr. pt. 3, Zool. p. 580 (*natata*).
1883—Eaton—Revis. Monogr. p. 67. Figs.
1908—Needham—Rep. Mich. Geol. Survey; p. 261 .... Mich.
1913—Clemens—Canad. Ent. 45: 332 .... Ont.
1915—Clemens—Marine and Fish. Sess. Pap. 39b, p. 116.
1920—Needham—Bull. Bur. Fish. 36: 283. (Figs. listed as *varia*) .... N.Y.
1930—Ide—Canad. Ent. 62: 206.
1932—Neave—Canad. Field-Ent. 36: 54 .... Man.

Length: body 11–13; wing 12–14 mm.

A dark brown species; both wings with numerous dark blotches and many margined cross veins; tails dark reddish brown.

Head deep blackish brown; antennae pale brown, tips whitish. Thorax deep blackish brown; a large pale patch on prothoracic pleura, mesothoracic pleura marked with smaller pale areas. Median portion of anterior and posterior margins of pronotum narrowly pale, also a narrow pale streak laterad of mesonotal scutellum and pale median area on prosternum. Base of fore leg pale; coxa, trochanter, femur and tibia dark red-brown, joinings blackish; tarsus paler red-brown, each joint darker apically. Middle and hind femora pale reddish brown, tibiae and tarsi yellowish with smoky tinge; all joinings darker.

Fore wing much as in *E. varia*, but less tinged with brownish apically, many more cross veins heavily margined, several additional brown clouds. Longitudinal veins dark brown, cross veins blackish, those in costal, subcostal and radial spaces widely margined except at apex of last two; many in disc likewise margined, also several near outer margin. Stripe at region of bulla deep brown in color, slightly wider than in *varia*; in apical third of wing beyond this are two dark brown clouds, at fork of radial sector and at fork in next space below, nearer margin. Cloud toward margin from inner end of bullar stripe larger than in *varia*, as is also that at base of long intercalary of media; between this and bullar stripe, another small cloud. Veins of hind wing also dark brown, those of basal subcostal space and many in disc dark-margined, the latter forming a number of small dark spots. Outer margin rather widely tinged with light brown (fig. 79).

Abdomen pale brown, with blackish brown markings. On each side of each tergite is a large irregular blotch occupying most of lateral area, and based on anterior margin; lateral and posterior edges of each blotch rather deeply emarginate, so as to leave small pale areas; lateral edges black. Posterior margins of basal and middle tergites narrowly pale brown, also median area, which is narrow on basal and apical tergites, widest on middle ones. On tergites 7–9 a pair of black submedian streaks, very short on 7. Sternites 1–6 with dark lateral triangles, widest posteriorly; 7–9 with dark longitudinal streaks laterally. A pair of small dark spots on each side of median line, on each sternite, also short dark oblique submedian streaks from anterior margin. Middle area of sternite 9 pale. Forceps base and forceps smoky brown; second forceps joint relatively short, as compared with *E. varia*. Penes broader than in *varia*; yellowish brown (fig. 82). Tails deep yellow to reddish brown, joinings darker.

This species is darker than any other in our fauna. Distinguished from *E. varia* as indicated above, also by its much darker body, legs and tails. In the Cornell collection are specimens from Ithaca, N. Y. and vicinity (V. Knox, C. N. Hardy, A. C. Hodson, *et al*); Ann Arbor, Mich. (Ward's Nat. Sci. Estab.); Burt Lake, Douglas Lake, and Walnut Lake, Mich.; Cass Lake and Muskoda, Minn. (L. A. Carruth); Sacandaga, N. Y. (V. S. L. Pate); Lake George, N. Y.; Pt. Elgin, Ont. (Mrs. E. G. Grieve); Union Springs, N. Y. (R. Matheson); Yellow Bay, Mont. (R. T. Young); Maligne L., Alta. (J. C. Bradley); Ile Bizard (Mrs. E. G. Grieve); Chicago, Ill. (P. S. Chapman); Great Falls, Mont. (C. C. Albright); Buffalo, N. Y. (C. Betten); N. Fairhaven, N. Y.; St. Cloud, Minn. (S. W. Friedrich, C. R. Yeager). We have seen others from Millport, O.; Long Lake, Ont.; Moose Factory, Hudson Bay; Westmoreland Co., Pa. (Carnegie Museum); from St. Paul, Minn. (Univ. of Minnesota); and from Coolin, Id. and W. Yellowstone, Wyo. (Calif. Acad. Science).

Ephemera triplex sp. nov.

Length: body 11; wing 10½–11 mm.

A rather small pale species; wings largely pale, with purplish bullar stripe and two other small clouds only.

Head yellowish; frontal shelf hyaline, frontal portion of carina purplish; basal joints of antenna purplish brown at apex. Vertex tinged faintly with reddish brown, and with purplish shading near bases of ocelli, which are dark-ringed. Thorax light yellowish brown; a broad purplish black discontinuous stripe on each side of pronotum, the posterior margin narrowly same color, pale at median line. Mesonotum tinged with reddish brown, on postero-lateral areas; a purplish black V-shaped marking anterior to scutellum. Purplish transverse marks on metanotum. Pleura with reddish brown streaks and triangle anterior to middle leg, often a streak anterior to hind leg. Fore leg light reddish brown; apex of femur, base and apical third of tibia, basal tarsal joint and tarsal joinings, purplish brown; each tarsal joint darker in apical portion. Middle and hind legs yellowish, coxae and trochanters largely reddish brown, as are apex of femur, base and apex of tibia, and tarsal joinings.

Wings hyaline; faint yellowish brown tinge in stigmatic area of fore wing, otherwise pale. Costa, subcosta and radius of fore wing yellowish, all other longitudinal veins silvery white. Cross veins of fore wing, except in anal region, purplish black, rather fine, those of first three costal spaces and a few in disc, narrowly dark-margined. A curved stripe at bulla, extending backward to space above media; slightly wider in third space and at fork of posterior branch of radial sector. A small cloud posterior to this on media, nearer the margin; another at base of median intercalary. Basal subcostal cross veins of hind wing faintly yellowish, all other veins colorless.

Abdomen pale yellowish white; tergites with dark longitudinal pencilings as in *E. blanda*; often with crescent-shaped lateral blotches bordering outer stripe. Lateral portion of posterior margin of each tergite and sternite narrowly purplish. Sternites marked as in *blanda*. Outer margin of forceps base may be marked with purplish. Genitalia quite similar to *E. varia* and *blanda*; 2nd joint of forceps less bowed than in *varia*. Tails yellowish white, joinings purplish brown; in places, alternate joinings wider.

Female largely pale yellowish; markings as in male, but less distinct; thorax and legs paler.

Holotype—Male imago, Marlinton, W. Va.; July 30, 1930 (J. G. Needham). No. 1346.1 in Cornell University collection.

Allotype—Female imago, Pigeon River, W. Va.; July 7, 1930 (J. G. N.). No. 1346.2 in Cornell collection.

Paratypes—5 male imagos, 5 female imagos; Wardensville, W. Va., July 21, and Fairmont, W. Va., Aug. 1, 1931 (P. N. Musgrave). No. 1346.3–12 in Cornell Collection.

Ephemera varia Eaton

1861—Hagen—Syn. Neurop. Ins. N. Amer. p. 38 (*decora*).
1875—Hagen—Ann. Rep. U. S. Geol. and Geogr. Survey of Terr. pt. 3: Zool. p. 578 (*decora*).
1883—Eaton—Revis. Monogr. p. 69 (renamed). Fig.
1901—Needham—Bull. N. Y. St. Mus. 47: 429. Figs., nymph described .... N. Y.
1907—Needham—Bull. N. Y. St. Mus. 124: 193.
1908—Needham—Rep. Mich. Geol. Survey 1908; p. 262 .... Mich.
1911—Morgan—Ann. Ent. Soc. Amer. 4: 100. (*simulans*).
1920—Needham—Bull. Bur. Fish. 36: 283 (*nec.* figs. of *varia*) .... N. H.
1925—McDunnough—Trans. R. S. C. 19: 208 .... Ont.

Length: body 11; wing 12 mm.

A rather pale species; fore wing tinged with brownish except in anal region, hind wing pale; no dark blotch in apical third of wing beyond bullar stripe.

Frontal portion of head reddish, carina black; basal antennal segments black apically, filament brownish, tips pale. Vertex red-brown, blackish between and at bases of ocelli. Thoracic notum light red-brown. A wide blackish longitudinal stripe on each side of pronotum. Mesonotal scutellum pale; dark brown area anterior to this;

middle portion of mesonotum paler yellowish, also a pale median line; antero-lateral margins brown. Metanotum pale, with transverse dark band near center. Pleura yellowish; red-brown oblique streak before middle and hind leg. Sternum pale yellowish, anterior portion of mesosternum marked with brown. Coxa and trochanter of fore leg yellowish; femur rather olive brown, yellowish at base, apex darker; base and apex of tibia widely brown, remainder yellowish, apical half brown-tinged; tarsus yellowish brown, short basal joint, apical half of distal joint and all joinings, dark brown. Middle and hind legs creamy white, claws faintly tinged with brownish.

Wings hyaline; fore wing tinged with brownish except in anal region and near base, where wing membrane and veins are pale or colorless; stigmatic area most deeply tinged. Longitudinal veins yellowish to yellow-brown. Cross veins blackish, those of basal costal and subcostal spaces margined narrowly with dark brown, many in the disc with wider marginings. A wide dark purplish band at bulla, slightly curved, and extending backward as far as space above anterior branch of media, wide enough to cover three cross veins. Posterior to this and nearer the margin, a smaller dark purple cloud; another at basal end of long intercalary of media. No cloud in apical third of wing beyond bullar stripe, though a few cross veins in third space are margined. Hind wing pale; costa and subcosta yellowish, basal cross veins in these spaces pale purplish; all other veins colorless, none dark-margined. Outer margin may be faintly yellow.

Abdominal tergites largely purplish grey; lateral margins, postero-lateral angles and mid-dorsal area pale yellowish white. On tergites 2–9, a pair of purplish black oblique submedian streaks extend backward and laterad from anterior margin on each side, conjoined posteriorly; inner of these not extending beyond middle of tergite, outer a little longer. Between these and along outer margin, whitish spaces. Blackish markings along pleura much as in *E. blanda.* Posterior margins of tergites, and middle portion of sternites, dark purplish. Sternites pale yellowish white. A black longitudinal streak on each side, and a dark mid-ventral streak on sternites 8 and 9, as in *E. blanda.* Forceps base yellowish brown, forceps yellowish; second joint longer than in *E. simulans*; penes rather narrower (fig. 82). Tails yellowish to greyish yellow, joinings dark brown.

The darker abdominal tergites and more extensive blotches in fore wing distingish this species from the allied *E. blanda.* Venter of abdomen paler, wings less extensively dark-margined, body paler than *E. simulans.* Nymph shown in fig. 80.

Specimens of this species in the Cornell collection are from Ithaca, N. Y. and vicinity (V. Knox, A. C. Hodson, C. N. Hardy, *et al*); Bethany, W. Va. (J. G. Needham); Cayuta, N. Y. (O. R. Smith); Danby, N. Y. (R. C. Tasker); Hinckley and Canajoharie, N. Y. (H. K. Townes, Jr.). We have seen others from Cheat Mt., W. Va. (Carnegie Museum); Allegheny State Park, N. Y. (Miss E. W. Eno); Saranac Inn and Johnstown, N. Y. (N. Y. S. Museum).

Genus **Pentagenia** Walsh

1863—Walsh—Proc. Ent. Soc. Philad. 2: 196.
1883—Eaton—Revis. Monogr. p. 75.
1920—Ulmer—Stett. Ent. Zeit. 81: 109.
1920—Needham—Bull. Bur. Fish. 36: 282.
1933—Spieth—J. N. Y. Ent. Soc. 41: 347.

Rather large mayflies, wing 14 to 15 mm. in length; a broad dark band dorsally on abdomen. Eyes of male quite large, approximated apically. Posterior margin of head of female almost straight between the eyes. Fore leg of male very short relatively, only slightly longer than head and thorax combined. Fore tibia 1⅓ to 1½ times as long as femur; tarsus slightly longer than femur. Basal tarsal joint short; second subequal to fifth; third and fourth progressively shorter. Hind leg very similar to fore leg; distal tarsal joint about as long as the three preceding joints together. Proportions of joints of leg of female much as in hind leg of male. Wings very much as in Hexagenia, except that the cells in the costal angulation of the hind wing may be divided by a cross line into two series. Penes simple, rod-like, divergent apically, united only at base. Subanal plate of female with a V-shaped median notch on apical margin; a short projection at each outer corner, directed outward and backward. Tails three in both sexes, middle one very short.

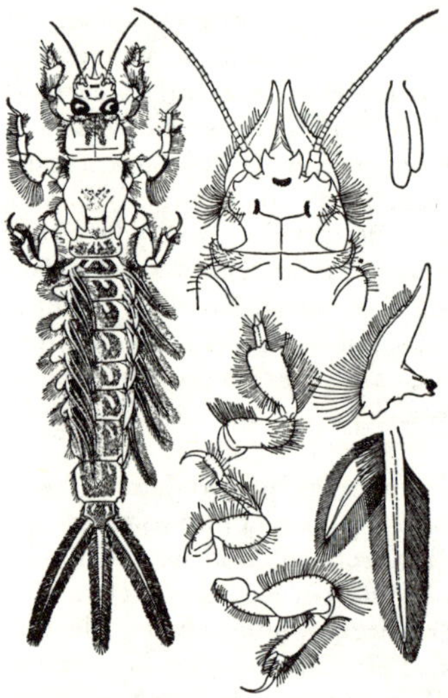

FIG. 83—Nymph of *Pentagenia vittigera* (drawing by C. H. Kennedy).

Frontal process on head of nymph bifid. Mandibular tusks relatively stout, crenate on outer (upper) margin. Maxillary palp very stout, 3-jointed; second and third joints fringed with long hairs. Galea-lacinia wide, somewhat conical. Labial palp 2-jointed; distal joint long, clavate, covered densely with long hairs; several long spines near tip (fig. 83).

Wings and genitalia of the imago are figured by Eaton, Needham and Spieth; the two last mentioned also give nymphal structures.

Genotype—*P. vittigera* (in Palingenia), Walsh.

Distribution—Nearctic; central and southern United States.

KEY TO SPECIES OF PENTAGENIA

Male adults

1—Venation colorless except on costal margin; usually four dark spots in fore wing in region of bulla.......... **vittigera**

—All veins pale brown; no dark spots on fore wing.......... **robusta**

Pentagenia robusta McDunnough

1926—McDunnough—Canad. Ent. 58: 185.......... Ohio

Length: body 17–18; wing 14½–15½ mm.

Eyes slightly larger than in *P. vittigera;* thorax and broad dorsal abdominal stripe deeper, shinier blackish brown; broken brown lateral lines on abdominal sternites.

Eyes larger than in *P. vittigera*, general appearance more robust. Thoracic notum deep shiny blackish brown; only a narrow lateral yellow margin on prothorax. Pleura tinged with brown, sternum largely brown. Fore tibia and tarsus light brown, darker at joints. No traces of black dots on the fore wing; both longitudinal and cross veins distinctly pale brown. The broad dorsal band on the abdominal tergites is deep shiny black-brown. The abdominal sternites are marked with broken brown lateral lines, as in Ephemera; sternites 8 and 9 with considerable brown shading. Tails distinctly brown, joinings narrowly pale; forceps tinged with brown. In other respects, similar to *vittigera.*

This species differs from *P. vittigera* as indicated above; the main differences being the lack of black spots in forewing and the pale brown venation; deeper and shinier color of thorax and abdominal dorsum; dark lateral streaks on sternites; tails darker. The type material was from Cincinnati, Ohio, in the Ohio River. We have seen no specimens of this species.

Pentagenia vittigera Walsh

1862—Walsh—Proc. Adad. Nat. Sci. Philad. p. 373 (Palingenia) .... Ill.
1863—Walsh—Proc. Ent. Soc. Philad. 2: 197 (*vittigera*), 198 (*quadripunctata*).
1863—Hagen—Proc. Ent. Soc. Philad. 2: 174 (Palingenia).
1868—Eaton—Ent. Mo. Mag. 5: 85.
1871—Eaton—Trans. Ent. Soc. London p. 63 (*vittigera*); 64 (*quadripunctata*).
1881—Eaton—Revis. Monogr. p. 76 (*vittigera*); 77 (*quadripunctata*).
1894—Banks—Ent. News 5: 178 (*quadripunctata*) .... Kans.
1920—Needham—Bull. Bur. Fish. 36: 282. Figs. (*quadripunctata* = *vittigera*) .... Ia.
1926—McDunnough—Canad. Ent. 58: 185.

Length: body 17–18; wing 14½–15½ mm.

Mesonotum and dorsal abdominal stripe duller and not quite as deep in color as in *P. robusta;* venation pale except on costal margin; no black lines on sternites.

Face yellowish; ocelli heavily black-ringed at base; antennae pale reddish, tip of filament white; eyes of living insect "yellowish on upper ¾ and dark ferruginous on lower ¼" (Walsh). In alcohol, eyes appear dark bluish grey. Pronotum yellowish, with a median purplish brown band. Mesonotum purplish to blackish brown; several black longitudinal streaks on anterior and median areas; sutures narrowly black; paler areas above wing root and on each side anterior to scutellum. Metanotum purplish brown, posterior margin black. Pleura yellowish; a dark oblique streak anterior to middle and hind leg. Sternum pale yellowish. Legs yellowish. Fore tibia purplish at tip (according to Walsh, at base also); tarsal joinings same color, also apical margins of distal tarsal joints of middle and hind legs. "Wings hyaline; veins and cross veins fine, subequal, hyaline, except on the costa, where they are coarse, the first vein fuscous at base, yellowish at tip; the second and third yellowish throughout; the costal cross veins fuscous at base, becoming yellowish towards the tip of the costa; in the hind wings two costal veins, with their connecting cross veins, are pale fuscous" (Walsh). In specimens we have examined, all veins of hind wing are colorless. In addition, there may be four dark dots on the fore wings, more or less distinct in different individuals, sometimes obsolescent. Of these Walsh says, under *P. quadripunctata;* "four distinct fuscous dots, each surrounded by a slight cloud and conspicuous to the naked eye, on the front wings, extending in a slightly curved line from the middle of the costa to the center of the disc; *viz.* one on the second vein of the costa, and one on the fourth, sixth and ninth veins respectively from the costal margin."

Lateral margins of abdominal tergites yellowish white; sternites pale yellowish white, unmarked, 8 and 9 rather deeper yellow. Dorsally a wide purplish black median band occupies a large portion of each tergite, its lateral margin irregular, widest near each anterior margin. Within this dark band pale oblique submedian streaks extend backward from the anterior margin to about the middle of each tergite; very faint traces of a narrow pale mid-dorsal line may be present. Genitalia and tails wholly pale, unmarked. Genitalia as in fig. 78.

This is a more slender species than *P. robusta*, with paler venation and tails, and no dark ventral markings. There are specimens of this species in the Cornell collection from Fairport and Muscatine, Ia. (H. E. Schraedick); Hannibal, Mo. (V. S. L. Pate); College Station, Austin and Bryan, Texas (V. A. Little, H. J. Reinhard and H. B. Parks); Monteagle, Tenn. (A. G. Richards); and Hope, Ark. Ward's Nat. Sci. Estab.). We have seen others from Ramsey Co., Minn. (Univ of Minnesota), and Manhattan, Kans. (State College, Kans.)

Genus **Hexagenia** Walsh

1863—Walsh—Proc. Ent. Soc. Philad. 2: 197.
1883—Eaton—Revis. Monogr. p. 48.
1920—Ulmer—Stett. Ent. Zeit. 81: 108.
1920—Needham—Bull. Bur. Fish. 36: 278.
1921—Ulmer—Arch. Naturg. 87, Abt. A; 233.
1924—McDunnough—Canad. Ent. 56: 90.
1931—Traver—Ann. Amer. Ent. Soc. 24: 591.
1933—Spieth—J. N. Y. Ent. Soc. 41: 347.

Mayflies of moderate to large size, wings ranging from 10 to 20 mm. in length. Eyes of male moderate to large in size; may be approximated apically, or separated by space less than diameter of one eye. In both sexes, upper portion of eye differs in color from lower portion. Posterior margin of head slightly emarginate; anterior margin of pronotum tends to cover posterior part of head. Fore leg of male usually somewhat shorter than body. In genotype, fore tibia slightly more than twice length of femur; tarsus about 1⅓ times as long as tibia. Basal fore tarsal joint very short; 2nd joint longest, almost ½ as long as tibia; 3rd, 4th and 5th joints progressively shorter. In hind leg of male, femur slightly shorter than tibia; in female, these joints are subequal. In both sexes, three basal tarsal joints very short, fourth joint as long as these three combined. No crowding of cross veins at bulla, as in Ephemera. Cross veins in disc of one or both wings may be dark-margined. Penes variable in shape (see fig. 84), usually separated near base. Subanal plate of female only slightly produced apically. Tails two, in both sexes.

Frontal process on head of nymph entire; its apical margin may be curved, truncate, conical or rounded. Mandibular tusks long, slender, smooth; upcurved, tips divergent. Maxillary palp and galea-lacinia much as in Ephemera. Labial palp 2-jointed.

Wings and genitalia of the imago, and mouthparts and gills of the nymph are figured by Eaton, Needham and Spieth; nymphal structures and genitalia by Traver; genitalia by Ulmer (1921) and McDunnough.

Genotype—*H. bilineata* (in Baetis), Say.

Distribution —Nearctic, Neotropical, African; in Nearctic, generally distributed.

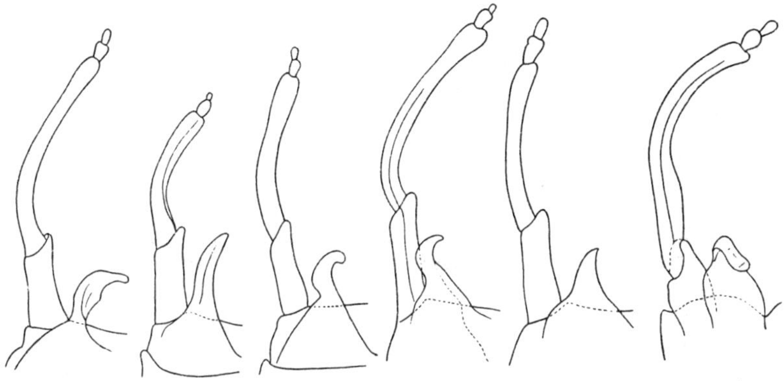

FIG. 84—Male genitalia of six species of Hexagenia.

KEY TO THE SPECIES

Male adults

1—Wings largely suffused with reddish brown; tips of penes recurved (see fig. 84). .... **recurvata**
—Wings in large part hyaline; tips of penes not recurved 2
2—Penes short, broad, bluntly pointed; outer fourth of hind wing deep purple-brown .... **atrocaudata**
—Penes not as above; dark outer margin of hind wing, if present, not as wide as above 3
3—Penes long, erect, almost straight; prominent yellow abdominal markings **rigida**
—Penes not as above; abdominal markings variable 4
4—Penes slender, somewhat beak-shaped; abdomen banded longitudinally with dark and light areas .... **bilineata**
—Penes more or less hook-like; abdominal markings not as above 5
5—Abdomen largely pale yellowish white; no prominent dark markings except on ganglionic areas .... **pallens**
—Abdomen variable in color, but not as above 6
6—Abdomen largely yellow, marked with purplish red; a median ventral band on thorax and abdomen .... **venusta**
—Abdomen variable in color; no such mid-ventral band 7
7—Outer margin of hind wing distinctly darker than adjacent area 8
—Outer margin of hind wing not darker than adjacent area 15
8—Abdominal tergites reddish or blackish brown; ventral triangles attain anterior margin 9
—Abdominal tergites whitish or yellowish, may be pink-tinged; ventral markings variable 10
9—Abdominal tergites reddish brown; pale postero-lateral triangles present. **occulta**
—Abdominal tergites blackish brown; no pale postero-lateral triangles. **viridescens**
10—Abdomen yellow, marked dorsally with brown or blackish 11
—Abdomen whitish or pinkish, dorsal markings reddish or purplish brown 13
11—Both wings tinged with very pale brownish grey; narrow grey median line on tergites 1–6 .... **munda**

—Wings not as above; wide dark median streak on all tergites 12

12—Costal margin of fore wing reddish brown; a few small dark dots in disc of hind wing .... **orlando**

—Costal margin of fore wing almost as transparent as area behind it; no spots in wing .... **limbata**

13—Small species, wing under 13 mm. in length; ventral triangles seldom attain anterior margin .... **elegans**

—Larger species, wing 14–19 mm. in length; ventral triangles variable 14

14—Wing 14 mm. in length; ventral triangles attain anterior margin .... **mingo**

—Wing 17–19 mm. in length; ventral triangles seldom attain anterior margin .... **weewa**

15—Large species, wing 19 mm. or more in length; abdominal tergites yellowish with black-brown markings .... **affiliata**

—Medium-sized species, wing not over 17 mm. in length; abdominal tergites variable 16

16—Pale submedian areas of middle tergites surrounded by dark color 17

—Pale submedian areas of middle tergites not surrounded by dark color 18

17—Wing 14 mm. in length; posterior margins of sternites pale next to pleural fold .... **mingo**

—Wing 17 mm. in length; posterior margins of sternites deep rose-colored next to pleural fold .... **rosacea**

18—Complete ventral triangles on apical sternites, traces on basal and middle ones .... **marilandica**

—No such ventral triangles .... **carolina**

VERIFICATION TABLE

Adults

Species	*Wing Length*	*Dk. marg. hd. wing*	*Abdomen color*	*Abd. Marks*	*Ventral triang.**	*Distr.*
affiliata	19½	No	Dk. br.	Yellow	No	E
atrocaudata	19½	Yes	Blackish	Oliv.	Yes	E
bilineata	17–19	Yes	Purp. bl.	Whit.	Yes	C
carolina	14–17	No	Creamy	Lav.	None	S E
elegans	10–12	Yes	White	Red. br.	No	S E
limbata	13–15 ?	Yes	Yellow	Brown	?	B.C.
marilandica	16	No	Creamy	Dp. rose	No	Md.
mingo	14½	Var.	Creamy	Red.-br.	Yes	W. Va.
munda	14	Yes	Yel.-br.	Black	No	S E
occulta	17	Var.	Red.-br.	Yel. red	Yes	E
orlando	12–14½	Yes	Yellowish	Dk. br.	Yes	Fla.
pallens	16	No	Whitish	Wanting	None	Kan.
recurvata	15–18	Yes	Yel. br.	Yellow	None	E
rigida	16	Yes	Brown	Yellow	None	E
rosacea	17	No	Whitish	Dp. rose	Yes	N.C.
venusta	13–16	Var.	Yellow	Purp. red	None	S W
viridescens	18	Yes	Black-br.	Pale	Yes	E
weewa	17–19	Yes	Whitish	Purp. br.	No	Fla.

* Ventral abdominal triangles attain anterior margins.

KEY TO THE SPECIES OF HEXAGENIA

Nymphs

1—Frontal process angular; first rudimentary gill simple................ **recurvata**
—Frontal process not angular; first rudimentary gill bifid....................... 2
2—Frontal process truncate....................................... **atrocaudata**
—Frontal process not truncate.. 3
3—Frontal process conical.. 4
—Frontal process dome-shaped or rounded.................................. 6
4—Body of female 32 to 36 mm. in length; of male, 23 to 28 mm.......... **carolina**
—Body of female 21 to 26 mm.; of male, 15 to 23 mm......................... 5
5—Mandibular tusk sharply upturned in apical half....................... **rigida**
—Mandibular tusk more gradually upturned apically................... **bilineata**
6—Body of female nymph 25 to 27 mm.; of male, 16 to 21 mm.; gills dark purple-grey .... **mingo**
—Body of female nymph 21–32 mm.; of male, 18 to 29 mm.; gills brownish.. **occulta**

VERIFICATION TABLE

Nymphs

Species	*Length of body, male*	*Length of body, female*	*Frontal process*	*First gill*
atrocaudata	23–28	27–35	Truncate	Bifid
bilineata	15–22	21–26	Conical	Bifid
carolina	23–28	32–36	Dome-shaped	Bifid
mingo	16–21	25–27	Dome-shaped	Bifid
occulta	18–29	21–32	Dome-shaped	Bifid
recurvata	20–25	25–29	Angular	Simple
rigida	18–23	21–26	Conical	Bifid

Hexagenia affiliata McDunnough

1927—McDunnough—Canad. Ent. 59: 118. Fig...Ont.
1931—Traver—Ann. Ent. Soc. Amer. 24: 611–13.

Length: body 19–22; wing 18–20 mm.

Abdominal tergites largely blackish brown, pale yellow lateral triangles surrounded by dark color; ventral triangles do not attain anterior margin; penes of *limbata* type.

Frontal portion of head yellowish white; antennae pale brown; vertex purplish. Upper portion of eyes with a greenish tinge; separated by a space about equal to one diameter of eye. Pronotum purplish brown, the lateral and median areas paler. Meso- and metathorax red-brown; scutellum marked with purplish; an oblique dark red streak on pleura anterior to middle leg; posterior portion of prosternum purplish. Fore leg reddish brown; extreme basal portion of middle tarsal joints may be slightly paler. Coxae and trochanters of middle and hind legs pale reddish; femora, tibiae and tarsi yellow; claws, distal tarsal joints and tarsal joinings purplish brown. Wings hyaline; costal margin of fore wing tinged with reddish brown, paler at apex. Veins dark purplish brown; a few cross veins in and behind costal margin of fore wing and in disc of hind wing narrowly margined; in the latter, a few small dark spots. No dark outer margin on hind wing.

Abdominal tergites largely blackish brown; yellow postero-lateral triangles and somewhat irregular submedian yellow areas, one on each side of each tergite, not based on anterior margin as in *rigida,* but wholly surrounded by dark color. Posterior margins of middle tergites may be very narrowly paler, much less evident than in *rigida.* Sternites yellowish white; mid-ventral line purplish black over ganglionic areas. A semi-circular reddish brown patch based on the posterior margin of each basal and middle sternite extends forward only as far as the middle of the segment, not attaining the anterior margin; on the apical sternites these patches are triangular and extend somewhat further forward. Reddish oblique submedian streaks from the anterior margin are more or less evident. Tails broadly and irregularly banded with yellow and brown. Genitalia largely dark red-brown; penes of the *H. limbata* type; distal forceps joint very small, third one quite well developed.

The abdominal pattern and lack of dark outer margin on the hind wing, distinguish these species from others having the same type of genitalia. There are specimens of this species in the Cornell collection from Severn, Ont. (J. McDunnough); others which seem to belong here are from Cross L., Minn. and North Bay, Ont. (D. Dunavan), and Lake George, N. Y.

Hexagenia atrocaudata McDunnough

1920—Needham—Bull. Bur. Fish. 36: fig. 64 (*bilineata,* form *munda*).
1924—McDunnough—Canad. Ent. 56: 92. Fig. . . . . Ont., Va.
1927—McDunnough—Canad. Ent. 59: 117. Fig.
1931—Traver—Ann. Ent. Soc. Amer. 24: 611–13. Figs. . . . . N. Y., Ind., W. Va.

Length: body 22; wing 20 mm.

A large dark species; abdomen blackish brown with olivaceous areas; outer margin of hind wing widely purplish brown; penes short, broad, bluntly pointed.

Head light red-brown; carina and frontal margin pale, vertex yellow. Eyes deep orange; very large, almost contiguous above. Pronotum blackish, median and lateral areas paler. Meso- and metanota and pleura dark reddish to olive brown; a paler reddish median streak on mesonotum; yellowish markings on pleura below wing roots, above leg bases. Sternum deep reddish to purplish brown. Fore legs deep reddish black, middle tarsal joints pale yellowish basally. Middle and hind legs greenish yellow; distal tarsal joint, tarsal joinings and claws purplish black. Wings hyaline; all veins blackish. Costal margin of fore wing "heavily tinged with brown in the apical half and between veins 2 and 3 in the basal half" (McDunnough). Cross veins in costo-apical space tend to anastomose; only a few cross veins in either wing are narrowly dark-margined, no conspicuous dark spots. Outer margin of hind wing with a very wide purplish black border.

Abdominal tergites dark blackish brown; 1 and 2 largely dark, only a narrow paler median line. Median yellowish olive triangular areas, widened posteriorly, on tergites 3–6; on 7 and 8, a pale median streak at anterior margin only. Pale antero-lateral triangles on tergites 3–10, separated from pale postero-lateral triangles by the blackish-brown lateral stripes, on the middle tergites. "The triangular dorsal portion of segment 10 is dull brown, edged with black and with geminate black medio-dorsal streaks" (McD.). Median triangles on sternites dark brown, the base of each extending over most of the posterior margin of the sternite; the apices attain the anterior margins. A small dark spot in the antero-lateral angle of each sternite, so that only lateral triangles of the pale olivaceous yellow background remain. Near center of each sternite are two small pale dots, one on each side. Genitalia largely dark red-brown, with a few yellow markings; penes distinctive, being broad, short and bluntly pointed (see fig. 84). Tail joints blackish brown, each with a narrow yellow basal ring. See frontispiece.

The characteristic genitalia separate this species from all others in the genus. There are specimens of this species in the Cornell collection from E. Aurora, Hagaman, Rochester and Ithaca, N. Y.; Pendleton Co., Greenbrier River, New River, Athens and Smoke Hole, W. Va.; Howe, Ind. (E. B. Williamson); N. Tazewell, Va.; Chambersburg, Pa. (V. Argo); Ashe Co., N. C.; Ann Arbor and Lansing, Mich.; Kingston, Ringgold and Rome, Ga. (P. W. Fattig).

Hexagenia bilineata Say

1824—Say—West. Quart. Rep. 2, p. 303 (Baetis)..Minn.
1859—LeConte—Complete writings of T. Say, 1: 203 (Baetis).
1862—Walsh—Proc. Acad. Nat. Sci. Philad. p. 373 (Palingenia)..............................Ill.
1863—Walsh—Proc. Ent. Soc. Philad. 2: 189 (Palingenia); 199 (transferred to Hexagenia).
1863—Hagen—Proc. Ent. Soc. Philad. 2: 174 (Palingenia).
1871—Eaton—Trans. Ent. Soc. London, p. 66. Figs.
1883—Eaton—Revis. Monogr. p. 50..La., Texas
1920—Needham—Bull. Bur. Fish. 36: 278 (in part) figs. 1–7 (in part), 9–11, 13–14, 62.
1921—Ulmer—Arch. f. Naturg. 87: 233–39. Figs.
1924—McDunnough—Canad. Ent. 56: 90.
1927—McDunnough—Canad. Ent. 59: 116. Figs.......................................Miss. R.
1931—Traver—Ann. Ent. Soc. Amer. 24: 611–13. Figs.

Length: body 18–21; wing 17–19 mm.

A large species; abdomen appears alternately light and dark, longitudinally; outer margin of hind wing dark-bordered; penes characteristic, tips slender, tapering.

Head light reddish brown, frons pale; bases of antennae purplish brown; blackish spot between eye and antenna. Eyes dark grey above; large, separated by space equal to about half of one eye diameter. Pronotum pale yellow laterally, median area tinged with purplish grey; longitudinal stripes wide, purplish brown. Meso- and metathorax light red-brown; mesonotum greenish yellow, blackish markings on scutellum; pale yellowish area on pleura between wing roots and bases of legs. Posterior portion of prosternum purplish black. Foreleg long, in some specimens almost equal to body in length; tarsal joints relatively long. Dark purplish brown, tarsus duller and slightly paler than femur and tibia. Middle and hind legs pale red-brown; tibiae tinged with purplish; hind femur often greenish yellow; tarsal joints very narrowly pale greyish at base. Wings hyaline, faintly grey-tinged; costal margin of fore wing tinged with reddish brown, basal costal space largely hyaline. Longitudinal veins deep red-brown, cross veins black. Many cross veins on costal margin and disc of fore wing, and in most of hind wing, rather widely dark-margined; in disc of hind wing, several small dark spots. Outer margin of hind wing rather widely bordered with purplish brown.

Abdominal tergites largely purplish black; posterior margins narrowly whitish; pleural fold and areas adjacent to it, whitish. Rather large linear or triangular pale submedian patches on each tergite, surrounded by dark ground color. Viewed from side, abdomen appears banded longitudinally with alternate light and dark stripes. Sternites largely occupied by purplish brown median triangles, apices of which attain anterior margin. In some specimens, these widen out into a continuous dark median band leaving only crescentic pale lateral areas; in others, triangles are represented by lateral dark boundaries only, median area paler. Ganglionic areas purplish black; pale oblique submedian streaks and central pale dots on each sternite, in dark specimens. Tails deep purplish brown, a very narrow pale ring at base of each joint.

Genitalia pale red-brown shaded with purplish grey; distal joints of forceps slender. Penes slender, tapering to a point; directed inward (see fig. 84).

The distinctive penes and the longitudinally banded appearance of abdomen distinguish this from other species. Specimens of this species in the Cornell collection are from Fairport, Ia.; Lucedale, Miss. (H. Dietrich); Waco, Tex. (F. F. Bibby); Knoxville, Tenn. (Miss L. Shipe); St. Cloud, Minn. (C. R. Yeager); Cloverport and Lexington, Ky. (H. Garman); Gadsden, Ala. (H. Dietrich); Rome, Tombigbee River, Marietta and Atlanta, Ga. (P. W. Fattig). Many of the Georgia specimens are unusually small.

Hexagenia carolina Traver

1931—Traver—Ann. Ent. Soc. Amer. 24: 601, 616. Figs...N. C.
1932—Traver—J. Elisha Mitchell Sci. Soc. 47: 103.

Length: body 20½–24; wing 14½–17 mm.

A large pale species; abdominal stripes rose or lavender; a few small spots in hind wing only.

Eye golden brown above; head creamy white; purplish spot near base of eye on each side, another near carina. Pronotum creamy white, lateral stripes purplish. Remainder of thoracic notum and pleura yellowish; brownish markings. Median anterior portion of mesosternum and ventral aspect of middle coxa purplish; sternum elsewhere creamy white. Fore femur light rose; tibia deeper rose; basal tarsal joint and all but extreme base of distal joint, claws and tips of middle tarsal joints, purplish; remainder of tarsus creamy. Fore leg rather short, about half as long as body. Middle and hind legs creamy white; tarsal joinings narrowly purplish, also claws and apical half of distal joint. Wings hyaline; costal border of fore wing yellowish brown; costa, subcosta and radius golden brown, other longitudinal veins and cross veins purplish brown except in anal region. In hind wing, subcosta and radius purplish brown, other longitudinal veins yellowish white; cross veins purplish black, a few in disc faintly margined to form one or two very small dark spots, not conspicuous. Outer margin not darkened.

Abdomen creamy white on 1–7, 8–10 yellowish; posterior margin of each creamy white. Faint grey median dorsal streak on tergites 1–6; on 7–10 this median streak is wider and darker grey. Lateral oblique stripes grey-rose to lavender; on tergite 1, occupy most of lateral space, and joined at anterior margin to median line; on 2–6, similar but narrower lines connect stripe to median line anteriorly, on 6 joined also on posterior margin; 7–9, marks appear as lateral arms from median stripe. Ganglionic areas of sternites 6–10 purplish brown. Faint yellowish median streak on 8 and 9, and a faint triangle on 7, otherwise unmarked. Genitalia of the *carolina* type; very similar to *limbata* type, but penes more erect, slightly more slender (see fig. 84). Second joint of forceps rather swollen distally; two distal joints of moderate size; penes and forceps marked with purplish. Tails reddish brown; each joint darker apically, joinings very dark, basal end pale whitish; tails creamy white at extreme tips.

The type material, from the locality of Greensboro, N. C., is in the Cornell collection, also other specimens from Dutchman's Creek, N. C. (J. W. Bailey) and Atlanta, Ga. (P. W. Fattig).

Hexagenia elegans Traver

1931—Traver—Ann. Ent. Soc. Amer. 24: 594. Fig. . . . . N. C.
1932—Traver—J. Elisha Mitchell Sci. Soc. 47: 103.

Length: body 12–17; wing 10½–12 mm.

A rather small slender species; abdomen whitish, with purplish red lateral stripes; triangles on middle and apical sternites do not reach beyond middle; dark spots in both wings; hind wing dark on outer margin.

Eye golden tan above. Vertex creamy, flesh-pink near posterior margin. Prothorax flesh-colored, lateral stripes mahogany-red. Remainder of thoracic notum creamy white; pleura similar, with few reddish markings. Dark purple spot posteriorly on prosternum; mesosternum anteriorly, and metasternum wholly dark red-brown; remainder of mesosternum paler shade of same color. Fore femur and tibia deep rose-red; basal and distal joints of tarsus, and distal half of middle joints, purplish brown; basal half of middle joints whitish. Fore leg rather short, little more than half as long as body. Middle and hind legs white; claws, distal tarsal joint except at base, apical margin of third joint, and joinings of others purplish brown. Wings very iridescent; costal border of fore wing dark brown; longitudinal veins of fore wing, of costal margin of hind wing, and all cross veins, purplish black. Many veins in both wings margined, forming many dark spots in each wing. Outer margin of hind wing prominently purplish brown.

Abdomen whitish; sternites 1–6 flushed with pink; segments 7–10 distinctly yellowish. Faint grey median line on tergites 1–6, becoming broader and reddish brown on 7–10. Lateral stripes on tergites vary from lavender to purplish rose; connected on posterior margin to median stripe. Ventral triangles indicated by outer dark stripes only, except on apical sternites, where triangle is fully colored; these do not reach beyond middle of each sternite except on basal segments, where they may almost attain anterior margin. Genitalia of *carolina* type; two distal joints extremely small. A rose-colored spot on each side of purplish brown forceps base; forceps and penes with purplish brown markings. Tails dark brown on apical portion of each joint, darker at joining; basal half of each joint greyish white. In some joints the joint is wholly dark except for narrow basal ring.

The type material, from the Chowan River near Winton, N. C. (Miss L. E. Brett), is in the Cornell collection. Quite typical specimens are present also from the Etowah River at Rome and Cartersville, Ga. (P. W. Fattig). Other specimens, also collected by Prof. Fattig at Jonesboro, Athens, Atlanta and Rome, Ga., are allied to *H. elegans*, but somewhat larger; the abdominal color pattern is more extensive and darker in color. These may represent a new species, but for the present are not described as such. Dark spots are present in both fore and hind wings, even more prominent than in typical *elegans*. Still other specimens from Washington, D. C. (V. Argo), lack the dark wing spots, but are otherwise similar to *elegans*. We have seen a specimen from Edenton, N. C. (C. S. Brimley).

Hexagenia limbata Guerin

1829—Guerin—Iconog. Reg. Anim. 2, pt. 1; pl. 60, fig. 7–9; 3, p. 384 (Ephemera). . . . . N. Amer.
1832—Gray—Grif. Cl. Ins. 2: pl. 94, fig. 7.
1842—Rambur—Hist. Nat. d. Ins. Neurop., p. 295, pl. 8, fig. 2 (Ephemera).
1843—Pictet—Hist. Nat. Neurop. 2: 146, pl. 12 (Palingenia).
1853—Walker—List Neurop. Ins. Brit. Mus. pt. 3: 548 (Palingenia).
1861—Hagen—Syn. Neurop. Ins. N. Amer. p. 43 (Palingenia).
1862—Walsh—Proc. Acad. Nat. Sci. Philad. p. 373 (Palingenia). . . . . Ill.

1863—Walsh—Proc. Ent. Soc. Philad. 2: 197.
1868—Eaton—Ent. Mo. Mag. 5: 85.
1871—Eaton—Trans. Ent. Soc. London, p. 65, pl. 1.
1876—Provancher—Natur. Canad. 8: 265 (Palingenia)..Que.
1877—Provancher—Fn. Ent. Canad. 2. fasc. 1: 82 (in Palingenia).
1883—Eaton—Revis. Monogr. p. 50 (*bilineata*, in part) p. 55, (*variabilis*).
1921—Ulmer—Arch. f. Naturg. 87: 233–39. Figs.
1924—McDunnough—Canad. Ent. 56: 90.
1927—McDunnough—Canad. Ent. 59: 117..B. C.
1931—Traver—Ann. Ent. Soc. Amer. 24: 611–13.

Length: body 19; wing 18.5 mm.

A pale yellow species; abdominal tergites with distinct yellow areas and brown markings; outer margin of hind wing dark-bordered.

Thorax reddish brown; blackish lateral stripes on pronotum. Legs yellow; fore femur and tibia blackish. Wings hyaline, with no conspicuous spots; anterior margin of fore wing not much darkened; outer margin of hind wing brown. Abdomen yellow marked with brown; mid-dorsal line of tergites widely dark; lateral stripes begin at antero-lateral angles and extend backward to meet median stripe at posterior margin. Tail joints pale yellowish basally, dark brown distally. Genitalia of the *limbata* type (see fig. 84, *H. occulta*).

Specimens in the Cornell collection which we had previously considered to be of this species are probably pale forms of *H. occulta*; a specimen from the Museum of Comparative Zoology (given to our collection) from British Columbia, and labelled *limbata*, seems likewise to be *occulta*. This specimen was selected by Eaton as *limbata*, but is distinctly flushed with reddish brown on the abdominal tergites. The above description is drawn from the original from Rambur, and from notes by Drs. Ulmer and McDunnough.

Hexagenia marilandica Traver

1931—Traver—Ann. Ent. Soc. Amer. 24: 599..Md.

Length: body 18–19; wing 15–16 mm.

A rather large species; abdomen whitish, dorsum tinged with pinkish, lateral stripes purplish rose, no mid-dorsal stripe on middle tergites.

Head greyish white; eye slate grey above; space between eyes equal to one diameter of eye. Pronotum tinged with yellowish; lateral stripes rather broad, purplish brown. Mesonotum creamy white, bordered with light brown; metanotum creamy yellow, marked with lavender. Pleura whitish. Prosternum white, two purple streaks between legs. Anterior portion of mesosternum dark purplish rose; posteriorly light tan with lateral lavender marks. Metasternum white, bordered with purplish. Fore coxa and trochanter purplish; femur and tibia rose-red, joinings darker. Basal and distal tarsal joints, distal third of middle joints, and claws, purplish brown. Middle and hind legs creamy white; coxa of middle leg purplish red anteriorly; claws, apical two-thirds of distal tarsal joints, and joinings of fourth and fifth joints, purplish brown. Costal margin of fore wing brown, darker in distal half but paler again at apex. Longitudinal veins brownish rose; cross veins purplish black; a few cross veins in each wing very narrowly margined, may form a few very small dark spots in disc of hind wing. Hind wing not darker on outer margin.

Abdomen creamy white with pinkish flush; 8–10 yellowish. Greyish mid-dorsal stripe on tergites 6–10, 1–6 with only a very fine grey line. Lateral stripes seem to form a continuous zigzag pattern on each side; purplish rose in color. Posterior margins of tergites 7–9 widely dark. A purplish mark on ganglionic area of each sternite;

on sternites 8 and 9, greyish rose triangles extend to about the middle of the segment. On basal and middle sternites, no complete triangles, but short oblique streaks on posterior margin at each side mark the outer boundaries; on 6 and 7, short grey lines extend from these to mid-line at center. Basal joints of tails yellow in basal half, dark red-brown distally. In remainder of tail, each joint dark red-brown, darker distally, with a narrow yellow basal ring. Genitalia of the *carolina* type; basal forceps joint relatively long, the two distal joints very small and poorly developed. Forceps base, penes and forceps amber, marked with purplish brown.

The type material, taken at Conococheague Pk. near Hagerstown, Md. (V. Argo), is in the Cornell collection. We have also several males of this species (V. Argo) from Washington, D. C. Several males taken at Jonesboro, Rome and Atlanta, Ga. (P. W. Fattig) seem to belong here. Prof. Fattig likewise collected many male specimens from Toccoa Falls, West Point and Jonesboro, Ga., which are larger and stouter than typical specimens of *marilandica*, the bodies 21 to 22 mm. in length; eyes larger than in *H. marilandica*; abdominal markings reddish brown, quite similar in pattern to that species. They may represent a new species intermediate between *carolina* and *marilandica*.

Hexagenia mingo Traver

1931—Traver—Ann. Ent. Soc. Amer. 24: 597..W. Va.

Length: body 17–17½; wing 14–14½ mm.

Abdomen whitish or pinkish, lateral stripes reddish brown; mid-dorsal dark stripe on all tergites; ventral triangles reach anterior margin.

Head light tan; creamy yellow markings on vertex. Eye creamy to light tan above; rather large, separated by space less than diameter of eye. Pronotum light tan, lateral stripes wide, dark red-brown. Mesonotum light tan with dark brown markings; metanotum light brown. Pleura whitish, washed with light brown on mesothorax. Prosternum pale anteriorly, dark red-brown posteriorly; center of mesosternum light red-brown, margins dark brown; metasternum dark brown. Fore coxa and trochanter brown; femur and tibia mahogany brown, joinings darker; in holotype, tarsus entirely dull brown, in other specimens the basal part of the middle joints often paler. Middle and hind legs canary yellow; distal tarsal joints, claws and tarsal joinings purplish brown. Costal margin of fore wing reddish brown; apex and base of costal space light tan. Longitudinal and cross veins of both wings dark purplish; several cross veins in each wing somewhat margined, tending to form a few small spots in hind wing. Outer margin of hind wing narrowly dark in some specimens, not in others.

Abdomen whitish or with a pinkish flush; wide dark mid-dorsal stripe on all tergites, becoming wider at posterior margins. Tergites 1 and 2 largely purplish brown. Lateral stripes arise at antero-lateral margins and extend to median stripes at posterior margin; these are quite wide and occupy most of each side of tergite. Reddish brown ventral triangles extend forward to the anterior margin on sternites 1–8; ganglionic area of each somewhat darker in color. Two small clear round spots near center of each sternite, one on each side. Basal joints of tails yellowish brown in basal half, somewhat darker brown distally; nearer the tip, joints wholly brown except for narrow yellow basal ring. Genitalia intermediate between *limbata* and *carolina* types; forceps base yellow, two purplish brown spots at anterior margin; forceps and penes yellow marked with purplish brown.

The type material, taken at Millwood, W. Va. (J. G. Needham), is in the Cornell collection. In addition, we have a number of male specimens from Washington, D. C.

(V. Argo). Several males taken at Lake Worth near Albany, Ga. (P. W. Fattig), resemble *H. mingo* in abdominal markings and size, but have large dark dots in the hind wings and a few small ones in the fore wings, as in *elegans*. These may represent a new species.

Hexagenia munda Eaton

1883—Eaton—Revis. Monogr. p. 53. . . . . N. C.
1927—McDunnough—Canad. Ent. 59: 118.

Length: body 18; wing 14 mm.

Abdomen light yellowish, tergites with pitch-black U-shaped markings; triangles on sternites do not reach beyond middle of each segment; wings tinged with very light brownish grey, no large dark spots in wings.

Thoracic notum light raw-umber; a broad blackish brown stripe on each side of pronotum, extending backward on mesonotum beyond wing roots. Prothoracic pleura yellowish. Metanotum piceous in part. Anterior portion of mesosternum, and a spot at joining of meso- and metasternum, piceous. Fore coxa yellowish; anterior portion of middle coxa piceous. Fore legs from trochanter to basal joint of tarsus pitch-brown, trochanter and femur at base paler on inner side; tarsal joints 2–4 "pitch-brown distally, but warm sepia grey in their proximal halves." Middle and hind legs pale yellowish; distal tarsal joint, apical margin of fourth joint, and claws, deep blackish brown. "Wings transparent, almost uniformly tinted with extremely light bistre grey; the marginal and submarginal areas of the fore wing distinctly bistre-grey; the terminal margin of the hind wing very narrowly tinged with light greyish; longitudinal neuration bronzy raw-umber brown, becoming light yellowish at the wing roots, and in part appearing black in certain lights; cross veinlets black, bordered narrowly with blackish in the same parts of the wing as *bilineata*" (Eaton).

"Abdomen very light yellow ochraceous, marked with pitch-black above, and light burnt umber beneath" (Eaton). Tergites 1–7 with broad lateral stripes from the antero-lateral angles, "rounded off posteriorly just before the distal border of the segment"; these are joined together on posterior and anterior margins by narrow bands of same color as stripes, so that markings appear like "the letter U closed by a line across the top, with arms of equal thickness" (Eaton). Median area of 1–6 pale except for a narrow greyish line; tergite 7 with a broad dark median stripe. On tergites 8 and 9 the dark median line is present as on 7; but lateral stripes do not reach anterior margin. Sternites 1–7 with markings arising from posterior margins, "semielliptical in the hinder, but in the form of a small segment of a circle in other segments"; these do not extend beyond the middle of each sternite. Sternites 8–10 "bright light brown ochreous." A black mid-ventral streak at anterior margin of each sternite. Tails "intense bistre-brown, lighter close to their insertion, their joinings very light"; some joints lighter at anterior margins.

The type specimen in the museum at Cambridge has genitalia of the *limbata* type. In the Cornell collection there is a single male, from Two Run Creek, Kingston, Ga. (P. W. Fattig), which Dr. Banks has compared with the type. It is slightly larger than the type, the wing being 16 mm. in length; fore coxa tinged with purplish brown; bands on middle tergites not completely united anteriorly; otherwise similar to description. A specimen from Ann Arbor, Mich., is doubtfully placed here. From the Chattahoochee River at Atlanta, Ga. (P. W. Fattig) we have a third specimen. This is certainly a near ally of *H. munda*, but differs in having the dark mid-dorsal stripe present and equally wide on all tergites, so that the lateral pale patches are separated; the wings are much more strongly tinged with brown, the hind wing more distinctly brown-margined, than in the typical *munda*. Until more material is available, we shall not describe this as a new species.

Hexagenia occulta Walker

1853—Walker—List Neurop. Ins. Brit. Mus. pt., 3: 564 (Palingenia; also *Baetis angulata*)...Canada
1901—Needham—N. Y. St. Mus. Bull. 47: 427, pl. 16 (*variabilis*)...N. Y.
1907—Needham—Rep. Mich. Geol. Survey, p. 262 (*variabilis*)...Mich.
?1911—Morgan—Ann. Ent. Soc. Amer.; 4: 99 (*variabilis*).
?1925—Needham—Ent. News 36: 112 (*bilineata*).
1927—McDunnough—Canad. Ent. 59: 119...Man.
1931—Traver—Ann. Ent. Soc. Amer. 24: 611–13.
1932—Neave—Contr. Canad. Biol. and Fish. 7: 7. Figs.
1932—Neave—Canad. Field-Nat. 46: 54.

Length: body 17–18; wing 18–19 mm.

A reddish brown species, extremely variable as to amount of reddish suffusion; outer margin of hind wing usually dark-bordered; ventral triangles attain anterior margin.

Frontal portion of head yellowish; vertex suffused with yellow; remainder of head dark brown. Eyes yellowish brown above; separated by space less than one eye diameter; relatively smaller than in *H. viridescens*. Pronotum dark red-brown; median line narrowly pale, lateral margins yellow; a large pale area above fore coxa. Meso- and metathorax reddish brown, depth of color variable; mesonotum reddish yellow, scutellum and region anterior to it yellowish. A large pale area on pleura before hind wing root; a smaller pale area below root of fore wing. Sternum dark purplish brown. Fore leg dark red-brown; femur and tibia brighter than tarsus; middle tarsal joints sometimes slightly paler at bases. Fore leg long, only slightly shorter than body. Middle and hind legs yellow with greenish tinge; claws, distal tarsal joint except at base, and distal margins of other joints, light purplish brown. Wings hyaline; costal margin tinged with reddish to purplish brown, deepest in stigmatic area, palest in basal costal space. Veins purplish black; several to many cross veins on costal margins and in discs of both wings margined, those in disc of hind wing forming small dark spots. Outer margin of hind wing usually with a well-defined purplish brown border.

Abdominal tergites largely reddish brown, amount of reddish suffusion very variable. Pleural fold usually yellow in whole or in part; yellow postero-lateral triangles and yellowish submedian streaks or narrow triangles are typically present; the former may be much reduced; the latter are usually surrounded by dark color. Large red-brown triangles occupy most of each sternite, their apices attaining the anterior margin; ganglionic areas purplish black; pale oblique submedian streaks at anterior margin, and two pale central dots, in each sternite. Lateral dark streaks may be present, often connected to central triangles; remaining areas yellowish. Tail joints yellow basally, deep brown apically; joinings appear as yellow rings. Genitalia yellowish red; second joint of forceps whitish basally; penes of *limbata* type (see fig. 84).

This species differs from *H. limbata* in the reddish suffusion of the yellow areas; dark specimens may be readily confused with *H. viridescens*, but the eyes of the latter are relatively larger and darker above.

Specimens of this species in the Cornell collection are from Glenwood Creek, Minn. (R. O. Christenson); North Bay, Ont. (D. Dunavan); St. Cloud, Minn. (G. Friedrich); Larimore, N. D. (L. A. Carruth); L. Winnipeg, Man. (F. Neave); Lansing, Mich.; Willowick, Mentor Park and Cedar Point, O. (J. R. Traver); Burt Lake,

Mich.; Canajoharie, N.Y. (H. K. Townes, Jr.): Cranberry Lake and Cayuga Lake, N.Y. We have seen others from Long Lake, Ont. (Carnegie Museum); Mound and St. Paul, Minn. (Univ. of Minnesota).

Hexagenia orlando Traver

1931—Traver—Ann. Ent. Soc. Amer. 24: 608..Fla.

Length: body 14–18; wing 11–14 mm.

A rather small species; abdomen yellowish, lateral stripes dark brown; a few small dark spots in disc of hind wing, its outer margin very narrowly purplish.

Head largely purplish; eye golden brown above; eyes rather large, separated by space less than one diameter of eye. Pronotum light tan, usually purplish in median area; lateral stripes dark purplish brown. Meso- and metanota yellowish brown; brown spot in center of posterior portion of former, and lateral and posterior margins golden brown with creamy markings; creamy markings centrally and posteriorly on metanotum. Pleura yellowish brown with creamy markings. Median area of meso- and metasterna pale reddish brown; laterally golden brown. Fore coxa and trochanter dark brown; femur and tibia dark red-brown, joinings darker; basal and distal tarsal joints, and distal portions of middle joints, purplish brown. Middle and hind legs yellow; joinings of tibia, tarsus and all tarsal joints, as well as claws and distal tarsal joints, purplish brown. Costal margin of fore wing reddish brown. Longitudinal veins of both wings golden brown, cross veins purplish black; a few cross veins of fore wing narrowly margined, but no dark spots; in hind wing, a few small dark spots in disc, the outer margin very narrowly purplish.

Abdomen yellow to light tan, deep yellow apically. A greyish or blackish mid-dorsal line on all tergites. On 1–7 the oblique lateral stripes arise at the antero-lateral angles, extending backwards to the median stripe on the posterior margin; except on 7, these stripes are narrower posteriorly. On the basal and middle tergites, narrow dark stripes extend along the anterior margin; posterior margins dark brown; hence the pale areas on each side are largely surrounded by dark margins. Ganglionic area of each sternite purplish brown; on sternite 9, a median orange triangle attains the anterior margin; the lateral margins of similar triangles are faintly indicated by slightly wavy brown lines on sternites 1–8. Sternite 1 may show two pale round spots near the center, one on each side; the following sternites may show traces of these pale dots. Basal tail joints light brown distally, yellow at base; at tip, joints light brown, joinings black; in remainder of tail, joints dark red-brown, joinings darker, and a narrow yellow ring at base. Genitalia of the *carolina* type, quite slender. Forceps base and basal joint of forceps amber brown with creamy markings; distal joints light purplish brown.

The type material, taken at Orlando, Fla. (M. D. Leonard), is in the Cornell collection. Other specimens from Kissimmee, Fla. (P. N. Musgrave), are probably of this species, although exhibiting some variations. We have also many male specimens from Spring Creek near Decatur, Ga. (J. C. Bradley), which bear a rather close resemblance to this species. They differ from typical *H. orlando* in these points: size slightly larger; median area of pronotum pale greyish, hyaline; vertex of head pale brown; thoracic sternum largely deep purplish; many cross veins of both wings distinctly margined, many large dark spots in disc of hind wing; ventral triangles more distinct. For the present we regard these as forms of *orlando*.

Hexagenia pallens sp. nov.

Length: body 19; wing 16 mm.

A pale whitish species, without prominent dark markings; longitudinal veins pale; hind wing not dark-margined; penes of *limbata* type.

Male—Head purplish red, frontal margin pale; bases of antennae deep rose-colored. Eyes large, separated by a space less than one eye diameter; in alcoholic specimen, purplish black above and below. Pronotum yellowish white, lateral stripes pale purplish rose. Meso- and metathorax flesh-colored, with distinct yellowish tinge. Pale purplish red oblique streak anterior to wing roots, and on pleura anterior to each middle and hind leg; a purplish spot at joining of pro- and mesosternum. Fore coxa yellowish with pinkish tinge; trochanter pale rose; femur and tibia deep purplish red; basal and distal joints of tarsus, and claws, deep purplish; distal portion of middle joints greyish lavender, joinings purplish, basal portions greyish white. Middle and hind legs pale yellowish; claws, distal tarsal joints and joinings of 2nd and 3rd joints pale brown with rose tinge. Wings hyaline, pale. Costal margin of fore wing faintly yellowish; costa, subcosta and radius deep yellowish amber, subcosta tinged with brown. Other longitudinal veins of fore wing, and subcosta of hind wing, pale yellowish; those of hind wing behind subcosta white. Cross veins deep blackish purple; in costal margin of fore wing rather coarse, margined with purplish; elsewhere fine, rather inconspicuous, only a few in disc of each wing very narrowly margined, not forming dark spots. Outer margin of hind wing not dark-bordered.

Abdomen semi-hyaline, pale yellowish white, the two apical segments opaque, more yellow. Faint semi-hyaline purplish shading laterally on basal and middle tergites; a faint brownish median streak on 9; no other dorsal markings. On apical sternites, a very small pale purplish blotch in each antero-lateral angle; ganglionic areas purplish on all sternites; median area of 1–3 faintly shaded with purplish, the posterior margin of 1 somewhat darker; no other markings. Tails reminiscent of *H. venusta*; in middle area, groups of two or three joints all rose-colored except for narrow basal pale ring on each; between these dark joints, a pale joint, rose-colored at apex only. In basal and apical areas, joints rather uniformly rose-colored apically and pale at base. Genitalia pale yellowish white; tips of forceps purplish, the distal joints rather short, second joint rather long; penes of the *limbata* type.

Female very similar to male. Pale yellowish white, egg masses yellow-orange. Foreleg largely pale yellowish white; femur streaked longitudinally with deep rose, purple at apex; tibia purplish rose at base, blackish at apex; middle tarsal joints pale, joinings narrowly purplish.

Holotype—Male imago; Manhattan, Kansas, taken at trap lantern during summer of 1932, by Prof. R. C. Smith. No. 1345.1 in Cornell collection.

Allotype—Female imago; same data. No. 1345.2 in Cornell collection.

Paratypes—29 female imagos, same data. No. 1345.3–31 in Cornell collection.

This pale species might be considered an albinistic form of *H. venusta*. It is much paler than any other species of our fauna except *H. carolina* in which the abdomen is semi-opaque; the eyes are much smaller and further separated in that species, the thorax and abdomen dorsally with distinct darker markings, the penes less curved at apex.

Hexagenia recurvata Morgan

1913—Morgan—Ann. Ent. Soc. Amer. 6: 395. Fig. .... N. Y.
1920—Needham—Bull. Bur. Fish. 36: 279. Figs.
1921—Ulmer—Arch. f. Naturg. 87: 239.
1931—Traver—Ann. Ent. Soc. Amer. 24: 611–13, 615. Figs. .... Mass., W. Va., Mich.

Length: body 18–20; wing 15–18 mm.

Wings heavily tinged with dark reddish brown; abdomen pale yellowish brown, dark markings few and inconspicuous; tips of penes narrow, recurved.

Head red-brown; carina, basal antennal joints and tip of filament paler. Eyes yellowish above; separated by space equal to about half diameter of one eye. Pronotum light red-brown; lateral margin narrowly darker; lateral stripes very narrow, indistinct. Meso- and metathorax deep reddish to blackish brown; faint yellowish markings along postero-lateral margin of mesonotum and on a small area anterior to scutellum. Creamy white markings on pleura below wing roots and around bases of legs. Ventral area of fore coxa pale reddish; upper part of coxa, trochanter and femur yellowish red; tibia pale red-brown; tarsus greyed lavender, the claws, tip of distal joint and bases of middle joints narrowly paler. Middle and hind legs yellowish; coxae red-brown; tarsi very faintly grey-tinged; tarsal joinings narrowly purplish grey. Wings heavily tinged with reddish or purplish brown, leaving hyaline areas only at base of each wing, and beyond disc and near apex of fore wing. Veins blackish; cross veins of both wings rather widely margined with dark red-brown, in dark areas. Outer margin of hind wing not darker than adjacent areas.

Abdomen pale yellowish brown; segment 1 light red-brown; sternites slightly paler than tergites. Pleural fold and posterior margins of all segments paler; tergites with very faint grey mid-dorsal line, narrow on basal ones, somewhat wider apically. Faint semi-opaque yellowish triangles based on posterior margins may be visible. A brownish line on each spiracular area. On each sternite, an oblique brownish submedian streak from anterior margin on each side, and two dark dots near center, one on each side. No other dark markings. Tails light purplish brown; joinings darker, each preceded by a narrow paler ring. Forceps base, basal forceps joint and penes pale reddish; remainder of forceps greyed lavender. Penes distinctive, the distal third of each narrowed and recurved. Basal forceps joint short and stout, second joint relatively long and incurved apically; distal joints normally well developed (see fig. 84).

The distinctive recurved tips of the penes, and the brown-tinged wings reminiscent of *Ephemera guttulata* distinguish this from other species of the genus. Nymph shown in fig. 47.

There are specimens of this species in the Cornell collection from McLean, Freeville and Michigan Hollow, N. Y.; Conesus Lake, N. Y.; Pine Lake, Herkimer Co., N. Y.; Passadumkeag Bog, Me. (A. E. Brower); Marquette Co., Mich.; Elk Garden and Wardensville, W. Va.; Granby, Mass.

Hexagenia rigida McDunnough

1883—Eaton—Revis. Monogr. pl. 7, fig. 11b (*bilineata*).
1920—Needham—Bull. Bur. Fish. 36: figs 61, 65 (*bilineata* and form *falcata*).
1924—McDunnough—Canad. Ent. 56: 90. Fig. . . . . Ont., Que., Man., Vt., Pa.
1927—McDunnough—Canad. Ent. 59: 118. Fig.
1931—Traver—Ann. Ent. Soc. Amer. 24: 611–13.
1932—Neave—Contr. Canad. Biol. and Fish. 7: 20. Figs.
1932—Neave—Canad. Field-Ent. 46: 54.

Length: body 19–20; wing 16–18 mm.

Abdominal tergites largely brown with yellow submedian and lateral blotches; usually a wavy brown lateral stripe on each side of venter, no dark triangles on middle sternites; penes long, almost straight.

Head yellow; vertex suffused with reddish brown. Eye olive brown above; eyes separated by space about equal to one eye diameter. Pronotum pale yellow, lateral stripes wide, purplish black. Meso- and metanota light reddish brown, the posterior portions yellowish; tips of scutella blackish; pleura light red-brown. Sternum reddish to purplish brown, deepest on posterior of prosternum. Ventral portion of fore coxa deep purplish black; trochanter, femur and tibia reddish brown; tarsus may be reddish brown except base of second joint, or each middle joint may be yellowish basally. Middle and hind legs pale yellow; tarsal joinings and last tarsal joint purplish brown. Wings hyaline; costal margin of fore wing light olive brown; all veins blackish brown. Cross veins of costal margin heavily black; a few nearer the disc may be narrowly margined; in hind wing, several in disc usually margined, forming several small blackish spots. Outer margin of hind wing typically with narrow dark margin, but this may be lacking.

Abdominal tergites mainly brownish; yellowish triangular submedian blotches and smaller blotches next to lateral margin, on each tergite; the yellow areas may be quite bright in color, or more or less obscured by a reddish tinge. Posterior margins usually narrowly pale. Sternites pale yellow; a purplish black blotch in each anterolateral angle, an oblique lateral stripe attached to its posterior end, so that lateral stripes appear to extend "from the central area of one segment backwards to the middle of the lateral margin of the following segment" (McDunnough). Ends of these stripes usually connected, so that a more or less continuous wavy lateral line is formed. Mid-ventral line deep brown on apical segments; often traces of dark line here on other segments; apical sternites with median dark triangles which do not attain the anterior margin. Forceps brown, second joint yellow. Penes rather long, almost straight (see fig. 84). Tails pale yellow, each joining narrowly purplish brown.

The long straight penes are distinctive. There are specimens of this species in the Cornell collection from Knowlton, Que. (G. H. Fisk), sent by Dr. McDunnough; Fairport, Ia.; Burt Lake, Mich,; Willowick, O. (J. R. Traver); North Bay, Ont. (D. Dunavan); Buffalo, N. Y.

Hexagenia rosacea Traver

1931—Traver—Ann. Ent. Soc. Amer. 24: 607.. N. C.

Length: body 20; wing 17 mm.

Abdomen whitish; lateral stripes deep rose-colored; median dark stripe on all tergites; hind wing not dark on outer margin, and with no dark spots.

Head greyish white; carina rose-colored distally; eye creamy above, quite large; space between eyes less than one diameter of eye. Pronotum greyish white, lateral stripes purplish red. Mesonotum whitish with creamy markings and rose-brown border; metanotum yellow with rose markings. Pleura white, with lateral rose stripes to base of middle legs. Prosternum white, a faint lavender streak posteriorly; mesosternum deep rose, posterior margin white; metasternum purplish rose. Venter of fore coxa in distal half, trochanter and femur rose-colored; tibia deep rose; basal and distal tarsal joints, and distal third of middle joints, deep purplish rose. Middle and hind legs white; tarsal joinings narrowly purplish, as are claws and distal two-thirds of fifth joint. Wings very iridescent; costal margin of fore wing light brown in basal half, dark brown distally. Costa, subcosta and radius golden brown, other longitudinal veins dark purple. Cross veins purplish black; no dark spots in either wing. Hind wing not darker on outer margin.

Abdomen white on segments 1–6; 7–10 yellow. A rather wide median dorsal stripe on all tergites; greyish brown on 1–5, reddish brown apically; progressively wider from 7 to 9. Tergites 1 and 2 dull rose grey except for white mark near pleural fold, one on each side near anterior margin, and pinkish crescent posteriorly at median line. Tergites 2–7 with deep rose oblique lateral stripes from antero-lateral angles to posterior margin, joining median stripe; on anterior margin a narrow streak extends halfway to median stripe. Posterior margins of 4–9 deep rose-colored. A dark rose streak marks ganglionic area on each sternite; triangles deep rose, the middle portion of each pale, so as to appear on middle sternites as two small lateral triangles; these do not attain the anterior margin. Posterior margins of sternites 2–10 deep rose-colored. Basal joints of tails deep rose distally, yellow at base, reddish black at joinings; in middle area, joints dark brown in distal two-thirds, yellow at base, joinings darker; in distal portion, joints dark brown except for narrow light ring at base. Genitalia of the *carolina* type; golden yellow, marked with reddish brown.

The type specimen, taken at Penrose, N. C., is in the Cornell collection. We have seen no others. This species has a shorter fore femur and more extensive color pattern than *H. carolina*; the second tarsal joint of the fore leg is shorter than in *H. marilandica*, in which species the mid-dorsal stripe is lacking on basal tergites. From specimens of *H. mingo* in which the dark outer margin of the hind wing is lacking, it is distinguished only by the larger eye and more extensive rose border on posterior margin of sternites.

Hexagenia venusta Eaton

1883—Eaton—Revis. Monogr. p. 54. . . . . Texas, Utah L.
1894—Banks—Ent. News 5: 178. . . . . Kans.
1921—Ulmer—Arch. f. Naturg. 87: 235, 237, 239.
1931—Traver—Ann. Ent. Soc. Amer. 24: 611, 612.

Length: body 15–18; wing 15 mm.

A yellowish species marked with purple-madder; longitudinal veins yellow, except on costal border; no large dark spots in veins.

Head yellow; second basal joint of antenna, and markings at base of and between ocelli, purplish brown, Thorax yellow; purplish red stripe on each side of pronotum, extending back on mesonotum to wing roots; an oblique purplish streak on pleura anterior to middle leg. Sternum largely deep purplish red, including median posterior area of prosternum; posterior portion of mesosternum paler than other portions, dark and widest posteriorly on metasternum. Fore legs madder-brown or purple-brown, tibia brighter than femur, tarsus dull; tarsal joints 1 and 5 wholly dark, others pale whitish in basal portion. Fore leg rather short, somewhat more than half as long as body. Middle and hind legs yellowish white; claws, all but extreme base of distal tarsal joint, and tip of third tarsal joint purplish brown. Wings hyaline, tinted very faintly with pale yellowish; costo-apical region and to a lesser extent the whole costal margin tinged distinctly with raw-umber. Costa, subcosta and radius of fore wing reddish brown (Eaton says, dark pitch-brown); other longitudinal veins in both wings yellowish. Cross veins blackish; those in costal border margined with dark brown, most extensively in stigmatic area; a few in disc of each wing very narrowly margined, but no dark spots in either wing. Outer margin of hind wing purplish or blackish brown, often in alternate wide and narrow areas.

Abdomen whitish yellow, semi-hyaline, on segments 2–7; 8–10 opaque, somewhat deeper yellow. Lateral somewhat triangular purplish red patches are present, one on each side, on all tergites except the last two, "the triangles truncate posteriorly, and in segments 1–4 reaching from base to joining, but in the others not starting from the base of the segment" (Eaton). A broad median streak on tergites 8 and 9, not reaching the anterior margin, on 8; rather purplish grey. A purplish red mid-ventral streak on all sternites, connected with dark metathoracic sternum; widest on basal sternites, becoming a mere line apically. Ganglionic area of each sternite blackish. Within the purplish band, except on sternites 8 and 9, a whitish dot on each side of median line, near center of segment; a similar pair of white dots on metathoracic sternum. In antero-lateral angle of each sternite, a small purplish triangle. Tails pale yellow, irregularly marked with purple-madder, sometimes at joinings, sometimes at apical margins of joints, at intervals an entire joint being dark. In female, tails wholly light yellowish, unmarked. Genitalia of the *limbata* type; pale yellow, marked with purple.

Specimens of this species in the Cornell collection are from Waco, Texas (F. F. Bibby); Huntsville, Tex. (V. A. Little); Austin, Tex. (H. B. Parks); Passagoula River, Miss. (H. Dietrich); Abilene, Kans. (V. S. L. Pate); Lawrence, Kans.; College Station, Tex. (H. J. Reinhard); Ames, Ia. (H. A. Scullen); Manhattan, Kans. (R. C. Smith); Stillwater, Sand Creek and Alfalfa Creek, Okla. (R. D. Bird); Decatur Lake, Ill.; Winter Haven, Tex. (S. E. Jones); Columbia, Mo.; Bourbon Co., Lonn Co. and Crawford Co., Kans. (R. H. Beamer). We have seen others from Mineola, Mo. (Ward's Nat. Sci. Establishment).

Hexagenia viridescens Walker

1853—Walker—List Neurop. Ins. Brit, Mus. pt. 3; p. 550 (Palingenia)........................Ont.
1927—McDunnough—Canad. Ent. 59: 118. Fig.
1931—Traver—Ann. Ent. Soc. Amer. 24: 611–13.

Length: body 19–20; wing 18–18 mm.

A large dark species; submedian pale markings on tergites reduced to narrow streaks; outer margin of hind wing dark-bordered; ventral triangles attain anterior margin.

Frontal portion of head yellowish; antennae, vertex and occiput purplish brown. Eyes deep brown above; large, separated by space less than one eye diameter. Pronotum deep purplish brown; lateral areas behind and above fore coxa narrowly yellowish. Meso- and metathorax purplish brown with a distinct reddish tinge; sternum and an oblique streak on pleura anterior to middle leg and below wing roots somewhat darker. Fore leg dark red-brown; middle tarsal joints very slightly paler basally. Fore leg long, almost equal to body in length. Middle and hind legs yellowish with a distinct reddish tinge, most evident at apices of femora; claws, distal tarsal joint and distal third of other joints, blackish brown. Wings hyaline; costal margin of fore wing reddish to olive brown; all veins deep purplish brown. A few margined cross veins in disc of hind wing tend to form small dark spots. Outer margin of hind wing with a rather wide purplish brown border.

Abdominal tergites largely deep purplish or blackish brown; posterior margins, small postero-lateral triangles and narrow submedian dashes on each tergite pale yellowish. Pleural fold yellowish white. Sternites paler reddish brown, each with a dark red-brown median triangle, the apex of which attains the anterior margin. Mid-ventral line purplish black on apical sternites, and at ganglionic areas on middle ones. Tails deep red-brown, each joint with a narrow yellowish ring at base. Genitalia deep purplish brown; penes of the *limbata* type.

The very limited pale dorsal markings, dark-margined hind wing and ventral triangles which attain anterior margin distinguish this species from others with similar genitalia except *H. occulta*; dark specimens of the latter species may have very similar maculation. The larger eyes, dark above, and slightly larger body size of *H. viridescens* are means of separation.

Specimens of this species in the Cornell collection are from Severn, Ont. (C. H. Curran), sent by Dr. McDunnough; Douglas Lake and Burt Lake, Mich.; North Bay, Ont. (D. Dunavan).

Hexagenia weewa Traver

1931—Traver—Ann. Ent. Soc. Amer. 24: 605...Fla.

Length: body 22–24; wing 17–19 mm.

A large species; abdominal stripes purplish brown; outer margin of hind wing dark, several dark spots in disc.

Head yellowish white; eyes purplish grey above, large, separated by space less than one diameter of eye. Pronotum light tan; lateral stripes purplish brown, extensions to wing base brown. Mesonotum creamy white bordered with brown. Metanotum yellowish. Pleura yellowish, shaded with brown on mesothorax. Posterior portion of prosternum, anterior of mesosternum and entire metasternum dark purplish brown; posterior of mesosternum lighter shade of same color. Ventral portion of fore coxa, trochanter, femur, tibia and basal tarsal joint mahogany brown; basal half of middle tarsal joints yellow, distal half purplish brown, as is the distal joint. Middle and hind legs yellowish white; very narrow purple line at each tarsal joining, distal tarsal joints purplish brown except at base. Costal margin of fore wing reddish brown. Costa, subcosta and radius brown, other longitudinal veins of both wings purplish brown. Cross veins purplish black; no spots in fore wing, a few small ones in disc of hind wing; outer margin of hind wing purplish.

Ground color of abdomen white, sometimes with faint lavender tinge; segments 8–10 pale yellowish. Well-defined mid-dorsal streak on all tergites; on 7–9 wide, dark purple; on 1–6 narrower, greyish purple, bifid anteriorly. Lateral oblique streaks purplish brown; triangular on 1 and 2, the base on pleural margin, and narrow extensions to median line an anterior margin. On middle tergites these marks arise from antero-lateral angles and do not reach the posterior margin; on 3–7 appear as elongated streaks with free ends pointed; on 8 and 9 the streaks are outward extensions of the median line, not attaining the anterior margin. Ganglionic area of each sternite purplish black; median triangles lavender brown, narrow on sternites 8 and 9, attaining anterior margin only on the basal segments; a small clear round spot on each side near center of each sternite. Basal tail joints light brown, dark brown at joinings; other joints dark red-brown, somewhat darker distally, each with a narrow yellow ring at base. Forceps base purplish brown; penes streaked with purplish brown; penes of the *carolina* type, second joint of forceps unusually long, the two distal joints unusually well developed. Entire body of female clear canary yellow.

The type specimens of this species taken at Chipola Lake, Fla. (J. G. Needham) are in the Cornell collection; we have seen no others.

Subfamily POTAMANTHINAE

Pronotum of male only slightly wider than long, somewhat emarginate on posterior margin. Fore leg of male almost as long as body; femur three-fourths as long as tibia, which is nearly as long as tarsus. Claws blunt, similar, on fore leg of male, all others in both sexes dissimilar. First anal vein of fore wing forked near margin; not attached to margin by cross veins, but several isolated cross veins lie between it and margin. Posterior branch of media more strongly curved than anterior branch of cubitus. Costal projection of hind wing rather obtuse; inner fork of radial sector usually shorter than its stem; media forked ¼ distance from base. Tails three, in both sexes. Forceps 3-jointed, first joint longest. Nymph not fossorial.

Genus **Potamanthus** Pictet

1843—Pictet—Hist. Nat. 2, Ephem. Neurop. p. 208.
1871—Eaton—Trans. Ent. Soc. London, p. 76 (restricted).
1883—Eaton—Revis. Monogr. p. 78.
1920—Ulmer—Stett. Ent. Zeit. 81: 110.
1920—Needham—Bull. Bur. Fish. 36: 287.
1927—Argo—J. N. Y. Ent. Soc. 35: 322.
1933—Spieth—J. N. Y. Ent. Soc. 41: 345.

Pale mayflies of medium size, wing ranging from 7 to 13 mm. in our species. Eyes of male more or less remote, separated by distance equal to one, or in some species two diameters, of eye, which varies in size in different species. Ocelli quite prominent, median smaller than laterals. Posterior border of head somewhat emarginate. Pronotum of male only slightly wider than long, somewhat emarginate at middle of posterior margin; narrowed slightly toward anterior margin. Pleura somewhat expanded, so that a small part is visible from above, on each side of pronotum. Fore leg of male not quite as long as body. Femur about ¾ the length of tibia, tarsus slightly longer than tibia. Five tarsal joints on fore leg, basal one very short; second longest, other joints becoming progressively shorter toward tip; claws similar, blunt. Femur of hind leg 1¼ times as long as tibia; tarsus and tibia subequal; four tarsal joints; claws of middle and hind legs of male, and all legs of female, dissimilar.

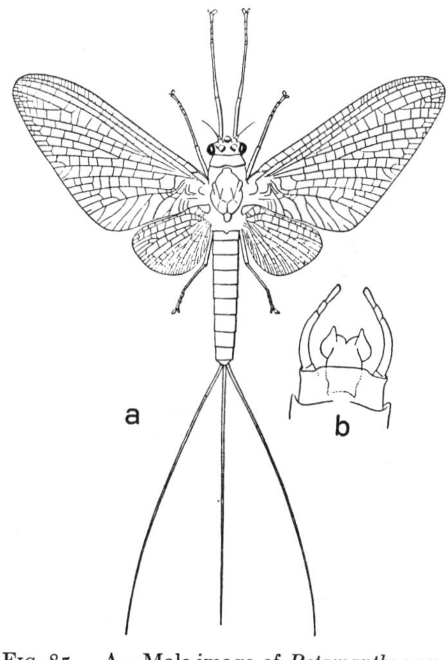

FIG. 85. A—Male imago of *Potamanthus sp.* (drawing by C. H. Kennedy). B—Genitalia of same.

First anal vein of fore wing forked near margin, not attached to margin by cross veins; however, several isolated cross veins, free at basal end, lie between it and margin. Posterior branch of media much more strongly curved at base than is anterior branch of cubitus. Costal projection of hind wing rather obtuse; inner fork of radial sector shorter than or equal to its stem; median vein forked $\frac{1}{4}$ from base (see fig. 85a).

Pleural fold of abdominal segments not expanded laterally; no postero-lateral spines. Tails three, in male and female; middle tail somewhat shorter than laterals, especially in male. Forceps base of male slightly emarginate on posterior margin, between forceps limbs. Forceps apparently 3-jointed, no short basal joint, so that the first joint is longest; second and third together not equal to first; third may be slightly longer than second. Penes united; a rounded lateral and a rather acute outer apical lobe on each side (see fig. 85b). Posterior margin of 9th sternite of female prolonged backward to form a broad extension partially covering the 10th sternite, its posterior margin slightly emarginate.

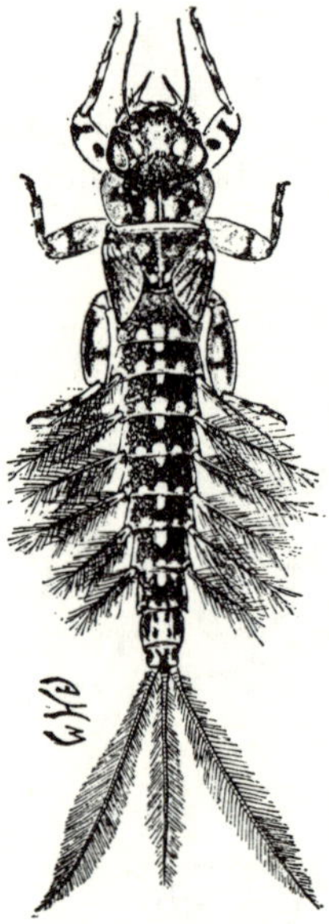

FIG. 86. Nymph of *Potamanthus sp.* (from Morgan).

Nymph, a sprawling form; not fossorial (see fig. 86). Frontal margin of head somewhat rounded. Mandibular tusks variable in length, in different species; widest at base, tapering toward apex from about $\frac{1}{4}$ of distance from base. Short spines borne on dorsal surface and a few on each lateral margin, also an outer fringe of hairs, on enlarged basal portion only; remainder of tusk bare. Antero-lateral portion of pronotum expanded, so that sclerite is narrowest at posterior margin; anterior border emarginate on each side of median line. Maxillary and labial palps each 3-jointed. Labrum wider than long. Femora flattened, tibiae and tarsi slender; a long curved apical spine at tip of fore tibia. Fore leg longest in some species; in others, the second exceeds it in length. Gills lateral in position. Gill on segment 1 rudimentary; on 2–7 paired, each division slender and pointed, margins fringed with long hairs. Tails three; bare at extreme base and tip, margined with hairs elsewhere.

Wings and genitalia of the imago and structures of the nymph are figured by Eaton (1883), by Needham (nymph, as Polymitarcys), and by Spieth; also by Argo (1927), who gives some account of habits. Argo resolved a long-standing confusion of the nymphs of this genus with those of Ephoron, and pointed out the true differential characters by which the two may be distinguished.

Genotype—*P. luteus* (in Ephemera), Linnaeus.

Distribution—Holarctic; in Nearctic, central southern and eastern.

KEY TO SPECIES

Adults

1—Cross veins of fore wing wholly pale, in both sexes** 2
—Cross veins in all or part of fore wing brownish or blackish, in female or in both sexes 6
2—A lateral reddish or fuscous spot on each abdominal tergite, eyes of male small .... **rufus**
—No such lateral spots on tergites; eyes of male variable in size 3
3—Vertex and mesonotum pale yellowish white; eyes of male large .... **diaphanus**
—Vertex reddish or brownish, usually mesonotum also; eyes of male variable 4
4—Abdominal tergites 1–5 reddish; eyes of male small .... **myops**
—Abdominal tergites wholly pale; eyes of male variable 5
5—Red median stripe on vertex, pronotum and mesonotum .... **medius**
—Vertex fuscous; red median stripe on pronotum .... **inequalis**
6—Cross veins in portion at least of fore wing dark in female, wholly pale in male .... **verticis**
—Cross veins dark in both sexes 7
7—Wing of male 7–8 mm. in length; vertex, mesonotum and abdomen wholly pale .... **neglectus**
—Wing of male 13 mm. in length; vertex, median area of mesonotum and lateral streaks on abdomen reddish .... **distinctus**

VERIFICATION TABLE

Adults

Species	*Wing*	*Eyes,* ♂	*Cross-veins* ♂	*Cross-veins* ♀	*Abd. mks**	*Distr.*
diaphanus	9½	Large	Pale	Pale	None	N. Y.
distinctus	11	Moderate	Dark	Dark	Stripes	E
inequalis	11½	Small	Pale	——	None	N. Y.
medius ♀	10½–11½	——	——	Pale ?	None	Kan.
myops	12½–13	Small	Pale	Pale	None	Ill.
neglectus	8–9	Small	Dark	Dark	Spots	E
rufus	13½–14	Small	Pale	Pale	Spots	N. Y.
verticis	8–10	Large	Pale	Dark	Spots	C

Potamanthus diaphanus Needham

1907—Needham—N. Y. St. Mus. Bull. 124: 193 .... N. Y.
1920—Needham—Bull. Bur. Fish. 36: 287.
1926—McDunnough—Canad. Ent. 58: 186.
1927—Argo—J. N. Y. Ent. Soc. 35: 320–2.

Length: body 8–10; wing 9½ mm.

A pale species; wings of male wholly pale, tail joinings not darker; no abdominal markings.

Head and thorax pale yellowish white. Eyes and ocelli black. "Tips of femora, tarsi and tibiae very faintly infuscated, a subapical inferior spot on the fore tibia

** Female of *inequalis* and male of *medius* not known.
* Markings on sides of abdomen.

being more distinct" (Needham). Wings hyaline whitish. Abdomen pale yellowish white, the tergites "with a satiny sheen." Tails white, joinings not noticeably darker.

According to Argo, the wings of both sexes are wholly pale, and the eyes of the males rather large, separated by a distance not greater than diameter of one eye. We have seen no specimens that are known to be of this species. The type material, from Buffalo, N. Y., cannot be located. The species seems to be distinct from others in the genus.

Potamanthus distinctus sp. nov.

Length: body 11; wing 11 mm.

Eyes of male moderately large; cross veins of both wings black; vertex, middle of pronotum, fore femur and tibia reddish; a continuous wide lateral purplish red abdominal stripe.

Male—Vertex of head red, except a narrow yellow strip next to eye; bases of antennae reddish, filament pale. Ocelli pale greyish flushed with pink; dark-ringed at base. Eyes black, moderately large, separated by a space slightly greater than one diameter of eye. Pronotum yellowish, with wide red median dorsal stripe. Mesonotum yellowish red, scutellum paler. Pleura and sternum yellowish, somewhat flushed with pink. Metanotum yellowish white. Fore femur quite deep red, extreme base yellowish; tibia reddish, deeper in basal half, tinged with yellow apically, purplish black at tip; short basal tarsal joint purplish black, other joints yellow faintly tinged with reddish, claws and joinings blackish brown. Middle and hind legs yellow, claws tinged with fuscous. Wings hyaline; main longitudinal veins of costal margin yellow, other longitudinals pale yellowish white. Cross veins of both wings, except outer margin of hind wing and anal region of both wings, black.

Abdomen pale yellowish. A wide continuous purplish red stripe occupies lateral areas of abdominal tergites; posterior margins narrowly darker, within this stripe. Sternites unmarked. Genitalia yellowish. Tails yellowish, joinings distinctly reddish brown.

Female very similar to male. Wings faintly amber-tinged; cross veins black, in both wings. All markings as in male.

Holotype—Male imago (dried); Willoughby, Ohio, July 1, 1934 (J. R. Traver); No. 1348.1 in Cornell collection.

Allotype—Female imago (dried), same locality, Aug. 9, 1934 (J. R. T.); No. 1348.2 in Cornell collection.

Paratypes—2 female imagos; E. Canada Creek, Little Falls, N. Y.; (H. K. Townes, Jr.); July 24, 1934; No. 1348.3–4, Cornell collection.

This species is most closely related to *P. rufus*, but smaller, with black cross veins in wings of both sexes. Slightly larger than *verticis*, from which it is distinguished by the continuous red abdominal stripe and the black cross veins in wings of male. From two localities in West Virginia (Justice, Smoke Hole) we have specimens quite similar to this species, but with tails wholly pale, and reddish markings of vertex and pronotum not evident. We have seen other specimens from Herkimer, Kast Bridge, and Little Falls, N. Y. (H. K. Townes, Jr.).

Potamanthus inequalis Needham

1908—Needham—N. Y. St. Mus. Bull. 134: 74. . . . . N. Y.
1926—McDunnough—Canad. Ent. 58: 186.
1927—Argo—J. N. Y. Ent. Soc. 35: 320 (*inequalis* = *diaphanus*).

Length: body 11; wing 11½ mm.

Head fuscous, remainder of body pale; tail joinings faintly darker; wings wholly pale.

Head fuscous to purplish brown. Eyes of male small. Thorax pale yellowish; median reddish streak on pronotum, also on anterior portion of mesonotum of one specimen. Legs yellowish white. For tibia infuscated at tip, also tarsal joinings. All joints of fore tarsus faintly smoky. Fore femur with faint reddish tinge. Wings hyaline, whitish, "with pinkish iridescence" (Needham); all veins wholly pale. Abdomen translucent, yellowish white; apical tergites "of a dull satiny whiteness" (Ndm.). Very faint opaque brownish areas on each tergite in same position as distinct dark lateral spots on some species of this genus. Tails white, joinings very faintly darker.

Above description drawn from the original, with additional notes on three male imagos from the New York State Museum, (taken June 29, 1909, Schenectady, N. Y.), which are almost certainly of the type material. We do not agree with Argo that this species is synonymous with *P. diaphanus*; however, it may well be a synonym of *myops*. Until we have been able to study undoubted material of the latter, we retain *inequalis* as a valid species.

Potamanthus medius Banks

1908—Banks—Trans. Amer. Ent. Soc. 34: 259 .... Kans.
1926—McDunnough—Canad. Ent. 58: 186.
1927—Argo—J. N. Y. Ent. Soc. 35: 321.

Length (female): body ?; wing 10½–11½ mm.

Vertex and prothorax with broad median stripe; tails joinings brownish; only female is known.

Head and thorax pale yellowish; a "broad median reddish stripe on prothorax and vertex of head, and less distinctly on mesothorax; the part on head is broader then elsewhere, and on the face it is broadened to reach the eyes, or nearly so" (Banks). Eyes of female "scarcely if any more than two diameters apart." Fore femur and tibia slightly infuscated at apex. Tails narrowly brownish at joinings; median tail "rather smaller than the others."

We have a single female from Manhattan, Kans., which is slightly larger than indicated for this species. It has the reddish vertex and median pronotal stripe, however, as well as brownish tail joinings. Fore femur tinged with pale reddish, darker at apex; apex of tibia and joinings of fore tarsus rather purplish red. Wings wholly pale, no dark cross veins; no dark markings on abdomen. For the present, we are holding this specimen as a probable *medius*.

Potamanthus myops Walsh

1863—Walsh—Proc. Ent. Soc. Philad. 2: 207 (Ephemera) .... Ill.
1871—Eaton—Trans. Ent. Soc. London, p. 71 (Ephemera).
1883—Eaton—Revis. Monogr. p. 72 (Ephemera).
1926—McDunnough—Canad. Ent. 58: 186 (transferred to Potamanthus).
1927—Argo—J. N. Y. Ent. Soc. 35: 320, 322, 323.

Length: body 12–13; wing 12½–13 mm.

Eyes of male small; wings wholly pale, in both sexes; abdominal segments 1–5 ferruginous, but no lateral markings.

Walsh's description is largely a comparison with *P. flaveola* (= *verticis* Say), in which he states the size of male (did not have female); eyes "separated by a space 2½ times their own diameter"; no lateral fuscous markings on abdomen, but one of his two specimens had abdominal segments 1–5 ferruginous, "6 and 9 paler, 7 and 8 yellowish"; the entire abdomen of the other specimen was yellowish. Dr. McDunnough states that the wings of the female have no black cross veins; the comparison with *flaveola*, by Walsh, would indicate wholly pale wings in male also. One would expect, further, that the vertex would be ferruginous; thorax pale ferruginous; legs yellowish, apical portion of fore femur reddish, tips of tibia and tarsal joinings fuscous; abdomen yellowish; tails whitish, joinings fuscous.

There are several male specimens in the Cornell collection from Elkhart, Ind., in a rather poor state of preservation, which probably belong here. These males agree well with the above account, except for the tails, which are wholly pale, joinings not darker.

*Potamanthus neglectus sp. nov.

1927—Argo—J. N. Y. Ent. Soc. 35: 320, 322. Figs. (*verticis*) .... Md.

Length: body 8–8; wing 8–9 mm.

A very slender pale species; eyes of male very small; cross veins in wing disc black, in both sexes; tail joinings blackish.

Male—Eyes very small, separated by a space 2½ to 3 times diameter of eye. Ocelli white, black-ringed at base. Vertex of head reddish. Thorax whitish, with median reddish stripe. Fore legs yellowish; femur purplish at apex, tibia black-tipped, tarsal joinings fuscous. Middle and hind legs whitish. Wings hyaline, all longitudinal veins pale whitish. Cross veins in disc of each wing black; those of costal margin and outer margin of both wings, and posterior third of fore wing, colorless. Abdomen pale creamy white, segments 2–7 semi-hyaline, apical segments opaque, somewhat yellowish; reddish brown lateral spots as in *verticis*. Genitalia creamy white. Tails whitish, joinings distinctly but narrowly black.

Female very similar to male; cross veins of both wings black, as in male, including a few in basal half of costal margin of fore wing; fore femur less distinctly ferruginous at tip.

Holotype—Male imago. Ithaca, N. Y., July 2, 1932, (Miss C. N. Hardy) No. 1349.1 in Cornell collection.

Allotype—Female imago; same locality, Aug. 2, 1934 (Miss A. M. French). No. 1349.2 in Cornell collection.

Paratypes—No. 1342-3—25 in Cornell Collection. 4 male imagos, 19 female imagos; Ithaca, N. Y. (Miss C. N. Hardy), July 2, 1932, and Conococheague Pk. near Hagerstown, Md. (V. Argo), July 22, 1926.

It is this species which Argo considered to be *P. verticis* Say. It agrees well with the original description of that species, drawn from a female specimen. We are following McDunnough, however, in considering *flaveola* a synonym of *verticis*, mainly because specimens fitting that description occur in Indiana and Illinois, while the present species is more eastern in distribution. The males of *verticis* have larger eyes than *neglectus*, and wholly pale wings; we can point to no means of separating the females of the two species. In this species as well as in others of the genus, the reddish markings of vertex, thorax and fore legs fade out almost completely in alcohol.

Potamanthus rufus Argo

1927—Argo—J. N. Y. Ent. Soc. 35: 323. Fig. (*rufous*) .... N. Y.

Length: body 15; wing 13½–14 mm.

Vertex and thorax reddish; fuscous spots on sides of abdominal tergites; tail joinings reddish; wings wholly pale, in both sexes.

* This name may fall as a direct synonym of *P. verticis*.

Eyes of male small, separated by a space equal to 2½ diameters of eye; eyes of female small, separated by at least three eye diameters. Vertex of head, and antennae, reddish; thorax likewise ferruginous. Fore femur and tibia "strongly ferruginous" at tips; tarsal joinings fuscous. Wings wholly pale, without darker cross veins, in both sexes. Abdomen yellowish, with "fuscous spots on sides of segments." Tails with joinings "strongly ferruginous" in both sexes. Fore femur and tibia of female reddish throughout their lengths. Female said to be "more strongly ferruginous" than male. (We have taken the liberty of Latinizing the name of this species.)

In the Cornell collection are several much-faded specimens of this species, from McLean and Corning, N. Y., which may be part of the type lot. Others which seem similar are from Ithaca, N. Y. We have seen a specimen from Canajoharie, N. Y. (H. K. Townes, Jr.).

Potamanthus verticis Say

1839—Say—J. Acad. Nat. Sci. Philad. 8: 42 (Baetis) .... Ind.
1853—Walker—List Neurop. Ins. Brit. Mus. 3: 562.
1859—LeConte—Complete Writings of T. Say, 2: 412.
1862—Walsh—Proc. Acad. Nat. Sci. Philad. p. 377 *Ephemera flaveola*) .... Ill.
1863—Hagen—Proc. Ent. Soc. Philad. 2: 178 (*Ephemera flaveola*).
1869—LeConte—Descript. of Ins. of N. Amer. by T. Say, p. 412.
1871—Eaton—Trans. Ent. Soc. London, p. 70 (*Ephemera flaveola*).
1883—Eaton—Revis. Monogr. p. 71 (*Ephemera flaveola*).
1926—McDunnough—Canad. Ent. 58: 186 (*flaveola* = *verticis*).

Length: body 7½–9½; wing 8–9 mm.

Vertex reddish; wings wholly white in male, those of female with blackish cross veins; lateral brownish markings on abdomen; tails fuscous at joinings.

Say's description, evidently drawn from a female specimen, calls for a ferruginous vertex; two reddish stripes on thorax, "confluent before, and becoming obsolete behind"; cross veins (nervures) of wings black "except those of the margin"; legs white, fore femur reddish at tip, tip of tibia and tarsal joinings fuscous; joinings of tails black. Additional data from Walsh's description of *flaveola*, as follows. "Orbits of ocelli and basal joints of antennae a little dusky"; antennal filament pale. Eyes of male large, separated by space equal to one diameter of eye. Thorax pale ferruginous. Legs yellowish; apical half of fore femur ferruginous; tips of tarsi of fore and hind legs fuscous. Costal margin of fore wing with slightly yellowish tinge; costa, subcosta and radius coarse, yellowish, other veins in wing of male fine, hyaline. Wing of female with fuscous cross veins in both wings, "except at the tip and along the posterior margin." Abdominal tergites "with a lateral fuscous vitta on joints 3–7 interrupted at the sutures." Tails whitish, joinings fuscous.

We follow McDunnough in considering *flaveola* Wlsh. a synonym of *verticis* Say. Identifications of *verticis* by Hagen, Eaton, and Banks, who placed the species in the Heptagenine group, we think are therefore incorrect. The species which Argo held as *verticis*, we are treating as *neglectus*. In specimens we have seen, which seem to be of this species, the lateral abdominal markings consist of a rounded spot on each marked tergite, above which is usually a convex mark.

Specimens in the Cornell collection which seem to be of this species are from Knoxville, Tenn. (Miss L. Shipe); Rockford, Ill. (Miss R. Marshall); Buffalo, Ithaca and Fulton, N. Y.

ADDENDUM

In a recent article (Canad. Ent. 67: 117–122), Dr. Ide has described and figured nymphs of three species of Potamanthus,—*rufus*, *flaveola*, and *walkeri*,—and imagos likewise of his new species *walkeri*. This species will fall, in our key to the imago, with *verticis* Say, from which it may be separated by its slightly larger size and entirely pale abdomen.

Subfamily **CAMPSURINAE**

Eyes of male, small, remote. A triangular median projection on anterior portion of pronotum. Legs of both sexes weak and aborted, except fore legs of male. Wings translucent. Cross veins much less numerous than in Ephemerinae and Ephoroninae. Costal strip of fore wing compressed and continued around tip of wing. No marginal intercalaries. Outer fork (Of) of Rs (R_{4+5}) very deep, deeper than fork of R_{2+3}. Triple triad reduced, its distal fork (3) wanting. Vein M forked very near the base; its posterior branch shorter than the intercalary. Cu_1 with a sigmoid curvature. Two cubital intercalaries, attached basally to Cu_1. Forceps 2-jointed; second joint very long and slender.

Genus **Campsurus** Eaton

1868—Eaton—Ent. Mo. Mag. 5: 83.
1883—Eaton—Revis. Monogr. p. 38.
1920—Ulmer—Stett. Ent. Zeit. 81: 106.
1920—Ulmer—Festsch. f. Zsch. 25: 17.
1924—Needham & Murphy—Bull. Lloyd Lib. 24: 13.
1933—Spieth—J. N. Y. Ent. Soc. 41: 347.

Yellowish or whitish mayflies of moderate size,—wing 12 to 15 mm., in our species. Eyes of male small, remote, separated by a distance greater than diameter of eye. Middle ocellus rather small, lateral ones large, prominent. Pronotum much wider posteriorly than at anterior margin; its anterior margin prolonged forward in median area as a somewhat triangular projection under which posterior portion of head is concealed. Fore leg of male short, not as long as abdomen. Coxae rather elongate, closely approximated. Femur slightly shorter than tibia; tarsus about one and a third times as long as tibia. Basal tarsal joint rounded, short; other four joints subequal to one another, each fully three times length of basal one; each is narrowed slightly at base. Claws dissimilar, one considerably the longer, in length almost equal to distal tarsal joint; weak, slender. Middle and hind legs reduced to mere stubs; all legs of female aborted. See male imago, fig. 43.

Venation less compact than in Ephoron, cross veins distinctly less numerous, so that space between any two of these in middle area of wing is about as wide as long. Main veins of costal margin of fore wing usually purplish, others paler. No marginal intercalaries, nor cross veins very near outer margin. Posterior branch of radial sector forked very close to base. Between branches of cubital vein are two long intercalaries; these may be joined to one another basally or enter anterior branch of cubitus close together. Hind wing relatively wider for its length than in Ephoron. Costal projection obtuse, very near base. Median vein forked about halfway from base to margin.

Tails two, in both sexes. Pleural fold not noticeably expanded. Posterior margin of 9th sternite of female almost straight. Forceps arise from short basal projections set directly on posterior margin of 9th sternite of male, so that a true forceps base is lacking. Forceps two jointed, if this basal portion be considered part of forceps. Lateral slender spine-like processes may arise from base of forceps. Penes variable in size and shape, usually decurved, and generally not united on median line.

Wings and genitalia of the imago are figured by Eaton (1883), Needham and Murphy, and Spieth. Figures of nymphal structures appear in Ulmer (Festsch. f. Zschokke) and Needham and Murphy (fig. 87).

Genotype — *C. latipennis* (in Palingenia), Walker.

Distribution—Nearctic, Neotropical.

FIG. 87. Nymph of *Campsurus sp.*

KEY TO SPECIES OF CAMPSURUS*

Male imagos, except *puella*

1—Fore leg entirely blackish; obscure fuscous markings on abdominal segments along pleural fold . . . . **decoloratus**

—Fore femur pale, tibia and tarsus purplish or blackish; abdomen not marked as above 2

2—Pronotum and dorsum of abdomen more or less shaded with grey or blackish brown 3

—Pronotum and dorsum of abdomen with no such dark shading . . . . **primus**

3—Meso- and metanota reddish brown, pleura and sternum somewhat paler; pronotum with a blackish brown lateral areas . . . . **incertus**

—Meso- and metathorax pale yellowish brown; pronotum clouded with grey . . **puella**

VERIFICATION TABLE

Adults

Species	*Wing length*	*Head*	*Meso-notum*	*Fore femur*	*Lat.*** *str.*	*Distr.*
decoloratus	14½	Br. black	Yel. br.	Black	Present	Tex.
incertus	12	Purp. bl.	Red-br.	Pale	Absent	Ga.
primus	12	Yellow	Yellow	Yellow	Absent	Ill.
puella	?	?	?	Pale	?	La.

* The species *primus* and *incertus* would fall in the genus Tortopus Ndm. & Murph.; as we are not yet certain that Tortopus is a valid genus, we retain the above-mentioned species in Campsurus.

** Lateral fuscous stripes on abdominal tergites.

Campsurus decoloratus Hagen

1861—Hagen—Syn. Neurop. Ins. N. Amer. p. 43 (Palingenia)............................Mexico
1871—Eaton—Trans. Ent. Soc. London, p. 65 (Hexagenia).
1883—Eaton—Revis. Monogr. p. 41.
1924—McDunnough—Canad. Ent. 56: 7 Fig..Texas

Length: body 16; wing 14½ mm.

Mesothorax yellowish fuscous; fore legs blackish; abdomen yellowish, with lateral obscure fuscous stripes.

Head brownish-black, antennae pale. Mesothorax yellowish fuscous. Fore legs blackish, middle and hind legs yellowish. Wings hyaline, yellowish; subcosta fuscous, other veins yellowish. Abdomen yellowish, with lateral obscure fuscous stripes. Tails yellowish. (Description taken from Hagen and Eaton). Above description is probably of a female specimen.

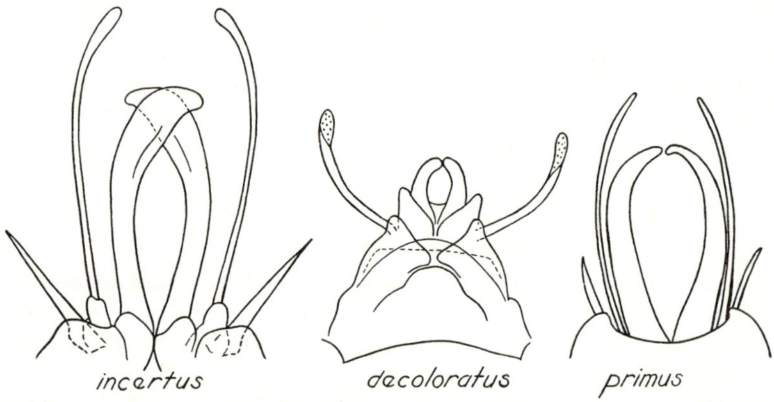

FIG. 88. Male genitalia of three species of Campsurus (*primus*, after McDunnough).

There are several specimens of Campsurus in the Cornell collection from Weslaco, Texas, (S. W. Clarke), and College Station, Tex. (F. F. Bibby), all males, which have blackish fore legs, reddish brown thorax; costal margin and three main veins, also costal cross veins, pale purplish, other veins yellowish. Body 13–14 mm., wings 11–12 mm. Genitalia similar to that figured by Dr. McDunnough, of a specimen from Brownsville, Tex., and referred to *decoloratus* (see fig. 88). Abdomen yellowish, the slightly expanded pleural fold somewhat reddish. (Are these the fuscous stripes of Hagen?) We are holding these males as *decoloratus*. Other species of Campsurus in our fauna have the fore femora pale.

Campsurus incertus sp. nov.

Length: body 11–12; wing 12 mm.

Thoracic notum largely reddish brown; fore femur pale; median dorsal area of abdomen more or less shaded with purplish grey.

Male—Basal joint of antenna purplish or brownish; filament pale. Ocelli pale, widely ringed with purplish black at base. Vertex of head reddish brown tinged with purplish or blackish in region of ocelli. Triangular forward-projecting portion of pronotum brownish black, in holotype, its anterior margin pale hyaline, in male paratypes. Remainder of pronotum rather bright yellowish, the lateral areas shaded with greyish, and with a blackish brown rectangular area of each side of median line, widened beyond center, obsolescent posteriorly; posterior margin very narrowly purplish on each side, not in middle. Median area of mesonotum yellowish to yellowish red, elsewhere reddish brown; sutures and margins blackish, also a diffuse V-shaped mark anterior to scutellum, each arm continued forward as a diffuse submedian stripe. Metanotum pale reddish brown, central area blackish. Pleura and sternum reddish yellow to pale red-brown. Fore femur whitish; tibia and tarsus purplish, tarsus paler apically; claws silvery. Middle and hind legs yellowish. Wings hyaline whitish; costa of fore wing greyish, subcosta and radius rather pale purplish; no dark cross veins nor darker shading on costal margin. Basal portion of costa and subcosta of hind wing grey; all other veins of both wings colorless.

Abdomen yellowish. Median area of tergites shaded with purplish grey, pale dorsally, darker on apical tergites. On basal tergites, mid-dorsal area pale, grey shading laterally and on posterior margins; distinct dark transverse triangle on each side of tergite 1, near median line. A diffuse darker submedian line on middle tergites, paler shading on each side. In male paratypes, grey shading much less distinct and considerably paler than in holotype. Genitalia quite similar to those of C. *primus* but the penes distinctly wider apically than at base (see fig. 88) which is the reverse of condition found in *C. primus*. Tails whitish, joinings somewhat opaque.

Female very similar; in some specimens, the dark markings are much more prominent and deeper in color than in the male, the dorsal marking becoming a wide blackish band; thoracic markings and vertex of head more blackish. In others the shading is not deeper nor more pronounced than in male. All legs yellowish. In fore wing, longitudinal veins from radius to anal region pale greyish lavender.

Holotype—Male imago; Spring Creek, Ga. (J. C. Bradley); Aug. 1931. No. 1347.1 in Cornell University collection.

Allotype—Female imago; same data. No. 1347.2 in Cornell collection.

Paratypes—2 male imagos, 55 female imagos; Eufaula, Ala., Sept. 4, 1931, Macon, Ga., Aug. 1932 (P. W. Fattig). No. 1347.3–59, Cornell University collection.

The dark median shading on pronotum and abdominal tergites seems to distinguish this species from others in our fauna, except from *puella,* in which the thorax is paler. It is close to *primus,* and apparently even more closely allied to *puella,* which is imperfectly known. It is quite possible that it may prove to be synonymous with *puella.*

In addition to the type specimens, we have others from Albany and Macon, Ga. (P. W. Fattig).

Campsurus primus McDunnough

1924—McDunnough—Canad. Ent. 56: 7. Fig. .... Ill.
1924—Needham and Murphy—Bull. Lloyd Lib. 24: 15 (In key; also footnote).
1926—McDunnough—Canad. Ent. 58: 185 (in discussion of *C. puella*; under *Ephoron leukon*).

Length: body 10; wing 12 mm.

Fore femora yellow; no dark shading on costa; penes long and broad.

Head pale yellowish, antennae whitish. Thorax somewhat deeper yellow. Fore femur yellow, tibia and tarsus blackish. Wings hyaline, venation pale; costa of fore wing with no shading of dark color. A long thin process projects from base of each forceps, about one-third as long as the forceps. Penes almost as long as forceps; broad and long, somewhat incurved apically. Tails whitish.

This species is somewhat smaller than *C. decoloratus* and lacks the dark femur, yellowish wings and dark shading of costal margin of that species. The genitalia are quite distinct from *decoloratus*, as represented by specimens from Texas. Dr. McDunnough states that *primus* may prove to be synonymous with *puella*. We are considering it to be a distinct species; we have seen no specimens. Genitalia very similar to *incertus* except for shape of penes (see fig. 88).

Campsurus puella Pictet

1843—Pictet—Hist. Nat. Neurop. 2, Ephem.; p. 145 (Palingenia)................................La.
1853—Walker—List Neurop. Ins. Brit. Mus.; p. 548 (Palingenia).
1861—Hagen—Syn. Neurop. Ins. N. Amer. p. 40 (Palingenia).
1871—Eaton—Trans. Ent. Soc. London, p. 58 (? Campsurus).
1873—Hagen—Trans. Ent. Soc. London, p. 391 (*Polymitarcys alba*).
1883—Eaton—Revis. Monogr.; p. 47 (*Polymitarcys alba*).
1926—McDunnough—Canad. Ent. 58: 185 (transferred to Campsurus).

Length: body 10–11?; wing 11–12? mm.

Pictet's very brief description of an incomplete female of this species, from New Orleans, is mainly a comparison with *Polymitarcys virgo*; the lengths of body and wing indicated above are of a male specimen of *P. virgo*, as recorded by Eaton.

Pictet says this species differs from *P. virgo* in its much shorter thorax, arched at the middle and prolonged forward in a sharp point (on which structural character it has been placed in Campsurus); the ocelli are black, but the head is not black-tinged, as in *P. virgo*; the feet are brownish at the extremity and femora are pale; network of wings is less compact and less white. Descriptions of *P. virgo* indicate that the body of that species is whitish; pronotum clouded with grey, tinged with brownish grey on anterior margin; remainder of thorax pale yellowish brown; abdominal tergites more or less clouded with grey, especially on posterior margins; wings white, with costa, subcosta and radius of fore wing grey; tails white.

We are holding the name *puella*, in the hope that future collecting in the vicinity of New Orleans may result in the capture of imagos of both sexes, and thus make it possible to determine the relation of this to other species in the group. *Campsurus primus* or *Campsurus incertus* may prove to be snyonyms of *puella*, if this species can be correctly identified.

Subfamily **NEOEPHEMERINAE**

Pronotum much wider than long. Claws dissimilar on all legs, in male and female. Wings hyaline. Basal costal cross veins weak or vestigial; stigmatic cross veins distinctly aslant, tending to anastomose. First anal vein attached to hind margin by one to three cross veins, the basal of these being the coarser. Posterior branch of median vein hardly more strongly curved at base than is anterior branch of cubitus. Two long cubital intercalaries, the posterior one deeply forked, attach Cu_1 to the hind margin. Costal projection of hind wing acute. Long marginal intercalaries. Pleural fold of abdominal segments expanded laterally, segments 8 and 9 bearing short postero-lateral spines. Tails three, in both sexes. Penes short, dome-shaped, separated only by a V-shaped median cleft on apical margin. Basal joint of forceps relatively long, more than one-half the length of second joint.

KEY TO GENERA OF SUBFAMILY NEOEPHEMERINAE

Adults

1—First anal vein of fore wing attached to hind margin by a single cross vein at right angles to it; fork of M in hind wing longer than its stem.... **Neoephemera**

—First anal vein of fore wing attached to hind margin by two or three cross veins; fork of M in hind wing subequal to its stem...................... **Oreianthus**

Genus **Neoephemera** McDunnough

1925—McDunnough—Canad. Ent. 57: 168.

Eyes of male large, almost contiguous apically. Posterior margin of head of female very slightly convex. Fore leg of male short; tarsus subequal to tibia; tibia 1¼ to 1½ times as long as the femur. In fore leg of female, tibia very slightly shorter than femur; tarsus slightly longer than tibia; apical joint of tarsus longest, equal to 2nd and 3rd joints combined. In hind leg of male, tarsus slightly shorter than femur; tibia about 1¼ times as long as femur; distal tarsal joint as long as the three preceding joints combined. Apical margin of prosternum very slightly extended in middle area, between legs. Basal costal cross veins obsolescent, those in stigmatic area strongly anastomosed. First anal vein of fore wing simple, "connected with hind margin by only a single cross vein, running almost at right angles to it from approximately its center and cutting off a large triangular space which may or may not contain a single short marginal intercalary" (McD.). Inner fork of radial sector in hind wing much longer than its stem. Median vein forked near base, so that the fork is longer than its stem. Forceps 4-jointed, the two apical joints very short, subequal. Subanal plate of female appears almost straight on apical margin. Nymph not known.

Genotype—*N. bicolor* McD.

Distribution—Nearctic; eastern Canada.

Neoephemera bicolor McDunnough

1925—McDunnough—Canad. Ent. 57: 168..Que.

Length: body 9; wing 10 mm.

This species is the sole representative of the genus.

Head ruddy brown; margin of eyes pale ochreous. Thorax reddish brown; scutellum and anterior portion of mesonotum paler. Legs pale yellowish; fore tarsi slightly smoky. Wings hyaline; all veins colorless. Abdominal tergites reddish brown; posterior margins narrowly pale yellowish; a broad pale yellow median stripe, which becomes wider posteriorly on each tergite, is widest on 7 to 9, where only lateral patches of the brown color remain. Sternites pale, with lateral brownish shading; posterior margins narrowly whitish; on each sternite "just beneath the flange" is a blackish dot. Forceps pale yellowish; tails whitish. Genitalia shown in fig. 78.

The type specimen was collected at Laprairie, Quebec. We have seen specimens of this species, of both sexes,from the type locality, which were kindly loaned us by Dr. McDunnough; he likewise presented a male imago to the Cornell collection.

Genus **Oreianthus** Traver

1871—Joly—Bull. Soc. Hist. Nat. Toulouse 4: 147 (Caenis).
1881—Eaton—Ent. Mo. Mag. 17: 196 (Caenis).
1884—Eaton—Revis. Monogr. p. 138 (?Tricorythus).
1917—Lestage—Ann. Biol. Lacust. 8: 370 (Caenis).
1921—Rousseau—Les larv. et nymph. aquat. ins. Europe, p. 241 (Tricorythus).
1931—Traver—Canad. Ent. 62: 103.
1932—Traver—J. Elisha Mitchell Sci. Soc. 47: 111.

Characters of imago refer to female only, as male is not known. Eyes small, far apart, set out on angles of head. Median ocellus smaller than lateral ones. A distinct median carina on vertex. Head rather small. Thorax and abdomen robust; body slightly widest across mesothorax. Posterior margin of pronotum quite deeply excavated in median area. Prosternum produced into a median carina; a smaller and more rounded projection is present also on anterior portion of mesosternum. Foreleg not quite half the length of body; tibia slightly longer than femur; tarsus about ½ as long as tibia. Hind leg shorter than others, the tibia somewhat shorter relatively. In each leg, the distal tarsal joint is much longer than any of the others. Long intercalary between branches of media

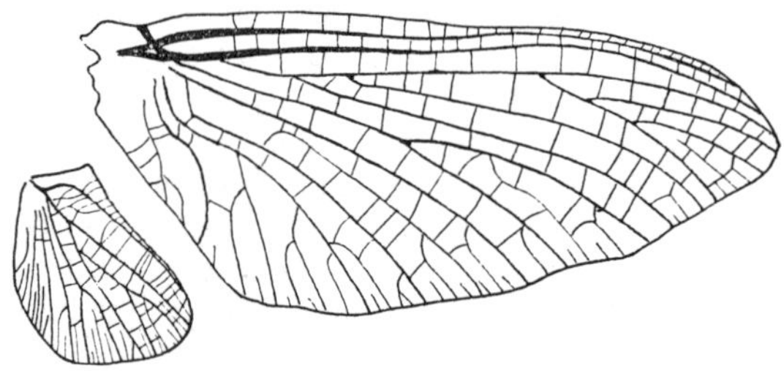

FIG. 89. Wings of *Oreianthus purpureus*.

extends fully ¾ the distance to the base. Cu_1 joined to hind margin by two or three intercalaries; the most anterior of these may appear to be a marginal fork of the anal vein. In hind wing, the inner fork of the radial sector is approximately as long as its stem. Median vein forked about halfway to margin. (See fig. 89).

Abdomen stout. Pleural fold on segments 1–9 developed into expanded plates, the postero-lateral angles of which, on 8 and 9, bear short spines. The median portion of posterior margin of 9th sternite extends backward as a blunt projection.

Head of nymph almost twice as wide as long. Second joint of antenna almost as long as first, but more slender. Prothorax slightly longer than head at middle line, considerably longer on lateral margins, as antero-lateral angle is produced forward in a large blunt spine; lateral margins developed into thin expanded areas. A semicircular lobe at antero-lateral margin of mesonotum. Body widest at greatest width of mesothorax, just anterior to base of third legs. Short median backward-projecting spine on metanotum. Traces of a median anterior projection to mesosternum. Legs long, finely hairy throughout length; hind legs longest, reaching to tip of abdomen. Femora compressed; a forward-projecting flange on dorsal edge apically, and a V-shaped excavation on ventral edge. Flange at tip of tibia on ventral margin. Tibiae and tarsi less definitely compressed, more rounded. Femur slightly longer than tibia; tarsus of fore leg more than half, of hind leg less than half, the length of tibia. Claws simple, not denticulate. On abdominal segment 1, a small upturned gill filament reminiscent of Caenis. Gills on second segment elytroid plates, grown together on median line, and covering all following pairs of gills; in shape, very like those of Caenis. Median spine on tergites 1 and 2; tergites 6–9 with a distinct median carina, on which is borne a short spine, except on tergite 9. Lateral margins of segments 3–9 prolonged into flattened processes. very pronounced on 6–9; blunt backward-projecting spines on 3–9, becoming progressively longer posteriorly. Abdomen gradually narrowed back of 6th segment. Tails three. See fig. 90.

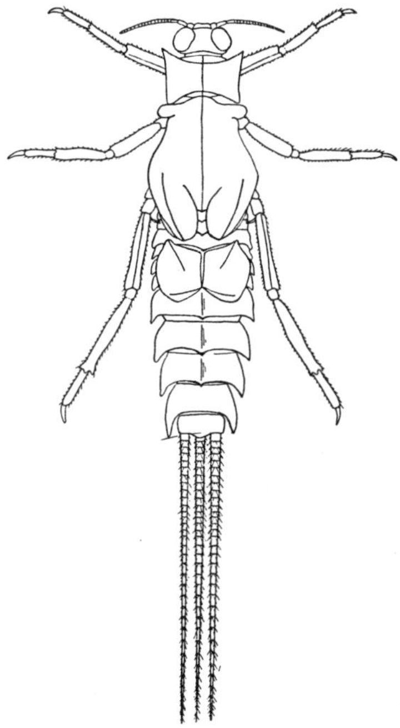

FIG. 90. Nymph of *Oreianthus purpureus*.

Genotype—*O. purpureus* Traver.

Distribution—Holarctic; in Nearctic, southeastern United States.

Oreianthus purpureus Traver

1931—Traver—Canad. Ent. 63: 103. Figs. .... N. C.
1932—Traver—J. Elisha Mitchell Sci. Soc. 47: 111. Figs.

Length (female): body 16; wing 17 mm.

General color purplish; mesonotum light brown; venation purplish. Only the female imago and nymph of this species are known.

Head of female purplish-brown; ocelli white, narrowly dark-ringed at base; white areas at bases of antennae; paler areas lateral of each eye, on vertex; antennae purplish brown; eyes dark grey. Pronotum purplish brown. Mesonotum yellowish brown, becoming bright reddish brown posteriorly; yellow-brown portion with narrow darker sutures, and brownish-red shading along each, also at anterior margin. Metanotum light purplish brown. Pleura pale brownish; purplish brown markings, most evident as oblique streaks anterior to middle and hind leg. Prosternum and anterior portion of mesosternum light grey with purplish mottling; remaining portion purplish brown. Fore and middle legs deep purplish brown; femur narrowly pale grey at base; tibia pale yellow-brown basally and at extreme apex; a yellow mark on inner margin of each tarsal joining. Third leg light brown; all joinings purplish brown. Wings hyaline; all veins deep purplish, cross veins finer than longitudinals; costa, subcosta and radius thicker and darker than others; 9 to 10 basal costal cross veins, the basal two or three of these very faint, others somewhat better developed. In the costo-apical space the cross veins are fine, quite faint, becoming almost invisible at apex of this space; a tendency for some of these to anastomose; these veins are 20 or more in number, considerably slanted and somewhat irregular.

Abdomen reddish purple, many segments considerably mottled with fine paler marks, especially on sternites, which thus appear paler than tergites; tergites 2 and 3 also much mottled. Pleural fold and area immediately adjoining it, on tergites and sternites, blackish; a paler streak on each tergite near posterior margin; on basal sternites, a pale spot in this dark area, next pleural fold. Overlapping portions of tergites appear darker; intersegmental areas of sternites paler. On basal tergites, a blackish diffuse mid-dorsal line; this becomes wider and more diffuse on 6, not present on other segments, but a narrow pale median hair-line is present on middle ones. Tracheal branchings faintly greyish white. Tails very dark purplish brown or black; 2nd, 3rd, 5th and 6th basal joints shorter than others.

Besides the type material, we have seen a female imago from Greenville, S. C. (H. K. Townes, Jr.). Nymphs of this species from Rock Creek, Ga. (P.W. Fattig) are in the Cornell collection; also nymphs of two other species of this genus; one from Sweetwater Branch, Liberty Co., Fla. (J. G. Needham), another from Upatoi Creek, Ga. (P. W. Fattig). The reference of Oreianthus to the family Baetidae, in Traver 1932, should be disregarded.

Family **HEPTAGENIIDAE**

Eyes of male simple; variable in size in the different genera. Hind tarsus with five freely movable joints. Venation plentiful, in both wings. Two parallel pairs of cubital intercalaries, unattached basally, of which the pair nearer the hind angle is the longer. Outer fork of Rs in hind wing present, except in Arthroplea. Forceps usually 4-jointed (5 joints, in Arthroplea).

The nymphs are sprawling forms, of protective coloration; bodies more or less flattened, heads strongly depressed; eyes lateral in position. Labrum variable in shape, but always considerably wider than its length. Mandibles slender in basal half. Canines two in number, the outer always longer than the inner; both are distinctly scoop-shaped, their margins more or less crenate; one or two sharp spines may be borne at tip of inner canine. Lacinia present or wanting; in the latter case, a row of hairs is often present in region of lacinia. Maxillary palp two-jointed; distal joint often thickly set with spines in apical portion. A fringe of long hairs is present on inner margin of galea-lacinia, usually also a second row parallel to these, a short distance back from margin. Upper margin or crown may likewise bear hairs or spines, or both. Labial palp two-jointed; distal joint thickly set with spines in apical portion. Glossae and paraglossae variable in shape and relative size in the different genera.

Gills are present on segments 1–7 of the abdomen. Each pair consists of a dorsal lamellate and a ventral fibrillar portion; however, the latter may be wanting or very much reduced. In several genera, the fibrillar portion has pushed upward so that it appears to lie dorsal of the plate-like part. The genus Anepeorus is not yet known in the nymphal state.

We recognize a single subfamily, the Heptageninae. Its characters are those of the family. The ten North American genera may be characterized as follows:

KEY TO GENERA OF HEPTAGENINAE

Adults

1—Outer fork (Of) of Rs (R_{4+5}) of hind wing wanting; male forceps 5-jointed **Arthroplea**
—Outer fork of Rs of hind wing present; male forceps 4-jointed................2
2—Fore tarsus of male not more than ¾ as long as tibia; apical margin of forceps base deeply excavated...**Anepeorus**
—Fore tarsus of male distinctly longer than tibia; apical margin of forceps base not as above..3
3—Basal joint of fore tarsus of male ½ or less than ½ as long as second joint....4
—Basal joint of fore tarsus of male at least ⅔ as long as second joint, may equal or slightly exceed it in length...6
4—Stigmatic cross veins more or less anastomosed; basal joint of fore tarsus of male ⅙ to ¼ as long as second.................................**Rhithrogena**
—Stigmatic cross veins not anastomosed; basal joint of fore tarsus of male variable. 5

5—Basal joint of fore tarsus of male 1/3 to 1/2 as long as second; penes more or less distinctly L-shaped..**Stenonema**

—Basal joint of fore tarsus of male 1/6 to 1/3 as long as second; penes not distinctly L-shaped..**Heptagenia**

6—Basal joint of fore tarsus of male 2/3 to 3/4 as long as second joint.............7

—Basal joint of fore tarsus of male equal to or slightly longer than second joint....8

7—Stigmatic area of fore wing divided by a fine line into an upper and lower series of cellules; basal costal cross veins very weak......................**Cinygma**

—Stigmatic area of fore wing not divided as above; basal costal cross veins usually well developed..**Cinygmula**

8—Stigmatic cross veins of fore wing more or less strongly anastomosed, sometimes forming two series of cellules; basal costal cross veins weak.........**Ironopsis**

—Stigmatic cross veins of fore wing not anastomosed; basal costal cross veins variable..9

9—Basal costal cross veins of fore wing weak; stigmatic cross veins not slanting.**Iron**.

—Basal costal cross veins of fore wing strong; stigmatic cross veins distinctly slanting..**Ironodes**

VERIFICATION TABLE

Adults

Genus	Basal costal cross-veins	Stigmat. c. vs. Slant	Stigmat. c. vs. Anast.[1]	Claws[2] of ♂	Forc. joints.	♂ eyes contig.
Anepeorus[3]	Rather weak	Yes	Part.	Similar	4	No
Arthroplea[4]	Rather weak	No	No	Dissim.	5	Almost
Cinygma	Weak	Yes	Fully	Dissim.	4	No
Cinygmula	Strong	Yes	No	Var.	4	No
Heptagenia	Var.	No.	No	Dissim.	4	No
Iron	Weak	No	No	Var.	4	Usually
Ironodes	Strong	Yes	No	Dissim.	4	No
Ironopsis	Weak	Yes	Fully	Dissim.	4	Yes
Rhithrogena	Rather weak	Yes	Part.	Dissim.	4	Usually
Stenonema	Strong	Var.	No	Dissim.	4	No

KEY TO THE GENERA

Nymphs

1—Fibrillar portion of each pair of gills wanting; second joint of maxillary palp at least four times as long as galea-lacinia..........................**Arthroplea**

—Fibrillar portion of each pair of gills present; second joint of maxillary palp much shorter..2

2—Tails two..3

—Tails three..5

3—A triad of stout spines present at tip of galea-lacinia; no submedian spines on abdominal tergites..4

—No such triad of spines on galea-lacinia; a pair of submedian spines present on abdominal tergites..**Ironodes**

[1] Stigmatic cross veins anastomosed partially (Part) or fully, forming two complete rows of cells.

[2] Fore claws, male tarsus.

[3] Fore tarsus of male shorter than tibia.

[4] Second branch of radial sector (Of) not forked.

4—Median row of long hairs on abdominal tergites; front of head slightly emarginate .... **Ironopsis**

—No such median row of hairs on abdominal tergites; front of head not emarginate .... **Iron**

5—Gills of seventh pair reduced to a single slender tapered filament or spine; tracheae, if present in this filament, without lateral branches .... **Stenonema**

—Gills of seventh pair flat and plate-like, quite similar to preceding pairs, tracheae with lateral branches always present in seventh pair 7

6—Fibrillar portion of gills reduced to a few tiny threads; front of head distinctly emarginate .... **Cinygmula**

—Fibrillar portion of gills well developed on segments 1–6, may be wanting on segment 7; front of head entire or very slightly emarginate 7

7—Gills of first and last pairs much enlarged, converging beneath body of nymph .... **Rhithrogena**

—Gills of first and last pairs not as large as middle pairs; all gills directed laterally, not converging beneath body 8

8—A chitinized area present on each mandible below molar region; inner canine about ½ as long as outer .... **Cinygma**

—No such chitinized area on mandible; inner canine at least ¾ as long as outer .... **Heptagenia**

VERIFICATION TABLE

Nymphs

Genus	*Canines*[1]	*Gal.-lac.*[2] *armature*	*Fibril. gills*	*Lat. spines*	*Tails*
Arthroplea	2:3	Absent	Absent	Present	3
Cinygma	1:2	Hairs	Ventral	Absent	3
Cinygmula	1:4	Spines	Ventral	Short	3
Heptagenia	3:4+	Spines	Ventral	Var.	3
Iron[3]	1:2	Absent	Dorsal	Short	2
Ironodes[4,5]	3:4+	Hairs	Dorsal	Short	2
Ironopsis[3,4]	2:3	Hairs	Dorsal	Short	2
Rhithrogena	1:2	Spines	Dorsal	Short	3
Stenonema	7:8+	Both	Ventral	Var.	3

[1] Ratio of length of inner canine to outer.
[2] Hairs or spines on crown of galea-lacinia.
[3] Triad of stout spines at tip of galea-lacinia.
[4] Median row of long hairs on tergites.
[5] Submedian spines on tergites.

Genus **Stenonema** Traver

1885—Eaton—Revis. Monogr. p. 262–3 (*generis incerti*); p. 278 (Ecdyurus, in part); p. 298 (Heptagenia, in part).
1905—Needham—Bull. N. Y. St. Mus. 86: 54 (Heptagenia).
1910—Banks—Canad. Ent. 42: 197 (Heptagenia).
1913—Clemens—Canad. Ent. 45: 247 (Heptagenia).
1915—Clemens—Marine & Fish. Sess. Pap. 39b; 131 (Heptagenia).
1920—Ulmer—Stett. Ent. Zeit. 81: 141 (Heptagenia, in part); 143 (Ecdyonurus, in part).
1924—McDunnough—Canad. Ent. 56: 117 (Ecdyonurus).
1933—Traver—J. Elisha Mitchell Sci. Soc. 48: 173.
1933—Traver—J. N. Y. Ent. Soc. 41: 113.
1933—Spieth—J. N. Y. Ent. Soc. 41: 330 (Ecdyonurus).

Mayflies of small to moderate size, wings ranging in length from 6 to 16 mm. Variable in color. Eyes of male moderately large in most species, but not closely approximated apically. Posterior border of head of female slightly emarginate. Posterior margin of pronotum quite deeply excavated in median area. Fore leg of male as long as or slightly longer than body. In fore leg of male, tibia about 1⅓ times as long as femur; tarsus about 1⅓ times as long as tibia. Basal fore tarsal joint ⅓ to ½ as long as the second, which is subequal to third; fourth slightly shorter; fifth subequal to first. In fore leg of female, femur subequal to or slightly shorter than tibia; tarsus slightly shorter than tibia. Basal fore tarsal joint fully ½ as long as second. In hind leg of both sexes, femur subequal to or slightly longer than tibia; tarsus about ½ as long as tibia. Basal, second and third tarsal joints usually subequal; fifth as long as the three preceding joints together. Claws dissimilar on all tarsi. Typical Heptagenine venation. Basal costal cross veins well developed, strong. In some species of the *tripunctatum—pulchellum* group, cross veins at and below bulla are considerably crowded. In species of the *interpunctatum* group, a short black transverse dash is present at bulla, in third interspace, and the hind wing is dark on outer margin. Stigmatic cross veins may be very slightly aslant, usually show no tendency to anastomosis, although a few may be forked. Cross veins in costal strip and at bulla may be dark-margined.

Forceps four-jointed. Penes more or less L-shaped. In *interpunctatum* group, spines are usually present on lower lateral margins. A pair of stout median spines is present, smaller lateral and apical ones occur in some species. Subanal plate of female well developed, extending beyond 10th segment; its apical margin entire, truncate or very slightly emarginate in middle area.

Three groups of nymphs are recognizable on structural characters. In all groups, the gill of the 7th pair is thread-like or spine-like, unlike those of the preceding pairs; a single trachea with no lateral branches may be present in this filament. Gills of the three types shown in Pl. IX. Front of head may be slightly emarginate. Inner canine of mandible about ⅞ as long as outer one. Nymphs of the *tripunctatum* group have pectinate claws. Lamellate portions of gills 1–6 rounded at apex; 7th gill fringed with long hairs, containing a single forked trachea. Postero-lateral spines are present on middle and apical abdominal segments. Upper margin of galea-lacinia bears both hairs and spines. In some species of the *pulchellum* group the claws are pectinate, in others no pectinations occur. Apex of gills truncate; 7th gill fringed laterally and containing no trachea. Postero-lateral spines may be present on middle abdominal segments. Upper margin of galea-lacinia as in preceding group. Nymphs of the *interpunctatum* group have no pectinations on the claws. Apex of gills distinctly pointed; 7th gill may have a few fine hairs along the margin, and contains an unbranched trachea. No postero-lateral spines on middle abdominal segments. Upper margin of galea-lacinia bears a row of pectinate spines. Tails three, in all groups.

Genotype—*S. tripunctatum* (in Heptagenia), Banks.

Distribution—Nearctic; general.

KEY TO SPECIES

Male adults

1—Genitalia of the *interpunctatum* or *carolina* types (see fig. 91); basal cross veins of costal and subcostal spaces, and in next space below, thickened, often margined at center (or in costal space, near costal margin); in third space below bulla, these veins usually connected by a black dash.....................2

—Genitalia not as above; cross veins in above-mentioned spaces, if thickened, are widened equally, not more at center than elsewhere; no such black dash in third space below bulla..13

2—Genitalia of *carolina* type; no distinct spines along outer lateral margin of penes; cross veins in 3rd space below bulla not connected by a black dash........3

—Genitalia of the *interpunctatum* type; distinct spines present on outer lateral margin of penes; black dash usually present below bulla..................4

3—Eyes relatively small and remote, separated by a space slightly more than diameter of eye; no dark markings on face, no stigmatic dots.........**carolina**

—Eyes larger, of normal size, separated by space not greater than half of diameter of eye; dark markings on face, stigmatic dots present.............**candidum**

4—No black dash below bulla in fore wing; stigmatic dots present, black mark below antenna, dark lateral mark on pronotum and dark oblique streaks on pleura ....**frontale**

—Black dash present below bulla in fore wing, connecting 2 to 4 cross veins; stigmatic dots and markings on face and thorax variable.................5

5—Dark species; abdominal tergites 2–7 shaded heavily with grey, blackish or reddish black, except on lateral margins...............................6

—Paler species; abdominal tergites 2–6 usually wholly pale except for narrow dark posterior margins (greyish shading may be present dorsally on basal tergites).8

6—No distinct oblique dark streaks on pleura anterior to middle and hind legs, although dark markings are present above coxae; distinct dark mid-ventral markings on sternites 1–6..................................**gildersleevei**

—Distinct oblique dark streaks present on pleura anterior to middle and hind legs; no such distinct dark markings on sternites............................7

7—Basal joint of fore tarsus almost or fully ½ as long as second joint; continuous dark transverse line on frontal margin of head.....................**ohioense**

—Basal joint of fore tarsus slightly more than ⅓ but less than ½ of second; dark marks on face not connected to form a continuous line...........**canadense**

8—Stigmatic dots present; dark oblique streaks present on pleura, anterior to middle and hind legs...9

—Stigmatic dots wanting; usually no such dark oblique streaks on pleura, although reddish shading may be present in this area...........................11

9—A small black mark at postero-lateral angle of pronotum, in addition to usual black lateral streak; stigmatic dots large, distinct; wing 10–11 mm. in length ....**majus**

—No such additional dark mark at postero-lateral angle of pronotum; stigmatic dots small, indistinct, often diffuse or reduced to a mere streak; wing 8–9 mm. in length 10

10—Mesonotum yellow, often with dark median band, anterior margins dark; greyish geminate median line on abdominal tergites, grey tinge may spread over dorsal portions of these segments . . . . . **proximum**

—Mesonotum dark red-brown, no darker anterior margin; faint narrow grey geminate line may be present on basal tergites only . . . . . **conjunctum**

11—Black mark on face at base of antenna; oblique black lateral streak on pronotum; length of fore tarsal joint variable 12

—Usually no such dark marks on face or pronotum, if present much reduced; basal joint of fore tarsus about ½ as long as second . . . . . **heterotarsale**

12—Basal joint of fore tarsus short, about ⅓ as long as second; median band wholly wanting on hind femur . . . . . **interpunctatum**

—Basal joint of fore tarsus longer, about ½ as long as second; faint dark median band on hind femur . . . . . **pallidum**

13—Cross veins at and below bulla distinctly crowded; in first three spaces these are arranged as 2, 2, 3,—3, 3, 3,—or 3, 4, 4 14

—Cross veins at and below bulla much less crowded; in first three spaces these are arranged as 1, 1, 1—1, 2, 1—1, 2, 2,—occasionally 1, 2, 3—or 2, 2, 2 21

14—Hind wing distinctly dark on outer margin; cross veins at bulla arranged as 3, 3, 3, or 3, 4, 4 15

—Hind wing not dark on outer margin; number of cross veins at bulla variable . . 17

15—Abdominal segments 2–7 hyaline, white; on posterior margin of each tergite are two lateral and one median short dark transverse mark; wing 9 mm. in length . . . . . **scitulum**

—Abdominal segments 2–7 yellowish brown tinged with darker brown; markings on posterior margins of tergites variable; wing 10–14 mm. in length 16

16—Three dark transverse dashes on posterior margin of each tergite, as in *scitulum* . . . . . **femoratum**

—Wide black posterior margin on each tergite; sagittate dark mark on each, at median line . . . . . **pudicum**

17—Cross veins at bulla arranged as 3, 3, 3 or 3, 4, 4,; 2 short dark lateral dashes and one median dash on posterior margin of each tergite 18

—Cross veins at bulla somewhat less crowded, usually arranged as 2, 2, 3; uninterrupted dark line on posterior margins of tergites 19

18—Abdominal segments reddish brown; basal joint of fore tarsus about ⅓ as long as the second . . . . . **birdi**

—Abdominal segments 2–7 pale whitish or yellowish; basal joint of fore tarsus about ½ as long as second . . . . . **tripunctatum**

19—Posterior ½ or ⅔ of abdominal tergites 1 to 8 largely occupied by a wide dark brown band; dark lateral markings on sternites 20

—Abdominal tergites 1 to 8 largely whitish; sternites immaculate . **rubromaculatum**

20—All tibiae distinctly reddish; basal joint of fore tarsus less than ⅓ of second; dark lateral patches on sternites large, conspicuous . . . . . **vicarium**

—Tibiae light amber, not red-tinged; basal joint of fore tarsus slightly more than ⅓ of second; dark lateral patches on sternites smaller . . . . . **rivulicolum**

21—Abdominal tergites largely yellowish to reddish brown 22

—Abdominal tergites 1–8 largely pale yellowish white 23

22—Abdominal tergites reddish brown; no white lateral border on 9 and 10; sternites more or less tinged with reddish or purplish . . . . . **ithaca**

—Abdominal tergites yellowish brown; lateral margins of 9 and 10 distinctly bordered with alabaster white; sternites pale smoky amber.......... **fuscum**

23—Dark stigmatic dots present, or an oblique dark streak in stigmatic area..... 24

—No such dark stigmatic markings.......... 31

24—Oblique purplish black streak in stigmatic area, in place of usual dark dot. **bellum**

—Dark dot present in stigmatic area.......... 25

25—Outer margin of hind wing distinctly brownish.......... **integrum**

—Hind wing without dark outer margin.......... 26

26—Tails wholly pale, not darker at joinings; dark posterior margin on each tergite reduced to a small dark transverse dash at median line..... **mediopunctatum**

—Tails whitish, distinctly darker at joinings; dark posterior margins on tergites somewhat more extensive.......... 27

27—Thoracic notum pale yellowish brown; wing 8–11 mm. in length.......... 28

—Thoracic notum deep brown, may be almost blackish; wing 7–8 mm. in length. 30

28—Dark posterior margins of abdominal tergites considerably reduced, evident in dorsal portion only; thorax distinctly tinged with pinkish or purplish. **annexum**

—Dark posterior margins of abdominal tergites much better defined; thorax not tinged with pinkish or purplish.......... 29

29—Basal abdominal tergites more or less tinged with purplish grey; no red or brown stain on stigmatic area of fore wing.......... **varium**

—No such greyish shading on basal abdominal tergites; stigmatic area of fore wing with a distinct reddish stain.......... **nepotellum**

30—Thorax distinctly tinged with reddish; scutellum white-tipped; abdominal segments slightly tinged with smoky.......... **rubrum**

—Thorax blacker, without red shading; scutellum broadly white; abdominal segments whiter, without smoky tinge.......... **pulchellum**

31—Small species, wing 6 mm. in length; distinct dark outer margin on hind wing .... **exiguum**

—Larger species, wing 8–11 mm. in length; hind wing not dark-margined...... 32

32—Dark posterior margins of tergites reduced to two short lateral transverse dashes; tails not darker at joinings.......... **bipunctatum**

—Dark posterior margins of tergites more extensive; tails may be darker at joinings.......... 33

33—Thoracic notum yellowish; tails distinctly darker at joinings; median and apical purplish red bands on all femora.......... **luteum**

—Thoracic notum dark olive brown to red-brown; tails may be faintly darker at joinings, or wholly pale; no median band on middle or hind femur........ 34

34—Basal joinings of tails purplish, others faintly dark-tinged; posterior margins of tergites rather wide, red-brown; thorax with distinct reddish markings; scutellum whitish.......... **placitum**

—Tails wholly pale, not darker at joinings; posterior margins of tergites narrowly red-brown in dorsal portion; thorax and abdomen with less red shading than above; scutellum white-tipped only.......... **terminatum**

VERIFICATION TABLE: (the *interpunctatum* group)

Male adults

Species	*Wing length*	*Bl. mks. on face*	*Pleural*[1] *streaks*	*Stigmat. dots*	*Tergites 2 to 6*	*Distr.*
canadense[2]	9	Dash	Present	Obscure	Dark	E
candidum[3]	8½	Dash	Absent	Present	Pale	Ohio
carolina[3]	11	None	Absent	Absent	Pale	E
conjunctum[4]	8	Dash	Present	Present	Pale	Ohio
frontale[3]	8	Dash	Present	Present	Pale	E
gildersleevei	10–11	Line	Absent	Obscure	Dark	Ohio
heterotarsale	11	None	Absent	Absent	Pale	E
interpunctatum[2]	7–7½	Dash	Absent	Absent	Pale	E
majus	10–11	Dash	Present	Present	Pale	E
ohioense	9–11	Line	Present	Present	Dark	E
pallidum	7½–8	Dash	Absent	Absent	Pale	N.C.
proximum	9	Dash	Present	Present	Obscured	N.Y.

[1] Oblique dark streaks on thoracic pleura.
[2] Basal fore tarsal joint very short, about ⅛ of second.
[3] No black transverse dash in wing below bulla.
[4] Second joint of fore tarsus about 2½ times as long as first.

VERIFICATION TABLE: (the *tripunctatum-pulchellum* group)

Male adults

Species	*Wing length*	*Bu. c. vs.*[1] *crowded*	*Hd. wing*[2] *dk. tip.*	*Stigmat. dots*	*Tergites 2 to 7*	*Distr.*
annexum	10–11	Slight	No	Present	Whitish	S E
bellum	7	Slight	No	Streaks	Whitish	S E
bipunctatum	8	Very sl.	No	Absent	Whitish	E
birdi	12	Much	No	Absent	Brown	Okla.
exiguum	6	Not	Yes	Absent	Whitish	S E
femoratum	10–12	Much	Yes	Obscure	Yel. br.	E
fuscum	13	Very sl.	No	Absent	Brown	E
integrum	6	Mod.	Yes	Present	Whitish	C
ithaca	11–12	Slight	No	Obscure	Red br.	E
luteum	10½	Very sl.	No	Absent	Whitish	N E
mediopunctatum	10	Slight ?	No	Present	Whitish	N E
nepotellum	8–10	Slight	No	Present	Whitish	E
placitum	9½	Very sl.	No	Absent	Whitish	E
pudicum	13–14	Much	Yes	Obscure	Olive br.	S E
pulchellum	7–8	Very sl.	No	Present	Whitish	E
rivulicolum	13	Mod.	No	Absent	Brown	N E
rubromaculatum	8–10	Mod.	No	Present	Whitish	E
rubrum	8	Very sl.	No	Present	Whitish	E
scitulum	9	Much	Yes	Absent	Whitish	Okla.
terminatum	8–10	Very sl.	No	Absent	Whitish	C
tripunctatum	12–13	Much	No	Obscure	Whitish	E
varium	10–11	Slight	No	Present	Whitish	S E
vicarium	13–16	Mod.	No	Absent	Dk. rd. br.	E

[1] Cross veins crowded at and below bulla.
[2] Hind wing dark on outer margin.

KEY TO THE SPECIES

Nymphs

1—Gills on segments 1–6 rounded at apex; 7th gill fringed laterally, contains a single trachea 2
—Gills on segments 1–6 truncate or pointed at apex; 7th gill variable 4
2—Frontal border of head slightly emarginate; ventral abdominal markings indistinct, on sternites 5–9 only . . . . . **femoratum**
—Frontal border of head not emarginate; ventral abdominal markings distinct, consist of two rows of dark spots on sternites 2–9 3
3—Small pale spot present at median area of frontal margin of head . . . **tripunctatum**
—No such small pale spot at median area of frontal margin . . . . . **birdi**
4—Gills on segments 1–6 pointed at apex; 7th gill contains a single trachea 5
—Gills on segments 1–6 truncate at apex; 7th gill without a trachea 15
5—Small species, 6–7 mm. in length . . . . . **pallidum**
—Larger species, 8½–13 mm. in length 6
6—Pale longitudinal streak on each side of dark median stripe on abdominal tergites almost or completely continuous 7
—Pale submedian streak not continuous, more or less interrupted 11
7—Pale median spot on frontal margin of head . . . . . **frontale**
—No such pale median spot on frontal margin 8
8—Pale spot on head near median ocellus shaped somewhat like fluer-de-lis design . . . . . **heterotarsale**
—Pale spot on head near median ocellus more or less triangular, not as above 9
9—Tergite 10 dark except for narrow pale anterior margin; dark median spot on tergite 9 distinctly diamond-shaped . . . . . **conjunctum**
—Tergite 10 with prominent pale submedian streaks; median dark spot on tergite 9 usually lens-shaped, not as above 10
10—Lateral spines on segments 8 and 9 subequal, almost twice as long as their width at base . . . . . **ohioense**
—Lateral spine on segment 8 shorter than that on 9, about as long as its width at base . . . . . **proximum**
11—Pale submedian streaks very narrow, confined to anterior half of each tergite, so that abdomen does not appear striped . . . . . **carolina**
—Pale submedian streaks wider and more extensive on each tergite, so that abdomen appears striped 12
12—Pale median spot on frontal margin of head . . . . . **majus**
—No such pale median spot on frontal margin 13
13—Posterior margins of tergites narrowly blackish; ventral markings absent or indistinct . . . . . **gildersleevei**
—Posterior margins of tergites not blackish; ventral markings distinct 14
14—Median dark stripe on tergites almost continuous, very wide on tergites 1–7, narrower on tergites 8–10 . . . . . **canadense**
—Median dark stripe on tergites 4, 8 and 9 reduced to mere trace, so that these segments appear largely pale . . . . . **candidum**
15—Abdominal segments with a rather wide blackish band on posterior margin, dorsally and ventrally 16
—Abdominal segments without such conspicuous dark posterior bands dorsally and ventrally 19
16—Claws with two small pectinations on lateral margin near apex . . . . . **luteum**

—Claws without such pectinations 17

17—Large species, body 18–20 mm. in length .... **vicarium**

—Smaller species, 10–12 mm. in length 18

18—Entire posterior half of sternite 9 dark .... **rivulicolum**

—Sternite 9 with 2 dark lateral marks, sometimes a median mark also; not as above .... **fuscum**

19—Dorsum of abdomen more or less unicolorous, without distinct color pattern ... 20

—Dorsum of abdomen with rather distinct pattern of pale and dark areas 22

20—Usually no distinct dark markings on venter of abdomen; claws not pectinate .... **annexum**

—Distinct dark markings present on venter of abdomen; claws pectinate 21

21—Dark ventral markings consist of 4 to 6 dark dots across sternite, and dark lateral streaks .... **rubromaculatum**

—Dark ventral markings less extensive; a dark U-mark on sternite 9, median dark patch on 8 .... **rubrum**

22—Postero-lateral spines present on segments 3–9; body 12–14 mm. in length .... **pudicum**

—Postero-lateral spines present in 5–9 or 7–9 only; body usually smaller (7–12 mm.) 23

23—Dark ventral markings on segments 8 and 9, or 9 only 24

—Dark ventral markings occur also on middle sternites 26

24—A dark lateral mark on each side of sternite 9; two short submedian streaks may also be present 25

—Median dark crescent-shaped mark on sternites 8 and 9, submedian streaks and dark dots on other sternites .... **varium**

25—Tergite 7 largely pale in median area, except for narrow dark markings on anterior margin .... **pulchellum**

—Tergite 7 largely dark; two narrow pale submedian streaks and narrow pale line next to posterior margin .... **terminatum**

26—Postero-lateral spine on segment 9 equals or exceeds half the length of lateral margin of segment 10 .... **nepotellum**

—Postero-lateral spine on segment 9 slightly shorter, not quite half the length of lateral margin of segment 10 .... **ithaca**

VERIFICATION TABLE

Nymphs of the interpunctatum group

Species	*Body length*	*Pale[1] spot*	*Pale[2] streaks*	*Ventr.[3] marks*	*Dk. marg.[4] tergites*	*Spine 8 vs. 9*
canadense	9	No	No	Yes	No	Equal
candidum	8½	No	No	Yes	No	Shorter
carolina	10–11	No	No	No	No	Longer
conjunctum	11	No	Yes	Yes	No	Equal
frontale	9–10	Yes	Yes	Yes	No	Equal
gildersleevei	11–13	No	No	No	Yes	Equal
heterotarsale[5]	9–10	No	Yes	Yes	No	Equal
majus	11–13	Yes	No	Yes	No	Equal
ohioense	10	No	Yes	Yes	No	Equal
pallidum	6–7	No	No	Yes	Yes	Shorter
proximum	10	No	Yes	Yes	No	Shorter

1 Median pale spot on frontal margin of head; this character is not always constant.
2 Continuous pale submedian streaks on tergites.
3 Ventral abdominal markings.
4 Abdominal tergites with dark posterior margins.
5 Assuming that *affine* equals *heterotarsale*.

VERIFICATION TABLE

Nymphs of the tripunctatum group.

Species	*Body length*	*Frons emarginate*	*Median pale spot on frons*	*Ventral marks on*
birdi	11–12	No	No	2–9
femoratum	9–12	Yes	No	5–9
tripunctatum	9–12	No	Yes	2–9

VERIFICATION TABLE

Nymphs of the pulchellum group

Species	*Body length*	*Claws[1] toothed*	*Dorsal[2] pattern*	*Margins[3] dark*	*Lat.[4] sp. on*	*Spine 8 vs. 9*
annexum	10–11	No	No	No	5–9	Longer
fuscum	10–12	No	Yes	Yes	3–9	Shorter
ithaca	10–11	No	Yes	No	7–9	Longer
luteum	10	Yes	Yes	Yes	?	?
nepotellum	8–10	No	Yes	No	5–9	Longer
pudicum	12–14	No	Yes	No	3–9	Longer
pulchellum	7–8½	Yes?	Yes	No	7–9	Longer
rivulicolum	10–11	No	Yes	Yes	3–9	Shorter
rubrum	8–9	Yes	No	No	7–9	Longer
rubromaculatum	9–10	Yes	No	No	?	Longer
terminatum	9	No	Yes	No	7–9	Longer
varium	9–12	No	Yes	No	5–9	Longer
vicarium	18–20	No	Yes	Yes	3–9	Shorter

1 Two small pectinations near apex; a difficult character to determine.
2 Distinct dorsal pattern on abdomen.
3 Wide dark posterior margins on tergites and sternites.
4 Postero-lateral spines on abdominal segments.

Stenonema annexum Traver

1933—Traver—J. Elisha Mitchell Sci. Soc. 48: 179. Fig. . . . . N. C.

Length: body 8½–9; wing 10–11 mm.

A species of *pulchellum* group, allied to *S. rubromaculatum*; spiracular dots present; tails darker at joinings.

Head pale brownish to clay-colored; an indistinct narrow dark line across center of median carina; antennae faintly dusky; vertex dull purplish brown, often with a dark brown spot on each side. Thorax pale brownish to clay-colored, distinctly tinged with pale purplish or pinkish. Dark purplish brown median geminate line on pronotum, its anterior and posterior margins narrowly purplish black. Purple streak on pleura, extending down on base of fore leg. Ochreous markings on pleura. A wide paler median stripe on mesonotum, including scutellum; latter does not appear as distinctly whitish, however, as in some other allied species, but is tinged with pinkish. Posterior portion of metanotum, especially laterally, much darker than anterior, this dark area appearing as a brown band at base of paler abdomen. A brown area on pleura between bases of fore and middle legs; a wide oblique dark streak back of fore

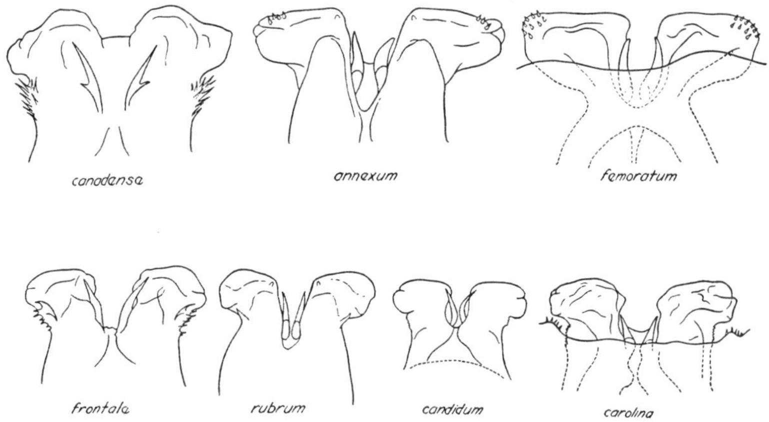

FIG. 91. Male genitalia of seven species of Stenonema.

leg. Purplish black markings on fore coxa, on lateral margins; middle and hind coxae with similar black markings, and a dark spot above each; tinged with reddish brown. Sternum clay-colored; prosternum and median area of mesosternum tinged with brown. Fore femur dusky, distinctly yellow-tinged; median and apical purplish bands; tibia widely black at apex; tarsal joinings yellowish brown, claws dark brown at base. Basal joint of fore tarsus about ½ the length of the second. Middle and hind legs yellowish white; femora with median and apical bands; tibiae narrowly black at apices; tarsal joinings as in fore leg. Wings hyaline; longitudinal veins of costal margin yellow, others somewhat yellowish; cross veins dark reddish brown, heavier than longitudinals, humeral vein especially thickened; 5 to 6 cross veins in basal costal space, about 11 beyond bulla; 2 cross veins in costal space, at bulla; 2 or 3 below it, in subcosta. Stigmatic area with reddish stain in subcostal space, in most specimens. Veins of hind wing almost colorless.

Abdominal segments 1–6 and basal portion of 7 hyaline whitish; very narrow dark posterior margins on tergites, most evident in dorsal area, appearing as dark mid-dorsal dashes. Distinct dark stigmatic dots present on segments 2 –8, quite faint on 2 and 8, prominent on 3–7. Segments 8 –10 opaque, also apical portion of 7; pale reddish yellow; posterior margins of tergites pale purplish brown, most evident on 10. Genitalia of *rubromaculatum* type, slightly less L-shaped than in *S. rubromaculatum* (see fig. 91). Tails white; at base, a few joinings pale, beyond this, alternate joinings narrowly purplish black.

Allied to *S. rubromaculatum* and *S. nepotellum*. The purple-pink flush of the thorax seems distinctive; apical tergites more extensively marked than in former, femora more distinctly banded than in latter species. The type material, from Goshen Swamp, N. C., taken by Prof. Needham, is in the Cornell collection. Nymphs differ from those of its allies in almost total lack of ventral markings. We have also a single specimen of this species from Barnwell, S. C. (D. Dunavan), and have seen others from the vicinity of Greenville, S. C. (H. K. Townes, Jr.).

Stenonema bellum Traver

1933—Traver—J. Elisha Mitchell Sci. Soc. 48: 202..N. C

Length: body 6½; wing 7 mm.

A small species of *pulchellum* group, allied to *S. integrum*; dark streaks in spiracular areas; hind wing not dark-bordered; tails dark-ringed.

Head pale whitish; a small black dot at inner lower corner of eye, opposite antenna; narrow reddish line below antenna; antennae wholly pale. A V-shaped reddish mark on vertex between lateral ocelli; on each side of head near posterior margin, a large dusky triangular spot. Thorax creamy white; an oblique purplish red streak laterally on pronotum, its posterior end continued laterally as a reddish line parallel to hind margin; a black line in postero-lateral angle. Oblique purplish black line on pronotal pleura extending to coxa of fore leg. Anterior and posterior to middle and hind legs, and extending up between them, are blackish pencilings on pleura. Oblique purplish marks laterally on metanotum. Legs pale yellowish white, femora yellowish; fore femur with tinges of rose; all femora with median and apical purplish rose bands. Apex of fore tibia faintly dusky; claws pale brownish yellow; at base of each tibia, a purplish rose streak. Wings hyaline. Costa, subcosta and radius yellow, other longitudinal veins pale. Cross veins of fore wing dark brown, paler in anal region. No reddish stigmatic stain. A slight crowding of cross veins at bulla, 2 in each of first three spaces. Little tendency for serial arrangement of cross veins, as in *S. integrum*. Hind wing wholly pale, veins colorless, outer margin not darkened.

Abdominal segments 1–7 hyaline white. Posterior margins of all tergites narrowly purplish black, not restricted to dorsal area. Extending forward from this dark margin on each side is a purplish black oblique streak, in place of the more usual spiracular dot. A bifurcate purplish streak on posterior portion of mid-dorsal line on tergites 3 and 6, and traces of this dark shading on 2 and 7; a black median dot at posterior margin of 4. Segments 8–10 opaque, yellowish. Genitalia much as in *S. integrum*; forceps relatively longer. Tails white; alternate joinings narrowly purplish black.

The type material of this species, taken at Penrose, N. C., is in the Cornell collection. We have seen no other specimens. This species is allied to *S. integrum* and *bellum*; may be distinguished as indicated under those species.

Stenonema bipunctatum McDunnough

1926—McDunnough—Canad. Ent. 58: 191 (Ecdyonurus) . . . . Ont.
1933—Traver—J. Elisha Mitchell Sci. Soc. 48: 175.

Length: body 7; wing 8 mm.

A member of *pulchellum* group; no spiracular dots; two black transverse dashes on posterior margins of tergites; tails white, not darker at joinings.

Frontal portion of head whitish; vertex olive- to reddish brown. Thoracic notum dark sepia brown, especially on posterior portion and on "lateral flanges" at bases of middle legs. Legs pale; fore legs somewhat tinged with smoky; tip of fore tarsus blackish; basal fore tarsal joint almost two-thirds the length of the second. All femora with a reddish spot at apex; median reddish band also on fore and middle femora. Wings hyaline; no shading in stigmatic area. Veins fine, black; cross veins in costo-apical region slightly thickened.

Abdominal segments 2–7 hyaline white; posterior margin of each tergite with two small black transverse dashes dorsally on the posterior margin, most evident on 4–7. No dark spiracular dots. Segments 8–10 opaque, whitish; posterior half of tergite 8, and most of 9 and 10 with bright brown shading. Forceps whitish; genitalia of *pulchellum* type. Tails whitish, not darker at joinings.

The lack of spiracular dots and wholly white tails, as well as the two dark dashes in place of the usual dark posterior margins of tergites, distinguish this species from others closely allied to it. We have a few male imagos of this species from Rockford, Ill.

Stenonema birdi sp. nov.

Length: body 11; wing 12–13 mm.

A species of the *tripunctatum* group; closely allied to *S. femoratum* but lacking the dark border on hind wing.

Male imago.—Head and thorax as in *femoratum*. Eyes very slightly smaller and set back somewhat further on angles of head; bluish black in holotype, greyish blue in some of paratypes (in alcohol). Legs pale reddish brown, somewhat darker than in *femoratum*, but with similar markings. Basal joint of fore tarsus slightly more than ⅓ but distinctly less than ½ the length of the second (in *femoratum*, slightly more than ½). Wings rather narrower in proportion to length than in *femoratum*. Venation slightly darker brown; cross veins much less distinctly margined. Cross veins crowded at bulla in first three spaces, but much less crowded behind third place than in that species; no faint dark band in this region. Basal costal cross veins 4 to 5 in number, tending to be slightly curved (not so in *femoratum*). Stigmatic cross veins rather more numerous, aslant, occasionally one or two are forked; this area stained as in *femoratum*. Hind wing *not darkened* at apex nor along outer margin.

Abdominal segments red-brown, darker than in *femoratum*; sternites little if any paler than tergites. Marked very similar to *femoratum*, but no obscured dark stigmatic spots. Tracheae quite distinctly outlined in greyish black. Pleural fold narrowly pale. Tails light reddish brown, darker at joinings, alternate joinings wider. Genitalia very much as in *femoratum*.

Female very similar to male. Eyes greyish (in alcohol). Opaque pinkish-white patches on thoracic pleura and sternum, sometimes indicated in male, much more extensive and more distinct. Sternites 8 and 9 almost wholly opaque pinkish white; markings of this color on lateral portions of tergites 8–10. Wing 15–17 mm. in length.

Holotype—Male imago. Comanche Co., Oklahoma, Feb. 28, 1932 (R. D. Bird, W. C. C.). No. 1344.1 in Cornell University collection.

Allotype—Female imago; same data. No. 1344.2 in Cornell collection.

Paratypes—13 male imagos, 16 female imagos; same location, Feb. 18–28, 1932 (R. D. Bird, W. C. C.). No. 1344.3–31 in Cornell collection.

Other imagos not included among the types are from McCurtain Co., Oklahoma, May 6, 1931 (Lois Bird). Nymphs taken at the same time and place as the type material, but not reared, are tentatively associated with the above imagos. These nymphs resemble *tripunctatum* in dorsal and ventral maculation; frontal margin of head not emarginate, as in *femoratum*; no pale spot on frontal margin.

Stenonema canadense Walker

1853—Walker—List Neurop. Ins. Brit. Mus. p. 569 (Baetis)............Canada.
1861—Hagen—Syn. Neurop. Ins. N. Amer. p. 47 (Baetis).
1871—Eaton—Trans. Ent. Soc. London, p. 138 (Ecdyonurus).
1885—Eaton—Revis. Monogr. p. 278 (Ecdyonurus).
1910—Banks—Canad. Ent. 42: 200 (Heptagenia).
1913—Clemens—Canad. Ent. 45: 258 Figs. (Heptagenia)............Ont.
1915—Clemens—Contr. Canad. Biol., p. 140 (Heptagenia). Figs.
1925—McDunnough—Trans. R. S. C., p. 222 (Ecdyonurus)............Que.
1930—Ide—Canad. Ent. 62: 230 (Ecdyonurus).
1933—Traver—J. Elisha Mitchell Sci Soc. 48: 175.
1933—McDunnough—Canad. Ent. 65: 43 (in discussion of *Ecdyonurus heterotarsale*).

Length: body 8–9; wing 9 mm.

A dark reddish brown species of *interpunctatum* group; black markings on face and pronotum, and oblique black streaks on pleura; abdominal tergites largely dark red-brown.

Head dark red-brown; a black band on face below each antenna, but not continuous across face; a dark spot on median carina; a small black dot near inner margin of eye, posterior margin of head black, the dark line thickest at median line. Pronotum brown; median line, posterior margin at middle area, and oblique lateral streaks, black. Mesonotum deep brown; scutellum narrowly tipped with pale yellowish. A faint purplish red streak anterior to wing root. Pleura yellowish, sutures dark red-brown; an oblique prominent black streak below each wing root, anterior to middle and hind legs. Between base of fore and middle leg, a darker brown area. Sternum red-brown, a blackish area laterally on mesosternum, below base of middle leg. Femora greenish yellow, fore femur slightly tinged with red-brown; median and apical black bands on all femora. Fore tibia yellowish brown, black at apex; may be very slightly longer than femur; tarsus smoky brown, basal joint more than ⅓ but less than ½ length of second. Middle and hind tibiae and tarsi rather yellowish; claws and tarsal joinings brownish, tarsi may be faintly smoky. Costo-apical area of fore wing tinged with brownish, with a faint reddish tinge at stigma. Longitudinal veins of costal margin yellowish brown; others black-brown; cross veins black, those of costal margin somewhat thickened; cross veins basad of bulla narrowly dark-margined, in radial space more widely margined; at bulla 3 widely margined veins are usually connected by a black dash. Hind wing narrowly blackish on outer margin.

Abdominal tergites 1–9 largely blackish brown; mid-dorsal line and posterior margin darker; pale submedian streaks, lateral margins, and antero-lateral triangles remain; these areas yellowish. Extent of pale submedian marks apparently somewhat variable. No dark spiracular dots apart from general dark areas. Tergite 10 somewhat paler, reddish brown. Sternites 1–7 yellowish, posterior margins faintly and narrowly pale brown; some specimens show some brownish shading except on lateral margins, but no distinct dark mid-dorsal band. Forceps smoky; penes pale brown; of the *interpuncatum* type (see fig. 91). Tails pale yellowish white on basal half, dark smoky apically; joinings dark brown. Distinguished from *S. gildersleevei* as indicated under that species.

There are specimens of *canadense* in the Cornell collection from the Ottawa Golf Club, Que., sent by Dr. McDunnough; from Hull, Que.; Montreal and Ile Bizard, Que. (Mrs. E. G. Grieve); Port Elgin, Ont. (Mrs. Grieve); Niagara Falls, N. Y. (C. Betten); Juanita Island, L. George (J. G. Needham); Ft. Jackson and Hopkinton, N. Y. (L. Harper); St. Cloud, Minn. (G. W. Friedrich); Canajoharie, N. Y. (H. K. Townes, Jr.).

Stenonema candidum sp. nov.

Length: body 7½; wing 8½ mm.

A species of the *interpunctatum* group; no dark bar in wing below bulla; no dark oblique streaks on pleura; stigmatic dots present, prominent; genitalia of *carolina* type.

Male imago.—Eyes greenish, in living insect; of usual size, not small as in *S. carolina*. Head yellowish; black dot on edge of median carina, a short black dash below each antenna; posterior margin dark brown, two small dark dots above it. Prothorax yellowish; black oblique lateral streak on pronotum, also short dark mark near postero-lateral margin. Mesonotum dark red-brown; scutellum yellowish. Pleura and sternum yellowish; faint greyish pencilings above leg bases; no oblique dark streaks. Legs yellowish. Fore femur distinctly greenish, other femora more faintly so, in living insect. Blackish median and apical bands on fore and middle femora; median band lacking on hind femur. Apex of fore tibia, and tarsal joinings, blackish, apical joint of fore tarsus deep smoky. Basal fore tarsal joint fully ½ as long as the second. Wings as usual in this group, but with no black bar joining the margined cross veins below bulla. Very faint brownish stigmatic stain. Basal costal and subcostal cross veins slightly thickened but not margined.

Abdominal segments 2–6, and base of 7, hyaline; whitish. Stigmatic dots purplish black, very distinct. Posterior margins of tergites narrowly blackish, with rather distinct black dot at median line. Segments 8–10, and apical portion of 7, opaque; tergites pale red-brown, sternites yellowish. Tails white, not darker at joinings. Penes as in fig. 91; resemble *carolina* in almost total loss of spines on lower lateral margin.

Female similar to male, except for usual differences. Thorax and abdominal segments yellowish. Median band on hind femur represented by a dark streak near upper margin. Narrow continuous dark line across frontal margin of head. Nymph very similar to that of *S. frontale* (see Pl. XXXIV). Differs in the following respects: no pale median spot on frontal margin of head; dark lateral markings on abdominal tergites more extended, joined laterally by a narrow dark bar to median dark areas, on tergites 2–3 and 5–7; postero-lateral spine on segment 9 appears very slightly longer. Length, 8½ mm.

Holotype—Male imago. Willoughby, Ohio, July 23, 1934 (J. R. Traver). No. 1342.1 in Cornell University collection.

Allotype—Female imago; same data. No. 1342.2 in Cornell collection.

Paratypes—Male imago, reared; Pleasant Valley, near Willoughby, O., Aug. 24, 1933. No. 1342.3 in Cornell collection.

This species is superficially very close to *S. frontale*, but in genitalic type it is allied to *S. carolina*.

Stenonema carolina Banks

1914—Banks—Proc. Acad. Nat. Sci. Philad. 66: 616 (Heptagenia) . . . . N. C
1925—McDunnough—Trans. R. S. C. 19: 222 (Ecdyonurus) . . . . Tenn., Que.
1933—Traver—J. Elisha Mitchell Sci. Soc. 48: 176, 177, 197. Fig.

Length: body 10; wing 10½–11½ mm.

A yellowish species, of *interpunctatum* group; no dark marks on face; lateral dark mark on pronotum; no spiracular dots; abdominal segments dark on posterior margins; no spines on lower lateral portion of penes; eyes rather small and remote.

Head and thorax pale yellowish or reddish yellow; no distinct dark marks on head, but a reddish median line on vertex, and posterior margin reddish. Antennae dusky. Eyes relatively small, remote. A rather faint dark oblique streak on each side of pronotum. Femora yellowish, sometimes with greenish tinge; median and apical dark bands on fore and middle legs, apical dark mark on hind leg, sometimes a faint trace of dark median spot. Tibiae and tarsi yellowish, more or less tinged with smoky; apices of tibiae and tarsal joinings darker. Basal fore tarsal joint ½ or slightly more than ½ the length of second. Wings hyaline; longitudinal veins amber brown, cross veins rather darker brown. Costal cross veins before and in region of bulla with a dark spot at the costal margin, instead of being uniformally margined; radial cross veins before bulla and two or three at bulla dark-margined in usual manner, but not connected by a dark dash. Apical portion of costal margin yellowish brown. Outer margin of hind wing narrowly dark-bordered.

Abdominal segments 1–7, sometimes basal portion of 8, more or less hyaline, with faint greyish or greenish tinge. Posterior margin of each segment, dorsally and ventrally, with a narrow but distinct dark border, so that abdomen appears annulate; these dark margins deeper in color on tergites, especially in dorsal area. No dark spiracular markings. Segments 8–10 opaque, tinged with reddish brown. Forceps brown; penes quite similar to those of *S. frontale*, being narrower and more rounded apically than in *S. interpunctatum*; differ from both of these in the lack of distinct short spines on lower lateral portion, only a slight roughening of the margin being evident (see fig. 91, *carolina* type). Tails rather deep smoky brown basally, somewhat paler brown apically; joinings opaque, appear very faintly darker.

This species seems distinct from all others described in this genus, by reason of the lack of lateral spines on the penes together with the yellowish dark-banded abdomen. There are specimens in the Cornell collection from North Carolina, including a male taken by Dr. Beutenmuller at Black Mt. in 1912; from Ringwood and Rockland Co., N. Y., from Hopkinton, N. Y. (L. Harper); Wardensville, W. Va. (J. R. T.); Mt. LeConte, Tenn. (J. T. Needham); Rocky Bottom and Clemson College, S. C. (D. Dunavan); Greenville Co., S. C. (H. K. Townes, Jr.). The nymph described in a previous paper (Traver, 1933) as being of this species was later found to have been erroneously associated. The real nymph of *S. carolina* occurs in material collected near Swannanoa, N. C. The dorsal pale abdominal markings are inconspicuous, and ventral markings apparently lacking.

Stenonema conjunctum sp. nov.

Length: body 8; wing 9 mm.

A yellowish species of the *interpunctatum* group; black marks on face and pronotum; spiracular dots faint, diffuse, as is median band of hind femur.

Head yellowish; black spot on median carina; one at inner corner of eye; black dash below antenna. Antennal filament dusky, tip paler. A small black dot near each eye, on vertex, halfway to posterior margin; reddish shading near posterior margin. Pronotum yellowish; an oblique black streak on each side. No dark mark at postero-lateral angle; often a small black dot on posterior margin at median line. Mesonotum reddish brown; scutellum pale, the lateral areas anterior to it darker brown; faint darker median and anterior pencilings. Pleura yellowish; a rather prominent purplish black oblique streak below roots of fore and hind wings, anterior to bases of middle and hind legs; greyish shading above leg bases. Sternum yellowish. Legs yellowish, femora greenish yellow; black median and apical bands on fore and middle femora, median band on hind leg fainter, rather diffuse, may even be wanting. Longitudinal black pencilings on all femora. Fore tibia black at apex; other tibiae narrowly dark brown apically; all tarsi faintly smoky. Basal joint of fore tarsus slightly more than ⅓, but distinctly less than ½ the length of second. Wings hyaline; cross veins dark-margined as in *S. interpunctatum*; a black dash at bulla in radial space. Costa, subcosta and radius yellow in basal two-thirds, brownish apically; other longitudinal veins dark brown, cross veins black. Apical portion of costal margin tinged with yellowish brown, extending around tip of wing. 3rd and 4th marginal intercalaries at wing tip thickened, black. Outer margin of hind wing dusky.

Abdominal segments 2–8 semi-hyaline, yellowish white. Posterior margins of tergites narrowly purplish black. Spiracular dots small, diffuse, not prominent as in *S. frontale*; may be absent on basal segments. Occasional specimens have very slight greyish shading medially and laterally on basal tergites, but much less distinct than in *S. proximum*. Segments 9 and 10 opaque; tergites usually rather dark red-brown, posterior margin of 10 yellowish. Sternites yellowish. Forceps yellowish, penes flesh-colored, with the inner apical portion rather elevated, as in *interpunctatum*. Tails pale whitish, joinings narrowly reddish at base, brownish apically. (Male).

Holotype—Male imago (reared); Pleasant Valley, near Willoughby, O., Aug. 24, 1932. (J. R. Traver). No. 1339.1 in Cornell University collection.

Allotype—Female imago (reared); same location, Aug. 22, 1933; No. 1339.2 in Cornell collection.

Paratypes—10 male imagos, 15 female imagos, same location; August and early September, 1929 to 1933 (J. R. T.). No. 1339.3–27 in Cornell collection.

This species is distinguished from *interpunctatum* by the oblique dark pleural streaks, the diffuse median band on hind femur and diffuse spiracular dots. From other pale species in this group it may be separated by the shorter basal joint of the fore tarsus. In addition to the type material and other specimens taken at same locality, we have specimens which seem to be of this species from Hagaman, W. Caneville, Kast Bridge, Ingham Mills, Dolgeville, Canajoharie, Kinderhook and Hoosick, N. Y. (H. K. Townes, Jr.); Black Lake, N. Y.; Kirtland and Leroy, O. (J. R. T.).

Stenonema exiguum Traver

1933—Traver—J. Elisha Mitchell Sci. Soc. 48: 201. Fig. . . . . N. C., Ga.

Length: body 5½; wing 6 mm.

A small species of *pulchellum* group, allied to *S. integrum*; no dark spiracular dots; hind wing dark-margined; tail wholly pale.

Head pale whitish; faint reddish dot at center of vertex, no other marks; antennal filament dusky at base. Thorax pale creamy white, immaculate. Legs pale creamy white. Median and apical rose-colored or lavender bands on all femora; rose-colored spot at apex of tibia on middle and hind legs. Tip of fore tibia purplish black; tarsal joinings narrowly dark. On all tarsi, claws and distal half of apical joint dusky. Basal joint of fore tarsus slightly more than one half the length of the second. Wings hyaline, iridescent. Costa, subcosta and radius yellowish brown; other longitudinal veins pale yellow; humeral cross vein purplish black; other cross veins prominent, very dark brown; 4 basal costal cross veins, 8–9 beyond bulla. No tendency toward crowding of veins at bulla, as in *S. integrum*; tendency for serial arrangement of veins below bulla less evident than in that species. Distinct reddish stain in stigmatic area. Outer margin of hind wing narrowly dark; all veins colorless.

Abdominal segments 1–7 hyaline white; posterior margins of all tergites narrowly purplish black in dorsal area. No dark spiracular markings. Segments 8–10 opaque creamy whitish. Genitalia very similar to those of *integrum*; forceps relatively longer than in that species. Tails white, not darker at joinings.

The type material, from Woodlawn, N. C. (J. R. Traver), and Atlanta and Rome, Ga. (P. W. Fattig) is in the Cornell collection. The wholly pale tails and lack of dark spiracular markings distinguish this species from *S. integrum*; it may be separated from *bellum* by the presence of the reddish stigmatic stain and dark-margined hind wing.

Stenonema femoratum Say

1823—Say—West Quart. Rep. 2: 162 (Baetis)..Ohio
1859—LeConte—Complete Writings of T. Say, 1: 171 (Baetis).
1861—Hagen—Syn. Neurop. Ins. N. Amer. p. 48 (Baetis).
1863—Hagen—Proc. Ent. Soc. Philad. 2: 169 (Baetis).
1925—McDunnough—Canad. Ent. 57: 190 (Ecdyonurus).
1933—Traver—J. Elisha Mitchell Sci. Soc. 48: 174.

Length: body 9–11; wing 10–12 mm.

A species allied to *S. tripunctatum*; abdominal tergites 2–6 yellowish brown, sternites yellow-white; black median spot and lateral dashes on tergites posteriorly; hind wing dark-tipped; cross veins crowded at bulla.

Head reddish brown; frontal margin paler. Purplish spot on median carina; a small black dot at lower corner of eye, another below antenna; purplish brown median spot on vertex, halfway to posterior margin. Antenna dark brown basally, tips pale. Thorax largely dark red-brown. Median area of pronotum blackish, likewise posterior margin; oblique purplish brown streak laterally, halfway to lateral margin. A wide purplish black streak extends down from base of fore wing across prothorax and fore coxa. Below each wing root, a similar dark oblique streak on pleura. Large areas on pleura anterior to wing root and below wings. Tip of mesonotal scutellum yellowish white; sternum paler red-brown. Legs yellowish. Fore femur tinged with yellow-brown; a small black spot at base; median purple-red band, apical dark brown one; tip of fore tibia and tarsal joinings blackish; distal tarsal joint and claws smoky brown. Basal joint of fore tarsus more than ½ length of second. Median and apical bands on middle and hind femora; tips of tibiae and tarsal joinings dark brown, last two tarsal joints tinged with smoky brown.

Wings hyaline, often with faint milky tinge. Cross veins crowded in first six spaces at bulla; usually 3 in costal space, and 4 to 5 in subcostal and radial spaces. These veins dark-margined, and wing membrane between them brown-tinged, forming a curved dark streak in wing, as in *S. pudicum*. Cross veins of costal margin thickened, many of them margined; humeral vein especially heavy; others in fore wing dark brown, many narrowly margined, but not as heavy as costals. Longitudinal veins of costal margin yellow, others in fore wing brown; in hind wing, yellow. Apex and upper portion of outer margin of hind wing heavily brown-tinged.

Abdominal tergites 2–6 yellowish brown dorsally, tinged with brown posteriorly and laterally; at median line of each tergite, on posterior margin, a black dot, and lateral dark streaks laterad of this, one on each side. Spiracular dots faintly visible, on 2–7. Sternites 2–6 yellowish, the antero-lateral and lateral margins narrowly brownish. Segments 7–10 rather deep red-brown. Genitalia quite similar to *S. tripunctatum*, but the penes with stronger median spines, the outer portions more wedge-shaped, the inner margins more distinctly incurved (see fig. 91) .Tails pale smoky at base, pale yellowish in middle area, dusky at tips; joinings purplish brown, alternate joinings wider.

In the Cornell collection there are reared specimens of this species from Kirtland and Euclid, Ohio. It is distinguished from *S. tripunctatum* by the darker abdominal tergites and dark-tipped hind wing. It is this species which Walsh confused with the form now known as *Siphloplecton interlineatum*; Eaton, following Walsh, placed the latter species in Siphlonurus, under the name *femoratus*. Other specimens in our collection are from Cabin Creek, Swamp Creek near Dalton, and Blairsville, Ga. (P. W. Fattig); Greenville Co., S. C. (H. K. Townes, Jr.); Montreal, Que. (Mrs. E. G. Grieve); Canajoharie, N. Y. (H. K. Townes, Jr.).

Stenonema frontale Banks

1910—Banks—Canad. Ent. 42: 199 (Heptagenia)... Mass.
1913—Clemens—Canad. Ent. 45: 259. Fig. nymph (Heptagenia)............................. Ont.
1915—Clemens—Contr. Canad. Biol., p. 141. Fig. (Heptagenia).
1930—Ide—Canad. Ent. 62: 230 (Ecdyonurus).
1930—McDunnough—Canad. Ent. 62:61 (Ecdyonurus).
1933—McDunnough—Canad. Ent. 65: 43 (in discussion of *Ecdyonurus heterotarsale*).
1933—Traver—J. Elisha Mitchell Sci. Soc. 48: 175.

Length: body 8; wing 9 mm.

A pale species of *interpunctatum* group; black markings on face and pronotum black streaks on pleura, dark spiracular dots.

Head yellowish; a black spot under each antenna, another on median carina, and a smaller dot near inner margin of each eye. Vertex somewhat reddish brown; posterior margin darker, apparently a dark spot on each side. Pronotum yellowish to light brown; an oblique black streak on each side; a small dark spot often present on postero-lateral corner. Mesonotum dark reddish brown; scutellum yellowish. Pleura yellowish; two oblique black streaks, one below root of each wing, anterior to bases of middle and hind legs. Sternum yellowish. Femora yellowish, each with median and apical blackish bands. Tibiae whitish; fore tibia blackish at apex, very slightly longer than femur. Fore tarsus greyish white, joinings slightly darker; basal joint almost ½ length of second. Tarsi on middle and hind legs pale, smoky at tips, joinings smoky. Wings hyaline; apical portion of costal margin brownish. Longitudinal veins brown, cross veins black. A few cross veins in costal and radial areas narrowly black-margined; under the bulla, two or three are more broadly margined, but with no dark bar connecting them. Wings rather long and narrow; about 6 basal costal cross veins and 13 or 14 beyond bulla.

Abdominal segments 1–7 whitish, hyaline; posterior margins of tergites rather widely blackish brown; distinct large dark spiracular dots. Tergites 8–10 reddish brown, sternites yellowish. Forceps pale; penes rather flesh-colored; apical portion of penes less elevated at inner margin than in *S. interpunctatum*, so that these structures appear shorter and more rounded (*frontale* type; see fig. 91). Tails pale at base, tip pale smoky; joinings very faintly darker.

PLATE XXXIV.

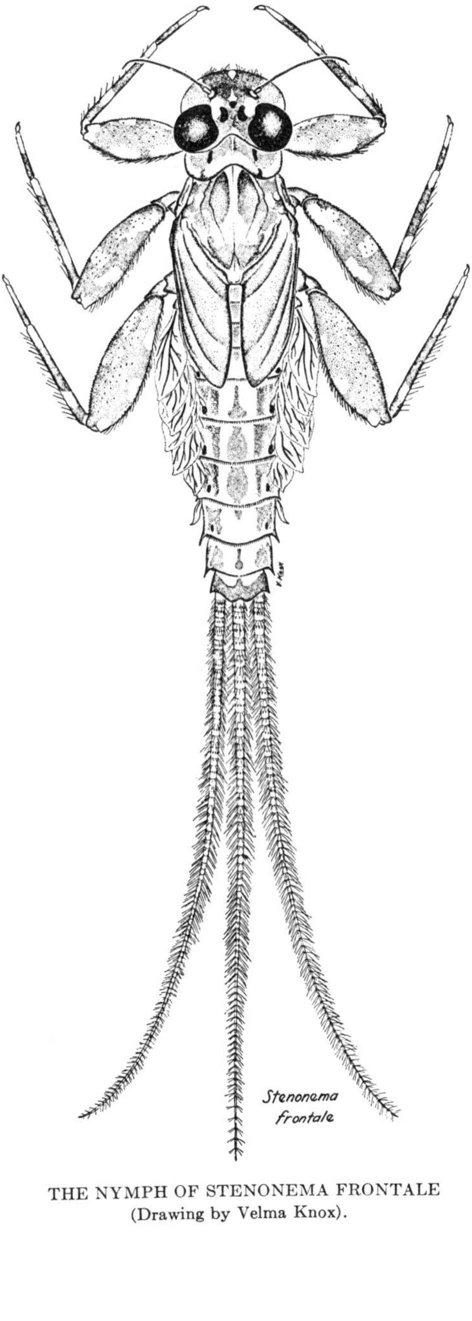

THE NYMPH OF STENONEMA FRONTALE
(Drawing by Velma Knox).

We have studied a nymph and reared imago from Go Home Bay, Ont. (G. S. Walley), kindly lent us by Dr. McDunnough, which is similar to the species considered by Clemens to be *frontale* (see nymph, Pl. XXXIV). We accept this determination, and the above description is based in part on the specimens mentioned. The species is rather closely allied to both *conjunctum* and *majus*. From the former it is distinguished by the more prominent spiracular dots and longer fore tarsal joint in the male. The larger size of *majus* and the more extensive dark markings separate that species from *frontale*. The very similar *candidum* differs in type of genitalia.

Stenonema fuscum Clemens

1913—Clemens—Canad. Ent. 45: 254. Fig. (Heptagenia) .... Ont.
1915—Clemens—Contr. Canad. Biol., p. 136. Fig. (Heptagenia).
1925—McDunnough—Trans. R. S. C. 19: 222 (Ecdyonurus)................................ Que.
1933—McDunnough—Canad. Ent. 65: 35 figs. (Ecdyonurus)................................ N. B.
1933—Traver—J. Elisha Mitchell Sci. Soc. 48: 174.

Length: body 10; wing 13 mm.

A species of the *vicarium* group; no crowding of cross veins at bulla; stigmatic area amber, no red stain; abdominal tergites 1–7 normally brown, little contrast with darker posterior bands.

Eyes large; pearly grey in living insect. Epistoma of head pale smoky hyaline; smoky brown streaks on each side of carina; pale yellowish anterior to ocelli; bases of antennae pale smoky, ruddy basal rings. Vertex rather bright brown, yellowish shading behind lateral ocellus. Prothorax largely smoky. Mesonotum normally deep brown; on rear half of scutellum and in depression anterior to scutellum, whitish shading. Mesonotum sometimes much paler and whitish areas more extensive; in others, color is darker and little pale shading remains. Metanotum largely deep brown; anterior projection paler. Anterior to wing roots a reddish streak is split by a pale line. Pleura and sternum largely clouded with brownish; an oblique yellowish streak from fore wing root to base of fore leg, similar one at root of hind wing. All coxae smoky brown, streaked laterally and at base with blackish. Legs light amber; fore femora suffused with smoky; median and apical purple red bands on all femora; all tarsal joinings blackish, also claws. Basal joint of fore tarsus usually slightly more than one-third the length of second. Wings hyaline; costa, subcosta and radius smoky amber, other longitudinals blackish; cross veins rather fine, blackish; no crowding of veins at bulla. Stigmatic area slightly suffused with amber, but no reddish stain.

Abdominal tergites normally appear an uncontrasted brown; the dull smoky amber of each segment extending into forward-projecting indistinct streaks (like inverted question marks) from the deeper brown posterior margin, so that there is no sharp line of demarcation between the two. Dark streak on mid-dorsal line. Dark posterior border occupies about ⅓ of the tergite. In pale specimens, a tendency to a rather more banded appearance; in dark forms, whole area may appear dark brown. Tergites 8–10 opaque; suffused with rather bright brown; lateral edges of 9 and 10, sometimes of 8 also, distinctly bordered with alabaster white. Sternites dull pale smoky amber; immaculate; apical sternites more yellowish. In dark specimens, traces of brown triangular shading on each side of median line. Forceps and tails very pale smoky amber; tail joinings narrowly brown. Genitalia quite similar to those of *pulchellum*.

This species is closely allied to *S. rivulicolum*; distinguished from that species by the lack of contrast between ground color and margin of anterior tergites, presence of pale lateral margin on tergites 8–10, absence of brown patches in postero-lateral angles of sternites, absence of reddish stain in stigmatic area.

Reared specimens of this species from Ithaca, N. Y. (Y. C. Hsu), Willoughby, O. (J. R. Traver); and Cayuta, N. Y. (O. R. Smith), are in the Cornell collection.

Stenonema gildersleevei sp. nov.

Length: body 10–11; wing 10½–11 mm.

A large species of *interpunctatum* group; abdominal tergites largely purplish black; fore tarsus shorter than or about as long as tibia, basal joint almost ½ of second.

Head reddish yellow; usually a continuous black transverse stripe across face below antenna, from inner corner of eye (in one paratype, no continuous stripe, but black dash below antenna, black spot on carina). Antennal filament dusky, tip paler. A black spot on each side of vertex; a small reddish median spot; dark shading along median line and posterior margin. Pronotum largely purplish black in median area, including posterior margin; lateral areas yellowish, margined laterally and posteriorly with faint dusky lines; a wide black oblique mark on each side. A short black transverse streak at base of fore coxa. Mesonotum reddish brown; a darker median stripe, distinctly purplish black in anterior and posterior portions, including scutellum; antero- and postero-lateral margins of mesonotum, and a median black triangle occupying most of metanotum, also blackish; above root of hind wing, a rather large yellow spot. Pleura yellowish, tinged with reddish brown; purplish black markings above leg bases and anterior and posterior to middle and hind coxae; no prominent oblique streaks anterior to legs, as in *S. canadense*. Sternum yellowish.

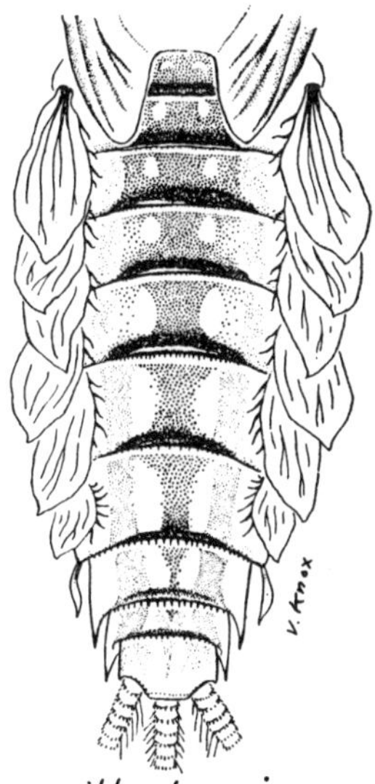

FIG. 92. Nymph of *Stenonema gildersleevei;* dorsal view of abdomen (drawing by Velma Knox).

Legs yellowish; femora deeper yellow; tarsi often tinged with smoky. Wide median and apical purplish black bands on all femora (may be fainter on hind leg). Apex of fore tibia blackish; apical portion of each tarsal joint deeper smoky; apices of middle and hind tibiae, and tarsal joinings, narrowly dark brown; distal tarsal joints deep smoky apically. Basal fore tarsal joint almost ½ the length of the second; in one specimen, however, as short as in *S. canadense.* Fore tarsus relatively short, usually about as long as tibia (may be slightly shorter). Wings hyaline, but with faint milky tinge. Longitudinal veins of costal margin yellowish, others amber to light reddish brown. Cross veins purplish black; those of basal costal border narrowly margined as usual; at bulla, 2 or 3 in radial space connected by a black dash. A faint brownish stain in stigmatic area, extending around tip of wing. Veins in anal region of fore wing, and in posterior half of hind wing, finer and paler; outer margin of hind wing narrowly dusky.

Abdominal segments 1–6 and base of 7, semi-hyaline, pale yellowish white; tergites largely occupied by a wide median purplish black area, leaving only the lateral areas and very narrow submedian streaks pale. Posterior margins and mid-dorsal line deeper in color; diffuse spiracular marks are present, but often fused with the dark median area so as to be indistinct. A purplish black transverse streak on each side of mid-ventral line of basal and middle sternites, on posterior margin; on basal segments, a diffuse purplish grey triangle extends backward obliquely from this dark streak. Segments 8–10, and apical portion of 7, opaque, deep yellow, tergites often shaded with reddish brown; posterior margins and median line on 8 and 9 blackish, with faint lateral shading; 10 pale except for faint median and submedian streaks and a narrow dark mark on posterior margin at median line. Genitalia yellowish to pale reddish brown; penes much as in *S. frontale.* Tails pale yellowish; joinings opaque; on some specimens, basal joinings faintly brownish.

Female larger than male, somewhat paler, but with same dark markings; abdominal tergites deep red-brown. Dorsal view of abdomen of nymph shown in fig. 92.

Holotype—Male imago, reared from nymph; Kirtland, Ohio, in a small spring-fed stream from Gildersleeve Mt.; Aug. 31, 1930. No. 1337.1 in Cornell collection.

Allotype—Female imago, reared; same location; Sept. 3, 1932; No. 1337.2 in Cornell collection.

Paratypes—3 male imagos, 5 female imagos, reared; same location; late August to early September, 1929, 1930, 1932. Nos. 1337.3–10 in Cornell collection.

This species is very similar to *S. canadense* in the imago stage; apparently distinguished by the relatively shorter fore tarsus with slightly longer basal joint; lack of prominent dark oblique streaks on pleura; paler tails with less conspicuous joinings; more definite mid-ventral markings. The size is slightly larger; the penes show slight differences. Distinguished from *ohioense* as indicated under that species.

Stenonema heterotarsale McDunnough

1933—McDunnough—Canad. Ent. 65: 42 (Ecdyonurus)............................Que., Ont., Ill.
1933—Traver—J. Elisha Mitchell Sci. Soc. 48: 184 (*affine*)..................................N. C.

Length: body 9; wing 11 mm.

A species of the *interpunctatum* group; basal fore tarsal joint about half the length of second; no dark markings on face or pronotum; no spiracular dots; no median band on hind femur; abdomen pale dorsally.

Eyes light green, in living insect. Head pale yellow; anterior and posterior margins semihyaline. Brown shading on vertex, "deepest in a triangular patch covering posterior area, becoming almost purple-black" (McD.). Pronotum light yellow, without dark markings. Mesonotum largely brown; posterior portion pale hyaline yellowish; scutellum "tipped with opaque ochreous." Metanotum light yellowish ochre. Pleura yellowish; anterior to wing roots is a reddish streak; deeper shading between fore and middle legs. Sternum yellowish. Legs light yellow, fore femur more distinctly yellow; blackish median and apical bands on fore and middle femora, distinct on fore leg but faint on middle femur; no median band on hind femur, apex blackish. Tip of fore tibia and all claws smoky. Basal joint of fore tarsus about ½ the length of the second. Wings much as in *interpunctatum,* except that the stigmatic area of fore wing is shaded with light amber and lacks the reddish tinge of that species.

Abdominal segments 2–7 hyaline, yellowish white; posterior margins of tergites with a fine blackish band. No dark spiracular dots. Segments 8–10 opaque; tergites shaded with ruddy brown, narrowly in median area of 8, on all of 9 and 10; lateral area of 8 pale yellowish; 8 and 9 with blackish posterior margins. Sternites wholly pale; 8 and 9 almost alabaster white. Genitalia very similar to *interpunctatum,* "but appear to have the latero-apical margin of the penes somewhat more oblique than in this species" (McD.). Tails pale, very faintly darker at joinings.

The species from North Carolina described as *S. affine* by Traver (1933) seems to be *heterotarsale*, described a few months before. We therefore sink *affine* to that species. We have, in addition to the material from North Carolina, a series of imagos of both sexes which Dr. McDunnough very kindly sent to the Cornell collection.

This species may be distinguished from *S. interpunctatum* by the absence of black markings on face and pronotum, and the longer basal fore tarsal joint of the male. Occasional specimens show traces only of such dark markings. From *pallidum* it may be separated by its larger size, as well as by the same features which distinguish it from *interpunctatum*.

Stenonema integrum McDunnough

1924—McDunnough—Canad. Ent. 56: 9 (*Heptagenia integer*)................................ Ill.
1933—Traver—J. Elisha Mitchell Sci. Soc. 48: 175.

Length: body 5; wing 6 mm.

A small species of *pulchellum* group; spiracular dots present; tails dark-ringed; outer margin of hind wing dark.

Head and thorax yellowish or whitish clay-colored; antennae pale whitish; ocelli black-ringed at base. Fore leg yellowish, femur with faint reddish tinge; tibia black at tip; tarsal joinings narrowly blackish. Basal tarsal joint fully one-half, often slightly more than one-half of second joint. All femora with median and apical reddish bands, sometimes indistinct on hind leg. All claws dark purplish. Wings hyaline; longitudinal veins fine, pale yellowish. Cross veins "thicker and distinctly dark, comparatively few in number, especially in the basal area, only two or three on costa before the bulla, where there is a group of four or five; below this the cross veins of the radial sector and median areas form an irregular line across the wing, preceded by long elongate cells" (McD.). A distinct reddish stain in stigmatic area, in subcostal space. Veins of hind wing hyaline; outer margin narrowly but distinctly brownish.

Abdominal segments 1–7 hyaline white; posterior margins of tergites very narrowly blackish dorsally; traces of a purplish black streak on posterior half of mid-dorsal line, on tergites 3 and 6. A short black mark (rather a streak than a dot) on each tergite laterally, at end of black border on posterior margin. Segments 8–10 opaque, yellowish white. Genitalia rather similar to *pulchellum* type, but lateral portions of penes shorter and rounder, the whole being less L-shaped in appearance. Tails whitish, joinings narrowly dark brown.

Distinguished from *S. pulchellum* and its immediate allies by the dark-margined hind wings and the fewer cross veins, which tend to be crowded at bulla and arranged in series beyond and below this point. Distinguished from its near ally, *exiguum*, by the presence of dark spiracular marks and dark tail joinings; also from *bellum* by the dark margin of the hind wing and the reddish stigmatic stain. We have specimens of this species from Lawrence, Kans., which Dr. McDunnough kindly sent to the Cornell collection. Other specimens which are very similar although not identical in all points are from Austin, Texas; also from Ringgold and Blairsville, Ga. (P. W. Fattig).

Stenonema interpunctatum Say

1839—Say—J. Acad. Nat. Sci. Philad. 8: 41 (Baetis)..Ind.
?1843—Pictet—Hist. Nat. d. Ins. Neurop. 2: 186. Fig. (in part, *Baetis flaveola*)................Tenn.
1853—Walker—List. Neurop. Ins. Brit. Mus. pt. 3: 562 (Baetis).
1859—LeConte—Complete writings of T. Say, 2: 411 (Baetis).

1861—Hagen—Syn. Neurop. Ins. N. Amer. p. 44 (Baetis).
1862—Walsh—Proc. Acad. Nat. Sci. Philad. p. 372 (Palingenia)............................... Ill.
1863—Walsh—Proc. Ent. Soc. Philad. 2: 190 (Palingenia).
1863—Hagen—Proc. Ent. Soc. Philad. 2: 177 (Palingenia).
1871—Eaton—Trans. Ent. Soc. London, p. 142.
1885—Eaton—Revis. Monogr. p. 267 (in Heptagenia).
1910—Banks—Canad. Ent. 42: 200 (in Heptagenia).
1921—Ulmer—Arch. f. Naturg. Hft. 6: 242 (Heptagenia) (*flaveola* = *interpunctatum*).
1933—McDunnough—Canad. Ent. 65: 42 (in discussion of *Ecdyonurus heterotarsale*)..... Kans., Ont.
1933—Traver—J. Elisha Mitchell Sci. Soc. 48: 175, ? 178................................? N. C.

Length: body 7½; wing 7½–8½ mm.

A species of *interpunctatum* group, having a black dash in fore wing near bulla; black markings on face and pronotum; basal fore tarsal joint about ⅓ of second; no dark spiracular markings; no median dark band on hind femur.

Head yellowish green; black dash below each antenna and black markings on carina. Ocelli black-ringed at base; black line through eye; antennal filament blackish. Pronotum yellowish, a black lateral streak on each side. Mesonotum brownish. Pleura and sternum yellowish, without distinct dark markings. Legs yellowish or greenish yellow; femora of fore and middle legs with median and apical black bands, hind femur black apically, not in middle. Basal joint of fore tarsus about ⅓ as long as the second. Wings hyaline; longitudinal veins yellowish; cross veins in anterior half of fore wing black in costal margin, dark brown behind this region. Say states, "wings on the anterior margin greenish." Cross veins in basal costal, subcostal and radial spaces margined with black; at bulla, 2 in the radial space with a black dash connecting them at about the center. Cross veins behind costal margin in apical portion beyond disc tend to be arranged in series. At apex of wing, marginal intercalaries in 2nd and 3rd spaces often somewhat darker and more thickened than others. Veins in anal region of fore wing and in most of hind wing paler. Outer margin of hind wing with narrow but distinct black border, the marginal intercalaries also dark in this area. Stigmatic area with reddish tinge, according to Dr. McDunnough.

Abdominal segments 1–7 hyaline whitish; posterior margins of tergites narrowly blackish; no dark spiracular dots. Segments 8–10 opaque; tergites "more or less obviously ferruginous" (Say). Genitalia of *interpunctatum* type (see fig. 91, *canadense*); a pair of stout median spines; shorter spines on outer margin of penes below the expanded upper portion. Tails immaculate, according to Say; pale whitish.

The short basal joint of the fore tarsus distinguishes this species from the allied *S. heterotarsale* and *S. pallidum*. Specimens in the Cornell collection from Knoxville, Tenn. (Miss Lula Shipe); from Goshen, O. (C. R. Crosby); and from Ringgold, Swamp Creek, Rock Creek and Rome, Ga. (P. W. Fattig), seem to be of this species. None of these, however, show a reddish stigmatic stain; several have very faint spiracular marks on tergites 6 to 8; several have a distinct brownish tinge in costal spaces at extreme apex of wing and continued around wing tip. The species referred to by Drs. Needham and Clemens under this name are probably not the true *interpunctatum*.

Stenonema ithaca Clemens & Leonard

1905—Needham—Bull. N. Y. St. Mus. 86: 54. Fig. nymph (*Heptagenia interpunctatum*)....... N. Y.
1924—Clemens & Leonard—Canad. Ent. 56: 17 (Heptagenia).
1930—Ide—Canad. Ent. 62: 229.. Ont.
1933—McDunnough—Canad. Ent. 65: 23 (in discussion of *nepotellum*)......................... Que.
1933—Traver—J. Elisha Mitchell Sci. Soc. 48: 178, 199.................................. N. C.

Length: body 9–10; wing 11–12 mm.

A reddish brown species; penes distinctly L-shaped; no spiracular dots; tails pale, joinings darker.

Head red-brown; purplish band across middle area of carina; black rings at bases of antennae. Thorax deep red-brown; pleura somewhat paler than notum and sternum. Median area and posterior margin of pronotum blackish. A purplish black streak laterally on prothorax extending down across fore coxa; similar dark streak on pleura anterior to wing roots. Dark purplish or reddish brown markings on pleura below root of each wing. Coxae deep red-brown, apical and lateral margins and a small apical spot reddish black. Posterior portion of mesosternum darker than remainder of sternum. Fore femur light reddish brown, with median and apical dark purplish red bands; tibia yellowish red, blackish at apex; tarsus yellowish with smoky tinge, joinings narrowly darker. Basal joint of fore tarsus somewhat less than ½ the length of the second. Middle and hind legs yellowish, sometimes with faint reddish tinge; femoral bands as on fore leg; tips of tibiae and tarsal joinings reddish brown. Wings hyaline; longitudinal veins yellowish to reddish brown, cross veins dark red-brown; veins of costal margin slightly thickened, cross veins not dark-margined. Only a very slight crowding of cross veins at bulla; usually 1 in costal space, 2 or 3 in each of two following spaces. Costo-apical region somewhat opaque, may be faintly brown-tinged; a faint reddish or brownish stain occasionally present in stigmatic area, typically without such a stain. Veins of hind wing paler and finer.

Abdominal tergites 1–8 largely occupied by wide reddish to purplish brown bands on the posterior margin, from which deep shading of same color extends forward almost to the anterior margins. A mid-dorsal streak of same dark color extends the length of the abdomen, sometimes divided by a very narrow pale median line. Near each spiracle is a pale irregularly oval space surrounded by purplish grey markings. No dark spiracular dots. A pale yellowish area remains on each side of each tergite on anterior margin; pleural fold usually narrowly pale. Sternites greyish, more or less tinged with pale reddish or purplish; posterior margins narrowly purplish grey, so that abdomen appears somewhat annulate. Traces of brownish triangles in postero-lateral areas may be present. Segments 9 and 10, often the posterior portion of 8, opaque, bright reddish with distinct yellow tinge; tergites with usual dark markings. Penes distinctly L-shaped; of *rubromaculatum* type. Tails pale greyish or yellowish white; joinings narrowly deep purplish brown, alternate joinings wider.

This species is darker red-brown than *S. fuscum* or *rivulicolum*, which it somewhat resembles; abdominal sternites dark-tinged and with darker posterior margins; the penes are also more sharply L-shaped. There are many reared specimens of this species in the Cornell collection from Ithaca, N. Y., and vicinity; others are from the mountain region of North Carolina (J. R. Traver); Willoughby, O. (J. R. T.); Allegheny State Park, N. Y. (R. E. Coker); Ft. Jackson, Hopkinton and Nassau, N. Y. (L. Harper); Smoke Hole, W. Va. (J. R. T.); Flat Creek, Rome, Ingham Mills, Dolgeville and Canajoharie, N. Y., and Greenville Co., S. C. (H. K. Townes, Jr.); Clemson College, S. C. (D. Dunavan); Blue Ridge, Bear Foot Creek and Carroll Creek, Ga. (P. W. Fattig).

Stenonema luteum Clemens

1913—Clemens—Canad. Ent. 45: 252. Fig. (Heptagenia) .... Ont.
1915—Clemens—Contr. Canad. Biol. p. 135. Fig. (Heptagenia).
1933—McDunnough—Canad. Ent. 65: 34. Fig. (Ecdyonurus) .... Que.

Length: body 9–10½; wing 10½ mm.

A large pale species of *pulchellum* group; no spiracular dots; stigmatic area of wing light amber; tails smoky at joinings.

Eyes pale green, in living insect. Head pale yellowish white anterior to ocelli; carina with faint oblique reddish streak; antennae narrowly red-ringed at base; Vertex yellowish to yellow-brown near ocelli; reddish brown oblique streaks posteriorly; a red-brown spot may be present behind median ocellus, and a narrow line at margin of eye. Pro- and mesonotum very pale ochreous; scutellum broadly creamy white; pleura yellowish white; an area between fore and middle legs and a portion of mesosternum tinged with light brown. Anterior portion of metanotum creamy; light brown posteriorly, this area appearing as a dark basal band on the pale abdomen. Coxae streaked with smoky on anterior and posterior margins; a faint pinkish patch behind posterior black streak. Legs pale amber. Fore femur with smoky tinge; median and apical purplish red bands on all femora. Tip of fore tibia, claw and tarsal joinings smoky. Basal fore tarsal joint rather short, slightly less than ½ the length of second. Wings hyaline; stigmatic area of fore wing light amber with slight reddish tinges. Costa, subcosta and radius dark smoky amber; other veins blackish; cross veins well developed.

Abdominal segments 1–7 hyaline, whitish yellow; dorsal posterior margins of tergites crossed by narrow black lines. No spiracular dots. Segments 8–10 opaque, creamy white; tergites 8 and 9, sometimes anterior portion of 10, shaded with pink. Forceps pale; genitalia of the *pulchellum* type. Tails pale, joinings smoky.

This species is distinguishable from allied forms by reason of its large size, pale coloration and lack of spiracular dots. Specimens of this species in the Cornell collection are from Port Elgin, Ont., and Ile Bizard, Que. (Evelyn G. Grieve). The previous reference to *luteum* as a probable member of the *tripunctatum* group (J. Elisha Mitchell Sci. Soc. 48: 174) was based on a mis-determination of this species.

Stenonema majus sp. nov.

Length: body 9½–10½; wing 11 mm.

A large rather pale species of *interpunctatum* group; black marks on face and pronotum, wide oblique pleural markings; black spiracular dots; median band on hind femur.

Head yellowish; a black dash below antenna; black spot on median carina, and very small dark spot at inner margin of each eye. Two short black marks on each side of vertex near eye, small median reddish dot; dusky shading near posterior margin. Pronotum yellowish; posterior margin blackish in median area, connected at each end to the wide black lateral streak; an incomplete black longitudinal streak above fore coxa, appearing as a black spot at postero-lateral corner of pronotum and black streak anteriorly. Mesonotum dark red-brown; scutellum and postero-lateral margins yellowish; anterior to scutellum, a darker brown area. On each side of mesonotum anteriorly, a longitudinal black stripe. Pleura yellowish; a wide oblique purplish black streak below roots of fore and hind wings, anterior to middle and hind legs. Purplish black shading posterior to middle and hind coxae, and wide purplish shading anterior to hind coxa. Antero-lateral and postero-lateral areas of metonotum with considerable blackish shading. Sternum yellow.

Legs yellowish white, femora deeper yellow. Purplish black median and apical bands on all femora, median band on hind femur sometimes rather faint. Fore tibia blackish at apex; basal tarsal joint fully ½ the length of the second. Tibiae of middle and hind legs sometimes faintly dusky in middle area; tarsi pale, joinings scarcely darkened; claws dusky. Wings hyaline; longitudinal veins amber yellow, cross veins black; cross veins margined as usual, but near bulla those in costal space are thickened near costa. Black dash at bulla, on 1, 2 or 3 cross veins. Costal margin amber-tinged, faintly at base, more deeply apically, continuing around tip of wing. Outer margin of hind wing dusky.

Abdominal segments 1–7 hyaline, pale yellowish white; posterior margins of tergites purplish black, rather wider on basal than on apical segments; middle sternites may be faintly dark-shaded posteriorly at median line. Basal tergites sometimes with faint greyish streak on mid-dorsal line, and with faint lateral shading. Spiracular dots large, black, prominent; at or near the ends of the dark bands. Segments 8–10 opaque; 8 and 10 largely yellow, tergite 9 distinctly reddish. Forceps pale, apical joinings dusky; penes flesh-colored, quite similar in structure to those of *S. frontale.* Tails pale whitish, narrowly brown at joinings. Female similar to male but paler, with usual sex differences.

Holotype—Male imago, reared from nymph; Cascadilla Cr., Ithaca, N. Y., July 16, 1932 (J. R. Traver). No. 1340.1 in Cornell collection.

Allotype—Female imago, reared; same location, July 20, 1932. No. 1340.2 in Cornell collection.

Paratypes—4 male imagos, 5 female imagos, same location; July 20–23, 1932; No. 1340.3–11 in Cornell collection.

The large size, extensive black markings of pronotum and pleura, together with the long basal joint of the fore tarsus, the prominent spiracular dots and structure of genitalia, distinguish this from other allied forms. Besides the type material, we have other specimens of this species from Ithaca and vicinity; Old Forge, N. Y. (J. G. Needham); and from Canajoharie and Oriskany, N. Y. (H. K. Townes, Jr.).

Stenonema mediopunctatum McDunnough

1926—McDunnough—Canad. Ent. 58: 191 (Ecdyonurus)....................................Ont.

Length: body 9; wing 10 mm.

A species of *pulchellum* group; thorax almost wholly deep black brown; spiracular dots usually present; tails not darker at joinings.

Frontal portion of head pale whitish; vertex sepia brown with ruddy shading next the eyes. Thorax, dorsally and ventrally, deep black brown, except a whitish patch on pleura anterior to wing roots and slight paler shading below wing roots. Posterior portion of scutellum with faint paler shading; a small whitish median dot anterior to scutellum, and two small white lateral streaks. Legs whitish; median and apical purplish bands on femora; fore tarsus black-tipped at apex; basal joint of fore tarsus more than half as long as second. Wings hyaline; veins fine, black, cross veins in costo-apical region slightly thickened.

Abdominal segments 2–7 hyaline white; scarcely a trace of darker posterior margins, only a small dark transverse dash on posterior margin of each, at mid-dorsal line. Black spiracular dots present on segments 4–7 (may be obsolete). Segments 8–10 opaque, whitish; tergite 8 in posterior half, and all of 9, shaded with light sepia brown. Forceps whitish; genitalia of the *pulchellum* type(?). Tails wholly white, not darker at joinings.

The very dark thorax, short dark median dash on tergites and fine veins distinguish this species from its near allies. The specimen which we had tentatively placed in this species previously, and to which reference was made in the J. Elisha Mitchell Sci. Soc. 48: 174, belongs to the *tripunctatum* group, but we doubt that it is the true *mediopunctatum;* we are therefore considering this species as a member of the *pulchellum* group, as Dr. McDunnough stated. We have not seen any specimen known to be of this species.

Stenonema nepotellum McDunnough

1933—McDunnough—Canad. Ent. 65: 20. Figs. . . . . Que., Ont., Ill.

Length: body 8; wing 8–10 mm.

A species of the *pulchellum* group; spiracular dots present; tails dark-ringed; reddish stain in stigmatic area of fore wing.

Head very similar to *S. rubromaculatum*. Mesonotum ochre-brown; antero-lateral margins narrowly whitish. Scutellum whitish, also lateral margins to base of fore wing. Metanotum largely ochre-brown; median elevation whitish. Pleura yellowish-white, with brown shading as in *rubromaculatum*; anterior to wing roots shaded with light reddish. Legs very much as in *rubromaculatum*; posterior margin of fore coxa narrowly black; less reddish shading laterally than in that species; purplish bands on femora also less distinct, may be wanting on hind leg. Wings as in *rubromaculatum*.

Abdominal segments 1–7 hyaline, pale yellowish white; posterior margins of tergites somewhat more distinctly black than in *rubromaculatum*; black spiracular dots well defined. Segments 8–10 opaque white; tergites largely suffused with pinkish brown; on anterior margin of 8, slight pale submedian wedge-shaped marks; posterior margin of 10 also white. Genitalia of *rubromaculatum* type. Tails pale; joinings narrowly brown.

In the imago stage, this species is very close to *S. rubromaculatum* and *S. annexum*. may be distinguished as indicated under those species. Ventral markings of nymph are distinct from either of these, and very similar likewise to *S. ithaca*. In the Cornell collection are specimens of this species, many of which have been reared from nymphs, taken at Ithaca, N. Y., and in the vicinity of Willoughby, Ohio, from the Chagrin River. A specimen taken at Ft. Jackson, N. Y. by Mr. L. Harper, probably belongs here. It may be this species which Prof. Needham has listed as *S. pulchellum* (see references under that species). We have seen other specimens from Ohio Pyle, Pa. (Carnegie Museum); Kast Bridge, Middleville, Chatham Center, Prattsville, Canajoharie and Newport, N. Y. (H. K. Townes, Jr.).

Stenonema ohioense sp. nov.

Length: body 8–10; wing 9–10 mm.

A rather dark species of *interpunctatum* group; basal joint of fore tarsus about ½ length of second; spiracular dots distinct; tails darker at joinings.

Head yellowish red; continuous black transverse line across face from inner margin of eye, below antenna. Antennal filament dusky, tip paler. A black spot on each side of vertex, and submedian black streaks continued backward to meet the blackish posterior margin. Pronotum reddish brown in median area, yellowish laterally; anterior and posterior margins black, the ends of the latter continuous with the oblique black streaks on each side; just before the posterior margin, several other shorter blackish transverse marks. A black mark in postero-lateral corner. Mesonotum dark red-brown; a median strip, and a narrow streak along lateral margin, brighter red. Median area at anterior and posterior portions distinctly blackish, this area taking the form of a dark triangle anterior to scutellum, which is not paler than areas around it. Metanotum dark red-brown, its central area with a large triangle outlined prominently in black, the base extending across middle of segment, its apex on anterior margin. Pleura yellowish red; a short purplish black mark anterior to middle coxa, faint dark marks at base of hind coxa. Sternum reddish yellow.

Fore legs greenish yellow with slight smoky tinge. Median and apical blackish femoral bands. Tips of tibiae blackish; tarsal joinings narrowly dark brown. Basal joint of fore tarsus almost or fully ½ as long as second joint. Middle and hind legs yellowish white; apical purplish black femoral bands prominent, median bands fainter; tips of tibiae and tarsal joinings brownish. All claws dusky; faint blackish longitudinal pencilings on all femora. Wings hyaline; costa, subcosta and radius yellowish brown, other longitudinal veins red-brown, cross veins slightly darker; all veins of anal region of fore wing and in hind wing finer and slightly paler. Cross veins in costal border dark-margined as usual; two or three radial veins at bulla more widely margined, often (but not always) connected by black dash. Apical portion of costal margin tinged with greenish amber, becoming brownish around tip of wing. Outer margin of hind wing narrowly purplish brown.

Abdominal segments 1–6, and tergite 7, semi-hyaline yellowish white; tergites 1–5 rather heavily shaded with purplish black in median area, the median geminate streak and posterior margins deeper blackish. Lateral margins and somewhat crescentic submedian streaks remain pale. Dark shading less extensive on 6; confined to median line and posterior margin, on 7. Sternites very faintly brownish at middle of posterior margins. Segments 8–10, and sternite 7, opaque, deep yellowish, tergites sometimes with a reddish tinge. Posterior margins of 8 and 10, and median line of 8, faintly dusky. Distinct black spiracular dots on segments 2–8. Forceps dusky; penes pale yellowish, very similar in structure to *S. frontale.* Tails pale yellowish basally, joinings narrowly dark brown; apical portions smoky brown, joinings not distinct.

Female paler, but with similar dark markings.

Holotype—Male imago, reared from nymph; East Creek, near Painesville, Ohio, Aug. 17, 1932. No. 1338.1 in Cornell collection.

Allotype—Female imago, reared; same location, Aug. 18, 1932. No. 1338.2 in Cornell collection.

Paratypes—10 male imagos, 6 female imagos, same location, Aug. 15 to 18, 1932, and Paine Creek, Leroy, Ohio, Aug. 7, 1933. No. 1338.3–18 in Cornell collection.

Besides the type series from Ohio, we have other specimens, many of them reared, from Ithaca, Cascadilla Creek, and Ringwood, N.Y. This species may be distinguished from *S. gildersleevei* by the less extensive dark markings on pleura and tergites, and the relatively longer male fore tarsus. The basal joint of the fore tarsus is longer than in *S. canadense,* and the genitalia show slight differences.

Stenonema pallidum Traver

1933—Traver—J. Elisha Mitchell Sci. Soc. 48: 181 .... N. C.

Length: body 6½–7; wing 7½–8 mm.

A small pale species of *interpunctatum* group; black markings on face and pronotum; basal joint of fore tarsus slightly less than half of second; no dark spiracular dots; faint median band on hind femur.

Head pale yellowish; a small reddish spot at center of vertex; black dash below each antenna; base of antennal filament dusky, tip pale. Thoracic notum yellow to yellowish brown; pleura slightly deeper yellow. A black longitudinal streak on each side of pronotum, no other dark markings. Sternum yellowish, with deeper orange area at each side near base of hind leg. Legs yellowish white; fore femur yellowish brown, with conspicuous median and apical purplish black bands. Similar but paler dark bands on other femora. Fore tibia purplish black at apex; tarsal joinings of fore leg, claws, and tips of distal tarsal joints of other legs, faintly brownish. Basal joint of fore tarsus half the length of the second. Wings hyaline; markings as in *S. interpunctatum;* fourth apical intercalary narrowly darker than others; 4 to 5 basal costal cross veins, 2 oblique ones at bulla, about 12 beyond. Cross veins in anal region of wing rather pale.

Abdominal segments 1–7 hyaline or semi-hyaline, whitish. Rather wide purplish black posterior margins on tergites; no dark spiracular dots. Segments 8–10 opaque, tergites faintly yellowish, sternites whitish. Genitalia very similar to those of *S. interpunctatum.* Tails pale, not darker at joinings.

This species is very similar to *interpunctatum,* and may in fact prove to be synonymous with it. For the present we hold it to be a distinct species, because of the faint but distinct dark median band on hind femur, the longer basal joint of the fore tarsus, and the paler thoracic notum and apical abdominal tergites. The basal joint of the fore tarsus, while not as long relatively as in *heterotarsale,* is yet rather intermediate in length between that species and *interpunctatum.* There is also the possibility that *pallidum* may be synonymous with *flaveola* Pict. (the female subimago, which Hagen and Ulmer have both considered to be *interpunctatum;* see references under that species). The type material of *pallidum* is in the Cornell collection.

Stenonema placitum Banks

1910—Banks—Canad. Ent. 42: 199. Fig. (Heptagenia)..N. Y.
1926—McDunnough—Canad. Ent. 58: 193 (in discussion of *Ecdyonurus rubrum*)..........Ont., Que.
1933—Traver—J. Elisha Mitchell Sci. Soc. 48: 175.

Length: body 8½; wing 9½ mm.

A reddish species of *pulchellum* group; no spiracular dots; no median band on middle and hind femora; basal joinings of tails darkened.

Frontal portion of head yellowish, tinged with reddish around bases of antennae and ocelli; ocelli narrowly black-fringed at base; vertex reddish brown, a creamy white area just back of ocelli; posterior margin reddish. Thoracic notum dark olive brown with reddish tinges; scutellum largely whitish. Metanotum red-brown, scutellum pale. Pleura yellowish white; a distinct deep rose patch anterior to base of fore wing; rose markings anterior to middle coxa. A black spot on coxae of middle and hind legs, another just above it. Reddish brown shading laterally on pronotum, and between bases of fore and middle leg; opaque creamy streak below wing roots. Sternum yellowish white. Fore leg yellowish; femur tinged with olive brown, tarsus slightly smoky. Very faint median band on femur, apex blackish; tip of tibia, claws and tarsal joinings blackish. Basal tarsal joint slightly less than ½ length of second. Middle and hind legs whitish; no median femoral bands; tips of joints and apical half of distal tarsal joints slightly darkened. Wings hyaline; costa, subcosta and radius yellowish brown; other veins dark red-brown, cross veins very slightly heavier than longitudinals. Veins of hind wing paler, brownish anteriorly. A faint reddish brown stain in stigmatic area. Space basad of humeral cross vein somewhat more darkened than in *S. terminatum.*

Abdominal segments 1–7 whitish hyaline; posterior margins of tergites, especially dorsally, rather widely red-brown; joinings of sternites opaque. No dark spiracular dots. Segments 8–10 opaque; tergites rather bright red-brown, sternites yellowish. Genitalia yellowish white; of *pulchellum* type. Tails whitish; basal joinings light purplish red; in remainder of tail, joinings opaque, faintly darker than joints.

Distinguished from the allied *terminatum* by the more extensive reddish markings and darker basal tail joinings, also by wider pale area on scutellum. Absence of spiracular dots and median femoral bands on middle and hind legs separate it from other allied species. There is a specimen of this species in the Cornell collection from Ile Bizard, Quebec, taken by Dr. Evelyn G. Grieve.

Stenonema proximum sp. nov.

Length: body 9–9½; wing 10 mm.

A yellowish species of *interpunctatum* group; black dots on face, black streaks on pronotum, also oblique black streaks on pleura; spiracular dots diffuse, usually present; greyish median streak on tergites 1–4.

Head yellowish; a black spot below each antenna, a very small dark spot near inner margin of each eye; usually no dark spot on carina, which is whitish. Antennal filament dusky, tips paler. Reddish shading at bases of ocelli, sometimes faintly near posterior margin of head; two short black streaks near each eye, on vertex, also a very small reddish median dot. Pronotum yellowish; an oblique black lateral streak quite prominent; a small black dot on posterior margin at median line, sometimes another anterior to this. Mesonotum yellowish, sometimes tinged with reddish; median area may be slightly darker. Anterior to wing roots, a purplish black streak, along margin of mesonotum. Pleura and sternum similar in color to mesonotum; a prominent purplish black oblique streak below roots of fore and hind wings, anterior to middle and hind legs, and between this and base of middle leg, another shorter dark streak (in some specimens, these streaks are rather faint). Legs yellowish, with faint greenish tinge; median and apical black bands prominent on fore and middle femora, median band on hind femur diffuse, may be wanting; longitudinal black pencilings on all femora. Tip of fore tibia black; tarsus faintly smoky; basal joint fully ½ the length of the second. All tarsal joinings faintly darkened. Wings hyaline; costa, subcosta and radius yellowish brown, other longitudinal veins darker brown, cross veins black; those of basal spaces in costal margin narrowly dark-margined. At bulla, a short black dash, often on one or two cross veins only. Apical portion of costal margin tinged with yellowish or greenish yellow. Outer margin of hind wing dusky.

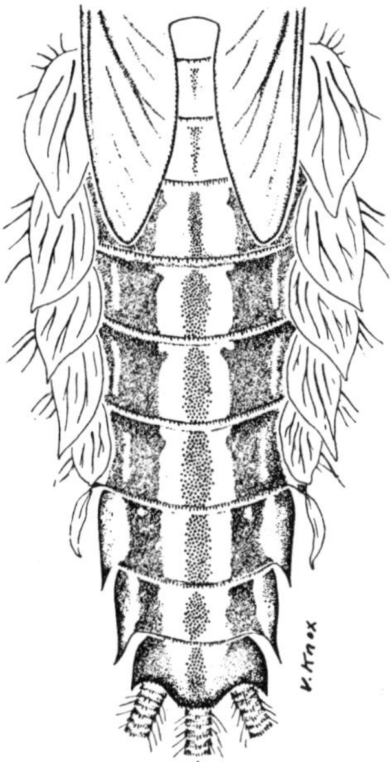

FIG. 93. Nymph of *Stenonema proximum;* dorsal view of abdomen (drawing by Velma Knox).

Abdominal segments 1–7 yellowish, often with greenish tinge (whitish, in alcohol); posterior margins of tergites purplish black. Diffuse blackish spiracular dots are usually present, not very prominent; may be wanting on very pale specimens. Posterior margins of sternites may be faintly grey-tinged in median area. Mid-dorsal line of tergites 1–4, sometimes also 6–8, with a distinct greyish streak, may appear geminate; on basal tergites also, faint lateral greyish tinges may be present. Segments 8–10 opaque, deeper yellow, may be tinged with reddish dorsally; traces of grey shading in spiracular areas (at ends of dark posterior bands). Genitalia very similar to *interpunctatum*; forceps yellowish, the penes flesh-colored. Tails yellowish at base, often becoming dusky in middle and apical portions; joinings faintly darker. (Male.) Dorsal abdominal pattern of nymph as in fig. 93.

Holotype—Male imago (pinned)—reared from nymph—taken near White Church, N. Y., in vicinity of Ithaca; June 13, 1933 (J. R. Traver). No. 1341.2 in Cornell University collection.

Allotype—Female imago—(pinned)—same data; June 13, 1933; (J. R. T.). No. 1341.2 in Cornell collection.

Paratypes—7 male imagos, 6 female imagos; same locality, June 10–20, 1933 (J. R. T.). No. 1341.3–15 in Cornell collection.

The longer basal joint of the fore tarsus distinguishes this species from *S. interpunctatum* and *conjunctum*; the dark marks on the face and pronotum, oblique pleural streaks and diffuse spiracular marks separate it from *heterotarsale*; by the two latter characters it is separable also from *pallidum*. In details of the structure of the penes it differs from other pale species of this group.

Besides the type material, there are other specimens in the Cornell collection which seem to be of this species, from Ithaca, N. Y.; Honeoye Falls, N. Y.; Lake Forest, Ill. (J. G. Needham).

Stenonema pudicum Hagen

1861—Hagen—Syn. Neurop. Ins. N. Amer. p. 39 (Ephemera)................................D. C.
1885—Eaton—Revis. Monogr. p. 280 (*Ecdyonurus vicarius*).
1925—McDunnough—Canad. Ent. 57: 191 (not synonym of *vicarium*)........................N. C.
1933—Traver—J. Elisha Mitchell Sci. Soc. 48: 174, 187, 200.

Length: body 12–12½; wing 13–14 mm.

A rather large species, allied to *S. vicarium*; hind wing dark-margined; cross veins crowded at bulla; sagittate dark mid-dorsal streak on tergites.

Head reddish brown; from lower corner of eye a transverse purple-black line passes under antenna and across carina; orange-red shading below antenna; ocelli black-ringed at base. Vertex with blackish shading between ocelli; oblique streaks from median line to posterior margin. Pronotum deep red-brown; geminate purplish-black median streak, darker anteriorly; posterior margin black in median area; purplish streak from margin to trochanter of fore leg. Mesonotum yellowish brown, red-brown anteriorly; scutellum and adjacent areas blackish brown; margins and sutures red-brown. Metanotum dark red-brown. A large pale area anterior to wing roots; postero-lateral area of pronotum pale. Pleura dark red-brown, intersegmental areas yellowish; sutures narrowly blackish. Small purplish spots on each side of middle and hind coxae. Sternum dark red-brown, pale areas around leg bases. Coxae and trochanters of all legs red-brown. Fore leg yellowish to olive brown; wide purple-brown median and apical bands on femur; apex of tibia black; claws, distal tarsal joint, and tarsal joinings narrowly brown. Basal joint of fore tarsus more than ½, sometimes almost ⅔ length of second. Middle and hind. legs paler yellowish olive; purplish median and apical bands on femora prominent; tarsal joinings deep brown, tarsi may be slightly smoky. Wings hyaline, but with very faint milky tinge. Many cross veins crowded together at bullar region, in first 6 to 8 spaces, and margined with dark brown, forming a dark curved streak in wing. Costa, subcosta and radius thickened, amber or deep yellow; other longitudinal veins of fore wing dark red-brown, finer. Cross veins purplish black, thicker than longitudinals; many in costal border and in disc are dark-margined. Distinct reddish brown stain in stigmatic area, in costal and subcostal spaces. Hind wing with a rather wide purplish outer margin; veins yellow-brown anteriorly, pale yellow posteriorly. See drawing of wing, fig. 29.

Abdominal segments yellowish olive tinged with reddish or purplish brown; tergites darker than sternites, which are distinctly yellowish. On each tergite a rather wide purplish black posterior margin, widest on anterior segments, where half to two-thirds of tergite may be dark. A sagittate dark mark at mid-dorsal line of each, based on posterior margin and reaching about to center of segment. Laterally, at end of each dark band, an irregular triangular mark of some color as band; apex of triangle extends toward center of segment; lateral areas in some specimens largely dark except for a single pale oval area. Pleural fold may be narrowly whitish. Narrow dark shading at anterior and lateral margins of sternites; ganglionic area whitish, other portions may be slightly grey-tinged. Genitalia pale brownish; penes much as in *tripunctatum* but apical margin more rounded, inner margin less distinctly incurved. Tails greyish white, joinings purplish brown; alternate joinings wider.

The crowding of cross veins at bulla seems to ally this species with *S. tripunctatum* and *femoratum*; the dark-margined hind wing separates it from the former, the sagittate mid-dorsal mark and uninterrupted dark posterior margins distinguish it from the latter species. However, the nymphal characters and genitalia are closer to *vicarium*, which lacks the dark-margined hind wing. Reared specimens of this species in the Cornell collection are from the mountains of North Carolina; a variant form from near Danbury, N. C.; other typical specimens from Ringwood, N. Y. We have also three other male specimens, from Clemson College and Greenville, S. C., taken by Prof. D. Dunavan and Mr. H. K. Townes, Jr., which lack the sagittate dark median mark; the posterior margins of tergites are likewise narrower. As nymphs of these are not known, we hold them under *S. pudicum*.

Stenonema pulchellum Walsh

1862—Walsh—Proc. Acad. Nat. Sci. Philad. p. 375 (in Palingenia)........................... Ill.
1863—Walsh—Proc. Ent. Soc. Philad. 2: 203 (in Palingenia).
1863—Hagen—Proc. Ent. Soc. Philad. 2: 177 (in Palingenia).
1871—Eaton—Trans. Ent. Soc. London; p. 141 (in Heptagenia).
1885—Eaton—Revis. Monogr. p. 299 (in Heptagenia).
? 1901—Needham—N. Y. St. Mus. Bull. 47: 420 (in Heptagenia)........................... N. Y.
? 1907—Needham—Rep. N. Y. St. Ent. p. 193 (in Heptagenia).
1913—Clemens—Canad. Ent. 45: 258 (as *Heptagenia luridipennis-nec* Burmeister). Fig........ Ont.
1915—Clemens—Contr. Canad. Biol. p. 139. Fig. (same as above).
1926—McDunnough—Canad. Ent. 58: 193 (in discussion of *Ecdyonurus rubrum*).
1933—McDunnough—Canad. Ent. 65: 23. Figs... Ia.
1933—Traver—J. Elisha Mitchell Sci. Soc. 48: 175.

Length: body 6–7; wing 7–8 mm.

A species of *pulchellum* group; dark spiracular dots present; scutellum broadly white; tails darker at joinings.

Eyes of living insect pearly white. Ocelli reddish, blackish at base; bases of antennae fuscous, tips pale. Vertex dark red-brown. Thoracic notum brownish black, with no reddish tinges; scutellum broadly white, not merely white-tipped, as in *S. rubrum*. Pleura pale, with yellowish brown and occasional pink tinges; a pinkish area anterior to wing roots. Central area of mesosternum brown. Fore legs pale yellowish; femur with median and apical darker bands; tips of tibia and tarsal joinings fuscous; basal tarsal joint fully one-half the length of the second. Middle and hind legs whitish, markings as on fore leg, but paler; median femoral band sometimes obsolescent. Wings hyaline; a pale brown (not reddish) stain in stigmatic area of fore wing.

Costa, subcosta and radius yellow, especially in basal two-thirds; all other veins reddish brown; cross veins somewhat heavier than longitudinals, which are quite fine. Cross veins more numerous than in *rubrum*. Veins in hind wing, behind costa, hyaline.

Abdominal segments 1–7 white, hyaline; posterior margins of tergites faintly darker; dark spiracular dots present, on segments 3–7. Segments 8–10 opaque; tergites dark ruddy, sternites paler. Genitalia of the *pulchellum* type (see fig. 91, *rubrum*); penes not as sharply L-shaped as in *rubromaculatum*. Tails whitish; joinings narrowly fuscous, in the basal area alternately wide and narrow.

The blacker thorax, with its lack of reddish shading, the white scutellum and purer white abdomen distinguish this species from the allied *rubrum*. The presence of spiracular dots separates it from *terminatum* and *placitum*. The smaller size, smaller eye, darker thorax and lack of red stigmatic stain distinguishes it from *rubromaculatum*. We have examined specimens of this species from Pleasant Valley, Ia., which Dr. McDunnough kindly lent us for study. The above description is drawn in part from these, as well as from Walsh's account and Dr. McDunnough's notes on the species.

Stenonema rivulicolum McDunnough

1933—McDunnough—Canad. Ent. 65: 40. Fig. . . . . Que.

Length: body 10; wing 13 mm.

A member of the *vicarium* group; slight crowding of cross veins at bulla; reddish stigmatic stain; pale and dark areas of tergites distinct.

Eyes of living insect pale grey with greenish tinge; smaller than in *S. fuscum*. Anterior portion of head very similar to *fuscum*; vertex deeper brown, intensified behind median ocellus and at middle of posterior margin. Mesonotum deep brown; pale areas around scutellum less distinct than in *fuscum*; posterior margin black-tinged. Metanotum deep brown. Pleura deeper brown; reddish anterior area as in *fuscum*, but pale streaks at bases of legs less distinct, deeper ochreous. Blackish shading above base of middle coxa, also tinges of red in this region. Legs much as in *fuscum*, but with black patches at bases of all legs; black streaks on coxae broader; femoral bands darker in color, deep purple to purplish brown, and noticeably wider. Basal fore tarsal joint somewhat longer, almost half the length of the second. Wings hyaline; a distinct reddish stain in stigmatic area; cross veins distinctly thicker and slightly margined with diffuse brownish in costal space; tendency for cross veins at bulla to be somewhat crowded.

Abdominal tergites much as in *fuscum*, but posterior borders wider, half the width of segment anteriorly, deep purplish brown; anterior margin of this band sharply defined, not diffuse as in *fuscum*. Pale areas light amber, less smoky than in that species. A longitudinal median band on tergites 2–9, purple brown, wider than in *fuscum*; widened toward posterior margin on each segment. Segments 8–10 opaque, shaded with light brown; lateral margins paler brown, but not distinctly white as in *fuscum*. Sternites 1–7 light yellowish, with small brown patches in postero-lateral angles of at least the anterior segments; sternites 8 and 9 shaded with light brown. Forceps pale smoky; genitalia as in *fuscum*. Tails smoky, darker at joinings.

Distinguished from its near ally *fuscum* as indicated above. Apparently intermediate between *fuscum* and the larger *vicarium*,. We have specimens of this species from Fairy Lake, Que. (including a paratype) which Dr. McDunnough kindly gave to the Cornell collection.

Stenonema rubromaculatum Clemens

1913—Clemens—Canad. Ent. 45: 256. Fig. (Heptagenia) . . . . Ont.
1915—Clemens—Contr. Canad. Biol., p. 138. Fig. (Heptagenia).
1930—McDunnough—Canad. Ent. 62: 61 (Ecdyonurus).
1930—Ide—Canad. Ent. 62: 230 (Ecdyonurus).
1933—McDunnough—Canad. Ent. 65: 16. Fig. (Ecdyonurus) . . . . Que., N. B., Mass., Ill., N. S.
1933—Traver—J. Elisha Mitchell Sci. Soc. 48: 175.

Length: body 8; wing 8–10 mm.

A member of *pulchellum* group; stigmatic area with reddish stain; spiracular dots present; tails dark-banded.

Eyes pale pearly grey, in living insect. Frons and carina whitish; a faint brownish oblique streak on each side of carina, and minute dark dot at its base. Bases of antennae and a lateral streak next eye on level with antenna rusty or tawny brown; antennae whitish. Vertex pale creamy; yellowish tinges around lateral ocelli and a light rusty or tawny brown spot behind median ocellus; two oblique tawny brown streaks extend backward from median line to eyes; posterior portion of head hyaline. Faint yellow-brown shading between eyes and lateral ocelli. Prothorax whitish; faint smoky shade on median area. Mesonotum light olive grey or pale clay; antero-lateral margins alabaster white; entire scutellum white, "lateral elevations and hollow behind the scutellum being tinged with pale sepia brown" (McD.). Pleura whitish; area anterior to base of middle leg tinged with brown; a black dot behind middle coxa, and brownish shading anterior to it. Anterior portion of metasternum, including median projection, alabaster white; posteriorly pale sepia brown; remainder of sternum whitish. Fore coxa tinged laterally with reddish brown; a fine black streak extends obliquely upward from this toward prothorax. Femur and tibia of fore leg subequal; very faintly amber-colored; femur with smoky shading and a median and apical purple-red band; tibia blackish at apex, which color extends over to base of tarsus. Tarsus pale; joinings and distal joint smoky; basal joint slightly less than half the length of second. Coxae and trochanters of middle and hind legs alabaster white, shaded with brown slightly at apex; bases of coxae with tawny brown shading, fainter on hind legs. Femora and tibiae very pale amber; faint median and apical femoral bands; femur slightly longer than tibia. Tarsi pale whitish; joinings and distal joint smoky. Wings hyaline; stigmatic area amber-tinged, a distinct purplish-red stain in subcostal space. Longitudinal veins deep smoky amber; cross veins blackish brown. Humeral cross vein heavily blackish; basal costal veins moderately fine and rather widely spaced, 3–5 of these. At bulla, 2 cross veins in costal space, 2 in subcosta, 3 below radius; 11–13 in costo-apical space. Cross veins very fine in anal area, in hind wings largely hyaline.

Abdominal segments 1–7 pale hyaline whitish; very narrow dark posterior margins on tergites in dorsal area; fine dark spiracular dots, 2–7. Segments 8–10 opaque alabaster white; tergites shaded with pinkish brown; only a narrow median pink streak on 8, no shading on posterior margin of 10. Forceps dull whitish; genitalia of *rubromaculatum* type (McDunnough's *tripunctatum* type), distinctly L-shaped, the apical margin almost straight, and forming a right angle with vertical margin. Tails whitish; joinings narrowly blackish brown.

This species is very close to *S. nepotellum* and to *annexum* in the imago stage, although readily separable on nymphal characters. Imago may be distinguished from former species by the duller and paler thorax, the narrower dark posterior margins of tergites, and the more restricted pinkish shading on tergite 8. The thorax is also paler than in *annexum*, and tergites 8–10 less extensively marked than in that species.

Color pattern of ventral surfaces of nymphs of these species is distinctive. We have seen no specimens known to be of this species.

Stenonema rubrum McDunnough

1913—Clemens—Canad. Ent. 45: 252 (*Heptagenia flavescens—nec* Walsh). Figs. .... Ont.
1915—Clemens—Contr. Canad. Biol. p. 134. Figs. (same as above).
1926—McDunnough—Canad. Ent. 58: 192 (*Ecdyonurus ruber*) .... Que.
1933—McDunnough—Canad. Ent. 65: 24.
? 1933—Traver—J. Elisha Mitchell Sci. Soc. 48: 190 .... N. C.

Length: body 7; wing 8 mm.

A rather small species of *pulchellum* group; dark spiracular dots present; faint reddish tinge in stigmatic area of wing; tail joinings purple-brown.

Frontal portion of head pale whitish; carina with slight reddish markings; vertex deep red-brown. Thoracic notum deep brown slightly tinged with reddish; pleura whitish, with red-brown markings anterior to wing roots and extending down to bases of middle legs; faint pinkish shading below wing roots. Mesonotal scutellum white-tipped; lateral area beneath scutellum, and anterior projection of metanotum, likewise pale. Sternum pale ochreous slightly yellow-tinged. Fore legs dull amber in color; coxa largely reddish; femur with median and apical purplish bands; tibia at apex, and tarsal joinings, blackish. Basal joint of fore tarsus rather more than half the length of the second. Middle and hind legs pale yellowish white; femora, tibiae and tarsi marked as on fore leg; a small reddish spot on each coxa. Wings hyaline; all veins dark, cross veins slightly thicker than longitudinal ones; a faint ruddy tinge in stigmatic area of fore wing.

Abdominal segments 1–7 pale hyaline slightly tinged with smoky; posterior margins of tergites narrowly blackish. Black spiracular dots present on segments 3–7. Tergites 8–10 deep ruddy brown; sternites opaque whitish. Genitalia of the *pulchellum* type (see fig. 91); forceps pale, tips slightly infuscated. Tails pale, joinings narrowly purplish brown.

The ruddier color of the thorax, the restricted whitish markings on the scutellum, and the smoky tinge of abdominal segments, distinguish this species from the allied *S. pulchellum.* The presence of dark spiracular dots separates it from *terminatum* and *placitum.* It may be this species which Dr. Banks identified as *terminatum* (1910, Canad. Ent. 42: 200). The North Carolina specimens (Traver '33) previously referred tentatively to this species may be merely a southern variety, or may represent a closely allied new species.

Specimens of this species in the Cornell collection are from Nassau, Hopkinton and Ft. Jackson, N. Y. (L. Harper); Canajoharie, N. Y. (H. K. Townes, Jr.); Ithaca, N. Y. (Miss C. N. Hardy); Kent Falls, Conn. (C. P. Alexander); Greenville Co., S. C. (H. K. Townes, Jr.); Cleveland, and Notala River, Ga. (P. W. Fattig).

Stenonema scitulum sp. nov.

Length: body 8–8½; wing 9 mm.

A species of the *tripunctatum* group; very close to *S. tripunctatum*, but with hind wing distinctly dark-margined; eyes large.

Male imago.—Eyes considerably larger relatively than in typical *tripunctatum*; almost contiguous apically. Head and thorax yellowish; short black submedian streaks on pronotum; blackish markings on pleura as in *tripunctatum.* Legs yellowish; median and apical bands on all femora, median band on hind femur may be very faint. Apex of fore tibia, all tarsal joinings and claws blackish. Basal joint of fore tarsus fully ½ as long as second. Wings very similar to *tripunctatum*, except that hind wing is distinctly blackish brown on outer margin, most evident at and below tip. Abdominal segments 2–7 hyaline, white; dark streaks on posterior margins of tergites as in *tripunctatum*; no stigmatic marks. Segments 8–10 opaque; tergites pale red-brown, sternites paler. Tails white, alternate joinings purplish black. Genitalia very similar to *tripunctatum*.

Female very similar to male. Wing 10–12 mm. in length.

Holotype—Male imago; Latimer Co., Oklahoma, June 10, 1931 (R. D. Bird). No. 1343.1 in Cornell University collection.

Allotype—Female imago, same data. No. 1343.2 in Cornell collection.

Paratypes—15 male imagos, 13 female imagos; same data. No. 1343.3–30 in Cornell collection.

This species might have been considered merely a very small pale variety of *tripunctatum*, except for the relatively larger eyes and the dark-tipped hind wing.

Stenonema terminatum Walsh

1862—Walsh—Proc. Acad. Nat. Sci. Philad. p. 375 (Palingenia) .... Ill.
1863—Walsh—Proc. Ent. Soc. Philad. 2: 203 (Palingenia).
1863—Hagen—Proc. Ent. Soc. Philad. 2: 177 (Palingenia).
1871—Eaton—Trans. Ent. Soc. London, p. 141 (Heptagenia).
1885—Eaton—Revis. Monogr. p. 299 (Heptagenia).
1926—McDunnough—Canad. Ent. 58: 193 (discussion of *Ecdyonurus rubrum*).
1933—Traver—J. Elisha Mitchell Sci. Soc. 48: 175.

Length: body 6½–8; wing 8–10 mm.

A rather ruddy species of *pulchellum* group; no spiracular dots; no median band on middle and hind femora; basal fore tarsal joint about ⅓ of second; tails not darker at joinings.

Eyes light yellow-green, in living insect. Head yellowish, vertex slightly tinged with reddish; bases of ocelli black-ringed. Thoracic notum reddish brown; scutellum tipped with whitish. Pleura yellowish with light red-brown shading between fore and middle legs, anterior to hind coxa, and narrow red-brown lines below wing roots. Anterior portion of metanotum darker than posterior. Sternum yellowish. Fore legs yellowish; middle and hind legs yellowish white. Fore femur with rather indistinct median and apical ruddy bands; on middle and hind femora the median band is wanting (traces of it occasionally in middle femur). Tip of fore tibia blackish; tips of other tibiae, and all tarsal joinings, narrowly infuscated. Basal joint of fore tarsus less than half, sometimes only slightly more than ⅓ the length of the second. Wings hyaline; veins reddish brown except main longitudinal veins of costal margin, which are yellowish. Cross veins not noticeably heavier than longitudinals except in basal costal and subcostal spaces. Stigmatic area of fore wing slightly cloudy, usually not dark-stained. Veins of hind wing largely hyaline.

Abdominal segments 1–6, and basal half of 7, hyaline whitish. Dorsal portions of posterior margins of tergites narrowly reddish brown; no dark spiracular dots. Segments 8–10 and apical portion of 7 opaque, reddish brown, sternites paler than tergites. Genitalia pale yellowish white, of the *pulchellum* type. Tails whitish, not darker at joinings.

This species is most closely allied to *S. placitum*, from which it is distinguished by the narrower posterior margins of tergites, generally paler color with less reddish shading, and lack of darker rings at tail joinings (in *placitum*, basal joinings are purplish, others very faintly dark-tinged). The lack of dark spiracular dots and the shorter basal joint of fore tarsus separate it from *pulchellum* and *rubrum*, as well as the absence of a median femoral band on middle and hind legs. We have two male imagos of this species from Gimli, L. Winnipeg, Man., kindly given us by Mr. F. Neave. Other specimens which seem to belong here, although slightly smaller than the above, were taken by Prof. R. E. Rodock at Lewiston, Idaho.

Stenonema tripunctatum Banks

1910—Banks—Canad. Ent. 42: 199 (Heptagenia) .... N. Y., Wis.
1913—Clemens—Canad. Ent. 45: 253. Figs. (Heptagenia) .... Ont.
1915—Clemens—Contr. Canad. Biol. p. 137. Figs. (Heptagenia).
1930—Ide—Canad. Ent. 62: 230 (Ecdyonurus).
1933—McDunnough—Canad. Ent. 65: 19 (in discussion of *Ecdyonurus rubromaculatum*).
1933—Traver—J. Elisha Mitchell Sci. Soc. 48: 176.

Length: body 9–11; wing 12–13 mm.

Thorax largely brown; abdomen paler, one median and two lateral black transverse dashes on posterior margins of tergites; hind wing not dark-margined; penes distinctly L-shaped.

Head yellowish, with red-brown shading between ocelli and on vertex; a small dark dot on each side of median carina, slightly below level of antennae. Eyes large, almost contiguous apically. Thorax largely red-brown; a dark median streak on anterior portion of pronotum, sometimes divided by a pale line; an oblique lateral dark streak on pronotum, about halfway to lateral margin. A purplish black streak extends from base of fore trochanter to posterior margin of prothorax. Dark shading anterior to scutellum on mesonotum, and on postero-lateral margins; scutellum somewhat paler than remainder of mesonotum. Pleura pale red-brown; a prominent dark spot below fore wing above base of middle leg; a less conspicuous short dark streak below hind wing. Small dark dot on each trochanter. Metanotum largely dark red-brown. Sternum paler red-brown, anterior portion of mesosternum yellowish. Legs yellowish. Fore femur tinged with yellowish brown; median and apical dark bands; apex of tibia blackish; claw and distal tarsal joint smoky, tarsal joinings blackish. Basal joint of fore tarsus about ½ the length of the second. Middle and hind femora with dark median and apical bands; tips of tibiae and tarsal joinings dark brown; distal tarsal joints smoky. Wings hyaline; humeral cross vein thickened, blackish; costo-apical space with yellowish tinge, distinct purplish red stain in stigmatic area, in costal and subcostal spaces. Longitudinal veins yellowish; cross veins blackish brown; those in costal margin thickened, very prominent; in anal region and along outer margin of fore wing, finer, somewhat paler. Cross veins at bulla tend to become crowded, so that there are usually 3 in the costal, and 3 or 4 each in the subcostal and radial spaces at this point. In hind wing veins are fine, pale; no distinct dark outer margin as in the allied *femoratum* and in *scitulum*.

Abdominal segments 1–7 pale whitish or yellowish, sometimes with slight brownish tinge. On posterior margin of each tergite, a dark transverse streak at median line, and two similar dark lateral streaks; spiracular dots present, near end of lateral dark streak. Segments 8–10 similar to thorax in color. Genitalia of the *tripunctatum* type (see fig. 91, *femoratum*); penes almost straight on apical margin; inner margin incurved, median spines well developed. Tails pale, joinings purplish black; alternate nearer joinings wider.

The three black dashes on posterior margins of tergites distinguish this species from others except *S. femoratum*, *scitulum* and *birdi*. The first named species is much darker, with dark-tipped hind wings; *scitulum* is small and pale, with dark-tipped hind wings; *birdi* brownish, hind wing as in *tripunctatum*. What appears to be merely a color-phase of this species is quite common; in this phase, the thorax is pale creamy or clay-colored, abdominal segments 1–7 almost pure white. Other markings are similar to the darker phase just described. The specimen previously recorded from North Carolina (Traver 1933), as a probable *tripunctatum*, seems nearer *femoratum*, and may be a new species.

Specimens of this species in the Cornell collection are from Comanche Co., Okla. (R. D. Bird); Allegheny State Park, N. Y. (R. E. Coker); Saranac, N. Y. (C. R. Crosby, H. Dietrich); Port Elgin, Ont. and Montreal, Que. (Mrs. E. G. Grieve); Imboden, Ark. (B. Marshall); Austin, Tex. (H. B. Parks); Iowa City, Ia. (T. M. Parsons); Raquette Lake, N. Y. (R. C. Tasker); Farmington, Mo. (Miss L. C. Thomsen); Willoughby, O. (J. R. Traver); Ithaca, Oriskany, Rome and Ithaca, N. Y (H. K. Townes, Jr.). We have seen other specimens from Silver Islet and Long Lake, Ont.; Grant's Pass, Ore.; Mound and Ramsey Co., Minn.; and Rochester, N. Y.

Stenonema varium Traver

1933—Traver—J. Elisha Mitchell Sci. Soc. 48: 192..N. C.

Length: body 9–11; wing 10–11 mm.

A species of *pulchellum* group; spiracular dots present; tails dark-ringed; basal abdominal tergites more or less tinged with dull purplish grey.

Head pale yellowish brown; no markings on frons. Faint orange dot at center of vertex; a short dusky curved mark, quite faint, on each side of dot. Filament of antenna pale brown. Thorax yellowish brown; darkest on metathorax and on each side of mesonotal scutellum. Scutellum white, with narrow median brown line; whitish area anterior to it. Anterior and posterior margins of pronotum narrowly purplish black. Thoracic sutures very narrowly dark brown. Pleura with a broad whitish area anterior to wing roots, at base of fore leg and between middle and hind legs. Sternum similar to notum, darker than pleura. Coxa and trochanter pale, the areas around coxa brown; a purplish streak extends toward pronotum from base of fore leg. Femora pale yellowish; median and apical purplish brown bands. Tibiae paler than femora; apices purplish brown. Fore tarsus pale, in some specimens faintly dusky; basal joint one-half or slightly more than one-half of second. Middle and hind tarsi pale yellowish; all tarsal joinings narrowly brownish. A pale brown cloud on each wing, near point of attachment. Humeral vein heavily infuscated; costa and subcosta at base with broad brown bands. Costa, subcosta and radius yellow; other longitudinal veins yellow-brown. Cross veins dark red-brown, thicker than longitudinals. No crowding of veins at bulla. Stigmatic area opaque whitish, without red or brown stain. 4 to 5 basal costal cross veins, about 8 in costo-apical space. Veins in anal region of hind wing wholly hyaline.

Abdominal tergite 1 largely dull purplish grey; segments 2–6, and basal half of 7, semihyaline, white. Posterior margins of all tergites purplish black, becoming wider on anterior segments. Spiracular dark dots present on 2–7. On 2–5, a narrow purplish black transverse streak extends from spiracular dot laterally almost to median line, parallel to and near the anterior margin. On median line of tergites 1–3 a dull purplish grey streak, with lateral extensions on each side near anterior margin. These grey markings on tergites quite variable; in some, median line dark on tergites 1–8, and 1–4 may be heavily shaded with purplish grey; other specimens are quite pale. Apical half of segment 7, and 8–10 opaque, pale yellowish brown. Forceps silvery white; penes and forceps base pale yellow-brown. Genitalia intermediate between *pulchellum* and *rubromaculatum*. Tails white; joinings narrowly purplish brown, alternate joinings wider. Tips faintly dusky.

The type material, taken in the vicinity of Greensboro, N. C., is in the Cornell collection; also other specimens taken in Greenville Co., S. C. (H. K. Townes, Jr.).

Stenonema vicarium Walker

1853—Walker—List. Neurop. Ins. Brit. Mus. pt. 3: 565 (in Baetis)............... St. Lawrence Riv.
1861—Hagen—Syn. Neurop. Ins. N. Amer. p. 48 (in Baetis).
1871—Eaton—Trans. Ent. Soc. London p. 150. Figs.
1885—Eaton—Revis. Monogr. p. 280 (in part) (in Ecdyonurus).
1925—McDunnough—Trans. R. S. C. p. 222... Que.
1933—Traver—J. Elisha Mitchell Sci. Soc. 48: 174....................................... N. C.

Length: body 12–14; wing 13–16 mm.

A large species allied to *S. fuscum* and *S. rivulicolum*; abdominal tergites largely dark red-brown: dark lateral patches on basal and middle sternites.

Eyes large. Head reddish brown, frons paler; purplish streak on each side of median carina; small dark spot at lower corner of eye; red-brown shading on vertex around bases of ocelli, darker shading posteriorly; two darker oblique submedian streaks. Thorax reddish brown. Pronotum shaded with purplish black; purplish streak on prothorax extending down on to fore coxa. Mesonotum with yellowish tinge; tip of scutellum reddish yellow, four small yellow spots anterior to it. A yellowish area anterior to wing root; yellowish intersegmental areas on pleura; purplish black markings around bases of legs and below wing roots. Purplish marks above and laterad of middle and hind coxae; coxae shaded with blackish. Mesosternum rather dark brown. Fore femur yellowish brown, with wide and conspicuous reddish purple median and apical bands. Middle and hind femora yellowish; bands as in fore femur, but more reddish. All tibiae distinctly reddish; deeper in color basally, paler apically. Fore tarsus yellowish brown with smoky tinge; middle and hind tarsi yellowish red. Basal joint of fore tarsus quite short; less than 1/3 but slightly more than 1/4 of second. Wings hyaline. Longitudinal veins yellowish brown, cross veins purplish brown, much heavier than longitudinals. Cross veins somewhat crowded at bulla; usually 2 or 3 in costal and subcostal spaces, 3 or 4 in radial space; some in costal margin and disc are narrowly dark-margined. Veins in anal regions of both wings finer, somewhat paler. Our specimens (determined by Dr. McDunnough) show no reddish stigmatic stain. Walker's description calls for the "fore border pale brown for one-third of the length from the tip." Dr. McDunnough states that, as compared with *S. fuscum*, there is "a decidedly ruddier tinge" in this area. Hind wing *not* dark-tinged at apex or on outer margin.

Abdominal tergites 1–8 largely occupied by a wide dark reddish to purplish brown band, leaving only a narrow and irregular anterior margin, an oval area near spiracle, and submedian streaks from anterior margin, paler yellowish brown. A wide mid-dorsal geminate streak, and dorsal portion of posterior margin of each, blackish brown. Tergites 9 and 10 largely reddish yellow, median streak faint; little or no dark shading. Sternites creamy yellowish; basal and middle ones with conspicuous purplish brown lateral patches, based on posterior margin; from each, a triangular dark area extends forward to about the middle of the segment. On apical sternites, these areas faint or wanting. Genitalia much as in *pudicum*; distinctly L-shaped. Tails usually rather deep olive brown; joinings deep reddish to purplish brown, alternate joinings wider. See figures of nymph and imago on Pl. I.

This species is larger than *fuscum* or *rivulicolum*, and has a relatively shorter basal fore tarsal joint. The dark ventral markings are much more extensive than in the latter, while *fuscum* is immaculate ventrally. The red tibiae seem distinctive. There are many reared specimens of this species in the Cornell collection from Ithaca, N. Y., and vicinity; specimens from Spero, N. C., are probably pale forms.

Genus **Heptagenia** Walsh

1863—Walsh—Proc. Ent. Soc. Philad. 2: 197.
1881—Eaton—Ent. Mo. Mag. 18: 24 (restricted).
1885—Eaton—Revis. Monogr. p. 265.
1905—Needham—Bull. N. Y. St. Mus. 86: 52 (Ecdyurus).
1920—Ulmer—Stett. Ent. Zeit. 81: 141.
1924—McDunnough—Canad. Ent. 56: 116.
1933—Traver—J. N. Y. Ent. Soc. 41: 108.
1933—Spieth—J. N. Y. Ent. Soc. 41: 330.

Mayflies of variable size, ranging in length of wing from 5 to 15 mm. Coloration varying from creamy white with few or no dark markings, to forms in which the body is largely dark purplish to reddish brown. In most species, eyes of male large, but not contiguous apically. Posterior margin of head of female somewhat emarginate. Pronotum quite deeply excavated on posterior margin in median area. Fore leg of male somewhat longer than body; of female, about 3/4 as long as body. Fore femur of male subequal to or slightly shorter than tibia; tarsus usually about 1 1/4 times the length of tibia. Basal fore tarsal joint relatively short; in most species, 1/6 to 1/4 as long as the second; in other species, nearly 1/3 as long as second. Second joint equal to or slightly longer than third; fourth shorter than third; fifth longer than first, about 1/2 of fourth. Fore femur and tibia of female subequal; tarsus about 3/4 of tibia; basal tarsal joint about 1/2 as long as second, which is subequal to fifth; third shorter than second; fourth subequal to first. In hind legs of both sexes, tibia about 3/4 as long as femur; tarsus slightly more than 1/2 of tibia. Basal tarsal joint shorter than second in many species, longer than second in others. Fifth joint approximately equal to third and fourth combined; fourth joint usually shortest. Claws dissimilar on all tarsi. Typical Heptagenine venation. In some species, costal and subcostal cross veins of fore wing may be thickened or margined; in those of the *maculipennis* group (not including *umbratica*), cross veins in several spaces below the bulla are likewise margined and tend to be crowded together. No anastomosis of cross veins in stigmatic area.

Apical margin of forceps base of male variable in shape; generally slightly emarginate in median area, occasionally more deeply emarginate; in a few species, distinctly convex. Penes usually united in basal half or two-thirds; the free distal portions divergent apically, variable in shape. Generally each division of penes possesses an apical lobe, and a lateral lobe-like extension which often bears a prominent marginal spine. A pair of stout spines is typically present between the divisions of the penes; one to three pairs of more slender spines may occur on or near the inner margin of each division, near the apex. Forceps four-jointed; basal joint very short, second joint longest. In species of the *flavescens*, *pulla* and *maculipennis* groups, and in *H. umbratica*, the third joint is subequal to or very slightly longer than the distal joint, these two together not longer than 3/4 of the second. In species of the *lucidipennis* and *simplicioides* groups, and in *H. kennedyi*, the third joint is usually 1 1/2 to 1 3/4 the length of the distal joint, these two joints together being usually subequal to the second. Subanal plate of female well developed, its apical margin obtusely or retusely rounded.

Head of nymph flattened; often large for size of body. Thorax and abdomen slightly convex dorsally, flattened ventrally. Labrum more than four times as wide as its length. Inner canine of right mandible about 3/4 as long as outer one; one rather long and one shorter spine-like process at tip. On left mandible, inner canine often fully as long as outer one; two or three blunt teeth or projections at tip of each outer canine. In *marginalis* and *julia*, a single long slender spine on left mandible represents the lacinia; on right mandible, lacinia absent, a row of long hairs occurs in lacinial region. In species of the *maculipennis* and *lucidipennis* groups, lacinia is present on each mandible, composed of two or three long slender spines. Both joints of maxillary palp more slender than in Rhithrogena; many long hairs (not pectinate) near tip of distal joint. A series of pectinate spines along apical margin of galea-lacinia. Glossae of labium rather strongly arched away from each other; in *marginalis* and *julia*, tips convergent, not so in species of the other groups. Lateral portions of hypopharynx somewhat larger than central portion; each lateral portion is strongly outcurved, narrowest at tip, outer margin rather sinuate.

As indicated above, nymphs of this genus appear to fall into two main groups. In one of these, the *flavescens—pulla—elegantula** group (represented by *marginalis, julia* and *elegantula*), the pronotum is definitely widest at anterior margin; claws not pectinate (i.e., no spines laterally near tip); no postero-lateral spines on segment 8, usually none on segments 6 and 7; filamentous portion of 7th pair of gills present. In the *maculipennis—lucidipennis* group, the pronotum is widest at or near the middle; claws pectinate; postero-lateral spines present on segments 6–8; filamentous portion absent from gills of 7th pair. In all species studied, the lamellate portion of the 7th pair of gills is similar in form to the preceding pairs; smaller than the central pairs but larger than the first pair. Filamentous portions of gills well developed, ventrad of lamellate portions. Tails three, approximately equal in size and length. General color of abdominal tergites dark brownish or blackish, with pale submedian streaks on each tergite, which may coalesce to form larger pale median areas.

* In my paper on Heptagenine mayflies of North America (J. N. Y. Ent. Soc. 41, 1933), I used the terms "*inconspicua* group" and "*elegantula* group" for those which I here call, respectively, the *lucidipennis* group and *pulla* group.—J. R. T.

The nymph figured by Eaton on Pl. 60 of the Revisional Monograph is evidently not a true Heptagenia. However, the nymph figured on Pl. 61, and figs. 24 and 25 of Pl. 62, as Ecdyonurus and a nameless genus, are true Heptagenias.

Genotype—*H. flavescens* Walsh.

Distribution—Holarctic; in Nearctic, quite generally distributed.

KEY TO THE SPECIES

Male adults*

1—Genitalia of the *maculipennis* type, or modified form thereof (see figs. 96, *maculipennis*, and 95, *umbratica*) 2
—Genitalia not as above 9
2—Cross veins of costal and subcostal spaces, and of several spaces below bulla, rather heavily margined; in latter species, veins tend to be crowded together, as in *Stenonema tripunctatum;* veins in hind wing and anal half of fore wing may be pale 3
—Cross veins of fore wing not as above; veins in costal half of this wing brownish, all others pale; subapical spines on penes small . . . . . **umbratica**
3—Abdominal tergites largely dark brown; lower pair of subapical spines on penes very long and slender . . . . . **hebe**
—Abdominal tergites 2–6 largely or wholly whitish; may have lateral purplish streaks of blotches; lower pair of subapical spines shorter 4
4—Abdominal tergites 2–6 wholly whitish (posterior margins may be narrowly dark), without lateral dark streaks or blotches 5
—Dark lateral streaks or blotches present on tergites 2–6 8
5—Posterior margins of abdominal tergites narrowly blackish 6
—Abdominal tergites wholly pale, no dark posterior margins . . . . . **maculipennis**
6—Thoracic notum and pleura largely deep brown . . . . . **walshi**
—Thorax yellowish white 7
7—Wing 9 mm. in length; no dark markings on thoracic pleura above leg bases **thetis**
—Wing 7 mm. in length; dark markings present on thoracic pleura above leg bases . . . . . **juno**
8—Lateral markings on abdominal tergites 2 and 3 consist of large semi-rhomboidal patches; on 4–6, these are smaller and subtriangular . . . . . **aphrodite**
—Lateral markings on tergites reduced to oblique streaks or very small triangles . . . . . **minerva**
9—Genitalia of the type shown in fig. 94 . . . . . **elegantula**
—Genitalia not as above 10
10—Genitalia of the *pulla* type (see fig. 94) 11
—Genitalia not as above 14
11—Wing 7½–8 mm. in length . . . . . **julia**
—Wing approximately 11 mm. in length 12
12—A wide purplish brown median stripe occupies most of each abdominal tergite . 13
—Abdominal tergites 2–7 pale yellowish; posterior margins dark brown; pale brown submedian triangular patches on each, based on posterior margin . . . . . **adaequata**

*(Does not include *vitrea* Wlk. or *manifesta* Etn.)

13—Mesonotum brown with a deeper brown median stripe; no distinct ruddy tinge on abdominal sternites. . . . . **solitaria**

—No such dark median stripe on mesonotum, which is quite uniformly red-brown; abdominal sternites distinctly ruddy-tinged in median area. . . . . **pulla**

14—Genitalia of the *flavescens* type (see fig. 94). 15

—Genitalia not as above. 19

15—A wide reddish brown median dorsal stripe on abdominal tergites, no dark lateral oblique streaks; yellowish species. 16

—No such median dorsal stripe, but prominent blackish oblique lateral streaks on each side of each abdominal tergite; brown or pale whitish species. 17

16—Wing 9–10 mm. in length; apical costal cross veins pale, indistinct, basal ones black, distinct; tibiae usually reddish in basal and apical thirds. . . . **cruentata**

—Wing 12–15 mm. in length; apical costal cross veins distinct, red-brown, basal ones faint, yellowish; tibiae reddish at extreme tip only. . . . . **flavescens**

17—Dark bronzy-brown species; wing 11 mm. in length. . . . . **townesi**

—Pale whitish or yellowish species; wing length variable, 8 to 10 mm. or over. . 18

18—Wing 10 mm. or more in length; black transverse marks on mesonotum; black mid-dorsal line on tergites; tails dark. . . . . **marginalis**

—Wing 8 mm. in length; mesonotum unmarked; traces only of dark mid-dorsal line on tergites 6, 7, 9 and 10; tails whitish. . . . . **dolosa**

19—Genitalia of the *lucidipennis* type (see fig. 97). 20

—Genitalia not as above. 24

20—Wing 5 to 5½ mm. in length. 21

—Wing 7 mm. in length. 22

21—Dark areas on tergites contain a pale median line and pale submedian streaks; a slight hook or spine on lateral margin of penes, as in *maculipennis* group . . . . **rusticalis**

—No pale markings within the dark dorsal area on tergites; no such lateral hook on penes. . . . . **inconspicua**

22—Mesonotum and tergites 1–6 light brown; median area of each tergite pale, but no pale median streaks or lateral triangles. . . . . **horrida**

—Mesonotum and tergites 1–6 darker brown; no distinct pale median area, but often with pale median streaks and lateral triangles. 23

23—Distinct pale median area and submedian streaks within the dark area of tergites; pale postero-lateral triangles; apical margin of forceps base emarginate at center and with a prominent blunt projection on each side. . . . . **perfida**

—Pale markings on tergites very indistinct; apical margin of forceps base not as above. . . . . **lucidipennis**

24—Genitalia of the *simplicioides* type (see fig. 96). 25

—Genitalia not as above. 30

25—Abdominal tergites largely creamy white or pale yellowish, with no conspicuous darker markings. 26

—Abdominal tergites purplish rose or brown, or light clay color with conspicuous brown lateral patches. 27

26—Median spines on penes very long and twisted, the apical half capable of being turned at right angles to lower half; semi-tranluscent brownish patches, very inconspicuous, may be present on tergites; (see fig. 98). . . . . **rodocki**

—Median spines on penes not so long, very slightly twisted at base only; no such semi-translucent patches on tergites; (see fig. 96) . . . . **simplicioides**

27—Tergites purplish rose in color; median area pale hyaline with rose tinge; pale postero-lateral triangles; sternites whitish with faint rose tinge........**rosea**
—Abdominal segments not as above..28
28—Fore tarsus distinctly longer than tibia; abdominal tergites red-brown; a wide mid-ventral red-brown band on thorax and abdomen...........**rubroventris**
—Fore tarsus short, subequal to tibia; abdominal tergites red-brown, or clay-color with dark lateral patches; no such continuous dark mid-ventral band on thorax and abdomen..29
29—Abdominal segments red-brown; sternites rather paler than tergites......**otiosa**
—Abdominal segments light clay-color; prominent deep brown semi-rectangular lateral patches on tergites, lateral dark streaks on sternites..........**criddlei**
30—Pale creamy white species, with no dark median band on tergites; wing 7½ to 9 mm. in length; no lateral spines on outer margin of penes.............31
—A broad median purplish band on abdominal tergites, lateral margins pale yellowish; wing 5½ to 6 mm. in length; 5 to 8 small lateral spines on lower outer margin of penes..**spinosa**
31—Penes superficially resemble those of *maculipennis* group, but have much smaller median spines (see fig. 96); third joint of forceps considerably longer than distal joint, these two together almost as long as second joint.......**kennedyi**
—Penes superficially resemble those of *elegantula*, but have a single median pair of spines and no inner processes (see fig. 95); third joint of forceps only slightly longer than distal joint, these two together not as long as ¾ of second joint...**persimplex**

VERIFICATION TABLE

Adults

Species	*Wing length*	*Veins*	*Tergites 1 to 6*	*Dorsal marks*	*Genit. type of*	*Distr.*
adaequata	11	Black	Yellow	Lat. ptch.	Pulla	Alta.
aphrodite	7	Dark	Pale	Lat. ptch.	Maculip.	E
criddlei	7	Brown	Pale	Lat. ptch.	Simplic.	Wyo.
cruentata	9–10	Brown	Yellow	Med. bd.	Flaves.	C.
dolosa	8	Brown	Pale	Lat. str.	Flaves.	Ga.
elegantula	9	Black	Pale	None	Elegant.	W.
flavescens	12–15	Fusc.	Yellow	Med. bd.	Flaves.	C.
hebe	7	Dark	Brown	Dors. ptch.	Maculip.	E
horrida	7	Pale	Brown	None	Lucidip.	Ont.
inconspicua	5½	Pale	Yellow	Med. bd.	Lucidip.	Man.
julia	7½–8	Brown	Yellow	Med. bd.	Pulla	S E
juno	7	Pale	Whitish	None	Maculip.	E
kennedyi	9	Pale	Pale	None	Kennedyi	Calif.
lucidipennis	7	Dusky	Brown	None	Lucidip.	E
maculipennis	7–8	Dusky	Yellow	None	Maculip.	E
manifesta	5–7	Pale	Ferrug.	None	Lucid.?	Ill.
marginalis	10	Brown	Pale	Lat. str.	Flaves.	E
minerva	7	Dark	Yellow	Lat. str.	Maculip.	E
perfida	7	Pale	Brown	None	Lucidip.	Ont.
persimplex	7½	Pale	Pale	None	Persimpl.	C

VERIFICATION TABLE (*Continued*)

Adults

Species	*Wing length*	*Veins*	*Tergites 1 to 6*	*Dorsal marks*	*Genit. type of*	*Distr.*
pulla	11	Brown	Pale	Med. bd.	Pulla	E
rodocki	8	Pale	Yellow	Lat. ptch.	Simplic.	Ida.
rosea	7	Amber	Rose	None	Simplic.	Ore.
rubroventris	8	Brown	Brown	None	Simplic.	Calif.
rusticalis	5	Pale	Brown	None	Lucidip.	E
simplicioides	8	Pale	Whitish	None	Simplic.	Alta.
solitaria	11	Brown	Brown	Med. bd.	Pulla	Alta.
spinosa	5½–6	Pale	Whitish	Med. bd.	Lucidip.	N. C.
thetis	9	Pale	Whitish	None	Maculip.	N. C.
townesi	11	Brown	Brown	Lat. str.	Flaves.	S. C.
umbratica	8–9	Brown	Yellow	Lat. ptch.	Maculip.	Que.
vitrea	6	Brown	Brown	?	?	Hud. Bay
walshi	6	Black	Pale	None	Maculip.	E
otiosa	7	Dark	Brown	None	Simplic.	Ore.

KEY TO THE SPECIES

Nymphs

1—Pronotum widest at or near the middle; filamentous portion of seventh pair of gills wanting; claws with pectinations. 2

—Pronotum widest at anterior margin; filamentous portion of seventh pair of gills present; claws lacking pectinations. 9

2—Head very large for size of body, slightly wider than pronotum; head and thorax together almost or fully as long as abdomen; many small dark brown spots or granulations on frontal margin of head. 3

—Head of normal size, not wider than pronotum; head and thorax together not more than ¾ as long as abdomen; no such dark spots on frontal margin of head. 7

3—Tracheae of gills very indistinct; pale markings on tergites 7 and 8 not coalesced; no pale spots along frontal margin of head; lateral dark streaks on sternite 9, none on preceding sternites. . . . . **thetis**

—Tracheae of gills very distinct; pale markings on tergites 7 and 8 coalesce to form a large pale blotch; three pale spots on each side of frontal margin of head; a row of lateral dark streaks on each side of abdominal sternites. 4

4—V-shaped pale median triangle based on anterior margin of tergite 9 occupies most of middle of that tergite; its apex almost attains the posterior margin. . . 5

—Posterior margin of tergite 9 brown, anterior portion largely pale; no such pale median V-shaped area as above. 6

5—Pale areas on femora usually more extensive than dark areas; dark area on sternite 9 at each posterior lateral angle, sometimes smaller dark areas at anterior angles, but these not connected to form a continuous dark margin. . . . . . **hebe**

—Dark areas on femora usually more extensive than light areas; dark margin on sternite 9 almost or entirely continuous, not divided into posterior and anterior areas. . . . . **aphrodite**

6—Gills rather heavily shaded with purplish black, leaving paler areas at extreme tip, on inner median space next to main trachea, and near base of outer margin; dark markings at anterior margin of sternites 8 and 9, in median area . . . . . **juno**

—Gills may be tinged with grey, but are not heavily dark-shaded as above; no dark transverse markings on abdominal sternites . . . . . **maculipennis**

7—Femora largely pale, with 2 dark zigzag bands across middle area; tarsus pale in middle, dark at each end . . . . . **lucidipennis**

—Femora largely brown, with pale basal and distal spots and pale median zigzag band; tarsus wholly dark brown 8

8—Pale submedian markings on abdominal tergites obscured, tergites 8–10 largely dark . . . . . **simplicioides**

—Very distinct pale submedian streaks and round spots on abdominal tergites; large pale median area on tergites 8 and 9 . . . . . **rubroventris**

9—Oblique black line present on each side of each abdominal tergite; very short postero-lateral spines on segments 6 and 7 . . . . . **marginalis**

—No such oblique black lines on tergites; no postero-lateral spines on segments 6 and 7 10

10—Pronotum with wide pale median streak and pale lateral areas; pale submedian streaks coalesce on tergites 8 and 9 to form a large pale blotch on each . . . . . **elegantula**

—Pale lateral areas on pronotum more restricted; pale median areas may be present on tergites 4 and 8, but not on 9 11

11—Body 8–9 mm. in length . . . . . **julia**

—Body 12 mm. in length . . . . . **pulla**

VERIFICATION TABLE

Nymphs

Species	*Length of body*	*Pronotum widest at*	*Claw pectinate*	*Filaments on gill 7*	*Abd. spines on 8*	*Abd. spines on 6–7*
aphrodite	6–7	Middle	Yes	No	Yes	Yes
elegantula	8–10	Front	No	Yes	No	No
hebe	6–7	Middle	Yes	No	Yes	Yes
julia	7½–9	Front	No	Yes	No	No
juno	6–7	Middle	Yes	No	Yes	Yes
lucidipennis	7–8	Middle	Yes	No	Yes	Yes
maculipennis	6–7	Middle	Yes	No	Yes	Yes
marginalis	9½–11	Front	No	Yes	No	Yes
pulla	12	Front	No	Yes	No	No
rubroventris	9–10	Middle	Yes	No	Yes	Yes
simplicioides	8	Middle	Yes	No	Yes	Yes?
thetis	8–9	Middle	Yes	No	Yes	Yes

Heptagenia adaequata McDunnough

1924—McDunnough—Canad. Ent. 56: 120 . . . . . Alta.

Length: body 10; wing 11 mm.

Mesonotum pale ochreous tinged with deeper color; abdominal tergites 2–7 pale yellowish, posterior margins and submedian triangles brown; genitalia of the *pulla* type.

Head light ochreous; a black dash next to lower corner of each eye; two blackish spots on vertex near the eyes. Bases of antennae black-ringed. Reddish brown shading at base of median carina; a small spot of same color on each side of median ocellus. Thorax pale ochreous; mesonotum tinged with deeper color. Legs ochreous. A black streak at base of middle and hind coxae. Femora marked with median and apical purplish bands; fore tibia smoky at apex. Basal joint of fore tarsus ⅕ as long as the second; of hind tarsus, slightly shorter than second. Wings hyaline. Basal portions of subcosta and radius thickened, yellowish; remaining lontigudinal veins fine, blackish. Cross veins blackish, thicker than the longitudinals, especially well marked along the costa.

Abdominal segments 2–7 semitranslucent, very pale yellowish; tergites 1–8 with narrow dark brown posterior margins and paler brown submedian triangular patches on the posterior margin of each. Tergites 9 and 10 shaded with light brown; sternites pale yellowish. Genitalia of the *pulla* type. Tails pale, joinings very distinctly brown.

The brown triangular patches on the tergites distinguish this species from *H. elegantula* and from *H. solitaria.* Allied to *solitaria* but differing from *elegantula* in type of genitalia. The larger size and different genitalic type separate it from *H. criddlei,* which has rather similar abdominal markings. We have seen no specimens which seem to belong with certainty to this species.

Heptagenia aphrodite McDunnough

1926—McDunnough—Canad. Ent. 58: 194. . . . . .Ont.
1933—Traver—J. Elisha Mitchell Sci. Soc. 48: 143–4, 148. Fig. Nymph desc. . . . . N. C.
1933—Traver—J. N. Y. Ent. Soc. 41: 111.

Length: body 6; wing 7 mm.

Mesonotum pale smoky brown; cross veins of costal area and some in wing disc margined with dark brown; tergites 2–6 pale yellowish dorsally and laterally, with broad dark lateral patches between; genitalia of the *maculipennis* type.

Head pale yellow, shaded with orange or ruddy at base of ocelli and between these and eyes; anterior and posterior margins tinged with black. Pronotum pale yellow, tinged laterally with black. Mesonotum pale smoky brown, the lateral portion anterior to wings rather light ochreous; thus there appears to be a broad dark mid-dorsal stripe containing a faint pale central line; scutellum pale whitish, a smoky brown area directly anterior to it. Pleura light yellowish; a broad blackish stripe above the leg bases. Sternum light yellowish. Legs pale yellowish; fore femora (usually not others) with a faint dark apical streak; fore tibiae and tarsi tinged with smoky. Wings hyaline; longitudinal veins fine, dark; cross veins heavier, those of the costal and subcostal spaces heavily dark-margined. Usually two veins in each of the first 6 or 7 spaces below the bulla are faintly heavier; a faint brown cloud at the fork of the posterior branch of the radial sector, 3 to 4 cross veins in costal space before the bulla, about 9 to 11 beyond it.

Abdominal tergite 1 deep brown except for the pale yellowish anterior margin. On tergites 2 and 3 are broad dark brown lateral patches, semi-rhomboidal with concave lateral edges; these do not quite reach the anterior margins of the tergites, but are joined by a narrow dark band on the posterior margin of each; thus a pale yellowish dorsal patch (larger on 3 than 2) and pale lateral areas remain on each of these tergites. The dark lateral patches on tergites 3–6 are smaller, semi-triangular, and the pale areas therefore larger. Tergites 7–10 deep black-brown. Sternites pale yellowish. Forceps and tails pale. Genitalia of the *maculipennis* type; apical projection of penes, as well as apical spines, stronger than in *H. minerva*, but not as well developed as in *H. hebe*.

This species is rather intermediate between *minerva* and *hebe*, both in color characters and structural details of the penes. The lateral dark patches on the tergites are smaller than in *hebe*, but larger than in *minerva*. Specimens of this species in the Cornell collection are from Fairmount, W. Va. (P. N. Musgrave); N. Fk. Potomac River, W. Va. (J. R. Traver); Waynesville, Caraway Creek near Sophia, N. Fk. Swannanoa River, Laurel River and Big Almance Creek, N. C. (J. R. Traver); Ringgold, Atlanta, Cleveland, Rome, and Swamp Creek, Ga. (P. W. Fattig); St. Cloud, Minn. (G. W. Friedrich); Knoxville, Tenn. (Miss L. Shipe); Ithaca, N. Y. (Miss C. N. Hardy, A. C. Hodson); Corning, N. Y. (C. R. Crosby, P. J. Chapman); Hagaman, Canajoharie and Little Falls, N. Y. (H. K. Townes, Jr.).

Heptagenia criddlei McDunnough

1927—McDunnough—Canad. Ent. 59: 261 .... Wyo.

Length: body 7; wing 7 mm.

Mesonotum deep black-brown; deep brown lateral patches on tergites, lateral dark streaks on sternites; fore legs short, deep purple-brown; genitalia of the *simplicioides* type.

Head dark olive brown; tinges of paler ochreous behind ocelli and at bases of antennae. Thorax deep black-brown; sutures of pleura marked with ochreous. Fore legs very deep purple-brown, slightly paler at base; tibia slightly longer than femur, tarsus and tibia subequal; individual joints of tarsus short, basal joint about ⅓ as long as second, which is subequal to third and fourth. Middle and hind legs somewhat paler than fore legs; basal joint of hind tarsus distinctly longer than second. Wings hyaline, venation light brown; cross veins very fine, not at all thickened; costal cross veins before the bulla very thin and few in number.

Abdominal segments light clay-color. Tergites with deep brown semi-rectangular patches on each side, extending anteriorly and laterally from the posterior margin, but not attaining the anterior margin except on 1 and 2, where almost the entire segment is occupied by these patches. Sternites with lateral longitudinal brown streaks or bands "adjoining the suture" (next to pleural fold?); in our specimens, ganglionic areas outlined in dark brown. Forceps and tails dull smoky. Genitalia of the *simplicioides* type (see fig. 96).

This species is distinguished from others allied to it by the short fore tarsi of the male, the fine cross veins and the abdominal pattern; the latter distinguishes it from *H. otiosa*, its nearest ally, which is more uniformly red-brown in color. In the Cornell collection there are specimens from Canyon, Yellowstone Nat. Pk., Wyo. (N. Criddle), kindly presented by Dr. McDunnough; South Park, Colo. (E. G. Oslar); Waterton Lakes, Alta. (J. H. Pepper), also given by Dr. McDunnough; Flathead Lake and Glacier Nat. Pk., Mont. (R. T. Young); Yakima, Wash. and Ft. Collins, Colo. (M. D. Leonard).

Heptagenia cruentata Walsh

1863—Walsh—Proc. Ent. Soc. Philad. 2: 205 .... Ill.
1871—Eaton—Trans. Ent. Soc. London, p. 140.
1885—Eaton—Revis. Monogr. p. 300.
1924—McDunnough—Canad. Ent. 56: 118 (*reversalis*) .... Man.
1926—McDunnough—Canad. Ent. 58: 301 (in discussion of *perfida*).
1929—McDunnough—Canad. Ent. 61: 180 (*reversalis* = *cruentata*).
1931—McDunnough—Canad. Ent. 63: 92 (in discussion of *Pseudiron centralis*).

Length: body 7½–8½; wing 9–10 mm.

Allied to *H. flavescens*, but smaller; mesonotum and abdominal tergites with light brown dorsal band, tibiae usually twice-banded; genitalia of the *flavescens* type.

Head yellowish, vertex sanguineous; ocelli fuscous; antennal filament dusky, tip whitish. Eyes in living insect pale bluish grey, divided by a black line. Thoracic notum sanguineous or light brown. Legs yellow. Reddish median and apical bands on each femur, the area between these paler reddish. Usually all the tibiae have the basal and apical third pale reddish; but as Dr. McDunnough has indicated, these markings may be almost obsolescent. Claws and tarsal joinings of all legs dusky. Basal joint of fore tarsus about ⅓ of second; in hind leg, slightly shorter than second. "Wings hyaline; veins and cross veins moderate, the veins finer than the cross veins, all fuscous except at extreme base of wing and the basal two-thirds of the costa, where the veins are respectively hyaline and yellowish" (Walsh). According to the original description, all the costal cross veins, including the humeral vein, are of "uniform coarseness," but in specimens which Dr. McDunnough has of this species, the apical costal cross veins are pale and indistinct, the basal ones black and distinct. Costa uniformly tinged with yellow throughout its length; subcosta thickened and darker "for the length of half a millimeter" (Walsh) in the region of the bulla, as in *H. flavescens.*

Abdomen yellow; tergites with a median dorsal band of light brown to sanguineous, the apical tergites paler; the overlapping portions appear darker, presenting an annulate appearance. Forceps pale, tips brownish. Tails whitish, joinings dark brown; toward the base the dark joinings are alternately narrow and wide. Genitalia of the *flavescens* type (see fig. 94).

This species is closely allied to *H. flavescens*, but it is smaller, with yellow costa, and usually with distinctly banded tibiae. In *flavescens*, the fore tarsus is slightly longer than the tibia, while in *H. cruentata* it is said to be of the same length as tibia. We have seen but one specimen of this species, from Lincoln, Neb. (D. B. Whelan).

Heptagenia dolosa sp. nov.

Length: body 8; wing 8 mm.

A pale species closely allied to *H. marginalis*; genitalia of the *flavescens* type.

Male—Head pale yellowish white; frontal shelf hyaline, anterior margin narrowly purplish black; a faint line of same color on each side of carina below antenna. Base of antenna white; filament faintly dusky at base, tip pale. Four small inconspicuous spots on vertex; two on posterior margin, two anterior to these, one near each ocellus. Eyes very large, blue-black in alcohol. Thorax pale yellowish white. Lateral margin and an oblique lateral stripe on pronotum black; a black streak above fore coxa. Two large and one small black spots above middle coxa, arranged as in *H. marginalis;* a small black dot above and behind hind coxa, a similar dark dot on lateral margin of metanotum. Mesonotum unmarked. Legs pale whitish; extreme base of fore femur reddish brown; a narrow apical streak, a wide median band, and a streak along the ventral edge, deep purplish red. Apex of fore tibia, and distal tarsal joint, narrowly brown; basal tarsal joint with faint brown shading. Middle and hind femora pale basally, median bands pale and incomplete; apical streak often incomplete, apices of tibiae and tarsal claws purplish. Three main veins of costal margin heavy, yellowish; other veins fine, brownish: 4 to 5 costal cross veins before bulla, 12 to 15 beyond it.

Abdominal segments 1–8 white, hyaline; 9 and 10 opaque, creamy yellow. Posterior margin of each tergite narrowly purplish black; a blackish longitudinal streak on each side of each tergite, halfway between median line and pleural fold; a faint oblique dark mark at each spiracle; tergites 6 and 7, occasionally also 9 and 10, with traces of a faint dark mid-dorsal streak. Dark markings less prominent than in *H. marginalis*. Genitalia creamy white. Details of structure of penes differ somewhat from *marginalis* (see fig. 95). Second forceps joint somewhat shorter relatively than in that species. Tails wholly whitish.

Holotype—Male imago. Chattahoochee River, Atlanta, Ga., July 4, 1931. Collector, Prof. P. W. Fattig. No. 1324.1 in Cornell collection.

Allotype—Female imago; same location, June 10, 1932. (P. W. Fattig). No. 1324.2 in Cornell University collection.

Paratypes—3 male imagos; 4 female imagos; same data as holotype, also July 4–10 and Aug. 1–18; Lawrenceville, Ga., June 13, 1932. Same collector. No. 1324.3–9 in Cornell University collection.

The smaller size, paler color of tails, single median femoral band, lack of dark mesonotal markings and more restricted mid-dorsal markings on tergites, as well as details in the structure of the penes, distinguish this species from *marginalis* (as considered under that head).

Heptagenia elegantula Eaton

1885—Eaton—Revis. Monogr. p. 253 (Rhithrogena) .... Colo.
1914—Banks—Proc. Acad. Nat. Sci. Philad. p. 615 (*coxalis*).
1923—Dodds—Trans. Amer. Ent. Soc. 49: 102. Fig. (Rhithrogena).
1924—McDunnough—Canad. Ent. 56: 117: op. cit., p. 119 (*querula*) .... Man.
1927—Needham—Utah Agri. Exp. Sta. Bull. 201: 15. Fig. nymph (Rhithrogena) .... Utah
1933—Traver—J. N. Y. Ent. Soc. 41: 111.

Length: body 9–10; wing 9–12 mm.

A yellowish species; reddish brown dorsal stripe on mesonotum, black markings on pleura above coxae; genitalia as in fig. 94.

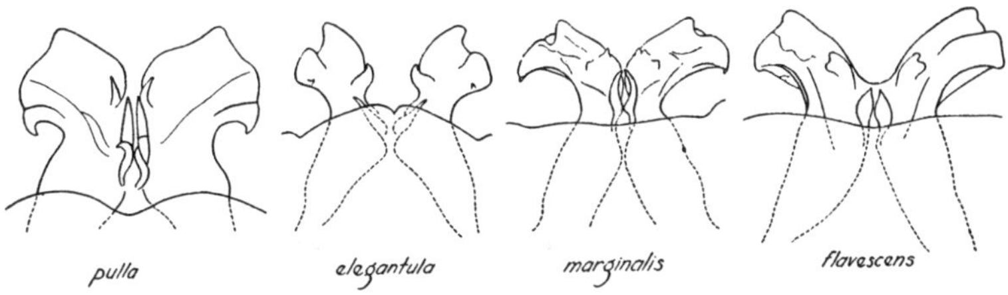

FIG. 94. Male genitalia of four species of Heptagenia.

Head pale yellowish; a reddish spot on frontal carina, and red shading on vertex between eyes and ocelli. Thoracic notum pale yellowish; a median reddish brown stripe on mesonotum, and pinkish area anterior to scutellum. Pleura and sternum likewise yellowish; pinkish areas on pleura below base of each wing; a black streak above base of middle and hind coxae, and a black spot posterior to hind coxae. Fore leg light bronzy brown; femur indistinctly banded at middle with light reddish brown, red-brown also at apex; apex of tibia, distal tarsal joint and tarsal joinings, also claws, dull reddish to blackish brown. Middle and hind legs yellowish amber; no median femoral band; tip of tibia not darkened. Basal joint of fore tarsus about ⅙ the length of the second joint. Wings hyaline. Costal margin of fore wing may be tinged with pale lemon yellow, with greenish or bistre-grey. Subcosta and radius yellowish amber in basal half; other longitudinal veins fine, pitch brown to black.

Humeral cross veins thickened posteriorly, blackish. Cross veins dark; those in costal space thicker than others, black; a well-marked black bulla.

Abdominal segments pale; 2–7 semi-hyaline, 8–10 opaque. Posterior margins of tergites 2–7 smoky greyish or brownish; tergites 8–10 tinged with pink. Sternites pale yellowish. Tails pale yellowish, joinings dark brown. Genitalia distinctive.

Specimens of this species in the Cornell collection are from Denver, Clear Creek and Ft. Logan, Colo. (E. G. Oslar); Lincoln, Neb. (D. H. Whelan); Johnson Co., Kans. (R. H. Beamer); Glacier National Pk., Mont. (A. S. Hazzard); Manhattan, Kans. (R. C. Smith); Logan River, Utah (J. G. Needham). We have seen others from Ft. Collins, Colo. (L. H. Sweetman), in the collection of the Colorado Agric. College, and from Platte Canyon, Colo. (E. G. Oslàr), in the Canadian Nat. Collection, loaned us by Dr. McDunnough.

Heptagenia flavescens Walsh

1862—Walsh—Proc. Acad. Nat. Sci. Philad., p. 373 .... Ill.
1863—Walsh—Proc. Ent. Soc. Philad. 2: 177, 205.
1868—Eaton—Ent. Mo. Mag. 5: 90.
1871—Eaton—Trans. Ent. Soc. London, p. 142. Fig.
1873—Hagen—Trans. Ent. Soc. London, p. 404.
1881—Eaton—Ent. Mo. Mag. 18: 24.
1885—Eaton—Revis. Monogr. p. 266.
1910—Banks—Canad. Ent. 42: 200 .... Minn.
1924—McDunnough—Canad. Ent. 56: 117. Fig .... Man.

Length: body 9–13; wing 11–15 mm.

A large yellowish species; a broad reddish brown dorsal band on abdominal tergites; genitalia of the *flavescens* type (see fig. 94).

Head yellowish; vertex ferruginous; antennal filament dusky, paler at tip; eyes bright greenish yellow, in living insect. Thorax largely ferruginous dorsally, the mesonotum shaded with blackish anteriorly, on scutellum and anterior to it. Pleura largely yellowish with reddish brown shading; sternum similar. Fore legs pale ferruginous; femora with a median and apical darker band; tibiae at apex, tips of tarsi and tarsal joinings, fuscous. Basal joint of fore tarsus about ¼ as long as the second. Middle and hind legs yellow; femora fuscous apically; tips of tarsi and tarsal joinings somewhat dusky. Basal joint of hind tarsus distinctly shorter than the second. Wings hyaline; a pale ferruginous cloud in the costo-apical region. "Veins and cross veins moderate, subequal, fuscous, except the three costal veins which are coarse, yellowish on their basal two-thirds, fuscous on their terminal one-third, where the cross veins also are coarse, the oblique basal cross vein being always very coarse; halfway to the tip the second costal vein is always thickened and obfuscated for the length of half a millimeter, sometimes obviously, sometimes indistinctly." (Walsh). Basal costal cross veins faint.

Abdomen yellowish, with a broad reddish-brown dorsal band on the tergites, becoming yellowish red on the apical ones. Lateral margins of all tergites pale, and on the basal and middle ones are pale submedian oblique markings from the anterior margin; posterior margins of tergites 1–6 darker. Sternites pale greenish or yellowish white, the apical ones more yellow. Segments 1–6 semi-hyaline, 7–10 opaque. Forceps pale, the tips fuscous; tails whitish; joinings reddish brown in distal portion, often pale at extreme base, and in intermediate portion alternately pale and reddish. Penes rather widely spreading apically; two large somewhat conical median spines.

This species is most nearly allied to *H. cruentata*, but it is considerably larger; the fuscous apical portion of the costal veins, the faint basal costal cross veins and coarse apical ones, and the lack of prominent banding on the tibiae distinguish it from that species. Specimens of this species in the Cornell collection are from Manhattan, Kans. (R. C. Smith, R. H. Painter); Fairport, Ia. (H. E. Schradieck); Atlanta, Ga. (P. W. Fattig); College Station, Tex. (V. A. Little). We have seen also a specimen from Ramsey Co., Minn., in the collection of the Univ. of Minn.

Heptagenia hebe McDunnough

1905—Needham—N. Y. St. Mus. Bull. 86: 51, 57. Fig. (*maculipennis*)........................N. Y.
1911—Morgan—Ann. Ent. Soc. Amer. 4: 101 (*Ecdyonurus maculipennis*).
1924—McDunnough—Canad. Ent. 56: 122..Que., Ont.
1926—McDunnough—Canad. Ent. 58: 195 (in discussion of *aphrodite*).
1930—Ide—Canad. Ent. 62: 229.

Length: body 6; wing 7 mm.

Mesonotum olive brown; cross veins of costal margin and some in disc of wing distinctly margined with dark brown; tergites largely deep brown, paler dorsally; genitalia of the *maculipennis* type (see fig. 97).

Frontal portion of head pale yellow; a dark streak along base of frontal shelf. Vertex deep brown, posterior margin darker. Antennal filament dusky at base, tip pale. Pronotum yellowish in median area, shaded heavily with blackish laterally, and with two short black lateral stripes. Mesonotum olive-brown, with slight ochreous shading; metanotum dark red-brown, with two dark transverse streaks near center. Pleura light yellow; a broad brownish lateral stripe above the bases of the legs. Sternum light yellow. Fore coxae and bases of femora yellowish; femora pale brownish apically, with a dark streak near apex on ventral edge; tibiae often yellowish basally, pale brown apically, tip darker brown; tarsi pale brown. Wings hyaline; costal border tinged with brown. Cross veins of the costal and subcostal spaces quite heavily margined with dark brown, especially before the bulla; at the bulla, several cross veins in the first 7 or 8 spaces are slightly margined, and at the fork of the posterior branch of the radial sector is a small brown cloud. Other cross veins of the disc more or less margined; all cross veins of fore wing dark brown, heavier than the paler brown longitudinal veins. 5 to 6 costal cross veins before bulla, about 12 beyond it. See frontispiece.

Abdominal tergites deep brown, the dorsal portions of 4–6 semihyaline whitish, often also base of 7; pale submedian streaks on other tergites; lateral margins pale smoky, on 2–7. Tergites 8–10 opaque, reddish brown, lateral margins yellowish. Sternites pale yellowish, 8 and 9 opaque. Forceps and tails pale yellowish. Penes of the *maculipennis*. Innermost pair of spines longer than in any other species of this group.

The dark abdominal tergites and the longer spines on the penes separate this species from others of the *maculipennis* group. It is probably most nearly allied to *H. aphrodite*. Specimens of this species in the Cornell collection are from Ithaca, N. Y., and vicinity (Miss C. N. Hardy, A. C. Hodson, J. G. Needham, V. Knox, H. K. Townes, Jr., J. R. Traver); Leroy and Kirtland, O. (J. R. Traver); Ft. Jackson and Hopkinton, N. Y. (L. Harper); Hagerstown, Md. (V. Argo); Knoxville, Tenn. (Miss, L. Shipe); Goshen, O. (C. R. Crosby); Kent Falls, Conn. (C. P. Alexander); St. Cloud, Minn. (G. W. Friedrich); Fairport, Ia. (H. E. Schraedick); Flat Creek, Dolgeville, Chatham Center, Canajoharie, Prattsville, Burtonsville and Mill Point, N. Y. (H. K. Townes, Jr.).

Heptagenia horrida McDunnough

1926—McDunnough—Canad. Ent. 58: 301. Fig. .... Ont.

Length: body 7; wing 7 mm.

Mesonotum light brown; abdominal tergites light brown, median area paler but with no mid-dorsal pale line; genitalia of the *lucidipennis* type.

Head light brown, extreme frontal area yellow; a round black spot below lateral ocellus and near the eye, as in *H. perfida*. Thoracic notum light brown, becoming deeper in the region of the scutellum; posterior edges of meso- and metanota blackish. Pleura and sternum light ochreous, not as yellow as in *perfida*; brown shading at bases of legs. Legs light ochreous; fore femur strongly tinged with reddish brown in the apical half, other femora with slight ruddy tinges in this area. Wings hyaline; longitudinal and cross veins pale. Abdominal tergites light brown, the median area paler and the lateral area darker, but the pale mid-dorsal streak and postero-lateral triangles of *perfida* are not present in this species. Posterior margin of each tergite dark brown; sternites light ochreous.

This species is most nearly allied to *H. perfida*, from which it may be distinguished as indicated above; also the apical margin of the forceps base is less deeply emarginate and lacks the lateral projections; details of penes somewhat different (see fig. 96, after McDunnough). Length of forceps joints as in that species. We have seen no specimens of this species.

Heptagenia inconspicua McDunnough

1924—McDunnough—Canad. Ent. 56: 118. Fig. .... Man.
1931—McDunnough—Canad. Ent. 63: 92 (in discussion of *rusticalis*).

Length: body 4; wing 5½ mm.

Mesonotum olive brown; abdominal tergites 2–7 pale yellow, with a median broad reddish brown band; genitalia of the *lucidipennis* type (see fig. 97).

Head light yellow; orange-brown shading on face next to eyes. Thoracic notum pale to dark olive brown; pleura light yellow. Legs pale yellow. Fore femora and tibiae smoky brown apically; basal tarsal joint about ⅓ of the second. Basal joint of hind tarsus longer than the second. Wings hyaline, venation pale; basal costal cross veins almost invisible. Abdominal segments pale yellow; 2–7 semihyaline, 7–10 opaque. A broad median band of light or ruddy brown, variable in intensity on different individuals, on tergites 2–7, the posterior margin of each tergite opaque, so that the abdomen appears annulate. Forceps and penes pale ochreous. Penes with a wide apical and a narrow lateral lobe, a single pair of large median spines, but none laterally or apically. Tails whitish, the basal joinings ruddy brown.

This species is apparently allied to *H. lucidipennis*, but it is considerably smaller. The abdomen would seem to be paler than in *H. rusticalis*, and it lacks the pale markings present in the dark dorsal areas of that species. There are specimens of this species in the Cornell collection from Treesbank, Man. (T. Criddle, R. M. White), and Wawanesa, Man. (R. M. White), kindly presented by Dr. McDunnough.

Heptagenia julia Traver

1933—Traver—J. Elisha Mitchell Sci. Soc. 48: 149. Fig. .... N. C.
1933—Traver—J. N. Y. Ent. Soc. 41: 111 (as sp? from N. C.).

Length: body 7–8½; wing 7½–8½ mm.

Allied to *H. pulla*, but smaller; mesonotum reddish brown; tergites 2 to 7 pale yellowish, with a median purplish brown band; genitalia of the *pulla* type.

Head pale yellowish; vertex tinged with reddish brown; narrow black line on face at corner of eye, below level of antenna base; antennae yellowish brown; ocelli with black bases. Pronotum light reddish brown, shaded with blackish brown in median area, especially on posterior margin; mesonotum dark reddish brown, scutellum very dark; metanotum similar. A conspicuous round black spot above fore coxa; black longitudinal streak on pleura above bases of legs, more or less discontinuous so as to form several oblique and some horizontal dark lines. Pleura light red-brown; sternum yellowish, posterior portion of mesosternum rather darker. Fore leg dusky; femur with indistinct median purplish band, purplish brown at apex; black streak distally on ventral margin. Tibia purplish brown at apex; tarsal joinings dark. Femora of middle and hind legs yellowish with faint dusky tinge, narrowly dark apically; tibiae and tarsi dusky. Middle coxa with one dark spot; usually two on hind coxa. Basal joint of fore tarsus about ¼ of second; of hind leg, distinctly shorter than second joint. Wings semi-hyaline, stigmatic area cloudy; costa, subcosta and radius yellowish, all other veins light brown; 4 to 5 costal cross veins basally, about 10 beyond bulla.

Abdominal segments 1–7 pale yellowish, semi-hyaline; 8–10 yellowish red, opaque. Posterior margins of tergites narrowly purplish black. A wide median purplish brown band on tergites 1–4, wider posteriorly on each tergite; on 5–7 the median band is present but narrower, confined to area each side of median line. Within this dark band are lunate submedian paler streaks; occasionally traces of a pale mid-dorsal line on anterior margins. A faint black line marks the stigmatic area, on each segment. Sternites unmarked.

Genitalia very similar to *H. pulla*, with penes differing slightly in detail from that species, the apical projections being more acute, and the two median pairs of spines almost alike in size and shape. Tails dusky at base, becoming pale yellowish brown distally; in some specimens almost wholly pale; joinings distinctly dull purplish, alternately wide and narrow in basal area.

The type material of this species is in the Cornell collection. It is a smaller species than any other of this group except *H. criddlei*, which has different abdominal markings; its nearest ally is *pulla*, distinguished from it is already noted. We also have specimens from Greenville, S. C. (H. K. Townes, Jr.).

Heptagenia juno McDunnough

1924—McDunnough—Canad. Ent. 56: 121. Fig. .... Que.
1926—McDunnough—Canad. Ent. 58: 194 (in discussion of *walshi*).

Length: body 6; wing 7 mm.

A pale whitish yellow species, abdominal tergites 2–7 faintly margined with brownish; only the basal costal cross veins dark-margined; genitalia of *maculipennis* type.

Head, thorax and abdomen pale whitish yellow. Lateral margin of prothorax blackish; a faint blackish streak on pleura above the leg bases, and a faint brownish shading below the wings. Legs pale yellow; all femora with a short black apical streak on the ventral edge. Wings hyaline. Longitudinal veins largely pale, as are also cross veins in the posterior (cubito-anal) half of the fore wing and all the hind wing. Basal costal cross veins blackish, rather heavily margined, also a few at bulla in each of the first three spaces; those of subcostal space slightly thickened in basal half, not margined distinctly. No cloud at fork of posterior branch of radial sector. There is no tendency for the dark marginings to coalesce, nor is there a distinct line of heavier cross veins below bulla.

Abdominal tergites with narrow dark posterior margins; no other markings. Segments 2–7 hyaline, 8–10 yellowish to light brown dorsally, yellowish ventrally. Genitalia and tails whitish. Penes much as in *H. maculipennis*, but with a single apical spine only (see fig. 95).

Specimens of this species in the Cornell collection from Newfield, N. Y., show several additional characters, not mentioned in the original description. Anterior

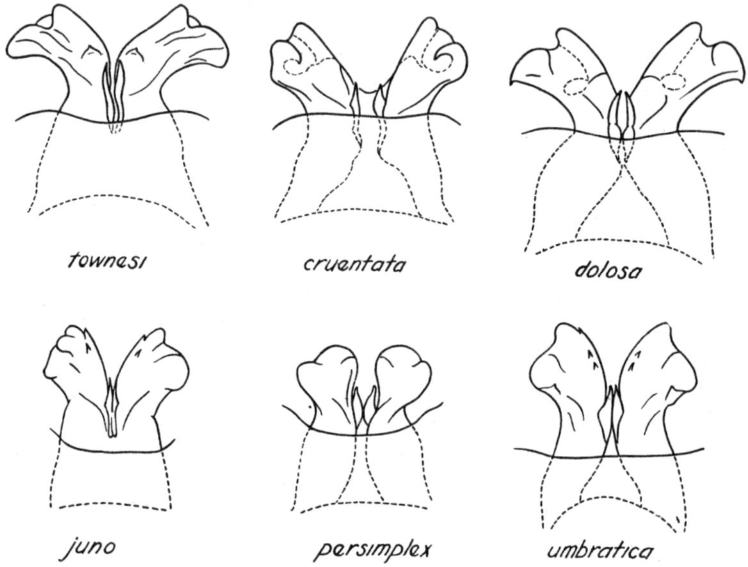

FIG. 95. Male genitalia of six species of Heptagenia.

portion of head and margin of frontal shelf shaded with deep rose-red, deepest on and near median carina. Bases of antennae pale; filaments deep smoky at base, tips pale. Faint purplish shading between eyes and lateral ocelli. Posterior margin of head narrowly blackish. Faint lateral streaks on pronotum, one on each side.

This species, allied to *H. maculipennis*, *H. walshi*, and *H. thetis* may easily be distinguished from the first two of these by the pale thorax; from *maculipennis* it is separable also because of the dark posterior margins of the tergites; tergites 2–7 are not tinged with smoky, as in *walshi*, and tergites 8–10 are much paler. The fore leg is longer than in the latter species. The smaller size, slight differences in structure of penes, and presence of dark marks on pleura separate it from *thetis*. We have specimens of this species also from Slaterville, N. Y., and have seen others from Stratford, N. Y. (H. K. Townes, Jr.) and Ohio Pyle, Pa. (Carnegie Mus. Coll.).

Heptagenia kennedyi McDunnough

1924—McDunnough—Canad. Ent. 56: 225. Fig. .... Calif.
1933—Traver—J. N. Y. Ent. Soc. 41: 110.

Length: body 7–8; wing 9 mm.

A pale species; penes with an erect apical lobe and two pairs of median spines, also small paired spines near tip of the longer of these.

Head pale; a few slight dark dots on clypeus. Thorax light brown; prothorax tinged with smoky posteriorly. Legs pale; no dark markings at bases of coxae or femora. Fore femora and tibiae tinged with black at apices; basal fore tarsal joint 1/6 as long as the second. Wings vitreous; venation colorless. Abdomen pale hyaline; tergites tinged with smoky along the posterior margins, giving the appearance of faint bands. Third joint of forceps considerably longer than the fourth joint, these two together as long as or longer than the second joint.

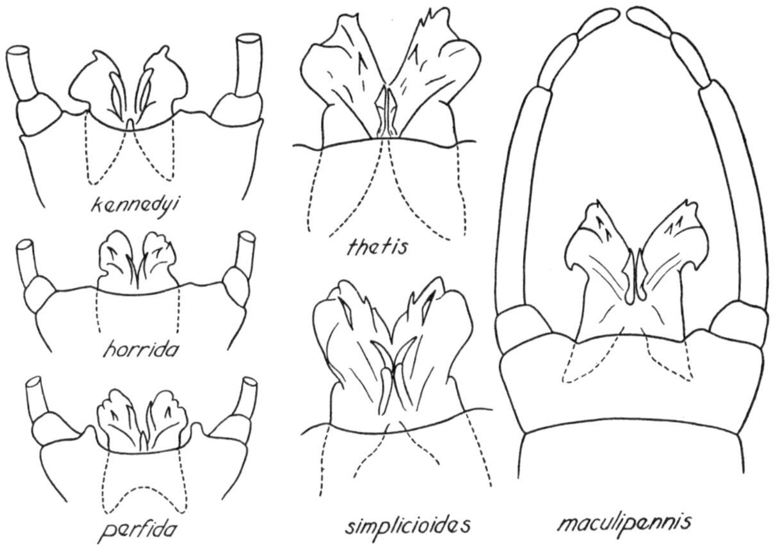

FIG. 96. Male genitalia of six species of Heptagenia (*kennedyi*, *horrida*, and *perfida*, after McDunnough).

The colorless veins and the long third joint of the forceps would seem to ally this species with the *inconspicua-lucidipennis* group, but the penes are different from others of this group, bearing a superficial resemblance (according to the published figure reproduced in fig. 96) to the *maculipennis* group. We have seen no specimens of this species.

Heptagenia lucidipennis Clemens

1913—Clemens—Canad. Ent. 45: 329 (Ecdyonurus). Fig., nymph............................Ont.
1915—Clemens—Marine and Fish. Sess. Pap. 39b; 118 (same as above).
1924—McDunnough—Canad. Ent. 56: 118.
1931—McDunnough—Canad. Ent. 63: 92 (in discussion of *rusticalis*).
1933—Traver—J. N. Y. Ent. Soc. 41: 111.

Length: body 6; wing 7 mm.

Mesonotum dark brown; abdominal tergites largely brown; genitalia of the *lucidipennis* type.

Frontal portion of head very slightly shaded with brown; vertex dark brown or reddish. Thoracic notum dark brown; pleura and sternum light yellowish. Femora of fore leg dark, those of the middle and hind legs yellowish. Basal joint of fore tarsus about 1/4 the length of the second; in the hind leg, the basal tarsal joint is distinctly longer than the second joint, all joints of this tarsus being relatively long for the size of the species. Wings hyaline; longitudinal veins slightly dusky, especially the costa and subcosta; cross veins wholly colorless. Abdominal tergites blackish brown, by the original description; Dr. McDunnough states that this species has paler coloration than *H. rusticalis*, and with less distinct markings. Sternites pale. Forceps base and penes yellow; forceps tinged with black. Tails pale, the basal portion tinged slightly with black. Penes with a prominent inner apical projection or lobe; on the ventral surface of each division of penis is a rather prominent spine (see fig. 97). In *H. rusticalis*, the inner apical lobe is not so well developed, and the spines on the ventral surface are usually lacking.

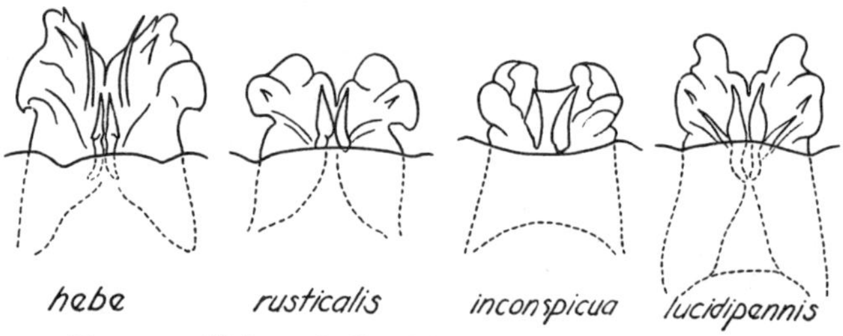

FIG. 97. Male genitalia of four species of Heptagenia.

There are specimens of this species in the Cornell collection from Port Elgin, Ont. (Mrs. E. G. Grieve); Willoughby, O. (J. R. T.); Ithaca, N. Y. (Miss C. N. Hardy); Ringwood, N. Y. (J. R. T.); and Danby, N. Y. (R. C. Tasker).

Heptagenia maculipennis Walsh

1863—Walsh—Proc. Ent. Soc. Philad. 2: 206 .... Ill.
1912—Wodsedalek—Biol. Bull. 22: 253.
1913—Clemens—Canad. Ent. 45: 329. Nymph fig .... Ont.
1915—Clemens—Marine Fish, Sess. Pap. 39b: 117. Nymph fig.
1924—McDunnough—Canad. Ent. 56: 120. Fig .... Man.
1933—Traver—J. N. Y. Ent. Soc. 41: 111.

Length: body 4½–6; wing 6½–8 mm.

A pale yellowish white species; mesonotum ferruginous; cross veins of costal border thickened and dark-margined; genitalia of the *maculipennis* type (see fig. 96).

Vertex of head ferruginous; bases of ocelli and filament of antenna dusky. Thorax ferruginous above, pleura with dark streak above leg bases, as in others of this group. Legs whitish, femora yellowish, with a short fuscous streak apically on ventral edge; tips of fore tibiae and all the claws fuscous. Wings hyaline; veins "fine and dusky, cross veins coarse and black," (Walsh)except in hind wing and cubito-anal region of fore wing, where they are wholly pale. Costal and subcostal cross veins of fore wing thickened, some margined and appearing to form dark spots. "At the base about 4 at the usual distance, the first in both rows and the next three in the costal row heavily bordered with black, then an open space, then on the middle of the costa a group of about 3 cross veins close together and bordered with black, then a space with fewer cross veins than usual, then at three-fourths the distance to the tip another similar group of 4 or 5 similarly bordered."

Basal abdominal tergites pale yellow, with no dark markings; "the terminal ⅔ of joint 6, the whole of 7 and 8 and the extreme base of 9, piceous" (Walsh); tergite

10 somewhat paler. Sternites pale yellowish white. Forceps and tails whitish. Two subapical spines on each division of penes, near inner margin.

The lack of dark posterior margins on the abdominal tergites distinguishes this species from the allied *H. juno*, *H. walshi*, and *H. thetis*. We have specimens of this species in the Cornell collection from Willoughby, Kirtland, Leroy and Painesville, O. (J. R. Traver); Knoxville, Tenn. (Miss L. Shipe); and Hannibal, Mo. (W. D. Sargent, V. S. L. Pate).

Heptagenia manifesta Eaton

1862—Walsh—Proc. Acad. Nat. Sci. Philad. p. 371 (*Baetis debilis*)............................Ill.
1863—Hagen—Proc. Ent. Soc. Philad. 2: 170 (*B. debilis*).
1871—Eaton—Trans. Ent. Soc. London, p. 130 (*Siphlonurus debilis*).
1885—Eaton—Revis. Monogr.; 253 (Rhithrogena; renamed).
1924—McDunnough—Canad. Ent. 56: 117, 119 (in discussion of *inconspicua*).

Length: body 4–7; wing 5–7 mm.

Mesonotum piceous; abdominal tergites ferruginous; venation pale hyaline, bases of three main costal veins fuscous at base; type of genitalia uncertain.

Eyes brown above, the lower portion fuscous. Thoracic notum generally piceous. Legs pale greenish; femora with a narrow median band, the 'knee' fuscous; apex of fore tibia, tips of tarsi, and joinings of fore tarsus, fuscous. Wings hyaline; venation pale greenish hyaline, except the basal third of costa, subcosta and radius, and occasionally their tips also, which are fuscous. Abdominal tergites ferruginous, the posterior half or third of each somewhat piceous. Forceps pale; tails uniformly whitish; penes blackish, according to Hagen.

Nothing is known of this species other than the above description, hence its affinities with others of the genus cannot at present be determined. Dr. McDunnough thinks it may be allied to *H. inconspicua*.

Heptagenia marginalis Banks

1910—Banks—Canad. Ent. 42: 198..Va., Pa.
1924—McDunnough—Canad. Ent. 56: 117.
1926—McDunnough—Canad. Ent. 58: 302 (in discussion of *solitaria*).
1933—Traver—J. Elisha Mitchell Sci. Soc. 48: 143, 146. Fig. Nymph desc...................N. C.
1933—Traver—J. N. Y. Ent. Soc. 41: 111.

Length: body 10; wing 10 mm.

A large pale species, with prominent black markings on head, thorax and abdominal tergites; genitalia of the *flavescens* type (see fig. 94).

Frontal margin of head widely blackish, the entire median area of frons shaded with same color; a black streak below antenna on each side of carina; an oblique dark streak on each side from corner of frons to base of antenna. A blackish line on posterior margin of head and a black spot on each side of vertex near ocelli. Base of antenna white; filament blackish basally, tip white. Thorax yellowish white. Lateral margin of pronotum, and a prominent lateral streak on each side, black; a black streak above and behind the fore coxa; a black spot on each side above middle coxa, a third spot behind and above latter of these; blackish markings above hind coxa. A blackish transverse mark on each side of median area of mesonotum, anterior to scutellum, which is greyish on the lateral margin. Legs pale, according to original description; femora somewhat darker at tips, not in middle; basal joint of fore tarsus of male shorter than distal joint. Wings faintly darker on costal border, especially in the costo-apical area. Venation brown, no cross veins margined; humeral cross vein pale. About 6 costal cross veins before bulla, 20–22 beyond it.

Abdomen white; segments 1–7 semi-hyaline, 8–10 opaque yellowish. Each tergite with a narrow black posterior margin, often a narrow band of blackish shading anterior to it; a broad black somewhat oblique streak on each side, between median line and pleural fold. A faint black mark at each spiracle, these marks connected by a dark hair line along the spiracular area. A narrow blackish streak along the mid-dorsal line, geminate on basal tergites. Sternites pale, unmarked; subanal plate of female rather darker. Genitalia pale, tips of forceps smoky; penes of *flavescens* type. Tails rather dark, pale smoky at base, becoming deeper smoky apically, joinings obscurely darker.

In specimens in the Cornell collection from Ft. Jackson, N. Y., and Sophia, N. C., the entire fore leg tends to be smoky brown; each femur has two conspicuous wide purplish bands, one near center and one subapical in position; tibiae brown at apex. First tarsal joint of male fore leg about ⅕ as long as second joint. The middle and hind legs of the North Carolina specimens also tend to be smoky brown on tibia and tarsus, femora yellowish. In the Ft. Jackson form, the middle and hind tibiae and tarsi are faintly marked with smoky. Since specimens from these localities agree in size and in other details with the original description, we are considering all to be of one species, with local variations in color.

In the Cornell collection are specimens of this species from many localities in the piedmont and mountain regions of North Carolina (J. R. Traver); Marlinton, W. Va. (J. G. Needham); N. Fk. Potomac River, W.Va. (J. R. T.); Ft. Jackson and Hopkinton, N. Y. (L. Harper); Willoughby, O. (J. R. T.); New River, W. Va. (J. G. N.); and Fairmount, W. Va. (P. N. Musgrave).

Heptagenia minerva McDunnough

1910—Banks—Canad. Ent. 42: 200 (*maculipennius*).
1924—McDunnough—Canad. Ent. 56: 121 .... Ont.
1926—McDunnough—Canad. Ent. 58: 194 (restricted); 195, (in discussion of *aphrodite*).

Length: body 6; wing 7 mm.

Face pale yellow with black markings along anterior margin; no ruddy shading on vertex. Thorax pale lemon yellow; a geminate brown mid-dorsal line on the anterior portion of mesonotum, and two small black dots laterally at base of scutellum. A broad blackish stripe on pleura above the leg bases. Legs pale yellow; fore femora with a median ruddy streak and an apical black longitudinal dash on the ventral edge; other femora with similar dark apical dash, but median band may be absent. Fore tarsus and apical half of fore tibia smoky. Basal tarsal joint ¼ to ⅙ of second. Wings hyaline; longitudinal veins fine, dark; cross veins heavier, margined in the costal and subcostal spaces, those at bulla with margins sometimes coalescing; veins below bulla and brown cloud on posterior branch of radial sector, as in *H. hebe* and *H. aphrodite* 3 to 4 costal cross veins before the bulla.

Abdomen pale yellowish white, tergites 8–10 tinged with ruddy brown. The lateral brown markings of the tergites are restricted to a series of lateral dashes or small triangular patches. Sternites unmarked. Forceps and tails pale. Apical projection of penes less prominent than in *H. aphrodite.*

The pale thorax with dark mid-dorsal line on mesonotum anteriorly, the two dark dots before scutellum, the ruddy median streak on fore femora and the much restricted lateral abdominal markings separate this species from the closely allied *H. aphrodite.* Specimens of this species in the Cornell collection are from Harpers Ferry and New Windsor, Md. (V. Argo).

Heptagenia otiosa McDunnough

1935—McDunnough—Canad. Ent. 67: 104. Fig.......................................Ore.

Length: body 7; wing 7 mm.

A dark-colored species of the *simplicioides* group, allied to *H. criddlei.*

Head and thorax rather dark reddish brown. A yellow streak on mesothorax anterior to wing root; two black dots on each coxa; a short black streak anterior to middle coxa. Scutella and areas on each side of these, blackish. Fore leg quite short; tarsus subequal to tibia; deep purplish brown, femur somewhat paler at base. Middle and hind legs reddish brown; tibia somewhat paler than femur; tarsus smoky brown. Wings hyaline; veins fine, dark brown. Cross veins in basal costal space and between bulla and stigmatic area few in number and very indistinct. Abdominal segments red-brown; tergites shaded with purplish brown in posterior two-thirds; sternites slightly paler than tergites. Genitalia of the *simplicioides* type; quite similar to *criddlei* but with larger subapical spines near inner margin, and an additional small apical spine near outer margin. In the type specimen which Dr. McDunnough lent us for study, the median spines were lacking (see fig. 98, "n. sp. McD.").

Distinguished from *criddlei* by the details of genitalic structure noted above, as well as by the darker abdomen which lacks the dark markings of that species. Type specimen from Maupin, Ore., in Canad. National Collection.

Heptagenia perfida McDunnough

1926—McDunnough—Canad. Ent. 58: 301. Fig.......................................Ont.

Length: body 7; wing 7 mm.

Mesonotum brown; abdominal tergites rather deep brown with pale median and submedian streaks; genitalia of the *lucidipennis* type (see fig. 96).

Eyes pale bluish white in living insect. Head yellowish tinged with ruddy; below each lateral ocellus near the margin of the eye is a round black spot. Thoracic notum brown; anterior sutures tinged with ruddy; scutellum paler. Pleura and sternum light yellow, pleura with faint ruddy shading at bases of legs. Fore legs light yellow at base; femora pale at extreme base, otherwise deep reddish brown; tibia smoky yellow with a black spot apically; tarsi smoky. Middle and hind legs pale yellowish; femora with ruddy tinges, tibiae with faint reddish streaks at base. Wings hyaline; venation pale, veins fine; costal cross veins indistinct except those of the apical area, which are 12–13 in number, simple and straight.

Abdominal tergites rather deep brown, seeming to be tinged with reddish in certain lights. A narrow pale mid-dorsal line and lunate submedian streaks on each tergite. The lateral margin of the dark area is somewhat oblique on each segment, leaving a pale postero-lateral triangle on each. Sternites pale yellowish. Tails pale, the basal joinings narrowly reddish brown. Third forceps joint much longer than the distal one, these two together fully as long as the second. Apical margin of forceps base distinctly emarginate in the median area, with a rather prominent blunt spine-like projection of each side. Penes with a rounded apical lobe as well as a lateral one; two pairs of median spines, and a lateral spine on each side. Evidently allied to *H. inconspicua.*

This species is larger than *H. rusticalis* or *H. inconspicua,* to which it is evidently allied. Differs in details of structure of penes from *H. horrida,* its closest ally; also from *H. lucidipennis.* We have seen no specimens.

Heptagenia persimplex McDunnough

1929—McDunnough—Canad. Ent. 61: 179. Fig. .... Ia.

Length: body 7; wing 7½ mm.

A small pale species; genitalia distinctive; it may be allied to *H. elegantula* (see fig. 95).

Eyes widely separated; vertex of head between eyes bright yellow, becoming pale creamy below antennae. Thorax creamy; yellowish tinges on pleura or bases of fore legs. Legs pale yellowish, shading into dirty whitish on tibiae and tarsi. Fore tibia smoky at apex; fore tarsus distinctly longer than tibia; basal tarsal joint normally about ⅓ the length of the second; second subequal to third, twice as long as fourth; fifth about as long as first. Considerable variation exists, in different specimens, however, as to the relative lengths of the fore tarsal joints. Wings hyaline; cross veins dark. 5 to 6 basal costal cross veins; stigmatic cross veins somewhat thicker than the others, but not as much so as in *Anepeorus simplex* which is superficially very similar to the present species.

Abdomen pale creamy; segments 1–7 hyaline, 8–10 opaque. Tails whitish. Apical margin of forceps base rather deeply excavated; penes slightly resemble those of *H. elegantula* on the one hand, and those of the *H. lucidipennis* group on the other. Forceps joints, however, as in the former species. We have specimens of this species in the Cornell collection from Hannibal, Mo. (W. D. Sargent) and Lincoln, Neb. (D. B. Whelan).

Heptagenia pulla Clemens

1913—Clemens—Canad. Ent. 45: 330 (Ecdyonurus). Fig. nymph. .... Ont.
1915—Clemens—Marine and Fish. Sess. Pap. 39b: 118 (Ecdyonurus; nymph fig.).
1924—McDunnough—Canad. Ent. 56: 120.
1930—Ide—Canad. Ent. 62: 229.
1933—Traver—J. Elisha Mitchell Sci. Soc. 48: 152. Fig. (in discussion of *julia*).
1933—Traver—J. N. Y. Ent. Soc. 41: 111.

Length: body 10–11; wing 10–11 mm.

Mesonotum light reddish brown; abdominal tergites largely dark reddish to purplish brown, sternites with median reddish shading; genitalia as shown in fig. 94.

Frontal portion of head pale; a brownish transverse streak across carina below bases of antennae. Vertex dark brown. Pronotum dark brown. Remainder of thoracic notum paler reddish brown; a purplish brown streak above the fore coxa, more or less continuous with one of two dark marks anterior to wing roots. Pleura yellowish; above and on each side of middle and hind coxae are two dark spots; faint dark markings between middle and hind legs. Sternum pale reddish brown. Fore femora and tibiae yellowish, tips reddish brown; in fresh specimens, both joints often washed with red-brown. Basal fore tarsal joint about ⅙ the length of the second. Tarsus whitish, joints faintly reddish. Middle and hind legs pale yellowish white; 'knees' reddish; claws and distal tarsal joints tinged with red. Basal tarsal joint of hind leg shorter than second one. Wings hyaline; longitudinal and cross veins brown, the latter somewhat thicker than longitudinals in anterior portion of fore wing. Stigmatic area somewhat opaque, clouded, 6 to 7 basal costal cross veins, about 15 beyond bulla.

Abdominal segments 1–6 semi-hyaline; 7–10 opaque. Tergites largely dark reddish to purplish brown, leaving the antero-lateral angles whitish. A narrow pale mid-dorsal line and pale lunate submedian marks on each tergite; posterior margins darker purplish brown. Apical tergites tinged with yellowish red. Sternites 1–6 pale yellowish white; posterior margins rather widely reddish, mid-ventral area tinged with pinkish or light red. Abdomen appears annulate, due to dark posterior margins of segments. Tails pale whitish, joinings distinctly dark red-brown. Penes rose-pink at tips.

The species is allied to *H. julia*, but it is considerably larger; the two pairs of median spines on penes are unequal in size, while in *julia* all four spines are approximately equal. The western *H. solitaria* is apparently very close to this species, also. Above description is drawn largely from specimens taken at N. Fairhaven, N. Y. (J. G. Needham) and another from Willowick, O. (J. R. Traver). Other specimens in the Cornell collection are from L. Winnipeg and Pigeon River. Man. (F. Neave); Buffalo, N. Y. (C. Betten); Ft. Jackson, N. Y. (L. Harper); and Slaterville, N. Y. (J. R. T.).

Heptagenia rodocki sp. nov.

Length: body 8; wing 8 mm.

Mesonotum pale reddish brown; semi-translucent brownish lateral patches on abdominal tergites and sternites; unique pair of long twisted spines in median portion of penes.

Male—Head pale reddish yellow; a small faint dark dot usually present on each side of median carina; eyes large, deep reddish purple (alcohol). Bases of antennae pale; filament smoky, tips pale. Thorax yellowish red to pale reddish brown; sutures very narrowly smoky, no other dark markings. Metanotum darker than other portions. Fore coxa and trochanter shaded faintly with purplish brown; fore leg pale reddish brown; femur usually darker brown at extreme base and apex; tibia blackish brown apically; tarsus smoky brown, distal joint and claw darker, joinings narrowly darker. Tibia slightly longer than femur; tarsus only slightly longer than tibia; basal joint of fore tarsus ¼ to ⅕ of the second joint. Middle and hind legs pale yellowish, coxae and trochanters often pale reddish; distal tarsal joint, claws and tarsal joinings dark smoky brown. Basal joint of hind tarsus distinctly longer than second joint. All femora more or less marked at base, near median area and longitudinally between these points, with opaque whitish patches, which in transmitted light appear black. Wings hyaline, venation pale; costa, subcosta and radius pale yellowish brown apically, other longitudinal veins very pale yellow; cross veins whitish, almost invisible, especially in basal costal space; 4 or 5 veins before bulla, 12–16 beyond it, in costal area, the latter faintly yellow-brown.

Abdomen pale yellowish; segments 2–7 semi-hyaline; 8–10 opaque, deeper yellowish red; joinings subopaque, appear darker in some lights. Tergites 2–7 with semi-translucent lateral patches, brownish, widely rectangular in shape, occupying most of each side, leaving a pale median line and pale lateral areas; these arise from or near the posterior margin and are darkest here; on some tergites, each patch may be divided anteriorly into 2 or 3 dark streaks. On some specimens, these patches are not present, and are difficult to see even when present. Sternites 2–7 with elongated lateral patches much like those of the tergites. Genitalia most resembling *H. simplicioides*. Third joint of forceps longer than terminal joint, these two together almost as long as second joint. Penes with a rather wide median and a narrow apical lobe on each side; a pair of apical spines, one near apical margin, on each side; median basal area occupied by two very long and twisted spine-like processes, which seem to be capable of rotation near the middle, so that the upper half may extend laterally at right angles to lower portion, as flattened blade-like structures (see fig. 98). These spines are apparently unique among the species in our fauna. Tails whitish, unmarked.

Female wholly pale yellowish; a narrow purplish line behind the eye. Base of fore leg pale reddish brown, all claws and distal joint of fore tarsus purplish brown.

Holotype—Male imago; Lewiston, Idaho, on store window, July 20, 1931; Prof. R. E. Rodock, collector. No. 1326.1 in Cornell University collection.

Allotype—Female imago; same data. No. 1326.2 in C. U. collection.

Paratypes—3 male imagos, same data; No. 1326.3–5 in C. U. collection.

This species, unique because of the peculiarly long and twisted median spines on the penes, seems to be allied to *H. simplicioides*, which lacks the darker abdominal markings, and to *H. rubroventris* in which the abdominal markings are wide median dorsal and ventral bands. The species is named for Prof. Rodock, who has sent us many specimens of mayflies from Idaho.

Heptagenia rosea sp. nov.

Length: body 7; wing 7 mm.

Mesonotum pale reddish brown; abdominal tergites largely purplish rose; genitalia of the *simplicioides* type.

Male—Head purplish; antennae purplish, pale at tips. Pronotum reddish brown washed with purplish. Mesonotum pale reddish brown; scutellum paler, but margined narrowly with purplish brown. Metanotum similar in color, its median projection pale; posterior portion darker. Pleura reddish brown, with no dark markings; sternum similar to mesonotum. Fore legs missing from single available specimen. Middle and

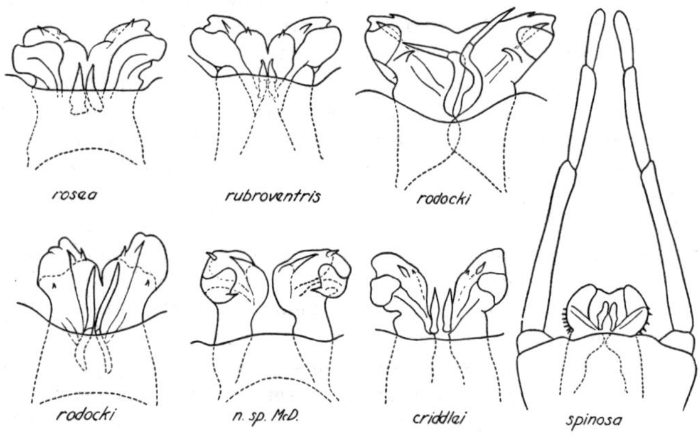

FIG. 98. Male genitalia of six species of Heptagenia (drawing designated n. sp. McD., is *H. otiosa*: published since this drawing was prepared).

hind legs yellowish amber; middle femur with a reddish apical streak on ventral edge; distal joints of all tarsi, claws and tarsal joinings smoky brown. Basal joint of hind tarsus slightly longer than second joint. Wings hyaline; longitudinal veins very pale yellowish to amber; cross veins colorless, almost invisible, except in costo-apical space. 4 to 5 basal costal cross veins, 10 to 12 beyond bulla.

Abdominal segments 1–6 semi-hyaline, apical segments opaque. Postero-lateral triangles and antero-median area of each tergite hyaline whitish, the latter tinged with rose; remainder of each tergite purplish rose, including the posterior margins. Intersegmental areas narrowly pale, so that abdomen appears somewhat annulate. Segments 7–10 washed with reddish brown; sternites 1–6 pale whitish tinged faintly with pink. Genitalia pale amber. Third joint of forceps considerably longer than the distal joint, these two together not quite as long as second joint. Penes with a median apical, apical and lateral lobes; two large median spines, unusually stout at base; on each side a small apical spine. Evidently allied to *H. simplicioides*, *H. rubroventris*, and *H. rodocki*, from all of which it is distinguished by the single apical spine on each side of penes, and by the color pattern of the abdomen. In general appearance, close to *H. perfida* and *H horrida*; also to *H. lucidipennis*, but differing in details of structure of penes. Tails pale amber in basal half, becoming pale smoky apically; joinings opaque, very faintly rose-colored. Penes as in fig. 98.

Holotype—Male imago—Gold Hill, Oregon, July 21, 1926; Mr. E. H. Nast, collector. In collection of California Academy of Science.

Heptagenia rubroventris sp. nov.

Length: body 7½–8; wing 8 mm.

Thorax yellow marked with reddish brown; abdomen largely reddish brown dorsally; a wide mid-ventral band of red-brown on thorax and abdomen; genitalia of *simplicioides* type (see fig. 98).

Male—Head yellowish. Face sprinkled finely with minute dark specks; area at base of antennae brown; brown median streak on vertex. Bases of antennae pale; filament dusky, tips pale. Pronotum yellowish, with a wide blackish median stripe. A wide purplish-black longitudinal stripe above base of fore leg extending backward along pleura above bases of other legs, most prominent on prothorax. Mesonotum yellow, washed heavily with reddish brown anteriorly and on entire posterior margin, including scutellum; a faint reddish-brown median band is more or less evident; antero-lateral margin purplish black. Metanotum dark red-brown. Pleura yellow marked with red-brown, black streak above legs bases. Sternum pale yellow, with a prominent median purplish or reddish brown band. Fore femur yellow at base, remainder bright reddish brown with purplish tinge; apex narrowly black. Tibia pale yellowish in basal portion, distinctly tinged with smoky apically; tarsus pale smoky grey, terminal joint and joinings narrowly darker. Basal joint of fore tarsus ¼ to ⅕ of second; tarsus distinctly longer than tibia. Middle and hind femora pale yellowish, marked only on dorsal apical margin with faint reddish tinge; tibiae pale, sometimes faintly smoky; terminal tarsal joint and claw purplish grey, other joints often tinged with smoky; joinings narrowly dark; basal joint as long as or very slightly longer than second joint. Wings with a distinct milky cloud in stigmatic area, faintly milky before bulla in first two spaces, sometimes entire wing with faint milky tinge. Costa, subcosta and radius yellowish; costa sometimes brownish beyond bulla. Other longitudinal veins fine, blackish brown; cross veins pale, almost invisible, especially in costal space before bulla, where there are usually 5 veins; 11 to 12 beyond bulla in costal space.

Abdominal segments 2–6 semi-hyaline; 7–10 opaque. A brown median band occupies all but extreme lateral area of all tergites; on 2–6 this is purplish brown with a tinge of red; on 7–10 bright red-brown, somewhat yellowish on 10. Within this dark band is a geminate darker median streak enclosing a narrow pale mid-dorsal line, and pale submedian lunate streaks on each side of the dark area. A dark ? mark separates the lateral edge of central dark band from the narrow pale lateral area above the pale pleural fold. Posterior margins of tergites 2–5 narrowly pale; of remaining tergites, dark red-brown. A wide purplish brown median band occupies most of sternites 1–5, is more restricted on 6 to 8, obsolescent on 9; lateral portions of 2–6 pale yellowish white; 7–9 yellow. Overlapping portions of sternites appear darker in area of dark median band. Within the dark band on each sternite are two pale dots, one on each side of median line near center of sternite, and two pale oblique submedian dashes from anterior margin, forceps base yellow; forceps smoky, darkest at tip of second joint. Penes much as in *simplicioides*.

Female very similar to male. Reddish brown areas at bases of ocelli; reddish patches between lateral ocelli and eyes, connected by a faint reddish transverse band; a wide reddish area at center of posterior margin and a median dark streak on vertex connecting this band and the transverse line. Dorsal abdominal band much more restricted than in male; ventral dark band present on thorax, almost obsolescent on abdomen. Cross veins of wing faintly dark except in costal area.

Holotype—Male imago. San Gabriel Canyon, San Gabriel Mts., Calif.; June 28, 1932. Mr. C. D. Michener, collector; No. 1323.1 in C. U. collection.

Allotype—Female imago; same data; No. 1323.2 in C. U. collection.

Paratypes—9 male imagos, 10 female imagos, same locality; June 2, 5, 8, 14, 24, 28; July 23, 1932; No. 1323.3–21 in C. U. collection.

The wide purple-brown mid-ventral band distinguishes this species from others previously described. In type of genitalia it is allied to *H. simplicioides*, also to *H. rosea*, *H. rodocki*, and *H. criddlei*. We have specimens of this fine species also from Waddell Creek, collected by Dr. David Shepherd and Dr. P. R. Needham. Nymphs taken with the type specimens, and one from Waddell Creek which was reared, are quite similar structurally to *H. lucidipennis*. Body 9 mm. in length; femora largely brown, with pale spots at base and apex and a pale median zigzag band; tarsus wholly dark brown.

Heptagenia rusticalis McDunnough

1931—McDunnough—Canad. Ent. 63: 92.......................................Que.

Length: body 5; wing 5 mm.

Mesonotum brown; abdominal tergites largely brown, paler laterally; genitalia of *lucidipennis* type.

Head brown between the eyes, becoming yellowish anteriorly; ochreous shading at bases of antennae. Pronotum brown, shading into yellowish on the lateral margins. Mesonotum brown, likewise metanotum. Pleura and sternum paler, dull yellowish; tinges of brown on pleura. Legs dull yellowish, like sternum in color; femora with slight tinges of brown, fore tarsus suffused with pale smoky. Basal joint of fore tarsus about ¼ of the second; on hind leg, basal tarsal joint longer than the second. Wings hyaline; longitudinal veins faint, cross veins almost invisible except those of the costo-apical space, which are 5 or 6 in number.

Abdomen largely brown dorsally, paler laterally; segments 1–6 semi-hyaline, 7–10 opaque and somewhat ruddier in color dorsally. Each tergite with a faint pale mid-dorsal line bordered by a pair of indistinct pale submedian streaks, not attaining the posterior margin, and most distinct on the anterior tergites. In transmitted light a small pale antero-lateral patch and a larger postero-lateral one are visible, separated by a narrow brown strip. Sternites dull dirty yellowish, tinged with ruddy on the posterior segments. Forceps and tails pale smoky. Apical inner projection of each division of penes less prominent than in *H. lucidipennis*; the lateral margin drawn out into a slight hook, much as in the *H. maculipennis* group, which character is not found in *lucidipennis* or *H. inconspicua*. Subapical spines on the ventral surface near outer margin usually small (see fig. 97).

This species, intermediate between *lucidipennis* and *inconspicua*, is smaller than the former, likewise somewhat darker in color and with clearer abdominal markings. It is considerably darker than the latter species, and differs in details of genitalic structure. Specimens in the Cornell collection are from Enfield and Taughannock Glens and Ithaca, N. Y. (J. G. Needham, V. Knox, C. N. Hardy, J. R. Traver), and from Willoughby, O. (J. R. T.).

Heptagenia simplicioides McDunnough

1924—McDunnough—Canad. Ent. 56: 120..Alta.

Length: body 7–8; wing 8 mm.

A pale yellowish species; genitalia of the *simplicioides* type (see fig. 96).

Head and thorax pale creamy to yellowish white. Clypeus slightly sprinkled with dark brown dots. Pinkish shading on pleura anterior to wing root and anterior to middle coxa. Legs pale whitish. Apex of fore tibia, basal fore tarsal joint, claws, distal joints and joinings of middle and hind legs deep smoky. Remaining joints of fore tarsus slightly smoky. Basal fore tarsal joint about ¼ as long as the second. Basal joint of third tarsus apparently variable, as it is less than the second in specimens from Waterton Lakes, but slightly longer than second on those from Yakima, Wash. Wings hyaline; both longitudinal and cross veins pale. Basal costal cross veins almost invisible, very few in number. 11 to 12 cross veins beyond bulla, those in stigmatic area quite regularly spaced; these, and cross veins of apical portion of subcostal space, slightly coarser than others, which are very fine and faint.

Abdominal segments pale creamy to yellowish white; 2–6 semi-hyaline, 7–10 opaque, the latter distinctly more yellowish. On tergites, the tracheae are outlined in pale grey. Tails pale yellowish white; a few of the basal joinings are pale reddish, others not darker than tail joints. Penes quite similar to *H. rodocki*, but the median spines shorter and less twisted; no lateral spines near margin are noticeable.

We have specimens of this species in the Cornell collection from Lethbridge, and Waterton Lakes, Alta. (J. H. Pepper) which were presented to the Cornell collection by Dr. McDunnough. Others from Yakima, Wash. (M. D. Leonard) seem to belong here.

Heptagenia solitaria McDunnough

1924—McDunnough—Canad. Ent. 56: 119 ....Alta.
1926—McDunnough—Canad. Ent. 58: 302..Colo.

Length: body 9; wing 11 mm.

Mesonotum with dark brown median stripe; costal margin of fore wing tinged with pale lemon yellow; cross veins brown; dark dorsal stripe on abdomen; genitalia of the *pulla* type; probably allied to that species.

Head light brown; black dash on frons next to eye, below base of antenna; vertex with ruddy tinge. Thoracic notum brown, with a deeper brown mid-dorsal stripe, most evident on anterior portion of mesonotum. Pleura pale ochreous brown; short black streaks just beyond bases of middle and hind coxae, and similar streaks above these, in line with wing roots; all of these dark markings may be obscure. Legs pale yellowish brown; all the femora, the fore tibia, and joints of the fore tarsus tipped with purplish. Basal joint of fore tarsus about ⅙ of the second; basal joint of hind tarsus slightly shorter than second. Wings hyaline; costal margin tinged with pale lemon yellow. Basal half of subcosta and radius thickened, pale yellowish; other longitudinal veins fine, deep brown. Cross veins thicker than longitudinals, especially along the costa; deep brown.

Abdominal segments very pale brownish. Tergites 2–7 with a dark purplish-brown median stripe enclosing a narrow pale mid-dorsal line; on posterior margins a narrow dark band of same color, with submedian broad dark bands projecting forward from it, but not attaining the anterior margin. The dark dorsal and submedian stripes may coalesce to form a simple dark shading, or with paler lunate marks submedially from the anterior margin. Tergites 8–10 opaque, light reddish brown; sternites paler. Forceps ochreous; tails dull ochreous, joinings narrowly brown. Genitalia of the *pulla* type; upper pair of median spines smaller than lower ones.

Larger and paler than *H. criddlei* another dark species of this group; apparently close to *H. pulla*, but rather darker in color markings, and with darker legs and tails. Specimens of this species in the Cornell collection from Glacier Nat. Pk., Mont. (A. S. Hazzard) and others from Colorado (G. S. Dodds) may belong here.

Heptagenia spinosa Traver

1933—Traver—J. Elisha Mitchell Sci. Soc. 48: 153. Fig. .... N. C.

Length: body 5–5½; wing 5½–6 mm.

A small species, distinct because of the lateral spines on penes, but evidently related to the *lucidipennis* group; mesonotum reddish brown, tergites with a broad median purplish dorsal stripe. Penes as in fig. 98.

Head pale yellow, no distinct markings; ocelli black-ringed at base; bases of antennae white, filament dusky. Thoracic notum dark reddish brown; pronotum purplish brown in median area and center of posterior margin; metanotum darker, with purplish posterior margin and transverse line near center. Pleura and sternum creamy yellowish white, unmarked. Legs pale whitish with a faint yellowish tinge, fore leg more yellow than others; a faint dark stripe on fore femur apically on ventral edge, and femoro-tibial joining pale purplish. Basal joint of fore tarsus ¼ to ⅓ of second; in hind leg, basal joint distinctly longer than second. Middle and hind femora with faint dark rings at apices. Wings hyaline, all veins colorless.

Abdominal segments 1–7 whitish, semi-hyaline; 8–10 opaque, yellow dorsally, creamy white ventrally. A broad median purplish band on tergites 1–8, the lateral margins remaining widely whitish. Within this dark band is a very narrow pale mid-dorsal line and whitish submedian streaks on each tergite. Posterior margins of tergites 1–8 narrowly dark purple. Median purplish band is lighter posteriorly, on tergites 1–5, then darker to center of tergite. Forceps faintly smoky; penes pale yellow, united and bearing several short spines on the lower lateral margins, as well as the usual two median blunt spines. Forceps long and slender, the two terminal joints together about as long as the second joint. Tails white, with yellowish tinge at base; in basal third, joinings narrowly purplish, pale in remaining portion.

The long terminal joints of the forceps, and the fact that the basal tarsal joint of the hind leg is distinctly longer than the second joint, would seem to ally this species with the *lucidipennis* group. The type material, from the Swannanoa River, N. C., is in the Cornell collection. We have seen no other specimens.

Heptagenia thetis sp. nov.

1933—Traver—J. Elisha Mitchell Sci. Soc. 48: 145. Nymph described (as *juno*) .... N. C.
1933—Traver—J. N. Y. Ent. Soc. 41: 111 (as *juno*).

Length: body 8–9; wing 9 mm.

A species of the *maculipennis* group; very similar to *H. juno*, but with no dark marks on pleura.

Male—Head pale yellowish white; a large smoky spot on each side of vertex between eye and lateral ocellus. Thorax yellowish; no dark longitudinal marks above leg bases, as in other species of this group. Legs pale yellowish; a short black streak apically on ventral margin of fore femur, not evident on other femora. Costa, subcosta and radius of fore wing pale amber; other longitudinal veins hyaline. Cross veins in anterior half of fore wing blackish; all others hyaline. Basal costal cross veins and a few in subcostal and third space slightly thickened; two cross veins at bulla in each of first three spaces; stigmatic cross veins finer and fainter than in *H. juno*, and more numerous (11 to 13 in *thetis*, 8 to 9 in *juno*). Abdomen pale whitish; segments 2 to 6 and basal portion of 7 semi-hyaline; apical segments opaque, light red-brown dorsally, somewhat paler ventrally. Posterior margins of tergites narrowly purplish black, most evident in dorsal portion. Tails whitish, unmarked. Genitalia much as in *juno*; penes with more distinct apical lobe, and longer spine on inner apical margin (see fig. 96).

Female very similar to male, aside from usual sex differences. Nymph 8 to 9 mm. in length (size erroneously stated in original description). No pale marks on frontal margin of head. Pale median marks on tergites 7 and 8 not coalesced. Lateral dark streaks on sternite 9 only; dark transverse marks at median line of anterior margin of sternites 8 and 9. Tracheae of gills very indistinct.

Holotype—Male imago, reared from nymph. The Cascades, near Danbury, N. C., May 4, 1930. No. 1322.1 in Cornell University collection.

Allotype—Female imago, reared; same data. No. 1322.2 in Cornell collection.

Paratypes—Two female imagos, one male subimago; same location; May 4 and 19, 1930. No. 1322.3–5 in Cornell collection.

Distinguished from *H. juno* by its larger size, as well as by features indicated above. The dark posterior margins of tergites separate it from *H. maculipennis*; from *H. walshi* it may be distinguished by the much paler mesonotum and pleura. The complete absence of dark markings on the pleura seems distinctive of this species.

Heptagenia townesi sp. nov.

Length: body 11; wing 11–12 mm.

A bronzy-brown species allied to *H. marginalis*; genitalia of the *flavescens* type (see fig. 95).

Male—Head deep red-brown on vertex, paler before the ocelli; anterior margin widely black, frontal shelf shaded with blackish. Posterior margin purplish black; a black median streak from middle of vertex to posterior margin, with a black submedian streak on each side. Antenna pale basally; filament smoky brown. Pronotum dark red-brown; lateral margin and a prominent lateral streak, black. A purplish streak above and behind fore coxa. Mesonotum deep red-brown, heavily shaded with blackish anteriorly, laterally above wing roots, and on entire posterior margin, including scutellum. Pleura paler red-brown, with wide yellowish areas anterior to and below wing roots. Blackish markings above leg bases as in *H. marginalis*. Sternum reddish brown, sutures narrowly darker. Fore leg dark bronzy brown; femur darker on each edge, with an indistinct median purplish band, and a semicircular dark mark at apex, as in *marginalis;* tip of tibia darker. Basal tarsal joint ¼ to ⅕ of the second. Tarsus about 1½ times as long as tibia. Middle and hind coxae, trochanter and femora pale whitish, the femora largely washed with purplish grey; margins and apical markings as in fore femur, also two indistinct longitudinal streaks; median band almost obsolescent. Tibiae and tarsi pale smoky brown, tarsal joinings darker.

Wings hyaline. Costa, subcosta and radius heavy, dark red-brown, almost black basally; cross veins in first two spaces similar, but not margined. All other veins finer, red-brown. Apical portion of costal space tinged with brownish; basal portion before humeral cross vein tinged with purplish. 5 to 6 cross veins in costal space before bulla, 19–22 beyond it.

Abdominal tergites 1–7 light yellowish to bronzy brown, semi-hyaline; 8–10 brighter with tinge of reddish, opaque. Posterior margins widely blackish; blackish wedge-shaped longitudinal stripes, one on each side, the wide part of each wedge directed forward. Traces of a blackish geminate mid-dorsal line, most distinct on 6 and 7, obsolescent on 4, 5, 8 and 9. Faint yellowish submedian triangles based on the posterior margin, are evident on some tergites. Sternites yellowish white, 9 red-brown; posterior margins greyish brown; on each sternite are two dark dots near the center, one on each side of median line, and two dark oblique submedian streaks from the anterior margins. Forceps and tails dark reddish to smoky brown. Penes quite similar to those of *H. marginalis;* second forceps joint relatively longer than in that species.

Holotype—Male imago; Greenville, S. C. June 2, 1932. Mr. H. K. Townes, Jr., collector. No. 1325.1 in C. U. collection.

Paratypes—3 male imagos, same data; No. 1325.3–5 in Cornell collection.

The slightly larger size, dark bronzy color and ventral abdominal markings distinguish this species from *H. marginalis;* it is considerably larger than the pale *H. dolosa*, another allied species. The species is named for Mr. H. K. Townes, Jr., who has sent us specimens of several species of mayflies from South Carolina.

Heptagenia umbratica McDunnough

1931—McDunnough—Canad. Ent. 63: 92 .... Que.

Length: body 7–8; wing 8–9 mm.

Mesonotum brown; abdominal tergites 2–6 pale yellow, posterior margins black, with brown lateral patches; genitalia as in fig. 95.

Frontal portion of head light yellow; frontal shelf semi-hyaline with a blackish streak; vertex brown. Thoracic notum and pleura brown; sternum light yellow. Legs pale yellowish, with scattered black streaks at their bases; fore femur slightly deeper yellow, fore tibia and tarsus smoky; basal joint of fore tarsus about 1/5 of the second, which is slightly shorter than the third. Wings hyaline; longitudinal and cross veins in costal half of fore wing brownish, all other veins pale; 3 to 4 basal costal cross veins.

Abdominal segments 2–6 light yellow, semi-hyaline; tergites narrowly black on the posterior margins, and with brown triangular submedian lateral patches based of the posterior margin, the apex rather blunt and not quite reaching the anterior margin. Segments 7–10 opaque; tergites brown, with the dark lateral patches faintly indicated; sternites shaded with brown. Forceps creamy, tinged with brown basally. Tails whitish, unmarked. Penes distinctly reminiscent of the *maculipennis* group, although the lack of shading on the cross veins separates it from species of that group. We consider it a probable modification of *maculipennis* and its allies.

We have three specimens of this species, which Dr. McDunnough kindly presented to the Cornell collection. These are from Knowlton and N. Hatley, Que. (J. McDunnough).

Heptagenia vitrea Walker

1853—Walker—List Neurop. Ins. Brit. Mus. p. 555 (Palingenia)....................... Hudson Bay
1871—Eaton—Trans. Ent. Soc. London, p. 143 (Heptagenia).
1885—Eaton—Revis. Monogr.; p. 254 (Rhithrogena).
1924—McDunnough—Canad. Ent. 56: 117 (Heptagenia).

Length: body 6; wing 6 mm.

Described from a female subimago; perhaps not even correctly placed as to genus. Its identity may never be correctly known; at least it cannot be determined at present. We leave the name as that of an undetermined species.

Body testaceous (Walker), light brown ochreous according to Eaton. Legs nearly the same color as the body, each femur with a median blackish grey band and light brown at the 'knee' (after Eaton). Wings light smoky grey, veins extremely light brown ochraceous (Eaton), testaceous according to Walker. Eaton's further additions to the original description are: fore tarsus light warm sepia-brown; 6 to 7 basal costal cross veins; 15–17 beyond the bulla, all simple and straight.

Heptagenia walshi McDunnough

1926—McDunnough—Canad. Ent. 58: 193.. Ont.

Length: body 6; wing 6 mm.

Mesonotum deep black-brown; tergites 1–6 pale, posterior margins dark; costal cross veins and some below bulla margined; genitalia of *maculipennis* type.

Head deep brownish, a pale yellow transverse band in the region of the antennae. Thorax deep blackish brown dorsally; slightly paler shading in region of scutellum. Pleura largely dark, sternum light yellowish. Legs light yellow; all femora with a fine dark streak apically; fore tibia brown at tip; fore tarsi slightly smoky, the basal joint about ¼ the length of the second. Wings hyaline; a faint milky tinge along the costa of the fore wing and in the basal half; veins and cross veins blackish. Basal costal cross veins, also those of subcosta in same region, and those below the bulla, margined broadly with black; some cross veins near apex of wing are also thickened; a slight tinge in the fork of the posterior branch of radial sector.

Abdominal segments 1–6 pale hyaline slightly tinged with smoky, the tergites with narrow dark brown posterior margins. Segments 7–10 opaque; tergites deep brown, sternites light ochreous. Forceps and tails pale.

This species may be separated from the allied *H. juno* by the deep brown thorax and slight smoky tinge on basal abdominal segments. The dark posterior margins of tergites 1–6 distinguish it from *H. maculipennis.* We have seen no specimens.

Genus **Cinygma** Eaton

1885—Eaton—Revis. Monogr. p. 247 (in part).
1920—Ulmer—Stett. Ent. Zeit. 81: 142 (in part).
1924—McDunnough—Canad. Ent. 56: 130 (in part).
1933—Traver—J. N. Y. Ent. Soc. 41: 120 (in part).
1933—McDunnough—Canad. Ent. 65: 75 (restricted). Figs.
1934—McDunnough—Canad. Ent. 66: 181 (correction of fig. of maxilla in previous paper).

Eyes of male imago much as in Cinygmula. Posterior margin of head of female only slightly emarginate. Fore leg of male about as long as body; femur only slightly shorter than tibia, which is about ½ as long as tarsus. Basal fore tarsal joint more than ½, usually ⅔ to ¾ as long as the second, which is subequal to third; fourth about ¾ as long as third;

fifth ½ of first. Basal joint of hind tarsus slightly longer than second; fifth joint only slightly longer than basal one. Claws dissimilar on all tarsi. Wings with typical Heptagenine venation; the stigmatic space is, however, divided by a fine line near the costa into a series of smaller costal and larger subcostal cells (see fig. 26b). Basal costal cross veins very weak. Posterior branch of media in fore wing forms almost a right angle at its union with the anterior branch. Forceps and forceps base much as in Cinygmula. Penes united almost to apex, much as in Heptagenia; a pair of small median spines present near base, and a smaller spine on each division somewhat below the apex. Posterior margin of subanal plate of female retuse.

Labial and maxillary palps of nymph much as in Heptagenia, but differing in shape of apical portion of former. Inner lobes (glossae) of labium distinctly triangular. Maxillae quite similar to those of Heptagenia. Upper portion of mandible much as in Heptagenia, "but possess, basad of the molar region proper, opposed lobes of a heavily chitinized nature" (McD.), evidently akin to those found in Ironodes and Ironopsis. Inner canine ⅔ to ¾ as long as outer one. Legs normal; claws with two small pectinations near apex. Gills present on segments 1–7; "oblong, ovate, with a well-developed bunch of filaments on underside at base and a strongly defined tracheal system, the first gill is decidedly smaller than the others and there is no trace of any ventral extension of either the first or seventh gill-pair, such as is found in Iron or Rhithrogena" (McD.). Postero-lateral margins of abdominal segments not produced into spinous processes. Tails three.

Genotype—*C. integrum* Eaton.

Distribution—Nearctic; western North America.

KEY TO THE SPECIES OF CINYGMA

Male Adults

1—Abdominal tergites 2–7 largely whitish, with dark posterior margins and lateral L-shaped marks or triangles in posterior portion; penes variable, in one species lyre-shaped .... 2.

—Abdominal tergites 2–7 with pale hyaline areas laterally and on each side of median line, elsewhere purplish brown; penes not lyre-shaped (see fig. 99) ... **integrum**

2—Penes distinctly lyre-shaped; geminate brown mid-dorsal line and brown lateral triangles in posterior half of tergites 2–7 .... **lyriformis**

—Penes not lyre-shaped, but very similar to those of *integrum*; no dark mid-dorsal line; lateral marks pale smoky, L-shaped .... **dimicki**

Cinygma dimicki McDunnough

1934—McDunnough—Canad. Ent. 66: 182 .... Ore.

Length: body 11; wing 11 mm.

Close to *integrum* in type of genitalia; abdominal tergites 1–7 largely pale yellowish.

Head pale yellowish; vertex tinged faintly with light brown. Pronotum pale ochreous with brown tinges; lateral margins blackish, and an oblique black streak on each side. A prominent black streak extends down on base of fore leg. Mesonotum pale brown; lateral margins and scutellum light yellow. Pleura light brown; a pale yellow patch "anterior to wing base, this patch bordered both dorsally and ventrally by distinct black streaks; black streaks at bases of both mid and hind legs. Metanotum pale brownish with a paler, more yellowish, median projection" (McD). Legs dull amber; two median purplish-brown bands on all femora; tibiae and tarsi tinged with smoky; a black streak on coxae. Wings hyaline; longitudinal and cross veins fine, dark.

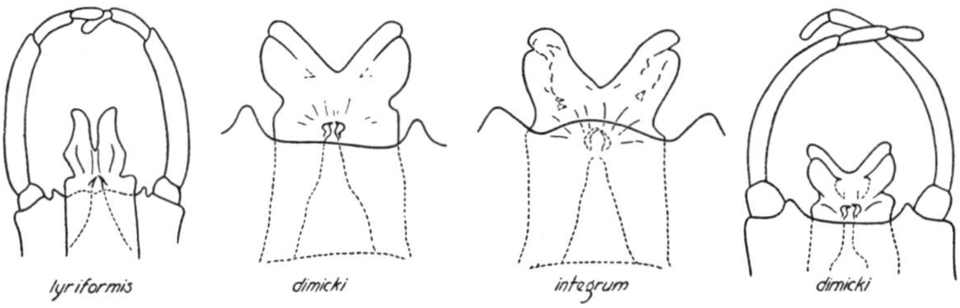

FIG. 99. Male genitalia of Cinygma (*lyriformis*, after McDunnough; *dimicki* with two enlargements).

Abdominal tergites 1–7 pale yellowish, hyaline; posterior margins narrowly black; a dark streak on posterior half of lateral margin, on each. Tergites 8–10 opaque, tinged with light brown. "On all but the first and the last two segments there are indistinct, lateral, L-shaped smoky marks with small comma-like streaks attached ventrad to the anterior ends of same" (McD). Sternites pale yellowish, 8 and 9 with faint tinges of brown. Forceps pale smoky. Tails deep smoky, joinings narrowly darker. Penes much as in *C. integrum*, but median spines smaller. Each division of the penes is evenly rounded at the apex, "without the slight tooth found in *integrum;* there is also less of a lateral excavation at the junction between the basal and apical sections" (McD.) (see fig. 99).

Quite distinct from *C. lyriformis* in details of structure of penes; distinguished from the closely-allied *C. integrum* as indicated above. There is a specimen of this species in the Cornell collection from Corvallis, Ore., taken by Prof. H. A. Scullen on May 31, 1931.

Cinygma integrum Eaton

1885—Eaton—Revis. Monogr.; p. 248. Fig. .... Wash., Ore.
1924—McDunnough—Canad. Ent. 56: 130.
1926—McDunnough—Canad. Ent. 58: 302. Fig. .... B. C.
1933—McDunnough—Canad. Ent. 65: 75. Fig. nymph parts.
1933—Traver—J. N. Y. Ent. Soc. 41: 121.
1934—McDunnough—Canad. Ent. 66: 181–2 (corrected figure of maxilla, nymph).

Length: body 11; wing 12 mm.

Head red-brown. Thoracic notum light red-brown; darker brown anteriorly and in postero-lateral areas. Pleura and sternum very similar in color, pleura with numerous blackish pencilings between wing roots and bases of legs. Fore leg yellowish to rather dark red-brown; two indistinct darker bands on femur, which is also narrowly darker at apex; apex of tibia blackish; basal tarsal joint somewhat more than ½ the length of the second. Middle and hind legs amber to light reddish brown; femora usually with two darker bands; in dark specimens, femora washed with purplish red. Wings hyaline; venation dark brown. A faint greyish tinge in stigmatic area and at base of fore wing. Cross veins in basal costal and subcostal spaces paler, rather indistinct; those in apical portion of costal space divided by a fine line near the costa into an upper series of small cells and a lower series of larger ones, as in *C. lyriformis.*

Abdominal segments 2–7 semi-hyaline. Tergite 2 largely purplish brown. Posterior margins of 3–7 dark purplish brown; geminate mid-dorsal line of same dark color, on each side of which is a whitish crescentic area. A wider dark purple-brown stripe between median line and pleural fold extends from posterior margin almost to anterior margin; between this dark stripe and pleural fold is another pale area. Sternites tinged with faint purplish brown. Segments 8–10 opaque; on tergites "the prevailing color becomes modified either with light burnt-umber or with opaque burnt-sienna or brown ochre, while the median stripe is posteriorly obliterated more or less in every segment, and in segments 8 and 9 the lateral stripes attain the base" (Eaton). Sternites light burnt-umber; no dark markings. Forceps and forceps base reddish brown. Tails light smoky greyish. Penes less distinctly lyre-shaped than in *lyriformis*, each division somewhat broader and notched at tip (see fig. 99).

The darker abdomen and differences in structure of penes serve to separate this species from *lyriformis*. Sepcimens of *integrum* in the Cornell collection are from Corvallis, Oregon (H. A. Scullen).

Cinygma lyriformis McDunnough

1924—McDunnough—Canad. Ent. 56: 226. Fig. (Ecdyonurus)............................Alta.
1926—McDunnough—Canad. Ent. 58: 302 (transferred to Cinygma; in discussion of *integrum*).
1933—McDunnough—Canad. Ent. 65: 75.
1933—Traver—J. N. Y. Ent. Soc. 41: 121.

Length: body 11½; wing 12 mm.

Head brown; clypeus pale; black shading at bases of ocelli. Thorax brown; posterior portion of mesonotum with slight ochreous shading, and yellow line along antero-lateral margin anterior to wing root. Pleura with pale shading at wing roots. Fore femur and tibia light brown; two rather indistinct darker brown bands on femur, apex also darker; base and apex of tibia narrowly blackish brown. Fore tarsus pale greyish, basal joint about ½ the length of the second. Femora of middle and hind legs similar to fore femur, tibiae paler. Wings hyaline; veins fine, light brown; cross veins very indistinct, almost colorless, in costal space except at apex of wing; a fine line parallel to and just below the costa divides this apical portion into an upper series of small costal cells and a lower subcostal series of much larger ones.

Abdominal segments 2–7 semi-translucent whitish; posterior margins of tergites deep brown. Traces of a geminate brown mid-dorsal line are present, and lateral triangular extensions from the posterior margin, halfway between mid-dorsal line and pleural fold, reach to about the middle of each tergite. Sternites with a mid-ventral row of small brown spots. Segments 8–10 opaque; tergites light brown, sternites tinged with ochreous brown. Forceps and tails dull brown. Penes distinctly lyre-shaped (see fig. 99).

Differs from *integrum* in the paler abdomen and in structural details of the penes. We have seen no specimens of this species.

Genus **Rhithrogena** Eaton

1881—Eaton—Ent. Mo. Mag. 18: 23.
1885—Eaton—Revis. Monogr. p. 250.
1905—Needham—N. Y. State Mus. Bull. 86: 59 (nymph).
1920—Ulmer—Stett. Ent. Zeit. 81: 142.
1926—McDunnough—Canad. Ent. 58: 195.
1933—Traver—J. N. Y. Ent. Soc. 41: 115.
1933—Spieth—J. N. Y. Ent. Soc. 41: 330.
1934—McDunnough—Canad. Ent. 66: 185.

Mayflies of medium to large size, usually dark brownish or reddish brown in color. Eyes of male large, usually contiguous apically. Posterior margin of head of female almost straight. Posterior margin of pronotum deeply excavated at median area. Fore leg of male somewhat longer than body; of female, slightly less than or about as long as body. Fore femur of male about $\frac{3}{4}$ the length of the tibia, which is about $\frac{3}{4}$ as long as tarsus. Basal fore tarsal joint relatively short, $\frac{1}{6}$ to $\frac{1}{4}$ the length of the second joint; third subequal to second, fourth about $\frac{3}{4}$ as long as third; fifth longer than first, about $\frac{1}{2}$ length of fourth. Fore tarsus of female little more than $\frac{1}{2}$ as long as tibia; femur about $\frac{3}{4}$ length of tibia. Basal fore tarsal joint of female about $\frac{1}{2}$ as long as second, usually subequal to fourth; second and fifth joints longest, third slightly shorter than second. Tibia of hind leg in both sexes about $\frac{1}{4}$ length of femur; tarsus $\frac{1}{2}$ or less than $\frac{1}{2}$ of tibia. Basal tarsal joint equal to or very slightly shorter than second; second, third and fourth subequal; fifth fully as long as the three preceding joints combined. Claws dissimilar on all tarsi. Typical Heptagenine venation. In North American species, cross veins of stigmatic area of fore wing are anastomosed, but this is not true of European species. Basal costal and subcostal cross veins weak. In hind wing, second fork of radial sector usually arises basad of the first fork.

Apical margin of forceps base of male usually slightly excavated at middle line, with a slightly rounded projection on each side, thus appearing undulate. Forceps four-jointed; basal joint very short, second long and slender, fourth slightly shorter than third; combined length of the two distal joints usually less than the second joint. Penes usually rod-shaped, separated from one another very near the base; in a few species, each penis division is shorter and stouter, broadly rounded at tip. Spines are usually present on penes, but in two species of the *jejuna* type these are wholly lacking. In the *anomala* type (nearest to the European *R. semicolorata* Curt., the genotype), a spatulate process is present on the ventral surface of each penis division, not quite halfway between base and apex; small spines are borne at the tip of this process; others are usually present on inner margin of each division just below apex. In the *brunnea* type, a strong lateral spine is present near the base of each penis division; smaller spines may occur on ventral or dorsal surface of each division, near inner margin, halfway between the lateral spine and apex; others are often present at or near the apex, usually lateral in position. In the *robusta* type (penis divisions short and broad) a strong apical spine is present on inner margin, while other spines occur as in the *brunnea* type. Subanal plate of female well developed, its apical margin entire, usually obtusely rounded.

Head of nymph flattened, slightly deflexed. Posterior margin straight, frontal and lateral margins rounded; frontal margin very slightly emarginate at median line. Head widest at level of anterior portion of eyes. Labrum slightly emarginate in median area of anterior margin; fully four times as wide as long. Outer canine at least twice as long as the inner; it slants inward obliquely, and is "sharply truncate at its outer base" (Eaton) opposite the inner canine. Lacinia wanting; a row of finely pectinate spines in the lacinial region. Proximal joint of maxillary palp dilated broadly near the base; distal joint "stoutly clavate, oblique at the extremity" (Eaton), and bearing many parallel series of pectinate spines. A series of pectinate spines is present on upper apical margin of galea-lacinia. Base and tip of distal joint of labium appear partially separated from the remaining portions. Glossae more or less conical, widely separated at base, usually slightly divergent apically. Median lobe of hypopharynx somewhat quadrangular; lateral lobes much narrower than median lobe, each dilated on outer margin near middle area.

Thorax and abdomen convex dorsally, flattened ventrally. Lateral margins of pronotum flare slightly near anterior border; posterior margin excavated in median area. Femur flattened; upper surface bears many minute spine-like processes; spines likewise present on margins, those on posterior margin longer. Tibia subequal to or slightly shorter than femur; usually with a few short spines on median portion of upper surface and along margins. Tarsus less than half the length of tibia; claw with a single stout spine about ⅓ from the base on inner margin, and 2 or 3 smaller pectinations laterally near apex. Gills borne on segments 1–7; each pair consists of a dorsal fibrillar and a ventral lamellate portion. Lamellate portions of first and last pairs meet beneath body of nymph; margins of intermediate pairs overlap one another, the free edge of each being deflected and pressed against the surface to which the nymph clings, thus forming an "adhesive disc" on ventral surface of body. Lateral margins of lamellae are not thickened, however, as in the genus Iron. Lamellae of first pair of gills largest, the anterior lobe much elongated; posterior lobe of gills of seventh pair elongated; intermediate pairs somewhat more oval in outline. Lateral margin of each abdominal segment bears a short anterior and a blunt posterior spine-like process near point of attachment of gill. Tails three; middle tail fringed with fine hairs on each side except in basal third; outer tails fringed in corresponding area on inner margin only.

Genotype—*R. semicolorata* (in Baetis), Curtis.

Distribution—Holarctic; in Nearctic, of general distribution.

KEY TO THE SPECIES*

Male Adults

1—Lateral spine or spatulate spine-bearing process present on each division of penes, near base . . . . . . 2.
—No such lateral spine or spine-bearing process on penes . . . . . . 14.
2—Strong lateral spine present on each division of penes, near base . . . . . . 3.
—Lateral spatulate process, bearing several short spines at tip, present on each division of penes, somewhat higher than spine of previous group . . . . . . 12.
3—Divisions of penes relatively short and stout, more or less truncate apically, bearing prominent spine on inner apical margin (see fig. 101, *robusta*) . . . . . 4.
—Divisions of penes relatively long and slender, not truncate apically; no such prominent spine on inner apical margin (see fig. 100) . . . . . . 5.
4—Wings distinctly tinted with brownish amber; tips of inner apical spine on penes recurved dorsally . . . . . . **brunneotincta**
—Wings not tinted; tip of apical spine on penes not recurved . . . . . . **robusta**
5—Small sharp spine present on latero-apical margin of each division of penes, somewhat as in Heptagenia (see fig. 101) . . . . . . **virilis**
—No such small sharp latero-apical spine present; latero-apical margin rounded . 6.
6—Spine or spines present on ventral or dorsal surface of each division of penes, near central portion; usually also finer spining near apex . . . . . . 7.
—Neither ventral nor dorsal spines present on any portion of divisions of penes . . . . . . **gaspeensis**
7—Ventral spine present on each division of penes, near inner margin, in central portion . . . . . . 8.
—No such ventral spine on divisions of penes; one to several spines on dorsal surface, in central portion . . . . . . 11.
8—Divisions of penes only slightly divergent apically . . . . . . 9.
—Divisions of penes strongly divergent apically . . . . . . 10.
9—Each division of penes broad at apex; latero-apical spining minute . . . . . **doddsi**
—Each division of penes narrowed at apex; latero-apical spines better developed . . . . . . **petulans**
10—Each division of penes narrowed at apex; 5–6 small apical ventral spines **morrisoni**
—Each division of penes broader at apex; midventral spine strong, apical spines less well developed . . . . . . **brunnea**
11—A single central dorsal spine on each division of penes; wing 11½ mm. in length . . . . . . **futilis**
—Five or six dorsal spines on each division of penes; wing 15 mm. in length . . . . . . **flavianula**
12—Abdominal tergites 2–7 smoky, with lavender tinge; longitudinal veins yellowish, cross veins colorless; tails not darker at joinings . . . . . . **fasciata**
—Abdominal tergites 2–7 reddish brown, usually with purplish tinge; longitudinal veins reddish brown, cross veins variable in color; tails distinctly darker at joinings . . . . . . 13.
13—Cross veins red-brown; tails smoky brown; wing 10 mm. in length . . . . . . **amica**
—Cross veins colorless; tails whitish; wing 7 mm. in length; body distinctly more reddish than in preceding species . . . . . . **anomala**
14—Penes strongly outcurved at tips; no small spines near apex of each division of penes . . . . . . 15.

* This is Dr. McDunnough's key to the *brunnea* group, with modifications.

—Penes very slightly or not at all outcurved at tips; several to many small apical spines (see fig. 101) . . . . **impersonata**

15—Penes distinctly narrowed apically; apical margin of forceps base projecting backward, with a distinct rounded median excavation (see fig. 102) . . . . **jejuna**

—Penes broad apically, not at all narrowed; apical margin of forceps base regularly undulate (see fig. 102) . . . . **undulata**

VERIFICATION TABLE

Adults†

Species	*Wing length*	*Mark on femur*	*Longit. Veins*	*Lat. spine on penes**	*Genitalia, type of*	*Distr.*
amica	10	Spot	Red-brown	Present	anomala	N. Y.
anomala	7	Spot	Red-brown	Present	anomala	N. E.
brunnea	13	Streak	Dp. brown	Present	brunnea	Calif.
brunneotincta	13	Absent?	Brown	Present	robusta	Que.
doddsi	11	Streak	Dk. brown	Present	brunnea	W
exilis♀	7	Spot	Colorless	Present	anomala?	N. C.
fasciata	6	Spot	Yellow	Present	anomala	N. C.
flavianula	15	Streak	Bl. brown	Present	brunnea	W
fuscifrons♀	7½	Spot	Yellow	Present	?	N. C.
futilis	11½	Streak	Dk. brown	Present	brunnea	W
gaspeensis	11–12	Streak	Brown	Present	brunnea	Que.
impersonata	10	Streak	Blackish	Absent	jejuna	Que.
jejuna	9–10	Streak	Dk. brown	Absent	jejuna	Que.
morrisoni	10½	Streak	Brown	Present	brunnea	W
petulans	11	Streak	Dk. red-br.	Present	brunnea	Calif.
robusta	14	Absent	Dp. brown	Present	robusta	W
uhari♀	7	Spot	Colorless	Present	?	N. C.
undulata	9	Streak	Brown	Absent	jejuna	W
virilis	15	Streak	Dk. brown	Present	brunnea	B. C., Alta.

KEY TO THE KNOWN SPECIES

Nymphs

1—Small nymphs, body 5 to 7 mm. in length . . . . 2.

—Larger nymphs, body 9 to 12 mm. in length . . . . 5.

2—Tergites 7–9 yellow, contrasting sharply with dark central and basal tergites . . 3.

—Tergites 7–9 not differing markedly in color from central tergites, although usually somewhat lighter . . . . 4.

3—Tergites 1–6 and basal half of 7 dark reddish brown; middle portion of head dark brown . . . . **fuscifrons**

—Tergites 1 and 2 pale yellow, 3–6 dark red-brown, 7 wholly pale; middle portion of head yellow or shaded with very pale brown . . . . **exilis**

4—Dorsum of abdomen light chestnut or cinnamon brown; no conspicuous markings . . . . **uhari**

—Dorsum of abdomen orange or orange-brown; tergites 4–8 dark brown basally, with pale posterior margins . . . . **fasciata**

† Males unless indicated.

* Including spatulate process.

5—Gills bright rose-pink in color; intermediate pairs bear each a rounded, ear-like erect dorsal lobe . . . . . **doddsi**
—Gills not as above; pale in color, with no such erect dorsal lobe . . . . . 6.
6—Nymph 12 mm. or more in length . . . . . 7.
—Nymph 9–10 mm. in length . . . . . **amica**
7—Gills colorless . . . . . **robusta**
—Gills tinged with purplish in dorsal half . . . . . **virilis**

VERIFICATION TABLE

Nymphs

Species	*Body*	*Gills*	*Tergites 1–6*	*Tergites 7–9*
amica	9–10	Greyish	Brown	Brown
doddsi	9–12	Rose-red	Green-brown	Green-brown
exilis	5–7	Whitish	Red-brown	Yellow
fasciata	5–7	Whitish	Orange-brown	Orange-brown
fuscifrons	5–7	Whitish	Brown	Yellow
robusta	12	Whitish	?	?
uhari	5–7	Greyish	Chestnut-br.	Chestnut-brown
virilis	12	Purplish	Brown	Brown

Rhithrogena amica sp. nov.

Length: body 9–10; wing 10 mm.

A medium-sized species of the *anomala* group; cross veins mainly brown, tails dark smoky brown.

Frontal portion of head greyish brown, vertex and median carina red-brown. Black semi-circular mark at base of antenna; fine dark hair-line between antenna and ocellus, from corner of eye. Bases of antennae pale; filament dusky, tip pale. Thoracic notum dark reddish brown; pronotum with darker shading laterally. Mesonotum may be yellow-tinged laterally. Pale spot anterior to scutellum; scutellum and all posterior portions of mesonotum blackish brown. Pleura paler than notum, more yellowish, with red-brown areas; purplish grey streak anterior to wing roots, purplish markings above and anterior to coxae. Sternum light red-brown, posterior portion of mesosternum deep red-brown. Fore legs olive brown, femur with reddish tinge; apex of femur and tibia blackish brown. Middle and hind legs yellowish; femora with reddish tinge; tarsi and apices of femora smoky brown. All femora with blackish spot at center, and usually with fine black longitudinal hair-line from base to apex. Small black spot at apex of each trochanter. Wings hyaline; basal half of fore wing may have very faint brownish tinge. Venation pale red-brown; basal costal and all of subcostal cross veins so faint as to be almost invisible. Humeral cross vein red-brown, pale at costal margin. Stigmatic area opaque whitish.

Abdominal tergites dark red-brown, 2–7 with purplish tinge; apical tergites brighter red-brown, less tinged with purple; lateral margins of 9 and 10 largely yellowish. Posterior margins narrowly darker brown; pleural fold slightly paler. Traces of pale mid-dorsal line and of short submedian streaks near center of each basal and intermediate tergite; pale marks at anterior and lateral margins (prominent in *R. anomala*), either absent or very faint. Sternites pale red-brown, more yellowish apically, usually semi-hyaline; sternite 9 yellow, outlined narrowly in dark brown. Genitalia red-brown, including forceps base; penes very similar to *R. anomala*, but slightly less outcurved at tips (see fig. 101). Tails dark olive to smoky brown, joinings narrowly darker.

Female very similar to male; head and thorax largely greenish-olive brown. Nymph blackish red-brown dorsally. Gills pale greyish. Central portion of basal and middle area of each femur largely occupied by a large pale somewhat cross-shaped mark. Tails red-brown.

Holotype—Male imago, reared from nymph. Ellis Hollow, N. Y., Ringwood Creek; May 16, 1933 (J. R. Traver). No. 1336.1 in Cornell University collection.

Allotype—Female imago, reared from nymph. Same data as holotype. No. 1336.2 in Cornell collection.

Paratypes—15 male imagos, 17 female imagos, reared; Ellis Hollow, May 12–20, 1933; Slaterville, N. Y., May 20–25, 1932; same collector. No. 1336.3–34 in Cornell collection.

This species is slightly larger and darker than the closely-allied *R. anomala*, and with less of a reddish tinge; the darker tails and brown cross veins also serve to distinguish it from that species. It is distinctly larger and darker than *R. fasciata*, which is also a member of this group.

Rhithrogena anomala McDunnough

1930—McDunnough—Canad. Ent. 60: 239 .... Que.
1933—Traver—J. Elisha Mitchell Sci. Soc. 48: 171 (in discussion of *fasciata*).
1933—Traver—J. N. Y. Ent. Soc. 41: 116.

Length: body 7; wing 7 mm.

A rather small reddish brown species; lateral spine-bearing projection on penes above base.

Head pale red-brown; clypeus pale hyaline; black semi-circular mark at base of antenna. Base of antenna pale; filament brown, tip pale. Thorax rather light red-brown, sometimes with yellowish tinge. Posterior and lateral margins of pronotum blackish. Tip of scutellum, lateral areas below it on each side, and small lateral patches anterior to it, purplish brown; two or three small yellowish patches anterior to scutellum. Blackish streak anterior to wing roots. Pleura paler than notum or sternum; large yellow areas near sternum anterior to middle and hind legs; narrow purplish black markings above bases of legs. Purplish intersegmental areas between pro- and mesosternum. Fore leg light reddish brown, tarsus more yellowish, may be faintly smoky; apex of femur and tibia deeper red-brown. Middle and hind legs yellowish, tarsi faintly smoky; femora red-tinged. Small black spot (not streak) at center of each femur; small black dot at apex of each trochanter; short black streak above middle and hind coxae. Wings hyaline; longitudinal veins red-brown, cross veins colorless almost invisible; humeral cross veins dark red-brown. Stigmatic area opaque whitish.

Abdominal tergites 2–7 deep reddish brown with purplish tinge. Pleural fold and intersegmental areas pale hyaline; paler semi-hyaline elongate areas on each side near pleural fold. Mid-dorsal line narrowly pale hyaline; short wide submedian streaks at anterior margins, and two small pale spots, one on each side of median line, in posterior half of each tergite (on basal ones, these may connect with submedian streaks). Tergites 8–10 brighter red-brown, 10 distinctly reddish. Sternites hyaline; very similar in color to tergites, but paler; apical sternites paler than basal and middle ones. Genitalia reddish brown. Penes outcurved at tip; 2 or 3 serrations near inner apical margin of each side, and a short wide lateral process halfway from base to apex, at apex of which are several short spines (see fig. 101). Tails whitish, joinings distinctly red-brown.

We have specimens of this species in the Cornell collection from the St. Regis River at Ft. Jackson, N. Y. (L. Harper), and from Saranac, N. Y. (C. R. Crosby, H. Dietrich). These are slightly larger than indicated for the type, but seem otherwise similar. The species is allied to both *fasciata* and *amica*, from which it may be distinguished by the detailed structure of the penes and the redder coloration of body.

Rhithrogena brunnea Hagen

1875—Hagen—Ann. Rep. U.S. Geol. & Geogr. Survey of Terr. 1873, pt. 3, Zool. p. 581 (Heptagenia) .... Calif.
1885—Eaton—Revis. Monogr. p. 253 (*hageni*)
1924—McDunnough—Canad. Ent. 56: 117, 225 (Heptagenia, in discussion of that genus and *R. flavianula*).
1926—McDunnough—Canad. Ent. 58: 195.
1933—Traver—J. N. Y. Ent. Soc. 41: 116.
1934—McDunnough—Canad. Ent. 66: 185.

Length: body 10; wing 13 mm.

A species of the *brunnea* group, penes distinctly outcurved at tip, with large ventral spine on inner margin.

Thoracic notum "bright brown ochreous, the peak of the mesonotum flavescent, a line adjacent to the tegulae and a streak below it and in front of these in a depression brown-black" (Eaton). Fore legs brown, tibia and tarsus blackish brown. Middle and hind legs yellowish brown, tarsi somewhat deeper brown. Femora marked in basal

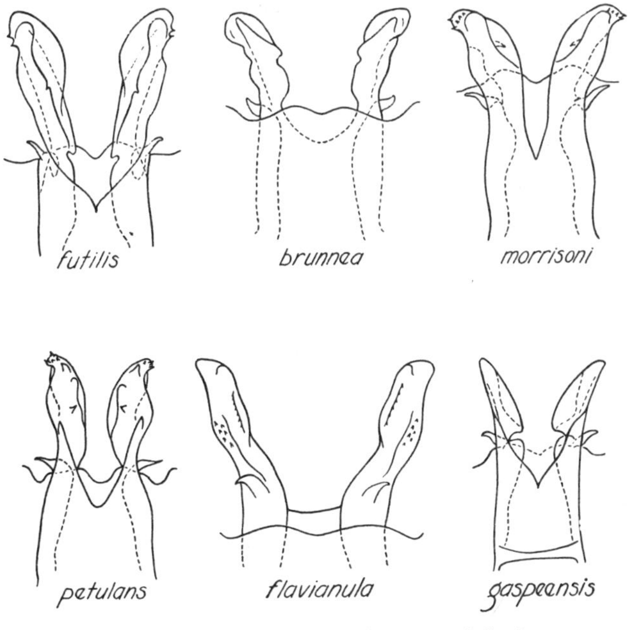

FIG. 100. Male genitalia of six species of Rhithrogena, of the *brunnea* group (after McDunnough, except *brunnea*).

area "with a longitudinal brown-black streak tapering at both ends" (Eaton). Wings hyaline; fore wings very faintly tinted with light brownish grey in stigmatic area. Venation deep brown, paler at wing bases. 4 or 5 basal costal cross veins before bulla; 5 to 7 simple cross veins and 11–14 anastomosing veins are beyond bulla. Abdominal tergites pitch-brown, posterior margins darker. Sternites reddish yellow, immaculate. Tails dark brown, joinings mostly opaque. Tips of penes distinctly curved outward; a basal lateral spine on outer margin, and a strong ventral spine on or near inner margin nearer apex. Apical spines weaker than in *R. morrisoni* (See fig. 100).

A larger species than *R. morrisoni*, differing in detailed structure of penes as indicated. The allied species, *R. virilis*, lacks the ventral median spine and possesses a small sharp spine on the latero-apical edge of each division of the penes, which is not present in *brunnea*. We have seen no specimens of the true *R. brunnea*. The identity of the nymph from Utah, referred to this species by one of us (Needham, Utah Agri. Exp. Sta. Bull. 201, 1927, p. 14), is doubtful.

Rhithrogena brunneotincta McDunnough

1933—McDunnough—Canad. Ent. 65: 280. Fig. .... Que.

Length: body 12; wing 13 mm.

A species allied to *robusta*; portions of both wings are distinctly tinted with brownish amber.

Head brown; paler shading in region of antennae. Thorax deep pitch-brown; mesonotum shaded with paler brown anterior to scutellum. A reddish shade anterior to wing roots; pleura and bases of legs shaded with ochreous brown. Fore legs deep smoky, femur with slight olive tinge. Middle and hind legs olive brown, femora and tarsi shaded slightly with smoky. Wings hyaline; basal two-thirds of fore wing and costal half of hind wing distinctly tinted with brownish amber. Veins and cross veins brown, fairly strong; cross veins in stigmatic area thickest, those in the costal space not much anastomosed, as is usual in other species of the genus.

Abdominal tergites deep brown; lateral margin and small antero-lateral and postero-lateral triangles lighter olive brown. Intersegmental areas on basal tergites narrowly pale, so that dorsum appears annulate. Sternites lighter olive brown, posterior ones distinctly brown. Forceps and tails deep smoky. (See fig. 102, after McDunnough).

This species is allied to the western *R. robusta*, from which it may be distinguished by the brown-tinted wings and structural details of the penes. In *robusta*, penes are sharply truncate at tip, the inner apical spine is not recurved, and lateral spines differ slightly from those of the eastern *brunneotincta*. We have seen no specimens of this species.

Rhithrogena doddsi McDunnough

1923—Dodds—Trans. Amer. Ent. Soc. 49: 104. Fig. (*hageni*) .... Colo.
1926—McDunnough—Canad. Ent. 58: 196 .... Alta.
1927—Needham—Bull. Utah Agri. Exp. Sta. 201: 13. Fig. nymph (*R. mimus*) .... Utah
1927—Needham—Canad. Ent. 59: 133 (nymph, *nec* imago; *R. mimus*).
1933—Traver—J. N. Y. Ent. Soc. 41: 116, 122.
1934—McDunnough—Canad. Ent. 66: 187. Fig. (in key, also in discussion of *futilis*).

Length: body 10; wing 11 mm.

A species of the *brunnea* group; penes broad at apex, almost erect.

Head deep brown; a pale band in region of antennae. Thorax brown; scutellum slightly paler; pleura mainly pale ochreous. Legs dark amber-brown. Fore tibia and tarsus smoky brown; a dark longitudinal streak on basal half of each femur, often entire femur appears shaded with black. Wings hyaline; venation fine, dark brown; basal costal cross veins obsolescent. Abdominal tergites brown; intersegmental areas paler. Sternites dull yellowish brown. Forceps and tails smoky brown. Penes almost straight, not outcurved at tip, wide apically; a ventral spine on each division, near inner margin, halfway between apex and lateral spine (see fig. 101).

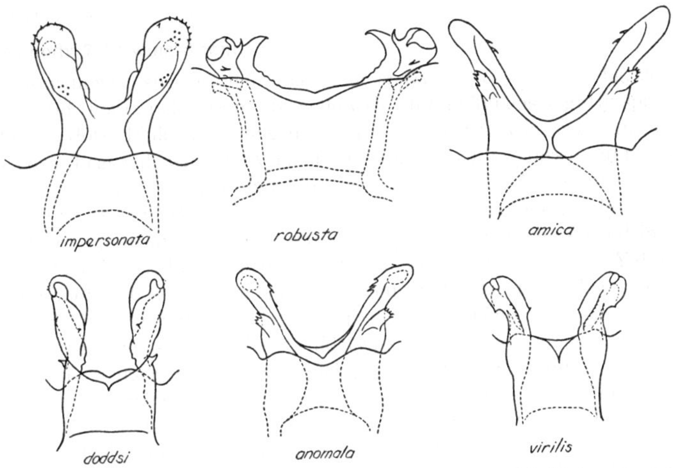

FIG. 101. Male genitalia of six species of Rhithrogena; (*doddsi* and *virilis*, after McDunnough.)

Very closely allied to *R. futilis*, from which it may be distinguished by the detailed structure of the penes, which are wider at tip and more erect than in that species; in *futilis* also, a small dorsal spine is present on each division of penes, but no ventral one. We have several of Dr. Dodds' specimens in the Cornell collection, also many specimens reared by one of us in Utah. The rose-red gills and small erect dorsal lobe on each of the intermediate pairs seems distinctive of the nymph of this species.

Rhithrogena exilis Traver

1933—Traver—J. Elisha Mitchell Sci. Soc. 48: 166..N. C.

Length (female); body 6; wing 7 mm.

A small pale species; male imago not known, genitalia of subimago seem to be of *anomala* type.

Body of female yellow. Black spot on each side of head, underneath clypeus; black half-ring at base of each antenna. Irregular purplish black lateral marks on pronotum. Faint darker markings on pleura; thoracic sternum unmarked. Legs yellow; a purplish black median spot at center of each femur. Wings semi-hyaline; venation colorless. Abdominal segments 2–6 flushed with pinkish orange; posterior margins of tergites 2–4 narrowly dull purple. Sternites pale, unmarked.

Tails white, unmarked.

Male subimago orange-brown; dark marks on pleura above leg bases; mesosternum dark-margined; brown streak runs forward from each side of scutellum; dark marks on each coxa and at each joining of fore leg; median femoral marks as in female. Genitalia not yet developed, but in one specimen seem to be of the *anomala* type.

Female very similar to *R. uhari*, but with paler abdominal tergites. Slightly larger and paler than *R. fuscifrons*. The type material of this species, from the vicinity of Black Mt., N. C., is in the Cornell collection. We have seen no other specimens.

Rhithrogena fasciata Traver

1933—Traver—J. Elisha Mitchell Sci. Soc. 48: 170. Fig. .... N. C.

Length: body 6; wing 6 mm.

A rather small pale species, allied in type of genitalia to *R. anomala*.

Head pale yellow; clypeus colorless, hyaline; small black spot at base of antenna; small black spot on each side beneath clypeus. Purplish area around base of antenna, which is whitish; base of filament smoky, tip white. Pronotum deeply emarginate

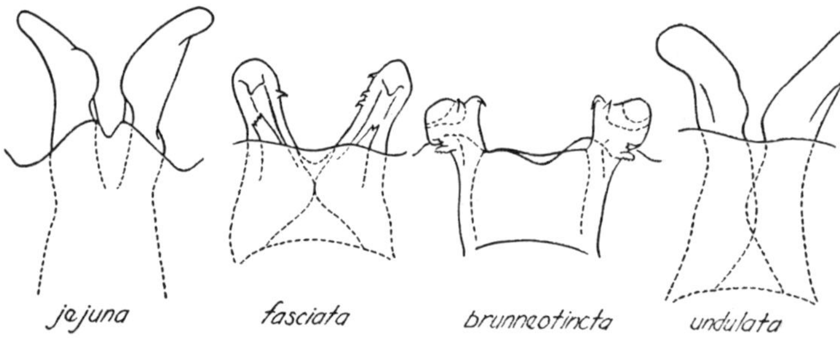

FIG. 102. Male genitalia of four species of Rhithrogena (*brunneotincta* after McDunnough).

posteriorly; yellowish orange. Mesonotum pale yellow, brown on each side of scutellum. Metanotum brown. Pleura pale with ochreous markings. Venter pale ochreous; posterior portion of mesosternum tinged with brown. Legs yellowish brown, femora deeper brown; round purplish spot at center of each femur. All joinings narrowly dark purplish. Wings semi-hyaline; longitudinal veins yellowish to yellow-brown, cross veins colorless. Abdominal segments smoky, some with ochreous tinge. Segments 1 and 2 pale, very faintly smoky; 3–6 deeper smoky, with distinct lavender tinge; 7–10 tinged with pale ochreous. Posterior margins of all tergites dull pale lavender, so that abdomen appears banded. Indistinct pale submedian dashes and pale median line on tergites 2–7. Sternites pale, immaculate; no dark banding. Genitalia pale brown; penes quite similar in form to *R. anomala* (see fig. 102). Tails whitish, not darker at joinings.

Rhithrogena flavianula McDunnough

1924—McDunnough—Canad. Ent. 56: 225. Fig. (Heptagenia)................................ Colo.
1926—McDunnough—Canad. Ent. 58: 195 (transferred to Rhithrogena).
1934—McDunnough—Canad. Ent. 66: 185 (in key).
1933—Traver—J. N. Y. Ent. Soc. 41: 116.

Length: body 14; wing 15 mm.

A large species of the *brunnea* group, having basal lateral spines on penes.

Clypeus dull grey, "crossed by a black band at its base" (McD.). Head ochreous at bases of antennae and around ocelli. Thorax deep brown. Anterior to the wing roots a pale ochreous patch, within which are two blackish streaks. Pleural sutures and bases of wings with ochreous markings. Fore leg blackish. Middle and hind legs golden brown; a black streak on basal half of each femur; tarsi blackish. Wings hyaline; stigmatic area tinged with pale amber. Venation blackish-brown. Abdominal segments deep reddish brown; lateral and posterior margins of tergites narrowly light yellow, so that abdomen appears annulate. Posterior sternites darker in color, the posterior margins yellow-banded as on the dorsum. Forceps deep brown. Tails blackish. Penes distinctly outcurved at tip, each bearing a series of very small spines near apex and on lateral margin. A large spine laterally at base on outer margin (see fig. 100).

The annulate abdomen, large size and genitalic structures should serve to distinguish this from other allied species. We have specimens of this species in the Cornell collection from the Yosemite Valley, Calif., and have seen others from the same location in the collection of the California Academy of Science. Two other specimens in our collection were taken in Colorado by Mr. E. J. Oslar (no other data).

Rhithrogena fuscifrons Traver

1933—Traver—J. Elisha Mitchell Sci. Soc. 48: 168..N. C.

Length (female): body 5; wing 7½ mm.

A small species, male not known, hence its relation to other species is doubtful.

Body of female light orange brown. Small black spot at base of antenna; antennae dark brown basally, tips paler; posterior margin of head brown. Posterior margin and lateral areas of pronotum tinged with purplish. Meso- and metanotal scutella, and their lateral margins, purple-tinged. Pleura paler than notum; a few purplish markings, and a dark streak above fore and middle legs. Sternum pale; posterior portion of mesosternum as dark as notum. Legs yellowish; tibiae and tarsi rather smoky. A dark purplish spot at center of each femur. All joinings dark brown. Wings semihyaline; longitudinal veins pale yellow, cross veins faint, paler. Abdominal tergites tinged with purplish brown, darker on 2–6; posterior margins very narrowly dark purplish. Sternites pale. Tails whitish, joinings very narrowly dark.

Allied to *R. uhari* and *R. exilis*, but with rather smaller body, darker color, and joinings of tails darker. The type material of this species, from several locations in the mountain area of North Carolina, is in the Cornell collection. We have seen no other specimens.

Rhithrogena futilis McDunnough

1928—McDunnough—Canad. Ent. 60: 9 (*doddsi*)..Alta.
1934—McDunnough—Canad. Ent. 66: 187. Fig..Wyo.

Length: body 11; wing 11½ mm.

A species of the *brunnea* group; very similar to *R. doddsi* in general appearance, but differing in details of structure of penes.

Description largely comparative. General color deeper brown than in *doddsi*, size simewhat larger. Anterior to the root of the fore wing "is a light ochreous patch tinged with ruddy-brown posteriorly and bisected by a blackish streak" (McD). Femora with dark longitudinal streaks. Abdominal segments with narrow pale yellowish intersegmental rings; on tergites "this pale area shows a faint smoky transverse streak" (McD). Penes narrower apically and somewhat more divergent than those of *R. doddsi*; no small ventral spine near center of each penis lobe, but a small dorsal spine in about the same position (see fig. 100, after McDunnough). "The minute apical spines on outer edge of each lobe which are present in most of the species of the group are somewhat better developed than in *R. doddsi*, but not as strong as in *R. morrisoni*" (McD). We have seen no specimens of this species.

Rhithrogena gaspeensis McDunnough

1933—McDunnough—Canad. Ent. 65: 279. Fig. .... Que.

Length: body 11; wing 11–12 mm.

A species of the *brunnea* group, having lateral spines on the penes.

Head deep brown, with blackish shading around bases of ocelli. Thorax deep pitch-brown; pleura paler brown, anterior to wing roots. Legs dull brown; a central longitudinal blackish streak on femora. Wings hyaline; veins fine, brown; extreme base of wing with a very faint brownish tinge. Abdominal segments dull brown; basal segments "slightly paler hyaline...near intersegmental areas" (McD)., but with no definite markings. Tails deep smoky.

Allied to *R. doddsi* and *R. morrisoni*, in type of male genitalia; differs from the former "in that the penis-lobes are much more narrowed apically and bend gently and evenly outward; there is further no trace of any other spining on either dorsal or ventral surfaces; in *morrisoni* the lobes are broader apically and more excurved and there is a very distinct, although small, spine on the inner ventral surface of each lobe" (McD). (See fig. 100, after McDunnough). We have seen no specimens of this species.

Rhithrogena impersonata McDunnough

1925—McDunnough—Canad. Ent. 57: 191 (Heptagenia). .... Que.
1926—McDunnough—Canad. Ent. 58: 195 (transferred to Rhithrogena).
1933—Traver—J. N. Y. Ent. Soc. 41: 116.

Length: body 9; wing 10 mm.

A species of the *jejuna-undulata* group, having no lateral spine basally on penes; small apical spines present on each division of penes.

Head and thorax deep brown; thoracic sternum paler, rather ochreous brown. Legs ochreous brown; femora "broadly shaded in their central portion with deeper brown, especially noticeable on the forelegs" (McD.). Wings hyaline; a slight brownish tinge at base. Venation strong, blackish. Abdominal tergites deep brown; sternites paler, ochreous brown. Forceps and tails dark brown. Penes with no lateral basal spines on outer margin; apices "broader and much less outcurved" (McD.) than in the allied *R. jejuna*, each division bearing several small apical spines (see fig. 101).

Distinguished from *R. jejuna* and *R. undulata* by the small apical spines on penes, and by the fact that the penes are barely or not at all outcurved at tips. We have a specimen of this species from Lachine, Que. (G. S. Walley), which Dr. McDunnough has kindly presented to the Cornell collection.

Rhithrogena jejuna Eaton

1853—Walker—List of Neurop. Ins. Brit. Mus. pt. 3; p. 568 (*Baetis fusca*). .... Ont.
1871—Eaton—Trans. Ent. Soc. London, p. 138. Fig. (*Heptagenia fusca*).
1885—Eaton—Revis. Monogr. p. 252 (re-named; transferred to Rhithrogena).
1924—McDunnough—Canad. Ent. 56: 117 (Heptagenia).
1925—McDunnough—Canad. Ent. 57: 191 (in discussion of *impersonata*).
1926—McDunnough—Canad. Ent. 58: 195.
1933—Traver—J. N. Y. Ent. Soc. 41: 116.

Length: body 8–9; wing 9–10 mm.

A species of the *jejuna-undulata* group, having no lateral spines near base of penes; tips of penes narrowed, distinctly outcurved.

Thoracic notum dull reddish black; pleura and sternum very similar. Legs sooty brown; fore tibia blackish at apex; all femora with a black streak longitudinally, which commences near base and is tapered at each end. Wings hyaline; membrane very faintly tinted with greyish, not noticeably darker in stigmatic area. Venation pitch-brown. About 3 basal costal cross veins and 19 beyond the bulla, many of which are anastomosed. Abdominal segments 2–5 "translucent burnt-umber brown" (Eaton); posterior margins of tergites darker; sternites slightly paler than tergites. Apical segments very similar in color, somewhat opaque. Forceps brown; tails reddish brown. Genitalia as in fig. 102.

This species is allied to *R. undulata* and *R. impersonata* in genitalic type. The lack of small apical spines and the outcurved tips of the penes separate this species from *R. impersonata*; the penes are likewise distinctly narrowed at the tips, by which character it may be distinguished from *R. undulata*. We have seen a specimen of this species, from the Museum of Comparative Zoology.

Rhithrogena morrisoni Banks

1924—Banks—Bull. Mus. Comp. Zool. 65: 424. Fig. (Epeorus)..............................Nev.
1925—McDunnough—Canad. Ent. 57: 191 (Heptagenia; in discussion of *impersonata*).
1926—McDunnough—Canad. Ent. 58: 195–6 (in discussion of Rhithrogena).
1933—McDunnough—Canad. Ent. 65: 280 (in discussion of *gaspeensis*).
1933—Traver—J. N. Y. Ent. Soc. 41: 116.
1934—McDunnough—Canad. Ent. 66: 188..Alta.

Length: body 9; wing 10½ mm.

A species of the *brunnea* group; humeral cross vein blackish; penes outcurved at tip.

Head and thorax dark brown. Pleura largely dark, but with several yellow spots. Fore legs brown; tips of tibiae darker. Middle and hind legs paler, femora with dark brown streaks. Wings hyaline; venation fine, brown. Humeral cross vein deep blackish. Stigmatic area tinted with smoky; cross veins mostly anastomosed or forking. Abdominal tergites dull brown, apical ones darker; posterior margins darker, lateral margins pale yellowish. Sternites dull yellowish, immaculate. Forceps dark brown. Tails dull yellowish to brownish, joinings not noticeably darker. Penes strongly divergent apically, narrowed toward the tip; a small but distinct ventral spine on each lobe, and 5 to 6 small spines ventrally near apex (see fig. 100, after McDunnough). Basal joint of forceps slightly swollen on inner margin.

Allied to *R. brunnea*, but with weaker mid-ventral spine on the penes, and stronger apical spines. We have seen no specimens of this species.

Rhithrogena petulans Seemann

1927—Seemann—J. Ent. and Zool. 19: 42 (Iron). Fig. (imago, *nec* nymph).................. Calif.
1933—Traver—J. N. Y. Ent. Soc. (transferred to Rhithrogena).
1934—McDunnough—Canad. Ent. 66: 185–6. Fig.

Length: body 10; wing 11 mm.

A species closely allied to *R. brunnea*; penes narrowed at apex, a small median ventral spine on each.

Frontal portion of head deep smoky grey; red-brown shading at bases of ocelli; vertex dark red-brown. Greyish streak across face in antennal region; antennae dark red-brown, tips paler. Pronotum dark olive-brown; darker brown shading along posterior margin. Mesonotum light red-brown with an olive tinge; median streak darker than area on each side. Pale spot anterior to scutellum; dark area on each side of this, from which a dark line extends forward above wing roots. Scutellum and all posterior portions of mesonotum, and posterior half of metanotum, blackish brown. Purplish streak anterior to wing roots. Intersegmental areas of pleura paler, yellowish brown. Narrow dark lines above leg bases. Sternum dark red-brown; anterior portion of mesosternum rather paler. Fore legs dark red-brown; middle and hind legs yellowish olive to yellow-brown. Black longitudinal streak, tapered at each end, on basal half of each femur. Coxae olive brown. Black spot at apex of each trochanter; apices of middle and hind femora, and tarsal joinings of these legs, deep brown; tarsi smoky brown. Wings hyaline; stigmatic area distinctly brown-tinged. Venation dark red-brown; longitudinal veins, especially in anterior half of fore wing, somewhat heavier than cross veins. Humeral cross vein deep purple-black. Basal costal cross veins well defined; about 6 before bulla, those in stigmatic area anastomosed as usual. Slight tendency for cross veins in 3rd and 4th spaces below bulla to be somewhat closer together than usual.

Abdominal segments 2–7 semi-hyaline, light purplish brown with reddish tinge; tergites darker than sternites. Intersegmental areas and lateral margins narrowly paler; posterior margins of tergites narrowly darker dorsally. Traces of a narrow pale mid-dorsal line, short submedian dashes near center of each tergite, and lateral faintly defined wedge-shaped areas based on anterior margin. Segments 8–10 opaque, brighter red-brown, with distinct yellowish tinge. Basal portion of forceps base pale; forceps deep brown, penes paler. Penes only slightly outcurved at apex; small apical spines rather well developed; small ventral spine near middle of each lobe; penes narrowed toward apex. (See fig. 100). Tails deep smoky brown; basal joinings darker, others somewhat opaque.

Specimens of this species in the Cornell collection taken by Mrs. T. M. Seemann in Cucumonga Canyon, Calif., March 29, 1923; others were taken by Mr. C. D. Michener in Icehouse Canyon, San Gabriel Mts., on Apr. 15, 1933. Description drawn largely from the latter specimens. Several others, taken by Prof. J. C. Bradley in the Santa Cruz Mts., Calif., in May, 1907, are apparently of this species.

Rhithrogena robusta Dodds

1923—Dodds—Trans. Amer. Ent. Soc. 49: 103. Fig. .... Colo.
1926—McDunnough—Canad. Ent. 58: 195.
1929—McDunnough—Canad. Ent. 61: 180. .... Alta.
1933—McDunnough—Canad. Ent. 65: 281 (in discussion of *R. brunneotincta*).
1933—Traver—J. N. Y. Ent. Soc. 41: 116.

Length: body 12; wing 14 mm.

A large dark brown species, allied to *R. brunneotincta*; dark mid-ventral ganglionic marks.

Head and thorax deep black-brown. A paler brown area anterior to wing roots; intersegmental areas of pleura pale brown. Fore leg dark brown; middle and hind legs slightly paler brown. No distinct dark streaks or dots on femora. Apices of femora, claws, tarsal joinings of middle and hind legs darker brown. Wings hyaline; stigmatic area distinctly brown-tinged, mainly in costal space, venation deep brown; longitudinal veins heavier than cross veins. Humeral cross veins light red-brown, pale at costal margin. Basal costal and subcostal cross veins quite pale and faint. Stigmatic cross veins of costa anastomosed as usual.

Abdominal segments 2–7 semi-hyaline; lateral and posterior margins of segments, and intersegmental areas, pale yellowish brown. Remainder of each tergite dark red-brown; of each sternite paler red-brown. Narrow paler mid-dorsal line and pale hyaline wedge-shaped submedian streaks near center of each tergite. Ganglionic areas of apical sternites blackish; these areas outlined in dark brown on basal and middle sternites. Segments 8–10 opaque, darker; almost black-brown. Genitalia dark red-brown. Penes widely divergent apically, tips roundly truncate. Three spines on apical portion; two on inner margin, of which the smaller is directed upward and the larger, its tip slightly curved, is directed inward; the one nearer the outer margin is short and stout. A long spine-like lateral process, its tip truncate, slightly curved outward, on outer margin of each side, above base (see fig. 101).

We have specimens of this species in the Cornell collection from the original material collected by Dr. Dodds. Other specimens are from Maligne Lake, Alta., taken by Prof. J. C. Bradley, July 4, 1915. The above description is drawn largely from the latter specimens. The brown-tinted wings of *R. brunneotincta*, as well as structural details of the penes, distinguish that species from *R. robusta*, to which it is closely allied.

Rhithrogena uhari Traver

1933—Traver—J. Elisha Mitchell Sci. Soc. 48: 165. . . . . N. C.

Length (female): body 6; wing 7 mm.

A small pale species; male imago not known, hence its relation to other species is doubtful.

Female imago pale yellowish. A black mark below antenna; faint brownish markings above frontal margin; base of antenna white; basal portion of filament dark, remainder light brown. Thoracic notum yellowish; metanotum yellow-brown, darker posteriorly. Small dark spot near center of lateral margin of pronotum. Faint purplish shading near wing bases; brown shading on each side of scutellum. Pleura paler than notum; purplish marks below wing roots; faint dark lines above bases of middle and hind legs. Sternum pale, mesosternum yellow. Legs pale whitish. Small round black spot at center of each femur. Tarsi smoky, with black spot dorsally at each joining. Wings semi-hyaline whitish, with faint milky tinge. Opaque cloud in stigmatic area. Venation whitish. Abdominal tergites light yellowish brown; 1–5 with purplish posterior margins; 6–10 more yellow, unbanded. Pleural fold whitish. Sternites pale yellow, immaculate. Tails pale yellow, joinings opaque.

Female very similar to *R. exilis* and *R. fuscifrons*; abdominal tergites rather darker than in the former, general color paler than the latter. The type material, from the Uharie Mts. in North Carolina, is in the Cornell collection. We have seen no other specimens.

Rhithrogena undulata Banks

1924—Banks—Bull. Mus. Comp. Zool. 65: 425. Fig. (Epeorus). . . . . N. M.
1925—McDunnough—Canad. Ent. 57: 191 (Heptagenia; discussion of *impersonata*).
1926—McDunnough—Canad. Ent. 58: 195 (transferred to Rhithrogena).
1933—Traver—J. N. Y. Ent. Soc. 41: 116.

Length: body 7½; wing 9 mm.

A species of the *jejuna-undulata* group, having no basal lateral spines on penes; tips of penes broad, distinctly outcurved.

Head and thorax dark brown; several yellow spots on pleura. Fore legs brownish; apex of tibia darker. Middle and hind legs pale; dark streaks on femora. Wings hyaline; venation brown. Stigmatic area tinged with smoky; cross veins in this region mostly simple. Abdominal tergites dark brown; lateral margins very narrowly pale. Sternites very pale. Forceps dark brown, tails paler brown. Apices of penes broad, distinctly outcurved; no basal lateral spine, nor small spines near apex (see fig. 102).

Allied to *R. jejuna* and *R. impersonata*. Tips of penes outcurved as in *R. jejuna*, but not narrowed as in that species; no small apical spines such as occur in *R. impersonata*. We have specimens of this species in the Cornell collection from Spring Creek, Wyo. (Mrs. E. B. Klots), and from Alsea, Ore. (H. A. Scullen); we have seen specimens from Greer, Ariz. (C. T. Vorhies), in the collection of the University of Arizona.

Rhithrogena virilis McDunnough

1928—McDunnough—Canad. Ent. 60: 9 (*brunnea*)..Alta.
1934—McDunnough—Canad. Ent. 66: 187. Fig..B. C.

Length: body 12; wing 15 mm.

A species of the *brunnea* group; distinct by reason of a small sharp spine on latero-apical margin of each lobe of penes.

Head brown; shaded with ochreous at bases of antennae; anterior portion largely smoky. Thoracic notum brown; scutellum and areas near it light yellowish brown. Ochreous shading on pleura, most distinct anterior to wing roots; a dark streak from wing base to pronotum. Median portion of sternum tinged with smoky. Fore legs deep smoky; middle and hind legs lighter brown; all femora with a longitudinal dark streak. Wings hyaline; costa faintly amber-tinged. Longitudinal and cross veins fine, dark brown; cross veins in stigmatic area anastomosed as usual. Abdominal segments brown, "irregularly paler along lateral edge and on posterior segments ventrally" (McD). Forceps and tails deep smoky. At the lateral margin of each lobe of the penes, at apex, is a small sharp spine, reminiscent of Heptagenia. (See fig 101).

A large species, distinct from others of the group as indicated above. We have seen no specimens.

Genus **Cinygmula** McDunnough

1885—Eaton—Revis. Monogr. p. 247 (Cinygma, in part).
1933—McDunnough—Canad. Ent. 65: 75. Figs. nymph.

Eyes of male imago moderately large, ovoid; not quite contiguous apically, but separated by a space less than 1/3 the shorter diameter of eye. Posterior margin of head of female slightly emarginate. Fore leg of male longer than body; of female, about as long as body. Fore femur of male about 3/4 as long as tibia, which is 2/3 to 3/4 the length of tarsus. Basal fore tarsal joint more than 1/2, usually 2/3 to 3/4 the length of the second. Basal joint of hind tarsus in both sexes slightly longer than the second; distal joint long, equal to first and second combined. Fore claws of male blunt, similar in *C. atlantica*, *C. confusa*, *C. mimus*, *C. ramaleyi* and *C. tarda*; but dissimilar in *C. hyalina* and *C. par*. Claws of middle and hind legs of male, and of all legs of female, dissimilar. Typical Heptagenine venation. In some species, wing membrane distinctly tinged with grey, yellow or amber; in others, wholly pale. Cross veins of stigmatic area usually not anastomosed; no distinct dividing line parallel to costal margin, as in Cinygma.

Forceps four-jointed, second joint longest; two apical joints together may equal the second; distal joint usually slightly shorter than third. Apical margin of forceps base may be slightly emarginate (*atlantica*, *ramaleyi*, *mimus*), or with a slight rounded projection at median area; a blunt projection on each side at inner margin of forceps. Penes separated almost to base; rod-like. A pair of short spines, often slightly outcurved, is usually present on inner margin near apex; often also a single spine is located laterally on each side, on lower outer margin. Subanal plate of female well developed, its posterior portion cleft at middle of apical margin by a V-shaped incision.

Frontal margin of head of nymph emarginate at median area, exposing the mouthparts when viewed from above. "Mouthparts distinctly reminiscent of those of Rhithrogena, notably the mandibles and the maxillary palpus" (McD.). Femora somewhat depressed. Claws with 3 to 4 pectinations. Gills present on segments 1–7; "pale, with only slight, indistinct tracheation and with the basal filaments *reduced to one or two* on the anterior gills; gill of segment 1 *largest,* heart-shaped, the ventral section tending to be slightly enlarged but not nearly to the extent found in Rhithrogena; remaining gills ovate, with that of segment 6 smallest and oblongovate" (McD.). Postero-lateral margins of abdominal segments only slightly produced. Tails three, as in Rhithrogena.

Genotype—*Cinygmula ramaleyi* (Ecdyurus) Dodds.

Distribution—Nearctic; western and northeastern North America.

KEY TO THE SPECIES OF CINYGMULA

Male Adults*

1—Wings distinctly tinged with yellow or amber. . . . . .2.
—Wings not tinged with yellow or amber; they may be faintly grey-tinted. . . . .7.
2—Spines present laterally on lower outer margin of penes, as well as subapically on inner margin. . . . . **confusa**
—Spines present subapically on inner margin of penes only. . . . . .3.
3—Two lobes at apex of each division of penes, the larger directed laterally; amber tinge quite evenly distributed over wing membrane. . . . . **ramaleyi**
—A single well-defined lobe on apex of each division of penes; amber tinge on wing membrane variable in extent. . . . . .4.
4—Cross veins of both wings, or those of costal half of fore wing only, lightly margined with smoky or brownish. . . . . .5.
—No cross veins margined in either wing. . . . . .6.
5—All cross veins of both wings faintly but broadly margined with pale smoky; no small ventral spine near apex of each lobe of penes; wings entirely amber-tinged. . . . . **reticulata**
—Cross veins of costal half of fore wing only, slightly margined with brown; small ventral subapical spine on each lobe of penes; wings tinged with amber at base and along costal margin of both wings. . . . . **gartrelli**
6—Wings uniformly suffused with light amber; a small ventral subapical spine on each lobe of penes. . . . . **uniformis**
—Amber tinge on wings distinctly deeper near base; no such ventral subapical spine on each division of penes. . . . . **mimus**
7—Wings faintly tinted with pale grey, distinctly brownish at extreme base; abdominal tergites 2–7 largely light brown. . . . . **par**

*Does not include *subaequalis.*

—Wings wholly pale, not as above; abdominal tergites 2–7 largely pale with darker markings..8.

8—Dark mid-ventral markings on abdominal sternites, at ganglionic areas.......9.

—No such dark mid-ventral markings.................................**atlantica**

9—No lateral spines on lower outer margin of penes; wing 7½ mm. in length..**tarda**

—Lateral spines present on lower outer margin of penes; wing 10 mm. in length...**hyalina**

VERIFICATION TABLE

Male Adults

Species	*Wing length*	*Wing tinted*	*Cr. veins margined*	*Ventral med. mks.*	*Distr.*
atlantica	8	No	No	None	E
confusa	10	Yes	No	Present	Alta.
gartrelli	10	Yes	Yes	Present	B. C.
hyalina	10	No	No	Present?	Alta.
mimus	10	Yes	No	None	W
par	9–10	Yes	No	Present	W
ramaleyi	10	Yes	No	Present	W
reticulata	9	Yes	Yes	Traces	Ore.
subaequalis	8	No	No	None	N. C.
tarda	7½	No	No	Present	Alta.
uniformis	10	Yes	No	Present	B. C.

KEY TO THE SPECIES

Nymphs

1—Small nymphs, 7 mm. in length; middle abdominal sternites largely brown, with pale lateral margins and submedian streaks; apical tergites not paler than preceding ones..**atlantica**

—Larger nymphs, 9–11 mm. in length; abdominal sternites pale, may have darker markings; apical tergites often distinctly paler than preceding ones.........2

2—Apical abdominal tergites white; femora brown, with a pale L-shaped longitudinal mark on basal half.....................................**ramaleyi**

—Apical abdominal tergites may be paler than preceding ones, but each has a narrow dark anterior margin; femora ochreous, with two broad brown bands...**uniformis**

Cinygmula atlantica McDunnough

1924—McDunnough—Canad. Ent. 56: 131. Fig. (Cinygma)................................N. S.
1933—Traver—J. N. Y. Ent. Soc. 41: 121 (Cinygma).
1933—McDunnough—Canad. Ent. 65: 75 (transferred to Cinygmula).

Length: body 7; wing 8 mm.

A rather small pale species; wings not tinted; spines on lower outer margin of penes.

Head yellowish brown; dark brown between eyes. Thorax light brown; lateral areas of pronotum, and postero-lateral areas of meso- and metanota, shaded with darker brown. Fore legs smoky brown. Middle and hind legs yellowish brown. Wings hyaline, not tinted; all veins fine, brown. Abdominal tergites 2–6 semi-hyaline, pale, with brown shading laterally. Tergites 2 and 3 pale in central median area only; tergites 4 to 6 largely pale, the brown shading restricted to postero-lateral areas. Submedian pale dashes, partially outlined in brown, are faintly indicated. Sternites pale ochreous, 2–6 semi-hyaline, 7–9 opaque. Tergites 7–10 opaque, almost wholly brown. Tails pale brown; penes smoky brown. A spine is present on lower outer margin of penes on each side, but none in inner margin (see fig. 103).

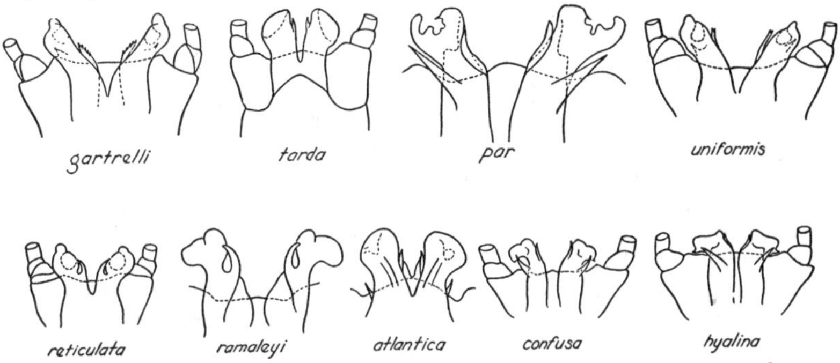

FIG. 103. Male genitalia of nine species of Cinygmula (after McDunnough, except *par*, *ramaleyi*, and *atlantica*).

This species is allied to *C. par*, by the structure of the genitalia, but it lacks the inner spine on the penes found in that species. The paler abdominal tergites and lack of mid-ventral dark markings in *C. atlantica* serve to distinguish it from *C. par*. We have reared specimens of this species in the Cornell collection from the vicinity of Ithaca, N. Y.

Cinygmula confusa McDunnough

1924—McDunnough—Canad. Ent. 56: 131. Fig. .... Alta.
1933—Traver—J. N. Y. Ent. Soc. 41: 121 (Cinygma).
1933—McDunnough—Canad. Ent. 65: 75 (transferred to Cinygmula).

Length: body 9; wing 10 mm.

A rather large species with amber-tinged wings; spines on inner and lower outer margins of penes; purplish mid-ventral patches on hyaline segments.

Head deep red-brown. Thorax light olive brown, the central portion of notum shaded with darker brown; purplish tinges on prothorax, and on pleura below the wings. Coxae of all legs light brown. Fore femur deep purplish, remainder of fore leg blackish; middle and hind legs light purplish brown. Wings "evenly suffused with deep amber color, with dark, clean cut venation" (McD.). Abdominal segments 2–7 hyaline. Posterior half of each tergite purplish brown; indistinct pale submedian streaks outlined in brown are most evident on tergites 2 and 3; on these tergites also, the brown band is wider than on the following segments. Sternites 2–7 with mid-ventral purplish patches, one on each segment. Segments 8–10 opaque, light brown; tergites with slight purplish shading; sternites somewhat paler than tergites. Forceps smoky brown basally; tails deep blackish brown. A spine on inner margin, another on lower outer margin of each side of penes (see fig. 103).

This species is allied to *C. hyalina* and *C. par*. The deep amber tinge of the wings distinguishes it from both of these species. The purplish tinge of the fore femur likewise separates it from *C. hyalina*; the fore leg of *C. par* is much lighter in color. We have seen no specimens of this species.

Cinygmula gartrelli McDunnough

1934—McDunnough—Canad. Ent. 66: 182. Fig. . . . . B. C.

Length: body 9; wing 10 mm.

A species of the *mimus-ramaleyi* group; wings amber-tinged at base, cross veins in costal half of fore wing lightly bordered with brown.

Eyes deep black-brown, not contiguous. Anterior portion of head sepia brown; antennal pits paler; blackish shading on each side of median carina. Vertex of head deep blackish. Thoracic notum sepia-brown; posterior portion of mesonotum blackish brown, "and with lighter shades on anterior projection of mesonotum and faintly cephalad of scutellum; pleura brown, with the membranous portions paler, a wine-colored patch anterior to base of fore wing from which a pale ochreous dash projects towards the pronotum; there are also faint ruddy tinges at bases of legs; sternum brown, shaded with black-brown" (McD.). Fore legs deep smoky brown; tibia slightly longer than femur and trochanter together. Middle and hind legs paler brown, femora with faint reddish tinge. Wings hyaline; tinged at base with brownish amber, "extending more faintly along costal margin of both wings; veins fine, brown; cross veins strong, brown, lightly bordered in the costal half of fore wing with brown shading which gives them a decidedly thickened appearance in this region" (McD).

Abdominal tergite 1 deep brown. Anterior half of tergites 2–7 pale, semi-hyaline yellowish brown; posterior portion of each deeper brown faintly tinged with reddish, the segments thus appearing banded. Tergites 8–10 opaque, rather evenly dark brown; pale lateral patches on 8 and 9. Sternites light yellowish brown; posterior margins, and often also the three apical segments, tinged faintly with reddish. "Ganglionic marks very faintly visible" (McD). Tails and forceps deep smoky. Penes quite similar to those of *C. mimus*, but with a small spine on the ventral surface of each division, near apex; median apical spines likewise slightly serrate on outer margin. (See fig. 103).

Differs from *C. mimus* as indicated above; allied also to *C. uniformis* which has the entire wings amber-tinged and no thickening of cross veins. Small spines near apex of penes are also present on this latter species. We have seen no specimens of this species.

Cinygmula hyalina McDunnough

1924—McDunnough—Canad. Ent. 56: 131. Fig. (Cinygma). . . . . Alta.
1933—Traver—J. N. Y. Ent. Soc. 41: 121 (Cinygma).
1933—McDunnough—Canad. Ent. 65: 75 (transferred to Cinygmula).

Length: body 9?; wing 10 mm.?

A species with untinted wings; spines on inner and lower outer margins of penes.

The description of this species is largely comparative, with only points of distinction between this and *C. confusa* being indicated. Said to be very similar to *confusa* in general appearance, but with hyaline wings, tinted very slightly in the stigmatic area only. Thorax somewhat deeper in color than in *confusa*; legs light to deep brown, but lacking the purple tinge of that species. Tergites 2 and 3 almost entirely brown, with faint traces only of the pale submedian streaks. Posterior margins of tergites brown, with less purplish tinges, and the "definition between the pale and dark areas is less distinct" (McD.). Sternites suffused with whitish, appearing more opaque than in *confusa*. Penes very similar in structure, but the basal spine on the lower outer margin is relatively longer and straighter, and the divisions of the penes relatively stouter, than in *C. confusa*. (See fig. 103).

This species is allied also to *C. par*, from which it may be distinguished by the paler wings and less extensive brown shading on abdominal tergites. We have seen no specimens of this species.

Cinygmula mimus Eaton

1885—Eaton—Revis. Monogr.; p. 249 (Cinygma)..Colo.
1924—McDunnough—Canad. Ent. 56: 131. Fig. (Cinygma)................................Alta.
1926—McDunnough—Canad. Ent. 58: 302 (in discussion of *Cinygma integrum*).
1927—Needham—Canad. Ent. 59: 133. Figs. (Rhithrogena, imago, *nec* nymph)............... Utah
1933—Traver—J. N. Y. Ent. Soc. 41: 121–2 (Cinygma).
1933—McDunnough—Canada. Ent. 65: 75 (transferred to Cinygmula).
1934—McDunnough—Canad. Ent. 66: 183–4 (in discussion of *gartrelli* and *reticulata*).

Length: body 9; wing 10 mm.

A rather large species, the wings tinged with yellowish brown; spine on inner margin of each side of penes only.

Head reddish to purplish brown. Thoracic notum light chestnut brown; remainder of thorax yellowish brown with reddish tinges; paler yellowish or creamy areas anterior to wing root, and below bases of wings on pleura. Legs light yellowish brown; tibio-tarsal joint and tarsal joinings of fore leg darker brown; claws of all legs brownish; tarsal joinings of middle and hind legs narrowly brown, distal tarsal joints tinged with light brown. Wings hyaline, the membrane tinged with yellowish brown, especially evident in hind wing and basal half of fore wing. Venation light yellowish brown, the cross veins often faintly red-brown. Abdominal segments 2–7 semi-hyaline, pale purplish to light madder brown; extreme anterior margin colorless, posterior margins narrowly darker. Sternites very similar to tergites, but somewhat paler; no dark mid-ventral markings. Segments 8–10 opaque, yellowish brown with a tinge of purple or madder; sternites more yellowish than tergites. Tails yellowish brown, often with faint red-brown tinge in basal portion; joinings opaque, faintly darker. Forceps light red-brown, penes yellow-brown; penes divergent almost from base, bearing on each side a single spine on inner margin. (See fig. 103, *gartrelli*).

The yellowish brown tinge of the wings, the absence of dark ventral markings and the details of structure of the penes, when taken together, distinguish this from other allied species. We have specimens of this species from Logan River, Utah (J. G. Needham); Moose, Wyo. (Mrs. E. B. Klots); Glacier National Park, Mont. Hazzard); and Lake Tahoe, Calif. (E. H. Nast).

Cinygmula par Eaton

1885—Eaton—Revis. Monogr. p. 249. Fig. (Cinygma)....................................Ariz.
1923—Dodds—Trans. Amer. Ent. Soc. 49: 106 (*Cinygma mimus*)..........................Colo.
1924—McDunnough—Canad. Ent. 56: 130, 132 (in discussion of *Cinygma confusa*).
1926—McDunnough—Canad. Ent. 58: 302 (in discussion of *C. integrum*).
1933—McDunnough—Canad. Ent. 65: 75 (transferred to Cinygmula).
1933—Traver—J. N. Y. Ent. Soc. 41: 121 (in Cinygma).

Length: 8; wing 9–10 mm.

A medium-sized species, wings tinged with light grey, darker at bases; dark mid-ventral ganglionic marks.

Thoracic notum light pitch-brown, "sometimes approaching light luteo-piceous anteriorly at the sides" (Eaton). Legs sooty to golden brown; base of fore femur and trochanter somewhat paler, base of first fore tarsal joint darker. Wings hyaline, "tinted with light bistre-grey throughout, but with a stronger tint than elsewhere" at the wing roots and in the stigmatic area; near wing roots the color becomes rather dark brown. Venation yellowish brown.

Abdominal segments 2–7 semi-hyaline; tergites largely light brown, the anterior margins, antero-lateral areas, narrow median line and submedian streaks being hyaline, "dirty-whitish." The three pale longitudinal streaks are outlined in darker color. Sternites paler, the ganglionic areas on 3–7 conspicuously marked with sepia-brown. Segments 8–10 opaque; tergites "Vandyke-brown," sternites "very light brownish ochre" (Eaton). Tails brown. A large spine on lower outer margin of penes on each side, a smaller spine on inner margin (see fig. 103).

This species is allied to both *C. confusa* and *hyalina*, also to the eastern *C. atlantica*. Distinguishing features are indicated under each of these species. We have a few of Dr. Dodds' specimens from Colorado, which evidently belong here; these are all imperfect, but the genitalia are as figured by Dr. McDunnough.

Cinygmula ramaleyi Dodds

1923—Dodds—Trans. Amer. Ent. Soc. 49: 101. Figs. (Ecdyonurus).......................... Colo.
1923—Dodds—Trans. Amer. Ent. Soc. 49: 109. Fig. (*Iron tollandi*).
1924—Dodds and Hisaw—Ecology 5: 145. Fig. nymph (Ecdyonurus).
1924—McDunnough—Canad. Ent. 56: 132. Fig. (Cinygma).................................. Alta.
1928—McDunnough—Canad. Ent. 60: 9, 239 (Cinygma) (*tollandi* = *ramaleyi*).
1929—McDunnough—Canad. Ent. 61: 178 (in discussion of *Cinygma tarda*).
1933—Traver—J. N. Y. Ent. Soc. 41: 120–1 (Cinygma).
1933—McDunnough—Canad. Ent. 65: 75 (transferred to Cinygmula). Figs. nymph parts.

Length: body 8; wing 10 mm.

A medium-sized species with amber-tinged wings; spines on inner margin of penes only.

Head purplish brown. Thorax rather bright red-brown; mesonotum yellowish brown, median stripe darker red-brown. Postero-lateral areas of mesonotum on each side of and posterior to scutellum, and lateral areas of metanotum, purplish brown. Narrow purplish markings above leg bases. Median area of mesosternum darker than pleura. Legs reddish brown; fore leg tinged with smoky. All femora with distinct reddish tinge; apex of fore tibia blackish; femoro-tibial joinings purplish brown, tarsal joinings narrowly dark brown, on all legs. Wings hyaline, distinctly amber-tinged throughout. Veins rather fine, reddish brown.

Abdominal segments 2–7 and basal portion of 8, semi-hyaline; hyaline areas of tergites light red-brown, the posterior half or two-thirds of each purplish brown. Tergite 2 almost wholly dark except for narrow hyaline anterior margin; ⅔ of tergite 3, and about ½ of tergites 4–7, dark-shaded, the dark color often concentrated in postero-lateral areas. Pale submedian streaks outlined in darker color are faintly indicated on basal tergites. Sternites greyish with tinge of pale brown; no distinct darker median markings on 2–6. Segments 9 and 10, and apical portion of 8, opaque; tergites red-brown, sternites paler. Ganglionic areas of sternites 7–9 opaque, pale red-brown. Forceps deep olive or smoky brown; penes light red-brown; a curved spine on inner margin of each side of penes (see fig. 103). Tails deep smoky brown.

Dr. Dodds' description gives no color characters; the above account is drawn from specimens taken by Prof. R. T. Young in Glacier National Park, Mont., in August, 1932. The species is allied to *C. mimus*, but differs in details of genitalic structure and in the more evenly distributed amber tinge of wings. It is larger than *C. tarda*, which species also lacks the tinted wings and differs in structure of the genitalia. In addition to the material mentioned above, we have some of Dr. Dodds' original specimens in the Cornell collection, which are too bleached to be of use in determining color characters.

Cinygmula reticulata McDunnough

1934—McDunnough—Canad. Ent. 66: 184. Fig. . . . . Ore.

Length: body 8; wing 9 mm.

A species of the *mimus-ramaleyi* group; wings suffused with pale amber, cross veins bordered with pale smoky.

Head and thorax brown. Lighter brown shading on mesonotum, anterior to scutellum; "pleura streaked with light ochreous; a faint ruddy patch anterior to base of fore wing" (McD). Legs pale brown; faint reddish tinges on all femora. "Wings entirely suffused with pale amber, veins and cross veins brown, the latter broadly but rather faintly bordered with pale smoky, giving a very characteristic checkered appearance to the entire wings, a feature not at all present in *C. mimus*" (McD). Anterior half of abdominal tergites 2–7 hyaline, posterior half "irregularly purple-brown (deeper in color than in *C. mimus*)" (McD). Tergites 8–10 opaque, purple-brown, lateral edges shaded with lighter brown, as is posterior margin of 10. Sternites much like tergites, but paler; forceps base light yellow-brown; ganglionic marks indicated as faint traces. Tails and forceps light smoky. Penes almost identical with those of *C. mimus* Etn., but the median apical spines are slightly larger. See fig. 103.

Distinguished from *C. mimus*, its nearest ally, by the deeper color of tergites, complete suffusion of wings with amber tinge, and the reticulated appearance of the wings. No small ventral spine near apex of penes, as in *C. uniformis* and *C. gartrelli*. We have a single specimen from Corvallis, Ore. (H. A. Scullen), which probably belongs here.

Cinygmula subaequalis Banks

1914—Banks—Proc. Acad. Nat. Sci. Philad. 66: 615 (Heptagenia). . . . . N. C.
1933—Traver—J. Elisha Mitchell Sci. Soc. 48: 176 (Stenonema).

Length: body ?; wing 8 mm.

Head reddish yellow; "a shining black ring at base of each ocellus, antennae pale" (Bks.). Thorax rather reddish yellow, posterior portion of notum darker. Legs pale, "a black dot at tip of femur and tibia I" (Bks.). Wings hyaline; venation mostly brown, no cross veins dark-margined; stigmatic area brownish yellow. "About five costals before bulla, fourteen beyond, a few near the middle are crossed" (Bks.). Hind wings not dark-margined. Basal joint of fore tarsus of male fully two-thirds as long as second; third equal to second; fourth "scarcely longer than first, the fifth not one-third of first." (Bks.). Abdomen pale; "segments dark on apical one-third or one-half above, last three segments brownish" (Bks.) Forceps pale. Tails faintly dark.

A sketch of the genitalia of the type specimen bears a remote similarity to the genitalia of *Cinygmula altantica*, which leads us to place the species tentatively in this genus. There is a possibility that this species and *C. atlantica* may be synonymous.

Cinygmula tarda McDunnough

1929—McDunnough—Canad. Ent. 61: 178. Fig. (Cinygma). . . . . Alta., Mont.
1933—Traver—J. N. Y. Ent. Soc. 41: 121 (in Cinygma).
1933—McDunnough—Canad. Ent. 65: 75 (transferred to Cinygmula).

Length: body 7; wing 7½ mm.

A small species with untinted wings; a spine on inner margin of penes only; dark mid-ventral ganglionic marks.

Head light wood-brown; bases of antennae ochreous; eyes dull greenish in living insect, almost contiguous. Often slight reddish or purplish shading next to the eyes. Thorax light brown; yellowish ochre shading on mesonotum anterior to scutellum, and on the anterior portions of the metanotum. Legs pale brown. Wings hyaline, untinted. Cross veins "in the basal half pale, indistinct, darker and better defined in the apical area, especially the costal ones" (McD.). Abdominal segments 2–7 hyaline; tergites light brown on posterior half or third, the postero-lateral corners darker; anterior portions of each tergite pale hyaline. Sternites pale with brown oval ganglionic markings. Segments 8–10 opaque; tergites light brown, sternites deep ochreous. Penes with a short spine on inner margins only (see fig. 103).
Tails dirty white.

This species is allied to *C. mimus* and *C. ramaleyi*, by the type of genitalia. Differs from both of these in structural characters of genitalia and in the untinted wings. We have seen no specimens of this species.

Cinygmula uniformis McDunnough

1934—McDunnough—Canad. Ent. 66: 183. Fig. .... B. C.

A species of the *mimus-ramaleyi* group; wings tinted throughout with light amber; cross veins not thickened.

Head smoky brown; antennal pits pale ochreous. Thoracic notum slightly paler brown than in *C. gartrelli*, the lighter brown shading anterior to the scutellum more extended. Pleura colored much as in *C. gartrelli*. Legs light brown; fore tibia considerably longer than trochanter and femur together, and longer than fore tibia of *C. gartrelli*. Middle and hind femora faintly tinged with reddish. "Wings *unicolorous light amber* with fine brown veins and crossveins" (McD). Abdominal tergites 2–7 with narrower brown posterior margins and much wider anterior hyaline areas than in *C. gartrelli*; no reddish tinges, as in that species; apical tergites opaque, "almost unicolorous sepia-brown with a fine, blackish, posterior border to 8" (McD). Sternites very pale ochreous; 8 and 9 with light yellow-brown shading; "ganglionic marks very faint, slightly ruddy" (McD). Tails and forceps light smoky. Penes with traces of small ventral spines near apex on each side, as in *C. gartrelli*; median spines not serrate on outer margin, otherwise very similar to that species (see fig. 103).

The uniform tinting of the entire wing, the presence of the small subapical spine on the penes, and the absence of thickened cross veins distinguish this species from the allied *C. mimus* and *C. reticulata*; separated from *C. gartrelli* as indicated above. We have seen no specimens of this species.

Genus **Iron** Eaton

1883—Eaton—Revis. Monogr. Pl. 24, fig. 44.
1885—Eaton—Revis. Monogr. p. 244.
1905—Needham—Bull. N. Y. St. Mus. 86: 53.
1920—Ulmer—Stett. Ent. Zeit. 81: 142 (also Epeorus, in part.)
1924—McDunnough—Canad. Ent. 56:129.
1933—Traver—J. Elisha Mitchell Sci. Soc. 48: 156.
1933—Traver—J. N. Y. Ent. Soc. 41: 116.
1933—Spieth—J. N. Y. Ent. Soc. 41: 330.

Mayflies of small to medium size, ranging in length of wing from 7 to 11 mm. Eyes of male large, usually contiguous apically. Posterior margin of head of female very slightly emarginate. Posterior margin of pronotum somewhat excavated in median area. Fore leg of male as long as or slightly longer than body. In the type species, fore femur of male about ⅔ as long as tibia, which is about ¾ the length of tarsus. Basal joint of fore tarsus as long as or slightly longer than 2nd joint; 3rd subequal to 2nd; 4th slightly shorter than 3rd; 5th about ½ as long as 4th. Fore femur of female only slightly shorter than tibia; tarsus subequal to femur. Basal fore tarsal joint fully ¾ of 2nd joint; 3rd and 5th subequal to 1st; 4th about ½ of 2nd. In hind leg of both sexes, femur about ¾ as long as tibia; tarsus about ½ as long as femur. Basal tarsal joint subequal to or very slightly longer than 2nd; 3rd and 4th progressively shorter; 5th about equal to 1st and 2nd combined. In most species, a prominent black mark is present at middle of each femur. Claws dissimilar on all tarsi, in both sexes, except in species of the *albertae* group, in which claws of male fore tarsus are similar and blunt. Typical Heptagenine venation. Cross veins of stigmatic area not anastomosed. In many species, basal costal cross veins weak and indistinct; in species of the *albertae* and *humeralis* groups*, the humeral cross vein is thickened and black in its posterior portion. Second fork of radial sector in hind wing arises basad of first fork.

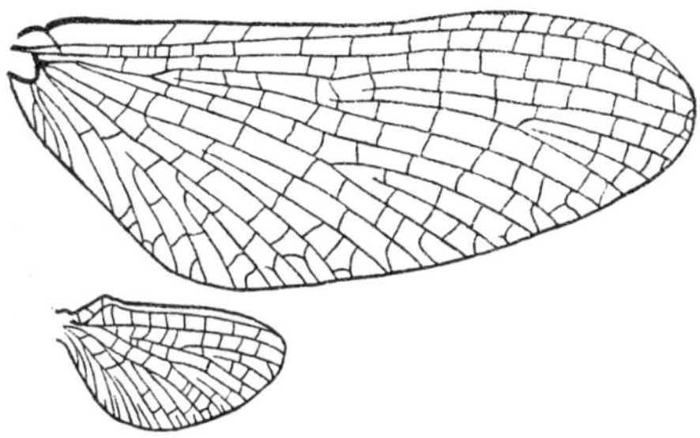

FIG. 104. Wings of *Iron fragilis* (drawing by Anna H. Morgan).

Apical margin of forceps base saliently produced, in species of the *longimanus* and *pleuralis* types, and in *I. deceptivus*; in the *humeralis* group this margin is more or less regularly undulate, while in the *albertae* group it may be slightly emarginate or somewhat produced. Forceps four-jointed, second joint longest. Fourth joint slightly shorter than third, these two combined being shorter than second joint. Penes united in basal ⅔ of their length (except in *I. deceptivus*). A pair of median spines is typically present; these are much reduced in the *humeralis* group, and wanting in *I. modestus* and in the *pleuralis* group. Long stout pick-shaped lateral spines occur in *I. pleuralis* and in *I. proprius*; shorter lateral spines are typically present in the other groups (absent in *I. modestus* and *I. deceptivus*). Apical margin of subanal plate of female emarginate in median area.

* In a previous paper (J. N. Y. Ent. Soc. 41: 118; 1933) I have used the term *nitidus* group; since *nitidus* Etn. has been transferred to another genus, I now refer to this as the *humeralis* group. The *albertae* and *pleuralis* groups are now recognized as distinct types. My reference to *pleuralis* in that paper applies to *fraudator* n. sp.; the reference to *modestus* should be disregarded, as further evidence indicates that I had mis-identified that species.—J. R. T.

Body of nymph flattened ventrally, slightly convex dorsally (see fig. 49). Frontal and lateral margins of head expanded, covering the mouthparts completely; relative extent of lateral and apical expansion and region of greatest width of head varies in different species. Labrum rather small, about 1½ to 2 times as wide as its length; antero-lateral margins rounded; a small rounded excavation at median line on anterior margin. Inner canine ½ to ¾ as long as outer one (see fig. 59). Lacinia absent, or represented by 2 or 3 short hairs; no fringe of hairs along lacinial region. A triad of stout spines, each slightly incurved and bearing at its apex a slender thread-like appendage, is present at inner apical margin of galea-lacinia. Near base of these spines, a group of 5–7 slender spines, in *longimanus* group; in *humeralis* group, these are usually absent. No row or series of spines along upper margin of maxilla, however, as in most genera of this family. Median portion of hypopharynx longer but somewhat narrower than lateral portions. Glossae of labium somewhat conical; may be slightly convergent apically. Lateral margins of pronotum slightly dilated at or anterior to middle. Leg joints fringed with hairs. Femoral flange (apical extension of femur) sharp-pointed in *humeralis* group, blunt at tip in other species examined. Claws with 3 to 6 pectinations laterally near tip.

Each abdominal segment bears a pair of spines at the postero-lateral angle. In the *humeralis* group, the outermost spine of each pair is long and slender, fully ½ as long as the length of its segment, and 3 to 4 times as long as its own width at base; the innermost spine is about ½ as long as the outer one. In the *longimanus* group, in *I. dispar* and also *I. sanctagabriel*, the outer spine is shorter and rather blunt, but not more than ¼ as long as the length of its segment, and about as long as its own width at base; inner spine about as long as outer one. Fibrillar portion of each gill appears dorsal in position; well developed. Lamellate portion thickened in a narrow band along outer margin, bearing many minute spines. Each gill overlaps the one behind it, and outer margins of all are pressed against surface to which nymph clings. Anterior lobe of lamellate portion of first gill better developed than others. In some species, these anterior lobes almost or completely meet beneath body of nymph; in other species, a long space intervenes between them. Gills of last pair usually approach one another beneath body. Tails two in number.

Genotype—*Iron longimanus* Eaton.

Distribution—Holarctic; in Nearctic, quite generally distributed.

KEY TO THE SPECIES

Male Adults*

1—Genitalia of the *albertae* type (see fig. 107); fore claws of male usually similar, blunt 2

Genitalia not as above; fore claws of male dissimilar 6

* Does not include *I. hesperus*, of which only the female is known.

2—No distinct blackish markings on thoracic pleura........................3
Thoracic pleura with distinct blackish markings........................4
3—Fore femur and tibia subequal; median spines on penes rather stout, relatively short; apical margin of forceps base somewhat excavated laterally (see fig. 107)........................**albertae**
—Fore femur shorter than tibia; median spines on penes longer and more slender; apical margin of forceps base slightly convex (see fig. 107).........**lagunitas**
4—Abdominal segments 2–7 yellowish, tergites may be faintly smoky; apical margin of the forceps base somewhat produced; venation variable...............5
—Abdominal segments 2–7 whitish, with pale brownish markings; apical margin of forceps base somewhat excavated laterally (see fig. 107); venation pale, most cross veins indistinct........................**youngi**
5—Venation brown, all cross veins distinct; penes as in fig. 107......**sancta-gabriel**
Venation pale, cross veins mostly indistinct; penes as in fig. 107.....**dulciana**
6—Genitalia of the *pleuralis* type (see fig. 106)........................7
Genitalia not as above........................8
7—Oblique lateral black streak on each tergite, cutting across postero-lateral angle; penes as in fig. 106........................**pleuralis**
—No such oblique lateral black streak on tergites........................**proprius**
8—Genitalia of the *longimanus* type (see fig. 105)........................9
Genitalia not as above........................13
9—Wing 7–8 mm. in length........................10
Wing 9–11 mm. in length........................11
10—Apical margin of forceps base regularly sinuate (see fig. 105); no oblique dark streak across postero-lateral angle of tergites........................**tenuis**
—Apical margin of forceps base extended forward more at median line than laterally; oblique black streak across postero-lateral angle of each tergite........................**fragilis**
11—Thorax and abdomen largely dark red-brown; venation dark brown..**fraudator**
Thorax and abdomen largely tinged with light red-brown; venation pale brown 12
12—Black posterior margins of tergites extend laterally to pleural fold.....**confusus**
Black posterior margins of tergites do not extend to pleural fold, but an oblique black line from this margin cuts across postero-lateral triangle to pleural fold........................**longimanus**
13—Genitalia of the *humeralis* type (see fig. 106, *punctatus*)..................14
—Genitalia not of the *humeralis* type........................18
14—Apical margin of forceps base as in fig. 105; venation light brown......**dispar**
—Apical margin of forceps base more regularly undulate, as in fig. 106; venation whitish or pale yellowish........................15
15—Wing 8–9 mm. in length........................16
—Wing 10–11 mm. in length........................17
16—Thorax blackish brown; no small lateral spines on outer margin of penes; eyes not contiguous apically........................**punctatus**
—Thorax pale yellowish brown; small lateral spines present on outer margin of penes, but not conspicuous; eyes contiguous apically...............**rubidus**
17—Thorax dull brown, abdominal tergites suffused with brown; penes longer and more widely separated apically than in following species...........**suffusus**
—Thorax yellowish, often with oblique dark markings on pleura; abdominal tergites typically much paler than above........................**humeralis**

18—Penes short and stout, spines lacking (see fig. 106); tails pale, joinings distinctly dark; femora with median and apical bands. . . . . **modestus**

—Penes longer and more slender, with long median spines (see fig. 105); tails dark, joinings not distinct; femora unmarked. . . . . **deceptivus**

VERIFICATION TABLE

Male Adults*

Species	*Wing length*	*Venation*	*Dk. mks. pleura*	*Tergites 2 to 7*	*Genital type of*	*Distr.*
albertae	11	Brown	No	Smoky	Albert.**	W
confusus	10–11	Pale br.	Few	Red-br.	Longim.	N. Y.
deceptivus	9–10	Dk. br.	No	Brown	Decept.	W
dispar	9	Lght. br.	Yes	Yellow	Humeral.	N. C.
fragilis	7	Pale	No	Yellow	Longim.	E
fraudator	10–11	Dk. br.	No	Dk. br.	Longim.	E
hesperus	12	Yellow	Yes	Pale	?	Wash.
humeralis	10–11	Whitish	Yes	Whitish	Humeral.	E
lagunitas	11	Lght. br.	No	Yellow	Albert.	Calif.
longimanus	11	Lght. br.	No	Brown	Longim.	W
modestus	7	Pale br.	No?	Pale	Modest.	Md.
pleuralis	10–11	Dk. br.	No	Dk. br.	Pleural.	N. Y.
proprius	10	Lght. br.	No	Pale br.	Pleural.	Colo.
punctatus	8	Pale	No	Whitish	Humeral.	E
rubidus	8–9	Yellow	Yes	Whitish	Humeral.	N. C.
sancta-gabriel	10½	Brown	Yes	Yellow	Albert.	Calif.
suffusus	11	Pale	No	Brown	Humeral.	E
tenuis	8	Pale	Yes	Whitish	Longim.	N. Y.
youngi	10–11	Pale br.	Yes	Whitish	Albert.	Ore.
dulciana	9½	Pale	Yes	Yellow	Albert.	B. C.

KEY TO THE SPECIES

Nymphs

1—Outer postero-lateral spines on middle abdominal segments long and slender, at least ½ the length of the respective segment, and at least three times as long as their own width at base; apical extension of femora 2 and 3 (flange) sharp-pointed at tip. 2

—Outer postero-lateral spines on middle abdominal segments shorter and rather blunt, not more than ¼ as long as the respective segment, and about as long as their own width at base; flange on femora 2 and 3 blunt at tip. 3

2†–Gills distinctly reddish; tracheae of gills not prominent, dull purplish in color; postero-lateral spines very long, slender and curved, being about 4 times as long as their width at base. . . . . **rubidus**

—Gills not red-tinged; tracheae of gills very distinct, blackish; postero-lateral spines somewhat shorter and straighter, being about 3 times as long as their width at base. . . . . **humeralis**

* Except *I. hesperus.*

** See figures under species whose names are here abbreviated.

† Nymphs of *suffusus* and *punctatus* probably belong here; these nymphs are not known to us.

3—Anterior lobes of first pair of gills not greatly prolonged, so that they lack much of meeting beneath body of nymph 4

—Anterior lobes of first pair of gills greatly prolonged, so that they meet or approximate meeting beneath body of nymph 7

4—Body of nymph small, usually 7 to 10 mm. in length 5

—Body of nymph larger, 11 to 14 mm. in length 6

5—Body of nymph 7–8 mm. in length; narrow pale median line on mesonotum, not widened at anterior margin . . . . . **tenuis**

—Body of nymph about 10 mm. in length; a wedge-shaped pale mark on median line of mesonotum, the widest portion near the anterior margin . . . . **deceptivus**

6—Distinct blackish markings on head, pronotum, and above leg bases; dorsum of abdomen deep red-brown with paler markings; sternum paler red-brown, the apical segments with dark lateral streaks . . . . . **sancta-gabriel**

—No such distinct dark markings on head and thorax; dorsum of abdomen chestnut brown with few darker markings; sternum yellowish, no dark lateral markings . . . . . **dispar**

7—Anterior lobes of first pair of gills meet beneath body of nymph 8

—Anterior lobes of first pair of gills approximate one another, but do not quite meet beneath body of nymph 9

8—Two large dark submedian spots at anterior margins of sternites 7 and 8, often also dark triangular spots on 9 near same region; two small dark dots often present near middle of sternites 7 and 8; narrow pale median line on mesonotum . . . . . **fragilis**

—Anterior margins of all sternites with a dark transverse streak on each side of median line, but not marked as above; two small dark dots near middle of sternite 9; no pale median line on mesonotum . . . . . **longimanus**

9—Two prominent dark submedian spots just anterior to middle area of each tergite; dark lateral marks on all sternites; 7–9 marked as in *fragilis*; in addition, dark submedian streaks on each sternite, as in *longimanus* . . . . . **fraudator**

—No such dark submedian spots on tergites; a lateral dark mark on each sternite, 7–9 marked as in *fragilis*, but no dark submedian streaks as in *fraudator* . . . . **confusus**

VERIFICATION TABLE

Nymphs

Species	*Body length*	*Teeth on claw*	*Femor. flange*	*Gills, 1st. pr.*	*Abd. spines*	*Distr.*
confusus	9–10	4	Blunt	Approx.	Short	N. Y.
deceptivus	10	?	Blunt	Remote	Short	W
dispar	11–13	4	Blunt	Remote	Short	N. C.
fragilis	7–8	4	Blunt	Meet	Short	E
fraudator	9–10	3–4	Blunt	Approx.	Short	E
humeralis	11	5	Sharp	Remote	Long	E
longimanus	7–10	3	Blunt	Meet	Short	W
rubidus	10	5	Sharp	Remote	Long	N. C.
sancta-gabriel	12–14	6	Blunt	Remote	Short	Calif.
tenuis	7	4	Blunt	Remote	Short	N. Y.

Iron albertae McDunnough

1924—McDunnough—Canad. Ent. 56: 129. Fig. . . . . Alta.
1929—McDunnough—Canad. Ent. 61: 177 (Epeorus) . . . . Wyo., Mont.
1933—Traver—J. N. Y. Ent. Soc. 41: 119 (Epeorus).

Length: body 10; wing 11 mm.

Eyes of living insect blue-grey. Thorax light brown; mesonotum with a central darker band which spreads out on posterior portion "where there is a median spot of pale ochreous" (McD.). No dark markings on pleura. Legs pale brown; all femora with dark spot near middle. Fore femur and tibia subequal. Tip of fore tibia blackish, other joinings dark brown. Wings hyaline; venation pale brown, fine, inconspicuous; humeral cross vein marked with blackish. Abdominal tergites 2–6 semi-hyaline, slightly smoky; posterior margins narrowly brown. Sternites 2–7 pale hyaline, with faint traces of dark spots at ganglionic areas. Segments 8–10 opaque; tergites brown, sternites creamy. Forceps creamy, brown at apex. Tails brown. Apical margin of forceps base somewhat excavated laterally. Divisions of penes not lobed, as in *I. longimanus*; laterally each division is prolonged into a beak-like process; see fig. 107. Median spines rather stout, slightly divergent apically; shorter than in *I. lagunitas*.

We have in the Cornell collection a paratype of this species which Dr. McDunnough very kindly gave us.

Iron confusus sp. nov.

Length: body 9–10; wing 10–11 mm.

A light reddish brown species, paler than *I. fraudator* but darker than *fragilis;* genitalia of the *longimanus* type.

Male imago (pinned specimen)—Head and thorax light red-brown. A dark spot above fore coxa; dark red-brown areas on pleura above and between middle and hind legs; very faint blackish pencilings above middle and hind coxae. Legs light red-brown; a conspicuous black mark at center of each femur; apex of fore tibia and all tarsal joinings narrowly dark brown. Wings hyaline; venation very pale brown; basal costal cross veins pale, indistinct.

Abdominal tergites more or less extensively tinged with light reddish brown; anterior margin pale yellowish; posterior margins narrowly but distinctly blackish. Sternites yellowish, more or less shaded with reddish in middle areas. Tails light yellow-brown, very indistinctly darker at joinings.

Female very similar to male, except for usual differences; venation somewhat darker brown. Postero-lateral spines of nymph short and blunt; femoral flange blunt; gills of first pair with large anterior lobes, which do not quite meet beneath body. No dark submedian spots on tergites. A dark lateral mark on each sternite; 7–9 with small dark submedian triangles at anterior margin.

Males preserved in alcohol differ from pinned specimens as follows: Head, thorax and abdomen considerably paler. Legs yellowish red; black femoral spot very distinct; fore tarsi yellowish. Abdominal segments 2–7 appear more yellowish, especially in anterior half; reddish shading on tergites less distinct, and almost lacking on sternites; black posterior margins of tergites even more distinct. Tergites 8–10 largely red-brown, sternites somewhat paler.

Holotype—Male imago (pinned). Slaterville, N. Y., May 20, 1934 (J. R. Traver). No. 1329.1 in Cornell University collection.

Allotype—Female imago (alcohol); same location, May 24, 1932 (J. R. T.). No. 1329.2 in Cornell collection.

Paratypes—31 male imagos, 13 female imagos (some pinned, others in alcohol, many reared); Slaterville, N. Y., May 23–29, 1932; Coy's Glen near Ithaca, N. Y., May 23–25, 1932. (J. R. T.). No. 1329.3–46 in Cornell collection.

This species is considerably larger than *I. fragilis*, and intermediate in color between this and *I. fraudator*. It has probably been confused with *I. fragilis*, in the past.

Iron deceptivus McDunnough

1923—Dodds—Trans. Amer. Ent. Soc. 48: Pl. 9. Fig. 22 (*longimanus*, late phase)..............Colo.
1924—McDunnough—Canad. Ent. 56: 132. Fig. (Cinygma)..................................Alta.
1929—McDunnough—Canad. Ent. 61: 177. (transferred to Iron).
1933—Traver—J. N. Y. Ent. Soc. 41: 118.

Length: body 8½–9; wing 9–10 mm.

A light brown species, very similar in general appearance to *Cinygmula ramaleyi;* genitalia distinctive.

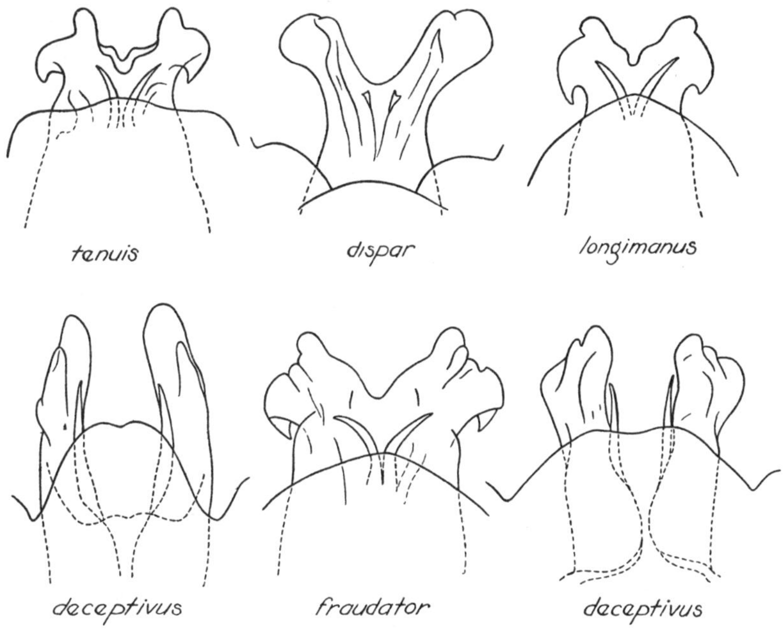

FIG. 105. Male genitalia of five species of Iron; two views of *I. deceptivus*.

Head and thorax reddish brown; lateral areas on each side of scutella of meso- and metathorax blackish brown. Legs light red-brown; apex of fore tibia, and all tarsal joinings, blackish brown; no markings on femora. Fore leg of male much longer than body; tibia slightly more than 1½ times as long as femur; first, second and third tarsal joints subequal one to another. Wings hyaline; venation dark brown, veins rather fine. Cross veins in costal space, except the 12 to 14 in the stigmatic area, very indistinct.

Abdominal segments 2–7 semi-hyaline; 8–10 opaque. Anterior portions of tergites pale brown; posterior portions shaded with purplish brown; the posterior margins and anteriorly-directed projections from them on each side, darker purplish brown. Sternites somewhat paler than tergites; small oval brown patches at ganglionic areas. Tails brown. Genitalia pale brown; in structure, unlike any other species of this genus. Penes appear to vary somewhat in shape according to their functional condition. As shown in the original figure, and in one slide lent us by Dr. McDunnough, penes are somewhat widened subapically, the lateral lobes very prominent. In other slides and specimens which Dr. McDunnough has permitted us to examine, the penes are long, slender, somewhat finger-like; the lateral lobes, folded flat, are barely noticeable. Median spines long, slender, slightly incurved. Apical margin of forceps base convex, projecting forward between forceps limbs; often very slightly emarginate at median line. We present figures of penes to show the two different phases indicated (see fig. 105, two views).

As noted, we have studied slides and specimens of this species, kindly loaned by Dr. McDunnough, who has given one male from the type locality to the Cornell collection. We have also a series of male imagos from Jocko Valley, near Arlec, Mont., in which the penes are long and slender, as in fig. 105. Head, thoracic pleura and abdominal tergites 8–10 with opaque pinkish white markings. A faint brown stain may be present in stigmatic area of wing. Posterior margins and lateral streaks on abdominal tergites darker, and contrasted more strongly with the pale anterior portion, than in type material observed. However, we are considering these to be *I. deceptivus.*

Iron dispar Traver

1933—Traver—J. Elisha Mitchell Sci. Soc. 48: 161. Fig....................................N. C.
1933—Traver—J. N. Y. Ent. Soc. 41: 119 (N. C. sp. No. 2).

Length: body 9; wing 9 mm.

Penes somewhat as in *I. humeralis,* but apical margin of forceps base different; veins brown in apical half of fore wing.

Head pale yellow-brown. Margin of clypeus, edge of median carina, and streak at inner corner of eye, purple. Small purplish dots between antennae and below base of each. Antennal filament dusky, the tip pale. Eyes much as in *I. rubidus.* Purple spot at median line on posterior margin of head. Thorax yellowish brown. Anterior margin of pronotum, also posterior margin at median line, purple; purplish submedian streak on each side. Median darker streak on mesonotum, bordered by pale yellow. Scutellum dark brown. Metanotum dark brown, paler laterally. Pleura yellowish; purple streak on anterior coxa and area above it; purplish marks and two dark spots above each of other coxae. Sternum pale yellow, mesosternum brownish. Fore femur yellow-brown; rather small median purple spot, pale brown apical mark. Tibia purplish brown, darker at apex and in basal third; tarsus pale yellow, joinings widely purplish; distal joint dusky. Basal joint of fore tarsus distinctly longer than second, which is subequal to third. Middle and hind legs pale yellowish white; purplish median spot on each femur; apices of tibiae and tarsal joinings narrowly darker. Venation in apical half of fore wing light brown; in basal half, veins almost colorless; humeral cross vein black next to subcosta. Apical subcostal cross veins of hind wing, and a few marginal intercalaries very pale brown, others colorless. Stigmatic area of fore wing opaque, whitish.

Abdominal segments 2–7 semi-hyaline, yellowish white; posterior margins of tergites purplish, this band widest at median line. Purplish black mid-dorsal line on posterior portion of tergites 1–9; lateral submedian purplish streaks on each, most extensive on tergites 1 and 2, also another series of streaks nearer the pleural fold.

Purplish mark near spiracular area. Segments 8–10 opaque, yellow. Sternites immaculate except for two short purple transverse bars on each side of median line at anterior margin of sternite 2. Tails dusky, faintly yellow-tinged; not darker at joinings. Penes of the *I. humeralis* type; no lateral spines near apex. Apical margin of forceps base as shown in fig. 105.

Distinguished from *I. punctatus*, which likewise lacks the lateral spines on the penes, by the paler head and thorax, more extensive abdominal markings and brown apical veins. The type material of this species is in the Cornell collection. We have likewise three female imagos from Greenville, S. C. (H. K. Townes, Jr.), which seem to be of this species.

Iron dulciana McDunnough

1935—McDunnough—Canad. Ent. 67: 103. Fig. .... B. C.

Length: body 9?; wing 9½ mm.

A species of the *albertae* group; blackish markings on thoracic pleura; apical margin of forceps base as in *I. sancta-gabriel.*

Head yellowish; antennae dusky. Red patches on vertex of head of female, between eyes and ocelli, and posterior to median ocellus; posterior margin of occiput blackish. Thoracic notum pale brown. Lateral and posterior margins of pronotum blackish. A reddish streak on each side of anterior portion of mesonotum. Posterior portions of meso- and metanota somewhat darker brown; dark median streak on mesonotum. Pleura and sternum yellow-brown. Creamy markings on pleura; a red-brown streak from pronotum to base of fore leg, another anterior to middle leg. Black dots on and above all coxae. Legs yellowish to olive brown. Tip of fore tibia, claws and tarsal joinings blackish brown. Median blackish spot on each femur, also a narrow red-brown apical band. Fore femur very slightly shorter than tibia. Wings hyaline. Venation yellowish in male, cross veins indistinct except those of stgimatic area, which are 9–11 in number. Humeral cross vein black near subcosta. In female, venation rather dark brown; veins fine but distinct.

Abdominal tergites yellowish to olive brown; posterior margins narrowly blackish. Sternites slightly paler. Tails yellowish to olive brown, joinings faintly darker. Genitalia as in fig. 107, "n. sp., McD."

The black thoracic markings ally this species to *I. sancta-gabriel* and *I. youngi*. Differences in detailed structure of the penes separate it from both species; the pale venation of the male also serves to distinguish it from the former species. The apical margin of the forceps base is much more produced than in *I. youngi*. We have seen the type specimens of this species, which are in the Canadian National Collection.

Iron fragilis Morgan

1905—Needham—N. Y. St. Mus. Bull. 86: 57. Figs (Iron sp?) .... N. Y.
1911—Morgan—Ann. Amer. Ent. Soc. 4: 104. Figs.
1924—McDunnough—Canad. Ent. 56: 129.
1933—Traver—J. N. Y. Ent. Soc. 41: 118–9.
1933—McDunnough—Canad. Ent. 65: 278 (*gracilis*) .... Que.

Length: body 7; wing 7 mm.

Genitalia of the *longimanus* type; closely allied to *I. tenuis*; a small pale species.

Eyes large, greyish; head parchment color. Basal segment of antenna white, remainder light brown. Ocelli with wide brown ring at base. Thoracic notum yellowish; lateral portions of pronotum and median area of mesonotum brown; subtriangular dark areas on each side of scutella of meso- and metanota. Pronotum deeply notched posteriorly. Pleura and sternum pale yellowish. Legs dull yellowish; a conspicuous dark brown spot near middle of each femur; claws and tarsal joinings brown. Fore legs longer than body. Wings hyaline, somewhat opaque in stigmatic area. Costa and subcosta almost colorless, other veins pale amber; cross veins rather finer than longitudinals. Humeral cross vein pale; veins of hind wing paler than fore wing.

Abdominal tergites 2–7 semi-hyaline, dull yellowish, sometimes with faint brownish tinge. Posterior margins narrowly purplish brown, the dark band slanting off obliquely toward spiracular area as it nears the pleural fold, much as in *I. longimanus*. Sternites paler yellowish. Segments 8–10 opaque, somewhat deeper in color. Traces of pale lunate submedian marks on middle tergites. Tails pale brownish at base, paler beyond; joinings indistinctly darker. Genitalia much as in *I. longimanus*; apical margin of forceps base less rounded at middle, somewhat angular. Median spines of penes more sharply outcurved toward apex, lobes of penes seem more prominent, because of deeper incisions between the apical and lateral portions.

There are specimens of this species in the Cornell collection from several localities in the vicinity of Ithaca, N. Y. This species is very closely allied to *I. tenuis* and *I. confusus*, with each of which it has probably been confused. Nymphs may be distinguished from each of the above species by the fact that the anterior lobes of the first pair of gills are very large and meet beneath the body of the nymph. All reared specimens we have seen which agree with the original description as to nymphal characters, are somewhat larger and have the posterior portion of each tergite more or less shaded with light purplish brown. The type material, other than a single vial of nymphs, is unfortunately missing.

Iron fraudator sp. nov.

1924—McDunnough—Canad. Ent. 56: 129 (*pleuralis*).
1925—McDunnough—Trans. R. S. C. 19: 221 (*pleuralis*)....................................Que.
1930—Ide—Canad. Ent. 62: 228. Fig. nymph (*pleuralis*)....................................Ont.
1933—Traver—J. N. Y. Ent. Soc. 41: 118–19 (*pleuralis*).
1933—McDunnough—Canad. Ent. 65: 278 (*pleuralis*).

Length: body 9–10½; wing 10–11 mm.

A dark reddish brown species, very similar in general appearance to *I. pleuralis*; genitalia of the *longimanus* type.

Male imago (pinned)—Head and thorax dark red-brown; reddish tinges around bases of antennae and on thoracic pleura. Pronotum heavily shaded with blackish; a black streak from lateral margin of pronotum downward to base of fore leg. A narrow pale yellowish streak anterior to wing roots; a few yellowish markings on pleura below wing roots and above bases of legs. Lateral areas on each side of scutella of meso- and metathorax blackish. Legs yellowish to olive brown; fore leg distinctly smoky. Prominent black spot near middle of each femur. Apex of fore tibia blackish. Distal joint of fore tarsus, and all joints of middle and hind tarsi, deep smoky to blackish brown. Wings hyaline; venation dark brown; faint brownish stain in stigmatic area. Basal costal cross veins rather indistinct; 12 to 14 in stigmatic area.

Abdominal tergites dark red-brown. Anterior margins, antero-lateral angles, an irregularly oval area near each spiracle, and submedian streaks based on anterior margin, yellow to pale yellowish brown. A pale narrow median line on tergites 2–7. Sternites paler; anterior margins yellowish, remaining portions flushed with pale red-brown. Tails dark brown. Forceps base light red-brown; forceps and penes darker. Penes very similar to *I. longimanus*, but with the lateral and apical lobes more distinct because more deeply dissected (see fig. 105).

Female very similar to male. Yellowish areas on pleura more extensive. Thoracic sternum largely yellowish, except posterior portion of mesosternum, which is red-brown. Basal costal cross veins more distinct than in male.

Specimens in alcohol usually appear somewhat paler than the description given for the holotype. Dark markings on posterior margins of tergites and around the pale oval spiracular area are more noticeable. Fore tibia usually about 1¼ times as long as femur; fore tarsus almost twice the length of the tibia, in some specimens, while in others it is not over 1½ times as long. In an occasional specimen the femora and tibiae are distinctly washed with red.

Nymph as described by Dr. Ide. Anterior lobes of first pair of gills do not quite meet beneath body of nymph. A pair of submedian blackish spots is usually present near middle of each tergite.

Holotype—Male imago (pinned), Ringwood, N. Y., May 7, 1933; (J. R. Traver), No. 1328.1 in Cornell University collection.

Allotype—Female imago, reared from nymph; Slaterville, N. Y., May 11, 1932 (J. R. T.). No. 1328.2 in Cornell collection.

Paratypes—22 male imagos, 11 female imagos. Of these, 15 specimens are pinned, others in alcohol; 4 males, 8 females, reared from nymph. Enfield, N. Y., May 2, 1933; Ellis Hollow, N. Y., May 5, '33; Slaterville, N. Y., May 11, '32; Ringwood, N. Y., May 7, '33. (J. R. T.). No. 1328.3–35 in Cornell collection.

This species is so similar in general appearance to *I. pleuralis* that we are unable to find any character other than genitalic structure by which to distinguish it. Further, the two species occur together; specimens of both have been taken in the same swarm.

Iron hesperus Banks

1924—Banks—Bull. Mus. Comp. Zool. 65: 424 (Rhithrogena)............................Wash.

Length (female): body 10; wing 12 mm.

Only the female imago of this species is known. Pale yellowish in color. Curved black mark on vertex between eye and lateral ocellus; black mark near posterior margin of head between eyes, a short extension from this along median line. Anterior and posterior margins of pronotum narrowly black; a black streak extends from edge of pronotum downward along edge of coxa and trochanter. Scutella of meso- and metanota darker than other portions of thorax. Pleura yellowish; a black spot anterior to middle coxa, and black mark at base of middle and hind coxae. Legs pale yellow; dark median and apical mark on each femur, the median spot large, black. Claws black. Basal joint of fore tarsus "is rather shorter than the second, which is barely longer than the third" (Banks). Basal joint of hind tarsus "a little longer than the second, and the second is a little longer than the third" (Banks). "Wings hyaline; venation yellowish, the cross veins more brownish; in costal space near base is a prominent black comma-shaped mark." One of us (Needham), after examining the type specimen, notes that the longitudinal veins appear brown, the cross veins somewhat heavier than the longitudinals. Cross veins of stigmatic area not anastomosed; this area rather narrow, about ⅔ as wide as space below it. About six basal costal cross veins, four beyond bulla anterior to stigmatic area.

Abdominal segments pale. At the spiracular area, two oblique black streaks, joined together at the ends, surround each spiracle, on segments 2–7. The enclosed pale area is largest on tergite 2, smaller on the following segments. Sternites immaculate; "last two segments fumose" (Banks.). Apical margin of subanal plate with a median notch.

The characters given in the original description, together with others noted in an examination of the type specimen, seem closer to the genus Iron than to Rhithrogena, in which genus the species was described. The species probably belongs in the *I. albertae* group; it may in fact be identical with some one of the species in that group which is now held under a different name.

Iron humeralis Morgan

1911—Morgan—Ann. Amer. Ent. Soc. 4: 105. Figs. (Epeorus)............................N. Y.
1924—McDunnough—Canad. Ent. 56: 129 (transferred to Iron).
1925—McDunnough—Canad. Ent. 57: 189. Fig..Ont.
1925—McDunnough—Trans. R. S. C. 19: 221..Que.
1929—McDunnough—Canad. Ent. 61: 177 (in discussion of *Epeorus albertae*).
1933—Traver—J. Elisha Mitchell Sci. Soc. 48: 158, 159 (in discussion of *rubidus*).
1933—Traver—J. N. Y. Ent. Soc. 41: 118-9.

Length: body 10; wing 10–11 mm.

A yellowish white species; venation pale; small lateral spines below apex of penes.

Head yellowish white; eyes olive-green in living insect, a brown band across lower portion; large, contiguous apically. Antennae brown; a brown line from base of antenna to inner margin of eye; median carina brown; ocelli ringed with olive at base. Thorax yellowish white. Posterior margin and postero-lateral corners of pronotum, and an oblique streak on each side, dark brown. Mesonotum yellowish; scutellum dark brown, with a pale area anterior to it, sometimes also lateral pale streaks on each side. Scutellum and lateral areas of metanotum dark brown. Pleura pale yellowish. A purplish brown streak extends downward across coxa from margin of pronotum. Purplish black markings below wing roots and above leg bases, consisting of dark lines and two or three dark spots on each line. Sternum yellowish; in many specimens, a transverse brown band extends across apical portion of mesosternum from point in front of middle leg; a similar dark band may extend across posterior portion of mesosternum from point anterior to hind leg on each side, widening on mesosternum to form a brown patch. Legs whitish; dark brown spot on each coxa; conspicuous dark median spot and apical band on each femur, most distinct on fore femur. Apex of fore tibia and fore tarsal joinings blackish. Basal fore tarsal joint about as long as second, which is subequal to third. Wings hyaline; venation almost colorless, although longitudinal veins may be faintly yellowish; humeral cross vein black next to subcosta. Stigmatic area opaque whitish.

Abdominal segments 2–6, often base of 7, semi-hyaline, whitish. Posterior margins of tergites narrowly brownish; on basal tergites, these bands are wider and extended forward slightly at each end to form irregular lateral patches (not present on all specimens). Traces of lateral brownish streaks on each tergite; median line blackish at posterior margin. Sternites pale, immaculate. Segments 8–10 and apical portion of 7 opaque, yellowish, often with tinge of reddish. Genitalia and tails pale whitish. Penes with small lateral spines below apex on each side (see fig. 107.)

This species is evidently quite variable in color, as we have specimens in which the abdomen is almost wholly white, few dark marks on thorax and no dark ventral thoracic bands; others in which the dark shading of the tergites is quite pronounced, thoracic markings prominent, legs greenish yellow with purplish red markings. We have seen none as dark as Dr. McDunnough has indicated, however. It is closely allied to *I. rubidus*, and *I. punctatus*, but it is larger than either of these. The latter also lacks the lateral spines on the penes. The paler thorax and relatively smaller genitalia distinguish it from *I. suffusus*. The species is common in the vicinity of Ithaca, N. Y. Specimens in our collection from Ft. Jackson and Nassau, N. Y. (L. Harper), are probably of this species. These are somewhat smaller than local specimens, with more prominent markings on thorax and abdominal tergites, but show no genitalic differences.

Iron lagunitas sp. nov.

Length: body 10; wing 11 mm.

Closely allied to *I. albertae,* but with slight differences in genitalic structure; fore femur shorter than tibia.

Male imago.—Head purplish brown. Thorax light red-brown, pronotum extensively shaded with purplish brown. Faint traces of a darker median stripe on mesonotum, as in *albertae.* Scutellum and lateral areas just anterior to it darker red-brown; outer margin of dark area, and median area anterior to scutellum, yellowish. Posterior margins and lateral depressions below scutella on meso- and metathorax also dark brown. No distinct blackish markings on pleura; faint purplish shading below wing roots and an oblique purplish streak anterior to root of fore wing. Legs olive brown. Fore femur somewhat yellowish near base, and with faint reddish tinge; tibia and tarsus smoky. Apex of fore tibia, tarsal joinings, claws and distal tarsal joint blackish. An indistinct purplish brown mark near middle of each femur; the area on each side of it is shaded with purplish, on middle and hind legs. Fore femur about ¾ the length of tibia. Fore claws similar, blunt. Wings hyaline; venation light brown. Humeral cross vein purplish black near subcosta, pale at costal margin. Basal costal cross veins rather weak, especially at costal margin. Cross veins between bulla and stigmatic area obsolescent. Stigmatic veins well developed, about 10 in number, not anastomosed. A faint brownish stain in stigmatic area.

Abdominal segments 2–7 semi-hyaline, yellowish; tergites more or less distinctly tinged with pale brownish. Posterior margins of tergites and lower edge of pleural fold darker brown. In one paratype, sternites likewise faintly shaded with brown. Segments 8–10 opaque, rather bright red-brown; sternites rather paler than tergites. Tails dark yellowish brown; very inconspicuously darker at joinings. Apical margin of forceps base very slightly convex. A distinct swelling, somewhat conical, near base of long joint of forceps, on inner margin. Penes very similar to *I. albertae,* but with longer and more slender median spines; lateral beak-like appendages on each division of penes slightly longer and more slender (see fig. 107).

Female very similar to male, aside from usual differences. Basal costal cross veins better developed.

Holotype—Male imago (originally pinned, now in alcohol); July 1, 1928, Lake Lagunitas, Calif.; Mr. E. H. Nast, collector. In collections of Calif. Acad. of Science.

Allotype—Female imago (pinned) same data. No. 1331.2 in Cornell University collection.

Paratypes—3 male imagos, 1 female imago, (pinned); 2 males (abdomens missing) in Cornell collection, No. 1331.3–5; 1 male, 1 female (abdomen of male missing) in collection of Calif. Acad. of Science.

Distinguished from *I. albertae* by the relatively longer fore tibia and by genitalic differences noted above. The lack of distinct blackish pleural markings separates it from *I. youngi* and *I. sancta-gabriel.*

Iron longimanus Eaton

1881—Eaton—Ent. Mo. Mag. 18: 26 (Epeorus, Colorado sp.)................................Colo.
1883–85—Eaton—Revis. Monogr. p. 245. Figs. (Epeorus, pl. 23; transferred to Iron, p. 245).
1923—Dodds—Trans. Amer. Ent. Soc. 49: 107. Fig. 21, *longimanus*; figs, 23–24, as Iron nymph No. 1.
1924—McDunnough—Canad. Ent. 56: 129.
1925—McDunnough—Trans. R. S. C. 19: 221 (in discussion of *pleuralis*).
1927—Needham—Utah Agr. Exp. Sta. Bull. 201: 15. Fig. nymph..............................Utah
1929—McDunnough—Canad. Ent. 61: 178 (in discussion of *deceptivus*).
1933—Traver—J. N. Y. Ent. Soc. 41: 118–9.

Length: body -10; wing 11 mm.

A rather pale brownish species; genitalia of the *longimanus* type (see fig. 105).

Head light reddish brown. Thoracic notum light yellowish to reddish brown ochre, pleura and sternum very similar in color. Fore femur and tibia "translucent raw umber" (Eaton); other femora similar, tibiae yellowish brown. All femora with a round black spot near the middle; fore tibia blackish at apex. Fore tarsus dull yellowish brown, middle and hind tarsi slightly paler in color; tarsal joinings may appear somewhat darker. Wings hyaline; venation light brownish amber, the humeral cross vein, base of subcosta and other longitudinal veins behind it, paler; veins in hind wing rather pale.

Abdominal tergites 2–9 "tinged extensively with light brownish" (Eaton); 9 and 10 often light brown-ochre. Tergites 2–7 semi-hyaline; posterior margins deep purplish brown, this band not extending around the postero-lateral angle, but a short dark oblique streak arising from this dark band runs to the spiracular area. Purplish brown shading along mid-dorsal line; median line narrowly pale, with darker streak on each side. Pale lunate submedian marks on each side lie within the shaded area. Segments 8–10 opaque. Sternites pale yellowish; nerve cord and ganglia withish, opaque; sometimes traces of pale submedian oblique streaks on basal and middle segments. Tails "light warm sepia brown" (Eaton). Apical margin of forceps base projects backward as a dome-shaped structure, concealing the bases of the penes. Each division of the penes has an erect apical lobe and a larger laterally-directed one, bearing a sharp lateral spine. A median pair of strong spines is present, their tips divergent.

In type of genitalia this western species is allied to the eastern *I. fraudator*, *I. fragilis* and *I. tenuis*. It is larger and somewhat darker in color than either of the two latter species. The lobes of the penes are likewise much less deeply cut than in any of the three. The darkest specimens we have seen are slightly paler than typical *fraudator*, which is likewise a little larger than *I. longimanus*. We have dark reddish-brown specimens of this species in the Cornell collection from South Park, Colo. (E. G. Oslar), and Maligne Lake, Alta. (J. C. Bradley). Paler specimens are from South Boulder Valley, Colo. (G. S. Dodds); Estes Park, Colo.; Glacier National Park, Mont. (A. S. Hazzard); Logan River, Utah, reared from nymph (J. G. Needham); and Moose, Wyo. (Mrs. E. B. Klots).

Iron modestus Banks

1910—Banks—Canad. Ent. 42: 202 (Epeorus)..Md., D. C.
1924—McDunnough—Canad. Ent. 56: 129 (transferred to Iron).

Length: body 6½; wing ?.

Penes unlike any other species in this genus; tails pale, alternate joinings distinctly dark.

Head pale. Thorax dark. Legs pale, all femora with median and apical brown marks; apex of fore tibia black. Basal fore tarsal joint "about as long as the second joint, the third as long as the second, the fourth much shorter, fifth one-half of the basal; the first, second and third together a little longer than the tibia" (Banks). Wings "scarcely darker in the apical costal area, longitudinal veins faintly brown, the cross veins darker brown, none marked with black, except the basal costal" (Bks.). 5 to 6 basal costal cross veins, 8 to 10 beyond bulla. Abdominal segments 2–8 pale, posterior margins of tergites (and sternites?) narrowly dark; "tip of the abdomen dark" (Bks.). Tails pale, alternate joinings dark.

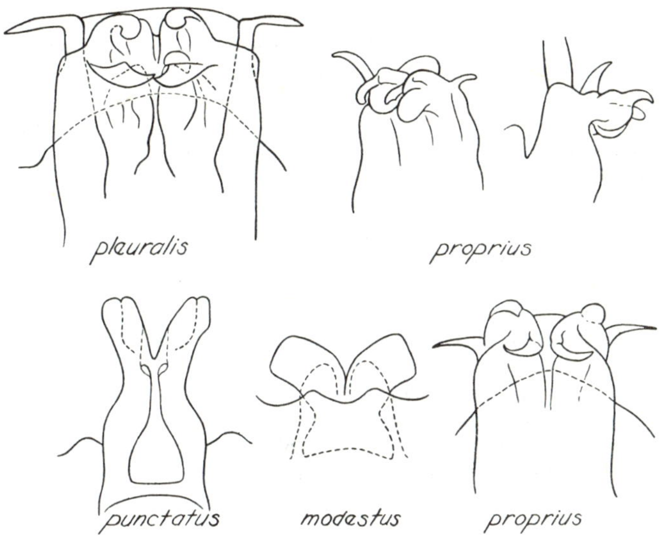

FIG. 106. Male genitalia of four species of Iron; three views of *proprius* (*punctatus*, after McDunnough).

Because of the length of the fore tarsal joints, this species would seem to belong either in this genus or in Eperous; genitalia, however, are quite distinct from any other known species in either genus, in our fauna. For the present we leave it in Iron. Reference to *I. modestus* in a previous paper (Traver, J. N. Y. Ent. Soc. 41: 118–9) was based on an error in determination, and should be disregarded. In fig. 106 is a sketch of the genitalia of the type specimen, prepared for Dr. McDunnough, which he permitted us to copy.

Iron pleuralis Banks

1910—Banks—Canad. Ent. 42: 102. Fig. (Heptagenia [Epeorus])............................N. Y.

Length: body 9–10, wing 10–11 mm.

A reddish brown species; cross veins of basal costal space indistinct; genitalia of *pleuralis* type, having a long prominent lateral spine on each side of penes.

A paratype of this species presented to the Cornell collection by Dr. Banks, is the basis for this description. Several other specimens in Dr. Banks' collection, taken at the same time and place and associated with *I. pleuralis*, prove to be *Rhithrogena jejuna*, as indicated by Dr. Banks' notes in his correspondence with us, and by a specimen which he kindly lent us for study. It is probable that the original description included both this species and *R. jejuna*.

Head of male reddish brown; no distinct markings. Thorax quite uniform dark red-brown; pronotum shaded with smoky brown except on anterior margin. No such distinct paler markings anterior to wing root and on pleura, as indicated in original description. Legs pale reddish brown. Fore leg longer than body. Basal fore tarsal joint subequal to second, third almost as long. A prominent black spot near median area of each femur. Apical margin of each trochanter very narrowly blackish; faint indications of a small darker spot at each apical angle. Basal joint of hind tarsus slightly longer than second joint. Middle and hind tarsi somewhat darker than their respective tibiae. Wings hyaline, seemingly with very faint yellowish tinge; hardly darker in stigmatic area. (Original description does not mention yellowish tinge, but states, "rather darker in costal area near the tip." Discrepancy may be due to fact that paratype has been in alcohol presumably since date of capture). Basal costal cross veins very faint; about 6 before the bulla. Four or five immediately following bulla are likewise very faint. Stigmatic cross veins heavier, distinct; about 14 in number. Venation light brown; longitudinal veins very slightly heavier than cross veins.

Abdominal segments 2–8 semi-hyaline; 9 and 10 largely opaque. Light reddish brown in color, with a yellow tinge; tergites barely darker than sternites. Tergites 2–7 with a paler brownish median area, short submedian streaks from anterior margin which may coalesce with the pale median streak; anterior margins and antero-lateral angles likewise paler brown. A distinct blackish streak at spiracular area, extending from posterior margin in a curve to the pleural fold, cutting off postero-lateral angle. Small pale brownish spot faintly indicated at ganglionic area of each sternite. Penes with two dorsally-directed lobes (best seen in side view) apically near median area; below these on dorsal side, a curved chitinized structure; on each side, a prominent laterally-directed spine-like process (see fig. 106). Distinctly different from genitalia of the *longimanus* type; adults may be readily confused with species of the latter type in their general appearance.

There are several local specimens in the Cornell collection, some pinned and others in alcohol, in which the genitalia are very similar to the above paratype of *I. pleuralis*. These have a somewhat paler mesonotum, and greyish areas anterior to wing root and on pleura, most evident in alcoholic specimens; more distinct black dots at apical angles of each trochanter. Wing with no yellow tinge, but slightly darker in stigmatic area; venation deeper brown; no pale median areas as on tergites, other pale areas as in Dr. Banks' specimen; tergites dark red-brown, distinctly darker than sternites, which are tinged with brown. Tails deep red-brown. We are considering these to be *I. pleuralis*.

Previous references to *I. pleuralis* in the literature by McDunnough, Ide, Traver and others seem to apply rather to *I. fraudator*, with which species the true *pleuralis* has been confused. The western *I. proprius* is the only other known species having genitalia of the *pleuralis* type; it differs from *I. pleuralis* in its somewhat smaller size and in details of the structure of the penes.

Iron proprius sp. nov.

Length: body 9–10; wing 10 mm.

Similar in general appearance to *I. longimanus*, but with genitalia of the *pleuralis* type.

The only specimens available are so faded from long immersion in alcohol that it is impossible to be certain of the original color. However, they seem to have been very similar to *I. longimanus*, from which species they are distinguishable mainly by the very different type of genitalia. A comparison of these with specimens of *longimanus* which likewise have been in alcohol for several years, shows that the dark spot on each femur is smaller and less distinct than in the latter species. The oblique black streak present on the postero-lateral angles of the tergites in *I. longimanus*, is apparently lacking in *I. proprius*. Genitalia very similar to *I. pleuralis*, but smaller; chitinized appendage attached to dorsal lobe of penes is narrower and more distinctly sickle-shaped than in that species. (See fig. 106).

Holotype—Male imago; Colorado (probably South Boulder Valley); no date. Collector, Prof. G. S. Dodds. No. 1330.1 in Cornell University collection.

Paratypes—Three male imagos, same data. No. 1330.3–5 in Cornell collection.

Iron punctatus McDunnough

1925—McDunnough—Canad. Ent. 57: 189. Fig. . . . . Ont., Que.
1929—McDunnough—Canad. Ent. 61: 177 (Epeorus).

Length: body 8; wing 8 mm.

Allied to *I. humeralis*, but smaller, with blackish-brown head and thorax; no lateral spines on penes.

Eyes smaller than in *humeralis*, not contiguous. Head and thorax deep blackish brown. An area posterior to scutellum, lateral edges of mesonotum, and pleural sutures, yellowish. Mesosternum shaded with ochreous between legs. Legs yellowish white; femora with dark median and apical marks; apex of fore tibia black. Wings hyaline; venation pale; humeral cross vein black-tinged. Abdominal segments 2–7 pale hyaline, posterior margins of tergites "very faintly and narrowly edged with blackish" (McD).; a distinct row of blackish dots at mid-dorsal line on posterior margin of each tergite; "in certain lights faint traces of a lateral row of small darkish spots is evident" (McD.). Segments 8–10 opaque; tergites with brown shading. Tails and forceps pale whitish. Genitalia shown in fig. 106.

Very similar to *I. humeralis*; distinguished by the smaller eyes, paler tergites, darker thorax and lack of lateral spines near apex of penes. A specimen in the Cornell collection taken at Buffalo, N. Y. by Prof. C. Betten, may be of this species.

Iron rubidus Traver

1933—Traver—J. Elisha Mitchell Sci. Soc. 48: 158. . . . . N. C.
1933—Traver—J. N. Y. Ent. Soc. 41: 119 (N. C. sp. No. 1).

Length: body 8–9; wing 8–9 mm.

Allied to *I. humeralis*; slightly smaller, eyes somewhat smaller, longitudinal veins faintly yellow-brown, lateral spines on penes less distinct.

Head pale yellowish; small lateral brown mark on carina, another at each corner of eye. Base of antenna pale, filament dusky at base, tip pale; eyes bright green in living insect, contiguous, but smaller than in *I. humeralis*. Posterior margin brown at middle line. Thorax pale yellow-brown. Lateral and posterior margins of pronotum brown; brown submedian streaks. Four small dark marks on mesonotum, between wings. Scutellum orange-brown, posterior margin black; lateral areas whitish. Posterior half of metanotum orange-brown, with purplish lateral shading. Pleura pale, with several distinct small purple-black marks on coxae and above each leg. Sternum pale yellowish. Legs pale yellowish; femora rather deeper yellow, each with dark median and apical marks; fore femur slightly dusky. Apex of fore tibia purplish. Basal joint of fore tarus very slightly shorter than second. Wings semi-hyaline;

stigmatic area opaque whitish; longitudinal veins faintly yellowish brown; cross veins hyaline, humeral cross vein black next to subcosta.

Abdominal segments 1–7 semi-hyaline, whitish. Posterior margin of each tergite narrowly purplish-black, paler at middle line. Mid-dorsal line, on tergites 3–9, purplish black in posterior half; on 1 and 2, a black median dot only. Purplish mark in stigmatic area of each tergite, and traces of dark lateral marks. Segments 8–10 opaque, tergites yellow-brown, sternites paler. Tails white, not darker at joinings. Penes much as in *I. humeralis*, but lateral spines near apex are much less prominent; second joint of forceps relatively longer than in that species.

Distinguished from *I. humeralis* as indicated; penes smaller than in *I. suffusus*, thorax paler than in that species or *I. punctatus*. The type material of this species is in the Cornell collection.

Iron sancta-gabriel sp. nov.

Length: body 10; wing 10½ mm.

A yellowish species, allied to *I. youngi*, also to *I. lagunitas* and *I. albertae*; dark markings on thoracic pleura and at postero-lateral angles of tergites.

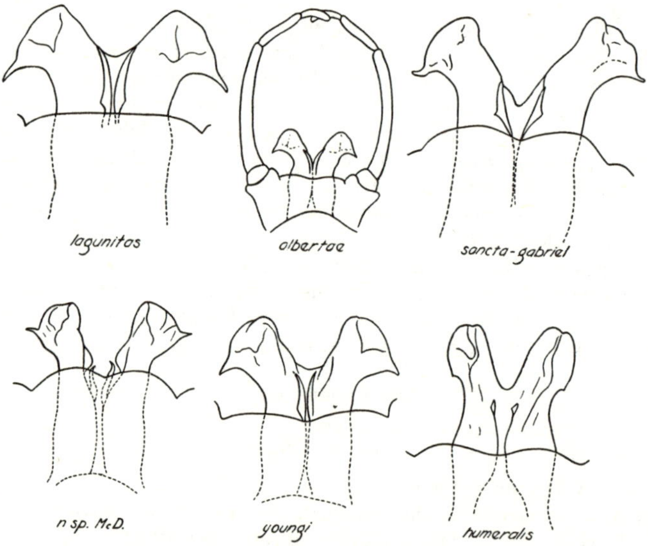

FIG. 107. Male genitalia of six species of Iron (drawing designated "n.sp. McD." is *I. dulciana* McD.: named after this drawing was prepared).

Male imago—Head yellowish. Thorax light red-brown, mesonotum quite distinctly deeper in color, especially anterior portion and lateral margins. Anterior and posterior margins of pronotum narrowly black; posterior portion shaded with blackish; two curved black submedian streaks terminate at median line near posterior margin. Posterior margins of meso- and metanota, including tips of scutella and depressions on each side of these, blackish. A blackish streak extends from lateral margin of pronotum to base of fore leg. A small black spot below root of fore wing; another anterior to middle coxa; two others, one above and one posterior to hind coxa. A small black dot on each coxa; a black streak on inner margin of each trochanter, also two minute dark dots at apical angles of each. Legs yellowish; fore tibia and all tarsi tinged faintly with olive brown. A round black spot at middle of each femur, also a subapical blackish band. Apex of fore tibia, all claws and tarsal joinings, blackish. Fore claws similar, blunt. Wings hyaline, appearing to have a very faint yellowish tinge. Costa, subcosta and radius yellowish brown, other longitudinal veins dark brown. Cross veins darker than longitudinals, those of costal margin almost blackish; slightly thicker than longitudinals. Humeral cross vein black in posterior half, pale next to costa. All costal cross veins well developed; 7 before bulla, 2 to 4 between bulla and stigmatic area, 9–10 stigmatic veins.

Abdominal segments 2–7 semi-hyaline, yellow; tergites faintly smoky. Posterior margins of tergites narrowly black. An oblique black mark cutting off postero-lateral angle of each, divides to enclose a whitish area at spiracle, as in *I. hesperus* and *I. youngi*. A black line in the same position on tergites 1 and 8. Segments 8–10 opaque, rather dark red-brown. Tails light red-brown with yellowish tinge; in basal portion, narrowly darker at joinings. Genitalia much as in *I. albertae* but differing in the forward-projecting apical margin of forceps base and the recurved lateral spines on each division of penes (see fig. 107).

Holotype—Male imago. San Gabriel Mts., Calif.; June 7, 1932. Mr. C. D. Michener, collector. No. 1332.1 in Cornell University collection.

A female subimago, taken in same location on May 21, is marked much like the male. An oblique dark mark on head between lateral ocellus and eye; posterior margin of head and extension along median line, dark brown. Nymphs taken on May 21 are rather similar to I. *dispar* in postero-lateral spines, femoral flanges and gills of first and seventh pairs. This species appears to be closely allied to *I. hesperus* of which only the female is known. It may indeed prove to be synonymous with that species, although there are several points on which the two do not agree.

We have several subimagos and a female imago, taken at Waddell Creek, Calif., by Prof. D. Shepherd, which are quite similar to *I. sancta-gabriel*, but with paler venation. Nymphs likewise have larger, more rounded gills. These may be the true *I. hesperus*.

Iron suffusus McDunnough

1925—McDunnough—Canad. Ent. 57: 189. Fig. .... Que.
1929—McDunnough—Canad. Ent. 61: 177 (Epeorus).

Length: body 11; wing 11? mm.

Similar to dark form of *I. humeralis* but with larger genitalia, and the penes more widely separated apically.

Larger than dark form of *I. humeralis*, and with larger eyes. Thorax dull brown, the posterior portions deeper brown. Legs pale yellowish; all femora with median and apical dark spots. Venation pale. Abdominal tergites suffused with brown, except for a double row of oval white spots. Tails pale; traces of brown rings at the basal joinings. Penes quite similar in shape to those of *I. humeralis*, but larger and more widely separated apically.

The larger size and slight structural differences in the penes seem to distinguish this species from *I. humeralis*, dark form. It is much darker than the form of *humeralis* which occurs near Ithaca, N. Y., from specimens of which the original description was drawn. We have a male imago from the collection of the New York State Museum taken at Corinth, N. Y., which seems to be of this species.

Iron tenuis sp. nov.

Length: body 7½–8; wing 8 mm.

A small pale species closely allied to *I. fragilis;* differs in slight details of genitalic structure, and in first pair of gills of nymph.

Head of male imago yellowish white; a narrow black line along margin of clypeus on each side of median carina. Thorax yellowish, somewhat parchment-colored. Posterior and lateral margins of pronotum very narrowly black; faint greyish shading on lateral areas. A narrow dark streak extends from lateral margin of pronotum down across prothorax to base of fore leg. Faint narrow darker markings above bases of middle and hind legs. Lateral areas on each side of scutella of meso- and metanota darker brownish. Legs pale yellowish; femora often with faint pinkish tinge. An elongated black spot (not rounded) near middle of each femur; a faint black hair-line extends from this to base of leg. Tip of fore tibia smoky brown; tarsal joinings of middle and hind legs very narrowly brown. Fore tarsus only slightly longer than tibia. Wings hyaline; longitudinal veins pale yellowish white; cross veins hyaline, almost invisible, particularly in basal costal space. Faint whitish cloud in stigmatic area; cross veins in this area faint, about 12 in number.

Abdominal segments 2–6, and basal portion of 7, semi-hyaline, very pale yellowish white. Segments 8–10 and apical portion of 7 opaque; pale reddish brown dorsally, paler ventrally. Posterior margins of all tergites very narrowly deep purplish brown. Tails whitish, very faintly brownish at extreme base. Genitalia as in fig. 105. Female not known.

The nymph is structurally quite similar to *I. fragilis*, *I. fraudator* and *I. confusus*. No dark submedian spots on tergites, as in *fraudator*. Tracheae of gills rather indistinct. Anterior lobes of first pair of gills quite short, extend forward only a short distance beyond base of leg, so that they do not meet, or even approximately meet, beneath body of nymph. Ventrally, dark brown submedian spots on anterior margin of sternites 7–9; often an oblique dark streak laterally on sternite 9. Body 7 to 8 mm. in length.

Holotype—Male imago, reared from nymph. Ringwood, N. Y. (near Ithaca), June 24, 1933; (J. R. Traver) No. 1327.1 in Cornell University collection.

Paratype—Male imago, reared; same data. No. 1327.3 in Cornell collection.

This small species seems to mature somewhat later than *I. fragilis*. The nymph is readily separated from that of *fragilis* by the structural characters of the first pair of gills. Penes of male somewhat more deeply and extensively lobed than in *I. fragilis*, especially on inner apical margin; apical margin of forceps base more regularly sinuate, its lateral projections almost on a level with the median one. It is quite possible that this species has been confused with *I. fragilis*, in the past.

Iron youngi sp. nov.

Length: body 10; wing 10–11 mm.

Allied to *I. albertae*, but with shorter fore femur, black spiracular markings and black markings on thoracic pleura.

Male imago—Head pale yellowish. Thorax fawn-colored. Anterior, posterior and postero-lateral margins of pronotum narrowly black; posterior portion somewhat smoky. Median olive brown streak on mesonotum. Tip of scutellum blackish, also posterior margins of mesonotum and depressions on each side of scutellum. Pale area anterior to scutellum; on each side of this, an olive brown area. Metathoracic scutellum and posterior margin of sclerite blackish. A brownish black streak extends downward from lateral margin of pronotum to base of fore leg. An oblique black streak on pleura anterior to root of fore wing; narrow black pencilings above, anterior and posterior to middle and hind coxae. A small black dot on each coxa. Legs yellowish. A prominent black spot at middle of each femur, and traces of a reddish brown or blackish apical band on each; this may be reduced to a small dark spot on ventral margin. Apex of fore tibia, all claws, and joinings of fore tarsus, blackish. Fore femur about ¾ as long as tibia. Femora may be faintly tinged with red. Wings hyaline; venation very pale brown. Basal costal cross veins indistinct, especially at costal margin; cross veins between bulla and stigmatic area obsolescent. Stigmatic cross veins well developed, slightly aslant, not anastomosed, about 14 in number. Humeral cross vein deep purplish black in posterior ⅔ of its length; pale next to costal margin.

Abdominal segments 2–7 semi-hyaline, whitish, faintly washed with pale brown. Posterior margins of tergites narrowly purplish black, most distinct dorsally. Median and lateral areas of tergites faintly brown-shaded; anterior margins and antero-lateral angles pale. A curved black streak, cutting across the postero-lateral angle on each side, divides into two branches at spiracle, so that a pale oval area is enclosed by black lines, as in *I. hesperus*. Ganglionic areas on sternites very faintly marked with brown. Abdominal segments 8–10 opaque, fawn-colored like thorax. Tails pale brown; apical portion of each joint is slightly deeper brown. Genitalia very much as in *I. albertae* (See fig. 107).

Holotype—Male imago; Upper Kintla Lake, Glacier National Park, Mont.; Aug. 16, 1933; (T. R. Young) No. 1333.1 in Cornell University collection.

Paratypes—4 male imagos, same data as holotype; 1 male imago, Cutright Cr., Glacier Nat. Pk.; Aug. 15, 1932; (A. S. Hazzard) No. 1333.3–7 in Cornell collection.

The black markings on thorax and spiracular areas of abdomen, and the more distinct median femoral spot, distinguish this species from *I. albertae* and *I. lagunitas*. The fore femur is also shorter than in the former species. The very indistinct. costal cross veins basad of the stigmatic area separate this species from *I. sancta-gabriel* to which it is closely allied. It seems to be allied also to *I. hesperus* but the dark posterior margins of the tergites and the faint brown markings at ganglionic areas of sternites should distinguish it from that species.

Genus **Ironodes** Traver

1927—Seemann—J. Ent. and Zool. 19: 42. Pl. 4, nymph figs. (as *Iron petulans*) . . . . Calif.
1933—Traver—J. N. Y. Ent. Soc. 41: 123 ('Nameless genus').
1935—Traver—Canad. Ent. 67: 32 (misprint Ironopsis).

Eyes of male small to moderate, usually not continuous apically. Posterior margin of head of female quite deeply emarginate. Fore leg of male slightly longer than body. Fore tibia fully 1½ times as long as femur; tarsus about 1½ times as long as tibia. Basal fore tarsal joint slightly longer than or subequal to second, which is subequal to third; fourth somewhat shorter than third, fifth about ½ the length of fourth. Fore femur of female slightly longer than tibia; tarsus subequal to femur. Basal fore tarsal joint about ⅔ as long as second, and subequal to fifth; third fully ¾ of second; fourth about ½ of first. In both sexes, hind tibia slightly shorter than femur. Hind tarsus ½ to ⅔ of tibia, in male;

only ½ as long as tibia in female. Basal joint of hind tarsus slightly longer than or subequal to second; third and fourth progressively shorter; fifth equal to third and fourth combined. Claws dissimilar on all tarsi. Typical Heptagenine venation. Basal costal cross veins strong; stigmatic cross veins simple, quite numerous, somewhat aslant. Posterior branch of media in fore wing forms almost a right angle at its union with anterior branch. Forks of radial sector in hind wing arise at about same point. Media of hind wing forked about ⅓ from base.

Apical margin of forceps base of male quite regularly undulate. Second forceps joint longest, often more or less bowed in apical third or near middle; third joint slightly longer than fourth, these two together not as long as second. Penes usually separated almost to base; somewhat compressed, elongate, usually divergent apically, and varying much in appearance according to position. A series of small spines is present on the ventral (or inner) surface of each division. See fig. 108. Apical margin of subanal plate of female slightly emarginate.

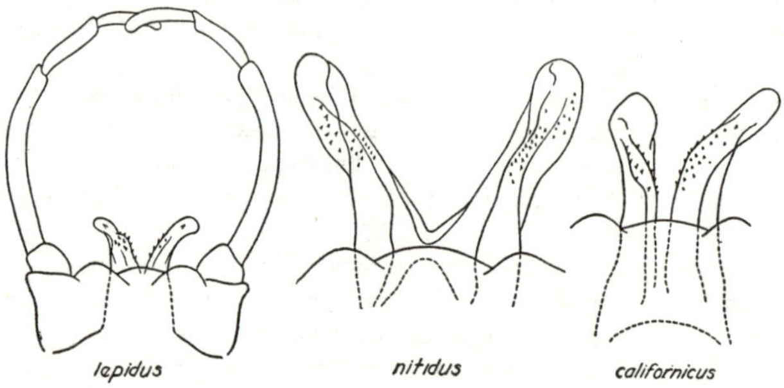

FIG. 108. Male genitalia of three species of Ironodes.

Body of nymph somewhat convex dorsally, flattened ventrally (see fig. 110, after Seemann). Lateral and frontal margins of head greatly expanded, covering all mouthparts. Frontal margin slightly emarginate at median line; densely fringed with hairs. Labrum 1½ to 2 times as wide as long; frontal margin widely emarginate in median area; antero-lateral angles very slightly rounded, lateral margins sloping toward base. Inner canine ¾ to ⅞ as long as outer one; a few hairs in lacinial region. Upper margin of mandible slopes sharply from base of canines to molar surface. A thickened area below molar surface, subparallel to it. Galea-lacinia of maxilla shaped much as in *pulchellum* group of Stenonema, but with a fringe of long hairs along margin next to palp and on upper margin; no triad of stout spines at tip, only a few slender straight spines. Second joint of palp clavate, its upper apical half bearing many series of short spines curved at tips. Glossae of labium slightly concave on their inner margin, strongly convex on outer margin; quite close together, slightly convergent apically. Median portion of hypopharynx much narrower than either lateral portion, likewise considerably shorter.

Pronotum emarginate on frontal margin, lateral margins slightly expanded; an emargination on posterior margin at median line and a smaller one on each side just within postero-lateral angle. A transverse ridge on pronotum, one-third of distance from anterior margin, in dorsal area; posterior two-thirds of this sclerite on a higher level than anterior portion. Two prominent flange-like projections on each coxa. Femoral flange rather short and blunt. Claws with pectinations near apex. Gill-bearing segments of abdomen bear very short blunt postero-lateral spines; these are wanting on apical segments. A row of submedian spines on each side of tergites 1–9, on posterior margins; each spine arises about one-third from anterior margin. Fibrillar portions of gills well developed, appear dorsal. Lamellate portions rather small, obovate; first pair smallest, seventh pair only slightly larger; intermediate pairs largest. Tails two.

Genotype—*I. nitidus* (in Iron) Eaton.

Distribution—Nearctic; western North America.

KEY TO SPECIES OF IRONODES

Male adults*

1—Large species, 11–14 mm. in length 2
—Smaller species, 8–9½ mm. in length 4

2—Abdominal segments yellowish brown; wing membrane very faintly brown-tinted, or wholly pale except on costal margin; mesonotum reddish or yellowish brown 3
—Abdominal segments brownish black; wing membrane faintly amber-tinted; mesonotum black-brown . . . . . . **flavipennis**

3—Mesonotum reddish brown; second joint of forceps strongly bowed; small ventral spines on penes numerous . . . . . . **californicus**
—Mesonotum yellow-brown; second joint of forceps less strongly bowed; small ventral spines on penes less numerous . . . . . . **nitidus**

4—Abdominal segments reddish brown except for rather wide pale anterior margin; no such subapical spines on penes, as in next species; stigmatic area of fore wing not darkened . . . . . . **arctus**
—Abdominal segments yellowish brown except for narrow pale anterior margin; small subapical spines on each division of penes; stigmatic area of fore wing brown-tinted . . . . . . **lepidus**

VERIFICATION TABLE

Male adults†

Species	*Wing length*	*Wings tinted*	*Stigmatic area*	*Tergites 2 to 7*	*Distr.*
arctus	9½	No	Not tinted	Red-brn.	Ore.
californicus	11–12	No	Brown	Yell.-brn.	Calif.
flavipennis	12–13	Amber	Smoky	Br.-black	B. C.
geminatus	8– 9½	Yellow	Yellow	Yellowish	W
lepidus	9– 9½	No	Brown	Yell.-brn.	Calif.
nitidus	11–14	Faint	Brown	Yell.-brn.	Pacific

* Does not include *geminatus*.

† Except *geminatus*.

KEY TO THE SPECIES

Nymphs

1—Body dark red-brown dorsally; apical third of each tarsus darker than remaining portions of legs 2

—Body darker, more blackish brown; all joints of legs concolorous, no portion darker than others . . . . **flavipennis**

2—Tracheae of gills distinct; no distinct dark lateral streaks on abdominal sternites . . . . **californicus**

—Tracheae of gills indistinct; distinct dark lateral streaks on basal and middle sternites . . . . **nitidus**

Ironodes arctus Traver

1935—Traver—Canad. Ent. 67: 33. Fig. . . . . Ore.

Length: body 9; wing 9½ mm.

Thorax darker red-brown, penes more slender than in allied species; abdominal segments reddish.

Eyes of male large, contiguous apically. Head purple-brown. Thorax dark red-brown; mesonotum almost black-brown. Paler red-brown areas on pleura; depressions on each side of scutellum black-brown. Fore legs red-brown; femora yellow-tinged; apex of tibia, and tarsal joinings, blackish. Middle and hind legs paler red-brown,

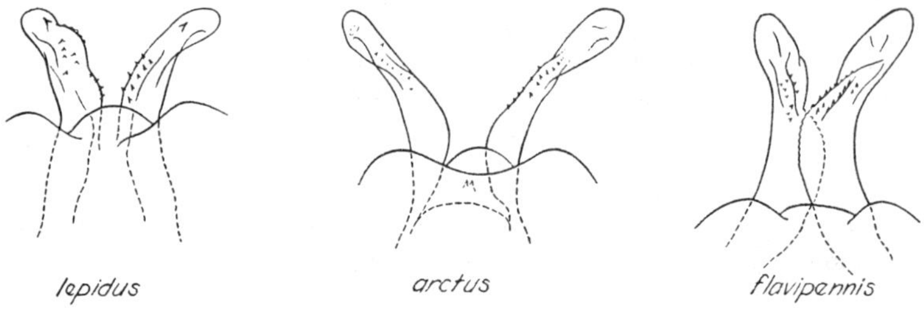

FIG. 109. Male genitalia of three species of Ironodes.

with yellow tinge; tarsi deep smoky; claws and all joinings darker. Wings hyaline; venation dark red-brown; stigmatic area *not* brown-tinged. 6 to 7 basal costal cross veins; about 9 stigmatic veins, simple, almost straight.

Abdominal segments 2–7 semi-hyaline, light red-brown; anterior fourth of each whitish hyaline, except sternite 7, so that abdomen appears annulate. Pleural fold red-brown. Posterior margins not noticeably darker or more opaque than preceding area, except slightly on dorsal portions of tergites. Segments 8–10 opaque, darker red-brown. Tails dark red-brown. Penes (see fig. 109) more slender than in *I. lepidus* and with no such subapical spines as in that species. Second joint of the forceps much as in *nitidus*. Two small spine-like structures between bases of penes may be artifacts.

The type material is in the Canadian National Collection; we have seen no other specimens. The distinct reddish tinge of the abdomen of these specimens may be due to too long a sojourn in the cyanide bottle.

Ironodes californicus Banks

1910—Banks—Canad. Ent. 42: 202 (Iron)..Calif.
1924—McDunnough—Canad. Ent. 56: 129 (Iron).

Length: body 10; wing 11–12 mm.

Very similar to *I. nitidus;* thoracic notum darker, more distinct dark markings on pleura.

Head light red-brown. Thorax pale brownish, according to original description. Specimens in our collection, determined by Dr. Banks as being of this species, have the notum and sternum distinctly red-brown, pleura paler. Quite distinct narrow black markings on pleura anterior to fore wing, above and anterior to middle and hind legs; coxae black-margined, with an oblique black streak across each; trochanters black-tipped. Depressions on each side of scutella dark brown. Legs light red-brown. Outer margins of femora narrowly darker brown; apex of fore tibia, tarsal joinings and claws dark brown. Wings hyaline, venation red-brown, somewhat more distinct than in *I. nitidus;* cross veins likewise more numerous than in that species. Tendency for costal and subcostal cross veins to be slightly heavier than others. Humeral cross vein, and extreme base of longitudinal veins of costal border, reddish. Stigmatic area faintly brown-tinged; remainder of costal margin even more faintly tinged with same color. 9–10 basal costal cross veins; 4 to 5 between bulla and stigma; about 14 in stigmatic area, most of these simple and distinctly aslant.

Abdominal segments 2–7 semi-hyaline, yellow-brown; posterior margins opaque, blackish on dorsal part of tergites. Tracheae faintly outlined in grey. Segments 8–10 opaque, reddish brown. Tails red-brown, somewhat darker basally. Penes as in fig. 108; small ventral spines somewhat more numerous than in *I. nitidus.* Second joint of forceps more strongly bowed near apex, and relatively somewhat longer than in *nitidus.*

This species is closely allied to *nitidus,* but is apparently distinguished from it as indicated above. We have specimens from Icehouse Canyon, San Gabriel Mts., Calif., taken April 15, 1933 (C. D. Michener). The nymph figured by Mrs. Seemann (J. Ent. and Zool. 19, 1910; Pl. 4) may be of this species (see fig. 110).

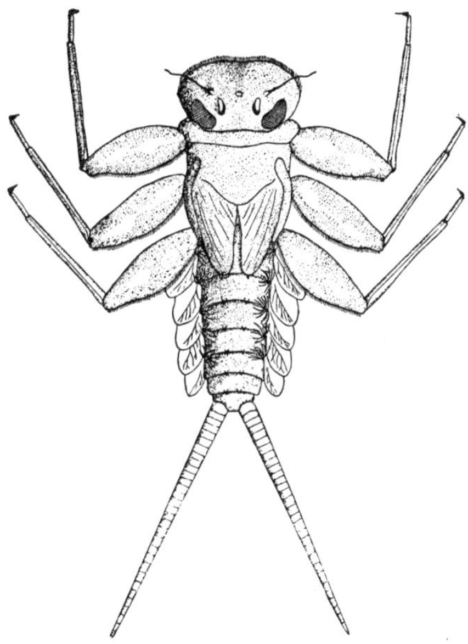

FIG. 110. Nymph of *Ironodes sp.* (from Seemann).

Nymphs taken in Mill Creek Canyon near Redlands, Calif., on May 19, 1932 (C. H. Abbott) may be of this species. Body dark red-brown dorsally, not as blackish as in *I. flavipennis;* similar thoracic markings, also a zigzag lateral line on the pronotum. Legs somewhat paler red-brown; femoral markings as in *flavipennis;* apical third of tarsus distinctly darker than other portions of legs. Markings on abdominal tergite as in *flavipennis*. Ventral markings also very similar, but in addition an oblique dark lateral line from anterior to posterior margin, on each division of thoacic sternum. Gills somewhat paler than in *I. flavipennis*, deep smoky with purplish tinge; tracheae distinct.

Ironodes flavipennis Traver

1935—Traver—Canad. Ent. 67: 33. Fig. .... B. C.

Length: body 9–10; wing 12–13 mm.

A dark brown species; wings of female distinctly amber-tinged, very faintly so in male.

Head dark red-brown; frontal portion paler, semi-hyaline; vertex and area around bases of ocelli deep red. Eyes large, almost contiguous apically. Thoracic notum and sternum blackish brown tinged with deep red; pleura paler red-brown. A conspicuous creamy patch anterior to base of fore wing, containing a narrow reddish streak. A distinct red spot above fore coxa, near end of creamy patch. Anterior to base of hind wing is a smaller creamy patch. A reddish oblique line anterior to base of middle leg; all coxae faintly red-tinged. Fore leg blackish brown with olive shading; femur somewhat paler than tibia and tarsus, tinged with reddish brown. Femora of middle and hind legs greenish yellow; tibiae dull olive with smoky tinge; tarsi deep smoky, almost blackish. Wings hyaline; venation very dark brown, except longitudinal veins of costal margin, which are amber-brown. Stigmatic area tinged with smoky brown; remaining portion of costal margin faintly amber-tinged in male, the remainder of wing likewise very pale amber. In female, entire wing quite uniformly tinted with rather deep greenish amber.

Abdominal segments 2–7 semi-hyaline, dark brown with faint red-brown tinge. Anterior margins slightly paler, more yellowish in some specimens. Posterior margins opaque, deeper brown. Pleural fold narrowly dark brown. Segments 8–10 opaque, deeper brown than preceding segments. Sternite 9 brighter red-brown. Tails and genitalia deep black-brown. Penes (see fig. 109) somewhat smaller than in *I. nitidus* the small ventral spines rather less numerous. Second joint of the forceps slightly more bowed in middle region than in *nitidus*. Nymph almost black-brown dorsally, yellowish ventrally; legs concolorous with body, a white mark basally on each femur. Gills purplish black, tracheae indistinct.

We have several specimens of this species, presented to the Cornell collection by Dr. McDunnough, from Summerland, B. C. (A. N. Gartrell).

? **Ironodes geminatus** Eaton

1885—Eaton—Revis. Monogr. p. 250.......... Wash., Calif., Colo.
1933—McDunnough—Canad. Ent. 65: 75.

Length: body ?; wing 8–9½ mm.

This species was described from male and female subimagos and a female imago. Description comparative, the species said to resemble *Heptagenia sulphurea* Mull. European and *H. elegantula* Etn., in the female sex. Vertex of head of female "suffused with light reddish brown, no black spots near the eyes either on the vertex or on the face, nor any at the bases of the coxae." Wings hyaline; membrane tinted uniformly with very pale yellowish. Venation brownish black, the main longitudinal veins being amber-colored near the base. Tails whitish, not darker at joinings.

Eaton placed this species in Cinygma; however, the characters given do not accord well either with *Cinygma integrum* and *C. lyriformis*, or with the seven species now placed in Cinygmula. It is possible that *geminatus* belongs in Iron or Epeorus; for the present we transfer it to Ironodes; we have seen no specimens that accord well with the original description.

Ironodes lepidus Traver

1935—Traver—Canad. Ent. 67: 35. Fig. .... Calif.

Length: body 8–9; wing 9–9½ mm.

Quite similar to *I. nitidus*, but smaller; a small spine-like structure is present near apex of each division of penes.

Frontal portion of head pale red-brown; a narrow creamy line below base of antennae; area around bases of antennae and ocelli, and vertex of head, quite bright red-brown. Thorax red-brown, olive-tinged; narrow yellowish streaks anterior to fore wing, below wing bases, and above bases of middle and hind legs. Very narrow blackish oblique line posterior to each coxa; apical margins of coxae and trochanters deep red-brown. Pronotum shaded with smoky brown. Legs light red-brown; tarsi smoky, or shaded with olive brown. Apex of fore tibia, tarsal joinings and claws somewhat darker. Wings hyaline; venation dark red-brown. Pale brownish tinge in stigmatic area. About 7 basal costal cross veins; 4 or 5 between bulla and stigma; 10 to 11 stigmatic veins, simple, very slightly aslant.

Abdominal segments 2–7 semi-hyaline; yellowish brown with slight olive tinge. Anterior margins narrowly pale hyaline on tergites and sternites, producing an indistinct annulate appearance. Posterior margins narrowly opaque; a blackish area on this margin, in dorsal portion of each tergite. Tracheae outlined in brownish black; a continuous dark wavy line along pleural fold. Segments 8–10 opaque, slightly deeper in color than preceding segments, and with reddish tinge; pleural fold narrowly creamy. Tails and genitalia reddish to olive brown. Penes (see figs. 108, 109) with a small spine-like structure near apex of each division, which is not present in other species of this genus. Numerous small spines on ventral surface.

Specimens in alcohol differ from the above description as follows: Thorax darker red-brown; yellow markings less distinct, blackish pencilings more pronounced; fore tarsus yellowish; pale anterior margins of abdominal segments less conspicuous. Four specimens of this species, from Silver Canyon, White Mts., Calif., are in the Cornell collection.

Ironodes nitidus Eaton

1885—Eaton—Revis. Monogr. p. 246. Fig. (Iron). .... Ore., Calif.
1924—McDunnough—Canad. Ent. 56: 129 (Iron).
1933—Traver—J. N. Y. Ent. Soc. 41: 118 (Iron).

Length: body 9–10; wing 11–14 mm.

A large yellowish brown species, slightly paler than the allied *I. californicus*.

Frontal margin of head yellowish brown, reddish brown elsewhere; a distinct reddish spot on vertex between eyes, sometimes with reddish shading also around bases of antennae and ocelli. Thorax light yellowish brown. Pronotum extensively shaded with smoky brown. Mesonotum not much darker than pleura; depressions on each side of scutellum dark brown. Many specimens have a creamy area anterior to base of fore wing, containing two red streaks; creamy streaks may be present also on pleura, below wing roots. Reddish areas on pleura above base of fore wing, above base of fore coxa, and anterior to bases of middle and hind coxae. Reddish shading on coxae. Fore femur and tibia brownish amber to light brown; tarsus "whitish ochre" (Eaton); apex of tibia, distal tarsal joint, claw and tarsal joinings darker brown.

Middle and hind legs pale yellowish; tibiae and tarsi tinged with olive brown. Bases of tibiae, tarsal joinings, distal tarsal joints and claws dark brown. Wings hyaline, iridescent; stigmatic area of fore wing "tinted with extremely light raw-umber; the remainder of the marginal and submarginal areas more faintly so tinted, and the other part of the wing imperceptibly so" (Eaton). Venation red-brown, quite distinct. 8–9 basal costal cross veins; about 4 veins between bulla and stigma; 12–14 stigmatic veins, mostly simple, distinctly aslant.

Abdominal segments 2–7 semi-hyaline, yellowish brown with distinct olive tinge; posterior margins opaque, darker, this opaque area filling the postero-lateral angles; on dorsal portion of each tergite, this margin is almost black. Anterior margins of sternites very narrowly whitish hyaline, most evident in antero-lateral angles. Segments 8–10 opaque; tergites yellowish white with faint red-brown tinge, or "light brown ochre, sometimes with the dorsal joinings light brown" (Eaton). Genitalia smoky brown; penes as in fig. 108. Second joint of forceps not quite as strongly bowed near apex as in *I. lepidus*, and much less bowed than in the allied *I. californicus*. Tails very pale yellow-brown beyond middle, slightly darker shade of same color basally.

Very similar to *I. californicus*, which we had first considered to be synonymous with it. Apparent differences between these species, other than noted above, are indicated in description of *californicus*. We have specimens of *I. nitidus* in the Cornell collection from Corvallis, Ore. (H. A. Scullen), and have seen others from Luckiamute River, King's Valley, Ore., lent us by Dr. McDunnough. Smaller specimens in our collection from Alsea, Oregon (H. A. Scullen), and others we have seen from Mary's Peak, Ore. (in Canadian National collection) appear to be similar except for size differences. Nymphs from Summerland, B. C., lent us for study by Dr. McDunnough, are very similar to those from California which we are tentatively calling *I. californicus*. The Summerland nymphs are slightly darker red-brown dorsally, tracheae of gills less distinct; abdominal sternites with lateral dark streaks.

Genus **Ironopsis** Traver

1935—Traver—Canad. Ent. 67: 36.

Large mayflies, purplish brown in color. Eyes of male large, contiguous apically. Posterior margin of head of female somewhat emarginate. Posterior margin of pronotum deeply excavated in median area. Fore leg of male considerably longer than body. Femur about 3/4 as long as tibia. Basal fore tarsal joint subequal to 2nd; 3rd slightly longer than 2nd; 4th subequal to 2nd; 5th about 1/2 as long as 4th. Fore femur of female almost as long as tibia; basal tarsal joint subequal to 3rd and 5th joints, about 3/4 as long as 2nd. In both sexes, basal tarsal joint of third leg subequal to 2nd; 3rd and 4th joints subequal, shorter than 2nd; 5th joint longest, equal to 2nd and 3rd combined. Claws dissimilar on all tarsi. Typical Heptagenine venation. Stigmatic cross veins more or less strongly anastomosed, with two distinct rows of cells in *I. permagnus*, and in some specimens of *I. grandis*. A rather long intercalary may be present between Cu_1 and the first set of the paired intercalaries. Forks of radial sector in hind wing arise at about the same point.

Apical margin of forceps base variable; in *I. grandis*, with rounded lateral projections; in *I. permagnus*, quite regularly undulate. Forceps four-jointed; the two distal joints together are longer than the long second joint. Penes rod-like, divergent at tips, and bearing a cluster of small spines on ventral surface some distance below apex. Apical margin of subanal plate of female with a V-shaped median cleft.

Nymph very similar to that of Iron (*longimanus* group). Frontal margin of head, and lateral margins anterior to eyes, much expanded; frontal margin very slightly emarginate at median line. Canines wider and blunter than in Iron; a row of hairs on each mandible, in lacinial region. Two thickened portions on mandible below molar region; one parallel to this region, the lower one on inner margin at right angles to molar surface. Glossae of labium somewhat conical, divergent apically, rather widely separated at base. A triad of stout spines at tip of galea-lacinia, as in Iron. Labrum deflexed; its antero-lateral margins angulate, not rounded as in Iron; lateral margins slope inward toward base; frontal margin almost straight, very slightly emarginate at median line. Fringe of long hairs on femur, tibia and tarsus; three lateral pectinations near apex of each claw. Antero-lateral margin of pronotum rounded. Extensions of postero-lateral margins of abdominal segments very short and blunt, much less developed than in any Iron nymph known. Gills of first pair very large; anterior lobes greatly expanded, meeting beneath body of nymph. Gills of intermediate pairs very well developed, thickened on outer margin and overlapping, as in Iron. Gills of seventh pair meet beneath body. Fibrillar portion of gills short, not prominent, appears dorsal. A prominent ridge of rather long hairs is present along the mid-dorsal line of all tergites.

Genotype—*I. grandis* (in Iron), McDunnough.

Distribution—Nearctic; western North America.

KEY TO THE SPECIES

Male adults

1—Wing of male 13–16 mm. in length; mesonotum light brown; small spines on ventral surface of penes relatively few in number .... **grandis**

—Wing of male 17–19 mm. in length; mesonotum black-brown; small spines on ventral surface of penes very numerous .... **permagnus**

Ironopsis grandis McDunnough

1924—McDunnough—Canad. Ent. 56: 130. Fig. (Iron) .... Alta.
1928—McDunnough—Canad. Ent. 60: 9.
1933—Traver—J. N. Y. Ent. Soc. 41: 118 (Iron).

Length: body 11; wing 13–16 mm.

Somewhat smaller than *I. permagnus;* thoracic notum paler; differs also in detailed structure of penes.

Head light brown. Smoky band across frontal margin of clypeus; paler shading around bases of antennae. Thorax light olivaceous brown, mesonotum somewhat reddish brown. Two black lines anterior to wing roots, enclosed in a light olive brown area, are very prominent; one of these extends down on prothorax to base of fore coxa. Obscure darker markings on pleura above bases of middle and hind legs. Fore leg typically blackish brown, paler at base of femur. In some specimens, femur and tibia are reddish brown, tarsus smoky to olive brown; femur shaded with olive brown, and with faint black pencilings. Middle and hind legs light olive to reddish brown; tarsi smoky, darker than tibiae; no darker markings. Wings hyaline; venation fine, dark. Stigmatic area faintly brown-tinged; cross veins very strongly anastomosed. Humeral cross vein faintly dark, especially near subcosta. About 11 costal cross veins before bulla.

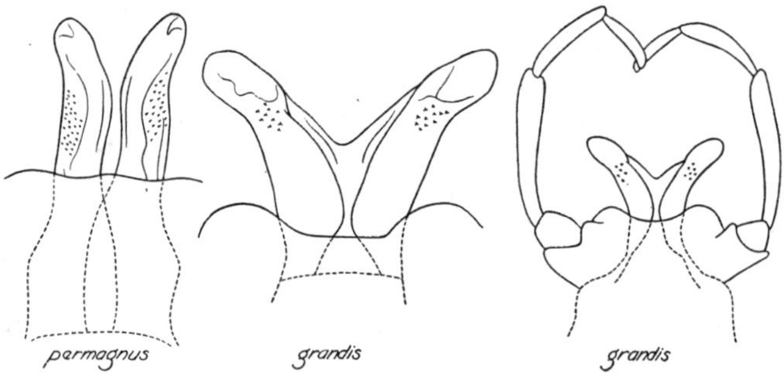

FIG. 111. Male genitalia of Ironopsis.

Abdominal tergites deep brown; lateral margins widely pale, yellowish; on tergites 4–8, pale yellowish submedian patches are based on the posterior margins; these patches are large on tergites 5–8, "occupying more than half the segmental area and separated by a narrow brown line from each other" (McD.). The anterior third of each tergite is occupied by the brown area, which "juts into the yellow patches dorsally as a dark triangle, laterally it separates these patches from the yellow lateral border and broadens out on the posterior portion of each segments, causing the dorsal edge of the yellow border to appear very irregular in outline" (McD.). Tergites 9 and 10 are largely yellowish. Sternites purplish brown, paler posteriorly. In some specimens, the pale areas on tergites are largely obscured by reddish shading. Forceps smoky brown, base light brown. Penes as in fig. 111. Tails deep blackish brown.

The paler mesonotum, the smaller size, and the relatively few spines on ventral surface of each division of penes, distinguish this species from *permagnus*. The apical margin of the forceps base is likewise more deeply emarginate in the median area. We have seen specimens of this species, kindly loaned by Dr. McDunnough, from Deep Creek, Peachland, B. C. and Mt. Apex, Hedley, B. C. (A. N. Gartrell), also from Waterton Lakes, Alta. (J. H. Pepper). Nymphs in the Cornell collection from Marble Fork, near Three Rivers, Calif. (J. C. Bradley) and Glacier National Park, Mont. (M. J. M.). are similar to those from Peachland, B. C., lent us by Dr. McDunnough. Structural features are given in the generic description. Body deep purplish brown; lateral margins of pronotum, and median area of tergites 1–4 and 8–9, largely yellowish. Median area of sternites mahogany brown, lateral areas pale; a series of dark lateral marks at edge of median dark area.

Ironopsis permagnus Traver

1935—Traver—Canad. Ent. 67: 37. Fig. .... Ore.

Length: body 13–15; wing 17–19 mm.

Larger than *I. grandis*, with darker mesonotum; spining on ventral surface of each division of penes much more extensive.

Head deep brown; no dark band on frontal margin; shaded with black around bases of antennae and sides of median carina. Thorax almost unicolorous black-brown. A single short black streak anterior to wing roots. Area anterior to wing roots often yellowish. Fore femur black-brown, paler at extreme base; tibia and tarsus blackish red-brown. Middle and hind legs light red-brown; femora shaded extensively with blackish; tarsi slightly deeper in color than tibiae. Coxae and trochanters of all legs black-tipped. Wings hyaline. Costa, subcosta and radius yellowish at extreme base, costa yellowish also in apical third; remainder of these veins thickened, dark reddish to purplish brown. Other longitudinal veins finer, paler red-brown; cross veins still finer, yellowish to light red-brown. 11 to 12 basal costal cross veins, very faint; humeral cross vein pale.

Abdominal tergites largely yellowish to olive brown. Large black-brown patches occupy tergites 2 and 3, leaving margins and median line pale. Tergites 4 and 5 with more restricted dark central areas, the margins irregular; lateral and antero-lateral margins pale, mid-dorsal line obscured. Tergites 6–8 with black-brown median triangle based on anterior margin, its apex extending backward along median line about ¾ the length of each tergite, leaving large pale postero-lateral triangles, with a small rounded brown mark in each. Tergites 9 and 10 more reddish brown; anterior margins narrowly black; black spot laterally on 9; median line and median area of posterior margin of 10 blackish. Sternites olive brown; basal ones shaded with red-brown. A rounded smoky spot near antero-lateral angle of each intermediate sternite, the two spots on each sternite connected by a smoky transverse band. Sternite 8 shaded with red-brown; 9 deeper brown, with greyish posterior and postero-lateral margins. Pleural fold greyish on all segments. Tails deep red-brown, almost black-brown at base; joinings narrowly paler. Forceps deep red-brown, paler apically. Penes light red-brown, with many small spines on ventral surface of each division (see fig. 111).

The type material, from Cloud Cap Inn, Mt. Hood, Ore. (H. A. Scullen), is in the Cornell University collection. We have seen no other specimens.

Genus **Anepeorus** McDunnough

1925—McDunnough—Canad. Ent. 57: 190.
1933—Traver—J. N. Y. Ent. Soc. 41: 122.

Eyes of male moderately large, not contiguous apically. Fore femur of male ¾ as long as tibia; tarsus about ⅔ as long as tibia. Basal tarsal joint more than ½ of the second, which is longer than the third; fourth longer than fifth, subequal to or slightly longer than basal one. Hind tibia slightly shorter than femur; basal tarsal joint subequal to second, distinctly longer than third and fourth joints, which are subequal to each other; fifth joint as long as second, third and fourth combined. Claws of fore leg of male equal, blunt; claws of middle and hind legs dissimilar. Wings with typical Heptagenine venation. Basal costal cross veins more or less weak; stigmatic cross veins strong, somewhat aslant, often forking. Forks of first and second branch of radial sector in hind wing arise at about same distance from margin. Median vein of hind wing forked about ⅓ of distance from base. Forceps four-jointed; forceps base deeply excavated on apical margin. Penes "united and broadly triangularly expanded at the base, apically forming two simple cylindrical lobes, separated by a V-shaped incision" (McDunnough). Nymph not known.

Genotype—*A. rusticus* McDunnough.

Distribution—Nearctic; southeast, central and western North America.

KEY TO THE SPECIES

Male adults

1—Abdominal segments 2–7 whitish hyaline; thorax yellow with pinkish tinge; legs white .... **simplex**

—Abdominal segments 2–7 semi-hyaline, brownish; thorax dull brown; fore legs brown, others ochreous .... **rusticus**

Anepeorus rusticus McDunnough

1925—McDunnough—Canad. Ent. 57: 190. Fig. .... Sask.
1933—Traver—J. N. Y. Ent. Soc. 41: 122.

Length: body 8; wing 9 mm.

A brownish species, thus distinct from the pale *A. simplex.*

Head brown; eye margins opposite antennae tinged with ochreous. Thorax dull brown; pale ochreous tinges on pleura, especially anterior to wing roots. Fore legs brown, the tarsus rather blackish; middle and hind legs pale ochreous. Wings hyaline. Longitudinal veins pale, cross veins darker; the latter are very fine except in apical portion of fore wing, where they are thickened and "slightly anastomosed along costa" (McD.). Abdominal tergites dull clay-brown; basal segments somewhat semi-hyaline; tergites 8–10 light ochreous brown. Faint traces of submedian and lateral rows of small dark brown spots are present. Sternites somewhat paler than tergites; each sternite with "two small central dots and narrow lateral oblique dashes." Forceps brownish ochreous; lateral apical portions of penes with fewer lobes than in *A. simplex.* Tails dirty whitish. Genitalia shown in fig. 112.

We have a specimen of this species, and have seen other specimens, from the Carnegie Museum collection, taken in Uinta Co., Utah, July 13, 1911. These are slightly smaller than indicated above, and the joinings of the abdominal segments, especially the basal ones, are opaque, presenting rather an annulate appearance; tails pale brownish.

Anepeorus simplex Walsh

1863—Walsh—Proc. Ent. Soc. Philad. 2: 204 (Heptagenia) .... Ill.
1871—Eaton—Trans. Ent. Soc. London: p. 141 (Heptagenia).
1885—Eaton—Revis. Monogr.: p. 300 (Heptagenia).
1929—McDunnough—Canad. Ent. 61: 178 (transferred to Anepeorus) .... Ia.
1933—Traver—J. N. Y. Ent. Soc. 41: 122.

Length: body 6–8; wing 7½–9½ mm.

A yellowish white species, as opposed to the general brown coloration of *A. rusticus.*

Head yellowish; eyes of living insect bright greenish yellow above and below; ocelli black-ringed at base. Thoracic notum flesh-colored; remainder of thorax yellow with distinct pinkish tinge. Legs white; femora yellowish; apex of fore tibia and tarsal joinings, and the "extreme tips of all the tarsi" (Walsh) brownish. Wings hyaline; all veins fine except cross veins in apical portion of costal space, which are coarse. Cross veins in basal portion of costal space very fine, indistinct, lacking in basal 1/5 of space, except for humeral cross vein. Veins in apical two-thirds of fore wing brownish; all others hyaline (Walsh says, all costal cross veins occasionally brownish). Abdominal segments 2–7 hyaline whitish, immaculate; segments 8–10 opaque, yellowish; in our Georgia specimen, tergites distinctly tinged with rose, especially on posterior margins of 9 and 10. Genitalia yellowish. Lateral apical portion of penes with two large and one smaller lobe on each side. Tails white, dusky at tips. Genitalia as in fig. 112.

The general pale coloration of this species distinguishes it at once from *A. rusticus*, the only other species thus far known in the genus. We have a single male specimen of this species from the Chattahoochee River at Atlanta, Ga., taken by Prof. P. W. Fattig.

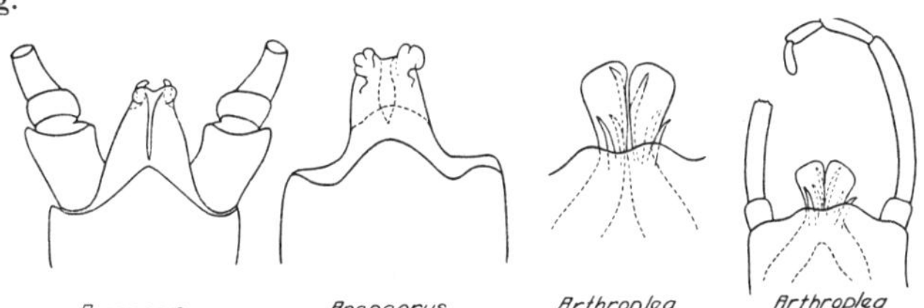

FIG. 112. Male genitalia (*rusticus* and *simplex*, after McDunnough; two enlargements of *bipunctata*).

Genus **Arthroplea** Bengtsson

1908—Bengtsson—Vet.-Akad. Arsbok: p. 239 (Arthroplea); p. 242 (Remipalpus, nymph).
1909—Bengtsson—Lunds Univ. Arssk. Afd. 2, B. 5; 17.
1920—Ulmer—Stett. Ent. Zeit. 81: 142.
1929—Blair—Ent. Mo. Mag. 65: 254 (Haplogenia). Figs.
1930—Bengtsson—Lunds Univ. Arssk. Afd. 2, B. 26: 24. Figs.
1930—Ide—Canad. Ent. 62: 42 (Cinygma, nymph). Figs.
1933—Traver—J. N. Y. Ent. Soc. 41: 122.

Eyes of male large, almost contiguous apically. Fore tibia almost as long as femur; tarsus twice the length of tibia. Basal fore tarsal joint about two-thirds the length of the second; third slightly longer than second; fourth equal to first, fifth about 1/2 of fourth. In female, basal fore tarsal joint subequal to second; third and fourth progressively shorter; fifth subequal to third. Tarsus slightly longer than femur, tibia about 3/4 as long as femur. In both sexes, hind tibia about 3/4 as long as femur; tarsus subequal to tibia. Basal tarsal joint somewhat longer than second; distal joint almost equal to joints two, three and four combined. Claws dissimilar on all tarsi. Fore wing with typical Heptagenine venation; basal costal cross veins rather weak; stigmatic cross veins almost straight, relatively few in number. In hind wing, second branch of radial sector (R_{4+5}) *not forked;* this character is unique among Nearctic Heptageniidae. Posterior margin of head of female almost straight. Posterior margin of pronotum deeply emarginate.

Forceps *five*-jointed; 2nd joint longest, fifth joint slender, nearly as long as fourth; the three distal joints together not as long as the second. Forceps base narrow at origin of forceps limbs, produced backward into two rounded lobes between the forceps, with a slight emargination at median line. Penes somewhat as in Cinygmula, but often lie so close together as to appear united; wide at base, narrowed about one-half the distance to the apex, above which point each division is somewhat expanded. A pair of long spines arise at narrowest portion, extending backward and divergent at tips; a smaller subapical spine on each division of penes. Subanal plate of female well developed, its apical margin rounded, entire.

Frontal margin of head of nymph somewhat emarginate at median area. Mandible rather as in Heptagenia, but with greater development of lacinia. Maxillary palp unique in its extreme length, being more than six times as long as the galea-lacinia; 2-jointed, second joint about twice the length of the basal one, "curved and provided with long moveable setae arranged in two rows, thus forming a basket-like apparatus"(Ide). Labium likewise unique; palp 2-jointed, second joint about 1½ times the length of the basal one, "concave along its medial surface and armed with long moveable bristles" (Ide). Glossae large, curved; paraglossae smaller, likewise curved; in appearance unlike any other known nymph of this family. Gills present on segments 1–7; single, pointed at tip, each with a distinct forward-projecting anterior lobe, less extensive on first gill; tracheae palmately branched. Postero-lateral spines present on abdominal segments 5–9, longest on 9. Tails three; laterals fringed on inner side only.

Genotype—*A. congener* Bengtsson.

Distribution—Palearctic; in Nearctic, occurs in eastern North America.

Arthroplea bipunctata McDunnough

1924—McDunnough—Occ. Pap. Boston Soc. Nat. Hist. 5: 76. Fig. (Cinygma)............Me., N. H.
1925—McDunnough—Trans. R. S. C.; p. 222 (Cinygma)....................................Que.
1930—Ide—Canad. Ent. 62: 42. Fig. nymph (Cinygma) p. 229 (transferred to Arthroplea).......Ont.
1933—McDunnough—Canad. Ent. 65: 75 (in discussion of Cinygmula).
1933—Traver—J. N. Y. Ent. Soc. 41: 122.

Length: body 6; wing 8 mm.

Head deep dark brown; clypeus hyaline. Thorax deep dark brown. Fore femur deep brown; tibia pale brown; tarsus dirty whitish. Basal fore tarsal joint about ⅔ the length of the second. Middle and hind legs dull yellowish; median and apical dark spots on femora; tarsal joinings and claws dark brown. Wings hyaline; a faint brownish stain at extreme base, much as in *Choroterpes basalis*. Longitudinal veins pale brown; cross veins colorless, indistinct except in the apical costal area. Abdominal tergites deep blackish brown; sternites somewhat paler, posterior margins narrowly dull whitish, "this color on the rear segments extending to the lateral portions of the dorsum" (McD.). Tails whitish, joinings narrowly dark brown. Genitalia as in fig. 112.

We have a pair of imagos of this species in the Cornell collection, from Lake Nipissing, Ont., which Dr. Ide kindly presented. We have seen no other specimens.

Family BAETIDAE

Mayflies of very diverse size and structure. Hind tarsi with three or four freely movable joints. In those groups in which five tarsal joints are present, the basal one of these is united immovably with the tibia. Eyes of the males are often divided. Venation may be greatly reduced, including the loss of the hind wing. Cubital intercalaries variable, but not occurring in two parallel pairs, except in the subfamily Metretopinae.

Nymphs likewise variable in structure and appearance; usually not flattened, often very perfectly stream-lined. Head not strongly depressed; eyes lateral in position. None are fossorial; some are sprawlers, others are climbers, still others are extremely agile swimmers.

KEY TO SUBFAMILY OF FAMILY BAETIDAE

Adults

1—Hind wings absent; outer fork of Rs (*Of*) of fore wing very deep...... **Caeninae**

—Hind wing present or absent; *Of* of forewing normal, or detached basally...... 2

2—1st anal vein of fore wing ends in outer margin; no cubital intercalaries........ **Baetiscinae**

—1st anal vein of fore wing ends in hind margin; cubital intercalaries present.... 3

3—*Of* of fore wing and posterior branch of M detached basally from their respective stems; eyes of male turbinate.......... **Baetinae**

—*Of* of fore wing and posterior branch of M normal, attached basally; eyes of male not as above.......... 4

4—*Of* of hind wing present (R_4 and R_5); tails 2.......... 5

—*Of* of hind wing absent (R_{4+5} separate); tails 3.......... 6

5—Cubital intercalaries of fore wing consist of a series of forking or sinuate veinlets attaching Cu_1 to hind margin.......... **Siphlonurinae**

—Cubital intercalaries of fore wing 2 to 4 in number, free basally, not as above.......... **Metretopinae**

6—Two pairs of cubital intercalaries, the anterior pair being the longer; 1st anal vein of fore wing attached to hind margin by a series of veinlets...... **Ametropinae**

—Two cubital intercalaries only; 1st anal vein not attached to hind margin as above.......... 7

7—Two short intercalaries in fore wing between median intercalary and posterior branch of M, also between the latter vein and Cu_1; male forceps with a single short terminal joint.......... **Ephemerellinae**

—No true intercalaries in fore wing in positions indicated above; male forceps with 2 (or 3) short terminal joints.......... **Leptophlebiinae**

VERIFICATION TABLE

Adults

	♂ *eyes*	*Of*[4]	*Cu. inter.*	*Hind wing*	*Of,*[4] *hd. w.*	*Costal angul.*	*Hind*[5] *tarsus*
Ametropinae[7]	entire	normal	4, long[2]	present	absent	acute	5
Baetinae[8]	turbin.	detach.	2, long	var.	absent	var.	4
Baetiscinae[8]	entire	normal	none	present	absent	obtuse	5
Caeninae[6], [7]	entire	deep	2, long	absent	———	———	4
Ephemerellinae[7]	grooved	normal	2, unequal	present	absent	low	4
Leptophlebiinae[7]	grooved	normal	2, short	present	absent	var.	4
Metretopinae[8]	entire[1]	normal	2 or 4	present	present	var.	5
Siphlonurinae[8]	entire[1]	normal	series[3]	present	present	var.	5

1—Horizontal line on surface marks out two areas.
2—Attached to Cu_2.
3—Forked, sinuate; attach Cu_1 to hind margin.
4—R_4 and R_5, outer fork (*Of*) of hind wing.
5—Number of joints of hind tarsus.
6—Marginal veinlets absent.
7—Tails three.
8—Tails two.

KEY TO SUBFAMILIES OF FAMILY BAETIDAE

Nymphs

1—Gills completely concealed beneath an enormously enlarged thoracic shield or carapace . . . . **Baetiscinae**

—Gills exposed; thoracic notum normal 2

2—Outer tails fringed on both sides 3

—Outer tails with heavy fringes on inner side only (may have a few short hairs on outer side) 5

3—Gills present on abdominal segments 1–7 . . . . **Leptophlebiinae**

—Gills absent from one or more of segments 1–7 4

4—Gills present on segments 3–7 or 4–7; in latter case, may be elytroid on segment 4, a rudimentary gill often present on segment 1 . . . . **Ephemerellinae**

—Gills present on segments 1–6 only, rudimentary on segment 1; gill on segment 2 elytroid, covering all those behind it . . . . **Caeninae***

5—Claws of middle and hind legs long and slender, about as long as the short tibia; claw of fore leg differs from others in structure 6

—Claws on all legs similar, sharp-pointed, much shorter than the tibia 7

6—Claw of fore leg bifid; no fleshy appendage attached to fore coxa . . . **Metretopinae**

—Claw of fore leg slender, curved, bearing several long spines; a fleshy appendage is attached to inner margin of fore coxa . . . . **Ametropinae**

7—Postero-lateral angles of the apical abdominal segments prolonged into thin flat lateral spines . . . . **Siphlonurinae****

—Postero-lateral angles of apical abdominal segments hardly more than acute, not prolonged into flat thin lateral spines . . . . **Baetinae**

* Oreianthus, of family Ephemeridae, has similar characters, but possesses two pairs of wing buds; in the Caeninae, hind wings are lacking.

** In Parameletus and some species of Ameletus, these spines are very short and weakly developed. The pincer-like process at tip of labial palp in Parameletus, and the series of pectinate spines on crown of maxilla in Ameletus, easily separate these two genera from all of the Baetinae.

VERIFICATION TABLE

Nymphs

Subfamily	All claws similar	Append.[1] f-coxa	Gills[4] on	Gill[2] elytr.	Outer[3] tails
Ametropinae	No	Yes	1–7	No	Inner
Baetinae	Yes	No	1–7	No	Inner
Baetiscinae[5]	Yes	No	1–6	No	Both
Caeninae	Yes	No	2–6	On 2	Both
Ephemerellinae	Yes	No	3 or 4, to 7	On 4, or O	Both
Leptophlebiinae	Yes	No	1–7	No	Both
Metretopinae	No	No	1–7	No	Inner
Siphlonurinae	Yes	Var.	1–7	No	Inner

1 Appendage on fore coxa.
2 Elytroid gill present.
3 Outer tails fringed on which side.
4 Well developed gills present an abdominal segments.
5 Thorax carapace-like, covering the gills.

Subfamily AMETROPINAE

Eyes of male entire. Tarsi five-jointed; basal joint of middle and hind tarsi more or less completely fused with tibia. Fore tarsus of male very long in proportion to the relatively short tibia. Claws dissimilar on all tarsi. Two pairs of cubital intercalaries in the fore wing; of these, the anterior pair is the longer, and attached basally to Cu_2. The 1st anal vein is attached to the hind margin by a series of straight and simple veinlets. Hind wing well developed, its costal angulation acute. In this wing, Of is wanting (R_{4+5}: not forked). Forceps four-jointed. Tails three, in both sexes, the middle tail almost as long as the laterals.

The nymphs are sprawling forms, in general appearance somewhat resembling those of the Metretopinae, but with distinct structural peculiarities which set them apart from all others. Our use of the term Ametropinae is much more restricted than Eaton's Section 13 (Revis. Monogr.; p. 230); it corresponds to a portion of Ulmer's Family Ametropodidae (Stett. Ent. Zeit. 81; 135).

Genus **Ametropus** Albarda

1878—Albarda—Ent. Mo. Mag. 15: 129.
1885—Eaton—Revis. Monogr. p. 231.
1920—Ulmer—Stett. Ent. Zeit. 81: 136.

Fore tarsus of male imago nearly five times as long as the tibia; first joint about one and one half times as long as the second and almost twice the length of the tibia; fourth joint equal to second; third and fifth progressively shorter. Fore tibia three fourths as long as the femur. Fore tarsus of female nearly three times as long as tibia, which is three-fifths the length of the femur. First tarsal joint slightly shorter than the tibia. The second branch of the median vein in the fore wing may be separated basally from its stem, appearing as a long intercalary. Cross veins of stigmatic area are simple, not anastomosing. Median vein of hind wing apparently forked about half way from the base (see fig. 113), and offset strongly to rearward.

Forceps base of male imago is deeply and roundly excavated, with an acute angle on each side. The basal joint of the forceps is relatively long, being almost two-thirds the length of the second; the two terminal joints are short and subequal to each other. The penes are united into a conic structure with a V-shaped apical cleft. Subanal plate of female likewise possesses a V-shaped median cleft on the apical margin.

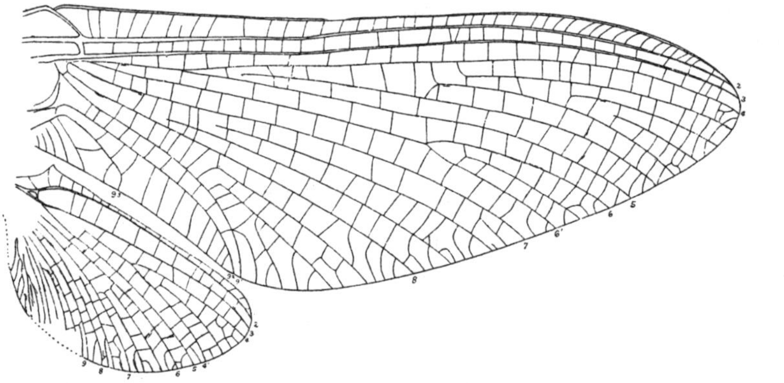

FIG. 113. Wings of *Ametropus fragilis*. (From Eaton).

Body of nymph somewhat like that of a Siphlonurine or Metretopine form. Head very small, the frontal margin cut away on each side almost to bases of antennae, so that the mouthparts are almost entirely exposed. Eyes small, set well forward on the antero-lateral angles of the head, anterior rather than lateral in position. Ocelli occupy an elevated area between and anterior to the eyes. A fringe of hairs on clypeus and tufts of hairs below each antenna. Labrum about twice as wide as long; deeply excavated on apical margin. Mandibles reminiscent of those of Siphlonurus, but with inner canine less well developed. Body of maxilla rather slender, rather similar to that structure in Hexagenia and Ephemera; 2 strong spines at tip. Maxillary palp 3-jointed; basal joint as wide as body of maxilla, and more than half as long. Middle portion of hypopharynx wider than long, excavated on apical margin; lateral portions extend laterad and forward of it. Labial palp 3-jointed, the basal joint very large, much dilated; distal joint short and conical, forming with the inner apical margin of the second joint a pincer-like process, as in Baetisca.

Pronotum slender, much wider than long. A peculiar membranous flap extends upward and backward from the posterior margin of the prosternum. Attached to the inner margin of the fore coxa is a peculiar membranous appendage, its apical portion densely clothed with hairs. Legs slender, fore leg shortest. Fore claw slender, slightly curved, bearing 4 or 5 long spines; similar spines on tarsus and tibia. Claws of middle and hind legs longer and straighter, each fully as long as the tarsus and about twice the length of the tibia. Abdomen very long, tapering rapidly beyond the seventh segment. Flattened lateral extensions present on segments 2–9; short postero-lateral spines on segments 8 and 9. Gills single, obovate; tracheae pinnately branched, but most of branches on the inner side. Tails three; outer ones heavily fringed on inner margin, and with a few short hairs on outer margin in distal portion.

Characters of the nymph are drawn from that of *A. albrighti*, which is unfortunately known in the nymphal stage only. However, the quite typical venation seems to leave no doubt as the proper placing of the species. This nymph is figured on Pl. XXXV.

Genotype—*A. fragilis* Albarda.

Distribution—Holarctic; in Nearctic, western only.

Ametropus albrighti sp. nov.

Length (female nymph, not fully mature); body 15–17; tails 6–8 mm.

Only the nymph of this species is known.

Entire body creamy white, with brown markings. A median brownish line on vertex and occiput of head. Brownish markings on posterior portion of pronotum, and on median and lateral areas of meso- and metanota, terminating in geminate median spots on the posterior margins of the latter segments. A broad dark brown band across the posterior portion of each abdominal tergite except the 10th; a dark median triangle extends forward from this band on each, but does not attain the anterior margin. Tails indistinctly darker at joinings; hairs on apical portions greyish brown. Structural characters as indicated in generic description, and as shown on Pl. XXXV.

Holotype—Female nymph (not fully mature); San Juan River, near Farmington, N. M., Oct. 14, 1934 (P. N. Albright). No. 1350.1 in Cornell University collection.

Paratypes—3 female nymphs, same locality, same collector, Nov. 1, 1933; 8 female nymphs, same data as holotype. No. 1350.3–13 in Cornell collection.

Several other nymphs of this species, collected by Mr. Albright in the same river, are in the Cornell collection. These nymphs were taken at an altitude of 5400 feet. Their habitat seems to be quite similar to that of Potamanthus.

Ametropus neavei McDunnough

1928—McDunnough—Canad. Ent. 60: 9 .... Alta.

Length (female): body 15: wing 15 mm.

A brown species; it is only known from a female imago.

Head light brown, darker around the ocelli. Prothorax dark brown, the lateral margins broadly pale. Mesonotum light brown, with darker shading anteriorly; deep blackish brown posteriorly; anterior to the scutellum is a pale spot. Femur and tibia of fore leg dark brown; tarsus pale, the first two joints somewhat smoky. Femora of middle and hind legs deep yellowish, each with a dark brown spot at the apex; tibiae and tarsi pale. Wings hyaline; cross veins pale and indistinct, longitudinal veins

PLATE XXXV

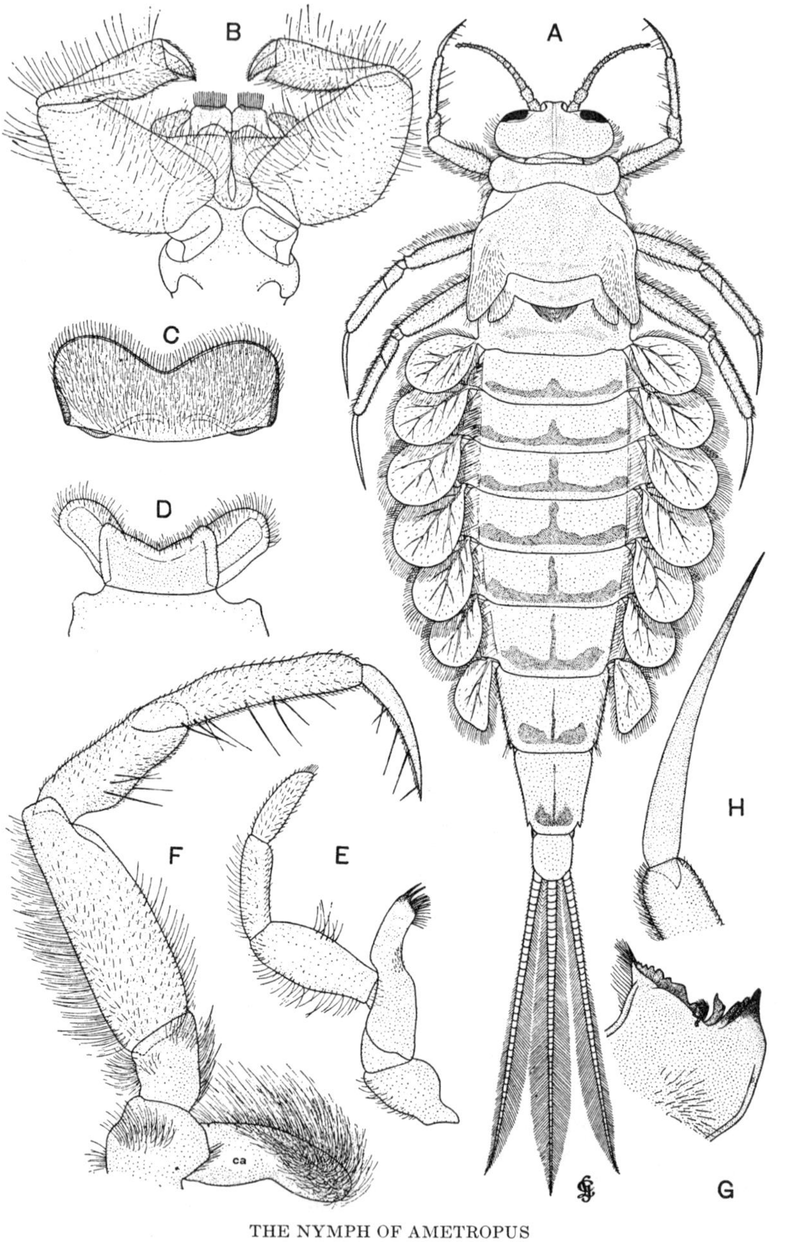

THE NYMPH OF AMETROPUS

A. Dorsal view of entire nymph; *B*, Labium; *C*, labrum; *D*, Hypopharynx; *E*, Maxilla; *F*, Fore leg, with *ca*, its coxal process of unknown function; *G*, Tip of mandible; *H*, Claw of hind foot (drawings by Chin-jén Luh).

brown. Abdominal tergites brown, with a broad pale area along the pleural fold. A double row of pale indistinct submedian spots is present; these are rather circular on the basal tergites, but elongate and tending to fork on the apical ones. Abdominal sternites light ochreous; sternites 1–8 are washed with brown along the lateral margins. A slight midventral V-shaped slit is present on the apical margin of the ninth sternite, forming two convex lobes which scarcely protrude enough to form a subanal plate. Tails three, equal in length; pale in color.

The holotype, a female imago, was taken at Old Fort Pt., Jasper Park, Alta., by Mr. F. Neave. We have seen no specimens of this species.

Subfamily **METRETOPINAE**

Tarsi five-jointed, the basal joint (except in the male fore tarsus) more or less completely fused with the tibia. First tarsal joint of middle and hind legs at least as long as the second, usually slightly longer; tibiae of middle and hind legs shorter than the femora. Claws unlike on all tarsi. Venation reminiscent of the Heptageniidae. One or two pairs of cubital intercalaries in the fore wing, which may or may not be proportioned as in that family. The first branch of the cubitus never joined to the hind margin by cross veins, as in the Siphlonurinae.

Forceps base more or less roundly excavated on the apical margin. Forceps four-jointed; second joint always the longest, the two terminal joints relatively short. Penes may be united to the apex, or divided halfway from the base into rod-shaped portions. Subanal plate of female variable. Tails two, in both sexes.

Nymphs (known for two of the genera only) are quite similar in body form to those of the Siphlonurinae.

This subfamily might well be placed in the Heptageniidae, if venation and genitalia only are considered. Since, however, the basal joint of the tarsus is not wholly free, but more or less fused with the tibia as in the Siphlonurinae, and the known nymphs are distinctly of the Siphlonurine type, it may equally well be placed in the Baetidae. It would seem to be a connecting link between the Heptageniidae on the one hand and the Siphlonurinae on the other. Ulmer (Stett. Ent. Zeit. 81; 135) includes the genus Metretopus in his family Ametropodidae.

KEY TO THE GENERA

Adults

1—A single pair of cubital intercalaries in fore wing; costal angulation of hind wing acute . . . . **Metretopus**

—Two pairs of cubital intercalaries in fore wing; costal angulation of hind wing obtuse 2

2—Fore tarsus of male fully three times the length of the tibia; basal joint of fore tarsus of male slightly longer than second; subanal plate of female obtuse . . . . **Siphloplecton**

—Fore tarsus of male only about twice as long as tibia; basal joint of male fore tarsus three-fourths of second; subanal plate of female excavated apically . . **Pseudiron**

VERIFICATION TABLE

Adults

Genus	*Occiput, female*	*Tars. vs. tibia*[1]	*Cu intercal.*	*Costal angul.*	*M_2 of hd. wg.*	*Subanal plate, ♀*
Metretopus	?	3:1	1 pair	Acute	Normal	Obtuse
Pseudiron	Convex	2:1	2 pairs	Obtuse	Angul.[2]	Excavated
Siphloplecton[3]	Concave	3:1	2 pairs	Obtuse	Rt. angle	Obtuse

KEY TO THE GENERA

Nymphs

1—Tarsi longer than tibiae; maxillary palp 2-jointed.................. **Metretopus**
—Tarsi subequal to tibiae; maxillary palp 3-jointed................ **Siphloplecton**

VERIFICATION TABLE

Nymphs

Genus	*Joints, max. palp*	*Tarsus vs. tibia*	*Divisions, fore claw*	*Gills*
Metretopus	2	Longer	Dissimilar	Single
Siphloplecton	3	Subequal	Similar	Double, 1–3[4]

Genus **Metretopus** Eaton

1901—Eaton—Ent. Mo. Mag. 2nd ser. 12: 253.
1909—Bengtsson—Lunds Univ. Arssk. Afd. 2, 5: 17.
1920—Ulmer—Stett. Ent. Zeit. 81: 136.

Medium-sized brownish mayflies. Fore tarsus of male imago about three times as long as the tibia, and twice the length of the femur. The first four tarsal joints are subequal to each other, the fifth about two-thirds the length of the first. In the fore leg of the female, the first joint is about one and one-third the length of the second; third and fourth progressively shorter, fifth equals the second. The tarsus is one and one-third times the length of the tibia, which is about equal to the femur in length. The middle and hind legs are very similar in proportion of joints to the fore leg of the female. Claws are dissimilar on all tarsi. In the fore wing, a single pair of intercalary veins is present between the branches of the cubitus; of these, the anterior vein is the longer. Cross veins in the stigmatic area tend to anastomose. Hind wing well developed, its costal angulation acute; the median vein forked near the base. See fig. 114, after Eaton.

The apical margin of the forceps base of the male is roundly excavated. Forceps four-jointed, the basal joint short and more or less completely fused with the second. Penes divided in the apical half into two

[1] Relative length of tarsus to tibia in fore leg of male imago.
[2] M_2 turns downward abruptly from main stem, for short distance.
[3] Short spines on fore tibia.
[4] In most species.

slender rod-like structures, each curving outward slightly, with a rounded lobe on the inner apical margin. (In *M. alter* Bgtssn., penes fused almost to the tip). Tails two in number, the median one represented only by a partly segmented rudiment.

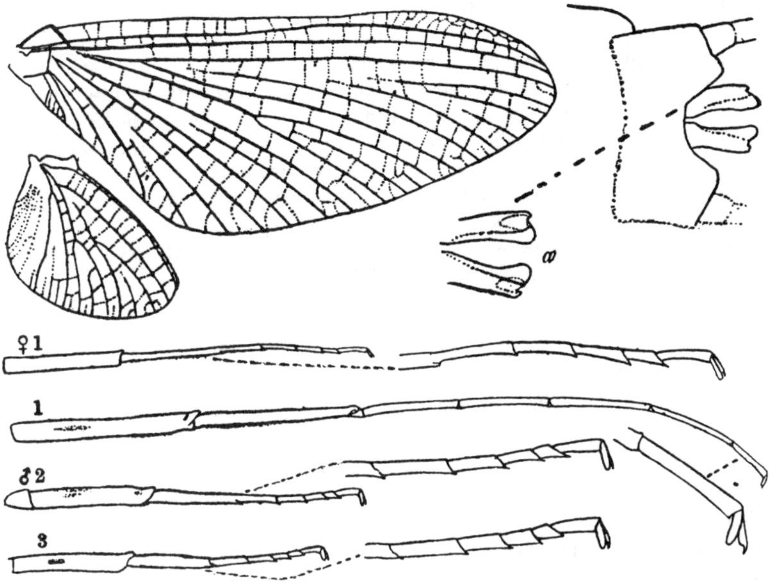

FIG. 114. Wings, legs and genitalia of *Metretopus norvegicus*. (From Eaton).

According to Bengtsson's description, the nymph seems to be quite similar to Siphloplecton, but with these differences. The gills, present on segments 1–7, are similar in form, and all are said to be single. The basal joint of the labial palp is rather longer and stouter than the second joint, its outer margin swollen; the second (end) joint is long and thin, thicker on the inner side. Tarsi plainly longer than the tibiae; femora almost as long as the tibiae and tarsi together. Claws of the middle and hind legs as in Siphloplecton, but the bifid claw of the fore leg consists of one rather short and thick division opposed to a longer and slenderer one. The maxillary palp is described as two-jointed.

Genotype—*M. norvegicus* Eaton.

Distribution—Holarctic; in Nearctic, apparently limited to western Canada.

Metretopus norvegicus Eaton

1901—Eaton—Ent. Mo. Mag. 37: 254. Fig. .... Norway
1909—Bengtsson—Lunds Univ. Arssk. N. F. 5: 16. Nymph described .... Norway
1925—McDunnough—Canad. Ent. 57: 187. .... Alta.
1928—Bengtsson—Tromso Mus. Arsh. Nath. Avd. 51: 15.

Length: body 10; wing 9 mm.

A brownish species, the middle abdominal segments of the male translucent; only species of the genus known in North America.

Thorax pitch-black above. Body somewhat discolored. Abdominal segment 1, and segments 7–10, bistre brown; the middle segments translucent. Posterior margins of tergites narrowly marked with umber brown. Mid-dorsal line of the same color, also "an oblique linear lanceolate streak on each side produced forwards from it, ****so as to constitute a tridentate marking, enclosing on each side of the dorsum a large almost right-angled triangular translucent space, rounded at the posterior angle, the shortest side of which is at the base of the segment. From the right angle of this triangle a fine tapering blackish streak proceeds towards the middle of the hypotenuse" (Eaton). Tails uniformly dark grey. The apical margin of the ninth sternite of the female is entire, and convex. The penes are divided half way to the base; each free portion is rather linear, and directed slightly outward, having a rounded lobe on the inner apical margin (see fig. 114).

Specimens of this species were taken at Slave Lake, Alta. (O. Bryant). We have not seen these, nor any European material.

Genus **Pseudiron** McDunnough

1931—McDunnough—Canad. Ent. 63: 91.

The genus was characterized from females only. Eyes small and closely appressed, as in Ametropus. On the posterior margin of the head is a rounded median projection, somewhat as in Isonychia. Wings with typical Heptagenine venation. The first tarsal joint in all the legs is slightly longer than the second; third subequal to the second; fourth one-half as long as the third; fifth equal to the first. In the hind legs the tarsus is fully as long as the tibia. All femora are considerably longer than the tibiae; claws dissimilar on all tarsi. The abdomen is very elongate. To these characters the following may be added. The posterior fork of the median vein in the fore wing is somewhat curved at its union with the main stem. The median vein in the hind wing is forked about one-fourth the distance from base to margin; the costal angulation is rounded. Tails two; a very minute rudiment of the middle tail may be present, but is unsegmented. Subanal plate excavated on apical margin.

The male imago of *meridionalis*, which seems best placed in this genus, shows the following additional characters. The first fore tarsal joint is about three-fourths as long as the second; third very slightly shorter than the second; fourth subequal to the first; fifth about one-half of the third. The fore tarsus is about twice the length of the tibia, which is slightly shorter than the femur. Middle and hind legs proportioned approximately as indicated for the female. Claws on all tarsi dissimilar. Genitalia reminiscent of *Metretopus norvegicus*.

In the characters of the male genitalia and the proportions of the tarsal joints of the middle and hind legs, this genus is close to Metretopus, but differs in possessing two pairs of cubital intercalaries in the fore wing, and in the proportions of the fore tarsal joints of the male. From Siphloplecton it is distinguished by the much elongated abdomen, the convex posterior margin of the head in the female, absence of spines on the fore tibia, proportions of the joints of the fore leg of the male, and the structure of the penes.

Genotype—*P. centralis* McDunnough.

Distribution—Nearctic; central and south-eastern United States.

Pseudiron centralis McDunnough

1931—McDunnough—Canad. Ent. 63: 91..Kans., Man.

Length (female): body 13; wing 13 mm.

A yellowish species; abdomen dorsally with a broad brown band; longitudinal veins pale, cross veins black, humeral cross vein pale. Only the female imago is known.

Head yellow. Basal antennal joints, the median portion of the vertex, and most of the space between and laterad of the lateral ocelli, light reddish. Pronotum ruddy brown, the lateral margins yellowish. Mesonotum light brown tinged with yellow; sternum yellow tinged with brown. Femora light amber, each with a median and an apical purplish red band; tibiae and tarsi pale yellowish. Wings hyaline, rather narrow. Longitudinal veins pale, the subcosta and radius amber-tinged. Humeral cross vein and those of the anal area pale; all others of the fore wing black, thickened in the basal costal and subcostal areas. Those of the stigmatic area are faint and evenly anastomosed. Abdominal tergites light yellow, the posterior margins brown; broad brown median patches, narrowed posteriorly, form a continuous dorsal band, its outer edge irregular, much like that of *Pentagenia vittigera*. The three apical tergites are almost wholly brown, and the postero-lateral margins of tergites 1–7 are deeper in color than other portions of the margin, thus presenting the appearance of a lateral row of brown spots. Sternites pale yellowish. The subanal plate is rather short, narrowed posteriorly to about half the width of the ninth sternite; the apical margin has a sharp V-shaped excavation. Tails pale whitish, joinings not darker.

The type material was taken at Lawrence, Kansas and Aweme, Man. Dr. McDunnough kindly lent us one of the paratypes for examination. In the Cornell collection are two females, one from Hannibal, Mo. (W. D. Sargent), the other from a trap lantern collection at Manhattan, Kans. (R. C. Smith), which are quite similar to the above description. Other females, one from Fairport, Ia. (H. E. Schraedick), the others from Manhattan, Kans., trap lantern collection, are paler and show some variation in color markings. We hold these tentatively under *P. centralis*, as probable variations of that species.

Pseudiron meridionalis sp. nov.

Length: body 13½; wing 10½ mm.

A reddish brown species; all veins reddish brown, the humeral cross vein very dark.

Male imago—Head yellowish, shaded with brown anterior to the ocelli, which are black-ringed at the bases. Below and laterad of each lateral ocellus is a small red mark. A purplish black streak below each antenna, near the corner of the eye; another on the median carina. Eyes deep bluish black. Pronotum yellowish red. Mesonotum reddish brown; scutellum and a median area near the anterior margin slightly paler. Posterior margins on each side of scutellum purplish black. Pleura yellowish, sutures reddish brown. Sternum pale reddish brown, the anterior median portion of the mesosternum yellowish. Femora reddish, each with three distinct dark reddish brown bands, located at base, apex and middle. Fore tibia pale reddish; darker red at the base, its apex blackish; tarsus yellow, the joinings and the entire distal joint dusky. Tibiae and tarsi of middle and hind legs yellow; tibiae reddish at extreme base; tarsal joinings and claws brownish. The first fore tarsal joint is about three-fourths the length of the second; third slightly shorter than the second; the fore femur is only slightly longer than the tibia.

Wings hyaline, venation reddish to purplish brown, the main veins of the costal margin (especially the subcosta) yellowish basally. A distinct reddish stain occupies the costal and subcostal spaces immediately before the bulla. Basal cross veins of the costal and subcostal spaces, including the humeral cross vein, are rather heavily margined; in the costal area, heaviest toward the subcosta. All cross veins somewhat heavier than the longitudinals. Those of the stigmatic area anastomose in the apical portion of that space, forming two rows of cells, of which those nearer the costa are smaller.

Abdomen reddish brown; tergites darker than the sternites, especially the three apical ones, which are also brighter. Tergites 1–4 narrowly pale on the posterior margins; 5 and 6 with distinct dark posterior margins, which are less evident on 7–9. All tergites are rather widely paler along the lateral margin, and have an indistinct wide paler median band. On each side of the median line, within this pale band, a slender darker triangle arises from the posterior margin, its apex almost attaining the anterior margin on the basal tergites, but extending only to the middle on the apical ones. On the anterior margin of each, at the mid-dorsal line, is a short black streak. Sternites faintly pinkish; unmarked, except for faint indications of a pale brown submedian triangle on each side, as on the tergites, and a narrow streak on each side. These are most clearly defined on the apical sternites. Tails pale whitish, faintly tinged with reddish at the extreme base; joinings not darker. Genitalia amber yellow. Penes rather similar to *Metretopus norvegicus*, but more widely separated apically; apical margin not lobed. Third joint of the forceps slightly longer than the fourth; the two terminal joints together are almost two-thirds as long as the second joint. Genitalia as in fig. 115.

Holotype—Male imago. Chattahoochee River, Atlanta, Georgia; (P. W. Fattig). No. 1371.1 in the Cornell University collection.

Genus **Siphloplecton** Clemens

1915—Clemens—Canad. Ent. 47: 258.

Rather large mayflies; mainly brown in color, with paler markings. Eyes of the male are not divided, and are contiguous apically; an oblique band of a different color crosses the lower portion, as in the Siphlonurinae. Eyes of female remote; occiput concave between the hind angle of the eyes. First joint of male fore tarsus is very slightly longer than the second; second and third are subequal, fourth slightly shorter; fifth about one-half the length of the first. Fore tibià shorter than femur; tarsus about three times the length of the tibia. A row of small spines is present on the dorsal surface of the tibia. In the middle and hind legs of both sexes, the basal joint is not separate from the tibia, its line of union being indicated by a lateral spine. This joint is about one and one-third the length of the second; the third and fourth joints are progressively shorter, the fifth about equal to the second. The tibia is about three-fourths as long as the femur, while the tarsus equals the femur or slightly exceeds it in length. Claws of all tarsi are dissimilar.

The wings are usually marked with dark clouds at the base and the bulla, and have the costal and discal cross veins infuscated; but in one species are wholly pale. The fore wing is nearly three times as long as wide; the hind wing is very well developed, the anal angle fully rounded; the costal angulation tends to be acute only in one species, *S. speciosum*. Between the branches of the cubitus are two pairs of intercalary veins, which may be free apically, or joined by cross veins to the cubital branches; of these, the anterior pair is shorter than the second, as in the Heptageniidae. In the aberrant species *speciosum*, the posterior vein of each pair is very short and weak, and may be joined by a cross vein to the anterior member. The stem of the posterior branch of the radial sector is slightly longer than the depth of its fork. The second branch of the median vein is strongly curved at the base. Cross veins of the stigmatic area are anastomosed, forming two rows of cells, those nearer the costa being the smaller. In the hind wing, the median vein is forked near the base.

The dorsum of the abdomen is variously marked and blotched with dark brown or black. The tails are two in number, with a segmented rudiment of the middle tail remaining. The forceps are four-jointed, the basal and terminal joints being short. The apical margin of the forceps base is only slightly excavated between the forceps limbs, so that the penes are partially concealed. The penes are united to form a rounded structure bearing minute spines on the lateral margins. The apical margin of the ninth sternite of the female is slightly produced and entire.

The nymph has a stream-line body strongly reminiscent of the Siphlonurinae; the abdomen is somewhat depressed. The labial palp is two-jointed; the second joint is expanded and truncate apically. Maxillary palp three-jointed; the second joint very short, the third equal to the first; the palp is longer than the body of the maxilla. The labrum narrows toward the base; its apical margin is almost straight. Mandibles are rather similar to Siphlonurus. The outer divisions of the hypopharynx are angulate laterally; the median division is shaped like that of Siphlonisca. Tibia and tarsus of the first leg are about equal in length, and together are slightly longer than the femur. The claw of the first leg is flattened and bifid, the two divisions being about equal in length. In the middle and hind legs, the combined tibia and tarsus do not exceed the femur in length; the tibia is shorter than the tarsus; the claw is very long and slender, sharp-pointed and without denticles, and nearly as long as the tarsus.

Gills are borne on segments 1–7; in most species, double on the first three segments. The outer gills are elongate oval in shape, cordate at the base on one side, the basal one slightly indented on the outer margin. Tracheation is pinnate. The postero-lateral angles of the abdomen bear well-developed spines. The three tails are approximately equal in length and size; the outer ones bear long hairs on the inner margin only, as in the Siphlonurinae.

Genotype—*S. basale* (in Baetis) Walker.

Distribution—Nearctic; eastern North America.

KEY TO THE SPECIES

Male adults

1—Cross veins of costal area and disc of fore wing brown and distinctly brown-margined; a dark cloud at bulla, more or less prominent....................2
—Cross veins of costal area and disc of fore wing (except humeral cross vein) pale, not dark-margined; no dark cloud at bulla........................**speciosum**
2—A short black line below the radius on the fore wing; cross veins of fore wing, other than those of the disc and the costal area, pale................**interlineatum**
—No such black line below the radius on the fore wing; all cross veins of the fore wing brownish..3
3—Abdominal tergites 2–6 heavily shaded with dark brown, and with black lateral pencilings; a lateral reddish brown mark on each sternite near the median line..**basale**
—Abdominal tergites 2–6 largely pale, the brown shading much reduced, no black pencilings; no such lateral marks on sternites ....**signatum**

VERIFICATION TABLE

Male adults

Species	*Wing length*	*Cloud at w. base*	*Cr. veins margined*	*Cloud at bulla*	*Tergites 3–7*	*Distr.*
basale	12–13	Yes	Yes	Yes	Brown	E
interlineatum	12–13½	Yes	In part	Yes?	Whitish	C
signatum	11½	Yes	Yes	Yes	Whitish	N. C.
speciosum	9–11	No	No	No	Whitish	Ga.

KEY TO THE SPECIES

Nymphs

1—Venter of abdomen marked with three longitudinal brown lines, one median, the others lateral; claws of second and third legs almost as long as their respective tarsi..**basale**
—Venter of abdomen not marked as above; claws of second and third legs ¾ to ⅘ the length of their respective tarsi, hence relatively shorter than above......2
2—Venter of abdomen marked with a single brown median line..........**signatum**
—Venter of abdomen marked with a very dark brown median line, and on each side near the pleural fold a row of black dots, two on each side of each sternite ....**speciosum**

Siphloplecton basale Walker

1853—Walker—List. Neurop. Ins. Brit. Mus. Pt. 3: 565 (Baetis)...........................Man.
1861—Hagen—Syn. Neurop. Ins. N. Amer. p. 50 (Baetis).
1871—Eaton—Trans. Ent. Soc. London p. 139. Fig. (Heptagenia).
1885—Eaton—Revis. Monogr. p. 298 (Heptagenia).

1913—Clemens—Canad. Ent. 45: 338. Fig. (*Siphlurus flexus*)................................Ont.
1915—Clemens—Canad. Ent. 47: 258. Fig. (transferred to Siphloplecton)....................Mich.
1915—Clemens—Contr. Canad. Biol. p. 124.
1924—McDunnough—Canad. Ent. 56: 128 (*flexus* =*basale*).
1930—Ide—Canad. Ent. 62: 227.
1932—Traver—J. Elisha Mitchell Sci. Soc. 47: 185..N. C.

Length: body 13–14; wing 12–13 mm.

A dark species; cross veins of costal area and disc of the fore wing heavily infuscated; a brown cloud at the base of each wing, and at the bulla of the fore wing. See frontispiece.

Frons pale; remainder of head blackish brown. Thorax deep black-brown, with a slight ruddy tinge. Fore legs reddish brown; a dark subapical femoral band, black knees, and darker tarsal joinings and tips of tibiae. Femora of middle and hind legs similar but paler; tibiae and tarsi paler, yellowish. A dark cloud at the base of each

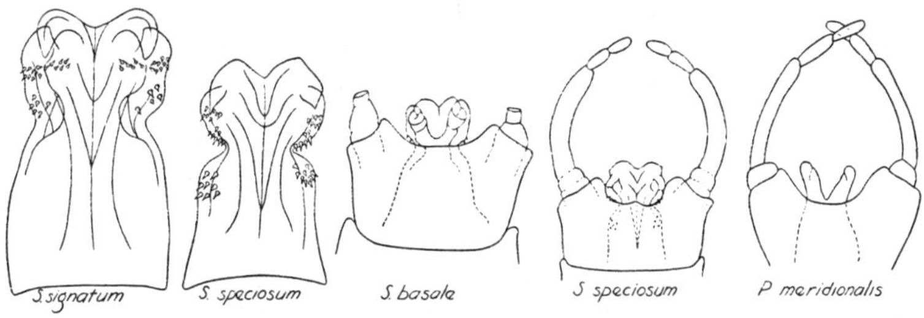

FIG. 115. Male genitalia of three species of Siphloplecton, and of one species of Pseudiron.

wing; cross veins of fore wing dark brown, those of the costal margin and the wing disc rather heavily margined; a brown cloud at the bulla, often a faint cloud in the stigmatic area, also at fork of hind branch of the radial sector. Wing of female with no such basal dark clouds. First segment of abdomen wholly dark brown; tergites 7–10 largely dark brown, 7 and 8 with pale antero-lateral angles, 10 pale laterally. Tergites 2–6 semi-hyaline, rather extensively shaded with dark brown on the posterior and lateral margins and the dorsal area; anterior margins, antero-lateral angles and lateral area pale. A curved line of black pencilings on each side of each tergite, and a dark brown oval patch on each side near the median line. Sternites 2–8 largely pale, semi-hyaline, with a reddish brown dot on each side of the median line, on each; sternite 9 dark brown with a pale streak on each side; ganglionic areas dark brown, most prominent on the apical segments.

Tails whitish, joinings ruddy brown. Genitalia largely pale reddish brown. Penes as in fig. 115.

The gills of the nymph are double on segments 1–3. On the venter of the abdomen are three uninterrupted longitudinal brown lines, one median and two lateral. The claws of the second and third legs are almost as long as the tarsi.

We have imagos of this species in the Cornell collection from Cranberry Lake, N. Y.; the campus of the Michigan Agricultural College, Lansing, Mich.; N. Fairhaven, N. Y., on the shore of Lake Ontario (J. G. Needham); Lake Nipissing, Ontario (F. P. Ide); and Spero, N. C. (J. R. Traver). We have likewise seen a specimen from Long Lake, Thunder Bay, Ont. Nymphs in the Cornell collection are from Douglas and Walnut Lakes, Michigan, and Old Forge, N. Y.

Siphloplecton interlineatum Walsh

1862—Walsh—Proc. Acad. Nat. Sci. Phil. p. 368 (*Baetis femorata*)............................Ill.
1863—Hagen—Proc. Ent. Soc. Phil. 2: 169.
1863—Walsh—Proc. Ent. Soc. Phil. 2: 190 (proposes name *interlineatum* for his *femorata*).
1924—McDunnough—Canad. Ent. 56: 129 (transferred to Siphloplecton)....................Man.

Length: body 12–13½; wing 12–13½ mm.

A paler species than *S. basale;* a black line below the radius at the bulla in the fore wing; cross veins in fore wing, other than the costals and discal veins, pale.

Eyes of live insect pearly white. Thorax blackish brown; prosternum often considerably marked with white; a broad conspicuous transverse whitish band between the hind coxae. Fore legs pale brown; the basal half of each femur may be brown; middle and hind legs paler. All femora with a broad dark band near the apex; tibiae darker at bases and tips; all tarsal joinings darker. Wings hyaline; cross veins of the costal margin and the disc fuscous, as are also the three longitudinal veins of the costal margin. All other veins of fore wing hyaline. A small brown spot is present at the fork of the hind branch of the radial sector, and sometimes a small cloud near the tip of the costa. The brown areas at the wing bases are more restricted than in *basale;* these are lacking in the female.

Abdominal tergites 1–5 whitish hyaline, the posterior margins and an oval spot on each side of the median line piceous, the dorsum slightly tinged with the same color. Apical tergites piceous, with lateral whitish areas. Sternites whitish hyaline. Tails whitish, the joinings fuscous; alternate joinings wider.

We have not seen a specimen of this species. Dr. McDunnough says of it, "From *basale* the species is distinguished by the paler abdomen and more restricted brown shading at the base of the wings in the male sex; the cross veins are pale except in the costal and subcostal regions and along the basal portion of the median vein."

Siphloplecton signatum Traver

1932—Traver—J. Elisha Mitchell Sci. Soc. 47: 186. Fig....................................N. C.

Length: body 14; wing 11½ mm.

In this species the wing is very similar to that of *S. basale*; the body differs from the latter in the much paler thorax and abdomen.

Frons greyish white; remainder of head pale brown. Pronotum dark brown, its posterior margin whitish. Mesonotum yellowish brown; scutellum margined with dark reddish brown; metanotum dark brown. Pleura yellowish, sutures brown. Fore femur light brown; tibia and tarsus yellowish white. Middle and hind legs pale yellowish brown; dark bands apically on all femora; tarsal joinings and tips of tibiae slightly darker. Abdominal tergites 1 and 2 largely brown; tergites 3–6 much paler than in *basale*, largely whitish and rather opaque. Brown markings on these tergites are reduced to a median triangle and lateral extensions of this, which do not quite attain the anterior margins; posterior margins are narrowly brown, and near the pale pleural fold is a brownish area. At the posterior margin of each tergite, a purplish mark on the median line and a lateral streak on each side of this. Tergites 7–10 largely dark reddish brown; pale antero-lateral areas on tergites 7 and 8 are more extensive than in *S. basale.* Ventrally wholly pale, except for a minute dark dot on the median line, at the anterior margin, and indistinct tan bars running forward from the posterior margin, parallel to the median line. Tails pale, joinings reddish brown. Genitalia light brown; penes as in fig. 115.

The nymph has double gills on segments 1–3; the claws of the second and third legs are slightly shorter relatively than in *S. basale;* ventral abdominal markings consist of a single brown median streak only.

The type material, from a lake near Liberty, N. C. (J. R. Traver) is in the Cornell collection. We have seen no other specimens.

Siphloplecton speciosum Traver

1932—Traver—J. Elisha Mitchell Sci. Soc. 47: 191. Fig. .... Ga.

Length: body 9–12; wing 9–11 mm.

A rather aberrant member of the genus; distinguished by the pale wings, and the very short second member of each pair of cubital intercalaries.

Frons pale. Remainder of head, and thorax, dark red-brown. On the thoracic sternum, a pale transverse band between the middle and hind pairs of coxae. Femora reddish brown, with dark reddish bands near the apex, less distinct on the fore legs; median area, and apices of middle and hind femora, pale. Fore tibia pale reddish brown, paler apically; tarsi pale, joinings narrowly darker. Middle and hind tibiae pale shaded medially with purplish red. Wings pale, hyaline; humeral cross vein dark purplish brown. All other cross veins wholly pale, except those of the basal costal space, which are pale brown next to the subcosta. Main longitudinal veins of the fore wing purplish brown at the extreme base, pale yellowish brown beyond; some almost colorless. No trace of the usual basal and bullar clouds. In the cubital space of the fore wing, each of the two pairs of intercalary veins consists of a moderately long anterior vein followed by a very short faint posterior one. Costal angulation of the hind wing tends to be acute.

Abdominal segment 1, and tergites 2, the apical half of 7, and 8–10, dark reddish brown; lateral areas of 7–10 pale. Tergites 3–6, and basal half of 7, largely pale, partially semi-hyaline; brown markings less extensive than in *S. basale.* Lateral and posterior brown margins rather wide on tergite 3; narrower and less strongly indicated on 4–6. An incomplete dark median line on each; on each side of this, a short dark streak, and another nearer the anterior margin, toward the pleural fold. Black pencilings on tergite 6 only. Sternites 2–9 wholly pale, except for brown oval patches on each side of the median line on sternite 2; greyish ganglionic areas on the three apical sternites; brown lateral shading on sternite 9. Tails white; joinings purplish black. Genitalia largely pale (see fig. 115).

Gills double on segments 1–3 of the nymph; claws of the second and third leg relatively shorter than in *S. basale;* venter of abdomen marked with a brown median longitudinal streak, and on each side near the pleural fold a row of black dots, two on each side of each sternite.

The type material, from Echeconee Creek, Ga. (P. W. Fattig) is in the Cornell collection. We have seen no other specimens.

Subfamily SIPHLONURINAE

Tarsi five-jointed; the basal joints of the middle and hind legs more or less completely fused with the tibiae, the five joints free and movable in the male fore tarsi. First joint of male fore tarsus at least ⅓ the length of the second joint, and may equal or exceed it in length. Claws variable in the different genera. The fore wing is narrowed just beyond the base, and widened somewhat at the anal angle; relatively long for its width;

the anterior branch of the cubital vein is attached to the inner margin of the wing by a series of cross veins, which are often forked and frequently sinuate. The hind wing is well developed, almost or fully as wide as long; anal region more or less well developed; costal projection usually somewhat obtuse, not especially prominent.

Forceps base usually very long, its apical margin, between the forceps limbs, more or less deeply emarginate. Forceps more or less distinctly four-jointed; basal joint short, usually (except in Isonychia) rather ringlike, often more or less fused with the long second joint, which is often strongly bowed; two (rarely three) terminal joints, which together are not longer than the second joint. Penes variable in different genera. Subanal plate of female wanting in Siphlonisca; in Isonychia and Ameletus, more or less deeply carved out on the apical margin; in the other genera, convex on this margin. Tails two, in both sexes.

Body of nymph stream-line in form, much compressed; legs relatively long, held at sides of body; claws (except in Isonychia) long and slender, not denticulate. Head inflexed (bent downward obliquely to the axis of the body). Gills relatively simple in structure; single, except in Siphlonurus and Isonychia; borne on segments 1–7. In Isonychia each pair is composed of a dorsal lamellate and a ventral fibrillar portion, while in Siphlonurus the two parts are similar and lamellate. Lateral margins of apical abdominal segments more or less expanded into flat lateral extensions, best developed in Siphlonisca, least so in Ameletus; a spine is present at the postero-lateral angle on each middle and apical segment. Tails three; middle tail heavily fringed on each side; lateral ones may have a few short hairs or spines on the outer margin, but are heavily fringed on the inner margin only.

KEY TO GENERA

Adults

1—Claws on each tarsus dissimilar; costal angulation of hind wing acute. . **Ameletus**
—Claws on each tarsus similar; costal angulation of hind wing obtuse or wanting. . 2
2—Hind tarsus longer than tibia (except in *marshalli*); all claws sharp-pointed . . . . **Siphlonurus**
—Hind tarsus shorter than or subequal to tibia; claws of middle and hind legs sharp-pointed, variable on fore leg of male. 3
3—Abdominal segments 5–9 dilated laterally into flat, broad expansions; no costal angulation on hind wing. . . . . **Siphlonisca**
—Abdominal segments 5–9 not dilated laterally, as above; costal angulation of hind wing present. 4
4—Media of hind wing simple, not forked; all claws sharp-pointed. . . . **Parameletus**
—Media of hind wing forked near outer margin; fore claws of male similar, blunt, in most species. . . . . **Isonychia**

VERIFICATION TABLE

Adults

Genus	*Tarsal claws*	*Spines on feet*	*Stig. c-veins*	*Fork of M.*	*H. w., c. angul.*	*Subanal plate*
Ameletus	Dissimil.	Few	Yes	Detached	Acute	Concave
Isonychia[1]	Similar	None	Slight.	Normal	Obtuse	Concave
Parameletus[2]	Similar	Present	Yes	Angled[5]	Obtuse	Convex
Siphlonisca[3]	Similar	Present	No	Normal	Absent	Absent
Siphlonurus[4]	Similar	Present	Slight.	Normal	Obtuse	Convex

[1] M of hind wing forked near margin. Fore tibia considerably longer than femur.
[2] M of hind wing simple, not forked.
[3] Broad lateral expansions on abdominal segments 5–9.
[4] Hind tarsus longer than tibia.
[5] Posterior branch of M forms almost a right angle at its union with stem.

KEY TO THE GENERA

Nymphs

1—Fore legs conspicuously fringed with long hairs; gill tufts present on bases of maxillae and fore coxae. . . . . **Isonychia**

—Fore legs without such conspicuous fringes; no maxillary or coxal gill tufts 2

2—Gill lamellae single on all segments. 3

—Gill lamellae double on segments 1 and 2, sometimes on other segments also . . . . **Siphlonurus**

3—Abdominal segments 5–9 greatly expanded laterally. . . . . **Siphlonisca**

—Abdominal segments 5–9 normal, not greatly expanded laterally. 4

4—A pincer-like process near tip of labial palp; no pectinate spines on crown of maxilla. . . . . **Parameletus**

—No such pincer-like process on labial palp; a fringe of pectinate spines on upper margin of maxilla. . . . . **Ameletus**

VERIFICATION TABLE

Nymphs

Genus	*Jts., mx. p.*	*Jts., lb. p.*	*Claws toothed*	*Coxal gills*	*Gills, on 1–2*	*Gills, on 3–7*	*Misc.*
Ameletus	3	3	No	None	Single	Single	a
Isonychia	2	2	Yes	Present	Double*	Double*	b
Parameletus	3	3	No?	None	Single	Single	c
Siphlonisca	3	3	No	None	Single	Single	d
Siphlonurus	3	3	No	None	Double	Var.	

* Upper and lower portions unlike.
a Long pectinate spines on crown of galea-lacinia.
b Long fringe of hairs, inner margin of fore leg.
c Pincer-like process on 2nd joint of labial palp.
d A midventral spine present on thorax.

Genus **Ameletus** Eaton

1885—Eaton—Revis. Monogr. p. 210.
1920—Ulmer—Stett. Ent. Zeit. 81: 134.

Medium-sized mayflies, usually brownish or yellowish brown in color. Male and female rather similar in coloration, although the abdomen of the female is often darker than that of the male. Eyes of male large, usually contiguous above; not divided. A transverse band of a paler color than the remainder of the eye may extend obliquely across the outer surface, as in Siphlonurus. Vertex and occiput of female usually with dark submedian marks. Hind margin of head slightly concave.

Fore leg of male about as long as the body. First fore tarsal joint is $\frac{1}{3}$ to $\frac{1}{2}$ as long as the second. Claws of all tarsi dissimilar, in both sexes. Hind tibia as long as, or slightly longer than, the tarsus. Fore wing relatively long and narrow, about three times as long as wide. Hind wing well developed; its costal angulation somewhat more acute than in Siphlonurus (see fig. 27a). The median vein of the hind wing is forked near the base. Venation typically dark; humeral cross vein strong and dark. Stigmatic cross veins usually forked and anastomosed. In some species, several to many of the cross veins in the costal half of the fore wing are brown-margined, thus causing the wing to have a speckled appearance.

Abdominal tergites usually with a rather wide dark posterior margin; the postero-lateral angles, and extensions along the pleural fold, are likewise dark. Ventrally, the ganglionic areas are often marked with black or brown; no other conspicuous markings.

The forceps base of the male is quite long, but usually relatively shorter than in Siphlonurus. It is usually rather deeply excavated on the apical margin between the forceps, and bears a tubercle or projection on each side of this excavation. The forceps are more or less distinctly four-jointed, the basal joint being short and ring-like, and often imperfectly separated from the long joint. The two terminal joints together are shorter than the long joint. Viewed from beneath, the tips of the penes are often visible, projecting beyond the forceps plate. The penes are long and slender, and may bear spines on the inner margin. The subanal plate of the female is rather broad, and slightly concave on its apical margin. Middle tail completely aborted.

Nymphs have a more perfect stream-line form than do nymphs of Siphlonurus (see fig. 45). Maxillary and labial palps each three-jointed, the former rather short and weak. The maxilla has a broad crown, upon which are borne many long pectinate spines (see fig. 66). Mandibles distinctly triangular, canines borne at one point of the triangle and the molar surface at another. Labrum a little longer than wide, emarginate on frontal border. Legs rather short, claws slender and not pectinate. Postero-lateral spines usually present on abdominal segments 4–9, but short and weak on segments 4–6. The lateral margins of these segments are only very slightly compressed. Gills single, simple in form, small and oval. The tails generally have a wide dark transverse band at the middle and a narrow dark band at the apex.

Genotype—*A. subnotatus* Eaton.

Distribution—Holarctic, Indo-australian; in Nearctic, quite generally distributed.

KEY TO THE SPECIES

Adults

1—Several cross veins in fore wing margined with brown, so that wing appears speckled 2
—No margined cross veins; wings not speckled 5
2—No dark marks on ganglionic areas of abdominal sternites . . . . **validus**
—Ganglionic areas of abdominal sternites marked with black or brown 3
3—Ganglionic areas black, heart-shaped . . . . **similior**
—Ganglionic areas may be black or brown; not heart-shaped 4
4—Medium-sized species, wing 10 mm. in length; wing of male amber-tinted . . . . **sparsatus**
—Larger species, wing 13 mm. in length; wing not amber-tinted . . . . . . **subnotatus**
5—Ganglionic areas of abdominal sternites with no dark markings 6
—Ganglionic areas marked with brownish black or orange 10
6—Wings distinctly amber-tinged; small species, wing 8 mm. in length . . . . **shepherdi**
—Wings not amber-tinged; larger species, wing 9 mm. or more in length 7
7—Both wings strongly and entirely tinged with umber brown . . . . . **vancouverensis**
—Wings not as above 8
8—Both wings suffused with brown at base . . . . **oregonensis**
—Both wings clear, no such basal suffusion 9
9—Abdominal tergites 2–7 largely light brown . . . . **aequivocus**
—Abdominal tergites 2–7 largely flesh-colored . . . . **lineatus**
10—Small species, wing 8 to 9 mm. in length 11
—Larger species, wing 10 to 16 mm. in length 15
11—Wings amber-tinged in basal costal area; a pair of wide overlapping spines near base of penes (see fig. 118) . . . . **browni**
—Wings and penes not as above 12
12—Ganglionic markings orange; longitudinal veins amber . . . . **cooki**
—Ganglionic markings dark brown or black; longitudinal veins brown to blackish 13
13—Abdominal tergites 2–7 deep brown; divisions of penes tend to converge apically (see fig. 116) . . . . **alticolus**
—Abdominal tergites 2–7 yellow-brown; divisions of penes divergent apically . . . 14
14—Penes without spines; long joint of forceps somewhat thickened at base and apex (see fig. 116) . . . . **ludens**
—Two small spines on inner margin of each division of penes; all forceps joints slender (see fig. 116) . . . . **celeroides**
15—Wings faintly smoky or grey-tinged 16
—Wings not as above 18
16—Very large species, wing 16 mm. in length; stigmatic area tinged with dark brown . . . . **exquisitus**
—Somewhat smaller species, wing 10–14 mm. in length; stigmatic area not dark-tinged 17

17—A single rather long slender spine attached to inner margin of each division of penes (see fig. 118); tails dark brown.............................**dissitus**

—Several small spines on inner margin of each division of penes (see fig. 116); tails whitish...**celer**

18—Abdominal tergites 2–7 yellow-brown; penes without spines (see fig. 116)..**velox**

—Abdominal tergites 2–7 greyish white; a single long slender spine on inner margin of each division of penes (see fig. 117)............................**vernalis**

VERIFICATION TABLE

Adults*

Species	*Wing length*	*Clouds in wing*	*Wings tinged*	*Abdom. tergites*	*Ganglionic marks*	*Distr.*
aequivocus	9	No	No	Light br.	None	Colo.
alticolus	8	No	No	Dp. brown	Dk. brown	Alta.
browni	8	No	Amber	Dp. yellow	Brown	Que.
celer	13–14	No	Smoky	Light br.	Dk. brown	B. C.
celeroides	8–8½	No	No	Light br.	Dk. brown	Alta.
cooki	8	No	No	Orange-br.	Orange	Mont.
dissitus	10–12½	No	Grey	Yellow	Dk. brown	Calif.
exquisitus	16	No	Grey	Yellow-br.	Brown	Ore.
lineatus♀	12–13	No	No	Flesh-col.	None	N. C.
ludens	9	No	No	Yellow-br.	Black	N. Y.
oregonensis	11	No	Brown	Yellow-white	None	Ore.
shepherdi	8	No	Amber	Yellow-red	None	Calif.
similior	9	Yes	No	Dp. brown	Black	Alta.
sparsatus	10	Yes	Amber	Yellow-br.	Black	W
subnotatus	13	Yes	No	Yellow-br.	Dk. brown	Colo.
validus	11	Yes	No	Pinkish-br.	None	Alta.
vancouverensis	10	No	Brown	Yellow-br.	None	B. C.
velox	13	No	No	Yellow-br.	Dk. brown	W
vernalis	12	No	No	Grey.-white	Brown	B. C.

KEY TO THE KNOWN SPECIES

Nymphs

1—Tracheation of gills plentiful, tracheae very distinct.........................2

—Tracheation of gills sparse, tracheae quite indistinct.........................3

2—A dark interior band of chitin across each gill about ¼ the distance from dorsal margin; tails largely blackish...**velox**

—No such dark interior band of chitin on gills; tails dark-banded in middle area ....**sparsatus**

3—Three distinct dark longitudinal stripes on venter of abdomen................4

—No such dark longitudinal stripes on venter of abdomen.....................5

4—Stripes wide, their edges not clearly defined; fused on last three sternites; no continuous pale line between each lateral stripe and the pleural fold.......**ludens**

—Stripes relatively narrower, edges clearly defined; fused only at posterior margin of sternite 9; a continuous white line present between each lateral stripe and pleural fold...**lineatus**

* Males, except *lineatus*.

5—A narrow dark interior band of chitin on each gill, much as in *velox*. 6
—No such dark interior chitinous band on gills. 7
6—Tails pale, not dark-banded; dark ganglionic marks present; gill tracheae very indistinct. . . . . **alticolus**
—Tails blackish in middle area; no dark ganglionic marks; gills tracheae moderately distinct, but sparse. . . . . **shepherdi**
7—Tails black in basal three-fourths; femur with faint dark median band and dark tip. . . . . **browni**
—Tails wholly pale, or with broad dark median band, but not as above; femur not banded as above. 8
8—Tails pale·ochreous; tarsi with deep brown distal bands only. . . . . **celer**
—Tails pale at base and apex, with broad dark median band; tarsi dark-banded basally as well as distally. . . . . **celeroides**

VERIFICATION TABLE

Nymphs

Species	*Body length*	*Tracheation*[1]	*Gill band*[2]	*Gangl. marks*	*3 abd.*[3] *stripes*	*Tails banded*	*Distr.*
alticolus	8–9?	No	Yes	Yes	No	No	Alta.
browni	9–10?	No	No	No?	No	Base	Que.
celer	10?	No	No	Yes	No	No	Alta.
celeroides	9?	No	No	Yes?	No	Mid.	B. C.
lineatus	12–14	No	No	No	Yes	Mid.	N. C.
ludens	10–11	No	No	Yes	Yes	Mid.	E
shepherdi	9	Mod.	Yes	No	No	Mid.	Calif.
sparsatus	10	Yes	No	Yes?	No	Mid.	W
velox	13?	Yes	Yes	?	No	No?	W

Ameletus aequivocus McDunnough

1934—McDunnough—Canad. Ent. 66: 164. Figs. . . . . Colo.

Length: body 9; wing 9 mm.

A brownish species allied to *A. cooki*; no dark ganglionic marks.

Head dark brown. Pronotum rather dark brown. Mesonotum paler brown; blackish brown shading on lateral areas and postero-laterally on each side of the scutellum. Metanotum brown, with blackish shading in posterior portion. Pleura and sternum deep brown; intersegmental areas paler brown. Fore femur and tibia deep brown with faint tinge of purplish; tarsus pale brown. Middle and hind legs pale brown, traces of reddish streaks on outer side of each femur. Wings hyaline; longitudinal veins fine, brown; cross veins indistinct, faintly brown-tinged; those of stigmatic area better defined, anastomosed regularly. Abdominal tergite 1 blackish brown; anterior margin and a large subtriangular lateral area on tergites 2–7 pale, semi-hyaline, other portions of these segments light brown; laterally the brown color extends forward almost to anterior margin. Sternite 1 light brown; 2–6 light yellowish; no dark ganglionic marks. Segments 8–10 opaque; tergites rather uniformly brown, paler brown lateral patches on 9 and 10. Sternites 7–9 light brown, posterior margin of 9 and forceps base rather paler. Tails dull pale brownish white.

[1] Refers to tracheae of gills, whether distinct or not.
[2] Dark band of chitin on interior, not edge, of gill.
[3] Three dark stripes on venter of abdomen.

Duller than the allied *A. cooki;* forceps base broader, penes closer together, and median spines longer than in that species (see fig. 118 from McDunnough). We have seen no specimens of this species.

Ameletus alticolus McDunnough

1934—McDunnough—Canad. Ent. 66: 163. Figs. .... Alta.

Length: body 8; wing 8 mm.

A rather deep brown species, allied to *A. velox*, *A. celer*, and *A. celeroides;* dark ganglionic areas present.

Head deep brown; areas laterad of antennae paler. Thorax deep brown. Postero-lateral angle of pronotum tipped with creamy. Antero-lateral margin of mesonotum creamy with a tinge of pink; "a creamy patch with slight pinkish shading anterior to wing base" (McD.). Creamy brown shading on pleura. Fore legs dark brown; middle

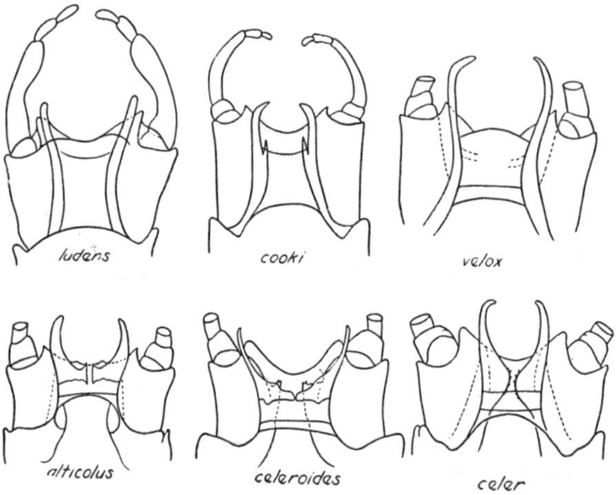

FIG. 116. Male genitalia of six species of Ameletus (*cooki*, *alticolus*, *celer*, and *celeroides*, after McDunnough).

and hind legs paler, creamy brown. Wings hyaline. Longitudinal veins fine, brown; cross veins indistinct, pale; stigmatic cross veins brown, finely anastomosed. Abdominal tergites "rather even, deep, shiny brown with little deepening of color along posterior margins of segments" (McD.). Sternites pale creamy; dark mid-ventral marks on ganglionic areas. Forceps "rather short and stumpy and the excavation of the apical margin of the forceps plate is much less deep" (McD.) than in *A. celer* and *A. celeroides;* divisions of the penes short and broad with no spines basally on ventral surface (see fig. 116 from McDunnough).

Distinguished from *A. celeroides* by the deep even brown of thorax and abdominal tergites, also in details of structure of penes. We have seen no specimens of this species.

Ameletus browni McDunnough

1933—McDunnough—Canad. Ent. 65: 278. Fig. .... Que.

Length: body 9; wing 8 mm.

A small clear-winged species; wings with a faint amber tinge, especially in the basal costal area.

Head and thorax deep mahogany brown. Median line of mesonotum broadly light yellowish brown; some oblique pale areas extend forward and downward from the wing roots and around the bases of the legs. Legs yellowish; femora tinged with ruddy brown; fore femur considerably darker than the others. Wings hyaline; a very faint amber tinge, most evident in the basal costal area; veins and cross veins fine, brown; cross veins in basal costal space indistinct. Abdominal tergites rather deep yellow; posterior margin of each tergite narrowly brown, and poorly defined brownish triangles extending forward from the end of each dark band above the pleural fold, their apices almost reaching the anterior margin. Sternites slightly paler yellow, shaded faintly with ruddy brown along the lateral margin; sternites 1–7 each with an oval brown ganglionic spot. Forceps and tails faintly smoky; tails with narrow dark joinings. Penes as in fig. 118.

Nymph light sepia brown, the head and thorax with pale streaks and lunate markings. Abdominal tergites marked with a wide pale median band, white submedian and antero-lateral dots and lateral oblique streaks; sternites pale brown, with traces of a whitish mid-ventral line, submedian dark streaks and a dark dot in the antero-lateral angle of each. Legs pale, with brownish black bands; gills rather small, oval, whitish, the edge darker, tracheation not visible. Tails blackish in the basal three-fourths, white beyond this, with a narrow dark tip.

This species, allied to *A. dissitus* and *A. cooki*, is distinguished from each of these by the finer details of the penes structure. It was described from specimens taken in the Gaspe Peninsula. We have seen no specimens.

Ameletus celer McDunnough

1934—McDunnough—Canad. Ent. 66: 161. Fig. . . . . Alta.

Length: body 13–14; wing 13–14 mm.

Allied to *A. velox*; abdominal tergites 2–7 pale yellowish brown, the posterior margins dark brown; dark mid-ventral ganglionic patches.

Head and thorax blackish brown. Postero-lateral margins of pronotum, and intersegmental areas of pleura, paler brown. Legs sepia-brown. Wings dull hyaline, very faintly tinged with smoky. Longitudinal veins fine, brown; cross veins very fine, light brown, rather faint. Abdominal tergites 2–7 semi-hyaline, pale yellowish brown; tracheae fine, numerous. Posterior margins dark brown; narrow brown submedian streaks extend forward from this band but do not attain the anterior margin. Tergites 8–10 opaque, rather deeper in color. Sternites paler than tergites; posterior margins not darker; "a median row of dark, oval ganglionic patches; area at base of forceps paler than preceding portion" (McD.). Tails whitish, faintly tinged with smoky; joinings narrowly blackish brown. Genitalia as in fig. 116. We have seen no specimens.

Ameletus celeroides McDunnough

1934—McDunnough—Canad. Ent. 66: 162. Fig. . . . . B. C.

Length: body 8; wing 8–8½ mm.

Allied to *A. velox* and *A. celer*; the abdomen is somewhat paler than in the latter; dark ganglionic patches present.

Very similar to *A. celer*, but much smaller. Description largely comparative. Thorax shiny blackish. Postero-lateral angle of pronotum creamy rather than brown, as in *celer*. Hind legs paler and duller than in *celer;* wings with no evidence of the smoky tinge found in that species; cross veins very fine, indistinct. Abdomen paler than in *celer;* the creamy shading "tends to replace the light brown shades on the abdomen, especially laterally on segments 6–8 and ventrally on segments 7 and 8; there is also a pale creamy line on the lateral edge of segment 9, this being sometimes found also on the two preceding segments" (McD.). No mention is made of ganglionic marks, hence their presence is assumed. Forceps joints "rather thin and fine and the penis-lobes are broadly divergent apically; the spining at the base ventrally seems to be limited to two small apical spines" (McD.); see fig. 116. We have seen no specimens of this species.

Ameletus cooki McDunnough

1929—McDunnough—Canad. Ent. 61: 174. Fig. .... Mont.
1931—McDunnough—Canad. Ent. 63: 89.

Length: body 7; wing 8 mm.

A small clear-winged species. Head blackish. Prothorax and postero-lateral portions of the meso- and metathorax also blackish. Mesothorax brown dorsally, with orange shading before the wing root and on and before the scutellum. Pleura and sternum largely blackish, shaded with orange at the bases of the legs, most prominent around the middle legs.

Fore legs blackish brown; middle and hind legs lighter brown, the basal joints tinged with orange-brown. Wings hyaline; longitudinal veins light amber, the cross veins pale and indistinct, except in the stigmatic area, where they are smoky.

Dorsum of abdomen orange-brown. Suffused brown shading along the lateral margin, tending to form obscure semi-triangular patches in the postero-lateral corners; sometimes extending narrowly along the posterior margin. Trachea of the anterior segments forming a black network. Sternites 2–6 dull hyaline whitish, with traces of orange on the ganglionic areas. Sternites 7 and 8 opaque, bright yellowish; sternite 9 largely blackish brown shaded with orange. Forceps base yellowish. Penes long and rather slender, each bearing a spine on the inner margin near the middle; penes curved only at tip (see fig. 116). Tails dirty whitish, the joinings dark.

Type material was taken at Bozeman, Mont., Aug. 7. We have no specimens of this species.

Ameletus dissitus Eaton

1885—Eaton—Revis. Monogr. p. 210. .... Calif.
1871—Eaton—Trans. Ent. Soc. London, p. 134. (*Isonychia manca*; genitalia mis-drawn).
1929—McDunnough—Canad. Ent. 61: 174. Fig.
1931—McDunnough—Canad. Ent. 63: 89.

Length: body 10–12½; wing 10–12½ mm.

A clear-winged species, in which the wings are faintly tinged with greenish grey. Body yellowish.

Head brownish. Thorax brown. Pronotum and posterior portion of metanotum blackish. Fore leg blackish brown; middle and hind legs yellowish to yellowish brown, the femora tinged with darker brown. Wings hyaline, very faintly tinged with greenish grey. Venation blackish brown. Stigmatic cross veins strongly anastomosed.

Abdomen bright yellowish to yellowish brown. Posterior margins of tergites 2–8 and the postero-lateral angles, blackish brown. Tergites 9 and 10 similarly marked, but somewhat darker in color. Sternites somewhat lighter in color than the tergites, "and seemingly marked each with a light pitch-brown blotch at the base in the middle" (Eaton). Forceps base light yellowish. Forceps blackish. Tails dark brown, joinings opaque.

The type material was from California, taken by Edwards and Osten-Sacken, in

April. A single male which we have seen, from the California Academy of Science, was taken in Alameda Co., Calif., on April 3, 1921. It differs slightly from the above description, in lacking the blackish ganglionic marks, but has instead, on each basal and middle sternite, a brown submedian streak. Genitalia as in fig. 118.

Ameletus exquisitus Eaton

1885—Eaton—Revis. Monogr. p. 212. . . . . Wash., Ore.

Length: body 14; wing 16 mm.

A large clear-winged species; the stigmatic area tinged with dark brown, wings very faintly tinged throughout with brownish grey.

Mesonotum yellowish brown. Metanotum blackish brown. Fore leg blackish brown, knee yellowish brown. Middle and hind legs yellowish brown to brown, the femora with an apical band. Wings hyaline, tinged faintly with brownish grey throughout; stigmatic area of fore wing tinged with dark brown. Venation dark reddish brown.

Tergites rich yellowish brown; the posterior margins, the postero-lateral angles, and a submedian streak, darker brown. Tergite 10 without these dark marks. Segments 2–6 largely semi-translucent; apical segments opaque. Sternites 2–6 light brownish grey; 7–9 yellowish brown. Ganglionic marks brownish. Forceps base dark brown on the lateral margins. Forceps blackish brown. Tails dark brown, darker near the base; joinings darker.

The large specimens, of which measurements are given above, were taken from Washington; a smaller specimen, measuring only 8 mm., wing 10 mm., from Mt. Hood, was also included by Eaton in this species. We have seen no specimens of this species.

Ameletus lineatus Traver

1932—Traver—J. Elisha Mitchell Sci. Soc. 47: 194. . . . . N. C.

Length: body 11; wing 11–12 mm.

A large pale species, clear-winged, and without ganglionic markings on the venter of the abdomen. Known only from females and nymphs.

Head yellowish brown. Prothorax dark yellowish brown, narrowly purplish on the posterior margin. Mesonotum clear greenish yellow, scutellum with blackish lateral margins. Pleura and sternum brown, the sternum dark brown on the posterior portions of the pro- and mesothorax. Legs brown. Wing with no clouds along the cross veins.

Abdominal tergites flesh-colored, margined laterally and posteriorly with orange-brown. Margin of pleural fold pale. Indistinct dark submedian streaks present. Sternites 1–6, and anterior half of sternite 7, flesh-colored; sternites 8–10, and the posterior half of 7, light olive brown. A purplish median mark on sternite 6, but no definite ganglionic marks on the other sternites. Submedian oval depressions present. Antero-lateral angles of sternite 8 whitish. Tails purplish brown, joinings darker.

The nymph has a cream-yellow thorax and abdomen, mottled and marked with dark brown. Legs banded. Tergites 6 and 7 largely cream colored; tergites 8–10 largely dark brown. Ventrally, three continuous dark brown stripes extend the length of the abdomen; one is median, the others lateral in position. Each dark lateral streak is separated from the lateral margin by a pale line. Lateral margins of segments 6–9 are narrowly flattened and thin; postero-lateral spines present on segments 2–9, well developed on 5–9. Tails whitish, with the two usual dark bands.

The type material, from the vicinity of Greensboro, N. C. is in the Cornell collection. The larger size and the apparent lack of dark ganglionic markings seem to distinguish this from the other two eastern species, *A. ludens* and *A. browni*.

Ameletus ludens Needham

1905—Needham—N. Y. St. Mus. Bull. 86: 19. Figs. .... N. Y.
1911—Morgan—Ann. Ent. Soc. Amer. 4: 117.
1913—Morgan—Ann. Ent. Soc. Amer. 6: 371.
1922—Clemens—Canad. Ent. 54: 77.
1924—Needham—Psyche 31: 308.

Length: body 8½; wing 9 mm.

A rather small clear-winged brownish species, with distinct blackish oval ganglionic marks on the venter of the abdomen.

Head of male blackish; of female, dark brown. Eyes of male honey-colored in life. Thorax reddish brown. Pronotum of female with paler median and lateral areas. Mesonotum yellowish to olive brown; scutellum reddish yellow; the area just anterior to it cream colored, and a pale cream colored patch on each lateral margin. Metanotal scutellum rather similarly marked. Anterior to the wing root is a pale cream colored area varied with a reddish tinge; similar patches on the pleura. Sternum blackish in the male, dark reddish brown in the female.

Fore legs dark reddish brown; middle and hind legs brown. Wings hyaline, veins blackish brown. Stigmatic cross veins tend to anastomose.

Abdomen of male semi-translucent except on the apical segments; pale yellowish brown above, paler beneath. Basal and apical segments washed with reddish brown. Rather wide dark brown bands on the posterior margins; postero-lateral angles dark brown. Sternites paler than tergites; apical sternites opaque reddish brown. Each ganglionic area is marked by a blackish oval spot, having a distinct black line through the center. Abdomen of female reddish brown dorsally, paler ventrally. Posterior margins, behind the dark bands, narrowly pale. Margin of pleural fold pale. Apical tergites and sternites often marked with creamy white areas which are tinged with red. Basal sternites pale, semi-translucent; apical sternite opaque.

Forceps blackish, tawny at the base. The long joint of the forceps is swollen basally and distally. The penes are long and slender, slightly divergent distally (see fig. 116). The subanal plate of the female is rather long, extending almost to the end of the tenth segment. Tails dark reddish brown basally, becoming somewhat paler distally. Joinings darker.

The mature nymph is dark reddish brown with cream colored markings. Legs pale, banded with brown. Abdominal tergites 8–10 very dark brown; basal and middle tergites varied with indistinct paler median, lateral and posterior markings. Dark brown oblique submedian marks present. Postero-lateral spines on segments 4–9, very small on 4 and 5. Ventrally, three very wide dark reddish brown bands occupy most of each sternite. These bands coalesce completely on sternites 8 and 9, and are more or less diffuse on the other sternites. No distinct pale streak separates the lateral dark line from the lateral margin of the sternite. Tails with the usual dark bands.

This species is of particular interest in that it ordinarily reproduces parthenogenetically. Although large numbers of nymphs and imagos have been examined, only two males have thus far been reported.

The species was described from specimens taken by Mr. D. B. Young at Newport, N. Y. It is quite common in the vicinity of Ithaca, N. Y., where the single adult male was taken by Dr. O. A. Johannsen on June 9, 1924. This specimen, and other local specimens of females and nymphs are in the Cornell collection. Nymphs which appear to be of this species were taken at Shinn's Run, near Shinnston, W. Va. (H. B. Bowman) on April 5, 1933.

Ameletus oregonensis McDunnough

1933—McDunnough—Canad. Ent. 65: 157. Fig. . . . . Ore.

Length: body 10; wing 11 mm.

Both wings of this species are suffused with brown at the base; genitalia are quite distinctive.

Head and thorax blackish; pleura shaded with paler ruddy tinges. Fore legs deep blackish brown. Middle and hind legs light yellowish brown, the femora largely suffused with reddish brown. Wings hyaline, the costal area faintly amber-tinged at

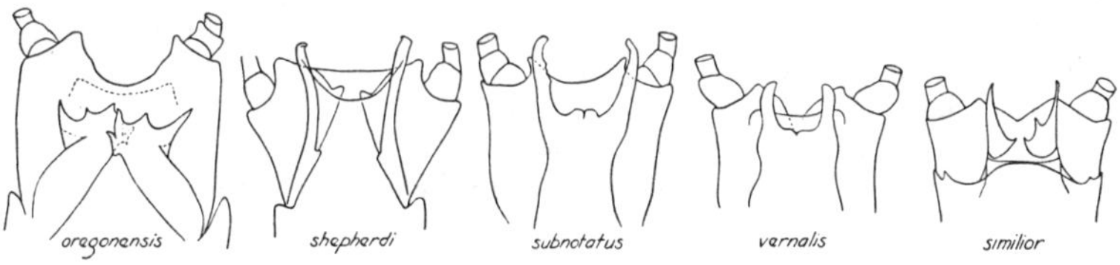

FIG. 117. Male genitalia of five species of Ameletus (*oregonensis*, *vernalis*, and *similior*, after McDunnough).

the base. The basal anal region of the fore wing is suffused with sepia brown, the stigmatic region similarly tinged. The whole basal area of the hind wing, extending along the costa half way to the apex, is prominently brown. Longitudinal and cross veins brown; cross veins numerous, those of the stigma anastomosed. Second abdominal tergite suffused with smoky brown; tergites 3-6 semi-hyaline, pale yellowish white; posterior margins, triangles in the postero-lateral corners, and two narrow submedian streaks, dark brown, which in certain lights appears reddish. Tergites 7–9 opaque, similarly marked, but the dark areas blackish brown. Tergite 10 largely brown, the posterior margin pale. Sternites 1–6 semi-hyaline, pale yellowish white; 7–9 opaque, suffused somewhat with brown. Forceps base pale, forceps deep smoky. Penes unique, each composed of a rather wide chitinous piece bearing at its apex three broad spines; penes short, not reaching to the apex of the forceps base (see fig. 117). Tails pale smoky, not darkened at the joinings. Nymph of this species is not known.

The type material, taken at Philomath, Oregon, is in the Canadian National Collection. We have specimens of this species in the Cornell collection from Philomath, Ore., and Waterton Lakes, Alta., presented by Dr. McDunnough.

Ameletus shepherdi Traver

1934—Traver—J. Elisha Mitchell Sci. Soc. 50: 238. Fig. . . . . Calif.

Length: body 8; wing 8 mm.

A small species with pale whitish abdomen and amber-tinted wings.

Head and thorax light reddish brown. Mesonotum yellowish, its anterior portion reddish margined with white; scutellum pale whitish, its postero-lateral margins dark reddish brown. Pleura yellowish red; sutures dark red-brown, intersegmental spaces whitish. Legs pale; tibia and tarsus of fore leg amber yellow, femur and middle and hind legs whitish. Wings hyaline, distinctly tinged with amber, most noticeably in the stigmatic area of the fore wing and the entire hind wing. Venation pale reddish brown to amber in the male, dark reddish brown in the female. Stigmatic cross veins anstomosed.

Abdominal segments 1, and 7–10, opaque, pale reddish to yellowish red; dorsum of posterior segments darker than venter. Intermediate segments whitish with faint tinge of pale yellowish red on the dorsum; semi-translucent, and with no distinct markings. Posterior margins and ganglionic areas opaque whitish. Dorsum of female pale reddish brown. Tails amber yellow, paler at tips. Genitalia amber yellow. Distal portion of penes directed outward. No spine, but a blunt thumb-like projection, on the inner margin of the penes. Tubercle on the inner distal margin of the forceps base blunt, quite prominent. Apical margin of forceps base very shallowly excavated (see fig. 117).

Nymph without distinct dark markings. Last two segments of abdomen dark reddish brown; posterior margins of all segments, also the lateral extensions, narrowly reddish brown. Postero-lateral spines on segments 2–9, very short on 2. A dark reddish brown band across the middle of the tail.

The type material was collected at Waddell Creek, Rancho del Oso, California, by Dr. David Shepherd. The specimens are in the collections of Stanford University and Cornell University.

Ameletus similior McDunnough

1928—McDunnough—Canad. Ent. 60: 9. . . . . Alta.
1929—McDunnough—Canad. Ent. 61: 174. Fig.

Length: body 9; wing 9 mm.

A medium-sized species, distinguished by the dark heart-shaped ganglionic marks on venter of abdomen. Wings with faint clouds.

Head and thorax shiny blackish brown, in the male; in the female, the thorax is rather bright reddish brown. Fore legs blackish brown; middle and hind legs lighter brown. Wings hyaline, veins dark, palest along the outer margin; some are faintly clouded with light umber brown, in the male.

Abdomen deep brown with lighter brown shading, the posterior margins definitely brown-ringed. Dark heart-shaped marks on ganglionic areas of sternites. Tubercles on the apical margin of the forceps base, between the forceps, rather poorly developed. Penes rather long and slender, only slightly incurved near the tip; each bears one or more spines on the inner margin (see fig. 117). Subanal plate of female rather small, covering less than half the tenth segment.

Type females were taken at Jacques Lake, Jasper Park, Alta. Sept. 4, 1926 (F. Neave). In 1928, a male specimen was taken in Waterton Lakes Park, Alta., in late August. We have no specimens of this species in our collection.

Ameletus sparsatus McDunnough

1931—McDunnough—Canad. Ent. 63: 89. . . . . Alta.

Length: body 8–9; wing 10 mm.

A medium-sized speckled-winged species, with dark ganglionic areas on sternites 7 and 8.

Head shining black. Pronotum black. Mesonotum reddish brown anteriorly, the middle area yellowish brown with deep blackish brown lateral margins. A yellowish brown area anterior to the wing root. Scutellum yellowish brown with a broad shiny black area on each side. Metanotum shining black, the anterior part shaded with brown. Pleura and sternum blackish brown, with lighter brown shading.

Legs of male missing. Fore leg of female blackish; middle and hind legs yellowish brown, the femora shaded with reddish black. Wings hyaline, tinged faintly with amber in the male. Veins and cross veins prominent, brown; cross veins in the costal half of the wing basad of the bulla margined with smoky brown, and forming small dark patches below the bulla and in the radial space between the bulla and the apex. Stigmatic cross veins strongly anastomosed.

Dorsum of abdomen yellowish brown on tergites 2–6, the apical tergites golden brown. Lateral and posterior margins of all tergites shaded rather broadly with darker brown. Sternites 2–6 dull smoky brown, apical sternites opaque yellowish brown; dark antero-median patches on sternites 7 and 8; sternite 9 with a broad deep brown lateral edging. Forceps base yellowish brown shaded with brown at the bases of the forceps and along the posterior margin. Forceps smoky brown. Resembling *A. dissitus* and *A. cooki* in structure of the penes (see fig. 118).

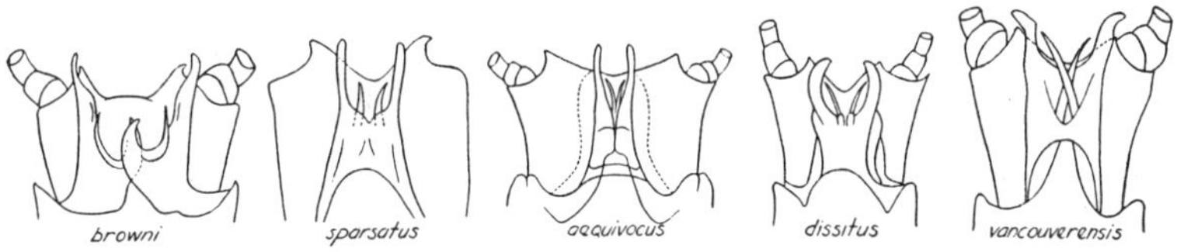

FIG. 118. Male genitalia of five species of Ameletus (after McDunnough, except *sparsatus*).

The type material came from Blairmore and Lundbreck, Alta. (J. H. Pepper). In the Cornell collection are several subimagos of both sexes and one male imago, all reared from nymphs, which seem to be of this species. These specimens are from the Yellowstone River in the Roosevelt Wild Life Forest (R. A. Muttkowski).

Legs of nymph banded. Abdomen orange-brown. Tergites 1, 7 and 10 largely pale except for a dark anterior margin. Tergite 9, and sternites 8 and 9, largely dark reddish to orange brown. Remaining tergites with large pale rather oval lateral spots; a pale median area bounded by dark brown submedian oblique streaks; anterior and posterior margins very narrowly black, especially in the postero-lateral angle. Sternites with similar marks, but the pale median streak usually wanting. Tails with a very wide median black band and a narrow black band at the apex.

Ameletus subnotatus Eaton

1885—Eaton—Revis. Monogr. p. 211 .... Colo.
1923—Dodds—Trans. Ent. Soc. Amer. 49: 105.
1928—McDunnough—Canad. Ent. 60: 9 .... Alta.
1931—McDunnough—Canad. Ent. 63: 89.

Length: body 11–12; wing 13 mm.

A large speckled-winged species, in which the ventral abdominal markings consist of dark dashes from the base of each segment, and two dark dots.

Head and thorax blackish brown. Fore leg dark reddish black; a black apical band on the femur. Middle and hind legs somewhat lighter brown; the femora banded apically with dark brown, the tarsi darker than the tibiae, especially toward the tips. Wings hyaline; both wings clouded with dark brown close to the wing roots. Venation blackish brown, many cross veins margined with brown. Small brown spots below the bulla on the radius, at the inner ends of the intercalaries of the radial sector, at the division of its outer fork, and at the inner end of the intercalary of that vein. Stigmatic cross veins branched and anastomosed, forming a rather dense network.

Abdomen pale yellowish brown dorsally; the posterior margins, the median stripe and the postero-lateral angles dark brown. Ventrally paler than the dorsum; sternites 2–6 marked with dark submedian dashes and a dark dot near the center on each side of the median line. Sternite 9 yellowish brown, a dark triangle on each side in place of the submedian streaks; lateral margins blackish brown. Forceps base shaded with dark brown. Forceps lighter brown. Tails blackish brown at the base.

The type material was taken in Colorado. In the Cornell collection are several male specimens collected by Dr. Dodds from Colorado, and two females from Pikes Peak, taken in June 1932 (R. J. Gilmore). Dr. Dodds' specimens differ slightly from the original description in possessing definite blackish ganglionic markings, which Eaton does not mention; also the dark lateral dashes on the sternites arise from the posterior margin of the middle and apical sternites, and the dark dots are not noticeable. In all other respects they match Eaton's description of the species. Genitalia of one of these specimens is shown in fig. 117.

Ameletus validus McDunnough

1923—McDunnough—Canad. Ent. 55: 50. Fig. .... Alta.
1928—McDunnough—Canad. Ent. 60: 9.

Length: body 10; wing 11 mm.

A pinkish brown species, with faint clouds on the fore wing and without dark ganglionic marks on the venter of the abdomen.

Head brownish. Thorax deep blackish brown, the mesothorax light brown on the anterior sutures and around the scutellum. Fore leg deep blackish brown; middle legs pinkish brown. Wings hyaline, veins dark; faintly clouded with light umber brown; palest along outer margin. Wing of female lacks the brown clouds.

Abdomen pinkish brown. Segments 1–6 semi-transparent, the sutures narrowly opaque; segments 7–10 opaque, segment 9 shaded laterally with smoky brown. Forceps base brown, with paler lateral margins and a yellow median area; posterior margin strongly excavated between the forceps; the tubercles on the apical margin well developed. Penes long and slender, curving outward near the tips. Subanal plate of female larger than in the allied species, *A similior*, and reaching nearly to the posterior margin of the tenth segment.

Type material was taken in September and October at Banff, Alta. (C. B. Garrett). We have seen no specimens of this species.

Ameletus vancouverensis McDunnough

1933—McDunnough—Canad. Ent. 65: 157. Fig. .... B. C.

Length: body 9; wing 10 mm.

The wings of this species are tinged throughout with umber brown; penes with two very long median spines.

Head and thorax deep mahogany brown. Mesothoracic scutellum light yellowish brown; two lateral arms extend forward from this area, thus forming with it a yellowish U-shaped patch. Pleura shaded with paler brown. Fore legs deep smoky brown; middle and hind legs umber brown, the femora with indications of a ruddy apical band. Wings tinged strongly and throughout with umber brown; stigmatic area even deeper brown; cross veins of the stigma anastomosed; both longitudinal and cross veins are dark brown, the cross veins very numerous. First abdominal segment deep mahogany brown; tergites 2–7 semi-hyaline, shiny light yellowish brown, the posterior margins narrowly dark brown, and the posterior half of each tinged with diffuse brownish shade. Tergites 8–10 are opaque, deep mahogany brown, tinged anteriorly with yellowish brown. Sternites 2–8 light ochreous, with a wash of dull brown; sternite 9 is rather bright yellow-brown in the central area, its base and lateral areas dark brown. Forceps base yellowish brown in central area, the apical and lateral margins mahogany brown; penes brown-tinged. Two long sharp spines are present between the lateral portions of the penes (see fig. 118). Tails deep smoky, joinings darker. The nymph is not known.

The type material of this species, from Vancouver Island, B. C., is in the Canadian National Collection. We have one specimen in the Cornell collection, from Courtenay, V. I., B. C. (J. D. Greyson) presented by Dr. McDunnough.

Ameletus velox Dodds

1923—Dodds—Trans. Ent. Soc. Amer. 49: 105. Figs. .... Colo.
1924—Dodds and Hisaw—Ecol. 5: 264, 139. Fig. nymph.
1927—Seemann—J. Ent. and Zool. 19: 47. .... Calif.
1928—McDunnough—Canad. Ent. 60: 9. .... Alta.
1929—McDunnough—Canad. Ent. 61: 174.

Length: body 12; wing 13 mm.

A large clear-winged species, with dark brown ganglionic markings.

Head brown, with a black stripe across the median carina. Thorax reddish brown. The red tinges are most evident on the lateral margins of the mesothoracic scutellum and an oblique streak across the pleura from below the wing root to a point anterior to the base of the middle leg. Fore leg reddish brown. Middle and hind legs yellowish. Wings hyaline. Longitudinal veins reddish brown, cross veins paler. No cross veins margined. Stigmatic cross veins forked and anastomosed.

Abdomen pale yellowish to reddish brown. Segments 2–7 semi-translucent, 7–10 opaque, tinged with reddish brown. Tergites narrowly pale on the posterior margins, smoky brown in the median area. A smoky lateral line lies above the pleural fold; antero-lateral angles pale. Tracheae tend to form a fine dark network on many of the tergites. Ganglionic areas of sternites marked with blackish brown; pale smoky sub-median marks may be visible. Tails reddish brown, narrowly darker at joinings. No long spine below the penes on the inner margin, as in *A. vernalis* (see fig. 116).

Color markings of the nymph are not given in the description, nor are the ventral markings shown in the figure.

The species was described from specimens taken in the South Boulder Valley, at Tolland, Colo., in July, 1915. Two male and one female specimens collected at that time by Dr. Dodds, are in the Cornell collection.

Ameletus vernalis McDunnough

1924—McDunnough—Canad. Ent. 56: 115. Fig. .... B. C.

Length: body 12; wing 12 mm.

A large clear-winged species with faint mid-ventral ganglionic marks on the abdomen.

Head and thorax deep blackish brown, the sutures yellowish. Fore femur deep blackish brown, a faint yellowish tinge at the base and a yellow line along the upper edge; tibia and tarsus black. Femora of middle and hind legs marked like fore femur, but slightly paler; tibiae and tarsi light brown. Wings hyaline, the fore wings shaded with brown on the apex and with a slight brown tinge at the base of R_3, where there is a group of 6 cross veins. Veins and cross veins brown except 3 or 4 costal cross veins beyond the bulla, which are pale and indistinct. Stigmatic cross veins anastomosed to form a series of smaller costal cells with larger ones below.

Abdominal segments 2–6 translucent, dull greyish white; segments 7–10 opaque yellowish. A dorsal series of deep purple-brown subtriangular patches is present on the tergites, the bases on the posterior margins, and with apices almost attaining the anterior margin. A similar series of lateral patches are brown in color, thus leaving pale anterior triangles between the darker ones. Tergite 9 is almost entirely brown. Sternites 2–6 show faint traces of mid-ventral brown spots connected by a fine line. Sternites 7 and 8 have brown lateral patches based on the posterior margins. Sternite 9 largely brown, yellowish on the posterior margin. Forceps deep smoky brown. Venter of female largely tinged with purplish brown. A long spine below the penes on the inner side (see fig. 115).

The type specimens were obtained at Oliver, British Columbia (G. B. Garrett). Two males of this species are in the Cornell collection, taken at Blewett, Wash. (M. D. Leonard).

Addendum

In a recent paper (Canad. Ent. 67: 100–103), Dr. McDunnough has given descriptions of the nymphs of three species of Ameletus, heretofore known only in the imago state, and of another species, *sparsatus*, a description of which had not before been published. Nymphs of *A. sparsatus*, *A. cooki*, and *A. validus*, belong together in what he terms the *validus* group, in which "the pattern on the dark [abdominal] segments consists of distinct pale oval spots and there is also a varying number of pale segments with only slight brown maculation" (McD.); femora pale, with a "*distinct brown quadrate or oval patch centrally* on the outer side" (McD.). In *sparsatus*, tergites 3–5, and 7, are largely brown, as are also sternites 3–9. Tergites 2, 3, and 4–8 of *cooki* are largely whitish, the sternites pale, those of segments 8 and 9 with wide transverse brown bands. In *validus*, the four median tergites 3–6 are brown with three oval pale patches on each; sternites "*almost entirely pale with faint brown streak*" laterally on 9.

The nymph of *A. similior*, on the other hand, belongs in the *velox* group; abdominal tergites largely brown without a definite color pattern, femora "pale brown without any dark central patch but usually a slightly paler tinge apically" (McD.). Abdominal tergites "largely pale brown with obscure and rather variable maculation"; "*segment 8 unicolorous deep brown with faint pale median line and small pale spot in antero-lateral corner*"; "segments 9 and 10 pale" (McD.). Sternites 3, 4 and 8 dark brown, others pale with faint dark lateral markings.

Genus **Siphlonurus** Eaton

1868—Eaton—Ent. Mo. Mag. 5: 89.
1885—Eaton—Revis. Monogr. p. 214 (Siphlurus).
1915—Clemens—Canad. Ent. 47: 245.
1919—Bengtsson—Lunds Univ. Arssk. Afd. 2, Bd. 5: 11 (Siphlurella = Siphlonurus in part).
1920—Ulmer—Stett. Ent. Zeit. 81: 135.
1923—McDunnough—Canad. Ent. 55: 46.
1933—Spieth—J. N. Y. Ent. Soc. 41: 327.

Medium to large-sized mayflies, usually strikingly marked, and often presenting an annulate appearance. Male and female very similar in color pattern. Head usually pale with a broad dark transverse band across the median carina, and black rings at the bases of the ocelli. Eyes of male large, usually contiguous apically; not divided. Eyes of both sexes typically have a transverse band extending obliquely across the outer surface, of a different color than the remaining portions of the eye. Vertex and occiput of female usually with a wide dark geminate submedian stripe. Hind margin of head slightly concave.

Vertex and pleura of thorax often marked with blotches of bright color, or with powdery white. Fore leg of male as long as or slightly longer than the body. First fore tarsal joint of male ¾ to ⅞ of the second, or subequal to it. First and second joints of fore tarsus possess blunt spine-like processes on the margins. Sharp spines are present on the margin of the fore tibia, and on margins and tips of tarsi of other legs. Claws of all tarsi are similar, sharp-pointed in both sexes. Hind tarsus longer than hind tibia.

Fore wing relatively long and narrow, about three times as long as wide (see fig. 24). Hind wing well developed, usually less than twice as long as wide. Costal angulation of hind wing not prominent, merely a rounded elevation. Median vein of hind wing forked about ⅓ of the distance from the base to the outermargin. Venation typically dark. Humeral cross vein strong and dark. Stigmatic cross veins frequently anastomosed.

FIG. 119. Male imago of *Siphlonurus alternatus*.

Markings of the abdominal tergites consist typically of dark and light somewhat triangular areas having the following arrangement. From a large dark area based on the posterior margin on each side, two triangles extend forward toward, but not reaching, the anterior margin. Pale triangular areas are thus left in the antero-lateral angles, and between the apices of the dark triangles, their bases on the anterior margin. Dark oval submedian marks are typically present; these are near the anterior margin, in the basal segments, but nearer the center in the apical ones. The venter of the abdomen is often strikingly marked. The color pattern consists of lateral blotches, oblique lateral lines, and median marks, often arranged in one or another of the following ways: 1), dark U-shaped marks; 2), lateral dark triangles, which may or may not be united anteriorly; 3), a continuous or discontinuous dark median line, with one or two pairs of small dots near it.

The forceps base of the male is very long, its apical margin between the forceps usually slightly concave. The forceps are more or less definitely four-jointed,—counting as the basal joint a ring-like enlargement at the base of the long joint, which is often imperfectly separated from it. The two distal joints together are not longer than the long joint. The inner apical margin of the short basal joint is often extended laterally beyond the base of the long joint. Penes very complex in structure, consisting usually of paired ventral and dorsal chitinized portions, of which one pair more or less completely overlies the other. Viewed from beneath, the penes are usually not visible above the apical margin of the forceps base. The subanal plate of the female is broad, and convex on its apical margin (in one species, extended into a median spine).

FIG. 120. Nymph of *Siphlonurus alternatus* (photo by Ann H. Morgan).

Nymphs have the compressed body and stream-line form typical of the subfamily (fig. 120). Maxillary and labial palps are each three-jointed. The maxilla is rather conical at the apex, bearing a few simple hairs and spines on its upper (outer) and inner margins. The mandibles are normal. Frontal margin of labrum entire, slightly concave. The claws are long and slender, not pectinate. Well-developed postero-lateral spines are borne on the thin flattened lateral margins of the abdominal segments. In some species, the gills are double on segments 1 and 2 only; in others, all seven pairs are double, one member of the seventh pair being very small. Gills large and plate-like, with abundant tracheation. Members of a pair are similar in form.

Genotype—*S. flavidus* (in Baetis), Ed. Pictet.

Distribution—Holarctic; in Nearctic, generally distributed.

KEY TO THE SPECIES

Adults

1—Hind wing blackish or brown, at least in the outer two-thirds of its length....2

—Hind wing not as above....3

2—Hind wing entirely coffee brown; cross veins in discs of both wings margined with dark brown....**marshalli**

—Hind wing blackish, at least in the outer two-thirds of its length; no margined cross veins....**mirus**

3—Hind wing of male orange-tinged; fore and hind wings of both sexes with a large dark brown blotch in the radial space.......................... **spectabilis**
—Wings not as above.. 4
4—Brown cloud in the axillary fold of the fore wing...................... **typicus**
—No such brown axillary cloud in fore wing...................................... 5
5—A continuous dark mid-ventral line on the abdominal sternites.............. 6
—No continuous dark mid-ventral line on sternites.............................. 9
6—Lateral margins of tergite 9 flaring, extending laterally beyond the margins of tergite 10; postero-lateral spine prominent.......................... **phyllis**
—Lateral margins of tergite 9 not flaring, hence tergite 9 not wider than tergite 10; spine inconspicuous.. 7

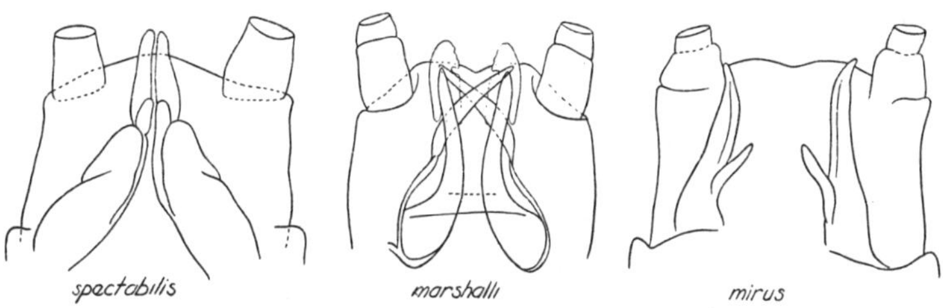

FIG. 121. Male genitalia of three species of Siphlonurus.

7—Ventral divisions of penes drawn out apically into a blunt projection; minute spines borne at apex along the inner margins.................. **marginatus**
—Ventral divisions of penes rounded apically; no apical spines, but 2 or 3 spines on the inner margins.. 8
8—Thorax and dorsum of abdomen shiny deep black-brown; thorax not marked with purplish rose.. **securifer**
—Thorax and dorsum of abdomen dark olive brown; thorax with conspicuous purplish rose markings... **decorus**
9—Prominent dark U-shaped marks on abdominal sternites (in *quebecensis*, more or less triangular laterally).. 10
—No dark U-shaped markings ventrally (if lateral triangles are present, these do not unite anteriorly).. 13
10—Markings of each sternite consist of dark brown lateral triangles united anteriorly at the median line; tails yellowish, joinings darker............. **quebecensis**
—Markings of sternites not triangular, but consisting of median U-shaped black marks; tails blackish at base, paler distally.............................. 11
11—A large rounded bulge on sternite 8 of the female; U-marks present on sternites 1–6 only.. **inflatus**
—No such large rounded bulge on sternite 8 of the female; U-marks present on sternites 1–9.. 12
12—U-marks on sternites 8 and 9 diffuse and poorly defined; a pale ruddy spot on the rear of the mesonotum.. **columbianus**
—U-marks clearly defined on sternites 8 and 9; no such ruddy mark on the mesonotum.. **occidentalis**
13—Very large species; wing of female 20 mm. in length.............. **luridipennis**
—Species smaller than above; wing of female not exceeding 16 mm. in length.... 14

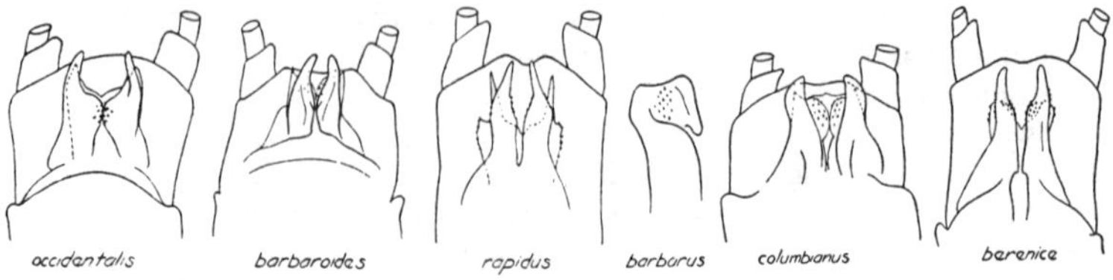

FIG. 122. Male genitalia of six species of Siphlonurus (*barbarus*, after McDunnough).

14—Oblique black lateral lines on abdominal sternites, not united anteriorly; two black dots near center of each sternite. . . . . **alternatus**

—Markings of abdominal sternites not as above. 15

15—Ventrally deep purplish brown, the posterior margins of the sternites yellowish; no dark pattern. . . . . **autumnalis**

—Paler ventrally, usually with darker markings. 16

16—Incomplete dark mid-ventral streak on abdominal sternites; no other markings . . . . **rapidus**

—No such incomplete dark mid-ventral marks; brownish lateral markings present, usually triangular, but not joined anteriorly. 17

17—Thorax and dorsum of abdomen pitch black, except for usual pale areas; ventral divisions of the penes prolonged into a point on the inner margin, spines on outer portion only (see fig. 122). . . . . **barbaroides**

—Thorax and dorsum of abdomen brown to reddish brown; ventral divisions of penes not as above. 18

18—Ventral divisions of penes finely serrate on outer and inner margins; no spines on central portion (see fig. 122). . . . . **berenice**

—Ventral division of penes not serrate on the margins, but bearing spines centrally (see fig. 122). . . . . **barbarus**

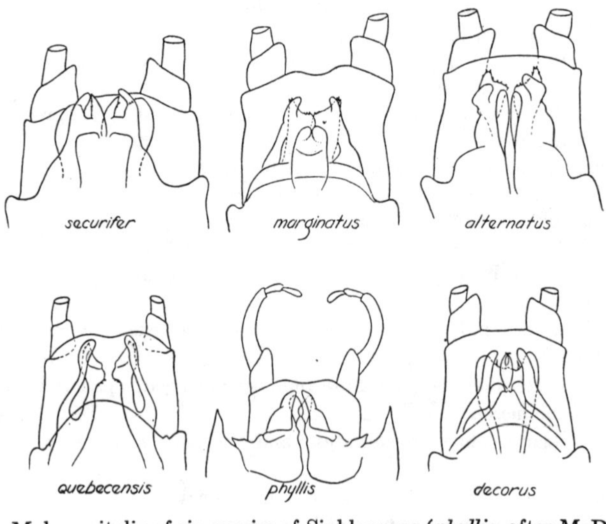

FIG. 123. Male genitalia of six species of Siphlonurus (*phyllis*, after McDunnough).

VERIFICATION TABLE

Adults[1]

Species	*Wing length*	*Black[2] U-marks*	*Two dk.[2] dots*	*Dark mid-ventral line*	*Forceps[3] angulate*	*Distr.*
alternatus	12	None	Present	Interrupted	Yes	E
autumnalis	13	None	None	None	No	Alta.
barbaroides	10–13	None	None	None	Yes	E
barbarus	11	None	None	None	Yes	E
berenice	10	None	None	None	Yes	E
columbianus	13	Present	Present	None	Yes	W
decorus	9–10	None	Present	Continuous	Yes	N. C.
inflatus ♀	14	Present	Present	None	?	Alta.
luridipennis	20	None	None	None	?	N. C.
marginatus	9	None	Present	Continuous	Yes	N. C.
marshalli[4]	13–14	None	None	None	No	Ark.
mirus[4]	11–13	None	Present	None	No	E
occidentalis	12–14	Present	Present	None	Yes	W
phyllis	13–14	None	Present	Continuous	No?	Canada
quebecensis	12–13	Partial	None	None	Slight	E
rapidus	12	None	None	Interrupted	Yes	E
securifer	11	None	Present	Continuous	Yes	E
spectabilis	13	Present	?	None	No	Calif.
typicus	10	Partial	None	None	?	Mass.

KEY TO THE SPECIES

Nymphs

1—Gills double on segments 1–7; one member of the 7th pair very small. 2
—Gills double on segments 1 and 2 only. 3
2—Abdominal sternites marked with an oblique black lateral line on each side, and two dark dots near center of each sternite. . . . . **alternatus**
—Abdominal sternites not marked as above. . . . . **phyllis**
3—Tails with dark transverse band somewhat beyond the middle, also a narrower dark apical band. 4
—Tails lacking the dark transverse bands. 6
4—A black mid-ventral line extends the length of the abdomen. . . . . **marginatus**
—No such black mid-ventral line on abdomen. 5
5—Sternites reddish brown, with pale apical and lateral margins and a wide pale submedian streak. . . . . **spectabilis**
—Sternites yellowish; apical segments with dark U-marks, others with broad lateral triangles united anteriorly. . . . . **quebecensis**
6—Legs banded; venter of abdomen with dark U-marks on each sternite. **occidentalis**
—Legs not banded; no such dark U-marks on the sternites. . . . . **mirus**

[1] Males, except *inflatus*.
[2] On abdominal sternites.
[3] Basal joint.
[4] Hind wing largely dark.

VERIFICATION TABLE

Nymphs

Species	*Body length*	*Legs banded*	*Gills, 3 to 6*	*Ventral marks*	*Tails banded*	*Distr.*
alternatus	12	Yes	Double	Lat. line	Middle	E
marginatus	9	Yes	Single	Med. line	Middle	N. C.
mirus	15	No	Single	Lat. triang.	No	E
occidentalis	13	Yes	Single	U-mks.	Tip	W
phyllis	12?	Yes	Double	?	Middle	Canada
quebecensis	12–15	Yes	Single	Lat. triang.	Middle	E
spectabilis	14	Yes	Single	Med. triang.	Middle	Calif.

Siphlonurus alternatus Say

1824—Say—West. Quart. Rep. 2: 304 (Baetis)....................................Northwest Terr.
1853—Walker—List Neurop. Ins. Brit. Mus. p. 567 (*Baetis annulata*)..........................N. Y.
1859—LeConte—Complete Writings T. Say, 1: 204 (Baetis).
1861—Hagen—Syn. Neurop. Ins. N. Amer. p. 49 (*B. alternatus*;) p. 48 (*B. annulatus*).
1862—Walsh—Proc. Acad. Nat. Sci. Phil. p. 369 (Baetis).. Ill.
1863—Hagen—Proc. Ent. Soc. Phil. 2: 169 (Baetis).
1863—Walsh—Proc. Ent. Soc. Phil. 2: 189 (Baetis).
1871—Eaton—Trans. Ent. Soc. London p. 127 (*Siphlonurus annulatus*); p. 129 (*S. alternatus*). Figs.
1876—Provancher—Naturl. Canad. 8: 267 (*B. femorata*)..Que.
1877—Provancher—Fn. Ent. d. Canad. 2, fasc. 1: 83 (*Baetis femorata*, also *Siphlurus alternatus*); 82, misprint, *S. alternans*.
1885—Eaton—Revis. Monogr. p. 219.
1901—Needham—N. Y. St. Mus. Bull. 47: 423. Nymph desc.
1907—Needham—N. Y. St. Mus. Bull. 124: 189.
1911—Morgan—Ann. Ent. Soc. Amer. 4: 116.
1915—Clemens—Canad. Ent. 47: 248. Fig., color pattern.
1930—McDunnough—Canad. Ent. 62: 61.
1930—Ide—Canad. Ent. 62: 228...Ont.
1930—Bengtsson—Lunds Univ. Arssk. N. F. Avd. 2, Bd. 26: 10 (*S. alternatus* placed in Siphlurella).

Length: body 12; wing 12 mm.

A rather large, somewhat pale species, with conspicuous ventral markings (see fig. 119).

Head whitish. Thorax yellowish brown. Wide reddish brown median stripe on metanotum and lateral margins of mesoscutellum. Scutella dark brown. Pleural sutures brown. Purplish red marks anterior to wing roots. Prosternum white, marked with two dark brown triangles. White band across anterior portion of mesosternum, enclosing a dark median spot.

Fore leg of male longer than body; yellowish. Apical dark band on femur; knee brown; apex of tibia, and tarsal joinings, brown. Femora of middle and hind legs whitish, with apical brown band. Tibiae and tarsi yellowish, joinings brown. Longitudinal veins of fore wing, and costal margin of hind wing, purplish black. Humeral cross vein purplish black. Costal cross veins before bulla pale, almost invisible; in stigmatic area pale brown, numerous, somewhat irregular, a few anastomosed. All other veins brown, the longitudinal veins all pale at the wing root.

Abdomen pinkish white, marked with chestnut brown. Posterior margins of tergites widely banded with brown; wide median brown streak; usual brown triangles and oval spots, the latter very prominent. Pale areas on anterior margin extensive. Tergites 8–10 with large powdery white areas. Ventrally, no lateral patch. Oblique lines extend from near the postero-lateral angle almost to the anterior margin near the median line, ending in a black dot. A pair of smaller black dots near center of each sternite, on each side of median line. A dark brown triangle on median line at the anterior margin. Sternites 7–9 of the female, and 9 of the male, largely yellow with reddish markings. Dark lateral mark on sternite 9 of the male.

Forceps yellowish. Penes yellow tipped with black. Tails reddish brown at the extreme base, becoming yellow and finally white in the distal portion. Joinings rather widely dark purplish brown. Penes as in fig. 123.

Nymph has double gills on all abdominal segments, one member of the 7th pair being very small (see 120). Femur with an apical brown band; tibia banded near base; tarsus twice banded. Usual black band across tail, beyond the middle. Ventral markings comparable to those of the imago.

This species, originally described from the middle west, is common throughout eastern North America. In the Cornell collection are specimens from Thuro, N. S. (R. Matheson); Ft. Jackson and Nassau, N. Y. (L. Harper); Saranac Lake and Old Forge, N. Y. (J. G. Needham); Mongo, Ind. (H. T. Spieth); also from the vicinity of Ithaca, N. Y.

Siphlonurus autumnalis McDunnough

1931—McDunnough—Canad. Ent. 63: 89. .... Alta.

Length: body 14; wing 13 mm.

A large dark species occurring in the late summer.

Head blackish. Pronotum deep smoky. Mesonotum dark brown, the scutellum blackish brown. Pleural sutures pale yellowish. Sternum blackish brown.

Fore leg deep blackish. Middle and hind legs dull yellowish with a smoky tinge; tarsi distinctly smoky. Wings hyaline; longitudinal and cross veins deep brown. Cross veins weak in the basal two-thirds of the costal space; in the stigmatic area heavier and strongly anastomosed.

Abdomen deep purplish brown dorsally, the pale anterior triangles tinged with yellow. Lateral margin narrowly yellowish, especially on the four apical segments. Ventrally deep purplish brown, the posterior margins narrowly yellowish. The median portion of the posterior half of sternite 9 shaded indistinctly with yellow, which extends over to the anterior margin of the forceps base. Remainder of forceps base, and forceps, deep smoky. The first joint of the forceps is rounded on its inner margin. The apical margin of the forceps base, between the forceps, is weakly excavated, and is on a level with the apex of the first joint of the forceps.

Type material from Blairmore, Alta. (J. H. Pepper). We have seen no specimens of this species.

Siphlonurus barbaroides McDunnough

1929—McDunnough—Canad. Ent. 61: 174. Fig., color pattern................................ Que.

Length: body 10–11; wing 10 mm.

A species much resembling *barbarus* and *columbianus*, but differing in finer details of structure of penes.

Thorax and dorsum of abdomen, except for usual pale areas, pitch black. Wings without the brownish tinge present in *columbianus*. Abdominal sternites paler than tergites. Lateral patches combine with oblique streaks to form lateral triangles. On sternites 1 and 2, these triangles are fused at the median line on the anterior margin, so that most of these segments are dark. Penes differ in details of structure from *barbarus* and *columbianus* (see fig. 122).

A specimen in the Cornell collection, from McLean, N. Y., is larger than the type, —wing and body each 12 mm. in length—but seems similar in other respects to the above description. In this specimen, a dark submedian dot on each side of each sternite is united with the dark lateral triangle.

Siphlonurus barbarus McDunnough

1924—McDunnough—Canad. Ent. 56: 225. Fig. .... N. Y.
1929—McDunnough—Canad. Ent. 61: 174–5. Fig.
1932—McDunnough—Canad. Ent. 64: 81 .... Que.

Length: body 10; wing 11 mm.

A medium-sized species resembling *S. berenice*, but with more clearly defined ventral markings.

Head brown, frontal margin pale. Thorax brown; pleural sutures pale yellowish. Legs pale yellowish brown, the fore legs somewhat darker. Tips of femora and tarsal joinings dark brown. Wings hyaline; veins brown.

Abdomen brown dorsally; anterior margin of each tergite semi-translucent, yellowish. Traces of dark dorsal and submedian longitudinal dashes visible on the posterior tergites. Pale yellowish ventrally. Lateral dark patches present. Sternite 9 largely dark brown. Forceps base pale yellowish; forceps smoky brown. Tails pale yellowish, joinings brown. Penes shown in fig. 122.

Type material was from Big Indian Valley, Catskill Mts., N. Y. (R. F. Pearsall). A single imperfect specimen from Quebec is in the Cornell collection.

Siphlonurus berenice McDunnough

1923—McDunnough—Canad. Ent. 55: 49. (Fig.) .... Que.
1924—McDunnough—Occas. Pap. Boston Nat. Hist. Soc. 5: 75 (*novangliae*—fig.) .... Conn.
1924—McDunnough—Canad. Ent. 56: 224 (*novangliae* = *berenice*).
1925—McDunnough—Trans. R. S. C.: p. 221 .... N. Eng., N. Atlantic States
1929—McDunnough—Canad. Ent. 61: 175 (color pattern).

Length: body 9–10; wing 10 mm.

A medium-sized brownish species; wings faintly tinted with brown.

Head yellowish. Thorax reddish brown; pleura tinged with yellow, sutures white. Reddish tinges at wing roots, on scutellum, and on mesosternum. Reddish yellow patch on each side of mesothoracic scutellum, also on metanotum.

Legs reddish brown, joinings darker. Wings faintly tinged with brown, especially on the costal margin. Veins dark reddish to purplish brown. Cross veins almost as heavy as longitudinal veins. Costal cross veins prominent; about 12 before the bulla, 22–26 beyond it. Those in stigmatic area are slightly curved, tend to anastomose, and some are incomplete, not reaching the subcosta.

Abdomen dark reddish brown dorsally, the apical tergites tinged with yellowish red. Posterior margins dark brown; usual dark lateral triangles and dark oval spots present. Pale areas on anterior margin somewhat reduced in size. Ventrally pale brown. Lateral dark patches present, united with oblique lines to form antero-lateral triangles; a pale oval space with a dark margin, in outer edge of this triangle. Sternite 1 dark brown on anterior margin. No dark median marks nor spots. Sternites 8 and 9 largely yellow with powdery white markings; dark lateral mark on sternite 9. Forceps and penes reddish brown; penes as in fig. 122. Tails dark reddish brown at base, becoming yellow in middle and distal portions. Joinings dark red-brown.

In the Cornell collection are many fine specimens of this species from Sunbury, Pa. (J. G. Needham), taken from a large swarm; Skinner, Me. (A. E. Brower); Liberty and Vandalia, N. Y.; and Kent Falls, Conn. (C. P. Alexander).

Siphlonurus columbianus McDunnough

1925—McDunnough—Canad. Ent. 57: 187. Fig. . . . . B. C.

Length: body 12; wing 13 mm.

A large species, much resembling *S. occidentalis* in its general appearance.

Head brown. Thorax deep brown. Mesothoracic scutellum blackish, with a paler ruddy spot directly anterior to it. Pleural sutures and wing bases yellowish. Fore legs deep blackish brown; middle and hind legs light yellowish brown. Femora marked with a slight purplish brown streak before the apex; tarsal joinings also purplish brown. Wings hyaline, with a faint brownish tinge in the stigmatic area. Veins entirely blackish.

Abdomen deep purplish brown dorsally, the pale anterior triangles light yellowish. Ventrally dull yellowish. The broad purplish brown oblique streaks are usually united on the anterior margin to form the blackish U-shaped marks. These are less distinct than in *occidentalis*, especially on sternites 8 and 9, where they barely join, are diffuse and poorly defined. Two tiny dark dots are usually present near the center of each sternite, within the dark U-mark. Segment 10 is wholly brown. Forceps dark with paler base. Penes as in fig. 122. Tails dark at base, becoming paler distally. Joinings brown.

Type material was from Agassiz, B. C. (R. Glendenning). In the Cornell collection are two specimens from the Yosemite Valley, Calif. (H. D. Michener, E. H. Nast); the latter specimen is from the California Academy of Science collection.

Siphlonurus decorus Traver

1932—Traver—J. Elisha Mitchell Sci. Soc. 47: 182. . . . . N. C.

Length: body 9–10; wing 9–10 mm.

A medium-sized species with conspicuous ventral markings.

Head brownish. Thorax dark olive brown. A large pale patch near the center of the pleura on each side. Pleural sutures pale; purplish rose markings on pleura and notum. A narrow pale transverse band crosses each division of the sternum.

Legs light brown, joinings darker. Femora with apical dark bands. Purplish rose spot on each coxa. Wings hyaline; longitudinal veins of fore wing purplish brown; all other veins pale brown. Costal cross veins before the bulla very pale, almost invisible. In the stigmatic area, pale but distinct; numerous, slightly curved, tending to anastomose. Stigmatic area, also the corresponding subcostal space, opaque white.

Abdomen dark olive brown dorsally, pale ventrally. Usual dark lateral triangles present on tergites; oval spots distinct. Mid-dorsal line pale; pale anterior triangles quite extensive. Ventrally, a blackish mid-ventral line extends continuously from the posterior margin of sternite 1 to the posterior margin of sternite 9, widened at each margin of each sternite. At center of each sternite, a small black dot on each side is very close to the dark median line. Anterior to this small spot, and laterad of it, is a larger dark brown oblong spot. The usual lateral patches are represented only by a narrow dark brown line parallel to the pleural fold. Sternites 7 and 8, and the postero-lateral angles of sternite 9, powder white.

Forceps and forceps base light brown with purplish rose markings. Penes shown in fig. 123. Tails dark brown basally, becoming light tan in the distal portion. Joinings dark purplish brown.

Type material from Goshen Swamp, N. C. (J. G. Needham), is in the Cornell collection.

Siphlonurus inflatus McDunnough

1931—McDunnough—Canad. Ent. 63: 90...Alta., B. C.

Length: body 14; wing 14 mm.

A large dark species, occurring late in the summer; known only from the female imago.

Head dull purplish on frontal portion, vertex dull yellow, with usual dark median streak. Thorax dull brown with blackish shading. Scutellum blackish, shaded with orange; antero-lateral portions of mesonotum with pale yellowish markings. Pleural sutures yellowish. Anterior portion of mesosternum pale yellowish, crossed by a purplish band; lateral yellow shading on posterior portion. Metasternum pale yellowish.

Legs dull yellowish, shaded rather heavily with purplish; fore tarsi deep smoky, others less dark. Wing hyaline, veins dark. Costal cross veins faint before the bulla. Stigmatic area with a faint purple tinge; cross veins weakly anastomosed.

Abdomen purplish brown dorsally; anterior triangles yellowish, rather extensive. Ventrally pale yellowish. Oblique bands broad, purplish black in color, and joined anteriorly to form U-shaped marks on sternites 1–6; two small black dots between the U-marks, on these segments. Posterior margin of sternite 7 broadly purplish black; only a pale oval central portion remains, surrounded by dark margins. Sternite 8 with a large rounded bulge, pale yellowish in color, bordered anteriorly and laterally by black stripes. A pale stripe extends backward from it to the posterior margin. Sternite 9 largely purplish black, with a narrow triangular median pale area. Tails blackish, somewhat paler distally.

Type material from Cameron Lake, Waterton Lakes Park and Blairmore, Alta., also Fernie B. C. (J. H. Pepper). We have seen no specimens of this species.

Siphlonurus luridipennis Burmeister

1839—Burmeister—Handb. d. Ent. 2: 801 (Baetis)..Carolina
1843—Pictet—Hist. Nat. Neurop. Ephem. 2: 192 (Baetis).
1853—Walker—List Neurop. Ins. Brit. Mus.; pt. 3: p. 563 (Baetis).
1861—Hagen—Neurop. Ins. N. Amer. p. 49 (*B. noveboracana*).
1926—Ulmer—Ent. Mittl. 15: 223 (not same as Eaton's *Ecdyonurus luridipennis*).
1926—McDunnough—Canad. Ent. 58: 190.

Length: body 16; wing 20 mm.

A very large dark species, known only from a female imago of the type material.

Head blackish. Frontal margin of clypeus yellowish. Usual dark transverse band across median carina, and dark median streak on vertex and occiput. Thorax rusty to chestnut brown. Pronotum brownish black, the lateral and posterior margins yellowish. Pleural and mesonotal sutures yellowish. A yellow area anterior to the scutellum. Sternum dark rusty brown, sutures yellowish.

Fore leg clear yellowish umber brown; femur darker toward the apex; knee somewhat clearer; tibia and tarsus light reddish brown. Middle and hind legs amber brown, joinings reddish brown. Wings hyaline, tinged faintly with grey, especially in the stigmatic area. Costa greenish yellow at the base. 11–13 costal cross veins before the bulla, 22–25 in the stigmatic area. The latter forked near the costa and anastomosed, forming a two-ranked network.

Abdomen smoky yellowish or greyish red dorsally, with diffuse dark shading, so that it appears very dark. Posterior margins of tergites dark, as well as the usual lateral triangles and oval spots. Median line also dark. Tergites 8–10 brighter in color, the oval spots smaller but more prominent. A short postero-lateral spine on segment 9.

Ventrally blackish, the apical sternites dark reddish brown. Posterior margin and anterior angles of sternites yellow; lateral margins reddish yellow; dark spots very indistinct. Tails broken.

The type material was taken by Mr. Zimmermann in Carolina. The above description is taken from Dr. Ulmer's account of the single remaining specimen, a female, in the Zoological Institute of the University of Halle.

Siphlonurus marginatus Traver

1932—Traver—J. Elisha Mitchell Sci. Soc. 47: 179. Figs. . . . . N. C.

Length: body 9–10; wing 9 mm.

Another medium-sized brownish species with a black mid-ventral line.

Head yellowish brown. No dark band across median carina, but a dark spot beneath each antenna. Thorax bright yellowish brown, the scutella darker. Pleural sclerites dark brown. Fore leg pale brown, tip of tarsus greyish white; femur with indistinct band. Middle and hind legs whitish; joinings brown; femora banded. Wings hyaline; longitudinal veins of fore wing brown, other veins pale yellow or whitish. Costal cross veins indistinct, almost invisible before the bulla; in stigmatic area, tend to anastomose. Stigmatic area, in costal and subcostal spaces, opaque white.

Abdomen olive brown dorsally, pale ventrally. Lateral triangles of tergites dark brown, much reduced in size; oval spots brown. Anterior pale triangles extended to occupy a large part of each tergite. Posterior margins dark brown. Pattern less distinct on apical tergites. Ventrally, a black median line extends continuously from sternites 1 to 9, widened near the posterior margin of each sternite. A small black dot on each side of median line, very near it, at the center of each sternite. Another oblique dark mark anterior to the dark dot, and laterad of it.

Forceps and forceps base yellowish brown. Penes as in fig. 123. Tails yellowish white, joinings dark reddish brown.

The nymph has double gills on segments 1 and 2 only. Legs yellowish; middle and hind femora with a brownish band, which is absent from the fore femur. Ventral markings are those of the imago. Usual black band across the tail beyond the middle.

The type material, from Liberty, N. C. (J. R. Traver), is in the Cornell collection.

Siphlonurus marshalli Traver

1934—Traver—J. Elisha Mitchell Sci. Soc. 50: 236. Fig. . . . . Ark.

Length: body 13–14; wing 13–14 mm.

Hind wings entirely stained with coffee brown, fore wing amber tinged in the basal half; cross veins in the discs of both wings heavily infuscated (see frontispiece). Hind tarsus barely longer than the tibia.

Head and thorax deep blackish brown; pleural sutures, areas anterior to the middle and hind legs, and a transverse band anterior to the hind legs, yellowish brown. Scutellum reddish. Fore leg dark reddish brown, tarsus slightly paler. Middle and hind legs yellowish, shaded somewhat with reddish brown. Hind tarsus but slightly longer than the tibia. Fore wing amber-tinged in its basal half; hind wing stained throughout with yellowish to coffee brown. Cross veins in the discs of both wings rather heavily infuscated with dark brown. In the fore wing, a dark zigzag band is formed at the bulla, extending across the wing to the first branch of the cubitus; between this and the wing base, in the radial space, another smaller cloud. Two darker clouds in the hind wing: at union of radius and its sector, and half-way from this to the margin. Abdomen dark reddish brown at base and apex, the middle segments paler; sternites yellowish, semi-hyaline in middle area. Middle tergites pale reddish brown, with wide dark posterior margins, dark reddish brown postero-lateral triangles, and a paler brown shading on the posterior portion of the ventral area. Tails deep reddish brown, almost black at the base.

Genitalia reddish brown. No swelling or extension on the inner apical margin of the first forceps joint. Penes seem to be of the type of *S. quebecensis* but differ markedly from that species in details of structure. See fig. 121.

The type material, from Imboden, Arkansas, was collected by Mr. Byron C. Marshall, for whom the species is named. Two specimens are in the Cornell collection.

Siphlonurus mirus Eaton

1885—Eaton—Revis. Monogr. p. 221.......N. H.
1915—Clemens—Canad. Ent. 47: 253. Figs.......N. Y.
1915—Alexander—MS. notes (in Clemens, Canad. Ent. 47: 254).
1932—Traver—J. Elisha Mitchell Sci. Soc. 47: 178.......N. C.

Length: body 10–14; wing 11–13 mm.

A large species, rather pale in color, with a conspicuous dark area on the hind wing.

Head yellowish brown. Thorax dark brown. Metanotum, except the areas on each side of the scutellum, yellowish to olive brown. Scutella yellowish. Pleural sutures cream colored. Pale bands anterior to wing roots; anterior to bases of middle and hind leg; and across the meso- and metasterna. Fore leg amber, the joinings and the two distal segments of the tarsus brown. Middle and hind legs yellowish; tarsi and all joinings brown, a diffuse smoky band on the femur. Fore wing hyaline, the longitudinal veins purplish black, those of the costal margin heavier. Cross veins light purplish black except in the costal space; these are pale and almost invisible. Stigmatic cross veins simple, almost straight, neither forking nor anastomosing. Hind wing dark brownish black, hyaline only at the base and in the basal half of the costal space. Cross veins in dark area margined with smoky brown. The extent of the hyaline space at the base of the wing varies with the sex, being more restricted in the male.

Abdomen yellowish brown, middle segments semi-hyaline. Posterior margins of tergites dark brown, widest at the middle line. The entire lateral margin also dark brown, except a pale line in the posterior angle. Dark lateral triangles reduced to small areas in the postero-lateral angle, and along the lateral margin; dark oval spots more or less apparent. Tergites 9 and 10 reddish brown. Ventrally yellowish brown on middle segments. Sternites 1 and 2, and the apical part of 3, dark reddish brown. Apical sternites may be largely yellow, with a dark brownish median line. Posterior half of each middle sternite shaded with brown, the extreme posterior margin whitish. A dark triangle runs forward from the posterior dark band, on each side of the median line, enclosing a small round spot. A dark oblique oval spot is often present, anterior to the center of each sternite, on each side. Mid-ventral line white in anterior half of each basal and middle sternite. Tails yellowish white, joinings reddish brown. Penes as in fig. 121.

The nymph has a conspicuous oblique black band on the front of the head, extending from below the antenna to the lower margin of the eye. Legs without dark bands; no dark transverse band across the tails. Gills double on segments 1 and 2 only. Ventral markings rather similar to those of the imago.

Specimens of this species in the Cornell collection are from Fishhouse and Sport Island, in the Sacandaga River (C. P. Alexander); Liberty, N. C. (J. R. Traver); Sandy Creek south of Newman, Ga. (P. W. Fattig); Greenville, S. C. (H. K. Townes,

Jr.); and nymphs from Highland, N. C. (J. G. Needham). Specimens from Sport Island and Liberty were reared from nymphs.

Siphlonurus occidentalis Eaton

1885—Eaton—Revis. Monogr.; p. 218 .... Colo., Wash.
1875—Hagen—Ann. Rep. U. S. Geol. and Geogr. Survey of Terr.; 281 (*H. brunnea*, in part).
1915—Clemens—Canad. Ent. 47: 248. Fig. Color pattern, nymph.
1923—Dodds—Trans. Amer. Ent. Soc. 49: 104. Fig.
1924—Dodds and Hisaw—Ecol. 5: 138, 264. Fig. nymph.
1928—McDunnough—Canad. Ent. 60: 9 .... Alta.
1930—Bengtsson—Lunds Univ. Arssk. N. F. 26: 10 (transferred to genus Siphlurella).
1931—McDunnough—Canad. Ent. 63: 89, 90.

Length: body 11–13; wing 12–14 mm.

A large brown species with conspicuous dark U-shaped marks ventrally.

Head yellowish brown. Thorax yellowish to reddish brown. Pronotum dark brown, grey laterally. Scutellum and lateral margin of mesonotum dark reddish brown; a yellow area anterior to scutellum, also to scutellum of metanotum. Antero-lateral portion of mesonotum and pleural sutures, yellowish, often with orange markings. White bands extend across the meso- and metasterna.

Fore leg dark reddish brown. Middle and hind legs yellow, the femora distinctly shaded with reddish brown; tips of tarsi brownish; joinings narrowly darker reddish brown. Wings hyaline, veins purplish black; the longitudinal veins heavier than the cross veins. A thickened area at the bulla, on the subcosta, radius and first division of the radial sector. Fore wing with a narrow brown shading at the base. Costal and subcostal cross veins pale. Stigmatic cross veins numerous, often forking near the costal margin.

Abdomen yellowish to yellowish brown, semi-hyaline in the pale areas, with reddish to purplish brown markings. All segments pale on the posterior margin. Median portion of each tergite yellowish brown; dark lateral triangles prominent, but confined to the postero-lateral angles and the lateral margin; dark oval spots conspicuous reddish brown. Apical tergites marked with yellow, often with powdery white areas. Ventrally, no dark lateral patches, but the oblique lines are broad and dark, united anteriorly to form a conspicuous U-mark on each sternite. Near the posterior margin a brown transverse line (may be crescentic) may almost or entirely unite the oblique dark bands. Two small dark dots usually present within the dark U-mark, anterior to the center.

Forceps base and forceps yellowish brown. Penes reddish brown (see fig. 122). Tails dark purplish brown at base, yellowish or whitish in middle and distal portions. Basal joinings faintly darker than segments.

The nymph of this species has double gills on segments 1 and 2 only. The legs are banded. The tails are dark-tipped, but not crossed by a median dark band.

Specimens of this species in the Cornell collection are from Glacier National Park, Mont. (A. S. Hazzard); Trinidad, Wash. (M. D. Leonard); Gothic, Colo. (J. D. Bowen); Yellowstone Park, Wyo. (P. R. Needham); Manitou, Colo., from the California Academy of Science collection (E. C. VanDyke); Harney Co., Ore. (W. T. Chamberlain); Moose and Centennial, Wyo. (Mrs. E. B. Klots, O. O. Blake); Logan River, Utah (J. G. Needham); Santa Fe, N. M. (P. B. Powell); South Boulder Valley, Colo. (G. S. Dodds); Florissant, and Twin Lakes, Colo. (T. D. A. Cockerell); and Pike's Peak, Colo. (R. J. Gilmore).

Siphlonurus phyllis McDunnough

1923—McDunnough—Canad. Ent. 55: 49. Fig. .... Alta.
1930—McDunnough—Canad. Ent. 62: 61. Figs., nymph described .... Que.

Length: body 12–13; wing 13–14 mm.

A rather large species having a conspicuous dark ventral stripe.

Head and thorax blackish brown. Thorax with yellowish shadings on the anterior portions. Legs dull olive brown; femora with apical brown bands. Wings hyaline, faintly tinged with brown in the basal area. Veins dark.

Abdomen dark brown dorsally. Dark oval spots and lateral triangles present. The pale anterior triangles are yellow, this color extending along the lateral margin. Ventrally yellowish. A brown mid-ventral stripe is present, and two small brown dots in each sternite, touching the median stripe. Anterior to the small dots, and laterad of them, are two large brown dots. Light brown shading along the lateral margin, in place of the usual lateral patches. Sternite 9 largely pale yellow, diffusely shaded with brown in the median area; a deep brown lateral streak on each side. Postero-lateral edge of sternite 9 produced into a prominent sharp tooth or spine. Forceps base squarely truncate on its apical margin. Forceps pale yellowish. Tails whitish. Genitalia as in fig. 123.

The nymph of this species has double gills, on each gill-bearing segment of the abdomen. The legs are banded, and the tail has the usual black band beyond the middle.

The type material was from Banff, Alberta, (C. B. Garrett). We have seen no specimens of the species.

Siphlonurus quebecensis Provancher

1876—Provancher—Naturl. Canad. 8: 267 (*Baetis canadensis*) .... Que.
1877—Provancher—Fn. Ent. d. Canad. 2: 84 (Baetis, Cloe and Heptagenia).
1877—Hagen—MS., Fn. Ent. d. Canad. 2: 82 (corrects Cloe to Heptagenia; species placed in Siphlurus).
1915—Clemens—Canad. Ent. 47: 250 (*triangularis*) Color pattern, nymph structures figured..... N. Y.
1925—McDunnough—Trans. R. S. C. p. 221.
1925—McDunnough—Canad. Ent. 57: 186 (*triangularis* = *quebecensis*).
1929—McDunnough—Canad. Ent. 61: 175 (color pattern of venter of abdomen).
1930—Ide—Canad. Ent. 62: 228 .... Ont.
1932—McDunnough—Canad. Ent. 64: 81.
1932—Traver—J. Elisha Mitchell Sci. Soc. 47: 179 .... N. C.

Length: body 11–13; wing 12–13 mm.

A handsome dark species, with more or less complete dark ventral U-shaped marks.

Head yellowish. Compound eyes purplish grey. Thorax reddish brown; pinkish markings on meso- and metanota; sutures of pleura widely cream colored. Fore leg of male light reddish brown. Femur with a dark ring at basal joining, and a brown band near apex; apex yellowish. Tibia and tarsus concolorous, brown. Middle and hind legs yellowish; base of femur, apical femoral band, knee and joinings of other segments brown. Last three tarsal joints, and claws, washed with brown.

Wings hyaline. Veins dark purplish brown, except stigmatic cross veins, which are pale brown, almost obsolete along the subcosta. Stigmatic area opaque white. Stigmatic cross veins numerous, anastomosing at the middle of their length. A thickened spot at the bulla, on subcosta, radius and radial sector.

Abdomen dark reddish brown dorsally, with usual pale lateral triangles and dark oval marks. Tergites 8–10 brighter red in color, and with larger pale areas, which are often powdery white. Ventrally whitish, subhyaline. Each sternite typically has the usual dark lateral patches and oblique dark streaks, the latter joined at the anterior margin to form the U-shaped dark mark. Sternites 1–7 pale in the median area, on the posterior margin and the antero-lateral angle. Sternites 8 and 9 largely dark except for a pale median streak in the posterior half. Genitalia pale reddish brown. Penes distinctive in form (see fig. 123). Tails pale yellowish white. Each joining rather widely ringed with purplish brown.

Nymphs have double gills on segments 1 and 2 only. Femur banded near the apex, as in the imago. Tibia with a basal band, tarsus twice banded. Ventral pattern of the abdomen comparable to that of the imago. A wide black band across the tails at about ⅔ of their length.

The Cornell collection includes a fine series of imagos from Stamford, Conn. (S. W. Bromley); dark specimens from Chelsea, Que. (W. T. M. Forbes); Ft. Ann, N. Y. (P. R. Needham); Bar Harbor, Me. (A. E. Brower); Ft. Jackson, N. Y. (L. Harper); Greenville, S. C. (H. K. Townes, Jr.); many pale specimens from the vicinity of Greensboro, N. C. (J. R. Traver); and specimens from the vicinity of Ithaca, N. Y.

Siphlonurus rapidus McDunnough

1924—McDunnough—Occas. Pap. Boston Nat. Hist. Soc. 5: 75. Fig. .... N. H., Mass.
1929—McDunnough—Canad. Ent. 61: 175. .... Que.

Length: body 10; wing 12 mm.

A medium-sized light brown species, having a discontinuous mid-ventral brown stripe.

Head grey to yellowish. Thorax dull yellowish to olive brown. Mesonotum yellowish, shaded with reddish brown on the median area. A pale spot before the scutellum. Scutellum reddish brown. Purplish streak anterior to wing roots. Pleural sutures yellowish. Pale bands across the meso- and metasterna. Fore leg yellowish brown, the tarsi yellowish; joinings purplish brown. Middle and hind legs yellowish, the femora washed with purplish brown; joinings purplish brown. Wings hyaline, very faintly tinged with brown. Stigmatic area opaque whitish. Veins dark purplish brown. All costal cross veins distinct, brown; about 11 before the bulla, 21–23 beyond it, in the stigmatic area. The latter veins are occasionally forked and anastomosed near the costal margin.

Abdomen purplish brown dorsally. The usual dark lateral triangles present, more or less diffuse; dark oval spots small, indistinct. Pale anterior triangles semi-hyaline, whitish; much reduced in size on the basal and apical tergites. Ventrally pale yellowish white, with a discontinuous dark purplish brown mid-ventral streak. Sternites 1 and 2 largely purplish brown. Forceps base yellowish white, forceps and penes brown. Tails yellowish white, joinings purplish brown. Genitalia as in fig. 122.

The Cornell collection includes specimens from Ft. Ann, N. Y. (P. R. Needham); Kent Falls, Conn., and Amherst, Mass. (C. P. Alexander); Lansing, Mich.; and Stamford, Conn. (S. W. Bromley).

Siphlonurus securifer McDunnough

1926—McDunnough—Canad. Ent. 58: 301. .... Ont.

Length: body 11; wing 11 mm.

A dark brown species with a dark mid-ventral line.

Thorax shiny deep blackish brown. Fore legs dull brown, with a faint brown apical band on the femur; tarsal joinings dark brown. Middle and hind legs yellowish brown; femoral band more distinct than on fore leg; tarsal joinings brown. Wings hyaline; all veins dark.

Abdomen deep shiny blackish brown dorsally. The anterior pale triangles are pale yellow. Ventrally pale yellow, with a continuous mid-ventral dark brown line. On each side of this line, near the center of each sternite, the usual small dark dot, and the usual lateral row of larger brown spots anterior to and laterad of the small dark dots. Forceps smoky brown. Tails dull smoky, the joinings brown. Penes as in fig. 123.

The type specimen was taken at Joe Lake, Alqonquin Park, Ont. (F. P. Ide). In the Cornell collection there is a male subimago with apparently similar genitalia, taken at Amherst, Mass. (C. P. Alexander). This specimen is considerably paler than indicated in the original description.

Siphlonurus spectabilis Traver

1934—Traver—J. Elisha Mitchell Sci. Soc. 50: 233. Fig. .... Calif.

Length: body 14; wing 13 mm.

A large pale species with speckled wings; hind wing of male orange-tinged; subanal plate of female running out to a point at the middle of the apical margin.

Head whitish, with usual dark band across the median carina. No oblique colored band across the eye. Thorax yellowish brown; lateral and anterior margins of mesonotum reddish brown. A purplish streak anterior to the wing roots; an oblique reddish brown streak across the mesopleura, from the base of the fore wing. Fore coxa with two black longitudinal streaks; middle and hind coxae with brown patches. Transverse dark band across mesosternum.

Fore leg shorter than body. Legs whitish, the fore femur and tibia, and all trochanters, yellowish. Fore femur reddish brown at apex, likewise fore tibia. Fore wing with a dark brown patch below the bulla in the radial space, and a dark triangle in the space of the sector below it. Several cross veins on costal margin and in wing disc margined. Yellowish at base, in anal region. Hind wing orange-tinged in male only. Large brown wedge-shaped spot in the radial space and the space of the sector below it; cross veins in the disc, and in the spaces beyond the radial fork, margined. Widened in the anal region, somewhat as in Parameletus.

Abdomen yellowish white, markings brown. Usual dark lateral triangles present; restricted to the postero-lateral angles and the lateral margin. Oval dark submedian spots rather indistinct. A longitudinal streak between these spots and the lateral triangles. Pale anterior triangles distinct in the antero-lateral angle, diffuse near median line. Tergites 7–10 reddish brown; a postero-lateral spine present on tergite 9. Ventrally, the oblique bands are broad, deep purplish brown in color, and united anteriorly to form dark U-marks. Sternite 9 purplish except for yellow lateral margins.

Forceps base and forceps purplish; penes reddish brown. No prominence on the inner margin of the forceps near the base; no clear separation of the basal segment from the long segment. Penes somewhat as in *alternatus*, but with long narrow forward-projecting processes in addition (see fig. 121). Apical margin of forceps base slightly concave between the forceps. Subanal plate of female extended backward on its apical margin into a prominent median spine.

Nymph has double gills on the first two abdominal segments only. Legs banded. Tergites 6 and 9 reddish. Ventrally, does not show pattern of imago, but has a dark triangle at the median line at the anterior margin of each sternite. Usual dark streak across the tails beyond the middle, but rather diffuse; orange-brown in color. Some of the type specimens are in the Cornell collection; others are at Stanford University.

Siphlonurus typicus Eaton

1885—Eaton—Revis. Monogr. p. 222..Mass.

Length: body 9–10; wing 10–11 mm.

A brownish species having a small brown cloud in the axillar fold of the fore wing.

Thorax deep yellowish brown. Fore femur and tibia reddish brown, tibia redder at tip; tarsus whitish, washed with grey distally; joinings blackish. Middle and hind femora pale reddish brown, tarsi whitish, tinged with brown; joinings light reddish. Wings hyaline, with a small brown cloud in the axillar fold of the fore wing. Veins blackish; the costa, subcosta and radius brownish black, as are also the bases of the main veins.

Abdomen dark brown dorsally, the apical segments washed with yellow. Dark brown median streak, the usual dark lateral triangles, and the pale anterior triangles on each tergite. Ventrally pale, sternites 8 and 9 yellowish to reddish brown. Faint U-shaped marks may be present, or fine lines converging toward the bases of the sternites. Forceps base pale in the middle area, light brown laterally. Forceps light brown. Tails greyish, the joinings blackish.

The type material was from Denham and Worcester, Mass. We have no specimens of this species, nor do we know of any record of it since Eaton's Monograph.

Genus **Isonychia** Eaton

1871—Eaton—Trans. Ent. Soc. London; p. 134.
1881—Eaton—Ent. Mo. Mag. 18: 21 (Chirotonetes, n. n.).
1881—Eaton—Ent. Mo. Mag. 17: 192 (Jolia).
1883—Eaton—Revis. Monogr. p. 42 (Jolia).
1885—Eaton—Revis. Monogr. p. 203 (Chirotonetes).
1905—Needham—Bull. N. Y. St. Mus. 86: 28 (Jolia = Chirotonetes).
1917—Clemens—Univ. Toronto Stud. Biol. Ser. 17 (Chirotonetes).
1920—Ulmer—Stett. Ent. Zeit. 81: 134 (Chirotonetes).
1923—McDunnough—Canad. Ent. 55: 46 (Isonychia valid).
1931—McDunnough—Canad. Ent. 63: 157.
1932—Traver—J. Elisha Mitchell Sci. Soc. 47: 200.
1933—Spieth—J. N. Y. Ent. Soc. 41: 329.

Medium to large-sized mayflies, usually reddish to purplish brown in color; male and female very similar. Eyes of male large, contiguous apically, not divided; usually marked with an obliquely transverse band of a different color, which is less conspicuous than in Siphlonurus. Head with an angulate backward projection at the inner corner of each eye; the posterior margin between these two projections is slightly raised, straight or very slightly emarginate. Remnants of gill tufts persist in the imago at the sides of the vestigial maxillae and on the bases of the fore coxae. Spinous median projections are present at the anterior margins of the meso- and metasterna, and near the middle of the mesosternum just anterior to the transverse groove between the middle legs. Of these, the first two are inconspicuous, the latter often rather prominent, sometimes appearing to be formed of two closely associated projections.

Fore leg of male somewhat shorter than the body. First fore tarsal joint usually subequal to or slightly longer than the second joint. Fore tarsus usually equals or slightly exceeds the tibia in length; rarely it is somewhat shorter. Fore tibia usually one-third longer than the femur.

In males, the claws of the fore tarsi are inflated and blunt, in most species; claws of the middle and hind legs of males, and of all tarsi of the females, are similar, sharp-pointed, on each tarsus. No spines or other projections are found on the legs. Hind tarsus about ¾ as long as the tibia. Fore leg red or brown; middle and hind legs pale. Fore wing 2 to 2½ times as long as wide (see fig. 28). Hind wing well developed, especially in the anal region, thus rather resembling Parameletus. Usually slightly longer than wide. Costal angulation obtuse. The median vein is forked near the outer margin.

The pleural fold of the 9th segment is slightly expanded laterally, bearing at its apex a short spine. Subanal plate of female usually decurved; more or less deeply excavated on the apical margin. Its outer angles acute when the excavation is deep, rounded when it is shallower. Forceps base of male of two distinct types. In species of the first group, it is deeply excavated apically, so that the penes project far beyond its margin. In species of the other group it is but shallowly excavated, so that all but the tips of the penes are concealed, when viewed from below. Dorsad of the forceps base, a muscular basal structure on each side seems to be the true basal joint of the forceps. The second and longest joint of the forceps is attached to this muscular base; it is somewhat enlarged near the base, but without an obvious suture at this point. The terminal joints together are usually shorter, or at most as long as, the long second joint. Penes are of four distinct types, within the genus. A rudimentary stub of the middle tail may be present.

The subimago is browner and duller than the imago. The wings are subhyaline, smoky brown in color, paler near the base; cross veins margined with brown. The eyes of the male are not quite contiguous above the head.

Nymphs possess a median frontal carina below the middle ocellus (see fig. 48). Mouthparts are very hairy. The labrum is somewhat quadrangular, covered with many stout bristles on the upper surface. Labial and maxillary palps are each two-jointed; in each, the basal joint is short, the distal joint wide and long, and set with many long bristle-like hairs. The crown of the maxilla is somewhat triangular, with an acute point at the apex of the outer margin. Gill tufts are present on the bases of the maxillae. The middle division of the hypopharynx is wide, slightly convex on the upper margin; the lateral divisions are rounded. Canines of the mandibles are quite incurved; the lacinia is prominent.

The fore tibia has a conspicuous apical spine, usually about one-half as long as the tarsus. On the femur, tibia and tarsus of the fore leg are fringes of long spine-like hairs, usually tawny orange in color; also few to many shorter stout spines. The flange at the apex of the femur is quite prominent. Gill tufts are present on the bases of the fore coxae. Claws are

PLATE XXXVI.

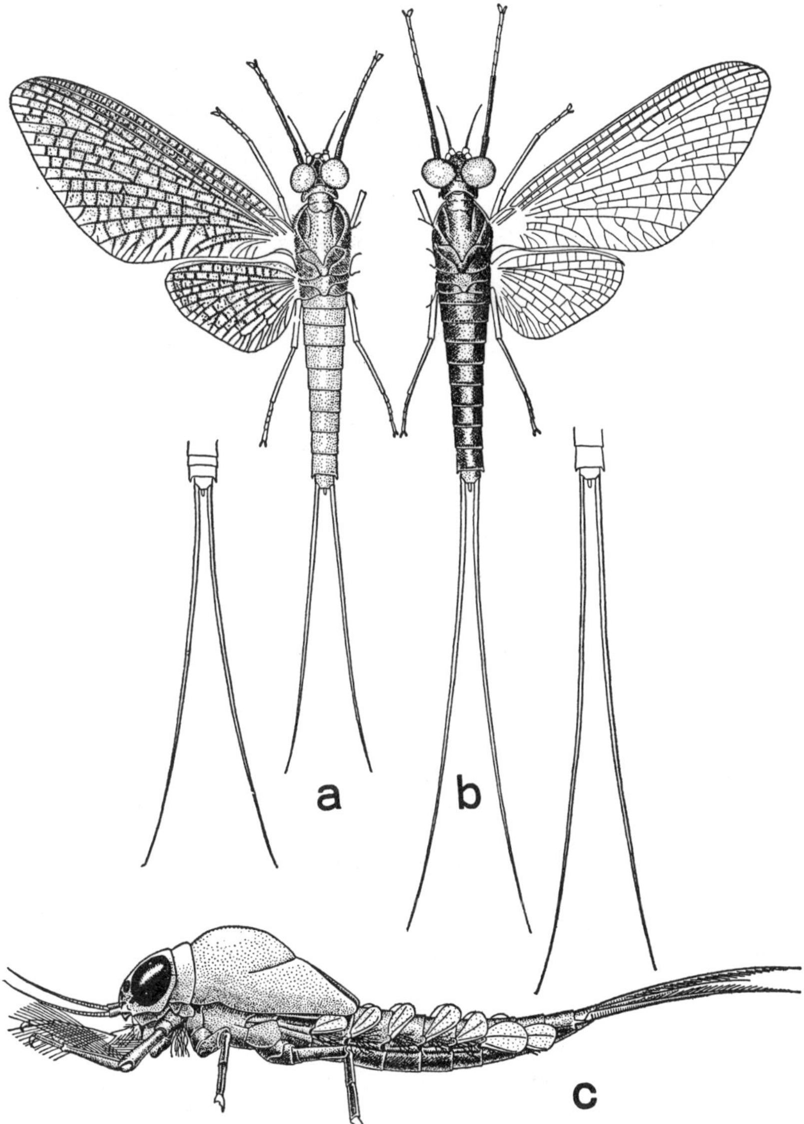

ADULTS AND NYMPH OF ISONYCHIA

a, male subimago; **b**, male imago; **c**, nymph of *I. sicca* (drawings by C. H. Kennedy).

rather stout, denticulate on the inner margin. The lateral extensions of the abdominal segments are noticeable only on the apical ones; on segments 8 and 9, these end in long sharp spines; shorter spines are also present on segments 5–7. The gills are similar in structure on each of segments 1–7; each pair consists of a basal filamentous portion covered by a dorsal plate-like portion. The dorsal plate bears small spines on its margin. The tails are banded near the center by a wide dark stripe, and near the tip by a narrower one; between these dark bands, the tails are pale.

Spieth (1933) considers this genus to be of the family Heptageniidae, and separates it from Siphlonurus and its allies, with which opinion we cannot concur.

Genotype—*I. ignota* (in Baetis), Walker.

Distribution—Holarctic; in Nearctic, rather widely distributed.

KEY TO THE SPECIES*

Adults

1—Forceps base of male and subanal plate of female deeply excavated apically....2
—Forceps base of male and subanal plate of female only shallowly excavated apically....17
2—Male genitalia of the *sicca* type (see fig. 126)....3
—Male genitalia not of the *sicca* type....6
3—Venation almost colorless; fore tibia pale in the middle and black at each end....**pictipes**
—Venation dark; tibia wholly dark....4
4—First joint of fore tarsus pale except the apex; remaining joints light brown....5
—Entire fore tarsus pale reddish black....**campestris**
5—Fore tarsus about ¾ as long as the tibia; cross veins of fore wing fine, brownish....**sicca**
—Fore tarsus usually fully as long as the tibia; cross veins of the fore wing heavy, black....**manca**
6—Male genitalia as shown at left in fig. 126....**diversa**
—Male genitalia of the *albomanicata* type (see fig. 124)....7
7—Venation pale (colorless to very pale brown)....8
—Venation dark (dark reddish or purplish brown to black)....15
8—Fore tarsus largely pale....9
—Fore tarsus largely smoky brown....11
9—Each joint of the fore tarsus shaded with smoky at the apex; abdomen bright reddish....**rufa**
—Fore tarsus usually wholly pale; abdomen not bright reddish....10

* *I. bicolor* Wlk. (based on female subimago) not included, as its actual identity cannot be known.

10—Abdomen flesh-colored to light orange; margin of pleural fold pale....**pacoleta**

—Abdomen dull rose red; margin of pleural fold blackish;......... **albomanicata**

11—Wing of male 13–14 mm. in length................................**sadleri**

—Wing of male not exceeding 12 mm. in length..........................12

12—Abdomen pale reddish, slightly duller and paler ventrally; wing 9–10 mm. ..**circe**

—Abdomen darker red, not duller nor paler ventrally; wing 10–12 mm........13

13—Thorax brownish red, abdomen deep wine red.......................**harperi**

—Thorax and abdomen both deep red..................................14

14—Thorax and abdomen dark wine red; apical segments of abdomen not brighter than the preceding segments....................................**fattigi**

—Thorax and abdomen brighter red; apical segments of abdomen brighter than the preceding segments....................................**christina**

15—Costal margin of fore wing with a roseate tinge, cross veins margined with reddish brown..**velma**

—Wing not as above..16

16—Abdomen intense reddish purple; pale yellowish triangles occupy each antero-lateral angle of the abdominal segments........................**intermedia**

—Abdomen deep dark red, brighter apically; no yellowish antero-lateral triangles ....**matilda**

17—Venation wholly pale..**arida**

—Venation dark..18

18—Tails pale (may be brown at base), distinctly ringed at each joining with red or brown..19

—Tails dark brown or black; joinings may be obscurely darker...............22

19—Abdomen deep red; mid-dorsal stripe obsolescent....................**georgiae**

—Abdomen yellowish with dark purplish brown markings; wide pale mid-dorsal stripe present, margined by dark submedian streaks....................20

20—Tarsus and distal portion of tibia of middle leg, and last two tarsal joints of hind leg, purplish brown...**notata**

—Middle and hind legs pale yellowish, except for purplish markings on claws and tarsal joinings...21

21—Wing of male 11½ mm.; wings not tinged with brown................**obscura**

—Wing of male 10 mm.; fore wing of male tinged with brownish in the basal and central areas...**annulata**

22—Venter of abdomen largely dark purplish brown.........................23

—Venter of abdomen largely light orange...............................**aurea**

23—Incurved portion of penes bearing 5–8 teeth or spines....................24

—Incurved portion of penes bearing but 2 spines (see fig. 126)............**thalia**

24—Incurved portion of penes bearing 7–8 spines; distal outer angle acute, with 2 small spines (see fig.125) ....**serrata**

—Incurved portion of penes bearing 5 spines; distal outer angle rounded, without spines (see fig. 125)...**similis**

VERIFICATION TABLE

Adults

Species	*Wing of ♂*	*Forceps base*	*Subanal pl. of ♀*	*♂ Fore tarsus*	*Venation*	*Dorsal stripe*	*Distr.*
albomanicata	11	Deep	Deep	Very pale	Pale	Yes	E
annulata	10	Shal.	Shal.	Pale	Brown	Yes	N. C.
arida	10½–12	Shal.	Shal.	Lght. br.	Pale	No	C
aurea	13–15½, ♀	?	Shal.	Dk. brown	Dk. br.	Obs.**	N. C.
campestris	9–11?	Deep	Deep	Lght. red. blk.	Dark	?	Alta.
christina	10	Deep	Deep	Smoky	Pale	Obs.	NE
circe	9–10	Deep	Deep	Pale*	Pale br.	No	Ga.
diversa	9	Deep		Brown	Pale	Yes	Tenn.
fattigi	10½–11½	Deep	Deep	Brown	Pale br.	Obs.	Ga.
georgiae	10½–12	Shal.	Shal.	Dk. brown	Dk. br.	Obs.	Ga.
harperi	12	Deep	Deep	Brown	Pale br.	Obs.	E,C
intermedia	12	Deep		Red-br.	Brown	Inter.	Ariz.
manca	11	Deep	Deep	Lght. br.	Dk. br.	No	Tex.,Ia.
matilda	11	Deep	Deep	Pale br.	Dk. br.	No	N. Y.
notata	12, ♀		Shal.	Dk. smoky	Brown	Yes	N. C.
obscura	11½	Shal.		Dk. br.	Dk. br.	Obs.	N. C.
pacoleta	10–11	Deep	Deep	Pale	Pale	Obs.	N. C.
pictipes	9½	Deep	Deep	Pale*	Pale	No	Ga.
rufa	10	Deep	Deep	Pale*	Pale	No	C
sadleri	13–14	Deep	Deep	Brown	Pale br.	No	N. Y.
serrata	13	Shal.	Shal.	Blackish	Brown	Obs.	N. C.
sicca	9–11½	Deep	Deep	Smoky br.	Brown	Obs.	C
similis	13	Shal.	Shal.	Blackish	Brown	Yes	N. C.
thalia	11	Shal.	Shal.	Grey	Dk. br.	Yes	N. C.
velma	15	Deep	Deep	Reddish	Dk. br.	Yes	Calif.

KEY TO THE KNOWN SPECIES

Nymphs

1—Basal joint of antenna pale....................2
—Basal joint of antenna dark....................4

2—Gills pale yellowish brown; pale mid-dorsal stripe the length of the abdomen, may be interrupted at posterior margin of each tergite....................3
—Gills purplish; pale mid-dorsal line on basal segments only....... **albomanicata**

3—Two diffuse dark spots in outer margin of each gill; pale mid-dorsal line interrupted at posterior margin of each tergite.................... **christina**
—A single diffuse dark spot in outer margin of each gill; pale mid-dorsal line continuous, not interrupted.................... **sadleri**

4—Antennae crossed by a dark transverse band about three joints wide, one-third of distance from base....................5
—Antennae without such a dark transverse band....................9

5—Tibial spine of fore leg ¾ as long as the fore tarsus; 7–9 denticles on each claw..6
—Tibial spine of fore leg ½ or less than ½ as long as the fore tarsus; 5–6 denticles on each claw....................8

* Joinings darker. **Obsolescent.

6—Mid-dorsal line wide, continuous the length of the abdomen; gills dark purplish .... **annulata**
—Mid-dorsal line wide or narrow, interrupted at the posterior margin of each tergite; gills greyish lavender........ 7
7—Mid-dorsal line usually present on basal segments only; 30–35 spines on fore tibia........ **notats**
—Mid-dorsal line continuous the length of the abdomen; 35–40 spines on fore tibia........ **similia**
8—Nymph dark red-brown; fore leg predominantly brown with narrow yellow markings........ **serrata**
—Nymph bronze; fore leg predominantly yellow with brown markings..... **thalia**
9—Gills dark purplish red; 5–6 denticles on each claw........ **matilda**
—Gills greyish lavender; 8–9 denticles on each claw........ 10
10—12–15 spines on fore tibia; tergite 10 entirely reddish brown........ **pacoleta**
—35–40 spines on fore tibia; tergite 10 yellowish in the basal third........ **aurea**

VERIFICATION TABLE
Nymphs

Species	*Body length*	*Antenna banded*	*Spur tib. I*[1]	*Spines on tib. I*	*Teeth on claw*	*Gill color*	*Distr.*
albomanicata[3]	12–15	No	½	10–13	8–9	Dk. purp.	E
annulata	9–11½	Yes	¾	20–25	7–9	Dk. purp.	N. C.
aurea	16–17	No	½[2]	40	8–9	Grey-lav.	N. C.
christina[3]	12–14	No	½	10–12	7–8	Pale br.	E
matilda	12–14	No	½	14–18	5–6	Purp.-rd.	N. Y.
notata	11–13	Yes	¾	30–35	8	Grey-lav.	N. C.
pacoleta	10–15	No	½	12–15	8	Grey-lav.	N. C.
sadleri[3]	15	No	½	12	7–8	Yel.-br.	N. Y.
serrata	14–16	Yes	½	25	5–6	Dk. purp.	N. C.
similis	12–15	Yes	¾	35–40	7–9	Grey-lav.	N. C.
thalia	12–13	Yes	½	20–25	5–6	Purp.	N. C.

Isonychia albomanicata Needham

1905—Needham—N. Y. St. Mus. Bull. 86: 31. Figs. (Chirotonetes)........ N. Y.
1911—Morgan—Ann. Ent. Soc. Amer. 4: 116 (Chirotonetes).
1913—Morgan—Ann. Ent. Soc. Amer. 6: 386, 391, 400 (Chirotonetes).
1913—Clemens—Canad. Ent. 45: 338 (Chirotonetes)........ Ont.
1917—Clemens—Univ. Toronto Biol. Series 17: 1–43. Figs. (Chirotonetes).
1923—McDunnough—Canad. Ent. 55: 47 (Chirotonetes becomes Isonychia).
1931—McDunnough—Canad. Ent. 63: 161 (*albomanicata* = *bicolor*).
1932—Traver—J. Elisha Mitchell Sci. Soc. 47: 203. Fig........ N. C.

Length: body 11–12; wing 11 mm.

A large species with pale fore tarsi, black posterior and lateral margins on the abdominal segments, and pale venation.

Head yellowish; frontal margin narrowly black. A longitudinal black mark beside each eye; antenna pale. Head of female with more or less definite reddish brown submedian streaks extending from the ocelli almost to the posterior margin. Angle be-

[1] Length of terminal spur on front tibia in proportion to length of fore tarsus.
[2] Somewhat less than ½.
[3] Basal joint of antenna pale (dark, in other species).

tween the eye and the posterior ocellus reddish brown. Well developed black spots in the corners of the occiput.

Thorax brownish, darker on the scutella of the meso- and metanotum, and on the sternum. No definite markings in the male. Thoracic notum of female yellowish; pleura often marked with rose or purplish, especially before the wing roots, and on the coxae and antecoxal pieces.

Femur and tibia of fore leg red; the femur paler at the base and along the basal part of the outer surface. Tibia blackish toward the apex, narrowly paler at the base; tarsus white, sometimes very faintly tinged with yellowish or pale smoky. Middle and hind legs wholly pale whitish. Wings subhyaline; venation wholly pale in male; in female, very pale brownish. Fore leg of male about as long as the body; tarsus slightly longer than tibia.

FIG. 124. Male genitalia of three species of Isonychia (*arida*, after McDunnough)

Abdomen dull rose red, not paler ventrally; apical segments brighter. Posterior and lateral margins of segments narrowly black, fading to smoky in the middle of the sternites. A pale mid-dorsal stripe extends the length of the abdomen. Segment 10 yellowish red. On the middle sternites, a pale lateral streak may border the dark pleural fold. Tails white, slightly tinged with yellowish or yellowish red at the base. In some specimens, a few joinings near the base may be faintly reddish; tails otherwise unmarked. Forceps and forceps base yellow to pale yellowish brown; the latter deeply excavated on its apical margin (see fig. 124). Subanal plate of female likewise deeply excavated apically, its outer angles acute.

The type material was taken at Ithaca, N. Y. (J. G. Needham). In the Cornell collection are other reared specimens from the vicinity of Willoughby, Ohio (J. R. Traver); from several places near Ithaca, N. Y. (C. N. Hardy, J. R. Traver); from the south fork of the Potomac River at the Smoke Hole, W. Va., and from the vicinity of Greensboro, N. C. (J. R. Traver). Other specimens not reared are from Fairport, Ia. (H. E. Schraedick); Lansing, Mich.; New Windsor, Md. and Chambersburg, Pa. (V. Argo); Ft. Jackson and Hopkinton, N. Y. (L. Harper).

This species may be distinguished from *christina*, with which it occurs locally, by the pale frons, the much paler fore tarsi, the dull rose of the abdomen (in *christina*, the abdomen is bright red), and by the lack of a red stain at the tail bases. It seems best to keep these two species separate, although further study may prove that but one very variable species is involved.

Isonychia annulata Traver

1932—Traver—J. Elisha Mitchell Sci. Soc. 47: 204. Figs. .... N. C.

Length: body 10; wing 10 mm.*

A rather small species, the abdomen largely whitish with purple brown markings, the tails conspicuously ringed at the joinings.

* Length of wing of male erroneously given as 13 mm., in original description.

Head yellowish brown. Antenna light brown. Head of female yellowish; corners of occiput blackish; a narrow black line between eye and lateral ocellus; a black spot on face below lateral carina, another at lower angle of eye. Pronotum whitish with purplish markings. Mesonotum yellowish brown anteriorly and at center, reddish laterally and on the scutellum. Pale area anterior to wing roots. Pleura and sternum whitish marked with reddish brown. Fore femur pale basally, distal half and all but the base of the tibia light red-brown. Tarsus very pale brownish, distal joint and claws purplish grey. Middle and hind legs yellowish white; claws and tarsal joinings faintly purplish grey. Fore wings with faint brownish tinge in basal and central area; veins dark brown, rather paler apically. Stigmatic area opaque whitish.

Abdomen creamy white, markings purplish. A wide mid-dorsal pale stripe is bordered by a purplish submedian streak on each side. Posterior margins of tergites purplish black; lateral margin narrowly blackish. A lateral purplish triangle on each side, based on the posterior margin, does not quite attain the anterior border. An irregular purplish mark occupies most of the postero-lateral angle. Sternite 1 entirely brownish; other sternites pale. Submedian triangles, based on the posterior margin, extend nearly to the anterior margin, and may be united basally. A purplish black line parallels the pleural fold from the center to the posterior margin of each sternite; a black dash is at its anterior end.

Tails white, tinged with brown at base; joinings purplish red. Forceps base and forceps pale brown. Forceps base very weakly excavated apically; penes of the *arida* type (see fig. 126). Subanal plate of female very weakly excavated on its apical margin; the outer angles rounded.

The type material was taken at Big Alamance Creek near Greensboro, N. C. (J. R. Traver), and is in the Cornell University collection.

Isonychia arida Say

1839—Say—J. Acad. Nat. Sci. Phil. 8: 42 (Baetis)..Ind.
1853—Walker—List Neurop. Ins. Brit. Mus. p. 562 (Baetis).
1859—LeConte—Complete Writings of T. Say, 2: 412 (Baetis).
1861—Hagen—Syn. Neurop. Ins. N. Amer. p. 46 (Baetis)................................D. C.
1862—Walsh—Proc. Acad. Nat. Sci. Phil. p. 370 (Baetis)....................................Ill.
1863—Hagen—Proc. Ent. Soc. Phil. 2: 170 (Baetis).
1863—Walsh—Proc. Ent. Soc. Phil. 2: 191-2 (Baetis).
1871—Eaton—Trans. Ent. Soc. London, p. 129 (Siphlurus).
1885—Eaton—Revis. Monogr. 206: (Chirotonetes). Fig.
1894—Banks—Ent. News 5: 178..Kan.
1931—McDunnough—Canad. Ent. 63: 159. Fig.

Length: body 9–13; wing 10–12 mm.

A species having pale yellowish areas in the antero-lateral angles of each abdominal segment, and pale venation.

Head light reddish to brownish. In female, bright yellow, the corners of the occiput blackish, and a black line between the eye and the lateral ocellus. Thorax dark reddish brown, very shiny; pleura a paler shade of the same color, sternum somewhat darker. In the female, a yellow lateral line extends across the pleura from the wing roots. Fore leg of male about as long as the body. Femur dark reddish brown, paler at the base; tibia and tarsus brown, darker at the joinings. Tibia may be paler in the middle, dark at either end; basal halves of the tarsal joints may be paler. Middle and hind legs yellow, tips of the tarsi a little darker. Wings hyaline, veins pale; in female, may be very faintly brownish.

Abdomen deep reddish brown, little if any paler ventrally. Each tergite and each sternite marked with yellow areas in the antero-lateral angles; in addition, another yellow patch is present on the anterior margin of each tergite, nearer the median line. There is no pale mid-dorsal stripe. Tergite 10 is deep bright orange.

Tails yellowish white, unmarked. Forceps light yellowish brown. Forceps base very slightly excavated apically. Penes separated by a shallow rounded notch; the outer angles acute. Below these is a toothed flap or projection, folded back onto the body of the penes (see fig. 124). Subanal plate of female likewise but slightly excavated on its apical margin, its outer angles rounded.

The type material was taken in Indiana, about the middle of June. We have seen a male and female imago of this species, pinned, which were kindly loaned by the U. S. National Museum.

This is the only known species, having the forceps base of the male only shallowly excavated, in which the venation is wholly pale.

Isonychia aurea Traver

1932—Traver—J. Elisha Mitchell Sci. Soc. 47: 224..N. C.

Length: body 16–17; wing 13–15½ mm. (female).

A large yellowish orange species, known from the nymph and female imago only; venation dark brown.

Head greenish yellow, vertex washed with orange. Keel of median carina purplish red. Antenna light grey basally, filament darker grey. Posterior border of occiput, and its corners, purplish. Thorax greenish yellow, the scutella dark reddish brown. Purplish grey markings on pleura anterior to the wing roots and above the hind leg. Sternum light yellow; brown spot near center of mesosternum; three whitish transverse bands between legs. Fore femur very light brown at base, the distal half dark red-brown. Tibia and tarsus very dark purplish brown. Middle and hind legs greenish yellow; claws and three distal joints of tarsus dark grey; first joint washed with grey at apex. Venation dark purplish brown; many stigmatic cross veins are anastomosed.

Abdomen light orange, brighter dorsally. Posterior margins and postero-lateral angles of tergites purplish brown. Brown crescentic submedian marks on 3–9. Ventrally, a deep orange triangle, based on the posterior margin, attains the anterior border; located medially. Tails very dark reddish brown basally, slightly paler brown beyond base. Joinings slightly darker. Subanal plate shallowly excavated apically, its outer angles rounded.

The type material, from a tributary of the Pigeon River at Hazelwood, N. C. (J. R. Traver) is in the Cornell collection.

Isonychia bicolor Walker

1853—Walker—List Neurop. Ins. Brit. Mus. p. 552 (Palingenia)..........................Canada
1861—Hagen—Syn. Neurop. Ins. No. Amer. p. 43 (Palingenia).
1871—Eaton—Trans. Ent. Soc. London, p. 128 (Siphlurus).
1885—Eaton—Revis Monogr. p. 221 (Siphlurus).
1931—McDunnough—Canad. Ent. 63: 161 (*albomanicata* = *bicolor*).

Length: body 11; wing 14 mm.

Only the female subimago of this species is known with certainty.

Vertex of head, and thoracic notum, yellowish brown. Thorax ventrally with a black spot on each side. Fore femur dark reddish brown; tibia rather less red, more blackish; tarsus very light yellowish white. Middle and hind legs yellow to pale amber, the tibiae and tarsi whitish; claws pitch brown. Wings as in all subimagos of this genus. Abdomen brownish, paler beneath. Tails light yellowish brown.

The type, a female subimago, was taken at St. Martin's Falls, Albany River, Hudson's Bay, by Dr. Barnston. Since there are certain points about the description of this specimen which do not agree well with subimago females of *albomanicata* Ndm. it seems better to retain the latter as a valid species, rather than to reduce it to synonymy with *bicolor*, as Dr. McDunnough has suggested. The above description is drawn from Walker, Hagen and Eaton.

Isonychia campestris McDunnough

1931—McDunnough—Canad. Ent. 63: 161 (as variety of *sicca*)............................Alta.

Length: body 9–10?; wing 9–11? mm.

Somewhat paler than *sicca*; venation dark.

Head of female pale yellowish, with traces, more or less distinct, of brown submedian stripes on the vertex and occiput. Blackish brown spot between eye and lateral ocellus, but no dark mark in corners of occiput. Thorax of male light brown, shaded somewhat with reddish and blackish brown on the scutellum. Fore leg entirely reddish black, tarsi somewhat paler. Middle and hind legs dull yellow, with slight red shading on the femora. Wings with fine veins, all dark-tinged.

Abdomen dorsally dull brown tinged with red, shaded laterally with deeper brown in the form of poorly defined dark patches; lateral margin pale brown. Light brown ventrally, rather paler than the dorsum. Penes similar in form to those of *sicca*. Subanal plate of female rather deeply excavated apically, its outer angles acute.

Type material was taken at Medicine Hat, and at Lethbridge, Alta. (J. H. Pepper, J. McDunnough). We have seen no specimens of this species.

This species may be distinguished from *I. sicca* by its generally paler color and its light brown thorax (in *sicca*, the thorax is deep reddish black).

Isonychia christina Traver

1934—Traver—J. Elisha Mitchell Sci. Soc. 50: 240..N. Y.

Length: body 10; wing 10 mm.

A bright red species, with pale smoky fore tarsi and very pale brownish venation.

Head reddish. Vertex of female cream-colored, largely suffused with reddish. Wide reddish areas between the ocelli, the eye and the lateral ocellus, and a wide median geminate red stripe; corners of occiput purplish black, the posterior margin reddish. Thorax very dark red, pleura somewhat paler. Deep rose markings on the pleura and mesonotum, and a pale area anterior to the scutellum. Femur and tibia of fore leg deep reddish black, each paler basally. Tarsus pale smoky brown, faintly paler at the joinings and the base of the first joint. Claws marked with deep rose. Tarsus fully as long as the tibia. Middle and hind legs yellowish white; claws and last tarsal joints tinged with reddish. Coxae deep rose colored. Wings hyaline, the stigmatic area opaque white. Venation very pale brownish; each longitudinal vein outlined by two narrow dark lines.

Abdomen intense red dorsally and ventrally, apical segments brighter. Posterior margins of all segments black. Pleural fold margined by an interrupted black line, parallel to which dorsally is another black line. Paler markings on tergites almost obsolescent. Ganglionic areas of the sternites paler, margined by dark oblique submedian streaks. Tails white, washed with brownish red at the base; 4 to 6 of the basal joinings are usually red, all others colorless. Genitalia reddish, forceps smoky brown, the terminal joint pale. Forceps base and subanal plate of female deeply excavated apically. Penes of the *albomanicata* type.

The type material was taken at Ithaca, N. Y., by Miss Christine Hardy. Other specimens in the Cornell collection from St. Lambert, Quebec, and from Ottawa, Ont. (J. McDunnough, F. P. Ide, G. S. Walley) seem to be of this species. It has doubtless been confused with *I. albomanicata.* Further study of these two species may show them to be but one species with a wide range of variation. However, as the nymphs show several points of difference, the two are best kept separate for the present. The abdomen of *christina* is bright red, that of *albomanicata* dull rose red; the fore tarsi are much paler and the frons of the head unmarked with red, in *albomanicata.*

Isonychia circe Traver

1934—Traver—J. Elisha Mitchell Sci. Soc. 50: 242. .... Ga.

Length: body 9½–10½; wing 9–10 mm.

A rather pale reddish species, with smoky fore tarsi and pale venation.

Head yellowish. A black spot between the eye and the base of the antenna. Antenna pale, tinged with red, its filament dusky. Eyes purplish. Head of female flesh-colored, with indications of a faint dusky pattern near the middle of the vertex, from which extensions may be traced forward and backward. A reddish line along the inner margin of the eye; space between eye and lateral ocellus reddish; corners of occiput black. Thorax reddish, tinged with yellow on the sides of the scutella, the pleura, and a spot anterior to the mesonotal scutellum. Scutellum, sternum and a small area in the middle of the anterior portion of the mesonotum dark reddish brown. Fore leg somewhat shorter than the body; tarsus usually equal to tibia; tibia relatively short, not much longer than the femur. Femur dark red, paler at base; tibia dark reddish black. Tarsus smoky brown; basal two-thirds of the first joint, basal half of the second, and a narrow margin at base of third, pale yellowish. Middle and hind legs clear pale yellow; claws faintly smoky at tip; coxae marked with rose. Wings hyaline. Stigmatic area somewhat opaque, whitish. Venation very pale yellowish.

Abdomen pale red, slightly duller and paler ventrally. Basal segments darker red; apical segments brighter and paler, often tinged with yellow and with powdery white areas; opaque. Middle segments semi-translucent. Posterior margins of all segments distinctly purplish black, these bands wider on the tergites. Pleural fold pale in the basal half of each segment, smoky apically; a small black mark at the stigma. Mid-dorsal line obscurely pale; narrow at base, wider apically. Opaque powdery white areas may margin it, on the apical tergites. Tails white, a few basal joints faintly yellowish. Two or three basal joinings bright red; all others white. Genitalia amber, sometimes tinged with rose. Forceps base rather deeply excavated apically. Penes of the *I. albomanicata* type; rather broad, and turning outward apically. Subanal plate of female with a moderately deep apical excavation; its outer angles subacute.

The type material was collected at Atlanta, Ga., from the Chattahoochee River, from the Alcova River near Monroe, Ga., and from the Apalachee River, also near Monroe, Ga. (P. W. Fattig). These specimens are in the Cornell University collection.

This species may be distinguished from *I. pacoleta,* which it most closely resembles, by the smaller size and the distinctly redder tinge of thorax and abdomen. In the female, the dorsum of the abdomen is brownish, the venter dull rose. In both sexes, the fore tarsus is smoky brown; the fore tarsus of *I. pacoleta* is whitish.

Isonychia diversa Traver

1934—Traver—J. Elisha Mitchell Sci. Soc. 50: 244. Fig....................................Tenn.

Length: body 9; wing 9 mm.

A small brownish species, with pale wings and unique genitalia.

Head and thorax dark reddish brown, brightest on the notum; pleura somewhat paler. Fore femur dark red, becoming blackish at the apex. Tibia blackish. Tarsus pale reddish brown, the basal half of the first and second joints yellowish. Fore tarsus as long as the tibia. Middle and hind legs whitish; femora slightly tinged with reddish, claws pale smoky. Wings hyaline, venation wholly pale.

Abdomen smoky brown with a reddish brown tinge. Tergites 9 and 10 are dark red-brown, sternites 8–10 bright reddish. Middle sternites paler than tergites, somewhat translucent. Posterior margins of all segments conspicuously dark. Pleural fold pale at margin; above this pale line is a dark strip at the center, and a dark mark in the stigma. Mid-dorsal line pale, widened posteriorly. Dark brown submedian streaks margin it, and are in turn margined by paler translucent strips, and these by a darker line. Submedian streaks obscure basally, distinct apically. Postero-lateral angles dark brown; antero-lateral angles paler, translucent. Sternites marked as are the tergites, with a wider median line. Tails white, unmarked.

Forceps base reddish; deeply excavated on apical margin. Forceps pale. Penes reminiscent of Siphloplecton in their form. Wide and united at the base, they taper suddenly, but again curve outward to form more or less rounded apical lobes, separated by a V-shaped notch. Each has a slight indentation on the apical margin (see fig. 126). Second joint of forceps at least a third longer than the terminal joints combined. Basal joint rather long and slender.

A single male imago of this species is known. This specimen was taken at Knoxville, Tenn. (G. Ainslee). It is in the Cornell collection.

Isonychia fattigi Traver

1934—Traver—J. Elisha Mitchell Sci. Soc. 50: 245...Ga.

Length: body 10–11½; wing 10½–11½ mm.

A dark red species with smoky fore tarsi and pale venation.

Front of head translucent; carina, frontal margin and bases of antennae marked with red. A black spot between the eye and the antenna; blackish shading between ocelli. Head of female whitish. A geminate rose streak on the middle of the vertex; a large red spot between the eye and the lateral ocellus; corners of occiput black, and black shading along the entire posterior margin. Thorax reddish brown; mesonotal scutellum, metanotum and sternum of male blackish. Pleura paler, with rose markings, which are more extensive on the female. Femur and tibia of fore leg reddish black; tarsus smoky brown, somewhat darker at apex of each joint. Fore tarsus of female pale at base of each joint, reddish apically. Middle and hind legs yellowish white; coxae rose colored; distal joint and joinings of tarsus reddish. Wings hyaline; venation of male pale, of female very pale brown.

Abdomen dark wine red. Posterior and lateral margins of segments rather widely blackish. Obscure paler mid-dorsal line, and traces of obscure submedian and lateral streaks. Not paler ventrally, but sternites rather translucent at base. Apical segments brighter red. Tails white; at base tinged with red, joinings reddish, elsewhere pale. Forceps base reddish brown; deeply excavated on its apical margin. Forceps reddish brown, terminal joints pale. Penes of the *albomanicata* type. Subanal plate of female rather deeply excavated apically, its outer angles subacute. Lateral extension of the 9th segment translucent.

The type specimens of this handsome species are from Swamp Creek, Dalton, and from the Apalachee River near Monroe, Georgia. All were collected by Prof. P. W. Fattig, for whom the species is named. Specimens are in the Cornell collection.

The thorax and abdomen of this species are dark wine red instead of the brighter red of *I. christina* which it somewhat resembles; in the latter species also, the apical segments of the abdomen are distinctly brighter in color than the preceding segments, which is not the case in *I. fattigi.*

Isonychia georgiae McDunnough

1931—McDunnough—Canad. Ent. 63: 159. Fig. .... Ga.

Length: body 10 mm.; wing 10 mm.

A reddish species with dark venation; fore tarsi purplish brown; tails distinctly ringed.

Head pale, with purplish red markings on carina, frontal margin and bases of antennae. Antenna reddish, pale at tip. Eyes purplish. Vertex red. Head of female yellowish, with a ruddy geminate median line and ruddy shading at the bases of the ocelli. A black line along the inner margin of the eye; corners of the occiput blackish. Thorax dark red; pleura somewhat paler. Anterior portion of mesonotum brighter red;

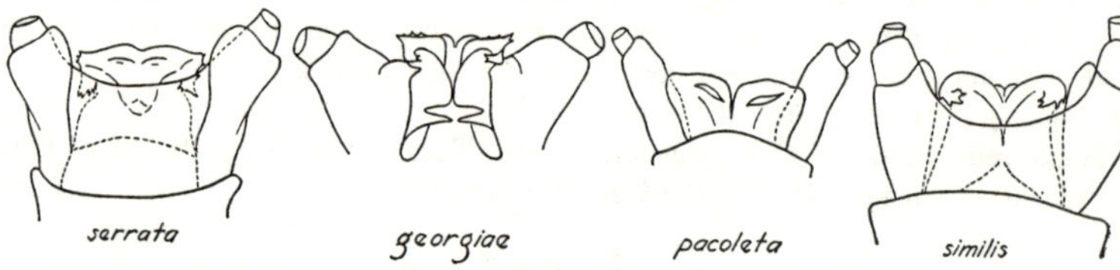

FIG. 125. Male genitalia of four species of Isonychia (*georgiae*, after McDunnough).

scutellum, metanotum and sternum reddish black. Fore leg dark purplish brown, the femur paler at base and shaded with ruddy apically. Tarsus only slightly paler than the dark tibia, and equalling it in length. Middle and hind legs yellowish white; coxae red. Distal joint of tarsus, tarsal joinings and claws reddish brown. Wings hyaline. Veins of the costal border purplish brown, all others blackish.

Abdomen dark wine red, brighter ventrally; basal portions of segments somewhat translucent. Apical segments suffused with bright red. Segments 1 and 2 reddish black. Posterior margins of all segments widely black; a black line margins the pleural fold. No pale median line. Faint traces of dark submedian lines. Lateral dark lines well developed, between these and the pleural fold. Antero-lateral angles and backward extensions from these somewhat translucent on the middle tergites; on tergites 6–9 large yellowish white powdery patches suffused with red occupy these areas. Tergite 10 and sternites 6–9 are largely bright red. Tails brownish basally, suffused with purplish brown; whitish distally. All joinings widely dark purplish brown.

Genitalia dark reddish brown. The long second joint is about equal in length to the combined terminal joints. Forceps base only slightly excavated apically, almost concealing the penes. These are quite similar to *I. arida*, but differ in finer details of structure (see fig. 125). Subanal plate of the female very shallowly excavated apically, its outer angles rounded.

The types of this species were taken in Rabun Co., Ga. (W. T. Davis). We have specimens in the Cornell collection from Towns Creek, near Cleveland, Ga., and from Carroll Creek near Blairsville, Ga. (P. W. Fattig).

Isonychia harperi Traver

1934—Traver—J. Elisha Mitchell Sci. Soc. 50: 246. . . . . N. Y., Ohio

Length: body 12 mm.; wing 12 mm.

A rather dark red species; fore tarsus reddish brown, venation pale.

Head reddish. Head of female yellowish; a large red spot between the eye and the lateral ocellus; corners of occiput black, posterior margin and a geminate median streak blackish. Notum and sternum of thorax reddish black, pleura paler, rather sharply distinct from the dark notum. Scutella, the mesosternum posteriorly, and a small area on anterior of the mesonotum, very deep reddish black. A rose spot anterior to the mesonotal scutellum. Fore tarsus slightly longer than the tibia. Femur dark red, paler at base; tibia deep reddish black. Tarsus reddish brown, the distal joints and the basal third of the first joint somewhat paler. Middle and hind legs greenish white, claws reddish. Wings hyaline, iridescent. Venation very pale yellowish; each longitudinal vein outlined by two narrow dark lines. Stigmatic area opaque whitish.

Abdomen dark wine red; all segments slightly paler basally, so that abdomen appears annulate. Apical segments brighter red, opaque; middle segments semi-translucent. Posterior margins of all segments darker; wider and darker on the tergites. Two dusky lines mark the pleural fold, one on the margin and the other dorsal to this; often a pale streak lies next to the pleural fold ventrally. A short black dash marks each stigma. Pale dorsal markings usually obscolesent. Tails deep reddish brown at the base, white in middle and at tips. All joinings colorless. Forceps base and penes reddish brown; forceps deep smoky brown, the terminal joint pale. Forceps base and subanal plate of female deeply excavated apically. Penes of the *albomanicata* type.

The type specimens were taken at Ft. Jackson, N. Y., from the St. Regis River (L. Harper). Other specimens in the Cornell collection which seem to be of this species are from Hopkinton, N. Y. (L. Harper); Cascadilla Creek, Ithaca, N. Y., Wilseyville, N. Y., Kellogg Creek near Painesville, Ohio (J. R. Traver); and from Brainard, Scoharie, Hagaman and Chatham Center, N. Y. (H. K. Townes, Jr.).

Isonychia intermedia Eaton

1885—Eaton—Revis. Monogr. p. 207. . . . . Ariz.

Length: body 12; wing 12 mm.

A dark purplish red species, having pale antero-lateral triangles on the abdominal segments; venation dark brown.

Thoracic notum intense raw-umber brown, somewhat paler before the scutellum of the mesonotum. Fore leg about $\frac{2}{3}$ as long as the body. Femur and tibia dark brown, femur rather paler at base; tarsus light brownish red, the first joint yellowish. Middle and hind legs light yellow, the claws and tarsal joinings narrowly purplish brown. Wings hyaline, venation dark brown; large longitudinal veins paler toward the wing base.

Abdomen intense purplish red. Tergites with yellowish white triangles occupying the antero-lateral angles, and between these and the median line another pale streak at the anterior margin. These spots increase in size posteriorly. Smaller pale triangles occupy the antero-lateral angles of the sternites. Sternite 10 and the forceps dull yellowish brown, washed with reddish purple; forceps base purplish red. Apical margin of the forceps base deeply excavated as in *I. sicca.*

The type material was from Arizona. We have seen no specimens of this species.

Isonychia manca Eaton

1871—Eaton—Trans. Ent. Soc. London p. 134. Fig. .... Tex.
1885—Eaton—Revis. Monogr. p. 206 (Chirotonetes). Fig. .... Mont.
1931—McDunnough—Canad. Ent. 63: 160. Fig. (as variety of *sicca*) .... Kans.

Length: body 11; wing 11 mm.

A reddish species having dark venation.

Head reddish brown; antenna brown. Head of female yellowish, shaded with brown. A double median brown stripe is usually present on the vertex and occiput; a blackish triangular patch between the eye and the lateral ocellus; the corners of the occiput have a small dark spot. Thorax reddish brown, the scutella and the sternum almost blackish, the notum more reddish. Fore leg longer than in *I. sicca*, the fore tarsus as long as the tibia. Femur and tibia of fore leg reddish black; tibia black at the apex, the femur somewhat paler basally; tarsus brownish or smoky, the first joint pale in its basal two-thirds. Middle and hind legs yellow, femora sometimes shaded with reddish; last tarsal joint rather smoky. Wings hyaline, venation dark. The cross veins of the fore wing are distinctly heavier and darker than in *sicca*, except in the stigmatic area, where the cross veins are pale in male specimens, as in *sicca.*

Abdomen reddish to reddish brown, distinctly redder than in *sicca*, and not much paler ventrally. Posterior margins of all segments, and the pleural fold, darker brownish to black. Faint traces of a paler median line, and of darker submedian streaks margining it. Tails pale, tinged with ruddy at the base, where the joinings are red. Other joinings not marked. Forceps base reddish, forceps smoky brown; the former deeply excavated apically. Penes of the *sicca* type, but rather broader and longer than in *I. sicca*, the upper margin less distinctly truncate. Subanal plate of female rather deeply excavated apically, its outer angles acute.

The specimen on which the original description was based was a female imago from Texas, collected by Mr. Belfrage. We have specimens in the Cornell collection from Manhattan, Kans. (R. H. Painter); from College Station, Texas (J. C. Gaines, H. J. Reinhard); from Austin, Tex. (H. J. Parks, Jr.); and from San Angelo, Tex. (S. E. Jones). The specimens tentatively referred to *manca* (Traver, 1933, J. Elisha Mitchell Sci. Soc. 47: 204) do not belong to this species; their position is doubtful.

The fore leg of the male of this species is relatively longer than in *sicca*, the tarsus usually being as long as the tibia. The cross veins of the wings are also much thicker and blacker, and the penes stouter and longer.

Isonychia matilda Traver

1934—Traver—J. Elisha Mitchell Sci. Soc. 50: 249 .... N. Y.

Length: body 12; wing 12 mm.

A dark red species; fore tarsus and tails dark reddish brown, venation dark brown.

Head dark reddish black. Frontal portion of head of female pale brown, darker at bases of antennae and ocelli. Vertex and occiput greyish; dark reddish brown between bases of ocelli; a narrow purplish brown line next to the inner margin of the eye; the corners of the occiput purplish brown, also the posterior margin. Notum and sternum of thorax deep reddish black; pleura reddish brown. Creamy white areas suffused with deep rose are on the pleura above each leg, and before the wing root. Fore tibia deep reddish black; femur very dark red, paler basally; tarsus entirely dark reddish brown. Tarsus slightly longer than the tibia; entire fore leg ⅚ as long as the body. Middle and hind legs yellowish white, tinged with pale reddish, the femora with indistinct dusky pencilings; claws and last tarsal joint dusky. Wings hyaline, iridescent. Venation dark reddish brown; stigmatic area, and sometimes the apical third of the fore wing, tinged faintly with brownish. Cross veins of the stigma numerous, often forked and anastomosing.

Abdomen deep red dorsally and ventrally, the apical segments brighter red. Basal and middle segments with a narrow pale translucent area anteriorly; remainder of each segment somewhat semi-translucent. Apical segments opaque. A double black line margins the pleural fold, and is paralleled dorsally by a third dark penciling. Only faint traces of a paler mid-dorsal line and of submedian streaks are present dorsally, but on the venter the pale mid-ventral streak and submedian marks are quite evident. Narrow black lateral streaks occur on the middle tergites. Tails dark reddish brown, paler brown apically; joinings slightly paler. Genitalia dark reddish brown. The combined terminal joints of the forceps are approximately equal to the long second joint. Forceps base deeply excavated apically, as is the subanal plate of the female. Penes of the *albomanicata* type.

The type specimens of this species were taken at Slaterville and Wilseyville, N. Y. (J. G. Needham, J. R. Traver). Other specimens from the vicinity of Ithaca, N. Y. are likewise in the Cornell collection.

Isonychia notata Traver

1932—Traver—J. Elisha Mitchell Sci. Soc. 47: 210..N. C.

Length: body (female) 13; wing 12 mm.

A yellowish orange species with light purplish venation, known from the nymph and the female imago only.

Head (of female) greenish yellow; antenna purplish. A black spot below the lateral carina on each side, another below each eye. Dark purplish spots in corners of occiput and a narrow black line between the eye and the lateral ocellus. Thorax yellowish with a greenish tinge. Mesonotal scutellum lavender; pleura marked with lavender; metanotal scutellum purplish. A dark brown mark on each side of median line of metanotum. Sternum brown, purplish posteriorly; mesosternal spinous process dark brown; two dark transverse bars between the fore legs. Fore leg brown, the tarsus rather paler brown; femur and tibia washed with purple. Middle and hind legs greenish white; tarsus and distal fourth of tibia of middle leg, and last two tarsal joints of hind leg, dark olive brown. Wings hyaline, stigmatic area opaque whitish. Venation light purplish rose.

Abdomen yellowish orange dorsally, yellow ventrally. A wide pale median dorsal streak extends the length of the abdomen, margined by dark submedian streaks. Lateral dark streaks parallel the submedian streaks. Posterior margins of all segments dark purplish brown. Ventrally dark submedian streaks margin a pale median line. A purplish red line parallels the pleural fold posteriorly on each sternite; at its anterior end is a dark purple spot. Tails light purplish brown, paler distally; joinings conspicuously dark purplish brown. Subanal plate only slightly excavated at its apical margin, its outer angles rounded.

The type material was taken at Bald Creek in Yancey Co., N. C. (J. R. Traver). These specimens are in the Cornell collection.

Isonychia obscura Traver

1932—Traver—J. Elisha Mitchell Sci. Soc. 47: 217. Fig. .... N. C.

Length: body 12; wing 11 mm.

A dark purplish brown species, with dark venation. Head and thorax dark purplish brown. Antenna dark purplish; purplish spots below the antennae. Purplish red spot at center of frontal margin. Lateral margins of pronotum and marks on pleura, somewhat paler. Lavender streak anterior to wing roots. Scutella black. Fore leg very dark purplish brown, tarsus slightly lighter than femur and tibia; base of femur light brown. Middle and hind legs white with greenish tinge; tarsal joinings narrowly purplish grey. Wings hyaline, venation dark purplish. Stigmatic area rather opaque white. Several stigmatic cross veins are forked.

Abdomen purplish brown, lighter in color ventrally. The anterior portion of each tergite is light brown, and a pale median dorsal streak extends the length of the abdomen. Posterior margins of all segments widely dark purplish brown. Dark submedian streaks margin the pale mid-dorsal area in the anterior portion of each tergite; from the dark posterior margin arise dark lateral streaks, rather diffuse. Pleural fold margined with purple. Ventrally, sternites 2–6 purplish rose, 7–9 a lighter shade of the same color. A wide cream margin parallels the pleural fold, and a narrow median pale line is bordered by wide purplish submedian streaks. A dark purplish brown streak, at its anterior end a dark dot, lies close to the pleural fold in the pale lateral area. Tails very dark purplish brown in the basal half, light grey distally; joinings reddish lavender. Forceps base dark reddish brown; forceps purplish brown. Forceps base very shallowly excavated apically; penes of the *arida* type (see fig. 124).

The type material, a single male imago, was taken at Penrose, N. C. (J. R. Traver). This specimen is in the Cornell collection.

Isonychia pacoleta Traver

1932—Traver—J. Elisha Mitchell Sci. Soc. 47: 218. Figs. .... N. C.

Length: body 11–12; wing 10–11 mm.

A yellowish orange species with pale venation. Head yellowish brown. Frontal margin narrowly blackish; a black spot at lower corner of each eye, and another below the frontal margin. Antenna greyish white. Pro- and metanota yellowish brown; mesonotum yellow, a reddish brown area near anterior margin. Scutella dark reddish brown. Pleura cream yellow with lavender markings. Sternum reddish brown. In female, corners of occiput purplish; no dark mark between eye and lateral ocellus. Fore femur and tibia purplish brown; the femur pale at base and in center along outer margin almost to the apex, the tibia narrowly pale at the extreme base. Tarsus white, faintly marked with lavender. Middle and hind legs white with yellow tinge; tips of claws and tarsal joinings faintly purplish grey. Wings subhyaline, veins entirely pale; stigmatic area opaque, some cross veins forked.

Abdomen light orange to bronze, the first 3 or 4 segments washed with rose; posterior margins of all segments purplish brown, bands on the tergites wider. Very narrow pale mid-dorsal line, bordered on posterior tergites with faint and diffuse submedian streaks. Pleural fold greyish white with a faint lavender streak. On sternites 2–7, indistinct light transverse dots may be evident. Tails white, very faintly flushed with flesh color or pale lavender at base; no darker joinings. Forceps base and forceps yellowish. Forceps base deeply excavated at its apical margin, as is also the subanal plate of the female; outer angles of the latter acute. Penes resembling those of *I. albomanicata*, but considerably broader (see fig. 125).

The type material, which is now in the Cornell collection, was taken near Tryon, N. C. in the Pacolet River (J. R. Traver). The species is larger and paler than *circe* and with the fore tarsi whitish instead of smoky brown.

Isonychia pictipes Traver

1934—Traver—J. Elisha Mitchell Sci. Soc. 50: 250 .... Ga.

Length: body 8–9; wing 8–10 mm.

A small pale species, with bicolored fore tibia and pale venation. Head pale reddish. Frontal margin narrowly black. A black spot between the eye and the antenna. Head of female flesh-colored; no reddish median stripe; a faint reddish line follows the inner margin of the eye. A large black spot in the corner of the occiput, and a smaller dark dot nearer the median line. Thorax reddish tinged with yellow, which is most evident on the lateral portion of the anterior half of the mesonotum and the sides of the metanotum. Scutella and sternum darker reddish brown. Fore femur madder red, paler at base, margined with black apically. Tibia bicolored; pale yellowish in the center, the basal third and a narrower band at the apex dark reddish brown. Tarsus pale smoky brown; the basal half of the first and second joints, and the basal third of the third joint, pale. Fore tarsus slightly longer than tibia. Middle and hind legs whitish tinged with amber; claws dusky. Wings hyaline, venation wholly pale.

Abdomen pale reddish, paler ventrally. Basal segments darker, apical segments brighter red. Prominent purplish black bands on the posterior margins. Stigma marked by a short black line. A dusky streak closely parallels the pleural fold on each tergite. An obscure paler mid-dorsal line may be present. Tails pale yellowish white, wholly pale at tip. Joinings pale. Extreme basal joints may be faintly flushed with pinkish. Genitalia yellow, often tinged with red. Forceps base deeply excavated apically; its lateral arms rather wide. Penes very similar to *I. sicca*. Subanal plate of female deeply excavated apically, its outer angles acute. This plate and the lateral extension of the 9th segment translucent.

The type material was collected in Williamson Swamp Creek near Bartow, from the Apalachee River north of Monroe, and from the Alcova River near Monroe, Ga. (P. W. Fattig). The specimens are in the Cornell collection.

The strikingly marked fore tibia should readily distinguish this species from *sicca*, *manca*, and *campestris*, which species it resembles in type of genitalia.

Isonychia rufa McDunnough

1931—McDunnough—Canad. Ent. 63: 162. Fig. .... Iowa, Kans.

Length: body 10; wing 10 mm.

A rather bright red species, with pale venation. Eyes ruddy chestnut in living male. Head light brown, bases of the ocelli black-ringed. Head of female pale yellow, shaded with brown around the bases of the ocelli and between the eyes and the lateral ocelli. Thorax ruddy brown. Coxae and femora of fore legs deep reddish; tibiae blackish; tarsi dull yellowish white, the apical portion of each joint shaded with smoky. The tarsus is as long as the tibia; the whole leg is somewhat longer than in *sicca*. Middle and hind legs rather deep yellowish. A small black lateral spot on the occiput. Wings hyaline, with pale venation in the male; in the female, the cross veins may be faintly tinged with brown.

Abdomen bright reddish dorsally, paler and duller ventrally. No blackish shading except a narrow posterior margin on each segment. Tails whitish yellow; traces of reddish rings in the basal joinings. Forceps pale reddish brown. Forceps base deeply excavated on its apical margin. Lateral arms of the forceps base rather broad, the inner margins wide and flaring, somewhat angulate apically. Penes of the *albomanicata* type. Subanal plate of the female deeply excavated apically, its outer angles acute.

This species was described from specimens taken at Davenport and Pleasant Valley, Ia. (G. S. Walley), and Lawrence, Kans. (W. J. Brown, Mrs. L. W. Brown). We have specimens in the Cornell collection from Fairport. Ia. (H. E. Schraedick); Lawrence, Kans.; Bryan and College Station, Tex. (V. A. Little); Manhattan, Kans. (R. C. Smith); and from Goshen, O. (C. R. Crosby). Other very similar specimens are from New Windsor, Md. (V. Argo), and Lucedale, Miss. (H. Dietrich).

Isonychia sadleri Traver

1934—Traver—J. Elisha Mitchell Sci. Soc. 50: 251 .... N. Y.

Length: body 14–16; wing 13–14 mm.

A large red species; fore tarsi smoky brown, venation pale.

Head pale reddish. A black spot between eye and base of antenna. Median carina, frontal margin and base of antenna reddish; antennal filament pale. Ocelli ringed at base with dark red-brown. Eyes purplish, transverse band pale. Head of female yellowish, with median reddish geminate streak; red areas between eyes and lateral ocelli; occipital corners black. Thorax very dark reddish brown; scutella, median area of mesonotum and mesosternum, dark reddish black. Pleura rather paler brown, with deep rose markings. Narrow rose markings on sides of mesonotum. Fore leg shorter than body; tarsus fully as long as tibia. Femur very dark red-brown, paler basally. Tibia deep red-black. Tarsus dark smoky brown; fourth joint and bases of first and fifth joints tend to be paler. Middle and hind legs yellowish white; claws, distal tarsal joint and tarsal joinings reddish; coxae rose-colored. Wings hyaline, venation very pale yellowish brown. Longitudinal veins each outlined by two parallel brownish lines. Stigmatic area opaque whitish.

Abdomen intense dark red, brighter apically. All segments with rather wide black posterior margins, those of tergites widest. Traces of a paler mid-dorsal line are most evident on apical segments; reddish submedian streaks may margin it apically, becoming smoky brown on middle and basal tergites. Pleural fold margined with black, a narrow paler area at each end of each tergite. Tails white, tinged with reddish brown basally. 6 or 8 basal joinings red, all others colorless. Genitalia reddish brown. Forceps base deeply excavated apically; its lateral arms rather wider than in *I. albomanicata*. Penes of the *albomanicata* type. Subanal plate of female rather deeply excavated apically, its outer angles subacute.

The type specimens were taken at the Fish Hatchery of Cornell University, on Cascadilla Creek, at Ithaca, N. Y. (W. O. Sadler), and are in the Cornell collection

Isonychia serrata Traver

1932—Traver—J. Elisha Mitchell Sci. Soc. 47: 222. Figs .... N. C.

Length: body 14; wing 13 mm.

A species having purplish brown markings on a bronze abdomen, and dark venation.

Head brown. A black mark below the frontal margin. Reddish purple spot on each side of median carina; another on outer side of base of antenna. In female, purplish markings at the corners of the occiput and between the eye and the lateral ocellus.

Pronotum brown. Mesonotum yellowish brown margined with cream color; a yellowish spot anterior to the scutellum, which is dark brown. Metanotum and sternum dark brown. Pleura pale with brown and purplish markings. Fore leg purplish; femur grey at base and apex; tarsus light greyish purple. Distal half of last joint of fore tarsus and base of claws purple. Middle and hind legs white, last tarsal joint and claws purplish grey. Venation dark purple brown.

Abdomen bronze dorsally, the posterior half of each tergite purplish brown; wide dark posterior margins. A paler mid-dorsal line is present, bordered by oblique dark submedian streaks. Pleural fold and postero-lateral angles very dark brown. Ventrally dark purplish brown, except for the pale antero-lateral angles, a very narrow pale posterior margin bordered by the dark central color, an oblique pale submedian dash and four pale dots arranged transversely across each sternite near the center. On sternites 6–9, the lateral margins are also pale. Narrow dark bar parallels the pleural fold posteriorly on each sternite; at its anterior end is a dark dot. Tails very dark brown, slightly paler distally. In basal portion, joinings slightly paler brown. Forceps and forceps base yellowish or bronzy brown. Forceps base very slightly excavated on its apical margin; penes of the *arida* type, the incurved lower portion bearing 7–8 saw-like teeth (see fig. 125). Subanal plate of female almost straight across the apical margin, its outer angles rounded.

The type material was taken at Allen's Creek near Waynesville, N. C. (J. R. Traver); these specimens are in the Cornell collection.

Isonychia sicca Walsh

1862—Walsh—Proc. Acad. Nat. Sci. Phil.; p. 371 (in Baetis)....Ill.
1863—Hagen—Proc. Ent. Soc. Phil. 2: 170 (in Baetis).
1863—Walsh—Proc. Ent. Soc. Phil. 2: 191–2, (in Baetis).
1871—Eaton—Trans. Ent. Soc. London; p. 130 (in Siphlurus).
1885—Eaton—Revis. Monogr.; p. 208 (in Chirotonetes) (*siccus*, in part; fig. probably inaccurate).
1918—Needham—Bur. Fisheries Bull. 36: 275. Fig.
1930—Ide—Canad. Ent. 62: 227....Ont.
1931—McDunnough—Canad. Ent. 63: 159. Fig....Ia.

Length: body 8½–10; wing 9–11½ mm.

This species has short fore legs and dark venation; the abdomen is paler and redder ventrally.

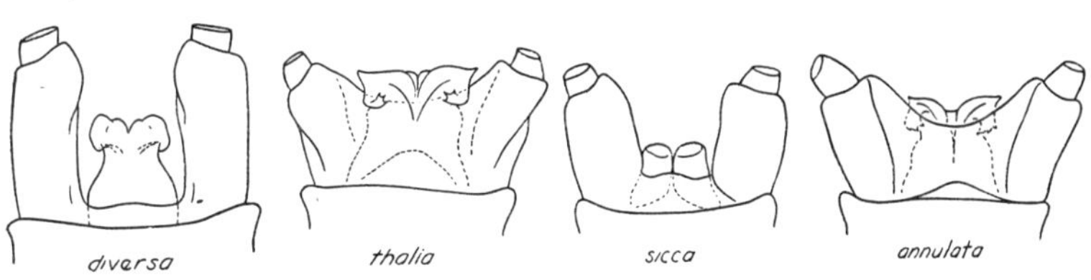

FIG. 126. Male genitalia of four species of Isonychia.

Head light reddish. Antennae pale. Head of female light yellowish, shaded with light brown; a sub-triangular reddish patch between the eye and the lateral ocellus; a small blackish spot in the corners of the occiput. Thorax reddish black, the pleura reddish. Fore leg shorter than in other allied species, the fore tarsus about ¾ the length of the tibia (in some specimens almost as long as tibia). Fore leg dark reddish to reddish black; tip of tibia black; tarsus smoky, the basal joint whitish except in the apical third. Middle and hind legs clear light yellow, sometimes tinged with reddish; tips of tarsi faintly smoky. Wings hyaline, with fine veins distinctly dark-tinged.

Abdomen reddish black dorsally, slightly paler and redder ventrally. Pleural fold also paler and redder than the tergites. The posterior margins are narrowly darker, and longitudinal dark dashes are usually present in the spiracular area. Faint traces of a paler mid-dorsal line and of smoky submedian streaks may be present. Tails whitish or greenish white, the basal joinings definitely reddish; joinings in remainder of tail faintly darker. Forceps base and basal joint of forceps reddish; forceps smoky; forceps base deeply excavated apically. Penes short and quite distinctly truncate apically (see fig. 126). Subanal plate of female rather deeply excavated on its apical margin, its outer angles acute.

The species was described from specimens taken near Rock Island, Ill. We have specimens in the Cornell collection from Fairport, Ia. (H. E. Schraedick), and have seen others from Lincoln, Neb. (D. B. Whelan), in the collection of the Nebraska Agricultural College.

This species may be distinguished from *I. manca* by the shorter fore leg and the somewhat paler venation (brownish in *sicca*, blackish in *manca*.) The deep reddish black of the thorax separates it from *campestris*, in which the thorax is light brown.

Isonychia similis Traver

1932—Traver—J. Elisha Mitchell Sci. Soc. 47: 213. Figs. .... N. C.

Length: body 12; wing 12½ mm.

Very similar in appearance to *I. serrata*, but differing in the detailed structure of the penes.

Head yellowish brown. Median carina and frontal margin purplish red. Black bar below frontal margin. Head of female yellowish white, with faint shading of brown in the corners of the occiput, and a very narrow dark line beside the eye near the lateral ocellus. Thorax dark reddish brown; margins of mesonotum, and a spot just anterior to the scutellum, pale yellowish. Pleura pale with lavender markings. Sternum very dark brown. Fore leg dark purple, femur somewhat lighter purple at base. Middle and hind legs yellow; claws and tarsal joinings purplish grey. Wings hyaline, venation dark purple. Stigmatic area faintly opaque, its cross veins tending to fork and anastomose.

Dorsum of abdomen bronze; the posterior margins and postero-lateral angles widely dark purplish brown, also the submedian lines bordering the paler mid-dorsal line, and a lateral line extending from the dark margin. A bronze triangle is thus left in the antero-lateral angle, a larger one nearer the median line, and the pale median streak. The anterior half of each sternite bronzy, its posterior half dark purplish brown. Four paler median dots on sternites 2–5. A dark purple spot in each antero-lateral angle near the pleural fold, and purplish shading posterior to it. Tails purplish rose, darker at base. Joinings dark purplish brown. Forceps and forceps base yellowish brown. Forceps base only slightly excavated apically. Penes of the *arida* type; the incurved portion bears 2 large and 2 small teeth (see fig. 125). Subanal plate of female slightly excavated on its apical margin, rather than almost straight as in *I. serrata*. Outer angles rounded.

The type material was taken at Cedar Creek, a tributary of the Tuckaseegee River, near Glenville, N. C. (J. R. Traver). These specimens are in the Cornell collection, as are also several female imagos taken in Greenville Co., S. C. (H. K. Townes, Jr.).

Isonychia thalia Traver

1932—Traver—J. Elisha Mitchell Sci. Soc. 47: 208 (*georgiae*)....N. C.
1934—Traver—J. Elisha Mitchell Sci. Soc. 50: 253. Fig.

Length: body 12½; wing 12 mm.

A reddish brown species with dark venation.

Head greyish white; median carina red-purple. Black spot below lateral carina, on each side. Head of female yellow, the corners of the occiput dark purple. Notum yellow, borders light brown; mesonotal scutellum purplish. Pleural light brown with pale markings, washed with lavender anterior to wing roots. Reddish brown on sternum. Fore femur white at base, purple distally; tibia white at base, remainder greyish lavender; tarsus pale greyish lavender. Middle and hind legs whitish; claws and last tarsal joints pale smoky.

Tergites reddish brown. Median pale streak extends the length of the abdomen; pale also laterally, next to pleural fold. Pleural fold narrowly dark purple. Dark submedian and lateral streaks, somewhat interrupted, on each side. Ventrally rather purplish brown, with pale mid-ventral streak bordered by darker submedian streaks. Purple line parallels the pleural fold. In female, abdomen marked with purplish rose. Tails dull purplish brown, joinings somewhat darker. Penes of the male of the *arida* type (see fig. 126). Subanal plate of the female very slightly excavated on its apical margin, its outer angles rounded.

The type material, from the Davidson River, N. C. (J. R. Traver) is in the Cornell University collection. We have seen no other specimens. In type of genitalic structure, this species is close to *I. georgiae*, with which it was originally confused; it is larger than that species, and differs in color of abdomen.

Isonychia velma Needham

1932—Needham—Canad. Ent. 64: 273. Fig....Calif.

Length: body 14; wing 16 mm.

A large reddish plumbeous species with bicolored legs and fine dark spots on the wings.

Head reddish brown; antennae and median carian darker. In the female, the top of the head is largely reddish, the median stripe paler, the corners of the occiput darker. Median carina purplish red, likewise the frontal margin. Thorax reddish brown; many sclerites are edged with whitish, and the notum is more uniformly brown. Fore legs reddish, the tibia darker, the femur paler at the base. Middle and hind legs reddish on the coxae, the outer half of the femora and the tarsi; basal half of femora, and tibiae, pale yellow. Fore wing with a roseate tinge along the costal border; a touch of the same color at the extreme anal angle of both wings. Hind wings of male with a faint brown border around the outer margin, which is almost obsolescent in the female. Longitudinal veins brown. Cross veins of the subcostal and radial spaces of the fore wing, and the costal cross veins more faintly, bordered with reddish brown to form diffuse darker spots. Remainder of wing hyaline.

Abdomen reddish plumbeous with obscure pale mid-dorsal and mid-lateral stripes margined by brighter plumbeous, forming interrupted lines. These are darker and confluent on segment 9. "The large thin flat lobes on the side of segment 9, and the apical ventral pair at its apex, are translucent." (Needham). Posterior margins of all segments dark. Pleural fold margined above and below by a darker strip. Sternites marked with an obscure paler mid-ventral stripe, margined posteriorly by dark submedian streaks based on the posterior margin. Tails of the male whitish, at base tinged with red. In the female, the tails are yellowish, and more distinctly red at the base. Forceps base and forceps of male brown, as are also the penes. The latter are somewhat of the *albomanicata* type; the forceps base deeply excavated apically. Subanal plate of female very deeply excavated at its apical margin, the outer angles acute.

The type material, collected at Putah Creek near Maticello, Calif., and from Russian River and Cloverdale, Calif. (J. C. Bradley). is in the Cornell collection. We have seen a pinned female of this species from Sobre Vista, Sonoma Co., Calif. (J. A. Kusche) from the collection of the California Academy of Science.

Genus **Siphlonisca** Needham

1908—Needham—N. Y. St. Mus. Bull. 134: 71.
1915—Clemens—Canad. Ent. 47: 254.

Large brownish mayflies, male and female very similar. Only one species, the genotype, is known. Eyes of male large, contiguous apically;

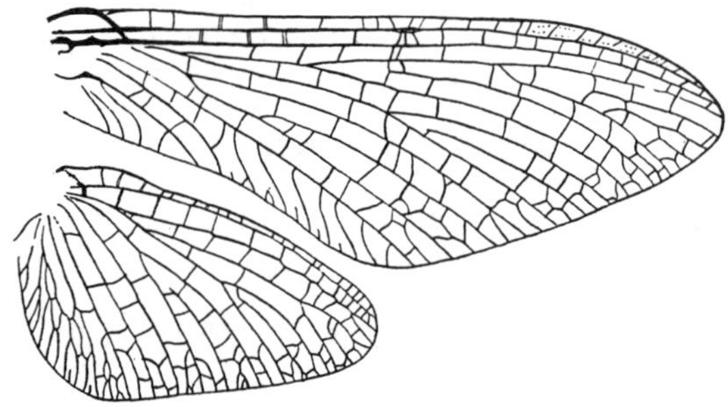

FIG. 127. Wings of *Siphlonisca aeroaromia*.

not divided. A mid-ventral spine present on the mesosternum; a similar spine on the metasternum. Fore legs of male somewhat shorter than the body. First fore tarsal joint of male somewhat longer than the second joint, which is subequal to the third and fourth; fifth slightly shorter than the fourth. Blunt spine-like processes present on the fore tibia and on all the fore tarsal joints. Hind tarsus slightly shorter than the tibia; distal joint as long as any two of the other joints. Claws on all tarsi similar, sharp-pointed, in both sexes. Fore wing about there times as long as wide (see fig. 127). Hind wing well developed; no costal angulation. Slightly concave on the costal margin, one third of the distance out from the base. First fork of R_s very long-stalked. Median vein of hind wing forked about one third of the distance from the base.

Conspicuous wide lateral expansions on the margins of abdominal segments 5–9 (see fig. 21). Forceps base of the male about as long as wide; rather acutely excavated on its apical margin between the forceps. On each side of this excavation is a blunt projection. Apically the lateral portions of the forceps base, and hence the bases of the forceps, are quite widely divergent. Forceps four-jointed. Basal joint short and ring-like; long joint somewhat bowed; fourth joint slightly shorter than the third. The two terminal joints together are about half as long as the long joint. The ventral division of the penes is drawn out apically into a long slender point, in the single known species; the dorsal division consists of two long slender pointed structures on each side, of which the outermost is the longer. Subanal plate of female completely lacking, since the 9th ventral sternite is not prolonged. A very short partially segmented rudiment of the middle tail remains.

The head of the nymph is relatively small for the size of the body. Maxillary and labial palps are each three-jointed. The tip of the maxilla is rather conical, bearing simple hairs and a few spines on the inner apical margin. Mandibles somewhat wedge-shaped, the canines relatively large and strong. Labrum about twice as wide as long; its front margin entire, slightly convex. The middle lobe of the hypopharynx is wide, entire and slightly concave on its apical margin.

Spines are present on the meso- and metasterna of the thorax, as in the imago. Legs rather short; claws slender, not pectinate. Lateral margins of abdominal segments 5–9 widely flaring, as in the imago. Gills all single; very broad and plate-like, rather cordate at the base, with abundant palmate tracheation. Middle tail very slightly shorter than the outer ones.

Genotype—*S. aerodromia* Needham.

Distribution—Nearctic; known only from state of New York.

Siphlonisca aerodromia Needham

1908—Needham—N. Y. St. Mus. Bull. 134: 71. Figs. . . . . N. Y.
1915—Clemens—Canad. Ent. 47: 254. Nymph.
1915—Alexander—MS. notes (in Clemens, Canad. Ent. 47: 255)

Length: body 19; wing 17 mm.

Head brownish; median carina yellow, margined laterally and frontally with black; an oblique black streak below each antenna. Thorax reddish brown. The meso- and metathoracic scutella, the sternum, and oblique streaks on the pleura, very dark red-brown. Fore leg of male reddish brown; tarsi paler than femora and tibiae. Middle and hind legs yellowish. Wings hyaline, strongly tinged with amber, which is most distinct in the male. Venation of the fore wing very dark red-brown; of most of the hind wing, black. Humeral cross vein strong, dark. A few costal and subcostal cross veins before the bulla are narrowly brown-margined. Cross veins of the first three spaces grouped in twos or threes at the bulla. Stigmatic cross veins simple, straight, about 14 in number; not forked nor anastomosed.

Abdomen pale yellowish brown with brown markings. Middle segments semi-translucent; apical segments opaque, washed with yellow. Tergites dark red-brown on the posterior margins, postero-lateral angles, and along the lateral margins; the flaring lateral extensions also dark. Median line pale greyish, narrow on basal tergites, widening progressively on those behind. A wide dark submedian stripe bounds it on either side. Middle sternites narrowly pale on the posterior margins. A dark apical cross band on each sternite, next to the pale posterior margin. Median area washed with reddish brown; bounded on each side by a rather wide darker brown submedian streak arising from the posterior margin, and extending forward almost to the anterior border. Each streak ends in an oval dark brown spot.

Forceps base yellowish, with two longitudinal reddish brown stripes. Forceps dark red-brown, penes slightly paler red-brown (see fig. 128). Tails white, unmarked; may be faintly brownish at extreme base.

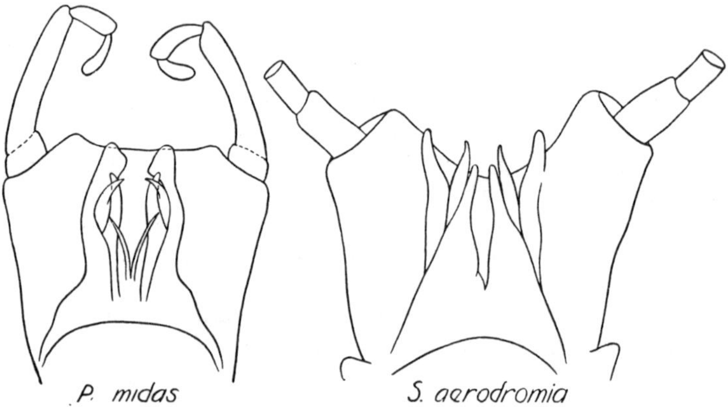

FIG. 128. Male genitalia of *Siphlonisca aerodromia* and *Parameletus midas*.

Nymph reddish brown. Legs not banded. Gills cover most of the dorsum of the abdomen; these are pale greyish white with purplish tracheation. Dorsum of abdomen darker along the median line and along a line parallel to the lateral margin, inside the line of the gills. Ventrally, two lateral dark brown longitudinal stripes extend the length of the abdomen, near the lateral margin. Tails brown, tips white; not banded.

The type specimens of this species, from Sport Island, Sacandaga River, N. Y. (C. P. Alexander), also others from the same location, are in the Cornell collection.

Genus **Parameletus** Bengtsson

1908—Bengtsson—Vet. Akad. Arsbok. 6: 242.
1909—Bengtsson—Lunds Univ. Arssk. N. F. Afd. 2, v. 4: 13 (as Potameis).
1909—Esben-Peterson—Deutsche Ent. Zeitschr. p. 554 (as Sparrea).
1910—Aro—Piirteitä päiväkorennoisten elämäntavoisla ja kehityksestä. Viipuri (as Palmenia).
1920—Ulmer—Stett. Ent. Zeit. 81: 134 (Potameis = Sparrea).
1923—McDunnough—Canad. Ent. 55: 48 (as Siphlonuroides).
1930—Bengtsson—Lunds Univ. Arsskr. N. F. 26: 13 (Potameis = Sparrea = Palmenia = Parameletus).
1932—McDunnough—Canad. Ent. 64: 81 (Siphlonuroides = Parameletus).

Medium-sized brownish mayflies, male and female very similar. Eyes of male large, contiguous apically; not divided. Fore leg of male fully as long as the body. First fore tarsal joint equal to second, which is subequal to the third. First joint of fore tarsus and fore tibia (see fig. 41b) thickly set with blunt spine-like processes; spines at tips of tarsal joints, middle and hind legs. Hind tarsus about as long as the tibia. Claws of all tarsi similar, sharp-pointed, in both sexes. Fore wing about 2½ times as long as wide (see fig. 129). Hind wing better developed in the anal region than in other genera of this subfamily; less than twice as long as wide. Median vein of hind wing *not forked.*

Forceps base of male slightly longer than wide; more or less shallowly excavated on its apical margin. Very blunt projections on each side of this excavation. Forceps more or less distinctly four-jointed; basal joint incompletely separated from the long joint. Penes consisting of the usual ventral and dorsal portions. Subanal plate of the female quite broad, its apical margin convex. A short 5-jointed rudiment of the middle tail remains.

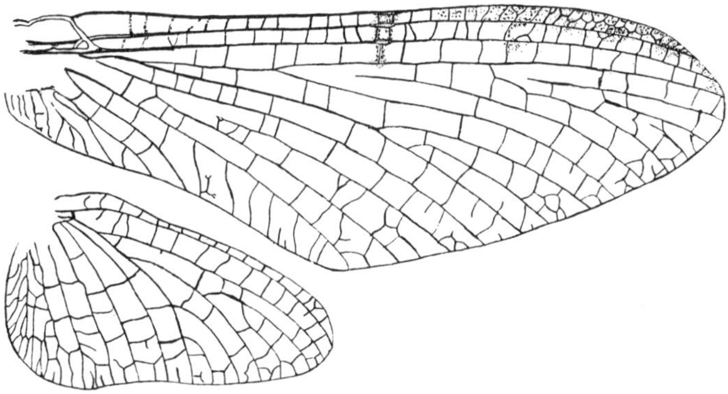

FIG. 129. Wings of *Parameletus midas.*

Maxillary and labial palps of the nymph each three-jointed; the second joint of the latter with a thumb-like process on its inner apical margin, much as in Baetisca. Maxilla rather conical apically, bearing hairs on the inner margin. Gills single on all segments; 7th pair smallest. Simple in form, plate-like, each about ⅔ as broad as long; tracheation pinnate. Postero-lateral spines on 5–9, rather weakly developed.

Genotype—*P. chelifer* Bengtsson.

Distribution—Holarctic; in Nearctic, known only from eastern Canada.

KEY TO SPECIES OF PARAMELETUS

Adults

1—Fore wing clouded with brownish on inner two-thirds of its length....... **croesus**
—Fore wing with no such brown cloud; costal margin may be shaded with brown .... **midas**

Parameletus croesus McDunnough

1923—McDunnough—Canad. Ent. 55: 48. Fig. (Siphlonuroides)............................... Ont.
1932—McDunnough—Canad. Ent. 64: 81 (Siphlonuroides = Parameletus).

Length: body 12–13; wing 13 mm.

Head of male yellowish; ocelli black-ringed; median carina marked with black. Head of female with a dark frontal band and dark submedian streaks on vertex and occiput. Thorax dull brown. Mesonotum shaded with yellowish anteriorly, especially along the median line. A broad black oblique line on pleura from base of fore wing forward to the lateral margin of the prothorax. Scutella blackish, the surrounding areas shaded with pale orange. Fore legs deep brown; femur paler at apex and along the upper margin. Middle and hind legs dull olive brown with darker tarsi. Wings hyaline, venation dark. Fore wings clouded with brownish over the inner two-thirds of their length, except for the costal half of the extreme base, which is pale. Brownish shading along the costa of the hind wing. Female with no brown shading on the wing.

Abdomen dull brown dorsally, with pale antero-lateral patches, triangular in shape. On the basal segments these pale areas are indistinct and semi-transparent; on the apical segments they are bright yellow and opaque, being especially prominent on tergite 10. Subdorsal dark streaks present, more or less obscure on the basal segments. Sternites yellowish. A large brown triangular patch occupies most of each sternite except the anterior margin and the antero-lateral angles. Sternite 9 brown except for a small yellow anterior triangle. Forceps base and forceps brown. Forceps base quite deeply excavated apically. Tails black-brown with darker joinings.

Type material was taken at Ottawa. Ont. (J. McDunnough). We have seen no specimens of this species.

Parameletus midas McDunnough

1923—McDunnough—Canad. Ent. 55: 49. Fig. (Siphlonuroides)..............................Ont.
1930—McDunnough—Canad. Ent. 62: 61. Nymph. (Sparrea sp.).
1932—McDunnough—Canad. Ent. 64: 81 (Siphlonuroides = Parameletus).

Length: body 10; wing 10 mm.

Head of male blackish. Head of female yellowish with a brown band across the median carina, and brown submedian streaks on the vertex and occiput. Thorax olive brown, shaded with yellowish olive anteriorly on the median line and on the mesothoracic scutellum. Thoracic pleura of female with yellow marks and blotches. Legs olive brown, the fore legs darker than the middle and hind pairs. Wings hyaline, venation dark. A faint trace of brown shading on the fore wing between the subcosta and radius basally; a brown dot at the base above the inner margin, and a faint shading of brown in the stigmatic area. A small brown cloud at the bulla. The entire costal margin may appear shaded with brown, especially in the female. Stigmatic cross veins in the apical portion of the costal space tend to anastomose.

Abdomen brown dorsally, the posterior margins blackish. Sternites duller and paler brown, with faint traces of small lateral dark spots on each. In female, each sternite is narrowly marked with yellowish. Forceps base brown, very slightly concave on the apical margin. Forceps brown, penes dark red-brown. Ventral division of the penes rounded at the apex, extending beyond the tips of the dorsal division; the latter composed of two rather slender curved structures, pointed at the tips, and converging apically (see fig. 128). Tails brown with darker joinings.

The type material was collected at Ottawa, Ont. (J. McDunnough). In the Cornell collection are several specimens presented by Dr. McDunnough, collected at Thunder River, Que. (W. J. Brown).

SUBFAMILY LEPTOPHLEBIINAE

Tarsi four-jointed, except in the male fore tarsus, where a very short additional basal joint is present. Claws dissimilar on all tarsi in Nearctic genera. Hind tibia usually longer than the femur; tarsus shorter than tibia. Wings four, in the Nearctic genera. In the fore wing, there are no free intercalaries between the long intercalary of media and the posterior branch of that vein; nor between this latter branch and the anterior branch of cubitus. The branches of the cubital vein are usually separated from one another at the base. Two to four rather long intercalaries are present between the two branches of cubitus. Hind wing variable in the different genera; with or without a costal projection.

Forceps three-jointed (occasionally a small fourth apical joint may be present; we are unable to distinguish the basal fourth joint, so-called, in Choroterpes). Of these, the basal joint is always much longer than either of the other two. Subanal plate of female with a median cleft, more or less deeply excavated according to the genus. Tails three; in Blasturus, the median tail is shorter and weaker than the lateral ones.

Nymphs rather slender; in some genera, somewhat depressed in the head and thoracic regions, in others the entire body is more or less compressed. Claws denticulate. Gills present on segments 1–7; lamelliform, or with the outer margins bearing long fringes which are united at the base. The members of each pair of gills are similar in form; they are present on segments 2 or 3 to 6, or 2–7. In some genera the gills on all segments are similar to one another in form. Postero-lateral spines are present on segments 9, or 8 and 9; in two genera, these spines occur on all but the basal segments.

KEY TO THE GENERA

Adults

1—Hind wing much reduced in size, venation rather scanty; a distinct costal angulation halfway to apex. 2
—Hind wing somewhat larger, venation less reduced; no costal angulation, costal margin slightly concave at center. 6
2—Costal angulation of hind wing obtuse; no sag in Of of fore wing. 3
—Costal angulation of hind wing acute; a distinct sag usually evident in Of of fore wing. 4
3—M of hind wing forked; fore tarsus of male shorter than tibia. . . . . . . **Thraulodes**
—M of hind wing not forked; male fore tarsus subequal to tibia. . . . . **Choroterpes**
4—Forceps base of male entire; apical margin of subanal plate of female only slightly cleft. . . . . **Thraulus**
—Forceps base of male rather deeply cleft, as is the apical margin of subanal plate of female. 5
5—In hind wing, subcosta ends in margin at outer side of costal angulation; two end joints of forceps short, together not ¼ length of preceding joint. **Habrophlebiodes**
—Mhind wing, subcosta extends almost to apex of wing; two end joints of forceps together equal in length the preceding joint. . . . . **Habrophlebia**
6—Middle tail shorter and weaker than laterals. . . . . **Blasturus**
—Middle tail approximately equal to laterals in length and thickness. 7
7—Penes separated almost to base; a long flap-like appendage, narrowed distally, is attached near apex of each, and extends inward and downward between lobes of penes. . . . . **Leptophlebia**

—Penes usually not as above; usually two sets of appendages, an upper short pair extending laterally, and a lower longer pair (the reflexed spur) extending backward, often obliquely. . . . . **Paraleptophlebia**

VERIFICATION TABLE

Adults

Genus	♂ Tars. I vs. tib.	Sag[5] in Of	Fk. of M hd. wing	Costal angul.	Forceps base	Subanal[4] plate
Blasturus[1]	Longer	Present	Present	Absent	Cleft	Deep
Choroterpes	Subequal	Absent	Absent	Obtuse	Entire	Slight
Habrophlebia[2]	Longer	Present	Absent	Acute	Cleft	Deep
Habrophlebiodes	Subequal	Present	Absent	Acute	Cleft	Deep
Leptophlebia	Longer	Present	Present	Absent	Cleft	Deep
Paraleptophlebia	Var.	Present	Present	Absent	Cleft	Deep
Thraulodes	Shorter	Absent	Present	Obtuse	Entire	Slight
Thraulus[3]	Shorter	Present	Absent	Acute	Entire	Slight

[1] Middle tail much shorter than laterals.
[2] Two end joints of forceps together as long as preceding joint.
[3] Long slender projections on dorsal aspect of forceps base between forceps and penes.
[4] Relative depth of cleft in apical margin.
[5] Refers to the asymmetry of this fork (Of) of Vein Rs.

KEY TO THE GENERA

Nymphs

1—Gills of first pair differing in form from those of the following pairs. 2
—Gills of first pair similar in form to those of the following pairs. 6
2—Gills filamentous; each gill, on segments 2–7, consists of two clusters of slender filaments. . . . . **Habrophlebia**
—Gills lamelliform; on segments 2–7 not as above. 3
3—Entire inner and outer margin of each gill on the middle segments finely dissected, forming a deep fringe of delicate filaments. . . . . **Thraulus**
—Inner and outer margins of gills on middle segments not as above; each gill of these pairs with a long terminal extension. 4
4—Gill on segment 1 single, unbranched; terminal extension of upper gill of each middle pair (sometimes of lower gill also) rather broad and spatulate **Choroterpes**
—Gill on segment 1 definitely bifid except at base; terminal extension of each gill on middle segments very slender. 5
5—Lamella of each gill on the middle segments with one or two blunt lobes at the base of the terminal extension; double row of denticles on inner margin of each claw (upper row near tip only). . . . . **Blasturus**
—Lamella of each gill on the middle segments without such lobes at base of terminal extension. . . . . **Leptophlebia***
6—Lateral spines present on abdominal segments 2–9; gills diminishing in size to rearward. . . . . **Thraulodes**
—Lateral spines present on abdominal segments 8 and 9, or on 9 only; gills not noticeably diminishing in size to rearward. 7
7—Spine on 9th segment not more than ¼ the length of that segment; labrum with

* *L. gracilis* as in Blasturus.

shallow indentation only on fore border; spinules on posterior margins of tergites 1–10 .... **Paraleptophlebia**
—Spine on 9th segment long and slender, ½ as long as the length of that segment; labrum rather deeply indented on fore margin; spinules on posterior margins of tergites 7–10 only .... **Habrophlebiodes**

VERIFICATION TABLE

Nymphs

Genus	*Gills alike on*	*Gill form*	*Maxil. spines*	*Teeth on claw*	*Lateral spines on*
Blasturus	2–6	Broad, lobed	4–7	2 rows	8+9
Choroterpes	2–7	Broad, lobed	15	1 row	3–9
Habrophlebia	1–7	Fimbriate	5	1 row	8+9
Habrophlebiodes[3]	1–7	Lanceolate	5–6	1 row	8+9
Leptophlebia	2–7	Broad, not lobed[2]	5–6	1 row[2]	8+9
Paraleptophlebia	1–7	Lanceolate	4–6	1 row	8+9[4]
Thraulodes	3–7[1]	Lanceolate	16–18	1 row	2–9
Thraulus	2–7	Deep fringe	16–18	1 row	8+9

Genus **Leptophlebia** Westwood

1840—Westwood—Introd. mod. classif. ins., 2: 31.
1881—Eaton—Ent. Mo. Mag. 17: 193 (restricted).
1885—Eaton—Revis. Monogr. p. 91.
1917—Bengtsson—Ent. Tidskr. 38: 177 (Euphyurus).
1917—Lestage—Ann. Biol. Lacust. 8: 327.
1920—Ulmer—Stett. Ent. Zeit. 81: 115.

This, the type genus of the subfamily, is poorly represented in North America. *L. marginata* Linn. is well known in Europe, and was once reported by Dr. A. S. Packard from Labrador. We associate with it here the two species that are most like it in form of male genitalia. These two, however, do not conform in all characters. Most of the North American species that have been described under this generic name will here be found in Paraleptoplebia.

The distinctive characters seem to be these: three long subequal tails; posterior branch of the radial sector in the fore wing sagged to rearward beyond its fork; costal margin of hind wing with no angulation but with a slight concavity near the middle; penes much as in Blasturus; male fore tarsus longer than tibia. Wings as in fig. 130.

The nymphs are quite slender. The gills are rather broadly lanceolate and rather abruptly narrowed to a long tail-like tip, but with no excised lateral lobes (except in *L. gracilis*) at the base of this slender portion. There are well developed lateral spines on abdominal segments 8 and 9, and the transverse apical border of all the segments is beset with a dense row of marginal spinules. The tarsal claws have (except in *L.*

[1] Broader on 1 and 2.
[2] Except in *L. gracilis*.
[3] Has row of spinules on posterior margins of abdominal tergites.
[4] On 8 and 9, or on 9 only.

gracilis) a single inferior row of denticles (lacking the upper short row of Blasturus). Mouthparts as in the subfamily.

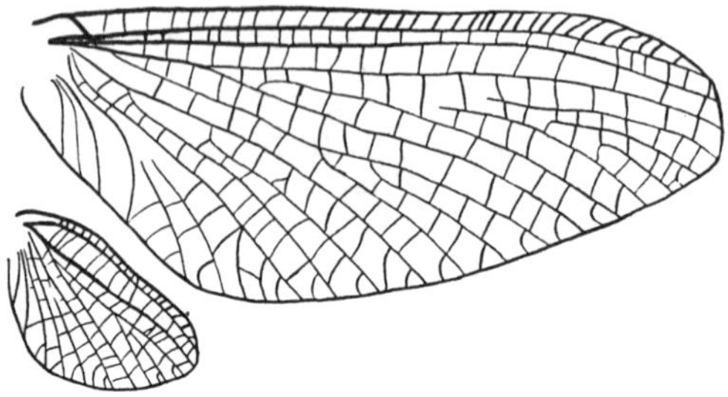

FIG. 130. Wings of *Leptophlebia johnsoni*. (From Eva Gordon.)

The characters here given are those of the European species of the genus. Of the two Nearctic species which we are placing tentatively in Leptophlebia, the nymph of *L. johnsoni* is not known, and that of *L. gracilis* agrees with those of the genus Blasturus as regards the characters of gills and claws. Characters used in the table are drawn from *L. marginata* Linn., a European species.

Genotype—*L. vespertina* (in Ephemera), Linnaeus.

Distribution—Holarctic; in Nearctic, known in east and southeast only.

Leptophlebia gracilis Traver

1932—Traver—J. Elisha Mitchell Sci. Soc. 47: 133 (Blasturus)............................N. C.

Length: body 6–7; wing 5½–6½ mm.

This is a dainty, pale brown, yellow-legged species, with the middle abdominal segments whitish, cross-barred with pale brown. Head in life orange, including the upper eye. Lower eye and basal rings of ocelli blackish.

Thorax reddish brown or fawn colored with paler marginal lines on the prothorax, in the notal groove, before the wing roots, and in the pleural sutures. Legs pale yellowish, darker on front femora; all with pale tarsi. Fore legs a little longer than the body. Wings subhyaline, tinged with brown toward the front by reason of the heavier veins in that portion of the wing, wholly hyaline toward the hind angle. Costal cross veins wanting before the bulla; those in the stigmatic area not very numerous, strongly aslant, and occasionally anastomosing.

Abdomen slender, pale reddish brown toward the ends, whitish in the middle portion, with dorsal cross bands of fawn color overspreading the apical half of segments 3 to 6. On the ventral side the yellow ganglia show through the translucent skin. The spiracles and the extreme apical dorsal margins of segments 8 to 10 are hair-lined in black. Tails subequal, whitish, with joinings of segments narrowly ringed with reddish. Forceps of the male four-jointed, white; penes red, rather narrowly linear and straight with a deep parallel-sided cleft between, each with a minute and very inconspicuous terminal tooth, and with a long slender reflexed spur that reaches proximally nearly to the end of the cleft (see fig. 131).

In the female the entire dorsal surface is light reddish brown with the abdominal tergites slightly darker. Legs yellow beyond the pale reddish brown femora.

The subimago in both sexes has wings, feet, and tails of a dull, smoky-brownish color.

This is a species of doubtful position in the system. The nymph is like Blasturus in form of gills and in having a second row of denticles on the tarsal claw, and for that reason was placed in Blasturus in the original description. The adult is like Leptophlebia in wing venation and male genitalia, hence the species is transferred to Leptophlebia.

The type material of this species, taken near Greensboro, N. C. (J. R. Traver) is in the Cornell collection.

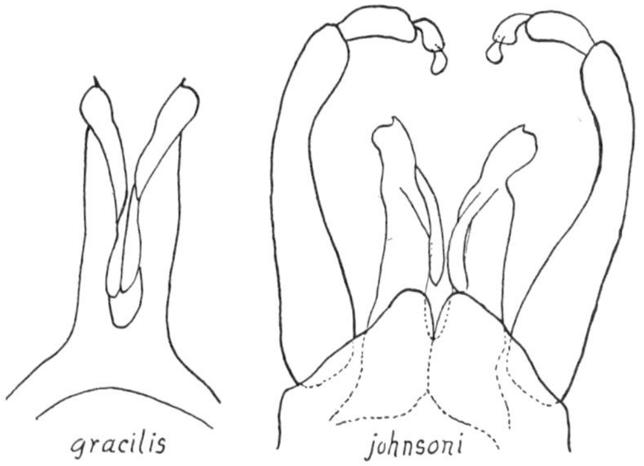

FIG. 131. Male genitalia of two species of Leptophlebia.

Leptophlebia johnsoni McDunnough

1924—McDunnough—Occas. Pap. Soc. Nat. Hist. Boston 5: 73 (fig.)........................N. H.
1925—McDunnough—Trans. R. S. C. p. 209..Que.
1925—McDunnough—Canad. Ent. 57: 169.
1930—McDunnough—Canad. Ent. 62: 55..Ont.

Length: body 9; wing 8–9 mm.

This is a handsome, shining brown species with whitish mid-abdomen, conspicuously ringed tails and brown-tipped wings; head, thorax, and basal segment of abdomen shining blackish-brown above, paler on the sides; fore legs brownish, middle and hind legs paler especially toward the tarsi. Wings hyaline except for a faint brown apical cloud that is heaviest on the stigmatic region. Costal cross veins obsolete, except in the stigmatic region where they are numerous, oblique, sometimes forked. The subcostal cross veins in this region are few and strong and erect.

Abdomen white on segments 2 to 7 with black dots on the spiracles and yellow ganglia showing through the skin beneath; on the lateral margins there is sometimes a faint wash of brown and there is sometimes a still more faint cloud of brown on the mid-dorsum, especially on segment 7; segments 8–10 are brown above with chalky white streaks in old males at the sides, that extend rearward on the base of the forceps.

Forceps white, four-jointed, the three small terminal joints successively more minute, and more or less deciduous. Penes long, yellow, their tips surpassing the principal bend of the forceps, separated by a Y-shaped notch; the reflexed spurs are long, at first convergent then bent rearward and parallel to their tips; the apical notch between the triangular divisions of the 9th sternite is unusually narrow (see fig. 131). The tails are distinctly ringed with rich red brown on the joinings of all the segments.

The female is paler than the male on head, thorax and legs, darker on the abdomen, where the middle segments are only a little paler than the end ones, and that mainly on the ventral side. The 9th abdominal sternite is produced beyond the tip of the 10th in a transparent triangular lobe that is divided by a median apical excision in equilateral-triangular form.

The type was from Jaffrey, N. H. Dr. Ide had the species from Kearney, and Sand Lake, Ont. Specimens in the Cornell collection are from Cranberry Lake, N. Y. (P. R. Needham); Stamford, Conn. (S. W. Bromley); and Amherst, Mass. (C. P. Alexander). Like *gracilis*, its position in the group is doubtful.

ADDENDUM

In a recent article (Canad. Ent. 67: 122, 1935), Dr. Ide has described the nymph of *L. johnsoni*. This nymph is very similar to that of *L. gracilis*, and it may well be that the latter species is a synonym of the former, as Drs. Ide and McDunnough suggest.

Genus **Paraleptophlebia** Lestage

1871—Eaton—Trans. Ent. Soc. London p. 77 (Leptophlebia, in part).
1881—Eaton—Ent. Mo. Mag. 17: 193 (Leptophlebia, in part).
1884—Eaton—Revis. Monogr. p. 91 (Leptophlebia, in part).
1917—Bengtsson—Ent. Tidskr. 38: 176 (Leptophlebia).
1917—Lestage—Ann. Biol. Lacust. 8: 340.
1920—Ulmer—Stett. Ent. Zeit. 81: 116.
1930—Ide—Canad. Ent. 62: 207 (Leptophlebia).
1933—Gordon—Bull. Brooklyn Ent. Soc. 28: 116 (Leptophlebia).
1933—Spieth—J. N. Y. Ent. Soc. 41: 342 (Leptophlebia).
1934—Traver—J. Elisha Mitchell Sci. Soc. 50: 189.

Mayflies of small to medium size—wing of male 5–12 mm. in length—in which the male and female are markedly different in color. Males are usually brownish or blackish on head, thorax and tip of abdomen; legs usually yellowish or brownish, typically without bands on the femora. The middle abdominal segments are sometimes dark brown; in other species, whitish, with or without obscure brownish markings. Tails three, about equal in length and thickness; slightly longer than the body. Females are typically reddish or reddish brown in color, never with whitish middle abdominal segments.

The wings are hyaline or subhyaline, in some species tinted throughout with amber. The posterior branch of the radial sector is sagged to rearward. The hind wing has no costal angulation, but is somewhat concave near the center of the costal margin. The median vein is forked. There are no intercalaries in the cubital fork, nor between cubitus and media (see fig. 132).

This genus differs from Leptophlebia principally in the structure of the male genitalia. The penes, while showing a great variation in form, usually bear one or two pairs of appendages. The more apical of these we term the lateral apical processes; the lower pair, the reflexed spurs. Forceps base much as in Leptophlebia. Forceps normally three-jointed, but may become four-jointed either through the cutting off of a short basal ring-like segment from the long joint, or by the addition of a third short terminal joint. The subanal plate of the female is deeply cleft on its apical margin.

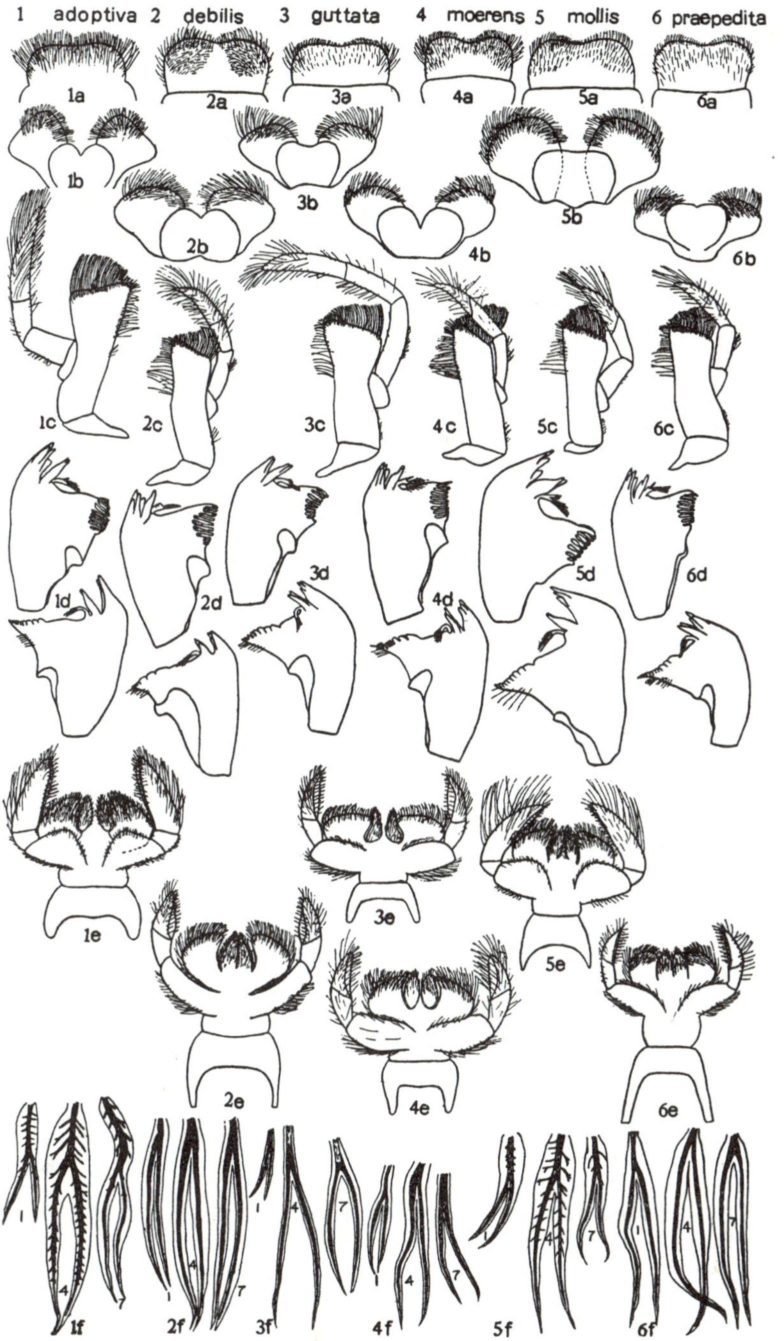

EASTERN SPECIES OF PARALEPTOPHLEBIA, NYMPHAL STRUCTURES

a, labrum; *b*, hypopharynx; *c*, maxilla; *d*, mandibles; *e*, labium; *f*, gills from segments 1, 4, and 7 (from Gordon).

Subimagos rather paler in body than the imagos, with pale smoky wings. Marks on the dorsum of the thorax rather resemble those of the subimago of Blasturus.

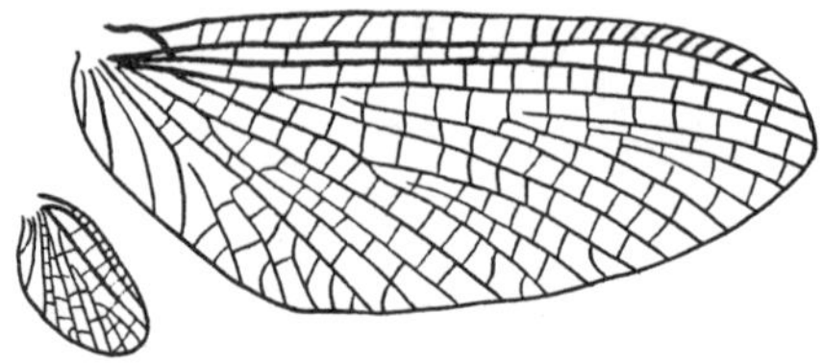

FIG. 132. Wings of *Paraleptophlebia moerens*. (From Eva Gordon.)

The nymphs are slender, the entire body usually compressed. In some species, however, the head is somewhat depressed. Claws are slender and pectinate. Gills similar in form on all segments, narrowly lanceolate, double on all segments. Mouthparts as in Blasturus and Leptophlebia; 4–6 large pectinate spines below the crown of the maxilla. Spinules present on the apical margins of tergites 1–10. Lateral spines on segments 8 and 9 only; in some species, present only on segment 9.

Genotype—*P. cincta* (in Leptophlebia), Eaton.

Distribution—Holarctic; in Nearctic, general throughout North America.

KEY TO THE SPECIES

Adults

1—Middle abdominal segments predominantly white 2
—Middle abdominal segments predominantly brown 18
2—Middle abdominal segments wholly white (may have dark stigmatic marks) . . 3
—Middle abdominal segments clouded or bordered with black or brown 6
3—No dark marks on abdominal spiracles 4
—Dark marks present on abdominal spiracles 5
4—Reflexed spur present (see fig. 133) . . . . **vaciva**
—Reflexed spur absent (see fig. 133) . . . . **mollis**
5—Penes not separated distally; no reflexed spur . . . . **swannanoa**
—Penes separated distally by a rounded apical notch; reflexed spur present . **guttata**
6—Long joint of forceps with a large rounded superior basal dilation 7
—Long joint of forceps more regularly tapering to base 9
7—Tips of penes with a pair of convergent wide lobes projecting within the apical notch; wing 10 mm . . . . **bicornuta**
—Tips of penes with no such lobe within the apical notch; wing 6–8 mm 8
8—Brown apical cross bands broad and with detached spots; wing 8 mm . . . **debilis**
—Brown apical cross bands narrow marginal lines, with no detached spots; wing 6–7 mm . . . . **moerens**
9—Penes with reflexed spur 10
—Penes with no reflexed spur; sometimes with apical teeth 15
10—Penes lyre-shaped; reflexed spur not visible from above . . . . **volitans**
—Penes not lyre-shaped; reflexed spur visible from above 11
11—Divisions of 9th sternite produced apically into long acuminate processes . **falcula**

—Divisions of the 9th sternite with no such acuminate apical processes.......12

12—Middle abdominal segments heavily shaded and clouded with brown; tails ringed with brown at base.......13

—Middle abdominal segments white with narrow black apical markings, not shaded with brown; tails concolorous.......14

13—Reflexed spur rounded at tip, not wider at base than elsewhere; cleft between penes relatively shallow (about ½ the length of the reflexed spur)...**compar**

—Reflexed spur acute at the tip, tapering from a wide base; cleft between penes relatively deep (fully as deep as the length of the reflexed spur)...**californica**

14—Black streak in postero-lateral corners of tergites 2–6; remainder of apical margin pale.......**strigula**

—Entire apical margin of each middle abdominal segment narrowly black..**sculleni**

15—Penes with apical teeth.......16

—Penes without apical teeth.......17

16—Black transverse dash in postero-lateral corners of tergites 2–6; remainder of apical margin pale.......**pallipes**

—Entire apical margin of each middle abdominal segment narrowly black.......**temporalis**

17—Apical process of penes prolonged laterally into a beak-like appendage..**assimilis**

—No such lateral beak-like appendage at apex of penes; penes notched on inner and outer margins near apex.......**heteronea**

18—Long joint of forceps marked off by a short basal subjoint.......**praepedita**

—Long joint of forceps not marked off by such a short basal subjoint.......19

19—Long joint of forceps with a large rounded superior basal dilation; wing 12 mm.......**packii**

—Long joint of forceps with no such basal dilation; wing not over 10 mm.......20

20—Penes with the reflexed spur.......21

—Penes lacking the reflexed spur.......26

21—Femora distinctly ringed with brown.......**bradleyi**

—Femora not ringed with brown.......22

22—Reflexed spur of penes acute at apex.......23

—Reflexed spur of penes rounded at apex.......24

23—Reflexed spur widest at center, its width at this point approximately equal to ½ its length.......**invalida**

—Reflexed spur widest near base, gradually tapering toward the tip; its greatest width not more than ⅓ of its length.......**gregalis**

24—Apical notch between the penes rounded, not longer than wide.......25

—Apical notch between the penes elongate, more than twice as long as wide...26

25—Apical notch between the penes almost closed at the top by the approximated inner margins of the penes.......**associata**

—Apical notch between the penes as wide at the top as at the middle; inner margins of the penes not approximated.......**clara**

26—Reflexed spur widest near the center; legs brownish.......**ontario**

—Reflexed spur not wider at center than elsewhere; legs pale.......**georgiana**

27—Apical notch between the penes a narrow cleft; oval clear space present on each side of penes below apical process.......**adoptiva**

—Apical notch between the penes wider, rounded or V-shaped; no such clear space on each side of the penes below apical process.......28

28—Wings subhyaline with reddish veins; penes separated apically by a shallow rounded notch. . . . . **rufivenosa**

—Wings deep amber-colored; penes very long and slender, separated almost to the base by a deep V-shaped notch. . . . . **brunneipennis**

VERIFICATION TABLE

Male Adults

Species	*Wing length*	*White tergites*[1]	*Forceps lobe*[2]	*Reflexed spur*[3]	*Distr.*
adoptiva	7–8	None	Absent	Absent	E
assimilis	7	Obsolete	Absent	Absent	SE
associata	8	None	Absent	Present	Calif.
bicornuta	10	Obsolete	Present	Present	W
bradleyi	9	None	Absent	Present	Ga.
brunneipennis	8	None	Absent	Absent	Wash.
californica	8	Obsolete	Absent	Present	Calif.
clara	7	None	Absent	Present	Calif.
compar	7½	Obsolete	Absent	Present	Calif.
debilis	8–9	Obsolete	Present	Present	E
falcula	7	Obsolete	Absent	Present	Ore.
georgiana	5	None	Absent	Present	Ga.
gregalis	7½–9	None	Absent	Present	W
guttata	5½–6½	Present	Absent	Present	E
heteronea	6–7	Obsolete	Absent	Absent	W
invalida	7	None	Absent	Present	B. C.
moerens	6–7	Obsolete	Present	Present	E
mollis	8	Present	Absent	Absent	E
ontario	6–7	Absent	Absent	Present	Ont.
packii	12	Absent	Present	Present	Utah
pallipes	7	Obsolete	Absent	Absent	W
praepedita	5–6	Absent	Absent	Present	E
rufivenosa	7–10	Absent	Absent	Absent	W
sculleni	6½	Obsolete	Absent	Present	Ore.
strigula	7	Obsolete	Absent	Present	Ont.
swannanoa	5¼	Present	Absent	Absent	SE
temporalis	7–8	Obsolete	Absent	Absent	W
vaciva	8	Present	Absent	Present	Ore.
volitans	5–6	Obsolete	Absent	Present	E

KEY TO THE SPECIES

Nymphs

1—Mandible with a forward-projecting tusk longer than the head. 2

—Mandible with no such long tusk. . . . . 3.

2—Tracheae of gills with many small lateral branches; tusk almost twice the length of the body of the mandible. . . . . **packii**

[1] Middle abdominal segments.

[2] A conspicuous lobe on the basal joint of the forceps.

[3] Appendage dependent from the penis tip.

—Tracheae with few or no lateral branches; tusk only slightly longer than body of mandible . . . . **bicornuta**

3—Mandibles very large, visible from above for most of their length; tips extend slightly beyond head . . . . **swannanoa**

—Mandible smaller, only a small portion visible from above; tips not extending beyond head 4

4—Tracheae of gills with conspicuous lateral branches 5

—Tracheae of gills without such conspicuous lateral branches 6

5—Color pale brown; canines of mandibles not strongly directed inward, a line along upper edge of molars passing through bases of canines . . . . **adoptiva**

—Color uniform dark brown; canines of mandibles strongly directed inward, a line along upper edge of molar passing well below bases of canines . . . . **mollis**

6—Second and third joints of maxillary palp together are 1⅓ to 1½ times the length of the first joint 7

—Second and third joints of maxillary palp together about equal the first joint in length 9

7—Postero-lateral angles of gill-bearing segments with a black streak along margin . . . . **strigula**

—Postero-lateral angles of gill-bearing segments with a black spot on the margin . . 8

8—Abdominal tergites largely dark brown, with narrow pale submedian streaks at anterior margin; legs wholly pale . . . . **guttata**

—Abdominal tergites largely yellowish with irregular brown markings; a dark brown "knee spot" on each leg, and partial dusky band near apex of femur . . . . **ontario**

9—Legs pale, barred with darker brown . . . . **debilis**

—Legs uniform brown, or nearly so 10

10—Gills hairy . . . . **volitans**

—Gills not hairy 11

11—Segments of abdomen nearly twice as wide as long . . . . **moerens**

—Segments of abdomen nearly equal in length and width . . . . **praepedita**

VERIFICATION TABLE

Nymphs

Species	*Body length*	*Head above*	*Lat. abd. spines on*	*Tracheal branchlets*[1]	*Distr.*
adoptiva	8–10	Spotted	9	Present	E
debilis	7–8	Brown	8, 9	Absent	E
guttata	6½–7	Brown	9	Absent	E
moerens	7½–8	Spotted	8, 9	Absent	E
mollis	6½–7½	Spotted	9	Present	E
ontario	6½–7	Brown?	8, 9	Absent	E
packii	9	Brown	8, 9	Present	W
praepedita	6½	Pale line	8, 9	Absent	E
strigula	6½–7½	Brown?	?	Absent	E
swannanoa	4½–5½	Brown	9	Present	S
volitans	6½	Spotted	8, 9	Absent	E

[1] Lateral dark branchlets on gill tracheae.

Paraleptophlebia adoptiva McDunnough

1929—McDunnough—Canad. Ent. 61: 169 (fig.) .... Que.
1930—Ide—Canad. Ent. 62: 209 (fig. nymph) .... Ont.
1933—Gordon—Brooklyn Ent. Soc. Bull. 28: 124 (figs) .... N. Y.

Length: body 7; wing 7–8 mm.

This is a blackish species with a brown abdomen that is only a little paler in the middle. Top of head and thorax shining black, only the sutures on the sides paler; dorsum of the first abdominal segment also blackish. Legs brown including the tarsi, paler brown beyond the knees. Wings subhyaline with brown veins and darker brown wing roots. Costal cross veins faint and rather numerous; those of the stigmatic region long, sinuate and crowded, but little conjoined. The abdomen is nearly uniform brown, only the base of segments 3–5 showing a suggestion of pale rings; and the apical segments are only a little darker on the dorsum. Forceps yellowish brown; the 2nd joint twice as large as the 3rd. Penes appressed, half as long as the first joint of the forceps. The sperm ducts terminate before the apex between a rounded inner lobe and an outer triangular tooth that is turned sharply laterally; below this the outer margin of the penis is suddenly widened and the basal portion parallel-sided (see fig. 135). Tails uniform pale brown.

The types of this species came from Kazubazua, Que. (G. H. Fiske). Specimens in the Cornell collection are from Ithaca, Salmon Creek, Harford, Freeville, and McLean, N. Y.

Paraleptophlebia assimilis Banks

1914—Banks—Proc. Acad. Nat. Sci. Phil. 66: 614. Fig. .... N. C.

Length: body 7; wing 7 mm.

Head and thorax blackish brown; dark red-brown markings on thoracic pleura. Fore legs yellowish brown; middle and hind legs yellow. Fore leg longer than body. Wings hyaline; longitudinal veins of the costal margin, and apical portions of other longitudinals, yellowish brown. Cross veins almost invisible, except in the stigmatic area, where they are simple, straight and about 17 in number. No cloud in the stigmatic space.

First segments of abdomen very dark reddish brown. Segment 2, the apical two-thirds of segment 7, and segments 8–10, also reddish brown, somewhat paler than the first segment. Segments 3–6, and basal portion of 7, white, semi-hyaline, marked with brown. Posterior margins of these segments widely banded dorsally with reddish to purplish brown. The postero-lateral angles filled with a triangular patch of the same color. Traces of a dark streak on the posterior portion of the median line may be present. Two small dark spots in the pleural fold on each segment, near the anterior and posterior margins. Pleural fold narrowly margined with brown. Mid-ventral ganglionic areas marked with orange-brown.

Forceps base pale brown, forceps pale. Tails yellowish white, joinings faintly darker. The narrow basal joint of the forceps mentioned by Dr. Banks, similar to that of *P. praepedita*, seems not to be a true joint. A slide of the genitalia of one of the paratypes was kindly loaned us by Dr. McDunnough, who obtained the specimen from Dr. Banks; the sketch (see fig. 134) is made from that slide. The lateral apical processes of the penes are large beak-like structures; no reflexed spurs are present.

The type material was taken at Black Mt., N. C., from the north fork of the Swannanoa River, in May. We have several fine specimens of this species in the Cornell collection from Greenville, S. C. (H. K. Townes, Jr.).

Paraleptophlebia associata McDunnough

1924—McDunnough—Canad. Ent. 56: 221. Fig. . . . . Calif.

Length: body 8; wing 8 mm.

A dark brown species, wings amber-tinted.

Head and thorax deep black-brown. Fore leg dark brown, middle and hind legs somewhat paler brown. Wings hyaline, tinged with amber color. Longitudinal veins light brown. Stigmatic area definitely brownish.

Abdomen brown; the anterior segments semi-translucent, the three posterior segments ruddier and opaque. Each segment is rather paler on the anterior margin. Posterior and lateral margins of the tergites blackish brown. Tails yellowish brown, darker at the base; joinings darker brown. Forceps light brown. Apical notch between penes lobes rounded, almost closed at the top by the approximated inner apical margins of the penes. Reflexed spur rather short,—not as long as in *P. compar*,—and rounded at the apex.

The types of this species were obtained by Dr. C. H. Kennedy from Santa Cruz Is., Calif. One male imago taken by Dr. J. C. Bradley at Eelton, Santa Cruz Mts., Calif., is in the Cornell collection.

Paraleptophlebia bicornuta McDunnough

1926—McDunnough—Canad. Ent. 58: 229. Fig. . . . . Alta.

Length: body 9; wing 10 mm.

This is a species with almost as much brown as white on the middle abdominal segments. Head and thorax blackish brown above, a little paler on the sides and beneath. Fore legs brown with paler tarsi; middle and hind legs paler beyond the brown femora. Wings hyaline with brown longitudinal veins; costal cross veins obsolete except in the stigmatic area where they are numerous, long, strongly aslant and sometimes forked.

Abdomen dark brown on the first segment above, lighter on the second, and reddish brown on segments 8 to 10; 3 to 7 whitish with brownish cloud on the posterior half of the dorsum of each. A black dot on the stigma and a diffuse roundish, brown spot above it. Mid-ventral areas also faintly tinged with brown, especially segment 7. Sternite 1 wholly brown; 8 and posterior half of 9 a bright reddish brown. The basal joint of the forceps is brown; the distal joints are white; at the base of the former is a conspicuous lobe above, similar to that in *P. debilis*. Penes broadly separated by a very wide U-shaped notch; dilated and truncate at tips and bearing just before the tips on the dorsal side a pair of long flat lobes that converge by their rounded ends within the U-shaped cleft. Reflexed spur bifid, almost crescentic in form, attached by the back of this crescent; the outer horn longer (see fig. 135). Tails pale brown.

The types were from Alberta. Several male specimens from Montana (R. T. Young) and others from Glacier National Park (A. S. Hazzard) are in the Cornell collection.

Paraleptophlebia bradleyi Needham

1932—Needham—Canad. Ent. 64: 274. . . . . Ga.

Length: body 8; wing 9 mm.

A dark brown species with yellowish wings, the femora black-ringed.

Head and thorax brown. Legs yellowish brown, with black rings on the apical third of all the femora. Wings hyaline, tinged with yellow. Veins brown. Costal cross veins numerous and distinct; stigmatic cross veins crowded and irregular, and anastomosed to form two rows of cells in the costal space, the outermost row being half the width of the inner one. Entire stigmatic area, and the radial space below it, semi-opaque, greyish.

Abdomen brown, darker at the ends. Tergites 1–8 with a narrow pale mid-dorsal line. On each side of this line are two rows of pale triangular spots, confluent on the basal margin. Ventrally paler, somewhat yellowish; ganglia not marked. Genitalia pale brown. Basal joint of the forceps slowly tapering and sinuately curved; the two small terminal segments dilated internally and subtriangular. Three short terminal joints are present, on one specimen. Penes long, tapering, separated by a V-shaped cleft for three-fourths of their length. Reflexed spur curved, spatulate, acute at the tip, reaching to the level of the bottom of the cleft (see fig. 134).

Type material was collected at Billy's Lake, in the Okeefinokee Swamp, Ga., in late December, 1914, by Dr. J. C. Bradley. This material is in the Cornell collection.

Paraleptophlebia brunneipennis McDunnough

1924—McDunnough—Canad. Ent. 56: 221 .... Wash.

Length: body 7; wing 8 mm.

A dark reddish brown species with amber wings.

Head and thorax blackish. Fore legs deep blackish brown, the tarsi somewhat paler. Femora of middle and hind legs brown; tibiae somewhat paler, more golden brown. Wings unicolorous deep amber; cross veins very numerous. About 14 costal cross veins before the bulla, and 23 or more beyond it; many of the latter are anastomosed.

Abdomen deep reddish brown, the last sternite and the forceps base blackish. Forceps paler brown. Penes united only at the extreme base, long and slender and bent outwards slightly at the tip; appear distinctly U-shaped (see fig. 135). Tails pale smoky brown.

The type material was taken at Rock Lake, Wash. (A. L. Melander). Dr. McDunnough has kindly lent us a slide of the genitalia of one of the paratypes, from which the sketch is made.

Paraleptophlebia californica Traver

1934—Traver—J. Elisha Mitchell Sci. Soc. 50: 195. Fig .... Calif.

Length: body 8; wing 8 mm.

A rather pale reddish brown species.

Head reddish brown; median carina and bases of antennae yellow. An oblique black line on each side of the face. Thorax reddish; mesonotum bright reddish brown. Mid-dorsal line of prothorax black; posterior margin white. Sternum and pleural sutures yellow with reddish brown marginings. Legs pale yellowish white; femora tinged with yellow. Wings hyaline. Longitudinal veins of the costal border yellowish; all others pale, except at extreme base, which are yellow. Stigmatic cross veins simple, curved, a few forked near the costal margin.

Segment 1 of abdomen largely opaque; 2–7 semi-hyaline; pale yellowish brown with smoky markings. Tergites 6 and 7, and segments 8–10, opaque reddish brown. Posterior margins of all segments smoky brown, this band darkest in the postero-lateral angles. Two oblique streaks on each side of each tergite, from the antero- and postero-lateral angles. Mid-dorsal line pale, bordered by purplish submedian streaks. A dark line on the pleural fold.

Genitalia golden brown; tips of penes purplish brown. Forceps widened gradually toward the base. Penes separated by a rounded notch, into which projects a short blunt process from each division of the penes. Reflexed spur widest at base, its tip acute (see fig. 134). Tails pale smoky, darker at base.

The type material, taken at Waddell Creek, Rancho del Oso, Calif. (D. Shepherd) is in the Cornell collection.

Paraleptophlebia clara McDunnough

1924—McDunnough—Canad. Ent. 56: 221 (*gregalis*).
1933—McDunnough—Canad. Ent. 65: 155. Fig. .... Calif.

Length: body 6; wing 7 mm.

A brown species; longitudinal veins light brown.

Head and thorax brown. Fore legs brown. Wings hyaline; cross veins pale, almost invisible; longitudinal veins light brown. Abdominal tergites deep blackish brown, the apical segments somewhat lighter brown. Blackish markings in the region of the spiracles. Sternites pale brownish; 1-7 hyaline, 8 and 9 opaque, deeper brown. Penes as in fig. 135.

A single male specimen of this species, from the type locality, Los Gatos, Santa Clara Co., Calif., is in the Cornell collection.

Paraleptophlebia compar Traver

1934—Traver—J. Elisha Mitchell Sci. Soc. 50: 193. Fig. .... Calif.

Length: body 7½; wing 7½ mm.

A rather pale brown species, resembling *P. californica* in coloring, but like *P. associata* in genitalia.

Head reddish yellow. Thorax reddish brown, the pleural sutures pale. A pale spot anterior to the mesothoracic scutellum. Notum tinged with yellow. Purplish black pencilings on the pronotum, and in an oblique line along the pleura from the base of the fore wing. Black marks at the base of the hind coxa. Legs pale whitish faintly tinged with yellow. Apical end of fore femur and base of fore tibia reddish brown. Wings hyaline. Longitudinal veins in anterior half of fore wing pale brown; other veins colorless. Stigmatic cross veins simple, somewhat slanting, 8–10 in number.

Abdomen pale with smoky markings. Segments 1–5, and the basal portion of 6, whitish and semi-hyaline. Apical segments opaque, yellowish. Posterior margins of tergites with a broad purplish grey band; lateral margins with faint smoky clouds. A small dark blotch in the postero-lateral angles. A purplish grey band, with a narrow black penciling, along the pleural fold. Faint indications of a narrow dark geminate mid-dorsal line on basal and apical tergites. A purplish mark on the posterior margin of each sternite, at the postero-lateral corner only. Tails greyish at base, silvery white at tips. Joinings in basal half purplish grey.

Forceps base pale reddish brown. Forceps brown at base, pale distally. Penes very similar to *P. associata*, but with longer reflexed lobes (see fig. 135).

Type material, taken at Waddell Creek, Rancho del Oso, Calif. (D. Shepherd) is in the Cornell collection.

Paraleptophlebia debilis Walker

1853—Walker—List Neurop. Ins. Brit. Mus. p. 569. (Baetis) .... N. S.
1861—Hagen—Syn. Neurop. Ins. N. Amer. p. 46.
1884—Eaton—Revis. Monogr. p. 98.
1907—Needham—N. Y. St. Mus. Bull. 124: 189 (*mollis*) .... N. Y.
1919—Morrison—Canad. Ent. 51: 143 (*mollis*). (fig).

1920—Ulmer—Arch. Natg. 85, Abt. A, Hft. 11: 27 (*separata*).
1925—McDunnough—Canad. Ent. 57: 169 (*separata* =*debilis*)............................ Que.
1925—Needham—Ent. News 36: 112 (*mollis*).
1926—McDunnough—Canad. Ent. 58: 299 (fig.).
1928—McDunnough—Canad. Ent. 60: 8... Alta.
1930—McDunnough—Canad. Ent. 62: 55... Ont.
1930—Ide—Canad. Ent. 62: 210 (fig. nymph).
1932—McDunnough—Canad. Ent. 64: 78.
1932—Traver—J. Elisha Mitchell Sci. Soc. 47: 113.................................... N. C.
1933—Gordon—Brooklyn Ent. Soc. Bull. 28: 122 (figs).

Length: body 8; wing 8–9 mm.

This is a handsome, clear-winged, rich brownish, late-season species. Head above, thorax, and both ends of abdomen deep rich brown. Legs pale. Wings hyaline with pale veins. Costal cross veins obsolete except in the stigmatic area where they are numerous, simple, curved, and strongly aslant, and where the proximal end (next to subcosta) of each is obsolescent.

Abdomen white on segments 2 to 7, with brownish purple half-rings across the apices of the segments. A brown point on the spiracle, and a rounded brown spot before each spiracle above the lateral margin. Ganglia marked with yellow. Forceps with a large, ventral, basal lobe-like dilation. Penes separated by a U-shaped notch; each dilated at its outer end, narrowing the notch. On the outer side at the end there is a recurved lateral tooth. The reflexed spur is twice as long as the cleft is deep; it is arcuate, outcurving and acute at the apex (see fig. 134). Tails white.

This is a common species in the northeastern states. The dates of local material in the Cornell collection are: Ithaca, from the 11th of October to the first of November; McLean, N. Y., October; Old Forge, N. Y., August 1st; Potter's Mills, Pa., October 31st (S. C. Bishop) and Bear Gully Creek, Willoughby, O. (J. R. Traver) August 25th. Western specimens are from Jasper Park, Alta. (Miss E. G. Fisher) and Blue L., Grand Coulee, Wash. (M. D. Leonard).

Paraleptophlebia falcula Traver

1934—Traver—J. Elisha Mitchell Sci. Soc. 50: 190. Fig.................................... Ore.

Length: body 7; wing 7 mm.

This is another bi-colored, white-winged, pale-legged species with a pair of dark spots on the dorsum of the 2nd abdominal segment. Head, thorax, and end segments of the abdomen reddish brown above; paler beneath. Legs pale with brown knees. Wings white, hyaline, with only a touch of brown on the root of the radial vein and no darkening of the stigmatic area. Costal cross veins obsolete, extremely pale even in the stigmatic area where they are few and straight and simple; in the succeeding subcostal space they are obsolete and there is no touch of milky whiteness in the stigmatic area.

Abdomen brown on the end segments, paler beneath and at the tip; segments 3 to 6 whitish; 2 and 7 less so. There is a pair of round brown spots on the dorsum of 2, and segment 7 is covered above by a broad triangular cloud. A faint touch of brown lies like a shadow across the middle of the intervening white segments. Spiracles marked with distinct black dots. The apices of the triangular divisions of the 9th sternite are very peculiar; they are prolonged in linear strap-like form that attains the farthest level of the penial notch. Basal segment of forceps white, widened in the middle and tapering near the tip. Penes separated by an open U-shaped cleft a little deeper than wide; each terminates in a flat, triangular process. The reflexed spurs are sickle-shaped, at first bending inward until the two almost touch, then widely outcurving to the side (see fig. 134). Tails white.

A single male specimen from Corvallis, Ore. (H. A. Scullen) is in the Cornell collection.

Paraleptophlebia georgiana Traver

1934—Traver—J. Elisha Mitchell Sci. Soc. 50: 191. Fig. . . . . Ga.

Length: body 6; wing 6 mm.

This is a small rather plain brownish mayfly with the middle abdominal segments black edged above in both male and female. Head above and upper division of the compound eyes brown, lower division black.

Thorax brownish, lighter on the sides. Legs whitish, a little tinged with amber on the femora. Wings of the male whitish hyaline; of the female tinged with light amber and with darker veins. Their costal cross veins are lacking except in the stigmatic area where they are few and simple.

Abdomen brown above, darkened across the ends of the middle segments so as to appear cross-banded; paler beneath. There are brown rings on the joinings of a few of the basal segments of the tails. The long joint of the forceps of the male is very slightly swollen at one-third its length and but little tapered beyond. The third joint is half as long as the second. The U-shaped cleft between the penes is somewhat widened downward and a little narrowed at the extreme outer end by the short hyaline lobes at the orifice (see fig. 133). Brown apices, obliquely truncate ending externally in a short tooth. Reflexed spur much longer than the middle cleft. Tip flattened, subclavate in outline, widened toward the free end. Lobes of the tenth sternite elongate triangular, separated by a rather narrow V-shaped cleft and rounded on the tips. The type material, collected from Swamp Creek, Dalton County, Ga. (P. W. Fattig), is in the Cornell University collection.

Paraleptophlebia gregalis Eaton

1884—Eaton—Revis. Monogr. p. 99. . . . . Ore.
1924—McDunnough—Canad. Ent. 56: 221. . . . . Calif.
1926—McDunnough—Canad. Ent. 58: 297.

Length: body 8; wing 8–9 mm.

A reddish brown species with the thorax blackish above. Fore legs reddish brown except the tarsi which are paler, as also are middle and hind legs. Wings hyaline with the longitudinal veins a little tinged with brown as are also the cross veins of the stigmatic area. The latter are irregular, somewhat branching, twelve to sixteen in number.

Abdomen brownish with darker lines across the joinings of the middle abdominal segments. Tails dingy whitish, becoming ferruginous basally. The long joint of the forceps is conical at base and parallel sided for half its length, then slowly tapering to the apex. The third joint is about half as long as the second. The broadly U-shaped notch between the penes is almost closed by a pair of hyaline, oblong, broad, flat, obliquely converging lobes that extend most of their length beyond the chitinized tip. They thus give the cleft a flask-shaped outline (see fig. 133). External to these the apex is obliquely truncate, the shorter external angle ending in a sharp backwardly-directed tooth. The reflexed spur is hyaline, semi-elliptical with a straightish external and a convex internal margin and rather acute apex. It is a little longer than the middle cleft is deep. The lobes of the divided tenth sternite are very broadly triangular and rounded on the ends.

We have seen a single male specimen from Yorkville, Calif. (E. P. Van Duzee) from the collection of the California Academy of Sciences. It is slightly smaller than the measurements given by Eaton, the fore wing being but 7½ mm. in length.

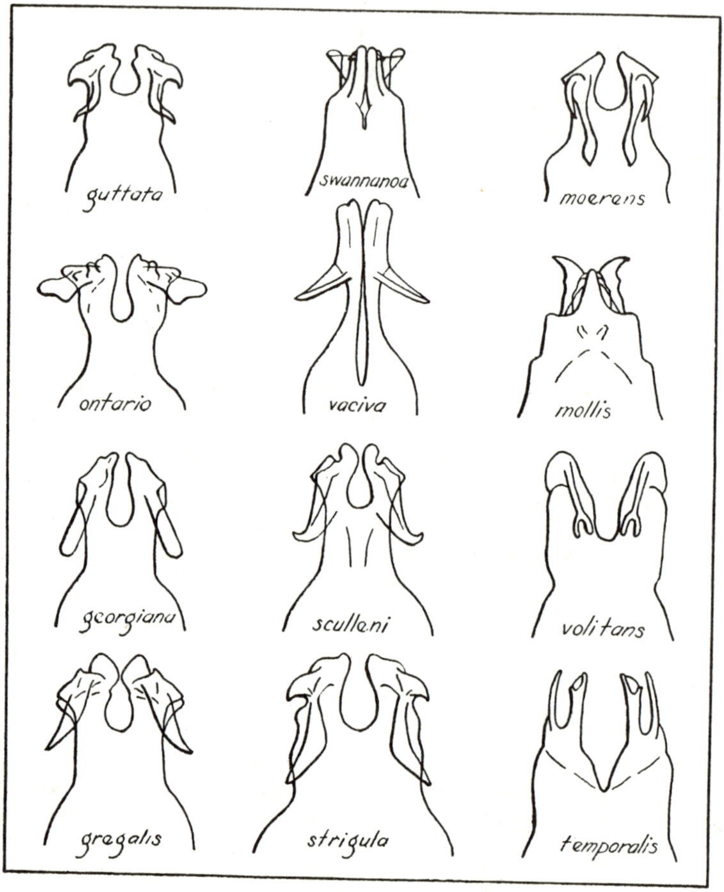

FIG. 133. Male genitalia of twelve species of Paraleptophlebia.

Paraleptophlebia guttata McDunnough

1924—McDunnough—Canad. Ent. 56: 95 (fig.) .... Que.
1925—McDunnough—Trans. R. S. C. Sect. V: 209, 223.
1930—Ide—Canad. Ent. 62: 210 (fig. nymph) .... Ont.
1932—Traver—J. Elisha Mitchell Sci. Soc. 47: 113 .... N. C.
1932—McDunnough—Canad. Ent. 64: 209, 210.
1933—Gordon—Brooklyn Ent. Soc. Bull. 28: 126 (figs) .... N. Y.

Length: body 5–6; wing 5½–6½ mm.

This is a reddish brown species with small paired brown spots on the white middle segments of the abdomen. Top of head and thorax and end segments of abdomen shining blackish brown. Legs whitish with a touch of pale brown on the knees. Wings whitish including the stigmatic area and its cross veins. Costal cross veins obsolete except in the stigmatic area where they are about 12 in number; the middle ones longer, curving and sometimes forking.

Abdomen white on segments 2 to 7 with a row of little roundish brown spots in the spiracular area toward the hind angles of the segments and with a faint suggestion of clouds in the middle of the dorsum. Ganglia yellowish. Forceps white; the long basal segment not tapering so much as in other species but more parallel-sided or even a little widened in the middle. Penes broadly separated for half their length by a U-shaped notch, rounded on the distal ends with a large, recurved, external hook at the tip. The reflexed spur is long, white, and outcurved sinuously like a goat's horn (see fig. 133). Tails white.

The types were from Covey Hill, Que. (C. H. Curran). This species is apparently common in the eastern United States. In the Cornell collection there are specimens from Coy Glen and Van Buskirk Glen (Eva Gordon); Taughannock Falls (K. Jordan); Ringwood and Ithaca (P. R. Needham); Salisbury, Stratford, Dolgeville, Flat Creek and N. Blenheim, N. Y. (H. K. Townes, Jr.). West Virginia specimens are from McLaughlin Spring near Marlinton, Briggs Run near Smoke Hole, Wardensville, Elk Garden and Valley Head (J. G. Needham). Other specimens were taken at the Pigeon River near Waynesville, N. C. (J. R. Traver).

This species is very like *P. mollis* in coloration, save for the spots on the white of the abdomen, and like *P. moerens* in form of genitalia.

Paraleptophlebia heteronea McDunnough

1923—Dodds—Trans. Ent. Soc. Amer. 49: 104 (*pallipes*)....................Colo.
1924—McDunnough—Canad. Ent. 56: 93 (fig.)....................Alta.

Length: body 5–6; wing 6–7 mm.

This is a blackish species with brown-and-white ringed abdomen. The top of head and thorax are rich reddish brown, the sides duller. The fore legs are brown, with paler tarsi and the other legs are wholly pale. The wings are whitish with cross veins absent from most of the costal space, with only a few irregular ones in the stigmatic region. The basal segment of the abdomen and the three apical ones are brown. The middle segments are paler, whitish at the base with the rings broader on segments 4 and 5 and disappearing on 7, and the black spiracles have a broken line connecting them. Basal segment of the forceps rather broad, then nearly parallel-sided to the middle, then strongly tapering; the 3rd segment is nearly as long as the 2nd but much more slender. The penes are separated by a V-shaped notch. The sperm ducts terminate in a projecting tube that is longer than the spines at its outer side, and that has its outer cavinate margin revolute (see fig. 134).

The type material came from Waterton Lakes, Alta. and a paratype has been kindly loaned us by Dr. J. McDunnough. We have also in the Cornell collection specimens taken by Dr. Dodds near Boulder, Colo., in South Boulder Valley between 5800 and 10,580 feet.

Paraleptophlebia invalida McDunnough

1926—McDunnough—Canad. Ent. 58: 298 (fig.)....................B. C.

Length: body 6–7; wing 7 mm.

This is an all brown species, with faintly-ringed abdomen. Thorax shining brown above and beneath a little paler on the sides before the wing roots and about the leg bases. Legs nearly uniform, paler brown only a little darker on the fore femora and knee joints. Wings hyaline, the membrane iridescent, the three long veins of the front brown, the others nearly colorless, as are all the cross veins except those of the stigmatic area. These are about ten in number, simple and oblique.

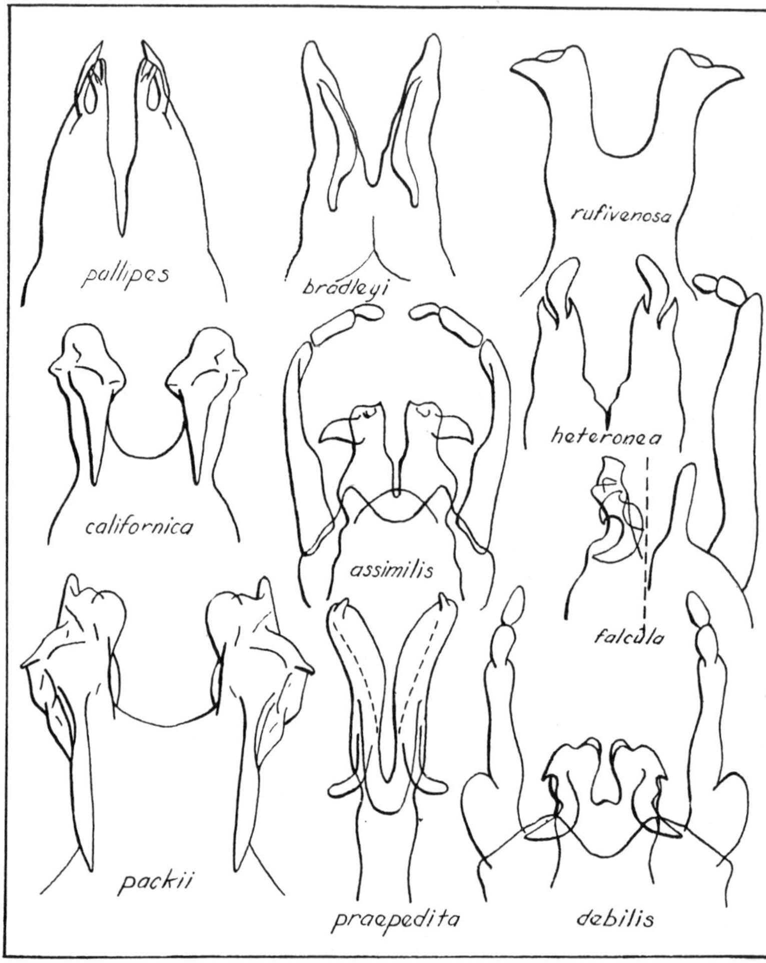

FIG. 134. Male genitalia of ten species of Paraleptophlebia.

Abdomen brown, only a little paler below, and narrowly on the joinings of the middle segments. Segments 7 to 10, darker; forceps and penes paler. The cleft between the penes is linear U-shaped. The pale end lobe and the external end tooth are minute, and the reflexed spur is short, with an outcurving acute point (see fig. 135). Tails very pale brown, with very faint indication of darker joinings on basal segments.

Type material was taken at Departure Bay, B. C., and slides made from it were kindly loaned us for study by Dr. McDunnough.

Paraleptophlebia moerens McDunnough

1924—McDunnough—Canad. Ent. 56: 94 (fig.) .... Que.
1925—McDunnough—Trans. R. S. C. p. 209.
1930—Ide—Canad. Ent. 62: 211.. Ont.
1932—Traver—J. Elisha Mitchell Sci. Soc. 47: 112.. N. C.
1933—Gordon—Brooklyn Ent. Soc. Bull. 28: 126 (figs) .... N. Y.

Length: body 6; wing 6 mm.

This is another small, clear-winged species with whitish, cross-lined abdomen. Head and thorax dark brown above. Legs pale. Wings whitish with a milky stigmatic area covering costal and subcostal interspaces. Costal cross veins almost wholly obsolete even in the stigmatic area where their faint vestiges are numerous, simple, and considerably aslant.

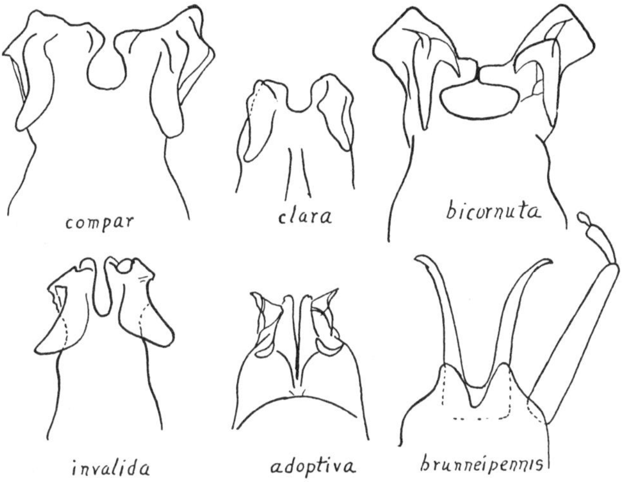

FIG. 135. Male genitalia of six species of Paraleptophlebia.

Abdomen brown on the ends; segments whitish on 3 to 6; less so on 2 and 7. The dorsal apical cross-bands of the white segments are well marked, as are also the lateral, longitudinal, spiracular lines which meet them at the hind angle of the segments. Forceps with a large basal lobe-like dilatation of its first segment. Penes separated for half their length by an oval notch, the aperture of which is closed by a lobe-like prolongation inward of their distal ends. Outer angle of the tips broadly rounded. The proximal portion of its margin is prolonged forward in an incurved spine; attached to this spine is the usual reflexed spur which is very slender (see fig. 133). Tails white.

This is an early season species. Specimens in the Cornell collection are from "The Glen," Ithaca, N. Y. (Eva Gordon); drift net collection near Ithaca (P. R. Needham); Greensboro, N. C. (J. R. Traver).

Paraleptophlebia mollis Eaton

1871—Eaton—Trans. Ent. Soc. London p. 88 (fig.).
1861—Hagen—MS. (*Cloe*).
1884—Eaton—Revis. Monogr. p. 97 (fig.).. N. H., N. Y., Mass., N. C.
1921—Ulmer—Arch. Natg. Abd. 6, Heft 6: 254 (fig.)

1924—McDunnough—Canad. Ent. 56: 95 (fig.)..Que., Ont.
1925—McDunnough—Trans. R. S. C.; p. 209.
1930—Ide—Canad. Ent. 62: 209 (fig. nymph).
1933—Gordon—Brooklyn Ent. Soc. Bull. 28: 127 (figs.)

Length: body 8; wing 8 mm.

This is a white-winged, white-legged, white-tailed species with dark brown thorax and brown tipped white abdomen. Head, thorax, and basal segment of the abdomen brown, darker above and beneath, with paler sutures at the sides. All appendages white. Wings white with whitish veins and with scarcely any darkening of the stigmatic area. Costal cross veins almost undiscoverable except in the stigmatic area where they are rather crowded, strongly aslant, sinuous, and sometimes forked. Segments 2 to 7 of the abdomen entirely translucent white, with no markings on either ganglia or spiracles. Segment 8 is pale brown, 9 and 10 are somewhat darker brown. Forceps and tails white. Penes pale brown, separated by a narrow cleft; each tip bears externally a thin, flat, triangular lobe that terminates laterally in a sharp, upcurving thornlike point (see fig. 133).

This is a common species in the northeastern states and has been collected at numerous localities near Ithaca, N. Y., including Varna, Harford, Ringwood, Etna, North Spencer, Tarbell and Slaterville Springs (P. R. Needham, *et al*). Other specimens in the Cornell collection are from the St. Regis River near Ft. Jackson, N. Y. (L. Harper); Hinckley, Salisbury and Rome, N. Y. (H. K. Townes, Jr.); and Kent Falls, Conn. (C. P. Alexander).

Paraleptophlebia ontario McDunnough

1926—McDunnough—Canad. Ent. 58: 299 (fig.)..Ont.
1933—Gordon—Brooklyn Ent. Soc. Bull. 28: 124.

Length: body 6; wing 7 mm.

This is another small clear-winged species of rich red-brown color and bi-colored abdomen. Head, thorax and both ends of abdomen brown. Legs pale brown with only the front femur a little darker. Wings hyaline with the radial vein pale brown, darker at the root. Costal cross veins obsolete, nearly so even in the stigmatic area where they are simple and but little curved.

Middle abdominal segments obscurely paler across the base of the dorsum and on the sides beneath the mid-lateral carina; end segments darker above. Genitalia and tails somewhat brownish at the base. Penes separated by a rather narrowly U-shaped cleft, truncate on the apex in a sinuous line that ends externally in a recurved tooth. Reflexed spurs very short, hardly longer than the depth of the notch, curved and blunt at the tip (see fig. 133).

The type material was taken at Jordan, Ont. (G. S. Walley). In the Cornell collection are reared specimens of this species from the vicinity of Willoughby, O. (J. R. Traver), and one male from Cinnamon Lake, N. Y. (C. R. Crosby), as well as one specimen from the type locality, presented by Dr. McDunnough.

Paraleptophlebia packii Needham

1927—Needham—Canad. Ent. 59: 44 (fig.)..Utah

Length: body 9; wing 12 mm.

This is our largest species of the genus. The sexes are alike in color, all brown, deepening to shining black on top of thorax and at both ends of abdomen, paler on appendages. Wings subhyaline with brown veins, and with a brown tint in the membrane at the roots. Costal cross veins numerous but weak as far as the stigmatic area where they are crowded; their tips strongly curved outward and often forked. Sometimes they anastomose forming a few small cells in an outer series next the costa. Vein Cu_2 unattached at base.

The forceps has a rounded basal lobe-like dilatation. The penes are separated by a broad U-shaped notch as wide as deep; the terminal orifice is guarded by a rounded inner lobe and an outer triangular tooth. The reflexed spur is short, lance-like or blade-like, and acute at the apex. The subanal plate in the female is divided into two elongate triangular lobes that extend rearward beyond the abdomen more than the length of the 9th segment (see fig. 134).

The species is known only from the north fork of Ogden River, Utah, whence it was collected by the late Dr. H. J. Pack and reared by him on the first of October 1926.

Paraleptophlebia pallipes Hagen

1875—Hagen—Ann. Rep. U. S. Geol. Survey p. 582. . . . . Calif.
1884—Eaton—Revis. Monogr. p. 98 (*memorialis*).
1926—McDunnough—Canad. Ent. 58: 296 (fig.) . . . . B. C., Colo.
1927—Seemann—J. Ent. and Zool. 19: 45 (*rufivenosa*) . . . . Calif.
1928—McDunnough—Canad. Ent. 60: 8. . . . . Alta.

Length: body 6; wing 7 mm.

This is a species with rather highly contrasting colors. Top of head and thorax are black, the tip of the abdomen is brown, the intervening part is white. The legs are pale brownish with white tarsi. The wings are whitish, a little milky along the costa the entire length but more so in the region of the stigma, where cross veins are very obscure, few in number, straight and unbranched. The extreme wing base is tinged with brown.

Abdominal segments 3 to 6 are white with touches of brown on the border behind the spiracles and on the spiracles themselves. The ganglia show through and are yellow. Segment 2 is pale brown and 7 to 10 are darker.

The basal joint of the forceps is somewhat irregularly tapering to its apical 4th and then widened slightly to the tip. The 3rd segment is nearly as long as the 2nd but much more slender. The penes are separated by a goblet-shaped notch. The sperm duct ends in a slender terminal tube behind which externally stands a long, sharp, curving spine directed posteriorly, its length more than half the depth of the notch and its tip surpassing the tip of the sperm duct (see fig. 134).

Specimens in the Cornell collection are from Corvallis, Ore. (H. A. Scullen); Pikes Peak, Colo. (R. J. Gilmore); San Dimas Canyon near Claremont, Calif. (Mrs. T. M. Seemann); and Yosemite Valley, Calif. (F. Schulthess).

Paraleptophlebia praepedita Eaton

1884—Eaton—Revis. Monogr. p. 99 (fig.) . . . . Mass.
1905—Needham—N. Y. St. Mus. Bull. 86: 49 (fig.) . . . . N. Y.
1911—Morgan—Ann. Ent. Soc. Amer. 4: 109.
1919—Morrison—Canad. Ent. 51: 144 (fig.)
1925—McDunnough—Trans. R. S. C. Sect. V, p. 209. . . . . E. Can.
1930—Ide—Canad. Ent. 62: 211.
1933—Gordon—Brooklyn Ent. Soc. Bull. 28: 129 (figs).

Length: body 5; wing 5–6 mm.

This is an all brown species without color pattern, with only varying depths of tint. Top of head and thorax and base of abdomen dark brown, with the sides and the middle of the abdomen and all the appendages paler brown. Fore legs darker than the others; all legs darker at the knees. Wings subhyaline with brown veins; costal cross veins few and faint, those of the stigmatic area short, simple and regular.

Abdomen paler on the middle segments darkening toward both ends. Spiracles indicated by minute and faint brownish dots. The 9th sternite is very long, one-half longer than wide, carrying the forceps far to rearward. The first segment of the forceps is divided by a basal constriction which sets off a sub-segment one-fourth as long as wide. Penes as long as the forceps, separated at half their length by a narrow, V-shaped cleft, flattened horizontally in their apical third and outcurving, and bearing a minute posteriorly-directed triangular tooth at the tip. At two-thirds of their length there arises a pair of reflexed spurs that curve outward and upward to rounded tips. These spurs are about one-third as long as the main segment of the forceps (see fig. 134). Tails nearly uniform pale brown.

This species is rather widely distributed in the eastern United States. Specimens in the Cornell collections are from North Spencer, Harford, and Ithaca, N. Y. (Eva Gordon); Ithaca, N. Y. (Anna H. Morgan); McLean, N. Y. (Mrs. Harold Morrison); Sherburn, Mass. (C. A. Frost); LaGrange, Ind. (E. B. Williamson); and from Walnut Lake and Maple River, Mich. (J. G. Needham).

Paraleptophlebia rufivenosa Eaton

1884—Eaton—Revis. Monogr. p. 99 .... Ore.
1924—McDunnough—Canad. Ent. 56: 221 .... B. C.

Length: body 7; wing 7 mm.

This is a brownish species without distinct color pattern. The head and thorax and tip of the abdomen are dark brown; middle abdominal segments, legs, and tails paler brown. Wings with reddish longitudinal veins and with the cross veins also tinged except in the basal 3rd; most heavily in the stigmatic region, where the cross veins are oblique, rather crowded, and some of them forked. 9th sternite narrowly divided by a V-shaped notch. Forceps tapering rather regularly from base to apex, slightly corrugated internally; the 3rd segment as large as the 2nd. Penes separated for about half their length by a V-shaped notch that is rounded proximally; their outer margins parallel almost to the tip where a triangular tooth projects laterally. There is no reflexed spur and there is no apical tooth (see fig. 134).

We have a single male specimen in the Cornell collection from Grangeville, Idaho (J. M. Aldrich).

Paraleptophlebia sculleni Traver

1934—Traver—J. Elisha Mitchell Sci. Soc. 50: 189. Fig. .... Ore.

Length: body 6½; wing 6½ mm.

This is another brown and white, clear-winged species with black-ringed abdomen. Thorax brown above with paler sutural lines on the sides. Legs pale. Wings hyaline. Costal cross veins obsolete except in the creamy stigmatic area which covers costal and subcostal interspaces. Stigmatic cross veins few, simple, slightly curved.

Abdomen brown on segments 1 and 8 to 10; light brown on 2 and 7; white on 3 to 6, and on the base of 7, with blackish apical half-rings and lateral stripes that are darker at the postero-lateral angles of the segment. Ganglia and venter of segments 7 to 9 yellowish. The basal segment of the forceps tapers suddenly at the base, then very slightly to its tip. The penes are separated by an oval U-shaped notch, longest at the inner angle which projects rearward in a rounded lobe. Lateral to this lobe the apex is obliquely truncate, ending laterally in a minute recurved hook; the reflexed spurs are elongate, sickle-shaped and incurving at the tip (see fig. 133).

The type, a single male specimen from Corvallis, Ore. (H. A. Scullen), is in the Cornell collection.

Paraleptophlebia strigula McDunnough

1932—McDunnough—Canad. Ent. 64: 209. . . . . Ont.

Length: body 7; wing 7 mm.

This species has a pale abdomen shaded with brown; it is close to *P. guttata*. Head and thorax deep blackish, the pleura tinged with brown. Legs whitish. Fore femora tinged with brown throughout; middle and hind femora brown-tinged apically, with the knee brown. Wings hyaline, veins pale. Head of female red-brown, vertex shaded with black. Thorax deep red-brown. Legs pale amber. Longitudinal veins faintly amber-colored.

Abdomen of male whitish hyaline on segments 2–6 and the anterior portion of 7; tergite 2 with considerable dark shading. Apical segments opaque, deep chocolate brown. A short black transverse streak on the postero-lateral corner of segments 2–6, which often has some slight brown suffusion anterior to it, distinguishes this species from *P. guttata*; in the latter, black lateral dots are present instead of the black streaks. Tails and forceps whitish, faintly brown-tinged at the base. Penes very similar to *guttata*, but convergent, not divergent apically; the beak-like lateral projections are somewhat shorter and the ribbon-like reflexed spurs longer. Abdomen of female deep red-brown (see fig. 133). The subanal plate with a deep narrow U-shaped apical excavation.

Type material was taken at Dundas, Ont. by Mr. G. S. Walley, who reared several specimens from the nymphs. A paratype, taken by Mr. W. J. Brown at the same time and place, was kindly given to the Cornell collection by Dr. McDunnough.

Paraleptophlebia swannanoa Traver

1932—Traver—J. Elisha Mitchell Sci. Soc. 47: 113. Fig. . . . . N. C.

Length: body 5; wing 5¼ mm.

This is another brown and white species with all appendages white. It is a little smaller than *P. mollis* and emerges a little later in the season. It differs from that species in having the abdominal spiracles marked with brown dots. The penes are appressed (thus eliminating an intervening cleft), and the thin, lateral lobes at the apex are broader and more extensively up-reared, with the sharp external points wider apart (see fig. 133).

The typical locality was Swannanoa River, N. C. (J. R. Traver). Other specimens in the Cornell collection are from Waynesville, Selica and Davidson River, N. C. (J. R. Traver); Greenville Co., S. C. (H. K. Townes, Jr.); and Timpson Creek, Ga. (P. W. Fattig). Mandibles of the nymph are shown in fig. 136.

Paraleptophlebia temporalis McDunnough

1926—McDunnough—Canad. Ent. 58: 297. Fig. . . . . B. C.

Length: body 6; wing 7–8 mm.

This is another brownish, pale-legged species with diffuse pale basal rings on the middle abdominal segments. Head and thorax brown above, with the sides and the terminal abdominal segments paler brown. Legs pale brown, whitish beyond the knees. Wings subhyaline with pale brown veins. Cross veins nearly obsolete in most of the costal space; single and straight and spaced well apart in the stigmatic region. The brownish rings upon segments 2–7 of the abdomen are very diffuse and ill-defined and only slightly darker on segments 2 and 7 than on the intervening segments. The black of the spiracle is extended rearward in an **R**-shaped line covering the postero-lateral angle. The basal segment of the forceps is tapered to three-fourths its length. There is a wide V-shaped notch between the two penes. The sperm duct ends in a long straight, tubular process on the outer sides of which stands a long spirally twisted spine (see fig. 133).

This species differs from *P. pallipes* in that the above mentioned spine lies in the same plane as the tubular process, not curving upwards as in that species.

The typical locality was Alta Lake, Mons, B. C. (J. McDunnough). In the Cornell collection there is a specimen of this species from Covallis, Ore. (H. A. Scullen).

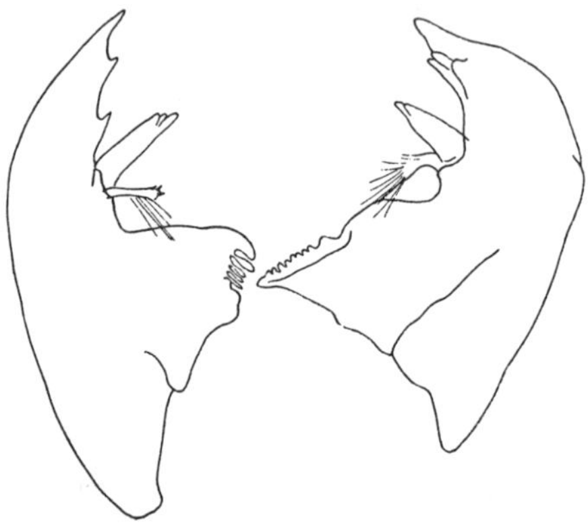

FIG. 136. Left and right mandibles of nymph, of *Paraleptophlebia swannanoa.*

Paraleptophlebia vaciva Eaton

1884—Eaton—Revis. Monogr. p. 97.......... Ore.
1926—McDunnough—Canad. Ent. 58: 296.

Length: body 7½; wing 8 mm.

This is a graceful black and white species with faintly tinted wings. Head, thorax, front coxae, and the end segments of the abdomen are shining brownish black. Fore legs tawny, with touches of brown on trochanter, knees, and both ends of tarsi. Other legs whitish with a touch of brown at the knees and tips of tarsi. Wings subhyaline, with the long veins of the costal strip brown, as are also the widely spaced and regular cross veins of the stigmatic area.

Abdomen strikingly bicolored. Segment 1 black, 2 to 7 translucent white with conspicuous brown marks on the ganglia beneath and with faint mid-dorsal clouds above; these become darker on the 6th segment and spread diffusely over more than half of the 7th. Segments 8 to 10 are blackish above and brownish beneath. Forceps pale brown. Penes long, separated by a narrow cleft, tapering in the basal half, then irregularly cylindric, bearing externally a visor-like backwardly-directed barb half their own length from the tip (see fig. 133). Tails whitish.

The typical locality was Mt. Hood, Ore. We have specimens in the Cornell collection from Crater Lake, Ore. (H. A. Scullen).

Paraleptophlebia volitans McDunnough

1924—McDunnough—Canad. Ent. 56: 95. Fig.......... Que.
1930—Ide—Canad. Ent. 62: 207. Nymph, figs.

Length: body 5; wing 5–6 mm.

A species with pale abdomen marked with purplish brown.

Head and thorax shiny brownish black. Legs dull whitish. Wings hyaline, veins pale. About 16 stigmatic cross veins; simple, slightly curved, not anastomosed. Fore wing narrowly brownish at the extreme base.

Abdomen hyaline white on segments 3–7; tergite 2 dark brown; segments 8–10 dark blackish brown, opaque. Tergites with narrow brown bands on the posterior margins, and irregular purplish brown markings dorsally and laterally on 5 and the apical segments. A series of short, lateral, longitudinal dark dashes above the spiracular line, each on the posterior margin of the tergite. Sternite 2 faintly shaded with brown; the apical sternites lighter brown than the corresponding tergites. Ganglionic patches outlined in ruddy brown. Forceps white. Penes separated for about half their length by a V-shaped notch; rather lyre-shaped in general appearance. The reflexed lobes are short, horse-shoe-shaped at the tip, and not visible from above (see fig. 133).

Type material was taken from the Ottawa Golf Club (R. Ozburn), and from Algonquin Park, Ont. (J. McDunnough). In the Cornell collection are specimens from Joe Lake, Algonquin Pk., Ont. (F. P. Ide); from Cacapon River, W. Va., (J. G. Needham); and from Mossy Creek, Fort Valley, Ga. (P. W. Fattig).

Genus **Blasturus** Eaton

1881—Eaton—Ent. Mo. Mag. 17: 193.
1884—Eaton—Revis. Monogr. p. 100.
1920—Ulmer—Stett. Ent. Zeit. 81: 116.
1932—Traver—J. Elisha Mitchell Sci. Soc. 47: 120.
1933—Spieth—J. N. Y. Ent. Soc. 41: 342.

These are brown or reddish brown or blackish mayflies of rather large size (fore wing 8–12 mm.) with pale middle and hind legs, and darker fore legs, and without femoral color bands (see fig. 12). The knee caps are narrowly cross-lined with black. The wings are subhyaline, sometimes clouded with brown, generally a little darker toward the costa, with numerous irregular and more or less anastomosing cross veins in the stigmatic area. The posterior branch of the radial sector sags heavily to rearward. The front margin of the hind wing is without costal angulation.

The abdomen is lighter in color than the thorax, especially on the middle segments, and in the brown of its dorsum there is a narrow mid-dorsal pale line, with a pair of obliquely placed pale marks at the front of each segment beside it. The paler joinings of the segment often give the abdomen an annulate appearance. The tails are three, as in allied genera, but the middle one is greatly reduced, often becoming but half as long as the others. They are generally ringed with darker color on the joinings of this segment.

The forceps of the male is occasionally four-jointed by the development of an extra little joint at the tip. Like the two normal small apical joints, it may readily become detached. The penes are rather long. They are separated by a more or less U-shaped notch. They have a weak little acuminate spine projecting from the tip of each, and they bear a large flattened and curved reflexed spur that reaches well down to the base of the middle notch. The subanal plate of the female is prolonged and deeply emarginate.

The subimago is lighter in body color and darker in wings than the imago. On the top of the brown thorax is a middorsal pale line narrow in front, suddenly widened to an I-shaped mark along the middle prominences and entirely surrounded by another pale line in the oval notal furrow. The crest of the scutellum also is broadly pale, and there are pale streaks across the metanotum. Middorsal pale line and the oblique pair of darker and some lateral pale markings are always present on the abdomen. The wings are smoky brown. The tails are much shorter than in the imago, but the middle tails less reduced in proportionate length.

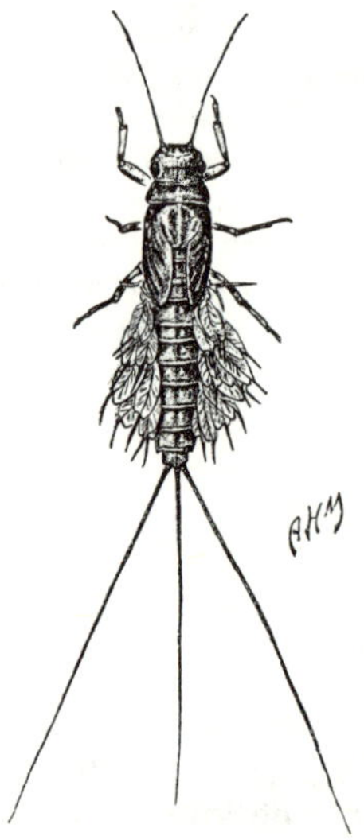

FIG. 137. Nymph of *Blasturus cupidus*. (From Morgan.)

Nymphs of Blasturus (see fig. 137) are chiefly distinguished by the breadth of the tracheal gills and by the form of the emargination of their tips. They are double and lamelliform on all seven segments, narrowly so on segment one. On segments 1 and 7 their margins are entire; on the intervening segments they are deeply notched on one or both sides of a long tail-like terminal filament. The species differ in the depth of these notches and in the angulation of the tips of the adjacent lobes. They differ also in the relative size of the lateral spines on abdominal segments 8 and 9. The claws in this genus bear beside the usual inferior row of denticles, here very numerous and small, an upper, more lateral, shorter row of fewer denticles, that vary in number with the species. Mouthparts and other appendages are much as in Leptophlebia, the middle tail hardly differing from the lateral ones in the nymph.

Spieth (1933) figures the fore wing of some Ephemerine genus, probably Hexagenia, (fig. 31), as that of Blasturus. His statements regarding the venation of the fore wing of Blasturus, very evidently based on this figure, are therefore incorrect.

Genotype—*B. cupidus* (in Ephemera), Say.

Distribution—Nearctic; generally distributed.

KEY TO THE SPECIES

Adults

1—Outer ⅓ to ⅔ of tip of fore wing with a distinct brown cloud or irregular oblique band 2
—Outer portion of fore wing with no such dark marking 4

2—A dark oblique band across outer portion of fore wing, leaving outer margin pale . . . . . **vibex**
—Outer ⅓ to ⅔ of tip of fore wing covered by a brownish cloud which extends fully to the outer margin 3
3—Wing 10–11 mm. in length; cloud extends across outer ⅓ of wing tip . . . . **nebulosus**
—Wing 13 mm. in length; cloud extends across outer ⅔ of wing tip, also outer portion of hind wing . . . . . **pacificus**
4—A brown stain in stigmatic area of fore wing 5
—Stigmatic area of fore wing not noticeably darker than adjacent areas 6
5—Sternite 9 distinctly dark brown; tails whitish to light tan . . . . . **cupidus**
—Sternite 9 not darker than preceding segments; tails brown . . . . . **gravastellus**
6—Penes of the "scarf" type (see fig. 138, *grandis*); appendages extend backward to base of notch between divisions of penes 7
—Penes of the "hooded" type (see fig. 138, *cupidus*); appendages do not attain the notch between divisions of penes 8
7—Large species, wing 10–12½ mm. in length; fore femur yellow . . . . . **grandis**
—Smaller species, wing 7½–9 mm. in length; fore femur orange to reddish brown . . . . . **intermedius**
8—Middle tail almost ⅔ as long as the laterals; fore femur yellowish to pale brown, tails whitish . . . . . **austrinus**
—Middle tail not more than ⅓ as long as the laterals; fore femur red-brown, tails red-brown . . . . . **collinus**

VERIFICATION TABLE

Male Adults

Species	*Wing length*	*Cloud on wing*	*Fore femur*	*Tails*	*Tail joinings*	*Middle** tail*	*Distr.*
austrinus	8–10½	None	Yel. br.	whitish	Dk. br.	⅔	SE
collinus	10	Wing roots	Rd. br.	Rd. br.	Dk. br.	⅓	SE
cupidus	10–12	On stigma	Rd. br.	Lt. tan	Obscure	⅓	E
grandis	10–12½	None	Yel.	Whitish	Rd. br.	½	N. C.
gravastellus	8–12	On stigma	Brown	Brown	Obscure	⅓ to ½	W
intermedius	7½–9	None	Or.br.	Brownish	Dk. br.	⅔	N. C.
nebulosus	10–11	⅓ of tip	Pale br.	Greyish	Dk. br.	¼	E
pacificus	13	⅔ of tip	Dk. br.	Brown	Obscure	¼	Ore.
vibex	10–12	Streak	Rd.br.	Brown	Black	⅓	Ore.

KEY TO THE SPECIES*

Nymphs

1—Legs distinctly banded 2
—Legs not banded, or with indistinct bands 4
2—Body 9–16 mm. in length; 6 to 7 large pectinate spines below crown of maxilla . . . 3
—Body 7–9 mm. in length; 4 or 5 large pectinate spines below crown of maxilla . . . . . **intermedius**
3—Medium-sized species, body 10–12 mm. in length; black markings on pronotum . . . . . **austrinus**

** In proportion to length of laterals.
* Does not include *pacificus;* most of the gills are wanting, in our specimens of this species.

—Large species, body 12–16 mm. in length; no black markings on pronotum. **grandis**

4—Upper portion of third pair of gills with median extension of main trachea as long as or longer than gill plate proper; lateral lobes on lower portion unequal in length. . . . . **nebulosus**

—Upper portion of third pair of gills with median extension of main trachea shorter than gill plate proper; lateral lobes on lower portion variable. 5

5—Lateral lobes at tip of lower gill of third pair unequal in length; 5 to 6 large pectinate spines below crown of maxilla. . . . . **collinus**

—Lateral lobes at tip of lower gill of third pair approximately equal in length; 6 to 7 large pectinate spines below crown of maxilla. 6

6—Postero-lateral spines on abdominal segments 8 and 9 about equal in length, the latter extending beyond center of 10th segment. . . . . **austrinus**

—Postero-lateral spine on segment 8 shorter than that on 9, the latter rather blunt, not extending to center of 10th segment. . . . . **cupidus**

VERIFICATION TABLE

Nymphs

Species	*Body length*	*Pect. sp.*[1] *on max.*	*Legs banded*	*Spine on 8 vs. 9*
austrinus	10–12	6–7	Var.	Subequal
collinus	8½–10½	5–6	No	Subequal
cupidus	10–11	6–7	No	Shorter
grandis	12–16	6–7	Yes	Shorter
intermedius	7–9	4–5	Yes	Shorter
nebulosus	10–11	6–7	No	Shorter
pacificus	15	5–6	No	Shorter

Blasturus austrinus Traver

1932—Traver—J. Elisha Mitchell Sci. Soc. 47: 129. Figs. . . . . N. C.

Length: body 9–11; wing 8–11 mm.

This is a shining brown species with pale legs and conspicuously ringed tails. Antennae brown, ringed with yellow at the base. Thorax above shining brown with a yellow streak in the notal furrow on each side. Wings hyaline with brown veins and scarcely any darkening of the stigmatic area. Legs yellow to pale yellowish brown, a little darker on the front of the femur and with narrow brown cross marks on both ends of it. Abdomen yellowish brown, paler beneath, darker at the ends above, and with pale rings on the joinings of the middle segments. These become obscured in old specimens. Mid-dorsal pale line evident, interrupted on the front half of each tergite. Paired oblique marks short, subtriangular, more or less obscured. Genitalia pale brown in ground color. Tails whitish, joinings rich dark brown.

The type material, from the vicinity of Greensboro, N. C. (J. R. Traver), is in the Cornell collection. This species is smaller than *B. grandis*, and much resembles *B. cupidus*. It has, however, more conspicuously ringed tails, almost no darkening of the stigmatic area, and the 9th sternite in the male is not conspicuously darker than preceding segments.

[1] Number of pectinate spines below crown of maxilla.

Blasturus collinus Traver

1932—Traver—J. Elisha Mitchell Sci. Soc. 47: 124. Fig. .... N.C.

Length: body 10; wing 9½–10 mm.

Besides the male subimago from which the original description was drawn, a male imago and a male subimago are now known, from South Carolina. The nymph, male subimago and female imago are fully described and illustrated in the paper cited above. The following characters of the male imago are drawn from the holotype—a subimago old enough so that the subimaginal cuticle is partly shed and the mature coloration is well advanced—and supplemented by reference to the South Carolina male.

The thorax is very dark brown above, dark reddish brown below, and somewhat lighter on the sides. The brown pronotum is paler on its hinder margin. The legs are nearly uniform yellowish to reddish brown. Wings hyaline, with the long veins of the costal strip brown, darkened toward the wing roots, and with the faintly brown irregular cross veins of the stigmatic area extensively conjoined.

Abdomen nearly uniform brown above, hardly paler on the middle segments, but distinctly paler beneath. Dorsum without median pale line, but with narrow double mid-dorsal black line; the paired oblique marks are almost obsolete. The spiracles are distinctly marked with black. Tails dark reddish brown, distinctly ringed with darker brown on the joinings of the segments, the rings alternately wider and narrower on the basal third. Genitalia yellowish.

The type material was collected near Highlands, N. C. (J. G. Needham); these, with other specimens from Greenville Co., S. C. (H. K. Townes, Jr.), are in the Cornell collection.

Blasturus cupidus Say

1823—Say—West. Quart. Rep. ii; p. 163 (Ephemera) .... Ohio
1853—Walker—List Neurop. Ins. Brit. Mus. p. 538, 553 (*hebes, concinnus, pallipes*) .... N. S., N. F.
1859—LeConte—Complete Writings of T. Say, p. 172.
1861—Hagen—Neurop. Ins. N. Amer. p. 47: 51 (Potamanthus).
1862—Walsh—Proc. Acad. Nat. Sci. Phil. p. 372 (Potamanthus) .... Ill.
1863—Hagen—Proc. Ent. Soc. Phil. p. 172 (Potamanthus).
1871—Eaton—Trans. Ent. Soc. London p. 89 (fig.; Leptophlebia).
1872—Packard—Guide to study of insects; p. 595 (Potamanthus).
1881—Eaton—Ent. Mo. Mag. 17: 193 (*concinnus*).
1884—Eaton—Revis. Monogr. p. 101 (fig.) .... Canada, N. Y., N. H., N. C.
1889—Packard—Psyche 5: 168 (Ephemera) .... R. I.
1903—Berry—Amer. Nat. 37: 25 (nymph fig.) .... N. J.
1905—Needham—N. Y. St. Mus. Bull. 86: 40 (fig. gills).
1907—Needham—N. Y. St. Mus. Bull. 124: 189.
1911—Morgan—Ann. Ent. Soc. Amer. 4: 107.
1913—Clemens—Canad. Ent. 45: 333 .... Ont.
1925—Traver—Canad. Ent. 57: 211.
1925—McDunnough—Trans. R. S. C. p. 210 .... Que.
1930—Neave—Ecol. 6: 568 .... Man.
1930—Ide—Canad. Ent. 62: 211.
1930—McDunnough—Canad. Ent. 62: 54.
1932—McDunnough—Canad. Ent. 64: 78.
1932—Traver—J. Elisha Mitchell Sci. Soc. 47: 120, 121, 122 (figs.)

Length: body 9–11; wing 10–12 mm.

This is a blackish species with subhyaline wings that are opaque and stained with brown in the stigmatic area. Antennae wholly brown. Legs brown, darker on the first femora with knee caps narrowly black. Wings hyaline with brown veins that are distinctly darkened in the stigmatic area where the cross veins are irregular and somewhat anastomosing.

Abdomen dark brown above becoming blackish toward the ends with the apex of the tenth segment yellowish. Minute black elongate dashes mark the spiracles at the side margin. Under parts fawn color with the ninth sternum distinctly darker brown. The pale mid-dorsal line of the abdomen is obscure or wanting. The paired oblique spots are more or less crescentic and suffused with blackish in old specimens.

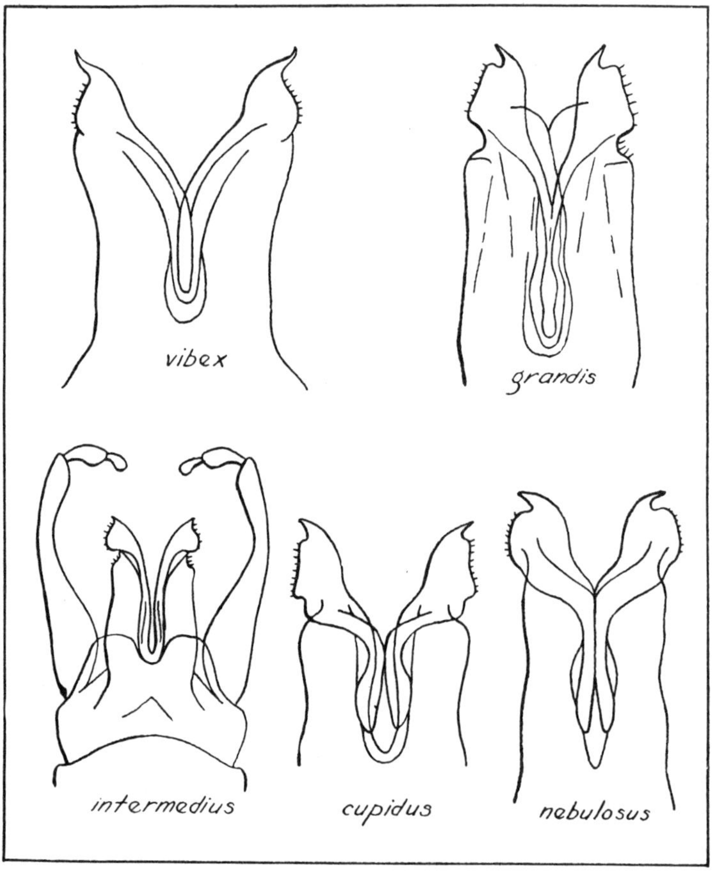

FIG. 138. Male genitalia of five species of Blasturus.

The tails are obscurely ringed with brownish bands on the joinings of the segments; for a distance beyond the base the rings are alternately wide and narrow. Penes as in fig. 138.

This is a variable species of wide distribution in the northern and eastern parts of the United States. We have reared specimens in the Cornell collection which Dr. Osgood Smith obtained at Catatonk Creek near Candor, N. Y. Other specimens are from Willoughby, Ohio, (J. R. Traver); these correspond well with Say's original description. Many reared specimens also from the vicinity of Ithaca, N. Y., and a fine series of imagos from Amherst, Mass. (Inez Williams, C. P. Alexander).

Blasturus grandis Traver

1932—Traver—J. Elisha Mitchell Sci. Soc. 47: 127 .... N. C.

Length: body 11–13; wing 10–13 mm.

This is a rather large species of lighter brown color, alternately yellowish brown. Thorax above shining brown with yellow streaks in the notal furrow each side. Legs dull yellow with the tarsi a little paler and the usual lines on the knee caps very narrow. Wings yellowish in front and toward the outer margin; hyaline at the center and toward the rear; hind wings white.

Abdomen pale yellowish brown, paler beneath especially in the middle of the segments. Segments 1 to 6 have the mid-dorsal pale line almost entire, and the paired oblique marks confluent with the pale rings at the base of the segments. The latter give the abdomen a distinctly ringed appearance. The dorsum of 7 to 10 brown, practically without color pattern. Tails white, distinctly ringed with reddish brown. Genitalia as in fig. 138.

This species is larger and paler than others of the genus; the "scarf" type of penes is found only in this species and in *B. intermedius*. The type material, from the vicinity of Greensboro, N. C. (J. R. Traver) is in the Cornell collection.

Blasturus gravastellus Eaton

1884—Eaton—Revis. Monogr. p. 102 .... Mont.

Length: body 8–10; wing 8–12 mm.

This is a blackish-brown-legged species with subhyaline yellowish veined wings; antennae brown. Thorax black above and below with paler streaks in the lateral sutures. Legs wholly brown, especially the fore legs. Wings with brownish veins except in the hyaline basal portion; the regular cross veins of the stigmatic area somewhat opaque.

Abdomen brown, paler beneath, the color of the under surface being fairly uniform from end to end; on the dorsal side the middle segments are of a lighter brown with distinct rings of paler color on their joinings. The mid-dorsal pale line is very faint, but broadly dilated at the hind margin of the middle segments. The paired oblique spots are obscure, linear, confluent in front with the pale intersegmental rings. Genitalia of the male yellowish, darker toward the apex. Tails brown, ringed obscurely with somewhat darker brown, and the rings of nearly uniform appearance.

This is a Rocky Mountain species very similar to *cupidus*. It lacks the brown color on the venter of the ninth abdominal segment. The stigma is less opaque and the rings of the tails are more even.

We have seen three specimens of males from the collection of the Colorado Agricultural College, bearing the label "Hayden, Colorado, June 13, '07."

Blasturus intermedius Traver

1932—Traver—J. Elisha Mitchell Sci. Soc. 47: 136. Figs .... N. C.

Length: body 9; wing 8 mm.

A rather small species, distinguished from all others except *B. grandis* by the distinctive "scarf" type of penes.

Frontal margin of head red-brown; vertex and notum of thorax blackish. Sternum deep reddish black. Pleura dark red-brown with black markings. Fore leg red-brown (orange, in recently transformed specimens), tarsus slightly paler than femur and tibia. Venation orange to reddish brown; no stain at wing roots or in stigmatic area. Abdominal tergites dark red-brown, sternites more yellowish. Ganglionic areas orange; sternite 9 reddish brown. Usual pale oblique submedian streaks on tergites, dark markings at spiracles, and faint dark oblique submedian streaks on sternites. Middle tail fully ⅔ the length of the laterals. Tails dark red-brown, joinings obscurely darker. Penes much as in *B. grandis*, with the appendages extending backwards almost to end of notch (see fig. 138).

In addition to the type material, taken near Buies Creek, N. C. (J. R. Traver), we have in the Cornell collection several male imagos reared from nymphs. These were taken from Turnbull Swamp, Braden Co., N. C. (J. G. Needham), in March, 1934.

Blasturus nebulosus Walker

1853—Walker—List Neurop. Ins. Brit. Mus. p. 554 (Palingenia)....................... Hudson Bay
1861—Hagen—Syn. Neurop. Ins. N. Amer. p. 52 (Potamanthus).
1862—Walsh—Proc. Acad. Nat. Sci. Phil. p. 372 (*Potamanthus odonatus*)........................ Ill.
1863—Walsh—Proc. Ent. Soc. Phil. p. 193, 194 (Potamanthus).
1863—Hagen—Proc. Ent. Soc. Phil. p. 171 (Potamanthus).
1871—Eaton—Trans. Ent. Soc. London, p. 89 (Leptophlebia).
1884—Eaton—Revis. Monogr. p. 103.
1913—Clemens—Canad. Ent. 45: 333.. Ont.
1930—Ide—Canad. Ent. 62: 211.
1932—Traver—J. Elisha Mitchell Sci. Soc. 47: 120, 121, 122 (fig.).

Length: body 10–12; wing 10–11 mm.

This is a brownish black species with ringed abdomen and a pale brown cloud covering the tip of the fore wing. Face brown including antennae.

Thorax shining brown above and below, with somewhat paler sides, that area traversed by pale sutures. Fore legs reddish brown. Middle and hind legs yellowish. Wings hyaline with brown veins, and with a brownish cloud covering the apical third of the fore wing. Its straight inner margin lies one or two cells beyond the posterior fork of the radial sector. Costal cross veins faint except in the stigmatic area where they are crowded and irregular and slightly anastomosing. Hind wings wholly hyaline.

Abdomen brown, paler beneath, a little darker on segments 1 and 2 and again on 8 and 9; 10 a little paler than 9. Middle pale line very fine and partly interrupted. Oblique marks crescentic and isolated. Joinings of the middle segments whitish, giving the abdomen an obscurely ringed appearance. Tails pale brown, ringed with dark brown on the joinings. Middle tail one-fourth as long as the others.

Genitalia yellowish amber. Penes separated by a goblet-shaped cleft. Reflexed spur strongly bent and its dilated tip extending below the level of the cleft between the penes. Lobes of the forceps base separated by a sharply V-shaped cleft and subtruncate on their apices (see fig. 135).

This species seems to differ from *cupidus* by the apical brown cloud on the fore wings and by the more nearly uniform coloration of the abdominal segments. The alternating darker and lighter rings on the tails are very pronounced. The depth of color of the apical cloud of the fore wing varies markedly in different specimens. In those which have been in alcohol for several years, the brown cloud may be almost obliterated.

The more typical of the specimens we have seen were collected at Bar Harbor, Me. (A. E. Brower). The westernmost specimens are from Colorado. Two males from the collection of the Colorado State College of Agriculture were collected at Hayden, Colo. Another male was collected by Dr. Elsie B. Klots at Moose, Wyo. We have specimens from a number of New York localities. Several males with darker brown tails, but otherwise similar, were taken at Williamsburg, Va., from Matoaka Lake on the campus of the College of William and Mary (Minnie M. Shreve, Elizabeth L. Purnell, Ellis M. Schwab).

Blasturus pacificus McDunnough

1933—McDumough—Canad. Ent. 65:155. Fig. .... Ore.

Length: body 13; wing 13 mm.

This is a rather large blackish species with brown wings. Head and thorax above shining black; sides of the latter paler in the sutures. Fore legs nearly uniform dark brown; middle and hind legs pale brown. Wings brown at the roots, and with a brown cloud covering more than the apical half of the fore wings and about half of the hind wings. The inner margin of the brown in the fore wings is angulated at the outer fork, behind which it is abruptly widened. The cross veins in the clear area before it are pale brown, even in the costal strip. In the stigmatic area they are numerous, irregular, and in the middle portion, anastomosing at midway their height. Costal and subcostal cross veins in the hind wing are about equal in number. All longitudinal veins are brown.

Abdomen brown, darker on both ends, hardly paler beneath but distinctly paler on the joinings of the middle segments, giving this part a ringed appearance. Tails nearly uniform brown. Middle tail one-fourth as long as the others.

Genitalia yellowish brown, of the usual form with small apical tooth of the penes directed laterally, and the long arcuate reflexed spur as long as the cleft between the penes is deep.

The male subimago has shorter tails (17 mm.) with the middle one less reduced (8 mm.), and with little indication of the basal hyaline area of the wings, these being brown, and only a little paler toward the anal angle. The brown of the body is paler, with the scutellum mostly yellowish and with yellow lines on most of the sutures of the mesothoracic dorsum. On each abdominal segment is a pair of basal oblique dashes, plainest toward basal segments, with a faint pale median longitudinal line between them. The female imago is similar to the male.

In the Cornell collection are several specimens, including one male imago, several subimagos and several partly transformed nymphs, from Corvallis, Ore. (M. F. Canova). Another male imago is also from Corvallis, Ore. (H. A. Scullen).

Blasturus vibex Traver

1934—Traver—J. Elisha Mitchell Sci. Soc. 50: 197.. Ore.

Length: body 13; wing 10–12 mm.

This is a dark brown species with an oblique streak of brown across the fore wing. Face yellowish with brown sutures. Antennae brown. Top of head blackish.

Thorax dark brown with pale streaks in the lateral sutures. Fore legs brown, the femora a little reddish. Middle and hind legs yellowish brown. Wings hyaline with brown veins and with an oblique streak that overspreads the stigmatic area and, narrowing to rearward, extends across the base of the bisector of the outer fork almost to the hind margin. Costal cross veins faint except in the stigmatic area, where they are crowded and irregular, a few of them forking or anastomosing. Hind wing hyaline.

Abdomen brown, darker on the ends, with obscure pale lines on the lateral margins and on the joinings of the middle segments. Tails brownish, narrowly and regularly ringed with black; middle tail a third as long as the others. Genitalia yellowish brown. Apical tooth of penes acuminately pointed and inclined obliquely to rearward. Reflexed spur rather slender, its tip barely reaching to the level of the bottom of the cleft between the penes. The latter is deeper than the cleft between the obliquely triangular lobes of the 9th sternite (see fig. 138.)

A female tentatively placed in this species is similar to the male but a little paler on fore legs and tails, and lacks the brown streak across the fore wings, having only the stigmatic area a little brownish. The subanal plate slightly surpasses the tenth segment and ends in two broadly rounded lobes separated by a similarly rounded apical notch.

In the Cornell collection are two male imagos collected at Corvallis, Ore. (H. A. Scullen), and one female imago from the same locality (J. Hudson). Dr. McDunnough considers this a pale form of *B. pacificus*, rather than a distinct species.

Genus **Choroterpes** Eaton

1881—Eaton—Ent. Mo. Mag. 17: 194.
1884—Eaton—Revis. Monogr. p. 104.
1920—Ulmer—Stett. Ent. Zeit. 81: 116.
1933—Spieth—J. N. Y. Ent. Soc. 41: 342 (Thraulus).

These are small mayflies (wing 5 to 7 mm.) with blackish or brownish bodies and brownish or reddish wing roots. The upper eyes are very

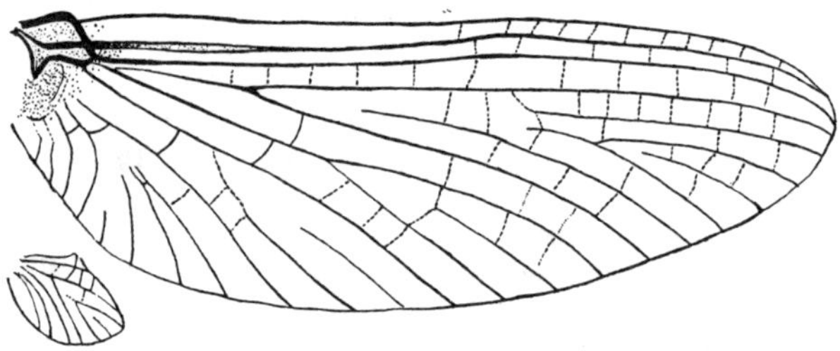

FIG. 139. Wings of *Choroterpes basalis.*

broad. The tarsal claws are dissimilar. The longitudinal wing veins are brownish toward the front, those toward the rear are indistinct and wellnigh invisible as are all the cross veins. The pale cross veins in the stigmatic area are few and distant and simple. In the anal angle of the fore wing between veins Cu_1 and Cu_2 are four intercalaries of which 1 and 3 are long and 2 and 4 are short; they are variously joined by cross veinlets. The costal margin of the hind wing is produced near the middle in a bold prominence, obtusely rounded at its tip, and the subcostal vein reaches the front margin just beyond this hump (see fig. 139.)

The abdomen is greyish or blackish brown, usually with a narrow median pale line, and with variable hair-lines of black on the lateral margin just below the spiracles, often with paler paired light spots or streaks between. The venter is pale. An undivided forceps base is set off from the sternum of the 9th segment by a deep transverse groove. Forceps three-jointed, the long joint greatly widened at the base, the two end joints short, subequal, the sutures separating them incomplete on the inner side. Penes long, simple, lacking reflexed spurs and apical processes, appressed or slightly separated by a narrow V-shaped cleft. Tails three, subequal, hardly longer than the body.

The female is similar, a little paler, and with less differentiation in color between thorax and abdomen. The 9th abdominal sternite is produced into an ovate lobe that is only a little emarginate at its tip.

The subimagos are similar, with a better development of color pattern on the thoracic dorsum, and with uniformly pale greyish wings.

The nymph is depressed of body, as befits the lotic habitat. The antennae are rather small, hardly longer than the head is wide. The labrum is widened forward and deeply emarginate on its front border. The maxillae are wider distally than in Blasturus, being squarish beyond the muscle attachment level. The pectinate spines on the lacinial border number 12 to 16, with an especially prominent one at the end of the row. The glossae of the labium are small, closely appressed and end in a transverse row of stiff bristles.

The gills are double (see fig. 65), except that of segment 1, which is single and linear-lanceolate. They are rather broadly lamelliform, and the two leaves of each gill are unlike in the extent and depth of the emarginations at the tip. The tail-like apical portion beyond the lateral excisions is more or less lanceolate in form according to the species. The tarsal claws are armed beneath with a single row of more than a dozen minute blunt denticles.

Spieth (1933) figures the wings and genitalia of this genus (figs. 29, 30 and 63) as Thraulus. It is very probable that his figures of the nymphal structures of Choroterpes are correct; the parts figured are very similar in both genera.

Genotype—*C. picteti* Eaton.

Distribution—Holarctic, Neotropical, Indo-australian. In Nearctic, quite generally distributed.

KEY TO THE SPECIES

Male adults

1—Abdomen whitish; no color on wing membrane . . . . **nanita**

—Abdomen grey to greyish brown above, paler ventrally; a distinct brownish stain on wing membrane, at least near the base 2

2—Reddish brown stain covers proximal half of fore wing and most of hind wing; tails whitish, joinings darker at base only 3

—Wings brown-stained at extreme base only; tails grey or brown, all joinings darker 4

3—Thorax reddish brown; longitudinal veins brown at base only, paler beyond . . . . **ferruginea**

—Thorax blackish; longitudinal veins brown for entire length . . . . **basalis**

4—Swelling at base of long joints of forceps large and bulbous; wing 5–6½ mm. in length........5
—Swelling at base of long joint of forceps tapering gradually; wing 7–9 mm. in length........6
5—Sternite 7–9 largely white; abdomen appears distinctly white-ringed **albiannulata**
—Sternite 9 entirely brown, a brown ganglionic mark on 8; abdomen less dinstinctly ringed........**terratoma**
6—Fore femur pitch-brown; abdominal tergites paler at joinings and on median line, but with no distinct pale pattern........**inornata**
—Fore femur yellowish in part; abdominal tergites 2–9 with distinct pattern of pale spots........**oklahoma**

VERIFICATION TABLE

Male adults

Species	*Wing length*	*Stain on membrane*	*Abd. terg. 2 to 6*	*Tails*	*Base of forceps*	*Distr.*
albiannulata	5–6	Limited	Grey	Brown	Bulbous	Alta.
basalis	5–6	Extensive	Grey-brown	White	Tapered	E
ferruginea	6	Extensive	Grey-brown	White	Tapered	E
inornata	7–9	Limited	Dark grey	Grey	Tapered	Ariz.
nanita	7	Absent	Whitish	White	Tapered	Tex.
oklahoma	7½	Limited	Grey-brown	Grey	Tapered	Okla.
terratoma	6½	Limited	Grey-brown	Brown	Bulbous	Calif.

Choroterpes albiannulata McDunnough

1924—McDunnough—Canad. Ent. 56: 221........Alta.

Length: body 6; wing 5–6 mm.

This is a blackish species with white-ringed abdomen. Head blackish above with antennal flagellum pale. Thorax shining black above and below, paler around wing and leg bases; the mesepisternal area with coppery reflections. Legs dull grey, almost sooty black on the front femora of the male. Tarsi paler. Wings hyaline, a little greyish at the stigma, with hardly a suggestion of darkening at the extreme base. Veins pale brown, darker from the radius forward. All cross veins faint, those of the stigmatic area simple, erect, and distant.

Abdomen brownish at the ends, grey on segments 2 to 6, paler beneath. The joinings of the middle segments are narrowly white above and below, giving the abdomen a distinctly ringed appearance. There is a row of diffuse blackish dots on the abdominal spiracles. Forceps, penes, and tails greyish, the last narrowly ringed with brown on the joinings of the segments.

Described from a specimen kindly given to the Cornell collection by Dr. J. D. McDunnough, from Lethbridge, Ont. (J. H. Pepper).

This species is near *C. inornata*, but slightly smaller, with distinctly less color in the wings.

Choroterpes basalis Banks

1900—Banks—Trans. Ent. Soc. Amer. 26: 248 (Leptophlebia)........Canada, Pa.
1905—Needham—N. Y. St. Mus. Bull. 86: 38 (nymph)........N. Y.
1911—Morgan—Ann. Ent. Soc. Amer. 4: 109.
1913—Clemens—Canad. Ent. 45: 334........Ont.

Length: body 4–5½; wing 5–6 mm.

This is a blackish species with very dark grey-brown abdomen and a deep brown stain on the fore wing bases, extending well outward along the subcosto-radial inter-space. Head brown above, including the antennae. Thorax blackish above and below, brown at the sides, with obscurely pale sutures. Legs tawny yellow, with submedian blackish rings on the femora, and with a wider mark of the same color on the knees. Tarsi whitish. Wings heavily stained with brown at base, the membranes being lightly tinged over the proximal half of the fore wing and nearly the whole of the hind wing. The longitudinal veins are darker than in the other species.

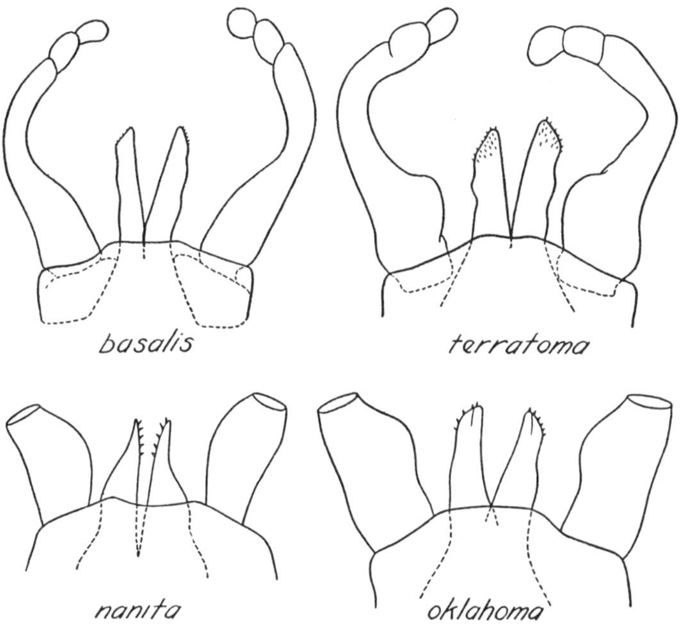

FIG. 140. Male genitalia of four species of Choroterpes.

Abdomen dark greyish brown or blackish above, pale beneath. In fully mature males of this species, the pale markings of the abdomen are well nigh obliterated; the middorsal line is very narrow and two pairs of spots are very faint. The black hair-line on the lateral margin is interrupted only on the joinings of the segments. In old specimens there tends to appear a parallel, more obscure black line close above it and the two are sometimes conjoined on each segment before the spiracle. The ventral surface is pale with a large blackish U-mark covering the margins of the 9th sternite except in the rear. Forceps white beyond their enlarged base. Penes brownish (see fig. 140). Tails white.

The type material was from Sherbrooke, Canada, and Montgomery Co., Pa. In the Cornell collection are specimens from Ithaca, N. Y., among which are several reared from nymphs; other reared specimens are from LeRoy and Kirtland, O. (J. R. Traver).

Choroterpes ferruginea Traver

1907—Needham—N. Y. St. Mus. Bull. 124: 192 (*basalis*)................................ N. Y.
1925—Needham—Ent. News 36: 113 (*basalis*).
1934—Traver—J. Elisha Mitchell Sci. Soc. 50: 205..N. Y.

Length: body 5½; wing 6 mm.

This is a reddish brown species with grey-brown abdomen and reddish wing bases. The head is brown above with paler antennae. Thorax nearly uniform reddish brown. Legs yellowish, becoming whitish on the tarsi; a distinct brown spot on the middle of the femur and a touch of the same color on the knee. Wings subhyaline to whitish, with a diffuse reddish brown tint covering a considerable area of the fore wing and most of the hind wing. Longitudinal veins near the costal margin brown, elsewhere invisible, as are all the cross veins.

Abdomen grey-brown, becoming darker and reddish at the ends; pale translucent beneath except on the end segments. A very narrow mid-dorsal pale line that becomes obsolete at the rear is overspread on segment 4 and 5 by broad pale middorsal triangles; smaller triangles in the same position on 3 and 6. A continuous black line follows the lateral margin on segments 2 to 8, with a parallel interrupted pale line in the spiracular area immediately above it. Venter of 9 reddish at the sides. Forceps pale. Penes reddish. Tails white, sometimes faintly ringed with reddish brown toward the base; often wholly pale.

The type material was collected at Wintergreen Point, Old Forge, N. Y. (J. G. Needham). Other specimens in the Cornell collection are from Commissioner's Island and Juanita Island, Lake George, N. Y.

Choroterpes inornata Eaton

1892—Eaton—Biol. Centr. Amer. Neuropt. p. 6. Figs.................................Mex., Ariz.

Length: body 6–9; wing 7–9 mm.

This is a blackish species with brown wing-roots and grey abdomen. Head blackish above with the tips of the antennae paler. Prothorax rufescent brown, very broad and with a broad quadrangular middorsal excavation of the hind lobe. Synthorax shining black, paler before the wing roots and around the leg bases. Legs all brown, darker on the front femora of the male. Wings subhyaline, with a darker costal strip due to the dark brown coloration of the first three longitudinal veins. The membrane is brown at the base, dark brown in the humeral cell, and faint for a distance beyond it. Cross veins very faint, even in the stigmatic area.

Abdomen dark grey, darker and brownish on the end segments, dirty white beneath. The paler middle segments are narrowly white across their joinings, and a very faint pale hair-line traverses them middorsally. There is a row of minute blackish marks on the spiracles.

Described from a N. Sonoran specimen from the type collection, kindly loaned by Dr. J. D. McDunnough.

Choroterpes nanita Traver

1934—Traver—J. Elisha Mitchell Sci. Soc. 50: 203. Fig.....................................Tex.

Length: body 7; wing 7 mm.

This is a dainty brown species with sooty blackish, sinuous cross bands on the dorsum of the whitish abdomen and with ringed tails. Head brown above, paler in front, antennae brown with paler basal segments.

Thorax brown, paler around the leg bases and in the rear of the mesoscutellar hump, and with interrupted hair lines of black on the lower margin of the pleural sclerites. Legs white with only hair lines of black on the anterior face of the femora. Wings white with only a few of the stronger veins brown (especially vein R and its basal connections), there being no color in the wing membrane. Cross veins hyaline and almost invisible even in the stigmatic area.

Abdomen whitish with a double row of broad, diffuse cross-laid ?-shaped sooty black markings separated by a wide middorsal pale line. Segments 9 and 10 darker, especially the side of the 9th sternite. Tails white, distinctly ringed with black, alternate segments more narrowly. Forceps white, penes brown (see fig. 140).

Subimago paler, with faint, smoky-greyish wings and yellowish crown. The rich brown thorax shows a middorsal yellow stripe, narrowed to a hair line at the front, and spear-pointed behind. Conjoined with this toward the rear is an oval line of yellow in the notal furrow, below which the brown of the thorax is darker—almost black—before the wing roots.

The type material, one adult male and one male subimago, taken at Austin, Texas (H. J. Parks, Jr.), is in the Cornell collection.

Choroterpes oklahoma Traver

1934—Traver—J. Elisha Mitchell Sci. Soc. 50: 204. Fig. . . . . Okla.

Length: body 7; wing 7½ mm.

This is a very dark brown species, with grey-brown mottled abdomen. Head dark brown above with paler brown antennae. Thorax shining blackish brown above, with a very narrow pale line in the notal furrow, and a wider pale strip just beneath it before the wing roots. Sides and sterna also dark brown. Legs brown and dull yellow, with whitish tarsi. The brown about equals the yellow in extent, is very diffuse, showing little cross-banding, but is darker on the longitudinal carinae. Wings subhyaline, darker in the subcostal space, and with brown longitudinal veins in the costal half, behind which they are transparent, as are all cross veins.

Abdomen greyish brown, paler beneath, darker on the end segments, and with a pretty pattern on segments 2 to 9 of large pale middorsal V marks through which runs the narrow middorsal pale line conjoining them. On the lateral margin of segments 2 to 8 there is a minute white spot surrounded in front by a blackish crescent. Tails greyish, the segments paler toward the base of each, obscurely darkened on the joinings, and so, faintly ringed. Sides of the 9th sternite marked with an elongate brown triangle; remainder white, including the forceps. Penes brown (see fig. 140).

The type material, two males from Murray County, Okla. (C. Sandoza), are in the Cornell collection.

Choroterpes terratoma Seemann

1927—Seemann—Pomona Jour. Ent. and Zool. 19: 46. . . . . Calif.

Length: body 6½; wing 6½ mm.

This is a blackish species with greyish abdomen. Head black above and in front, including basal segments of antennae; flagellum of antennae grey.

Thorax dull black, paler in the sutures, narrowly so in the notal furrow. Legs black and yellow, more black than yellow, except on tibiae of middle and hind legs, the usual middle and apical bands on the femora tending to run together. Wings subhyaline with the longitudinal veins dark brown except at the anal angle, where they are transparent, as are all cross veins.

Abdomen dark greyish-brown above, paler beneath. Dorsum with a narrow median pale line that is overlaid and widened by a series of inverted V-marks on the pale joinings of the middle segments. Side margins of segments 2 to 8 with a series of black marks shaped like fish hooks, pointed to rearward. Tails greyish brown, the joinings of the segments toward the base ringed with darker brown. Genitalia as in fig. 140.

The female is similar in coloration, only a little paler, with some yellow markings on the head (on median and occipital lines and around the eyes), and on the sides of the prothorax. The triangular prolongation of the 9th sternite projects far beyond the tip of the 10th segment and is notched at the tip.

Besides the type material which is in the Cornell University collection, we have two adult males, one female and one nymph from San Gabriel Canyon near Claremont, Calif. (C. D. Michener); also, slide mounts of two adult females collected at Cloverdale, Calif., (J. C. Bradley).

Genus **Habrophlebiodes** Ulmer

1919—Ulmer—Arch. Naturg. 85: Abt. A, Hft. 11: 39.
1919—Morrison—Canad. Ent. 51: 139 (Leptophlebia).
1920—Ulmer—Stett. Ent. Zeit. 81: 118.

These are slender, brownish, clear-winged mayflies with three tails that are a little longer than the body. The cross veins of the wings are so

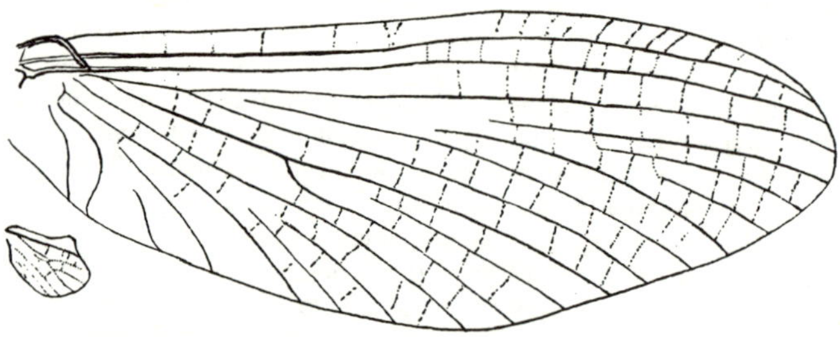

FIG. 141. Wings of *Habrophlebiodes betteni.*

colorless as to be scarcely visible, but the longitudinal veins are brownish. The costal margin of the hind wing is sharply angulated beyond the middle, and the subcosta ends in the margin at the outer side of the angulation (see fig. 141).

The prolongation of the 9th abdominal sternite is deeply cleft, and the forceps are three-jointed; the penes are closely appressed and bear divergent and decurving reflexed spurs that are subclavate at their tips. The subanal plate of the female is deeply cleft, and the parts of the 7th and 8th sternites adjacent to the openings of the oviduct are prolonged into a sort of ovipositor (Morrison, 1919). See fig. 20.

Nymphs are very similar to those of Paraleptophlebia, having narrowly lanceolate gills very much resembling those of *P. adoptiva.* The anterior margin of the labrum is somewhat more deeply cleft than in Paraleptophlebia. Nymphs of Habrophlebiodes are distinguished from all others of the subfamily Leptophlebiinae by the absence of spinules from the posterior margins of tergites 1–6; on tergites 7–10 these spinules are well developed.

Genotype—*H. americana* (in Habrophlebia), Banks.

Distribution—Nearctic; in east and southwest.

KEY TO THE SPECIES

Male adults

1—Apical extensions of forceps base broadly rounded at tips; all abdominal segments with conspicuous wide dark bands on posterior margins, continuous around body.. **annulata**

—Apical extensions of forceps base more or less acuminate at tips; posterior dark bands, if present on abdominal segments, much less conspicuous............ 2

2—Sternites paler than tergites, 3–5 whitish; legs pale yellowish white... **americana**

—Sternites not paler than tergites, middle sternites whitish along basal margins only; legs pale brown.. **betteni**

VERIFICATION TABLE

Adults

Species	*Wing length*	*Legs*[1]	*Sternites of 3 to 5*	*Tip of forceps base*	*Tails*	*Distr.*
americana	5	Whitish	Whitish	Acuminate	Whitish	E
annulata	4½	Red-brown	Pale br.	Blunt	?	Okla.
betteni	5½	Pale brown	Brown	Acuminate	Brown	E

Habrophlebiodes americana Banks

1903—Banks—Ent. News 14: 235. Fig. (Habrophlebia).................................. N. J.
1903—Berry—Amer. Nat. 37: 26. Nymph.
1919—Ulmer—Arch. Naturg. 85, Abt. A, Heft 11: 39 (transfer to new genus).
1925—McDunnough—Trans. R. S. C. p. 210 (*Choroterpes betteni* = *H. americana*).
1930—Ide—Canad. Ent. 62: 211.. Ont.

Length: body 5; wing 5 mm.

Head and thorax reddish brown. Legs whitish. Fore femur brownish, darker at the apex; hind femur with a dusky apical band and faint indications of a median band. Wings hyaline. Longitudinal veins of the costal margin, and the bases of the other longitudinals, very pale brown. All other veins colorless.

Middle abdominal segments more or less semi-transparent, the venter paler than the dorsum. Tergites obscured with purplish brown markings, on the posterior and lateral margins, the median line, and oblique lateral streaks from the postero-lateral angles. Pale submedian anterior triangles are thus left. Basal and apical tergites opaque, dark red-brown; the apical sternites much paler than the tergites, and with a yellowish tinge. Middle strenites largely pale. Each is marked with a purplish brown band extending transversely across the sclerite from the postero-lateral angle, leaving a narrow median pale space on the posterior margin, and wide pale areas anteriorly. On the apical segments, this band is much reduced in size. Ganglionic areas opaque white on sternites 2–6; on 7–9, marked with pale orange-brown. Forceps base pale

[1] Except fore femur.

brown; forceps yellowish white (see fig. 142). Tails white, narrowly darker at the joinings.

The species was originally described from specimens reared by Mr. Edward Berry at Passaic, N. J. We have specimens in the Cornell collection from the vicinity of Ithaca, N. Y. (P. R. Needham); Ringwood, N. Y. (Eva Gordon); the Cascades near Danbury, N. C., Wayah Creek near Franklin, and Penrose, N. C. (J. R. Traver); from Greenville, S. C. (H. K. Townes, Jr.); and from Marlinton, W. Va. (J. G. N).

Comparison of these specimens with the type material of *betteni* Ndm., shows that the latter is a valid species, and not, as Dr. McDunnough has suggested, a synonym of *americana* Bks.

Habrophlebiodes annulata Traver

1934—Traver—J. Elisha Mitchell Sci. Soc. 50: 199. Fig. .... Okla.

Length: body 4½; wing 4½ mm.

This is a slender, reddish-brown species, with black-ringed abdomen. Head and thorax black above, reddish brown on the sides and beneath. Femora rufous on the fore legs, brownish on the others; all legs becoming pale straw-yellowish on the tarsi. Wings hyaline with brown longitudinal veins and with no visible cross veins except in the stigmatic area, where they are very few (6 to 8), simple, strongly aslant, and incomplete at their subcostal ends.

Abdomen distinctly ringed with black all around the apical margin of the abdominal segments. Pattern almost lacking, but there is an obsolete middorsal pale hair line, bordered by equally obscure lines of black, and there is another obscure black hair line on the lateral margin. Midventral areas, paler brown. The prolonged 9th sternite is deeply divided by a narrow cleft into two very blunt-tipped lobes. Forceps red-brown. Penes tawny (see fig. 142). Tails lacking in present material. Two males collected in Oklahoma (R. D. Bird), the type material, are in the Cornell collection.

Habrophlebiodes betteni Needham

1907—Needham—N. Y. State Mus. Bull. 124: 194 (? Choroterpes) .... N. Y.
1919—Morrison—Canad. Ent. 51: 139 (fig. nymph, imago).
1925—McDunnough—Trans. R. S. C. p. 210 (*C. betteni* = *H. americana*) .... Que.

Length: body 5½; wing 5½ mm.

This is an all-brown mayfly that is pale only on the appendages. Head and thorax shining blackish brown above, a little lighter on the sides. Legs brownish, much paler beyond the knees. Front femora all very dark, the others lighter, obscurely twice-banded with blackish. Wings subhyaline with brown longitudinal veins and very faint cross veins. Costal cross veins wanting before the bulla, in the costal area few (7 to 10), simple, oblique, and entire.

Abdomen nearly uniform brown above and below with a pair of obscure, dorsal, basal, paler triangles on each of the middle segments. Prolongation of the 9th sternite deeply divided by a wide V-shaped notch into two rather acutely triangular lobes. Forceps brownish, the main joint triangular lobes. Forceps brownish, the main joint regularly tapering. Penes yellowish with the clavate and divergent reflexed spurs still paler (see fig. 142). Tails pale brown, narrowly ringed with darker brown.

The type material was taken by Dr. C. Betten at Hamburg, N. Y. In the Cornell collection are other specimens from McLean, Harford, Salmon Creek at North Lansing, and Ringwood, N. Y.

Some fine specimens collected by Mr. L. Harper at Big Hollow Brook and the St. Regis R. at Hopkinton, N. Y. and at Boyden Brook in St. Lawrence Co., are darker than the type specimens or the local material. The veins are more distinctly brown, especially at the wing base and in the stigmatic area; the pale areas of the abdominal tergites are almost obsolescent; and a wide pale semi-hyaline band parallels the pleural fold ventrally.

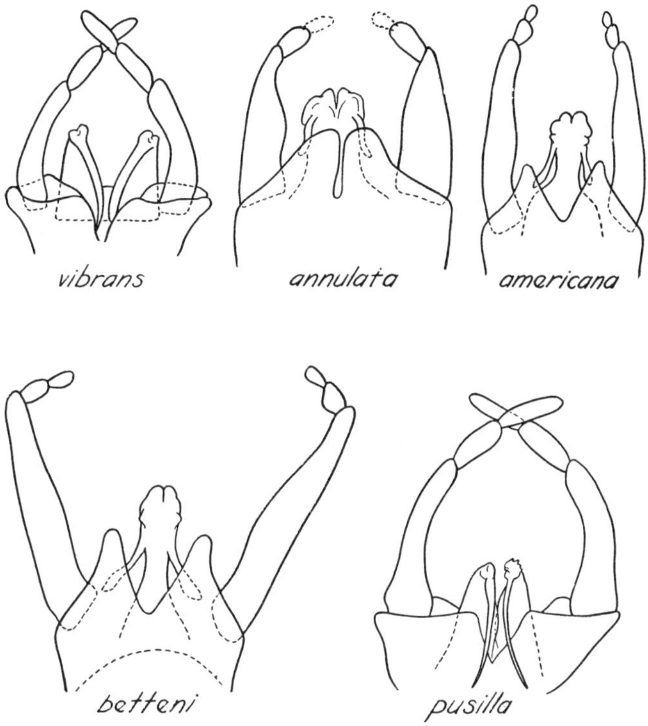

FIG. 142. Male genitalia of *Habrophlebia vibrans*, and *pusilla;* also of *Habrophlebiodes annulata, americana*, and *betteni*.

Genus **Habrophlebia** Eaton

1881—Eaton—Ent. Mo. Mag. 17: 195.
1884—Eaton—Revis. Monogr. p. 114.
1920—Ulmer—Stett. Ent. Zeit. 81: 117.

These are dainty little clear-winged mayflies, with three tails of which the middle one is the longest. The fore legs of the male are longer than the body, and in them the tibia is nearly twice as long as the femur. The fore wings have few and simple cross veins in the stigmatic area and none before the bulla. The hind wings are minute, with a sharp angulation near the middle of the costal margin. Subcosta meets the wing margin between this projection and the wing apex. There are few cross veins, and these are mainly in the costal and subcostal interspaces. See fig. 143.

The abdomen is slightly tapering, with the 9th sternite transversely divided at the base of the prolonged portion (base of forceps), doubly

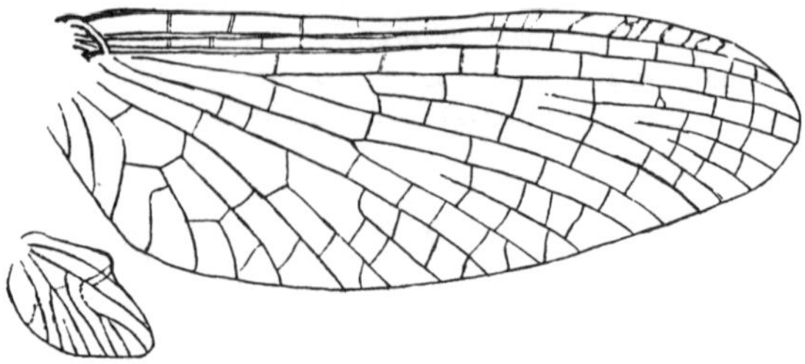

FIG. 143. Wings of *Habrophlebia vibrans.*

margined at the end, the margins completely encircling the base of the forceps. The main joint of the forceps is but little longer than the two terminal joints taken together. The reflexed spur of the penes is very long and slender, its tip resting when retracted in a cleft in the lower marginal carina of the 9th sternite. The subanal plate of the female is deeply excavated on its apical margin.

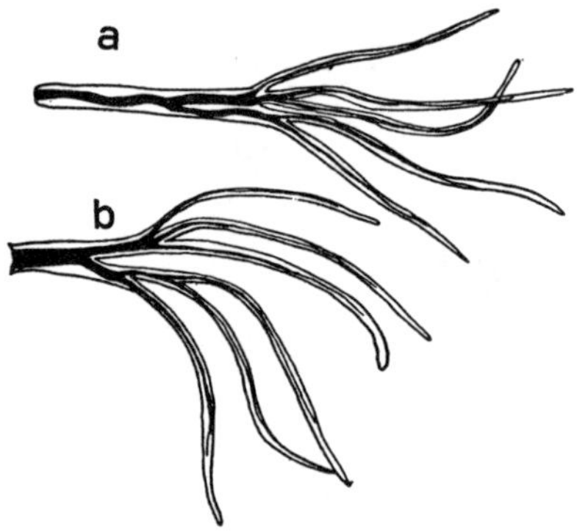

FIG. 144. Gills of *Habrophlebia pusilla: a,* 1st gill; *b,* 4th gill.

The subimago rather closely resembles the imago, but the wings are dull subhyaline and color patterns are less developed.

The nymph of this genus is most easily recognized by the form of its gills (see fig. 144). These are double and narrowly lamelliform at the base, thus conforming to the subfamily type, but they quickly break up into a number of linear filaments, outspread at the sides of the abdomen. The filaments vary in number, and are more numerous in the upper than in the lower divisions of each gill.

Genotype—*H. fusca* (in Ephemera), Curtis.

Distribution—Holarctic; in Nearctic, known only from the east and southeast.

KEY TO THE SPECIES

Male adults

1—Cleft between the penes very narrow; reflexed spur almost as long as the main joint of the forceps...**pusilla**

—Cleft between the penes wider at the apex than half the length of the reflexed spur; the latter considerably shorter than the main joint of the forceps.....**vibrans**

Habrophlebia pusilla Traver

1932—Traver—J. Elisha Mitchell Sci. Soc. 47: 118. Fig.....................................N. C.

Length: body 4; wing 4½ mm.

Head and thorax reddish brown. Antennae greyish beyond their brown bases. Fore femur light reddish brown; tibia and tarsus yellowish, a reddish brown apical band on the tibia. Middle and hind legs whitish, darker at joinings of segments. Wings hyaline, venation wholly pale. Abdominal tergites reddish brown; sternites much paler. There are large paired white patches on tergites 3 to 6, and whitish lines parallel the side margins of segments 7 and 8. Tails yellowish white, Penes longer than in *H. vibrans*, with the cleft between them very narrow (see fig. 142).

The type specimen, a male subimago reared from the nymph, taken at the Cascades near Danbury, N. C. (J. R. Traver), is in the Cornell collection; likewise a male imago from Greenville, S. C. (H. K. Townes, Jr.). We are at present unable to place definitely some larger, darker, but very similar specimens from Greenville, S. C. (H. K. Townes, Jr.) and the Great Smoky Mts., N. C. (P. N. Musgrave).

Habrophlebia vibrans Needham

1907—Needham—N. Y. St. Mus. Bull. 124: 192 (figs.)....N. Y.
1914—Banks—Proc. Acad. Nat. Sci. Phil. 66: 614 (*jocosa*)....N. C.
1925—McDunnough—Trans. R. S. C. p. 210 (*jocosa* = *vibrans*).

Length: body 5; wing 5 mm.

This is a dainty little clear-winged mayfly with a line of large whitish triangles on the back of the abdomen, giving it a serrated appearance when viewed from the side. Head brown above, paler on the middorsal line. Antennae pale beyond the swollen basal segment. Eyes globular, not quite touching above. Synthorax fawn color, prothorax darker brown. Legs pale, darker and distinctly brownish on front femora and tibiae especially toward their distal ends. Wings whitish hyaline with only a tint of amber in the veins toward the base. Cross veins of the costal strip nearly wanting before the bulla, and only a few (8–10) short simple parallel ones in the stigmatic area.

Abdomen brown at the ends with large whitish crescents covering half of the dorsum of the purplish middle segments and with the apical segments tawny. Beneath, it is paler brown at the ends, becoming whitish under the middle segments. Forceps pale. Penes elongate, conic, separated by a deep V-shaped cleft, without apical teeth but with a very long gently outcurving spine-like reflexed spur that is as long as the penes (see fig. 142). Penes light brown. Tails whitish, with darker joinings on a few basal segments, best marked at apex of the second joint.

Besides the type material from Bald Mountain Pond outlet near Old Forge, N. Y., taken from a swarm (J. G. Needham), there are other specimens in the Cornell collection from Ft. Ann, N. Y. (P. R. Needham) and Hopkinton, N. Y. (L. Harper).

Genus **Thraulodes** Ulmer

1919—Ulmer—Arch. Naturg. 85: 33.
1920—Ulmer—Stett. Ent. Zeit. 81: 116.

These are rather small mayflies with long tails, and well preserved cross veins in the fore wings. The upper eyes are very broad and meet above the head. The legs are moderate, the fore tarsus of the male being shorter than either tibia or femur. The costal cross veins of the fore wing are well preserved in the stigmatic area, but may be wanting basad of the bulla. In both wings a line of cross veins joins the principal forks, and in the hind wing there are no other cross veins. The costal angle of the hind wing is prominent rounded at the tip, and situated at the middle of the costal border. Subcosta of hind wing only slightly shortened, and extending well beyond the costal angulation. Media of the hind wing is *forked,* a character which separates this genus from others of this subfamily which possess the costal angulation. See fig. 145.

The forceps are three-jointed, with the last two segments very small. forceps base entire and prolonged a little in the middle. Penes usually bear each a long slender sharp-pointed reflexed spur. Subanal plate of female only slightly concave on the apical margin. Tails three, much longer than the body, in the male.

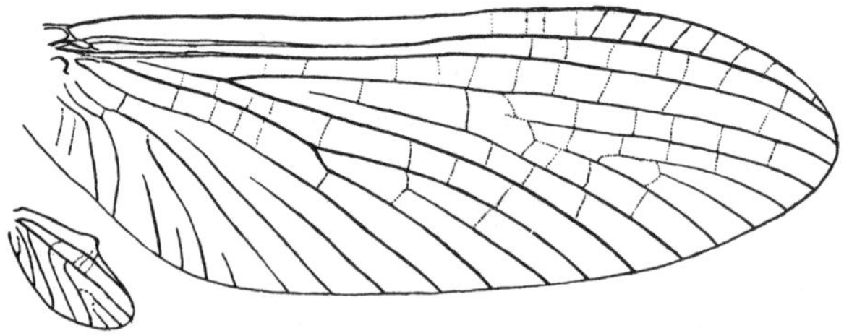

FIG. 145. Wings of *Thraulodes speciosus.*

Gills of the nymph are lanceolate, widest on segments 1 and 2. Spinules are present on the apical margin of abdominal tergites 1–10; postero-lateral spines are borne of segments 2–9. 16–18 large pectinate spines are present below the crown of the maxilla, which is broadly rounded and set with great numbers of more slender pectinate hairs on its crown. Maxillary palp three-jointed, the outer apical margin of the second joint prolonged into a slender process which lies appressed against the side of the short third joint. Labial palp three-jointed, the last joint very short. Outer margin of mandible strongly bent inward at each end; canines long and slender, lacinia prominent. Labrum twice as wide as long.

Spieth (1933, J. N. Y. Ent. Soc. 41, fig. 55) figures, as the fore wing of Ephemerella, a wing which is probably that of a species of Thraulodes.

Genotype—*T. laetus* (in Thraulus), Eaton.

Distribution—Neotropical, Nearctic; in the latter region, of occasional occurrence in southwestern United States.

Thraulodes speciosus Traver

1934—Traver—J. Elisha Mitchell Sci. Soc. 50: 201. Fig. .... Ariz.

Length: body 8; wing 9 mm.

This is a beautiful fawn-colored species with long ringed tails, and cross veins of the fore wings heavily marked with brown. Head blackish with the very large and attingent upper eyes fawn-colored, and flagella of antennae pale. Thorax fawn-colored with blackish lines in the pleural sutures, and especially around the leg bases. Legs honey yellow, with the fore femur heavily streaked with blackish, the other femora with heavy brown apical cross-bands; both ends of fore tibiae and the tips of all tarsi touched with brown. Wings hyaline with pale brown longitudinal veins and dark brown cross veins over most of their area. The costal cross veins are much heavier before the bulla than in the stigmatic area. Little brown spots surround the breaks of the bulla. Two rather well defined lines of more heavily marked cross veins traverse the basal half of the wing; one from the bulla backward through the posterior branch of the radial sector, to the median vein; the other parallel to it and nearer the wing base. In the hind wing the costal angulation is prominent, rounded on end, and about midway the wing length. Only two or three cross veins just behind the outer end of the hump are heavily marked with brown.

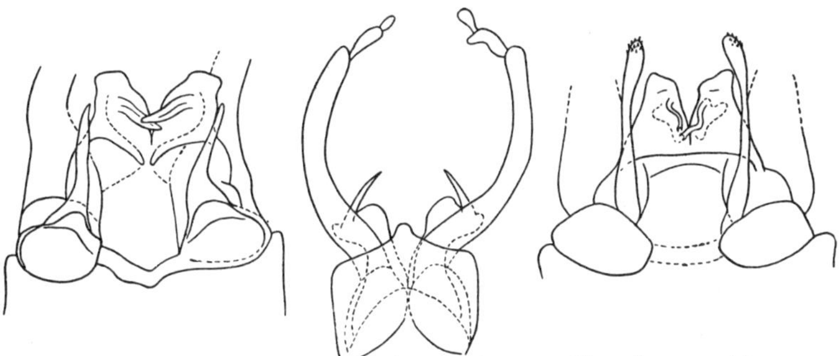

Thraulus albertanus *Thraulodes speciosus* *Thraulus presidianus*

FIG. 146. Male genitalia of Thraulus and Thraulodes.

Abdomen brown at base and rufous at tip with intervening segments 3 to 6 pale and strongly marked with apical cross bands of brown. Through the translucent venter. the nerve ganglia show plainly. The last ganglion is heavily pigmented with black. Tails white, heavily ringed with blackish on the joinings of the segments. Forceps hyaline, three-jointed beyond a bulbous segment-like basal enlargement that is covered by the 9th sternite, the last segment half as large as the one before it, both very small. Penes honey yellow, separated by a V-shaped cleft, each constricted in the middle then dilated to a broad doubly rounded apex with a long straight stout spine arising at the end. See fig. 146.

We have seen two male and two female specimens that were collected at an altitude of 8,000–9,000 feet in Rustler Park in the Chiricahua Mountains of Arizona (J. A. Kusche); the allotype and one paratype are in the Cornell collection: holotype in the Calif. Academy of Sciences.

Genus **Thraulus** Eaton

1881—Eaton—Ent. Mo. Mag. 17: 195.
1884—Eaton—Revis. Monogr.; p. 106.
1920—Ulmer—Stett. Ent. Zeit. 81: 117.

These are rather small mayflies, of very pretty coloration. The upper eyes are very broad, contiguous above the head. The legs are of moderate length, with claws of each tarsus unlike. Hind wings are small with a rather acute and very prominent angulation on the middle of the costal margin, and with very few (5–9) cross veins (see fig. 147). The male forceps is three-jointed, its two terminal points very short, hardly longer than broad, its base not bifid. Between the prolongation of the 9th sternite (forceps base) and the penes there is a pair of processes of variable form according to the species, that project to rearward underneath the penes. Tails three, equal, a little longer than the body. The subanal plate of the female is slightly concave on its apical margin.

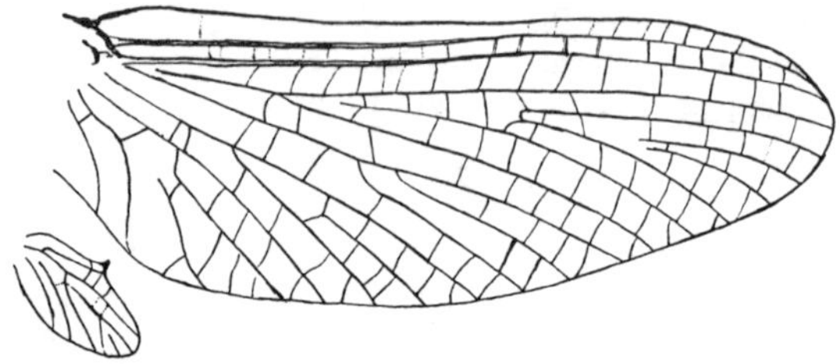

FIG. 147. Wings of *Thraulus presidianus.*

The subimago is similar but with greyish wings.

The nymph is easily recognized by its gills, which are double, on the first abdominal segment deeply forked with two slender linear divisions, on succeeding segments oval, lamelliform, and fringed around the entire margin with linear fringes that are as long as the body of the gill is wide.

As noted in the discussion of the genus Choroterpes, the wings and genitalia figured by Spieth (1933, J. N. Y. Ent. Soc. 41, figs. 29, 30, 33) as of the genus Thraulus are really of the genus Choroterpes.

Genotype—*T. bellus* Eaton.

Distribution—Holarctic, Neotropical, Indo-australian; in Nearctic, known only from western North America.

KEY TO THE SPECIES

Adults

1—Basal half of fore wing and most of hind wing smoky; wing 10 mm. in length .... **albertanus**

—Wings wholly hyaline, with a tinge of amber at the extreme base; wing 7 mm. in length........ **presidianus**

Thraulus albertanus McDunnough

1931—McDunnough—Canad. Ent. 63: 82........ Alta.

Length: body 8; wing 10 mm.

Head and thorax deep black; membranous portions of the pleura slightly tinged with ruddy. Coxae and trochanters of legs black. Femora reddish brown, the bases and apices tipped with black; tibiae and tarsi pale, with a brownish tinge; a black ring at the base of the tibia. Fore wings semi-hyaline, the basal half strongly tinged with smoky; veins black, cross veins numerous and fine, black except the basal costal cross veins, which are obsolescent. Hind wings largely smoky; the blunt costal projection is before the middle of the wing.

Abdomen dorsally dirty whitish, marked with broad slightly oblique bands of deep smoky; pale areas are thus left of the anterior and posterior margins, giving the abdomen a ringed appearance. Sternites more evenly smoky black, ringed with whitish. Forceps dull yellowish; two fine long spines project backwards from the forceps base (see fig. 146). Tails whitish.

Type material was obtained at Medicine Hat, Alta. (J. H. Pepper). We have a specimen of this species in the Cornell collection, and have seen others from the Carnegie Museum, taken in Uinta Co., Utah.

Thraulus presidianus Traver

1934—Traver—J. Elisha Mitchell Sci. Soc. 50: 199. Fig....................................Tex.

Length: body 6; wing 7 mm.

This is a clear-winged blackish-brown species with nearly white abdomen. Head very dark brown including the very broad and contiguous upper eyes, and the basal segment of the antenna.

Thorax brown, darker around the margins of the dorsum, with three subparallel obscure pale lines on the mesoscutum that converge into a pale streak on the front of the scutellum. Sides and leg bases paler (remainder of legs lacking). Wings hyaline, with a tinge of amber on their extreme bases, that in the fore wing diffuses outward along the radial vein toward the bulla. Beyond this, the longitudinal veins are but little more evident than are the faint cross veins. In the stigmatic area the cross veins are weak, simple, and scarcely aslant.

Abdomen faintly brownish or yellowish-white above and below, with a wash of brown on the dorsum of segments 1 and 7 to 10, and with a touch of the same on the lateral margins of some of the intervening segments. Forceps broken off the one known specimen. The pair of internal processes that overlie their base in this species are very long, linear and slender, longer than the penes, so that their finely denticulate tips project into view to rearward. Penes short, separated by a V-shaped notch; in form pyramidal, their apices truncated, their reflexed spurs minute, slender, awl-shaped, their incurving tips lying across the base of the interpenial cleft (see fig. 146).

The type, a single male specimen collected at Presidio, Tex. (E. L. Thomas), is in the Cornell collection.

Subfamily **BAETISCINAE**

Tarsi five-jointed, the basal joint immovable and fused with the tibia, except in the fore leg of the male; claws of male fore tarsus similar, blunt; claws on other legs dissimilar. Fore wing with posterior branch of median vein extending almost to the wing base, and detached basally from its stem; the two branches of cubitus appear to arise independently of one another; the first anal vein ends in the outer margin, and is attached to the anal angle and the inner margin by cross veins. Cross veins numerous in both wings. Hind wing large, almost circular in outline, venation copious; marginal intercalaries very long. Sixth abdominal segment considerably longer than any of the others. Forceps three-jointed; penes triangular, appearing to be united into a single conical structure. Tails two, in both sexes.

Nymph possesses a mesonotal shield which covers most of the dorsum of the thorax, abdominal segments 1–5, and the basal half of segment 6; apically this shield fits into a pyramidal elevation on tergite 6. Gills, on segments 2–6, completely concealed beneath this shield. Labial palp with a pincer-like structure distally. Legs rather short; tibia very short; femur equal to tibia and tarsus combined; claws long and slender, not denticulate. Tails three, fringed with hairs on both sides except at the base.

Genus **Baetisca** Walsh

1862—Walsh—Proc. Acad. Nat. Sci. Philad. p. 378.
1864—Walsh—Proc. Ent. Soc. Philad. 3: 200.
1880—Joly—Bull. Soc. Angers, 8 & 9: 157.
1920—Ulmer—Stett. Ent. Zeit. 81: 130.
1931—Traver—J. N. Y. Ent. Soc. 39: 45.
1932—McDunnough—Canad. Ent. 64: 213.
1933—Spieth—J. N. Y. Ent. Soc. 41: 357.

Mayflies of medium size, with robust bodies, the abdomen tapering in its posterior half (see frontispiece). Both sexes are brownish in the imago stage, without prominent markings. Eyes of male large, not divided; almost contiguous on the upper surface. Eyes of female rather small, remote, set far out on the postero-lateral angles of the head; median ocellus slightly smaller than the lateral ones. Posterior margin of head very slightly emarginate at the median line, and with very slight backward prolongations on each side of this emargination.

Mesothorax unusually large and chunky, pro- and metathoracic regions relatively narrow. Mesonotal scutellum extends backward as a prominent spinous process well beyond the posterior margin of the metanotum. The posterior margin of the prosternum bears two spinous projections, one on each side of the median line. Fore leg of male about as long as the body; five free tarsal joints, the basal of these being the longest, almost 1½ as long as the second; second slightly longer than the third, which is subequal to the fourth; fifth slightly shorter than the fourth. Fore tibia about ¾ the length of the femur, and only slightly longer than the basal tarsal joint; tarsus nearly 2½ times as long as the tibia. Claws of male fore tarsus similar, blunt. Tarsi of middle and hind legs likewise five-jointed; basal joint immovable and fused so completely with the tibia that only a spur on the inner margin marks its position. Basal tarsal joint slightly longer than the second; second and third subequal, fourth slightly shorter; distal joint longest, almost equal in length to the preceding joints combined; claws dissimilar. Tibia of hind leg about ½ as long as the tarsus and ¾ the length of the femur. Legs of female similar to hind leg of male.

Fore wing long and wide, its greatest width being near the base in a line drawn directly through the anal angle. Anterior branch of radial sector very weak at the extreme base. Posterior branch of median vein and the median intercalary extend almost to the base of the wing; the posterior branch is detached basally from the median stem. The two branches of cubitus appear to arise independently of one another; parallel at base, diverging slightly toward the outer margin. First anal vein ends in the outer margin, and is attached to the anal angle and the outer margin by a series of straight cross veins, which in the middle area lie parallel to one another. Second and third anal veins short, ending in the inner margin near the base. Cross veins numerous and quite regularly arranged; short marginal intercalaries in most of the interspaces; all except those of the anal region are very fine and almost invisible. Hind wing almost circular in outline, with a blunt costal projection quite near the base. Venation copious, the anal region in particular being well developed. Median vein forked about one-third the distance from the base. Cross veins numerous, but none very near the outer margin. Very long marginal intercalaries occur, usually in groups of 3 or 4; one or two long median ones and two shorter laterals, in each interspace.

Abdomen relatively short and chunky, tapering quite sharply in its apical half. Segment 1 almost completely fused with the thorax; segments 2–5 rather short and subequal; segment 6 very long fully twice the length of 5; 7 longer than 5 but considerably shorter than 6; 8 and 9 progressively shorter than 7, each being longer ventrally than dorsally; 10 very short. Segment 9 bears rather long postero-lateral spines. Near the middle of the 6th tergite is a peculiar transverse elevation. Tails 2, in both sexes, the median one represented by a short conical unsegmented rudiment; lateral tails quite short, usually not as long as the body. Forceps three-jointed; basal joint moderately long and rather stout, divergent apically; second joint longer and more slender, sharply bowed, and widened more or less at the base, where it bears a rounded inward projection; distal joint short and conical. Penes triangular in shape, tapering to a point distally, and pressed so close together as to appear united into a single conical organ. Apical margin of 9th sternite of female extended backward to form a bifid subanal plate, with a deep V-shaped median cleft.

The wings of the subimago are in large part smoky or blackish brown, with clear hyaline areas around most of the cross veins, and with irregularly arranged more extensive hyaline areas near the base and across the disc. The depth and amount of dark color on the wings varies with the species, and to some extent with individuals in a given species.

The stout-bodied nymph (see fig. 61) is unique among North American genera is possessing a large mesonotal shield or carapace which completely covers the dorsum of the meso- and metathorax and abdominal segments 1–5, as well as the basal half of segment 6. Its apical margin fits into a pyramidal elevation on tergite 6; the gills, borne on segments 2–6 only, are completely enclosed by this shield as in a gill chamber. Only one other genus, the European Prosopistoma, has such a thoracic shield. Dorsal and lateral spines are often but not always, present on this shield. The head frequently bears more or less prominent frontal projections, and spinous projections from the anterior angles of the genae. The second joint of the labial palp bears a spine-like projection on its inner margin, which with the distal joint forms a pincer-like structure. Postero-lateral spines, more or less well developed, are usually present on abdominal segments 6–9. Tails 3; rather short, approximately equal in length, bearing a fringe of hairs on both sides except at the extreme base. They are usually held erect or reflexed over the abdomen. Legs rather short; femur approximately equal to tibia and tarsus combined; tibia very short, ⅓ to ¼ the length of the tarsus. Claws rather long and slender, curved, sharp at the tip; not denticulate.

Genotype—*B. obesa* (in Baetis), Say.

Distribution—Nearctic; eastern North America.

KEY TO THE SPECIES

Male adults

1—Both fore and hind wings flushed with reddish or orange, at least in the basal third 2
—Wings not as above 3
2—Wing 9–10 mm. in length, flushed with orange; legs greenish white .... **carolina**
—Wing 8–8½ mm. in length, flushed with reddish; legs yellowish brown .... **rubescens**
3—Longitudinal veins wholly brown .... **laurentina**
—Longitudinal veins wholly or in large part pale (may be yellowish on costal margin, or dark near extreme base) 4
4—Longitudinal veins wholly pale; dorsum of abdomen dark brown, venter yellowish white; wing 8 mm. in length .... **lacustris**
—Longitudinal veins of costal margin yellowish, radius piceous at extreme base; dorsum and venter of abdomen pale reddish brown; wing 10 mm. in length .... **obesa**

VERIFICATION TABLE

Male adults

Species	*Wing length*	*Wing-base reddish*	*Veins (mainly)*	*Legs*	*Abdom. tergites*	*Abdom. sternites*	*Tails[1] ringed*	*Distr.*
carolina	9–10	Yes	Brown	Gr. white	Purp. br.	Lt. tan	Faintly	E
lacustris	8	No	Pale	Yel. white	Brown	Yel. white	No	Gt. Lakes
laurentina	10	No	Brown	Dull amber	Deep br.	Pale br.	Yes	NE
obesa	10	No	Pale	Gr. yel.	Red-br.	Red-br.	Yes	E
rubescens	8–8½	Yes	Brown	Yel. br.	Red. br.	Red-br.	No[2]	Que.

KEY TO THE SPECIES

Nymphs

1—Mesonotal shield with well-developed spines,—dorsal, lateral, or both 2
—Mesonotal shield with neither dorsal nor lateral spines .... **callosa**
2—Both dorsal and lateral spines present, on mesonotal shield 3
—Dorsal spines lacking, lateral spines present 4
3—Frontal projections prominent .... **obesa**

[1] Tail joinings darker, at least in basal half.
[2] Tails opaque at joinings,—not dark-ringed.

—Frontal projections reduced to small blunt tubercles. . . . . **laurentina**
4—Genae produced into spine-like processes; frontal projections wanting; lateral spines of mesonotal shield rather short. . . . . **carolina**
—Genae not produced into spine-like processes; two small rounded frontal projections; lateral spines of mesonotal shield very long and sharp. . . . . **lacustris**

VERIFICATION TABLE

Nymphs

Species	*Body length*	*Frontal process*	*Genae*[1] *produced*	*Dorsal spines*[2]	*Lateral spines*[2]	*Distr.*
callosa	3½–4	Short	No	No	No	E
carolina	7–10	Absent	Yes	No	Yes	E
lacustris	8–10?	Short	No	No	Yes	Gt.Lakes
laurentina	6–8?	Short	No	Yes	Yes	NE
obesa	8–10?	Prominent	Yes	Yes	Yes	E

Baetisca callosa Traver

1931—Traver—J. N. Y. Ent. Soc. 39: 57. Fig. nymph. . . . . W. Va.
1932—McDunnough—Canad. Ent. 64: 213. . . . . Que.

Length (nymph): body 3½–4½ mm. (immature specimens).

Only the nymph of this species is known.

The nymph lacks both the dorsal and lateral spines on the mesonotal shield. Frontal projection short, genae not produced into spine-like processes.

From other nymphs of this genus thus far described, it is distinguishable at once by the total lack of both dorsal and lateral mesonotal spines. It is not probable that these spines would be present in the fully mature nymph, although only immature specimens are known as yet.

The type specimens of this species are in the Cornell collection. We have other specimens from Mossy Vly, near Speculator, N. Y. (V. S. L. Pate).

Baetisca carolina Traver

1931—Traver—J. N. Y. Ent. Soc. 39: 51. Figs. . . . . N. C.
1932—Traver—J. Elisha Mitchell Sci. Soc. 47: 112.
1932—McDunnough—Canad. Ent. 64: 213. . . . . Que.

Length: body 8½–10; wing 9–10 mm.

Abdominal tergites of male imago light purplish brown, sternites light tan; fore wing in basal third, and hind wing entirely tinged with orange-brown; longitudinal veins brown.

Head and pronotum light tan, mesonotum paler, with a greenish tinge; scutellum tipped and margined with black; pleura light brown. Thoracic sternites yellowish tan, anterior portion of mesosternum brown. Legs greenish white; tips of claws and all joinings pale brown. Wings hyaline, somewhat iridescent. Basal third of fore wing, and all of the hind wing, tinged with orange-brown. Longitudinal veins brown; cross veins very faint, pale. Abdominal tergites 1–6 light purplish brown, darker anteriorly; sternites light tan. Tergites 7–10 brownish; sternites 8 and 9 light yellow, 6 and 7 brownish; pleural fold purplish brown. Brown lateral dots and transverse dashes on

[1] Genae produced forward into spine-like processes.
[2] On mesonotum.

sternites 1–5. Forceps yellowish white, brown at tip; tails white, slightly brown at base, joinings in basal half light brown. Genitalia as in fig. 148.

The nymph (see fig. 61) lacks dorsal spines on the mesonotal shield, but short lateral spines are present; the genae are produced into upcurved spines; the frontal projection is wholly wanting.

From *B. rubescens*, which likewise has the wings flushed with a reddish color, this species may be separated by its larger size and pale whitish legs; also the wings are tinged with orange rather than reddish. The type specimens of this species are in the Cornell collection. Other specimens are from the vicinity of Morgantown, W. Va. (P. N. Musgrave).

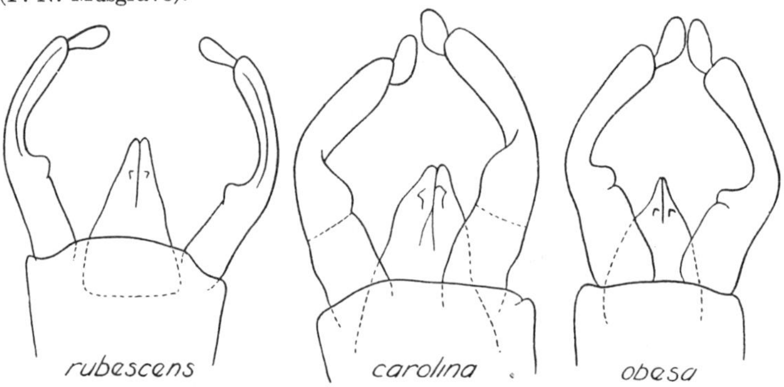

FIG. 148. Male genitalia of three species of Baetisca.

Baetisca lacustris McDunnough

1913—Clemens—Canad. Ent. 45: 333 (*obesa*)....................................Georgian Bay
1915—Clemens—Marine Fish. Sess. Pap. 39b: 119 (*obesa*).
1932—McDunnough—Canad. Ent. 64: 214. Fig. nymph....................Ont., Man., Gt. Lakes

Length: body 7; wing 8 mm.

Abdominal tergites of male imago brown, sternites pale yellowish white; venation wholly pale.

Head and thorax brown; thoracic sternum pale brown. Legs pale yellowish white. Wings hyaline, venation wholly pale. Abdominal tergites brown; sternites pale yellowish white, contrasting strongly with the dark upper surface. Forceps yellowish white. Tails whitish, not darker at joinings.

The nymph has lateral but no dorsal spines on the mesonotal shield; the genae are not produced into spines; the frontal projections are small and rounded. The lateral mesonotal spines are very long and sharp.

This species is distinguished from others in the genus by the wholly pale venation and the striking contrast in color between the dorsal and ventral surfaces of the abdomen. We have specimens of this species in the Cornell collection from Willowick and Mentor Park, Lake Co., Ohio. on the shore of Lake Erie (J. R. Traver).

Baetisca laurentina McDunnough

1932—McDunnough—Canad. Ent. 64: 214. Fig. nymph..........................Que., Ont., N. B.

Length: body 8; wing 10 mm.

Abdominal tergites of male imago deep brown, sternites paler brown; longitudinal veins brown.

Head and thorax deep brown; lateral flange-like edge of mesonotum paler; thoracic sternum deep brown. Legs dull amber, the fore legs deeper in color than the second and third pairs. Wings hyaline; longitudinal veins brown; cross veins pale, almost invisible except in the anal region. Abdominal tergites deep brown: sternites paler brown. Forceps dull dirty amber colored; tails pale brownish, the joinings darker.

The nymph possesses both dorsal and lateral spines on the mesonotal shield; the frontal projections are reduced to very small blunt tubercles; the genae are very slightly produced only.

The dark longitudinal veins separate this species from *obesa*, in which the veins are largely pale. We have seen no specimens of this species.

Baetisca obesa Say

1839—Say—J. Acad. Nat. Sci. Phil. 3: p. 378 (Baetis)....................................Ind.
1843—Pictet—Hist. Nat. Neurop. 2: 195 (Baetis).
1853—Walker—Cat. Neurop. Ins. Brit. Mus. p. 563 (Baetis).
1859—LeConte—Works of T. Say, 2: 412 (Baetis).
1861—Hagen—Syn. Neurop. Ins. N. Amer. p. 45 (Baetis).
1862—Walsh—Proc. Acad. Nat. Sci. Phil. p. 378 (transferred to new genus Baetisca)............Ill.
1863—Walsh—Proc. Ent. Soc. Phil. 2: 187.
1864—Walsh—Proc. Ent. Soc. Phil. 3: 200. Fig. nymph.
1868—Eaton—Ent. Mo. Mag. 5: 89.
1870—Packard—Guide to study of insects, p. 595. Fig.
1871—Eaton—Trans. Ent. Soc. London, p. 101. Figs. (wing misdrawn).
1880—Joly—Bull. Soc. d'Etud. Sc. Angers, p. 157. Fig. nymph. (translation of Walsh, 1864).
1881—Vayssiere—Ann. d. Sc. Nat. Zool. 11: 4, 5. Fig. nymph.
1882—Vayssiere—op. cit. 13: 72: 98, 99–103. Fig. nymph.
1885—Eaton—Revis. Monogr. p. 228. Figs......................................Mich., Calif.
1905—Needham—Bull. N. Y. St. Mus. 86: 30. Figs.....................................N. Y.
1922—Murphy—Bull. Lloyd Library 22: 16
1931—Traver—J. N. Y. Ent. Soc. 39: 61.
1932—McDunnough—Canad. Ent. 64: 213, 214.

Length: body 7–8; wing 9–10 mm.

Abdominal tergites of male imago pale reddish brown; venation yellowish on costal margin, otherwise pale.

Head ferruginous; antenna similar in color, the filament generally pale. Thorax ferruginous; sternum "paler behind, especially the space between the posterior coxae" (Walsh). Legs pale greenish yellow; in the fore leg, the apex of the femur, tips and joinings of the tarsi, are slightly fuscous. Middle and hind legs paler, only the tips of tarsi cloudy. Wings hyaline; longitudinal veins of the costal margin yellowish, radius piceous at the extreme base, "a few of the principal veins slightly tinged with fuscous" (Walsh); all other veins pale. Cross veins almost invisible. Abdominal tergites pale reddish brown, the posterior margins whitish; sternites somewhat paler. Tails whitish, the joinings fuscous, especially in the basal portion. Forceps pale, sometimes reddish toward the tip. Genitalia as in fig. 148.

The nymph has both dorsal and lateral spines on the mesonotal shield; the genae are somewhat produced into spine-like processes; the frontal projections are very prominent and well developed.

The paler venation separates this species from the allied *B. laurentina.* We have specimens in the Cornell collection from the M.A.C. campus at Lansing, Mich.; Douglas Lake, Mich.; Elkhart, Ind.; Three-Mile Is., L. Winnepesaukee, N. H.; and a nymph from Sandy Creek, Ga. (P. W. Fattig).

Baetisca rubescens Provancher

1876—Provancher—Natur. Canad. 8: 267. . . . . Que.
1878—Provancher—Pet. Faun. Canad., Neurop. p. 84 (*Cloe unicolor*); p. 82 (3), (*Cloe rubescens*).
1885—Eaton—Revis. Monogr. p. 169 (transferred to Baetis).
1921—McDunnough—Canad. Ent. 53: 118 (transferred to Baetisca).
1931—Traver—J. N. Y. Ent. Soc. 39: 45.

Length: body 6–7; wing 8–8½ mm.

Abdominal tergites of male imago reddish brown, sternites paler reddish to yellowish brown; both wings flushed with reddish in the basal portions; longitudinal veins brown. (See frontispiece).

Head and thorax rather dark reddish brown. Legs yellowish brown; all joinings of fore leg narrowly reddish brown. Wings hyaline; the basal third and most of the costal margin of the fore wing, and most or all of the hind wing, flushed with reddish; in the hind wing this color is deepest at the base, fading gradually toward the outer margin. Longitudinal veins brown; cross veins pale, almost invisible. Abdominal tergites reddish brown, darker next to the pale pleural fold; sternites paler reddish to yellowish brown. Posterior margins of all segments rather dark reddish brown, most evident on the tergites. Forceps and tails pale yellowish; tail joinings slightly opaque, but not darker than the joints. Genitalia as in fig. 148.

The smaller size and darker legs separate this species from *B. carolina*, which has the wings flushed with orange rather than with reddish.

We have specimens of this species in the Cornell collection from St. Lambert, Que. (W. J. Brown), kindly presented by Dr. McDunnough. The above description is drawn in large part from these specimens.

Addendum

A new species of Baetisca,—*B. bajkovi* Neave,—has been described recently from Manitoba by Dr. Neave (Internat. Rev. Hydrobiol, Hydrogr. 31: 166, 1934). A figure of the nymph is given; this is the only stage known. Nymph allied to *lacustris*, but with "well-developed frontal tubercles and genal projections" (Neave).

Subfamily **EPHEMERELLINAE**

Tarsi four-jointed, except in the fore leg of the male, where a very short basal joint is present. Claws dissimilar on all legs. Venation quite similar to that of Blasturus, but with two or more free intercalaries between the bisector of the median fork and the posterior branch of media; also between media and the anterior branch of cubitus; no 'sag' in the posterior branch of the radial sector. Vein Cu_2 strongly angulate before the middle. Hind wing quite well developed.

Genus **Ephemerella** Walsh

1862—Walsh—Proc. Acad. Nat. Sci. Philad. p. 377.
1884—Eaton—Revis. Monogr. p. 124.
1905—Needham—Bull. N. Y. St. Mus. 86: 42.
1909—Bengtsson—Lunds Univ. Arssk. Afd. 2, B. 5: 6 (Chitonophora =Ephemerella in part).
1920—Ulmer—Stett. Ent. Zeit. 81: 119.
1930—Walley—Canad. Ent. 62: 13.
1931—McDunnough—Canad. Ent. 63: 32, 190.
1931—Steger—Psyche 38: 27.
1932—Traver—J. Elisha Mitchell Sci. Soc. 47: 142.
1933—Spieth—J. N. Y. Ent. Soc. 41: 351.

Mostly small to moderate-sized mayflies, although a few species are large. Variable in color and markings; most of the species are brownish. Eyes of males divided; the small basal portion usually darker in color than the large distal portion. Eyes usually closely approximated above. Eyes of female smaller, not divided; set well out on the sides of the head. Occipital margin usually more or less straight between the eyes. Pronotum usually with a median carina.

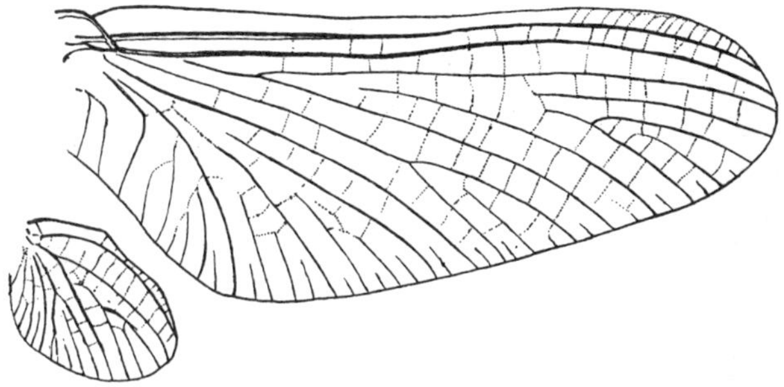

FIG. 149. Wings of *Ephemerella cornuta.*

Second and third tarsal joints of the male fore leg frequently subequal; fourth joint about three-fourths of the third; distal joint rather less than one-half of the fourth. In the hind tarsus of both sexes, the three basal joints are rather short and approximately subequal, while the distal joint is almost as long as the three basal ones combined. Femora and tibiae variable as to their relative length, in different species; third tarsus shorter than tibia. Claws dissimilar on all tarsi in both sexes.

Hind wing with a slight rounded costal elevation near the base, beyond which is a very shallow depression in the costal margin. Median vein of hind wing forked slightly less than half way from the base to the margin (see fig. 149). Forceps three-jointed, the second joint longer than either the basal or the distal one. Penes variable in shape in the different groups; with or without minute spines. Apical margin of the forceps base entire. Apical margin of the subanal plate of the female usually obtusely rounded, but in some species emarginate.

Nymphs of this genus are very variable in form of body; several lines of adaptation seem to have been followed by different species. Some are slender stream-lined forms; others are flattened ventrally, with dorsum arched, and with broad flattened femora. Many have developed prominent dorsal tubercles or spines on head, thorax, legs and abdomen. Some species are smooth, others are densely covered with hairs. The mouthparts are quite constant throughout the genus, especially within the different groups. The labrum is wider than long, sometimes slightly emarginate on the anterior border. The size and direction of slant of the mandibular canines varies considerably. Maxillae are more or less conical at the apex, with one or more broad terminal spines. The maxillary palp may be rudimentary, or even entirely wanting in some species; at its greatest development, it does not extend beyond the galea-lacinia, and is usually much shorter. The labial palp is also relatively short; the four apical divisions of the labium are frequently short, conical and almost equal. The tarsal claws of most species bear several denticles.

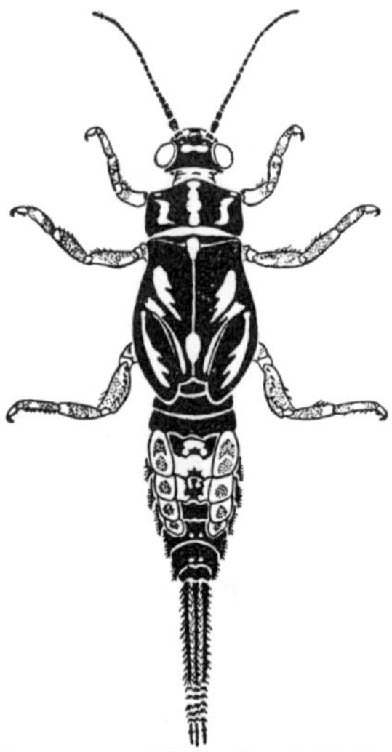

Fig. 150. Nymph of *Ephemerella serratoides* (drawing by Evelyn George).

Lateral extensions are usually present on the segments of the abdomen, bearing postero-lateral spines of greater or less development. Paired dorsal spines are present in many species. Gills are borne on segments 3–7 in many species, but on 4–7 in others. In the latter case, the gill on segment 4 may be operculate or semi-operculate, covering all or parts of the succeeding pairs of gills; when this gill is fully operculate, a short gill rudiment is usually present on segment 1 also. In all species, the gills are wholly dorsal in position. The tails are usually about as long as the body, and are often hairy, or with whorls of spines at the joinings. Many species of this genus hold the tails upright or arched upward over the abdomen.

On the basis of genitalic structure and parallel nymphal structures, several groups are recognizable within this genus, as indicated in the following keys and tables.

Spieth (1933) presents an incorrect figure of the fore wing (fig. 55) of this genus; it is quite probable that his figure is really that of Thraulodes. We do not agree with his conclusion that Ephemerella and Tricorythodes are closely allied.

Genotype—*E. excrucians* Walsh.

Distribution—Holarctic; in Nearctic, very generally distributed.

Groups of species within the genus Ephemerella

We recognize seven groups of species within this genus. The most homogeneous and distinct of these is the *bicolor* group, containing 10 species. Penes of male united, swollen at base but not at apex, bearing no spines; a distinct tubercle is present basally between the divisions of the penes. Nymphs bear a rudimentary gill on segment 1; none on 3; gill on 4 operculate. 9th segment of abdomen longer than the 8th; prominent dorsal abdominal spines; lateral margins of abdominal segments much flattened and expanded, bearing prominent postero-lateral spines.

The *hecuba* group contains but the single species, *hecuba*. Imago of this species, described by Shepherd (Canad. Ent. 61: 262) would appear to be in reality a subimaginal form. In this, the gills of the nymph are persistent in the winged form, as occurs occasionally in other species, in the subimago stage. A spine-like tooth at apical margin of fore trochanter; abdominal segments 9 and 10 bear postero-lateral spines. Head of nymph very wide on frontal margin; gills on 4–7 only, operculate on 4; dorsal spines on tergites 2–6 only. Lateral margins of segments 2–9 produced into prominent flattened teeth. A sharp spine at apex of each femur; maxillary palp present.

The heterogeneous *simplex* group contains 3 species. Penes of male broadened more at apex than base, and bearing minute papillae laterally near apex. Nymphs have gills on segments 4–7 only, that on 4 semi-operculate; maxillary palp present; abdominal segment 9 not longer than 8.

In the *serrata* group, with 11 species, the penes of the male are broadened at base as much or more than at apex; barb-like lateral dorsal spines, one on each side, are usually present. Nymphs have gills on segments 3–7, none operculate; no spines on anterior margin of fore femur; a whorl of spines at each tail joint; maxillary palp weak or wanting.

To the *fuscata* group belong 21 species. In the male genitalia, the apical forceps joint is at least twice as long as wide; the long second joint may be distinctly bowed or irregularly swollen. The highly specialized nymphs often bear spines on occiput and dorsum of thorax and abdomen; in many species, the fore femora also bear spines or tubercles on the anterior margin. Gills present on segments 3–7; maxillary palp present.

The heterogeneous *needhami* group contains 9 species. Lateral apical margins of penes project forward or outward as distinct processes, usually rather long and slender; apical notch between divisions of penes rather deep; penes may bear several to many spines. Nymphs bear gills on segments 3–7. Femora with only short and inconspicuous spines on margins, usually none on upper surface of fore femur. Dorsal abdominal spines may be present; maxillary palp present; tails may have whorls of spines at each joining, in the basal half.

Probably the most generalized forms are the 11 species in the *invaria* group. Penes much as in the preceding group, but with lateral apical processes shorter, blunt, and directed inward; apical notch between divisions of penes more shallow. Spines always present on upper part of penes. Nymphs have gills on segments 3–7. Femora bear rather long spines on posterior margins; on upper surface of fore femur, a transverse band of spines near apical end. Maxillary palp present; short dorsal abdominal spines may be present. No whorls of spines at tail joinings.*

* Names have been suggested for certain of these groups, as follows: for the *hecuba* group, **Timpanoga** (monotypic) by the senior author (Ann. Ent. Soc. Amer. 20: 108, 1927); for the *fuscata* group, **Drunella,** type *E. grandis* (N. Y. St. Mus. Bull. 86: 42, 1905), and **Eatonella,** type *E. doddsi* (Ann. Ent. Soc. Amer. 20: 108, 1927), by the senior author; for the *needhami* group, **Chitonophora,** type *E. aurivillii*, by Bengtsson in Lunds Univ. Arssk. (N. F.) 5: 6 (of reprint), 1909. A proper application of these names awaits a better definition of the groups of species, based on more complete knowledge of them in all stages, and especially of the type species of Ephemerella, *E. excrucians*.

KEY TO THE GROUPS OF SPECIES OF EPHEMERELLA

Adults

1—Spine-like tooth at apex of fore trochanter........................**hecuba** group
—No such distinct spine at apex of fore trochanter............................2
2—Third joint of the forceps at least twice as long as wide......................3
—Third joint of forceps scarcely longer than wide..............................4
3—Third joint of forceps four to five times as long as wide; second joint not strongly bowed, but swollen basally, slender apically; brownish species. . **attenuata** group
—Third joint of forceps at least twice as long as wide; second either strongly bowed or irregularly swollen; blackish species........................**fuscata** group
4—Penes united, swollen basally, but not apically; no spines present on penes; a distinct tubercle present between them at the base.............**bicolor** group
—Penes united, somewhat swollen apically; spines usually present on penes; with or without a tubercle between the bases of the penes..........................5
5—Lateral apical margins of the penes projecting forward as distinct processes; usually several to many spines on the penes..............................6
—Lateral apical margins of the penes rounded, and with no forward-projecting processes; spines, if present, barb-like and dorso-lateral in position..........7
6—Lateral apical processes of the penes short, blunt and directed inward; the apical notch between the penes relatively shallow....................**invaria** group
—Lateral apical processes of the penes rather long and slender, directed forward or outward; the apical notch between the penes deeper..........**needhami** group
7—Penes broadened more at apex than at base; minute papillae borne near the apex on each side..**simplex** group
—Penes broadened as much or more at base than at apex; no such minute papillae near the apex; usually with barb-like lateral dorsal spines, one on each side ....**serrata** group

Nymphs

1—Gills present on abdominal segments 3–7....................................2
—Gills absent on segment 3, present on 4–7.................................5
2—Teeth or spines generally present on the anterior margin of the fore femur; when absent, head tubercles present and spines of abdomen extremely large........ ....**fuscata** group
—Teeth or spines wanting on the anterior margin of the fore femur.............3
3—A whorl of spines at each tail joining; maxillary palp weak or wanting......... ....**serrata** group
—Tails fringed with long hairs in the apical portion, may have whorls of spines at the base; maxillary palp normally developed..............................4
4—Rather long spines along the posterior margins of the femora, and on the upper surface of the fore femur, where they are arranged in an irregular transverse band near the apical end; tails never with whorls of spines at the base. **invaria** group
—Only short and inconspicuous spines along the margins of the femora; usually no spines on the upper surface of the fore femur, but if present, generally distributed and not in a transverse apical band; tails may have whorls of spines in the basal half...**needhami** group
5—Knee spines present...**hecuba** group
—Knee spines absent...6

6—Gill rudiment present on segment 1; gill on segment 4 operculate; maxillary palp wanting; segment 9 longer than segment 8 .... **bicolor** group
—Gills on 4–7 only; gill on segment 4 usually semi-operculate; maxillary palp present; segment 9 not longer than segment 8 .... **simplex** group

KEY TO THE SPECIES OF THE *BICOLOR* GROUP

Adults

1—Wing under 8 mm. in length 2
—Wing 8 mm. or more in length 6
2—Tails dark-ringed 3
—Tails wholly pale, dark rings wanting .... **coxalis**
3—Abdomen pale reddish brown or yellowish dorsally; usually with blackish markings 4
—Abdomen dark brown dorsally; blackish markings usually present 5
4—Ganglionic areas of the middle abdominal sternites dark-ringed; usual curved row of dark dots present, may be rather indistinct .... **doris**
—Ganglionic areas not darkened; the curved row of dark dots wanting ..... **bicolor**
5—A distinct brown spot at apex of femur; tail rings blackish .... **minimella**
—Only an indistinct streak near apex of femur; tail rings reddish brown .... **aestiva**
6—Femora, tibiae, thoracic venter and abdomen thickly sprinkled with fine black dots .... **lutulenta**
—No such sprinkling of fine black dots on legs, thorax and abdomen 7
7—A ventral globular swelling present on the penes halfway from base to apex .... **prudentalis**
—No such globular swelling on the penes 8
8—Ganglionic areas of the middle abdominal sternites brownish 9
—Ganglionic areas not distinctly darkened .... **verisimilis**
9—Abdomen reddish brown ventrally .... **funeralis**
—Abdomen pale ventrally .... **temporalis**

VERIFICATION TABLE

Adults of the *bicolor* group

Species	*Wing length*	*Tergites 1 to 6*	*Sternites 1 to 6*	*Ganglia dark*	*Bl. dots, sternites*	*Distr.*
aestiva	6	Dk. br.	Pale	No	Present	E
bicolor	6–7	Red. br.	Pale	No	Absent	E
coxalis	7	Blackish	Pale	Yes	Absent	E
doris	7½	Yellow	Pale	Yes	Indis.	SE
funeralis	8–10	Dp. br.	Red-br.	Yes	Var.	E
lutulenta	10	Bl. br.	Pale	No	Indis.	E
minimella	6	Dk. br.	Pale br.	No	Present	E
prudentalis	8	Red-br.	Pale	No	Present	Que.
temporalis	8–10	Dp. br.	Pale	Yes	Present	E
verisimilis	8–9	Dp. br.	Brown	No	Indis.	Que.

KEY TO THE SPECIES OF THE *BICOLOR* GROUP

Nymphs

1—Occipital tubercle wanting or very rudimentary 2
—Occipital tubercle better developed 3

2—Row of dorsal spines suddenly widened on tergite 5; spines relatively small, especially on tergites 5–7 . . . . . **bicolor**
—Row of dorsal spines not suddenly widened on tergite 5; spines much larger and more upright, especially on tergites 5–7 . . . . . **minimella**
3—Distance between the dorsal spines on tergites 5–7 considerably less than the length of each of these tergites at the median line 4
—Distance between the dorsal spines on tergites 5–7 about equal to the length of each of these tergites at the median line 6
4—Occipital tubercles large, rather conical; rows of dorsal spines subparallel 5
—Occipital tubercles smaller, rounded; rows of dorsal spines convergent to rearward . . . . . **prudentalis**
5—Lateral margin of prothorax distinctly concave; femora relatively short and chunky, the fore femur only twice as long as wide . . . . . **doris**
—Lateral margin of the prothorax almost straight; femora longer and narrower, the fore femur about three times as long as wide . . . . . **temporalis**
6—Postero-lateral spines on segments 2 and 3 poorly developed, the spine on 3 not longer than its width at the base 7
—Postero-lateral spines on segments 2 and 3 well developed, the spine on 3 being fully twice as long as its width at the base 8
7—Rows of dorsal spines divergent to rearward; hind femur relatively long and narrow . . . . . **verisimilis**
—Rows of dorsal spines subparallel; hind femur shorter and chunkier, its length about twice its width . . . . . **aestiva**
8—Postero-lateral spine on segment 9 incurved, the lateral margin of this segment only slightly convex . . . . . **funeralis**
—Postero-lateral spine on segment 9 straight, never incurved; the lateral margin of this segment strongly convex 9
9—Dorsal spine on tergites 1–3 broad, short and blunt; length of body slightly over 8 mm . . . . . **coxalis**
—Dorsal spine on tergites 1–3 long, slender, erect, finger-like; length of body 9–12 mm . . . . . **lutulenta**

VERIFICATION TABLE
Nymphs of the *bicolor* group

Species	*Body length*	*Occip. tuberc.*	*Spines,[1] rows*	*Spines[1] on 1 to 3*	*Spine of 3, L>W[2]*	*Width of rows on 5 to 7[3]*
aestiva	7	Mod.	Parallel	Long	No	Equal
bicolor	6–8	Reduc.	Diverge	Short	No	Equal
coxalis	8½	Reduc.	Diverge	Short	Yes	Equal
doris	7–8½	Strong	Parallel	Long	Yes	Less
funeralis	8½–9½	Mod.	Diverge	Long	Yes	Equal
lutulenta	9–12	Reduc.	Diverge	Long	Yes	Equal
minimella	6	Reduc.	Diverge	Long	No	Equal
prudentalis	8–9	Mod.	Converge	Short	No	Less
temporalis	9–11	Strong	Parallel	Long	Yes	Less
verisimilis	8–9	Mod.	Diverge	Long	No	Equal

[1] Rows of dorsal abdominal spines from front to rear.
[2] Postero-lateral spine on segment 3 longer than wide.
[3] Width between dorsal abdominal spines on 5–7 less than or equal to length of their respective tergites in median line.

KEY TO THE SPECIES OF THE *SIMPLEX* GROUP

Adults

1—Third joint of the forceps very long, about one-half the length of the second joint; fore tarsus of male not longer than the tibia; tails white........... **attenuata**
—Third joint of the forceps short; minute papillae present on the penes; fore tarsus of male at least one-third longer than the tibia; tails smoky.......... **simplex**

Nymphs

1—Tubercles present on the occiput, pronotum and mesonotum.......... **attenuata**
—No tubercles on head or thorax.. 2
2—No dorsal abdominal spines; no denticles on the claws; body very hairy.. **simplex**
—Dorsal spines present; 8–9 denticles on each claw; body not conspicuously hairy .... **margarita**

KEY TO THE SPECIES OF THE *SERRATA* GROUP

Adults

1—Middle tail fully twice the length of the outer ones.............. **heterocaudata**
—Middle tail not noticeably longer than the outer ones....................... 2
2—Abdominal segments 1–6 whitish, semi-hyaline.......................... **frisoni**
—Abdominal segments 1–6 not as above.. 3
3—Wing 8 mm. in length; tails smoky, longitudinal veins pale grey-brown.... **tibialis**
—Wing not exceeding 7 mm. in length; tails pale, longitudinal veins pale........ 4
4—Tails not darker at joinings.. 5
—Tails distinctly darker at joinings.. 6
5—Abdominal sternites with no dark markings.......................... **sordida**
—Abdominal sternites each with a curved row of small dark dots.......... **molita**
6—Abdominal sternites with no darker markings....................... **deficiens**
—Abdominal sternites with darker markings..................................... 7
7—Ventral abdominal markings consist of dark ganglionic patches and a wide lateral streak near each pleural fold; wing 7 mm. in length............... **micheneri**
—Ventral abdominal markings not as above; wing 5–6 mm. in length........... 8
8—Ventral abdominal markings consist of a lateral row of dark dashes on each side, and a curved row of small dark spots between.................. **serratoides**
—Ventral abdominal markings consist of the lateral row of dark dashes only; no dark spots between.. **serrata**

VERIFICATION TABLE

Adults of the *serrata* group

Species	*Wing length*	*Venation*	*Dk. apic. mks., fem. 2 and 3*	*Ventral abd. mks.*	*Tails ringed*	*Distr.*
deficiens	5	Pale	Present	Absent	Yes	E
frisoni[1]	5	Pale	Absent	Absent	No	Ill.
heterocaudata[2]	7	Smoky	Absent	Present	Yes	W
micheneri	7	Pale	Present	Present	Yes	Calif.
molita	6–7	Pale	Present	Present	No	NE
serrata	6	Pale	Present	Present	Yes	E
serratoides	5½–6	Pale	Present	Present	Yes	E
sordida	6	Pale	Present	Absent	No	E
tibialis	8	Smoky	Present	Absent	Yes	W

[1] Abdominal segments 1–6 white. [2] Middle tail very long.

KEY TO THE SPECIES OF THE *SERRATA* GROUP

Nymphs

1—Maxillary palp wanting. 2
—Maxillary palp present; may be very weak. 3
2—Dorsal abdominal spines present. . . . . **teresa**
—Dorsal abdominal spines absent. . . . . **deficiens**
3—Maxillary palp represented only by a very short rudiment. . . . . **cognata**
—Maxillary palp better developed. 4
4—Thoracic tubercles present. . . . . **serrata**
—Thoracic tubercles absent. 5
5—Middle tail almost three times as long as the outer ones. . . . . **heterocaudata**
—The three tails approximately equal in length. 6
6—Long hairs on head, legs and thorax; lateral extensions of abdomen very well developed, postero-lateral spines on 9 extending beyond margin of 10. . **sordida**
—No such conspicuous long hairs; lateral extensions of the abdomen poorly to moderately developed, the spine on 9 not extending beyond the margin of 10. . 7
7—Dorsal spines present on tergites 2–9; no ventral markings. . . . . **tibialis**
—Dorsal spines present on 3–8, best developed on 4–7; dark ventral markings present. 8
8—Ventrally a lateral series of dark curved marks and dark median blotches; several dark bands across the tails. . . . . **micheneri**
—Ventrally a lateral series of dark streaks, and between these a curved row of four dark dots; tails black at base, the tips pale. . . . . **serratoides**

VERIFICATION TABLE

Nymphs of the *serrata* group

Species	*Body length*	*Max. palp*	*Teeth on claw*	*Dors. abd. spines*	*Distr.*
cognata	7–8	Minute	9–10	Present	Calif.
deficiens[1]	5	Absent	8–9	Absent	E
heterocaudata[2]	6–7	Present	7	Present	Wyo.
micheneri	6½–8	Present	7–8	Present	Calif.
serrata[3]	5	Present	3–4	Present	E
serratoides[4]	5–6	Present	7	Present	E
sordida[1]	4–5½	?	?	Present	E
teresa	5–7	Absent	9–10	Present	Calif.
tibialis[1]	7–8	Present	7	Present	W

KEY TO THE SPECIES OF THE *FUSCATA* GROUP*

Adults

1—Large species, wing 11 to 15 mm. in length. 2
—Smaller species, wing 6 to 9 mm. in length 7
2—Second joint of forceps strongly bowed, with a deep constriction near the middle. . . . . **doddsi**
—Second joint of forceps not as above. 3
3—Very large species, wing 15–18 mm. in length. 4

[1] No dark ventral markings. [2] Middle tail very long. [3] Tubercles on dorsum of thorax.
[4] 4 dark spots on each sternite, as well as usual dark lateral markings.
* Does not include *E. depressa*, of which the male imago is not known.

—Moderate-sized species, wing 12 mm. in length .... **coloradensis**
4—Wings distinctly tinged with light amber .... **flavitincta**
—Wings not as above 5
5—Abdominal tergites 2–6 yellowish brown, with large median quadrate black patches and lateral circular black ones; wing 16–18 mm. in length .... **ingens**
—Abdominal tergites 2–6 black or deep purplish brown, margins paler; wing 15 mm. in length 6
6—Long joint of forceps considerably wider at base than at apex ...... **autumnalis**
—Long joint of forceps not wider at base than at apex .... **grandis**
7—Small species, wing 6 to 7 mm. in length 8
—Somewhat larger species, wing 8 to 9 mm. in length 10
8—Distinct purplish rose markings on head, thorax, legs and abdomen; tails orange-brown at base .... **conestee**
—Head, thorax and abdomen blackish brown; no such purplish markings; tails whitish 9
9—Second joint of forceps much inflated, with a deep constriction near the middle. **lata**
—Second joint of forceps only slightly inflated; no such deep constriction, **cornutella**
10—Femora conspicuously sprinkled with black dots 11
—Femora with no such sprinkling of black dots 12
11—Abdominal sternites each with a curved row of four dark dots; tails smoky brown .... **tuberculata**
—Abdominal sternites with no such curved row of dark dots; tails yellowish. **fuscata**
12—Second joint of forceps rather strongly bowed, with a rather deep constriction near middle; venter of abdomen not darker than dorsum 13
—Second joint of forceps only slightly bowed; no such deep constriction; venter of abdomen ruddier than the dorsum .... **flavilinea**
13—Hind femur only slightly longer than tibia; tails yellowish .... **cornuta**
—Hind femur as long as tibia and tarsus combined; tails purplish brown. **longicornis**

VERIFICATION TABLE

Adults of the *fuscata* group

Species	*Wing length*	*Venation*	*2 forc.*[1] *bowed*	*Dk. mks.*[2] *sternum*	*Tails*	*Distr.*
autumnalis	15	Dk. brown	No	Yes	Black	B. C.
coloradensis	12	Dk. brown	No	Yes	Brown	W
conestee	7	Yellowish	No	No	Brown	N. C.
cornuta	9	Pale	Yes	Var.	Yellow	E
cornutella	7	Pale	Mod.	No	White	E
depressa	10	Pale	?	No	White	Ont.
doddsi	14	Dk. brown	Yes	Yes	Brown	W
flavilinea	8	Pale	No	Yes	Black	W
flavitincta	18	Blackish	No	Yes	Black	Ore.
fuscata	8	Yellowish	No	Yes	Yellow	E
grandis	15	Dk. brown	No	Yes	Brown	W
ingens	16–18	Bl. brown	No	Yes	Black	B. C.
lata	6–7	Pale	Yes	No	White	E
longicornis	8–9	Pale br.	Mod.	Yes	Brown	N. C.
tuberculata	9	Pale br.	No	Yes	Brown	E

[1] 2nd joint of forceps strongly bowed.
[2] Abdominal sternites almost wholly dark, or with distinct dark markings.

KEY TO THE SPECIES OF THE *FUSCATA* GROUP

Nymphs

1—Teeth or spines present on the anterior margin of the fore femur............2
—No teeth or spines on the anterior margin of the fore femur................13
2—Occipital tubercles present...3
—Occipital tubercles wanting..7
3—Tubercles present on the pronotum and mesonotum........................4
—No tubercles on pronotum or mesonotum....................................5
4—Lateral extension of the 8th abdominal segment flaring widely at about the middle of its length; no fringe of hairs on the posterior margins of tergites 8 and 9...**allegheniensis**
—Lateral extension of the 8th segment not flaring widely, as above; the posterior margins of tergites 8 and 9 fringed with long hairs..............**tuberculata**
5—Frontal horns present; a frontal shelf covers the mouthparts..........**conestee**
—No frontal horns; frontal shelf incomplete, not covering the mouthparts......6
6—"Thumb" at distal end of fore tibia long and sharp; abdominal spines moderate; femora slender...**coloradensis**
—"Thumb" at end of fore tibia short and blunt; abdominal spines very short; femora stouter..**flavilinea**
7—Frontal shelf notched to receive the bases of the antennae..................8
—Frontal shelf not notched..10
8—Short dorsal spines present...**wayah**
—Dorsal spines wanting..9
9—A complete "adhesive disc" present on the venter of the abdomen; no pronotal tubercles; length 11½–13 mm..................................**doddsi**
—Only slight indications of such a ventral disc; short tubercles present on the pronotum; length 9 mm.......................................**fuscata**
10—Frontal horns rudimentary, not extending as far as the margin of the frontal shelf...**lata**
—Frontal horns conspicuous, usually projecting beyond the frontal shelf......11
11—"Thumb" at distal end of fore tibia short and blunt, not turning outward; frontal horns so strongly curved as to appear somewhat semilunar........**cornutella**
—"Thumb" of fore tibia longer, extending at least to middle of tarsus; sharp at tip; frontal horns less strongly curved.................................12
12—Frontal horns less than half as long as the interval between them; 'thumb' of fore tibia not extending beyond middle of tarsus..................**cornuta***
—Frontal horns more than half as long as the interval between them; 'thumb' of tibia extending beyond middle of tarsus.........................**longicornis**
13—Dorsal abdominal spines forming a rather regular series....................14
—Dorsal abdominal spines on tergites 8 and 9 very much enlarged...........17
14—Submedian mesothoracic tubercles well developed; occipital tubercles prominent 15
—No submedian mesothoracic tubercles, only an elevated submedian ridge; occipital tubercles smaller...16
15—Posterior pronotal spine of submedian pair long; lateral margins of abdominal segments less convex than in following species; body 16 mm. in length...... ....**flavitincta**

* Here belongs also *E. depressa*, which we are unable to separate satisfactorily from *E. cornuta*.

—Posterior pronotal spine of submedian pair distinctly shorter than above; lateral margins of abdominal segments distinctly convex; body 13–15 mm. in length .... **glacialis**

16—Dorsal abdominal spines of 8th and 9th pairs tend to converge slightly at apex; tibiae distinctly longer relatively than in the following species........ **ingens**

—Abdominal spines of 8th and 9th pairs straight or slightly divergent apically; tibiae shorter than above............ **grandis**

17—Submedian spines of pronotum rather slender, sharp-pointed, 2 in each group; submarginal pronotal spines sharp at tips, well developed................ 18

—Submedian spines of pronotum rather large and blunt, 3 in each group; submarginal pronotal spines short and blunt............ 19

18—Twin pairs of pronotal spines subequal; dorsal spine on 8th abdominal segment bent strongly backward and downward............ **autumnalis**

—Twin pairs of pronotal spines unequal, the anterior twice as high as posterior; spine on 8th segment not strongly bent, as above................. **spinifera**

19—Posterior pair of submedian pronotal spines small, subequal; lateral margin and postero-lateral spine on segment 9 practically straight............. **yosemite**

—Posterior pair of submedian pronotal spines unequal, the outer one longer; lateral margin of segment 9 flared anteriorly, its postero-lateral spine directed slightly outward............ **proserpina**

VERIFICATION TABLE

Nymphs of the *fuscata* group

Species	*Body length*	*Frontal horns*	*Occip. tuberc.*	*Pronot. spines*	*Mesonot. spines*	*Teeth, fem. I*
allegheniensis	8–8½	Yes	Yes	Yes	Yes	Yes
autumnalis[1]	11	No	Yes	Yes	Yes	No
coloradensis[1]	11	No	Low	No	No	Yes
conestee	5–7	Yes	Yes	No	No	Yes
cornuta[2]	10	Yes	No	No	No	Yes
cornutella[2]	6–8	Yes	No	No	No	Yes
depressa[2]	9	Yes	No	No	No	Yes
doddsi[2]	11½–13	No	No	No	No	Yes
flavilinea[1]	9–10	No	Yes	No	No	Yes
flavitincta[1]	16	No	Yes	Yes	Yes	No
fuscata[3]	9	Indic.[4]	No	Yes	No	Yes
glacialis[1]	13–15	No	Yes	Yes	Yes	No
grandis[1]	12½	No	Yes	Yes	Yes	No
ingens[1]	13–15	No	Yes	Yes	Yes	No
lata[2]	7	Yes	No	No	No	Yes
longicornis[2]	8–10	Yes	No	No	No	Yes
proserpina[1]	20	Indic.	Yes	Yes	Yes	No
spinifera[1]	11	No	Yes	Yes	Yes	No
tuberculata	10½	Yes	Yes	Yes	Yes	Yes
wayah	6–7	Indic.	No	No	No	Yes
yosemite[1]	13	Indic.	Yes	Yes	Yes	No

[1] No frontal shelf. [2] Dorsal abdominal spines wholly wanting.
[3] Dorsal abdominal spines very short. [4] Indications of very short frontal horns.

KEY TO THE SPECIES OF THE *NEEDHAMI* GROUP

Adults

1—Spines present on the penes 2
—No spines on the penes 3
2—Wing 11 mm. in length; venation pale brown, tails dark smoky brown . . . . **aurivillii**
—Wing 9 mm. in length; venation hyaline, tails whitish . . . . . **septentrionalis**
3—Wing 6–8 mm. in length; fore leg brown 4
—Wing 10–11 mm. in length; fore leg yellow . . . . . **euterpe**
4—Venation pale 5
—Venation dark purplish brown . . . . . **maculata**
5—Abdominal tergites deep reddish to blackish brown, tinged with wine-red; tails deep smoky brown . . . . . **needhami**
—Abdominal tergites olive brown with purplish brown markings; tails whitish . . . . . **catawba**

VERIFICATION TABLE

Adults

Species	*Wing length*	*Fore leg*	*Venation*	*Spines on penes*	*Tails*	*Distr.*
aurivillii	11	Yel. br.	Pale br.	Yes	Dk. smoky	G
catawba	7	Red-brown	Pale	No	Whitish	Calif.
euterpe	10–11	Yellow	Red-br.	No	Purp. br.	Calif.
maculata	8	Brown	Purp. br.	No	Dark br.	E
needhami	6	Smoky br.	Smoky	No	Dk. smoky	E
septentrionalis	9	Yellow	Pale	Yes	Whitish	N. C.

KEY TO THE SPECIES OF THE *NEEDHAMI* GROUP

Nymphs

1—Paired dorsal abdominal spines absent 2
—Paired dorsal abdominal spines present 5
2—Posterior margins of middle tergites prolonged slightly at the median line; legs very long and slender . . . . . **septentrionalis**
—Posterior margins of middle tergites with no such median prolongations; legs not unusually long and slender 3
3—Genae produced laterally to form an incomplete frontal shelf; length 8–11 mm. . . 4
—Genae not produced laterally; length 5–7 mm. . . . . . **catawba**
4—Legs banded; venter of abdomen marked with reddish lateral triangles; length 10–11 mm. . . . . . **euterpe**
—Legs not banded; ventral markings consist of dark bands on the anterior margins and lateral extensions from these; length 8–9 mm. . . . . . **maculata**
5—Dorsal spines long, conspicuous 6
—Dorsal spines short, inconspicuous 7
6—Dorsal spines divergent, curved at the tips; no pale dorsal stripe . . . . . **hystrix**
—Dorsal spines not divergent, the tips not curved; a pale median dorsal stripe is usually present . . . . . **needhami**
7—Lateral extensions of the abdomen inconspicuous; legs not banded; no ventral markings; length 7–8 mm. . . . . . **angusta**
—Lateral extensions of the abdomen well developed, conspicuous; legs banded; dark ventral markings present; length 9–13 mm. 8

8—Length 11½–13 mm.; no lateral prolongation of the genae; legs rather long and slender; pale dorsal stripe present. . . . . **concinnata**
—Length 9–10 mm.; genae slightly produced laterally; legs not unusually long and slender; no pale dorsal stripe. . . . . **aurivillii**

VERIFICATION TABLE

Nymphs of the *needhami* group

Species	*Body length*	*Frontal shelf*	*Legs banded*	*Teeth on claws*	*Dorsal spines*	*Pale dors. stripe*	*Tails*	*Distr.*
angusta	7–8	No	No	6–7	Present	Absent	Pale	Utah, Wyo.
aurivillii	9–10	Faint	Yes	9–10	Short	Absent	Yellowish	G
catawba	5–7	No	Yes	10–11	Absent	Absent	Brownish	N. C.
concinnata[1]	11½–13	No	Yes	8–9	Faint	Present	Yellow	Mont.
euterpe	10–11	Yes	Yes	4–5	Absent	Absent	Dk. brown	Calif.
hystrix	10	No	No	7–8	Present[2]	Absent	?	Mont.
maculata	8–9	Yes	No	3–4	Absent	Absent	Yellowish	Calif.
needhami	9	No	Var.	10	Present	Present	Yellowish	E
septentrionalis[1]	9	No	Yes	11–12	Absent	Present	Yellowish	E

KEY TO THE SPECIES OF THE *INVARIA* GROUP

Adults

1—Second joint of forceps distinctly swollen apically. 2
—Second joint of forceps not distinctly swollen apically. 9
2—Tails not dark-ringed; a pale yellowish white species. . . . . **dorothea**
—Tails distinctly dark-ringed; darker species. 3
3—24–26 spines on the penes, of which 8 to 10 are at the apex. . . . . **mollitia**
—7 to 20 spines on the penes; number of spines at the apex variable, but fewer than above. 4
4—Venation yellowish brown to brown. 5
—Venation paler hyaline. 6
5—Longitudinal veins yellowish to yellowish brown, cross veins pale; posterior margins of abdominal tergites dark brown, intersegmental areas pale. . **vernalis**
—Longitudinal and cross veins reddish brown; posterior margins of tergites smoky, paler than above, abdomen not so distinctly ringed in appearance. . **subvaria**
6—No spines at apex of penes. . . . . **fratercula**
—Usually 2 to 4 spines present at apex of penes. 7
7—Abdomen dull purplish brown ventrally. . . . . **infrequens**
—Abdomen pale yellowish white ventrally. 8
8—Several lateral spines on the penes; dorsal spines at the apex of the V, and a single ventral spine at the apex, on each side. . . . . **invaria**
—Lateral, dorsal and ventral spines as above, and in addition, two small clusters of ventral spines below the dorsal ones. . . . . **rotunda**
9—Thorax and abdominal tergites dark reddish to blackish brown. 10
—Thorax and abdominal tergites yellowish to olive-brown. . . . . **semiflava**
10—A brownish black spot at base of fore wing; thoracic notum deep blackish brown. . . . . **inermis**
—No such brownish spot at base of fore wing; thoracic notum reddish brown . . . . **excrucians**

[1] Legs very long and slender. [2] Very long, curved.

VERIFICATION TABLE

Adults of the *invaria* group

Species	*Wing length*	*Eyes*	*Venation*	*Swollen forceps*[1]	*Spines on penes*	*Distr.*
dorothea	7–8½	Red	Pale	Yes	12–14	E
excrucians	7–8½	Egg yel.	Pale	No	8	C,E
fratercula	8	Red-brown	Pale	Yes	18–20	Que.
inermis	6–8	?	Pale	No	8–10	W
infrequens	9	Light or.	Pale	Yes	14–16	W
invaria	8–9	?	Pale	Yes	12–14	E
mollitia	10	Salmon	Yellowish	Yes	24–26	Calif.
rotunda	10	Orange	Pale	Yes	12–18	E
semiflava	8	Mustard yel.	Pale	No	?	Ont.
subvaria	10	Red-brown	Brown	Yes	10	E
vernalis	8–10	Orange	Yel. brown	Yes	7–10	N. C.

KEY TO THE SPECIES OF THE *INVARIA* GROUP

Nymphs

1—Dorsal spines wholly wanting..2
—Dorsal spines present, but may be minute..............................5
2—Pronotum with distinct pale markings; pale submedian spots present on the posterior margins of the abdominal tergites..............................3
—No such distinct pale markings on the pronotum; no pale submedian spots on the posterior margins of the tergites..............................**infrequens**
3—Lateral extensions of the abdominal segments marked with a dark median transverse band; tails banded throughout their length..............................4
—Lateral extensions of the abdominal segments with no such dark median band; tails banded only at the base..............................**mollitia**
4—Pronotum pale at each corner only; no conspicuous pale blotches on the middle abdominal tergites..............................**dorothea**
—Entire lateral margin of pronotum pale; conspicuous pale blotches on the middle abdominal tergites..............................**inermis**
5—Dorsal spines well developed; no pale submedian spots on the posterior margins of the abdominal tergites..............................**subvaria**
—Dorsal spines very short, may be almost obsolescent; pale median spots usually present on the posterior margins of the tergites..............................6
6—Dorsal spines short, but apparent; color dark with a sprinkling of pale dots; never with dark transverse bands on the notum..............................7
—Dorsal spines much reduced, in some specimens almost obsolescent; often with brown transverse streaks across the pro- and mesonota..............................**invaria**
7—Median line of pronotum pale, usually also the median line of mesonotum, dorsum and venter of abdomen; postero-lateral spines not incurved at the tips; 8–10 denticles on each claw..............................**vernalis**
—No such pale median lines on thorax and abdomen; postero-lateral spines slightly incurved at the tips; 5–7 denticles only, on each claw..............................**rotunda**

1 Second joint of forceps distinctly swollen at apex.

VERIFICATION TABLE

Nymphs of the *invaria* group

Species	*Body length*	*Teeth on claw*	*Dorsal abd. spines*	*White dors.*[1] *spots*	*Dk. band, lat. ext.**	*Distr.*
dorothea[2]	7	6–7	Absent	Var.	Present	E
inermis[2]	6–8	8	Absent	Present	Present	W
infrequens	8–9	5–6	Absent	Absent	Absent	W
invaria	8–9	5–6	Reduced	Var.	Present	E
mollitia	8–10	8–10	Absent	Present	Absent	Calif.
rotunda	10	5–7	Short	Present	Present	E
subvaria	9–10	?	Present	Absent	?	E
vernalis	8–11	8–10	Short	Present	Present	N. C.

Mr. P. N. Albright has studied the gills of the Ephemerella nymphs of the Cornell collection and has prepared the following key based on gill structure. Living material was used for comparison in the case of a few local species. Only alcoholic specimens were in the main available, and in many cases not enough of these were obtained to allow corrections for individual variation. Nevertheless, the key well shows the extraordinary variety occurring within the comparatively simple Ephemerella gill-pattern.

A KEY TO THE SPECIES OF EPHEMERELLA NYMPHS

Based on Gill Structure

By Philip N. Albright

1—Gill on fourth segment elytroid 2
—No gill elytroid 11
2—Gill on segment four covers the next one for about one-third of its length . . . . . . **attenuata**
—Gill on segment four covers the next one for about three-fourths of its length . . . 3
—Gill on segment four almost covers the next one 5
—Gill on segment four completely covers the others 8
3—Pigmentation on the upper plate of gill on segment four is light or wanting . . . . . . **hecuba**
—Pigmentation of the upper plate of gill on segment four is heavy 4
4—Gill on segment 7 is minute . . . . . . **margarita**
—Gill on segment 7 is larger, about one-third the length of the others . . . . **simplex**
5—Gill on segment 7 is minute 6
—Gills are gradually reduced in size from the 4th to the 7th segments 7
6—Under plate of gill on fourth segment is cleft from the tip about three-fourths of its length . . . . . . **aestiva**
—Under plate of gill on segment four is cleft to its base . . . . . . **minimella**
7—Under plate of gill on fourth segment is deeply bilobed, pigmentation of upper plate is slight and more confined to the apex . . . . . . **funeralis**

[1] White submedian spots on apical margin of each tergite, at base of spine, if spine is present.
[2] Postero-lateral spines on segments 6–8 shorter than wide; in others, length equals width.
* A dark band crossing the lateral extensions of the abdominal segments.

—Under plate of gill on fourth segment is deeply bilobed, pigmentation of the upper plate is general and heavy . . . . **verisimilis**
—Under plate of gills on segment is merely notched at its apex . . . . **temporalis**
8—Gill on segment seven is much smaller than the others 9
—Gills on segments gradually become smaller from the fourth to the seventh . . . . **lutulenta**
9—Lobes of under plate of gill on segment 4 are mostly confined to the upper surface of the plate . . . . **coxalis**
—Lobes of under plate of gill on segment 4 are large and numerous and found on both surfaces of the plate 10
10—Gill on segment seven is minute . . . . **doris**
—Gill on segment 7 is about one-fourth the length of those on previous segments . . . . **bicolor**
11—Gill on segment three overlaps the one following one-half its length or more . . . 12
—Gill on segment three overlaps the one following one-third its length or less . . . 21
12—Pigmentation of upper plate of gill on segment three slight or none 13
—Pigmentation of upper plate of gill on segment three heavy 15
13—Secondary lobes of under plate of gill on segment three are linear in shape . . . . **micheneri**
—Secondary lobes of under plate of gill on segment three are broad; oval to orbicular 14
14—Secondary lobes of under plate are directed apically . . . . **aurivillii**
—Secondary lobes of under plate are directed laterally . . . . **mollitia**
15—Upper plate of gill on segment three just covers the lower plate 16
—Upper plate of gill on segment three extends beyond the lower plate one third to one-half the length of the upper plate 18
16—Secondary lobes of under plate of gill on segment three are not compactly arranged . . . . **allegheniensis**
—Secondary lobes of under plate of gill on segment three are compactly arranged 17
17—Secondary lobes of under plate of gill on segment three are directed apically . . . . **cornutella**
—Secondary lobes of under plate of gill on segment three are directed laterally . . . . **vernalis**
18—Main lobes of under plate of gill on segment three are two in number and unequal 19
—Main lobes of under plate of gill on segment three are two in number and equal 20
19—Secondary lobes of under plate of gill on segment three are directed laterally . . . . **serratoides**
—Secondary lobes of under plate of gill on segment three are directed apically . . . . **needhami**
20—Under plate of gill on segment three very heavily pigmented . . . . **longicornis**
—Under plate of gill on segment three with very slight or no pigmentation **rotunda**
21—Pigmentation of upper plate of gill on segment three none or slight 22
—Pigmentation of upper plate of gill on segment three heavy and either general or in a definite area 28
22—Gill on segment seven less than one-third the length of that on segment three . . 23
—Gill on segment seven more than one-half the length of that on segment three . . 25
23—Upper plate of gill on segment three only slightly longer than the lower . . **cornuta**
—Upper plate of gill on segment three one-fourth or more longer than the lower . . 24

24—Base of upper plate of gill on segment three is lobed............... **infrequens**
—Base of upper plate of gill on segment three is obliquely rounded...... **angusta**
—Base of upper plate of gill on segment three is truncate............. **maculata**
25—Secondary lobes of under plate of gill on segment three are hair-like or finger-like .... **serrata**
—Secondary lobes of under plate of gill on segment three are broad (oval or elliptical to spatulate)..26
26—Terminal lobes of under plate of gill on segment three converge and cross each other.. **lata**
—Not so..27
27—Under plate of gill on segment three cleft two-thirds of its length or less.. **catawba**
—Under plate of gill on segment three cleft three-fourths of its length or more .... **dorothea**
28—Gill on segment seven is minute...29
—Gill on segment seven larger, about one-fifth the length of the gill on segment three..33
29—Pigment of upper plate of gill on segment three is in a three lobed area (side lobes may be small)...30
—Pigment of upper plate of gill on segment three is general and shows no lobing.31
30—Under plate of gill on segment three cleft at least two-thirds of its length...... .... **concinnata**
—Under plate of gill on segment three cleft at most one-half of its length.. **inermis**
31—Upper plate of gill on segment three only one-third (or less) longer than the under plate.. **cognata**
—Upper plate of gill on segment three two-thirds (or more) longer than the under plate..32
32—Secondary lobes of under plate of gill on segment three are numerous, small, closely placed together, and appear of many layers................. **euterpe**
—Secondary lobes of under plate of gill on segment three are not close together and of only two layers.. **spinifera**
33—Pigmentation of upper plate of gill on segment three is of only one lobe or a central area...34
—Pigmentation of upper plate of gill on segment three is in a well-defined two- or three- lobed area...38
34—Secondary lobes of under plate of gill on segment three with a basal thumb-like lobe.. **grandis**
—Secondary lobes not so formed..35
35—Main lobes of under plate of gill on segment three are unequal.............36
—Main lobes of under plate of gill on segment three are equal or nearly so....37
36—Upper plate of gill on segment three is elliptical...................... **invaria**
—Upper plate of gill on segment three is elongate-cordate........ **septentrionalis**
37—Secondary lobes of under plate of gill on segment three are lanceolate.. **conestee**
—Secondary lobes of under plate of gill on segment three are elliptical to spatulate .... **flavilinea**
38—Secondary lobes of under plate of gill on segment three are finger-like.......39
—Secondary lobes of under plate of gill on segment three are broad oval or spatulate 40
39—Upper plate of gill on segment three covers those on the following segments one-third their length.. **bispina**

—Upper plate of gill on segment three covers those on the following segments not over one fifth of their length . . . . **deficiens**

40—Layers of secondary lobes of under plate of the gill on segment three apparently three . . . . **coloradensis**

—Layers of secondary lobes on under plate of gill on segment three are two . . **doddsi**

Ephemerella aestiva McDunnough

1931—McDunnough—Canad. Ent. 63: 64. Figs. . . . . Que.

Length: body 6; wing 6 mm.

A member of the *bicolor* group, very close to *E. bicolor*. Imagos are difficult to separate from that species. Nymph has moderately prominent occipital tubercles in both sexes; rows of dorsal spines subparallel; spines on tergites 1–3 moderately long, narrow and tapering bluntly.

Imago very similar to a large specimen of *E. bicolor*. Thorax somewhat darker brown than in that species. Legs light yellow. A dark spot on the side of each coxa; traces of dark median banding on the middle and hind femora, and a slight dark terminal streak on each. The first seven abdominal tergites are an even dark brown; the apical three, bright brown. A paler mid-dorsal stripe, enclosing a narrow dark median line, and margined by dark submedian streaks, may be present. Sternites 1–7 subhyaline, the basal three tinged with brown. Sternites 8 and 9 are whitish, opaque, and shaded with light brown. The usual curved row of four dark dots is more or less well developed. Tails pale, the joinings reddish brown. Genitalia of the usual type in this group.

The nymph has occipital tubercles rather well developed in both sexes. The femora are relatively short and chunky, the posterior margin of the hind femur being only slightly more than twice the greatest width. Rows of dorsal tubercles subparallel; those on tergites 1–3 are moderately long, rather narrow, blunt and tapering. The spines on segments 2 and 3 are poorly developed, being however slightly stronger than in *E. bicolor*. The margins of the lateral extensions bear many spinules, but are practically without the long hairs found in *bicolor*. General color dark brown, sprinkled with pale dots; traces of a pale mid-dorsal line margined with dark submedian streaks usually present. Legs dark brown with pale banding, which is reduced to small oval spots on the femora. Operculum sprinkled with pale dots, and with a pale median oval patch. Short dark lateral dashes and the usual curved row of four dark dots are usually present on the abdominal sternites. Tails dark brown, banded alternately with yellow.

The species was described from material taken in Quebec. In the Cornell collection are several nymphs which are tentatively referred to this species, and a few reared imagos. The nymphs are from Enfield Glen, Six Mile Creek and Wilseyville Creek, in the vicinity of Ithaca, N. Y.; Elmsdale N. S. (A. H. Leim); the Buckhannon River in W. Va. (Mrs. A. L. Burckmyer); Town Creek near Ellijay, Ga. (P. W. Fattig); and a stream near Reidsville, N. C. (J. R. Traver). The reared material is from Spencer, N. Y. (A. Louise Steger).

Ephemerella allegheniensis Traver

1934—Traver—J. Elisha Mitchell Sci. Soc. 50: 221. . . . . W. Va.

Length (nymph): body 8½ mm.

A species of the *fuscata* group, allied to *E. tuberculata*. Only the nymph is known. Separated from the latter species by the greater development of the occipital and thoracic tubercles, the widely flaring lateral margin of segment 8, and the lack of long hairs on tergites 8 and 9. See fig. 51.

Short blunt frontal horns are present, and a frontal shelf which is slightly notched laterally (see fig. 151e). Occipital tubercles high and prominent. Tubercles present also on the pro- and mesonota. Fore femur with numerous spines on the anterior margin and a few on the posterior. All femora flattened, the fore femur widest. Tibial 'thumb' well developed. Fore tarsus distinctly widened apically. Dorsal abdominal spines present, best developed on tergites 5–7. Gills borne on segments 3–7. No fringe of hairs on the posterior margin of tergites 8 and 9, as in *tuberculata*. The lateral extension of segment 8 flares outward widely at about the middle of its length. General color (in life) greenish brown with black markings. A transverse black band between the eyes; black streaks on the occipital tubercles; wide and prominent black markings on pro- and mesonota; black lateral patches on abdominal tergites 3, 4, 6 and 7. Legs yellowish with indistinct brown markings. Venter of abdomen yellowish; a lateral row of black dashes on each side; black submedian streaks; and a curved row of four black dots. Tails yellowish to reddish brown; near the middle, 4 or 5 narrow bands of white.

The type material of this species, in the Cornell University collection, is from the Cacapon River, W. Va. (J. R. Traver); and the Potomac River, Md., near Harpers Ferry (V. Argo).

Ephemerella angusta Traver

1934—Traver—J. Elisha Mitchell Sci. Soc. 50: 211 .... Utah, Wyo.

Length: body (nymph) 7–8 mm.

A slender species probably allied to the *needhami* group; only the nymph is known; lateral extensions of the abdomen inconspicuous; dorsal spines present.

Head and thorax smooth. Maxillary palp weak but perfect in form. Antennae pale. General color pale reddish brown. Wing pads and the mesonotum anterior to them dark brown in mature nymphs. Legs pale, with no conspicuous markings. 6 to 7 denticles on each claw. Hind leg rather long.

Abdomen relatively long and slender; middle segments slightly wider than the basal and apical ones. Lateral extensions inconspicuous. Postero-lateral spines on segments 4–9. Gills borne on segments 3–7. Dorsal spines present on tergites 2–8; short and inconspicuous on 2 and 3, well developed on 5–7. Anterior margin of each tergite pale. Posterior margin often narrowly dark brown; remainder reddish brown. Rather large lateral patches may be present next to the gills on the middle segments. No definite ventral markings. Tails pale, almost bare at base and tip, and with only a few short spines in the middle area; in mature nymphs, with narrow dark half-rings at the base.

The type material was taken in the Provo River and the North Fork of the Ogden River, Utah (J. G. Needham). Other specimens in our collection are from the Yellowstone National Park, Wyo. (J. T. Needham, R. A. Muttkowski).

Ephemerella attenuata McDunnough

1925—McDunnough—Canad. Ent. 57: 42 .... Que
1930—McDunnough—Canad. Ent. 62: 55.
1931—McDunnough—Canad. Ent. 63: 209. Figs. .... Ont.

Length: body 6; wing 6 mm.

A species considered to be of the *simplex* group; fore tarsus of male comparatively short, third joint of forceps long, about one-half the length of the second; nymph rather slender, with occipital tubercles and tubercles on the pro- and mesonota; gills on segments 4–7 only; dorsal spines present.

Eyes deep brown. Head and thorax deep blackish brown. Legs pale yellowish white; fore legs darker, the tibia and tarsus largely smoky. The middle and hind femora may show an apical brownish spot. Fore tibia about one and one-half times the length of the femur, and about as long as the entire tarsus. Hind tibia twice as long as the tarsus. Wings hyaline; venation hyaline, cross veins faint except in the stigmatic area, where they are strongly anastomosed. Abdomen dark brown dorsally, the three posterior segments paler brown. Paler ventrally, the eighth sternite quite pale creamy brown. Forceps whitish, shaded at base with light brown; third joint very long, about one half the length of the second. Penes united except for a slight V-shaped apical cleft; widened apically, narrowed again at the extreme apex (see fig. 157). Tails white.

Nymph rather slender; abdomen somewhat flattened. Occipital and prothoracic tubercles present, also a single median anterior and a similar posterior mesothoracic tubercle. The lateral extensions of the abdominal segments are prolonged into shorter and blunter postero-lateral spines than in *E. simplex;* the lateral margin of segment 8 is distinctly and characteristically sinuate. Dorsal spines are present, most distinct on segments 4–7. Gills present on segments 4–7 only. General color light brown. A few blackish spots are present on the anterior portions of the pro- and mesonota, and one dark spot near the base of the wing pads. Legs slender, pale; each femur is marked with a narrow broken median dark band; tibia with a median black ring, and a broader black band at the base of the tarsus. 8 to 9 denticles on each claw. Pale submedian dorsal stripes may be present, also a median dark streak on the apical tergites; on the anterior margins of the first five tergites are lateral pale dots; the lateral extensions are pale, with a median brown band. Ventrally a lateral row of dark dashes is present on each side, a curved row of four small dark dots across each sternite, and sometimes dark median blotches on the anterior margin of each. Tails rather short, pale in immature nymphs, with a dark band before the apex; in mature nymphs the tails are reddish at the base, and may be crossed by two dark bands beyond the middle. Basally they bear short spines; apically each is fringed with long fine hairs.

The type material was taken near Hull, Que. We have specimens in the Cornell collection from this locality and from S. Bolton, Que. (G. S. Walley), which were kindly sent us by Dr. McDunnough; also a series of nymphs from Catoctin Creek, a tributary of the Potomac River, (V. Argo).

Ephemerella aurivillii Bengtsson

1908—Bengtsson—K. Sven. Vet. Akad. Arsbok. 6: 243 (Chitonophora).
1908—Eaton—in Esb. Pet. Tromso Mus. Aarsh. 25: 149 (*aronii*).
1909—Bengtsson—Lunds Univ. Arssk. N. F. Afd. 2, V: 8.
1924—McDunnough—Canad. Ent. 56: 223. Fig. (*norda*)............................Alta., Alaska
1930—Bengtsson—Lunds Univ. Arssk. N. F. Afd. 26: 3.
1930—Walley—Canad. Ent. 62: 17. Fig. (*aronii*).. Que.
1930—McDunnough—Canad. Ent. 62: 55 (*aronii*).
1930—Ide—Canad. Ent. 62: 211 (*aronii*)..Ont.
1930—Lestage—Bull. and Ann. Soc. Ent. Belg. 70: 204 (Chitonophora).
1931—McDunnough—Canad. Ent. 63: 197.
1932—McDunnough—Canad. Ent. 64: 79.

Length: body 10; wing 11 mm.

A species of the *needhami* group; venation pale brown; spines present on the penes. Nymph with short dorsal spines.

Head brown. Thorax deep reddish brown; lateral margins of the mesonotal scutellum blackish brown; three small pale spots anterior to the scutellum. Pleura with yellowish interspaces. Fore leg dark yellowish brown, slightly tinged with smoky;

minute spines on the fore tibia. Middle and hind legs deep yellow; a small dark blotch may be present at the apex of the femur; claw, and shading on the distal joint of the tarsus, reddish. Wings hyaline, the longitudinal veins pale brown, deepest on the costal margin of the fore wing. Stigmatic area opaque whitish, the cross veins anastomosed.

Posterior two-thirds of each abdominal tergite deep smoky brown; anterior portions paler; apical tergites reddish brown. Traces of a dark median line, partially geminate, and of indistinct darker lateral streaks may be present. Pale yellowish brown ventrally, the basal sternites faintly suffused with purplish. Ganglionic areas pale. Traces of the usual submedian oblique dashes at the anterior margin, and of the small dark dot central dots. Genitalia pale brownish. Second joint of the forceps somewhat swollen at base and apex; third joint short, conical. Penes separated apically by a deep V-shaped cleft; many small spines are present along the inner apical margin (see fig. 155). Tails deep smoky brown; joinings not distinct.

Body of nymph yellowish to yellowish brown, with dark reddish brown markings. Head smooth; a tendency toward a slight lateral development of the genae to form a very shallow frontal shelf, much less complete than in *E. maculata* or *E. euterpe*. In some specimens, the face is brown and the vertex mottled. Thorax smooth. Pronotum indistinctly marked with darker brown; an incomplete dark line along the anterior margin. Dark brown markings also on the anterior margin of the mesonotum. Femora pale brownish, pale apically; fore and middle femora with a pale band near the apex; short spines on the posterior margins, and on the upper surface of the fore femur, near the apex. Tibiae and tarsi pale, each with two brownish bands.

Abdomen light reddish brown dorsally, the middle tergites with dark reddish brown lateral patches on each side above the gills. Gills present on segments 3–7. Lateral extensions of the abdomen very well developed, each bearing a postero-lateral spine; pale brownish with a darker spot on the margin of each. Inconspicuous dorsal spines present on segments 2–8; each bears minute spinules. Venter pale; a series of dark lateral streaks on each side. Tails pale, a few dark bands at base and apex only.

This species, originally described from Europe, has been recorded in North America from Alaska, Alberta and eastern Canada. In the Cornell collection are specimens of nymphs and imagos from Bradore Bay, Que. (W. J. Brown), presented by Dr. McDunnough. Other nymphs, somewhat larger and darker, are from the Joyce Kilmer Preserve, Pa. (A. Louise Steger).

Ephemerella autumnalis McDunnough

1934—McDunnough—Canad. Ent. 66: 158. Fig. . . . . B. C.

Length: body 11; wing 15 mm.

A species of the *fuscata* group, allied to *E. spinifera*.

Eyes deep reddish brown. Head black; pits at bases of antennae "and a small area posterior to mid-ocellus pale yellowish. Pronotum black with yellowish shades laterally" (McD.) at bases of fore legs. Mesonotum black; pale yellowish brown markings as indicated; a "lateral streak extending forward from base of wing, the lateral edge of the anterior projection, the lateral posterior edges reaching caudad to beginning of scutellum and relieving in this region a black dorsal band as broad as and extending over scutellum" (McD.). Metathorax black; slight yellowish shading on scutellum, intersegmental areas of pleura and around bases of legs. Fore femur and tibia light amber, "heavily suffused with blackish, the pale color only showing at bases of femora and at joint" (McD.); tibia subequal to femur; tarsus light smoky amber. Middle and hind femora and tibiae light amber, the femora suffused with blackish in distal third; tibia as long as femur and trochanter together; tarsi and claws dull smoky. Wings hyaline, tinged faintly with smoky in stigmatic area. Veins and cross veins rather fine, dark; stigmatic cross veins paler.

Abdominal tergites 1 and 2 almost wholly black; 3–7 blackish, anterior margins narrowly pale smoky, these are broader at median area of tergite 7. Traces "of a geminate dark median line filled with pale color; on segments 5–7 there is a tendency for the black area to break up into three patches, due to paler submedian prongs projecting from the posterior pale area into and almost separating the dark area; laterally there are faintly pale comma-shaped marks in the dark areas" (McD.). Tergites 8 and 9 all black except a narrow pale posterior margin; 10 pale, "narrowly dark on anterior border with median and lateral prongs projecting backward into the pale area" (McD.). Sternites 2–7 with a narrow pale anterior margin and wider pale posterior margin, blackish between these pale bands; two pale submedian lunate marks on anterior margin, two small dots near center of segment, and a larger lateral pale dot or dash, on dark portion of each sternite. Sternites 8 and 9 black; posterior margin of 9 pale. Forceps blackish; yellowish brown apically; base pale yellowish brown. Tails black.

Nymph quite similar to *E. spinifera*, but differing in these points: twin spines on pronotum subequal; postero-lateral spines of segments 8 and 9 shorter, most evident of 9; long dorsal abdominal spines on segment 8 "more strongly bent backward and downward" (McD.). The species matures fully a month later than *spinifera*. We have seen no specimens.

Ephemerella bicolor Clemens

1913—Clemens—Canad. Ent. 45: 336. Fig. .... Ont.
1915—Clemens—Cont. Canad. Biol. p. 123. Fig.
1925—McDunnough—Trans. R. S. C. 19: 212. .... Que.
1930—McDunnough—Canad. Ent. 62: 57. Fig.
1931—McDunnough—Canad. Ent. 63: 61. Fig. .... N. B.

Length: body 5–7; wing 6–7 mm.

A member of the *bicolor* group; imago paler than other related species. Nymph variable in both size and color; occipital tubercles wanting in the male, much reduced in the female; rows of dorsal spines divergent posteriorly, with an abrupt increase in the space between them on tergite 5.

Head of male light brown. Thorax dark reddish brown. Head of female pale yellow. Legs pale yellow; coxae tinged with brown, and with black spots. Tinges of red may be present near the apex of each femur. Wings hyaline; venation hyaline. Stigmatic cross veins anastomosed. Abdominal tergites pale reddish brown with a distinct tinge of olive-brown. Blackish streaks near the pleural fold, and in some specimens a paler mid-dorsal line margined by interrupted blackish submedian streaks. Ventrally pale, the basal sternites slightly tinged with brown. Tracheae appear as fine dark lines. Tails pale yellowish white, the joinings purplish red.

The occipital tubercles are wanting in the male nymph, and much reduced in the female. The rows of dorsal spines are distinctly divergent toward the rear, with a rather abrupt increase in the space between them on tergite 5. The spines on tergites 1–3 are short and chunky, and barely tapering. Postero-lateral spines on segments 2 and 3 are very poorly developed; these spines are well developed on the posterior segments, tipped with black and gracefully curved. The margins are set with short spinules, between which are long hairs. Color very variable, as is also the size. Pale specimens are strikingly marked with dark brown on the thorax and the posterior abdominal tergites. Legs distinctly banded. An oval patch is often present near the middle of the operculum. The usual four dots in a curved row, and the dark lateral dashes, are present in most specimens. Tails brownish, with indications of pale cross bands. Fore leg as in fig. 62e.

The species was described from material taken at Georgian Bay, Ont. We have imagos in the Cornell collection from St. Lambert, Que. (G. S. Walley), and from Lachine, Que. (F. P. Ide) which Dr. McDunnough very kindly sent us. Nymphs which appear to belong to this species are from Cayuga Flats, Beebe Lake and Six Mile Creek, in the vicinity of Ithaca, N. Y.; Slaterville Springs, N. Y.; Keuka Lake, N. Y. (C. R. Crosby); Grass River near Canton, N. Y.; and Elmsdale, N. S. (A. H. Leim).

Ephemerella catawba Traver

1932—Traver—J. Elisha Mitchell Sci. Soc. 47: 53, 176. Figs. . . . . N. C.

Length: body 6½–7; wing 7 mm.

A species of the *needhami* group; usually without spines on the penes; venation pale.

Head light brown; eyes light orange. Thorax reddish brown. Pronotum margined laterally with pale yellow, mesonotum blotched with yellow. Mesonotal scutellum and metanotum darker brown. Pleura brown, with yellow markings. Sternum brown; metasternum shaded with purple anteriorly. Fore leg light reddish brown, claw and last two tarsal segments greyish. Middle and hind legs white; claws and tarsi washed with light brown. Wings hyaline; veins colorless, except the three main veins of the costal margin, which are tinged with yellow. Stigmatic area opaque, cross veins anastomosed.

Abdomen olive brown dorsally; tergites 1–6 with purplish brown markings. Tergites 7–9 light chestnut brown; purplish brown submedian lines on 7. Anterior margins of tergites yellowish, posterior margins dark brown. Median line faintly greyish white; pleural fold greyish white. Sternites 1–6 greyish white with lavender tinge; sternite 1 chestnut brown on anterior and lateral margins. Faint darker submedian dashes, and dots near the median line. Tails smoky white at base, silvery white beyond; joinings purple. Penes as in fig. 155.

Nymph slender, smooth; olive green to brown, with or without white patches on abdominal tergites 4–7. Thoracic notum and legs mottled with yellow and brown. Fore femur smooth; two dark bands. No dorsal spines on abdomen. The flattened lateral margins are prolonged into sharp spines, yellow in color. Ventrally paler than above, usually without markings. Tails yellowish to reddish brown, with several narrow yellow cross bands. Usually found in moss on rocks, in rapid water.

The type material, from several locations in North Carolina, is in the Cornell University collection. This species was considered as belonging in the *invaria* group, at the time it was described. It seems as closely allied to the *needhami* group, however, and we are now transferring it to that group.

Ephemerella cognata Traver

1934—Traver—J. Elisha Mitchell Sci. Soc. 50: 231. . . . . Calif.

Length (nymph): body 7–8 mm.

A member of the *serrata* group, of which only the nymph and female subimago are known. Head and thorax of nymph smooth; dorsal spines present; postero-lateral spines longer than in *E. tibialis*. Abdominal sternites of female with black lateral marks and darkened ganglionic areas.

Mandibles of nymph well developed. Maxillary palp vestigial, represented only by a minute stub. No head or thoracic tubercles, but a slight elevation on each side of the pronotum on the posterior margin. Legs reddish brown; apex of tibia and a median tarsal band pale. 9 to 10 denticles on each claw. Gills present on tergites 3–7. Lateral extensions of abdomen, and postero-lateral spines, better developed than in the related species *tibialis*. True dorsal spines present on tergites 4–7; on 4 these are wider apart than elsewhere. General color dark reddish brown. Ventral abdominal markings consist of dark ganglionic blotches and a series of dark lateral streaks. Tails yellow at base, alternately brown and yellow beyond the middle. Tails of subimago white, joinings purplish black.

We have nymphs of this species from Waddell Creek and San Antonio Canyon, Calif. (D. Shepherd, J. G. Needham) and a reared female subimago from the first location.

Ephemerella coloradensis Dodds

1923—Dodds—Trans. Amer. Ent. Soc. 49: 97. Figs. .... Colo.
1927—Needham—Ann. Ent. Soc. Amer. 20: 113. .... Utah, Wyo., N. M.
1927—Needham—Utah Agr. Exp. Sta. Bull. 201: 8. Fig.
1929—McDunnough—Canad. Ent. 61: 170. Fig. .... Alta.
1930—Walley—Canad. Ent. 62: 19. Fig. .... Mont.

Length: body 12; wing 12 mm.

A species of the *fuscata* group; closely allied to *E. flavilinea*, from which the imago may be distinguished by the dark venation and the straighter second forceps joint, which is not swollen apically. Nymph with very low occipital tubercles; tibial 'thumb' longer and sharper at the tip than in *flavilinea;* middle and hind legs relatively more slender; dorsal spines more prominent.

Imago dark brown, without striking color markings. Thorax dark brown. Venation dark purplish brown. Abdomen rather uniform dark brown; pleural fold pale; margins of segments pale, appearing as light bands when the body is extended. Tails dark brown, deeper brown at the base. Apical margin of forceps base almost straight between the bases of the forceps limbs. Second joint of forceps moderately and regularly curved, not noticeably swollen near the apex; somewhat thinner near the middle than elsewhere.

Nymph without frontal horns or frontal shelf. Head roughened; occipital tubercles present, but very low and inconspicuous. No thoracic spines. Anterior margin of the fore femur with teeth or spines. Tibial 'thumb' quite long, sharp-pointed, relatively straight. Legs stout, but relatively more slender than in *E. flavilinea;* all femora somewhat flattened, the posterior margins with many hairs and a few short spines. Gills borne on segments 3–7. Dorsal spines present on tergites 1–9, somewhat longer on tergites 6–8 than on the basal segments. General color light brown, mottled with paler areas, especially on the legs. Tails each with two dark and two paler bands.

We have nymphs of this species from Gothic, Colo. (J. D. Bowen); Glacier Nat. Pk., Mont. (A. S. Hazzard, *et al*); Yellowstone Pk., Wyo. (R. A. Muttkowski); Teton Pk., Utah (A. S. Hazzard); Logan River, Utah (J. G. Needham); and the Big Blackfoot River, Potomac, Mont.

It is probable that Eaton's Ephemerella Nymph No. 1, from Washington Territory (figured on Pl. 38 of the Monograph, Figs. 1–10) is this species.

Ephemerella concinnata Traver

1934—Traver—J. Elisha Mitchell Sci. Soc. 50: 219. .... Mont.

Length (nymph): body 11½–13 mm.

A member of the *needhami* group, known in the nymphal stage only; close to *septentrionalis*, but with somewhat shorter legs; very inconspicuous dorsal spines.

Nymph dark reddish brown in color, with a wide pale median stripe almost the entire length of the body. Head smooth. Dark blackish brown anterior to the ocelli; vertex and occiput yellow, with a dark brown mark like three sides of a square enclosing a pale space, in which are other dark marks near the median line. Antennae white at base; a dark brown band just beyond the base; remainder yellowish white. Indistinct darker spots on each side of the pronotum, and anterior to the wing roots on the mesonotum. A wide median yellowish stripe extends the length of the thorax dorsally; median line pale, margined on each side by a narrow dark brown submedian streak. Pleura yellow with dark brown markings. Venter yellow, the median sclerites dark reddish brown. Legs long and slender, the hind leg longest; tarsus about ⅔ the length of the tibia; tibia and femur subequal. Short spines on the upper surface and along the posterior margin of each femur; longer spines on the anterior margin and at apex of each tibia, and on the anterior margin of each tarsus. Long hairs on the posterior margins of each joint. Femora dark brown, pale at apex; several rather large pale blotches at base and near the middle of the upper surface. Tibiae yellow, with two dark bands; tarsus dark brown, yellow at each end. Each claw bears 8 to 9 denticles on its inner margin.

Abdomen dark reddish brown dorsally; a wide pale median stripe is present on tergites 1–7. Within this pale streak lies the dark median line, narrow on the basal tergites, wider on tergites 6 and 7. Rather inconspicuous dorsal spines are present on tergites 2–8; each appears as a slight elevation upon which are borne numerous short spinules. Lateral extensions of the abdominal segments are quite prominent; each bears a rather long postero-lateral spine. Gills borne on segments 3–7; pale, each with a darker flask-shaped median mark. Tergites 8–10 largely dark reddish brown. Ventral markings consist of a lateral series of dark marks in each side; short oblique dark submedian marks at the anterior margin; and the usual two dark central dots. Tails yellow, marked with five or six darker bands. Basally a whorl of spines at each joining; distally the tails are fringed with rather long hairs.

This species is represented by two nymphs only, taken in Glacier National Park, Mont. (A. S. Hazzard). The specimens are in the Cornell collection.

Ephemerella conestee Traver

1932—Traver—J. Elisha Mitchell Sci. Soc. 47: 168. Figs. . . . . N. C.

Length: body 6; wing 7 mm.

A species of the *fuscata* group; by the nymphal structures allied to *E. tuberculata*.

Abdomen of male imago tinged with orange or rose color; body and legs with purplish rose markings; tails brown. Nymph has very short frontal horns, occipital but no thoracic tubercles, and dorsal abdominal spines.

Eyes yellowish to orange. Head light brown tinged with purple. Pronotum largely purplish rose; a prominent median carina. Mesonotum light olive brown; metanotum dark brown; both with purplish tinges. Pleura light olive brown, the sutures pale. Two rose colored streaks anterior to the wing roots. Fore femur greyish, streaked with purplish rose; tibia and tarsus dark olive brown. Middle and hind legs light olive brown, joinings and claws dark brown. Wings hyaline, veins pale yellowish; a purplish brown stain at base of fore wing. Abdominal tergites 1–3 deep purple with rose tinges; 4 and 5 orange marked with deep purplish red; 6–9 similar to 4 and 5 but paler. Anterior and middle tergites are darker on each side of mid-dorsal line, halfway to the pleural fold, and at the center of the pleural fold. Sternites greyish lavender with rose tinges; 7–9 yellowish. Tails orange brown at base, shading into smoky brown; joinings slightly darker. Forceps smoky brown; penes deep rose.

The frontal horns of the nymphs are very short; the frontal shelf is slightly notched laterally (see fig. 158b). A pair of short blunt tubercles on the occiput. No thoracic tubercles, but the dorsal surface is irregular. Many spines on the anterior margin of the fore femur, and a few on the posterior; tibial 'thumb' short, blunt. Dorsal spines present on tergites 2–7. Gills borne on segments 3–7. Head brown mottled with yellow and darker brown; thorax yellow with irregular brown mottling and streaks. Legs yellowish with indistinct brown bands. Tergites 1–4 brown mottled with yellowish laterally; 5 and 6 yellow; 7 and 8 yellow mottled with brown; 9 and 10 dark brown. Yellowish to yellowish brown ventrally; sternite 9 dark brown. Tails yellow to yellowish brown; 4 to 7 narrow whitish bands.

The type material, from the mountains of North Carolina, is in the Cornell collection (J. R. Traver). Other nymphs are from Forney's and Deep Creek, in the Great Smoky Mts., and from Mt. Leconte, Tenn. (J. G. Needham).

Ephemerella consimilis Walsh

1862—Walsh—Proc. Acad. Nat. Sci. Phil. p. 378.......... Ill.
1931—McDunnough—Canad. Ent. 63: 187.
1932—Traver—J. Elisha Mitchell Sci. Soc. 47: 147.

Length: body 7; wing 6½ mm.

This species, very inadequately described from a single male, cannot even tentatively be placed as to its possible relationship to other species in the genus. The type specimen has been destroyed, the nymph is not known. Until such time as much material can be collected from the type locality and carefully checked with the description of this species, its identity cannot be determined.

Ephemerella cornuta Morgan

1911—Morgan—Ann. Ent. Soc. Amer. 4: 114. Figs.......... Conn.
1931—McDunnough—Canad. Ent. 63: 211. Figs.......... N. H., Que.
1931—Steger—Psyche 38: 28. Figs.......... N. Y.
1932—Traver—J. Elisha Mitchell Sci. Soc. 47: 164.......... N. C.

Length: body 9; wing 9 mm.

A species of the *fuscata* group allied to both *E. cornutella* and *E. longicornis*; third forceps joint of the male relatively longer, and the second joint more strongly bowed, than in its immediate allies. Nymph has curved frontal horns, a frontal shelf, and spines on the anterior margin of the fore femur.

Head and thorax blackish brown. Pleural sutures, an area anterior to the wing root, and the antero-lateral margins of the mesonotum paler, with ruddy tinges. Femur and tibia of fore leg blackish brown, tarsus smoky brown. Middle and hind legs yellowish amber to olive brown. Hind femur very slightly longer than the tibia. Axillary cords extend backwards slightly beyond the scutellum, ending in blunt lobes. Wings hyaline, venation pale; the main veins of the costal margin often tinged with yellow basally, becoming yellowish brown apically. A brown spot at the base of the subcosta in the fore wing. Abdomen reddish brown dorsally and ventrally, the posterior segments darker; middle segments often with very narrow pale margins. Pleural fold may be narrowly pale. In some specimens there are faint traces of pale mid-dorsal and submedian streaks, pale ventral ganglionic areas and short brown submedian oblique streaks from the anterior margins. Tails pale yellowish. Our local form is slightly larger and paler than the one from Quebec. Specimens of the variety from North Carolina have wider pale areas on the abdomen; the abdominal tergites have darker transverse median bands and dark posterior margins; venation pale brown; tails purplish brown; size slightly larger than the New York form.

Head of nymph with curved frontal horns and a frontal shelf; no occipital tubercles. Thorax smooth. No dorsal abdominal spines. Gills borne on segments 3–7. General color reddish brown, the thorax often mottled with brown. Anterior margin of fore femur bears teeth or spines; the tibial spine or 'thumb' curves outward gradually from the base, and does not extend beyond the middle of the tarsus. Tails pale, with a single dark basal ring.

In the Cornell collection we have imagos of this species reared from local nymphs (A. Louise Steger); a pair from Knowlton, Que., which Dr. McDunnough kindly sent us; and reared specimens (known as variety A) from North Carolina, (J. R. Traver). Nymphs in our collection include the type material from Salisbury, Conn. (Lucy Smith); specimens from the vicinity of Ithaca, N. Y.; and from several localities in North Carolina (J. R. Traver).

Ephemerella cornutella McDunnough

1931—McDunnough—Canad. Ent. 63: 82.......... Que.
1931—McDunnough—Canad. Ent. 63: 211. Figs.......... N. H., N. B., N. S.
1932—Traver—J. Elisha Mitchell Sci. Soc. 47: 163.......... N. C.

Length: body 7; wing 7 mm.

A species of the *fuscata* group; very close to *E. cornuta*, but considerably smaller.

Head and thorax deep blackish brown. Pleura tinged with reddish below the bases of the wings. Legs light brown, tinged with ruddy basally. Wings hyaline, venation pale; cross veins almost invisible. Abdominal tergites smoky brown; posterior margins darker brown; dark brown oval patches, rather indistinct, situated laterally. Sternites light olive brown; posterior sternites reddish brown, opaque. Forceps dull brownish, the third joint relatively shorter than in *cornuta*. Penes as in fig. 153. Tails dull whitish, the joinings faintly reddish brown.

Nymph very much resembles a small *cornuta*, but is darker in color, varying from dark reddish brown to brownish black. The frontal horn is considerably more strongly incurved. The tibial 'thumb' of the fore leg is short and blunt, not curving outward as in *cornuta*. Other structural characters as in *cornuta*. A dark basal spot is present on the tibia, also a dark median band; tarsus dark basally. Tails yellowish brown.

We have imagos of this species in the Cornell collection from Knowlton, Que. (G. S. Walley), sent us by Dr. McDunnough; reared specimens from Black Mt., the Catawba River at Andrews Geyser, and a tributary of the Laurel River, N. C. (J. R. Traver); and other reared specimens from Salmon Creek at North Lansing, N. Y. Nymphs in our collection are from many localities in the mountains of North Carolina (J. R. Traver); from Ithaca, N. Y., and vicinity, by various collectors; McLaughlin Springs, W. Wa. (J. G. Needham); and the Chestatee River near Cleveland, Ga. (P. W. Fattig).

Ephemerella coxalis McDunnough

1926—McDunnough—Canad. Ent. 58: 186.......... Que.
1931—McDunnough—Canad. Ent. 63; 37. Fig.......... Ont.

Length: body 7; wing 7 mm.

A species of the *bicolor* group. First seven abdominal tergites heavily shaded with light brown and black; black spots on coxae, trochanter and femur. Nymph (tentatively associated) has the occipital tubercles rather low; dorsal spines on 1–3 broad and blunt.

Eyes ruddy brown. Head yellow. Thorax brown, brighter on the posterior part of the mesonotum; pleura with yellow shading. Legs rather bright yellow; two black spots on each coxa, and an apical black spot on each trochanter and femur, most distinct on the middle and hind legs. Wings hyaline; venation hyaline. The first seven abdominal tergites are heavily shaded with light brown and black; on tergites 1–4 only pale median patches are left unmottled. In each pale patch is a black median mark, and dark submedian shading. Tergites 5–7 have larger pale areas in which the dorsal streak is lacking, but the submedian streaks are prominent. Black streaks in the region of the spiracles. Tergites 8–10 are light brown with submedian dark marks anteriorly, and lateral dark streaks. Pale yellowish white ventrally; sternites 1 and 9 are shaded with brown. Ganglionic areas brown. Tails white, unmarked.

A nymph tentatively associated with the above imago is paler in color than other related species, with the head tubercles, antennae and dorsal spines prominently white. Head and thorax are sprinkled with pale dots; abdomen is light brown dorsally, with yellow and black lateral markings and traces of dark submedian marks along the rows of dorsal spines. Legs pale yellowish, the light brown bands with numerous pale spots. Occipital tubercles not prominent. The rows of dorsal spines diverge to rearward; those on tergites 1–3 are broad, short and blunt, and bend backward gently. The postero-lateral spines on segments 2 and 3 are moderately well developed; those on segments 4–7 are long, narrow and incurved. The operculum usually has a pale oval patch near the middle; the joint is pale; dark longitudinal veining is faintly present. Venter of abdomen pale, with the usual curved row of dark dots. Tails light brown, with only very faint traces of pale cross bands.

We have imagos of this species in the Cornell collection, presented by Dr. McDunnough, from Vaudreuil, Que. (G. S. Walley). Nymphs and several reared specimens which are close to *coxalis* are from the Uharie River near Farmer, N. C. (J. R. Traver); Asheboro, N. C. (J. G. Needham); and Chickamauga Creek, Ga. (P. W. Fattig).

Ephemerella deficiens Morgan

1911—Morgan—Ann. Ent. Soc. Amer. 4: 111. Figs. . . . . N. Y.
1925—McDunnough—Trans. R. S. C. 19: 212. Figs. (*atrescens*) . . . . Que.
1925—McDunnough—Canad. Ent. 57: 43 (*atrescens*) . . . . Ont.
1929—McDunnough—Canad. Ent. 61: 170 (*atrescens* = *deficiens*).
1930—Ide—Canad. Ent. 62: 211.
1931—Steger—Psyche 38: 30.
1931—McDunnough—Canad. Ent. 63: 205. Fig. . . . . N. B., N. S., N. H.
1932—Traver—J. Elisha Mitchell Sci. Soc. 47: 148, 177 . . . . N. C.

Length: body 5; wing 5 mm.

A member of the *serrata* group; distinguished from *E. sordida*, in the male imago, by the deep blackish brown abdominal tergites, the prominent dark apical spot on the middle and hind femora, and the blackish apical portion of the penes.

Head and thorax blackish brown. Eyes reddish brown, antennae brown. Vertex of head in female deep wine-red. Pronotum brown; meso- and metathorax brown, likewise the pleura, all with blackish markings. Sternum brown. Coxae brown; legs nearly white, the femora with a prominent dark brown spot apically, especially prominent on the middle and hind legs. Wings hyaline, veins pale; brown shading at base of wings. Abdominal tergites deep blackish brown; sternites paler. Tails pale, the joinings narrowly but distinctly purplish brown. Forceps pale; penes whitish basally, blackish apically (see fig. 156).

Nymph blackish (rarely dull brown), usually with a pale median thoracic stripe dorsally. Head and thorax smooth; maxillary palp entirely wanting; no dorsal abdominal spines. Lateral extensions of the abdominal segments and postero-lateral spines only moderately developed; tips of the spines whitish. Gills present on segments 3–7. Each claw with 8–9 denticles. Tails yellowish brown; a wide dark median band crosses the two outer ones; at each joining, a circle of prominent spines. Femora dark brown, tibiae and tarsi yellowish with wide brown bands. Two pale spots laterally on each side of the pronotum.

We have specimens of this species from Ithaca and vicinity (Anna H. Morgan, A. Louise Steger); Lost River, the north fork of the Potomac River at Smoke Hole, and Wardensville, W. Va. (J. G. Needham, J. R. Traver); several stations in North Carolina (J. R. Traver); Nassau, N. Y. (L. Harper); and the Mashpee River, Cape Cod, Mass. (W. D. Sargent).

Ephemerella depressa Ide

1930—Ide—Canad. Ent. 62: 212. Figs. .... Ont.
1931—McDunnough—Canad. Ent. 63: 212.

Length: body 6–7 mm.; wing 10–11 mm.

A species of the *fuscata* group; very close to *E. cornuta* (perhaps it is synonymous with that species). Male imago not known. Female separated from that of *cornuta*, in original description, by the usual but not constant fusion of vein M and Cu_1 in the fore wing. Nymph said to be distinguished from that of *E. cornuta* by the shape of the maxillary palp and the fore femur. From the illustrations, we can find no good distinguishing character for the nymphs of the two species. Further, the venational character mentioned is one which occurs occasionally in several species of this group, and is apparently constant in none. More material is necessary to determine the validity of this species.

Ephemerella doddsi Needham

1884—Eaton—Revis. Monogr., Pl. 39, figs. 1–22; p. 132. (Nymph No. III) .... Colo.
1923—Dodds—Trans. Am. Ent. Soc. 49: 95. Figs. (*Drunella grandis*).
1927—Needham—Ann. Amer. Ent. Soc. 20: 111 .... Mont., Calif., Utah, Wash.
1927—Needham—Utah Agr. Exp. Sta. Bull. 201: 7. Fig.
1928—McDunnough—Canad. Ent. 60: 8 .... Alta.
1930—Walley—Canad. Ent. 62: 14. Fig.

Length: body 13; wing 14 mm.

A member of the *fuscata* group; allied to *E. fuscata* in type of genitalia. Superficially very like *E. grandis*, but without the wide dark femoral bands, and with darker tarsi; genitalia distinct. Nymph without occipital or dorsal tubercles; no dorsal spines; anterior margin of fore femur with spines; an adhesive disc on venter of abdomen.

Head and thorax dark reddish brown; pleural sutures paler. Fore femur dark purplish brown; tibia slightly paler than femur; tarsus yellowish brown, paler at the tip. Middle and hind legs yellowish; femora shaded with brown, but without the prominent wide dark apical bands of *grandis;* the three basal tarsal joints shaded with purplish black. Wings hyaline; venation dark brown. Abdominal segments each with a broad median transverse band of dark brown, somewhat paler on the dorsal areas of the tergites; anterior and posterior margins widely pale. Pleural fold pale. All pale areas relatively wider than in *grandis*. Tails dark brown at the base, becoming pale at the tips. Second joint of the forceps strongly bowed, deeply constricted at the middle.

Nymph (see fig. 50) without frontal horns, but with a broad frontal shelf, notched widely to receive the bases of the antennae (see fig. 151 f). No occipital nor thoracic tubercles. Anterior margin of the fore femur with spines or teeth; smaller spines on its posterior margin, and on both margins of the other femora as well. All femora somewhat flattened. Tibial 'thumb' about one-third as long as the tarsus; strong, rather sharp at the tip; directed outward. No dorsal abdominal spines. A broad ventral adhesive disc is present on the abdomen, extending from sternite 1 to 9; long hairs form its outer margin, short hairs occupy the inner space. General color yellowish to tan, with darker areas on eyes, wings, gills, bases of tarsi and tails; abdominal tergites shaded indefinitely with brown. Tails yellowish, darker at bases; tips light brown.

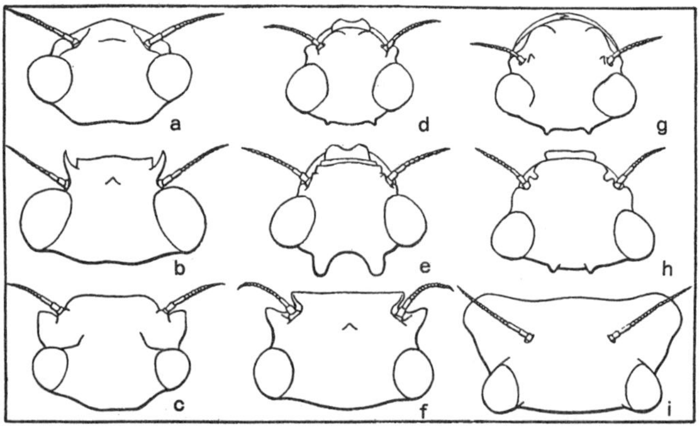

FIG. 151. Heads of Ephemerella nymphs (drawings by W. D. Sargent). *a, maculata; b, longicornis; c, fuscata; d, flavilinea; e, allegheniensis; f, doddsi; g, grandis; h, tuberculata; i, hecuba.*

We have reared imagos of this species from Logan River, Utah (J. G. Needham); and a single specimen from Wenatchee, Wash. (M. D. Leonard). Nymphs are from Logan River, Utah (J. G. Needham); Glacier Nat. Pk., Mont. (A. S. Hazzard); Gothic, Colo. (J. D. Bowen); Keweah River, Calif. (C. Juday); Estes Pk., Colo. (P. W. Claassen); Santa Fe, N. M. (P. B. Powell); Yosemite, Calif. (P. R. Needham); Yellowstone Pk., Wyo. (R. A. Muttkowski); Centennial, Wyo. (R. J. Gilmore); Teton Pk., Utah (A. S. Hazzard); Marble Fks., Calif. (J. C. Bradley); Ashford, Wash. (Hortense Butler); Missoula, Mont.; and Potomac, Mont. (F. E. Barry).

Ephemerella doris Traver

1934—Traver—J. Elisha Mitchell Sci. Soc. 50: 208.................................... N. C., Ga.

Length: body 7; wing 7½ mm.

A species of the *bicolor* group, close to *E. temporalis*. Imago paler and more yellowish than in the latter species. A distinct concavity of the lateral margin of the pronotum of the nymph seems a distinctive character.

Eyes orange. Base of antenna pale, filament brown. A dark spot below each antenna. Thorax yellowish brown; a few dark dots on the pronotum, and a black submedian spot on each side of the metanotum. Legs pale yellowish; a dark spot on each coxa. Apex of fore tibia, tarsal joinings and claws light reddish brown. Wings

PLATE XXXVIII

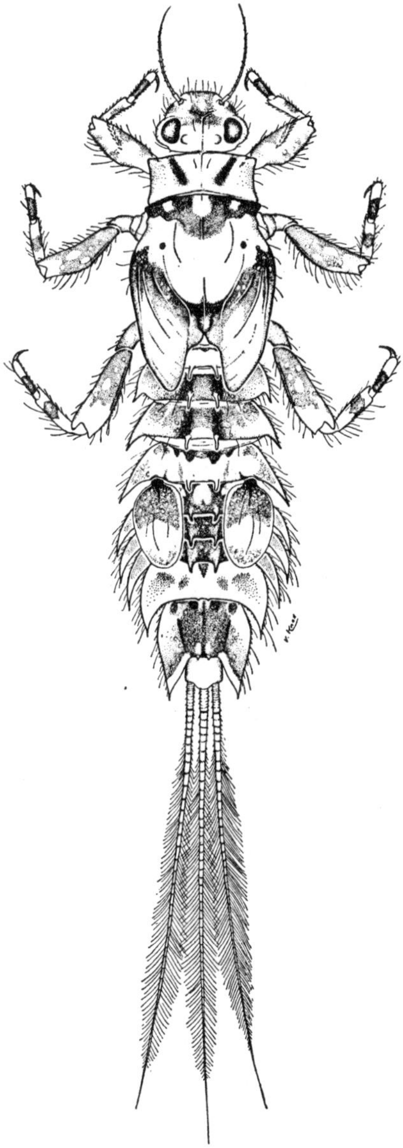

NYMPH OF *Ephemerella doris*
(Drawing by Velma Knox).

hyaline; venation hyaline; a faint brownish tinge at the base of the fore wing in the anal region, and a short dark streak at base of costa and radius. Stigmatic cross veins anastomosed. Abdomen yellowish, paler ventrally. Posterior margins of the basal and middle tergites are brownish red. Distinct dark submedian streaks on tergites 2, 3, and 6–7; faint indications of these on the other tergites. Tracheae dusky; black streaks in the region of the stigma. On tergites 8 and 9, a small black spot is on each side of the median line at the anterior margin; tergite 10 whitish. Basal and middle sternites have the usual curved line of dark dots, which are less apparent apically. Ganglionic areas black-margined. Tails pale, joinings purplish black. Penes as in fig. 157.

Nymph variable as to color pattern (see Pl. XXXVIII). In one phase, marked conspicuously with alternating areas of yellowish white and blackish brown. Lateral margins of the pronotum distinctly concave. Femora relatively short and broad, the hind femur being only twice as long as broad. Postero-lateral spines well developed on segments 2–9; on segment 3, about twice as long as the width at the base; on 4–7, long and more slender than in *E. temporalis*. Dorsal abdominal spines on tergites 1–3 quite long, blunt, and rather finger-like; bent backwards in the posterior half. Dorsal spines on 8 and 9 very small, but better developed than in *temporalis*. Operculum blackish brown, sprinkled with pale dots, as are the dark brown areas in other parts of the body. Rows of dorsal spines are subparallel, although the members of each pair on tergites 1–3 diverge slightly at the tips.

The type material was from the Piedmont area of North Carolina, in the Uharie River (J. R. Traver); from Lake Waccamaw in the coastal plain region of North Carolina (J. G. Needham); and from the Withlacoochee River near Macon, Ga. (P. W. Fattig). These specimens are in the Cornell University collection.

Ephemerella dorothea Needham

1907—Needham—N. Y. St. Mus. Bull. 124: 190. Fig. ....N. Y.
1925—McDunnough—Trans. R. S. C., p. 212. ....Que.
1931—McDunnough—Canad. Ent. 63: 194. Fig. ....N. B., N. H., Tenn., N. C.
1931—Steger—Psyche 38: 31.
1932—Traver—J. Elisha Mitchell Sci. Soc. 47: 148, 176.

Length: body 5–6 mm.; wing 7–8½ mm.

A small very pale species, of the *invaria* group; the second joint of the forceps enlarged distally; 12–14 spines on the penes.

Eyes of male bright red; the head of the female is often reddish on the vertex. Thorax and legs pale yellowish. Wings hyaline, iridescent, the venation wholly pale. Abdomen pale yellowish; in the female, often suffused with ruddy on the dorsum. Tails white, without dark markings. Genitalia yellowish, the second joint of the forceps enlarged distally. Spines on the penes number from 12 to 14, and are usually ventral and lateral in position (see fig. 152). (Dr. McDunnough reports specimens in which apical spines are present, but this is not true of any of our material).

The small yellowish brown nymph is speckled with pale dots. The lateral extensions of the abdomen are poorly developed, the postero-lateral spines short except on segments 8 and 9. There are no dorsal spines on the abdomen. Claws are strongly curved; each bears 8–9 denticles. Tails pale, rather darker at the tips; may be faintly banded.

The type material, reared by Prof. Needham at Old Forge, N. Y., in July, 1905, is in the Cornell University collection. We have also reared specimens from various localities in the vicinity of Ithaca, N. Y. (A. Louise Steger, P. R. Needham); and from Big Alamance Creek, Danbury, Caraway Creek, and Flat Creek at Black Mt., N. C.

(J. R. Traver). Imagos not reared are from Waterbury, Vt.; Covey Hill, Que. (G. S. Walley); Stamford, Conn. (S. W. Bromley); and the Beaverkill R., N. Y. (P. R. Needham). Nymphs are from Greenville, S. C. (H. K. Townes, Jr.); from the Great Smoky Mts., Tenn. (P. W. Musgrave); from Wardensville, W. Va. and Hillsville, Va. (J. R. Traver); and from Joyce Kilmer Stream, Pa. (A. Louise Steger).

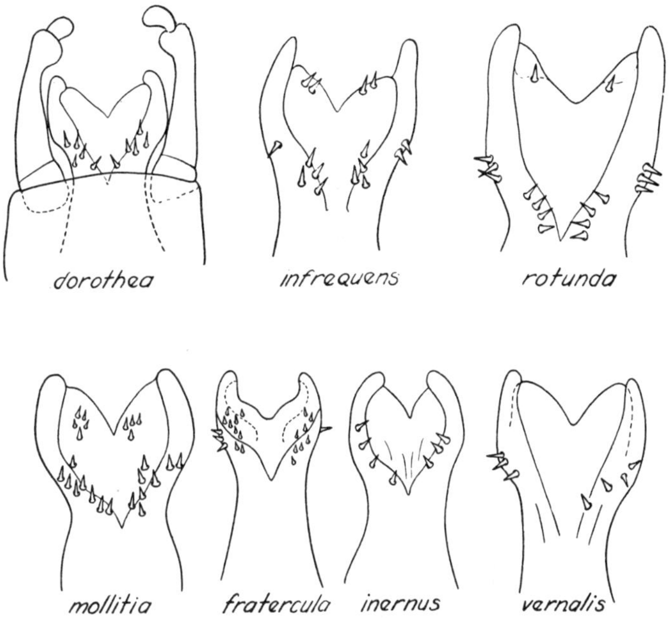

FIG. 152. Male genitalia of seven species of Ephemerella, of the *invaria* group. (Figure of *fratercula*, after McDunnough.)

Ephemerella euterpe Traver

1934—Traver—J. Elisha Mitchell Sci. Soc. 50: 213 .... Calif.

Length: body 10–11; wing 10–11 mm.

A member of the *needhami* group; venation light reddish brown; genitalia without spines; prominent dark ventral markings in nymph and imago; nymph without dorsal spines.

Head dark brown, frontal margin paler; eyes light orange. Thorax reddish brown. Pronotum shaded with smoky brown; mesonotum margined with darker brown, and with narrow median and submedian streaks on the scutum; scutellum outlined with black, its lateral margins dark red-brown. Two dark streaks on the pleura. Sternum light reddish brown with darker brown areas; ganglionic areas blackish. Legs yellowish. Fore femur tinged with smoky at base and apex; a small dark reddish brown spot at the apex. Distal joint of fore tarsus, and claws and tarsal joinings of all tarsi, pale smoky. Very small spines on the fore tibia. Wings hyaline; stigmatic area opaque whitish, the cross veins anastomosed near the costal margin. Longitudinal veins light reddish brown, those of the anal area wholly pale, as are all the cross veins.

Abdomen dorsally light reddish brown tinged with olive, the apical segments darker. Anterior margins narrowly dark, posterior margins pale; pleural fold pale, paralleled above by a dark line containing a dark stigmatic dot and a small pale oblong area. Traces of a pale median line and of pale submedian streaks on the basal tergites. Venter pale yellowish white; anterior margin of each apical sternite, and broad lateral triangles on all the sternites, deep purplish brown. Sternites 7–9 largely dark; ganglionic areas blackish brown. Tails dark purplish brown basally, somewhat paler distally; joinings not distinct. Genitalia reddish brown; terminal joint of forceps short, thick, conical. Penes separated apically by a V-shaped cleft; no spines are present (see fig. 155).

Nymph light reddish brown; a few indistinct darker markings on the thorax; legs fringed with hairs; tips of femora and tibiae, both ends of the tarsi, and claws, dark reddish brown; 4 to 5 pectinations on each claw. Apical tergites distinctly reddish. Broad lateral triangles mark each sternite, as in the imago. Tails blackish brown at the base, reddish brown distally. Two or three narrow dark bands near the base. Beyond the basal third, tails fringed with rather long hairs. Head and thorax smooth. Genae produced laterally and anteriorly to form frontal shelf. Gills on segments 3–7. Lateral extensions of the abdomen quite prominent; each bears a postero-lateral spine. No dorsal spines.

The type material, taken at Waddell Creek, Rancho del Oso, Calif. (D. Shepherd), is in the Cornell collection and in the collection of Stanford University.

Ephemerella excrucians Walsh

1862—Walsh—Proc. Acad. Nat. Sci. Phil. p. 377. . . . . Ill.
1884—Eaton—Revis. Monogr. p. 130.
1925—McDunnough—Canad. Ent. 57: 169. Fig. . . . . Ont.
1930—Ide— Canad. Ent. 62: 211.
1931—McDunnough—Canad. Ent. 63: 192.

Length: body 5½–7½ mm.; wing 7–8½ mm.

A member of the *invaria* group; second joint of forceps not swollen apically; nymph unknown.

Eyes (in living male) egg-yellow. Vertex of head, and antennae, reddish brown. Dorsum of thorax reddish brown. Legs pale yellow; first tarsal joint of fore leg, all tarsal joinings of fore leg, and tips of all the tarsi, cloudy. Wings hyaline; venation wholly hyaline; a slight yellowish tinge on the costa of the fore wing.

Dorsum of abdomen reddish brown, sometimes almost reddish black. Tails whitish, the joinings brown except near the tips. Second joint of forceps little if any swollen apically; no spines at the apex of the penes.

The type material, from the vicinity of Rock Island, Ill., has been destroyed, but other specimens in the Museum of Comparative Zoology at Cambridge, Mass. were sent by Walsh to Hagen. One of these has been selected by Dr. Banks as the lectotype. Specimens in the Cornell collection from the Upper St. Lawrence River, from Knowlton, Quebec, and from Fredericton, N. B., although considerably paler, are tentatively placed in this species by Dr. McDunnough; he has kindly sent us several specimens collected at Horning's Mills, Ont. (F. P. Ide).

Ephemerella flavilinea McDunnough

1926—McDunnough—Canad. Ent. 58: 188. . . . . Alta.
1929—McDunnough—Canad. Ent. 61: 169. Fig. . . . . Mont., Wyo.
1930—Walley—Canad. Ent. 62: 18. Fig.

Length: body 6½; wing 8 mm.

A species of the *fuscata* group; very close to *E. coloradensis*. The imago is distinguished from the latter species by the pale venation, and rather sharp constriction of the second joint of the forceps, near the apex. Nymph with shorter abdominal spines, stouter legs and a shorter tibial 'thumb.'

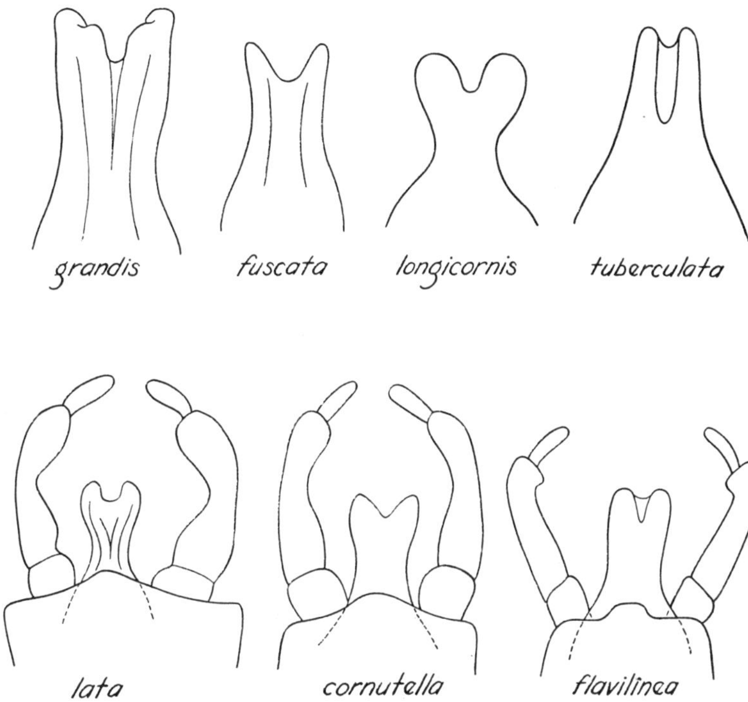

FIG. 153. Male genitalia of seven species of Ephemerella, of the *fuscata* group.

Head and thorax deep ruddy brown. Femur and tibia of fore leg deep blackish brown; tarsi dull whitish. Tibia once and a half as long as the femur, and slightly shorter than the tarsus. Middle and hind legs are dull yellowish; a reddish brown apical patch on each femur on the inner side. Wings hyaline, venation pale. Abdominal tergites deep ruddy brown; sternites more reddish than the dorsal surface. A pale yellowish line along the pleural fold, which tends to extend upward at the edge of each segment, especially on the posterior ones. Tails blackish at the base, paler toward the tips. Second joint of the forceps with a rather sharp constriction near the apex; the tip bent inward (see fig. 153).

Nymph without frontal horns or a frontal shelf (see fig. 151 d). Occipital tubercles present, but no thoracic spines. Anterior margin of the fore femur with teeth or spines. Dorsal abdominal spines present on 2–9; very small on 9; somewhat shorter than in *coloradensis*. Tibial 'thumb' shorter than in *coloradensis;* fore tibia more slender; middle and hind legs stouter. General color variable, ranging from light to dark brown; pale specimens appear distinctly mottled. Femora with three pale indistinct areas; tibiae with distinct pale bands at each end; tarsi banded in the middle. Tails pale; narrow blackish bands beyond the middle and at the tips.

A pair of imagos of this species from Lundbreck, Alta. (J. H. Pepper) were presented to the Cornell collection by Dr. McDunnough. Nymphs in this collection are from Yellowstone Park (R. A. Muttkowski); Glacier National Park (A. S. Hazzard); and Moose, Wyo. (Elsie B. Klots).

Ephemerella flavitincta McDunnough

1934—McDunnough—Canad. Ent. 66: 155. . . . . Ore.

Length: body 15; wing 18 mm.

A species of the *fuscata* group, allied to *E. glacialis*; wings are tinged with light amber.

Head black; pits at bases of antennae yellowish. Thorax shiny black. Intersegmental areas of pleura dull yellow-brown; brighter yellow around bases of legs. Fore legs deep black, tarsi slightly paler. Coxae and trochanters of middle and hind legs blackish, ring of membrane between them yellowish. Femora largely black, yellowish on basal margin. Tibiae and tarsi rather bright yellow; distal tarsal joint tinged with smoky. Wings hyaline, "tinted with light amber, the color paling apically; veins and cross veins strong, blackish," (McD.), stigmatic cross veins rather faint. No fusion between posterior branch of median and Cu_1.

Abdominal tergites black; posterior margins narrowly pale yellowish, "most noticeably in lateral angles and on segments 8 and 9, the other portions being somewhat dulled (at least in specimen examined) with smoky brown." (McD.). Sternites black; posterior margins of sternites 1–7 rather bright yellow; sternites 8 and 9 wholly black. Forceps entirely black. Tails black, becoming pale smoky apically.

Nymph very similar to that of *E. glacialis*, but larger than that species, with broader head and better developed occipital tubercles; posterior spine of submedian spine on pronotum much longer than in *E. glacialis*.

The large size and the amber-tinged wings of the imago should distinguish this species from all others in the genus. We have seen no specimens.

Ephemerella fratercula McDunnough

1925—McDunnough—Trans. R. S. C. 19: 213. Fig. . . . . Que.
1931—McDunnough—Canad. Ent. 63: 197.

Length: body 7; wing 8 mm.

A species of the *invaria* group; numerous ventral spines in the apical area of the penes; venation wholly pale. Nymph unknown.

Eyes reddish brown. Head light yellow; vertex shaded with brown. Prothorax smoky brown. Mesonotum light yellow; pleura and sternum deeper yellow; brown shading at bases of wings. Legs light yellow; femur of fore leg deeper yellow than tibia and tarsus, which are slightly darker than the middle and hind legs. Wings hyaline; venation wholly hyaline; cross veins indistinct except in the stigmatic area.

Dorsum of abdomen deep smoky brown on the basal and middle tergites, lighter yellowish brown apically; basal tergites deeper smoky than middle ones. Posterior margins narrowly blackish brown. Venter light yellow, the apical sternites opaque and slightly deeper yellow; ganglionic areas smoky brown; no reddish brown shading, as in *E. subvaria*. Forceps as in *E. invaria;* numerous ventral spines on the apical area of the penes (see fig. 152). Tails pale, the joinings narrowly dark brown.

The species is known only from the type locality, Covey Hill, Que. We have one specimen taken at Covey Hill (G. S. Walley), which was kindly sent us by Dr. McDunnough.

Ephemerella frisoni McDunnough

1927—McDunnough—Canad. Ent. 59: 10.......... Ill.
1931—McDunnough—Canad. Ent. 63: 206.

Length: body 5; wing 5 mm.

A member of the *serrata* group; male imago distinguished by the pale whitish anterior abdominal segments. Nymph unknown.

Head and thorax deep blackish brown. Mesothorax broad; axillary cords barely extending beyond the tip of the scutellum; pleura with lighter brown tinges. Fore femora light sepia brown; tibiae and tarsi dull whitish; tibiae slightly shaded at base with sepia, and rather long, being about equal in length to the entire tarsus and more than twice the length of the femur. Middle and hind legs whitish; femora tinged with brown; tibiae about as long as the femora. Wings hyaline, all veins pale; a brown tinge at the extreme base. Abdominal segments 1–6, and the anterior half of segment 7, dull whitish, semi-hyaline; segments 8–10, and the posterior half of 7, pinkish brown, opaque; paler ventrally. On the anterior segments traces of a fine ruddy dorsal line are evident; also "lateral rows of small ruddy spots or streaks situated one on each segment in the central area" (McDunnough). Forceps pale with tinge of brown. Between the forceps at the base is a prominent tubercle, as the *bicolor* group; however, there are small lateral barbs apically on the penes. Tails white, unmarked.

This species is known only from the type specimens; we have not seen these.

Ephemerella funeralis McDunnough

1925—McDunnough—Trans. R. S. C. 19: 210. Fig.......... Que.
1931—McDunnough—Canad. Ent. 63: 39. Figs.......... Ont.

Length: body 7½–9; wing 9 mm.

A species of the *bicolor* group, quite close to *E. temporalis*, but considerably darker. Occipital tubercles of nymph quite well developed. Rows of dorsal spines divergent to rearward; spines on tergites 1–3 very erect, thin and tapering.

Eyes orange. Head yellowish brown; a black spot below each antenna, next to eye. Vertex shaded with blackish. Head of female deep yellow, vertex brown. Thorax reddish brown. A median depressed O-shaped area on the pronotum, sometimes outlined in rose color; on the anterior margin the lateral carinae around this depression run up into a spine. A rose-colored mark near the lateral margin on each side; blackish pencilings near the median line. Mesonotum yellowish laterally; median line dark red; a median geminate black streak just anterior to the scutellum. A blackish streak before each wing root, surrounded by a yellow area. Coxae brown-tinged, with a purplish black transverse streak. Fore leg amber, other legs paler yellow; claws and tips of last tarsal joint dusky. Wings hyaline; venation pale, the main veins of the costal margin yellowish.

Abdominal tergites reddish brown, 8–10 darker. Posterior margins, pleural fold, lateral streaks, a narrow median line, subdorsal streaks and stigmatic marks blackish brown. A pale yellowish area on each side of the dark mid-dorsal line. Sternites somewhat paler reddish brown. Oblique black dashes and dark dots distinct. Ganglionic areas blackish brown, on middle sternites. Tails yellowish brown at the base, whitish apically; joinings purplish black. Genitalia dark reddish brown.

Occipital tubercles of the nymph well developed, rounded. The rows of dorsal spines diverge distinctly to rearward. The spines on tergites 1–3 are thin, tapering and very erect. The lateral extensions and postero-lateral spines are better developed in this species than in any other of the *bicolor* group; the spine on segment 9 is fully twice as long as its width at the base. The lateral margins of segment 9 are subparallel, the spines at the tip being distinctly incurved. General color brown, rather unicolorous; markings of the femora very indistinct; operculum unmarked or with faint pale dots. The usual curved row of dark dots on the sternites is absent, but the short dark oblique dashes are present. Tails unicolorous brown; joinings of a few of the basal segments darker brown.

Although Dr. McDunnough secured only female nymphs and imagos, we have reared rather a long series of both sexes from nymphs taken at Freeville and McLean, N. Y., which he has kindly identified as *E. funeralis*. In these specimens the usual dark markings of the abdominal sternites are present and distinct. The lack of a ventral globular swelling on the genitalia distinguishes this species from *E. prudentalis*; aside from its darker coloration it may readily be confused with *E. temporalis*. Nymphs in our collection are from the vicinity of Ithaca, N. Y.; Watkins Glen, N. Y. (A. Louise Steger); Sand River, near Buckhannon, W. Va. (Mrs. A. L. Burckmyer); Stephens Co., Ga. (P. W. Fattig); Great Falls, Va. (J. R. Greeley); Kirtland, O. (J. R. Traver); and Stump House Mt., S. C. (J. G. Needham).

Ephemerella fuscata Walker

1853—Walker—List. Neurop. Ins. Brit. Mus. 3: 570 (in part) (Baetis)..................Hudson Bay
1861—Hagen—Neurop. Ins. N. Amer., p. 47 (Baetis).
1884—Eaton—Revis. Monogr., p. 129 (*walkeri*).
1905—Needham—N. Y. St. Mus. Bull. 86: 43. Fig. (in part, adult *nec* nymph, *bispina*).........Ind.
1931—McDunnough—Canad. Ent. 63: 214. Figs. (*bispina* =*fuscata*).................... Que., N. B.

Length: body 8; wing 8 mm.

A blackish, yellow-legged species; imago very similar to *E. tuberculata*, but smaller, more slender, with fewer dark femoral dots, and no ventral dark dots. Nymph very close to *E. wayah*, but distinguished by the presence of a lateral prothoracic tubercle, wider femora and shorter dorsal spines.

Head and thorax deep blackish brown. Fore femur and tibia deep brown; tarsus deep amber brown. Middle and hind legs yellowish, the femora heavily shaded with ruddy brown. All femora sprinkled rather sparsely with black dots. Wings hyaline; venation pale yellowish. Abdomen deep smoky brown; middle segments semi-translucent, apical segments opaque, deeper brown. Basal and middle sternites somewhat paler than the tergites; dark triangles in the antero-lateral angles, but no curved row of dark dots. Tergites rather unicolorous, but somewhat paler laterally. Tails yellowish, joinings unmarked. Genitalia very similar to *E. tuberculata* (see fig. 153).

Head of nymph with broad frontal shelf, notched on each side to receive the bases of the antennae (see fig. 151 c). No occipital tubercles; head somewhat roughened. A single small prothoracic tubercle on each side near the middle of the lateral margin. Anterior margin of fore femur with spines. All femora very much flattened, the posterior margin with a wide backward extension. Head, legs and body very hairy. Dorsal spines very minute. Ventral surface of abdomen rather less hairy than in *E. wayah;* no adhesive disc. General color pale greyish to deep olive brown; thorax, tibiae and tarsi with suffused darker markings. Usually a mid-dorsal row of dark blotches on the abdomen; ventrally, a lateral row of dark streaks on each side, but no curved row of dark dots between, as in *E. tuberculata*. Tails short, pale.

We have imagos of this species in the Cornell collection from Elkhart, Ind. (R. J. Weith); and from Vaudreuil, Que. (G. S. Walley) kindly presented by Dr. McDunnough. Nymphs from Cascades Point, Que., were likewise given us by Dr. McDunnough.

Ephemerella glacialis Traver

1884—Eaton—Revis. Monogr., p. 131; figs. 11–15, Pl. 38 (Nymph No. 2)....................Wash.
1930—Walley—Canad. Ent. 62: 17 (in discussion of *E. grandis*).
1934—Traver—J. Elisha Mitchell Sci. Soc. 50: 207..Mont.

Length: body (nymph); 14–15 mm.

A species of the *fuscata* group; closely allied to *E. grandis*. Distinguished from that species by the greater development of the occipital, submedian mesothoracic and submarginal prothoracic tubercles. Only the nymph is known.

Head with no frontal shelf nor true frontal horns. Occipital tubercles quite prominent; conical, directed backward. Submarginal tubercles of the pronotum larger than in *grandis;* anterior to each, a very small spine. A large submedian tubercle, and posterior to it two smaller ones, of which that nearer the median line is very minute. Mesothoracic tubercles near the anterior margin much as in *E. grandis;* submedian tubercles well developed (in *grandis*, there is an elevated submedian ridge only); median posterior tubercle larger than in *grandis*. Anterior margin of the fore femur unmodified. Legs much as in *grandis*, but appear more slender, especially as to the femora of the middle and hind legs. Tibiae pale at each end; apical half of each tarsus pale. Gills borne on segments 3–7. Dorsal spines prominent, very similar to *grandis;* spines on 8 and 9 somewhat larger relatively. General color deep blackish brown in maturer forms, usually with a pale median dorsal stripe the length of the body.

The type material, from Midvale Creek, Glacier National Park, Mont. (A. S. Hazzard, J. E. H., M. J. M.) is in the Cornell University collection.

Ephemerella grandis Eaton

1884—Eaton—Revis. Monogr. p. 128..Colo., Wyo.
1927—Needham—Ann. Ent. Soc. Amer. 20: 108. Fig..Utah
1927—Needham—Utah Agr. Exp. Sta. Bull. 201: 8. Fig.
1930—Walley—Canad. Ent. 62: 17.

Length: body 14; wing 15 mm.

A species of the *fuscata* group. Male imago very similar to that of *E. doddsi*, but distinguished by the greater amount of dark coloring on the femora and the abdominal tergites, the pale middle and hind tarsi, and the structure of the forceps. Nymph very close to *E. glacialis*, but with much lower submedian mesothoracic spines and occipital tubercles.

Head and thorax dark reddish brown. Pleural sutures and an area anterior to the wing roots pale. Scutellum bordered with blackish brown; its spine quite slender and long. Fore legs brownish, tarsi somewhat paler. Middle and hind legs pale yellowish white. All femora washed with purplish brown, and with wide dark brown subapical bands. Wings hyaline; longitudinal veins dark purplish brown. Abdominal segments dark purplish brown with wide pale margins, so as to appear conspicuously ringed. Pleural fold whitish. Segments 8–10 largely dark; 8 and 10 with pale posterior margins. Tails deep purplish brown at base, becoming paler towards the tips. Second joint of the forceps relatively straight, not much swollen; third joint rather short (see fig. 153).

Nymph with no frontal horns nor frontal shelf (see fig. 151 g). Occipital tubercles present, not as well developed as in *E. glacialis*. A pair of submedian prothoracic tubercles, the anterior of each pair being the larger; laterad of each pair is a slight rounded elevation. A conical tubercle on the margin of the prothorax on each side; anterior to this, a slight elevation. On each side of the median line of the mesonotum, on the anterior margin, is a low rounded tubercle, and a single median tubercle is situated near the posterior margin. No true submedian tubercles, only a slightly elevated ridge at the usual location of these. Dorsal spines present on segments 2–9; erect, slender, finger-like, directed backwards toward the tips; slowly and gradually increasing in size to rearward. Spines on 8 and 9 not conspicuously larger than those preceding. No teeth on the anterior margin of the fore femur. Gills borne on segments 3–7. Legs yellowish, banded with dark brown. General color of body light reddish brown; dark lateral patches on the middle abdominal tergites; V-shaped dark submedian marks on the anterior margins of same tergites. A lateral row of dark marks on each side of the venter, connected by a wide dark band across the middle of each sternite. Tails reddish at base, alternately red and yellowish beyond.

We have reared imagos of this species in the Cornell collection from Logan, Utah (J. G. Needham). Nymphs are from Estes Park, Colo.; and Pecos, N. M. (T. D. A. Cockerell).

Ephemerella hecuba Eaton

1884—Eaton—Revis. Monogr. p. 133. Pl. 40. Figs. 1–17.................Colo.
1905—Needham—N. Y. St. Mus. Bull. 86: 46. Fig.................N. M.
1927—Needham—Ann. Ent. Soc. Amer. 20: 115.................Wyo., Utah
1927—Needham—Utah Exp. Sta. Bull. 201: 9. Fig.
1929—Shepherd—Canad. Ent. 61: 260. Figs.................Calif.
1930—Walley—Canad. Ent. 62: 14. Fig.

Length: body 15; wing 13 mm.

A unique species considered to be in a group by itself. Persistent respiratory gills borne on segments 4–8 of the imago, and prominent postero-lateral spines on segments 9 and 10; nymph has an entire broad frontal shelf, gills present on tergites 4–7 only, lateral extensions of the abdomen on segments 2–9 produced into long slender spines, dorsal spines obsolescent.

Head mottled brown, margins whitish; eyes light brown. Basal segments of antennae light yellow, remainder brown. Pronotum widest near the anterior margin, the antero-lateral angles rounded; mottled brown, the carinae reddish. Anterior median portion of mesonotum brown; posterior portion reddish brown. Metanotum light yellow. Fore coxa light yellow; trochanter reddish brown, a spine-like tooth at its distal end; femur dark brown, its margins blackish; tibia curved, notched basally on the anterior margin; first tarsal joint one-third as long as the second. Wings semihyaline, venation distinct.

Tergites 1–4 dark yellow, with transverse reddish stripes which widen laterally and posteriorly into comma-shaped figures; on either side of the median line are light yellow circular spots increasing in size and distinctness posteriorly. Tergites 5–10 almost entirely covered by reddish bands posteriorly and by the circular whitish spots anteriorly. The anterior and lateral margins of each sternite are occupied by prominent reddish bands; a semilunar yellowish strip is present along the posterior margin of each. Forceps light yellow; second joint constricted near the apex, third joint short, emarginate posteriorly. Apical portion of the penes separated by an irregularly V-shaped notch.

Body of nymph much flattened, hairy; head small in comparison with other parts of the body (see fig. 151 i); widest in the region of the 5th abdominal segment. Head light brown, widest across the broad entire frontal shelf. Maxillary palp well developed; mandible heavy, canines moderately well developed. Pronotum widest at the posterior margin; on each side near the middle of the lateral margin is a short tubercle; another shorter tubercle is present on each side of the median line near the posterior margin. Legs hairy; a short conical spine or tubercle is present on the dorsal surface of the middle coxa. Femora bear at the apices an acute spine. Traces of two blackish bands on each femur, and one near the middle of each tibia. Claws without pectinations. Basal half of tarsus blackish. Abdomen stout, much flattened; the lateral extensions prolonged into sharp saw-tooth-like postero-lateral spines. Widest at the fifth segment. Gills present on segments 4–7 only. The gill on 4 operculate. Very faint traces of inconspicuous dorsal spines may occur on tergites 4 to 6. The median line is dark brown on tergites 3 and 4; tergites 5–9 show faint submedian brownish areas, and reddish lateral streaks next to the gills. Sternites marked with wide reddish to brownish bands occupying the anterior half and much of the lateral portion of each. Faint indications of pale submedian oblique marks at the anterior margin. Postero-lateral spines reddish brown at tips, paler just before the tips; pale at bases, and crossed beyond the base by a wide brownish band. Tails with one or two very narrow blackish rings near the base, and a wide band of black across all three at about two-thirds of their length; tips narrowly black.

The species was described from specimens taken in Colorado. We have specimens of nymphs in the Cornell collection from Pecos, N. M. (T. D. A. Cockerell); from the Provo River at Heber, Utah (J. G. Needham); and from California, locality not given (Mrs. S. T. Yeams). As indicated in discussion of groups of the genus, it is probable that the description given by Dr. Shepherd for the imago applies rather to the subimaginal stage. We have presented this account, however, as he gives it.

Ephemerella heterocaudata McDunnough

1929—McDunnough—Canad. Ent. 61: 170..Wyo., Mont.
1930—Walley—Canad. Ent. 62: 15. Fig.

Length: body 6½; wing 7 mm.

A species tentatively referred to the *serrata* group; imago olive brown with black markings; middle tail more than twice as long as the outer ones. Nymph with well-developed dorsal abdominal spines; head and thorax smooth; black abdominal markings as in the imago; middle tail much longer than the outer ones.

Eyes dark reddish brown. Thorax dark olive brown, the pronotum with considerable black shading. Median and lateral sutures of the mesonotum black-marked. A yellowish patch within which is a black line lies anterior to the wing base; the black line continues to the fore coxa. Mesosternum blackish brown with olive brown shading. Legs deep olive; fore femur and tibia tinged with black; fore tarsus about equal in length to the tibia, and both of these joints longer than the femur. Wings hyaline; longitudinal veins somewhat smoky, cross veins pale. Abdominal tergites olive brown; the first four tergites with broad black bands anteriorly, which bands are reduced to semicircular dorsal patches on the posterior tergites. A series of lateral black patches is also present, which tend to fuse with the dorsal bands on the anterior segments. Sternites pale olive; an irregular mid-ventral black band and similar black lateral bands. Forceps smoky, the third joint short. Penes united, forming an upcurved rod. See fig. 156. Tails smoky, the middle one twice as long as the outer ones.

Head and thorax of nymph smooth. Legs rather short; femora somewhat flattened. Lateral extensions of the abdominal segments practically wanting, the postero-lateral spines not produced. Gills present on tergites 3–7. Dorsal spines prominent, strong; incurved, becoming progressively longer and further apart from tergite 2 to 7, but shorter and closer together on tergites 8–9. General color dark brown, the head blackish; legs without pale bands. Abdominal sternites with a mid-ventral reddish brown streak and on each side a lateral row of brownish dashes. Tails brownish, the joinings darker; middle tail almost three times the length of the outer ones.

We have two male imagos of this species, in the Cornell collection, taken in the Yellowstone Park, Wyo. (C. R. Crosby).

Ephemerella hystrix Traver

1927—Needham—Ann. Ent. Soc. Amer. 20: 116. . . . . Mont.
1934—Traver—J. Elisha Mitchell Sci. Soc. 50: 212.

Length: body (nymph) 10 mm.

A species of the *needhami* group, of which only the nymph is known; distinctive by reason of the very long slender divergent dorsal spines. See fig. 53 b.

Head of nymph smooth; thorax smooth. Femora slightly flattened; spines on posterior margins. Tibiae and tarsi fringed with hairs on the posterior margin. 7 to 8 denticles on the inner margin of the claw. Color of entire body light reddish brown. Joinings of the legs, the apex of the tarsus, the claw, and small spots on the coxa and trochanter, dark reddish brown.

Gills present on segments 3–7. On tergites 2–9 are long slender dorsal spines, curved near the tips; each bears a series of spinules near its tip, on the anterior margin. Spines on segment 2 are relatively short, and close together near the median line; behind this segment the spines become gradually longer and further apart; spines on 6 and 7 are longest, subequal to one another. Spines on 8 and 9 are slightly convergent, and shorter than those on 6 and 7. Spines on the middle segments tend to turn outward apically. Short lateral extensions present on the abdominal segments, bearing short postero-lateral spines on segments 4–7; these angles somewhat incurved and rounded on 8 and 9. Basal tergites pale reddish brown, becoming progressively darker in color. Traces of a dark median line on 7 and 8, and of dark submedian streaks on the preceding tergites; large dark lateral patches beneath the gills and on tergites 8 and 9. Each sternite marked with a broad dark reddish brown lateral triangle on each side, the two triangles connected on the anterior margin. Ganglionic areas pale on sternites 1–6; on 7–8, dark brown.

A single specimen, the holotype, was taken in Big Blackfoot River, Potomac, Mont. (F. T. Barry). It is in the Cornell collection. This is the nymph which Prof. Needham described but did not name, in his paper on the Rocky Mountain species of Ephemerella.

Ephemerella inermis Eaton

1884—Eaton—Revis. Monogr. p. 127. . . . . Colo.
1924—McDunnough—Canad. Ent. 56: 223.
1927—Needham—Utah Agr. Exp. Sta. Bull. 201: 9. Fig.
1928—McDunnough—Canad. Ent. 60: 238.
1930—Walley—Canad. Ent. 62: 15. Fig. . . . . Wyo.

Length: body 5–6; wing 6–8 mm.

A species of the *invaria* group, in which the second joint of the male forceps is not swollen apically; venation pale; nymph without dorsal spines, mottled irregularly with pale areas.

Thoracic notum polished; deep blackish brown in color. Fore leg dull greenish grey; the coxa, trochanter and knee lighter; tarsus greyish white; femur light yellowish amber. Femora and bases of tibiae of the middle and hind legs yellowish amber; distal portion of tibiae and tarsi dirty whitish; the tip of the last tarsal joint, the claws and all the tarsal joinings, light reddish brown. Wings hyaline; venation very pale yellowish to hyaline. A light blackish brown spot at the extreme base of the fore wing.

Dorsum of abdomen dark reddish brown to blackish brown, the last three tergites yellowish to reddish; lateral margins light yellowish; the joinings opaque. Ventrally paler; genitalia light yellowish. Second joint of the forceps not swollen at the apex. Usually there are no apical spines on the penes, and but 4 or 5 spines on each side, at some distance from the apex (see fig. 152). Tails greyish white, the joinings opaque.

The nymph is rather dark reddish brown in color; the head and thorax are irregularly mottled with light and dark areas, which are sometimes rather extensive on the pronotum. The lateral margins of the pronotum are pale. Legs blackish brown, the femora with three large pale areas; tibiae pale at apex and near the base; tarsi pale distally. Spines on the joints are very similar to *E. mollitia*. Lateral extensions of the abdominal segments very well developed. Dorsal spines absent. A rather large pale spot on each side of the median line, on the posterior margin of each tergite. On the middle tergites the median line is often pale, sometimes wholly whitish. Rather extensive pale areas are often present just above the gills on these tergites, most pronounced on tergites 5 and 6. These markings are variable in different specimens. Tails yellowish brown, with several median and apical dark bands. Ventrally yellowish brown, the apical sternites and the lateral margins of the middle ones blackish brown. Traces of dark submedian oblique dashes on the anterior margins, and of dark dots nearer the center of the sternites, may be present.

The species was described from specimens taken at Denver, Arkansas Canyon and Colorado Springs, Colo. It has been reported also from Yellowstone Lake, Wyo. We have specimens of this species in the Cornell collection from that location, kindly sent us by Dr. McDunnough. Other specimens which seem to be of this species are from Glacier National Park, Mont. (A. S. Hazzard); from Tioga and Merced Rivers in the Yosemite Park, (P. R. Needham); Spring Creek and Powder Canyon, Wyo. (Elsie B. Klots); Logan River and Provo River, Utah (J. G. Needham); Centennial, Wyo. (R. J. Gilmore); Snake River, Wyo. (Elsie B. Klots); Reno, Nev. (E. P. Van Duzee); Jemez Springs, N. M. (Helen E. Murphy); Pecos, N. M. (T. D. A. Cockerell); and Pleasant Valley, Ore. (M. D. Leonard). It is probably this species which Dr. Dodds described as Ephemerella No. 2.

Ephemerella infrequens McDunnough

1924—McDunnough—Canad. Ent. 56: 223. Fig. .... Alta., Calif., Utah
1927—Needham—Ann. Ent. Soc. Amer. 20: 114. Fig. (*inermis*).
1928—McDunnough—Canad. Ent. 60: 238 (Needham's *inermis* = *infrequens*).
1930—Walley—Canad. Ent. 62: 12.

Length: body 8; wing 9 mm.

A species of the *invaria* group; second joint of forceps swollen at apex; nymph without dorsal spines, and with no pale spots on posterior margins of tergites.

Eyes of living male light orange. Head light brown, clypeus and bases of antennae pale yellow. Pronotum and the lateral edges of the mesonotum anterior to the wing base purplish brown. Mesonotum deep olive brown, becoming reddish brown posteriorly; two slightly raised blackish lines anterior to the scutellum. Bases of wings and the pleural sutures tinged with yellowish brown. Prosternum yellowish; remainder of sternum brown. Legs light yellow, the fore tibiae and tarsi rather duller than the femora; tarsal joints slightly marked with brown. Wings hyaline; venation pale; cross veins indistinct, especially in the basal portion of the costal space.

Abdomen light reddish brown dorsally; posterior margins smoky. A narrow yellowish line along the pleural fold. Dull purplish brown ventrally, shading into yellowish brown on the apical sternites. The middle sternites show traces of pale semitranslucent areas along the anterior margins. Forceps yellowish, penes similar, with a darker transverse band near the base. Tails yellowish at the base, whitish beyond; joinings narrowly deep brown. Usually four apical spines on the penes, and four or five others lower down, in two irregular groups (see fig. 152).

The nymph is dark reddish to blackish brown in color. Antennae dark brown at base, remainder yellowish. Vertex of head, each side of the pronotum, and the mesonotum anterior to and between the wing roots, indistinctly mottled with darker brown; these mottlings are much less apparent than in the very similar nymph of *E. inermis*. Legs reddish brown; femora blotched irregularly with yellowish; tibiae and tarsi each pale-banded near the base and again at the apex. Femora with the usual posterior spines; fore femur with spines as usual on the upper surface near the apex. 5 to 6 denticles on each claw. No dorsal abdominal spines. Lateral extensions present, but slightly less developed than in *E. inermis*, and lacking the dark mark across the middle, which is present in the latter species. No pale spots at the posterior margins of the tergites, near the median line, such as are present in *E. inermis* and *E. mollitia*. A pale mid-dorsal line is usually present. Pale lateral areas, if present, are inconspicuous and hidden beneath the gills; often these are entirely wanting. Venter without conspicuous markings. Tails reddish brown, crossed by several yellowish bands.

The type material was taken at Waterton Lakes, Alta., and Lake Tahoe, Calif. We have specimens in the Cornell collection from Seton Lake, Lillooet, B. C., sent to us by Dr. McDunnough; from Logan River and Box Elder Canyon, Utah, (J. G. Needham); from Glacier National Park (A. S. Hazzard); and from Moose, Wyo. (Elsie B. Klots).

Ephemerella ingens McDunnough

1934—McDunnough—Canad. Ent. 66: 154. Fig. . . . . B. C.

Length: body 13–15; wing 16–18 mm.

A species of the *fuscata* group, allied to *E. grandis*.

Eyes deep blackish brown. Head black. Pits of antennae and a small area posterior to the middle ocellus pale yellowish. Pronotum black. Mesonotum black, lateral posterior margins shaded slightly with brown. "Membranous area between base of wing and fore coxa pale yellowish, shaded with light brown, and containing a blackish streak" (McD.). Metathorax black; slight yellow shading on median area of notum, and on pleura between sclerites and around leg bases. Fore legs blackish; edges of femur near base with pale smoky yellow tinge. Coxae and femora of middle and hind legs blackish; femora and trochanters tinged with dull yellowish at base and on edges. Tibia as long as femur and trochanter together, dull yellowish. Tarsus smoky yellow, often distinctly smoky. Wings hyaline. Veins and cross veins black-brown; stigmatic cross veins somewhat indistinct. No fusion of posterior fork of media with Cu_1.

Abdominal tergites smoky yellow or yellowish brown on lateral and posterior margins. Tergites 2–7 largely "occupied by a large quadrate median black patch and lateral circular patches of same color, indistinctly spearated by bands of dull yellow-brown; on the rear segments the black areas more or less coalesce" (McD.). Sternites with broader light yellow-brown posterior margins which are more distinct than those of the tergites; remainder of each sternite blackish with faint reddish tinge; "faint traces of small oblique submedian pale streaks on anterior margin" (McD.). Forceps dull brown, tails black.

Nymph very similar to *E. grandis*. The tibiae, especially of hind legs, relatively longer than in that species; occipital tubercles likewise longer. Rows of dorsal abdominal spines somewhat closer together, and spines on 7 and 8 larger in proportion to preceding pairs, than in *grandis*. We have seen no specimens of this species.

Ephemerella invaria Walker

1853—Walker—List Neurop. Ins. Brit. Mus. 3: 568 (Baetis)............................Hudson Bay
1884—Eaton—Revis. Monogr. p. 129.
1925—McDunnough—Trans. R. S. C. 19: 212. Fig..Que.
1930—Ide—Canad. Ent. 62: 211..Ont.
1931—McDunnough—Canad. Ent. 63: 196. Fig.

Length: body 8; wing 8–9 mm

A species of the *invaria* group; genitalia much as in *E. rotunda;* venation pale; nymph smaller than *rotunda*, the dorsal spines almost obsolescent.

Eyes of male probably orange, in life. Head and thorax light reddish brown; pronotum shaded with smoky. Mesosternum rather deeper reddish brown. Legs pale whitish. Wings hyaline; venation pale; stigmatic area opaque whitish. Dorsum of abdomen light smoky to purplish brown; basal tergites darker than the apical ones. Posterior margins rather dark smoky brown; pleural fold yellowish. Ventrally pale yellowish white, the middle sternites semi-translucent; apical sternites opaque, yellowish. Tails white; joinings purplish black, very distinct. Genitalia very similar to *rotunda*, but with somewhat fewer spines on the penes.

The nymph is smaller than *rotunda;* the dorsal abdominal spines are very poorly developed, and may be almost obsolescent. The color varies from pale yellowish brown heavily sprinkled with brown dots, to specimens which are almost wholly brown with yellowish brown dots. In the paler forms, the brown dots may be distributed rather uniformly, or may be arranged in bands across the anterior portion of the pronotum and between the roots of the wing pads on the mesonotum. Large pale lateral patches may be present on tergites 5 and 6; these patches may run together to form an almost wholly white dorsal area. The pale dots at the location of the dorsal spines are often present, as in *E. rotunda.*

The type material, described from St. Martin's Falls, Albany River, Hudson Bay, is in the British Museum. Dr. McDunnough has kindly sent to the Cornell collection specimens taken at Kazubazua, Que., in which the dorsum of the abdomen is light smoky brown; also other specimens from St. Lambert, Que. (G. S. Walley, W. J. Brown), in which the basal and middle abdominal segments are almost wholly pale. We have nymphs of the yellowish brown color form from Candor, N. Y.

Ephemerella lata Morgan

1911—Morgan—Ann. Ent. Soc. Amer. 4: 112. Figs....................................Mo., N. Y.
1926—McDunnough—Canad. Ent. 58: 187 (*inflata*)..Que.
1931—McDunnough—Canad. Ent. 63: 211. Figs. (*inflata* = *lata*).
1932—Traver—J. Elisha Mitchell Sci. Soc. 47: 176. Fig...................................N. C.

Length: body 6–7; wing 6–7 mm.

A species of the *fuscata* group; rather closely allied to *E. cornutella*, from which it may be separated by the more inflated second joint of the forceps, which is likewise rather sharply constricted near the middle. Frontal horns of nymph very short; prothorax and 8th tergite usually whitish in whole or part; no dorsal spines.

Head and thorax deep blackish brown. Pleura and lateral areas of the mesosternum shaded with dull olive brown. Fore femur and tibia rather dark brown, about equal in length; coxa rather olive brown; tarsus pale, whitish. Middle and hind legs paler, the femora pale olive brown, the tibiae and tarsi lighter; an indistinct reddish longitudinal streak dorsally on the femora. Hind femur about as long as tibia and tarsus combined. Wings hyaline; venation pale, the main veins of the costal margin tinged with yellow; a small brownish spot at the base of the fore wing. Abdomen somewhat paler blackish brown dorsally, almost unicolorous. Indications of a pale mid-dorsal line on the anterior tergites. Sternites paler, dull olive brown. Second joint of forceps strongly inflated at each end and sharply constricted near the middle; penes united almost to the apex, the apical notch shallow and rounded (see fig. 153). Tails dull whitish.

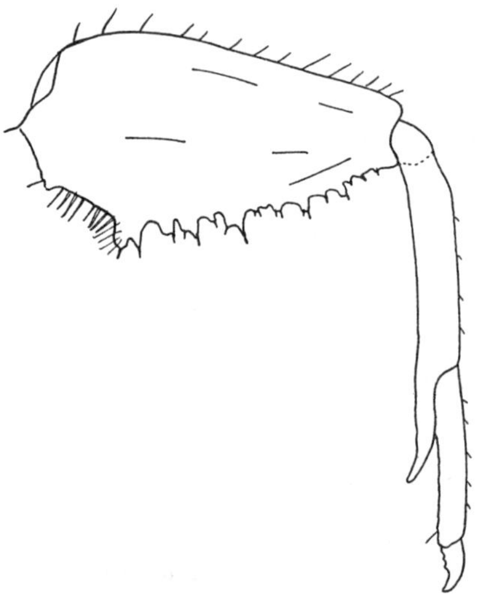

FIG. 154. Left fore leg of nymph of *Ephemerella longicornis.*

Nymph brownish black; prothorax and 8th tergite usually conspicuously white, but the amount of pale color seems to be variable. Frontal horns present but very short, not reaching to the margin of the frontal shelf. Anterior margin of fore femur with teeth or spines; posterior margin rather widely expended. Tibial 'thumb' stout, almost straight, extending about to the middle of the tarsus. No dorsal spines (although mentioned in the original description, the type material and all other specimens examined show no such spines). Tails light brown; basal joinings dark brown.

We have in the Cornell collection the nymphs from which the original description was drawn; these are from Sandy River, Me., and Fall Creek, Ithaca, N. Y. (Anna H. Morgan). Others are from Davidson River and Flat Creek at Black Mt., N. C. (J. R. Traver). Reared imagos are from Davidson River, N. C. (J. R. Traver); and from Fulford, Que. (L. J. Milne). The latter were sent us by Dr. McDunnough.

Ephemerella longicornis Traver

1932—Traver—J. Elisha Mitchell Sci. Soc. 47: 164. Figs.................................N. C.

Length: body 8–9; wing 8–9 mm.

A species of the *fuscata* group; a very close ally of *E. cornuta.* Hind femur rather longer relatively than in *cornuta;* tarsi and tails darker; pleural fold rather widely pale. Nymph distinguished from *cornuta* by the longer frontal horns and the longer and more curving tibial spine.

Head largely brown. Thorax dark olive brown dorsally. A round pale spot on each side of the median line of the pronotum, which is also pale laterally. A greyish streak anterior to the wing roots; pleura with paler markings. Fore femur yellowish brown, shaded with dark brown; tibia and tarsus dark brown, last tarsal joint paler brown. Femora and tibiae of middle and hind legs yellow margined with brown; tarsi dark brown, paler on the distal joints. Hind femur as long as tibia and tarsus combined. Wings hyaline, longitudinal veins pale yellowish brown; a brown spot at base of fore wing. Abdomen dark reddish brown dorsally; a pale submedian spot on each side of the mid-dorsal line, on the anterior margins of the middle tergites. Pleural fold rather widely pale. Tergites darker brown next to the pleural fold and on the posterior margins. Light yellowish brown ventrally; on each sternite a dark brown lateral streak lies next to the pleural fold. Postero-lateral spine on segment 9 longer than that of *cornuta*. Tails purplish brown. Third joint of forceps relatively shorter, and second joint less strongly bowed, than in *E. cornuta;* penes spread more widely at apex (see fig. 153).

Nymph quite similar to that of *E. cornuta*. Light reddish in color, the thorax often checkered and mottled with contrasting light and dark areas. Abdomen rather lighter in color than thorax. Frontal horns very long and relatively straight; their length is more than half the width of the space between them (see fig. 151 b). Prominent spines on the anterior margin of the fore femur; its posterior margin almost straight. The 'thumb' of the fore tibia is longer than in *cornuta*, extending beyond the middle of the tarsus; it likewise curves outward sharply in its distal portion (see fig. 154). Postero-lateral spines of the abdominal segments are somewhat longer than in *cornuta*. Abdomen yellow ventrally, with a row of dark marks on each side, near the pleural fold. Sternites 2–8 hairy, also the posterior margins of the legs. Legs yellowish with indistinct reddish bands.

The type material of this species, taken in the Appalachian region of North Carolina, is in the Cornell collection.

Ephemerella lutulenta Clemens

1913—Clemens—Canad. Ent. 45: 335. . . . . Ont.
1913—Clemens—Canad. Ent. 45: 336 (*lineata*).
1915—Clemens—Cont. Canad. Biol., p. 121 (in part). Fig. nymph; p. 122 (in part nymph, *lineata*).
1924—McDunnough—Occas. Pap. Boston Nat. Hist. Soc. 5: 74.
1925—McDunnough—Trans. R. S. C. 19: 212. . . . . Que.
1931—McDunnough—Canad. Ent. 63: 32. Figs. . . . . N. B.

Length: body 8–9; wing 10 mm.

A large species of the *bicolor* group; imago distinguished by the sprinkling of very small dark dots on legs, thorax and abdomen. The male nymph lacks occipital tubercles; dorsal spines on tergites 1–3 are long, erect and finger-like, the rows of spines divergent to rearward.

Face dark brown; a reddish grey streak on the median carina, and two lateral streaks to bases of antennae. Thorax dark reddish brown. Legs greenish yellow, claws brown. Wings hyaline; venation hyaline. Abdomen blackish brown dorsally, segments 8 and 9 rather paler in color. Posterior margins of tergites narrowly darker reddish brown. Pleural fold blackish brown on the margin; a black streak in each stigmatic region. Legs, venter of thorax and the entire abdomen sprinkled with very fine dark dots. Tails reddish brown near the base, paler distally; joinings dark brown.

In the male nymph the occipital tubercles are practically wanting, but are much better developed in the female. Rows of dorsal spines diverge to rearward. The spines on tergites 1–3 are long, erect and rather finger-like, and bend backward; on 8 and 9 they are almost obsolete. The postero-lateral spines are well developed, the spine on segment 3 being fully twice as long as its width at the base; the spines on segments 4–8 are very long and curved, bearing spinules and hairs on the margins. The operculum is veined longitudinally and usually bears several to many small pale dots. The color varies from dirty brown to a dark blackish brown; legs brown with pale bands, the femora with pale spots. Tails rather unicolorous brown, the cross bands of paler color less distinct than in *E. temporalis*. An inhabitant of rather large bodies of water.

The species was described from material taken at Georgian Bay, Ont. We have imagos of this species in the Cornell collection, sent us by Dr. McDunnough, from Gauvreau Lake, Que. Nymphs in this collection are from the outlet of Lake Mendota, Wis.; from Tri Lakes, Ind.; Conesus Lake, N. Y.; a tributary of the Cape Fear River, N. C. (J. R. Traver); and from Little Flat Creek near Compton, Tenn. (Lula Shipe).

Ephemerella maculata Traver

1934—Traver—J. Elisha Mitchell Sci. Soc. 50: 216. Fig. .... Calif.

Length: body 8; wing 8 mm.

Head blackish brown; frontal margin pale. Eyes dull orange. Thorax very dark olive brown. Mesonotal scutellum and its lateral margins, the lateral margins of the

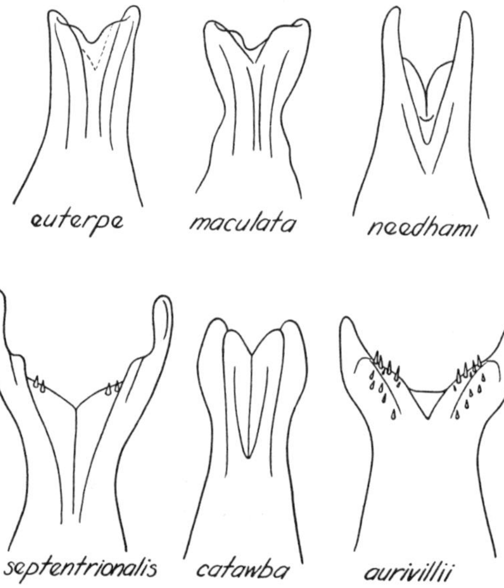

FIG. 155. Male genitalia of five species of Ephemerella, of the *needhami* group.

mesonotum, and most of the sternum, blackish brown. Ganglionic areas blackish. Pleura dark brown with yellowish interspaces. Fore femur yellowish at base, the apex brown; tibia deep olive brown; tarsus paler smoky brown. Middle and hind legs yellowish white, with dark pencilings on the femora at base and apex; tibiae brownish at base; tarsi pale smoky brown. Wings hyaline; longitudinal veins dark purplish brown, cross veins of the fore wing finer but distinctly dark. A dark reddish brown spot at the base of the subcosta and radius of fore and hind wings, and a dark streak at the base of the anal region of the fore wing. Stigmatic area opaque whitish; cross veins anastomosed.

Abdomen smoky to olive brown dorsally; tergites 1–2 and 7–8 shaded with blackish brown, middle tergites reddish brown in the median area. A dark brown line above the pale pleural fold, containing a dark stigmatic spot and a dark crescent, and two or three pale areas. Base of tergite 9, and most of 10, yellowish. Posterior margins narrowly pale. Ventrally pale yellowish white with wide brownish black anterior margins and lateral triangles on each sternite; ganglionic areas blackish; sternites 8 and 9 largely brownish black. Tails black at base, deep brown distally; joinings indistinctly darker. Genitalia dark smoky brown. Third joint of forceps short, stout, conical. Penes separated at apex by a wide V-shaped cleft; no spines are present (see fig. 156).

Nymph yellowish to pale yellowish brown with brown markings. Head smooth (see fig. 151 a). Genae produced anteriorly and laterally to form a frontal shelf overhanging the mouthparts. Thorax smooth; dark brown spots and pencilings on each side of the pronotum, and anterior to the wing roots on the mesonotum. Legs yellowish white; tibiae marked narrowly at each end with dark reddish brown; apex of tarsus and tip of claw dark red. Femora moderately flattened; long hairs on the posterior margins of femora and tibiae. 3 to 4 short blunt denticles on each claw. Well developed lateral extensions on the abdominal segments, each bearing a postero-lateral spine. No dorsal spines. Gills present on segments 3–7. Tergites 1–2, 5–6 and the median and lateral areas of 7, shaded with reddish brown; dark lateral patches on these tergites. Ventrally yellowish, with brownish black anterior margins and lateral triangles on all sternites, as in the imago. Ganglionic areas darkened. Tails pale yellowish white, tips dark brown; two dark reddish brown bands at the base.

The type material, consisting of male and female imagos and nymphs taken at San Gabriel Canyon in the San Gabriel Mts., Calif. (C. D.. Michener), and at Waddell Creek (D. Shepherd), is in the Cornell University collection.

Ephemerella margarita Needham

1927—Needham—Ann. Ent. Soc. Amer. 20: 114. . . . . Utah
1927—Needham—Utah Agri. Exp. Sta. Bull. 201: 9. Fig.
1930—Walley—Canad. Ent. 62: 14. Fig. . . . . Wyo., Mont., Alta.
1931—McDunnough—Canad. Ent. 63: 210. Fig. . . . . N. H.

Length: body 9½ mm.

A species of the *simplex* group, of which only the nymphal stage is known; dorsal abdominal spines present, but no occipital nor thoracic tubercles.

Head and thorax of nymph smooth. Legs pale; tibiae and tarsi banded at the middle, tarsi also at the tips. 8 to 9 denticles along the inner margin of each claw, the distal one being longest. General color olive brown with blackish markings. Lateral extensions of the abdominal segments are well developed, and terminate in long postero-lateral spines on segments 4 to 9; abdomen broadest at segment 7. Gills present on segments 4 to 7 only, semi-operculate on segment 4. Rather short dorsal spines are present on tergites 3 to 9. Two color phases are known. In the best marked phase, a median blackish spot is present on tergites 2 and 3, and 5 and 6; apical margin of 8 with a black streak; tergite 9 largely blackish except for the lateral extensions and two pale submedian apical dots. Lateral extensions pale, each crossed near the middle by a broad brown band. Brownish ventrally; sternite 8 largely pale yellowish; sternite 9 blackish brown. Tails yellowish white, with a broad black band across the middle and a narrow black band at the tip.

The species was described from Provo River, Utah. We have specimens from this locality; also from Box-Elder Canyon, and Weber River at Winship, Utah (J. G. Needham).

Ephemerella micheneri Traver

1934—Traver—J. Elisha Mitchell Sci. Soc. 50: 227. Fig. . . . . Calif.

Length: body 6–7; wing 7 mm.

A member of the *serrata* group; yellowish with brownish markings; tails white, joinings dark; two or three blunt processes on the apical margin of each division of the penes. Nymph with well developed maxillary palp, smooth head and thorax, and dorsal spines.

Eyes orange. Head yellowish shaded with brown. Thorax yellowish, with brownish black markings; a dark band and streak on the prothoracic pleura; scutum, scutellum and median line of mesonotum margined with dark brown; two lateral dark streaks and three median dots on metanotum. Two blackish streaks on the pleura anterior to the middle and hind legs; coxae black-marked; median areas of the pro-

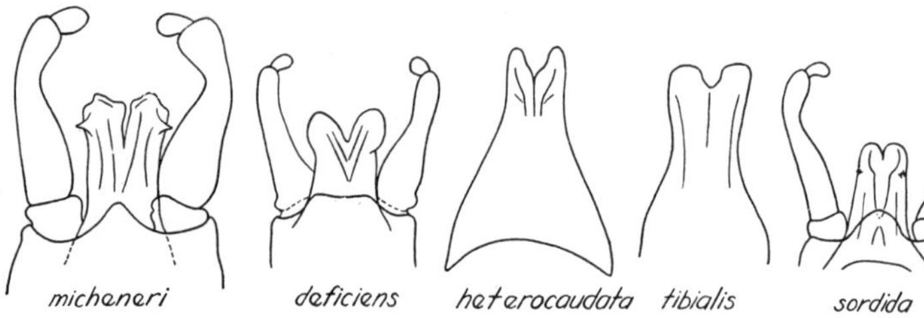

FIG. 156. Male genitalia of five species of Ephemerella.

and mesosterna outlined in black, ganglia of the meso- and metasterna darkened. Legs yellowish; fore tibia and tarsus pale smoky. A narrow reddish brown apical band on each femur. Claws and basal tarsal joints smoky. Wings subhyaline to hyaline; venation pale. A purplish streak on the membrane of the fore wing near the base, below the costa; a brownish spot at base of costa and subcosta.

Abdominal tergites brownish; 2–8 rather heavily shaded with reddish to olive brown laterally. Posterior margins pale; brownish black stigmatic marks; pleural fold pale. Sternites yellowish; ganglionic areas purplish black; a reddish brown streak or triangular area borders the pleural fold, on each side. Genitalia yellowish. Two or three blunt processes are present on the apical margin of each division of the penes (see fig. 156). Tails white, joinings purplish black.

Head and thorax of nymph smooth; maxillary palp rather well developed. Thoracic ganglia and the anterior margins of the pro- and mesosterna blackish. Legs pale yellowish; a black spot at base of tarsus; apex of tarsus, and claw, blackish brown. 7 to 8 denticles on each claw. Dorsal spines present on tergites 2–8; short on 2 and 8, longest on 5–7. Extensions of abdominal segments and the postero-lateral spines short; no spine on segment 8. General color yellowish to yellowish brown, the basal and middle tergites shaded with black. Blackish lateral patches are often present on the middle tergites. Ganglionic areas of the sternites are darkened; a double row of curved brown marks is present on each side. Tails alternately yellow and brownish black.

The type material of this species, in the Cornell collection, is from San Gabriel Canyon, in the San Gabriel Mts., Calif. (C. D. Michener).

Ephemerella minimella McDunnough

1931—McDunnough—Canad. Ent. 63: 63. Figs. . . . . Que.

Length: body 6; wing 6 mm.

A member of the *bicolor* group, rather close to *E. bicolor*. Thorax and abdomen of imago darker brown than *bicolor*, the legs brighter yellow, and with a distinct brown spot apically on each femur. Nymph with very weak postero-lateral spines on segments 2 and 3; occipital tubercles wanting, in the male; dorsal spines long, thin and more upright than in *bicolor;* rows of spines divergent, but not suddenly widened on tergite 5.

Eyes orange. Head yellow with brown shading at the bases of the ocelli. Thoracic notum very dark brown, with yellow shading laterally and in the sutures. Sternum rather bright brown. Legs quite bright yellow; coxa tinged with brown, and with a black spot; near the apex of each femur is a distinct brown spot. Wings hyaline. Abdominal tergites 1–7 dark brown, 8–10 with paler shading; tergites 1–4 tinged slightly with smoky subdorsally. A narrow brown dorsal line is present, and a pair of blackish subdorsal dots near the anterior margins, which are fainter on the posterior tergites. On tergites 8 and 9 the dorsal line and the subdorsal dots coalesce. Short oblique black streaks and longitudinal black lines are present at the stigmatic region. Ventrally pale brown, the last two sternites creamy. On sternites 2–7, the usual curved row of dark dots is distinct. Tails pale, joinings blackish.

Occipital tubercles are wanting in the male nymph. Postero-lateral spines are very poorly developed on segments 2 and 3. Dorsal spines longer, thinner and more upright than in the closely related *bicolor,* especially those on tergites 5–7, which are twice as long as in *bicolor.* The rows of spines are evenly divergent to rearward, and not suddenly widened on tergite 5. Tails not banded.

The type material was taken at Knowlton, Que., and is in the Canadian National collection. We have a single nymph from Burncoat Swamp, N. C. (J. G. Needham), which is doubtfully referred to this species.

Ephemerella molita McDunnough

1930—McDunnough—Canad. Ent. 62: 57..Que., N. S.
1931—McDunnough—Canad. Ent. 63: 206.

Length (female): body 6; wing 6–7 mm.

This species, known only from two type females, is tentatively placed by Dr. McDunnough in the *serrata* group.

Head of female dull yellowish, shaded with brown around the ocelli and near the eyes; posterior to the ocelli in the central area are numerous irregular black pencilings and two small brown tubercles, the latter closer together than the rear ocelli. Vertex smooth. Prothorax blackish, its posterior margin almost straight, with a small wart-like excrescence on each side. Mesothorax dark brown with paler brown shading. Pleura and sternum deep brown shaded with ruddy. Legs pale yellowish. Femora rather short; traces of a brown band at apex of each. Tarsi about two-thirds the length of the tibiae. Wings hyaline, all veins pale. Both wings are shaded distinctly with brown at the base, this brown area on the fore wing extending well beyond the humeral cross vein.

Abdominal tergites deep brown, the three posterior segments brighter and paler. Pleural fold shaded with blackish. Sternites paler, especially posteriorly. On each is a curved row of four very small black dots. Subanal plate yellowish; flatly scoop-shaped in appearance. Tails dull whitish, unmarked. We have seen no specimens of this species.

Ephemerella mollitia Seemann

1927—Seemann—J. Ent. and Zool. 19: 44. Figs..Calif.

Length: body 10; wing 10 mm.

A species of the *invaria* group; spines on the penes more numerous than in any other species of this group; venation pale. Nymph without dorsal abdominal spines.

Eyes salmon pink in life. Antennae purplish black at base, second joint pale; filament dark brown (in alcoholic specimens, rather pale.) Thorax deep yellow, shading into deep orange on the median line, the lateral margins of the mesonotum above the wing roots, and the lateral margins of the scutellum. Scutellum margined narrowly with brown. Anterior and posterior margins of pronotum, the median line, and markings on each side of the median line, purplish black. Deep pinkish to orange at the base of each leg. Legs pale (in life the fore leg may have had a ruddy tinge); tarsal joinings of the fore leg narrowly brownish. Wings hyaline; venation pale yellowish to hyaline.

Basal and middle abdominal tergites deep reddish; posterior margins rather widely purplish black, the anterior margins paler (probably the "white or translucent joinings" of the original description). Apical tergites yellowish, tinged with reddish laterally. Pleural fold pale on the extreme margin; an interrupted dark line parallels it closely on each tergite. Ventrally paler, the basal and middle sternites tinged with reddish. Forceps smoky brown, penes pale. Tails white, the joinings black. 24 to 28 spines on the penes; an apical cluster consisting of 4 to 6 spines on each side, and others irregularly arranged further back toward the base (see fig. 152).

Nymph reddish brown; very similar to *E. inermis*, but with fewer bands on the tails. Vertex and occiput of head, the pronotum on each side, and the mesonotum anterior to and between the wing roots, mottled with irregular dark reddish brown spots, which are surrounded by a yellowish ground color. Lateral margins of the pronotum pale yellowish. Genae show a slight tendency toward lateral development. Legs pale reddish to greyish brown; femora indistinctly mottled. A brownish band at the base of tibia and tarsus; another dark band beyond the middle of the tibia. Short spines on the posterior margins of the femora, and the anterior margins of tibiae and tarsi; shorter spines on the upper surface of the fore femur near the apex. 8–10 denticles on each claw. Lateral extensions of the abdominal segments well developed. Dorsal spines absent. Two pale spots on the posterior margin of each tergite, one on each side of the median line. Faint indications of a pale median line and of pale submedian streaks, on the middle tergites. Larger pale areas are usually present laterally, just above the gills, on tergites 3–7; most extensive on tergite 5. Tails yellowish. In mature nymphs, each joining in the basal half appears black-ringed, due to the subimaginal tail within. In younger nymphs, tails pale except for one or two dark basal rings.

The type material, from San Dimas, San Antonio and Cucamonga Canyons, Calif. (Mrs. T. M. Seemann) is in the Cornell University collection. Other specimens are from Bishop, Calif. (O. R. Smith); Tuolumne River, Calif. (P. R. Needham); Sequoia Nat. Park., Calif. (J. C. Bradley); and Farmington, N. M. (P. N. Albright).

Ephemerella needhami McDunnough

1905—Needham—N. Y. St. Mus. Bull. 86: 47. Figs. (in part, male imago, *excrucians*)......... N. Y.
1905—Needham—N. Y. St. Mus. Bull. 86: 43 (in part, nymph, *nec* imago, *bispina*)............. Ind.
1925—McDunnough—Canad. Ent. 57: 171. Fig.. Que.
1931—McDunnough—Canad. Ent. 63: 202. Figs.

Length: body 6–6½; wing 6–6½ mm.

A rather small, dusky species, having a brownish area at the base of the fore wing; venation faint smoky, brownish at base; no spines on penes. Nymph distinguished by the erect dorsal spines on segments 2–8, parallel on 3–8.

Head dark reddish brown; frons yellowish; antennae reddish brown; eyes bright orange-red. Thorax deep red-brown, the pronotum largely black; mesonotum brownish black. Sternum somewhat paler, tinged with yellowish. Fore leg smoky brown, the femur slightly paler than tibia and tarsus, and with a small reddish apical spot. Tibia about twice as long as the femur. Middle and hind legs yellowish, the femora usually with a longitudinal ruddy streak and a reddish apical spot. Wings hyaline; a brownish spot at the base of the fore wing; longitudinal veins faintly smoky; Sc and R of the fore wing brownish at extreme base. Stigmatic area opaque whitish; cross veins anastomosed near the costal margin.

Abdomen deep reddish to blackish brown dorsally, usually tinged with wine-red. Posterior margins of tergites blackish brown. Traces of pale brownish submedian streaks on each side of the dark median line may be present. Pleural fold blackish brown. Venter usually, but not always, distinctly tinged with wine-red. Dark markings consist of two short oblique submedian streaks at the anterior margin, and two small dark dots nearer the center of each sternite. Genitalia light reddish brown. Second joint of the forceps not swollen at the apex. Penes separated apically by a wide rather V-shaped notch; no spines are present (see fig. 155). Tails deep smoky brown; joinings not distinct.

The nymphs are rather variable in color pattern, ranging from dark reddish brown to almost black; usually but not always, a broad pale median line is present on thorax and abdomen. Within this pale stripe may be a dark median line; the pale stripe itself is bordered by darker stripes. The prothorax is pale on the lateral margins; on the mesothorax, between the wing buds, are often two small pale rounded spots. Head and thorax smooth. Tergites 2–8 each bear erect dorsal spines, shortest on tergite 2, where the spines are also closest together. On the following tergites the spines are roughly parallel, not diverging as in *E. hystrix*. Each spine bears several minute spinules on its upper surface. Lateral extensions of the abdomen only moderately developed, and with short postero-lateral spines. Venter marked as in the imago, with dark submedian oblique streaks at the anterior margin and two dark dots nearer the center of each sternite. Tails pale yellowish, with one or two dark bands at the base and apex only.

The species was described from Laprairie, Que. In the Cornell collection we have the specimens described as *E. excrucians* and *E. bispina*, Needham, 1905. Other specimens are nymphs from Elmsdale, N. S. (A. H. Leim); and imagos taken at Ithaca, N. Y. (W. O. Sadler), and at Mattawamkeag, Me. (A. E. Brower).

Ephemerella proserpina Traver

1934—Traver—J. Elisha Mitchell Sci. Soc. 50: 223..Calif.

Length (nymph): body 20 mm.

A species of the *fuscata* group, allied to *E. spinifera* and *E. yosemite*. Distinguished from both of these species by its very large size, and the inequality of the submedian tubercles on the pronotum. Only the nymphal stage is known.

Only very slight indications of frontal horns are present; there is no frontal shelf. Occipital tubercles are prominent. The submedian tubercles of the pronotum are unequal, the posterior pairs shorter than the anterior. Three pairs of submedian tubercles are present on the pronotum, two pairs behind the first pair; the submarginal tubercles are smaller even than in *E. yosemite*. Submedian mesonotal tubercles much as in *yosemite*, but a true median tubercle is present in this species, near the posterior margin. Anterior margin of the fore femur without teeth or spines. Fore tibia and tarsus rather stout; tibial 'thumb' blunt, moderately long. Middle and hind legs relatively longer and more slender than in *yosemite;* femora not flattened. Dorsal spines present on tergites 2–9; erect and closest together on 2; 3–7 slope backward slightly; rows divergent on 3–5, subparallel on 6 and 7. Spines on 8 and 9 at least twice as long and large as on the preceding tergites; that on 9 more slender than the 8th one, straight, and directed backward. Gills present on tergites 3–7. General color very dark reddish brown. Tails with narrow yellow bands except at the base.

The type, a single female nymph, was taken by Mr. Charles Michener in Mill Creek Canyon, the San Bernardino Mts., Calif. This specimen is in the Cornell collection.

Ephemerella prudentalis McDunnough

1931—McDunnough—Canad. Ent. 63: 40. Figs. . . . . Que.

Length: body 8; wing 8 mm.

A member of the *bicolor* group, related to *E. temporalis*. There is a globular swelling on the penes halfway from base to apex. Occipital tubercles of nymph moderately developed. Rows of dorsal spines convergent to rearward. Dorsal spines on tergites 1–3 blunt and finger-like, barely curved.

Eyes orange. Head yellowish brown; a curved black streak beneath each antenna. Thorax reddish brown, the pleura and the anterior portion of the mesonotum yellowish; several black dots on the pronotum. Metasternum purplish red. Fore leg amber yellow, a brown spot at the apex of the tibia. Middle and hind legs paler yellow; claws and tarsal joinings narrowly reddish brown. Wings hyaline; venation hyaline. On each main vein of the fore wing, and on median vein of the hind wing, is a short brown basal streak; on each wing also, a faint basal brownish stain. Abdominal tergites reddish brown, shading into yellowish apically. Posterior margin of the basal and middle tergites, and the pleural fold, blackish brown. Short dark submedian streaks or dots on the anterior margin of each tergite. Black stigmatic dashes. Ventrally paler, the basal sternites faintly washed with yellowish brown. Black dots and dashes distinct. Tails yellowish white, joinings purplish black.

Occipital tubercles are present in nymphs of both sexes, but are only moderately developed. The rows of dorsal spines are widest apart on the basal tergites, and converge to rearward. The spines on tergites 1–3 are finger-like and taper bluntly, but are shorter and less curved than in *E. temporalis*. Postero-lateral spines on segments 2 and 3 are only moderately developed. In color the nymph resembles a dark *temporalis;* in some specimens a pale dorsal stripe is present. The hinge of the operculum is pale, and the operculum itself is sprinkled with pale dots. Tails not noticeably banded, but each joining is darker.

The type material was from Knowlton, Que. We have specimens of imagos from this locality in the Cornell collection, which Dr. McDunnough very kindly sent us.

Ephemerella rotunda Morgan

1905—Needham—N. Y. St. Mus. Bull. 86: 45. Fig. (*Ephemerella sp.*) . . . . N. Y.
1911—Morgan—Ann. Ent. Soc. Amer. 4: 113. Fig.
1924—Needham—Psyche; 31: 309 (*feminina*).
1926—McDunnough—Canad. Ent. 58: 188 (*feminina* =*rotunda*).
1931—Steger—Psyche 38: 32. Fig.
1931—McDunnough—Canad. Ent. 63: 195. Figs. . . . . Que., Ont.
1932—Traver—J. Elisha Mitchell Sci. Soc. 47: 176 (also 149, in discussion of *inconstans*) . . . . N. C.

Length: body 10; wing 10 mm.

A species of the *invaria* group; venation pale; spines on the penes more numerous than in *E. invaria* and *E. subvaria;* nymph dark brown with many fine pale dots, the dorsal spines better developed than in *invaria.*

Eyes probably orange in life. Head yellowish, shaded with reddish on the occiput. Antennae pale. Thorax light reddish yellow; pronotum shaded with purplish black. A dark brown area on each side of the mesonotum and two narrow blackish submedian streaks anterior to the scutellum. A dark brown line above each of the middle and hind legs; pleural sutures narrowly brown. Legs pale whitish. Wings hyaline; venation very pale yellowish to hyaline; stigmatic area rather opaque white.

Abdominal tergites pale yellowish, each with a broad purplish brown transverse band across the middle. On tergites 1–2 and 8–9, this dark band occupies most of the sclerite. Pleural fold pale. Traces of a pale median line, most distinct on the middle tergites. Ventrally yellowish white, sometimes with faint indications of yellowish brown transverse bands across the middle of each sternite, similar to those on the tergites. Oblique submedian streaks at the anterior margin of each sternite; may be rather indistinct. Tails white, basal joinings distinctly purplish black; beyond the middle these are but faintly indicated. Spines on the penes rather more numerous than in *E. invaria;* two small clusters of ventral spines are present, which are not found in *invaria.* Two to four apical spines are usually present (see fig. 152).

The nymph is larger than *E. invaria* or *E. subvaria;* it is dark reddish brown in color, sprinkled with numerous very small pale dots. Each corner of the pronotum is pale. There are usually other smaller pale areas on each side of the pronotum, including a crescentic mark and two small round spots between this and the median line. Eight to ten somewhat larger pale areas are usually present on the mesonotum, between and anterior to the wing roots. Rather large pale lateral areas may be present on tergites 5 and 6; these may be connected along the anterior margin. Near the middle of each lateral extension is a dark bar. Venter somewhat paler than the dorsum; darkest on the apical sternites and the lateral portions of the middle ones. Traces of a pale median triangle at the anterior margin may be present on the middle and apical sternites. Tails yellowish brown; crossed by two or three dark bands near the middle, and another apically. White spots are present on each tergite, at the base of each dorsal spine. These spines are better developed than in *invaria,* but less prominent than in *subvaria.*

The species was described from four specimens taken in the vicinity of Ithaca, N. Y. All of these have been destroyed. However, we have many specimens in the Cornell collection, both of nymphs and imagos, including much reared material, from Ithaca and nearby localities. Two female imagos sent by Dr. McDunnough, taken at Knowlton, Que. (G. S. Walley), are paler than our local material. We have also several nymphs of this species from Scott's Creek near Sylva, N. C. (J. S. Gutsell).

Ephemerella semiflava McDunnough

1926—McDunnough—Canad. Ent. 58: 300 .... Ont.
1931—McDunnough—Canad. Ent. 63: 192.

Length: body 7; wing 8 mm.

A species of the *invaria* group; second joint of forceps not swollen apically; nymph unknown.

Eyes of living male mustard yellow. Head yellowish, tinged with reddish brown near the ocelli. Thorax olive brown shaded with yellow; light yellowish brown near the scutellum; reddish brown shading anterior to the wing roots. Blackish transverse streak near the posterior margin of the prothorax; pleura and sternum light yellowish brown; an oblique stripe of pure yellow on the pleura from the base of the fore wing to the base of the fore leg. Legs light yellow, becoming whitish on the fore tarsi. Fore tibia about one and one-half times as long as the femur; tarsus about twice the length of the tibia. Wings hyaline; venation hyaline.

Abdomen yellowish olive brown; lateral rows of indistinct oblique brown stripes, more or less fused on tergites 8–10. Tergites 2 and 3 faintly blackish on the posterior margins. Tails white, with faint brownish joinings at the base. Genitalia very similar to *E. excrucians.* It is quite possible that this species may be synonymous with *excrucians.*

The type material, from Tilsonburg, Ont., is in the Canadian National Collection at Ottawa. We have no specimens of this species.

Ephemerella septentrionalis McDunnough

1925—McDunnough—Canad. Ent. 57: 171. Fig. . . . . Ont.
1931—Steger—Psyche 38: 34. Figs. . . . . N. Y.
1931—McDunnough—Canad. Ent. 63: 201. Fig. . . . . Que.

Length: body 8; wing 9 mm.

A species of the *needhami* group; femora and tibiae of imago very long as compared with tarsi; legs of nymph very long and slender; short spines on inner apical margin of penes.

Head brownish; eyes light orange-red. Thorax yellowish to olive brown dorsally; ventrally pale yellowish with reddish tinges; pleura yellow, tinged with ruddy. Legs very long; femur and tarsus of hind leg subequal in length, each being more than twice as long as the tarsus. Pale yellowish, the femora brownish at the apex. Wings hyaline, venation entirely colorless. Abdominal tergites 1–7 dull olive to smoky brown; traces of black lateral dashes often present; posterior margins narrowly darker. Tergites 8–10 bright brown with a reddish tinge. Sternites pale yellowish, hyaline except the apical segments, which are opaque and somewhat darker. An interrupted median brownish stripe may be present, sometimes extending forward on to the thoracic sternum; four dark dots on each sternite, arranged in a curving line. A small dark spot may be present at the middle of the posterior margin.

Tails yellowish white, joinings rather faintly reddish brown. Second joint of forceps somewhat swollen on the inner apical margin. Penes separated apically by a deep rounded notch; several short spines on the margin of this notch, and on each side a larger toothlike process which projects into the notch (see fig. 155).

The yellowish brown nymph is remarkable for the extreme length and slenderness of the legs, which are fringed with rather long hairs. Body yellowish, thickly sprinkled with very small brown dots. Head smooth; a tuft of hairs on each side of the occiput. Thorax smooth. Head shaded with brown along the median line; prothorax pale on median line, which is bounded on each side by a dark brown stripe. Legs yellowish; femora with three bands of dark brown; tibiae and tarsi banded near the middle, this band often lacking on the hind tibia. 11 to 12 denticles on the inner margin of each claw. Lateral extensions present on the abdominal segments; each bears a posterolateral spine. Gills present on segments 3–7. No dorsal spines, but the posterior margin is produced backward at the median line. A narrow brown median stripe the length of the dorsum, and traces of pale lateral stripes. Ventral markings consist of two dark oblique dashes on the anterior margin, and two small dark dots nearer the center of each sternite. Tails light brown at base, yellow apically; in the middle area, alternately banded with black and yellow. A whorl of spines at each joining, in the basal portion; apically without these spines, and margined with longer hairs.

We have imagos and nymphs of this species in the Cornell collection, kindly sent by Dr. McDunnough, from Knowlton, Que. (G. H. Fisher, L. J. Milne, G. S. Walley).

Other reared specimens are from North Harford, N. Y. (A. Louise Steger). Imagos are from Old Forge, N. Y. (J. G. Needham), and Ft. Jackson, N. Y., the St. Regis River (L. Harper).

Ephemerella serrata Morgan

1911—Morgan—Ann. Ent. Soc. Amer. 4: 109. Figs.....N. Y., Mass.
1931—McDunnough—Canad. Ent. 63: 207. Fig.....Que.
1932—Traver—J. Elisha Mitchell Sci. Soc. 47: 143.....N. C.

Length: body 5; wing 6 mm.

A brightly colored species; eyes of male orange-red, body brighter brown than *E. deficiens* or *E. sordida.* Nymph with feeble prothoracic spines or warts; no curved row of dark spots on each sternite, as in *E. serratoides.*

Eyes of living male orange-red; head brown, lighter in front. Antennae pale brown. Thorax yellowish dorsally, shaded with brown. Posterior margin of mesonotum margined with brown; pleura with brown shading. Sternum pale with median and lateral dark brown areas. Legs pale; femora banded with brown apically. Wings hyaline, tinged with brown at the base; veins pale. Abdomen brown dorsally, pale ventrally, with no curved row of dark dots such as occur in *E. serratoides.* Tails pale, the basal joinings brown.

Head of nymph somewhat roughened on the vertex, but with no distinct occipital tubercles. Feeble wart-like processes on each side of the median line of the prothorax. Dorsal spines represented by wart-like spicule-bearing processes. Gills present on tergites 3–7. Postero-lateral spines become increasingly acuminate to rearward, that on the 9th segment extending beyond the end of segment 10. Lateral margins of abdominal segments bear conspicuous spinules. Tails pale, with a median band of dark brown. Each claw with 3–4 denticles. General color brown; legs and venter paler; tergites 5–6 with brown pencilings. Abdominal sternites marked with a lateral row of dark dashes, but with no curved row of dark dots between these rows.

We have imagos of this species in the Cornell collection, reared by Dr. A. H. Morgan from Sheffield, Mass., and Ithaca, N. Y. Nymphs in this collection are from Ithaca and vicinity; Catoctin Creek, Md. (V. Argo); Lost River, W. Va. (J. G. Needham); Wilmington, N. Y. (P. R. Needham); also from three stations in the Appalachian region of North Carolina: Laurel River, Wayah Creek and a tributary of the Pigeon River at Waynesville (J. R. Traver). Mention of these was accidentally omitted from *Mayflies of North Carolina* (*J. Elisha Mitchell Sci. Soc.* 47: 143, 1932).

Ephemerella serratoides McDunnough

1930—McDunnough—Canad. Ent. 62: 55 (*serrata*).....Que.
1931—McDunnough—Canad. Ent. 63: 83.
1931—McDunnough—Canad. Ent. 63: 207. Figs.
1932—Traver—J. Elisha Mitchell Sci. Soc. 47: 143.....N. C.

Length: body 5; wing 5½–6 mm.

A species of the *serrata* group; imago very similar to *E. serrata,* but distinguished by the presence of four tiny dark dots in a curved row across each abdominal sternite. Nymph with a smooth head and prothorax; dorsal spines present; a curved row of dark dots on each abdominal sternite.

Eyes of male orange-red; head yellowish between the eyes. Vertex of head in female light yellow with blackish shading. Pronotum of male brown with smoky shading. Mesonotum, pleura and sternum rather a bright brown with paler shading. Legs dull yellow; middle and hind tibiae each with an apical reddish black spot. Wings hyaline; all veins pale. Abdominal tergites olive to reddish brown with indistinct darker shading in the form of a broken median line and rows of submedian and lateral spots. Sternites pale reddish brown, becoming yellowish brown posteriorly. Sternites 2–8 each bear a curved row of four dark dots across the center. Forceps dull yellowish brown. Tails whitish; basal joinings reddish.

Head and thorax of nymph entirely smooth, and without long hairs; brown, sprinkled with pale dots. Dorsal spines represented by wart-like spicule-bearing processes, as in *E. serrata*. Lateral margins of the abdominal segments bear spicules; the postero-lateral spines are somewhat more pronounced than in *serrata*, but not as well developed as in *E. sordida*. Abdominal tergites are quite evenly brown; tergites 5 and 6 paler laterally, and with indications of a dark broken median line and similar lateral dark lines. Sternites paler, each with a curved row of four dark dots; on each side a lateral row of dark dashes as in *serrata*. About 7 denticles on each claw. Tails blackish brown at base, the tips pale.

We have nymphs of this species in the Cornell collection from Elk Garden, the Bluestone River and the Cacapon River, W. Va., and from the Great Smoky Mts. (J. G. Needham); the Potomac River, Md., near Harper's Ferry (V. Argo); from Wardensville, W. Va., likewise from two stations in the Appalachian region of North Carolina: Bald Creek and the Ocona Lufty River (J. R. Traver). Reference to these stations was accidentally omitted from *Mayflies of North Carolina*.

Ephemerella simplex McDunnough

1925—McDunnough—Canad. Ent. 57: 41 .... Que.
1930—Ide—Canad. Ent. 62: 211 .... Ont.
1931—McDunnough—Canad. Ent. 63: 208. Figs .... N. B.
1932—Traver—J. Elisha Mitchell Sci. Soc. 47: 177 (sp. No. 1) .... N. C.

Length: body 6; wing 6 mm.

A small, dark colored species; penes ending in two broad lobes, with minute papillae at the apices; nymph somewhat flattened, hairy, without occipital, thoracic or dorsal abdominal spines; gills on segments 4–7.

Eyes deep brown; head deep blackish brown. Thorax deep blackish brown; lateral margin of the mesonotum, and the dorsal transverse suture, pale brown. Mesosternum blackish brown in the central portion, light brown laterally between the middle and hind legs. Fore leg blackish; middle and hind legs dull yellowish brown; femur and tibia of hind leg subequal, tarsus two-thirds of the tibia. Wings hyaline; longitudinal veins brownish; cross veins apparent only in the stigmatic area. Abdominal tergites blackish brown, almost unicolorous, the lateral margins slightly paler; tergites 8 and 9 shaded with paler brown. Dull yellowish brown ventrally; sternite 9 shaded with blackish brown, as are also the forceps. Penes curve upward; each ends in a broad lobe, with a slight ventral heart-shaped depression just basad of the lobes; minute papillae are borne near the apex of each lobe (see fig. 157). Tails dull smoky.

The nymph is rather broad and flat and very hairy. Pale olivaceous to deep brown in color, shaded variously with dark brown on the abdominal tergites. No occipital nor thoracic tubercles, nor dorsal abdominal spines are present. Genae considerably produced laterally. Femora rather chunky, the fore femora bearing a series of long hairs transversely across the middle. Femora with two dark bands; tibiae banded apically, tarsi with a median dark band. No denticles on the claws. Lateral extensions of the abdomen rather well developed, and bearing postero-lateral spines on segments 4–9. Gills borne on segments 4–7 only, that on segment 4 being rather strongly operculate. Markings of the tergites consist of interrupted dark submedian and lateral streaks which may broaden and coalesce. Sternites pale; unmarked, or with lateral rows of dark dashes. Tails pale, the joinings faintly ringed; with weak whorls of spinules and some fine fringing hairs apically.

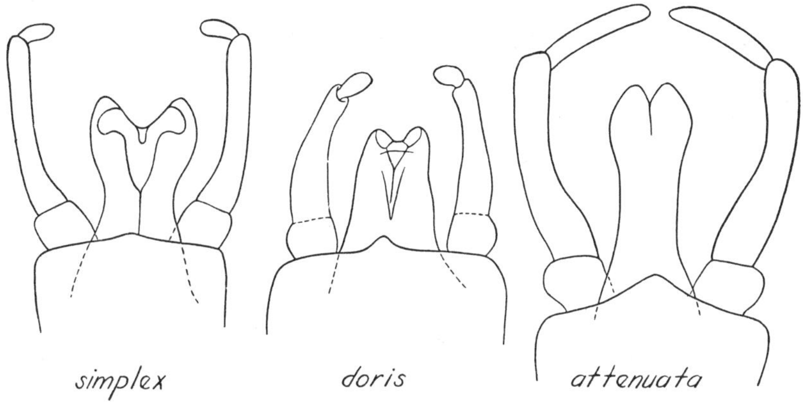

FIG. 157. Male genitalia of three species of Ephemerella.

The species was described from material taken at Laprairie, Que. We have imagos of this species in the Cornell collection from Cascades Pt., Que., which Dr. McDunnough kindly sent us; and a series of males from North Fairhaven, N. Y., on Lake Ontario (J. G. Needham, *et al*). Nymphs in our collection are those sent by Dr. McDunnough from Cascades Pt., and others from several localities in North Carolina (J. R. Traver).

Ephemerella sordida McDunnough

1925—McDunnough—Canad. Ent. 57: 42 .... Que.
1930—McDunnough—Canad. Ent. 62: 55 ....
1930—Ide—Canad. Ent. 62: 212 .... Ont.
1931—McDunnough—Canad. Ent. 63: 205. Figs.

Length: body 5; wing 6 mm.

A species of the *serrata* group; male imago duller in color than *E. deficiens* with tails unmarked. Head and thorax of nymph hairy; lateral extensions of abdominal segments prominent, the postero-lateral spines long and flat; no dark ventral markings.

Eyes deep blackish brown. Head and thorax blackish brown; pleura with lighter brown shading; lateral extensions of mesosternum paler. Head of female "light ruddy ochreous" (McD.). Legs pale yellowish, fore legs somewhat tinged with smoky. Middle and hind femora with indications of brown spots apically. Fore tibia twice the length of the femur, and slightly longer than the entire tarsus. Wings hyaline, all veins pale. Abdominal tergites dull blackish, the three posterior segments brownish. Sternites paler, rather smoky, becoming opaque and pale grey posteriorly. Genitalia and tails whitish; tails unmarked. Penes as in fig. 156.

Head and thorax of nymph smooth, but quite hairy. Dorsal spines represented by the usual wart-like processes. Postero-lateral spines longer than in any other species of this group; the spine on segment 9 extends well beyond the 10th segment. Gills on segments 3–7. Legs hairy, the anterior margin of the femora with a few spinules. General color pale yellowish brown to olive; no ventral markings. Femora usually dark with pale bands; tibiae pale with a wide median dark band; tarsi dark basally. Tails pale, without dark markings.

We have three male imagos in the Cornell collection, taken by Mr. G. S. Walley at Knowlton, Que., which Dr. McDunnough very kindly sent to us.

Ephemerella spinifera Needham

1927—Needham—Ann. Ent. Soc. Amer. 20: 110....................................Mont., Utah
1927—Needham—Utah Agr. Exp. Sta. Bull. 201: 8. Fig.
1930—Walley—Canad. Ent. 62: 17. Fig..Wyo.
1934—McDunnough—Canad. Ent. 66: 156.

Length (nymph): body 11 mm.

A species of the *fuscata* group, allied to *E. proserpina*, *E. yosemite*, and *E. autumnalis;* distinguished from the last by the suddenly enlarged spines on 8 and 9; from the two former species by the unequal size of the submedian pronotal spines.

No frontal horns nor frontal shelf. Occipital tubercles prominent. Submarginal spines of the pronotum sharp-pointed, well developed; a pair of submedian spines on each side, unequal in size, the anterior of these being twice as high as the posterior. A pair of small tubercles near the median line, on the anterior margin of the mesonotum; two pairs of submedian mesonotal spines, those of the posterior pair being the larger; and a median tubercle-like elevation near the posterior margin. A median tubercle on the prosternum. Dorsal spines very prominent; those on segments 2–7 subequal, the pairs on 8 and 9 fully four times the size of the preceding ones. No teeth or spines on the anterior margin of the fore femur; tibial 'thumb' rather slender, sharp-pointed, not reaching to the middle of the tarsus. General color pale brownish to reddish brown; ventrally yellowish, the abdomen with an indistinct row of darker dashes on each side. Tails yellowish.

The type material, from Montana and Utah, is in the Cornell collection; this consisted of immature nymphs only. Mature specimens, as taken from the Gardiner River in Yellowstone National Park and figured by Walley (1930), differ in some slight points from the original description.

Ephemerella subvaria McDunnough

1925—McDunnough—Trans. R. S. C. 19: 213 (dark form of *invaria*)..........................Que.
1930—Ide—Canad. Ent. 62: 211 (dark form of *invaria*)..Ont.
1931—McDunnough—Canad. Ent. 63: 84, 194.

Length: body 9; wing 10 mm.

A species of the *invaria* group; close to *E. invaria*, but distinguished by the prominent dark veins and cross veins; nymph rather similar to *E. rotunda*, but with distinct dorsal spines, and often paler in color.

Eyes of living male bright red brown. Head deep brown. Thorax deep brown, with reddish tinges on the pleura. Legs deep dull amber; fore femur tinged with smoky, middle and hind femora often with traces of reddish median shading. Wings hyaline; longitudinal and cross veins distinctly light brown; no cross veins in the basal two-thirds of the costal and radial spaces.

Abdomen light reddish brown dorsally; shaded laterally with smoky; posterior margins smoky. Ventrally paler reddish brown, the apical sternites somewhat darker. Lateral margins of segments yellowish. Forceps dull yellowish. Penes very similar to *E. invaria;* apical spines usually absent. Tails pale, yellowish at base; joinings distinctly smoky brown.

Nymph slightly smaller than *E. rotunda*, and paler in color; tergites 6 and 7 often almost entirely pale, while 5 has only a dark median subtriangular patch; tergites 8 and 9 always dark. Dorsal spines present on abdomen, much better developed than in *rotunda;* lateral extensions much as in *rotunda*. Pale dorsal spots at base of spines not present.

The type material, from Cascades Point, St. Lawrence River, and from other localities in Quebec, is in the Canadian National Collection. We have specimens of this species from Cascades Point (G. S. Walley), which Dr. McDunnough kindly sent to the Cornell collection. It is possible that certain local specimens, from the vicinity of Ithaca, N. Y., may be of this species.

Ephemerella temporalis McDunnough

1924—McDunnough—Occas. Pap. Boston Nat. Hist. Soc. 5: 74. Fig. .... Ont.
1925—McDunnough—Trans. R. S. C. 19: 212. .... Que.
1930—McDunnough—Canad. Ent. 62: 58. Fig.
1931—McDunnough—Canad. Ent. 63: 35. Figs. .... N. B.

Length: body 8; wing 8–10 mm.

A member of the *bicolor* group, allied to both *E. prudentalis* and *E. doris*. No minute black dots on legs and abdomen, as in *E. lutulenta*. Occipital tubercles of nymph strongly developed in both sexes, distinctly conical in shape; rows of dorsal spines subparallel, those on tergites 1–3 long, tapering, finger-like and bent backward.

Eyes reddish yellow. Head deep yellowish. Thorax bright brown, tinged with olivaceous dorsally. A paler orange-yellow streak anterior to the wing roots; pale creamy shading on the pleura at the wing bases. Legs yellow, shading into dirty white on the tibiae. Claws and joints of the tibiae blackish; a dark dot apically on the femora. Wings hyaline; venation hyaline. On the fore wing, the costa is slightly yellowish at the base. Abdominal tergites 2–7 deep brown, 8 and 9 reddish brown, 10 yellowish brown. Posterior margins of tergites 2–4 blackish. On the basal tergitess the faint brown mid-dorsal line is partially geminate. A small subdorsal dark dot on the anterior margin of each tergite. Faint dark markings in the region of the spiracles; lateral margins of 8 and 9 narrowly yellowish. Pale yellowish white ventrally; basal sternites tinged with brown. The usual curved row of dark dots is distinct, and the ganglionic areas of sternites 4–7 are brown. Tails whitish, joinings dark brown.

The occipital tubercles of nymphs of both sexes are very strongly developed and distinctly conical in shape. The rows of dorsal spines are subparallel, sometimes slightly divergent on tergite 4. These spines on tergites 1–3 are long and tapering, finger-like and bent backward. The postero-lateral spines on segments 2 and 3 are well developed, that on segment 3 being almost twice as long as its width at the base. The general color is very dark brown, with a heavy sprinkling of minute pale dots. A pale dorsal stripe is present in some specimens. The operculum is sprinkled with pale dots, and may have two pale patches along the joint. Ventrally the curved row of dark dots and the dark lateral dashes are distinct. Tails distinctly banded, alternately dark brown and pale yellow.

The species was described from material taken at Ottawa, Ont. We have imagos in the Cornell collection from Knowlton, Que., Lake Nipissing, Ont., the mouth of the Nieve River, and Fredicton, N. B., presented by Dr. McDunnough. Other imagos, reared, are from Green Lake, N. Y., and from Hamburg Lake and a small lake near Liberty, N. C. (J. R. Traver). Nymphs are from the forestry farm at the University of Michigan; Washtenaw Creek, Mich., Douglas Lake, Mich.; Spencer Lake, N. Y.; and Oochee Creek, Ga. (P. W. Fattig). Imagos tentatively referred here are from Amherst, Mass. (C. P. Alexander); Catatunk, Me. (A. E. Brower); and Ft. Anne, N. Y. (P. R. Needham).

Ephemerella teresa Traver

1934—Traver—J. Elisha Mitchell Sci. Soc. 50: 2?0. . . . . Calif.

Length (nymph): body 5–7 mm.

A species of the *serrata* group; seemingly very close to *E. micheneri*. Dorsal abdominal spines present. Only the nymphal stage is known.

Head and thorax smooth. Maxillary palp wanting. Median line of pronotum elevated anteriorly, depressed posteriorly; bordered by lateral carinae. Legs pale; apices of joints reddish brown. 9–10 denticles on each claw. Thoracic ganglia darkened. A purplish line along the costal margin of the developing wing. Lateral abdominal extensions and postero-lateral spines only moderately developed. Dorsal spines present on tergites 4–7; short and furthest apart on 4; longer and subequal on 5–7. Gills borne on tergites 3–7. General color pale reddish brown; darker lateral patches present on tergites 4–8, and dark dots on 9. Sternites yellowish; ganglionic areas often darkended, and a row of dark lateral dashes on each side, as in *E. serrata*. Tails pale yellowish, with a whorl of dark reddish brown spines at each joining.

We have the type specimens of this species in the Cornell collection, from San Antonio Canyon, Calif. (Mrs. Theresa Robinson Seemann).

Ephemerella tibialis McDunnough

1924—McDunnough—Canad. Ent. 56: 224. . . . . Alta.
1926—McDunnough—Canad. Ent. 58: 186.
1929—McDunnough—Canad. Ent. 61: 170. Fig. . . . . Mont.
1930—Walley—Canad. Ent. 62: 18, fig.

Length: body 7½; wing 8 mm.

A member of the *serrata* group; abdominal segments of male translucent whitish on the anterior margins; tails deep smoky; fore tibia twice as long as the femur. Nymph with the usual wart-like dorsal spines; lateral extensions of the abdomen not prominent; head and thorax smooth; tails distinctly banded.

General color of male imago deep purplish brown. Thorax almost blackish dorsally. Fore femur pale smoky yellowish, tibia deep smoky, tarsus paler than the tibia; tibia fully twice as long as the femur. Middle and hind legs dull yellowish, the femora with reddish bands apically. Wings vitreous; venation pale grey-brown, the cross veins rather more distinct than is usual for this group. Anterior margins of abdominal segments 3–8 narrowly whitish, translucent. Sternites with indications of dark ganglionic blotches. Forceps reddish basally, blackish distally. Penes blackish (see fig. 156). Tails deep smoky, slightly darker at the joinings.

Head and thorax of nymph smooth. Legs rather short. Lateral extensions of the abdominal segments not prominent. Dorsal spines represented by the usual wart-like processes. Gills present on tergites 3–7. A slender blackish nymph, sometimes with a narrow mid-dorsal pale line on thorax and abdomen. A small pale spot at the base of the femur ventrally; an apical pale band and an incomplete subapical pale mark also. A pale band near each end of the tibia; tarsus pale at base and apex. Tails alternately ringed with brown and pale areas.

We have a pair of imagos from Banff, Alta., taken by Dr. McDunnough, who kindly gave them to the Cornell collection. Other imagos in our collection tentatively referred to this species are from Yellowstone Park, Wyo. (C. R. Crosby), and from the San Gabriel Mts., Calif. (C. D. Michener). Nymphs are from Moose, Wyo. (Elsie B. Klots); and Yellowstone Park, Wyo. (P. R. Needham).

Ephemerella tuberculata Morgan

1911—Morgan—Ann. Ent. Soc. Amer. 4: 112. Figs..N. Y.
1931—McDunnough—Canad. Ent. 63: 213. Figs..Ont., Que.
1932—Traver—J. Elisha Mitchell Sci. Soc. 47: 175. Fig..N. C.

Length: body 9; wing 9 mm.

A species of the *fuscata* group; by the nymphal structures, closely allied to *E. conestee*, and *E. allegheniensis*. Imago very similar to *E. fuscata*, but distinguished from it by the dark ventral abdominal markings and the greater number of black spots on the femora. The lateral extension of the 9th segment of the nymph is not flaring, as in *E. allegheniensis:* the thoracic tubercles distinguish it from *E. conestee.*

Head and thorax dark reddish to blackish brown; pale areas anterior to and laterad of the scutellum, and along the antero-lateral margin of the mesonotum. Pleura with paler shading. All coxae and trochanters yellowish, the former sometimes with one or more dark dots. Fore femur and tibia blackish; tarsus pale to smoky brown. Middle and hind femora yellowish, shaded with olive brown; tibiae and tarsi pale yellowish. All femora heavily sprinkled with black dots. Wings hyaline; longitudinal veins smoky to pale reddish brown. Tergites 1 and 7–10 largely dark reddish to blackish brown; 7 and 8 may have a pale median streak between the dark lateral areas. Middle tergites somewhat paler, the middle areas yellowish brown, the posterior margins and lateral areas dark brown; traces of a broken dark mid-dorsal line. Pleural fold pale. Sternites 1–8 light yellowish brown, the anterior ones sometimes tinged with reddish and washed with blackish. Sternite 9 is deep blackish brown. Each pale sternite has a blackish lateral streak near the pleural fold, and a curved row of four dark dots between these, across the middle of the segment. Forceps smoky brown, penes blackish (see fig. 153). Tails smoky brown.

The nymph has very short frontal horns, and a frontal shelf which is slightly notched to receive the bases of the antennae (see fig. 151 h). Occipital tubercles present; tubercles also on the pro- and mesonota. Femora flattened, the fore femur widest and bearing teeth or spines on the anterior margin. Tibial 'thumb' about half as long as the tarsus, rather stout, almost straight (see fig. 62 f). Dorsal spines present on abdominal tergites. Gills borne on segments 3–7. Posterior margins of tergites 8 and 9 are fringed with long hairs. General color (in life) dull brownish green, often with a prominent whitish dorsal stripe. Paler ventrally, with a row of indistinct dark marks on each side of the abdominal sternites, somewhat as in *E. serrata*. Tails yellow at base, with one brown band; beyond the middle, alternately yellow and brown.

We have one male imago from Quebec, in the Cornell collection, collected by Dr. McDunnough, and other reared imagos from the mountains of North Carolina (J. R. Traver). Besides the nymphs from North Carolina, we have one from the St. Regis River, N. Y.

Ephemerella unicornis Needham

1905—Needham—N. Y. St. Mus. Bull. 86: 45..Ind.
1931—McDunnough—Canad. Ent. 63: 187.

Length: body 5; wing 5 mm.

The identity of this species, described from a single male subimago, cannot at the present time be correctly determined. When more material from the type locality is available, such determination may be made possible. The character by which this species was distinguished, "an erect conic tubercle upon the front margin of the middle lobe of the mesothorax," is said by Dr. McDunnough to "occur sporadically in several species very similar in the subimago stage." The type specimen is in the Cornell collection.

Ephemerella verisimilis McDunnough

1930—McDunnough—Canad. Ent. 62: 57. Fig. . . . . Que.
1931—McDunnough—Canad. Ent. 63: 65. Figs. . . . . Ont., N. B.

Length: body 7; wing 8 mm.

A species of the *bicolor* group, intermediate between *E. bicolor* and *E. temporalis;* abdominal sternites darker and legs deeper yellow than in these species. Occipital tubercles of nymph well developed; rows of dorsal spines evenly divergent to rearward.

Eyes orange. Head reddish brown with purplish brown shading around the ocelli. Thorax deep brown. Legs rather bright yellow, with traces of orange patches at the base, the middle and the apex of each femur. Wings hyaline, venation hyaline. Abdominal tergites deep brown, with no indication of yellow or orange shading. Abdominal sternites darker than in *bicolor* or *temporalis*. Faint indications of the usual dark dots and oblique streaks. Tails smoky, the joinings narrowly darker.

The occipital tubercles of the nymph are quite well developed in both sexes. The rows of dorsal spines are gradually and evenly divergent to rearward; those of tergites 1–3 are slightly longer and more erect than in *E. bicolor*, the tips more pointed. The postero-lateral spines on segments 2 and 3 are considerably better developed than in *bicolor*. Tails brown with no pale bands.

The type material was taken at Bradore Bay, Que. Dr. McDunnough has kindly sent to the Cornell collection imagos from that locality (W. J. Brown). Doubtful specimens in our collection include material taken at Old Forge Pond (J. G. Needham), and nymphs from Mashpee River, Mass. (W. D. Sargent).

Ephemerella vernalis Banks

1914—Banks—Proc. Acad. Nat. Sci. Phil. 66: 614. . . . . N. C.
1925—McDunnough—Trans. R. S. C. 19: 213.
1931—McDunnough—Canad. Ent. 63: 195 (*rotunda*).
1932—Traver—J. Elisha Mitchell Sci. Soc. 47: 149. Figs. (*inconstans*).

Length: body 8–9½; wing 8–10 mm.

A species of the *invaria* group; the second joint of the forceps enlarged distally; venation yellowish to yellowish brown; usually 7 spines only, on penes.

Head light brown; eyes light orange. Thorax brown; pronotum often with a median pale streak; pleura marked with pale yellow; mesonotal scutellum margined with dark reddish brown. Ventrally brown, darkest at the center of the mesosternum. Fore leg yellowish to light brown; purplish streaks on distal portion of femur. Middle and hind legs very similar. Venation very light yellowish brown, the main veins of the costal border yellowish basally. Stigmatic area opaque.

Abdominal tergites 2–7 light purplish brown with a narrow median whitish line; 2–4 washed with brownish, 5–7 paler; posterior margins dark brown, postero-lateral angles and intersegmental rings greyish white; a purplish line margins the pale pleural fold. Tergites 8–10 washed with yellowish brown; a median light brown line is present, and a brownish patch on each side; one or more short dark dots near each stigma. Sternites 1–5 greyish, lavender tinged; 6–9 yellowish brown; on each, a brown line margins the pleural fold, less distinct on 6–9. Posterior margins and ganglionic areas whitish. Tails greyish to light brown, joinings purplish brown. Penes with fewer spines than *E. rotunda*, usually only 7, occasionally 10; of these, two may be apical in position, but more often there are no apical spines (see fig. 152).

Nymphs resembling *E. rotunda*, but less robust. Vary in color from light yellowish brown to dark mahogany; body finely dotted and mottled with numerous small yellowish dots. Legs barred with yellow. A median yellowish dorsal streak is often present on the dorsum of the abdomen. Minute paired dorsal spines are present, and a whitish spot marks the base of each spine. Lateral extensions of the abdomen are more rounded on the outer margin than in *rotunda;* end spines not incurved; each lateral extension is crossed by a black streak. Indications of a yellowish mid-ventral line and a dark lateral line near the margin; posterior segments usually darker than the preceding ones.

This species differs from *rotunda* as mentioned; spine on the penes fewer in number than in *E. invaria;* venation paler than in *E. subvaria*. Examination of the genitalia of a male paratype sent us by Dr. Banks establishes the fact that the species described from North Carolina as *E. inconstans* is in reality the true *vernalis;* we cannot agree with Dr. McDunnough that *vernalis* is a synonym of *rotunda*.

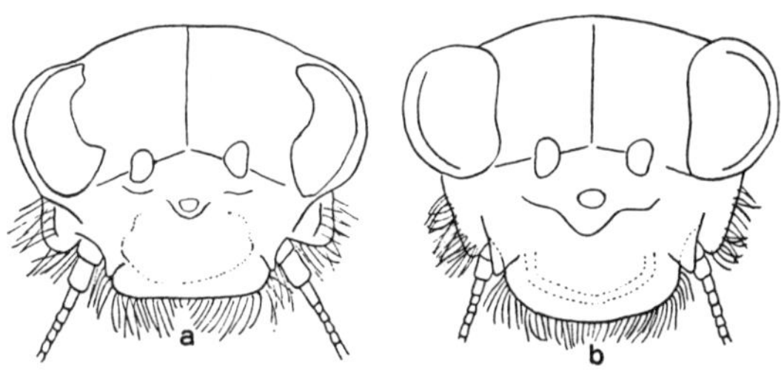

FIG. 158. Heads of Ephemerella nymphs; a, *wayah;* b, *conestee.*

Ephemerella wayah Traver

1932—Traver—J. Elisha Mitchell Sci. Soc. 47: 172. Figs. . . . . .N. C.

Length (nymph): body 6–7 mm.

A member of the *fuscata* group, and closely related to *E. fuscata*. Distinguished from this species in the nymphal stage by the lack of prothoracic tubercles, better developed abdominal spines, slightly narrower femora, and somewhat longer tibial thumb. Only the nymph and female subimago are known.

Head of nymph (see fig. 158 a) with a broad frontal shelf, widely notched on each side to receive the bases of the antennae. Frontal horns barely indicated. No occipital tubercles. All femora much flattened, but without the wide prominent posterior extension of *E. fuscata;* fore femur with spines on the anterior margin. Tibial spine almost one-third as long as the tarsus; blunt, stout, better developed than in *fuscata* (see fig. 159). Legs and entire body very hairy. No prothoracic tubercles, as in *fuscata*. Dorsal spines present,—not conspicuous, but better developed than in *fuscata*. Sternite 9 and the lateral margins of 3–8, very hairy, forming a very incomplete adhesive disc. General color (in life) grey-green; yellowish brown in alcohol. Marked irregularly with dark brown on legs, head, thorax and abdomen; tergites 8–10 usually yellowish. Abdominal sternites yellow; a lateral row of dark marks on each side, much as in *E. serrata*.

The type material, from the Appalachian region of North Carolina, is in the Cornell University collection.

Ephemerella yosemite Traver

1934—Traver—J. Elisha Mitchell Sci. Soc. 50: 225..Calif.

Length (nymph): body 13 mm.

A member of the *fuscata* group, allied to *E. spinifera*. It is distinguished from the latter species by the larger size, the blunter submedian prothoracic tubercles, which are of unequal size, the smaller and blunter lateral prothoracic tubercles, and the shorter and thicker leg joints. Only the nymph is known. See fig. 53 a.

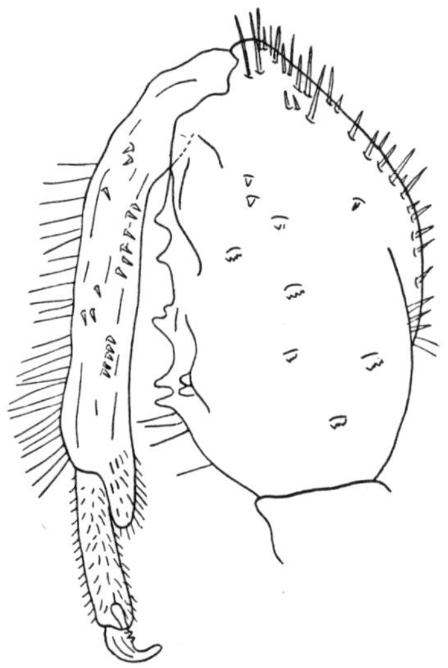

FIG. 159. Right fore leg of nymph of *Ephemerella wayah*.

Very slight indications only of frontal horns. No frontal shelf. Occipital tubercles present; also two lateral prothoracic tubercles, and two pairs of submedian ones; two pairs of mesothoracic tubercles and a median posterior elevation. Ventrally, a tubercle on the prosternum. No teeth or spines on the anterior margin of the fore femur. Joints of the fore leg relatively thick and wide; tibial 'thumb' moderately long, blunt. Dorsal spines on segments 2–7 are quite prominent, pointed at the tips, and slant backward slightly. Spines on 8 and 9 show a sudden increase in size, being more than twice as high as the preceding pairs, and correspondingly larger. Those on 8 are erect, but the pair on 9 slant backward. Gills borne on segments 3–7. General color blackish brown with yellow markings. Thorax dark brown; median posterior area of mesonotum yellow, sprinkled with fine dark dots. Venter yellowish brown with dark dots. Femora dark brown with several irregular pale areas; tibiae blackish, the ends yellow, a dark median tarsal band. Abdominal tergites blackish brown, tergite 8 largely yellow. Abdominal sternites dull reddish brown with numerous fine dark dots; posterior margins darker; ganglionic areas faintly darker; a lateral row of dark streaks on each side. Tails blackish brown with narrow yellow bands; usually 3 dark segments to 1 pale one.

The type material of this species, in the Cornell collection, was taken by Dr. P. R. Needham from the Merced River, Yosemite Valley, Calif., and the Feather River at Chester, Calif.

ADDENDUM

Since this manuscript was prepared, Dr. McDunnough has published (Canad. Ent. 67: 95–104, 1935) an article on western species of Ephemeroptera, in which appear descriptions of two new species of Ephemerella, and much additional knowledge regarding three previously described species. One of the two new species, *Ephemerella lapidula*, falls into the *fuscata* group; it is closely allied to *E. flavilinea*, the male imago being somewhat larger than typical *flavilinea* specimens from Alberta, and with blacker shading on femora and abdominal sternites. The distal joint of the forceps is "distinctly longer than in *flavilinea*." The nymph (described by Walley, Canad. Ent. 62: 18, 1930, as *flavilinea*) is best distinguished from true *flavilinea* by the "shape of the dorsal abdominal tubercles," which "are slightly longer, especially on the eighth segment, distinctly more pointed apically and *noticeably projecting upward* above the plane of the segment" (McD.). Type locality, Firehole River, Yellowstone Park, Wyo.

The second new species, *Ephemerella columbiella*, is closely allied to *heterocaudata*. The male imago is somewhat smaller than the latter species (wing length 6 mm., in *E. columbiella*), the dark abdominal markings being very similar to those of the nymph, and rather less intense in depth of color than in *E. heterocaudata*. In the nymph, the dorsal tubercles are much smaller than in the allied *heterocaudata;* the lateral tails are relatively shorter, being only about ⅓ the length of the long middle tail; and the dark abdominal markings are rather more extensive, with heavier lateral dark stripes. Type locality, Trepanier Cr., Peachland, B. C.

The true nymph of *Ephemerella flavilinea* is described (McD., 1935, p. 95), from material taken recently in several localities in British Columbia. In this species, the dorsal abdominal tubercles "are *much flattened* and scarcely elevated at all from the segmental plane; they are quite broad at the base, tapering sharply to a blunt, rounded apex, and are feebly granulate" (McD.).

A re-description of *Ephemerella hystrix* (on p. 98), adds considerable useful information regarding this species; the lateral tails are slightly shorter than the middle one, for which reason Dr. McDunnough thinks this species will prove to be allied to *heterocaudata* rather than to the *needhami* group, when the imago is known.

A reared male imago of *Ephemerella hecuba* is described (p. 99), and the genitalia figured. Dr. McDunnough questions whether Shepherd had the true *hecuba*, and indicates that the so-called imago described by him was in reality a subimago. Distal joint of forceps short; penes shaped much as in the *invaria* group, but with no short spines; body 12 mm., wing 12 mm.

Subfamily **CAENINAE**

Relatively small mayflies, varying in color from pale whitish to deep brownish black. Eyes of males simple, not divided nor grooved, widely remote from one another. Lateral ocelli unusually large and prominent. Tarsi 4-jointed, except in fore leg of male, where an additional short basal joint is present. Hind wing wholly wanting. Fore wing ample, very well developed, especially in the cubito-anal region. Wings often semi-opaque; usually the costa and subcosta are tinged with purplish except at the tip; outer and hind margins ciliate even in the imago. No marginal intercalaries. Cross veins few, or at most not very numerous. Forceps base may be entire, sometimes excavated on apical margin; or bear a median and two lateral divisions on its apical margin. Forceps and penes variable.

The nymphs are sprawling forms, more or less densely covered with fine hairs. Body tapers toward the tail. Lateral margins of abdominal segments 2–9 more or less produced into flattened processes bearing postero-lateral spines. Gills present on 2–6; usually also a filamentous rudiment on segment 1; gill on segment 2 elytroid, almost completely covering those behind it. Tails three; hairs quite evenly distributed on both margins of each.

KEY TO THE GENERA

Adults

1—Posterior branch of M in fore wing, and intercalary between branches of this vein, much shorter than anterior branch, not extending to base of wing; forceps 3-jointed . . . . **Tricorythodes**

—Posterior branch of M in fore wing, and intercalary between branches of this vein, as long as anterior branch, extending to base of wing; forceps 1-jointed . . . 2

2—Prosternum twice as wide as its length, fore coxae therefore are widely separated; second joint of antenna three times as long as basal joint . . . . **Brachycercus**

—Prosternum two or three times longer than broad, coxae therefore much closer together; second joint of antenna not more than twice as long as basal joint . . . . **Caenis**

Nymphs

1—Gills on segments 2–6 single; operculate gill on 2 quadrate, those on 3–6 with deeply fringed margins 2

—Gills on segments 2–6 double; operculate gill on 2 triangular, those on 3–6 not fringed, margins entire . . . . **Tricorythodes**

2—Three prominent tubercles on head; maxillary and labial palps 2-jointed . . . . **Brachycercus**

—No tubercles on head; maxillary and labial palps 3-jointed . . . . **Caenis**

Genus **Tricorythodes** Ulmer

1892—Eaton—Biol. Centr. Amer. Neur. p. 12 (Tricorythus).
1919—Ulmer—Arch. Naturg. 85: 51.
1920—Ulmer—Stett. Ent. Zeit. 81: 122.
1933—Spieth—J. N. Y. Ent. Soc. 41: 353 (Tricorythus).

Mayflies of small to moderate size, wing length ranging from 3 to 7 mm. General structure much as in Caenis and Brachycercus, although most species are blackish or dark brown in color. Female often paler than male. Eyes of both sexes, and ocelli as well, very widely separated, as in the subfamily. Posterior margin of head of female almost straight between the eyes.

Fore leg of male long and slender, almost or fully as long as body, somewhat shorter than wing. In other respects, this genus is close to Tricorythus, which is not represented in our fauna. Fore femur ⅖ as long as tibia, which is 1½ times as long as the tarsus. Basal fore tarsal joint very short; 2nd joint longest, remaining joints progressively shorter; 2nd joint as long as femur, at least twice the length of 3rd joint, or as long as joints 3–5 combined. Middle and hind femora somewhat longer than their respective tibiae, and 2½ times the length of the tarsi. Terminal tarsal joint longest, equal to other three combined; 3rd joint very oblique at apex. Claws of fore leg of male similar, blunt; on other legs, and on all tarsi of female, dissimilar. Posterior branch of media of fore wing, and long intercalary between the two branches of that vein, much shorter than anterior branch, not extending up to the base. Cross veins more numerous than in Caenis; usually several to many in each intercalary space. Subcosta and radius usually purplish black except at apex.

Tails three, in male and female; those of the male about three times as long as the body. Forceps base entire, but more or less deeply excavated in median area of apical margin. Penes united for about $\frac{3}{4}$ of their length, forming a somewhat conical structure; distal portions separated from one another by a rounded or V-shaped notch. Forceps 3-jointed; basal joint widest, somewhat shorter than the 2nd; an ovoid swelling at base of 2nd joint on inner margin, remainder of joint relatively long and slender, tapering distally. Apical joint very short, more or less ovoid. Posterior margin of subanal plate of female entire, may be slightly obtuse; plate well developed.

General appearance of nymph much like that of Caenis. Apical margin of labrum rather deeply notched at median line. Canines of mandibles rather wide, lobed or toothed apically, inner canine as long as outer one. Lacinia well developed. Maxilla rather conical at apex, palp very much reduced. Glossae and paraglossae of labium fused almost to apex, much as in Ephemerella. Palp relatively small, 3-jointed; apical joint very slender, basal one considerably longer than 2nd and 3rd combined. Median portion of hypopharynx obtuse at apex; about as wide as each lateral lobe, but shorter. Pronotum relatively longer in proportion to width than in Caenis; femora rather wider in median area; claws slightly longer, not pectinate. Apparently no gill rudiment on segment 1. Gill on segment 2 consists of a large, elytroid, somewhat triangular, upper lamella and a bifid, smaller lower portion; one division of the latter is long and slender, the other somewhat ovoid. Each gill on segments 3–6 consists of two plate-like divisions with entire margins, not fimbricate as in Caenis and Brachycercus. A whorl of spines at each tail joint. Nymph dwells amid fine sand and gravel on stream beds, or in moss and other plant growths on large stones.

Spieth (1933) places this genus in the same family with Ephemerella, removing it entirely from relationship with Caenis and Brachycercus. Although there are certain points of likeness between Tricorythodes and Ephemerella as regards nymphal structures, the venation of the imago is clearly much nearer to the other genera of the Caeninae than to Ephemerella. We consider Tricorythodes an aberrant member of the Caeninae.

Genotype—*T. explicatus* (in Tricorythus), Eaton.

Distribution—Nearctic; widely distributed in North America.

KEY TO THE SPECIES

Adults

1—Femora heavily sprinkled with numerous black dots........................2
—Femora not as above........................3
2—Head and pronotum blackish, femora smoky grey........................**atratus**

—Head and pronotum pale brown, middle and hind femora yellowish..... **allectus**
3—Small species, wing 3–3½ mm. in length.................................. 4
—Somewhat larger species, wing 5–7 mm. in length.......................... 6
4—Dorsum of abdomen yellowish white, dark markings much restricted; black mid-ventral markings.. **texanus**
—Dorsum of abdomen red-brown or dull black, in former case with extensive black markings; no such dark mid-ventral marks.............................. 5
5—Mesonotum and dorsum of abdomen dark red-brown; femora light red-brown .... **minutus**
—Mesonotum and dorsum of abdomen black; fore femur black, others reddish black .... **stygiatus**
6—All femora deep pitch-brown; fore tibia light brown, others darker brown.... .... **explicatus**
—All femora reddish brown; tibiae greyish or whitish......................... 7
7—Median whitish blotches on tergites 3–8; tibiae largely smoky grey........ **fictus**
—Tergites 3–8 largely black in median area; tibiae whitish.................. **fallax**

VERIFICATION TABLE

Adults

Species	*Wing length*	*Pronot.*	*Mesonot.*	*Fem. 2, 3*	*Excavat., forc. base*	*Distr.*
allectus	3½–4	Pale br.	Brown	Yellow	Deep	E
atratus	4–4½	Black	Bl. br.	Smoky	Deep	Que.
explicatus	6–7	Black	Black	Bl. br.	Shallow	Tex.
fallax	6	Dp. br.	Bl. br.	Red-br.	Shallow	W
fictus	5–5½	Bl. br.	Red-br.	Red-br.	Mod.	Okla.
minutus	3½	Bl. br.	Red-br.	Lght. br.	Shallow	Utah
stygiatus	3½	Black	Black	Red-bl.	Mod.	Que.
texanus	3	Yellow	Red-br.	Yellow	Mod.	Tex.

Tricorythodes allectus Needham

1905—Needham—Bull. N. Y. St. Mus. 86: 47. Figs. (? Caenis)............................ N. Y.
1906—Cockerell—Univ. Colo. Studies 3: 136 (Tricorythus).
1911—Morgan—Ann. Ent. Soc. Amer. 4: 115 (Tricorythus).
1919—Ulmer—Arch. Naturg. 85: 51 (transferred to new genus).
1931—McDunnough—Canad. Ent. 63: 264.

Length: body 3½–4½; wing 3½–4 mm.

Thoracic notum dark red-brown; femora and tibiae with many black flecks and streaks.

Head and pronotum pale brown; pronotum with blackish lateral markings and minute dark flecks over entire surface. Remainder of thorax darker red-brown. Fore femur pale red-brown with yellowish areas; middle and hind femora similar in some specimens, in others largely yellowish; all are heavily flecked and speckled with black. Fore tibia of male smoky grey, paler next to dark basal band; other tibiae whitish with similar dark bands at base, and variably flecked with black in median area. In specimens preserved in alcohol, a reddish median band may be present. All tibiae pale, in female specimens. Fore tarsus pale smoky; others whitish, distal joints pale smoky. Ground-work of abdominal tergites pale yellowish brown; in most specimens, more or less heavily overlaid with blackish except on lateral areas of middle segments;

anterior and posterior margins of tergites may be narrowly black. Short black streaks on lateral margins of middle and apical tergites. Sternites paler, with faint red-brown tinge; indistinct grey marks near pleural fold, and obscure smoky tinges on median and lateral areas. Tails pale, tinged and margined with blackish in the basal areas. Genitalia as in *T. atratus*.

Above description drawn largely from four specimens mounted in Canada balsam; these were taken at Ithaca, N. Y., and are evidently type material. It is quite possible that *T. atratus* is merely a dark form of the above species, but for the present we retain both names. The paler head, pronotum and femora of *T. allectus* seem to be the main distinguishing features. Other specimens in the Cornell collection which we are holding as *allectus* are from Ithaca and Old Forge, N. Y. Quite similar are several from Pigeon River and Wardensville, W. Va.

Tricorythodes atratus McDunnough

1923—McDunnough—Canad. Ent. 55: 39 (Tricorythus)....................................Que.
1931—McDunnough—Canad. Ent. 63: 265. Fig.

Length: body 2–3; wing 4–4½ mm.

Very close to *T. allectus*, but with darker head, pronotum and femora; same black speckling on femora and middle of tibiae.

Head dull blackish, slightly reddish between eyes; eyes and posterior ocelli black. Pronotum blackish, with black lateral patches. Remainder of thorax deep black-brown; paler areas on pleura at bases of legs and at wing roots. Coxae and trochanters brownish, shaded faintly with smoky. "Femora smoky grey, overlaid to a varying extent with black streaks and dots, tending to coalesce to form an apical band which at times (especially in alcohol material) shows a ruddy tinge" (McD.). Fore tibia

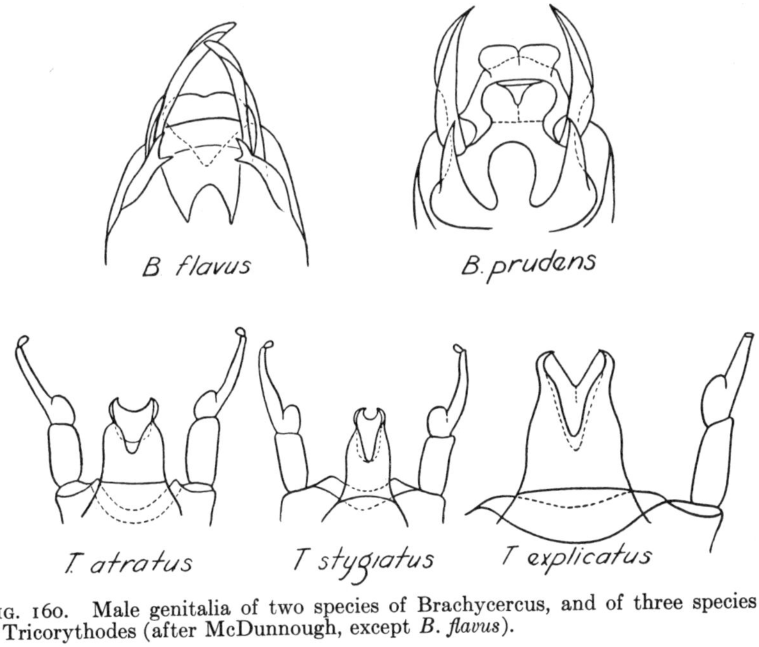

FIG. 160. Male genitalia of two species of Brachycercus, and of three species of Tricorythodes (after McDunnough, except *B. flavus*).

brownish; middle and hind tibiae dull whitish; each has a black spot at base, and blackish markings near the middle, which may have a reddish appearance. Fore tarsus pale smoky; middle and hind tarsi whitish. Abdomen of a dirty grey ground color, more or less completely overlaid with blackish dorsally, except on the lateral areas of middle segments. Anterior and posterior margin of each tergite with narrow black streaks; a row of small black spots near pleural fold on each side. Sternites dull smoky grey. Tails pale; basal area tinged faintly with smoky grey. Apical margin of forceps base deeply excavated in median area (see fig. 160).

This species is doubtfully distinct from *T. allectus*, but would appear to have much darker head, pronotum and femora. We have seen no specimens.

Tricorythodes explicatus Eaton

1892—Eaton—Biol. Cent. Amer. Neurop. p. 11. Fig. (Tricorythus)........................Mexico
1903—Banks—Proc. Ent. Soc. Wash. 5: 238 (Tricorythus)................................Ariz.
1920—Ulmer—Arch. f. Naturg. 85: 51 (transferred to Tricorythodes).
1931—McDunnough—Canad. Ent. 63: 267. Fig..Tex.

Length: body 4?; wing 5–7 mm.

Body pitch-black; femora deep pitch brown, tibiae brownish, tarsi yellowish white; forceps base of male with a very shallow apical excavation.

Head and eyes dull black; ocelli shining milky-white, dark-ringed at base; posterior tubercles shiny. Thorax dull black; pronotum with a hourglass-shaped lateral shiny brownish-black patch. Femora of all legs deep pitch-brown; the fore tibia 2½ mm. long, light brown; middle and hind tibiae shaded with pitch-brown, becoming yellowish-white apically; all tarsi yellowish white. Wings whitish, hyaline; venation indistinct; subcosta and radius blackish except at base. Abdomen wholly dull black, with satiny lustre. Forceps whitish; penes (see fig. 160) shiny blackish brown, base paler, and bent upward at right angles; forceps base very shallowly excavated apically. Tails whitish, the edges and joinings black.

In size and type of forceps base, this species is allied to *T. fallax*, but is distinguished from it by the wholly blackish thorax and abdomen, the darker middle and hind femora, the brownish tibiae and the pale whitish tails. It is larger than *T. fictus*, which is likewise much paler in body color and exhibits a somewhat deeper excavation of the forceps base. We have a series of specimens of this species in the Cornell collection from Ft. Davis, Texas, (H. A. Scullen). Among these are a few females, which agree well with Eaton's description. It seems quite possible that the specimens recorded from Alberta by Dr. McDunnough may represent a closely allied species.

A series of males from Therma, N. M., Cimarron Canyon (A. B. Klots, R. H. Post), may belong here, but are in some respects intermediate between *T. explicatus* and *T. fallax*. The size is smaller than in either species—wing length 5 mm.—; in depth of body color and deep brown femora they resemble *explicatus;* but the fore tibiae are wholly pale, sternite 9 is largely whitish except at base, and the tails are deep smoky grey.

Tricorythodes fallax sp. nov.

1931—McDunnough—Canad. Ent. 63: 265 (in discussion of *allectus*); 268 (under *explicatus*)........ ....Wyo., Sask., N. Y., Que., Ont.

Length: body 5; wing 6 mm.

Mesonotum deep blackish brown; middle and hind femora pale reddish brown. fore tibia and tarsus white; tails smoky grey; forceps base very shallowly excavated,

Male—Head deep mahogany brown, often with a pale spot in middle of vertex; ocelli whitish, black-ringed at base; eyes black; antennae pale, the second joint brown basally, filament brownish at extreme base. Pronotum deep mahogany brown; postero-lateral corners pale whitish. Mesonotum blackish brown; antero-lateral margins and sutures anterior to scutellum pale yellowish; scutellum blackish. Pleura deep red-brown, with rather extensive pale areas below and behind the wing roots and around the leg bases. Prosternum whitish; meso- and metasterna pale brown in median area, remainder deep mahogany brown. Coxae and trochanters of all legs deep red-brown, with slight blackish markings. Fore femur dark red-brown, a white area on outer side near base, dorsal and ventral edges black; 'knee' black; tibia and tarsus wholly white except terminal tarsal joint which is pale smoky; apex of tibia narrowly black. Middle and hind femora pale reddish brown, often with whitish areas near base; 'knees' black; tibiae and tarsi whitish, each tibia with a smoky ventro-basal and dorso-apical dark spot. Wings hyaline, whitish; subcosta and radius purplish black, other longitudinal veins finer but of same dark color.

Abdomen dark reddish brown. Tergites 1, 2, 8 and 9 almost wholly black; tergites 3–7 black in dorsal area, brown laterally, with variable amount of blackish shading; tergite 10 reddish brown. Posterior margins of all tergites and sternites narrowly pale, giving abdomen an annulate appearance. Heavy black markings, somewhat irregular, along spiracular area of basal and middle tergites. Hyaline, oblique, short dashes on each side of sternites next to pleural fold; mid-ventral area faintly greyish black; a pair of small hyaline dots anterior to center of each sternite, one on each side of median line. Sternite 9 brown at base and apical margin of forceps base, remainder white. Forceps white basally and at tip, second joint smoky; penes as in *T. explicatus*. Forceps base very shallowly excavated apically, as in *explicatus*. Tails rather deep smoky grey, somewhat paler apically; joinings hyaline, base and apex of each joint darker.

Holotype—Male imago; Waddell Creek, Calif., July, 1931 (D. Shepherd). No. 1365.1 in Cornell University collection.

Paratypes—25 male imagos, same data; No. 1365.3–27 in Cornell collection.

Similar to *T. explicatus* in size and in type of forceps base; differing from that species in the paler femora, wholly white fore tibia, paler abdomen which likewise appears annulate, and the smoky grey tails. Distinguished from *T. fictus* as indicated under that species. We have a series of specimens of both sexes from Zillah, Wash. (M. D. Leonard), which are smaller than the California specimens, and with thoraces deeper black, but otherwise quite similar; the females have extensive yellow areas on the thoracic pleura, the lateral and ventral areas of the abdomen, as in the specimens from Yellowstone Park mentioned by Dr. McDunnough. As we have no females of the California species, we do not include the Washington specimens among the types.

We are placing tentatively in this species the eastern forms referred to by Dr. McDunnough as a variety of *T. explicatus*. We have several series of these from Ithaca, N.Y., and vicinity. They differ from the California specimens in the more extensive pale thoracic areas, blackish median and brown edge-markings of the femora, the abdomen not at all annulate, no dusky mid-ventral markings, tails paler grey with blackish joinings, presence of a pale mid-dorsal abdominal streak, and paler venation. These may prove to be a species distinct from any thus far described.

Tricorythodes fictus sp. nov.

Length: body 5; wing 5–5½ mm.

A dark red-brown species, heavily shaded with smoky black; abdomen with dorsal markings and intersegmental rings whitish; forceps base with a moderately deep apical excavation.

Male—Vertex of head quite bright red-brown; posterior ocelli whitish, black-ringed at base; eyes black; antennae tinged faintly with reddish basally, filament pale. Thorax deep red-brown; pronotum with black margins, black median line with a pale area on each side of it, heavy black shading laterad of pale area and beyond this a large patch of brown. Scutellum, except tip, lateral sutures and a V-mark anterior to it, black; tip of scutellum pale grey. Anterior to and below wing roots and around the base of middle leg are pale areas; rather extensive black markings on pleura. Prosternum largely pale, with heavy smoky shading. Ganglionic areas of thoracic sternum pale greyish outlined with black. Femora red-brown; dorsal and ventral edges black; a blackish longitudinal streak on outer surface, and extensive whitish areas on the dorsal side of this, in basal half; middle and hind femora heavily shaded with greyish black, becoming reddish brown apically. Extreme base of tibia black; a narrow pale area beyond this, remainder of segment smoky grey. Tarsi whitish, terminal joint of fore tarsus smoky brown; others faintly tinged with red-brown. Wings hyaline whitish; subcosta and radius greyish in basal two-thirds; basal costal space semi-opaque, tinged with grey; longitudinal veins of anterior half of wing dark, paler toward margin.

Reddish brown ground color of abdomen almost wholly obscured by general and extensive greyish-black shading. Intersegmental areas pale. Tergites 3–8 with median whitish areas on the posterior margins, most extensive on 8; each pale area bordered anteriorly and laterally by a black line, which then extends laterad along posterior margin to pleural fold; on median line of each tergite is a pale hyaline spot, largest on 6–8. Tergite 9 blackish except for median pale spot; 10 reddish brown with black median streak. Pleural fold pale, heavily marked with black. Sternites with pale intersegmental rings widest in median area; a mid-ventral blackish spot on each; short black lateral dashes next to pleural fold. Forceps base smoky black, with narrow whitish median streak; apical margin with a moderately deep excavation. Forceps pearl-grey; penes pale brown. Tails rather deep smoky grey at base, paler distally; basal joinings blackish.

Female with pale areas on vertex of head; more extensive areas on thorax dorsally and laterally; abdomen rather light red-brown, the blackish markings deeper in color but less widely distributed; intersegmental areas not so noticeably pale.

Holotype—Male imago; Murray Co., Oklahoma; March 20, 1932 (A. Sandoz). No. 1367.1 in Cornell collection.

Allotype—Female imago; same data. No. 1367.2 in Cornell collection.

Paratypes—19 male, 19 female imagos; same data; No. 1367.3–40 in Cornell collection.

The whitish blotches on the abdominal tergites, the duller black of the abdomen and the smoky grey tibiae, distinguish this species from *T. fallax*. It is much smaller than *T. explicatus* with paler red-brown of thorax and abdomen, the latter marked with whitish; the forceps base is excavated more deeply than in either of these species.

Tricorythodes minutus sp. nov.

1926—Needham—Bull. Utah Agri. Exp. Sta. 201: 12 (*explicatus*)............................Utah

Length: body 2½; wing 3½ mm.

A small reddish brown species, quite similar in general appearance to *T. fallax;* forceps base rather shallowly excavated; middle and hind legs wholly red-brown.

Male—Head quite bright reddish on vertex; posterior margin blackish; antennae pale reddish; eyes black; posterior ocelli whitish, black-ringed at base. Thorax deep red-brown; pronotum shaded heavily with blackish in median and lateral areas. Mesonotum deeper in color; a narrow pale yellowish streak on sutures anterior and laterad of scutellum; pale areas on pleura limited to regions below and anterior to wing roots and around leg bases. Femora of all legs, and tibiae and tarsi of middle and hind legs, light reddish brown; fore tibia and tarsus whitish; 'knee' of fore leg darker red-brown. Wings whitish, hyaline; venation pale except for purplish subcosta and radius.

Abdomen reddish brown. Tergites 1, 2, 8 and 9 largely blackish; intermediate tergites reddish brown laterally, with a narrow red-brown median line, on each side of which is a heavy blackish shading, leaving the posterior margins narrowly pale. Sternites apparently unmarked except for a large black lateral triangle on each side of sternite 9. Tails pale smoky at base, pale distally; joinings and lateral margins narrowly blackish, in basal portion. Forceps base reddish brown, darker on apical margin, which is quite shallowly excavated. Penes light reddish; forceps very pale smoky.

Holotype—Male imago. Provo River, Utah· July 22, 1926: (J. G. Needham). No. 1366.1 in Cornell collection.

Paratypes—11 male imagos; same data; No. 1366.3–13 in Cornell collection.

Distinguished from *T. stygiatus* by its paler color; the lack of black freckling on femora and tibiae and the reddish brown of the abdomen separates it from *T. allectus* and it is much darker in color than *T. texanus*. The shallow excavation of the apical margin of the forceps base also distinguishes it from all of the above species which approach it in size.

Other specimens in the Cornell collection, from the mouth of the Logan River, Utah, and Logan River at the sugarworks (J. G. Needham), are probably of the same species. Several of these were reared from nymphs.

Tricorythodes stygiatus McDunnough

1931—McDunnough—Canad. Ent. 63: 267. Fig. ....Que.

Length: body 2½–3; wing 3½ mm.

Mesonotum shiny blackish; thoracic sternum and abdomen dull blackish; hind femora deep rufo-piceous, tibiae largely same color.

Head dull black, the posterior tubercles shiny black; eyes and lateral ocelli black; basal joint of antenna black, paler apically. Median area of pronotum dull black, with faint ruddy tinge in the mid-dorsal portion; shiny black laterally. Mesonotum shiny black, the scutellum with ruddy tinges; pleura shiny black; anterior to the wing roots and around the bases of the legs are dull areas. Sternum dull black. Coxa and trochanter of fore leg black; femur blackish with a pale hyaline streak longitudinally; tibia and tarsus whitish, tinged faintly with ruddy; tibia 1½ mm. in length. Coxae and trochanters of middle and hind legs blackish, femora shiny rufo-piceous; tibiae largely rufo-piceous, becoming paler apically; tarsi whitish. Wings whitish hyaline; subcosta and radius darker. Abdominal tergites completely suffused with dull blackish, tinged slightly with ruddy in the median area, especially on the basal tergites. Sternites dull blackish; lateral margins somewhat shiny; posterior margins faintly whitish. Genitalia dull whitish (see fig. 160). Tails blackish basally, becoming pale smoky apically.

The black body and deep reddish brown femora separate this from other small species in the genus. We have seen no specimens of this species.

Tricorythodes texanus sp. nov.

Length: body 2½–3; wing 3 mm.

Thorax yellowish olive; abdomen pale yellowish white, with blackish markings dorsally, and a mid-ventral row of dark dots; legs largely pale; forceps base with a moderately deep apical excavation.

Male—Head yellowish; vertex sometimes tinged with brown; eyes black; lateral ocelli pearl-grey, heavily ringed with black at base; antennae pale. Thorax yellowish to yellowish olive; mesonotum reddish brown with an olive tinge in the median area; scutellum pale with a few faint blackish markings. A few fine blackish markings on pleura above bases of legs. Blackish oblique lateral dashes on mesosternum, anterior to base of middle leg; large ganglionic areas of thoracic sternum usually faintly smoky and narrowly outlined in blackish; a transverse black mark between anterior coxae. Legs largely pale yellowish white; fore coxae, trochanters and femora yellowish tinged with smoky; dorsal and ventral edges of femora black, often with a blackish streak dorsally near apex; 'knee' brownish. Fore tibia silvery, tinged with smoky in median area; tarsus silvery white. A dark dash on coxae of middle and hind legs, and a black ventro-apical dot on trochanters. Femora with a narrow longitudinal blackish line near the middle of the outer surface, often broken up into shorter dashes. A small black spot dorsally at base of each tibia. Wings hyaline, whitish; subcosta and radius faint purplish except at wing tip.

Abdomen pale yellowish white, the apical segments deeper yellow. Tergites 1 and 2 almost wholly obscured by smoky shading; remaining tergites marked with smoky or blackish dorsally, the dark areas tending to be confined to a mid-dorsal black streak and two wider smoky lateral streaks, with a pale area between. Lateral margins quite widely pale; posterior margins narrowly pale; a series of short black lateral dashes in spiracular area. Sternites with similar short lateral dashes next to the pleural fold, and small blackish mid-ventral spots, one on each basal and middle sternite, more or less well developed; these are smallest on basal sternites, largest on sternite 6, obsolescent on apical ones. These spots may be much reduced in size, or even wholly obsolescent, in some specimens. Genitalia silvery white. Forceps base with a moderately deep apical excavation; basal half of penes covered by this plate. Tails silvery white, sometimes with faint smoky edges at extreme base.

Female slightly larger than male; body almost wholly yellowish; mesonotum dull greyish olive; lateral streaks on middle and hind femora more prominent than in male, tending to form a disconnected purplish spot near base. Tergites 1 and 2 yellowish in median area, with wide lateral oblique dark markings; mid-ventral markings of thorax and abdomen larger and more prominent than in male. Mid-dorsal streak becomes wider apically and lateral streaks on tergites narrower, in both sexes.

Holotype—Male imago, pinned (now in alcohol); Devil's River, Texas, July 2, 1917 (C. U. Biological Expedition). No. 1364.1 in Cornell collection.

Allotype—Female imago, pinned; same data; No. 1364.2 in C. U. collection.

Paratypes—6 male imagos, 15 female imagos, many pinned; same data; No. 1364.3–23 in Cornell collection.

The pale color of legs and body of this small species should separate it from others of the genus. The apical margin of the forceps base is less deeply excavated than in the blackish *T. stygiatus*, however it is not as shallow as in the reddish brown *T. minutus*.

Genus **Brachycercus** Curtis

1834—Curtis—London Edin. Philos. Mag. s. 3, 4: 122.
1839—Burmeister—Handb. d. Ent. 2: 796 (Oxycypha).
1884—Eaton—Revis. Monogr. p. 141 (Caenis, in part).
1917—Bengtsson—Ent. Tidskr. p. 186 (Eurycaenis).
1920—Ulmer—Stett. Ent. Zeit. 81: 122 (Eurycaenis).
1924—Lestage—Ann. Soc. Ent. Belg. 65: 61.

Small mayflies, similar to Caenis in venation and general appearance. Second joint of antenna considerably elongated, fully three times the length of the basal joint. Prosternum twice as wide as its length, so that the fore coxae are widely separated from one another. Fore tibia of male about one and one-third times as long as the femur; tarsus shorter than tibia; leg as long as or slightly longer than the body. Femora of middle and hind legs about equal in length to tibiae and tarsi combined. First five segments of abdomen relatively short; segments 7–10 longer. Forceps base bears on each side of its posterior margin a prolongation, at the tip of which is a blade-like process tapering to a point distally; a median somewhat ovoid process extends almost as far backward as the lateral processes. Penes much as in Caenis; forceps one-jointed, blade-like processes, somewhat twisted, tapering apically to sharp points. Posterior margin of 9th sternite of female convex.

Nymphs very similar to those of Caenis, but with a prominent conical tubercle arising from the base of each ocellus; a short tubercle may also be present on the lateral margin of the pronotum. Head and pronotum narrower than in Caenis. Labial palp two-jointed; labrum about 2½ times as broad as long. Maxillary palp two-jointed. Fore legs relatively short; middle and hind legs longer than in Caenis, the femora relatively longer and more slender; claws longer, more slender and less curved apically than in Caenis. Lateral curved blade-like projections of middle abdominal segments much more prominent.

Genotype—*B. harrisella* Curtis.

Distribution—Holarctic, African; in Nearctic, known only from eastern North America.

KEY TO THE SPECIES

Adults

1—Small species (wing 3 mm. in length) 2
—Larger species (wing 4½ to 5 mm. in length) 3
2—Meso- and metanota straw to canary yellow; fore femur light purplish brown; abdominal tergites creamy white, 8–10 with a purplish black mid-dorsal streak .... **flavus**
—Meso- and metanota light brown; fore femur pale smoky; abdominal tergites yellowish white, immaculate .... **prudens**
3—Mesonotum dark red-brown, metanotum purplish; tergites 1–6 purplish grey, margins lavender; posterior margins of sternites purplish .... **nitidus**
—Mesonotum pale brown; metanotum similar; tergites 1–6 pale, posterior margins blackish; dark lateral spots on some of sternites .... **idei**

VERIFICATION TABLE

Adults

Species	*Wing length*	*Meso-notum*	*Fore femur*	*Terg. 1 to 6*	*Dk. mks., sternum*	*Distr.*
flavus	3	Yellow	Purp. br.	Cream.	None	La.
idei	4½	Pale br.	Blk. br.	Pale	Lat. sp.	N. C.
nitidus♀	5	Dk. br.	Oliv. br.	Grey	Post. mg.[1]	Ont.
prudens	3	Brown	Smoky	Yellow	None	Sask. Kans.

KEY TO THE SPECIES

Nymphs

1—Tubercles present on the lateral margin of the pronotum, one on each side, anterior to center; legs not banded.................................... **nitidus**

—No such prothoracic tubercles; legs distinctly dark banded, each joint with a single dark submedian band.................................... **lacustris**

Brachycercus flavus sp. nov.

Length: body 3; wing 3 mm.

Mesonotum straw to canary yellow; fore femur light purplish brown; abdomen pale whitish, tergites 8–10 with a black mid-dorsal streak.

Male—Head pale red-brown; a narrow black median line on vertex, and a transverse black line between ocelli; eyes black, lateral ocelli grey; antennae whitish. Pronotum yellowish, shaded with faint purplish brown in median area, around base of leg, and on the lateral and posterior margins. Mesonotum straw to canary yellow, deeper yellow posteriorly; median line, and suture on each side of this, narrowly blackish brown, most evident near center of sclerite. Metanotum bright yellow, shaded with purplish brown in the posterior half. Pleura yellow; a smoky mark below each wing; antero-lateral area above sternum shaded with purplish brown. Sternum yellow; anterior margin and antero-lateral areas of meso-sternum shaded with purplish brown. Fore coxa and femur light purplish brown; tibia pale smoky at base, becoming silvery white apically; tarsus silvery white. Middle and hind legs wholly whitish. Wings semi-hyaline whitish; subcosta and radius shaded with purplish, which in the region of the bulla forms a faint wash in the costal space, leaving hyaline areas at base and apex of this space; all longitudinal veins in anterior half of wing purplish, the color fading out toward the margin; other veins colorless.

Abdomen hyaline, pale creamy whitish; basal tergites shaded with smoky in the median area; tergites 8–10, and posterior half of 7, with a narrow purplish black mid-dorsal streak. No other markings. Tails white. Genitalia as in fig. 160.

Holotype—Male imago, pinned (now in alcohol); Sabine River Ferry, La. opp. Orange, Tex.; June 20, 1917. (C. U. Biological Expedition). No. 1363.1 in Cornell collection.

Paratypes—5 male imagos; same data. No. 1363.3–7 in Cornell University collection.

This species, most closely allied to *B. prudens*, may be distinguished from it by the bright yellow of the thorax, the slightly darker fore femur, the black mid-dorsal streak on the apical abdominal tergites, and the slightly longer male forceps. It is smaller than other described species in our fauna.

[1] Posterior margins of sternites purplish.

Brachycercus idei Lestage

1930—Ide—Canad. Ent. 62: 218. Fig. (*Eurycaenis pallidus;* n. preoccupied)....................Ont.
1931—McDunnough—Canad. Ent. 63: 264 (in discussion of *Eurycaenis prudens*).
1931—Lestage—Bull. Ann. Soc. Ent. Belg. 71: 119 (n.n., for *pallidus* Ide).

Length: body 3½; wing 4½ mm.

Mesonotum pale brown; fore femur dark black-brown; abdominal tergites pale, the posterior margins black.

Eyes dark; lateral ocelli transparent; vertex pale in the median area, dark cinnamon brown toward the eyes. Pro- and mesonota pale brown, pleura paler; a narrow dark transverse line between the fore coxae; anterior portion of mesosternum dark piceous; metasternum paler, with a dark piceous transverse band. Fore legs longer than body; femora dark blackish brown. Middle and hind legs entirely pale; femora about as long as tibiae and tarsi combined. Wings whitish, especially toward the posterior margin; subcosta and radius dark piceous in the basal two-thirds of their length. Abdominal tergites 4–10 pale, the posterior margins blackish. Sternites pale, with dark lateral spots on some of them. Tails whitish. Forceps longer than in *B. prudens.*

This species is larger than *B. prudens* or *B. flavus,* and has longer forceps, as well as black posterior margins on the abdominal tergites which are not present in those species. The thorax and abdominal tergites are paler, the fore femur darker, than in *B. nitidus,* of which only the female is known. Its relation to *B. lacustris,* known in the nymphal stage only, cannot at present be determined. We have seen no specimens of *B. idei.*

Brachycercus lacustris Needham

1908—Needham—Rep. Mich. Geolog. Survey, p. 263 (*Ephemerella sp.*).....................Mich.
1918—Needham—N. Y. St. Coll. Forest. Tech. Pub. 9: 249 (Caenis).........................N. Y.
1932—Traver—J. Elisha Mitchell Sci. Soc. 47: 139 (in discussion of *nitidus*).

Length (nymph): body 7–7½ mm.

Only the nymph of this species is known; it may be that this is really the nymph of *B. idei.*

Nymph apparently without tubercles on the prothorax, such as are present in *B. nitidus.* General color pale. A pair of zigzag bands on the vertex, from eyes to middle of occiput. Prothorax wholly pale. Mesonotum darker at anterior and posterior margins; curved black median pencilings. Legs pale, each joint with a single brown submedian band. "Basal black mid-dorsal triangles" on "segments in front and rear"; (Ndm.); middle segments obscurely brownish. Anterior margin of each segment, dorsally and ventrally, brownish, paler in the median area; a pair of black dashes on the posterior portion of tergite 10. Tails wholly pale, but with faintly brown joinings in the distal portion, at least of some specimens.

The absence of prothoracic tubercles and the paler color, as well as the banded legs, distinguishes this nymph from that of *B. nitidus.* The nymph of *B. idei* is not known. The large size would seem to preclude the possibility that this is the nymph of *B. flavus* or *B. prudens.* The specimens of this species taken at Oneida Lake, N. Y. (F. C. Baker), are in the Cornell collection.

Brachycercus nitidus Traver

1932—Traver—J. Elisha Mitchell Sci. Soc. 47: 139. Fig. (Eurycaenis)........................N. C.

Length (female): body 4½; wing 5 mm.

Mesonotum dark reddish brown; fore femur light olive brown; abdomen purplish grey, tergites 7–10 yellowish.

The male imago is not known. Head of female light yellowish brown; median carina and anterior portion of vertex purplish black; eyes black; ocelli pearl-grey, black-ringed at base; antennae whitish. Pronotum dark purplish brown, paler laterally. Mesonotum dark reddish brown, a paler strip on each side of the median line; sutures blackish. Metanotum purplish. Pleura light brown; darker red-brown between the middle and hind legs. Sternum light brown shaded with purple; sutures purplish black; anterior portion of mesonotum heavily shaded with purple. Fore femur longer than tibia and tarsus combined; light olive brown, margined with dark brown; tarsus whitish, tibia same color as femur. Femora of middle and hind legs similar to first; tibiae yellowish; tarsi whitish. Wings greyish white; subcosta and radius purplish except at tip; longitudinal veins of anterior portion of wing purplish grey basally.

Abdominal tergites 1–3 purplish black; 4–6 purplish grey in median area, greyish white laterally; posterior margins lavender. Tergites 7–9 greyish yellow; posterior margins and median line greyed lavender; 10 yellowish brown, purplish medially and on posterior margin. Sternites 1–6 purplish, finely mottled with very tiny pale dots; posterior margins darker purple; two larger light spots, indistinct, near center of sternite 1; lateral margins greyish lavender. Sternites 7–9 yellowish; 8 with brown shading. Tails greyish white, tinged faintly with lavender at base.

Nymph with a prominent tubercle on each side of the pronotum, situated anterolaterally. Thoracic notum dark reddish brown, with blackish pencilings. Legs yellowish, with no darker markings. Abdomen reddish brown dorsally, somewhat paler ventrally. Tails yellow at base, amber brown distally.

This species is larger than *B. flavus* and *B. prudens*; likewise it is much darker in color. It is darker than *B. idei*, another large species, and with different markings. The nymph differs from *B. lacustris* in possessing tubercles on the prothorax; the color is darker, color pattern different; tibiae and tarsi unbanded. The type specimens of this species are in the Cornell collection.

Brachycercus prudens McDunnough

1931—McDunnough—Canad. Ent. 63: 264 (Eurycaenis). Fig. .... Sask., Kans.

Length: body 3; wing 3 mm.

Mesonotum bright brown; fore femur pale smoky; abdomen wholly pale yellowish white.

Eyes black; antennae pale; head lightly suffused with pale brownish. Pronotum pale yellowish white, the median area tinged faintly with smoky posteriorly. Mesonotum rather bright brown, the median area with slight sepia shading; lateral and posterior portions paler. Metanotum light brown. Pleura pale, the episternal area brown-tinged. Coxae and femora of fore legs pale smoky; tibiae and tarsi whitish. Middle and hind legs wholly whitish. Wings semi-hyaline whitish, subcosta and radius dark except at the tip. Abdomen entirely pale yellowish white. Forceps faintly tinged with smoky, tails white. Forceps shorter than in *B. idei* or *B. flavus* (see fig. 160).

This species is closely allied to *B. flavus*, from which it may be separated by the bright brown mesonotum, the shorter forceps, and the lack of a dark mid-dorsal streak on the apical abdominal tergites. It is smaller than either *B. nitidus* or *B. idei;* tergites paler than in the former, and lacking the black posterior margins of the latter species. We have seen no specimens of *B. prudens.*

Genus **Caenis** Stephens

1835—Stephens—Ill. Brit. Ent. 6: 61.
1884—Eaton—Revis. Monogr. p. 141 (in part).
1920—Ulmer—Stett. Ent. Zeit. 81: 121.
1923—Campion—Ann. Mag. Nat. Hist. 11, s. 9, 64; 515 (Ordella).
1924—Lestage—Ann. Soc. Ent. Belg. 65: 61 (Ordella).
1931—McDunnough—Canad. Ent. 63: 254.
1933—Spieth—J. N. Y. Ent. Soc. 41: 356.

Small mayflies, ranging in wing length from 2 to 5 mm. (see frontispiece). Male and female very similar in coloration. Abdomen usually pale yellowish white, the tergites often more or less suffused with blackish. Second joint of antenna not more than twice the length of the basal joint. Posterior margin of head almost straight. Pronotum deeply excavated posteriorly in the median area. Prosternum narrow, two or three times longer than broad, so that the fore coxae are more or less closely approximated. Fore tibia of male usually about twice the length of the femur; tarsus shorter than tibia; second joint longer than any of the others, distal joint shorter than the fourth. Fore leg of male usually about as long as body. Middle and hind femora about 1⅓ times as long as their respective tibiae; tarsi slightly shorter than tibiae. Claws as in Tricorythodes.

Posterior branch of media in the fore wing, and the long intercalary between the two branches of this vein, as long as the anterior branch, and extending up to the base of the wing (see fig. 31). Cross veins of wing arranged singly, usually no intercalary region with more than one cross vein; 2 or 3 cross veins only, in radial region. Subcosta and radius blackish or purplish except at apex. Short lateral filaments extend backward from the postero-lateral margins of the middle abdominal segments. Tails three; relatively very long in male, often three times the length of the body; middle tail may be slightly the longest. Forceps base entire, its apical margin slightly convex. Penes united, broad and plate-like; forceps 1-jointed, generally sharp-pointed at the tips. Posterior margin of subanal plate of female entire. Subimagos have more prominently ciliated wing margins and longer lateral filaments on abdomen.

Head and pronotum of nymph (see fig. 46) narrower than mesonotum; relatively wider than in Brachycercus. Labrum less broad relatively than in that genus, its apical margin slightly concave; not excavated at median line. Mandibles somewhat as in Tricorythodes; canines less broad than in that genus, but distinctly wider than in Brachycercus. Maxilla conical at apex; palp 3-jointed, distal joint longest; entire palp fully twice the length of the galea-lacinia. Glossae and paraglossae as in Brachycercus,—well differentiated, not fused as in Tricorythodes. Palp 3-jointed; basal joint stoutest, the two distal joints much less slender than in Tricorythodes. Middle lobe of hypopharynx wider than in that genus, otherwise similar. Fore leg relatively longer than in Brachycercus; claws slender, somewhat curved apically, not pectinate. A well-developed filamentous gill rudiment on segment 1. The operculate gill on segment 2 is single; large and quadrate, its outer apical angles somewhat rounded. Each of the gills on segments 3–6 is single, deeply fimbricate on the outer ⅔ of its entire margin. Postero-lateral spines on middle segments somewhat more prominent than in Tricorythodes; not upcurved as in Brachycercus, and much less sharply toothed than in that genus. A whorl of spines is present at each tail joint. These nymphs dwell in mud and silt, as do those of Brachycercus.

Genotype—*C. macrura* Stephens.

Distribution—Holarctic, Indo-Australian; in Nearctic, widely distributed.

KEY TO THE SPECIES

Adults

(After McDunnough, 1931, with additions and modifications.)

1—No dark stigmatic dots or streaks on abdominal tergites....................2
—Dark stigmatic streaks present on at least some of the abdominal tergites....4
2—Head and thorax dark; two lateral bands of blackish shading on tergites..**tardata**
—Head and thorax paler; tergites almost or entirely immaculate..............3
3—Mesonotum reddish brown; fore leg whitish except for a purplish spot on knee ....**anceps**
—Mesonotum pale yellowish; fore femur much tinged with smoky........**ridens**
4—Stigmatic dark streaks on the posterior tergites only; head pale.........**hilaris**
—Stigmatic streaks present on the anterior tergites; head at least partially dark-shaded....................5
5—Hind femora with a dark apical band....................6
—Hind femora with the dark band reduced to a dorsal streak.................7
6—Femora sprinkled with black dots....................**punctata**
—Femora without such a sprinkling of dark dots....................**diminuta**
7—Head crossed by a band of purplish or ruddy shading, leaving the posterior portion paler....................8
—Head entirely suffused with purplish or blackish....................11
8—Mesonotum light yellow-brown; size small (3 mm.); male forceps of moderate length....................**jocosa**
—Mesonotum deeper brown; size larger (wing 4 mm.); forceps moderate to long..9
9—Hind femora and tibiae sprinkled with black dots; male forceps of moderate length....................**delicata**
—Hind femora and tibiae without such a sprinkling of black dots; male forceps longer....................10
10—Mesonotum light brown; male forceps very long, thin and straight....**forcipata**
—Mesonotum deep brown; male forceps not so long and slightly curved..**latipennis**
11—Size small (3 mm.); male forceps short....................**amica**
—Size large (4–5 mm.); male forceps moderate, curved................**simulans**

VERIFICATION TABLE

Adults

Species	*Wing length*	*Rear of head*	*Dk. mk.,*[1] *fem. III*	*Spirac.*[2] *marks*	*Bl. terg.*[3] *1 to 6*	*Forceps*	*Distr.*
amica	2	Black	Streak	1–7	Yes	Short	E
anceps	2	Pale	Dot	None	No	Short	E
delicata	4	Pale	Triang.	1–9	Yes	Mod.	Okla.
diminuta	3	Black	Band	1–7	Yes	Short	E
forcipata	4	Pale	Triang.	1–7	Yes	Long	E
hilaris	2½	Pale	Streak	7–9	No	Short	E
jocosa	3	Pale	Streak	1–2, 5–6	Yes	Mod.	E
latipennis	4	Pale	Streak	1–7	Yes	Long.	N
punctata	4	Purpl.	Band	1–9	Yes	Short	E
ridens	2½	Pale	Absent	None	No	Short	Kans.
simulans	3½–4	Black	Streak	1–7	Yes	Mod.	N
tardata	3	Black	Streak	None	Yes	Short	N W

Caenis amica Hagen

1861—Hagen—Syn. Neurop. Ins. N. Amer. p. 55. .... Md.
1884—Eaton—Revis. Monogr.; p. 147 (*diminuta*, in part).
? 1907—Needham—N. Y. St. Mus. Bull. 124: 192 (*diminuta*, in part)........................ N. Y.
1931—McDunnough—Canad. Ent. 63: 256. Fig.

Length: body 2; wing 2 mm.?

Allied to *C. diminuta*, but the black apical mark on the middle and hind femora is a streak only, not a band.

Head fulvous; with black bands. Antennae white. Prothorax fulvous, banded with black; mesothorax brassy fulvous. Legs whitish, the fore legs much longer; cinereous, the femora obscurer; middle and hind femora with a black spot above upon the apex, this spot not continued downward to form a band, as in *C. diminuta;* on the outer side of the trochanter is a black streak. Wings semi-hyaline, whitish-grey, "the two costal veins black" (Hagen). Abdomen pale; tergites shaded with grey, the posterior margins and the sides marked with black. Tails white. Forceps short, much as in *diminuta*, but somewhat more curved than in the latter species, and the tips more finely pointed (see Pl. XXXIX).

Distinguished from *C. diminuta* by the lack of an entire black band on the hind femora, by the brighter brown of the mesonotum, and the black streak on the trochanter.

Specimens in the Cornell collection which seem very similar to this species are from Bar Harbor, Me. (A. E. Brower); Saranac Inn, N. Y. (J. G. Needham); Justice and Ripley, W. Va. (J. G. Needham); and Hamburg Lake, near Greensboro, N. C. (J. R. Traver).

Caenis anceps sp. nov.

Length: body 2; wing 2 mm.

Allied to *C. ridens*, but smaller, with brownish thorax and wholly pale abdomen; forceps short.

---

[1] Dark mark on hind femur.
[2] Spiracular markings on abdominal segments.
[3] Blackish color suffuses abdominal tergites.

PLATE XXXIX.

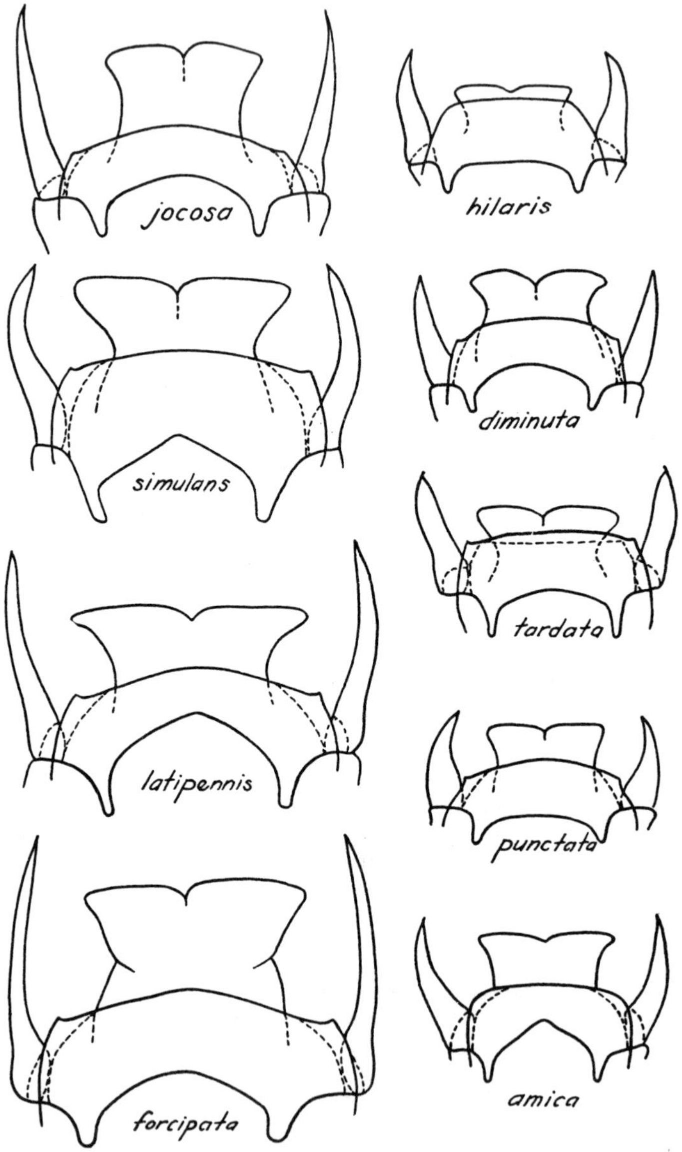

GENITALIA OF CAENIS
(after McDunnough).

Male imago.—Head pale; a purplish black transverse band along the front margin of the vertex; another band, slightly paler and incomplete on the median line, between the eyes; posterior margin narrowly purplish black. Antennae pale. Pronotum pale, with faint darker shading laterally. Mesonotum light reddish brown, rather bright; median line very narrowly darker; faint darker shading on the postero-lateral margins of the scutellum. Pleura and metanotum paler reddish brown; blackish pencilings on the pleura anterior to the wing root, below the wing, and above the bases of the middle and hind legs. Sternum yellowish. Wings semi-hyaline whitish, the subcosta and radius purplish for most of their length. Legs whitish; fore femur with a darker band near the apex, fore tibia darker basally, so that the 'knee' appears purplish to dark smoky; middle and hind legs wholly pale, except that a very tiny black dorso-apical dot is present, in most specimens, on the femora. Abdomen wholly whitish, somewhat yellow-tinged apically; without stigmatic markings or darker shading on any segment. Forceps and tails whitish; forceps relatively short.

Female similar to male, except that the mesonotum is paler reddish brown, the median line not darker; pleura paler, more yellowish, than in male; middle and hind femora often lack the black dorso-apical dot.

Holotype—Male imago; Varna, N. Y., July 15, 1931 (J. R. Traver). No. 1368.1 in Cornell collection.

Allotype—Female imago, same data; No. 1368.2 in Cornell collection.

Paratypes—40 male imagos, 7 female imagos; same data. No. 1368.3–49 in Cornell collection.

This species is very close to *C. ridens;* it may in fact prove to be only an eastern form of that species. However, it is smaller than typical *ridens*, as represented in our collection by a female from Lawrence, Kans., which was sent us by Dr. McDunnough; there are more numerous blackish markings on the pleura; the mesonotum is reddish brown in both sexes, and in the female sex is not darker on the median line; the fore leg is much paler; also a dark dot is often present near the apices of the middle and hind femora. For this reason, we are considering these eastern specimens as a distinct species.

In addition to the type material, there are many other specimens of this species in the Cornell collection, from Ithaca and McLean, N. Y. We have seen others from Canajoharie, N. Y. (H. K. Townes, Jr.).

Caenis delicata sp. nov.

Length: body 4½; wing 4 mm.

A species with moderately long forceps; mesonotum reddish brown; head flesh-colored with purplish markings.

Head of male flesh-colored, more or less shaded with purplish, usually leaving a paler area next to the posterior margin; a black line margins the vertex in front, and the posterior border; a wide purplish band extends across between the eyes. Eyes black; ocelli whitish, black-ringed at base; antennae pale, a purplish longitudinal streak on the outer side of the basal joint. Pronotum pale; anterior and lateral margins strongly, and posterior margin more faintly, bordered with blackish; median area more or less suffused with purplish; two black dots near the posterior margin, one on each side of the median line; antero-lateral corner and extreme lateral margin purplish black. Mesonotum reddish brown; a streak on each side of median line darker; faint dark markings bordering the scutellum; a faint dark V-mark anterior to it; a dark streak inward from the wing root, another anterior to wing. Metanotum reddish brown, postero-lateral corners yellowish; considerable blackish shading in the antero-lateral area; lateral and posterior margins black. Pleura similar in color to mesonotum; deeper brown above episternum; anterior margin blackish; extensive black markings above the bases of the legs and below the wing roots. Prosternum pale, with lavender markings; remainder of sternum pale in median area, brownish laterally.

Fore legs rather widely separated. A black antero-dorsal dot on fore coxa; fore femur purplish black on outer side, paler near middle, a black streak on the dorsal edge; tibia and tarsus whitish, faintly tinged with smoky. Middle and hind legs whitish; a black streak on coxa and trochanter; femur marked apically with a black triangle, its base toward apex of joint. Femora rather thickly speckled, especially in basal half and on dorsal edge, with small purplish black dots, which often tend to form a dark longitudinal streak in the middle of the outer surface. Dorsal edge of tibia with three or more black dots or short streaks. Wings hyaline whitish; subcosta and radius purplish; several longitudinal veins behind these are faintly lavender basally.

Abdomen yellowish white; apical tergites yellowish. Tergites 1–6 and antero-lateral portion of 7, shaded with greyish lavender or purplish, deepest in color laterally; median area of tergites 1 and 2 usually pale, often cut by a dark median line; lateral margins pale. Posterior margins narrowly purplish black, paler in the median area; faint stigmatic streaks on tergites 1–6; heavy markings on 7 and 8, somewhat fainter on 9. Tergites 8–10 pale except for purplish lateral markings, forming with the lateral dark streaks an ovoid dark area around the stigma. A median and two lateral dark dots on 9; 3 longitudinal median streaks and two lateral dark dots on 10. Sternum with a pair of dark dots, obliquely arranged, very evident on each side of sternites 6–8; one or two dark dots on each side of the anterior sternites; a black transverse dash on each side of the anterior margin of 9. Forceps faintly smoky, tails whitish. Forceps of moderate length; stout at base, the apical two-thirds slender; tip sharp-pointed; slightly curved.

The females are larger than the males, bodies 6 to 7 mm. in length; tergites 1–6 largely suffused with purplish, the posterior margins pale; dark dots on middle and hind femora more numerous and more widely scattered than in the male.

Holotype—Male imago. Murray Co., Okla.; April 2, 1931. (R. D. Bird). No. 1369.1 in the Cornell University collection.

Allotype—Female imago; same data. No. 1369.2 in Cornell collection.

Paratypes—15 male imagos, 12 female imagos, same data; No. 1369.3–29 in Cornell collection.

This species is allied to *C. jocosa* in the type of the genitalia, but differs in its much larger size, and in the presence of dots on femur and tibia. In the latter respect it differs from all other described species of our fauna except *C. punctata*, from which it is separable by reason of the longer forceps and lack of a dark femoral band, which is reduced to a dark triangle.

In addition to the type material, there are other specimens in the Cornell collection from Murray and Cleveland Co., Okla, (R. D. Bird); Manhattan, Kans. (R. C. Smith); and Medora, Kans. (R. H. Painter).

Caenis diminuta Walker

1853—Walker—List Neurop. Ins. Brit. Mus. pt. 3: 584. . . . . Fla.
1861—Hagen—Syn. Neurop. Ins. N. Amer. p. 55.
1871—Eaton—Trans. Ent. Soc. London, p. 95.
1884—Eaton—Revis. Monogr. p. 147.
1931—McDunnough—Canad. Ent. 63: 257. Fig. . . . . Ont.

Length: body 3; wing 3 mm.

A small species, with a dark band on the hind femora; blackish head; forceps short.

Head blackish over a pale ground color; antennae pale. Pronotum pale, with heavy blackish suffusion; laterally paler; often with a narrow black line along the lateral and posterior margins. Mesonotum deep bronzy brown, slightly tinged with smoky; scutellum paler, with black shading. A black transverse band on the metanotum. Pleura the same color as the mesonotum; black shading anterior to the wing roots, and a few black streaks below these areas. Sternum pale with a slight brownish tinge. Coxae and femora of the fore legs smoky, a blackish apical band on the femora; tibiae and tarsi paler, only slightly tinged with smoky. Middle and hind legs whitish; coxae tinged with smoky, and with a black spot laterally; a distinct black streak dorsally at the apices of the femora, this streak continuing downward on the outer side to form a blackish apical band, a fine black line may be present basally on the dorsal edge of each femur. Wings semi-hyaline, whitish; subcosta and radius dark.

Abdomen pale yellowish white; tergites 1–6 shaded heavily with blackish; tergites 1, 2, and the base of 3 pale in the median area, the mid-dorsal line black; posterior margins of 1–6 black, the lateral margins pale, with distinct black stigmatic dots. Tergites 7 and 8 largely pale, the black shading consisting only of two faint submedian patches, a black stigmatic dot, and the black posterior margin. Tergite 9 with a deeper yellow tinge dorsally; shaded heavily with blackish laterally; often a dark dot in the yellow area. Tergite 10 yellowish. Sternites pale yellowish; 1–6 with fine black streaks in the antero-lateral areas; 6–9 with more or less distinct black lateral dots, arranged obliquely, the posterior dot nearest the median line; these dots may be visible at times on the basal sternites also. Forceps and tails pale. Forceps short, rather stubby (see Pl. XXXIX).

The black apical band of the hind femora, the small size, the dark head and dark brown mesonotum distinguish this species from others with short forceps.

Specimens of this species in the Cornell collection are from Gainesville, Fla. (T. H. Hubbell); Spring Creek, Ga. (J. C. Bradley); and Ocala, Fla. (M. D. Leonard).

Caenis forcipata McDunnough

1907—Needham—Bull. N. Y. St. Mus. 124: 192 (*diminuta*, in part) . . . . N. Y.
1931—McDunnough—Canad. Ent. 63: 257 . . . . Ont., Que.

Length: body 4; wing 4 mm.

A species with long slender forceps; mesonotum light brown. A purplish or blackish band across the head between the eyes; head paler before the posterior margin. Eyes black, antennae pale. Pronotum pale yellowish, the median area with blackish shading; traces of a narrow black border line, most evident on the lateral margins. Mesonotum dark yellowish brown; scutellum paler, shaded slightly with black; a dark streak anterior to the wing root. Pleura similar to mesonotum, but with paler areas anterior to and below the wing roots; around the bases of the legs are some dark streaks. Sternum pale, faintly tinged with brown. Fore legs smoky at base; coxae and femora shaded with smoky, the femora with a black apical spot dorsally; tibiae and tarsi pale, faintly tinged with smoky. Middle and hind legs pale yellowish, a black streak on the outer side of each coxa and a black dot on the postero-ventral edge; a black streak on the outer side of the trochanter; a dorso-apical triangular mark on each femur, the apex pointing forward. Wings semi-hyaline whitish, the subcosta and radius dark, as usual.

Abdominal tergites 1–6 pale yellowish, heavily suffused with blackish; pale only in small lateral areas around the black stigmatic dots, and on the median area of 1 and 2, which show traces of a dotted black mid-dorsal line. Tergite 7 shaded anteriorly with black in the submedian areas; tergites 8–10 pale yellowish. Stigmatic dots present on tergites 1–7, very heavy on 7, obsolescent on 8–10; tergite 10 with three black streaks dorsally. Sternites pale yellowish; only faint traces of the obliquely-arranged dark dots on 6 to 8. Forceps and tails pale. Forceps very long and slender (see Pl. XXXIX).

This species is larger than *C. jocosa*, and has a darker mesonotum, as well as longer forceps; it is distinguishable from *C. latipennis* by the longer, thinner forceps, which are not curved as in *C. simulans*. Specimens in the Cornell collection from Old Forge, N. Y. (J. G. Needham) may be of this species.

Caenis hilaris Say

1839—Say—J. Acad. Nat. Sci. Philad. 8: 43 (Ephemera).................................... Ind.
1853—Walker—List Neurop. Ins. Brit. Mus. pt. 3: p. 583.
1859—LeConte—Complete works of T. Say, 2: 413 (Ephemera).
1861—Hagen—Syn. Neurop. Ins. N. Amer., p. 54.
1862—Walsh—Proc. Acad. Nat. Sci. Philad., p. 381.. Ill.
1863—Hagen—Proc. Ent. Soc. Philad. 2: 179.
1871—Eaton—Trans. Ent. Soc. London, p. 96.
1884—Eaton—Revis. Monogr., p. 147.
1931—McDunnough—Canad. Ent. 63: 255................................Ia., Md., D. C., Kans.

Length: body 2–2½; wing 2½ mm.

A small species with short forceps; mesonotum pale yellowish; head whitish.

Head of female whitish, of male sometimes shaded with light purplish red; eyes black, ocelli pale with black basal rings; antennae slightly smoky beyond the basal joint. Pronotum whitish; antero-lateral margins may be tinged with blackish; near the posterior margin are two small black dots, one on each side of the median line. Mesonotum pale yellowish; a faint blackish V-mark anterior to the scutellum. Pleura pale, with some short blackish streaks at the bases of the legs and the wing roots. Sternum pale. Legs white; fore femora shaded with smoky apically, fore tibiae faintly smoky; tibia twice the length of the femur, tarsus shorter than tibia. On the dorsal edge of each femur near the apex is a fine short black streak. Wings semi-hyaline whitish, subcosta and radius purplish black for most of their length.

Abdomen whitish. Tergites 1 and 2 shaded faintly with blackish; a fine short blackish streak laterally on the margins of segments 7–9. Sternites white, unmarked. Forceps and tails whitish. Forceps short, narrowed evenly toward the tip; the forceps base covers most of the penes (see Pl. XXXIX).

The spiracular streaks on segments 7–9 only, distinguish this species from others allied to it; the dark femoral dashes separate it from *C. ridens*, and the paler mesonotum and slight shading on tergites 1 and 2 separate it from *C. anceps*.

Specimens of this species in the Cornell collection are from Quicksand, Ky. (C. R. Crosby); Latimer Co., Okla. (R. D. Bird); Manhattan, Kans. (R. C. Smith); Marlinton, Wardensville, Justice and Smoke Hole, W. Va. (J. G. Needham); Fairmount, W. Va. (P. N. Musgrave); Lake Forest, Ill. (J. G. Needham); Lucedale, Miss. (H. Dietrich); Hagerstown, Md. (V. Argo); Stamford, Conn. (S. W. Bromley); and Canajoharie, N. Y. (H. K. Townes, Jr.).

Caenis jocosa McDunnough

1908—Needham—Rept. Mich Geolog. Survey, p. 263 (*diminuta*)........................Mich.
? 1911—Morgan—Ann. Ent. Soc. Amer. 4: 115 (*hilaris*)........................N. Y.
1931—McDunnough—Canad. Ent. 63: 260. Fig........................O., Ill., Ont., Que., N. B.

Length: body 3; wing 3 mm.

A small species, forceps moderately long; mesonotum light yellowish brown.

Head pale yellowish; a pale purplish tinge between the eyes, but paler again on the posterior margin; eyes black, antennae pale. Pronotum pale yellowish; on each side of the median line, on the posterior margin, is a blackish dot; area between and anterior to these dots faintly ruddy-tinged; slight black streaks on the lateral margin. Mesonotum light yellowish brown; scutellum paler, tinged faintly with smoky; anterior to the scutellum is a faint smoky median V; a dark lateral streak anterior to the wing root. A transverse blackish shade on the median area of the metanotum. Pleura pale yellowish, shaded lightly with yellow-brown, most intense in the region of the episternum. Sternum pale yellowish. Fore legs tinged with blackish basally; femora smoky, with a black streak apically on the dorsal side; tibiae rather more than twice the length of the femora, pale smoky; tarsi pale. Femora of the middle and hind legs pale yellowish; black dot on coxa, a faint black streak on trochanter; femur with a strong black dorso-apical streak, below which on the outer side is a small black dot. Wings semi-hyaline whitish; subcosta and radius purplish black.

Abdomen pale yellowish. Dorsal areas of tergites 1–6 with blackish suffusions, leaving as pale areas a lateral margin cut by a dark line along the posterior margin of each. Segments 1, 2, 5 and 6 with faint black stigmatic dots. Tergite 7 pale except for two submedian dark patches, and a large, well-defined stigmatic dot. Tergites 8–10 pale, 8 with faint black shading laterally; a faint dark median dot on 9; two blackish median longitudinal streaks, rather slight, on 10. Sternites pale yellowish; 1–5 with faint transverse black streaks laterally; 6–8 with two very faint obliquely-arranged dark dots on each side, the anterior dot near the lateral margin, the posterior one halfway between the median line and the lateral margin. Forceps pale, tails whitish. Forceps moderately long, sharp-pointed (see Pl. XXXIX).

The extent of black suffusion and markings varies considerably in different specimens; in pale specimens, the head may be almost wholly pale, the prothorax pale except for the two dark dots, the abdominal tergites with the dark shading very faint or almost wanting, except the stigmatic dot on 7. The somewhat darker thorax, longer fore legs and much longer forceps distinguish this species from *hilaris:* the smaller size and shorter forceps, as well as the paler thorax, separate it from *forcipata* and *latipennis*.

Specimens of this species in the Cornell collection are from McLean and Ithaca, N. Y.; Thousand Island Park, St. Lawrence River (Mrs. M. R. Miller); St. Cloud, Minn. (J. Shadlack); Lake Minnewaska, Minn.; Walnut Lake, Mich. (J. G. Needham); and Cleveland Co., Okla. (R. D. Bird).

Caenis latipennis Banks

? 1905—Needham—N. Y. St. Mus. Bull. 86; pl. 11 (*diminuta*)........................N. Y., Wash.
1907—Banks—Ent. News 18: 14.
1931—McDunnough—Canad. Ent. 63: 262........................Sask., Ont.

Length: body 4?; wing 4–4½ mm.

A species with long forceps; mesonotum dark brown, head paler posteriorly. A purplish or blackish band across the head between the eyes; head paler before the posterior margin. Pronotum pale yellowish, with blackish shading in the median area. Mesonotum shining brown, darker than in *C. forcipata;* pleura paler brown. Legs pale; darker markings at the apices of the femora. Fore tibia plainly more than twice the length of the femur. Wings semi-hyaline whitish, the subcosta and radius very dark and prominent. Abdominal tergites suffused heavily with blackish or blackish brown, much as in *C. forcipata,* but the dark color extends over the posterior tergites more than in that species. Sternites paler. Tails pale, "faintly annulate with brown" (Banks). Forceps long and sharp-pointed, but shorter relatively than in *forcipata* (see Pl. XXXIX).

This species, allied in type of forceps to *C. forcipata,* may be distinguished from it by the darker brown of the mesonotum, the more extensive dark shading of the abdominal tergites, and the relatively shorter forceps. The eastern specimens which Dr. McDunnough is holding under this name are paler as regards the thorax and dorsum of abdomen. The figure given as *C. diminuta* (by Needham, 1905) may have been of such a specimen. We have seen no specimens of *latipennis.*

Caenis punctata McDunnough

1901—Needham—N. Y. St. Mus. Bull. 47: 426 (*diminuta*)................................N. Y.
1925—Needham—Ent. News 36: 113 (*diminuta*).
1931—McDunnough—Canad. Ent. 63: 259. Fig....................................Que., Ont.

Length: body 3; wing 3 mm.

A species with relatively short forceps; mesonotum yellow-brown; head shaded with purplish.

Head with purplish shading on a pale ground color, the whole vertex appearing somewhat ruddy; eyes black, antennae pale. Pronotum pale yellowish; median portion shaded heavily with blackish, lateral areas paler; margins with a strong black border. Mesonotum yellowish brown; posterior portion paler, with black streaks and dots; a black V-mark anterior to the scutellum, outlining the sutures. Black shading on the metanotum. Pleura yellowish brown; a blackish streak and shading anterior to the wing root; some black shading over the episternum; heavy black streaking around the bases of the legs and below the wing roots. Sternum pale. Fore legs with blackish shading basally; coxa and trochanter mostly pale with slight black streaks; femur deep smoky, a blackish band at the apex and traces of a median one; tibia and tarsus pale whitish, faintly tinged with smoky. Middle and hind legs whitish; femora with a slight sprinkling of small black dots, and a dark apical band, also traces of a median band, especially on the middle leg; coxae and trochanters each with a black spot; hind tibia with an occasional black dot basally. Wings semi-hyaline whitish, subcosta and radius dark.

Abdomen pale yellowish, the tergites suffused with blackish, much as in *C. diminuta;* median areas of tergites 1 and 2, and 7–10, pale; traces of a black dotted mid-dorsal line on 1–3; black stigmatic dots or dashes on each segment; a median black dot on tergite 9, and several minute black dashes on 10. Sternites pale; a pair of prominent black dots, obliquely arranged, on each side of sternites 7–9, the outer dot visible also on the preceding sternites; black streaks along the lateral margins. Forceps and tails pale. Forceps short, broad at the base, and narrowed considerably about halfway to the tip, so that the inner margin is strongly curved (see Pl. XXXIX).

The black apical femoral band is similar to that of *C. diminuta;* it is separable from that species by reason of the black dots on the femora, the darker thorax, and the curved forceps. It is close to *C. delicata*, but smaller; the latter species likewise lacks the black apical band on the femur, which is reduced to a streak on the dorsal area.

Specimens of this species in the Cornell collection are from Saranac Inn, N. Y. (P. R. Needham); Thousand Island Park, St. Lawrence River (Mrs. M. R. Miller); Juanita Is., L. George, N. Y. (J. G. Needham).

Caenis ridens McDunnough

1931—McDunnough—Canad. Ent. 63: 256. . . . . Kans.

Length (female): body 2½–3; wing 2½ mm.

This species, known only in the female sex, has a pale yellow mesonotum and pale head, with no dark stigmatic dots.

Head of female whitish; between the rear ocelli is a slightly smoky shading; eyes black, antennae pale. Pronotum whitish with considerable smoky shading. Mesonotum pale yellowish, tinged faintly with brownish, this color rather deeper than in *C. hilaris;* mid-dorsal line faintly dark; posterior portion shaded with smoky; a slight dark streak inward from the wing roots. Two faint dark transverse streaks near the posterior margin of metanotum. Pleura wholly pale except for a small curved dark mark anterior to the base of the middle leg and a slight deepening of color anteriorly. Sternum pale. Fore legs at base, and entire fore femur tinged considerably with smoky; tibia and tarsus paler; width between the leg bases greater than in *hilaris*. Middle and hind legs wholly whitish, unmarked. Wings semi-hyaline whitish, subcosta and radius purplish.

Abdominal tergites wholly white, except for slight smoky shading laterally on 1 and 2. No traces of dark lateral streaks or stigmatic dots on any segment. Sternites white, unmarked. Tails white, the middle one longer than the laterals.

This species, allied to both *C. hilaris* and *C. anceps*, is separable from the former by reason of the lack of lateral black dashes on the tergites and the dorsal marks on the femora; from *anceps* it is distinguished by the paler mesonotum, fewer dark markings on the pleura, the lack of femoral markings, and the presence of slight dark shading on tergites 1 and 2 laterally.

A paratype of this species, from Lawrence, Kans., was kindly presented to the Cornell collection by Dr. McDunnough. Other specimens which seem to belong here are from Walnut Lake, Mich., and Lake Forest, Ill. (J. G. Needham); and Port Elgin, Ont. (Mrs. E. G. Grieve).

Caenis simulans McDunnough

1926—Needham—Bull. Lloyd Lib. 27: 90. . . . . N. Y.
1931—McDunnough—Canad. Ent. 63: 263. Fig. . . . . Que., Ont., N. B., Sask., Alta.

Length: body 3½–4; wing 3½–4 mm.

A species with forceps of moderate length; mesonotum bronzy-brown; head blackish. See frontispiece.

Head entirely suffused with blackish above; eyes black, antennae pale. Pronotum pale yellowish; median area and antero-lateral margin strongly suffused with blackish; margins narrowly black. Mesonotum bronzy-brown, darker than in *C. jocosa;* scutellum paler, faintly shaded with black. Metanotum pale shaded with black. Pleura bronzy-brown, paler around the bases of the legs; a few faint black streaks; a fine black line on the anterior margin. Sternum light yellowish. Fore legs with blackish shading at the bases; femora blackish, with a mere trace of a dark apical spot; tibiae and tarsi faintly smoky. Middle and hind legs pale yellowish; slight black dots and streaks on coxa and trochanter; dorso-apical black streak on femur fainter than in *jocosa*. Wings hyaline whitish; subcosta and radius purplish.

Abdominal tergites pale yellowish, shaded with black as in *C. jocosa;* black stigmatic dots on segments 1–7; a broken black mid-dorsal line on the first two or three tergites; tergites 8 and 9 deeper yellowish, with traces of a dark median dot on 9 and dark median longitudinal streaks on 10. Sternites pale; a pair of obliquely arranged dark dots on sternites 6–8; faint black lateral dashes on the basal sternites; and a small round brownish mid-ventral spot on each segment. Forceps faintly smoky; tails pale. Forceps of moderate length, distinctly curved, and slightly thicker than in *C. jocosa* (see Pl. XXXIX).

This species, allied to *C. jocosa*, may be distinguished from it by the larger size, lack of black spots on the pronotum, presence of brownish mid-ventral spots, darker mesonotum and more curved forceps. The species appears to be quite variable as to markings, dark forms having faint black lateral dashes on tergites 8–10; in some, the mid-ventral spots may be wanting. Western specimens which seem to belong here are somewhat larger than the eastern ones, with darker mesonotum and wider-spread blackish suffusion of the tergites.

Specimens of this species in the Cornell collection are from McLean and Ithaca, N. Y. (J. G. Needham, J. R. Traver); Saranac, N.Y. (J. G. Needham); Nassau, N.Y. (N. Y. St. Mus. Coll.); Douglas Lake, Mich. (J. G. Needham); Seton L., Minnetonka, Minn.; Little Wall Lake, Ia. (Phoebe Hall); Flathead Lake, Mont. (R. T. Young); and Shuswap L., B. C. (W. A. Clemens).

Caenis tardata McDunnough

1931—McDunnough—Canad. Ent. 63: 260. . . . . .Alta., Sask.

Length: body 3; wing 3 mm.

A species with short forceps; mesonotum deep brown, head blackish.

Head shaded heavily with blackish; eyes black, antennae pale. Pronotum pale with blackish shading, deepest in the dorsal area; traces of a narrow black line on the posterior margin dorsally, none on the lateral margins. Mesonotum deep brown; faint blackish markings on sutures and lateral margins; black shading on the scutellum. Pleura deep brown; a pale area anterior to the wing root is margined dorsally by a black streak; dark streaks also around the bases of the legs and along the anterior margin of the episternum. Sternum pale, tinged slightly with brown. Fore legs shaded with deep brown at the base; femora smoky, with a darker spot apically; tibiae and tarsi pale. Femora of middle and hind legs pale whitish; black spot on the outer side of each coxa, another at base on the inner side; a fine black dorso-apical streak on each femur, which may be reduced until it is almost invisible. Wings semi-hyaline whitish, subcosta and radius dark.

Abdominal tergites pale yellowish white; broad lateral blackish shading on tergites 1–6, separated by a pale median area; lateral margins pale, no stigmatic dots; posterior margins narrowly dark. Tergites 7 and 8 entirely pale; faint dark shading on 9 and 10, and a tiny median dot on 9. Sternites wholly pale yellowish white, sternite 9 slightly deeper in color, often with a faint smoky transverse streak in the median area near the posterior margin. Forceps and tails pale. Forceps short, rather blunt at the tip (see Pl. XXXIX).

The deep brown mesonotum and the blunt apex of the forceps distinguishes this species from others having short forceps. We have seen no specimens of this species.

Subfamily **BAETINAE**

In this subfamily, the diversity in size and appearance of the upper and lower divisions of the eyes of the male reaches its maximum development (fig. 16). The upper portion is known as the turbinate eyes; these are set on a more or less cylindric stalk, the facets occurring on the flattened upper surface only. Facets of turbinate and lower eyes differ in size, those of the turbinate portion being the larger. The middle ocellus is smaller than the lateral ones.

Venation is greatly reduced in this group. In the fore wing, the posterior branches of media and cubitus are detached basally from their respective stems, thus appearing as long intercalaries. Cross veins in the basal costal space are often wholly wanting. Behind the subcosta, the other cross veins, relatively few in number, are usually arranged in several more or less definite series across the wing. Short marginal intercalaries are well developed, occurring singly or in pairs in each interspace between main veins or between long intercalaries. Except in the genus Callibaetis, the number present in each interspace is constant for a given genus. The hind wing is much reduced in size, considerably longer than wide; or it may be wholly wanting. Venation consists of 2 to 3 longitudinal veins, with or without a few cross veins between. A costal projection is usually present near the base of this wing. In those genera in which the hind wing is absent or very greatly reduced, the anal portion of the fore wing is better developed than in genera with larger hind wings.

The middle and hind tarsi are four-jointed; the basal one of these joints tends to become partially fused with the tibia. In the fore leg of the male, five joints are present, the basal one of these very short. Claws dissimilar on all tarsi.

Tails two. Penes internal; in some forms, a 'penis cover,' single or divided, is present between the basal joints of the forceps.

Nymphs of most species are slender, stream-lined forms (see fig. 162), the head flexed downward, as in the Siphlonurinae. Legs rather long and slender. Postero-lateral margins of abdominal segments usually without backward-projecting spines, except in Callibaetis, where there is a slight development of them. Tails two or three in number; the outer ones with a thick fringe of hairs on the inner margin, few or none on the outer margin; middle tail fringed on both sides. Gills present on abdominal segments 1–7, single or with a ventral portion which on the more apical segments may consist of a minute recurved flap only. Gills lamellate.

KEY TO THE GENERA

Adults (males)

1—Hind wing present....................2
—Hind wing absent....................5
2—Fore wings usually with numerous costal cross veins before the bulla; hind wings

with a moderate number of cross veins, at least in the costal region. . **Callibaetis**
—Fore wings without costal cross veins before the bulla; hind wings with no cross veins, or with a very few of them. 3
3—Marginal intercalaries of fore wing occur singly. . . . . **Centroptilum**
—Marginal intercalaries of fore wing occur in pairs. 4
4—Hind wing reduced to a mere thread, with no costal projection, and only occasional traces of a single vein; fore tarsus of male one-half to two-thirds the length of the tibia. . . . . **Heterocloeon**
—Hind wing considerably larger, costal projection usually (not always) present; at least two longitudinal veins; fore tarsus of male at least as long as the tibia . . . . **Baetis**
5—Marginal intercalaries of fore wing occur singly. 6
—Marginal intercalaries of fore wing occur in pairs. . . . . **Pseudocloeon**
6—Basal joint of hind tarsus very long, equal in length to the three other joints together; 2nd joint of forceps short and broad, its inner margin with a prominent wide extension. . . . . **Neocloeon**
—Basal joint of hind tarsus only moderately long, equal in length to the third and fourth joints together; 2nd joint of forceps longer and narrower, conical, tapering apically, with no extension on its inner margin. . . . . **Cloeon**

VERIFICATION TABLE

Adults (males)

Genus	*Hind wing*	*Marginal veinlets*	*H. w., costal angulation*	*Fore tarsus to tibia*	*1st joint* tars. III*
Baetis	Present	Paired	Var.	Longer**	Short
Callibaetis	Present	Var.	Present	Longer**	Mod.
Centroptilum	Present	Single	Present	Longer**	Long
Cloeon	Absent	Single	———	Longer	Mod.
Heterocloeon	Present	Paired	Absent	Shorter	Short
Neocloeon	Absent	Single	———	Longer**	Long
Pseudocloeon	Absent	Paired	———	Subequal	Short

KEY TO THE GENERA

Nymphs

1—Gill lamellae double (often with a mere recurved flap on ventral or dorsal surface) on abdominal segments 1–6. 2
—Gill lamellae single on all abdominal segments. 4
2—Tracheae of gill lamellae pinnately branched; second pair of wing buds present. . 3
—Tracheae of gill lamellae palmately branched; second pair of wing buds absent . . . . **Cloeon**
3—Double portion of gill a flap on the ventral surface; maxillary palp 2-jointed . . . . **Callibaetis**
—Double portion of gill a flap on the dorsal surface; maxillary palp 3-jointed. . . . . **Centroptilum**
4—Tails two. 5
—Tails three. 7

* Short—1st = 2nd + 3rd; Mod.—1st = 3rd + 4th; Long—1st = 2nd + 3rd + 4th.
** Or at least subequal.

5—Second pair of wing buds present (in Heterocloeon, very minute)............6
—Second pair of wing buds absent............................**Pseudocloeon**
6—Gills largely blackish, with a narrow pale border................**Heterocloeon**
—Gills not as above (*bicaudatus*, and the *propinquus* group)...............**Baetis**
7—Middle tail shorter and weaker than the outer ones; distal joint of labial palp rounded...**Baetis**
—Middle tail practically as long and stout as outer ones; distal joint of labial palp dilated apically...8
8—Second pair of wing buds present; gill lamellae symmetrical, with normal pinnate branching...**Centroptilum**
—Second pair of wing buds absent; gills asymmetrical, pinnately branched, but all branches on inner side...................................**Neocloeon**

VERIFICATION TABLE

Nymphs

Genus	*Joints max. palp*	*Joints lab. palp*	*Distal jt. lab. palp*	*Claws toothed*	*Wing bud of III*	*Gills on 1–6*	*No. of Tails*
Baetis	2 or 3	3	Rounded	Yes	Present	Single	2 or 3
Callibaetis	2	3	Conical	Yes	Present	Double	3
Centroptilum	3	3	Dilated	No	Present	Var.	3
Cloeon[1]	3	3	Conical	Yes	Absent	Double	3
Heterocloeon	2?	3?	?	?	Present	Single	2
Neocloeon[2]	3	3	Dilated	No	Absent	Single	3
Pseudocloeon	2	3	Rounded	Yes	Absent	Single	2

Genus **Callibaetis** Eaton

1881—Eaton—Ent. Mo. Mag. 17: 196.
1885—Eaton—Revis. Monogr. p. 191.
1900—Banks—Trans. Amer. Ent. Soc. 26: 248.
1920—Ulmer—Stett. Ent. Zeit. 81: 126.
1933—Spieth—J. N. Y. Ent. Soc. 41: 336.

Mayflies of small to medium size, the fore wing ranging in length from 6½ to 12 mm. (Nearctic species). See frontispiece. Bodies and usually also the legs of both sexes more or less thickly sprinkled with small reddish or dark brown dots, which on the body often appear to be set in slight depressions. Bodies of males usually darker than the females. The wings of the females are generally ornamented with brownish along the costal margin and sometimes also in the disc, while those of the males may be wholly hyaline, or with variable amounts of dark color, always less than in the females. Turbinate eyes of male moderately large in size, rather lengthily oval (see fig. 16a). Eyes of female relatively large, but not set high on the side of the head. Posterior margin of head of female slightly emarginate at the median line, on each side of which is a slightly rounded lobe directed backwards.

Fore leg of male slightly shorter than the body. Fore tibia about 1 and

[1] Gill tracheation palmate; in other genera, pinnate.
[2] Tracheal branches on one side of gill only.

⅓ times as long as the femur; tarsus very slightly longer than tibia. In the female, the fore tibia is almost as long as the femur; tarsus slightly shorter than tibia. Hind tibia of male ½ to ¾ the length of the femur,

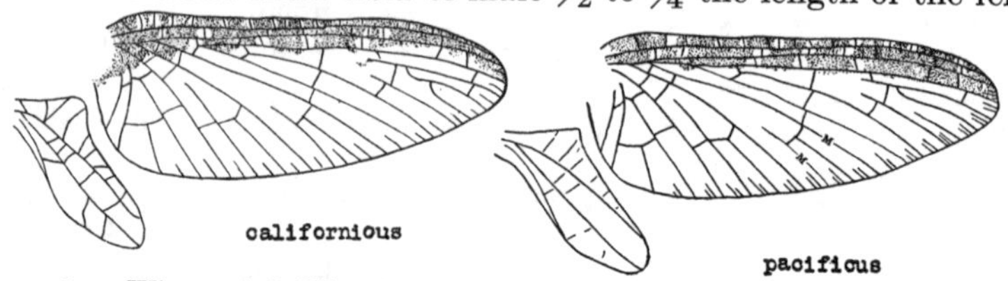

FIG. 161. Wings of *Callibaetis californicus* and *C. pacificus* (from Seemann.)

the tarsus very slightly shorter than the tibia. Basal tarsal joint relatively short, not longer than the 2nd and 3rd joints combined. Hind leg of female very similar to that of the male.

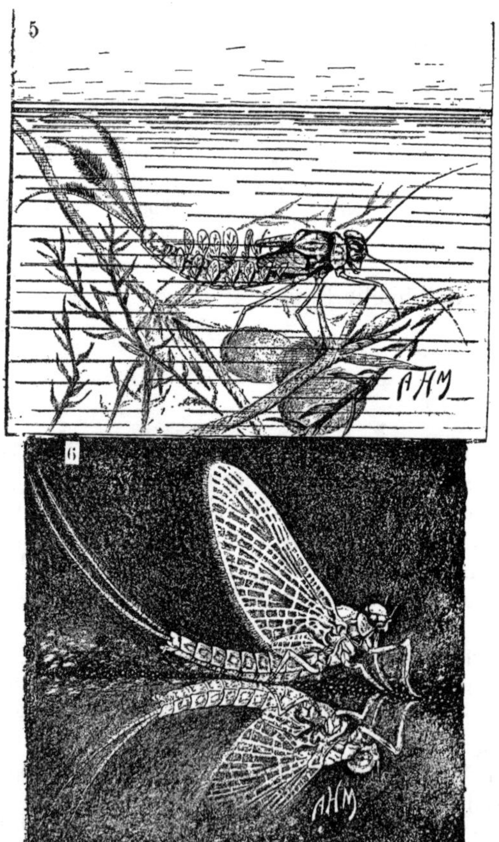

FIG. 162. *Callibaetis fluctuans* (from Morgan): *5*, nymph; *6*, subimago.

Fore wing of many species with more numerous cross veins than in the other genera of this subfamily; in others, however, these are few in number and arranged in the usual irregular series (see fig. 161). Marginal intercalaries occur singly in some species, in pairs in other species, the number apparently not always constant within a single species. In most species, intercalaries near the tip and the anal region are paired, although those near the middle of the outer margin may occur singly. In most species, cross veins are present in the basal costal space. Hind wing relatively large, and widened at the base; the costal projection is large and very obtuse. Three longitudinal veins and several well-developed cross veins are present, also a few intercalaries between 2nd and 3rd veins.

Basal joint of forceps stout, slightly longer than wide; sometimes slightly produced inwards on its inner margin. Second joint slender, conical, tapering toward the apex, where it is practically fused with the third joint. Third joint usually bowed, narrowest near its union with the 2nd joint; widened apically, but narrowed again at the apex. Distal joint generally clavate, about twice as long as wide. A small rounded or conical 'penis cover' is present between the basal forceps joints. Posterior margin of the 9th sternite of the female almost straight, barely if at all produced backwards. Subimago as in fig. 162.

Maxillary palp of nymph 2-jointed; barely as long as the body of the maxilla, but extending slightly beyond the tip of the galea-lacinia. Labial palp 3-jointed, the distal joint rather short and conical. Canines of the mandible rather long and slender. Claws long and slender, denticulate; each is about ½ to ¾ the length of its respective tarsus. Gills rather large, more or less pointed at the tip, irregularly subovate; double on segments 1–6, or 1–7. On segments 1 and 2, the ventral portion of each gill is rather well developed, sometimes bearing a small secondary flap; on the remaining gills, the ventral portion consists merely of a small recurved flap. Tracheation is pinnate. Tails three, about equal in length and thickness. See fig. 162.

Genotype—*C. pictus* (in Baetis), Eaton.

Distribution—Nearctic, Neotropical; in Nearctic, generally distributed.

KEY TO THE SPECIES

Adults*

1—Cross veins of fore wing behind the vitta (or behind the fourth longitudinal vein) relatively few in number,—only about 15 to 25; usually none very near the hind margin; usually a single irregular row of them across the wing........2

—Cross veins of fore wing behind the vitta much more numerous,—about 35 to 60 of them; usually several are close to the hind margin; two or more very irregular rows of them across the wing........8

2—Intercalaries of middle area of outer margin of fore wing usually occur in pairs. 3

—Intercalaries of middle area of outer margin usually occur singly........5

3—Wing 5½ to 7 mm. in length; in wing of female, several brownish clouds on veins behind vitta, also on hind margin; cross veins brown in male........**pictus**

—Wing 7–9 mm. in length; no such distinct brown clouds behind vitta, in wing of female; cross veins in wing of male pale or brownish........4

4—Abdominal tergites 5, 6 and 9 of male with large pale areas; in female, tergites 1, 4, 7 and 8 almost wholly dark brown, others pale with dark markings; cross veins brown in male wing........**pacificus**

—Abdominal tergites 7 and 9 of male with large pale areas; in female, all tergites pale brown; cross veins in wing of male pale........**signatus**

5—Tails distinctly dark brown at the joinings........6

—Tails wholly pale, or pale yellowish only at joinings........7

6—Basal costal space in fore wing hyaline, costa unmarked; hind wing relatively

* Does not include *fuscus* or *doddsi*.

short and wide, more than ½ as wide as its greatest length; cross veins of hind wing incomplete, pigmented . . . . **floridanus**

—Basal costal space, to beyond the middle, with pale narrow bands on cross veins, brown between veins; hind wing relatively narrow and long, less than ½ as wide as its greatest length; cross veins of hind wing complete, pale. **californicus**

7—Basal costal space of fore wing of female interrupted brown and white; posterior margin of vitta practically straight; legs greyish brown, femora with many minute dots . . . . **montanus**

—Basal costal space of fore wing of female hyaline; posterior margin of vitta irregular, often with 6 or 7 large obtuse lobes; legs whitish, femora barely or not at all dotted . . . . **fluctuans**

8—Intercalaries of the middle area of the fore wing usually occur in pairs 9

—Intercalaries of the middle area of the fore wing usually occur singly 17

9—Longitudinal veins of fore wing brown in places, or alternately dark and light throughout 10

—Longitudinal veins of fore wing wholly dark or light, not alternately dark and light as above 15

10—Wing of male 7 to 8 mm. in length 11

—Wing of male longer, 9 to 12 mm. in length 14

11—Basal costal space in fore wing of female hyaline . . . . **coloradensis**

—Basal costal space in fore wing of female brown 12

12—Vitta of fore wing of female blackish brown; male wing may be slightly pigmented 13

—Vitta of fore wing of female light brown; male wing not pigmented . . **ferrugineus**

13—Vitta in wing of female a broad almost continuous band; no clouds in disc or along outer margin of wing . . . . **americanus**

—Vitta in wing of female relatively narrow, divided into several lobes which are widely separated by pale areas; small clouds usually present on veins and along outer margin . . . . **carolus**

14—Vitta in wing of female, and slight markings along costal margin of male, dark brown; tails distinctly darker at joinings; many clouds along veins and outer margin of wing of female . . . . **hageni**

—Slight markings along costa of male wing blackish; female not known; tails yellowish, apparently not darker at joinings . . . . **nigritus**

15—Wing of male entirely pale, no vitta; longitudinal veins pale . . . . **pallidus?**

—Wing of male with a brownish vitta; longitudinal veins brown 16

16—Costal space in fore wing of male wholly hyaline; space between subcosta and fourth vein with a light brown or reddish brown vitta extending from base halfway to tip; 2 or 3 pale dots in this area; wing (male) 8 mm. in length . . . . **semicostatus**

—Costal space in fore wing of male interrupted brown and white; vitta of male wing not so sharply delimited; wing (male) 9–10 mm. in length . . . **skokianus**

17—Wing of female 6½ to 7 mm. in length; hind margin of vitta sinuate, with four projections; clouds along the outer margin . . . . **pretiosus**

—Wing of female 8–9 mm. in length; vitta not as above; clouds may be present or absent 18

18—Clouds present in disc and along outer margin of wing; vitta of female relatively narrow, its lobes separated by wide pale areas . . . . **carolus**

—No clouds in disc or along outer margin of wing; vitta of female a broad band, almost continuous . . . . **americanus**

VERIFICATION TABLE

Adults

Species	*Wing ♂ length*		*Cross* veins*	*Marginal veinlets*	*Clouds ♀ on veins*	*♂ Wing pigmented*	*Tails*	*Distr.*
americanus	8–9	♀	Many	Var.	No	?	Ringed	Colo.
californicus	7–8		Few	Single	No	No	Ringed	Calif.
carolus	8		Many	Var.	Yes	Slightly	Ringed	Calif.
coloradensis	7½	♀	Many	Paired?	No	?	Smoky?	Colo.
doddsi	6½		Few	?	?	No	?	Colo.
ferrugineus	7		Many	Paired	No	No	Mkd.?	E
floridanus	7	♀	Few	Single	No	?	Mkd.	Fla.
fluctuans	6½–7½	♀	Few	Single	No	No	Pale	E
fuscus	10		Many?	?	?	Yes	?	Colo.
hageni	12	♀	Many	Paired	Yes	Slightly	Ringed	G ?
montanus	8	♀	Few	Single	No	?	Ringed	SW
nigritus	9–9½		Many	Paired?	No	Slightly	Pale?	Colo.
pacificus	7½–8		Few	Paired	No	No	Ringed	Calif.
pallidus	9½	♀	Many	Paired	No	?	Pale	Colo.
pictus	5½–7		Few	Paired	Yes	No	Ringed	SW
pretiosus	6½–7	♀	Many?	Single	Yes	?	Ringed	Va.
semicostatus	8		Many?	Paired	No	Yes	Pale	Man.
skokianus	9–10		Many	Paired	No	Yes	Pale	E
signatus	7–7½		Few	Paired	No	No	Ringed	Ariz.

KEY TO THE KNOWN NYMPHS**

1—Broad dark bands across tail, at least one-third the width of the hairy portion . . . 2
—Tail wholly pale, or with a narrow dark band near the tip only 3
2—Dark longitudinal markings on each edge of femur, near apex; abdominal tergites 2, 3, 5 and 6 with darker markings than others . . . . **montanus?**
—No such dark markings near apex of femur; basal and intermediate tergites quite uniform in coloration . . . . **californicus**
3—Femur with a broad pre-apical band; body 10–11 mm. in length . . . . **hageni**
—No such dark femoral band; body ranging in length from 9 to 13 mm. 4
4—Body 9 mm. in length . . . . **pacificus**
—Body 12–13 mm. in length . . . . **skokianus**

Callibaetis americanus Banks

1900—Banks—Trans. Amer. Ent. Soc. 26: 250 . . . . Wash., Colo., ? N. H., Ont.
1930—Ide—Canad. Ent. 62: 226.

Length: body 7½; wing 8–9 mm. (female).

A very dark brown species; cross veins of fore wing numerous; marginal intercalaries seemingly variable, single or double; male not known.

Head and thorax very dark brown, almost black. Femora thickly dotted with brown; tarsi marked with brown. Many cross veins in the fore wing; in the type specimens, the marginal intercalaries, as studied by one of us (J. G. N.), were very

* Few = 15 to 25; Many = 35 to 60.

** The description of *fuscus* by Dodds is not sufficiently specific to permit incorporating the species in a key.

difficult to see, appearing to be absent altogether from some spaces, single in others, and in still other spaces occurring in pairs. "Vitta of wing dark brown, broader at base, hind edge quite even, with three principal indentations, usually quite narrow, costal area with hyaline spots, no clouds along the veins" (Banks). Cross veins mostly white; longitudinal veins alternately brown and pale. Hind wing relatively long and narrow, the costal border emarginate immediately behind the basal projection; cross veins moderately numerous; usually two or more marginal intercalaries; base of this wing often brown. In the fore wing, the costal brace is pale in the middle, brown at each end; on each side there is a pale area on the costal margin, a brown cloud next to subcosta. Abdomen very dark brown, almost blackish. Tails darker at the joinings.

We have three female specimens in the Cornell collection which agree well with Dr. Banks' description of this species. Two of these, from Gothic, Colo., were collected by Mr. J. D. Bowen; the third was taken by Prof. R. T. Young at Upper Kintla Lake, Glacier National Park, Mont. In these, the hind margin of the wing is alternately dark and light; the marginal intercalaries, while small and rather difficult to see, are mostly single; in one specimen, several interspaces are without an intercalary, and in at least two spaces the veins are paired. One of Dr. Banks' paratypes, from Colorado, which he kindly sent to our collection, is much paler, and has several small but distinct clouds along the veins; also the stippling on the femora is almost wholly wanting. It seems probable that this represents another, and as yet undescribed, species allied to *C. americanus.* The eastern species listed here may not be the true *americanus.*

Callibaetis californicus Banks

1900—Banks—Trans. Amer. Ent. Soc. 26: 249. . . . . Calif.
1927—Seemann—J. Ent. and Zool. 19: 48. Figs.

Length: body 6–8; wing 7–8 mm.

A reddish brown species; wing of male not pigmented.

Head and thorax of male dark reddish brown. Fore femur reddish brown; tibia paler red-brown, darker distally, blackish at femoro-tibial joining; tarsus yellowish, joinings narrowly brownish. Middle and hind legs yellowish white; femora with a prominent pre-apical reddish brown band, and streaked longitudinally with red-brown in distal half; 6 to 8 round reddish brown dots along the middle area of each. Tibiae blackish at femoral joining; distal tarsal joint red-brown, other joints reddish apically; joinings narrowly red-brown. Head and thorax of female paler red-brown except the postero-lateral areas of mesonotum; stippled with reddish brown. Fore leg yellowish white, the dark reddish brown femoral band very prominent; apex of tibia, and all of the tarsus except the basal joint, reddish.

Wings hyaline; cross veins few in number (18 to 22 beyond first branch of radius) marginal intercalaries single (see fig. 161); longitudinal veins pale yellowish brown in both sexes; cross veins yellowish in wing of male, and in female wing beyond the disc; those in disc of female wing snowy white. Wing of male not pigmented. Vitta in wing of female rather dark reddish brown, including the basal costal space; cross veins fenestrate with hyaline, those in the costal space to beyond the middle being "short bands from vein to costa," rather than the usual round or oval spots. Hind margin of vitta "with about 8 or 10 indentations," those in the basal half quite wide. Hind wing rather narrow for its length, and narrowed toward the tip; reddish brown at extreme base. Cross veins only moderately numerous; usually no marginal intercalary; the costal border distinctly emarginate immediately behind the basal projection.

Abdomen of male reddish brown, paler yellowish brown ventrally; tergites with paler lateral and posterior margins; anterior margin of each very narrowly darker; an indistinct darker line marks the spiracular area; traces of a pale mid-dorsal line. On each tergite and sternite, a pair of reddish brown oblique submedian streaks from anterior margin about to the middle of the segment; one or two dark dots in a line with each dark streak, immediately posterior to it. Median and lateral areas of each segment with red-brown stippling, the dots not very close together. Abdomen of female paler, flesh-colored, but similarly marked. Forceps reddish brown at base, yellowish apically. Tails whitish, joints relatively very long; alternate joinings purplish brown.

We have reared specimens of this species in the Cornell collection from Chino, Calif., (Mrs. Theresa M. Seemann); also several from Laguna Beach, Calif. (J. G. Needham).

Callibaetis carolus sp. nov.

Length: body 8; wing 8 mm.

A brown species; cross veins in fore wing numerous; marginal intercalaries variable, single or paired; wing of male usually very slightly pigmented.

Turbinate eyes of male orange-brown. Head red-brown. Basal joints of antenna pale, black apically; filament smoky brown, pale at extreme base and at tip. Thorax red-brown, notum reddish brown; pleura with paler yellowish areas, some of which are stippled. Fore leg brownish; in some specimens, femora yellowish except at apex, tarsi distinctly pale; in others, entire leg smoky to pale red-brown; femur with distinct pre-apical dark band. Middle and hind legs whitish or yellowish white; faint traces of a dark band near apex of femur, and occasional dark longitudinal streaks; stippling, if present, very faint; femoro-tibial joint and tarsal joinings darker; tarsi more or less shaded with light brown. Wings hyaline, iridescent; cross veins numerous; marginal intercalaries variable, usually paired in male, often single in female. Wing of male usually with faint blackish-brown markings along the costa, quite variable in extent; extreme apical portion of costal space shaded with brown. In well marked specimens, longitudinal veins are brown with many pale areas; cross veins pale, outer margin brown.

Head and thorax of female yellowish brown; two longitudinal brown streaks on vertex, with a darker dot at center on each side, another at posterior margin; dark streaks extend backward as submedian marks on pronotum. Thoracic notum yellowish in anterior and central portions; middle of scutellum yellow; postero-lateral margins of mesonotum dark red-brown. Legs whitish or yellowish, tarsi usually wholly brownish; femora usually with some reddish brown stippling. Vitta of wing blackish-brown, including basal costal space except basad of costal brace; this brace is brown at each end and pale in the middle, as in *C. americanus*, and on each side near the subcosta is a brown cloud; costal space interrupted brown and white; vitta rather narrow, quite variable, but its posterior margin usually with four distinct lobe-like extensions, separated by wide pale areas; the most apical of these extensions has three or more small secondary lobes. Cross veins in vitta fenestrate with hyaline. Hind margin of wing alternately brown and white; small clouds often present along some of the veins, also on margin around the intercalaries. Longitudinal veins as in male.

Abdominal tergites of male rather dark red-brown, the apical margins darker; dark submedian streaks, usually separated by a narrow pale mid-dorsal line; dark line along pleural fold; stippling moderately dense and regular. On each side of the dark submedian streaks is a paler yellowish streak, which on the apical tergites may extend over most of the lateral area. Sternites paler, yellowish brown, with the usual dark submedian streaks; stippling somewhat finer than on dorsum; apical sternites often opaque creamy or yellowish white. Forceps and tails yellowish or whitish, tail joinings distinctly but narrowly blackish. Abdomen of female somewhat paler, otherwise similar.

Holotype—Male imago. Big Bear Lake, San Bernardino Mts., Calif.; Aug. 24, 1932; Mr. Chas. D. Michener, collector. No. 1358.1 in Cornell collection.

Allotype—Same data, Aug. 25, 1932. No. 1358.2 in Cornell collection.

Paratypes—50 male imagos, 12 female imagos; same location; Aug. 24 and 25, 1932. No. 1358.3–64 in Cornell collection.

In addition, there are several other specimens taken at the same time and place, not included among the types. This species we name in honor of the collector. It is quite close to *C. nigritus*, but smaller than the types of that species, and differing in having the tails well marked at the joinings. We describe this as a new species, because of the abundance of material at hand, although it may fall as a synonym of *nigritus* when more is known of that species in both sexes. The female from Dr. Banks' collection, listed there under *C. americanus*, but having distinct clouds in the wing, probably belongs here, as well as two other females in our collection, from Clear Creek, Colo., collected by Mr. E. J. Oslar.

Callibaetis coloradensis Banks

1900—Banks—Trans. Amer. Ent. Soc. 26: 250 .... Colo.

Length: body 7; wing 7½ mm. (female).

A dark brown species, the body almost blackish; cross veins of fore wing numerous.

Head and thorax of female dark brown. Femora finely stippled with brown; tips of tarsi brown. Cross veins of wing numerous; marginal intercalaries paired ?. Vitta dark brown, "broken up into three principal spots, one, the largest, apical, one pterostigmatical, one before the middle, and a smaller one at base; basal costal space hyaline; the longitudinal veins are in places slightly marked with brown" (Banks). Most of the cross veins are white. Hind wing rather long and moderately narrow, the costal border somewhat emarginate behind the basal projection; marginal intercalaries long, more like longitudinal veins than intercalaries, 3 or 4 in number (from MS notes by J. G. N. on the type specimen); cross veins tend to be arranged in groups of two. Abdomen dark brown, almost blackish; shorter proportionately "than in other species of this section" (Banks). Tails missing from type specimen, not mentioned in description.

We have several male specimens in the Cornell collection from Pike's Peak, Tabernash, and Estes Park, Colo., which agree well with Dr. Banks' description of the female of this species. As there are no females associated with the males, we are uncertain of their identity, but are holding them tentatively as *C. coloradensis*. In these males, the entire outer margin of the fore wing is dark brown; longitudinal veins alternately dark and light, as in *C. americanus;* cross veins mostly pale; marginal intercalaries brown, occurring in pairs; base of hind wing dark brown; tails (in most specimens) pale smoky, very slightly darker or at least opaque at the joinings.

Callibaetis doddsi nom. nov.

1923—Dodds—Trans. Amer. Ent. Soc. 49: 100. Fig. (*vitreus* Dodds; *nec* Navas 1918) .... Colo.

Length: body 6½; wing 6½ mm.

A brownish species; cross veins of fore wing few in number; wing of male not pigmented.

Thorax light brown; sutures pale. Wings hyaline, fore wing not pigmented in male (female not known). Cross veins of fore wing relatively few in number; intercalaries not mentioned. Hind wing likewise unpigmented; cross veins few; only one marginal intercalary. Abdominal tergites 2–6 light grey-brown, 7–10 red-brown. Sternites paler, the dark freckling becoming less dense apically to the 6th and 7th sternites; sternites 8 and 9 without freckles. Each sternite is marked with a pair of dark submedian streaks. Basal joint of forceps irregular on the inner margin, and with a slight rounded protuberance at the inner apical margin. Second joint with distinct 'wrinkles' near the base on the inner margin.

This species, described from a single male, is apparently allied to *C. pictus*, *C. pacificus*, and *C. signatus;* however, the number of marginal intercalaries in each interspace of the fore wing is not mentioned, and we have not seen the type specimen. The name *vitreus*, which Dr. Dodds used for this species, being preoccupied (Navas, 1918, for a Neotropical Callibaetis), we re-name this species *doddsi*, in honor of the describer.

Callibaetis ferrugineus Walsh

1861—Hagen—Syn. Neurop. Ins. N. Amer. p. 53 (*Cloe undatus*).
1862—Walsh—Proc. Acad. Nat. Sci. Philad. p. 379 (Cloe).......... Ill.
1871—Eaton—Trans. Ent. Soc. London, p. 124.
1871—Hagen—op. cit. p. 402 (*B.* ? *undatus*, in part).
1873—Hagen and Eaton—Trans. Ent. Soc. London, p. 402 (in part).
1885—Eaton—Revis. Monogr. p. 193.
1900—Banks—Trans. Am. Ent. Soc. 26: 250.......... N. Y., Mich.
1907—Needham—Mich. Geol. Survey, p. 262.

Length: body 7½–9½; wing 7–8½ mm.

A reddish species; cross veins of fore wing numerous; marginal intercalaries paired; wing of male not pigmented. See frontispiece.

The two basal joints of the antenna each slightly fuscous at the tip; antennal filament whitish at base, fuscous at tip, in most specimens. Head and thorax ferruginous. "Anterior mesothoracic lobe subtruncate, the corners rounded." Thoracic sternum pale, generally speckled with reddish brown. Legs pale yellowish; tips of the tibiae, the tarsal joinings as well as the tips of each joint, and the basal tarsal joint of the fore leg, fuscous. Wings hyaline; costa a little yellowish; veins of the costal margin yellowish, and a few of the other longitudinal veins often tinged with fuscous; veins and cross veins moderate, subequal, hyaline. There are numerous cross veins in the hind wing. Abdominal tergites densely freckled with reddish brown, sometimes almost piceous; venter paler, reddish white, less densely freckled. Forceps and tails whitish.

According to Dr. Banks, the costal band on the wing of the female is light brown, covering the basal costal space; many of the cross veins fenestrate with hyaline; posterior margin with three principal indentations and two smaller ones near the tip; veins brown in parts. The femora are said to be thickly spotted with light brown, and tarsi and tails marked with brown. From the original description, however, the tails would seem to be unmarked, and femoral freckles, if present, inconspicuous.

We have specimens in the Cornell collection, mostly males, which seem to be of this species, from the following localities. Walnut Lake, Mich. (J. G. Needham); Raquette Lake, N. Y. (R. C. Tasker); Hagerstown, Md. (V. Argo); Bar Harbor, Me. (A. E. Brower); Fariport, Ia. (H. E. Schraedick); Green Lake, Wis.; and Ithaca, N.Y.

In all of these specimens, the tails are more or less distinctly darker at the joinings, the femora with few or no freckles. From Ithaca, also, we have several specimens, presumably of a related but undescribed species, which are considerably paler, a little larger, eyes orange-red, tails wholly white, unmarked, femora with a few brownish-red freckles.

One of us (Needham) examined the type, a male specimen, in the Museum of Comparative Zoology; the following additional descriptive points are taken from his notes.

"Femora pale, brown on tips of femur, tibia and tarsus of fore leg; on hind leg, tips of all tarsal joints brownish. Middle abdominal segments with large deep punctures as thick as holes in a colander—8 or 9 of these holes in a row across the top of each tergite; pale submedian longitudinal streaks on the dorsum of the abdomen. Costal margin of hind wing straight on each side of the projection. In female, costal brace brown, on fore wing, bordered on each side with hyaline. Punctures on abdomen, and hind wing, as in male."

Hind wing rather narrowed toward the tip; apparently with two to four well-developed intercalaries. Wing of type 7 mm. in length.

Callibaetis floridanus Banks

1900—Banks—Trans. Amer. Ent. Soc. 26: 249...Fla.

Length: body (female) 6; wing 7 mm.

A uniformly ferruginous species; cross veins of fore wing few in number; marginal intercalaries single; male not known.

Thorax ferruginous. Femora thickly dotted with brown; 10 to 15 large dots on the middle line of each, as well as many smaller ones on each side. Tarsal joints brown apically. Cross veins of fore wing few, only about 20 behind the vitta; marginal intercalaries, in the type specimen, occur singly except at the tip. Costal brace red-brown, with a clear space immediately beyond; this is followed by U-shaped bars of brown (the latter statement is taken from MS notes on the type specimen, by J. G. N.). "Costa of wing unmarked; basal costal region hyaline, vitta pale brownish, broader at base, with about seven or eight small narrow indentations on hind edge, otherwise the edge is very regular" (Banks). Cross veins in middle of wing brownish. Hind wing relatively short for its width, the costal margin nearly straight on each side of the costal projection. Cross veins rather few in number; usually one well-developed intercalary only; veins 2 and 3 convergent toward the tip. Abdomen uniformly ferruginous; tails pale, narrowly brown at the joinings.

Callibaetis fluctuans Walsh

1862—Walsh—Proc. Acad. Nat. Sci. Philad. p. 379 (Cloe).....................................Ill.
1863—Hagen—Proc. Ent. Soc. Philad. 2: 178 (Cloe).
1871—Eaton—Trans. Ent. Soc. London, p. 122 (Baetis).
1885—Eaton—Revis. Monogr. p. 193 (*ferrugineus*).
1900—Banks—Trans. Am. Ent. Soc. 26: 249...D. C.
1913—Morgan—Ann. Ent. Soc. Amer. 6: pl. 43...N. Y. ?

Length: body (female) 6–7; wing 6½–8 mm.

A brownish white species; cross veins of fore wing relatively few in number; marginal intercalaries single; wing of male not pigmented.

Head and thorax brownish white; vertex of female sometimes ferruginous, "with a double longitudinal carina." Antennal filament dusky, sometimes pale at the base, A double light brown band on the thoracic notum, confluent behind. Legs whitish, the tips of the tarsi cloudy. Wings hyaline, iridescent; veins rather coarse near the base; longitudinal veins generally brown, and occasionally edged with brown toward the base. Cross veins fine, always hyaline. In the wing of the female, the space between costa and subcosta is hyaline; on the costal margin are 15 to 18 small brown spots, some of these confluent. A light brown band is present behind the subcosta, containing about 14 round hyaline spots—"some of them confluent before or behind with the hyaline part of the wing." Posterior margin of this brown vitta variable, irregular; sometimes with 6 or 7 "large obtuse teeth." Marginal intercalaries occur singly. Cross veins behind the vitta relatively few in number (14 to 18, according to Walsh).

Abdomen, both tergites and sternites, generally brownish-white, sometimes varied with brown. Two of Walsh's specimens were pale brown, the 6th segment brown. Tails whitish. (Original description, of female only). The wing of the male (if we are correct in the identification of this species) is unpigmented. In both sexes, brown submedian curved marks are present on tergites and sternites.

One of us (Needham) has examined the type specimen (a female, labeled Rock Is., Ill., in Hagen's handwriting) at the Museum of Comparative Zoology. From his notes we quote the following:

"Tails white. Costal band continuous, fenestrate on cross veins except at bulla behind subcosta. Maculate line of brown from front of mesonotum through the wing base. Stigmatic cross veins simple, few, more or less erect. Vitta at base extends back to 1st anal vein. Femora finely dotted (not very plain)."

Banks (Trans. Amer. Ent. Soc. 26:249) apparently has some other species confused with *fluctuans*, for he says of it, "Setae narrowly annulate with brown; vitta of wing yellowish brown, including the basal costal space"; "about 18 cross veins beyond vitta, all brown." These statements do not coincide with Walsh's description.

We have three specimens in the Cornell collection from Fairport, Ia., two of them females, which seem to agree well with Walsh's description; also two females from Ithaca, N. Y. From Ithaca and vicinity we have many specimens of a somewhat larger size, very similar in maculation of body and wings, but with all except the basal cross veins distinctly brown.

Callibaetis fuscus Dodds

1923—Dodds—Trans. Amer. Ent. Soc. 49: 100 .... Colo.

Length: body 8; wing 10 mm.

A brown species; cross veins numerous (?) in the fore wing; wing of male pigmented.

Thorax brown, with some pale areas; some of these are dark-freckled. Femora of middle and hind legs distinctly banded; fore femur less distinctly so. Fore wing in both sexes "with liberal pigmented areas at front margin as far back as the radius; cross veins all along the costal space." Numerous cross veins in the hind wing, which is pigmented at the base, as far as the costal projection. Abdomen of male "with light ground color and dark stippling, about equally dense on all segments." Irregular light and dark areas on the tergites; a dark stripe along the pleural fold; on each sternite a pair of large dark submedian markings. Female described as very similar to male in coloration, except for usual sex differences.

We have not seen the type specimens of this species. In size, it seems close to *C. hageni*, but its relation to other species in the genus cannot be determined until more is known about it than is stated in Dr. Dodds' description. In the Cornell collection are several male specimens from Yellowstone Lake, Wyo., Lapwai, Idaho (V. Argo), and Scott Lake, Ore. (H. A. Scullen, J. E. Davis), which we are holding tentatively under this name. Fore leg brownish, middle and hind legs yellow; femora freckled, forming a dark pre-apical femoral band; base of tibia blackish, at joint; tarsi brownish. Longitudinal veins brown, cross veins pale; marginal intercalaries paired; tails yellowish, paler at tips, not distinctly darker at joinings. Wing about 9 mm. in length.

Callibaetis hageni Eaton

1861—Hagen—Syn, Neurop. Ins. N. Amer. p. 50 (*Baetis tessellata*).
1871—Eaton—Trans. Ent. Soc. London, p. 84 (in discussion of *Leptophlebia colombiae*; *B. tessellata*).
1873—Hagen—Trans. Ent. Soc. London, p. 395 (*B. tesselleta*).
1885—Eaton—Revis. Monogr. p. 192. Figs. (re-named; *tessellata* preoccupied)..........Wash., Calif.
1900—Banks—Trans. Amer. Ent. Soc. 26: 250 (*tessellata*).
1930—McDunnough—Canad. Ent. 62: 59. Nymph..Que.

Length: body 12; wing 12 mm. (female).

Body piceous; cross veins of fore wing numerous; marginal intercalaries paired; male wing very slightly pigmented.

Body piceous; in female, the dorsal sutures of the thoracic notum are paler, yellowish; darker in the male, polished. Legs brown, in both sexes; femora "densely, minutely and inconspicuously dotted" (Eaton). Cross veins numerous in fore wing; intercalaries (in an Oregon specimen we have examined) mostly in pairs, occasionally single. Vitta of female wing quite wide, dark brown, its hind margin with about 5 lobes. According to Dr. Banks, the second lobe extends back on wing further than in other species of this group. Cross veins in the vitta fenestrate with hyaline; behind the vitta, along the veins, are about 3 rows of brown spots or clouds, "following and lying between the three main series of cross veins" (from MS notes by J. G. N.); the middle row of these spots is largest. Basal costal space included in vitta; longitudinal veins interrupted brown and white; outer margin of wing alternately brown and white; intercalaries brown, sometimes in a faint brownish cloud; cross veins mostly white. Stigmatic cross veins few, aslant; in the female fenestrate, more or less so in the male, but fainter in that sex. Wing of male (notes on a specimen in Museum of Comp. Zool.) almost clear—with two narrow brown widened portions on the costa; stigma slightly brown-tinged.

Abdomen "discolored" (in our Oregon specimen, deep reddish brown, almost blackish dorsally, only slightly paler ventrally); thickly dotted above and below. Tails white, joinings blackish.

We have seen specimens of this species, in the collection of the Univ. of Minnesota, from Portland, Ore.; in the collection of the Calif. Acad. of Science, from Logan Pk. Inn, Colo., Cannon Beach and Gold Hill, Ore., and San Francisco, Calif.; and in the collection of the Colorado Agr. College, from Ft. Collins, Colo. In the Cornell collection are specimens from Berkeley, Calif.; Cheyenne, Wyo. (Fanny T. Hartman); Yellow Bay, Mont. (R. T. Young); Yakima, Wash. (M. D. Leonard); Klamath Falls, Ore. (M. P. Canova); San Joaquin River, Calif. (C. D. Michener); and many specimens from Ithaca, N. Y. and vicinity, which seem to be the eastern form of this species. In these specimens, the wings range in size from 8½ to 12 mm. in length; in the eastern forms, the legs and sometimes the venter of the abdomen are paler than in the western form, but they appear similar in other respects.

Callibaetis montanus Eaton

1885—Eaton—Revis. Monogr. p. 196. Fig...Cent. Amer.
1900—Banks—Trans. Am. Ent. Soc. 26: 249...Mexico

Length: body (female) 6; wing 8 mm.

A brownish yellow species; cross veins of fore wing few in number; marginal intercalaries occur singly; male not known.

"Body discolored; notum testaceous, irrorated on each side with brown-ochre, and traversed lengthwise by a median brown-ochreous stripe intersected by a light yellowish line" (Eaton). Legs "somewhat of a dull light Vandyke-brownish grey, stippled or irrorated minutely and densely in the femora with very light burnt-umber brown"; tarsal joinings similar in color; distal joints of the tarsi darkened, claws black. Wings vitreous; cross veins of fore wing few in number, intercalaries single. Costal vitta light raw-umber brown in color, occupying the costal, subcostal and "next adjoining spaces," and widened at the base of the wing; this band varies in intensity apically and in the basal costal space. Veins similar in color, except the basal costal cross veins, which are whitish. 6 or 7 cross veins before the bulla; beyond it "about 7 rudimentary and 7 well-defined simple and almost straight" cross veins. Tails white; joinings and basal joints yellowish.

We have seen specimens that seem to be of this species, from Manhattan, Kans. (R. C. Smith); Tuscon, Ariz. (A. A. Nichol); from Sabinal, N. M. (C. W. Sabrosky, R. M. Painter); likewise reared females from Farmington, N. M. (P. N. Albright). Two specimens in the Cornell collection from Weslaco and Austin, Tex. (S. W. Clark, H. B. Parks, Jr.), seem to agree well with the description of *C. montanus*, as do two others from Sacramento, Calif. (Velma Knox). Associated with the Weslaco specimens are several males, similar in number of cross veins and intercalaries, but with wholly pale legs, the femora not at all stippled; claws blackish; eyes very large, deep orange in color; thorax red-brown; wing unpigmented, all veins whitish, stigmatic area opaque white; abdomen yellowish brown dorsally, with narrow pale mid-dorsal stripe and pale lateral margins; sternites pale yellowish or whitish, with usual dark submedian streaks. There is no certainty that these are males of *montanus*, however. Other apparently similar males are from San Angelo and El Paso, Texas; among specimens of this species collected at Mexico City, Mex. (G. W. P. Lassmann) is one male subimago similar to the Texas males.

Callibaetis nigritus Banks

1918—Banks—Bull. Mus. Comp. Zool. 62: 12 .... Colo.

Length: body ?; wing 9–9½ mm. (male).

Body dark brown; cross veins in fore wing numerous; marginal intercalaries occur mostly in pairs; wing of male very slightly pigmented.

The species was described from two male specimens. Body dark brown, thoracic notum blackish. Fore leg very dark, more or less blackish; middle and hind legs pale, the femora darker before the apex, tarsi wholly darker; femora very finely and equally freckled (this from the senior author's notes on the type specimens). Cross veins in fore wing numerous; intercalaries seem to occur mostly in pairs. Wing hyaline; longitudinal veins white with some black streaks; cross veins mostly pale; outer margin alternately dark and pale. "Costal area marked with black (not brown) spots, but not forming a continuous streak; various dots along costal margin, streaks along veins and larger spots beyond the middle, two being larger than the others" (Banks). Abdominal tergites with a darker median stripe, which "on the last few segments is pale in the middle." A pair of dark submedian streaks on each sternite. Forceps and tails pale yellowish.

We have in the Cornell collection duplicates of specimens in the collection of Oregon State College, which seem to agree with the above description, except for their smaller size. These are from Corvallis, Polina Lake, and Anthony Lake, Ore. (H. A. Scullen, Ruth J. Norberg). These may not represent the true *nigritus*, and are very close to the species from California we are naming *carolus*.

Callibaetis pacificus Seemann

1927—Seemann—J. Ent. and Zool. 19: 49. Figs. . . . . Calif.

Length: body 7–8; wing 7–8 mm.

Male dark brown, female pale tan; cross veins of fore wing few in number; marginal intercalaries paired; wing of male barely or not at all pigmented.

Head and thorax of male blackish brown; extensive pale areas on pleura and around bases of legs; scutellum pale in middle. Head and thorax of female pale tan, with flesh-colored tinge; anterior margin of mesonotum olive brown with pale median and lateral streaks; faint brownish stippling on some portions of pleura and sternum. Scutellum yellow. Fore leg of male rather dark olive to reddish brown; femur paler at tip, with a dark pre-apical band; tarsi paler olive brown; faint traces of a row of 6 to 8 round dark dots near middle of inner surface of femur. Middle and hind legs cream-colored to pale tan; apical margins of tibiae, a wide pre-apical femoral band, and part or all of each tarsal joint, dark brown; 10 to 12 faint darker dots along middle of each femur; tibiae likewise with faint dark stippling.

Wings hyaline; cross veins few in number (18 to 20 behind first branch of radius); marginal intercalaries paired. Longitudinal veins dark brown in both sexes; in female, four main veins of costal margin with numerous pale spaces. Cross veins dark brown in male, white in female. Wing of male wholly unpigmented, or with narrow brown markings on the costa only. Wing of female irregularly and variably mottled along the costal border with pale reddish brown (see fig. 161). Costal brace pale except at base; entire costal space usually interrupted brown and white; cross veins fenestrate irregularly with hyaline. Behind the costa, several brown clouds separated by hyaline areas; a small dark area at basal fork of radius; a paler and larger cloud at and on each side of the bulla; two or more clouds beyond the bulla, the outer one at or near the wing tip. Clouds may also be present in the wing disc. Costal angulation of hind wing rather acute; cross veins few; one long intercalary.

Abdominal tergites 5, 6 and 9 of male largely creamy white, except for brown oblique submedian streaks on the anterior margins; remainder of dorsum dark red-brown, each tergite narrowly greyish white on the lateral and posterior margins. Greyish median markings on basal tergites; two creamy submedian streaks the length of tergite 10. On each segment, a dark brown mark above the pleural fold; median part of posterior margin of each tergite dark brown. Sternites paler, the basal ones greyish tinged with pale brown, the apical ones creamy; minutely and densely stippled with red-brown; a pair of curved red-brown submedian marks on each. Abdomen of female light tan flushed with pinkish; tergites 7 and 8 largely dark brown; other tergites marked with dark brown submedian triangular blotches on anterior margin, dark streaks above pleura, and a lateral oblique streak from the posterior margin; tergite 4 with more extensive submedian brown markings. Tergites stippled coarsely with red-brown in the paler areas; sternites paler than tergites, finely stippled as in the male. Forceps smoky brown. Tails dark olive brown, somewhat paler distally; joinings blackish brown; alternate joinings wider, at the base. Tails of female slightly paler.

This species is closely allied to *C. pictus* and to *C. signatus*. We are unable to separate it satisfactorily from *pictus*, as we have seen no specimens of that species;

the larger size of *C. pacificus* may be distinctive; however, it is possible that *pacificus* is synonymous with *pictus*. From *C. signatus* it may be distinguished by the dark cross veins in the male wing and the more extensive pale areas on the abdominal tergites of that sex.

The type material of this species, from Claremont, San Dimas Canyon and the San Jacinto Mts., Calif., is in the Cornell collection. In addition, we have many specimens of both sexes from Pasadena, the San Bernardino Mts. and the San Gabriel Mts., Calif. (C. D. Michener); from Jemez Springs, N. M. (Helen Murphy); Yakima, Wash. (M. D. Leonard); and from Norman, Okla. (R. D. Bird). We have also seen specimens from the collection of the Univ. of Arizona, taken at Tuscon, Ariz.; and from the collection of the Calif. Acad. of Science, taken at Redlands, San Diego and Woodacre, Calif.

Callibaetis pallidus Banks

1900—Banks—Trans. Amer. Ent. Soc. 26: 251..Colo.

Length: body (female) 9; wing 10 mm.

A pale ferruginous species; cross veins of fore wing numerous; marginal intercalaries paired; male undescribed (wings unpigmented?).

Thorax pale ferruginous. Femora pale, not freckled; tarsi blackish. "Costa marked with brown and white; vitta pale ferruginous, extreme basal costal space hyaline, vitta with a number of hyaline spots, especially near the pterostigmatic region, hind edge with several broad indentations, especially on basal half" (Banks). Cross veins of fore wing numerous, most of them white; marginal intercalaries occur in pairs, in the type specimens. Abdomen pale ferruginous. Tails pale, unmarked.

We have one female specimen in the Cornell collection from Ft. Collins, Colo. (C. P. Gillete), which may be of this species. A male specimen, from the same locality, has the wings wholly pale, and agrees with the above description as to numbers of cross veins and marginal intercalaries in pairs. We are considering this as *C. pallidus*. The species reported as *pallidus* from Logan River, Utah (Needham, 1927), does not belong here, as the cross veins are fewer in number and the marginal intercalaries single; it is probably an undescribed species. We have also several males, from Boulder and Denver, Colo., and Tuscon, Ariz., which are near *pallidus*, but not this species.

Callibaetis pictus Eaton

1871—Eaton—Trans. Ent. Soc. London: p. 122. Fig. (Baetis)..........................**Calif., Tex.**
1881—Eaton—Ent. Mo. Mag. 17: 196.
1885—Eaton—Revis. Monogr. p. 195.
1900—Banks—Trans. Amer. Ent. Soc. 26: 249 (*undatus*).

Length: body 5–6½; wing 5½–7 mm.

A dark brown species; cross veins of fore wing relatively few in number; marginal intercalaries paired; wing of male not pigmented.

Thoracic notum of male pitch-brown; in female, dark yellowish. A row of 6 to 8 or more round spots on each femur, and a few dots, which are sometimes (especially on the hind legs) confluent in part; tarsal joinings, claws, and the "last 1–3 joints" blackish or reddish brown. Wings unpigmented in male; cross veins relatively few in number; marginal intercalaries in pairs; stigmatic area slightly reddish brown in the male. Venation pitch-brown, the four main veins of the costal margin white in places, in both sexes; cross veins dark in the male, white in the female. Outer margin of female wing white here and there. No cross veins in the basal costal space, in the male;

6–9 very faint veins in this space, in the female; 6–8 beyond the bulla, in male; 11–16 in this area, in the female, mostly simple, with no granulations.

In the fore wing of the female, the vitta consists of a light umber-brown band in the costal and subcostal spaces and along the wing base, the brown more or less regularly alternating with hyaline white areas; just beyond the bulla, in the costal space, the brown is more extensive for a short distance. Faint and rather irregular clouds of light brown are present in the disc of the wing and along the outer margin from the edge of the vitta almost to the anal veins. Hind wing very faintly colored at extreme base; cross veins relatively few in number; usually one rather long intercalary and one or two shorter ones. Wing relatively short for its width.

Body greatly discolored. Abdominal tergites "closely punctulate," sternites even more closely freckled; sternites and some portions of the tergites "rosy whitish-brown, with the punctulations dark red-brown; the remainder of the dorsum pitch-brown." Tails white or reddish white sometimes banded with grey; joinings black in the female, greyish in the male.

The evidence seem to us to be insufficient to warrant placing *pictus* in synonymy with *C. undatus*, as Dr. Banks has done. We consider each to be a valid species; the latter, insufficiently described, is not included here, as it has not been reported north of Mexico. In the Cornell collection we have specimens that may be *pictus*, from Mexico City, Mex. (G. W. P. Lassmann). This species is so closely allied to *pacificus*, separable from it mainly by its smaller size, that the latter may prove to be a synonym of *pictus*. We have many specimens held under the name *pacificus*.

Callibaetis pretiosus Banks

1914—Banks—Proc. Acad. Nat. Sci. Philad. 66: 615..Va.

Length: body ?; wing (female) 6½ mm.

A brown species; cross veins of fore wing rather numerous; marginal intercalaries occur singly; male not known.

Body of female brown. Two narrow white stripes on the thoracic notum. Legs pale, tips of tibiae dark, also the tarsal joints. Wings hyaline; veins and cross veins mostly white, but the longitudinal veins with some dark marks. Costal vitta consists of a "fairly broad brown stripe from base to tip, its hind margin sinuate with four projections, four hyaline spots in the subcostal area before the middle, about five pale spots on costal margin before the middle, and several more or less connected beyond." Along the hind margin are five or six faint irregular clouds. Hind wing much as in *C. floridanus*. Abdomen brown, with darker spots on the sides.

We have no specimens that seem to be of this species.

Callibaetis semicostatus Banks

1914—Proc. Acad. Nat. Sci. Philad. 66: 614. Fig..Man.

Length: body ?; wing 8 mm.

A brownish species; cross veins of fore wing moderately numerous; marginal intercalaries paired; wing of male pigmented.

Body of male brownish, with fine dots on many parts, particularly the venter. Some of the thoracic sutures are whitish. Legs pale, darker at tips of tarsi; femora without freckles. Wings hyaline. Longitudinal veins brownish, in the type specimens, although the original description states, "veins pale." Wing of male pigmented, the vitta bounded in front by the subcosta and behind "by the fourth vein"; it is light brown or reddish brown in color, and extends halfway from the base to the tip of the wing. Two or three pale dots are present in this vitta; sometimes there are also a few dark marks in the costal area. One of us (Needham) studied the type specimens, adding these notes; "Subcostal pale brown strip shows fenestra faintly near the base; vitta overspreads the stigmatic area more or less; stigmatic cross veins more or less irregular, very oblique." Hind wings fully twice as long as broad, angulate at the base. Abdominal sternites finely dotted. Tails white.

Only the male of this species is known; it is evidently allied to *C. skokianus* and *C. pallidus*. The smaller size and differences in vitta of male distinguish it from the former species. Its relation to *pallidus* cannot at present be determined; there is a possibility that these two species are in reality but male and female of a single species. From Estes Park, Colo., we have two males in the Cornell collection which seem to be of this species.

Callibaetis skokianus Needham

1903—Needham—N. Y. St. Mus. Bull. 68: 215..Ill.
1905—Needham—N. Y. St. Mus. Bull. 86: 40.
1930—Ide—Canad. Ent. 62: 227..Ont.

Length: body 9–10; wing 9–10 mm.

A large pale species; cross veins of fore wing numerous; marginal intercalaries paired; wing of male pigmented.

Turbinate eyes of male large, egg-yellow; head pale brownish, margins whitish. Vertex of female with two pale brown longitudinal streaks. Prothorax pale brown; median sutures of meso- and metanota pale, the former with a brown band on each side; brown spots on the sides inferiorly. Pleura irregularly speckled with brown. Legs yellowish white; femora dark at the apices, and with few or no freckles; tarsal joints dark apically, the distal one wholly brown. Costal band on fore wing of female reddish brown, covering the bases of all the veins proximally; narrowed regularly toward the apex, where it ends just before the wing tip; no posterior lobes; most of the cross veins fenestrate with hyaline except at the base; basal costal space interrupted brown and white. In the stigmatic area the band fades into a yellowish wash. In the male, this band is paler and usually ends just anterior to the yellowish stigmatic area. Cross veins very numerous, especially in the female; in this sex, about 70 are present behind costal vitta. Marginal intercalaries paired. Cross veins pale, longitudinal ones mainly brown. Stigmatic cross veins numerous, often anastomosed. Hind wing broadly rounded at tip, all veins brownish; usually a pair of well-developed marginal intercalaries.

Abdomen pale yellowish or flesh colored, thickly dotted with brown, tending to form a dark submedian stripe on each side of a paler mid-dorsal line. Above the spiracular line on each segment is a brown dash; a curved dark mark is present above the anterior end of each dash. Sternites paler than tergites, the dots more evenly distributed; a pair of brown submedian curved marks on each, those of 9 abbreviated. Forceps and tails white.

The type material of this species is in the Cornell collection. We have other specimens from Devil's Lake and Fargo, N. D., the latter collected by Mr. L. A. Carruth; one from Columbia, Mo.; also many from Ithaca, N. Y., which are slightly darker than the type specimens. We have seen other specimens, in the collection of the Univ. of Minnesota, collected at St. Paul, St. Peter and Lesueur Co., Minn.

Callibaetis signatus Banks

1918—Banks—Bull. M. C. Z. 62: 11..Ariz.

Length: body ?; wing 7–7½ mm.

A brownish species; cross veins of fore wing relatively few in number; marginal intercalaries paired; wing of male unpigmented.

Male dark brown, female pale brownish; scutellum of female pale in the middle. Legs pale, tarsi and tips of tibiae rather darker. Wings hyaline; longitudinal veins brown, cross veins hyaline, in both sexes. In wing of female, "various pale brown spots along the costal region," but these do not form a continuous streak; "several costal dots, a basal streak, one just before the middle of wing on subcosta and radius, and three or four beyond; several other brown spots." Wing of male unpigmented. In the type specimens, the basal costal space (female wing) is wholly pale; intercalaries of the fore wing are paired; cross veins relatively few in number. Hind wing shaped much as in *C. pacificus*. Abdomen of male nearly black dorsally, except for two large pale spots, "one over the penultimate segment, and the other on the third segment before the last"; venter finely dotted. Entire abdomen of female pale brownish, dotted. Tails pale, darker at the joinings.

This species is closely allied to both *C. pacificus* and *C. pictus;* in the former species, however, the pale spots on the abdomen of the male occur on tergites 5, 6 and 9 (we interpret the description of *C. signatus* as meaning that the pale spots are on tergites 7 and 9, although it may mean, on 6 and 9); and the cross veins in the disc, at least, of the fore wing of both sexes are brown. The wing of the female of *C. pictus* has brown clouds on the hind margin; the cross veins are pale in the female, dark in the male.

We have seen one male specimen from Sabinal, N. M. (C. W. Sabrosky), which seems to be of this species; the specimen is in the collection of State College, Kans. A female specimen in the Cornell collection, from Redlands, Calif. (Velma Knox), may also be of this species.

Genus **Baetis** Leach

1815—Leach—Brewst. Edinb. Encycl. 9: 137.
1840—Westwood—Introd. Mod. Classif. Ins. 2: 25 (Brachyphlebia).
1843—Pictet—Hist. Nat. 2. Ephem. Neurop. p. 239 (Cloe, in part).
1885—Eaton—Revis. Monogr. p. 156.
1920—Ulmer—Stett. Ent. Zeit. 81: 124.
1925—McDunnough—Trans. R. S. C. 19: 214.
1932—McDunnough—Canad. Ent. 64: 79.
1933—Spieth—J. N. Y. Ent. Soc. 41: 337.
1912—Bengtsson—Ent. Tidskr., p. 110 (Acentrella =Baetis, in part).

Small mayflies, the fore wing ranging in size (in Nearctic species) from 2½ to 8 mm.; most species range from 4 to 6 mm; wings of female slightly longer than in the male. Females more or less unicolorous, usually reddish or brownish, the head and thorax usually paler than in the male. Abdominal segments 2–6 of male imago usually hyaline or semihyaline; in some species, wholly or largely pale, in others olivaceous to deep brown. Turbinate eyes of male vary in size in different species. These sizes we designate (after McDunnough) as small, moderate, and large. The length of the stalk likewise varies, in different species. Eyes of female not divided; small, remote. Hind margin of head of female emarginate at the median line; on each side of this emargination is a slight outward prolongation.

Prothorax relatively small. Near the middle of the metanotum is a short protuberance. Fore leg of male about as long as the body. Fore tarsus as long as or slightly longer than the fore tibia. Basal joint very short; second usually slightly longer than the third; fourth and fifth progressively shorter. Fore tibia about one and one-third times as long as the femur. Basal joint of hind tarsus relatively short,—not longer than the second and third joints together. Hind femur and tibia subequal; tarsus shorter than tibia.

Marginal intercalaries of fore wing occur in pairs. Basal costal cross veins entirely wanting. Hind wing present; variable in size, usually rather well developed. In most (not all) species, a costal projection (acute or slightly curved) is present near the base of the wing. Longitudinal veins 2 or 3 in number; the 3rd, when present, usually much shorter than the others. In a few species only, there are more or less evident indications of cross veins between these.

Forceps four-jointed, the 2nd and 3rd joints more or less completely fused. We recognize two distinct types of genitalia on the basis of differences in the forceps joints, also various modifications of these two main types. In the *intercalaris* type (see fig. 164), the basal forceps joint usually bears a tubercle or projection on its inner apical margin; second joint relatively long and slender, conical, tapering apically; third joint quite long and slender, usually narrowest at its base. In the *moffati* type (see fig. 164), there is no true tubercle on the inner apical margin of the basal joint, although the inner margin may be prolonged inward; second joint usually shorter than above, distinctly cylindrical, not tapering apically; third joint rather less slender and not noticeably narrower at the base. In each type, the distal joint may be short and relatively stout, or rather long and slender, a condition which varies in different species.

The so-called 'modified *moffati*' type occurs in some of the species formerly placed in Acentrella, in which the costal projection of the hind wing is wholly wanting. In these, the inner margin of the second joint bulges inward quite sharply near the middle; the inner margin of the basal joint is distinctly prolonged inward; and a distinct 'penis-cover' is present. *B. spinosus*, in which the inner apical margin of the second joint bears a distinct inwardly-directed spine, may be considered a variant of the *moffati* type, also; while *B. frondalis* and *B. australis*, each with an excavation and spine situated between the bases of the forceps, represent another variation. A variation of the *intercalaris* type is seen in the *quilleri* group (*B. quilleri*, *B. endymion*, *B. erebus*), in which the penis-cover consists of two projecting rounded lobes, and the inner margin of the second joint is slightly tuberculate. In the female, there is little or no prolongation of the apical margin of the ninth sternite to form a subanal plate.

The nymphs have slender, stream-lined bodies; heads directed downwards; rather long and slender legs; claws with numerous fine denticles on the inner margins. A narrow but distinct median notch on the apical margin of the labrum; mandibles roughly triangular, the canines large with blunt teeth; maxillary palp two or three-jointed, usually not extending beyond the tip of the galea-lacinia; labial palp three-jointed, the distal joint distinctly rounded, short, never expanded truncate on the apical margin. Gills present on segments 1–7; single, on all segments; pinnately branched, usually obovate, but somewhat variable in shape in different species. In some species, the gills on segments 6 and 7 are narrow and lanceolate; in others, these gills are rounded and oval. In a few species (*B. bicaudatus*, and the *propinquus* group) there are but two tails. In all other known species, three tails are present, but the middle one is distinctly shorter and weaker than the two lateral ones.

Those species in which the costal projection of the hind wing is wholly wanting, we had previously placed in Acentrella. Since, however, there is but a slight distinction between wings of this type and those like *spinosus*, in which the costal projection is greatly reduced; and since, further the type of genitalia in this species is allied to the *moffati* type which is found in many species of Baetis, we do not now consider that these differences are worthy of generic rank.

Genotype—*B. bioculatus* (in Ephemera), Linnaeus.

Distribution—Holarctic, Neotropical, Indo-Australian; in Nearctic, generally distributed.

KEY TO THE SPECIES

Male adults*

1—Hind wing with three longitudinal veins of approximately equal length...... .... **thermophilos**

—Hind wing with two or three longitudinal veins, of which the first and second are approximately equal in length; the third, if present, is much shorter, ending near the middle of the hind margin.......... 2

2—Hind wing with but two longitudinal veins, the third wholly wanting.......... 3

—Third vein present in hind wing, although it may be very weakly developed... 14

3—Costal projection of hind wing present, although it may be poorly developed... 4

—Costal projection of hind wing entirely wanting.......... 9

4—Costal projection prominent, acute.......... 5

—Costal projection small, blunt, not prominent.......... 7

5—Costal projection not curved; fore wing about 6 mm. in length..... **bicaudatus**

—Costal projection distinctly curved; fore wing 3 mm. in length.......... 6

6—Abdominal tergites 2–6 dull brown; eyes small, dark reddish brown. . **akataleptos**

—Abdominal tergites 2–6 whitish hyaline; eyes large, blackish brown....... **harti**

7—A well-developed pointed projection on inner apical margin of second forceps joint.......... **spinosus**

—No such projection on inner margin of second forceps joint.......... 8

8—Abdominal tergites 2–6 largely shaded with purplish brown; excavation and spine between bases of forceps; fore wing 4½ mm. in length........ **australis**

—Abdominal tergites 2–6 wholly hyaline whitish, except for black stigmatic dots; no such excavation and spine between forceps bases; fore wing 3 mm. in length .... **pygmaeus**

9—Fore wing 7 mm. in length; venation yellowish brown; fore femur with reddish apical band.......... **amplus**

—Fore wing 6 mm. or less in length; venation pale; fore femur wholly pale, or banded apically with black.......... 10

* Omitting *unicolor, posticatus, minimus*.

10—Abdominal tergites 2–6 deep brown; wing 6 mm. in length............**frivolus**

—Abdominal tergites 2–6 whitish or yellowish, hyaline; may be tinged with smoky brown or marked with purplish red; wing not over 5½ mm..............11

11—Abdominal tergites 2 and 5, often other basal tergites as well, marked prominently with purplish red..................................**ephippiatus**

—No such purplish red markings on tergites..............................12

12—Abdominal tergites 2–6 tinged with smoky brown; all femora likewise tinged with smoky brown..**insignificans**

—Abdominal tergites 2–6 wholly whitish hyaline, without smoky tinges; fore femora may be smoky or pale reddish brown..........................13

13—Fore femur pale yellowish brown; a faint ruddy shading over spiracular area...**dardanus**

—Fore femur often pale ferruginous; a black dot at each spiracle......**propinquus**

14—Second longitudinal vein of hind wing forked about one-third the distance from base to margin; in this fork, one or more intercalaries..................15

—Second longitudinal vein of hind wing not forked.........................16

15—Distinct tubercle present on inner apical margin of basal forceps joint; blackish spiracular line present....................................**devinctus**

—Tubercle absent, or if present, very poorly developed; no dark spiracular line...**parvus**

16—Costal projection of hind wing prominent, distinctly curved; wing usually long and narrow, reminiscent of wing of Centroptilum (often 3 to 3½ times as long as wide); usually traces of cross veins..............................17

—Costal projection of hind wing variable, but not curved as above; wing relatively shorter and wider (usually not more than 2½ times as long as wide); usually no traces of cross veins..19

17—Abdominal tergites 2–6 dark reddish brown; venation light red-brown; tails dark red-brown basally, not darker at joinings.....................**erebus**

—Abdominal tergites 2–6 yellowish to dark olive brown; venation pale; tails pale, distinctly dark-ringed at basal joinings..............................18

18—Abdominal tergites 2–6 deep olive brown, sternites pale yellowish olive; hind wing very long and narrow...................................**quilleri**

—Abdominal tergites 2–6 yellowish to pale olive brown, sternites little if any paler; hind wing relatively shorter and wider........................**endymion**

19—Genitalia of the *moffati* type (2nd forceps joint cylindric; no tubercle at inner apical margin of basal forceps joint; see fig. 164)......................20

—Genitalia of the *intercalaris* type (2nd forceps joint conical, smaller toward apex; usually a tubercle is present at inner apical margin of basal forceps joint; see fig. 164)...29

20—Abdominal tergites 2–6 brownish (olivaceous to deep brown)..............21

—Abdominal tergites 2–6 hyaline whitish or yellowish.......................28

21—Between the bases of the forceps is a deep excavation, in which is situated a small sharp spine; third vein of hind wing short and faint, costal projection poorly developed..**frondalis**

—No such excavation or spine between the bases of the forceps; costal projection better developed; 3rd vein variable......................................22

22—Usually two well-developed intercalaries between veins 2 and 3 of hind wing..23

—Not more than one intercalary, sometimes none, between veins 2 and 3 of hind wing..25

23—Fore wing 5–6 mm. in length; traces of one or two intercalaries between veins 1

and 2 of hind wing; veins 1 and 2 of hind wing not distinctly convergent apically....24
—Fore wing 7 mm. in length; no intercalaries between veins 1 and 2 of hind wing; in hind wing, veins 1 and 2 are distinctly convergent apically.... **tricaudatus**
24—Wing 5 mm. in length; turbinate eyes moderate.... **incertans**
—Wing 6 mm. in length; turbinate eyes very large.... **vagans**
25—Fore wing about 5 mm. in length.... **adonis**
—Fore wing 6½ to 8 mm. in length....26
26—No intercalaries between veins 2 and 3 of hind wing; vein 3 very faint.. **moffati**
—One intercalary between veins 2 and 3 of hind wing; vein 3 variable....27
27—Vein 3 of hind wing faint; both margins of hind wing rounded, not parallel to one another.... **intermedius**
—Vein 3 of hind wing better developed; upper and lower margins of hind wing almost parallel; tails darker at joinings.... **parallelus**
28—Abdominal tergites 2–6 each with a prominent purplish red band on the posterior margin.... **piscatoris**
—Abdominal tergites 2–6 wholly pale except for dark stigmatic dots; no such dark bands on posterior margins.... **lasallei**
29—Tubercle on inner apical margin of basal joint of forceps wanting, or very poorly developed....30
—Tubercle on inner apical margin of basal forceps joint present, well developed..31
30—Abdominal tergites 2–6 yellowish white, shaded laterally and posteriorly with pale brown.... **levitans**
—Abdominal tergites 2–6 deeply tinged with maroon brown, except for pale lateral and anterior areas.... **pluto**
31—Abdominal tergites 2–6 brownish (olivaceous to deep brown)....32
—Abdominal tergites 2–6 whitish or yellowish white....33
32—Abdominal tergites 2–6 deep brown; fore wing 6 mm. in length; two intercalaries between veins 2 and 3 of hind wing; turbinate eyes large.... **brunneicolor**
—Abdominal tergites 2–6 olive brown; fore wing 4 mm. in length; no intercalaries in hind wing; turbinate eyes moderate.... **rusticans**
33—Turbinate eyes small....34
—Turbinate eyes large....35
34—Fore wing 5 mm. in length; eyes in living insect light brown; abdominal tergites 2–6 tinged with yellowish brown.... **flavistriga**
—Fore wing 4 mm. in length; eyes in living insect pale yellowish brown; abdominal tergites 2–5 yellowish, seldom with faint brownish tinge.... **nanus**
35—Abdominal tergites 2–6 ruddy brown on posterior margins, this color sometimes extending over the posterior half of each tergite.... **cingulatus**
—No such ruddy brown shading posteriorly on tergites 2–6....36
36—Intercalaries in first interspace of fore wing considerably longer than any of the other pairs; thorax shiny blackish.... **intercalaris**
—Intercalaries in first interspace of fore wing not longer than any of the other pairs; thorax yellowish or reddish brown....37
37—Intercalaries in second interspace of fore wing shorter than the 1st or 3rd pairs; abdominal tergites 2–6 not tinged with olivaceous; eyes light orange in living insect.... **pallidulus**
—Intercalaries in second interspace of fore wing not shorter than the first or third pairs; abdominal tergites 2–6 with faint olivaceous tinge; eyes red-brown in living insect.... **phoebus**

VERIFICATION TABLE

Adults*

Species	Wing length	Turbinate eye	Veins in hd. wing	Hd. w. costal projection	Abd. tergites 2 to 6	Genitalia, type of	Distr.
adonis	5	Large	3	Acute	Yel. brown	Moffati	Calif.
akataleptos	3	Small	2	Curved?	Brown	Intercal.	Alta.
amplus	7	Large	2	Absent	Yel. brown	Mod. mof.	N. C.
australis	4½	Mod.	2	Weak	Purp. br.	Moffati	N. C.
bicaudatus ♀	6½	?	2	Acute		?	W
brunneicolor	6	Large	3	Acute	Dp. brown	Intercal.	E
cingulatus	5½	Large	3	Acute	Yel. white	Intercal	E
dardanus	5½	Mod.	2	Absent	Whitish	Mod. mof.	Man.
devinctus	5	Mod.	3	Acute	Whitish	Intercal.	Calif.
endymion	5–5½	Mod.	3	Curved	Yel. brown	Intercal.	Okla.
ephippiatus	4–4½	Large	2	Absent	Yel. white	Mod. mof.	Ga.
erebus	5	Large	3	Curved	Red-brown	Intercal.	Ariz.
flavistriga	5	Small	3	Acute	Yel. brown	Intercal.	E
frivolus	6	Very large	2	Absent	Dp. brown	Mod. mof.	E
frondalis	5–6	Large	3	Weak	Dp. brown	Moffati	E
harti	3	Large	2	Curved	Whitish	Intercal.	Ill.
incertans	5	Mod.	3	Acute	Olive br.	Moffati	E
insignificans	4	Small	2	Absent	Pale	Mod. mof.	B. C.
intercalaris	5	Large	3	Acute	Whitish	Intercal.	E
intermedius	6½	?	3	Acute	Lght. br.	Moffati	Colo.
lasallei	5	?	3	Acute	Whitish	Moffati	Wis.
levitans	4½	Mod.	3	Acute	Yel. white	Intercal.	E
minimus ♀	7½	?	3	Acute		?	Colo.
moffati	7½	Mod.	3	Acute	Yel. brown	Moffati	W
nanus	4	Small	3	Acute	Yellowish	Intercal.	E
pallidulus	5	Large	3	Acute	Yellowish	Intercal.	Ill.
parallelus	8	?	3	Acute	Brown	Moffati	Nev.
parvus	5½	Mod.	3	Acute	White	Intercal	W, E ?
phoebus	5	Large	3	Acute	Whitish	Intercal.	E
piscatoris	6	Mod.	3	Acute	Yel. white	Moffati	Calif.
pluto	4½	Mod.	3	Acute	Brown	Intercal.	E
posticatus	8	?	?	?	Whitish	?	Ind.
propinquus	4–5	Large	2	Absent	Whitish	Mod. mof.	Ill.
pygmaeus	2½–3	Mod.	2	Acute	White	Intercal.	E
quilleri	5	Large	3	Curved	Olive br.	Intercal.	W
rusticans	4	Mod.	3	Acute	Olive br.	Intercal.	E
spinosus	5½	Large	2	Weak	Yel. white	Spinosus	W, E
thermophilos	5	?	3	Weak	Olive br.	?	W
tricaudatus	7	?	3	Acute	?	Moffati	Colo.
unicolor ♀	4½	?	?	?		?	D. C.
vagans	6–7	Very large	3	Acute	Olive br.	Moffati	E

* ♂s, unless otherwise indicated.

KEY TO THE KNOWN SPECIES*

Nymphs

1—Tails two 2
—Tails three; middle one weaker and usually considerably shorter than the outer ones 3
2—Tergites 1–9 dark red-brown; 10 yellow .... **amplus**
—Tergites 1–10 pale with dark markings; 3, 4 and 7 distinctly darker than others .... **bicaudatus**
3—Gills of segments 6 and 7 narrow, lanceolate .... **pygmaeus**
—Gills of segments 6 and 7 oval, rounded 4
4—Tails with a dark band beyond the middle .... **cingulatus**
—Tails with no such dark band 5
5—Middle tail about 1/6 the length of the outer ones .... **intermedius**
—Middle tail 1/2 or more of the length of the outer ones 6
6—Middle tail 3/4 to 5/6 the length of the outer ones .... **parvus, brunneicolor**
—Middle tails 1/2 to 2/5 the length of the outer ones 7
7—Tracheae of gills very indistinct .... **vagans**
—Tracheae of gills distinct .... **tricaudatus**

Baetis adonis sp. nov.

Length: body 5; wing 5 mm.

Abdominal tergites 2–6 of male imago light yellowish brown; genitalia of the *moffati* type.

Head and antennae light reddish brown. Turbinate eyes rather large, stalk of moderate height; stalk and upper surface orange in alcoholic specimens. Thorax reddish brown. A spot anterior to the scutellum, an area on each side of it, a line along the antero-lateral margin of the mesonotum, and the anterior portion of the metanotum, yellowish. Pleura rather extensively shaded with yellowish or olive brown. Margins of the scutellum, an area on mesonotum anterior to the yellowish portions, posterior of metanotum, and most of sternum, darker red-brown. Fore leg yellowish brown; often a darker area near apex of tibia. All coxae reddish brown. Middle and hind legs yellowish, tarsi and tarsal joinings reddish brown. Wings hyaline, venation yellowish brown. 6–7 cross veins in stigmatic area of fore wing; somewhat aslant, often with short horizontal veins between; granulations present. Marginal intercalaries quite short, usually none in the first interspace, often but one in the second. Hind wing with three veins, the third quite short, running close to hind margin and ending in the basal third of that margin. Tip of wing broadly rounded. Faint traces of an intercalary in some, between veins 2 and 3, and occasionally one between 1 and 2. Costal projection well developed.

Abdominal tergites 2–6 semi-hyaline, yellowish to olive brown, sometimes shaded with orange; posterior margins very narrowly pale. Sternites semi-hyaline, pale yellowish. Spiracular area unmarked. Segments 7–10 opaque, creamy yellowish, tergites shaded with reddish brown; posterior and lateral margins narrowly darker, especially on tergite 10. (In two of the paratypes, these segments are reddish brown without creamy shading). Tails yellowish white. Forceps yellowish or pale reddish brown. Genitalia of the *moffati* type; apical joint of forceps short.

Head of female yellowish; thorax and dorsum of abdomen yellowish to pale reddish brown, sternites somewhat paler. Legs yellowish, tarsi as in male. Tails yellowish,

* Adapted from McDunnough, Canad. Ent. 64: 79, 1932.

joinings opaque, appearing slightly darker. (One paratype wholly reddish brown, thorax with creamy markings).

Holotype—Male imago. San Gabriel Mts., Calif.; June 3, 1932. Chas. D. Michener, collector. No. 1353.1 in Cornell University collection.

Allotype—Female imago. Same data. No. 1353.2 in Cornell University collection.

Paratypes—9 male imagos; 1 female imago, same data, 1 female taken May 24, 1932, same location. No. 1353.3–13 in Cornell University collection.

In addition to the types, several specimens of both sexes which are apparently of this species were taken by the same collector on July 23, 1932, same location.

This species seems close to *intermedius*, but is considerably smaller; also, nymphs taken at the same location on the same date do not accord with Dr. Dodds' account of the nymph of that species. From *minimus* it is separable by reason of the shorter wing in relation to body length.

Baetis akataleptos McDunnough

1926—McDunnough—Canad. Ent. 58: 188. . . . . Alta.

Length: body 2½; wing 3 mm.

Abdominal tergites 2–6 of male dull brown; genitalia of the *intercalaris* type; hind wing two-veined, costal projection prominent.

Turbinate eyes of male imago small, oval in shape, and dark reddish brown. Thorax blackish brown. Legs pale smoky, the tarsi whitish. Wings hyaline, venation pale; cross veins in the stigmatic area of the fore wing few in number, not well developed, and without granulations. No intercalaries in the first interspace. Hind wing long and narrow; only two veins, the third one entirely lacking; costal projection prominent.

Abdominal tergites 2–6 semitranslucent, dull brown, sternites similar; posterior segments opaque, brighter brown. Genitalia very similar to *B. pygmaeus*, but with a finely setose projecting penis-cover, which is not present in that species, between the bases of the forceps. Related also to *B. harti*. Distinguishable from these species by reason of the brown basal segments of the abdomen.

We have seen no specimens of this species.

Baetis amplus Traver

1932—Traver—J. Elisha Mitchell Sci. Soc. 47: 233. Figs. (Acentrella). . . . . N. C.

Length: body 7; wing 7 mm.

Abdominal tergites 2–6 of male imago yellowish brown, the posterior margins hyaline whitish; hind wing two-veined, costal projection wanting.

Turbinate eyes large, almost circular; orange in color. A black mark across the face. Antennae light brown. Thorax reddish brown. Longitudinal dark brown streaks on the mesonotum, median and submedian in position. Metanotum distinctly reddish. Sternites light reddish brown. A red spot on the coxa of the fore leg; fore femur distinctly tinged with red, a band of this color near its apex. Fore tibia yellowish, tarsus yellowish white. Middle and hind legs yellowish, the femora light yellowish brown; tarsal joinings may be very narrowly reddish. Wings hyaline, venation yellowish brown. Cross veins of the stigmatic area of the fore wing 7–8 in number; only a slight tendency toward anastomosis; almost no granulations between them. Intercalaries absent from the first interspace, well developed in other spaces. Hind wing very small (1 mm. in length), long and narrow; two veins only, the third entirely absent. Costal projection entirely wanting.

Abdominal segments 2–6 (not 1–5, as stated in original description) yellowish brown, tinged with grey. The posterior margins rather widely opaque, giving this part of the abdomen an annulate appearance. Segments 7–10 light reddish brown, rather darker dorsally. Basal sternites slightly paler than their tergites. Genitalia of the modified *moffati* type, the second joint of the forceps being cylindrical. The distal forceps joint is relatively very long, fully three times as long as wide; it is almost one-half the length of the preceding joint.

The type material is in the Cornell collection. This species is very close to *B. frivolus*, but may be distinguished from it by the larger size, the color of the eyes and thorax, the paler abdomen and the distinctly brownish venation.

Baetis australis Traver

1932—Traver—J. Elisha Mitchell Sci. Soc. 47: 230. Figs. . . . . N. C.

Length: body 5; wing 4½–5 mm.

Abdominal tergites 2–6 of male imago purplish brown in lateral and posterior portions, the median and antero-lateral areas pale, semi-hyaline; genitalia of the *moffati* type; an excavation between the bases of the forceps, in which is a small spine.

Turbinate eyes large, oval, orange in color. Thorax deep ruddy brown. Intersegmental areas of the pleura yellowish. Anterior portion of mesosternum lighter yellowish brown. Legs pale yellowish white. Wings hyaline, venation pale. 6–8 stigmatic cross veins in the fore wing, not anastomosing; only the two most apical ones of these reach the subcosta. Stigmatic area somewhat opaque. No intercalaries in the first interspace. Hind wing rather long and narrow; the third vein wanting or obsolescent, mere traces of it sometimes seen at the extreme base. A very short intercalary may be present between veins 1 and 2. Costal projection much reduced.

Tergites 2–6 largely pale, semi-hyaline, in the dorsal area, with faint submedian brownish streaks; antero-lateral margins also pale. Remainder of each tergite purplish brown. Tergites 7–10 opaque, reddish brown, with a purplish tinge. Sternites greyish white, semi-translucent. A continuous double blackish line outlines the tracheae on each side; from it arise branches which extend on to the tergites and sternites. In the holotype, the upper of these is the longer. Between the bases of the forceps is an excavation, at the base of which is a short spine, much as in *frondalis;* a penis-cover also is present, over this excavation (see fig. 164).

While close to *B. frondalis*, this species is slightly more slender, and the basal abdominal tergites exhibit more pale areas. The excavation between the bases of the forceps is also less deeply carved out. The types are in the Cornell collection. We have also another male specimen from Mossy Creek, near Fort Valley, Ga. (P. W. Fattig). In the original description, it was stated that no spine was present in the excavation between the forceps bases. This was erroneous, as more careful study of the genitalia has shown that the spine was present, but displaced. It is very plainly present in the specimen from Georgia.

Baetis bicaudatus Dodds

1923—Dodds—Trans. Amer. Ent. Soc. 49: 110. Figs. . . . . Colo.

Length (female); body 4½; wing 6½ mm.

Only the female imago of this species is known. Body uniform light brown. Hind wing with two veins only; the third entirely wanting, according to the original description. Costal projection present, quite prominent.

The species has been reared, and a careful description of the nymph is given by

Dr. Dodds. It differs from all known nymphs except those of the *propinquus—frivolus* group, in having the middle tail represented only by a minute stub. Until males are reared, or correctly associated imagos of both sexes are collected from the type locality, the affinities of this species to others of the genus will not be known.

In the Cornell collection are several duplicate specimens of Dr. Dodds' Colorado species; among these is a vial of female imagos, which in color and size correspond well with the above description, also in the shape and costal projection of the hind wing, as figured. But in all of these, there is a faintly indicated third vein in the hind wing. We have not seen the type material of this species. A single female subimago from Cloverdale, Calif. (J. C. Bradley), seems very close to *B. bicaudatus*.

Baetis brunneicolor McDunnough

1925—McDunnough—Canad. Ent. 57: 173. Figs. .... Ont.
1930—Ide—Canad. Ent. 62: 221.
1932—McDunnough—Canad. Ent. 64: 79. Key to nymphs.

Length: body 6; wing 6 mm.

Abdominal tergites of male imago deep brown; genitalia of the *intercalaris* type, the tubercle on the inner margin of the first forceps joint well developed.

Turbinate eyes large, deep brown in color; the stalk rather short, and shaded with yellow. Head and thorax deep blackish brown. The lateral anterior margin of the mesonotum, the pleural suture and the lateral extensions of the mesosternum shaded with paler brown. Rear portion of the mesothorax with slight reddish brown markings. The median dorsal projection of the metathorax is cream-colored. Legs pale yellowish brown, the fore legs deeper in color, with smoky shading at the apex of the tibiae. Wings hyaline, venation pale. Strong granulations between the cross veins of the stigmatic area of the fore wing. Intercalaries of the first interspace much longer than those in the second space; all intercalaries well developed. Hind wing large and broad, the third vein well developed. Between the second and third veins are two marginal intercalaries; often another intercalary is present between veins 1 and 2.

Abdominal tergites 2–6 deep brown with faint ruddy tinges, and obsolescent pale submedian dashes anteriorly. Tergites 7–10 similar to the basal ones; sternites pale yellowish brown. Forceps of the *intercalaris* type, the inner tubercle of the first joint well developed.

We have seen no specimens of this species.

Baetis cingulatus McDunnough

1925—McDunnough—Trans. R. S. C. p. 216. Figs. .... Que.
1930—McDunnough—Canad. Ent. 62: 58.
1932—McDunnough—Canad. Ent. 64: 79. Key to nymphs.

Length: body 5; wing 5½ mm.

Abdominal tergites 2–6 of male imago pale yellowish white, narrowly ruddy on the posterior margins; genitalia of the *intercalaris* type.

Turbinate eyes large, deep brown in color. Face and antennae black-brown. Thorax deep olive brown. Lateral margin of mesonotum anterior to the wing roots frequently yellowish; scutellum whitish, often with ruddy tinges anterior to it; ruddy tinges on the pleura, the sutures pale. Anterior margin and projecting median portion of metanotum whitish. Legs pale yellowish; fore femora mostly pale brownish, as are the tips of the middle and hind ones. Wings hyaline, venation pale. First pair of intercalaries of the fore wing well developed, but not as long as in *B. intercalaris*, second pair shorter than the first or third pair. Hind wing slightly more than twice as long as its width. Third vein well developed, reaching to the middle of the hind margin. Intercalaries absent, or with a trace of one between the second and third vein.

Abdominal tergites 2–6 pale yellowish white, semi-translucent; the posterior margins narrowly ringed with reddish brown, which color may extend forward laterally over the posterior portion of each tergite. Segments 7–10 opaque, light reddish brown dorsally, the sternites whitish tinged with brown. Sternites 2–6 pale whitish. Tracheae blackish in the spiracular area, forming a more or less continuous wavy dark line. Tails and forceps white. Genitalia with a well-developed tubercle on the inner apical margin of the first joint.

The dark bands on the posterior margins of the basal abdominal tergites separate this species from the allied *B. phoebus*. In the Cornell collection are specimens that seem to be of this species, from Hopkinton and Nassau, N. Y. (L. Harper).

Baetis dardanus McDunnough

1923—McDunnough—Canad. Ent. 55: 41 .... Man.
1925—McDunnough—Canad. Ent. 57:174 (in discussion of *frivolus*), and 175 (in discussion of *spinosus*); 172 (in discussion of *propinquus*).
1926—McDunnough—Canad. Ent. 58: 300 (in discussion of *insignificans*).

Length: body 5; wing 5½ mm.

Abdominal tergites 2–6 of male imago hyaline whitish; genitalia of the modified *moffati* type; costal projection of hind wing entirely wanting; hind wing two-veined.

Turbinate eyes are slightly larger than in *B. pygmaeus*. Thorax dark shiny blackish brown, the lateral sutures marked with whitish. Legs pale yellowish. Wings hyaline. No costal projection near the base of the hind wing. Only two veins present in the hind wing. Abdominal segments 2–6 hyaline whitish; tergites 7–10 brownish olive, the sternites opaque white with brown shading. Faint ruddy shading along the pleural fold. Forceps whitish.

This species, according to Dr. McDunnough (Canad. Ent. 57: 172), is very close to *B. propinquus*, and may even be synonymous with it. We have seen no specimens of *B. dardanus*. It is related also to *B. insignificans*, from which it may be separated by the smoky brown shading on tergites 2–6 of that species. Its other allies, *B. frivolus* and *B. amplus*, are considerably larger; the basal tergites of the former are deep brown, of the latter yellowish brown.

Baetis devinctus sp. nov.

Length: body 5; wing 5 mm.

Abdominal tergites 2–6 of male imago hyaline whitish; genitalia of the *intercalaris* type, tubercle present on inner margin of basal forceps joint. Second vein of hind wing forked near base.

Head and antennae light red-brown. Turbinate eyes moderate in size; stalk yellowish, upper surface pale orange in alcoholic specimens. Thoracic notum rather dark reddish brown; margins of scutellum blackish. Pleura and sternum somewhat paler red-brown; intersegmental areas of pleura yellowish. Legs whitish; fore femora yellowish, other femora faintly yellow-tinged; coxae reddish brown. Wings hyaline, venation pale. A faint cloud in the stigmatic area of the fore wing; cross veins 6–8 in number; aslant, no apparent tendency to fork, sometimes incomplete toward the subcosta; between them, many granulations. Marginal intercalaries much as in *B. parvus*. Hind wing very similar to that of *parvus* (see fig. 163).

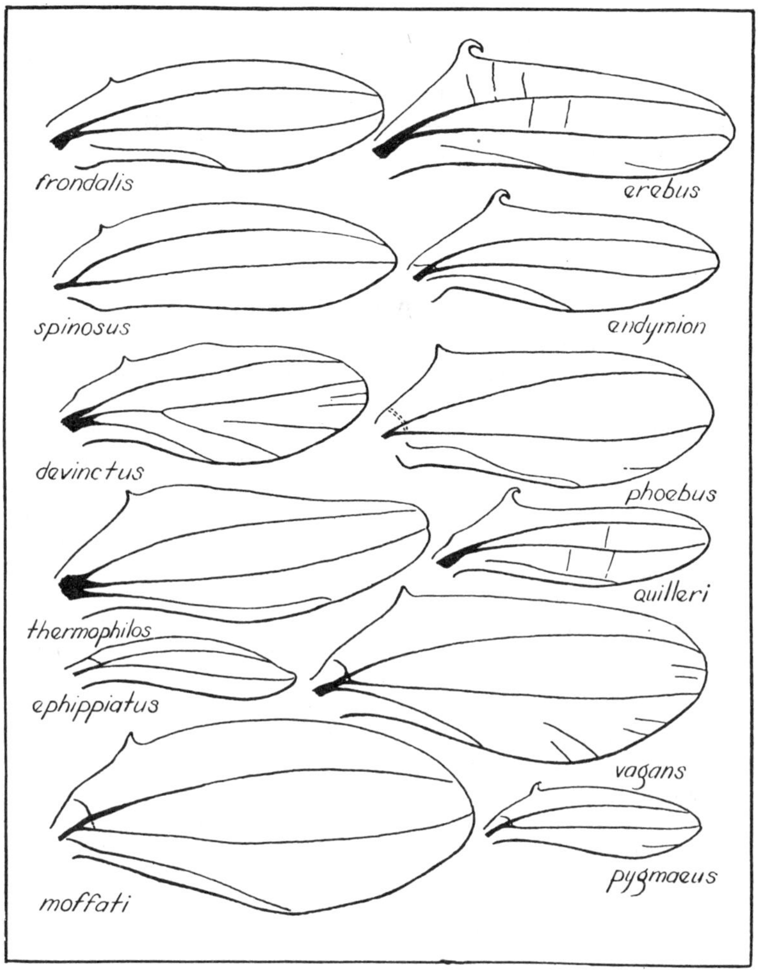

FIG. 163. Hind wings of twelve species of Baetis.

Basal segment of abdomen dark reddish brown dorsally, paler ventrally. Abdominal segments 2–6 hyaline, white. Spiracles outlined in black; geminate blackish tracings along the entire spiracular area; on the basal sternites especially, short oblique branches extend out from the main spiracular line. Segments 7–10 light reddish brown, opaque; sternites slightly paler than tergites; posterior margins of tergites and sternites darker red-brown, presenting an annulate appearance. Tails and forceps white. At the inner apical margin of the basal forceps joint is a distinct tubercle or inward projection; inner margin of this joint somewhat irregular. Apical joint of forceps relatively shorter than in *parvus*, being about one-third the length of the penultimate one (see fig. 164).

Head of female yellowish. Thorax reddish brown. Legs wholly pale whitish.

Abdomen yellowish, the tergites more or less washed with pale purplish or rose-tinged brown; intersegmental areas pale.

Holotype—Male imago. Waddell Creek, Calif.; no date; collector, Dr. David Shepherd. No. 1351.1 in the Cornell University collection.

Allotype—Female imago. Same data. No. 1351.2 in Cornell collection.

Paratypes—34 male imagos, 8 female imagos; same data. No. 1351.3–44 in Cornell collection.

This species is close to *B. parvus*, from which it may be distinguished by the tubercle on the inner margin of the basal forceps joint, the paler reddish brown thorax, and the blackish geminate spiracular line. The females are likewise much paler than in *parvus*.

Baetis endymion sp. nov.

Length: body 5; wing 5–5½ mm.

Abdominal tergites 2–6 of male imago semi-hyaline, yellowish to light olive brown; genitalia of the *intercalaris* type; hind wing of the *quilleri* type, three veins present.

Head light red-brown. Turbinate eyes moderate in size; in alcoholic specimens, pale orange with purplish shading. Thorax darker reddish brown; intersegmental areas of pleura paler; sutures of meso- and metanota blackish; scutellum rather paler reddish brown. Femur and tibia of fore leg and femora of middle and hind legs yellowish, with faint reddish brown shading especially at each end; other joints pale yellowish white. Wings hyaline, venation pale; a small reddish brown spot at extreme base of each wing. A small opaque whitish cloud in the stigmatic area of the fore wing; about 8 cross veins, somewhat aslant, often incomplete at the subcosta; granulations present. Marginal intercalaries rather well developed; one or two in the first interspace. Hind wing relatively shorter and wider than in *B. quilleri* (see fig. 163). Costal projection strongly curved, often wide at base. Third vein ending just beyond middle of hind margin; very faint traces of incomplete cross veins in each of the three spaces. Usually one intercalary between veins 2 and 3 (cross veins and intercalary, not shown in figure).

Abdominal segments 2–6, and basal half of 7, semi-hyaline; yellowish to light olive brown; sternites only slightly paler than tergites; posterior margins narrowly opaque whitish. Faint traces of a hyaline submedian streak on each side of mid-dorsal line, and of short oblique streaks from the anterior margin of each sternite. Spiracular area and spiracles prominently outlined by a blackish geminate line, from which arise lateral branches to each terglte and sternite. Apical half of segment 7, and segments 8–10 opaque, light reddish brown, posterior and lateral margins somewhat brighter reddish brown. Forceps pale red-brown basally, tips whitish. The rounded projection or tubercle on the inner apical margin of the basal forceps joint either entirely wanting or very poorly developed. Distal forceps joint long, fully half the length of the third joint (in one paratype, more than half as long). Two lobe-like processes (penis lobes?) project from the apical margin of the forceps base, between the basal joints of the forceps. Tails whitish; basal joinings narrowly reddish or purplish brown, other joinings opaque whitish.

Holotype—Male imago. Johnston Co., Oklahoma, March 19, 1932; collector, Mr. A. Sandoz. No. 1354.1 in Cornell collection.

Paratypes—16 males, same data; 1 male, same time and place, Prof. R. D. Bird, collector; 3 males, Murray Co., Okla., March 22, 1931, same collector. No. 1354.3–22 in Cornell collection.

The three males from Murray Co. are paler than the other paratypes, but similar in other respects.

This species is somewhat larger than *B. quilleri;* the hind wing is distinctly wider and shorter relatively than in that species or than in *B. erebus*. Its paler coloration distinguishes it from *erebus*, while the similarity in color of sternites and tergites separates it from *quilleri*, in which the sternites are considerably paler than the tergites.

Baetis ephippiatus sp. nov.

Length: body 4–4½; wing 4–4½ mm.

Abdominal tergites 2–6 of male imago pale yellowish white, semi-hyaline, marked with purplish red; hind wing two-veined, no costal projection; genitalia of the modified *moffati* type.

Head light reddish brown; antennae yellowish, a red-brown ring at apex of basal joint. Turbinate eyes large for size of insect, on high stalks; stalk yellowish flushed with pink, upper surface pale orange, in alcoholic specimens. Pronotum light red-brown. Mesonotum quite dark red-brown; median suture and margins of scutellum blackish; a pale line anterior to the wing root. Intersegmental areas of pleura yellowish red, remainder red-brown. Metanotum chestnut brown. Sternum red-brown, median area paler. Legs pale yellowish white; tarsi not clouded. In a few specimens, a narrow blackish ring near apex of femur, and a black hair-line the length of that joint. Wings hyaline, venation pale. A whitish cloud in stigmatic area of fore wing; 7–8 cross veins, aslant, often incomplete near subcosta; granulations present. Marginal intercalaries well developed; none in first interspace; those of second space short. Hind wing with no costal projection; small and narrow, leaf-like; two veins only (see fig. 163).

Abdominal segments 2–6 semi-hyaline, whitish with a faint yellowish tinge. Tergites 2 and 5 much marked with purplish red; geminate submedian streaks, and all except the anterior and lateral margins variably shaded with this color, in irregular saddle-shaped pattern. Often geminate streaks are present also on tergites 3, 4 and 6, with traces also of lateral shading. Posterior margins narrowly but distinctly purplish red (on tergites 2–4, these dark rings may be obsolescent). Spiracular line and spiracles outlined prominently by a geminate black line, with lateral branches on each segment. Segments 7–10 opaque; tergites reddish brown, 8 deeper in color than others, tinged with purplish; posterior margins darker. Sternites paler, yellowish; 8 usually almost wholly light reddish brown with purplish tinge; lateral margins of 9 brownish. Tails and forceps yellowish white. Genitalia of the modified *moffati* type; a rounded projection on inner apical margin of basal forceps joint; second joint swollen somewhat, near middle of inner margin; distal joint short. A truncate penis-cover is present between the basal joints (see fig. 164).

Holotype—Male imago. Mossy Creek, Ft. Valley, Ga.; June 1, 1931; collector, Prof. P. W. Fattig. No. 1356.1 in Cornell collection.

Paratypes—24 male imagos. Ft. Valley and Rome, Ga., June 1 and Aug. 3, 1931; same collector. No. 1356.3–26 in Cornell collection.

The saddle-shaped purplish red markings of tergites 2 and 5, as well as the dark posterior margins of all the tergites, distinguish this species from the closely allied *B. propinquus* and *B. dardanus*.

Baetis erebus sp. nov.

Length: body 4½–5; wing 5 mm.

Abdominal tergites 2–6 of male imago dark red-brown, semi-hyaline; genitalia of the *intercalaris* type; hind wing of the *quilleri* type.

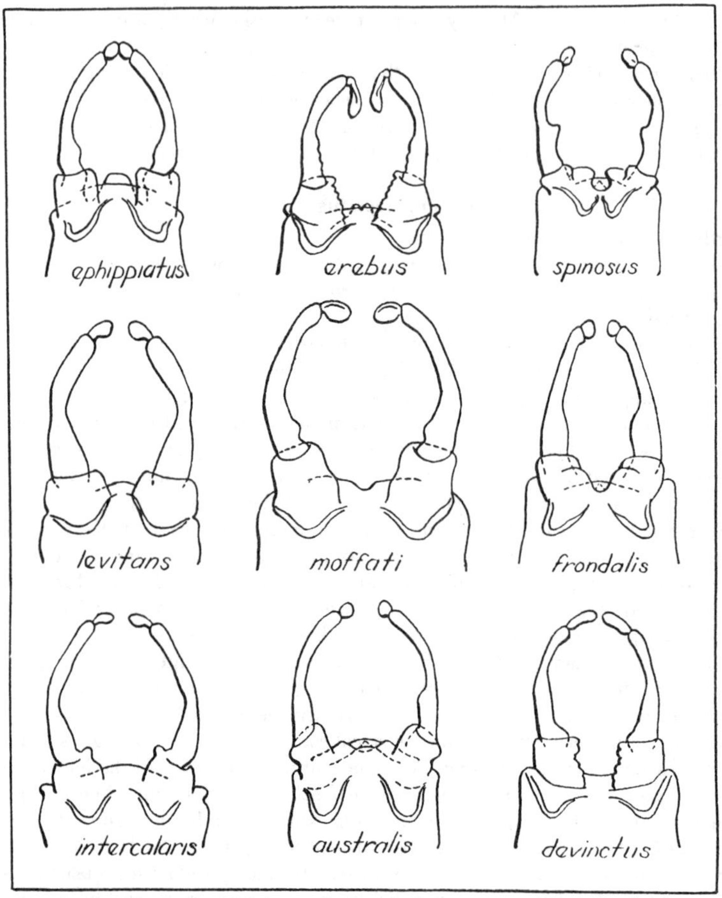

FIG. 164. Male genitalia of nine species of Baetis.

Head and thorax deep blackish red-brown. Turbinate eyes moderately large, but slightly smaller than in *B. quilleri;* orange-brown. Intersegmental areas of pleura paler red-brown. Fore leg wholly light reddish brown, tarsus slightly paler. Middle and hind legs yellowish; tarsi and sometimes bases of tibiae tinged faintly with reddish brown. Wings iridescent, hyaline; venation light red-brown; a small reddish brown spot at base of each wing. Small opaque whitish cloud in stigmatic area of fore wing. 5–8 stigmatic cross veins, somewhat aslant, incomplete toward the subcosta; granulations present between them. Marginal intercalaries rather short; none in the first interspace. Hind wing of the *quilleri* type, but relatively longer than in that species; costal projection rather fish-hook-like. Third vein ends just beyond middle of hind margin. Between veins 2 and 3 is a well-developed intercalary. Traces of cross veins in each of the three spaces, most evident between costal margin and 1st vein (see fig. 163).

Abdominal segments 2–10 rather dark reddish brown; segments 2–6 semi-hyaline, 7–10 semi-opaque; sternites only slightly paler than tergites. Posterior and lateral margins of tergites, and median posterior margins of sternites, narrowly darker and opaque. Traces of a pale mid-dorsal line and of short oblique darker dashes at the anterior margin of each tergite. Spiracular line and spiracles faintly outlined in dark brown. Lateral areas of sternites next to pleural fold, and in some specimens a similar area on the tergites paler, hyaline. Tails rather dark red-brown basally, the distal half whitish; joinings opaque. Forceps red-brown basally, tips yellowish. A tubercle at inner apical margin of basal forceps joint only slightly developed; inner margin of first and second joint irregular, roughened; distal joint very long, more than half the length of the third joint. Between the forceps bases are two small lobe-like projections (perhaps penis-covers): see fig. 164.

Two females taken at Tuscon, Ariz., are apparently of this species. Very similar to male, but all legs pale reddish brown; abdominal sternites yellowish red, paler than tergites. Wing 6 mm. in length.

Holotype—Male imago. Santa Catalina Mts., Bear Canyon, Ariz.; May 3, 1931. Collector, Prof. L. P. Wehrle. No. 1355.1 in Cornell collection.

Allotype—Female imago. Tuscon, Ariz., Jan. 15, 1921. In collection of Univ. of Arizona.

Paratypes—5 male imagos, same data as holotype; 1 female imago, same data as allotype; 1 male imago, Fish Creek, Ariz., May 5, 1918, collected by Prof. J. C. Bradley. No. 1355.3–9 in Cornell collection.

We have also two males taken at Austin, Texas, (H. B. Parks, Jr.), and 2 males taken at Phoenix, Ariz. (J. C. Bradley), which may be of this species.

The abdomen is distinctly more reddish in color and the hind wing relatively longer than in the allied *B. quilleri;* the dark reddish brown abdomen easily separates the species from the paler *B. endymion*, which is also of the *quilleri* type. The presence of three veins in the hind wing distinguishes all of these species from *B. akataleptos* and *B. harti*, which have but two veins.

Baetis flavistriga McDunnough

1921—McDunnough—Canad. Ent. 53: 120. Figs.. **Ont.**
1923—McDunnough—Canad. Ent. 55: 40. Fig
1925—McDunnough—Trans. R. S. C. p. 215 (key to males).

Length: body 4½; wing 5 mm.

Abdominal tergites 2–6 of male imago hyaline, tinged with yellowish brown; genitalia of the *intercalaris* type, the tubercle on the inner apical margin of the first forceps joint well developed.

Turbinate eyes small, oval, light brown with a yellowish tinge; stalk long, purplish brown, the upper third yellow. Face black; pale green markings around base of antennae and ocelli; antennae blackish. Thorax dark olivaceous; yellowish shading on the pronotum; an oblique yellow streak runs forward from the wing roots along the lateral margin of the mesonotum. Pleural sutures with yellow markings. Legs dirty whitish, femora pale yellow. Wings hyaline; venation pale. Intercalaries of the fore wing shorter than in *B. intercalaris*, particularly those of the first pair. Hind wing as in *B. nanus*.

Abdominal tergites 2–6 hyaline, tinged with pale yellowish brown; 7–10 light umber brown. The contrast between the basal and apical tergites is less marked than in *B. phoebus* and *B. intercalaris*. Sternites paler, the apical ones opaque. Slight black markings on the spiracular area. The tubercle on the first forceps joint well developed.

This species has been separated from the closely allied *B. nanus* by the color of the eyes in living insects, the darker thorax, and the browner basal abdominal tergites. As indicated under that species, *nanus* may fall as a synonym of *flavistriga*, when more material is available for study. Specimens of this species in the Cornell collection are from Ithaca, N. Y., and vicinity; also from Ft. Jackson, N. Y., the St. Regis River (L. Harper). Dr. McDunnough has presented a fine series from Ottawa, Ont., and the Golf Club, Que. (G. S. Walley, R. Ozburn).

Baetis frivolus McDunnough

1925—McDunnough—Canad. Ent. 57: 174. Figs. .... Que., Ont.
1926—McDunnough—Canad. Ent. 58: 300 (in discussion of *insignificans*).

Length: body 5; wing 6 mm.

Abdominal tergites 2–6 of male imago deep brown; genitalia of the modified *moffati* type, the fourth joint of the forceps very long; no costal projection on hind wing, which has two veins only.

Turbinate eyes very large, deep blackish brown, the edges paler. Head and thorax shiny blackish. Light brown markings on the postero-lateral edges of the prothorax, the antero-lateral edges of the mesothorax, the pleural sutures (these are tinged with reddish), the lateral edges and side projections of the mesosternum, and the postero-lateral margins of the mesonotum near the scutellum. Fore legs smoky brown; middle and hind legs pale yellowish. Wings hyaline, venation pale. Cross veins in the stigmatic area of the fore wing 6–7 in number, anastomosed. First intercalary in the first interspace rudimentary or lacking; all others well developed. Hind wing long and narrow, third vein entirely lacking. Costal projection entirely wanting.

Abdominal tergites deep brown, opaque. Sternites dull yellowish, becoming whitish posteriorly. Traces of a broken black spiracular line. Second joint of the forceps cylindrical, the fourth unusually long, being fully three times as long as wide, its tip knob-shaped. Between the bases of the forceps is a small raised plate.

This species, related, by reason of the complete lack of costal projection of the hind wing, to *propinquus*, *dardanus*, *insignificans* and *amplus*, may be distinguished from each of these by the deep brown color of the abdomen.

We have seen no specimens of this species.

Baetis frondalis McDunnough

1925—McDunnough—Canad. Ent. 57: 173, 174 (in discussion of *spinosus*) Figs. .... Que., Ont.

Length: body 5; wing 5–6 mm[2]

Abdominal tergites 2–6 of male imago deep brown; genitalia of the *moffati* type; a deep excavation between the bases of the forceps, in which is a short spine.

Head and thorax deep shiny black. Brown markings on the pleural sutures. Legs pale yellowish brown. Wings hyaline. Stigmatic cross veins of the fore wing few in number, not anastomosing; no intercalaries in the first interspace. Hind wing long and narrow, the margins sub-parallel; third vein reduced to a mere trace. Costal projection greatly reduced (see fig. 163).

Abdominal tergites 2–6 deep brown; the anterior margins, especially laterally, pale and partially semi-translucent. Tergites 7–10 opaque, deep brown. Sternites pale dull cream-colored, the opaque apical ones with partial brown shading. Tails and forceps pale. The third joint of the forceps is long and narrow, the apical one very short and somewhat truncate. Between the bases of the forceps is a deep excavation, covered by a small triangular plate; a small spine is situated at the base of the excavation (see fig. 164).

This species is very similar to *australis*, from which species it may be distinguished by the greater amount of brown shading on the basal abdominal tergites, and the somewhat darker legs. We have in the Cornell collection a fine series of males of this species, from Cascades Pt., Que., presented by Dr. J. McDunnough.

Baetis harti McDunnough

1924—McDunnough—Canad. Ent. 56: 7...Ill.
1926—McDunnough—Canad. Ent. 58: 188 (in discussion of *akataleptos*).

Length: body 2½; wing 3 mm.

Abdominal tergites 2–6 of male imago hyaline whitish; genitalia of the *intercalaris* type; only two veins present in hind wing, the costal projection very prominent.

Turbinate eyes large for size of insect, oval, blackish brown. Head blackish. Thorax deep black-brown. Legs pale yellowish. Wings hyaline (venation pale?). Intercalaries of the fore wing short but distinct; the first pair obsolescent. Hind wing narrow and long, with a prominent costal projection, much as in *B. quilleri*. Only two veins, the third entirely lacking.

Abdominal segments 2–6 hyaline whitish; segments 7–10 chocolate brown dorsally, the sternites slightly paler. Black dots along the spiracular area. Tails whitish. Genitalia of the *intercalaris* type.

This very diminutive species is distinguishable from *B. pygmaeus* by reason of the prominent costal projection of the hind wing; from *B. akataleptos*, which has a similarly prominent costal projection, it is separable by the pale basal abdominal segments, which in *akataleptos* are dull brown. We have seen no specimens of this species.

Baetis incertans McDunnough

1925—McDunnough—Trans. R. S. C. p. 220. Figs.......................................Que.
? 1932—Traver—J. Elisha Mitchell Sci. Soc. 47: 230, 231 (*Baetis sp.* No. 1, also *rusticans* ?)......N. C.

Length: body 4½; wing 5 mm.

Abdominal tergites 2–6 of male imago olive brown, the posterior margins deeper ruddy brown; genitalia of the *moffati* type.

Turbinate eyes moderate in size; red-brown. Head and thorax deep olive brown. Scutellum and posterior margins of the mesonotum only slightly paler; anterior projection of the metanotum whitish; pleura with ruddy brown tinges. Femora pale yellowish; tibiae and tarsi dirty whitish, a smoky apical spot on the fore tibia. Wings hyaline, venation pale. About 6 cross veins in the stigmatic area of the fore wing, with only slight granulations between them; veins not anastomosed. First pair of intercalaries obsolescent, second and third pairs rather short, subequal. Hind wing broad, more than twice as long as its width; third vein rather weak, running close to the hind margin. Between the 2nd and 3rd vein are two strong intercalaries.

Abdominal tergites olive brown; 2–6 semi-translucent and narrowly deeper ruddy brown on the posterior margins; 7–10 opaque. Sternites pale yellowish white, the apical ones opaque. Tails and forceps white.

The smaller size, smaller turbinate eyes, and abdomen more evenly brown in color, separate this species from *vagans*. The species reported from North Carolina, with subimago males only, as a possible *rusticans*, seems to be close to *B. incertans*, as is also *Baetis sp.* No. 1, reported in the same paper. Until more specimens can be obtained for study, we leave these doubtfully in *incertans*. We have seen a specimen of this species from Covey Hill, Que. (G. S. Walley), which Dr. McDunnough lent us for examination.

Baetis insignificans McDunnough

1926—McDunnough—Canad. Ent. 58: 300. Fig. .... B. C.

Length: body 4; wing 4 mm.

Abdominal tergites 2–6 of male imago pale, semi-hyaline, considerably tinged with smoky brown; genitalia of the modified *moffati* type; no costal projection on the hind wing, which has two veins only.

Turbinate eyes small, rather bright red. Thorax deep brown. The pleural sutures and the anterior projection of the metanotum are pale cream-colored. Fore femora dull smoky brown, tibiae and tarsi paler. Middle and hind legs pale ochreous, with smoky tinges on the femora. Wings hyaline, venation pale. Only two veins in the hind wing; the costal projection is entirely wanting. Hind wing narrow and leaf-like.

Abdominal segments 2–6 pale, semi-hyaline, the tergites considerably tinged with smoky brown. Segments 7–10 opaque; the tergites light sepia-brown, the sternites whitish. Tails white. The second joint of the forceps somewhat cylindrical; fourth joint of the forceps longer and narrower than in *B. dardanus*, but not as long as in *B. frivolus*. Between the bases of the forceps is a short projecting plate, not found in *dardanus*.

Aside from the genitalic differences noted, this species may be separated from *B. dardanus* by the smoky tinges of the basal abdominal tergites, which are also lacking in *B. propinquus*; *B. frivolus* and *B. amplus* are larger, darker species of the same group.

One male in the Cornell collection, from Tuttle, Idaho, seems to be of this species.

Baetis intercalaris McDunnough

1921—McDunnough—Canad. Ent. 53: 118. Figs. .... Ont.
1922—Murphy—Bull. Lloyd Libr. 22, Ent. Ser. 2: 42 (*propinquus*).
1923—McDunnough—Canad. Ent. 55: 39. Fig.
1924—McDunnough—Canad. Ent. 56: 8 (in discussion of *pallidulus*).
1925—McDunnough—Trans. R. S. C. p. 214 (key to males).

Length: body 4½; wing 5 mm.

Abdominal tergites 2–6 of male imago hyaline white; genitalia of the *intercalaris* type, the tubercle on the inner margin of the first forceps joint well developed.

Turbinate eyes large, lengthily oval, deep brown; upper part of stalk yellow. Face black; inner edge of ocelli, and a spot at base of eye, pale yellow-green. Antennae black, paler at tip; basal joint ringed with yellow at apex. Thorax shiny blackish; lateral margins of mesonotum and the scutellum tinged with dull olivaceous; brown shading on the pleural sutures. Legs dirty white, claws black. Wings hyaline, venation pale. Intercalaries of the fore wing long, those of the first interspace considerably longer than succeeding pairs. Hind wing well developed, at least two times as long as its width. Third vein strong, running rather close to the hind margin, and extending at least to the middle of that margin. Between the 2nd and 3rd veins, two intercalaries are usually present.

Abdominal tergites 2–6 hyaline white; 7–10 umber brown. Sternites whitish, the apical ones opaque. In the spiracular area, the tracheae are lightly marked with black. Tails white. The second forceps joint conical, a tubercle present on the inner apical margin of the first forceps joint (see fig. 164).

Distinguished from *B. flavistriga* by the paler basal abdominal tergites, the large turbinate eyes, and the very long first pair of intercalaries of the fore wing, the latter character also distinguishes it from *B. phoebus*.

This species is fairly common in the vicinity of Ithaca, N. Y., from which region we have a considerable number of specimens in the Cornell collection. Other specimens are from Old Forge, N. Y. (J. G. Needham); the Potomac River near Harpers Ferry, and Conococheague Park, Md. (V. Argo); Ft. Jackson, N. Y. (L. Harper); and Stamford, Conn. (S. W. Bromley).

Baetis intermedius Dodds

1923—Dodds—Trans. Amer. Ent. Soc. 49: 110. Figs. . . . . Colo.

Length: body 6; wing 6½ mm.

Abdominal tergites 2–6 of male imago light brown; genitalia of the *moffati* type.

Thorax dark brown. Abdomen light brown. (No other color characters given). Hind wing about twice as long as its width; third vein very faint, running close to the hind margin, and ending basad of the middle of that margin. One short intercalary between veins 2 and 3. Costal projection well developed.

We have seen no specimens of this species.

Baetis lasallei Banks

1924—Banks—Bull. Mus. Comp. Zool. 65: 425. Figs. . . . . Wis.

Length: body ?; wing 5 mm.

Abdominal tergites 2–6 of male imago whitish; genitalia apparently (judging by figure) of the *moffati* type.

Thorax brown, margin of mesonotum more yellow. Legs pale. Wings hyaline, venation pale. Intercalaries of fore wing moderately long. Hind wing large and broad, narrower toward each end, rounded at the tip. Third vein well developed; two long intercalaries between the 2nd and 3rd vein. Costal projection strong, and colored.

Abdominal segments 2–6 pale whitish, with a black dot on each side of each segment. Tergites 7–10 brown, sternites whitish. Tails pale. The second forceps joint rather cylindrical.

Said to be nearest to *B. phoebus*, from which it may be distinguished by the broader hind wing, which is also less pointed at the tip; the genitalia, however, seem not of the same type as *phoebus*. We have seen no specimens of this species. It is apparently close to *pallidulus*, but the presence of the two intercalaries in the hind wing would seem to distinguish it from that form.

Baetis levitans McDunnough

1925—McDunnough—Trans. R. S. C. pp. 216–219 (in discussion of *pluto*). Figs. . . . . Que.

Length: body 4; wing 4½ mm.

Abdominal tergites 2–6 of male imago pale yellowish white, with faint posterior and lateral brown shading; genitalia of the *intercalaris* type, the tubercle on the inner apical margin of the first joint scarcely or not at all developed (see fig. 164).

Turbinate eyes moderate in size; deep ruddy brown in color. Face and antennae black-brown. Thorax deep olive brown; lateral margins of the mesonotum, before the wing roots, often yellowish; scutellum dark, but a central and two lateral yellowish patches just anterior to it. Anterior projection of the metanotum whitish. Ruddy shading on the pleura, the sutures yellowish in part. Legs pale yellow. Femora more or less shaded with brown, which is continuous over the entire length of the fore femur reduced to a median and an apical spot on the middle femur, and on the hind one to an apical spot only. Wings hyaline, venation pale. First and second pairs of intercalaries of the fore wing rather short, frequently broken; the next two pairs longer. Hind wing more than two and one-half times as long as its width; narrower than in *B. cingulatus*, and more or less even in width throughout. Third vein well developed, running close to the hind margin. One intercalary may be present between the 2nd and 3rd vein.

Abdominal segments 2–6 pale whitish yellow, semi-translucent, the tergites shaded laterally and posteriorly with pale brown. Tergites 7–10 light russet brown; sternites opaque, creamy, with faint tinges of brown. Tracheae along the spiracular area strongly black, forming a wavy double black line with dorsal and ventral branches extending from it.

This species has larger eyes than *B. flavistriga*, which has quite similar abdominal markings. The lack of the tubercle on the first forceps joint, the smaller eyes and darker scutellum, separate it from *B. cingulatus*. In the Cornell collection are several specimens from Hopkinton and Ft. Jackson, N. Y., the St. Regis River, (L. Harper), which seem to be of this species.

Baetis minimus Dodds

1923—Dodds—Tans Amer. Ent. Soc. 49: 113. Figs. . . . . Colo.

Length (female): body 5; wing 7½ mm.

Only the female imago is known, hence the position of this species in reference to others of the genus is uncertain.

Species said to be much like *moffati*, but differing in the small size of the body in relation to the fore wing, and of the hind wing in relation to the size of the fore wing. Hind wing rather small (0.8 mm.), the third vein present but very faint; no intercalaries. Costal projection present near the base of the hind wing, moderately developed. No color characters are given.

We have seen no specimens that correspond with the above description.

Baetis moffati Dodds

1923—Dodds—Trans. Amer. Ent. Soc. 49: 112. Figs. . . . . Colo.
1924—McDunnough—Canad. Ent. 56: 8 (in discussion of *pallidulus*).
1925—McDunnough—Canad. Ent. 57: 173 (in discussion of *brunneicolor*).
1925—McDunnough—Trans. R. S. C., p. 220 (in discussion of *vagans*).

Length: body 7; wing 7½ mm.

Abdominal tergites 2–6 of male imago brownish; genitalia of the *moffati* type, see fig. 164, (second forceps joint cylindrical, no tubercle on the inner apical margin of the first joint).

No color characters are given in the original description, but from duplicate material of Dr. Dodds' in the Cornell collection (which specimens seem without doubt to be *moffati*), the following notes are presented.

Turbinate eyes somewhat shriveled, but appear to be of moderate size; reddish orange in color. Antennae pale reddish. Head and thorax deep red-brown, the meso- and metanota much darker red-brown, almost blackish on the margins of the mesonotal scutellum. Postero-lateral and lateral margins of the pronotum darker than central area. Antero-lateral margin of mesonotum, and several sutures of pleura and sternum, blackish. Fore leg reddish brown, tip of tarsus rather paler. Apical margins of coxa, trochanter and femur, and base of tibia, deep reddish black. Only stubs of middle and hind legs remain; these are paler reddish brown. Wings hyaline, venation pale yellowish brown in costal half of fore wing, elsewhere pale. Hind wing rather large; three veins, the third rather faint and running close to the hind margin to a little beyond the middle. Costal projection well developed. No intercalaries (see fig. 163).

Abdominal segments 2–6 light reddish brown, in part semi-translucent; tergites only slightly darker than sternites; intersegmental areas pale hyaline, giving the abdomen an annulate appearance. Segments 7–10 opaque, darker reddish brown, especially on the dorsum. Genitalia of the *moffati* type; basal joint of forceps large, with no tubercle on the inner apical margin; second joint swollen and cylindrical in shape, usually considerably shorter than the third more slender joint; fourth joint much shorter than the third, about as long as the second.

Besides the material collected by Dr. Dodds, presumably from Tolland, Colo. (exact location not given), there are several female imagos of this species in the Cornell collection from Pike's Peak, Colo. (R. T. Gilmore), and from Gothic, Colo. (J. D. Bowen).

Baetis nanus McDunnough

1923—McDunnough—Canad. Ent. 55: 42...Ont.
1925—McDunnough—Trans. R. S. C. p. 215 (key to male imagos).

Length: body 3; wing 4 mm.

Abdominal tergites 2–6 of male imago yellowish, semi-hyaline; genitalia of the *intercalaris* group, the tubercle on the inner apical margin of the first forceps joint well developed.

Turbinate eyes small, oval (smaller than in *B. flavistriga*); pale yellowish brown in living specimens, bright red with yellow around the rims, in dried ones. Thorax pale olivaceous; the posterior portions may be tinged with smoky; anterior sutures shaded with brownish. Legs dull hyaline whitish, except the fore femora, which are yellow. Wings hyaline, venation pale. Third vein of hind wing well developed, extending to or beyond the middle of the hind margin. Between veins 2 and 3, traces of an intercalary are often present.

Abdominal tergites 2–6 semi-hyaline, yellowish, sometimes tinged faintly with brown; 7–10 opaque, Isabella brown in color. Sternites pale dull ochreous, the lateral areas noticeably hyaline. Tracheae and spiracular line with streaks and dots of blackish. Genitalia of the *intercalaris* type. Tails whitish.

This species is very close to *B. flavistriga;* it may prove synonymous with that species. It has been distinguished from *flavistriga* by the color of the eyes in the living insect, the paler thorax, and the yellower appearance of the basal tergites. We have seen no specimens of this species.

Baetis pallidulus McDunnough

1924—McDunnough—Canad. Ent. 56: 8..Ill.

Length: body 4–5; wing 5 mm.

Abdominal tergites 2–6 of male imago very pale yellowish hyaline; genitalia of the *intercalaris* type, tubercle present on inner margin of basal forceps joint.

Turbinate eyes very large (larger than in *B. intercalaris*), circular; deep reddish in dried specimen, light orange in living insect. Thorax pale yellowish olivaceous to light red-brown; scutellum of the mesonotum creamy. Legs pale yellowish. Wings hyaline, venation pale. Marginal intercalaries of fore wing well developed; those of the first pair not longer than any of the others; the pair in the second interspace somewhat shorter than the 1st or 3rd pair. About 8 stigmatic cross veins; aslant, somewhat anastomosed; few or no granulations between them. Hind wing broad, very similar to that of *B. moffati;* usually one somewhat oblique intercalary between the 2nd and 3rd vein.

Abdominal segments 2–6 very pale yellowish, hyaline, not tinged with olivaceous. Tergites 7–10 light red-brown; sternites deep creamy, sometimes with faint reddish-brown tinge. A pale-centered black dot over each spiracle, and traces of a continuous dark spiracular line. Tails whitish. A tubercle present on the inner margin of the basal forceps joint.

Dr. McDunnough has very kindly sent several specimens of this species to the Cornell collection; these are from Jordon, Ont. (G. S. Walley). The species seems close to *B. phoebus*, but is apparently distinguishable from it by the larger eyes, which in life are light orange instead of red-brown; the basal abdominal tergites are likewise paler, and the 2nd pair of marginal intercalaries relatively shorter, than in that species.

Baetis parallelus Banks

1924—Banks—Bull. Mus. Comp. Zool. 65: 425. Figs. . . . . Nev.

Length: body ?; wing 8 mm.

Abdominal tergites 2–6 of male imago brownish; genitalia (judging by the figure) of the *moffati* type.

Head and thorax pale brown; margin of mesonotum hardly more yellow; remainder of mesonotum dark shining brown, sternum partly brown. Fore legs pale brownish, middle and hind legs pale. Wings hyaline, venation pale. Intercalaries of the fore wing moderately long. Hind wing elongate, the sides nearly parallel. Third vein quite well developed. One rather long intercalary between the 2nd and 3rd veins. Costal projection sharp.

Abdominal tergites 2–6 brownish; 7–10 darker brown. Sternites pale. Tails pale, darker at the joinings.

We have seen no specimens of this species.

Baetis parvus Dodds

1923—Dodds—Trans. Amer. Ent. Soc. 49: 113. Figs. . . . . Colo.
1925—McDunnough—Trans. R. S. C.; pp. 214, 215 (key to male imagos) . . . . Alta., ? Ont.
1928—McDunnough—Canad. Ent. 60: 8.
1930—Ide—Canad. Ent. 62: 221. Nymph fig. (Is this Dodds' *parvus*?).

Length: body 5½; wing 5–5½ mm.

Abdominal tergites 2–6 of male imago hyaline, whitish, unmarked; genitalia of the *intercalaris* type, no tubercle on inner apical margin of first forceps joint. Second vein of hind wing forked near base.

Head and thorax deep blackish brown; filament of antenna paler. Turbinate eyes moderate in size; both stalk and upper surface dark orange brown, in dried specimens. Paler red-brown areas on pleura below the wings; tip of scutellum paler brown. Legs whitish. Wings hyaline, venation pale. 5–8 cross veins in the stigmatic area of the fore wing, often with short horizontal veins between them, aslant, often incomplete on the subcostal margin, sometimes forking toward the subcosta. Usually one short intercalary in the first interspace; those of the second interspace slightly shorter than the following pairs. Hind wing unusually wide for its length; costal projection well developed. Three longitudinal veins, the second forked near the base; one long intercalary in this fork (usually also a pair of short intercalaries just below the anterior branch, not shown in Dr. Dodds' sketch; traces of another short one between the long intercalary and the posterior branch, in some specimens).

Basal tergite of abdomen wholly dark brown, sternite somewhat paler. Abdominal segments 2–6 hyaline white; usually no dark spiracular markings. Tergites 7–10 opaque, brown; wholly or in part suffused with deep rose-color. Sternites opaque creamy white, more or less shaded and marked with rose. Tails white. No tubercle or projection on the inner apical margin of the basal forceps joint. Apical joint of forceps fully half as long as the penultimate one.

The forked second vein of the hind wing distinguishes this species from all other described North American species excepting *B. devinctus;* the presence of a well-developed tubercle on the inner margin of the basal forceps joint, as well as the paler thorax, separates *devinctus* from *parvus*.

Specimens of this species in the Cornell collection are from Glacier National Park, Mont. (R. T. Young); Yellowstone Park, Wyo. (C. R. Crosby); and several fine specimens sent by Dr. McDunnough, collected at Waterton Lakes, Lundbreck and Blairmore, Alta. (J. H. Pepper). From Briggs Run at the Smoke Hole, W. Va., there are specimens of an allied species, much paler in color in both sexes. Until more material is available, we hold these under *B. parvus*.

Baetis phoebus McDunnough

1923—McDunnough—Canad. Ent. 55: 41...Ont.
1924—Banks—Bull. Mus. Comp. Zool. 65: 425 (in discussion of *lasallei*).
1925—McDunnough—Trans. R. S. C. p. 215 (key to male imagos).

Length: body 5; wing 5½ mm. (female?).

Abdominal tergites 2–6 of male imago hyaline whitish; genitalia of the *intercalaris* type, the tubercle on the inner apical margin of the first forceps joint well developed.

Turbinate eyes large, almost circular, with long stalk; in living specimens, red-brown with a narrow yellow rim. Head and thorax brown. Along the lateral margin of the mesonotum, forward from the wing roots to the anterior projection, is a yellow streak; yellowish brown shading on the posterior portion of the mesonotum and the anterior area of the metanotum. Pleural sutures marked more or less with ochreous. Legs very pale yellowish. Wings hyaline, cross veins pale. Intercalaries of the fore wing well developed, those of the first pair not longer than the following pairs. Hind wing broad, well developed; third vein extends rather beyond the middle of the hind margin; a short intercalary usually present between veins 2 and 3 (see fig. 163).

Abdominal tergites 2–6 hyaline, whitish, very faintly tinged with olivaceous; 7–10 bright ruddy brown. Sternites opaque whitish, variably tinged with brown. Slight blackish dots along the spiracular area. The tubercle on the first joint of the forceps well developed.

Close to *B. flavistriga*, but larger, and with larger turbinate eyes. Also distinguished by the paler basal abdominal tergites, which are much less shaded with olive than in that species. It is likewise very close to *B. pallidulus*, but with somewhat smaller eyes, which in life are red-brown, not light orange as in that species. We have seen no specimens.

Baetis piscatoris sp. nov.

Length: body 6; wing 6 mm.

Abdominal tergites 2–6 of male imago yellowish hyaline, a purplish red band on posterior margin of each; genitalia of the *moffati* type.

Head and antennae reddish brown. Turbinate eyes moderate in size, on a rather high stalk; stalk yellowish, upper surface light orange, in alcoholic specimens. Thorax rather dark reddish brown; a yellowish streak on mesonotum anterior to wing root, another along postero-lateral margin; a small yellow area anterior to the scutellum; margins of scutellum blackish. Intersegmental areas of pleura yellowish red. Fore legs light yellowish brown; middle and hind legs yellowish, joinings and tips of tarsi light brown. Wings hyaline, venation pale; longitudinal veins of costal margin faintly yellowish. A faint whitish cloud in stigmatic area of fore wing; 8–11 cross veins, aslant, sometimes incomplete, occasionally anastomosed; with granulations between them. Intercalaries of the first and second interspaces may be single or paired; shorter than the following pairs. Hind wing rather large; three longitudinal veins; costal projection well developed. Usually traces at least of two intercalaries between the 2nd and 3rd veins; often also a faint intercalary between veins 1 and 2.

Abdominal tergites 2–6 yellowish, hyaline; on the posterior margin of each is a prominent purplish-red band, not extending to the spiracular line. Tergite 2 is often slightly tinged with brownish. Sternites 2–6 paler yellowish. Spiracular area outlined by a geminate purplish line, from which on each segment arise lateral branches to tergite and sternite. Segments 7–10 opaque, light reddish brown, the margins narrowly darker; sternites only slightly paler than tergites. Tails pale yellowish brown, paler at tips. Forceps yellowish.

Female reddish or orange-brown; legs yellowish brown, tarsi somewhat darker. Venation light reddish brown. Abdominal sternites yellowish; tergites banded as in male. Spiracular markings prominent. Tails as in male.

Holotype—Male imago. Waddell Creek, Rancho del Oso, Calif.; no date. Dr. David Shepherd, collector. No. 1352.1 in Cornell collection.

Allotype—Female imago; same data. No. 1352.2 in Cornell collection.

Paratypes—5 male imagos, 3 female imagos; same data; No. 1352.3–10 in Cornell collection.

In type of genitalia, this species is allied to the pale *B. lasallei* and to the yellowish-brown *B. adonis*, from both of which it may be distinguished by the prominently banded abdomen; other species having the same type of abdomen are darker in coloration (*B. tricaudatus*, of the color of which we are not certain, is a larger species). *B. cingulatus*, which has a similarly-marked abdomen, has genitalia of the *intercalaris* type.

Baetis pluto McDunnough

1925—McDunnough—Trans. R. S. C.: 218. Figs. Que.
1928—McDunnough—Canad. Ent. 60: 239.
1930—McDunnough—Canad. Ent. 62: 58.
1930—Ide—Canad. Ent. 62: 221. Ont.

Length: body 4½; wing 4½ mm.

Abdominal tergites 2–6 of male imago largely maroon brown, with pale hyaline areas on the lateral and anterior margins; genitalia of the *intercalaris* type, having no tubercle on the inner apical margin of the first forceps joint.

Turbinate eyes moderate; deep black-brown. Head and thorax shiny blackish. Lateral margin of mesonotum, anterior to the wing roots, brown; a light brown spot on the antero-lateral corner of mesonotum; pleural sutures shaded with brown. Legs dull ochreous, tibiae and tarsi paler than the femora. Wings hyaline, venation pale. Heavy granulations present between the cross veins of the stigmatic area of the fore wing; cross veins anastomosed. First pair of intercalaries obsolescent, second pair rather short, third pair well developed, long. Hind wing almost three times as long as its width; third vein well developed, running beyond the middle of the hind margin.

Between veins 2 and 3 is one long intercalary, and a shorter somewhat obsolescent one. Traces of two short intercalaries between veins 1 and 2.

Abdominal tergites 2–6 semi-translucent, deeply tinged with maroon brown on most of each tergite; along the lateral and anterior margins, however, are pale hyaline areas. Tergites 7–10 opaque maroon brown. Sternites 2–6 pale hyaline; 7–10 opaque, whitish with brown tinges. Tracheae of the spiracular area prominently marked with black. Tails white. Forceps brown-tinged. There is no tubercle on the inner apical margin of the first forceps joint.

The maroon-brown of the abdomen and the blacker thorax, as well as the absence of the forceps tubercle, distinguishes this species from *B. rusticans*. The abdominal markings are darker than in *B. levitans*. We have seen no specimens of this species.

Baetis posticatus Say

1824—Say—Godman's West. Quart. Rep. 2: 162 (Cloeon) .... Pa.
1859—LeConte—Complete Writings of T. Say. 1: 172.
1861—Hagen—Syn. Neurop. Ins. N. Amer. p. 53 (Cloe).
1871—Eaton—Trans. Ent. Soc. London, p. 120 (Baetis).
1885—Eaton—Revis. Monogr. p. 169.

Length: body 8 mm.

Abdominal tergites 2–6 of male imago greenish white, hyaline.

Turbinate eyes reddish brown. Thorax black. Legs white, the fore femur obscure. Wings hyaline. Abdomen greenish white, hyaline, on segments 2–6; the apical segments fuscous. Tails white.

This species is little known up to the present time. Only the above incomplete description stands as a record. Perhaps further and more extensive collecting in the type locality may bring to light specimens of *B. posticatus*. Until then, it stands as a name only. We make no attempt to include this species in our keys.

Baetis propinquus Walsh

1862—Walsh—Proc. Acad. Nat. Sci. Phil., p. 380 (*Cloe vicina ?*; *nec. vicina* Hagen) .... Ill.
1863—Walsh—Proc. Ent. Soc. Phil. 2: 207, note 20; (*propinquus* for *C. vicina* Walsh).
1871—Eaton—Trans. Ent. Soc. London, p. 121 (transferred to Baetis).
1885—Eaton—Revis. Monogr. p. 169.
1925—McDunnough—Canad. Ent. 57: 172.

Length: body 4–5; wing 4½–5 mm.

Abdominal tergites 2–6 of male imago hyaline, whitish; genitalia of the modified *moffati* type; no costal projection on hind wing, which is two-veined.

Head and thorax piceous. Antennae fuscous, the tips paler. Turbinate eyes large. Legs pale; fore femur often pale ferruginous; tips of tarsi cloudy. Wings hyaline, venation pale. Hind wing long and narrow, with two veins only, the third being entirely wanting. No costal projection near the base of the hind wing. Abdominal segments 2–6 whitish, hyaline; tergites 7–10 piceous, sternites opaque whitish. Tails whitish, the basal joinings may be fuscous. A black dot near the spiracle, on each of the white segments.

The above description is taken from the original by Walsh, with additional data by Dr. McDunnough (Canad. Ent. 57: 172), regarding the specimen in the Museum of Comparative Zoology which he proposes to regard as the lectotype. The species, as thus defined, is very close to *B. dardanus*, which may prove to be a synonym of *B. propinquus*. We have examined a specimen from Shawneetown, Ill., which Dr. McDunnough sent us; in this specimen, all the legs are whitish except for the cloudy tarsi; the apical sternites are faintly tinged with pink, likewise the forceps.

Baetis pygmaeus Hagen

1861—Hagen—Syn. Neurop. Ins. N. Amer. p. 54 (Cloe)........................St. Lawrence R.
1863—Hagen—Proc. Ent. Soc. Phil. 2: 178 (Cloe).
1871—Eaton—Trans. Ent. Soc. London, p. 122.
1885—Revis. Monogr. p. 170.
1901—Needham—Bull. N. Y. St. Mus. 47: 421. Figs. (in part) Life history notes.............N. Y.
1905—Needham—Bull. N. Y. St. Mus. 86: 40.
1921—McDunnough—Canad. Ent. 53: 117 (*propinquus*, in part).
1922—Murphy—Bull. Lloyd Lib. Ent. Ser. 2: 42. Figs, life history notes.
1923—McDunnough—Canad. Ent. 55: 40. Fig. (*propinquus*).
1925—McDunnough—Trans. R. S. C. p. 214 (key to male imagos).
1925—McDunnough—Canad. Ent. 57: 172.
1930—McDunnough—Canad. Ent. 62: 58.
1930—Ide—Canad. Ent. 62: 221. Nymph, figs...Ont.
1932—McDunnough—Canad. Ent. 64: 79 (key to nymphs).

Length: body 3; wing 2½–3 mm.

Abdominal tergites 2–6 of male imago hyaline whitish; genitalia of the *intercalaris* type; hind wing with two veins only (see fig. 163).

Turbinate eyes moderately large, oval; deep brown; upper part of stalk yellow, basal part blackish. Face black; light green shading at base of antenna only. Antenna black, a white ring at the apex of the first joint. Thorax shiny black; faint pale markings along the pleural sutures. Legs white; fore femur slightly smoky. No granulations between the cross veins of the stigmatic area of the fore wing; no intercalaries in first interspace. Hind wing with two veins only, the third entirely wanting.

Abdominal segments 2–6 hyaline whitish; tergites 7–10 light sepia brown; sternites opaque white. Black dots over the spiracles. Tails white. A tubercle on inner apical margin of the first forceps joint is well developed.

The above description is based on the determination of *B. pygmaeus* by Dr. McDunnough (Canad. Ent. 57: 172), after careful examination of the few remaining fragments of Hagen's type, in the museum at Cambridge. Some of the literature listed above may not apply to the species as thus determined. In the Cornell collection there are specimens of this species taken by Dr. F. P. Ide at Coteau-du-lac, Que., which Dr. McDunnough has kindly presented, as well as several from Lachine and Ste. Annes, Que., and Ottawa, Ont. Other specimens are from Saranac Inn, N. Y. (J. G. Needham); the Smoke Hole, W. Va.; Conococheague Pk., near Hagerstown, Md. (V. Argo); and Ithaca, N. Y.

Baetis quilleri Dodds

1923—Dodds—Trans. Amer. Ent. Soc. 49: 112. Figs.......................................Colo.
1924—McDunnough—Canad. Ent. 56: 8 (in discussion of *harti*).

Length (female): body 4; wing 5 mm.

We are holding as *B. quilleri* one male and one female specimen from Tuttle, Idaho. Abdominal tergites 2–6 of male imago deep olive brown; genitalia of the *intercalaris* type; hind wing with curved costal projection as in Centroptilum, and with cross veins between 2nd and 3rd vein.

Head and thorax of male deep blackish brown. Turbinate eyes (dried) large, deep reddish brown. Pleura largely paler reddish brown. Fore and middle legs missing. Hind femur pale smoky brown; tibia and tarsus pale yellowish white. Wings hyaline, venation pale. 8–9 cross veins in stigmatic area of fore wing; somewhat aslant, complete, with faint granulations between them. Marginal intercalaries well developed; a single one only in the first interspace. Hind wing much as in Dr. Dodds' description, but with traces of one or more cross veins between veins 1 and 2, as well as a single well-defined one between veins 2 and 3 (see fig. 163). A faint marginal intercalary between veins 2 and 3, not seen in the female specimen. Abdominal tergites 2–10 deep olive brown, the basal ones semi-hyaline, posterior ones semi-opaque. Traces of darker oblique submedian dashes from the anterior margin of each tergite; at the end of each dash is a small dark dot. Sternites paler, yellowish olive, each with a distinct dark oblique submedian dash and dark dot near center of sternite on each side. Posterior margins of all segments opaque or slightly darkened, so that the abdomen appears annulate. Each spiracle marked with a small dark dot; a dark dash just above the spiracular line, which is faintly outlined in black. Tails missing. Genitalia imperfect, but apparently of the *intercalaris* type, quite similar to *B. erebus*.

Female somewhat paler reddish brown; abdominal sternites yellowish; tails yellowish, the joinings distinctly reddish brown.

Dr. Dodds' description, of the female only, is mainly diagnostic of the hind wing. Of this he says:

"Narrow, with prominent curved humeral point, three longitudinal veins; first and second distinct and arising close together, but not from a common stalk; third faint and near the hind margin, where it ends at about three-fourths length of wing. There are two faint cross-veins between second and third longitudinal veins."

Baetis rusticans McDunnough

1925—McDunnough—Trans. R. S. C. p. 217. Figs. . . . . Que.
1928—McDunnough—Canad. Ent. 60: 239 (in discussion of *pluto*).

Length: body 3½; wing 4 mm.

Abdominal tergites 2–6 of male imago olive brown, the posterior margins deeper ruddy brown; genitalia of the *intercalaris* type, the tubercle on the inner apical margin of the first forceps joint well developed.

Turbinate eyes moderate, red-brown. Head and thorax deep olive brown. Antero-lateral margins of the mesonotum paler, "at times terminating in a small yellowish spot on each side of the anterior tubercle"; posterior portion of mesonotum with lateral and central tinges of ruddy brown, the scutellum sometimes paler. Ochreous markings on the pleura. Anterior projection of metanotum whitish. Legs pale yellowish brown; femora, especially near the apex, tinged with deeper brown. Wings hyaline, cross veins slightly brown-tinged. Cross veins of the stigmatic area of the fore wing 4–6 in number, with slight granulations between them. First pair of intercalaries short or obsolescent, second pair shorter than the third. Hind wing about two and one-half times as long as its width; third vein extends weakly to the middle of the hind margin, at the base bent strongly toward vein 2. No intercalaries.

Abdominal tergites 2–6 semi-translucent, olive brown, the posterior margins narrowly deeper, ruddier brown; 7–10 opaque russet brown. Sternites 2–6 smoky hyaline; 7–9 whitish with brown tinges, the posterior margins edged with reddish brown. Tracheae of the stigmatic area form a black lateral line. Tails and forceps whitish. The tubercle on the inner apical margin of the first forceps joint is well developed.

Distinguished from *B. levitans* by the better development of the tubercle on the first forceps joint, and the much darker abdomen. The smaller size, and lack of intercalaries in the hind wing, separate it from *B. brunneicolor*. There is one specimen in the Cornell collection from Ft. Jackson, N. Y. (L. Harper) which seems to be of this species.

Baetis spinosus McDunnough

1925—McDunnough—Canad. Ent. 57: 174. Figs. . . . . Man.

Length: body 4½; wing 5½ mm.

Abdominal tergites 2–6 of male imago semi-translucent, whitish; only two veins in the hind wing; genitalia unique in possessing a strong pointed projection on the inner apical margin of the second forceps joint; excavation and spine between bases of forceps (see fig. 164).

Turbinate eyes deep red-brown, slightly smaller than in *B. intercalaris*. Head and thorax deep shiny black. The antero-lateral edges of the mesonotum, the pleural sutures, and the edges of the central portion of the mesosternum pale yellowish to yellowish brown. Legs white. Wings hyaline, venation pale. 8–9 cross veins in the stigmatic area of the fore wing; few or no granulations between them. No intercalaries in the first interspace. Hind wing long, leaf-like, the third vein entirely absent. Costal projection reduced to a mere trace (see fig. 163).

Abdominal tergites 2–6 semi-translucent, white or yellowish white. Tergites 7–10 deep chestnut or chocolate brown. Basal sternites whitish, semi-translucent; apical ones opaque white. Tails white. Faint black spiracular dots are present. A rounded projection on the inner apical margin of the first forceps joint; second joint somewhat cylindrical, its inner apical margin bearing a strong pointed projection. Apical joints much as in *B. frondalis*. Between the bases of the forceps is an excavation containing a sharp spine.

There are specimens of this species in the Cornell collection from Saranac Inn, N. Y. (J. G. Needham); from Atlanta, Ga., the Chattahoochee River (P. W. Fattig); and a paratype from Darlingford, Man., kindly presented by Dr. McDunnough. We have also examined specimens lent us by Dr. McDunnough, from Foster, Que., and the Rideau River, Ottawa.

Baetis thermophilos McDunnough

1926—McDunnough—Canad. Ent. 58: 188...Wyo.

Length: body 3½; wing 5 mm.

Abdominal tergites 2–6 of male imago dark olive brown, distinguished from all other known species by the very long third vein of the hind wing, which equals the other two veins in length.

Head and thorax shiny blackish. Legs smoky brown, with paler tarsi. Wings hyaline, venation pale. 7–8 stigmatic cross veins in the fore wing, without granulations. Hind wings long and narrow, and with three longitudinal veins; the third vein runs close to the hind margin, and equals the first two in length; between veins 2 and 3 is a single intercalary. The costal projection is rather blunt and short (see fig. 163).

Abdominal tergites 2–6 dark olive-brown, with pale narrow rings on the posterior margins. Tergites 7–10 opaque, brighter brown. Sternites pale ochreous. The bases of the forceps dusky, the two apical joints whitish; the last joint is narrow, and almost as long as the preceding one. Dr. McDunnough has presented to the Cornell collection a female of this species, from Upper Geyser Basin, Yellowstone Pk., Wyo. In this collection there are also several specimens from Eel River, Calif. (J. C. Bradley), which may belong here. Specimens of a closely-allied species, in which the long third vein is closer to the hind margin, are in this collection, from Cloverdale, Calif. (J. C. Bradley).

Baetis tricaudatus Dodds

1923—Dodds—Trans. Amer. Ent. Soc. 49: 111. Figs......................................Colo.

Length: body 7; wing 7 mm.

No color characters given. Genitalia of the *moffati* type.

Hind wing with three veins, the third running close to the hind margin to beyond the middle. Veins 1 and 2 converge distinctly toward the tip. Upper margin of hind wing almost straight, beyond the costal projection, which is quite well developed. Between veins 2 and 3 are two intercalaries, the second of which extends up from the hind margin toward the 2nd vein. The inner margin of the second (cylindrical) joint distinctly irregular and of a wrinkled appearance.

We have seen no male imagos of this species. Allied species, in which the upper margin of the hind wing is more rounded and the inner margin of the second forceps joint is smooth, occur in Logan, Utah, and Yellowstone Park. A single female *B. tricaudatus* in Denver, Colo. (E. J. Oslar) is in the Cornell collection. We have seen also a male subimago from Ft. Garland, Colo., which seems to be of this species; the abdomen of this specimen is dark brown dorsally, somewhat paler ventrally.

Baetis unicolor Hagen

1861—Hagen—Syn. Neurop. Ins. N. Amer. p. 54 (Cloe)................................D. C.
? 1862—Walsh—Proc. Acad. Nat. Sci. Phil., p. 380.
1863—Hagen—Proc. Ent. Soc. Phil. 2: 178.
1871—Eaton—Trans. Ent. Soc. London, p. 120 (transferred to Baetis).
1885—Eaton—Revis. Monogr. p. 169 (*hageni*).
1921—McDunnough—Canad. Ent. 53: 117.
1923—McDunnough—Canad. Ent. 55: 41.
1925—McDunnough—Canad. Ent. 57: 172 (in discussion of *brunneicolor*).

Length (female): body 4; wing 4½ mm.

This species, described from a female imago, has not yet been identified with certainty. Until such time as extensive collections of correctly associated individuals of both sexes can be made, near the type locality, the question of its identity must remain unsolved. We make no attempt to include this species in our key. There is no certainty that Walsh had the true *unicolor* Hagen, in his Illinois material. The following brief description of the female imago is taken from the original, by Hagen.

Body bronze-brown (brassy-brown, in the translation of Hagen's Synopsis.) Legs pale luteous. Wings hyaline. Tails white.

One of us (Needham) has examined the female type, labeled *Cloeon unicolor* Hagen, in the Museum of Comparative Zoology at Cambridge. In this specimen, the hind wing is long and narrow; veins 1 and 2 do not appear to converge towards the tip; the hind margin is upturned, making it impossible to determine the presence or absence of vein 3. Another specimen in the Museum, labeled *Cloe unicolor?* Hagen, from New Hampshire, is clearly a Callibaetis.

Baetis vagans McDunnough

1922—Murphy—Bull. Lloyd Lib. 22, Ent. Ser. 2: 40 (*posticatus*). Figs.......................N. Y.
1925—McDunnough—Trans. R. S. C. p. 219. Figs..Que.
1930—McDunnough—Canad. Ent. 62: 58.
1930—Ide—Canad. Ent. 62: 221..Ont.
1932—McDunnough—Canad. Ent. 64: 79. Key to nymphs.
1933—McDunnough—Canad. Ent. 65: 278...Gaspe

Length: body 6; wing 6 mm.

Abdominal tergites 2–6 of male imago dull olive brown; genitalia of the *moffati* type.

Turbinate eyes very large, deep red-brown. Head and thorax deep pitch brown; a pale yellowish spot is usually present on each side of the anterior tubercle of the mesonotum; scutellum pale brown; anterior to the scutellum, and on the postero-lateral margins, are light ruddy brown shadings. Paler brown shading on the pleural sutures.

Anterior margin of metanotum, and its anterior projection, distinctly whitish. Legs dull yellowish white; fore femur distinctly tinged with brown, and a small dark patch near the apex of the fore tibia. Wings hyaline, venation pale. Cross veins in the stigmatic area of the fore wing strongly anastomosed, forming an irregular network. First pair of intercalaries usually not well developed, may be broken; second and third pairs moderately long, subequal. Hind wing twice as long as its width. Third vein well developed, running to the middle of the hind margin. One or two intercalaries between veins 2 and 3, and traces of one or two intercalaries between veins 1 and 2 (see fig. 163).

Abdominal tergites 2–6 dull olive brown, paler on the anterior margins, semi-hyaline; a dark ruddy brown line marks the posterior margin, but does not extend along the lateral area. Tergites 7–10 opaque, ruddy or chocolate brown. Sternites 2–6 pale smoky semi-translucent; 7–10 pale whitish brown. Tails white. Tracheae in the spiracular area more or less marked with black. It is larger than the related species *B. incertans*, and with larger turbinate eyes.

This species is common in the vicinity of Ithaca, N. Y.; there are many specimens of it in the Cornell collection. These specimens are often somewhat larger than the type series, the fore wing often attaining 7 mm. in length. Dr. Murphy has published rather extensive notes on the life history of *B. vagans*, under the name of *B. posticatus*.

Genus **Centroptilum** Eaton

1869—Eaton—Ent. Mo. Mag. 6: 132.
1885—Eaton—Revis. Monogr. p. 174.
1920—Ulmer—Stett. Ent. Zeit. 81: 124.
1933—Spieth—J. N. Y. Ent. Soc. 41: 339.

Small mayflies, the fore wing ranging in size from 4 to 8 mm. (Nearctic species). Basal abdominal segments of the male imago are pale yellowish or whitish in most species, often with more or less extensive ruddy or brownish markings. Females paler, often light reddish brown. Turbinate eyes of male moderate to large in size, usually lengthily oval in shape (see fig. 16b). Eyes of female relatively large, and set rather high on the sides of the head. Posterior margin of head of female quite distinctly emarginate in the median area.

Fore leg of male about as long as the body. Fore tibia 1½ to 2 times the length of the femur; tarsus slightly longer than the tibia. In the female, the tarsus is nearly 9⁄10 the length of the tibia; tibia almost as long as the femur. In the hind leg of the male, the tarsus is about as long as the tibia, which is subequal to the femur. Basal tarsal joint quite long; usually equal in length to the 3rd and 4th joints together, sometimes slightly longer. Hind leg of female very similar to male. The marginal intercalaries of the fore wing occur singly. The small hind wing is narrow and elongate; costal projection curved; only two longitudinal veins; in most species, no traces of cross veins.

Basal joint of forceps rather stout, usually slightly wider than long. Second joint likewise rather stout, often prolonged inward on its inner margin. Third joint long and relatively slender, frequently somewhat bowed; usually dilated toward the apex. Distal joint variable; sometimes half as long as the third joint, but usually much shorter, clavate, two or three times as long as broad. A large 'penis-cover' is typically present between the basal forceps joints, which are relatively far apart; the shape of the apical margin of this cover varies in different species. Posterior margin of 9th sternite of female prolonged backwards very slightly in the median area.

Maxillary palp of nymph 3-jointed; fully as long as the body of the maxilla, and extending somewhat beyond the tip of the galea-lacinia. Labial palp 3-jointed, the distal joint distinctly dilated and truncate apically. Canines of the mandible long, slender, sharp-pointed, in the genotype; somewhat shorter and stouter in some other species. Claws long and slender; not denticulate, except in *C. elsa*; ½ to ¾ the length of their respective tarsi. Gills single on all segments, in the genotype and most other species; apparently variable in shape in different species; tracheation pinnate. Gills of *C. bellum* are reported (Ide—Canad. Ent. 62: 224, 1930) to possess a recurved dorsal flap,—*i.e.*, double instead of single. In this species also, the tracheae are branched on the inner side only, as in Neocloeon. Nymphs of *C. rivulare* and *C. ozburni* resemble *bellum*. Tails three, approximately equal in length and thickness.

Genotype—*C. luteolum* (in Ephemera), Müller.

Distribution—Holarctic, African; in Nearctic, quite generally distributed.

KEY TO THE SPECIES

Males*

1—Dorsum of abdomen largely bright ruddy, or with a prominent ruddy dorsal stripe 2

—Dorsum of abdomen not as above; tergites 2–6 yellowish or whitish, often shaded or margined with brown or ruddy 3

2—Abdominal tergites of female with a prominent ruddy dorsal stripe (male unknown) . . . . **infrequens**

—Abdominal tergites of male bright ruddy, except for a pale narrow lateral area; of female, pale yellowish tinged with brown posteriorly . . . . **quaesitum**

3—Abdominal tergites 7–10 opaque creamy whitish; inner margin of second forceps joint with an inward bulge or tubercle 4

—Abdominal tergites 7–10 yellowish brown, ruddy or chocolate brown, not as above; inner margin of second forceps joint with or without such a tubercle . . 6

4—Thoracic sternum and metanotum creamy white; fore wing 6 mm. in length . . . . **walshi**

—Thoracic sternum, and posterior half at least of metanotum, wood brown; wing 4 to 5 mm. in length 5

5—Fore wing 4 mm. in length; hind wing relatively narrow; tergite 9 not margined with pinkish; tubercle on second forceps joint rather poorly developed . . **album**

—Fore wing 5 mm. in length; hind wing relatively broader; lateral margin of tergite 9 pinkish; tubercle on second forceps joint strong and prominent . . . **convexum**

6—A sublateral row of four minute red transverse dashes on each side of tergites

* Except *infrequens*, of which only females are known.

2–6, on the intersegmental suture; tubercle present on second forceps joint .... **rufostrigatum**

—No such sublateral row of red dashes; tubercle on second forceps joint present or absent........ 7

7—Inner margin of second forceps joint distinctly and prominently tuberculate; distal joint of forceps very long........ 8

—Inner margin of second forceps joint may possess a distinct inward bulge, but is not tuberculate; distal joint usually shorter........ 9

8—Head and thorax shiny black; abdominal tergites 2–6 hyaline white; third joint of forceps distinctly bowed........ **conturbatum**

—Head and thorax olivaceous brown; abdominal tergites 2–6 yellowish tinged with brown; third joint of forceps not strongly bowed, but widened apically........ **asperatum**

9—A distinct tubercle is present between the bases of the forceps limbs........ 10

—No such tubercle between the bases of the forceps limbs........ 11

10—Tergite 10 yellowish; tergites 2–4 with narrow ruddy posterior margins; tergites 7–9 deep chocolate brown; wing 5 mm. in length........ **bifurcatum**

—Tergites 7–10 ruddy brown; posterior half of tergites 2–6 ruddy brown; wing 4 mm. in length........ **semirufum**

11—Wedge-shaped ruddy lateral markings (extensions of ruddy posterior bands) are present on tergites 2–6, above the spiracular line........ 12

—No such wedge-shaped ruddy markings on tergites 2–6........ 15

12—Mesothorax deep brown or blackish; femora usually with faint ruddy markings 13

—Mesothorax olivaceous to light reddish brown; femora with no such ruddy markings........ 14

13—Ruddy geminate dorsal line on tergites 2–6; eyes bright red, in dried specimen .... **intermediale**

—No such ruddy geminate dorsal line on tergites; eyes blackish brown, in dried specimen........ **simile**

14—Hind wing with a distinct marginal intercalary below the second vein, and with several distinct cross veins........ **venosum**

—Neither marginal intercalary nor cross veins present in hind wing........ **bellum**

15—Inwardly-directed projection present on inner margin of second forceps joint.. 16

—Such a projection wholly wanting, or very poorly developed........ 17

16—Third joint of forceps quite strongly bowed, widest at apex; distal forceps joint relatively long; hind wing wide (about 3 times as long as broad)........ **elsa**

—Third joint of forceps not distinctly bowed, not wider at apex than at base; distal forceps joint short; hind wing long and narrow (about 4 times as long as broad) .... **caliginosum**

17—Abdominal sternites 7–10 russet or chestnut brown, only slightly paler than the corresponding tergites........ **rivulare**

—Abdominal sternites 7–10 opaque, white, contrasting strongly with the red or brown of the corresponding tergites........ 18

18—Sternites 2–6 faintly but distinctly brownish on the posterior margins; tergites 7–10 chocolate brown; eyes blackish brown in dried specimen........ **ozburni**

—Sternites 2–6 with no such dark posterior markings; tergites 7–10 bright reddish brown; eyes bright red in dried specimen........ **fragile**

VERIFICATION TABLE

Male adults

Species	*Wing length*	*Eyes*[1]	*Meso-thorax*	*Red marks*[2] *on 2–6*	*Abd. terg. 7–10*	*Projection*[3] *on forceps*	*Distr.*
album	4	Or. br.	Yel. br.	Absent	Whitish	Present	E
asperatum	5	Yel. br.[5]	Oliv. br.	Absent	Oliv. br.	Present	Calif.
bellum	6	Red	Olivac.	Present	Red br.	Present	E
bifurcatum	5	Or. br.	Dp. oliv. br.	Absent	Choc. br.	Weak	W
caliginosum	6½	Bl. br.	Dp. brown	Absent	Choc. br.	Present	Que.
conturbatum	5	Bl. br.	Black	Absent	Choc. br.	Present	W
convexum	5	Chestnut	Brown	Absent	Whitish	Present	Ont.
elsa	5½	Pinkish[5]	Red-brown	Absent	Yel. br.	Present	Wyo.
fragile	5	Red	Bl. br.	Absent	Brt. br.	Absent	E
infrequens[4]	8♀	?	Yel. br.	?	Yellow	?	Man.
intermediale	6	Red	Dp. brown	Present	Maroon br.	?	Que.
ozburni	4½–5	Bl. br.	Blackish	Absent	Choc. br.	Weak	E
quaesitum[6]	7	Or. red[5]	Lit. br.	Absent	Brt. ruddy	?	Alta.
rivulare	5½–6	Orange	Bl. brown	Absent	Russet br.	Weak	Ohio
rufostrigatum	5	Bl. br.	Bl. brown	Absent	Choc. br.	Weak	E, C
semirufum	4	Red br.	Olivac.	Present?	Ruddy br.	?	NE
simile	5–6	Bl. br.	Blackish	Present	Ruddy br.	Weak	E
venosum	5½–6	Orange[5]	Red brown	Present	Lt. red br.	Present	Calif.
walshi	6	Reddish	Yel. br.	Absent	Creamy	Present	C

KEY TO THE KNOWN SPECIES

Nymphs

1—Gills double (possessing a small recurved dorsal flap) on some or all of segments 1–7 of the abdomen; tracheae mainly or wholly on the inner side of the gill. . . . 2

—Gills single (no flaps) on abdominal segments 1–7; tracheation of gill normally pinnate. 4

2—Tails dark brown distally, with a pale band about ⅔ the distance from the base. . 3

—Tails pale except for a very narrow dark apical mark, and a wider brown band about halfway from base to tip. . . . . **rivulare**

3—Abdominal sternite 9 marked with a broad brown band; other sternites distinctly brown on the posterior margins, often also with lateral brown markings **ozburni?**

—No such dark markings on abdominal sternites. . . . . **bellum**

4—Abdominal tergites wholly pale except for darker posterior margins; tracheation of gills very indistinct. . . . . **elsa**

—Abdominal tergites considerably marked and shaded with darker color; gill tracheation quite distinct. 5

5—Abdominal tergites 7–10 largely shaded with brown. . . . . **album**

—Abdominal tergites 7–10 largely pale. . . . . **convexum**

---

[1] Color in dried specimen, unless otherwise noted.
[2] Reddish wedge-shaped lateral markings, on tergites 2–6.
[3] Projection or tubercle on inner margin at 2nd forceps joint.
[4] Only female known.
[5] From alcoholic specimen.
[6] Abdominal tergites 2–10 bright ruddy.

Centroptilum album McDunnough

1926—McDunnough—Canad. Ent. 58: 189..Ont., Que.
1929—McDunnough—Canad. Ent. 61: 173 (in discussion of *conturbatum* and *walshi*).
1930—McDunnough—Canad. Ent. 62: 58. Nymph fig.
1932—McDunnough—Canad. Ent. 64: 80.
? 1932—Traver—J. Elisha Mitchell Sci. Soc. 47: 229..N. C.

Length: body 4; wing 4 mm.

Abdominal tergites 2–6 of male imago entirely hyaline.

Turbinate eyes dull orange-brown, in dried specimen. Head blackish brown; whitish ochreous tinges in median area behind ocelli. Pronotum blackish with paler patches laterally. Mesonotum pale ochreous with brown shading; scutellum and posterior portion almost entirely white, shaded faintly with pink. Metanotum deep brown; anterior margin white with pinkish shading and a median dark dot. Pleura and sternum mostly deep brown, with faint white shading at the intersegmental areas. Legs white. Wings hyaline, venation pale. Hind wing very narrow, much as in *C. fragile*, but not quite as long; costal projection prominent. Abdominal segments 2–6 entirely hyaline; segments 7–10 opaque whitish. Tails white. A slight inward bulge on the inner margin of the second forceps joint.

This species is allied closely to *C. convexum;* the hind wings are considerably more narrow than in that species, and the tubercle on the second forceps joint less well developed. We have seen no specimens of this species. The gills of the nymph are single on all segments; the dorsum of the abdomen is largely shaded with dark color.

Centroptilum asperatum sp. nov.

Length: body 5; wing 5 mm.

Abdominal tergites of male imago semi-hyaline, yellowish with a tinge of brown; one large and two very small protuberances on inner margin of second forceps joint; distal joint very long.

Head and thorax olivaceous brown. Antennae and turbinate eyes yellowish brown. Antero- and postero-lateral margins of mesonotum, a streak anterior to the wing root, and pleural markings, blackish brown; scutellum and a spot on each side of it, yellowish. Median area of metanotum yellowish. Posterior portion of sternum dark brown. a blackish line between this and mesosternum. Legs yellowish white. Wings hyaline, venation pale; 6 to 7 cross veins in stigmatic area, which is not clouded. 2 of these veins are forked near the subcosta, 2 are anastomosed. Hind wing rather wide for its length, much as in *C. elsa;* 2 longitudinal veins; costal projection rather short but strong (see fig. 165).

Abdominal segments 2–6 semi-hyaline, yellowish; tergites tinged with pale brown, especially along the spiracular area; a faint dark dash over each spiracle, and traces of a narrow broken dark spiracular line. Segments 7–10 opaque, olivaceous brown dorsally, yellowish white ventrally. Tails missing. Forceps somewhat as in *C. conturbatum*, but with a single larger median protuberance, beset with minute hairs, on the inner margin of the second joint; near the apex is a minute second protuberance, and a third one near the base (see fig. 166). Third (long) joint distinctly widened toward the apex; a slight protuberance near the base on the inner margin; not bowed. Distal joint relatively longer than in *C. conturbatum*, being more than half the length of the third joint. Plate between the forceps bases somewhat dome-shaped (in the single specimen we have, slightly concave on the posterior margin, but this may not be normal).

Holotype—Male imago. Big Bear Lake, San Bernardino Mts., Calif., Aug. 29, 1932; collector, Charles D. Michener. No. 1359.1 in Cornell University collection.

In general appearance, this species resembles *elsa*, but in type of genitalia is closer to *conturbatum*.

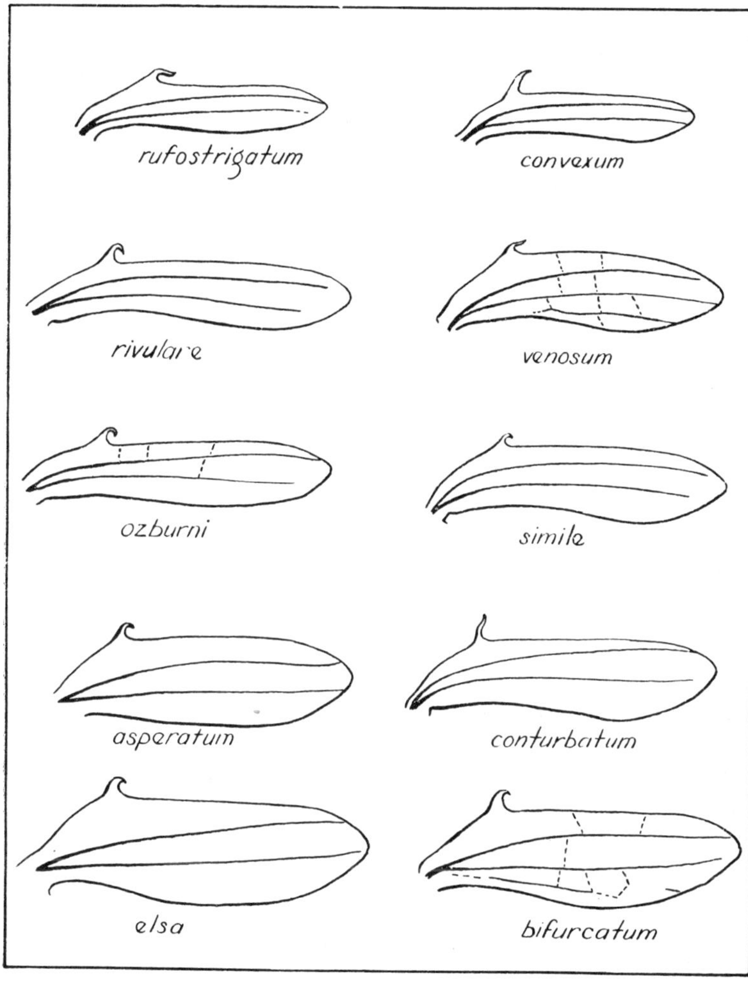

FIG. 165. Hind wings of ten species of Centroptilum.

Centroptilum bellum McDunnough

1924—McDunnough—Canad. Ent. 56: 96. Op. cit. 97(in discussion of *ozburni*)................Que.
1930—McDunnough—Canad. Ent. 62: 223. Nymph fig......................................Ont.
1931—McDunnough—Canad. Ent. 63: 88 (in discussion of *intermediale*).

Length: body 5; wing 6 mm.

Abdominal tergites 2–6 of male imago pale yellowish, the posterior margins narrowly ruddy brown.

Head and antennae dull blackish. Turbinate eyes deep reddish in dried specimen. Deep olive brown shading on the prothorax; mesonotum olivaceous shaded with variable amounts of yellowish subdorsally; lateral and posterior margins tinged with ruddy. Metanotum darker, deep olive brown; sutures darker; anterior dorsal portion shaded with ruddy. Legs yellowish white. Wings hyaline. Hind wing moderately broad, two longitudinal veins. Abdominal tergites 2–6 pale yellowish, semi-hyaline;

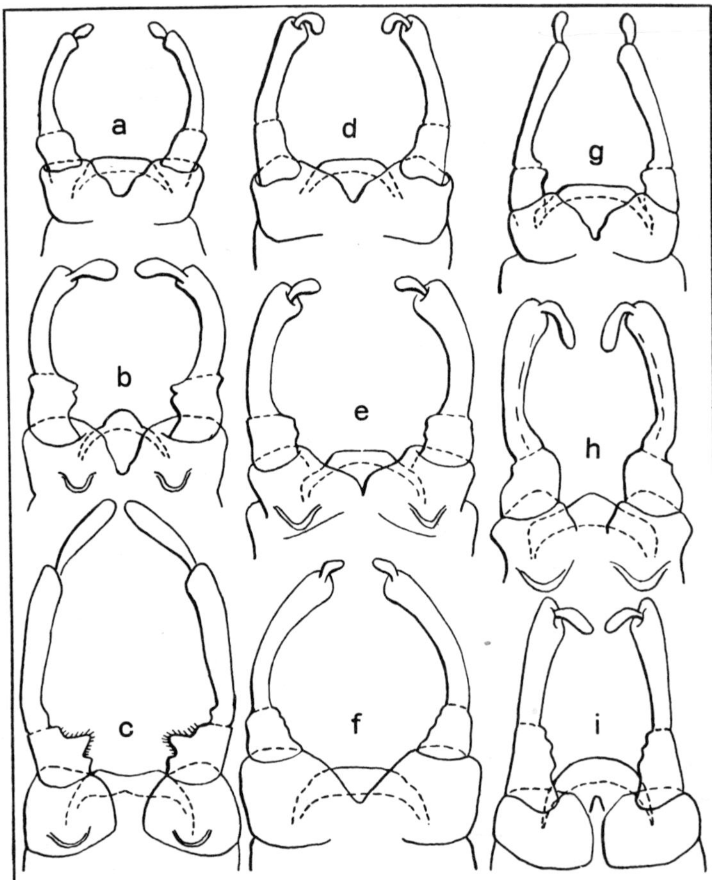

FIG. 166. Male genitalia of nine species of Centroptilum: *a, rufostrigatum; b, conturbatum; c, asperatum; d, fragile; e, venosum; f, rivulare; g, bellum; h, elsa; i, bifurcatum.*

on the posterior margin of each is a narrow ruddy brown band extending to the spiracular area, where it ends in a faint small reddish patch. Sternites 2–6 likewise pale, but with no ruddy posterior markings. Tergites 7–10 bright ruddy brown; sternites opaque yellowish shaded laterally with reddish, and with faint traces of ruddy posterior margins. Spiracular line blackish. Forceps and tails white. See figure 166.

Distinguished from *C. ozburni* by the paler thorax and the ruddier color of the posterior tergites. Hind wing relatively wider and posterior margins of tergites narrower than in the related *C. intermediale.* Closely allied also to *C. caliginosum*, but with more ruddy abdominal markings than in that species. There is one specimen of this species in the Cornell collection, kindly presented by Dr. McDunnough; it is from Lachine, Que. (F. P. Ide).

The nymph has double gills on the basal segments at least; abdomen marked dorsally with brown, segments 4, 5, 7, 8 and 10 being paler; tails dark distally, with a pale band about ⅓ from the tip.

Centroptilum bifurcatum McDunnough

1924—McDunnough—Canad. Ent. 56: 96 .... Alta.
1929—McDunnough—Canad. Ent. 61: 171. Fig.

Length: body 5; wing 5 mm.

Abdominal tergites 2–6 of male imago yellowish white, with traces of ruddy shading on the posterior margins of 2–4, and a faint reddish dorsal line. Hind wing relatively wide.

Head blackish, also the basal half of the antennae. Thorax deep olive brown. A pale yellow line extends forward from the wing base along the lateral margin of the mesonotum, ending in a small round yellowish patch. Posterior margin of mesonotum and pleural sutures yellowish. Legs pale yellowish white, femora somewhat deeper in color. Wings hyaline. Hind wing relatively wider than in most species of this genus; in some specimens, but not all, the lower of the two longitudinal veins may be faintly forked near its apex (see fig. 165).

Abdominal segments 2–6 hyaline, yellowish white; a faint reddish dorsal hair-line, and traces of ruddy shading on the posterior margins of tergites 2–4, extending to the spiracular line. Tergites 7–9 deep chocolate brown; 10 pale yellowish. Sternites 7–9 alabaster white. On the pale tergites, a broken blackish line marks the spiracular area. Forceps and tails whitish. Plate between forceps bases large, somewhat dome-shaped; a small pointed tubercle is borne on it (see fig. 166).

The greater width of the hind wing separates this species from *C. rufostrigatum*, which it resembles superficially. Three specimens in the Cornell collection, from Hollister and Tuttle, Idaho, seem to be of this species. In one of these only, the forking of the second vein in the hind wing appears. In each, however, there are faint traces of cross veins in the hind wing, 3 to 5 in each of the three spaces; in two of these, an interrupted third longitudinal vein occurs, its basal and apical portions missing. The second joint of the forceps is slightly irregular on the inner margin, but lacks a definite tubercle or projection as in *C. conturbatum*. In these specimens, the scutellum is prominently pale yellowish white. The presence of the small pointed tubercle between the forceps bases allied this species with *C. semirufum*, from which it may be distinguished by the much wider hind wing and the duller brown of the posterior tergites.

Centroptilum caliginosum McDunnough

1925—McDunnough—Canad. Ent. 57: 175 .... Que.

Length: body 6; wing 6½ mm.

Abdominal tergites 2–6 of male imago pale translucent with very faint tinge of brown.

Turbinate eyes deep blackish brown in dried specimen. Head and thorax deep brown. Lateral margin of mesonotum, the sutures, and a small patch anterior to the scutellum, paler brown. Legs dull whitish; fore legs tinged with smoky. Wings hyaline, venation pale; 6 to 7 cross veins in the stigmatic area of the fore wing; intercalary of first interspace missing. Hind wing long and narrow, the costal projection strong. Abdominal segments 2–6 pale translucent, the tergites very faintly tinged with brown, particularly along the lateral and posterior margins. Tergites 7–10 chocolate brown, sternites paler. Forceps and tails whitish. A well developed inward bulge is present on the inner margin of the second forceps joint.

This species, related to *C. rufostrigatum*, lacks entirely the reddish abdominal markings of that species, is also considerably larger, and has a much better developed bulge on the second forceps joint. It is distinguished from the allied *C. bellum* by the absence of ruddy markings on the tergites. We have not seen a specimen of this species.

Centroptilum conturbatum McDunnough

1929—McDunnough—Canad. Ent. 61: 171. Fig. .... Alta.

Length: body 5; wing 5 mm.

Abdominal tergites 2–6 of male imago hyaline white; inner margin of second forceps joint distinctly tuberculate; distal joint unusually long.

Turbinate eyes dark blackish brown, in dried specimen. Head and thorax shiny black; pleural sutures brownish. Legs white. Wings hyaline, venation pale (see fig. 165). Abdominal segments 2–6, and part of 7, hyaline white, sometimes with faint dusky shading posteriorly on the middle tergites. Tergites 8–10, and often part of 7, chocolate brown. Sternites white; 2–6 hyaline, 7–9 opaque. Tails white. Second joint of forceps distinctly tuberculate on its inner margin; third joint strongly bowed; distal joint unusually long (see fig. 166). The plate between the bases of the forceps is conical on its posterior margin.

There are specimens of this species in the Cornell collection from Moose, Wyo. (Elsie B. Klots), and from Waddell Creek, Calif. (D. Shepherd); most of the specimens from Waddell Creek are in the Stanford University Collection.

Centroptilum convexum Ide

1930—Ide—Canad. Ent. 62: 222. Figs. .... Ont.

Length: body 4; wing 5 mm.

Abdominal tergites 2–6 of male imago hyaline white, sometimes with a median pinkish shading.

Turbinate eyes pale chestnut brown in dried specimen. Head brown, with darker brown markings at bases of antennae and on median carina. Prothorax brown. Mesonotum brown, the pleura and scutellum marked with ochreous white. Metanotum pinkish on the anterior margin, brown in posterior half. Legs pale yellowish. Wings hyaline. Hind wing moderately broad, the costal projection prominent (see fig. 165). Abdominal segments 2–6 hyaline white, sometimes tinged faintly with pinkish in the median area. Segments 7–10 opaque white, the tergites with a pink flush, most noticeable on 9, where it appears as a pink lateral border. Tails white. Second joint of forceps with a distinct inwardly projecting tubercle.

From the closely-allied *C. album*, this species may be separated by the pink lateral margin on tergite 9, the somewhat darker thorax, the wider hind wing, and the greater development of the tubercle on the second forceps joint. It is somewhat smaller than *C. walshi*, and is distinguished from it by the brown thoracic sternum and posterior half of the metanotum, which areas are creamy white in *C. walshi*. In the Cornell collection there are specimens of this species from Horning's Mills, Ont., which Dr. Ide kindly presented.

The nymph is closely allied to *C. album*, with single gills on all segments, but the dorsal abdominal markings are more restricted, the last four segments being largely pale.

Centroptilum elsa sp. nov.

Length: body 5; wing 5–5½ mm.

Abdominal tergites of male imago yellowish with faint brown shading, semi-hyaline; a regularly rounded inward bulge on inner margin of second forceps joint; distal joint long.

Head red-brown; antennae pale, basal joints faintly tinged with red-brown. Turbinate eyes flesh-colored. Thorax red-brown; pleura and anterior portion of mesonotum shaded with yellowish; all sutures, pronotum, posterior portion of mesonotum, small sclerites in pleura below wing, posterior margin of metanotum and most of sternum, darker. Legs pale yellowish white. Wings hyaline, venation pale whitish; pale cloud in stigmatic area. 4 to 5 complete cross veins in this area; between these, several very incomplete ones, sometimes almost horizontal in position. Hind wing quite large, rather wide for its length; costal projction rather small; two longitudinal veins only (see fig. 165).

Abdominal segments 2–6 semi-hyaline; tergites yellowish tinged with brown, especially along the lateral margins; posterior margins scarcely darker than other parts; sternites yellowish white, the basal ones very faintly brown-tinged. Segments 7–10 opaque; tergites pale yellowish brown, the posterior margins very narrowly darker brown; sternites paler, a distinct brownish-red line along the pleural fold on sternite 9. Tails whitish. Plate between the bases of the forceps limbs distinctly convex on its posterior margin. Second joint of forceps with a regularly rounded inward bulge on the inner margin; outer margin distinctly curved, slanting inward at the apex; third (long) joint strongly bowed, widest near apex; distal joint long (see fig. 166).

Holotype—Male imago, reared from nymph; Moose, Wyo., July 30, 1929; collector, Dr. Elsie B. Klots. No. 13C0.1 in Cornell University collection.

Paratypes—2 male imagos, same data; No. 1360.3–4 in Cornell collection.

This species is named in honor of Dr. Elsie B. Klots, who collected and reared the nymphs. Several nymphs and subimagos of both sexes are also in the Cornell collection.

Head and thorax of nymph light red-brown. Abdomen yellowish; no distinct markings except the reddish brown posterior margins of the segments, most distinct dorsally. Legs yellowish; tarsi and claws tinged with reddish brown. Gills quite large and long; obovate; tracheae not distinct; apparently single on all segments. Tails yellowish, less stout than in the known eastern species, and with no thickened annuli. Middle tail slightly shorter than the laterals. Claws with a few indistinct denticulations; no short spines on lateral margins of segments 8–10, and no postero-lateral spines on any of the abdominal segments.

Centroptilum fragile McDunnough

1923—McDunnough—Canad. Ent. 55: 42 .... Ont.
1931—McDunnough—Canad. Ent. 63: 87 (in discussion of *intermediale*).

Length: body 4; wing 5 mm.

Abdominal tergites 2–6 of male imago whitish hyaline, slightly brown-tinged, the posterior margins faintly brown.

Turbinate eyes large, narrowly oval, rather kidney-shaped; pale yellow-brown in living insect, bright red in dried specimen. Face yellowish; antennae pale smoky, the bases white. In dried specimen, the face appears dark grey. Thorax shiny blackish brown; pale markings on the scutellum and adjoining areas; pleura brown, with pale markings near the wing bases. Legs white. Wings hyaline. Abdominal segments 2–6 hyaline whitish, slightly tinged with brown, the posterior margins faintly brown; slightly streaked with black, especially in the spiracular area. Tergites 7–10 rather bright brown; sternites opaque white. Second joint of forceps without a conspicuous inward bulge (see fig. 166).

This species may be distinguished from the allied *C. ozburni* by the bright red eyes in dried specimens (blackish brown, in *ozburni*), and the absence of dark posterior markings on the middle sternites. In the Cornell collection are specimens of this species from the Golf Club, Que. (G. S. Walley), which Dr. McDunnough very kindly gave us. Others which seem to belong here are from Ft. Jackson, N. Y. (L. Harper).

Centroptilum infrequens McDunnough

1924—McDunnough—Canad. Ent. 56: 98... Man.

Length (female): body 7; wing 8 mm.

A large species, of which only the female imago is known; distinctive by reason of the prominent ruddy dorsal stripe and large size.

Thorax of female light ochreous, shaded with creamy. Legs pale yellow. Wings hyaline, venation pale. Hind wing long and very narrow; two longitudinal veins, united for a short distance at the base, then running close to each other to beyond the costal projection, which is long and hook-like. Abdomen pale yellowish; last three segments opaque, others semi-hyaline. A broad and prominent ruddy stripe extends the entire length of the abdomen dorsally, deepest in color on tergites 7–9, and containing two pairs of minute hyaline submedian dots.

We have seen no specimens of this species. It is apparently related to *C. quaesitum*, but the female of that species has a pale yellowish abdomen, the posterior margins of the tergites tinged with brown.

Centroptilum intermediale McDunnough

1931—McDunnough—Canad. Ent. 63: 87... Que.

Length: body 6; wing 6 mm.

Abdominal tergites 2–6 of male imago pale whitish, hyaline; a geminate ruddy dorsal line; posterior margins broadly ruddy.

Turbinate eyes of dried specimen bright red. Thorax deep brown. Posterior portion of mesonotum, and the antero-lateral margin, with ruddy shading. Pleura brown, shaded with ruddy at base of wing and on the intersegmental areas. Sternum light sepia brown, the intersegmental membrane pale yellowish. Legs pale yellowish; femora show very faint traces of ruddy median bands. Wings hyaline, venation pale. Hind wing slightly longer and narrower than in *C. simile.*

Abdominal tergites 2–6 pale whitish, hyaline; a geminate ruddy dorsal line is present; posterior margins shaded broadly with ruddy, and with lateral wedge-shaped markings extending forward from these, halfway to the anterior margin. From near the apex of each wedge an oblique ruddy streak extends inward to the anterior margin; these streaks are most clearly marked on tergites 5 and 6, tending to break up into spots on tergites 2 to 4. Tergites 7–10 deep maroon brown. Sternites pale yellowish white; hyaline on 2–6, the apical sternites opaque with slight ruddy shading. Forceps and tails white.

The bright red eyes of dried specimens separate this species from the allied *C. simile,* in which the eyes are blackish brown; the thorax is also paler than in that species. From *C. bellum* it may be distinguished by the broader banding on the abdomen and the narrower hind wings. We have seen no specimens of this species.

Centroptilum ozburni McDunnough

1924—McDunnough—Canad. Ent. 56: 97...Que.

Length: body 4; wing 4½–5 mm.

Abdominal tergites 2–6 of male imago hyaline white, the posterior margins narrowly chocolate-brown.

Head and antennae blackish; tip of antennae white. Thorax unicolorous deep blackish dorsally, the pleura and sternum slightly more brownish; anterior edge of mesosternum tipped with pale ochreous. Legs whitish; traces of a ruddy median band on the femora. Wings hyaline. Hind wing long and narrow, two longitudinal veins (see fig. 165). Abdominal tergites 2–6 hyaline white; traces of an interrupted brown geminate mid-dorsal line; a narrow band of chocolate brown on the posterior margin of each, this band barely attaining the spiracular area. Tergites 7–10 chocolate brown, sternites alabaster white. Sternites 2–6 hyaline white, with very faint traces of brown posterior margins, most evident on 5 and 6. A blackish line marks the spiracular area. Tails white.

The dark thorax and the deeper brown of the posterior abdominal tergites distinguish this species from *C. bellum.* The darker color and more prominent banding of the abdomen separate it from *fragile.* There are specimens of this species in the Cornell collection from Lachine, Que. (F. P. Ide), which Dr. McDunnough was kind enough to send us. Others from Ringwood, N. Y., near Ithaca, seem to be of this species. The nymph of the Ringwood specimens is much like that of *C. bellum,* but has a broad band of brown on sternite 9, also lateral markings on some of the other apical sternites, and the sternites margined posteriorly with brown.

Centroptilum quaesitum McDunnough

1931—McDunnough—Canad. Ent. 63: 87...Alta.

Length: body 7; wing 7 mm.

Abdominal tergites 2–6 of male imago bright ruddy, except for a narrow pale lateral area.

Turbinate eyes light orange-red in dried specimen. Head light yellowish, as are the basal joints of the antennae; antennal filament smoky. Thorax rather light brown, the pleura with faint ruddy tinges; a yellow streak anterior to the wing root. Sternum yellowish, with ruddy brown shading. Femora pale amber; tibia and tarsus of fore leg with a faint smoky tinge; those of middle and hind legs pale whitish. Wings hyaline, venation pale. Hind wing rather narrow, with two longitudinal veins.

Abdominal tergites 2–10 bright ruddy, with the exception of a narrow pale lateral area; the ruddy color deepening on tergites 7–9, paler on 10. Sternites pale yellowish. Spiracular area marked with a black line, on tergites 2–6. Tails pale. Forceps likewise pale; between the bases of the forceps is a plate with a strongly convex posterior margin.

This species would seem to be related to *C. infrequens,* of which the female only is known. The narrower and longer hind wing of that species, as well as the ruddy dorsal band in the female, distinguishes it from *quaesitum*, in which the abdomen of the female is pale yellowish, the anterior tergites with brown posterior margins. We have seen no specimens of this species.

Centroptilum rivulare sp. nov.

Length: body 5½–6; wing 5½–7 mm.

Abdominal tergites 2–6 of male imago semi-hyaline, yellowish; anterior margins very narrowly brown.

Head reddish brown. Turbinate eyes quite large, rather kidney-shaped, bright orange margined with dark brown (alcoholic specimen). Thorax deep reddish to blackish brown; pleura and areas around the wing bases paler, yellowish brown. Legs yellowish or yellowish white, the fore leg rather more deeply tinged with yellow. Wings hyaline, venation pale. Stigmatic area of fore wing faintly opaque whitish; cross veins simple, slightly aslant, few in number; usually 5 to 6 only. Hind wing relatively long and narrow, much as in *C. ozburni;* costal projection strongly curved (see fig. 165).

Abdominal tergites 2–6 semi-hyaline, yellowish or yellowish white, often shaded faintly with olive brown; anterior margins very narrowly dark brown. Traces of pale brown submedian streaks on each; a pale brown dot anteriorly above the pleural fold; a short black oblique line at each spiracle; often a broken blackish spiracular line. Sternites 2–6 paler, more whitish than the tergites; unmarked. Segments 7–10 opaque, both tergites and sternites russet or chestnut brown, the latter only slightly paler than the tergites; anterior margins very narrowly darker. Forceps and tails whitish. Tubercle at inner margin of second forceps joint very poorly developed. 'Penis-cover' quite large, more or less truncate on the apical margin; distal joint of forceps short (see fig. 166).

Head of female flesh-colored; a dark brown median streak on vertex, extending forward between the ocelli. Mesothorax pale reddish brown; sutures narrowly darker, intersegmental areas pale. Legs yellowish white, tarsi often tinged with smoky brown. Venation pale yellowish brown. Abdomen pale reddish dorsally, venter yellowish. Broken black spiracular line present.

The nymph has small dorsal flaps on the basal pairs of gills.

Holotype—Male imago, reared; Kirtland, Ohio, Aug. 6, 1932; No. 1361.1 in Cornell collection.

Allotype—Female imago, reared; same data; No. 1361.2 in Cornell collection.

Paratypes—8 male and 13 female imagos, reared; Aug. 5–11, 1932; (J. R. Traver), same locality; No. 1361.3–23 in Cornell collection.

This species seems closely allied to *C. caliginosum* and *C. bellum;* the tubercle on the inner margin of the second forceps joint is less well developed, however, than in either of those two species. The reddish brown of the posterior sternites separates it from *caliginosum,* in which tergites 7–10 are chocolate brown, the sternites paler. From *bellum* it may be distinguished by the total lack of reddish markings on the basal abdominal tergites. The nymph resembles *bellum* and *ozburni* in its abdominal color pattern, but has paler tails and darker tarsi than either of those species; the sternites are pale, not marked with brown as in *ozburni.*

Centroptilum rufostrigatum McDunnough

1924—McDunnough—Canad. Ent. 56: 95. . . . . Man.
1925—McDunnough—Canad. Ent. 57: 176 (in discussion of *caliginosum*).

Length: body 4½–5; wing 5 mm.

Abdominal tergites of male imago pale yellowish white, each with a row of four minute red transverse dashes on the posterior margin on each side.

Face and thorax deep blackish brown. Thoracic sternum deep black brown; posterior margin of mesonotum and pleural sutures tinged with pale yellowish brown. Legs pale yellowish white. Wings hyaline. Hind wing long and narrow, two longitudinal veins (see fig. 165). Abdominal segments 2–6 pale yellowish white; each tergite with a lateral row of four minute red transverse dashes on each side, these dashes situated on the posterior margin of the tergite. Tergites 7–10 deep chocolate brown, sternites alabaster white; the pleural fold forms a sharp division between these two colors. A faint slightly curved hair-line of black marks the spiracular area, on the pale segments. Forceps and tails white. Inner margin of second joint of the forceps with a poorly developed inward bulge, much less apparent than in *C. caliginosum* (see fig. 166).

From the allied *C. caliginosum*, this species may be separated by the smaller size, and the lesser development of the bulge on the second forceps joint. The much narrower hind wing distinguishes it from *C. bifurcatum*. Dr. McDunnough has kindly sent specimens of this species to the Cornell collection; these are from Aweme, Man. (R. M. White), and Fredericton, N. B. (W. J. Brown). We have two specimens from Harpers Ferry, Md. (V. Argo) which may be of this species.

Centroptilum semirufum McDunnough

1885—Eaton—Revis. Monogr. p. 175 (under *luteolum*). . . . . Hudson Bay
1926—McDunnough—Canad. Ent. 58: 189. . . . . Ont.
1929—McDunnough—Canad. Ent. 61: 171 (in discussion of *bifurcatum*).

Length: body 4; wing 4 mm.

Abdominal tergites 2–6 of male imago ruddy brown on the posterior half or three-quarters, the remaining areas pale hyaline; a prominent tubercle between the bases of the forceps limbs.

Turbinate eyes red brown, in dried specimen; appear as flat discs, not shriveled longitudinally. Head pale yellowish below the antennae; vertex brown, with ruddy tinges around the ocelli. Thorax olivaceous brown; lateral margins of mesonotum paler; scutellum white, tinged anteriorly with ruddy. Anterior margin of metanotum pale; central area with strong ruddy tinges. Pleura brown, with small pale areas at bases of legs. Mesosternum pale yellowish. Legs yellowish white; femora with a faint ruddy streak in the median area long the dorsal edge. Wings hyaline, venation pale. 3–4 cross veins in the stigmatic area of the fore wing; hind wing narrow, two longitudinal veins; costal projection strong.

Abdominal tergites 2–6 shaded heavily with ruddy brown on the posterior half or three-quarters; anterior half of tergites 2 and 3 pale hyaline; pale area on 4–6 narrower, and divided by a fine oblique shade of brown into a large dorsal patch and a smaller one on the antero-lateral margin. Tergites 7–10 opaque ruddy brown. Sternites whitish; 2–6 hyaline, 7–9 opaque with faint ruddy shading laterally. A faint broken black line along the spiracular area. Forceps and tails white. Between the bases of the forceps limbs is a prominent tubercle.

The prominent ruddy banding of the abdomen and the brighter color of the posterior tergites distinguish this species from the allied *C. bifurcatum*. We have seen no specimens of *C. semirufum*.

Centroptilum simile McDunnough

1924—McDunnough—Canad. Ent. 56: 97 .... Que.
1925—McDunnough—Trans. R. S. C. p. 221.
1931—McDunnough—Canad. Ent. 63: 88 (in discussion of *intermediale*).

Length: body 4–6; wing 5½–6 mm.

Abdominal tergites 2–6 of male imago hyaline white, the posterior margins ruddy brown, with paler ruddy brown shading on the posterior third of each.

Head and thorax blackish. Lateral margin of mesonotum distinctly pale yellow-brown. Legs yellowish, the fore leg deeper in color than the others; ruddy marks at base, middle and apex of the femora; a dark hair-line connects these ruddy markings. Wings hyaline. Hind wings long and narrow (see fig. 165). Abdominal tergites 2–6 hyaline white; the posterior margin of each rather deep ruddy brown, with paler ruddy brown shading over the posterior third of each tergite, extending forward on each side above the spiracular line, almost to the anterior margin. Sternites 2–6 hyaline white, the posterior margins with traces of ruddy markings. Tergites 7–10 ruddy brown; sternites opaque whitish, 8 and 9 tinged with ruddy brown. Spiracular line black. Tails white.

This species is close to *C. ozburni*, but may be distinguished from it by the ruddier brown markings of the abdomen and the more extensive ruddy brown shading of the anterior tergites. The legs also are deeper in color, and more marked with ruddy color. There are several specimens of this species in the Cornell collection, which Dr. McDunnough kindly sent; these are from S. Bolton and Sweetsburg, Que. (J. McDunnough, G. S. Walley); and from Burke Falls, Ont. (F. P. Ide). We have others which seem to be of this species, from Ft. Jackson and Nassau, N. Y. (L. Harper).

Centroptilum venosum sp. nov.

Length (male): body 5½–6; wing 5½–6 mm.

Abdominal tergites 2–6 of male imago yellowish white, the posterior margins reddish brown; a distinct rounded bulge on inner margin of second forceps joint.

Turbinate eyes large; quite bright orange, usually paler along margin. Face light brown; paler areas around bases of antennae and on clypeal margin near the eye. Antennae pale brown, the second joint and tip of filament whitish. Pronotum reddish brown, its margins darker. Mesonotum reddish brown (yellowish, in pale specimens); sutures darker; usually a pale median area near the anterior margin; antero-lateral and postero-lateral margins creamy. Scutellum creamy white, more or less shaded with pinkish. Lateral areas below scutellum dark reddish brown. Faint greyish markings before the wing. Metanotum red-brown, the median projection and lateral and posterior margins darker, and with distinct red shading. Pleura and sternum yellowish to olive brown, sutures darker; yellowish areas below wing and around leg bases.

Legs yellowish white, fore leg slightly amber-tinged. Joinings sometimes narrowly darker; claws orange-brown. Wings hyaline; a slight cloud in stigmatic area. Longitudinal veins of fore wing yellowish brown; 6–10 cross veins in stigmatic space; marginal intercalary usually absent from first two interspaces. Hind wing wide for its length; two longitudinal veins and a distinct intercalary below the second vein, extending basad for half or more than half the wing length. Distinct cross veins are present; 3 to 5 in the costal space; 1 or 2 near the center of the space between veins 1 and 2; and 3 to 4 in the last space (see fig. 165). In the latter area, the cross veins may all extend from vein 2 to the intercalary, or the basal ones may reach the hind margin. Costal projection strongly curved.

Abdominal tergites 2–6 yellowish white, semi-translucent; posterior margins reddish or orange-brown. A wedge-shaped mark of the same color extends forward halfway to the anterior margin, above the spiracular area, at the end of each dark band, on each side. Indications of pale brown oblique submedian streaks on the anterior margin. An interrupted black line, almost straight, marks the spiracular area; near the posterior margin of each tergite, a short oblique branch extends downward, with a short red-brown dash at its extremity. Tergites 7–10 opaque, light red-brown, posterior margins darker; median area often marked with whitish or pinkish opaque streaks. Sternites yellowish white; 2–6 semi-translucent, the posterior margins sometimes faintly darker; 7–9 opaque, tinged more or less with pinkish, especially on the posterior margins. Tails whitish or yellowish white. Distal joint of forceps short; third (long) joint not strongly bowed; a distinct rounded bulge on inner margin of second joint. Plate between forceps bases broad, its posterior margin almost straight (see fig. 166).

Head and thorax of female yellowish brown; longitudinal reddish streaks on vertex; pronotum much marked with reddish brown. Abdominal tergites light reddish, yellower on apical tergites; posterior margins bright reddish or red-brown. Sternites pale reddish yellow. Black markings along the tracheae.

Holotype—Male imago. San Gabriel Canyon, San Gabriel Mts., California; June 24, 1932. Charles D. Michener, collector. No. 1362.1 in Cornell University collection.

Allotype—Female imago; same data. No. 1362.2 in Cornell collection.

Paratypes—8 males, 2 females; same data; June 24 to 26, 1932. No. 1362.3–12 in Cornell collection.

The broad hind wing with intercalary and cross veins distinguishes this species from *C. bellum;* the conspicuous abdominal banding separates it from *C. bifurcatum.* The distinctive venation of the hind wing should separate it from other species heretofore described.

In the Cornell collection are female specimens from Cobil's Canyon, near Claremont, Calif., reared by Mrs. Seemann (listed by her as Baetis No. 1; see *J. Ent. and Zool.* 19: 50, 1927); these have much larger hind wings than any species described from North America, being fully twice as long as those of *C. venosum.* Cross veins are also quite numerous and well developed. We are not dscribing this as a new species at the present time, as no males are known.

Centroptilum walshi McDunnough

1929—McDunnough—Canad. Ent. 61: 173. Fig. . . . . Ia., Ill.

Length: body 6; wing 6 mm.

Abdominal segments 2–6 of male imago hyaline white, with no markings except a broken black spiracular line.

Turbinate eyes pale yellow in living insect, bright reddish in dried specimen. Pronotum light ochreous brown; mesonotum similar, shading into creamy on the posterior third. Metanotum and all of the sternum creamy. Legs whitish. Wings hyaline, venation pale. Abdominal segments 2–6 hyaline white, with a broken black line along the spiracular area. Tergites 7–10 opaque, creamy, shaded with light purplish brown (perhaps due to discoloration); sternites similar, but with no dark shading. Tails whitish. Between the bases of the forceps is a plate having a convex posterior margin. The inner margin of the second forceps joint bears a distinct tubercle.

This species may be distinguished from the allied *C. album* by the stronger tubercle on the second forceps joint, and the pale metanotum and thoracic sternum; these areas are wood-brown in *album*, which is also considerably smaller in size.

One male and two females from Manhattan, Kans. (R. C. Smith), which seem to be of this species, are in the Cornell collection.

Genus **Heterocloeon** McDunnough

1925—McDunnough—Canad. Ent. 57: 175.

Small brownish mayflies, of which but a single species has been described. In this species, the basal abdominal segments of the male are pale, tinged with yellow and often with a brownish wash. Females paler than males; abdomen dirty olive green, basal segments tinged with yellowish. In both sexes, round reddish-brown spots are present on the mid-dorsal line near the anterior margin on tergites 2 and 3. Turbinate eyes of male large, oval. Posterior margin of head of female almost straight.

Fore leg of male short; when extended backwards, reaching only to the 2nd or 3rd abdominal segment. Fore tarsus quite short, only one-half to three-fourths as long as the tibia. Tibia one and one-third to one and one-half the length of the femur. In the hind leg, the tibia and femur are subequal; tarsus shorter than tibia; basal joint of tarsus relatively short, about as long as the second and third joints together. Marginal intercalaries of fore wing occur in pairs. Hind wing very greatly reduced in size, appearing as a mere thread, with traces of one vein only; costal projection wholly wanting.

Genitalia described under the single species; apparently characterized by the spinous projection on the inner apical margin of the third forceps joint (see fig. 167). Apical margin of 9th sternite of female extended backwards very slightly at the median line.

Nymph very similar to that of Pseudocloeon, having but two tails, and gills single on all segments. In the single species known, the tails are unbanded and the gills are largely blackish with a narrow pale border.

Genotype—*H. curiosum* McDunnough.

Distribution—Nearctic; eastern North America.

Heterocloeon curiosum McDunnough

1923—McDunnough—Canad. Ent. 55: 43 (Centroptilum).................................. Ont.
1925—McDunnough—Canad. Ent. 57: 175 (transferred to new genus).
1932—McDunnough—Canad. Ent. 64: 213. Nymph discussed.

Length: body 4½–5; wing 5–5½ mm.

Abdominal tergites 1–5 of male imago pale hyaline, shaded slightly with greenish yellow.

Head black. Turbinate eyes large, lengthily oval; sessile portion of stalk black, upper portion pale yellow; eyes deep brown. Thorax shiny black; pleura with brown shading below the wings. Legs whitish; fore femur blackish; coxae and claws brown. Wings hyaline, venation pale. 6–8 cross veins in the stigmatic area of the fore wing; somewhat aslant, some incomplete toward subcosta, and with short horizontal veins between several of them. Marginal intercalaries quite well developed except in the first interspace, where they may be wholly wanting, and between the anal veins. A reddish brown spot at base of fore wing. Hind wing a mere thread, much narrower than in Centroptilum; traces of a single vein may be present.

Abdominal segments 1–5 hyaline; tergites slightly shaded with pale greenish yellow, sometimes also tinged with light sepia brown. A small round red-brown spot is present on tergites 2 and 3, on the median line near the anterior margin; traces of similar spots may be present also on tergites 4 and 5. Tergites 6–10 light sepia brown; sternites shaded with opaque white. Spiracular area often marked with black. Forceps white. Tails white, the basal joinings red-brown. Between the basal joints of the forceps are two slight projections (penis-covers?) separated rather widely from one another. On the inner margin of the basal joint, near the base, is a slight protuberance. Second joint conical, tapering apically; third joint strongly bowed, a spinous projection at its inner apical margin; distal joint slightly bowed, about 2½ times as long as wide (see fig. 167).

There are specimens of this species in the Cornell collection from Montreal, Que. (Mrs. E. G. Grieve); Buffalo, N. Y. (C. Betten); Harpers Ferry, Md. (V. Argo); and Ft. Jackson, N. Y. (L. Harper).

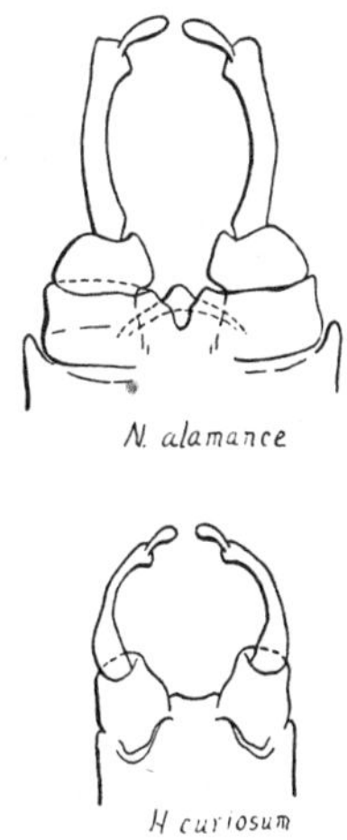

FIG. 167. Male genitalia of Neocloeon and Heterocloeon.

Genus **Pseudocloeon** Klapalek

1885—Eaton—Revis. Monogr. p. 179 (Cloeon, in part).
1905—Klapalek—Mitt. Naturh. Mus. Hamburg 22: 105.
1920—Ulmer—Stett. Ent. Zeit. 81: 125.
1931—McDunnough—Canad. Ent. 63: 85.
1932—McDunnough—Canad. Ent. 64: 212.
1933—Spieth—J. N. Y. Ent. Soc. 41: 338.

Small mayflies; fore wing ranging in size from 3½ to 6 mm. (Nearctic species). Basal abdominal segments of male range from whitish to dark olive brown. Females paler, yellowish to pale reddish brown. Turbinate eyes of male very large for the size of the insect (see fig. 16b). Eyes of female rather small (smaller than in the Cloeon group); widely separated. Posterior margin of head of female almost straight.

Fore leg of male about as long as the body. Fore tibia 1½ to 1⁵⁄₇ as long as the femur. Tarsus 1¼ the length of the tibia; second (longest) joint slightly longer than the third. In the hind leg, the tibia is very slightly longer than the femur, the tarsus slightly shorter than the tibia. Basal tarsal joint relatively short, about as long as the 2nd and 3rd joints combined. Both fore and hind legs of the female are very similar in proportionate lengths of joints to the hind leg of the male. In the fore wing (see fig. 26c), the marginal intercalaries occur in pairs; the hind wing is wholly wanting.

Basal joint of forceps rather large, usually slightly longer than wide; its inner margin may be somewhat prolonged inward. Second joint short and conical, tapering slightly toward the apex. Third joint longer and more slender, sometimes slightly bowed; usually not at all enlarged apically, even tending to become narrower at the apex. Distal joint at least twice as long as wide. A 'penis-cover' is usually present between the basal forceps joints. Apical margin of 9th sternite of female prolonged backwards very slightly in the median area.

Maxillary palp of nymph 2-jointed, rather short, extending only a very short distance beyond the tip of the galea-lacinia. Labial palp 3-jointed; the distal joint is rounded apically and set with short stout spines; basally it is more or less fused with the second joint, so that the palp often appears to be 2-jointed. Mandibular canines rather wide and stout, more or less even across the tops. Claws widest at base, tapering to slender tips; inner margin denticulate; each claw is about ⅓ as long as its tarsus. Gills single on all segments, obovate; tracheation pinnate. Tails two; the middle one represented by a minute unsegmented rudiment. In *P. carolina*, the body of the nymph is short and chunky, the thorax unusually wide; in other known species of nymphs, the body is more slender, of the type common to members of the subfamily Baetinae.

Genotype—*P. kraepelini* Klapalek.

Distribution—Nearctic, Neotropical, African, Indo-Australian; in Nearctic, generally distributed.

KEY TO THE SPECIES

Male adults

1—Abdominal tergites 2–6 brown..2
—Abdominal tergites 2–6 pale,—whitish or yellowish........................4
2—Thorax and abdomen dull olive brown or clay-colored; posterior tergites paler ochreous; fore femur dull whitish.................................**futile**
—Thorax and abdomen (including posterior tergites) clear shining brown or blackish; fore femur brownish..3
3—Dorsum of abdomen dark olivaceous brown; tails white, unmarked...**turbidum**
—Dorsum of abdomen much deeper polished brown; tails white, darker at joinings..**carolina**
4—Rather large ruddy lateral patches on abdominal tergites 2–6...............5
—No such large ruddy lateral patches, though small submedian reddish dots may be present..6
5—A small brown dot on mid-ventral line of each sternite near the center; no reddish markings on femora..**rubrolaterale**
—No such brown mid-ventral dots; middle and hind femora with a ruddy dash on the lower edge anteriorly, and ruddy spots at middle and apex......**parvulum**
6—Reddish or brownish spots at mid-ventral line, on at least some of the sternites..7
—No red or brown mid-ventral markings on the sternites....................9
7—A pair of submedian reddish dots present on tergites 2–6; a faint minute brown patch may be present above each spiracle...............................8
—No reddish submedian dots on tergites; on each, a faint blackish spiracular dot ....**punctiventris**

8—Femora, at least of middle and hind legs, distinctly banded at middle and apex with purplish red; a faint brown patch above each spiracle............ **virile**
—Femora pale, unbanded; no such brown patch above the spiracle....... **veteris**
9—Thorax reddish brown; 4 to 6 red dots and 2 red dashes on each of tergites 2 to 6.. **etowah**
—Thorax black or blackish brown; no such red dots and dashes on tergites 2 to 6. 10
10—Eyes red-brown in living insect 11
—Eyes yellowish green in living insect .... **chlorops**
11—Orange-brown bands on the dorsal portions of the posterior margins of tergites 2–6; apical tergites dull sepia brown........................... **cingulatum**
—Tergites 2–6 with no such bands on the posterior margins; apical tergites pale umber brown... **dubium**

VERIFICATION TABLE

Male adults

Species	*Wing length*	*Eyes*[2]	*Fore femur*	*Marks*[3] *on 2–6*	*Abd. terg. 2–6*	*Abd. terg. 7–10*	*Distr.*
carolina	4	Brown	Brown	No	Brown	Brown	E
chlorops[1]	4	Purplish	Smoky	No	Pale	Pale red-br.	Ont.
cingulatum	4	Dp. red-br.	Dp. smoky	No	Pale	Dp. sepia br.	Que.
dubium	4–4½	Red-br.	Pale red-br.	No	White	Umber br.	E
etowah	4	Orange[4]	Whitish	Yes	White	Red-brown	Ga.
futile	3½	Grn. br.	Whitish	No	Clay-br.	Ochreous	Alta.
parvulum	3½	Red.	Smoky	Yes	Whitish	Lt. choc. br.	N
punctiventris	4–4½	Lt. yel. br.	Smoky	No	Pale	Lt. red-br.	Ont.
rubrolaterale	4	Red-br.	Smoky	Yes	Yel. white	Sepia br.	Alta.
turbidum	4	Dp. red-br.	Dk. smoky	No	Dk. ol. br.	Dk. ol. br.	Alta.
veteris	6	?	Yellowish	Yes	Yel. white	Lt. choc. br.	C, S
virile	5½	Dp. red-br.	Smoky yel.	Yes	Pale	Choc. br.	Ont.

KEY TO THE KNOWN SPECIES*

Nymphs

1—Tails alternately banded with light and dark....................... **parvulum**
—Tails otherwise.. 2
2—Tails with median dark band.. 3
—Tails unbanded.. 5
3—Abdominal segments, especially in male, with strong color contrasts; segments 3, 4, 8–10 largely pale, others dark; 9 with a dark lateral dash .... **dubium**
—Abdominal segments largely dark; no lateral dash on 9.................... 4
4—Rear abdominal segments with traces of pale dorsal band, usual dark submedian and subventral dots distinct; anterior segments with pale submedian dots; medio-ventral dark dots of adult present..................... **punctiventris**
—Rear abdominal segments with dorsal pale band; dark dots obscure; pair of sub-

[1] Eyes yellowish green in living insect.
[2] Dried specimen, unless indicated.
[3] Red marks may consist of dots, dashes, streaks, or blotches.
[4] From alcoholic specimen.
* McDunnough's key (Canad. Ent. 64: 212, 1932) with slight changes in phraseology.

median pale spots on anterior portion of each segment; no medio-ventral dark dots . . . . **virile**

5—Short chunky species with broad thorax; long hairs on dorsal margin of femora and tibiae; first eight abdominal segments almost entirely dark . . . . **carolina**

—More elongate species; hairs on femora considerably shorter; abdomen paler with segments 4 and 8–10, particularly in male, frequently pale . . . . **cingulatum**

Pseudocloeon carolina Banks

1924—Banks—Bull. Mus. Comp. Zool. 65: 426 . . . . N. C.
1925—McDunnough—Trans. R. S. C. p. 221 . . . . Que.
1931—McDunnough—Canad. Ent. 63: 85. Key to nymphs.
1932—McDunnough—Canad. Ent. 64: 213. Key to nymphs.
1932—Traver—J. Elisha Mitchell Sci. Soc. 47: 227.

Length: body 4?; wing 4 mm.

Abdominal tergites 2–6 of male imago wholly dark polished brown.

Turbinate eyes brown in dried specimen. Thorax and femora clear polished brown, darker than in *P. turbidum.* Femora likewise polished brown. Wings hyaline, venation pale. Abdominal tergites 2–10 clear polished brown; sternites pale yellowish. Tails pale whitish, very slightly brownish at the joinings. Genitalia as in fig. 168.

The much darker brown of thorax and abdominal tergites, as well as the dark markings at the tail joinings, distinguishes this species from *turbidum.* There is a male imago in the Cornell collection from Buffalo, N. Y. (N. Y. State collection) which seems to be of this species. However, the middle and hind femora are paler than the original description would indicate. Nymphs of *carolina* were obtained in several mountain streams in western North Carolina.

Pseudocloeon chlorops McDunnough

1923—McDunnough—Canad. Ent. 55: 45 . . . . Ont.
1924—McDunnough—Canad. Ent. 56: 115 (in discussion of *turbidum*).
1931—McDunnough—Canad. Ent. 63: 86 (in discussion of *cingulatum*).

Length: body 3; wing 4 mm.

Abdominal tergites of male imago hyaline whitish with a faint yellowish tinge.

Turbinate eyes the size of those in *P: dubium;* in living insect, yellowish green with a purplish-brown band at the base of the stalk; in dried specimen, the eyes are deep purplish, with a wide pale red-brown outer margin. Head shiny black; bases of antennae and ocelli ringed with greenish. Thorax blackish, shining; lateral sutures (of pleura) pale; anterior portion of mesonotum shaded slightly with greenish. Legs whitish; fore femur smoky. Wings hyaline. Abdominal segments 2–6 hyaline, with faint yellowish tinge. Tergites 7–10 pale brown, ruddier than in *P. punctiventris;* sternites paler, more or less tinged with pale brown. A faint brown line marks the spiracular area. Tails white.

This species is very close to *P. dubium,* and may prove to be synonymous with it; the color of the eyes in the living insect is the main distinction. Of this Dr. McDunnough says; (it is) "a character which is not always to be relied upon." We have seen no specimens of this species.

Pseudocloeon cingulatum McDunnough

1931—McDunnough—Canad. Ent. 63: 85. Key to nymphs . . . . Que.
1932—McDunnough—Canad. Ent. 64: 213. Nymph included in key.

Length: body 4; wing 4 mm.

Abdominal tergites 2–6 of male imago pale hyaline, the dorsal portions of posterior margins narrowly orange-brown.

Turbinate eyes circular, deep red-brown, somewhat larger than in *P. dubium.* Thorax wholly deep blackish brown; membranous portions of pleura slightly tinged with ruddy. Fore femora deep smoky; tibiae and tarsi pale, slightly smoky. Middle and hind femora pale amber with a smoky tinge; tibiae and tarsi whitish. Wings hyaline, venation pale. Abdominal tergites 2–6 pale hyaline, the first two tergites faintly

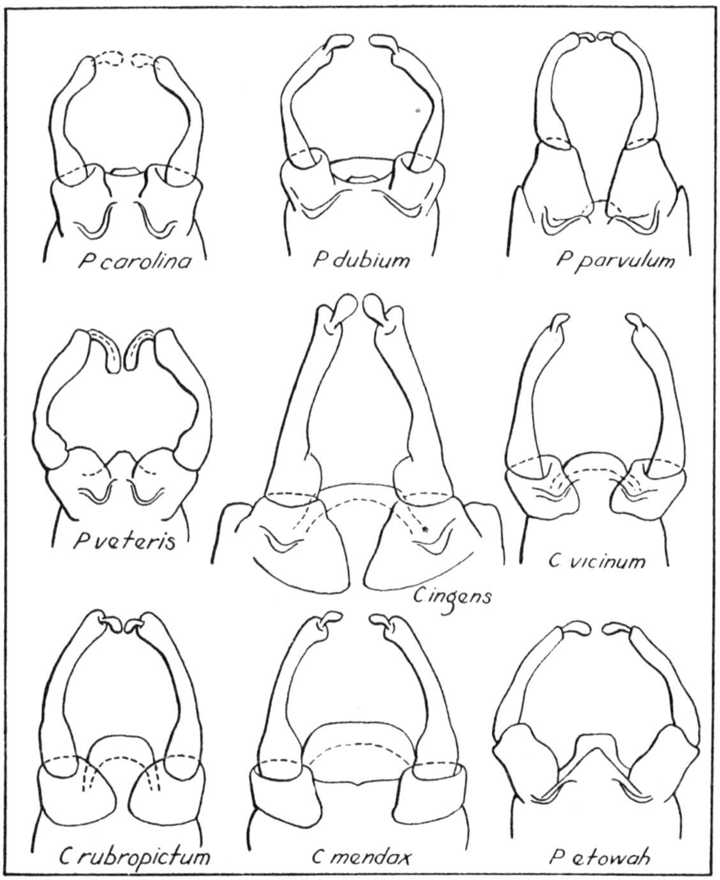

Fig. 168. Male genitalia of five species of Pseudocloeon, and of four species of Cloeon.

tinged with smoky; the dorsal portion of the posterior margins, especially of tergites 2 and 3, is finely orange-brown. Abdominal tergites 7–10 deep sepia-brown. Sternites 2–6 pale hyaline; 7–9 opaque, whitish tinged with sepia. Tails and forceps pale. Tracheae marked with black, forming a double black line along the spiracular area. Genitalia much as in *P. dubium.*

The orange-brown abdominal banding and the duller sepia brown of abdominal tergites 7–10 distinguish this species from *P. dubium;* it lacks the red submedian dots of *P. virile*, and the femora are not banded; the lack of a mid-ventral row of spots separates it from *P. punctiventris*. Dr. McDunnough has kindly sent to the Cornell collection specimens of this species, from the Golf Club and Laprairie, Que. (F. P. Ide, G. S. Walley).

Pseudocloeon dubium Walsh

1862—Walsh—Proc. Acad. Nat. Sci. Phil., p. 380 (Cloeon)....................................Ill.
1923—McDunnough—Canad. Ent. 55: 45 (in discussion of *punctiventris*, *chlorops*, and *Cloeon ingens*)....................................Ont.
1924—McDunnough—Canad. Ent. 56: 115 (in discussion of *turbidum* and *virile*).
Op. cit. p. 8 (in discussion of *veteris*).
1931—McDunnough—Canad. Ent. 63: 85 (in discussion of *cingulatum*); key to nymphs.
Op. cit. p. 87 (in discussion of *futile*).
1932—McDunnough—Canad. Ent. 64: 211 (in discussion of *parvulum*); key to nymphs.
1932—Traver—J. Elisha Mitchell Sci. Soc. 47: 227....................................N. C.

Length: body 2½–4; wing 3½–4½ mm.

Abdominal tergites 2–6 of male imago hyaline white, usually with black spiracular markings.

Turbinate eyes almost circular; ruddy brown in living specimens. Thorax shiny black. Legs pale. Wings hyaline. Abdominal segments 2–6 hyaline white; a dark dot over each spiracle, usually a pale hyaline center in each. Tergites 7–10 pale umber brown (piceous, in original description); sternites opaque whitish. Genitalia as in fig. 168.

The above description is taken from Walsh, and from Dr. McDunnough's account of the Ontario species he is calling *dubium*. In the type specimen at the Museum of Comparative Zoology, the cross veins of the stigmatic area are aslant, 5 to 6 in number, irregular and rather incomplete on the subcostal margin. In the Cornell collection is a series of males from Buffalo, N. Y., which match the above description except that there are no dark spiracular dots. Several nymphs which seem to be of this species are also in this collection, from the mountains of North Carolina. Dr. McDunnough has sent us several specimens from Rideau River, Ont. (J. McDunnough, G. S. Walley, L. J. Milne).

Pseudocloeon etowah sp. nov.

Length: body 4; wing 4 mm.

Abdominal tergites 2–6 of male imago semi-hyaline, white; marked with dots and dashes of red.

Head of male reddish brown; antennae very pale red-brown. Turbinate eyes very large, oval, light orange (alcoholic specimens); far apart at the anterior margins, closely approximated at the rear. Thorax reddish brown; postero-lateral margins of mesonotum narrowly pale yellowish; intersegmental areas of pleura, and an area anterior to wing root, yellowish. Legs whitish, coxae reddish brown. Wing hyaline, venation pale; main veins of anterior half of wing reddish brown at extreme base. A small faint whitish semi-opaque cloud in the stigmatic area; cross veins of this area few in number (about 4 to 6), simple, somewhat aslant; often with short horizontal vein traces between them. No intercalaries in first interspace; one or two, very short, in second space; other parts well developed.

Abdominal segments 2–6 white, semi-hyaline, a pair of small red submedian dots near the anterior margin of each tergite; another small red dot, or a pair of small dots close together, on each side, near middle of tergite, about halfway to the pleural fold; traces of geminate red mid-dorsal streaks, best developed on tergites 2, 3 and 6; and short red transverse dashes, one on each side of mid-dorsal line, near the posterior margin. Broken geminate blackish streaks mark the spiracular area. Tergites 7–10 reddish brown; sternites whitish or yellowish white. Forceps and tails white. Extending out between the bases of the forceps is a truncate 'penis-cover,' which by reason of its length and shape seems unique in this group (see fig. 168).

Head and pronotum of female flesh-colored; faint brownish markings on the vertex, and a brown transverse band across the pronotum. Remainder of thorax pale reddish brown; pale areas on mesonotum on each side of a brownish median stripe, and anterior to the scutellum; other large pale areas on the pleura. Dorsum of abdomen pale reddish, with traces of red markings as in the male; tracheae more or less outlined in black. Venter pale yellowish.

Holotype—Male imago; Rome, Ga., the Etowah River, Aug. 18, 1931; Prof. P. W. Fattig, collector. No. 1357.1 in Cornell collection.

Allotype—Female imago; Aug. 16, 1931; same data as holotype; No. 1357.2 in Cornell collection.

Paratypes—11 male imagos, 10 female imagos; same data as allotype; also Chickamauga River at Ringgold, Ga.; No. 1357.3–23 in Cornell collection.

The more numerous red dots and dashes on the tergites, the paler thorax, and the lack of ventral markings, distinguish this species from *P. virile*.

Pseudocloeon futile McDunnough

1931—McDunnough—Canad. Ent. 63: 86 . . . . Alta.

Length: body 3; wing 3½ mm.

Abdominal tergites of male imago clay-brown.

Turbinate eyes rather lengthily oval, well separated; light olive or greenish brown in dried specimen. Head brownish between the eyes; antennae pale smoky. Thorax deep clay-brown; traces of ruddy shading on the pleural sutures; scutellum of the mesonotum, and anterior portion of metanotum paler. Legs dull dirty whitish. Wings hyaline, venation pale. 4 or 5 oblique cross veins in the stigmatic area, with no granulations between them; paired intercalaries rather long and well developed except in the first and second interspaces. Abdominal tergites 2–6 clay-brown, very similar in color to the thorax; the triangular portion of tergite 1 paler, more ochreous. Tergites 7–10 paler, rather ochreous; traces of ruddy lateral shading, on some specimens; Sternites considerably paler, the apical ones shading into cream-color, with traces of ruddy streaks on the lateral margin. Tails pale; forceps pale creamy, widely separated at the base.

The shape and color of the eyes and the clay-brown of thorax and abdomen separate this species from *P. turbidum* and *P. carolina*. We have seen no specimens of this species.

Pseudocloeon parvulum McDunnough

1932—McDunnough—Canad. Ent. 64: 210 . . . . Ont., Alta.

Length: body 3; wing 3½ mm.

Abdominal tergites 2–6 of male imago dull hyaline whitish, suffused with a slight smoky or pale amber; large ruddy lateral patches on each side.

Turbinate eyes deep reddish in dried specimen and more oval than in *P. dubium*. Thorax blackish brown. Fore femur smoky, tinged slightly with amber; tibia and tarsus smoky white. Femora of middle and hind legs pale, very faintly tinged with amber, each with a ruddy dash anteriorly on the lower edge and a distinct ruddy lateral spot near the apex and well beyond the middle of the joint. Tibiae and tarsi whitish. Wings hyaline, tinged slightly with blackish brown at the extreme base; venation pale. A very few cross veins in the stigmatic area, no granulations between them. Abdominal tergites 2–6 dull hyaline whitish, suffused with a slight smoky or amber tinge. Traces of a fine ruddy mid-dorsal geminate line. Laterally on each tergite is a large diffuse ruddy or claret-colored patch; that on tergite 4 is less distinct, may even be obsolescent. Tergites 7–10 light chocolate brown; sternites opaque whitish tinged with brown, this being deepest on the lateral area of sternite 9. Basal sternites hyaline, faintly suffused with amber or smoky. Spiracular area marked with black. Forceps pale; basal joint suffused with light sepia brown; distal joint short and stubby, scarcely longer than broad (see fig. 168); in *P. dubium*, this joint is much longer. Tails whitish.

This species is smaller than any other thus far described, except *P. futile*, from which it is readily distinguished by the ruddy lateral abdominal patches and the dull hyaline of the basal abdominal segments. The ruddy lateral markings are less extensive than in *P. rubrolaterale*, which also possesses dark mid-ventral dots on the abdomen. Dr. McDunnough has kindly sent to the Cornell collection specimens of this species from Tillsonburg, Ont. (G. S. Walley).

Pseudocloeon punctiventris McDunnough

1923—McDunnough—Canad. Ent. 55: 45 (Cloeon)..Ont.
1924—McDunnough—Canad. Ent. 56: 115 (in discussion of *turbidum*).
1931—McDunnough—Canad. Ent. 63: 86 (in discussion of *cingulatum*).
1932—McDunnough—Canad. Ent. 64: 212. Key to nymphs.

Length: body 3–4; wing 4–4½ mm.

Abdominal tergites 2–6 of male imago whitish hyaline; a mid-ventral row of minute dots on posterior margins of sternites.

Turbinate eyes almost circular, larger than in *P. dubium;* light yellow-brown, stalk paler, with a purplish brown ring just above the base. Face blackish, antennae and base of ocelli tinged with pale yellowish green. Antennae black, the tips paler; pale ring at apex of first joint. Thorax blackish, shaded laterally with pale olive; slight greenish markings on anterior portion of mesonotum, its lateral margin, and on the scutellum; metanotal projection marked with greenish. Legs whitish; fore femur usually with smoky tinge. Wings hyaline. Abdominal segments 2–6 whitish hyaline; faint black dots over each spiracle. Tergites 7–10 light brown, ruddier than in *P. dubium;* sternites opaque white. On the posterior margin of each sternite, at the mid-ventral line, is a minute dot.

The lack of submedian red dots on tergites 2–6 and the paler apical tergites separates this species from *P. virile;* the absence of red patches on the tergites distinguishes it from *P. rubrolaterale;* while the presence of mid-ventral dots sets it apart from *P. dubium*. We have seen no specimens of this species.

Pseudocloeon rubrolaterale McDunnough

1931—McDunnough—Canad. Ent. 63: 86..Alta.
1932—McDunnough—Canad. Ent. 64: 211 (in discussion of *parvulum*).

Length: body 4; wing 4 mm.

Abdominal tergites 2–6 of male imago pale yellowish white; large lateral maroon patches on each.

Turbinate eyes of moderate size, almost circular; reddish brown in dried specimen. Head and thorax deep blackish; pleural membranes paler brown. Legs pale yellowish white; fore femur with smoky tinges. Wings hyaline, venation pale. Abdominal tergites 2–6 pale yellowish white; very distinct large lateral maroon or wine-colored patches on each; sternites pale, each with a small brown dot near the center, on the mid-ventral line. Tergites 7–10 sepia brown; lateral patches visible in certain lights; sternites paler brown, with mid-ventral dark dots. Tails and forceps pale.

The prominent wine-colored lateral patches on the tergites distinguish this species from all others thus far described. We have seen no specimens of this species.

Pseudocloeon turbidum McDunnough

1924—McDunnough—Canad. Ent. 56: 113.......................Alta.
1925—McDunnough—Trans. R. S. C., p. 221 (in discussion of *carolina*).
1931—McDunnough—Canad. Ent. 63: 87 (in discussion of *futile*).

Length: body 4; wing 4 mm.

Abdominal tergites 2–6 of male imago entirely dark olivaceous brown.

Turbinate eyes are large, circular; considerably larger than in *P. dubium;* reddish brown in living insect, deep red-brown to blackish in dried specimen. Thorax shining blackish; pleural sutures below the wings tinged with ruddy. All femora dark smoky brown; tibiae and tarsi pale yellowish white, those of the fore legs tinged with smoky. Wings hyaline. Abdominal tergites 2–10 entirely dark olivaceous brown; sternites pale yellowish white. Tails and forceps white.

This species is very similar to the eastern species *P. carolina,* but not so deep a brown in color of body. We have seen no specimens of this species.

Pseudocloeon veteris McDunnough

1924—McDunnough—Canad. Ent. 56: 8.......................Ill.

Length: body 4½; wing 6 mm.

Abdominal tergites 2–6 of male imago yellowish, semi-hyaline; submedian minute reddish dots in the central portion of each.

Turbinate eyes large, circular, much larger than in *P. dubium.* Thorax deep brown; pleural sutures and scutellum of the meso- and metanota paler. Legs yellowish. Wings hyaline. Abdominal tergites 2–6 semi-hyaline, yellowish; submedian rows of minute reddish dots in the center of each, one on each side of the median line. Tergites 7–10 light chocolate brown; sternites paler. On the pale sternites 2–6, a mid-ventral row of brown spots, one on the posterior margin of each sternite. Genitalia as in fig. 168.

The presence of the reddish dorsal dots on the abdomen separates this species from *P. punctiventris; P. virile,* which likewise has red dorsal dots, possesses red median dots ventrally only on sternites 5 and 6, and is also a smaller species. In the Cornell collection are specimens from Austin, Tex. (H. B. Parks, Jr.), which may be of this species; however, the brown ventral markings are not evident.

Pseudocloeon virile McDunnough

1923—McDunnough—Canad. Ent. 55: 46.......................Ont.
1924—McDunnough—Canad. Ent. 56: 116 (in discussion of *turbidum*).
1931—McDunnough—Canad. Ent. 63: 86 (in discussion of *cingulatum*).
1932—McDunnough—Canad. Ent. 64: 212. Key to nymphs.

Length: body 4; wing 5½ mm.

Abdominal tergites 2–6 of male imago hyaline; faint ruddy tinges on posterior margins; two small red submedian dots on anterior portion of each.

Turbinate eyes, in dried specimen, large, oval, deep reddish brown. Head blackish.

Antennae likewise blackish. Thorax shiny black; a pale yellow streak anterior to wing base, along antero-lateral margin of mesonotum; sutures of pleura and sternum slightly marked with paler. Fore femur and base of tibia smoky yellow; tarsus and remainder of tibia pale whitish. Middle and hind legs whitish; a median band and apex of femur pale purplish-red. Wings hyaline. Abdominal segments 2–6 whitish hyaline; tergites faintly tinged with ruddy on the segmental incisures; two minute red submedian dots on the anterior portion of each tergite. Tergites 7–10 chocolate brown; sternites pale, opaque, 7 and 8 with considerable ruddy shading. On sternites 5 and 6 there is a faint mid-ventral ruddy dot.

The presence of the dorsal submedian red dots and the mid-ventral dots on sternites 5 and 6 distinguishes this species from others allied to it. Dr. McDunnough has kindly sent to the Cornell collection specimens of this species from the Golf Club and Tillsonburg, Que. (G. S. Walley, R. Ozburn).

Genus **Neocloeon** Traver

1932—Traver—J. N. Y. Ent. Soc. 40:365. Figs.
1932—Traver—J. Elisha Mitchell Sci. Soc. 47: 112.

Small mayflies closely allied to Cloeon; fore wing 4½–6 mm. in length, in the single described species (see Pl. XL). Basal abdominal segments of male imago pale yellowish white. Turbinate eyes quite large, lengthily egg-shaped. Female pale orange or reddish brown; eyes rather large and set quite high on the sides of the head, as in Cloeon and Centroptilum. Posterior margin of head very slightly emarginate.

Proportional lengths of joints of fore leg in both sexes very similar to those of Cloeon.* In the third leg of the male, the tibia is ⅔ to ¾ the length of the femur; tarsus and tibia subequal; basal tarsal joint very long, equal in length to the other three joints combined (thus relatively longer than in Cloeon). In the female, the hind tibia is somewhat shorter proportionately, being only slightly more than ½ the length of the femur; tarsus very slightly shorter than tibia; basal tarsal joint as in the male. In the fore wing, the marginal intercalaries occur singly; the hind wing is wholly wanting.

Basal joint of forceps about 1⅓ times as wide as its length; the inner margin not contracted near the apex, as in Cloeon. Other joints as in the description of the single known species. Apical margin of 9th sternite of female almost straight.

Maxillary palp of nymph 3-jointed, the palp longer than the body of the maxilla, and extending well beyond the tip of the galea-lacinia. Labial palp 3-jointed, the distal joint quite widely dilated apically, especially on the inner margin, much as in Centroptilum. Canines of mandibles longer and more slender and tapering than in Cloeon. Claws very long and slender, not denticulate; almost ¾ as long as their respective tarsi. Gills single on all segments; the middle gills irregularly rounded in outline, others more or less straight on the outer (anterior) margin,

* In the original description, p. 366, the statements "tibia approximately ½ the length of the femur", and in regard to Cloeon, "fore tarsus of male nearly ⅙ as long as tibia", are errors.

lobed posteriorly; tracheation pinnate (in gill 4, approximating a palmate condition), all the main branches arising from the inner side. Tails three, approximately equal in length and thickness.

While there seem to be good characters on which to separate the male imagos and nymphs from those of Cloeon, we are unable as yet to find any character whereby females of these two genera may be distinguished. It is possible that when more species are known which would fall in the genus Neocloeon as here defined, it will become apparent that these should be classed as an aberrant group of the genus Cloeon, rather than placed in a separate genus. However, the nymphs resemble Centroptilum more closely than Cloeon.

Genotype—*N. alamance* Traver.

Distribution—Nearctic, eastern United States.

Neocloeon alamance Traver

1932—Traver—J. N. Y. Ent. Soc. 40: 365. . . . . N. C.
1932—Traver—J. Elisha Mitchell Sci. Soc. 47: 112.

Length: body 5–6; wing 5–6 mm.

Abdominal tergites 2–6 of male imago semi-hyaline, pale yellowish white marked with brown.

Head reddish brown. Turbinate eyes large, bright orange, somewhat egg-shaped. Pronotum olive brown; mesonotum bright red-brown, metanotum darker red-brown. A yellow line anterior to the wing root, and a yellowish brown area on mesonotum on each side of median sclerite. Pleura yellowish to olive brown; sternum red-brown shading into olive brown posteriorly. Legs whitish. Wings hyaline, venation pale. A very faint whitish cloud in the stigmatic area of the fore wing. 6–8 cross veins, aslant, usually complete; few or no granulations between them.

Abdominal tergites 2–6 semi-hyaline, faintly yellowish, marked and faintly washed with brownish. Posterior margin pale brown; an extension of this band forms triangular brown patches in the postero-lateral angles. A longitudinal submedian stripe, pale brown in color, margins the mid-dorsal line the length of each tergite, widening posteriorly. Between these stripes and nearer the mid-dorsal line are two short darker bars parallel to one another, from the anterior margin of about the first third of each tergite. Sternites 2–6 hyaline, pale yellowish white. Spiracular area marked with a broken brown line. Segments 7–10 semi-opaque; tergites olive to reddish brown, becoming darker posteriorly; sternites somewhat paler in color, 9 margined anteriorly and laterally with dark red-brown. Forceps and tails whitish. Genitalia reminiscent of *Centroptilum conturbatum;* second forceps joint wider than long, its inner margin prolonged inward in a somewhat ax-blade-shaped extension; third forceps joint with a slight swelling near the base on the inner margin, a more pronounced swelling apically, somewhat bowed in the middle; distal joint 2½ to 3 times as long as wide. Between the bases of the forceps is a conical penis-cover (see fig. 167).

The type material of this species is in the Cornell collection. We have seen no other specimens. However, we have nymphs and reared females of this genus from Kirtland, Ohio, which may represent another species. From Tallulah Falls, Ga., we have a single Neocloeon nymph.

PLATE XL.

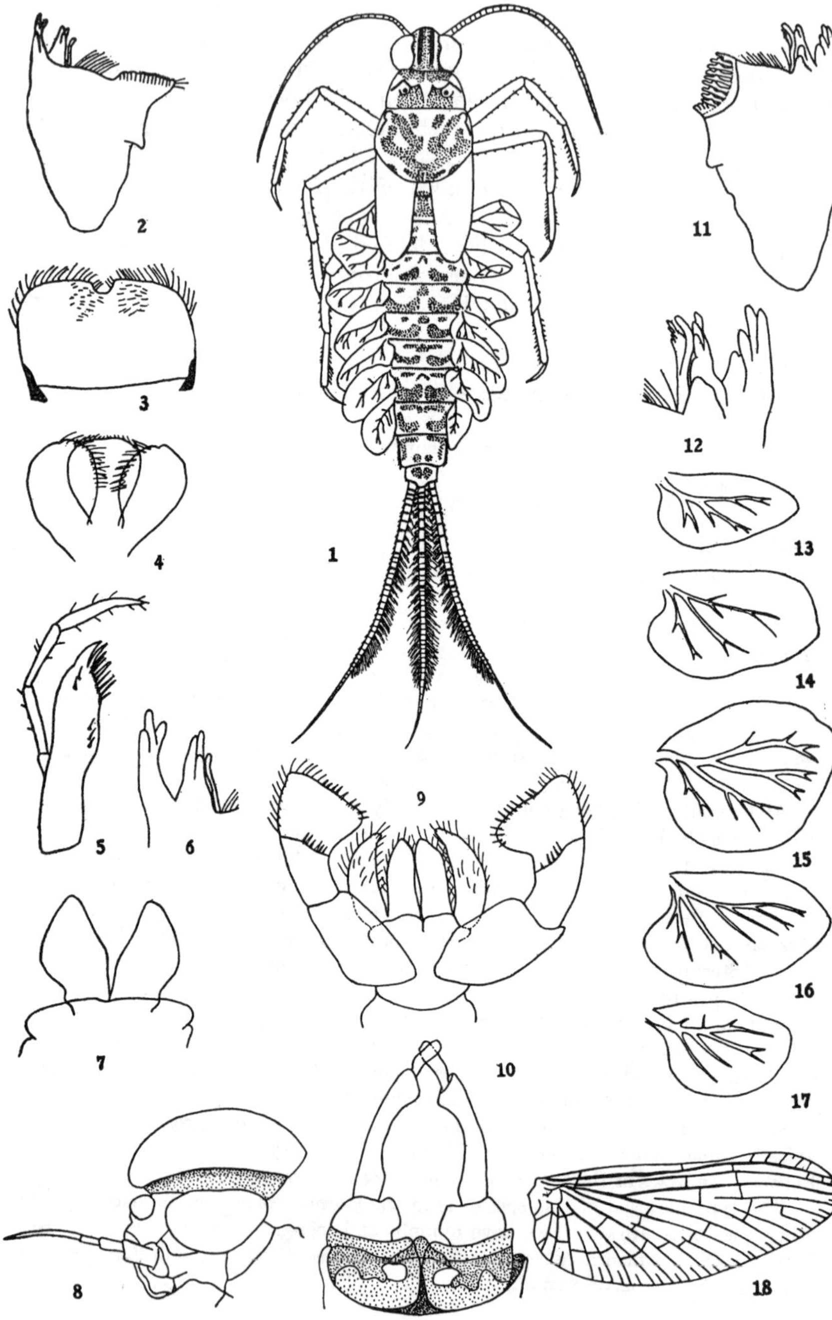

NEOCLOEON ALAMANCE

Fig 1. Nymph; Fig. 2. Right mandible; Fig. 3. Labrum; Fig. 4. Hypopharynx; Fig. 5. Right maxilla; Fig. 6. Canines of mandible; Fig. 7. Subanal plate of female imago, showing tenth sternal plates; Fig. 8. Head of male imago, side view; Fig. 9. Labium; Fig. 10. Male genitalia; Fig. 11. Left mandible; Fig. 12. Canines of mandible; Figs. 13-17. Gills of segments 1, 2, 4, 5, and 7; Fig. 18. Wings of male imago.

Genus **Cloeon** Leach

1815—Leach—Brewst. Edinb. Encycl. 9: 137.
1839—Burmeister—Handb. d. Ent. 2: 797 (Cloe, in part).
1843—Pictet—Hist. Nat. 2, Ephem. Neurop. p. 239 (Cloe, in part).
1861—Hagen—Syn. Neurop. Ins. N. Amer. p. 52 (Cloe, in part).
1885—Eaton—Revis. Monogr. p. 179.
1920—Ulmer—Stett. Ent. Zeit. 81: 125.
1933—Spieth—J. N. Y. Ent. Soc. 41: 341.

Small mayflies, length of fore wing (Nearctic species) ranging from 3 to 9 mm. Basal abdominal segments of male imagos range from whitish to dull olive brown; the paler species often have reddish markings on the tergites. Females paler than males; yellowish to pale orange or reddish brown. Turbinate eyes of male moderate to large in size. Eyes of female larger than in Baetis; much as in Centroptilum and Neocloeon. Posterior margin of head of female very slightly emarginate.

Fore leg of male slightly shorter than body. Fore tibia about 1⅓ times as long as femur; fore tarsus subequal to tibia. In female, femur and tibia subequal, tarsus "little more than ¾ as long as the tibia" (Eaton, p. 180). Basal joint of fore tarsus approximately equal in length to the 3rd and 4th joints combined. In hind leg of male, the basal tarsal joint is equal to the 3rd and 4th joints combined (shorter relatively than in Neocloeon); in female, almost as long as the three other joints combined. Hind tibia ¾ the length of the femur, in both sexes. In the genotype (Eaton, p. 180) the "hind tarsus of the male [is] about ¾ as long as the tibia"; but in *C. ingens*. the tibia is slightly shorter than the tarsus. In the female, the hind tarsus is slightly shorter than the tibia. In the fore wing the marginal intercalaries occur singly; hind wing wholly wanting.

Basal joint of forceps short and relatively stout, contracted suddenly near the apex on the inner margin; second joint somewhat conical, becoming narrower apically, and usually longer than wide; third joint rather long and slender, often dilated slightly at the apex; distal joint short and more or less clavate. A 'penis-cover' is usually present between the bases of the forceps; this structure varies somewhat in shape and size in different species. Apical margin of 9th sternite of female almost straight, very slightly extended backwards at the median line.

Maxillary palp of nymph 3-jointed; almost as long as the entire body of the maxilla, its tip extending well beyond the galea-lacinia. Labial palp 3-jointed, the apical joint somewhat conical, the outer margin extending out further than the inner one. Canines of mandibles rather short and blunt. Claws very slender at the tips, somewhat widened at the base; finely denticulate on the inner margin; usually not more than ⅓ the length of the tibia. Gills double on all segments or on segments 1–6; the larger division of each pair is irregularly rounded in outline, the tracheation more or less distinctly palmate. Tails three, approximately equal in length and thickness.

Genotype—*C. dipterum* (in Ephemera), Linnaeus.

Distribution—Holarctic, African, Indo-Australian, Neotropical; in Nearctic, occurs mainly in east and north.

KEY TO THE SPECIES**

Male adults

1—Abdominal tergites 2–6 brownish or reddish brown.......................... 2
—Abdominal tergites 2–6 pale whitish or yellowish.......................... 4
2—Wing 8–9 mm. in length; abdominal tergites 2–6 pale olive brown, thorax blackish or brownish black.......................... 3
—Wing 6–6½ mm. in length; abdominal tergites 2–6 pale ferruginous, thorax similar.......................... **mendax**
3—Legs yellowish; fore tibia smoky in basal third, a smoky dash before apex; hind tibia dusky at base; tergites 6 and 7 with lateral and dorsal ruddy shading .... **implicatum**
—Legs wholly smoky brown; no such ruddy shading on tergites.......... **ingens**
4—Ruddy or purplish brown lateral patches on abdominal tergites 2–6.......... 5
—Abdominal tergites without such ruddy or purplish lateral patches.......... 6
5—Wing 3 mm. in length; large purplish brown lateral patches on tergites 2–6, also geminate dorsal dashes of same color; black dashes on sternites........ **minor**
—Wing 6–6½ mm. in length; faint ruddy patches on tergites 2–6, above lateral margin; no dorsal dashes nor black dashes on sternites.............. **inanum**
6—Small submedian red dots on tergites 2–6 near the posterior margin, likewise above the spiracular line near center of each side.................... **rubropictum**
—No such red dots on tergites 2–6.......................... 7
7—Apical sternites pale fawn brown; short geminate ruddy mid-dorsal streaks on tergites 2 and 3; wing 4 mm. in length.......................... **insignificans**
—Apical sternites pale creamy; no such ruddy mid-dorsal streaks on tergites 2 and 3; wing 4½ to 5 mm. in length.......................... 8
8—Thoracic pleura, posterior margins of pro- and mesonota, and scutellum, pale creamy white; legs wholly pale whitish.......................... **simplex**
—No such pale creamy thoracic markings; fore legs fulvous at the base.... **vicinum**

VERIFICATION TABLE

Adult males*

Species	*Wing length*	*Meso-thorax*	*Abd. terg. on 2–6*	*Marks on*[1] *terg. 2–6*	*Abd. terg. 7–10*	*Distr.*
implicatum	8–9?	Bl. brown	Oliv. brown	Yes	Red-brown	Alta.
inanum	6½	Oliv. brown	Yellowish	Yes	Red-brown	Alta.
ingens	9	Blackish	Oliv. brown	No	Choc. brown	Alta.
insignificans	4	Brown	Hyaline	Yes	Fawn br.	Ont.
mendax	5–6	Ferruginous	Ferrug.	No	Ferruginous	E
minor	3	Bl. brown	Hyaline	Yes	Deep br.	Ont.
rubropictum	4	Dp. brown	Yel. white	Yes	Red-brown	E
simplex	5	Bl. brown	Whitish	No	Chestnut br.	Que.
triangulifer♀	5	Dp. brown	Brown	No	Brown	NE
vicinum	4½–5	Fulvous	Hyaline	No	Red-brown	E

** *C. triangulifer* not included.
* Female only available in *triangulifer*.
[1] Markings on tergites may be red or brown dots, mid-dorsal streaks, or lateral blotches.

KEY TO THE KNOWN SPECIES

Nymphs

1—Body 4–4½ mm. in length.......................................**rubropictum**
—Body 7½ mm. or more in length.......................................2
2—Body 7½ mm. in length; no ventral lobe on 7th gill; tails about ¾ as long as abdomen.......................................**mendax**
—Body 8½ mm. or more in length; a very small ventral lobe is present on 7th gill; tails as long as abdomen.......................................**ingens**

Cloeon implicatum McDunnough

1924—McDunnough—Canad. Ent. 56: 113.......................................Alta.
1930—McDunnough—Canad. Ent. 62: 59 (in discussion of *ingens*).

Length: body 8?; wing 9? mm.

Abdominal tergites 2–6 of male imago pale olive brown, with lateral ruddy shading; segmental incisures pale yellowish.

Turbinate eyes olive green in living insect; deep olive brown in dried specimen. Thorax dark blackish brown; sutures on the posterior portion of the mesonotum marked with ruddy brown. Pleura below the wings shaded with reddish and pale yellow. Sternum brown, paler than the notum. Legs pale yellowish; fore femur shaded with smoky; fore tibia smoky on the basal third, and with a smoky longitudinal dash before the apex, the joint dusky; hind tibiae slightly dusky at base. Wings hyaline. Abdominal tergites 2–7 pale olive brown, semi-hyaline, shaded laterally with ruddy, and on tergites 6 and 7 with ruddy dorsal shading also; segmental incisures pale yellowish. Tergites 8–10 bright ruddy brown. Sternites pale yellowish. A fine broken black line marks the spiracular area.

This species may be separated from *C. ingens*, which is closely allied to it, by the paler and redder coloration of the body. We have seen no specimens of this species.

Cloeon inanum McDunnough

1924—McDunnough—Canad. Ent. 56: 113.......................................Alta.
1925—McDunnough—Canad. Ent. 57: 186 (in discussion of *simplex*).

Length: body 6; wing 6½ mm.

Abdominal tergites 2–6 of male imago pale yellowish, with faint ruddy lateral patches.

Turbinate eyes dull orange in dried specimen. Thorax olivaceous brown; the posterior portions and the areas of the wing bases shaded with pale yellowish. Legs pale yellowish; fore femur and tibia with faint smoky tinges. Wings hyaline. Abdominal segments 2–6 semi-hyaline, pale yellowish dorsally; tergites with faint ruddy lateral patches above the spiracular line. Tergites 7–9 bright ruddy brown, 10 pale yellow; sternites pale yellowish. Forceps and tails pale yellowish.

From *C. mendax*, the only other described species of about the same size, *inanum* may be distinguished by the yellowish basal abdominal tergites and the olive brown of the thorax. We have seen no specimens of this species.

Cloeon ingens McDunnough

1923—McDunnough—Canad. Ent. 55: 44.......................................Alta.
1924—McDunnough—Canad. Ent. 56: 113 (in discussion of *implicatum*).
1930—McDunnough—Canad. Ent. 62: 59. Figs.......................................Que.
1932—McDunnough—Canad. Ent. 64: 80. Nymph described.

Length: body 8; wing 9 mm.

Abdominal tergites 2–6 of male imago dull olive brown, posterior margins narrowly black.

Head smoky, with ruddy tinges. Basal joints of antennae dusky. Turbinate eyes shriveled, discolored, in dried specimen, forming lunate lateral discs. Thorax shiny blackish; sutures tinged with ruddy brown; heavily shaded laterally with pale brown. Legs smoky brown. Wings hyaline. Abdominal tergites 2–6 semi-hyaline, dull olive brown, the posterior margins faintly blackish. Tergites 7–10 chocolate brown. Sternites dirty whitish; posterior ones opaque and shaded slightly with brown. The spiracular area is marked with black. Tails and forceps whitish; tails tinged at base with ruddy. Plate between bases of forceps limbs somewhat dome-shaped (see fig. 168).

This species may be distinguished from the allied *C. implicatum* by the smoky brown legs and the darker coloration of the body. There are specimens of *ingens* in the Cornell collection from Eustis, Me. (A. E. Brower). Another from Pingree Pk., Colo. (C. R. Crosby), seems also to belong here.

Cloeon insignificans McDunnough

1925—McDunnough—Canad. Ent. 57: 186. Fig. . . . . Ont.

Length: body 3; wing 4 mm.

Abdominal tergites 2–6 of male imago pale hyaline; short obsolescent ruddy geminate dorsal streaks on 2 and 3.

Eyes blackish brown. Head and thorax deep brown, almost unicolorous; sternum slightly paler. Legs whitish. Wings hyaline, venation pale. 5 to 6 cross veins in the stigmatic area; intercalaries absent from 1st and 2nd interspaces; 1st cross vein between radius and radial sector in a line with the second one. Abdominal segments 2–6 hyaline, wholly pale except for short ruddy mid-dorsal obsolescent streaks on tergites 2 and 3. Tergites 7–10 deep fawn brown, the sternites somewhat paler. Tails and forceps whitish. Plate between bases of forceps dome-shaped, its posterior margin convex.

This species may be separated from the allied *C. rubropictum* by the absence of red dots and black stigmatic markings, and the fawn-colored apical sternites. We have seen no specimens of this species.

Cloeon mendax Walsh

1862—Walsh—Proc. Acad. Nat. Sci. Phil., p. 381. . . . . Ill.
1885—Eaton—Revis. Monogr. p. 190.
1907—Needham—Bull. N. Y. St. Mus. 124: 190. . . . . N. Y.
1929—McDunnough—Canad. Ent. 61: 173 (in discussion of *Centroptilum walshi*).
1930—Ide—Canad. Ent. 62: 224. Nymph. Figs. . . . . Ont.

Length: body 4; wing 6–6½ mm.

Abdominal tergites 2–6 of male imago pale ferruginous; legs pale.

Antennal filament fuscous, pale at the tip. Head and thorax pale ferruginous; thoracic sternum pale greenish hyaline. Legs pale, the tips of the tarsi cloudy. Wings hyaline, venation hyaline. Abdominal tergites 2–10 pale ferruginous. Sternites pale greenish hyaline, the posterior segments opaque. Genitalia as in fig. 168.

Specimens in the Cornell collection taken by Prof. Needham at Old Forge and First Lake, N. Y. (J. G. Needham), seem to belong to this species. It may be distinguished from *C. ingens* and *C. implicatum* by its smaller size and ferruginous rather than olive brown abdominal tergites. In other species of which the male is known, abdominal tergites 2–6 are pale whitish or yellowish, not ferruginous.

Cloeon minor McDunnough

1926—McDunnough—Canad. Ent. 58: 190...Ont.

Length: body 3; wing 3 mm.

Abdominal tergites 2–6 of male imago hyaline; large lateral purplish brown blotches on each.

Head and thorax deep blackish brown. Anterior margin of metanotum tinged with purplish. Pleura and sternum largely dark; slight ruddy tinges anterior to the wing roots. Legs whitish; femora with traces of a ruddy median spot. Wings hyaline, venation pale. First cross vein of the radial sector considerably nearer the base of the wing than the following vein. Abdominal tergites 2–6 hyaline; geminate purplish brown dorsal dashes and large lateral blotches of the same color on each; on tergites 3 and 6 the lateral blotches coalesce dorsally, so that only the anterior portions of these are hyaline. Sternites 2–5 hyaline, each with a brown patch along the lateral margin on each side, and with transverse blackish dashes on the posterior margin, one on each side of the median line. Tergites 7–10 opaque, deep brown tinged with purplish; sternites light creamy brown, opaque; the posterior margin of sternite 9 shaded with purplish. Tails and forceps pale.

This species is distinct from all others thus far described, by reason of its small size and brownish abdominal blotches. We have seen no specimens of this species.

Cloeon rubropictum McDunnough

1913—Clemens—Canad. Ent. 45: 341 (*dubium*). Nymph described.
1923—McDunnough—Canad. Ent. 55: 43...Ont.
1925—McDunnough—Canad. Ent. 57: 186 (in discussion of *simplex* and *insignificans*). Fig.
1930—McDunnough—Canad. Ent. 62: 59...Que.
1931—McDunnough—Canad. Ent. 63: 88 (in discussion of *triangulifer*).
1930—Ide—Canad. Ent. 62: 226.

Length: body 3; wing 4 mm.

Abdominal tergites 2–6 of male imago yellowish white; small red submedian dots and traces of ruddy median line.

Turbinate eyes long and narrow, oval; in dried specimen, shriveling to form lunate discs, leaving central portion of head free. Head blackish brown. Thorax deep brown; margins of meso- and metanota and sutures slightly ruddy. Legs wholly pale yellowish white. Wings hyaline. Abdominal segments 2–6 pale yellowish white; a reddish inverted Y-shaped mark on tergite 2; tergites 3–6 with slight traces of a reddish median line, and small red submedian dots on the posterior margins; above the spiracular line and near the central area of each tergite is a faint red dot. Tergites 7–10 bright red-brown; sternites opaque whitish. A broken black line marks the spiracular area. Tails and forceps whitish. Plate between the bases of the forceps limbs a rectangular projection (see fig. 168).

The ruddy dots and broken black line of the abdominal tergites separate this species from *C. insignificans;* the darker thorax and ruddy markings distinguish it from *C. simplex*. Dr. McDunnough has very kindly sent to the Cornell collection specimens of this species, from the Golf Club and Ste. Anne's, Que. (F. P. Ide). Other specimens in this collection are from Willoughby, O. (J. R. Traver), and the vicinity of Ithaca, N. Y.

Cloeon simplex McDunnough

1925—McDunnough—Canad. Ent. 57: 185. . . . . Que.
1930—McDunnough—Canad. Ent. 62: 59.
1930—Ide—Canad. Ent. 62: 226. . . . . Ont.

Length: body 4½; wing 5 mm.

Abdominal tergites 2–6 of male imago pale whitish hyaline; traces of a black spiracular hair-line.

Turbinate eyes light green in living insect, pale orange-yellow in dried specimen, base deep brown. Head and thoracic notum dark blackish brown; posterior margin of pronotum creamy with a ruddy brown dot; antero-lateral margins of mesonotum paler brown, its posterior margin, the scutellum, and a small patch anterior to it, creamy. Anterior margin of metanotum creamy; pleura largely creamy, with tinges of ruddy brown. Prosternum whitish; meso- and metasterna largely brown, the central and lateral areas with creamy shading. Legs wholly pale whitish. Wings hyaline, venation pale; 6 to 7 cross veins in the stigmatic area, no granulations between them. Intercalaries of the 1st and 2nd interspaces wanting. First cross vein between radius and radial sector in a line with the following one.

Abdominal segments 2–6 pale hyaline whitish, with traces of a black hair-line along the spiracular area. Tergites 7–10 opaque, bright chestnut brown, with creamy lateral shading; posterior portion of 10 also creamy. Sternites wholly pale creamy. Tails and forceps white.

The paler eyes, pale thoracic markings and lack of red dots on the abdominal tergites distinguish this species from *C. rubropictum.* The wholly pale legs separate it from *C. vicinum,* in which the bases of the fore legs are fulvous. We have seen no specimens of this species.

Cloeon triangulifer McDunnough

1931—McDunnough—Canad. Ent. 63: 88. . . . . Ont.

Length (female): body 5; wing 5 mm.

Only the female imago of this species is known; distinct because of the triangular light brown lateral patches on sternites 2–7.

Head almost wholly light pinkish brown, slightly paler near the eyes. Thorax rather deep brown; sternum only slightly paler than notum; pleura tinged slightly with ruddy. Legs pale dull yellowish. Wings hyaline; venation faintly amber-tinged; cross veins almost invisible except in the stigmatic area. Abdominal tergites brown, faintly tinged with ruddy, especially on the posterior ones; sternites pale yellowish; on sternites 2–7 are triangular light brown lateral patches, based on the anterior margin and reaching almost to the posterior border; these patches are barely visible on sternites 8 and 9. Tails pale.

The darker thorax, especially the sternum, and the lateral brown triangles on the sternites, serve to separate the females of this species from those of *C. rubropictum.* We have seen no specimens of this species.

Cloeon vicinum Hagen

1861—Hagen—Syn. Neurop. Ins. N. Amer., p. 54. . . . . D. C.
1863—Hagen—Proc. Ent. Soc. Phil. 2: 178.
1871—Eaton—Trans. Ent. Soc. London, p. 107.
1885—Eaton—Revis. Monogr. p. 190.
1907—Needham—Bull. N. Y. St. Mus. 124: 182. . . . . N. Y.
1925—McDunnough—Canad. Ent. 57: 186 (in discussion of *simplex*).

Length (female): body 4; wing 4½ mm.

Abdominal tergites 2–6 of male imago whitish hyaline.

Turbinate eyes rufous. Thorax fulvous. Legs whitish, the fore leg fulvous at the base. Wings hyaline. In the type specimen of this species at the Museum of Comparative Zoology, there are about 5 cross veins in the stigmatic area, simple and almost straight, and not quite reaching the subcosta. Abdominal segments 2–6 whitish hyaline; the apical tergites fuscous, sternites pale. Tails whitish. In the type male, there is a slight inward-projecting angle at the apex of the basal forceps joint, on its inner margin.

In the Cornell collection is a single fragment of a male imago, minus head and legs, from the Cacapon River near Wardensville, W. Va., which shows similar cross veins in the stigmatic area of the wing, and a rather similar projection at the inner angle of the basal joint of the forceps (see fig. 168).

BIBLIOGRAPHY*

Adolph, G. E. 1879. Ueber Insectenflügel. Nova Acta der Ksl. Leop.-Carol. Deutschen Akademie der Naturf. 41: 215–291, pls. 27–32.

Allee, W. C. and Stein, E. R., Jr. 1918. Light reactions and metabolism in mayfly nymphs. J. Exp. Zool. 26: 423–458.

Argo, Virgil. 1927. The North American species of the genus Potamanthus, with a description of a new species. N. Y. Ent. Soc. 35: 319–328, pls. 14–15.

Banks, Nathan. 1892. (Ephemeridae in) A synopsis, catalogue and bibliography of the Neuropteroid insects of temperate North America. Trans. Amer. Ent. Soc. 19: 331–332, 345–348.

———. 1894. On a collection of Neuropteroid insects from Kansas. Ent. News 5: 178.

———. 1900. New genera and species of Nearctic Neuropteroid insects. Trans. Amer. Ent. Soc. 26: 245–251.

———. 1903. (Ephemeridae in) Neuropteroid insects from Arizona. Ent. Soc. Wash. 5: 238.

———. 1903. A new species of Habrophlebia. Ent. News 14: 235.

. 1907. A new mayfly of the genus Caenis. Ent. News 18: 13–14, 1 text fig.

———. 1908. Neuropteroid insects—Notes and descriptions. Trans. Amer. Ent. Soc. 34: 259.

———. 1908. A list of Neuropteroid insects from North Carolina. Proc. Ent. Soc. Wash. 9: 151.

———. 1910. Notes on our eastern species of the mayfly genus Heptagenia. Canad. Ent. 42: 197–202, text figs. 13–16.

———. 1914. New Neuropteroid insects, native and exotic. Proc. Acad. Nat. Sci. Philad. 66: 614–616, figs. 7–8, pl. 28.

———. 1924. Descriptions of new Neuropteroid insects. Bull. Mus. Comp. Zool. 65: 424–25, 4 text figs.

Bartram, J. 1750. A further account of the Libellae or mayflies. Phil. Trans. Royal Soc. London 10: 28.

———. 1750. Observations on a sort of Libella or Ephemeron. Phil. Trans. Royal Soc. London 10 (abridged): 843–5.

Bartram, W. 1791. Travels through North and South Carolina, Georgia, East and West Florida. Philadelphia, pp. 78–81.

Berry, E. W. 1903. New or hitherto unknown Ephemerid nymphs of the eastern United States. Amer. Nat. 37: 25–31, text figs. 1–4.

Bengtsson, Simon. 1914. Undersökningar ofver äggen hos Ephemeriderna. Ent. Tidskr.

———. 1930. Kritische Bemerkungen ueber einige nordische Ephemeropteren, nebst beschreibung neuer Larven. Lunds Univ. Arssk. N. F. Avd. 2, Bd. 26: 1–27, text figs. 1–50.

———. 1930. Beitrag zur Kenntnis der Ephemeropteren des Nordlichen Norwegen. Tromso Mus. Arsh. Naturh. Avd. 1, 51 (1928); 1–19, text figs. 1–11.

Bogoescu, Constantin. 1932. Contributiuni la Studiul morphologic si biologic al phanerelor larvelor de Ephemerine. Bucharest Inst. de Arte Grafice E. Marva. Bd. Princip. Mircea 10; 1–100, text figs. 1–105, 2 pls.

* See also the special bibliographies on thoracic structures appended to Chapters XI and XII.

Bowerbank, James. 1833. Observation on the circulation of the blood in insects. Ent. Mag. 1: 239–244, pl. 2.

Brauer, F., and Redtenbacher, J. 1888. Ein Beitrag zur Entwicklung des Flugelgeaders der Insecten. Zool. Anz. 11: 444–447.

Brown, W. A. 1934. (Ephemeroptera in) A contribution to the insect fauna of Timagami. Canad. Ent. 67: 211.

Burmeister, H. 1832. Handbuch der Entomologie, Vol. 1, Berlin. (Excellent résumé of works preceding it).

Carpenter, F. M. 1933. (Plectoptera in) The Lower Permian insects of Kansas. Part 6. Proc. Amer. Acad. Arts and Sci. 68, No. 11: 487–503, text figs. 26–29.

Chapman, R. N. 1918. The basal connections of the tracheae of the wings of insects; in Wings of Insects, Comstock; pp. 27–51, text figs. 17–35.

Chopra, B. 1927. The Indian Ephemeroptera. Rec. Ind. Mus. 29: 91–138, text figs. 1–18, pls. 8–10.

Claassen, P. W. 1922. The larva of a Chironomid which is parasitic on a mayfly nymph. Kans. Univ. Sci. Bull. 14: 395–405, 3 pls.

Clemens, W. A. 1913. New species and new life histories of Ephemeridae or mayflies. Canad. Ent. 45: 246–262, 329–341, pls. 5–7.

———. 1915. Rearing experiments and ecology of Georgian Bay Ephemeridae. Contr. Canad. Biol. 139: 114–143, pls. 13–18.

———. 1915. Mayflies of the Siphlonurus group. Canad. Ent. 47: 245–260, pls. 9–11

———. 1917. An ecological study of the mayfly Chirotenetes. Univ. of Toronto Stud., Biol. Ser. 17: 1–43; pls. 1–5.

———. 1922. The food of ciscoes in Lake Erie. Contr. Canad. Biol. 75–85.

———. 1928. Food of trout from streams of Oneida Co., N. Y. Trans. Amer. Fish. Soc. 58: 183–197.

Clemens, W. A., and Bigelow, N. K. 1922. The food of ciscoes (Leucichthys) in Lake Erie. Contr. Canad. Biol. 89–101.

Clemens, W. A., and Leonard, A. K. 1925. On two species of mayflies of the genus Heptagenia. Canad. Ent. 57: 17–18.

Cockerell, T. D. A., and Gill, Marie. 1906. Tricorythus, a genus of mayflies. Univ. of Colorado Stud. 3: 135–137.

Comstock, J. H. 1893. Evolution and taxonomy. In Wilder Quarter Century Book; 37–113, 3 pls.

———. 1918. (Ephemerida in) The Wings of Insects. Ithaca, N. Y. 214–223, text figs. 212–224.

———. 1925. (Ephemerida in) An introduction to entomology. Ithaca, N. Y. 308–313, text figs. 350–357.

Comstock, J. H., and Needham, J. G. 1898. The wings of insects. Amer. Nat. 33: 117–126.

Dodds, G. S. 1923. Mayflies from Colorado. Trans. Amer. Ent. Soc. 49: 93–114, pls. 8–9.

Dodds, G. S., and Hisaw, F. L. 1924-1925. Ecological studies of aquatic insects. Ecology 5: 137–148, 262–271. Ecology 6: 123–137, 380–390, 2 pls, 16 figs.

———. 1925. Ecological studies of aquatic insects. IV. Altitudinal range and zonation of mayflies, stoneflies and caddisflies in the Colorado Rockies. Ecology 6: 380–390, text figs. 1–3.

Drenkelfort, Heinrich. 1910. Neue Beiträge zur Kenntniss der Biologie und Anatomie von *Siphlurus lacustris* Eaton. Zool. Jahrb. Jena Abth. fur Anatomie Bd. 29, 527–617, pls. 40–42.

Eaton, A. E. 1871. A monograph on the Ephemeridae. Trans. Ent. Soc. London; 1–158, pls. 1–6.

———. 1881. An announcement of new genera of Ephemeridae. Ent. Mo. Mag. 17–18: 191–197, 21–27.

———. 1883–87. A revisional monograph of recent Ephemeridae or mayflies. Trans. Linn. Soc. London, Sec. Ser. Zool. 3: 1–352, pls. 1–65.

———. 1892. (Ephemeridae in) Biologia Centrali-Americana 38: 1–16, pl. 1.

Forbes, S. A. 1888. On the food relations of fresh-water fishes. Bull. Ill. St. Lab. Nat. Hist. 2: 475–538.

Gordon, Eva. 1933. Notes on the Ephemerid genus Leptophlebia. Bull. Brooklyn Ent. Soc. 28: 116–134, pls. 12–14.

Gros, A. J. 1923. Etudes sur les Primeres Stades des Ephemeres du Jura Francais. Am. Biol. Lacustre 12: 49.

Haase, Erich. 1891. Zur Entwicklung der Flügelrippen der Schmetterlinge. Zool. Anz. 14: 116–117.

Hagen, Hermann. 1861. (Ephemeridae in) Synopsis of the Neuroptera of North America, with a list of South American species. Smith. Misc. Coll. 38–55.

Handlirsch, Anton. 1906-08. (Jura formation Plectoptera in) Die fossilen Insekten und die Phylogenie der rezenten Formen. Leipzig. 600–604, 1228–9, pl. 46, figs. 26–33.

———. 1910. Ueber fossile Insekten. Ist Congress Internat. d'Ent., 2nd pt.; 177–184, pls. 6–10.

———. 1918. Fossile Ephemeridenlarven aus dem Buntsandstein der Vogesen. 'Verhandl.' der k.k. Zool.-Bot. Gesell. Wien. 1–3, 1 text fig.

Holste, Georg. 1910. Das Nervensystem von *Dytiscus marginalis*. Zeit. Wiss. Zool. 96: 419–476.

Hora, Sunder Lal. 1930. Ecology, bionomics and evolution of the Torrential Fauna, with especial reference to the organs of attachment. Phil. Trans. Royal Soc. London, ser. B, 218: 171–282, pls. 15–18.

Hsu, Yin-Chi. 1933. Some new morphological findings in Ephemeroptera. 5th Congr. Int. Ent. Paris 1932, 2: 361–368, 2 pls.

Ide, F. P. 1930. The nymph of the mayfly genus Cinygma Eaton. Canad. Ent. 62: 42–45, pl. 6.

———. 1930. Contributions to the biology of Ontario mayflies with descriptions of new species. Canad. Ent. 62: 204–213, pl. 17.

———. 1935. The effect of temperature on the distribution of the mayfly fauna of a stream. Univ. of Toronto Studies, Biol. Ser. 39, Pub. Ont. Fish. Res. Lab. 50: 9–76, pls. 1–10.

———. 1935. Post embryological development of Ephemeroptera (Mayflies). External characters only. Canad. J. Res. 12: 433–478, pls. 1–13.

———. 1935. Life history notes on Ephoron, Potamanthus, Leptophlebia and Blasturus with descriptions (Ephemeroptera). Canad. Ent. 67: 113–125, pls. 4–5.

Joly, Emile. 1876. Études sur l'embryogénie des Éphémères, notamment chez la *Palingenia virgo*. Mem. Acad. Sci. Toulouse; 243–254, pls. 1–2.

Jordan, D. S., and Evermann, B. W. 1904. American food and game fishes. N. Y.

Keating, W. H. 1824. Narrative of an expedition to the source of the St. Peter's River, Lake Winnipeek, Lake of the Woods, etc., in 1823, under S. H. Long, Philadelphia.

Kellogg, V. L. 1895. The Ephemeridae and venation nomenclature. Psyche 7: 311–315, text figs. 1–4.

Kennedy, C. H. 1926. The nymph of *Ephemera guttulata* Pictet with notes on the species. Canad. Ent. 58: 61–63, 1 text fig.

Kimmins, D. E. 1934. Notes on the Ephemeroptera of the Godman and Salvin collection, with descriptions of two new species. Ann. Mag. Nat. Hist. Ser. 10, 14: 338–353, figs. 1–17.

Lameere, A. 1917. Étude sur l'evolution des éphémères. Bule. Soc. Zool. France 42: 41–59, 61–81.

———. 1922. Sur la nervation aliare des insects. Bull. Class. des Sci. Acad. Roy. Belgique, ser. 5, 8: 38–149.

LeConte, John L. 1891. The complete writings of Thomas Say on the entomology of North America, I and II. A. E. Foote, Philadelphia.

Lestage, J. A. 1917. Contribution a l'étude des larves des Éphémères paléarctiques Ann. Biol. Lacustre 8: 213–456, text figs. 1–54.

———. 1918. Contribution a l'étude des larves Éphémères paléarctiques. Ann. Biol. Lacustre 9: 79–182.

———. 1924. Contribution a l'étude des larves des Éphémères: Ephemerelliden. Ann. Biol. Lacustre, 13: 227–302.

———. 1928. Les Ephémèroptères de la Belgique. Ann. Bull. Soc. Ent. Belg. 68: 251–264.

———. 1935. Contribution a l'étude des Ephéméroptères; IX.—Le Groupe Siphlonuridien. Bull. Ann. Soc. Ent. Belg. 75: 77–139, figs. 1–11.

Lubbock, John. 1866. On the development of *Chloeon* (*Ephemera*) *dimidiatum.* Trans. Linn. Soc. London 24: 61; and 25: 477.

Martynov, V. 1922. The interpretation of the wing venation and tracheation of the Odonata and Agnatha. Rev. Russe Ent. 18: 145–174, 3 text figs., 1 pl. (Translation in Psyche 37: 245–281).

Martynov, A. B. 1932. (Plectoptera in) New Permian Palaeoptera with the discussion of some problems on their evolution. Trav. Inst. Paleozool. Acad. Sci. Urss. 1: 10–17, text figs. 6–7.

McDunnough, J. 1923. New Canadian Ephemeridae with notes. Canad. Ent. 55: 39–50, text figs. 1–3.

———. 1924. New Ephemeridae from Illinois, Canad. Ent. 56: 7–9, 1 text fig.

———. 1924. New Ephemeridae from New England. Occas. Papers Boston Soc. Nat. Hist. 5: 73–76, pl. 6.

———. 1924. New Canadian Ephemeridae with notes, II. Canad. Ent. 56: 90–98, 113–122, 128–133, pls. 1, 3.

———. 1924. New North American Ephemeridae. Canad. Ent. 56: 221–226, pl. 5.

———. 1925. The Ephemeroptera of Covey Hill, Que. Trans. Roy. Soc. Canada 19: 207-224, pl. 1.

———. 1925. New Ephemerella species (Ephemeroptera). Canad. Ent. 57: 41–43.

———. 1925. New Canadian Ephemeridae with notes, III. Canad. Ent. 57: 168–176, 185–192, pls. 4–5.

———. 1926. Notes on North American Ephemeroptera with descriptions of new species. Canad. Ent. 58: 184–196.

———. 1926. New Canadian Ephemeridae with notes IV. Canad. Ent. 58: 296–303, pl. 3.

———. 1927. A new Ephemerella from Illinois (Ephemeroptera) Canad. Ent. 59: 10.

———. 1927. Notes on the species of the genus Hexagenia with description of a new species (Ephemeroptera). Canad. Ent. 59: 116–120, 1 text fig.

———. 1927. A new Heptagenia from the Yellowstone region (Ephemeroptera). Canad. Ent. 59: 261.

———. 1928. The Ephemeroptera of Jasper Park, Alta. Canad. Ent. 60: 8–10.

———. 1928. Ephemerid notes with description of a new species. Canad. Ent. 60: 238–240.

———. 1929. Notes on North American Ephemeroptera with descriptions of new species, II. Canad. Ent. 61: 169–180, 1 text fig., pls. 3–4.

———. 1930. The Ephemeroptera of the north shore of the Gulf of St. Lawrence. Canad. Ent. 62: 54–62, pls. 7–9.

———. 1931. The *bicolor* group of the genus Ephemerella with particular reference to the nymphal stages (Ephemeroptera). Canad. Ent. 63: 30–42, 61–68, pls. 2–5.

———. 1931. New species of North American Ephemeroptera. Canad. Ent. 63: 82–93.

———. 1931. The genus Isonychia (Ephemeroptera). Canad. Ent. 63: 157–163, pl. 8.

———. 1931. The eastern North American species of the genus Ephemerella and their nymphs (Ephemeroptera). Canad. Ent. 63: 187–197, 201–216, pls. 11–14.

———. 1931. New North American Caeninae with notes (Ephemeroptera). Canad. Ent. 63: 254–268, pls. 17–18.

———. 1932. Further notes on the Ephemeroptera of the north shore of the Gulf of St. Lawrence. Canad. Ent. 64: 78–81.

———. 1932. New species of North American Ephemeroptera II. Canad. Ent. 64: 209–215, 1 text fig.

———. 1933. Notes on the Heptagenine species described by Clemens from the Georgian Bay region, Ont. (Ephemerop.). Canad. Ent. 65: 16–24, 33–34, 2 text figs., pl. 1.

———. 1933. The nymph of *Cinygma integrum* and description of a new Heptagenine genus. Canad. Ent. 65: 73–76, pls. 2–3.

———. 1933. New species of North American Ephemeroptera III. Canad. Ent. 65: 155–158, 1 text fig.

———. 1933. New Ephemeroptera from the Gaspé Peninsula. Canad. Ent. 65: 278–281, 3 text figs.

———. 1934. New species of North American Ephemeroptera IV. Canad. Ent. 66: 154–164, 1 text fig., pls. 7–8; 181–188, 1 text fig., pl. 10.

———. 1935. Notes on western species of Ephemeroptera. Canad. Ent. 67: 95–104, figs. 1–4.

Morgan, Ann H. 1911. Mayflies of Fall Creek. Ann. Ent. Soc. Amer. 4: 93–119, pls. 6–12.

———. 1912. Homologies in the wing veins of mayflies. Ann Ent. Soc. Amer. 5: 89–106, pls. 5–9.

———. 1913. A contribution to the biology of mayflies. Ann. Ent. Soc. Amer. 6: 371–413, text figs. 1–3, pls. 42–54. (Extensive bibliography).

———. 1929. The mating flight and the vestigial structures of the stump-legged mayfly. *Campsurus segnis*, Ndm.

Morgan, Ann H., and Grierson, Margaret C. 1932. The functions of the gills in burrowing mayflies (*Hexagenia recurvata*). Physiol. Zool. 5: 230–245, pl. 1.

Morrison, Emily R. 1919. The mayfly ovipositor, with notes on Leptophlebia and Hagenulus. Canad. Ent. 51: 139–146, pls. 10–11.

Murphy, Helen E. 1922. Notes on the biology of some of our North American species of mayflies. Bull. Lloyd Library 22, Ent. Ser. 2: 1–46, pls. 1–7.

Neave, Ferris. 1929. (Ephemeroptera in) Reports of the Jasper Park Lakes investigations, 1925–26 IV. Aquatic insects. Contr. Canad. Biol. and Fish. N. S. 4; 187–189.

———. 1930. Migratory habits of the mayfly *Blasturus cupidus* Say. Ecology 11: 568–576, text figs. 1–3.

———. 1932; A study of the mayflies (Hexagenia) of Lake Winnipeg. Contr. Canad. Biol. and Fish. N. S. 7: 179–201, text figs. 1–10.

———. 1932. The mayflies of Lake Winnepeg. Canad. Field-Nat. 46: 54–55.

———. 1934. A contribution to the aquatic insect fauna of Lake Winnipeg. Int. Rev. Hydrobiol. 31: 157–170, figs. 1–3.

Needham, J. G. 1901. (Ephemeridae in) Aquatic insects in the Adirondacks. N. Y. S. Mus. Bull. 47: 418–429, pls. 15–16.

———. 1903. (Ephemeridae in) Aquatic insects in New York State. N. Y. S. Mus. Bull. 68: 213–17, text figs. 1–2.

———. 1905. (Ephemeridae in) Mayflies and midges of New York. N. Y. S. Mus. Bull. 86: 17–62, text figs. 1–14, pls. 4–12.

———. 1907. Additional data concerning New York mayflies. N. Y. S. Mus. Bull. 124: 189–194, pl. 10.

———. 1908. Notes on the aquatic insects of Walnut Lake. Rep. of Geol. Survey of Mich. for 1907. 259–263.

———. 1908. A peculiar mayfly from Sacandaga Park. Rep. State Ent. N. Y. 71–75.

———. 1908. Report of the Entomologic Field Station at Old Forge, N. Y. N. Y. S. Mus. Bull. 124.

———. 1918. A new mayfly, Caenis, from Oneida Lake, New York. N. Y. S. Coll. Forestry, Tech. Pub. 9: 249–251.

———. 1918. Aquatic insects. Ward and Whipple's Fresh Water Biology, pp. 867–946.

———. 1920. Burrowing mayflies of our larger lakes and streams. Bull. U. S. Bur. Fish. 36; 269–292, pls. 70–82.

———. 1924. The male of the parthenogenetic mayfly *Ameletus ludens*. Psyche 31: 308–310.

———. 1925. Some aquatic insects of Lake George. Ent. News 36: 112–114.

———. 1927. A Baetine mayfly nymph with tusked mandibles. Canad. Ent. 59: 44–47, 1 text fig.

———. 1927. The life history and habits of a mayfly from Utah. Canad. Ent. 59: 133–136, 1 text fig.

———. 1927. The Rocky Mountain species of the mayfly genus Ephemerella. Ann. Ent. Soc. Amer. 20: 107–117, 1 text fig.

———. 1932. Three new American mayflies (Ephemerop.). Canad. Ent. 64: 273–276, 1 text fig.

Needham, J. G., and Christenson, R. O. 1927. Economic insects in some streams of northern Utah. Bull. Utah Agr. Exp. Sta. 201: 6–16, text figs, 2–18.

Needham, J. G., and Lloyd, J. T. 1916. (Ephemerida in) Life of Inland Waters. Ithaca, N. Y.

Needham, J. G., and Murphy, Helen E. 1924. Neotropical mayflies. Bull. Lloyd Lib. 24, Ent. Ser. 4: 1–79, pls. 1–13.

Needham, J. G., and Needham P. R. 1927. A guide to the study of fresh water biology. New York and Albany; pp. 10–14, pls. 2 and 3.

Needham, P. R. 1928. A net for the capture of stream drift organisms. Ecology 9: 339–342.

———. 1930. Ecology of streams. The [Cold Spring Harbor] Biological Laboratory. 2, No. 3: 1, 3–5.

Packard, A. S. 1895. On a rational nomenclature of the veins of insects, especially those of the Lepidoptera. Psyche 7: 235–241.

Pearse, A. S. 1915. On the food of the small shore fishes in the waters near Madison, Wis. Bull. Wis. Nat. Hist. Soc. 13: 7–22.

Percival, E., and Whitehead, H. 1926. Observations on the biology of the mayfly *Ephemera danica* Muell. Proc. Leeds Phil. Soc. 1: 136–148, text figs. 1–3.

———. 1930. A Biological Survey of the river Wharfe. J. Ecol. 18: 286–302.

Phillips, J. S. 1930. A revision of New Zealand Ephemeroptera. Trans. N. Z. Institute 61: 271–390, pls. 50–67.

Pickles, Alan. 1931. On the metamorphosis of the alimentary canal in certain Ephemeroptera. Trans. Ent. Soc. London 79, Pt. II: 263–276, pls. 5–6.

Pictet, F. J. 1843. Histoire Naturelle, générale et particulière des Insectes Néuropterès. Famille des Éphémèrines. Geneva and Paris: 1–300, pls. 1–47. .

Riley, C. V. 1870. In Memoriam (Benj. D. Walsh). Amer. Ent. 2. No. 3: 65–68.

Reaumer, R. A. F. 1742. Memoires pour servir a l'histoire des insectes. Tome 6: 457–522, pls.

Redtenbacker, J. 1886. Vergleichende Studien ueber das Flügelgeader der Insecten. Ann. des K.K. naturh. Hof-museums, Bd. 1, Hft. 3: 153–231.

Ronalds, Alfred. 1877. The fly-fisher's entomology. 8th ed. London.

Say, Thomas. See J. L. LeConte.

Schoenemund, Eduard. 1930. Eintagsfleigen oder Ephemeroptera. Die Tierwelt Deutschlands und der angrenz. Meeresteile 19: 1–106, text figs. 1–186.

Scudder, S. H. 1882. A bibliography of fossil insects. Bull. Harvard Univ. 13.

———. 1890. The Tertiary insects of North America. Washington, Gov. print. off.

———. 1891. Index to known fossil insects of the world, including myriapods and arachnids. Bull. U. S. Geol. Survey 71: 1–744.

Seemann, Theresa M. 1927. Dragonflies, stoneflies, and mayflies of southern California. J. Ent. and Zool. 19: 40–51, 2 text figs., pls. 2–4.

Seiler Wilh. 1906. Beitraege zur Kenntniss der Ocellen der Ephemeriden. Zool. Jahrb. Anat. 22: 1–40.

Sellards, E. H. 1907. Types of Permian insects. Pt. II, Plectoptera. Amer. J. of Sci. 23: 345–355, text figs. 1–13.

Shafer, G. D. 1907. (*Callibaetis hageni* in). Histology and development of the divided eyes of certain insects. Proc. Wash. Acad. Sci. 8: 468–486, pls. 24–27.

Shepherd, David. 1929. *Ephemerella hecuba* Eaton; Description of various stages (Ephemerida, Baetidae). Canad. Ent. 61: 260–264, pls. 10–11.

Snodgrass, R. E. 1926. The Morphology of Insect Sense Organs and the Sensory Nervous System. Smithson. Misc. Coll. 77: 1–80.

Spieth, H. T. 1933. The phylogeny of some mayfly genera. J. N. Y. Ent. Soc. 41: 55–86, 327–390, pls. 16–24.

Spuler, A. 1892. Zur Phylogenie und Ontogenie des Flügelgeaders der Schmetterlinge. Zeit. wiss. Zool. 53: 597–649, 2 pls.

Steger, A. Louise. 1931. Some preliminary notes on the genus Ephemerella. Psyche 38: 27–35, 1 text fig.

Sulc, K., and Javrel J. 1924. Ueber Epoikische und Parastiische Chironomid Larven. Acta. Soc. Sci. Nat. Moravicae 1: 383–391, 19 text figs.

Swammerdam, Jan. 1737. (Ephemera in) Biblia naturae; sive Historia insectorum. Vols. 1 and 2, pls. 13–15.

Thoreau, Henry D. 1906. The writings of Thoreau edited by Bradford Torrey. Boston and New York.

Tillyard, R. J. 1923. The wing-venation of the order Plectoptera or mayflies. J. Linn. Soc. London-Zool. 35: 143–162, text figs. 1–10.

———. 1925. Kansas Permian insects. Amer. J. Sci. 9: 328–335, 3 text figs.

———. 1926. (Plectoptera in) Insects of Australia and New Zealand. Sidney. 57–64. text figs. E1–E6.

———. 1932. Kansas Permian insects, Pt. 15. The order Plectoptera. Amer. J. Sci. 23: 97–134, 237–272, 22 text figs.

Traver, J. R. 1925. Observations on the ecology of the mayfly *Blasturus cupidus*. Canad. Ent. 57: 211–218.

———. 1931. Seven new southern species of the mayfly genus Hexagenia, with notes on the genus. Ann. Ent. Soc. Amer. 24: 59–1620, 3 text figs, pl. 1.

———. 1931. The Ephemerid genus Baetisca. J. N. Y. Ent. Soc. 39: 45–66, pls. 5–6.

———. 1931. A new mayfly genus from North Carolina. Canad. Ent. 63: 103–109, pl. 7.

———. 1932. Mayflies of North Carolina. J. Elisha Mitchell Sci. Soc. 47: 85–236, pls. 5–12.

———. 1932. Neocloeon, a new mayfly genus (Ephemerida). J. N. Y. Ent. Soc. 40: 365–372, pl. 14.

———. 1933. Mayflies of North Carolina, Pt. III. J. Elisha Mitchell Sci. Soc. 48: 141–206, pl. 15.

———. 1933. Heptagenine mayflies of North America. J. N. Y. Ent. Soc. 41: 105–125.

———. 1934. New North American species of mayflies (Ephemerida). J. Elisha Mitchell Sci. Soc. 50: 189–254, pl. 16.

———. 1935. Two new genera of North American Heptageniidae (Ephemerida). Canad. Ent. 67: 31–38, 1 pl.

Ueno, Masuzo. 1928. Some Japanese mayfly nymphs. Mem. Coll. Sci. Kyoto Imper. Univ. Ser. B, 4: 19–63, text figs. 1–6, pls. 3–17.

———. 1931. Contributions to the knowledge of Japanese Ephemeroptera. Annot. Zool. Japan. 13: 189–231, text figs. 1–34, pls. 12–13.

Ulmer, Georg. 1920. Uebersicht ueber die Gattungen der Ephemeropteran, nebst Bemerkungen ueber einzelne Arten. Stett. Ent. Zeit. 81: 97–144.

———. 1921. Ueber einige Ephemeropteren-Typen älterer Autoren. Arch. f. Naturg. Abt. A, 6: 229–267, text figs. 1–21.

———. 1924. (Ephemeroptera in) Biologie der Tiere Deutschlands. Berlin. 1–40, text figs. 1–28.

———. 1925. (Ephemeropteren in) Beitrage zur Fauna sinica. Arch. f. Naturg. 91, Hft. 5: 86–110, text figs. 80–102.

———. 1932. Bemerkungen ueber die seit 1920 neu aufgestellten Gattungen der Ephemeropteren. Stett. Ent. Zeit. 93: 204–219.

———. 1932–33. Aquatic insects of China. Art. 6, Revised key to the genera of Ephemeroptera. Peking Nat. Hist. Bull. 7: 195–218, pls. 1–2.

Vayssière, Albert. 1882. Recherches sur l'organisation des larves des Éphémèrines. Ann. Sci. Nat. (6) Zool. 13, 1: 1–137, pls. 1–11.

———. 1934. Etude anatomique des larves nymphales des *Baetisca obesa* et *carolina*. Ann. Sci. Nat. Zool. (10) 17: 381–406, pls. 1–2.

Wagner, George. 1908. Notes on the fish fauna of Lake Pepin. Trans. Wiscon. Acad. Sci. Arts and Letters. 16: 23–37.

Walley, G. S. 1930. Review of Ephemerella nymphs of western North America (Ephemeroptera). Canad. Ent. 62: 12–20, pls. 1–3.

Walker, Francis. 1853. Catalogue of the species of Neuropterous insects in the collection of the British Museum Pt. III, Termitides and Ephemeridae; 533–585.

Walsh, Benj. D. 1862. List of the Pseudoneuroptera of Illinois, contained in the cabinet of the writer, with descriptions of over forty new species, and notes on their structural affinities. Proc. Acad. Nat. Sci. Phila. 13–14: 361–402.

———. 1863. (Ephemeridae in) Observations on certain N. A. Neuroptera, by H. Hagen, M. D., of Koenigsberg, Prussia; translated from the original French MS. and published by permission of the author with notes and descriptions of about twenty new N. A. species of Pseudoneuroptera. Proc. Ent. Soc. Phila. 2: 169–179, 188–210.

———. 1864. On the pupa of the Ephemerinous genus Baetisca Walsh. Proc. Ent. Soc. Phila. 3: 200–206, 1 text fig.

Walton, Izaak. 1861. The complete angler, or, The contemplative man's recreation. London.

Weiss, Harry B., and Ziegler, Grace M. 1931. Thomas Say, early American naturalist. Springfield, Baltimore. 14 pls.

Westwood, J. O. 1840 (Ephemeridae in) An introduction to the modern classification of insects. London: 24–34.

Wiebe, A. H. 1926. The first three larval stages of *Hexagenia bilineata* Say. Ohio J. Sci. 26: 267–275, pl. 1.

Williamson, Hugh. 1802. On the *Ephoron leukon,* usually called the White Fly of Passaik River. Trans. Amer. Philos. Soc. 5: 71–73.

Wodsedalek, J. E. 1911. Phototactic reactions and their reversal in mayfly nymphs. Biol. Bull. 21: 265–271.

———. 1912. Palmen's organ and its function in nymphs of the Ephemeridae, *Heptagenia interpunctata* (Say) and *Ecdyurus maculipennis* (Walsh). Biol. Bull. 22: 253–273, pls. 1–3.

———. 1912. Formation of associations in the mayfly nymphs *Heptagenia interpunctata* (Say). J. Animal Behav. 2: 1–19, text figs. 1–2.

———. 1912–13. Natural history and general behavior of the Ephemeridae nymphs *Heptagenia interpunctata* (Say). Ann. Ent. Soc. Amer. 5–6: 31–40.

Zimmerman, O. 1880. Ueber eine eigentümliche Bildung der Ruckengefässes bei einigen Ephemeridenlarven. Zeit. Wiss. Zool. 34: 404–406.

Addendum:

Barnard, K. H. 1932. South African May-flies (Ephemeroptera). Trans. R. Soc. of S. Afr. 20: 201–259, 48 figs.

INDEX

Whole figures, diagnostic figures, descriptions and definitions are indicated by **bold-face** type; exotic genera and species by (Ex.). Synonyms are printed in *italics*.

Q

R